Finance

Life Science

(continued on inside back cover)

MATHEMATICAL APPLICATIONS
for the Management, Life, and Social Sciences

MATHEMATICAL APPLICATIONS

for the Management, Life, and Social Sciences

Fifth Edition

Ronald J. Harshbarger
Georgia Southern University

James J. Reynolds
Clarion University of Pennsylvania

D. C. Heath and Company
Lexington, Massachusetts   Toronto

Address editorial correspondence to:

D. C. Heath and Company
125 Spring Street
Lexington, MA 02173

Acquisitions: Charles W. Hartford
Development: Philip Charles Lanza
Editorial Production: Janice Molloy
Design: Kenneth Hollman
Art Editing: Diane Grossman/Carol Reid
Production Coordination: Lisa Merrill/Charles Dutton
Permissions: Margaret Roll

PREFACE

To paraphrase English mathematician, philosopher, and educator Alfred North Whitehead, the purpose of education is not to fill a vessel but to kindle a fire. In particular, Whitehead encouraged students to be creative and imaginative in their learning and to continually form ideas into new and more exciting combinations. This desirable goal is not always an easy one to realize in mathematics with students whose primary interests are in areas other than mathematics. The purpose of this text, then, is to present mathematical skills and concepts and to apply them to ideas that are important to students in the management, life, and social sciences. We hope that this look at the relevance of mathematical ideas to a broad range of fields will help inspire the imaginative thinking and excitement for learning that Whitehead spoke of. The applications included allow students to view mathematics in a practical setting relevant to their intended careers. Almost every chapter of this book includes a section or two devoted to the applications of mathematical topics. An index of these applications on the inside covers demonstrates the wide variety used in examples and exercises. Although intended for students who have completed two years of high school algebra or its equivalent, this text begins with a brief review of algebra which, if covered, will aid in preparing students for the work ahead.

Pedagogical Features

Important pedagogical features that have been retained in this new edition are the following.

Intuitive Viewpoint. The book is written from an intuitive viewpoint, with emphasis on concepts and problem solving rather than on mathematical theory. Each topic is carefully explained, and examples illustrate the techniques involved. Many exercises stress computation and drill, but there are enough challenging problems to stimulate students in thoughtful investigations.

Flexibility. At different colleges and universities, the coverage and sequencing of topics may vary according to the purpose of this course and the nature of the student audience. To accommodate alternative approaches, the text has a great deal of flexibility in the order in which topics may be presented and the degree to which they may be emphasized.

Chapter Warm-ups. A Warm-up appears at the beginning of each chapter and invites students to test themselves on the skills needed for that chapter. The Warm-ups present several prerequisite problem types that are keyed to the appropriate sections in the upcoming chapter where those skills are needed. Students who have difficulty with any particular skill are directed to specific sections of the text for review. Instructors may also find the Warm-ups useful in creating a course syllabus that includes an appropriate scope and sequence of topics.

Applications. We have found that integrating applied topics into the discussions and exercises helps to provide motivation within the sections and demonstrate the relevance of each topic. In addition, we have found that offering separate lessons on applied topics such as cost, revenue, and profit functions brings the preceding mathematical discussions into clear and concise focus and provides a thread of continuity as mathematical sophistication increases. There are eleven such sections in this text. Overall, of the more than 5300 exercises in the book, more than 1900 are applied problems.

Objective Lists. Every section begins with a brief list of objectives that outlines the goals of that section for the student.

Procedure/Example and Property/Example Tables. Appearing throughout the text, these tables aid student understanding by giving step-by-step descriptions of important procedures and properties with illustrative examples worked out beside them.

Boxed Information. All important information is boxed for easy reference, and key terms are highlighted in boldface.

Review Exercises. At the end of each chapter, a set of Review Exercises offers the student extra practice on the topics in that chapter. These exercises are annotated with section numbers so that any student who has difficulty can turn to the appropriate section for review.

Changes in the Fifth Edition

The major goals of this revision were to include numerous real data problems, enhance the graphical and numeric approaches to problem solving, add other pedagogical features, lay a better groundwork for the presentation of the mathematics, and provide opportunities for the use of graphing and computer technologies for those instructors inclined to use them. These goals were achieved in the following ways.

1. We have made an extensive addition of applications involving real data and **mathematical modeling** in exercises and examples throughout the text. Over 220 problems among the exercise applications—about one of every eight applications—are developed around real data and contain source notes. Many sections include problems with functions that are modeled from real data, and some problems ask students to model functions from the data.

2. The pedagogy includes an **enhanced graphical and numerical viewpoint or approach,** which can be used effectively during instruction, whether or not the instructor chooses to require student use of graphics calculators or other graphing utilities. We have added a large number of graphs to our discussions, examples, and exercises that allow students to see relationships and that are designed to improve their ability to read and interpret graphs. In addition, the inclusion of real data problems provides increased opportunities to develop a numerical perspective in understanding and solving mathematical applications.

3. Special highlighted subsections that discuss the **use of graphing utilities** are added throughout the text. After the introduction to graphing with technology in new Section 1.3, Graphs and Graphing Utilities, discussion of the use of technology is placed in subsections and in examples within sections. These presentations are highlighted with a technology icon 📟 so that they may be emphasized or de-emphasized at the instructor's discretion. Some exercises that benefit from the use of technology—which includes graphics calculators, computer programs, and computer spreadsheets—are also highlighted with the technology icon. Of course, many additional problem investigations can also profit from the use of technology, at the option of the instructor. Technology is used to graph functions and to discuss generalizations, applications, and implications of problems being studied. It is also used for modeling functions from real data. The power that these tools bring to our investigations now permits the inclusion of problems that are larger in scope and that contain more data points.

4. Each section begins with a new **Application Preview** that establishes the context and direction for the concepts that will be presented. The preview motivates the mathematical discussion in the upcoming section and is then revisited in a worked-out example appearing later in the lesson.

5. New **Checkpoint** problems are introduced in each section. These in-lesson exercises ask questions and pose problems that allow students to check their understanding of the skills and concepts under discussion as they progress through the section. Solutions to Checkpoint problems appear before lesson exercises.

6. The **Consultant's Corner** feature has been expanded to appear at the end of every chapter. These case studies can be investigated with or without the use of graphing utilities and are excellent means of providing cooperative learning opportunities for students.

7. Exercise sets include **labeled applications** in which the subject matter of each problem is identified. Instructors and students can use these labels to easily select applications that are of special interest or are particularly suitable for their class.

8. A new end-of-chapter **Key Terms and Formulas** index precedes the Review Exercises in each chapter.

Other significant content improvements include the following.

CHAPTER *1* Linear Equations and Functions

The power and limitations of graphing technology are discussed, paving the way for future discussions involving tabular, graphical, and analytic methods of solving problems. The discussion of operations with functions, in Section 1.4, now includes composition of functions.

CHAPTER 2   Matrices

Section 2.4, Inverse of a Square Matrix, now includes a brief discussion of finding determinants for 2x2 matrices and using technology for finding determinants of larger square matrices. The discussion emphasizes the usefulness of technology in performing operations with large matrices and finding matrix inverses.

CHAPTER 3   Inequalities and Linear Programming

Section 3.3 now includes the use of technology to find the corners of feasible regions in linear programming. Two Consultant's Corners are included, one of which discusses sensitivity analysis. There is also an increased emphasis in converting real data tables and figures to matrices.

CHAPTER 4   Quadratic Functions, and
CHAPTER 5   Exponential Functions

These chapters now exhibit a much stronger modeling emphasis with significant additions of real data problems.

CHAPTER 6   Mathematics of Finance

An increased emphasis on the use of graphical comparisons provides a more visual exploration of functional relationships and the idea that "time is money." Also, compound interest problems are now given a more student-friendly pace, focusing in early exercises on identification of interest rate per period and number of periods.

CHAPTER 7   Introduction to Probability

The chapter has been rewritten and reorganized to present both the "set" and "event" notation styles/interpretations of probability theory. One-trial and multiple-trial experiments are now treated in separate lessons.

CHAPTER 8   Further Topics in Probability; Statistics

Chapter 8 has been rewritten and reorganized to extend the text's presentation of probability topics and emphasize a variety of probability distributions. The discussion of expected value has been relocated from Chapter 7 to new Section 8.2, Discrete Probability Distributions; Decision Making. New Section 8.4 presents a consolidated discussion of descriptive statistics concepts.

CHAPTER 9   Derivatives

Section 9.1, Limits, has been reorganized to strengthen the pedagogical links among the graphic, numeric, and symbolic approaches to limits. Examples and exercises throughout the chapter now place much greater emphasis on graphical relationships between functions and their derivatives, as well as greater emphasis on interpretation problems, both for limits and derivatives. The discussion in Section 9.2 now includes limits at infinity.

CHAPTER 10 Applications of Derivatives

There is now increased emphasis in Sections 10.1 and 10.2 on interpretation of the graphs of functions and their first and second derivatives and, in particular, what can be learned from the relationships among them.

CHAPTER 14 Functions of Two or More Variables

Section 14.2, Partial Differentiation, now combines coverage of first-order and higher-order partial derivatives into a single section. Section 14.5 now includes a graphical approach to solving constrained max-min problems.

Supplements

Instructor's Guide. This supplement contains answers to all even-numbered exercises in the text; full solutions to all even-numbered exercises; solutions to all Consultant's Corner problems; four forms of chapter tests for each chapter; answer keys to all test forms; and transparency masters for important tables and figures from the text.

Study & Solutions Guide, by Gordon Shilling. In addition to the solutions to all odd-numbered exercises in the text, this guide contains supplementary exercises that reinforce the concepts and techniques presented in the text. Answers to these problems are also provided.

ipsTest Computerized Testing. This computerized testing package, available in MS-DOS and Macintosh formats, combines algorithmic and fixed-item testing in one convenient software package. The carefully constructed algorithms use random number generators to produce different, yet equivalent, versions of each problem. A nearly unlimited number of different tests and quizzes can be generated quickly and easily. Questions may be output as either multiple choice or free response items. The Macintosh version features an editing function for changing existing problems or creating additional problems. In custom-designing tests, instructors can also draw from among the 2000 fixed items in the test bank, add their own test items to the bank, or edit existing items. Tests may be prepared using any combination of both algorithmic and fixed items from the software package. User notes and sample problem printouts are included with the program.

Test Item File. This is a printed version of all of the test items appearing in the computerized testing program. Fixed items are correlated to appropriate sections of the textbook. The Test Item File also includes sample problem printouts for each algorithmic testing item, organized by chapter.

BestGrapher, Version 4.0, by George Best. Available in MS-DOS and Macintosh formats, this tool software and its accompanying workbook emphasize the concepts of calculus. The software can be used for graphing, evaluating functions, and symbolic differentiation among other tasks. Graphing features now also include zoom, trace, and simultaneous plot capabilities. The workbook provides activities that encourage experimentation and discovery about functions, derivatives, areas, and other topics.

Acknowledgments

We would like to thank the many people who have helped us at various stages of revising this text. The encouragement, criticism, and suggestions that have been offered have been invaluable to us. Special thanks go to Dipendra Bhattacharya, Steve Gendler, Susie Lanier, Carol Nessmith, Sahib Singh, and Steve Wise for their contributions to this project.

For their reviews of draft manuscript and the many helpful comments that were offered, we would also like to thank Mitzi Chaffer, Central Michigan University; Florence Chambers, Joliet Junior College; R. D. Derderian, Providence College; Edward Gade, University of Wisconsin at Oshkosh; Dauhrice Gibson, Gulf Coast Community College; Betty L. Hawkins, Shoreline Community College; Rose Marie Kinik, Sacred Heart University; Vincent Koehler, Trinity College of Vermont; Lee H. LaRue, Paris Junior College; James Loats, Metro State College of Denver; Janice C. Lyon, Tallahassee Community College; Theodore F. Moore, Mohawk Valley Community College; James R. Nolan, Siena College; Todd D. Oberg, The University of Montana; Gordon Shilling, University of Texas at Arlington; Michel Smith, Auburn University; and Diane Van Nostrand, University of Tulsa.

We would also like to express our appreciation to the editorial staff at D. C. Heath for their continued enthusiasm and support.

Ronald J. Harshbarger
James J. Reynolds

CONTENTS

PART 1

ALGEBRA REVIEW

CHAPTER 0

ALGEBRAIC CONCEPTS

This chapter provides a brief review of the algebraic concepts that will be used throughout the text.

The review begins with sets and the real numbers, the number system used throughout the text. Special subsets of the real numbers, including intervals, are also discussed.

Exponents and radicals, the building blocks for algebraic expressions, are then introduced. Operations with algebraic expressions and factoring are reviewed and then followed by operations with, and simplification of, algebraic fractions.

You should already be familiar with the topics covered in this chapter, but it may be helpful to spend some time reviewing them. In addition, each chapter after this one opens with a warm-up page that identifies prerequisite skills needed for that chapter. If algebraic skills are required, the warm-up cites their coverage in this chapter. Thus, you will find that the following sections are a useful reference as you study later chapters.

0.1 SETS

We can begin to talk about sets by first looking at how blood is typed. The determination of blood types is based on the presence or absence of three antigens: the A antigen, the B antigen, and the antigen referred to as the Rh factor. Each different antigen combination determines a different blood type, and this scheme can be used to identify any individual as a member of a distinct group (or set) based on blood type (see problem 59 in the exercises). For example, a person belongs to the set of people with blood type AB^+ if that person's blood contains antigen A, antigen B, and a positive Rh factor.

A **set** is a well-defined collection of objects. We may talk about a set of books, a set of dishes, a set of students, or, as described above, a set of individuals with a certain blood type. There are two ways to tell what a given set contains. One way is by listing the **elements** (or **members**) of the set (usually between braces). We may say that a set A contains 1, 2, 3, and 4 by writing $A = \{1, 2, 3, 4\}$. To say that 4 is a member of set A, we write $4 \in A$. Similarly, we write $5 \notin A$ to denote that 5 is not a member of set A.

3

If all the members of the set can be listed, the set is said to be a **finite set.** $A = \{1, 2, 3, 4\}$ and $B = \{x, y, z\}$ are examples of finite sets. When we do not wish to list all the elements of a finite set, we can use three dots to indicate the unlisted members of the set. For example, the set of even integers from 8 to 8952 inclusive could be written as

$$\{8, 10, 12, 14, \ldots, 8952\}.$$

For an **infinite set,** we cannot list all the elements, so we use the three dots. For example, $N = \{1, 2, 3, 4, \ldots\}$ is an infinite set. This set N is called the set of **natural numbers.**

Another way to specify the elements of a given set is by description. For example, we may write $D = \{x: x$ is a Ford automobile$\}$ to describe the set of all Ford automobiles. $F = \{y: y$ is an odd natural number$\}$ is read "F is the set of all y such that y is an odd natural number."

EXAMPLE 1

Write the following sets in two ways.
(a) The set A of natural numbers less than 6.
(b) The set B of natural numbers greater than 10.
(c) The set C containing only 3.

Solution
(a) $A = \{1, 2, 3, 4, 5,\}$ or $A = \{x: x$ is a natural number less than 6$\}$
(b) $B = \{11, 12, 13, 14, \ldots\}$ or
$B = \{x: x$ is a natural number greater than 10$\}$
(c) $C = \{3\}$ or $C = \{x: x = 3\}$

Note that set C of Example 1 contains one member, 3; set A contains five members; and set B contains an infinite number of members. It is possible for a set to contain no members. Such a set is called the **empty set** or the **null set,** and it is denoted by $\varnothing$ or by $\{\quad\}$. The set of living veterans of the War of 1812 is empty because there are no living veterans of that war. Thus

$$\{x: x \text{ is a living veteran of the War of 1812}\} = \varnothing.$$

Special relations that may exist between two sets are defined as follows.

Relations Between Sets	
Definition	**Example**
1. Set X and Y are **equal** if they contain the same elements.	1. If $X = \{1, 2, 3, 4\}$ and $Y = \{4, 3, 2, 1\}$, then $X = Y$.
2. $A \subseteq B$ if every element of A is an element of B. A is called a **subset** of B. The empty set is a subset of every set.	2. If $A = \{1, 2, c, f\}$ and $B = \{1, 2, 3, a, b, c, f\}$, then $A \subseteq B$.
3. If C and D have no elements in common, they are called **disjoint.**	3. If $C = \{1, 2, a, b\}$ and $D = \{3, e, 5, c\}$, then C and D are disjoint.

In the discussion of particular sets, the assumption is always made that the sets under discussion are all subsets of some larger set, called the **universal set** *U*. The choice of the universal set depends upon the problem under consideration. For example, in discussing the set of all students and the set of all female students, we may use the set of all humans as the universal set.

We may use **Venn dagrams** to illustrate the relationships among sets. We use a rectangle to represent the universal set and closed figures inside the rectangle to represent the sets under consideration. Figures 0.1–0.3 show such Venn diagrams.

The shaded portion of Figure 0.3 indicates where the two sets overlap. The set containing the members that are common to two sets is said to be the **intersection** of the two sets.

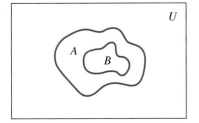

Figure 0.1
B is a subset of A; B ⊆ A.

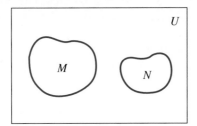

Figure 0.2
M and N are disjoint.

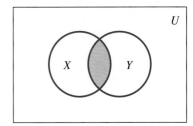

Figure 0.3
x and y are not disjoint.

Set Intersection   The intersection of *A* and *B*, written *A* ∩ *B*, is defined by

$$A \cap B = \{x: x \in A \text{ and } x \in B\}.$$

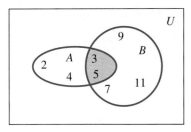

Figure 0.4

EXAMPLE 2
If *A* = {2, 3, 4, 5} and *B* = {3, 5, 7, 9, 11}, find *A* ∩ *B*.

Solution
A ∩ *B* = {3, 5} because 3 and 5 are the common elements of *A* and *B*. Figure 0.4 shows the sets and their intersection.

CHECKPOINT

Let *A* = {2, 3, 5, 7, 11}, *B* = {2, 4, 6, 8, 10}, and *C* = {6, 10, 14, 18, 22}. Use these sets to answer the following.
1. (a) Of which sets is 6 an element?   (b) Of which sets is {6} an element?
2. Which of the following are true?
   (a) 2 ∈ *A*   (b) 2 ∈ *B*   (c) 2 ∈ *C*   (d) 5 ∉ *A*   (e) 5 ∉ *B*

3. Which pair of A, B, and C is disjoint?

4. Which of A, B, and C are subsets of
   (a) the set P of all prime numbers?
   (b) the set M of all multiples of 2?

5. Which of A, B, and C is equal to
   $D = \{x \mid x = 4n + 2 \text{ for natural numbers } 1 \leq n \leq 5\}$?

The **union** of two sets is the set that contains all members of the two sets.

Set Union   The union of A and B, written $A \cup B$, is defined by

$$A \cup B = \{x\colon x \in A \text{ or } x \in B \text{ (or both)}\}.^*$$

Figure 0.5

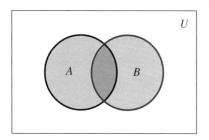

Figure 0.6

EXAMPLE 3
If $X = \{a, b, c, f\}$ and $Y = \{e, f, a, b\}$, find $X \cup Y$.

Solution
$X \cup Y = \{a, b, c, e, f\}$

EXAMPLE 4
Let $A = \{x\colon x \text{ is a natural number less than 6}\}$ and $B = \{1, 3, 5, 7, 9, 11\}$.
(a) Find $A \cap B$.
(b) Find $A \cup B$.

Solution
(a) $A \cap B = \{1, 3, 5\}$
(b) $A \cup B = \{1, 2, 3, 4, 5, 7, 9, 11\}$

We can illustrate the intersection and union of two sets by the use of Venn diagrams. Figures 0.5 and 0.6 show Venn diagrams with universal set U represented by the rectangles and sets A and B represented by the circles. The shaded region in Figure 0.5 represents $A \cap B$, the intersection of A and B, while the shaded region in Figure 0.6 represents $A \cup B$.

All elements of the universal set that are not contained in a set A form a set called the **complement** of A.

Set Complement   The complement of A, written A', is defined by

$$A' = \{x\colon x \in U \text{ and } x \notin A\}.$$

---

* In mathematics, the word *or* means one or the other or both.

Figure 0.7

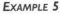

EXAMPLE 5

If $U = \{x \in N: x < 10\}$, $A = \{1, 3, 6\}$, and $B = \{1, 6, 8, 9\}$, find the following.

(a) A' (b) B' (c) $(A \cap B)'$ (d) $A' \cup B'$

Solution

(a) $U = \{1, 2, 3, 4, 5, 6, 7, 8, 9\}$ so $A' = \{2, 4, 5, 7, 8, 9\}$

(b) $B' = \{2, 3, 4, 5, 7\}$

(c) $A \cap B = \{1, 6\}$ so $(A \cap B)' = \{2, 3, 4, 5, 7, 8, 9\}$

(d) $A' \cup B' = \{2, 4, 5, 7, 8, 9\} \cup \{2, 3, 4, 5, 7\}$
$= \{2, 3, 4, 5, 7, 8, 9\}$

We can use a Venn diagram to illustrate the complement of a set. The shaded region of Figure 0.7 represents A' and the *un*shaded region of Figure 0.5 represents $(A \cap B)'$.

CHECKPOINT

Given the sets $U = \{1, 2, 3, 4, 5, 6, 7, 8, 9, 10\}$, $A = \{1, 3, 5, 7, 9\}$, $B = \{2, 3, 5, 7\}$, and $C = \{4, 5, 6, 7, 8, 9, 10\}$, find the following.

6. $A \cup B$ 7. $B \cap C$ 8. A'

EXAMPLE 6

The local newspaper of Oil City, Pennsylvania, is called *The Derrick*. On the morning of June 2,1994, *The Derrick* listed 17 stocks "of local interest." A prospective investor categorized these 17 stocks according to whether

- their closing price on June 1, 1994, was less than $40/share (set C)
- their price-to-earnings ratio was less than 20 (set P)
- their dividend per share was at least $1.50 (set D)

Of these 17 stocks

10 belonged to set P	6 belonged to both C and P
9 belonged to set C	7 belonged to both D and P
11 belonged to set D	3 belonged to all three sets
4 belonged to both C and D	

Draw a Venn diagram that represents this information. Use the diagram to answer the following.

(a) How many stocks had closing prices of less than $40 per share or price-to-earnings ratios of less than 20?

(b) How many stocks had none of the characteristics of sets C, P, or D?

(c) How many stocks had only dividends per share of at least $1.50?

(a)

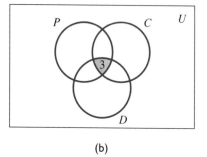

(b)

Figure 0.8

Solution

The Venn diagram for three sets has eight separate regions (see Figure 0.8 (a)). To assign numbers from our data, we must begin with some information that refers to a single region, namely that three stocks belonged to all three sets (see Figure 0.8 (b)). Because the region common to all three sets is also common to any pair, we can next use the information about stocks that belonged to two of the sets (see Figure 0.8(c)). Finally, we can complete the Venn diagram (see Figure 0.8(d)).

(a) We need to add the numbers in the separate regions that lie within $C \cup P$. That is, 13 stocks closed under $40 per share or had price-to-earnings ratios of less than 20.

(b) There is one stock outside the three sets C, D, and P.

(c) Those stocks that had only dividends of at least $1.50 per share are inside D but outside both C and P. There are three such stocks.

(c)

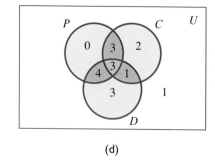

(d)

CHECKPOINT SOLUTIONS

1. (a) Sets B and C have 6 as an element.
   (b) None of A, B, or C has {6} as an element; {6} is itself a set and the elements of A, B, and C are not sets.

2. (a) True   (b) True   (c) False; $2 \notin C$
   (d) False; $5 \in A$ (e) True

3. A and C are disjoint.

4. (a) $A \subseteq P$
   (b) $B \subseteq M$ $C \subseteq M$

5. $C = D$

6. $A \cup B = \{1, 2, 3, 5, 7, 9\}$

7. $B \cap C = \{5, 7\}$

8. $A' = \{2, 4, 6, 8, 10\}$

EXERCISE 0.1

Use $\in$ or $\notin$ to indicate whether the given object is an element of the given set in the following problems.
1. x $\{x, y, z, a\}$
2. 3 $\{1, 2, 4, 5, 6\}$
3. 12 $\{1, 2, 3, 4, \ldots\}$
4. 5 $\{x: x$ is a natural number greater than 5$\}$
5. 6 $\{x: x$ is a natural number less than 6$\}$
6. 3 $\varnothing$

In problems 7–10 write the following sets a second way.
7. $\{x: x$ is a natural number less than 8$\}$
8. $\{x: x$ is a natural number greater than 6 and less than 10$\}$
9. $\{3, 4, 5, 6, 7\}$
10. $\{7, 8, 9, 10, \ldots\}$
11. If $A = \{1, 2, 3, 4\}$ and $B = \{1, 2, 3, 4, 5, 6\}$, is A a subset of B?
12. If $A = \{a, b, c, d\}$ and $B = \{c, d, a, b\}$, is A a subset of B?
13. Is $A \subseteq B$ if $A = \{a, b, c, d\}$ and $B = \{a, b, d\}$?
14. Is $A \subseteq B$ if $A = \{6, 8, 10, 12\}$ and $B = \{6, 8, 10, 14, 18\}$?

Use $\subseteq$ notation to indicate which set is a subset of the other in the following problems.
15. $C = \{a, b, 1, 2, 3\}, D = \{a, b, 1\}$
16. $E = \{x, y, a, b\}, F = \{x, 1, a, y, b, 2\}$
17. $A = \{1, \pi, e, 3\}, D = \{1, \pi, 3\}$
18. $A = \{2, 3, 4, e\}, \varnothing$
19. $A = \{6, 8, 7, 4\}, B = \{8, 7, 6, 4\}$
20. $D = \{a, e, 1, 3, c\}, F = \{e, a, c, 1, 3\}$

In problems 21–24, indicate whether the following pairs of sets are equal.
21. $A = \{a, b, \pi, \sqrt{3}\}, B = \{a, \pi, \sqrt{3}, b\}$
22. $A = \{x, g, a, b\}, D = \{x, a, b, y\}$
23. $D = \{x: x$ is a natural number less than 4$\}$, $E = \{1, 2, 3, 4\}$
24. $F = \{x: x$ is a natural number greater than 6$\}$, $G = \{7, 8, 9, \ldots\}$
25. From the following list of sets, indicate which pairs of sets are disjoint.

$A = \{1, 2, 3, 4\}$

$B = \{x: x$ is a natural number greater than 4$\}$

$C = \{4, 5, 6, \ldots\}$

$D = \{1, 2, 3\}$

26. If A and B are disjoint sets, what does $A \cap B$ equal?

In problems 27–30, find $A \cap B$, the intersection of sets A and B.
27. $A = \{2, 3, 4, 5, 6\}$ and $B = \{4, 6, 8, 10, 12\}$
28. $A = \{a, b, c, d, e\}$ and $B = \{a, d, e, f, g, h\}$
29. $A = \varnothing$ and $B = \{x, y, a, b\}$
30. $A = \{x: x$ is a natural number less than 4$\}$ and $B = \{3, 4, 5, 6\}$

In problems 31–34, find $A \cup B$, the union of sets A and B.
31. $A = \{1, 2, 4, 5\}$ and $B = \{2, 3, 4, 5\}$
32. $A = \{a, e, i, o, u\}$ and $B = \{a, b, c, d\}$
33. $A = \varnothing$ and $B = \{1, 2, 3, 4\}$
34. $A = \{x: x$ is a natural number greater than 5$\}$ and $B = \{x: x$ is a natural number less than 5$\}$

In problems 35–46, assume

$$A = \{1, 3, 5, 8, 7, 2\}, \qquad B = \{4, 3, 8, 10\},$$
$$C = \{2, 4, 6, 8, 10\},$$

and U is the universal set of natural numbers less than 11. Find the following.
35. A' 36. B' 37. $A \cap B'$
38. $A' \cap B'$ 39. $(A \cup B)'$ 40. $(A \cap B)'$
41. $A' \cup B'$ 42. $(A' \cup B)'$ 43. $(A \cap B') \cup C'$
44. $A \cap (B' \cup C')$ 45. $(A \cap B')' \cap C$
46. $A \cap (B \cup C)$

The difference of two sets, $A - B$, is defined as the set containing all elements of A except those in B. That is, $A - B = A \cap B'$. Find $A - B$ for each pair of sets in problems 47–50, if $U = \{1, 2, 3, 4, 5, 6, 7, 8, 9\}$.
47. $A = \{1, 3, 7, 9\}$ and $B = \{3, 5, 8, 9\}$
48. $A = \{1, 2, 3, 6, 9\}$ and $B = \{1, 4, 5, 6, 7\}$
49. $A = \{2, 1, 5\}$ and $B = \{1, 2, 3, 4, 5, 6\}$
50. $A = \{1, 2, 3, 4, 5\}$ and $B = \{7, 8, 9\}$

APPLICATIONS

51. **Sticker shock** The table on page 10 contains base prices for selected models of cars. Let A be the set of cars listed with a base price of at least \$20,000 in any year. Let B be the set of cars listed with a 1993 base price of at least \$20,000. Let C be the set of cars listed with an increase in the base price of at least 75% over 10 years.
(a) List the elements of A, B, and C.
(b) Is any of A, B, and C a subset of one of the others?
(c) Find $A \cap C$ and describe it in words.
(d) Write a verbal description for B'.
(c) Find $A' \cap C'$.

Entry-Level Base Prices: Then and Now

Model	1983	1988	1993	10-year Increase
Cadillac Seville	$21,440	$27,627	$36,990	73%
Chev. Camaro	$8,036	$10,995	$13,500	68
Chev. Corvette	$21,800	$29,480	$34,595	59
Ford Escort	$5,639	$6,586	$8,355	48
Ford Mustang	$6,727	$8,835	$10,719	59
Honda Accord	$8,345	$10,915	$13,950	67
Honda Civic	$4,899	$6,195	$8,400	71
Mazda RX-7	$9,695	$15,880	$32,500	235
Nissan Maxima	$10,869	$17,449	$20,960	93
Pont. Bonneville	$8,899	$14,099	$19,444	118
Toyota Celica	$7,299	$11,348	$14,198	95
Toyota Corolla	$6,138	$8,998	$11,198	82

Source: *Motor Trend,* February 1993, p. 89.

52. **Voting** The following persons were on the ballot for judge of River County.

Republicans: $A = \{$Robbins, Swartz, Gonzales$\}$

Democrats: $B = \{$Barth, McRoberts, Robbins$\}$

If only one person named Robbins is on both the Democratic and Republican ballot,
(a) list the elements of $A \cup B$ and describe this set.
(b) list the elements of $A \cap B$ and describe this set.
(c) If two judges are to be elected, write the set that contains all possible pairs of winners.

National health care Suppose that the table below summarizes the opinions of various groups on the issue of national health care. Use this table for problems 53 and 54.

| | Whites | | Non-whites | | |
Opinion	Rep.	Dem.	Rep.	Dem.	Total
Favor	100	250	30	200	580
Oppose	250	150	10	10	420
Total	350	400	40	210	1000

53. Identify the number of individuals in each of the following sets.
(a) Republicans and those who favor national health care
(b) Republicans or those who favor national health care
(c) White Republicans or those who oppose national health care

54. Identify the number of individuals in each of the following sets.
(a) Whites and those who oppose national health care
(b) Whites or those who oppose national health care
(c) Non-white Democrats and those who favor national health care

55. **Languages** A survey of 100 aides at the United Nations revealed that 65 could speak English, 60 could speak French, and 40 could speak both English and French.
(a) Draw a Venn diagram representing the 100 aides. Use E to represent English-speaking and F to represent French-speaking aides.
(b) How many aides are in $E \cap F$?
(c) How many aides are in $E \cup F$?
(d) How many aides are in $E \cap F'$?

56. **Advertising** Suppose that a survey of 100 advertisers in *U.S. News, These Times,* and *World* found the following.

14 advertised in all three
30 advertised in *Times* and *U.S. News*
26 advertised in *World* and *U.S. News*
27 advertised in *World* and *Times*
60 advertised in *Times*
52 advertised in *U.S. News*
50 advertised in *World*

Draw a Venn diagram representing this information and use it to answer the following.
(a) How many advertised in none of these?
(b) How many advertised only in *Times?*
(c) How many advertised in *U.S. News* or *Times?*

57. **College enrollments** Records at a small college show the following about the enrollments of 100 freshmen in mathematics, fine arts, and economics.

38 take math
42 take fine arts
20 take economics
4 take economics and fine arts
15 take math and economics
9 take math and fine arts
12 take math and economics but not fine arts

Draw a Venn diagram representing this information and label all the areas. Use this diagram to answer the following.
(a) How many take none of these three courses?
(b) How many take math or economics?
(c) How many take exactly one of these three courses?

58. **Survey analysis** In a survey of the dining preferences of 110 dormitory students at the end of the spring semester, the following facts were discovered about Adam's Lunch (AL), Pizza Tower (PT), and the Student Union.

   30 liked AL but not PT
   21 liked AL only
   63 liked AL
   58 liked PT
   27 liked the Union
   25 liked PT and AL but not the Union
   18 liked PT and the Union

Draw a Venn diagram representing this survey and label all the areas. Use this diagram to answer the following.
(a) How many liked PT or the Union?
(b) How many liked all three?
(c) How many liked only the Union?

59. **Blood types** Blood types are determined by the presence or absence of three antigens: A antigen, B antigen, and an antigen called the Rh factor. The resulting blood types are classified as follows.

   type A if the A antigen is present
   type B if the B antigen is present
   type AB if both the A and B antigens are present
   type O if neither of the A or B antigens is present

These types are further classified as *Rh-positive* if the Rh-factor antigen is present and *Rh-negative* otherwise.
(a) Draw a Venn diagram that illustrates this classification scheme.
(b) Identify the blood type determined by each region of the Venn diagram (such as A^+ to indicate type A, Rh-positive).
(c) Use a library or another source to find what percentage of the U.S. population has each blood type.

0.2 THE REAL NUMBERS

This text uses the set of **real numbers** as the universal set. We can represent the real numbers along a line called the **real number line.** This number line is a picture, or graph, of the real numbers. Each point on the number line corresponds to exactly one real number and each real number can be located at exactly one point. Thus two real numbers are said to be equal whenever they are represented by the same point on the number line. The equation $a = b$ (a equals b) means that the symbols a and b represent the same real number. Thus $3 + 4 = 7$ means that $3 + 4$ and 7 represent the same number. Table 0.1 lists special subsets of the real numbers.

Table 0.1 Subsets of the set of real numbers

	Description	*Example (some elements shown)*
Natural numbers	$\{1, 2, 3, \ldots\}$ The counting numbers.	
Integers	$\{\ldots, -2, -1, 0, 1, 2, \ldots\}$ The natural numbers, 0, and the negatives of the natural numbers.	
Rational numbers	All numbers that can be written as the ratio of two integers, a/b, with $b \neq 0$. These numbers have decimal representations that either terminate or repeat.	

Table 0.1 continued

Irrational numbers	Those real numbers that *cannot* be written as the ratio of two integers. Irrational numbers have decimal representations that neither terminate nor repeat.	
Real numbers	The set of all rational and irrational numbers (the entire number line).	

The properties of the real numbers are fundamental to the study of algebra. These properties follow.

Properties of the Real Numbers

Let a, b, and c denote real numbers.

1. Addition and multiplication are commutative.
$$a + b = b + a \qquad ab = ba$$

2. Addition and multiplication are associative.
$$(a + b) + c = a + (b + c) \qquad (ab)c = a(bc)$$

3. The additive identity is 0.
$$a + 0 = 0 + a = a$$

4. The multiplicative identity is 1.
$$a \cdot 1 = 1 \cdot a = a$$

5. Each element a has an additive inverse, denoted by $-a$.
$$a + (-a) = -a + a = 0$$

Note that there is a difference between a negative number and the negative of a number.

6. Each nonzero element a has a multiplicative inverse, denoted by a^{-1}.
$$a \cdot a^{-1} = a^{-1} \cdot a = 1$$

Note that $a^{-1} = 1/a$.

7. Multiplication is distributive over addition.
$$a(b + c) = ab + ac$$

Notice that property 5 provides the means to subtract by defining $a - b = a + (-b)$ and property 6 provides a means to divide by defining $a \div b = a \cdot (1/b)$. The number 0 has no multiplicative inverse, so division by 0 is undefined.

We say that a is less than b (written $a < b$) if the point representing a is to the left of the point representing b on the real number line. For example, $4 < 7$ because 4 is to the left of 7 on the number line. We may also say that

7 is greater than 4 (written $7 > 4$). We may indicate that the number x is less than or equal to another number y by writing $x \leq y$. We may also indicate that p is greater than or equal to 4 by writing $p \geq 4$.

EXAMPLE 1

Use $<$ or $>$ notation to write
(a) 6 is greater than 5.   (b) 10 is less than 15.
(c) 3 is to the left of 8 on the number line.
(d) x is less than or equal to 12.

Solution

(a) $6 > 5$ (b) $10 < 15$ (c) $3 < 8$ (d) $x \leq 12$

The subset of the real numbers consisting of all real numbers x that lie between a and b, excluding a and b, can be denoted by the *double inequality* $a < x < b$ or by the **open interval** (a, b). It is called an open interval because neither of the endpoints is included in the interval. The **closed interval** $[a, b]$ represents the set of all real numbers x satisfying $a \leq x \leq b$. Intervals containing one endpoint, such as $(a, b]$ and $[a, b)$, are called **half-open intervals.**

We can use $[a, +\infty)$ to represent the inequality $x \geq a$, and $(-\infty, a)$ to represent $x < a$. In each of these cases, the symbols $+\infty$ and $-\infty$ are not real numbers, but represent the fact that x increases without bound ($+\infty$) or decreases without bound ($-\infty$). Table 0.2 summarizes three types of intervals.

Table 0.2 Intervals

Type of Interval	Inequality Notation	Interval Notation	Graph
Open interval	$x > a$	(a, ∞)	
	$x < b$	$(-\infty, b)$	
	$a < x < b$	(a, b)	
Half-open interval	$x \geq a$	$[a, \infty)$	
	$x \leq b$	$(-\infty, b]$	
	$a \leq x < b$	$[a, b)$	
	$a < x \leq b$	$(a, b]$	
Closed interval	$a \leq x \leq b$	$[a, b]$	

CHECKPOINT

1. Evaluate the following, if possible. For any that are meaningless, so state.
   (a) $\dfrac{4}{0}$ (b) $\dfrac{0}{4}$ (c) $\dfrac{4}{4}$ (d) $\dfrac{4-4}{4-4}$

2. For (a)–(d), write the inequality corresponding to the given interval and sketc
   its graph on a number line.
   (a) $(1, 3)$ (b) $(0, 3]$ (c) $[-1, \infty)$ (d) $(-\infty, 2)$

Sometimes we are interested in the *distance* a number is from the origi
(0) of the number line, without regard to direction. The distance a number
is from 0 on the number line is the **absolute value** of a, denoted $|a|$. Th
absolute value of any nonzero number is positive, and the absolute value (
0 is 0.

EXAMPLE 2

Evaluate the following.
(a) $|-4|$ (b) $|+2|$ (c) $|0|$ (d) $\big|-5 - |-3|\big|$

Solution

(a) $|-4| = +4 = 4$ (b) $|+2| = +2 = 2$
(c) $|0| = 0$ (d) $\big|-5 - |-3|\big| = |-5 - 3| = |-8| = 8$

Note that if a is a nonnegative number, then $|a| = a$, *but if a is negativ*
then $|a|$ is the positive number $(-a)$. Thus

Absolute Value	$$|a| = \begin{cases} a & \text{if } a \geq 0 \\ -a & \text{if } a < 0 \end{cases}$$

In performing computations with real numbers, it is important to remen
ber the rules for computations with signed numbers.

Operations with Real (Signed) Numbers

Procedure	Example
1. (a) To add two signed numbers with the same sign, add their absolute values and affix their common sign.	1. (a) $(+5) + (+6) = +11$ $\left(-\dfrac{1}{6}\right) + \left(-\dfrac{2}{6}\right) = -\dfrac{3}{6} = -\dfrac{1}{2}$
(b) To add two signed numbers with unlike signs, find the difference of their absolute values and affix the sign of the number with the larger absolute value.	(b) $(-4) + (+3) = -1$ $(+5) + (-3) = +2$ $\left(-\dfrac{11}{7}\right) + (+1) = -\dfrac{4}{7}$

2. To subtract one signed number from another, change the sign of the number being subtracted and proceed as in addition.

3. (a) The product of two numbers with like signs is positive.

   (b) The product of two numbers with unlike signs is negative.

4. (a) The quotient of two numbers with like signs is positive.

   (b) The quotient of two numbers with unlike signs is negative.

2. $(-9) - (-8) = (-9) + (+8) = -1$
   $16 - (8) = 16 + (-8) = +8$

3. (a) $(-3)(-4) = +12$
   $\left(+\dfrac{3}{4}\right)(+4) = +3$

   (b) $5(-3) = -15$
   $(-3)(+4) = -12$

4. (a) $(-14) \div (-2) = +7$
   $+36/4 = +9$

   (b) $(-28)/4 = -7$
   $45 \div (-5) = -9$

When two or more operations with real numbers are indicated in an evaluation, it is important that everyone agree upon the order in which the operations are performed so that a unique result is guaranteed. The following **order of operations** is universally accepted.

1. Perform operations within parentheses.
2. Find indicated powers ($2^3 = 2 \cdot 2 \cdot 2 = 8$).
3. Perform multiplications and divisions from left to right.
4. Perform additions and subtractions from left to right.

EXAMPLE 3
Evaluate the following.
(a) $-4 + 3$ (b) $-4^2 + 3$
(c) $(-4 + 3)^2 + 3$ (d) $6 \div 2(2 + 1)$

Solution
(a) -1 (b) $-16 + 3 = -13$
(c) $(-1)^2 + 3 = 1 + 3 = 4$ (d) $6 \div 2(3) = 3 \cdot 3 = 9$

CHECKPOINT

True or false:
3. $-(-5)^2 = 25$
4. $|4 - 6| = |4| - |6|$
5. $9 - 2(2)(-10) = 7(2)(-10) = -140$

Operations with real numbers can be performed on a calculator. Calculator types vary widely, from those that perform only the four arithmetic operations to those that perform many operations automatically and those that have graphing capabilities and can be programmed.

The text assumes that you have a scientific or graphing calculator. A discussion of some of the capabilities of graphics calculators and graphing utilities may be found in Section 1.3.

CHECKPOINT SOLUTIONS

1. (a) Meaningless. A denominator of zero means division by zero, which is undefined.

   (b) $\frac{0}{4} = 0$. A numerator of zero (when the denominator is not zero) means the fraction has value 0.

   (c) $\frac{4}{4} = 1$

   (d) Meaningless. The denominator is zero.

2. (a) $1 < x < 3$

   (b) $0 < x \le 3$

   (c) $-1 \le x < \infty$ or $x \ge -1$

   (d) $-\infty < x < 2$ or $x < 2$

3. False. $-(-5)^2 = (-1)(-5)^2 = (-1)(25) = -25$. Exponentiation has priority and applies only to -5.

4. False. $|4 - 6| = |-2| = 2$ and $|4| - |6| = 4 - 6 = -2$.

5. False. Without parentheses, multiplication has priority over subtraction. $9 - 2(2)(-10) = 9 - 4(-10) = 9 + 40 = 49$.

EXERCISE 0.2

In problems 1–2, indicate whether the given expression is one or more of the following types of numbers: rational, irrational, integer, natural. If the expression is meaningless, so state.

1. (a) $\dfrac{-\pi}{10}$ (b) -9 (c) $\dfrac{9}{3}$ (d) $\dfrac{4}{0}$

2. (a) $\dfrac{0}{6}$ (b) -1.2916 (c) 1.414 (d) $\dfrac{9}{6}$

Which property of real numbers is illustrated in each part of problems 3 and 4?

3. (a) $8 + 6 = 6 + 8$

   (b) $5(3 + 7) = 5 \cdot 3 + 5 \cdot 7$ (c) $-e \cdot 1 = -e$

4. (a) $6(4 \cdot 5) = (6 \cdot 4)(5)$ (b) $-15 + 0 = -15$

   (c) $\left(\dfrac{3}{2}\right)\left(\dfrac{2}{3}\right) = 1$

Insert the proper sign $<$, $=$, or $>$ to replace $\Box$ in problems 5–10.

5. $-14 \,\Box\, -3$

6. $\pi \,\Box\, 3.14$

7. $0.333 \,\Box\, \dfrac{1}{3}$

8. $\dfrac{1}{3} + \dfrac{1}{2} \,\Box\, \dfrac{5}{6}$

9. $|-3| + |5| \,\Box\, |-3 + 5|$

10. $|-9 - 3| \,\Box\, |-9| + |3|$

In problems 11–22, evaluate each expression.

11. $-3^2 + 10 \cdot 2$

12. $(-3)^2 + 10 \cdot 2$

13. $\dfrac{4 + 2^2}{2}$

14. $\dfrac{(4 + 2)^2}{2}$

15. $\dfrac{16 - (-4)}{8 - (-2)}$

16. $\dfrac{(-5)(-3) - (-2)(3)}{-9 + 2}$

17. $\dfrac{|5 - 2| - |-7|}{|5 - 2|}$

18. $\dfrac{|3 - |4 - 11\,||}{-|5^2 - 3^2|}$

19. $\dfrac{(-3)^2 - 2 \cdot 3 + 6}{4 - 2^2 + 3}$

20. $\dfrac{6^2 - 4(-3)(-2)}{6 - 6^2 \div 4}$

21. $\dfrac{-4^2 + 5 - 2 \cdot 3}{5 - 4^2}$

22. $\dfrac{3 - 2(5 - 2)}{(-2)^2 - 2^2 + 3}$

23. What part of the real number line corresponds to the interval $(-\infty, \infty)$?

24. Write the interval corresponding to $x \geq 0$.

In problems 25–28, express each inequality or graph using interval notation.

25. $1 < x \leq 3$

26. $-4 \leq x \leq 3$

27.
    $\quad 0 \quad 2 \quad 4 \quad 6 \quad 8 \quad 10 \quad 12$

28.
    $\quad -2 \quad 0 \quad 2 \quad 4 \quad 6$

In problems 29–32, write an inequality that describes each interval or graph.

29. $(-\infty, 5]$

30. $(-2, \infty)$

31.
    $\quad 0 \quad 2 \quad 4 \quad 6 \quad 8$

32.
    $\quad 0 \quad 2 \quad 4 \; 5 \; 6 \quad 8$

In problems 33–40, graph the subset of the real numbers that is represented by each of the following and write your answer in interval notation.

33. $(-\infty, 4) \cap (-3, \infty)$

34. $[-4, 17) \cap [-20, 10]$

35. $x > 4$ and $x \geq 0$

36. $x < 10$ and $x < -1$

37. $[0, \infty) \cup [-1, 5]$

38. $(-\infty, 4) \cup (0, 2)$

39. $x > 7$ or $x < 0$

40. $x > 4$ and $x < 0$

In problems 41–46, use your calculator to evaluate each of the following. List all the digits on your display in the answer.

41. $\dfrac{-1}{25916.8}$

42. $\dfrac{51.412}{127.01}$

43. $(3.679)^7$

44. $(1.28)^{10}$

45. $\dfrac{-1}{(-0.3019)^5}$

46. $\dfrac{-1}{(-3.8)^{12}}$

APPLICATIONS

47. **Cigarette use** Using data from the National Institute on Drug Abuse, the percentage P of high school seniors who have tried cigarettes can be modeled by

$$P = 75.4509 - 0.706948t,$$

where t is the number of years past 1975. We can use this model to approximate the percentage of high school seniors in 1985 who had tried cigarettes by using $t = 10$ (because 1985 is 10 years after 1975).
   (a) Approximate the 1985 percentage.
   (b) Approximate the 1990 and 1995 percentages.
   (c) What seems to be happening to the percentage of high school seniors who have tried cigarettes?
   (d) While this model may accurately reflect cigarette use by high school seniors from 1975 to the present, it will certainly be inappropriate if $P < 0$. Why?
   (e) Find any value of t that makes $P < 0$. To what year does it correspond?

48. **Tax burden** According to data from the Interval Revenue Service, the per capita federal tax burden T (in hundreds of dollars) can be modeled by

$$T = 20.37 + 1.834t,$$

where t is the number of years past 1980.
   (a) What t-value represents the year 1990?
   (b) Approximate the per capita tax burden in 1990. Remember that T is in hundreds of dollars.
   (c) Estimate the per capita tax burden in the year 2000.

0.3 INTEGRAL EXPONENTS

If \$1000 is placed in a 5-year savings certificate that pays an interest rate of 10% per year, compounded annually, then the amount returned after 5 years is given by

$$1000(1.1)^5.$$

The 5 in this expression is an **exponent.** Exponents provide an easier way to denote certain multiplications. For example

$$(1.1.)^5 = (1.1)(1.1)(1.1)(1.1)(1.1).$$

An understanding of the properties of exponents is fundamental to the algebra needed to study functions and solve equations. Furthermore, the definition of exponential and logarithmic functions and many of the techniques in calculus also require an understanding of the properties of exponents.

For any real number a,

$$a^2 = a \cdot a, \quad a^3 = a \cdot a \cdot a, \quad \text{and} \quad a^n = a \cdot a \cdot a \cdot \ldots \cdot a \quad (n \text{ factors})$$

for any positive integer n. The positive integer n is called the **exponent,** the number a is called the **base,** and a^n is read "a to the nth power."

Note that $4a^n$ means $4(a^n)$, which is different from $(4a)^n$. The 4 is the coefficient of a^n in $4a^n$. Note also that $-x^n$ is not equivalent to $(-x)^n$ when n is even. For example, $-3^4 = -81$, but $(-3)^4 = 81$.

Some of the rules of exponents follow.

Positive Integer Exponents

For any real numbers a and b and positive integers m and n,

1. $a^m \cdot a^n = a^{m+n}$

2. For $a \neq 0$, $\quad \dfrac{a^m}{a^n} = \begin{cases} a^{m-n} & \text{if } m > n \\ 1 & \text{if } m = n \\ 1/a^{n-m} & \text{if } m < n \end{cases}$

3. $(ab)^m = a^m b^m$

4. $\left(\dfrac{a}{b}\right)^m = \dfrac{a^m}{b^m} \quad (b \neq 0)$

5. $(a^m)^n = a^{mn}$

EXAMPLE 1

Use properties of positive integer exponents to rewrite each of the following. Assume all denominators are nonzero.

(a) $\dfrac{5^6}{5^4}$ (b) $\dfrac{x^2}{x^5}$ (c) $\left(\dfrac{x}{y}\right)^4$ (d) $(3x^2y^3)^4$ (e) $3^3 \cdot 3^2$

Solution

(a) $\dfrac{5^6}{5^4} = 5^{6-4} = 5^2$ (b) $\dfrac{x^2}{x^5} = \dfrac{1}{x^{5-2}} = \dfrac{1}{x^3}$

(c) $\left(\dfrac{x}{y}\right)^4 = \dfrac{x^4}{y^4}$ (d) $(3x^2y^3)^4 = 3^4(x^2)^4(y^3)^4 = 81x^8y^{12}$

(e) $3^3 \cdot 3^2 = 3^{3+2} = 3^5$

For certain calculus operations, use of negative exponents is necessary in order to write problems in the proper form. We can extend the rules for positive integer exponents to all integers by defining a^0 and a^{-n}. Clearly $a^m \cdot a^0$ should equal $a^{m+0} = a^m$, and it will if $a^0 = 1$.

Zero Exponent   For any nonzero real number a, we define $a^0 = 1$. We leave 0^0 undefined.

In Section 0.2, we defined a^{-1} as $1/a$ for $a \neq 0$, so we define a^{-n} as $(a^{-1})^n$.

Negative Exponents

$$a^{-n} = (a^{-1})^n = \left(\dfrac{1}{a}\right)^n = \dfrac{1}{a^n} \qquad (a \neq 0)$$

$$\left(\dfrac{a}{b}\right)^{-n} = \left[\left(\dfrac{a}{b}\right)^{-1}\right]^n = \left(\dfrac{b}{a}\right)^n \qquad (a \neq 0, b \neq 0)$$

EXAMPLE 2

Write each of the following without exponents:

(a) $6 \cdot 3^0$ (b) 6^{-2}

(c) $\left(\dfrac{1}{3}\right)^{-1}$ (d) $-\left(\dfrac{2}{3}\right)^{-4}$

(e) $(-4)^{-2}$

Solution

(a) $6 \cdot 3^0 = 6 \cdot 1 = 6$ (b) $6^{-2} = \dfrac{1}{6^2} = \dfrac{1}{36}$

(c) $\left(\dfrac{1}{3}\right)^{-1} = \dfrac{3}{1} = 3$ (d) $-\left(\dfrac{2}{3}\right)^{-4} = -\left(\dfrac{3}{2}\right)^4 = \dfrac{-81}{16}$

(e) $(-4)^{-2} = \dfrac{1}{(-4)^2} = \dfrac{1}{16}$

Using the definitions of zero and negative exponents enables us to extend the rules of exponents to all integers and to express them more simply.

Rules of Exponents For real numbers a and b and *integers* m and n,

1. $a^m \cdot a^n = a^{m+n}$ 2. $a^m/a^n = a^{m-n}$ $(a \neq 0)$

3. $(ab)^m = a^m b^m$ 4. $(a^m)^n = a^{mn}$

5. $(a/b)^m = a^m/b^m$ $(b \neq 0)$ 6. $a^0 = 1$ $(a \neq 0)$

7. $a^{-n} = 1/a^n$ $(a \neq 0)$ 8. $(a/b)^{-n} = (b/a)^n$ $(a, b \neq 0)$

Throughout the remainder of the text, we will assume all expressions are defined.

EXAMPLE 3

Use the rules of exponents and the definitions of a^0 and a^{-n} to simplify each of the following with positive exponents.

(a) $2(x^2)^{-2}$ (b) $x^{-2} \cdot x^{-5}$ (c) $\dfrac{x^{-8}}{x^{-4}}$ (d) $\left(\dfrac{2x^3}{3x^{-5}}\right)^{-2}$

Solution

(a) $2(x^2)^{-2} = 2x^{-4} = 2\left(\dfrac{1}{x^4}\right) = \dfrac{2}{x^4}$

(b) $x^{-2} \cdot x^{-5} = x^{-2-5} = x^{-7} = \dfrac{1}{x^7}$

(c) $\dfrac{x^{-8}}{x^{-4}} = x^{-8-(-4)} = x^{-4} = \dfrac{1}{x^4}$

(d) $\left(\dfrac{2x^3}{3x^{-5}}\right)^{-2} = \left(\dfrac{2x^8}{3}\right)^{-2} = \left(\dfrac{3}{2x^8}\right)^2 = \dfrac{9}{4x^{16}}$

CHECKPOINT

1. Complete the following.

   (a) $x^3 \cdot x^8 = x^?$ (b) $x \cdot x^4 \cdot x^{-3} = x^?$ (c) $\dfrac{1}{x^4} = x^?$

   (d) $x^{24} \div x^{-3} = x^?$ (e) $(x^4)^2 = x^?$ (f) $(2x^4y)^3 = ?$

2. True or false:

   (a) $3x^{-2} = \dfrac{1}{9x^2}$ (b) $-x^{-4} = \dfrac{-1}{x^4}$ (c) $x^{-3} = -x^3$

3. Evaluate the following, if possible. For any that are meaningless, so state. Assume $x > 0$.
   (a) 0^4 (b) 0^0 (c) x^0
   (d) 0^x (e) 0^{-4} (f) -5^{-2}

EXAMPLE 4

Write $(x^2y)/(9wz^3)$ with all factors in the numerator.

Solution

$$\frac{x^2y}{9wz^3} = x^2y\left(\frac{1}{9wz^3}\right) = x^2y\left(\frac{1}{9}\right)\left(\frac{1}{w}\right)\left(\frac{1}{z^3}\right) = x^2y \cdot 9^{-1}w^{-1}z^{-3}$$
$$= 9^{-1}x^2yw^{-1}z^{-3}$$

EXAMPLE 5

Simplify the following so all exponents are positive.

(a) $(2^3x^{-4}y^5)^{-2}$ (b) $\dfrac{2x^4(x^2y)^0}{(4x^{-2}y)^2}$

Solution

(a) $(2^3x^{-4}y^5)^{-2} = 2^{-6}x^8y^{-10} = \dfrac{1}{2^6} \cdot x^8 \cdot \dfrac{1}{y^{10}} = \dfrac{x^8}{64y^{10}}$

(b) $\dfrac{2x^4(x^2y)^0}{(4x^{-2}y)^2} = \dfrac{2x^4 \cdot 1}{4^2x^{-4}y^2} = \dfrac{2}{4^2} \cdot \dfrac{x^4}{x^{-4}} \cdot \dfrac{1}{y^2} = \dfrac{2}{16} \cdot \dfrac{x^8}{1} \cdot \dfrac{1}{y^2} = \dfrac{x^8}{8y^2}$

CHECKPOINT SOLUTIONS

1. (a) $x^3 \cdot x^8 = x^{3+8} = x^{11}$ (b) $x \cdot x^4 \cdot x^{-3} = x^{1+4+(-3)} = x^2$

(c) $\dfrac{1}{x^4} = x^{-4}$ (d) $x^{24} \div x^{-3} = x^{24-(-3)} = x^{27}$

(e) $(x^4)^2 = x^{(4)(2)} = x^8$

(f) $(2x^4y)^3 = 2^3(x^4)^3y^3 = 8x^{12}y^3$

2. (a) False. $3x^{-2} = 3\left(\dfrac{1}{x^2}\right) = \dfrac{3}{x^2}$

(b) True. $-x^{-4} = (-1)\left(\dfrac{1}{x^4}\right) = \dfrac{-1}{x^4}$

(c) False. $x^{-3} = \dfrac{1}{x^3}$

3. (a) $0^4 = 0$

(b) Meaningless. 0^0 is undefined.

(c) $x^0 = 1$ when $x \neq 0$

(d) $0^x = 0$ since $x > 0$

(e) Meaningless. 0^{-4} would be $\dfrac{1}{0^4}$, which is not defined.

(f) $-5^{-2} = (-1)\left(\dfrac{1}{5^2}\right) = \dfrac{-1}{25}$

EXERCISE 0.3

Evaluate in problems 1–8. Write all answers without using exponents.

1. $(-4)^3$

2. -5^3

3. -2^4

4. $(-2)^5$

5. 3^{-2}

6. 6^{-1}

7. $-\left(\dfrac{3}{2}\right)^2$

8. $\left(\dfrac{2}{3}\right)^3$

In problems 9–16, use rules of exponents to simplify the expressions. Express answers with positive exponents.

9. $6^5 \cdot 6^3$

10. $\dfrac{7^8}{7^3}$

11. $\dfrac{10^8}{10^9}$

12. $\dfrac{5^4}{(5^{-2} \cdot 5^3)}$

13. $(3^3)^3$

14. $(2^{-3})^{-2}$

15. $\left(\dfrac{2}{3}\right)^{-2}$

16. $\left(\dfrac{-2}{5}\right)^{-4}$

In problems 17–20, simplify by expressing answers with positive exponents.

17. $(x^2)^{-3}$

18. x^{-4}

19. $xy^{-2}z^0$

20. $(xy^{-2})^0$

In problems 21–34, use the rules of exponents to simplify so only positive exponents remain.

21. $x^3 \cdot x^4$

22. $a^5 \cdot a$

23. $x^{-5} \cdot x^3$

24. $y^{-5} \cdot y^{-2}$

25. $\dfrac{x^8}{x^4}$

26. $\dfrac{a^5}{a^{-1}}$

27. $\dfrac{y^5}{y^{-7}}$

28. $\dfrac{y^{-3}}{y^{-4}}$

29. $(x^4)^3$

30. $(y^3)^{-2}$

31. $(xy)^2$

32. $(2m)^3$

33. $\left(\dfrac{2}{x}\right)^4$

34. $\left(\dfrac{8}{a^3}\right)^3$

In problems 35–46, compute and simplify so only positive exponents remain.

35. $(2x^{-2}y)^{-4}$

36. $(-32x^5)^{-3}$

37. $(-8a^{-3}b^2)(2a^5b^{-4})$

38. $(-3m^2y^{-1})(2m^{-3}y^{-1})$

39. $(2x^{-2}) \div (x^{-1}y^2)$

40. $(-8a^{-3}b^2c) \div (2a^5b^4)$

41. $\left(\dfrac{x^3}{y^{-2}}\right)^{-3}$

42. $\left(\dfrac{x^{-2}}{y}\right)^{-3}$

43. $\left(\dfrac{a^{-2}b^{-1}c^{-4}}{a^4b^{-3}c^0}\right)^{-3}$

44. $\left(\dfrac{4x^{-1}y^{-40}}{2^{-2}x^4y^{-10}}\right)^{-2}$

45. (a) $\dfrac{2x^{-2}}{(2x)^2}$

(b) $\dfrac{(2x)^{-2}}{(2x)^2}$

(c) $\dfrac{2x^{-2}}{2x^2}$

(d) $\dfrac{2x^{-2}}{(2x)^{-2}}$

46. (a) $\dfrac{2^{-1}x^{-2}}{(2x)^2}$

(b) $\dfrac{2^{-1}x^{-2}}{2x^2}$

(c) $\dfrac{(2x^{-2})^{-1}}{(2x)^{-2}}$

(d) $\dfrac{(2x^{-2})^{-1}}{2x^2}$

In many applications it is often necessary to write expressions in the form cx^n, where c is a constant. In problems 47–54, write the expressions in this form.

47. $\dfrac{1}{x}$

48. $\dfrac{1}{x^2}$

49. $(2x)^3$

50. $(3x)^2$

51. $\dfrac{1}{(4x^2)}$

52. $\dfrac{3}{(2x^4)}$

53. $\left(\dfrac{-x}{2}\right)^3$

54. $\left(\dfrac{-x}{3}\right)^2$

In problems 55–58, use a scientific calculator to evaluate the indicated powers.

55. 1.2^4

56. $(-3.7)^3$

57. $(-6)^5$

58. $(-.8)^{-9}$

APPLICATIONS

Compound interest If \$$P$ is invested for n years at rate i (as a decimal), compounded annually, the future value that accrues is given by $A = P(1 + i)^n$, and the interest earned is $I = A - P$. In problems 59–62, find A and I for the given P, n, and i.

59. \$1200 for 5 years at 12%

60. \$1800 for 7 years at 10%

61. \$5000 for 6 years at 11.5%

62. \$800 for 20 years at 10.5%

63. ***Investing*** Using data from the Investment Company Institute (published in *Investment Digest* of Valic Co., Summer 1992), the number of mutual funds N can be modeled with the formula

$$N = 4.8139(1.9387)^t,$$

where t is the number of decades past 1900.
(a) What t-value corresponds to 1950?
(b) Approximate the number of mutual funds in 1950.
(c) Approximate the number of mutual funds in 1990.
(d) Estimate the number of mutual funds in the year 2000.

64. ***Health care expenditures*** Using data from the Congressional Budget Office (reported in *Newsweek,* October 4, 1993), the national health care expenditure H (in billions of dollars) can be modeled with the formula

$$H = 45(1.094)^t,$$

where t is the number of years past 1960.
(a) What t-value corresponds to 1970?
(b) Approximate the national health care expenditure in 1970.
(c) Approximate the national health care expenditure in 1993.
(d) Estimate the national health care expenditure in 2005.

0.4 RADICALS AND RATIONAL EXPONENTS

A process closely linked to that of raising numbers to powers is that of extracting roots. From geometry we know that if an edge of a cube is x units, its volume is x^3 cubic units. Reversing this process, we determine that if the volume of a cube is V cubic units, the length of an edge is the cube root of V, denoted

$$\sqrt[3]{V} \text{ units.}$$

When we seek the **cube root** of a number such as 8 (written $\sqrt[3]{8}$), we are looking for a real number whose cube equals 8. Since $2^3 = 8$, we know that $\sqrt[3]{8} = 2$. Similarly, $\sqrt[3]{-27} = -3$ since $(-3)^3 = -27$. The expression $\sqrt[n]{a}$ is called a **radical,** where $\sqrt{}$ is the **radical sign,** n the **index,** and a the **radicand.** When no index is indicated, the index is assumed to be 2 and the expression is called a **square root;** thus, $\sqrt{4}$ is the square root of 4 and represents the positive number whose square is 4.

Only one real number satisfies $\sqrt[n]{a}$ for a real number a and an odd number n; we call that number the **principal nth root,** or, more simply, the **nth root.**

For an even index n, there are two possible cases:

1. If a is negative, there is no real number equal to $\sqrt[n]{a}$. For example, there are no real numbers that equal $\sqrt{-4}$ or $\sqrt[4]{-16}$ because there is no real number b such that $b^2 = -4$ or $b^4 = -16$. In this case, we say $\sqrt[n]{a}$ is not a real number.

2. If a is positive, there are 2 real numbers whose nth power equals a. For example, $3^2 = 9$ and $(-3)^2 = 9$. In order to have a unique nth root, we define the (principal) nth root, $\sqrt[n]{a}$, as the *positive* number b satisfying $b^n = a$.

We summarize this discussion as follows.

nth Root of a   The (**principal**) **nth root** of a real number is defined as

$$\sqrt[n]{a} = b \quad \text{only} \quad \text{if } a = b^n$$

subject to the following conditions:

	$a = 0$	$a > 0$	$a < 0$
n even	$\sqrt[n]{a} = 0$	$\sqrt[n]{a} > 0$	$\sqrt[n]{a}$ not real
n odd	$\sqrt[n]{a} = 0$	$\sqrt[n]{a} > 0$	$\sqrt[n]{a} < 0$

When we are asked for the root of a number, we give the principal root.

EXAMPLE 1

Find the roots, if they are real numbers.

(a) $\sqrt[6]{64}$ (b) $-\sqrt{16}$ (c) $\sqrt[3]{-8}$ (d) $\sqrt{-16}$

Solution

(a) $\sqrt[6]{64} = 2$ because $2^6 = 64$

(b) $-\sqrt{16} = -(\sqrt{16}) = -4$

(c) $\sqrt[3]{-8} = -2$

(d) $\sqrt{-16}$ is not a real number because an even root of a negative number is not real.

To perform calculus operations on expressions involving radicals, it is sometimes necessary to rewrite the radicals in exponential form with fractional exponents.

We have stated that for $a \geq 0$ and $b \geq 0$,

$$\sqrt{a} = b \quad \text{only if} \quad a = b^2.$$

This means that $(\sqrt{a})^2 = b^2 = a$, or $(\sqrt{a})^2 = a$. In order to extend the properties of exponents to rational exponents, it is necessary to define

$$a^{1/2} = \sqrt{a}, \quad \text{so that} \quad (a^{1/2})^2 = a.$$

Exponent 1/n   For a positive integer n, we define

$$a^{1/n} = \sqrt[n]{a} \quad \text{if } \sqrt[n]{a} \text{ exists.}$$

Thus $(a^{1/n})^n = a^{(1/n) \cdot n} = a$.

Since we wish the properties established for integer exponents to extend to rational exponents, we make the following definitions.

Rational Exponents   For positive integer n and any integer m (with $a \neq 0$ when $m \leq 0$):

1. $a^{m/n} = (a^{1/n})^m = (\sqrt[n]{a})^m$.
2. $a^{m/n} = (a^m)^{1/n} = \sqrt[n]{a^m}$

if a is nonnegative when n is even.

Throughout the remaining discussion we assume all expressions are real.

EXAMPLE 2
Write the following in radical form and simplify.

(a) $16^{3/4}$ (b) $y^{-3/2}$ (c) $(6m)^{2/3}$

Solution

(a) $16^{3/4} = \sqrt[4]{16^3} = (\sqrt[4]{16})^3 = (2)^3 = 8$

(b) $y^{-3/2} = \dfrac{1}{y^{3/2}} = \dfrac{1}{\sqrt{y^3}}$

(c) $(6m)^{2/3} = \sqrt[3]{(6m)^2} = \sqrt[3]{36m^2}$

EXAMPLE 3
Write the following without radical signs.

(a) $\sqrt{x^3}$ (b) $\dfrac{1}{\sqrt[3]{b^2}}$ (c) $\sqrt[3]{(ab)^3}$

Solution

(a) $\sqrt{x^3} = x^{3/2}$ (b) $\dfrac{1}{\sqrt[3]{b^2}} = \dfrac{1}{b^{2/3}} = b^{-2/3}$

(c) $\sqrt[3]{(ab)^3} = (ab)^{3/3} = ab$

As noted, our definition of $a^{m/n}$ guarantees that the rules for exponents will apply to fractional exponents. Thus, we can perform operations with fractional exponents as we did with integer exponents.

EXAMPLE 4
Simplify the following expressions.

(a) $a^{1/2} \cdot a^{1/6}$ (b) $a^{3/4}/a^{1/3}$ (c) $(a^3 b)^{2/3}$
(d) $(a^{3/2})^{1/2}$ (e) $a^{-1/2} \cdot a^{-3/2}$

Solution

(a) $a^{1/2} \cdot a^{1/6} = a^{1/2+1/6} = a^{3/6+1/6} = a^{4/6} = a^{2/3}$

(b) $a^{3/4}/a^{1/3} = a^{3/4-1/3} = a^{9/12-4/12} = a^{5/12}$

(c) $(a^3 b)^{2/3} = (a^3)^{2/3} b^{2/3} = a^2 b^{2/3}$

(d) $(a^{3/2})^{1/2} = a^{(3/2)(1/2)} = a^{3/4}$

(e) $a^{-1/2} \cdot a^{-3/2} = a^{-1/2-3/2} = a^{-2} = 1/a^2$

CHECKPOINT

1. Which of the following are *not* real numbers?

   (a) $\sqrt[3]{-64}$ (b) $\sqrt{-64}$

   (c) $\sqrt{0}$ (d) $\sqrt[4]{1}$

   (e) $\sqrt[5]{-1}$ (f) $\sqrt[8]{-1}$

2. (a) Write as radicals: $x^{1/3}$, $x^{2/5}$, $x^{-3/2}$

   (b) Write with fractional exponents: $\sqrt[4]{x^3} = x^?$, $\dfrac{1}{\sqrt{x}} = \dfrac{1}{x^?} = x^?$

3. Evaluate the following.

   (a) $8^{2/3}$ (b) $(-8)^{2/3}$

   (c) $8^{-2/3}$ (d) $-8^{-2/3}$

4. Complete the following.

   (a) $x \cdot x^{1/3} \cdot x^3 = x^?$ (b) $x^2 \div x^{1/2} = x^?$

   (c) $(x^{-2/3})^{-3} = x^?$ (d) $x^{-3/2} \cdot x^{1/2} = x^?$

   (e) $x^{-3/2} \cdot x = x^?$ (f) $\left(\dfrac{x^4}{y^2}\right)^{3/2} = ?$

5. True or false:

   (a) $\dfrac{8x^{2/3}}{x^{-1/3}} = 4x$

   (b) $(16x^8 y)^{3/4} = 12x^6 y^{3/4}$

   (c) $\left(\dfrac{x^2}{y^3}\right)^{-1/3} = \left(\dfrac{y^3}{x^2}\right)^{1/3} = \dfrac{y}{x^{2/3}}$

We can perform operations with radicals by first rewriting in exponential form, performing the operations with exponents, and then converting the answer back to radical form. Another option is to apply directly the following rules for operations with radicals.

Rules for Radicals	**Examples**

Given that $\sqrt[n]{a}$ and $\sqrt[n]{b}$ are real,*

1. $\sqrt[n]{a^n} = (\sqrt[n]{a})^n = a$ 1. $\sqrt[5]{6^5} = (\sqrt[5]{6})^5 = 6$

2. $\sqrt[n]{a} \cdot \sqrt[n]{b} = \sqrt[n]{ab}$ 2. $\sqrt[3]{2}\,\sqrt[3]{4} = \sqrt[3]{8} = \sqrt[3]{2^3} = 2$

3. $\dfrac{\sqrt[n]{a}}{\sqrt[n]{b}} = \sqrt[n]{\dfrac{a}{b}}$ $(b \neq 0)$ 3. $\dfrac{\sqrt{18}}{\sqrt{2}} = \sqrt{\dfrac{18}{2}} = \sqrt{9} = 3$

* Note that this means $a \geq 0$ and $b \geq 0$ if n is even.

Let us consider Rule 1 for radicals more carefully. Notice that if n is even and $a < 0$, then $\sqrt[n]{a}$ is not real, and Rule 1 does not apply. For example,

$$\sqrt{(-2)^2} \neq -2 \quad \text{since} \quad \sqrt{(-2)^2} = \sqrt{4} = 2 = -(-2).$$

We can generalize this observation as follows: If $a < 0$, then $\sqrt{a^2} = -a < 0$, so

$$\sqrt{a^2} = \begin{cases} a & \text{if } a \geq 0 \\ -a & \text{if } a < 0 \end{cases}$$

This means

$$\sqrt{a^2} = |a|$$

EXAMPLE 5
Simplify:
(a) $\sqrt[3]{8^3}$ (b) $\sqrt[5]{x^5}$ (c) $\sqrt{x^2}$

Solution
(a) $\sqrt[3]{8^3} = 8$ by Rule 1 for radicals
(b) $\sqrt[5]{x^5} = x$ (c) $\sqrt{x^2} = |x|$

Up to now, to *simplify* a radical has meant to find the indicated root. More generally, a radical expression $\sqrt[n]{x}$ is considered simplified if x has no nth powers as factors. Rule 2 for radicals ($\sqrt[n]{a} \cdot \sqrt[n]{b} = \sqrt[n]{ab}$) provides a procedure for simplifying radicals.

EXAMPLE 6
Simplify the following radicals; assume the expressions are real numbers.
(a) $\sqrt{48x^5y^6}$ $(y \geq 0)$
(b) $\sqrt[3]{72a^3b^4}$

Solution

(a) To simplify $\sqrt{48x^5y^6}$, we first factor $48x^5y^6$ into perfect square factors and other factors. Then we apply Rule 2.

$$\sqrt{48x^5y^6} = \sqrt{16 \cdot 3 \cdot x^4xy^6} = \sqrt{16}\,\sqrt{x^4}\,\sqrt{y^6}\,\sqrt{3x} = 4x^2y^3\,\sqrt{3x}.$$

(b) In this case, we factor $72a^3b^4$ into factors that are perfect cubes and other factors.

$$\sqrt[3]{72a^3b^4} = \sqrt[3]{8 \cdot 9a^3b^3b} = \sqrt[3]{8} \cdot \sqrt[3]{a^3} \cdot \sqrt[3]{b^3} \cdot \sqrt[3]{9b} = 2ab\sqrt[3]{9b}.$$

Rule 2 for radicals also provides a procedure for multiplying two roots with the same index.

EXAMPLE 7

Multiply the following and simplify the answers, assuming nonnegative variables.

(a) $\sqrt[3]{2xy} \cdot \sqrt[3]{4x^2y}$ (b) $\sqrt{8xy^3z}\,\sqrt{4x^2y^3z^2}$

Solution

(a) $\sqrt[3]{2xy} \cdot \sqrt[3]{4x^2y} = \sqrt[3]{2xy \cdot 4x^2y} = \sqrt[3]{8x^3y^2} = \sqrt[3]{8} \cdot \sqrt[3]{x^3} \cdot \sqrt[3]{y^2}$
$\qquad = 2x\sqrt[3]{y^2}$

(b) $\sqrt{8xy^3z}\,\sqrt{4x^2y^3z^3} = \sqrt{32x^3y^6z^3} = \sqrt{16x^2y^6z^2}\,\sqrt{2xz}$
$\qquad = 4xy^3z\sqrt{2xz}$

Rule 3 for radicals $(\sqrt[n]{a}/\sqrt[n]{b} = \sqrt[n]{a/b})$ indicates how to find the quotient of two roots with the same index.

EXAMPLE 8

Find the quotients and simplify the answers, assuming nonnegative variables.

(a) $\dfrac{\sqrt[3]{32}}{\sqrt[3]{4}}$ (b) $\dfrac{\sqrt{16a^3x}}{\sqrt{2ax}}$

Solution

(a) $\dfrac{\sqrt[3]{32}}{\sqrt[3]{4}} = \sqrt[3]{\dfrac{32}{4}} = \sqrt[3]{8} = 2$

(b) $\dfrac{\sqrt{16a^3x}}{\sqrt{2ax}} = \sqrt{\dfrac{16a^3x}{2ax}} = \sqrt{8a^2} = 2a\sqrt{2}$

Occasionally, we wish to express a fraction containing radicals in an equivalent form that contains no radicals in the denominator. This is accomplished by multiplying the numerator *and* the denominator by the expression that will remove the radical. This process is called **rationalizing the denominator.**

EXAMPLE 9

Express each of the following with no radicals in the denominator. (Rationalize each denominator.)

(a) $\dfrac{15}{\sqrt{x}}$ (b) $\dfrac{2x}{\sqrt{18xy}}$ $(x, y > 0)$

(c) $\dfrac{3x}{\sqrt[3]{2x^2}}$

Solution

(a) We wish to create a perfect square under the radical in the denominator.

$$\frac{15}{\sqrt{x}} \cdot \frac{\sqrt{x}}{\sqrt{x}} = \frac{15\sqrt{x}}{x}$$

(b) $\dfrac{2x}{\sqrt{18xy}} \cdot \dfrac{\sqrt{2xy}}{\sqrt{2xy}} = \dfrac{2x\sqrt{2xy}}{\sqrt{36x^2y^2}} = \dfrac{2x\sqrt{2xy}}{6xy} = \dfrac{\sqrt{2xy}}{3y}$

(c) We wish to create a perfect cube under the radical in the denominator.

$$\frac{3x}{\sqrt[3]{2x^2}} \cdot \frac{\sqrt[3]{4x}}{\sqrt[3]{4x}} = \frac{3x\sqrt[3]{4x}}{\sqrt[3]{8x^3}} = \frac{3x\sqrt[3]{4x}}{2x} = \frac{3\sqrt[3]{4x}}{2}$$

CHECKPOINT

6. Simplify:
   (a) $\sqrt[7]{x^7}$ (b) $\left[\sqrt[5]{(x^2 + 1)^2}\right]^5$
   (c) $\sqrt{12xy^2} \cdot \sqrt{3x^2y}$

7. Rationalize the denominator of $\dfrac{x}{\sqrt{5x}}$.

It is also sometimes useful, especially in calculus, to *rationalize the numerator* of a fraction. For example, we can rationalize the numerator of

$$\frac{\sqrt[3]{4x^2}}{3x}$$

by multiplying the numerator and denominator by $\sqrt[3]{2x}$, which creates a perfect cube under the radical:

$$\frac{\sqrt[3]{4x^2}}{3x} \cdot \frac{\sqrt[3]{2x}}{\sqrt[3]{2x}} = \frac{\sqrt[3]{8x^3}}{3x\sqrt[3]{2x}} = \frac{2x}{3x\sqrt[3]{2x}} = \frac{2}{3\sqrt[3]{2x}}.$$

CHECKPOINT SOLUTIONS

1. Only *even* roots of negatives are not real numbers. Thus $\sqrt{-64}$ and $\sqrt[8]{-1}$ are not real numbers.

2. (a) $x^{1/3} = \sqrt[3]{x}$, $x^{2/5} = \sqrt[5]{x^2}$, $x^{-3/2} = \dfrac{1}{x^{3/2}} = \dfrac{1}{\sqrt{x^3}}$

   (b) $\sqrt[4]{x^3} = x^{3/4}$, $\dfrac{1}{\sqrt{x}} = \dfrac{1}{x^{1/2}} = x^{-1/2}$

3. (a) $8^{2/3} = (\sqrt[3]{8})^2 = 2^2 = 4$
   (b) $(-8)^{2/3} = (\sqrt[3]{-8})^2 = (-2)^2 = 4$

   (c) $8^{-2/3} = \dfrac{1}{8^{2/3}} = \dfrac{1}{4}$

   (d) $-8^{-2/3} = -\left(\dfrac{1}{8^{2/3}}\right) = -\dfrac{1}{4}$

4. (a) $x \cdot x^{1/3} \cdot x^3 = x^{1+1/3+3} = x^{13/3}$ (b) $x^2 \div x^{1/2} = x^{2-1/2} = x^{3/2}$
   (c) $(x^{-2/3})^{-3} = x^{(-2/3)(-3)} = x^2$ (d) $x^{-3/2} \cdot x^{1/2} = x^{-3/2+1/2} = x^{-1}$

   (e) $x^{-3/2} \cdot x = x^{-3/2+1} = x^{-1/2}$ (f) $\left(\dfrac{x^4}{y^2}\right)^{3/2} = \dfrac{(x^4)^{3/2}}{(y^2)^{3/2}} = \dfrac{x^6}{y^3}$

5. (a) False. $\dfrac{8x^{2/3}}{x^{-1/3}} = 8x^{2/3} \cdot x^{1/3} = 8x^{2/3 + 1/3} = 8x$

   (b) False. $(16x^8y)^{3/4} = 16^{3/4}(x^8)^{3/4}y^{3/4}$
   $$= (\sqrt[4]{16})^3 x^6 y^{3/4} = 8x^6 y^{3/4}$$

   (c) True.

6. (a) $\sqrt[7]{x^7} = x$ (b) $[\sqrt[5]{(x^2+1)^2}]^5 = (x^2+1)^2$
   (c) $\sqrt{12xy^2} \cdot \sqrt{3x^2y} = \sqrt{36x^3y^3} = \sqrt{36x^2y^2 \cdot xy}$
   $$= \sqrt{36x^2y^2}\,\sqrt{xy} = 6xy\,\sqrt{xy}$$

7. $\dfrac{x}{\sqrt{5x}} = \dfrac{x}{\sqrt{5x}} \cdot \dfrac{\sqrt{5x}}{\sqrt{5x}} = \dfrac{x\sqrt{5x}}{5x} = \dfrac{\sqrt{5x}}{5}$

EXERCISE 0.4

Unless stated otherwise, assume all variables are nonnegative and all denominators are nonzero.

In problems 1–12, find the powers and roots, if they are real numbers.

1. $\sqrt{256/9}$ 2. $\sqrt{1.44}$

3. $\sqrt[5]{-32^3}$ 4. $\sqrt[4]{-16^5}$

5. $16^{3/4}$ 6. $32^{3/5}$

7. $(-16)^{-3/2}$ 8. $-27^{-1/3}$

9. $*\left(\dfrac{8}{27}\right)^{-2/3}$ 10. $\left(\dfrac{4}{9}\right)^{3/2}$

11. (a) $8^{2/3}$ (b) $(-8)^{-2/3}$

12. (a) $8^{-2/3}$ (b) $-8^{2/3}$

In problems 13–16, replace each radical with a fractional exponent. Do not simplify.

13. $\sqrt{m^3}$
14. $\sqrt[3]{x^5}$
15. $\sqrt[4]{m^2 n^5}$
16. $\sqrt[5]{x^3}$

In problems 17–20, write in radical form. Do not simplify.

17. $x^{7/4}$
18. $y^{11/5}$
19. $-(1/4)x^{-5/4}$
20. $-x^{-5/3}$

In problems 21–34, use the properties of exponents to simplify each expression so only positive exponents remain.

21. $y^{1/4} \cdot y^{1/2}$
22. $x^{2/3} \cdot x^{1/5}$
23. $z^{3/4} \cdot z^4$

24. $x^{-2/3} \cdot x^2$
25. $y^{-3/2} \cdot y^{-1}$
26. $z^{-2} \cdot z^{5/3}$

27. $\dfrac{x^{1/3}}{x^{-2/3}}$
28. $\dfrac{x^{-1/2}}{x^{-3/2}}$
29. $\dfrac{y^{-5/2}}{y^{-2/5}}$

30. $\dfrac{x^{4/9}}{x^{1/12}}$
31. $(x^{2/3})^{3/4}$
32. $(x^{4/5})^3$

33. $(x^{-1/2})^2$
34. $(x^{-2/3})^{-2/5}$

In problems 35–40, simplify each expression by using the properties of radicals. Assume nonnegative variables.

35. $\sqrt{64x^4}$
36. $\sqrt[3]{-64x^6 y^3}$
37. $\sqrt{128x^4 y^5}$
38. $\sqrt[3]{54x^5 x^8}$
39. $\sqrt[3]{40x^8 y^5}$
40. $\sqrt{32x^5 y}$

In problems 41–48, perform the indicated operations and simplify.

41. $\sqrt{12x^3 y} \cdot \sqrt{3x^2 y}$
42. $\sqrt[3]{16x^2 y} \cdot \sqrt[3]{3x^2 y}$
43. $\sqrt{63x^5 y^3} \cdot \sqrt{28x^2 y}$
44. $\sqrt{10xz^{10}} \cdot \sqrt{30x^{17}z}$

45. $\dfrac{\sqrt{12x^3 y^{12}}}{\sqrt{27xy^2}}$
46. $\dfrac{\sqrt{250xy^7 z^4}}{\sqrt{18x^{17} y^2}}$

47. $\dfrac{\sqrt[4]{32a^9 b^5}}{\sqrt[4]{162a^{17}}}$
48. $\dfrac{\sqrt[3]{-16x^3 y^4}}{\sqrt[3]{128y^2}}$

In problems 49–54, rationalize each denominator and then simplify.

49. $\sqrt{2/3}$
50. $\sqrt{5/8}$
51. $\dfrac{\sqrt{m^2 x}}{\sqrt{mx^2}}$

52. $\dfrac{5x^3 w}{\sqrt{4xw^2}}$
53. $\dfrac{\sqrt[3]{m^2 x}}{\sqrt[3]{mx^5}}$
54. $\dfrac{\sqrt[4]{mx^3}}{\sqrt[4]{y^2 z^5}}$

In calculus it is frequently important to write an expression in the form cx^n, where c is a constant. In problems 55–58, write each expression in this form.

55. $\dfrac{-2}{3\sqrt[3]{x^2}}$
56. $\dfrac{-2}{3\sqrt[4]{x^3}}$
57. $3x\sqrt{x}$
58. $\sqrt{x} \cdot \sqrt[3]{x}$

In calculus problems, the answers are frequently expected to be in a simple form with a radical instead of an exponent. In problems 59–62, write each expression with radicals.

59. $\dfrac{3}{2}x^{1/2}$
60. $\dfrac{4}{3}x^{1/3}$

61. $\dfrac{1}{2}x^{-1/2}$
62. $\dfrac{-1}{2}x^{-3/2}$

APPLICATIONS

Half-life In problems 63 and 64, use the fact that the quantity of a radioactive substance after t years is given by $q = q_0(2^{-t/k})$, where q_0 is the original amount of radioactive material and k is its **half-life** (the number of years for half the radioactive substance to decay).

63. The half-life of strontium-90 is 25 years. Find the amount of strontium-90 remaining after 10 years if $q_0 = 98$ kg.

64. The half-life of carbon-14 is 5600 years. Find the amount of carbon-14 remaining after 10,000 years if $q_0 = 40.0$ g.

65. *Population growth* Suppose the formula for the growth of the population of a city for the next 10 years is given by

$$P = P_0(2.5)^{ht},$$

where P_0 is the population of the city at the present time and P is the population t years from now. If $h = 0.03$ and $P_0 = 30,000$, find P when $t = 10$.

66. *Advertising and sales* Suppose it has been determined that the sales at Ewing Gallery decline after the end of an advertising campaign, with daily sales given by

$$S = 2000(2^{-0.1x}),$$

where S is in dollars and x is the number of days after the campaign ends. What is the daily sales 10 days after the end of the campaign?

67. *Company growth* The growth of a company can be described by the equation

$$N = 500(0.02)^{0.7^t},$$

where t is the number of years the company has been in existence and N is the number of employees. What is the number of employees when $t = 0$? (This is the number of employees the company has when it starts.)

68. *Company growth* For the company in problem 67, what is the number of employees when $t = 5$?

0.5 OPERATIONS WITH ALGEBRAIC EXPRESSIONS

In algebra we are usually dealing with combinations of real numbers (such as 3, 6/7, $-\sqrt{2}$) and letters (such as x, a, m). Unless otherwise specified, the letters are symbols used to represent real numbers and are sometimes called **variables.** An expression obtained by performing additions, subtractions, multiplications, divisions, or extraction of roots with one or more real numbers or variables is called an **algebraic expression.** Unless otherwise specified, the variables represent all real numbers for which the algebraic expression is a real number. Examples of algebraic expressions are

$$3x + 2y, \quad \frac{x^3y + y}{x - 1}, \quad \text{and} \quad \sqrt{x} - 3.$$

Note that the variable x cannot be negative in the expression $\sqrt{x} - 3$, and that $(x^3y + y)/(x - 1)$ is not a real number when $x = 1$, because division by 0 is undefined.

Any product of a real number (called the **coefficient**) and one or more variables to powers is called a **term.** The sum of a finite number of terms with nonnegative integer powers on the variables is called a **polynomial.** If a polynomial contains only one variable x, then it is called a polynomial in x.

Polynominal in x   The general form of a **polynomial in x** is

$$a_nx^n + a_{n-1}x^{n-1} + \cdots + a_1x + a_0,$$

where each coefficient a_i is a real number and where $i = 0, 1, 2, \ldots, n$. If $a_n \neq 0$, the **degree** of the polynomial is n, and a_n is called the **leading coefficient.** The term a_0 is called the **constant term.**

Thus $4x^3 - 2x + 3$ is a third-degree polynomial in x with leading coefficient 4. If two or more variables are in a term, the degree of the term is the sum of the exponents of the variables. The degree of a nonzero constant term is zero. Thus, the degree of $4x^2y$ is $2 + 1 = 3$, the degree of $6xy$ is $1 + 1 = 2$, and the degree of 3 is 0. The **degree of a polynomial** containing one or more variables is the degree of the term in the polynomial having the highest degree. Therefore $2xy - 4x + 6$ is a second-degree polynomial.

A polynomial containing two terms is called a **binomial,** and one containing three terms is called a **trinomial.** A single-term polynomial is a **monomial.**

Since monomials and polynomials represent real numbers, the properties of real numbers can be used to add, subtract, multiply, divide, and simplify polynomials. For example, we can use the Distributive Law to add $3x$ and $2x$.

$$3x + 2x = (3 + 2)x = 5x$$

Similarly, $9xy - 3xy = (9 - 3)xy = 6xy$.

Terms with exactly the same variable factors are called **like terms.** We can add or subtract like terms by adding or subtracting the coefficients of the variables.

EXAMPLE 1

Compute $(4xy + 3x) + (5xy - 2x)$.

Solution

$$(4xy + 3x) + (5xy - 2x) = 4xy + 3x + 5xy - 2x$$
$$= 9xy + x$$

Subtraction of polynomials uses the Distributive Law to remove the parentheses.

EXAMPLE 2

Compute $(3x^2 + 4xy + 5y^2 + 1) - (6x^2 - 2xy + 4)$.

Solution

Removing the parentheses gives

$$3x^2 + 4xy + 5y^2 + 1 - 6x^2 + 2xy - 4,$$

which simplifies to

$$-3x^2 + 6xy + 5y^2 - 3.$$

Using the rules of exponents and the Commutative and Associative Laws for multiplication, we can multiply and divide monomials, as the following example shows.

EXAMPLE 3

Perform the indicated operations.

(a) $(8xy^3)(2x^3y)(-3xy^2)$ (b) $-15x^2y^3 \div (3xy^5)$

Solution

(a) $8 \cdot 2 \cdot (-3) \cdot x \cdot x^3 \cdot x \cdot y^3 \cdot y \cdot y^2 = -48x^5y^6$

(b) $\dfrac{-15x^2y^3}{3xy^5} = -\dfrac{15}{3} \cdot \dfrac{x^2}{x} \cdot \dfrac{y^3}{y^5} = -5 \cdot x \cdot \dfrac{1}{y^2} = -\dfrac{5x}{y^2}$

Symbols of grouping are used in algebra in the same way that they are in the arithmetic of real numbers. We have removed parentheses in the process of adding and subtracting polynomials. Other symbols of grouping, such as brackets, [ ], are treated the same as parentheses.

When there are two or more symbols of grouping involved, we may begin with the innermost and work outward.

EXAMPLE 4

Simplify $3x^2 - [2x^2 - (3x^2 - 2x)]$.

Solution

$$
\begin{aligned}
3x^2 - [2x - (3x^2 - 2x)] &= 3x^2 - [2x - 3x^2 + 2x] \\
&= 3x^2 - [4x - 3x^2] \\
&= 3x^2 - 4x + 3x^2 \\
&= 6x^2 - 4x
\end{aligned}
$$

By the use of the Distributive Law, we can multiply a binomial by a monomial. For example,

$$x(2x + 3) = x \cdot 2x + x \cdot 3 = 2x^2 + 3x.$$

We can extend the Distributive Law to multiply polynomials with more than two terms. For example,

$$5(x + y + 2) = 5x + 5y + 10.$$

EXAMPLE 5

Find the following products.

(a) $-4ab(3a^2b + 4ab^2 - 1)$
(b) $(4a + 5b + c)ac$

Solution

(a) $-4ab(3a^2b + 4ab^2 - 1) = -12a^3b^2 - 16a^2b^3 + 4ab$
(b) $(4a + 5b + c)ac = 4a \cdot ac + 5b \cdot ac + c \cdot ac = 4a^2c + 5abc + ac^2$

The Distributive Law can be used to show us how to multiply two polynomials. Consider the indicated multiplication $(a + b)(c + d)$. If we first treat the sum $(a + b)$ as a single quantity, then two successive applications of the Distributive Law give

$$(a + b)(c + d) = (a + b) \cdot c + (a + b) \cdot d = ac + bc + ad + bd.$$

Thus we see that the product can be found by multiplying $(a + b)$ by c, multiplying $(a + b)$ by d, and then adding the products. This is frequently set up as follows.

Product of Two Polynomials	
Procedure	**Example**
To multiply two polynomials:	Multiply $(3x + 4xy + 3y)$ by $(x - 2y)$.
1. Write one of the polynomials above the other.	1. $3x + 4xy + 3y$ $\underline{ x - 2y}$
2. Multiply each term of the top polynomial by each term of the bottom one, and write the similar terms of the product under one another.	2. $3x^2 + 4x^2y + 3xy$ $\underline{ - 6xy - 8xy^2 - 6y^2}$
3. Add like terms to simplify the product.	3. $3x^2 + 4x^2y - 3xy - 8xy^2 - 6y^2$

> **EXAMPLE 6**
> Multiply $(4x^2 + 3xy + 4x)$ by $(2x - 3y)$.
>
> *Solution*
>
> $$4x^2 + 3xy + 4x$$
> $$\underline{2x - 3y}$$
> $$8x^3 + 6x^2y + 8x^2$$
> $$\underline{ - 12x^2y - 9xy^2 - 12xy}$$
> $$8x^3 - 6x^2y + 8x^2 - 9xy^2 - 12xy$$

Since the multiplications we must perform often involve binomials, the following special products are worth remembering.

◢ **Special Products**   A. $(x + a)(x + b) = x^2 + (a + b)x + ab$
B. $(ax + b)(cx + d) = acx^2 + (ad + bc)x + bd$

It is easier to remember these two special products if we note their structure. We can obtain these products by finding the products of the First terms, Outside terms, Inside terms, and Last terms, and then adding the results. This is called the FOIL method of multiplying two binomials.

> **EXAMPLE 7**
> Multiply the following.
> (a) $(x - 4)(x + 3)$ (b) $(3x + 2)(2x + 5)$
>
> *Solution*
>
> $$\text{First}\quad \text{Outside}\quad \text{Inside}\quad \text{Last}$$
> (a) $(x - 4)(x + 3) = (x^2) + (3x) + (-4x) + (-12) = x^2 - x - 12$
> (b) $(3x + 2)(2x + 5) = (6x^2) + (15x) + (4x) + (10)$
> $$= 6x^2 + 19x + 10$$

Additional special products are as follows:

Additional Special Products

C. $(x + a)^2 = x^2 + 2ax + a^2$ binomial squared
D. $(x - a)^2 = x^2 - 2ax + a^2$ binomial squared
E. $(x + a)(x - a) = x^2 - a^2$ difference of two squares
F. $(x + a)^3 = x^3 + 3ax^2 + 3a^2x + a^3$ binomial cubed
G. $(x - a)^3 = x^3 - 3ax^2 + 3a^2x - a^3$ binomial cubed

EXAMPLE 8

Multiply the following.

(a) $(x + 5)^2$ (b) $(3x - 4y)^2$ (c) $(x - 2)(x + 2)$
(d) $(x^2 - y^3)^2$ (e) $(x + 4)^3$

Solution

(a) $(x + 5)^2 = x^2 + 2(5)x + 25 = x^2 + 10x + 25$
(b) $(3x - 4y)^2 = (3x)^2 - 2(3x)(4y) + (4y)^2 = 9x^2 - 24xy + 16y^2$
(c) $(x - 2)(x + 2) = x^2 - 4$
(d) $(x^2 - y^3)^2 = (x^2)^2 - 2(x^2)(y^3) + (y^3)^2 = x^4 - 2x^2y^3 + y^6$
(e) $(x + 4)^3 = x^3 + 3(4)(x^2) + 3(4^2)(x) + 4^3 = x^3 + 12x^2 + 48x + 64$

CHECKPOINT

1. Remove parentheses and combine like terms. $9x - 5x(x + 2) + 4x^2$
2. Find the following products.
   (a) $(2x + 1)(4x^2 - 2x + 1)$ (b) $(x + 3)^2$
   (c) $(3x + 2)(x - 5)$ (d) $(1 - 4x)(1 + 4x)$

All algebraic expressions can represent real numbers, so the techniques used to perform operations on polynomials and to simplify polynomials also apply to other algebraic expressions.

EXAMPLE 9

Perform the indicated operations.

(a) $3\sqrt{3} + 4x\sqrt{y} - 5\sqrt{3} - 11x\sqrt{y} - (\sqrt{3} - x\sqrt{y})$
(b) $x^{3/2}(x^{1/2} - x^{-1/2})$ (c) $(x^{1/2} - x^{1/3})^2$ (d) $(\sqrt{x} + 2)(\sqrt{x} - 2)$

Solution

(a) We remove parentheses, then combine the terms containing $\sqrt{3}$ and the terms containing $x\sqrt{y}$.

$$(3 - 5 - 1)\sqrt{3} + (4 - 11 + 1)x\sqrt{y} = -3\sqrt{3} - 6x\sqrt{y}$$

(b) $x^{3/2}(x^{1/2} - x^{-1/2}) = x^{3/2} \cdot x^{1/2} - x^{3/2} \cdot x^{-1/2} = x^2 - x$

(c) $(x^{1/2} - x^{1/3})^2 = (x^{1/2})^2 - 2x^{1/2}x^{1/3} + (x^{1/3})^2 = x - 2x^{5/6} + x^{2/3}$

(d) $(\sqrt{x} + 2)(\sqrt{x} - 2) = (\sqrt{x})^2 - (2)^2 = x - 4$

   In later chapters we will need to write problems in a simplified form so that we can perform certain operations on them. We can often use division of one polynomial by another to obtain the simplification, as shown below.

Division of Polynomials

Procedure	Example
To divide one polynomial by another:	Divide $4x^3 + 4x^2 + 5$ by $2x^2 + 1$.
1. Write with both polynomials in descending powers of a variable. Include missing terms with coefficient 0 in the dividend.	1. $2x^2 + 1\overline{)4x^3 + 4x^2 + 0x + 5}$
2. (a) Divide the highest power of the divisor into the highest power of the dividend, and write this partial quotient above the dividend. Multiply the partial quotient times the divisor, write the product under the dividend, and subtract, getting a new dividend.	2. (a) $\begin{array}{r} 2x \\ 2x^2 + 1\overline{)4x^3 + 4x^2 + 0x + 5} \\ \underline{4x^3 + 2x } \\ 4x^2 - 2x + 5 \end{array}$
(b) Repeat until the degree of the new dividend is less than the degree of the divisor. Any remainder is written over the divisor and added to the quotient.	(b) $\begin{array}{r} 2x + 2 \\ 2x^2 + 1\overline{)4x^3 + 4x^2 + 0x + 5} \\ \underline{4x^3 + 2x } \\ 4x^2 - 2x + 5 \\ \underline{4x^2 + 2} \\ -2x + 3 \end{array}$ Degree $(-2x + 3) <$ degree $(2x^2 + 1)$ Quotient: $2x + 2 + \dfrac{-2x + 3}{2x^2 + 1}$

EXAMPLE 10

Divide $(4x^3 - 13x - 22)$ by $(x - 3)$, $\qquad x \neq 3$.

Solution

$$\begin{array}{r} 4x^2 + 12x + 23 \\ x - 3\overline{)4x^3 + 0x^2 - 13x - 22} \\ \underline{4x^3 - 12x^2 } \\ 12x^2 - 13x - 22 \\ \underline{12x^2 - 36x } \\ 23x - 22 \\ \underline{23x - 69} \\ 47 \end{array}$$

$0x^2$ is inserted so that each power of x is present.

The quotient is $4x^2 + 12x + 23$, with remainder 47, or

$$4x^2 + 12x + 23 + \frac{47}{x - 3}$$

CHECKPOINT

3. Use long division to find $(x^3 + 2x + 7) \div (x - 4)$.

CHECKPOINT SOLUTIONS

1. $9x - 5x(x + 2) + 4x^2 = 9x - 5x^2 - 10x + 4x^2$
   $$= -x^2 - x$$

   Note that without parentheses around $9x - 5x$, multiplication has priority over subtraction.

2. (a) $4x^2 - 2x + 1$

   $\underline{2x + 1}$

   $8x^3 - 4x^2 + 2x$

   $\quad\quad 4x^2 - 2x + 1$

   $\underline{8x^3 \quad\quad\quad\quad + 1}$

   (b) $(x + 3)^2 = x^2 + 2(3x) + 3^2 = x^2 + 6x + 9$

   (c) $(3x + 2)(x - 5) = 3x^2 - 15x + 2x - 10 = 3x^2 - 13x - 10$

   (d) $(1 - 4x)(1 + 4x) = 1 - 16x^2$ (Note that this is different from $16x^2 - 1$.)

3. $\begin{array}{r} x^2 + 4x + 18 \\ x - 4{\overline{\smash{\big)}\,x^3 + 0x^2 + 2x + 7}} \\ \underline{x^3 - 4x^2} \\ 4x^2 + 2x + 7 \\ \underline{4x^2 - 16x} \\ 18x + 7 \\ \underline{18x - 72} \\ 79 \end{array}$ $\left(\text{The answer is } x^2 + 4x + 18 + \dfrac{79}{x - 4}.\right)$

EXERCISE 0.5

For each polynomial in problems 1–4, (a) give the degree of the polynomial, (b) give the coefficient (numerical) of the highest degree term, (c) give the constant term, and (d) decide whether it is a polynomial of one or several variables.

1. $10 - 3x - x^2$
2. $5x^4 - 2x^9 + 7$
3. $7x^2y - 14xy^3z$
4. $2x^5 + 7x^2y^3 - 5y^6$

The expressions in problems 5 and 6 are polynomials (that is, they have the form $a_n x^n + a_{n-1} x^{n-1} + \cdots + a_1 x + a_0$).

5. For $2x^5 - 3x^2 - 5$,

   (a) $2 = a_?$
   (b) $a_3 = ?$
   (c) $-3 = a_?$
   (d) $a_0 = ?$

6. For $5x^3 - 4x - 17$,

    (a) $a_3 = ?$ (b) $a_1 = ?$

    (c) $a_2 = ?$ (d) $-17 = a_?$

In problems 7–10, evaluate each algebraic expression at the indicated values of the variables.

7. $4x - x^2$ at $x = -2$

8. $3x^2 - 4y^2 - 2xy$ at $x = 3$ and $y = -4$

9. $\dfrac{2x - y}{x^2 - 2y}$ at $x = -5$ and $y = -3$

10. $\dfrac{16y}{1 - y}$ at $y = -3$

In problems 11–18, simplify by combining like terms.

11. $(16pq - 7p^2) + (5pq + 5p^2)$

12. $(3x^3 + 4x^2y^2) + (3x^2y^2 - 7x^3)$

13. $(4m^2 - 3n^2 + 5) - (3m^2 + 4n^2 + 8)$

14. $(4rs - 2r^2s - 11rs^2) - (11rs^2 - 2rs + 4r^2s)$

15. $-[8 - 4(q + 5) + q]$

16. $x^3 + [3x - (x^3 - 3x)]$

17. $x^2 - [x - (x^2 - 1) + 1 - (1 - x^2)] + x$

18. $y^3 - [y^2 - (y^3 + y^2)] - [y^3 + (1 - y^2)]$

In problems 19–64, perform the indicated operations and simplify.

19. $(5x^3)(7x^2)$

20. $(-3x^2y)(2xy^3)(4x^2y^2)$

21. $(39r^3s^2) \div (13r^2s)$

22. $(-15m^3n) \div (5mn^4)$

23. $ax^2(2x^2 + ax + ab)$

24. $-3(3 - x^2)$

25. $(3y + 4)(2y - 3)$

26. $(4x - 1)(x - 3)$

27. $(1 - 2x^2)(2 - x^2)$

28. $(x^3 + 3)(2x^3 - 5)$

29. $(4x + 3)^2$

30. $(2y + 5)^2$

31. $\left(x^2 - \dfrac{1}{2}\right)^2$

32. $(x^3y^3 - 0.3)^2$

33. $(2x + 1)(2x - 1)$

34. $(5y + 2)(5y - 2)$

35. $(0.1 - 4x)(0.1 + 4x)$

36. $\left(\dfrac{2}{3} + x\right)\left(\dfrac{2}{3} - x\right)$

37. $(x - 2)(x^2 + 2x + 4)$

38. $(a + b)(a^2 - ab + b^2)$

39. $(x^3 + 5x)(x^5 - 2x^3 + 5)$

40. $(x^3 - 1)(x^7 - 2x^4 - 5x^2 + 5)$

41. $(18m^2n + 6m^3n + 12m^4n^2) \div (6m^2n)$

42. $(16x^2 + 4xy^2 + 8x) \div (4xy)$

43. $(24x^8y^4 + 15x^5y - 6x^7y) \div (9x^5y^2)$

44. $(27x^2y^2 - 18xy + 9xy^2) \div (6xy)$

45. $(x + 1)^3$

46. $(x - 3)^3$

47. $(2x - 3)^3$

48. $(3x + 4)^3$

49. $(0.1x - 2)(x + 0.05)$

50. $(6.2x + 4.1)(6.2x - 4.1)$

51. $(x^3 + x - 1) \div (x + 2)$

52. $(x^5 + 5x - 7) \div (x + 1)$

53. $(x^4 + 3x^3 - x + 1) \div (x^2 + 1)$

54. $(x^3 + 5x^2 - 6) \div (x^2 - 2)$

55. $x^{1/2}(x^{1/2} + 2x^{3/2})$

56. $x^{-2/3}(x^{5/3} - x^{-1/3})$

57. $(x^{1/2} + 1)(x^{1/2} - 2)$

58. $(x^{1/3} - x^{1/2})(4x^{2/3} - 3x^{3/2})$

59. $(\sqrt{x} + 3)(\sqrt{x} - 3)$

60. $(x^{1/5} + x^{1/2})(x^{1/5} - x^{1/2})$

61. $(2x + 1)^{1/2}[(2x + 1)^{3/2} - (2x + 1)^{-1/2}]$

62. $(4x - 3)^{-5/3}[(4x - 3)^{8/3} + 3(4x - 3)^{5/3}]$

63. (a) $(3x - 2)^2 - 3x - 2(3x - 2) + 5$

    (b) $(3x - 2)^2 - (3x - 2)(3x - 2) + 5$

64. (a) $(2x - 3)(3x + 2) - (5x - 2)(x - 3)$

    (b) $2x - 3(3x + 2) - 5x - 2(x - 3)$

APPLICATIONS

65. **Revenue** A company sells its product for $55 per unit. Write an expression for the amount of money received (revenue) from the sale of x units of the product.

66. **Profit** Suppose a company's revenue R (in dollars) from the sale of x units of its product is given by

$$R = 215x.$$

Suppose further that the total costs C (in dollars) of producing those x units is given by

$$C = 65x + 15,000.$$

(a) If profit is revenue minus cost, find an expression for the profit from the production and sale of x units.

(b) Find the profit received if 1000 units are sold.

67. **Investments** Suppose that you have $4000 to invest, and you invest x dollars at 10% and the remainder at 8%. Write an expression that represents the
    (a) amount invested at 8%.
    (b) interest earned on the x dollars at 10%.
    (c) interest earned on the money invested at 8%.
    (d) total interest earned.

68. **Medications.** Suppose that a nurse needs 10 cc of a 15.5% solution (that is, a solution that is 15.5% ingredient) of a certain medication, which must be obtained by mixing x cc of a 20% solution and y cc of a 5% solution. Write expressions involving x for
    (a) y, the amount of 5% solution.
    (b) the amount of ingredient in the x cc of 20% solution.
    (c) the amount of ingredient in the 5% solution.
    (d) the total amount of ingredient in the mixture.

69. **Package design** The volume of a rectangular box is given by V = (length) (width) (height). If a rectangular piece of cardboard that is 10 in. by 15 in. has a square with sides of length x cut from each corner (see Figure 0.9), and if the sides are folded up along the dotted lines to form a box, what expression of x would represent the volume?

Figure 0.9

0.6 FACTORING

We can factor monomial factors out of a polynomial by using the Distributive Law in reverse; $ab + ac = a(b + c)$ is an example showing that a is a monomial factor of the polynomial $ab + ac$. But it is also a statement of the Distributive Law (with the sides of the equation interchanged). The monomial factor of a polynomial must be a factor of each term of the polynomial, so it is frequently called a **common monomial factor.**

EXAMPLE I

Factor $-3x^2t - 3x + 9xt^2$.

Solution

1. We can either factor out $3x$ and obtain
$$-3x^2t - 3x + 9xt^2 = 3x(-xt - 1 + 3t^2)$$

2. or we can factor out $-3x$ (factoring out the negative will make the first term of the polynomial positive) and obtain
$$-3x^2t - 3x + 9xt^2 = -3x(xt + 1 - 3t^2).$$

If a factor is common to each term of a polynomial, we can use the above procedure to factor it out, even if it is not a monomial. For example, we can factor $(a + b)$ out of the polynomial $2x(a + b) - 3y(a + b)$. If we factor

$(a + b)$ from both terms, we get $(a + b)(2x - 3y)$. The following example demonstrates the **factoring by grouping** technique.

EXAMPLE 2

Factor $5x - 5y + bx - by$.

Solution

We can factor this polynomial by the use of grouping. The grouping is done so that common factors (frequently binomial factors) can be removed. We see that we can factor 5 from the first two terms and b from the last two, giving

$$5(x - y) + b(x - y).$$

This gives two terms with the common factor $x - y$, so we get

$$(x - y)(5 + b).$$

We can use the formula for multiplying two binomials to factor certain trinomials. The formula

$$(x + a)(x + b) = x^2 + (a + b)x + ab$$

can be used to factor trinomials such as

$$x^2 - 7x + 6.$$

EXAMPLE 3

Factor $x^2 - 7x + 6$.

Solution

If this trinomial can be factored into an expression of the form

$$(x + a)(x + b)$$

then we need to find a and b such that

$$x^2 - 7x + 6 = x^2 + (a + b)x + ab$$

That is, we need to find a and b such that $a + b = -7$ and $ab = 6$. The two numbers whose sum is -7 and whose product is 6 are -1 and -6. Thus

$$x^2 - 7x + 6 = (x - 1)(x - 6).$$

A similar method can be used to factor trinomials such as $9x^2 - 31x + 12$. Finding the proper factors for this type of trinomial may involve a fair amount of trial and error because we must find factors a, b, c, and d such that

$$(ax + b)(cx + d) = acx^2 + (ad + bc)x + bd.$$

Another technique of factoring can be used to factor trinomials such as those we have been discussing. It is especially useful in factoring more complicated trinomials, such as $9x^2 - 31x + 12$. This procedure for factoring second-degree trinomials follows.

Factoring a Trinomial

Procedure	Example
To factor a trinomial into the product of its binomial factors:	Factor $9x^2 - 31x + 12$.
1. Form the product of the second-degree term and the constant term.	1. $9x^2 \cdot 12 = 108x^2$
2. Determine if there are any factors of the product of step 1 that will sum to the middle term of the trinomial. (If the answer is no, the trinomial will not factor into two binomials.)	2. The factors $-27x$ and $-4x$ give a sum of $-31x$.
3. Use the sum of these two factors to replace the middle term of the trinomial.	3. $9x^2 - 31x + 12 = 9x^2 - 27x - 4x + 12$
4. Factor this four-term expression by grouping.	4. $9x^2 - 31x + 12 = (9x^2 - 27x)$ $\qquad\qquad\qquad\qquad + (-4x + 12)$ $\qquad = 9x(x - 3) - 4(x - 3)$ $\qquad = (x - 3)(9x - 4)$

In the example just completed, note that writing the middle term $(-31x)$ as $-4x - 27x$ rather than as $-27x - 4x$ (as we did) will also result in the correct factorization. (Try it.)

EXAMPLE 4
Factor $9x^2 - 9x - 10$.

Solution
The product of the second-degree term and the constant is $-90x^2$. Factors of $-90x^2$ that sum to $-9x$ are $-15x$ and $6x$. Thus

$$9x^2 - 9x - 10 = 9x^2 - 15x + 6x - 10$$
$$= (9x^2 - 15x) + (6x - 10)$$
$$= 3x(3x - 5) + 2(3x - 5)$$
$$= (3x - 5)(3x + 2).$$

We can check this factorization by multiplying.

$$(3x - 5)(3x + 2) = 9x^2 + 6x - 15x - 10$$
$$= 9x^2 - 9x - 10$$

Some special products that make factoring easier are the following.

Special Factorizations   The perfect-square trinomials:

$$x^2 + 2ax + a^2 = (x + a)^2$$
$$x^2 - 2ax + a^2 = (x - a)^2$$

The difference of two squares:

$$x^2 - a^2 = (x + a)(x - a)$$

EXAMPLE 5
Factor $25x^2 - 36y^2$.

Solution
Since the binomial $25x^2 - 36y^2$ is the difference of two squares, we get

$$25x^2 - 36y^2 = (5x - 6y)(5x + 6y).$$

These two factors are called binomial **conjugates** because they differ in only one sign.

EXAMPLE 6
Factor $4x^2 + 12x + 9$.

Solution
Although we can use the technique we have learned to factor trinomials, the factors come quickly if we recognize that this trinomial is a perfect square. It has two square terms, and the remaining term $(12x)$ is twice the product of the square roots of the squares $(12x = 2 \cdot 2x \cdot 3)$. Thus

$$4x^2 + 12x + 9 = (2x + 3)^2.$$

Most of the polynomials we have factored have been second-degree polynomials, or **quadratic polynomials.** Some polynomials that are not quadratic are in a form that can be factored in the same manner as quadratics. For example, the polynomial $x^4 + 4x^2 + 4$ can be written as $a^2 + 4a + 4$, where $a = x^2$.

EXAMPLE 7
Factor $x^4 + 4x^2 + 4$ completely.

Solution
Since the trinomial is in the form of a perfect square, letting $a = x^2$ will give us

$$x^4 + 4x^2 + 4 = a^2 + 4a + 4 = (a + 2)^2.$$

Thus

$$x^4 + 4x^2 + 4 = (x^2 + 2)^2.$$

> ### EXAMPLE 8
> Factor $x^4 - 16$ completely.
>
> **Solution**
> The binomial $x^4 - 16$ can be treated as the difference of two squares, $(x^2)^2 - 4^2$, so
> $$x^4 - 16 = (x^2 - 4)(x^2 + 4).$$
> But $x^2 - 4$ can be factored into $(x - 2)(x + 2)$, so
> $$x^4 - 16 = (x - 2)(x + 2)(x^2 + 4).$$

CHECKPOINT

1. Factor the following.
   (a) $8x^3 - 12x$ (b) $3x(x^2 + 5) - 5(x^2 + 5)$ (c) $x^2 - 10x - 24$
   (d) $x^2 - 5x + 6$ (e) $4x^2 - 20x + 25$ (f) $100 - 49x^2$

2. Consider $10x^2 - 17x - 20$ and observe that $(10x^2)(-20) = -200x^2$.
   (a) Find two expressions whose product is $-200x^2$ and whose sum is $-17x$.
   (b) Replace $-17x$ in $10x^2 - 17x - 20$ with the two expressions in (a).
   (c) Factor (b) by grouping.

3. True or false:
   (a) $4x^2 + 9 = (2x + 3)^2$ (b) $x^2 - x + 12 = (x - 4)(x + 3)$
   (c) $5x^5 - 20x^3 = 5x^3(x^2 - 4) = 5x^3(x + 2)(x - 2)$

A polynomial is said to be factored completely if all possible factorizations have been completed. For example, $(2x - 4)(x + 3)$ is not factored completely because a 2 can still be factored out of $2x - 4$. Confining our attention to factors with integer coefficients, we can factor a number of polynomials completely by using the following guidelines.

Guidelines for Factoring Completely	Look for: Monomials first. Then for: Difference of two squares. Then for: Trinomial squares. Then for: Other methods of factoring trinomials.

> ### EXAMPLE 9
> Factor completely $12x^2 - 36x + 27$.
>
> **Solution**
> $$12x^2 - 36x + 27 = 3(4x^2 - 12x + 9) \qquad \text{Monomial}$$
> $$= 3(2x - 3)^2 \qquad \text{Perfect square}$$

EXAMPLE 10

Factor completely $16x^2 - 64y^2$.

Solution

$$16x^2 - 64y^2 = 16(x^2 - 4y^2)$$
$$= 16(x + 2y)(x - 2y)$$

Note that factoring the difference of two squares immediately would give $(4x + 8y)(4x - 8y)$, which is not factored completely (since we could still factor 4 from $4x + 8y$ and 4 from $4x - 8y$).

CHECKPOINT SOLUTIONS

1. (a) $8x^3 - 12x = 4x(2x^2 - 3)$
   (b) $3x(x^2 + 5) - 5(x^2 + 5) = (x^2 + 5)(3x - 5)$
   (c) $x^2 - 10x - 24 = (x - 12)(x + 2)$
   (d) $x^2 - 5x + 6 = (x - 3)(x - 2)$
   (e) $4x^2 - 20x + 25 = (2x - 5)^2$
   (f) $100 - 49x^2 = (10 + 7x)(10 - 7x)$

2. (a) $(-25x)(+8x) = -200x^2$ and $-25x + 8x = -17x$
   (b) $10x^2 - 17x - 20 = 10x^2 - 25x + 8x - 20$
   (c) $ = (10x^2 - 25x) + (8x - 20)$
   $ = 5x(2x - 5) + 4(2x - 5)$
   $ = (2x - 5)(5x + 4)$

3. (a) False. $4x^2 + 9$ cannot be factored. In fact, sums of squares cannot be factored.
   (b) False. $x^2 - x + 12$ cannot be factored. We cannot find two numbers whose product is $+12$ and whose sum is -1.
   (c) True.

EXERCISE 0.6

Factor the following completely.

1. $9ab - 12a^2b + 18b^2$
2. $8a^2b - 160x + 4bx^2$
3. $4x^2 + 8xy^2 + 2xy^3$
4. $12y^3z + 4yz^2 - 8y^2z^3$
5. $5(y - 4) - x^2(y - 4)$
6. $2(x + 1) - 5x(x + 1)$
7. $6x - 6m + xy - my$
8. $x^3 - x^2 - 5x + 5$
9. $x^2 + 8x + 12$
10. $x^2 + 6x + 8$
11. $x^2 - x - 6$
12. $x^2 - 2x - 8$
13. $2x^2 - 8x - 42$
14. $3x^2 - 21x + 36$
15. $7x^2 - 10x - 8$
16. $12x^2 + 11x + 2$
17. $x^2 - 10x + 25$
18. $4y^2 + 12y + 9$
19. $49a^2 - 144b^2$
20. $16x^2 - 25y^2$
21. $2x^3 - 8x^2 + 8x$
22. $x^3 + 16x^2 + 64x$
23. $2x^2 + x - 6$
24. $2x^2 + 13x + 6$
25. $3x^2 + 3x - 36$
26. $4x^2 - 8x - 60$
27. $2x^3 - 8x$
28. $16z^2 - 81w^2$

29. $10x^2 + 19x + 6$

30. $6x^2 + 67x - 35$

31. $9 - 47x + 10x^2$

32. $10x^2 + 21x - 10$

33. $9x^2 + 21x - 8$

34. $9x^2 + 22x + 8$

35. $y^4 - 16x^4$

36. $x^8 - 81$

37. $x^4 - 8x^2 + 16$

38. $81 - 18x^2 + x^4$

39. $4x^4 - 5x^2 + 1$

40. $x^4 - 3x^2 - 4$

41. $10x^2 - 99x - 63$

42. $10x^2 - 27x - 63$

43. $10x^2 + 61x - 63$

44. $10x^2 + 9x - 63$

Use the following factorization formulas involving cubes to factor each expression in problems 45–52.

Factorizations with cubes

Perfect cube
$$a^3 + 3a^2b + 3ab^2 + b^3 = (a + b)^3$$

Perfect cube
$$a^3 - 3a^2b + 3ab^2 - b^3 = (a - b)^3$$

Difference of two cubes
$$a^3 - b^3 = (a - b)(a^2 + ab + b^2)$$

Sum of two cubes
$$a^2 + b^3 = (a + b)(a^2 - ab + b^2)$$

45. $x^3 + 3x^2 + 3x + 1$

46. $x^3 + 6x^2 + 12x + 8$

47. $x^3 - 12x^2 + 48x - 64$

48. $y^3 - 9y^2 + 27y - 27$

49. $x^3 - 64$

50. $8x^3 - 1$

51. $27 + 8x^3$

52. $a^3 + 216$

In problems 53–58 determine the missing factor.

53. $x^{3/2} + x^{1/2} = x^{1/2}(?)$

54. $2x^{1/4} + 4x^{3/4} = 2x^{1/4}(?)$

55. $x^{-3} + x^{-2} = x^{-3}(?)$

56. $x^{-1} - x = x^{-1}(?)$

57. $(-x^3 + x)(3 - x^2)^{-1/2} + 2x(3 - x^2)^{1/2}$
$$= (3 - x^2)^{-1/2}(?)$$

58. $4x(4x + 1)^{-1/3} - (4x + 1)^{2/3} = (4x + 1)^{-1/3}(?)$

APPLICATIONS

59. **Reaction to medication** When medicine is administered, the reaction (measured in change of blood pressure or temperature) can be modeled by (that is, described by)
$$R = \frac{cm^2}{2} - \frac{m^3}{3},$$
where c is a positive constant and m is the amount of medicine absorbed into the blood.* Factor the expression for the reaction.

60. **Sensitivity to medication** From the formula for reaction to medication given in problem 59, an expression for sensitivity S can be obtained, where
$$S = cm - m^2.$$
Factor this expression for sensitivity.

61. **Volume** Suppose that squares of side x are cut from four corners of an 8-by-8-inch piece of cardboard and an open-top box is formed (see Figure 0.10). The volume of the box is given by $64x - 32x^2 + 4x^3$. Factor this expression.

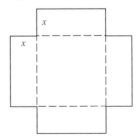

Figure 0.10

62. **Power in a circuit** Factor the following expression for the maximum power in a certain electrical circuit.
$$(R + r)^2 - 2r(R + r)$$

*Some Mathematical Models in Biology, R. M. Thrall, et. al., U.S. Department of Commerce, 1967.

0.7 ALGEBRAIC FRACTIONS

Evaluating certain limits and graphing rational functions require an understanding of algebraic fractions. The fraction 6/8 can be reduced to 3/4 by dividing both the numerator and denominator by 2. In the same manner, the algebraic fraction

$$\frac{(x + 2)(x + 1)}{(x + 1)(x + 3)}$$

can be reduced to

$$\frac{x + 2}{x + 3}$$

by dividing both the numerator and denominator by $x + 1$, if $x \neq -1$.

Simplifying Fractions   We *simplify* algebraic fractions by factoring the numerator and denominator and then dividing both the numerator and denominator by any common factors.*

EXAMPLE 1

Simplify $\dfrac{3x^2 - 14x + 8}{x^2 - 16}$, $x^2 \neq 16$.

Solution

$$\frac{3x^2 - 14x + 8}{x^2 - 16} = \frac{(3x - 2)(x - 4)}{(x - 4)(x + 4)}$$

$$= \frac{(3x - 2)\overset{1}{\cancel{(x - 4)}}}{\underset{1}{\cancel{(x - 4)}}(x + 4)}$$

$$= \frac{3x - 2}{x + 4}$$

We can multiply fractions by writing the product as the product of the numerators divided by the product of the denominators. For example,

$$\frac{4}{5} \cdot \frac{10}{12} \cdot \frac{2}{5} = \frac{80}{300},$$

which reduces to $\frac{4}{15}$.

---

* We assume all fractions are defined.

We can also find the product by reducing the fractions before we indicate the multiplication in the numerator and denominator. For example, in

$$\frac{4}{5} \cdot \frac{10}{12} \cdot \frac{2}{5},$$

we can divide the numerator and denominator by 5 and then by 4, giving

$$\frac{\overset{1}{\cancel{4}}}{\underset{1}{\cancel{5}}} \cdot \frac{\overset{2}{\cancel{10}}}{\underset{3}{\cancel{12}}} \cdot \frac{2}{5} = \frac{1}{1} \cdot \frac{2}{3} \cdot \frac{2}{5} = \frac{4}{15}.$$

Product of Fractions   We *multiply* algebraic fractions by writing the product of the numerators divided by the product of the denominators, and then reduce to lowest terms. We may also reduce prior to finding the product.

EXAMPLE 2
Multiply:

(a) $\dfrac{4x^2}{5y} \cdot \dfrac{10x}{y^2} \cdot \dfrac{y}{8x^2}$ (b) $\dfrac{-4x + 8}{3x + 6} \cdot \dfrac{2x + 4}{4x + 12}$

Solution

(a) $\dfrac{4x^2}{5y} \cdot \dfrac{10x}{y^2} \cdot \dfrac{y}{8x^2} = \dfrac{\overset{1}{\cancel{4x^2}}}{\underset{1\cdot1}{\cancel{5y}}} \cdot \dfrac{\overset{2}{\cancel{10x}}}{y^2} \cdot \dfrac{\overset{1}{\cancel{y}}}{\underset{2}{\cancel{8x^2}}} = \dfrac{1}{1} \cdot \dfrac{2x}{y^2} \cdot \dfrac{\overset{1}{1}}{\underset{1}{2}} = \dfrac{x}{y^2}$

(b) $\dfrac{-4x + 8}{3x + 6} \cdot \dfrac{2x + 4}{4x + 12} = \dfrac{-4(x - 2)}{3(x + 2)} \cdot \dfrac{2(x + 2)}{4(x + 3)}$

$$= \dfrac{\overset{-1}{\cancel{-4(x - 2)}}}{3\cancel{(x + 2)}} \cdot \dfrac{2\cancel{(x + 2)}}{\underset{1}{\cancel{4}}(x + 3)}$$

$$= \dfrac{-2(x - 2)}{3(x + 3)}$$

In arithmetic we learned to divide one fraction by another by inverting the divisor and multiplying. The same rule applies to division of algebraic fractions.

EXAMPLE 3

(a) Divide $\dfrac{a^2b}{c}$ by $\dfrac{ab}{c^2}$ (b) $\dfrac{6x^2 - 6}{x^2 + 3x + 2} \div \dfrac{x - 1}{x^2 + 4x + 4}$

Solution

(a) $\dfrac{a^2b}{c} \div \dfrac{ab}{c^2} = \dfrac{a^2b}{c} \cdot \dfrac{c^2}{ab} = \dfrac{\overset{a \cdot 1}{\cancel{a^2b}}}{\cancel{c}} \cdot \dfrac{\overset{c}{\cancel{c^2}}}{\underset{1 \cdot 1}{\cancel{ab}}} = \dfrac{ac}{1} = ac$

(b) $\dfrac{6x^2 - 6}{x^2 + 3x + 2} \div \dfrac{x - 1}{x^2 + 4x + 4} = \dfrac{6x^2 - 6}{x^2 + 3x + 2} \cdot \dfrac{x^2 + 4x + 4}{x - 1}$

$$= \dfrac{6(x - 1)(x + 1)}{(x + 2)(x + 1)} \cdot \dfrac{(x + 2)(x + 2)}{x - 1}$$

$$= 6(x + 2)$$

CHECKPOINT

1. Reduce $\dfrac{2x^2 - 4x}{2x}$.

2. Multiply: $\dfrac{x^2}{x^2 - 9} \cdot \dfrac{x + 3}{3x}$.

3. Divide: $\dfrac{5x^2(x - 1)}{2(x + 1)} \div \dfrac{10x^2}{(x + 1)(x - 1)}$.

If two fractions are to be added, it is convenient that each be expressed with the same denominator. If the denominators are not the same, we can write the equivalents of each of the fractions with a common denominator. We usually use the lowest common denominator (LCD) when we write the equivalent fractions. The **lowest common denominator** is the lowest-degree variable expression into which all denominators will divide. If the denominators are polynomials, then the LCD is the lowest-degree polynomial into which all denominators will divide. We can find the lowest common denominator as follows.

Finding the Lowest Common Denominator

Procedure	Example
To find the lowest common denominator of a set of fractions:	Find the LCD of $\dfrac{1}{x^2 - x}, \dfrac{1}{x^2 - 1}, \dfrac{1}{x^2}$.
1. Completely factor each denominator.	1. The factored denominators are $x(x - 1)$, $(x + 1)(x - 1)$, $x \cdot x$.
2. Write the LCD as the product of each of these factors used the maximum number of times it occurs in any one denominator.	2. x occurs a maximum of 2 times in one denominator, $x - 1$ occurs once, and $x + 1$ occurs once. Thus the LCD is $x \cdot x(x - 1)(x + 1) = x^2(x - 1)(x + 1)$.

The procedure for combining (adding or subtracting) two or more fractions is as follows.

Adding or Subtracting Fractions	
Procedure	**Example**
To combine fractions:	Combine $\dfrac{y-3}{y-5} + \dfrac{y-23}{y^2-y-20}$.
1. Find the LCD of the fractions.	1. $y^2 - y - 20 = (y-5)(y+4)$, so the LCD is $(y-5)(y+4)$.
2. Write the equivalent of each fraction with the LCD as its denominator.	2. The sum is $$\dfrac{(y-3)(y+4)}{(y-5)(y+4)} + \dfrac{y-23}{(y-5)(y+4)}.$$
3. Add or subtract as indicated by combining like terms in the numerator over the LCD.	3. $$= \dfrac{y^2 + y - 12 + y - 23}{(y-5)(y+4)}$$ $$= \dfrac{y^2 + 2y - 35}{(y-5)(y+4)}$$
4. Reduce the fraction, if possible.	4. $y - 5$ is a factor of the numerator, so the sum is $\dfrac{y+7}{y+4}$, if $y \neq 5$.

EXAMPLE 4

Add $\dfrac{3x}{a^2} + \dfrac{4}{ax}$.

Solution

1. The LCD is a^2x.

2. $\dfrac{3x}{a^2} + \dfrac{4}{ax} = \dfrac{3x}{a^2} \cdot \dfrac{x}{x} + \dfrac{4}{ax} \cdot \dfrac{a}{a} = \dfrac{3x^2}{a^2x} + \dfrac{4a}{a^2x}$

3. $\dfrac{3x^2}{a^2x} + \dfrac{4a}{a^2x} = \dfrac{3x^2 + 4a}{a^2x}$

4. The sum is in lowest terms.

EXAMPLE 5

Combine $\dfrac{y-3}{(y-5)^2} + \dfrac{y-2}{y^2-y-20}$.

Solution

$y^2 - y - 20 = (y - 5)(y + 4)$, so the LCD is $(y - 5)^2(y + 4)$. Writing the equivalent fractions, then combining them gives

$$\frac{y - 3}{(y - 5)^2} + \frac{y - 2}{(y - 5)(y + 4)}$$

$$= \frac{(y - 3)(y + 4)}{(y - 5)^2(y + 4)} + \frac{(y - 2)(y - 5)}{(y - 5)(y + 4)(y - 5)}$$

$$= \frac{(y^2 + y - 12) + (y^2 - 7y + 10)}{(y - 5)^2(y + 4)}$$

$$= \frac{2y^2 - 6y - 2}{(y - 5)^2(y + 4)}.$$

A fractional expression that contains one or more fractions in its numerator or denominator is called a **complex fraction.** An example of a complex fraction is

$$\frac{\dfrac{1}{3} + \dfrac{4}{x}}{3 - \dfrac{1}{xy}}.$$

We can simplify fractions of this type using the property $\dfrac{a}{b} = \dfrac{ac}{bc}$, with c equal to the LCD of *all* the fractions contained in the numerator and denominator of the complex fraction.

For example, all fractions contained in

$$\frac{\dfrac{1}{3} + \dfrac{4}{x}}{3 - \dfrac{1}{xy}}$$

have LCD $3xy$. We simplify this complex fraction by multiplying the numerator and denominator as follows:

$$\frac{3xy\left(\dfrac{1}{3} + \dfrac{4}{x}\right)}{3xy\left(3 - \dfrac{1}{xy}\right)} = \frac{3xy\left(\dfrac{1}{3}\right) + 3xy\left(\dfrac{4}{x}\right)}{3xy(3) - 3xy\left(\dfrac{1}{xy}\right)} = \frac{xy + 12y}{9xy - 3}.$$

EXAMPLE 6

Simplify $\dfrac{x^{-3} + x^2y^{-3}}{(xy)^{-2}}$ so that positive exponents remain.

Solution

$$\frac{x^{-3} + x^2 y^{-3}}{(xy)^{-2}} = \frac{\dfrac{1}{x^3} + \dfrac{x^2}{y^3}}{\dfrac{1}{(xy)^2}}, \quad \text{LCD} = x^3 y^3$$

$$= \frac{x^3 y^3 \left(\dfrac{1}{x^3} + \dfrac{x^2}{y^3} \right)}{x^3 y^3 \left(\dfrac{1}{x^2 y^2} \right)} = \frac{y^3 + x^5}{xy}$$

CHECKPOINT

4. Add or subtract:

   (a) $\dfrac{5x - 1}{2x - 5} - \dfrac{x + 9}{2x - 5}$

   (b) $\dfrac{x + 1}{x} + \dfrac{x}{x - 1}$

5. Simplify $\dfrac{\dfrac{y}{x} - 1}{\dfrac{y}{x} - \dfrac{x}{y}}$.

We can simplify algebraic fractions whose denominators contain sums and differences involving square roots by rationalizing the denominators. Using the fact that $(x + y)(x - y) = x^2 - y^2$, we multiply the numerator and denominator of an algebraic fraction of this type by the conjugate of the denominator to simplify the fraction.

EXAMPLE 7

Rationalize the denominators.

(a) $\dfrac{1}{\sqrt{x} - 2}$ (b) $\dfrac{3 + \sqrt{x}}{\sqrt{x} + \sqrt{5}}$

Solution

Multiplying $\sqrt{x} - 2$ by $\sqrt{x} + 2$, its conjugate, gives the difference of two squares and removes the radical from the denominator in (a). We also use the conjugate in (b).

(a) $\dfrac{1}{\sqrt{x} - 2} \cdot \dfrac{\sqrt{x} + 2}{\sqrt{x} + 2} = \dfrac{\sqrt{x} + 2}{(\sqrt{x})^2 - (2)^2} = \dfrac{\sqrt{x} + 2}{x - 4}$

(b) $\dfrac{3 + \sqrt{x}}{\sqrt{x} + \sqrt{5}} \cdot \dfrac{\sqrt{x} - \sqrt{5}}{\sqrt{x} - \sqrt{5}} = \dfrac{3\sqrt{x} - 3\sqrt{5} + x - \sqrt{5x}}{x - 5}$

CHECKPOINT

6. Rationalize the denominator $\dfrac{\sqrt{x}}{2\sqrt{x} - 3}$.

CHECKPOINT
SOLUTIONS

1. $\dfrac{2x^2 - 4x}{2x} = \dfrac{2x(x - 2)}{2x} = x - 2$

2. $\dfrac{x^2}{x^2 - 9} \cdot \dfrac{x + 3}{3x} = \dfrac{x^2 \cdot (x + 3)}{(x + 3)(x - 3) \cdot 3x} = \dfrac{x}{3(x - 3)} = \dfrac{x}{3x - 9}$

3. $\dfrac{5x^2(x - 1)}{2(x + 1)} \div \dfrac{10x^2}{(x + 1)(x - 1)} = \dfrac{5x^2(x - 1)}{2(x + 1)} \cdot \dfrac{(x + 1)(x - 1)}{10x^2}$

$$= \dfrac{(x - 1)^2}{4}$$

4. (a) $\dfrac{5x - 1}{2x - 5} - \dfrac{x + 9}{2x - 5} = \dfrac{(5x - 1) - (x + 9)}{2x - 5}$

$$= \dfrac{5x - 1 - x - 9}{2x - 5}$$

$$= \dfrac{4x - 10}{2x - 5} = \dfrac{2(2x - 5)}{2x - 5} = 2$$

(b) $\dfrac{x + 1}{x} + \dfrac{x}{x - 1}$ has LCD $= x(x - 1)$

$$\dfrac{x + 1}{x} + \dfrac{x}{x - 1} = \dfrac{x + 1}{x} \cdot \dfrac{(x - 1)}{(x - 1)} + \dfrac{x}{x - 1} \cdot \dfrac{x}{x}$$

$$= \dfrac{x^2 - 1}{x(x - 1)} + \dfrac{x^2}{x(x - 1)} = \dfrac{x^2 - 1 + x^2}{x(x - 1)}$$

$$= \dfrac{2x^2 - 1}{x(x - 1)}$$

5. $\dfrac{\dfrac{y}{x} - 1}{\dfrac{y}{x} - \dfrac{x}{y}} = \dfrac{\left(\dfrac{y}{x} - 1\right)}{\left(\dfrac{y}{x} - \dfrac{x}{y}\right)} \cdot \dfrac{xy}{xy} = \dfrac{\dfrac{y}{x} \cdot xy - 1 \cdot xy}{\dfrac{y}{x} \cdot xy - \dfrac{x}{y} \cdot xy}$

$$= \dfrac{y^2 - xy}{y^2 - x^2} = \dfrac{y(y - x)}{(y + x)(y - x)} = \dfrac{y}{y + x}$$

6. $\dfrac{\sqrt{x}}{2\sqrt{x} - 3} = \dfrac{\sqrt{x}}{2\sqrt{x} - 3} \cdot \dfrac{2\sqrt{x} + 3}{2\sqrt{x} + 3}$

$$= \dfrac{2x + 3\sqrt{x}}{4x - 9}$$

EXERCISE 0.7

Simplify the following fractions.

1. $\dfrac{18x^3y^3}{9x^3z}$

2. $\dfrac{15a^4b^5}{30a^3b}$

3. $\dfrac{x - 3y}{3x - 9y}$

4. $\dfrac{x^2 - 6x + 8}{x^2 - 16}$

5. $\dfrac{x^2 - 2x + 1}{x^2 - 4x + 3}$

6. $\dfrac{x^2 - 5x + 6}{9 - x^2}$

7. $\dfrac{6x^3y^3 - 15x^2y}{3x^2y^2 + 9x^2y}$

8. $\dfrac{x^2y^2 - 4x^3y}{x^2y - 2x^2y^2}$

In problems 9–38, perform the indicated operations and simplify.

9. $\dfrac{6x^3}{8y^3} \cdot \dfrac{16x}{9y^2} \cdot \dfrac{15y^4}{x^3}$

10. $\dfrac{25ac^2}{15a^2c} \cdot \dfrac{4ad^4}{15abc^3}$

11. $\dfrac{8x - 16}{x - 3} \cdot \dfrac{4x - 12}{3x - 6}$

12. $\dfrac{x^2 + 7x + 12}{3x^2 + 13x + 4} \cdot \dfrac{3x + 1}{x + 3}$

13. $(x^2 - 4) \cdot \dfrac{2x - 3}{x + 2}$

14. $\dfrac{4x + 4}{x - 4} \cdot \dfrac{x^2 - 6x + 8}{8x^2 + 8x}$

15. $\dfrac{x^2 - x - 2}{2x^2 - 8} \cdot \dfrac{18 - 2x^2}{x^2 - 5x + 4} \cdot \dfrac{x^2 - 2x - 8}{x^2 - 6x + 9}$

16. $\dfrac{x^2 - 5x - 6}{x^2 - 5x + 4} \cdot \dfrac{x^2 - x - 12}{x^3 - 6x^2} \cdot \dfrac{x - x^3}{x^2 - 2x + 1}$

17. $\dfrac{15ac^2}{7bd} \div \dfrac{4a}{14b^2d}$

18. $\dfrac{16}{x - 2} \div \dfrac{4}{3x - 6}$

19. $\dfrac{y^2 - 2y + 1}{7y^2 - 7y} \div \dfrac{y^2 - 4y + 3}{35y^2}$

20. $\dfrac{6x^2}{4x^2y - 12xy} \div \dfrac{3x^2 + 12x}{x^2 + x - 12}$

21. $(x^2 - x - 6) \div \dfrac{9 - x^2}{x^2 - 3x}$

22. $\dfrac{2x^2 + 7x + 3}{4x^2 - 1} \div (x + 3)$

23. $\dfrac{2x}{x^2 - x - 2} - \dfrac{x + 2}{x^2 - x - 2}$

24. $\dfrac{4}{9 - x^2} - \dfrac{x + 1}{9 - x^2}$

25. $\dfrac{a}{a - 2} - \dfrac{a - 2}{a}$

26. $x - \dfrac{2}{x - 1}$

27. $\dfrac{x}{x + 1} - x + 1$

28. $\dfrac{x - 1}{x + 1} - \dfrac{2}{x^2 + x}$

29. $\dfrac{4a}{3x + 6} + \dfrac{5a^2}{4x + 8}$

30. $\dfrac{b - 1}{b^2 + 2b} + \dfrac{b}{3b + 6}$

31. $\dfrac{3x - 1}{2x - 4} + \dfrac{4x}{3x - 6} - \dfrac{x - 4}{5x - 10}$

32. $\dfrac{2x + 1}{4x - 2} + \dfrac{5}{2x} - \dfrac{x + 4}{2x^2 - x}$

33. $\dfrac{1}{x^2 - 4y^2} - \dfrac{1}{x^2 - 4xy + 4y^2}$

34. $\dfrac{3x^2}{x^2 - 4} + \dfrac{2}{x^2 - 4x + 4} - 3$

35. $\dfrac{x}{x^2 - 4} + \dfrac{4}{x^2 - x - 2} - \dfrac{x - 2}{x^2 + 3x + 2}$

36. $\dfrac{1}{y^2 + 2y - 35} - \dfrac{y - 1}{y^2 - 4y - 5} + \dfrac{2}{y^2 + 8y + 7}$

37. $\dfrac{-x^3 + x}{\sqrt{3 - x^2}} + 2x\sqrt{3 - x^2}$

38. $\dfrac{3x^2(x + 1)}{\sqrt{x^3 + 1}} + \sqrt{x^3 + 1}$

In problems 39–48, simplify each complex fraction.

39. $\dfrac{3 - \dfrac{2}{3}}{14}$

40. $\dfrac{4}{\dfrac{1}{4} + \dfrac{1}{4}}$

41. $\dfrac{x + y}{\dfrac{1}{x} + \dfrac{1}{y}}$

42. $\dfrac{\dfrac{5}{2y} + \dfrac{3}{y}}{\dfrac{1}{4} + \dfrac{1}{3y}}$

43. $\dfrac{2 - \dfrac{1}{x}}{2x - \dfrac{3x}{x + 1}}$

44. $\dfrac{1 - \dfrac{2}{x - 2}}{x - 6 + \dfrac{10}{x + 1}}$

45. $\dfrac{\sqrt{a} - \dfrac{b}{\sqrt{a}}}{a - b}$

46. $\dfrac{\sqrt{x - 1} + \dfrac{1}{\sqrt{x - 1}}}{x}$

47. $\dfrac{\sqrt{x^2 + 9} - \dfrac{13}{\sqrt{x^2 + 9}}}{x^2 - x - 6}$

48. $\dfrac{\sqrt{x^2 + 3} - \dfrac{x + 5}{\sqrt{x^2 + 3}}}{x^2 + 5x + 4}$

In problems 49–54, rewrite each of the following so only positive exponents remain, and simplify.

49. (a) $(2^{-2} - 3^{-1})^{-1}$ (b) $(2^{-1} + 3^{-1})^2$

50. (a) $(3^2 + 4^2)^{-1/2}$ (b) $(2^2 + 3^2)^{-1}$

51. $\dfrac{2a^{-1} - b^{-1}}{(ab)^{-1}}$

52. $\dfrac{x^{-2} + y^{-2}}{x^{-2} - y^{-2}}$

53. $\dfrac{xy^{-2} + x^{-2}y}{x + y}$

54. $\dfrac{x^{-2} + xy^{-2}}{(x^2y)^{-2}}$

In problems 55 and 56, rationalize the denominator of each fraction and simplify.

55. $\dfrac{1 - \sqrt{x}}{1 + \sqrt{x}}$

56. $\dfrac{x - 3}{x - \sqrt{3}}$

In problems 57 and 58, rationalize the numerator of each fraction and simplify.

57. $\dfrac{\sqrt{x + h} - \sqrt{x}}{h}$

58. $\dfrac{\sqrt{9 + 2h} - 3}{h}$

APPLICATIONS

59. **Time study** Workers A, B, and C can complete a job in a, b, and c hours, respectively. Working together they can complete

$$\frac{1}{a} + \frac{1}{b} + \frac{1}{c}$$

of the job in one hour. Add these fractions to obtain an expression for what they can do in one hour working together.

60. **Focal length** Two thin lenses with focal lengths p and q and separated by a distance d have their combined focal length given by the reciprocal of

$$\frac{1}{p} + \frac{1}{q} - \frac{d}{pq}.$$

(a) Combine these fractions.
(b) Use the reciprocal of your answer in (a) to find the combined focal length.

Average cost A company's average cost per unit when x units are produced is defined to be

$$\text{Average cost} = \frac{\text{Total cost}}{x}.$$

Use this equation in problems 61 and 62.

61. Suppose a company's average costs are given by

$$\text{Average cost} = \frac{4000}{x} + 55 + 0.1x.$$

(a) Express the average cost formula as a single fraction.
(b) Write the expression that gives the company's total costs.

62. Suppose a company's average costs are given by

$$\text{Average cost} = \frac{40{,}500}{x} + 190 + 0.2x.$$

(a) Express the average cost formula as a single fraction.
(b) Write the expression that gives the company's total costs.

63. **Advertising and sales** Suppose that a company's daily sales volume attributed to an advertising campaign is given by

$$\text{Sales volume} = 1 + \frac{3}{t + 3} - \frac{18}{(t + 3)^2},$$

where t is the number of days since the campaign started. Express the sales volume as a single fraction.

KEY TERMS
& FORMULAS

Section	Key Terms	Formula
0.1	Sets and set membership	
	Natural numbers	$\{1, 2, 3, 4 \ldots\}$
	Empty set	$\varnothing$
	Set equality	
	Subset	$A \subseteq B$
	Universal set	U
	Venn diagrams	
	Set intersection	$A \cap B$
	Disjoint sets	$A \cap B = \varnothing$
	Set union	$A \cup B$
	Set complement	A'
0.2	Real numbers	
	Subsets and properties	
	Real number line	
	Inequalities	
	Intervals and interval notation	
	Absolute value	
	Order of operations	
0.3	Exponent and base	a^n has base a, exponent n
	Zero exponent	$a^0 = 1, a \neq 0$
	Negative exponent	$a^{-n} = \dfrac{1}{a^n}$
	Rules of exponents	
0.4	Radical	$\sqrt[n]{a}$
	Radicand, index	radicand $= a$; index $= n$
	Principal nth root	$\sqrt[n]{a} = b$ only if $b^n = a$ and $a \geq 0$ and $b \geq 0$ when n is even
	Fractional exponents	$a^{1/n} = \sqrt[n]{a}$ $a^{m/n} = \sqrt[n]{a^m} = (\sqrt[n]{a})^m$
	Properties of radicals	
	Rationalizing the denominator	
0.5	Algebraic expression	
	Variable	
	Constant term	
	Coefficient; leading coefficient	
	Term	
	Polynomial	$a_n x^n + \cdots + a_1 x + a_0$

Section	Key Terms	Formula
	Degree	
	Monomial	
	Binomial	
	Trinomial	
	Like terms	
	Distributive Law	$a(b + c) = ab + ac$
	Binomial products	
	Division of polynomials	
0.6	Factor	
	Common factor	
	Factoring by grouping	
	Special factorizations	
	Conjugates	$a + b; a - b$
	Quadratic polynomials	$ax^2 + bx + c$
	Factoring completely	
0.7	Algebraic fractions	
	Numerator	
	Denominator	
	Reduce	
	Product of fractions	
	Quotient of fractions	
	Common denominator	
	Lowest common denominator (LCD)	
	Addition and subtraction of fractions	
	Complex fraction	
	Rationalize the denominator	

REVIEW EXERCISES

0.1

1. Is $A \subseteq B$, if $A = \{1, 2, 5, 7\}$ and $B = \{x: x$ is a positive integer, $x \leq 8\}$?

2. Is it true that $3 \in \{x: x > 3\}$?

3. Are $A = \{1, 2, 3, 4\}$ and $B = \{x: x \leq 1\}$ disjoint?

In problems 4–7, use sets $U = \{1, 2, 3, 4, 5, 6, 7, 8, 9, 10\}$, $A = \{1, 2, 3, 9\}$, and $B = \{1, 3, 5, 6, 7, 8, 10\}$ to find the elements of the sets described.

4. $A \cup B'$ 5. $A' \cap B$

6. $(A' \cap B)'$ 7. Does $(A' \cup B')' = A \cap B$?

0.2

8. State the property of the real numbers illustrated in each case.

(a) $6 + \dfrac{1}{3} = \dfrac{1}{3} + 6$ (b) $2(3 \cdot 4) = (2 \cdot 3)4$

(c) $\dfrac{1}{3}(6 + 9) = 2 + 3$

9. Indicate whether each given expression is one or more of the following: rational, irrational, integer, natural, or meaningless.

(a) π (b) $0/6$ (c) $6/0$

10. Insert the proper sign $<$, $=$, or $>$ to replace each ☐.
    (a) π ☐ 3.14   (b) -100 ☐ 0.1   (c) -3 ☐ -12

For problems 11–18, evaluate each expression. Use a calculator when necessary.

11. $|5 - 11|$

12. $44 \div 2 \cdot 11 - 10^2$

13. $(-3)^2 - (-1)^3$

14. $\dfrac{(3)(2)(15) - (5)(8)}{(4)(10)}$

15. $2 - [3 - (2 - |-3|)] + 11$

16. $-4^2 - (-4)^2 + 3$

17. $\dfrac{4 + 3^2}{4}$

18. $\dfrac{(-2.91)^5}{\sqrt{3.29^5}}$

19. Write each inequality in interval notation, name the type of interval, and graph it on a real number line.
    (a) $0 \le x \le 5$ (b) $x \ge -3$ and $x < 7$
    (c) $(-4, \infty) \cap (-\infty, 0)$

20. Write an inequality that represents each of the following.
    (a) $(-1, 16)$ (b) $[-12, 8]$
    (c)

0.3

21. Evaluate each of the following without a calculator.

    (a) $\left(\dfrac{3}{8}\right)^0$

    (b) $2^3 \cdot 2^4$

    (c) $\dfrac{4^9}{4^3}$

    (d) $\left(\dfrac{1}{7}\right)^3\left(\dfrac{1}{7}\right)^{-4}$

22. Use the rules of exponents to simplify each of the following with positive exponents. Assume all variables are nonzero.

    (a) $x^5 \cdot x^{-7}$ (b) x^8/x^{-2} (c) $(x^3)^3$
    (d) $(y^4)^{-2}$ (e) $(-y^{-3})^{-2}$

For problems 23–28 rewrite each expression so only positive exponents remain. Assume all variables are nonzero.

23. $\dfrac{-(2xy^2)^{-2}}{(3x^{-2}y^{-3})^2}$

24. $\left(\dfrac{2}{3}x^2y^{-4}\right)^{-2}$

25. $\left(\dfrac{x^{-2}}{2y^{-1}}\right)^2$

26. $\dfrac{(-x^4y^{-2}z^2)^0}{-(x^4y^{-2}z^2)^{-2}}$

27. $\left(\dfrac{x^{-3}y^4z^{-2}}{3x^{-2}y^{-3}z^{-3}}\right)^{-1}$

28. $\left(\dfrac{x}{2y}\right)\left(\dfrac{y}{x^2}\right)^{-2}$

0.4

29. Find the following roots.
    (a) $-\sqrt[3]{-64}$ (b) $\sqrt{4/49}$

30. Write each of the following with an exponent and with the variable in the numerator.
    (a) $\sqrt{x}$ (b) $\sqrt[3]{x^2}$ (c) $1/\sqrt[4]{x}$

31. Write each of the following in radical form.
    (a) $x^{2/3}$ (b) $x^{-1/2}$ (c) $-x^{3/2}$

32. Rationalize each of the following denominators and simplify.

    (a) $\dfrac{5xy}{\sqrt{2x}}$

    (b) $\dfrac{y}{x\sqrt[3]{xy^2}}$

In problems 33–38, use the properties of exponents to simplify, so only positive exponents remain. Assume all variables are positive.

33. $x^{1/2} \cdot x^{1/3}$ 34. $y^{-3/4}/y^{-7/4}$ 35. $x^4 \cdot x^{1/4}$
36. $1/(x^{-4/3} \cdot x^{-7/3})$ 37. $(x^{4/5})^{1/2}$ 38. $(x^{1/2}y^2)^4$

In problems 39–44, simplify each expression. Assume all variables are positive.

39. $\sqrt{12x^3y^5}$

40. $\sqrt{1250x^6y^9}$

41. $\sqrt[3]{24x^4y^4} \cdot \sqrt[3]{45x^4y^{10}}$

42. $\sqrt{16a^2b^3} \cdot \sqrt{8a^3b^5}$

43. $\dfrac{\sqrt{52x^3y^6}}{\sqrt{13xy^4}}$

44. $\dfrac{\sqrt{32x^4y^3}}{\sqrt{6xy^{10}}}$

0.5

In problems 45–62, perform the indicated operations and simplify.

45. $(3x + 5) - (4x + 7)$
46. $x(1 - x) + x[x - (2 + x)]$
47. $(3x^3 - 4xy - 3) + (5xy + x^3 + 4y - 1)$
48. $(4xy^3)(6x^4y^2)$
49. $(3x - 4)(x - 1)$ 50. $(3x - 1)(x + 2)$
51. $(4x + 1)(x - 2)$ 52. $(3x - 7)(2x + 1)$
53. $(2x - 3)^2$ 54. $(4x + 3)(4x - 3)$
55. $(2x^2 + 1)(x^2 + x - 3)$ 56. $(2x - 1)^3$
57. $(x - y)(x^2 + xy + y^2)$ 58. $\dfrac{4x^2y - 3x^3y^3 - 6x^4y^2}{2x^2y^2}$
59. $(3x^4 + 2x^3 - x + 4) \div (x^2 + 1)$
60. $(x^4 - 4x^3 + 5x^2 + x) \div (x - 3)$
61. $x^{4/3}(x^{2/3} - x^{-1/3})$
62. $(\sqrt{x} + \sqrt{a - x})(\sqrt{x} - \sqrt{a - x})$

0.6

In problems 63–71, factor each expression completely.

63. $2x^4 - x^3$

64. $4(x^2 + 1)^2 - 2(x^2 + 1)^3$

65. $4x^2 - 4x + 1$ 66. $16 - 9x^2$ 67. $2x^4 - 8x^2$

68. $x^2 - 4x - 21$

69. $3x^2 - x - 2$

70. $12x^2 - 23x - 24$

71. $16x^4 - 72x^2 + 81$

72. Factor as indicated: $x^{-2/3} + x^{-4/3} = x^{-4/3}(?)$

0.7

73. Reduce each of the following to lowest terms.

(a) $\dfrac{2x}{2x + 4}$

(b) $\dfrac{4x^2y^3 - 6x^3y^4}{2x^2y^2 - 3xy^3}$

In problems 74–80, perform the indicated operations and simplify.

74. $\dfrac{x^2 - 4x}{x^2 + 4} \cdot \dfrac{x^4 - 16}{x^4 - 16x^2}$

75. $\dfrac{x^2 + 6x + 9}{x^2 - 7x + 12} \div \dfrac{x^2 + 4x + 3}{x^2 - 3x - 4}$

76. $\dfrac{x^4 - 2x^3}{3x^2 - x - 2} \div \dfrac{x^3 - 4x}{9x^2 - 4}$

77. $1 + \dfrac{3}{2x} - \dfrac{1}{6x^2}$

78. $\dfrac{1}{x - 2} - \dfrac{x - 2}{4}$

79. $\dfrac{x + 2}{x^2 - x} - \dfrac{x^2 + 4}{x^2 - 2x + 1} + 1$

80. $\dfrac{x - 1}{x^2 - x - 2} - \dfrac{x}{x^2 - 2x - 3} + \dfrac{1}{x - 2}$

In problems 81 and 82, simplify each complex fraction.

81. $\dfrac{x - 1 - \dfrac{x - 1}{x}}{\dfrac{1}{x - 1} + 1}$

82. $\dfrac{x^{-2} - x^{-1}}{x^{-2} + x^{-1}}$

83. Rationalize the denominator of $\dfrac{3x - 3}{\sqrt{x} - 1}$ and simplify.

84. Rationalize the numerator of $\dfrac{\sqrt{x} - \sqrt{x - 4}}{2}$ and simplify.

APPLICATIONS

85. *Job effectiveness factors* In an attempt to determine some off-the-job factors that might be indicators of on-the-job effectiveness, a company made a study of 200 of its employees. They were interested in whether the employees had been recognized for superior work by their supervisors within the past year, whether they were involved in community activities, and whether they followed a regular exercise plan. The company found the following.

    30 answered "yes" to all three
    50 were recognized and they exercised
    52 were recognized and were involved in the community
    77 were recognized
    37 were involved in the community but did not exercise
    95 were recognized or were involved in the community
    95 answered "no" to all three

(a) Draw a Venn diagram that represents this information.
(b) How many exercised only?
(c) How many exercised or were involved in the community?

Campaign Management

A politician is trying to win election to the city council, and as his campaign manager, you need to decide how to promote the candidate. There are three ways you can do so: you can send glossy, full-color pamphlets to registered voters of the city, you can run a commercial during the television news on a local cable network, and/or you can buy a full-page ad in the newspaper.

Two hundred fifty thousand voters live in the city, and 36% of them read the newspaper. Fifty thousand voters watch the local cable network news, and 30% of them also read the newspaper.

You also know that the television commercial would cost $40,000, the newspaper ad, $27,000, and the pamphlets mailed to voters, 90 cents each, including printing and bulk-rate postage.

Suppose the success of the candidate depends on your campaign reaching at least 125,000 voters, and because your budget is limited, you must achieve this goal at a minimum cost. What would your plan and the cost of that plan be?

If you need help devising a method of solution for this problem, try answering the following questions first.

1. How many voters in the city read the newspaper but do not watch the local cable television news?
2. How many voters read the newspaper or watch the local cable television news or both?
3. Complete the following chart indicating the number of voters reached by each promotional option, the total cost, and the cost per voter reached.

	Number of Voters Reached	Total Cost	Cost per Voter Reached
Pamphlet			
Television			
Newspaper			

4. Now, explain your plan and the cost of that plan.

LINEAR MODELS AND SPECIAL FUNCTIONS

Warm-up

Prerequisite problem type	for section	Answer	Section for review:

Evaluate:
(a) $2(-1)^3 - 3(-1)^2 + 1$
(b) $3(-3) - 1$
(c) $14(10) - 0.02(10^2)$
(d) $\dfrac{3-1}{4-(-2)}$ (e) $\dfrac{-1-3}{2-(-2)}$
(f) $\dfrac{3(-8)}{4}$ (g) $2\left(\dfrac{23}{9}\right) - 5\left(\dfrac{2}{9}\right)$

for section: 1.1, 1.2, 1.4, 1.5, 1.7, 1.8

Answer:
(a) -4
(b) -10
(c) 138
(d) $\dfrac{1}{3}$ (e) -1
(f) -6 (g) 4

Section for review: 0.2 Signed numbers

Locate on a number line:
(a) The integers from -1 to 3 inclusive

(b) The integers from -3 to 2 inclusive

for section: 1.2, 1.6, 1.8

Answer:
(a) a number line with points marked at $-3, -2, -1, 0, 1, 2, 3, 4$

(b) a number line with points marked at $-4, -3, -2, -1, 0, 1, 2, 3$

Section for review: 0.2 Real number line

(a) $\dfrac{1}{x}$ is *undefined* for which real numbers?

(b) $\sqrt{x-4}$ is a real number for which values of x?

for section: 1.4, 1.5, 1.6

Answer:
(a) Undefined for $x = 0$
(b) $x \geq 4$

Section for review: 0.2 Real numbers
0.4 Radicals

Identify the coefficient of x and the constant term for
(a) $-\frac{3}{2}x + 2$
(b) $9x$
(c) $7x - 300$

for section: 1.1, 1.6, 1.7, 1.8, 1.9

Answer:

	Coeff.	*Const.*
(a)	$-\frac{3}{2}$	2
(b)	9	0
(c)	7	-300

Section for review: 0.5 Algebraic expressions

Simplify:
(a) $4(-c)^2 - 3(-c) + 1$
(b) $[3(x + h) - 1] - [3x - 1]$
(c) $12\left(\dfrac{3x}{4} + 3\right)$
(d) $9x - (300 + 2x)$
(e) $-\frac{1}{5}[x - (-1)]$
(f) $2(2y + 3) + 3y$

for section: 1.1, 1.4, 1.5, 1.6, 1.7, 1.8, 1.9

Answer:
(a) $4c^2 + 3c + 1$
(b) $3h$
(c) $9x + 36$
(d) $7x - 300$
(e) $-\frac{1}{5}x - \frac{1}{5}$
(f) $7y + 6$

Section for review: 0.5 Algebraic expressions

CHAPTER 1

LINEAR EQUATIONS AND FUNCTIONS

A wide variety of problems from business, the social sciences, and the life sciences may be solved by using equations. Managers and economists use equations and their graphs to study costs, sales, national consumption, or supply and demand. Social scientists may plot demographic data or try to develop equations that predict population growth, voting behavior, or learning and retention rates. Life scientists use equations to model the flow of blood or the conduction of nerve impulses and to test theories or develop new ones by plotting experimental evidence.

In this chapter we begin by investigating linear equations and their graphs. We will also define the concepts of relation and function, introduce functional notation, and then relate all of these back to equations and graphs.

Numerous applications of mathematics are given throughout the text, but all chapters contain special sections emphasizing business and economics applications. In particular, this chapter introduces three important applications that will be expanded and used throughout the text as increased mathematical skills permit: national consumption; supply and demand as functions of price (market analysis); and total cost, total revenue, and total profit as functions of the quantity produced or sold (theory of the firm).

Linear equations are used by most businesses to predict such things as future revenues and costs. Linear functions are used in economics, biology, and sociology to relate data, and are also used to a large extent in accounting courses. Keynes used the linear function as the model for national consumption. In this chapter we discuss several important features of Keynesian analysis. We also formulate linear supply and demand functions and then determine the price and quantity at which market equilibrium occurs. We use linear revenue and cost functions to obtain profit functions and to find break-even points, and we solve problems involving linear equations from the social and life sciences.

1.1 SOLUTION OF LINEAR EQUATIONS IN ONE VARIABLE

❑ To solve linear equations in one variable
❑ To solve applied problems by using linear equations

APPLICATION PREVIEW

Woodwright Industries manufactures interior, raised panel doors with a hardwood veneer from a pressed mixture of scrap hardwood chips. They break even when total revenue *equals* total costs. If x represents the number of these doors that Woodwright sells and if total revenue is given by the expression $98x$ and total costs are given by $48x + 12,000$, then Woodwright breaks even when total revenue = total cost, that is, when

$$98x = 48x + 12,000.$$

In this section, we begin to study equality relationships such as this one.

An **equation** is a statement that two quantities or algebraic expressions are equal. The two quantities on either side of the equal sign are called **members** of the equation. For example, $2 + 2 = 4$ is an equation with members $2 + 2$ and 4; $3x - 2 = 7$ is an equation with $3x - 2$ as its left member and 7 as its right member. Note that the equation $7 = 3x - 2$ is the same statement as $3x - 2 = 7$. An equation with one literal number, such as $3x - 2 = 7$, is known as an equation in one variable. The literal number (x in this case) is called a **variable** because the value of the literal number determines whether the equation is true or not. For example, $3x - 2 = 7$ is true only for $x = 3$. Finding the value(s) of the variable(s) that make the equation true, that is, finding the **solutions,** is called **solving the equation.** The set of solutions of an equation is called a **solution set** of the equation. The variable in an equation is sometimes called the **unknown.**

Some equations involving variables are true only for certain values of the variables, while others are true for all values. Equations that are true for all values of the variables are called **identities.** The equation $2(x - 1) = 2x - 2$ is an example of an identity. Equations that are true only for certain values of the variables are called **conditional equations** or simply **equations.**

Two equations are said to be **equivalent** if they have exactly the same solution set. For example,

$$4x - 12 = 16$$
$$4x = 28$$
$$x = 7$$

are equivalent equations because they all have the same solution, namely 7. We can often solve a complicated linear equation by finding an equivalent equation whose solution is easily found. We use the following properties of equality to reduce an equation to a simple equivalent equation.

Properties of Equality	Examples
Substitution Property The equation formed by substituting one expression for an equal expression is equivalent to the original equation.	$3(x - 3) - \frac{1}{2}(4x - 18) = 4$ is equivalent to $3x - 9 - 2x + 9 = 4$ and to $x = 4$. We say the solution set is $\{4\}$, or the solution is 4.
Addition Property The equation formed by adding the same quantity to both sides of an equation is equivalent to the original equation.	$x - 4 = 6$ is equivalent to $x = 10$. $x + 5 = 12$ is equivalent to $x = 7$. (Subtracting 5 from both sides is equivalent to adding -5 to both sides.)
Multiplication Property The equation formed by multiplying both sides of an equation by the same nonzero quantity is equivalent to the original equation.	$\frac{1}{3}x = 6$ is equivalent to $x = 18$, and $5x = 20$ is equivalent to $x = 4$. (Dividing both sides by 5 is equivalent to multiplying both sides by $\frac{1}{5}$.)

If an equation contains one variable and if the variable occurs to the first degree, the equation is called a **linear equation in one variable.** The three properties above permit us to reduce any linear equation in one unknown to an equivalent equation whose solution is obvious. We may solve linear equations in one unknown by using the following procedure:

Solving a Linear Equation

Procedure	Example
To solve a linear equation in one unknown:	Solve $\dfrac{3x}{4} + 3 = \dfrac{2(x - 1)}{6}$.
1. If the equation contains fractions, multiply both sides by the least common denominator (LCD) of the fractions.	1. LCD is 12. $12\left(\dfrac{3x}{4} + 3\right) = 12\left(\dfrac{2(x - 1)}{6}\right)$
2. Remove any parentheses in the equation.	2. $9x + 36 = 4x - 4$
3. Perform any additions or subtractions to get all terms containing the variable on one side and all other terms on the other side.	3. $9x + 36 - 4x = 4x - 4 - 4x$ $5x + 36 = -4$ $5x + 36 - 36 = -4 - 36$ $5x = -40$
4. Divide both sides of the equation by the coefficient of the variable.	4. $\dfrac{5x}{5} = \dfrac{-40}{5}$ $x = -8$
5. Check the solution by substitution in the original equation.	5. $\dfrac{3(-8)}{4} + 3 = \dfrac{2(-8 - 1)}{6}$

EXAMPLE 1

Solve $2(3y - 1) = 4(y - 5)$ for y.

Solution

1. No fractions are involved.
2. $6y - 2 = 4y - 20$ Removing parentheses
3. $2y - 2 = -20$ Subtracting $4y$ from both sides
   $2y = -18$ Adding 2 to both sides
4. $y = -9$ Dividing both sides by 2
5. $2(-28) = 4(-14)$ Checking

EXAMPLE 2

The relation between degrees Celsius and degrees Fahrenheit is given by

$$F = \frac{9}{5}C + 32.$$

What Celsius temperature is equivalent to 176°F?

Solution

If $F = 176$, the equation is $176 = \frac{9}{5}C + 32$. Solving for C gives

$$880 = 9C + 160$$
$$720 = 9C$$
$$80 = C.$$

Thus 80°C is equivalent to 176°F.

CHECKPOINT

1. Solve for x.

   (a) $x + 9 = -2$ (b) $4x = 7$ (c) $-\dfrac{1}{3}x = 15$

   (d) $3(x - 7) = 19 - x$ (e) $4 - 5x = 4 + x$

2. Let $\dfrac{5(x - 3)}{6} - x = 1 - \dfrac{x}{9}$.

   (a) Find the LCD of the denominators.
   (b) Multiply both sides of the equation by the LCD.
   (c) Complete the solution.

EXAMPLE 3

Suppose that the relationship between a firm's profit P and the number x of items sold can be described by the equation

$$5x - 4P = 1200.$$

(a) How many units must be produced and sold for the firm to make a profit of $150?
(b) Solve this equation for P in terms of x.
(c) Find the profit when 240 units are sold.

Solution

(a) $5x - 4(150) = 1200$ (b) $5x - 1200 = 4P$

$$5x - 600 = 1200 \qquad\qquad P = \frac{5}{4}x - 300$$

$$5x = 1800$$
$$x = 360 \text{ units}$$

(c) $P = \dfrac{5}{4}x - 300$

$$P = \frac{5}{4}(240) - 300 = 0$$

We say that the firm breaks even (profit $P = 0$) when $x = 240$ units.

We now return to the Application Preview problem.

EXAMPLE 4

Suppose Woodwright Industries' total revenue from the sale of x interior, raised panel doors is given by

$$R = 98x,$$

and their total costs are given by

$$C = 48x + 12{,}000.$$

Woodwright breaks even when their total revenues equal their total costs. Find the number of doors that Woodwright must sell in order to break even.

Solution

Break-even occurs when $R = C$, so we must solve for x in

$$98x = 48x + 12{,}000$$
$$50x = 12{,}000$$
$$x = \frac{12{,}000}{50} = 240.$$

Thus, Woodwright Industries breaks even if 240 doors are sold.

With an applied problem, it is frequently necessary to convert the problem from its stated form into one or more equations from which the problem's solution can be found. The following guidelines may be useful in solving stated problems.

GUIDELINES FOR SOLVING STATED PROBLEMS

1. To begin, read the problem carefully to determine what you are to find. Use variables to represent the quantities to be found.

2. Reread the problem and use your variables to translate given information into algebraic expressions. Often, drawing a figure is helpful.

3. Use the algebraic expressions and the problem statement to formulate an equation (or equations).

4. Solve the equation.

5. Check the solution in the problem, not just in your equation or equations. The answer should satisfy the conditions.

EXAMPLE 5

A bicycle costs the wholesaler $28.00. What will the retailer sell it for if the wholesaler's markup is 20% of the wholesale selling price and the retailer's markup is 30% of the retail selling price?

Solution

We want to find the retailer's selling price. Let r = the retailer's selling price. Figure 1.1 illustrates the facts and how they relate to the variable r. We see that $0.70r$ represents both the retailer's cost and the wholesaler's selling price. Figure 1.1 also shows that

$$28 + 20\% \text{ (wholesale selling price)} = \text{wholesale selling price.}$$

Figure 1.1

That is,

$$28 + 0.20(0.70r) = 0.70r$$
$$28 + 0.14r = 0.70r$$
$$28 = 0.56r$$
$$50 = r.$$

Thus, the retailer's selling price is $50.

CHECKPOINT SOLUTIONS

1. (a) $x + 9 = -2$
$x = -2 - 9$
$x = -11$

   (b) $4x = 7$
$x = \dfrac{7}{4}$

   (c) $-\dfrac{1}{3}x = 15$
$x = (15)(-3)$
$x = -45$

   (d) $3(x - 7) = 19 - x$
$3x - 21 = 19 - x$
$4x = 40$
$x = 10$

   (e) $4 - 5x = 4 + x$
$-5x = x$
$0 = 6x$
$\dfrac{0}{6} = x$
$x = 0$

2. $\dfrac{5(x - 3)}{6} - x = 1 - \dfrac{x}{9}$

   (a) LCD of 6 and 9 is 18.

   (b) $18\left(\dfrac{5x - 15}{6}\right) - 18(x) = 18(1) - 18\left(\dfrac{x}{9}\right)$

   (c) $3(5x - 15) - 18x = 18 - 2x$
$15x - 45 - 18x = 18 - 2x$
$-45 - 3x = 18 - 2x$
$-63 = x$

EXERCISE 1.1

In problems 1–28, solve each equation.

1. $2x + 3 = 5$

2. $4x + 2 = 5$

3. $4x - 7 = 8x + 2$

4. $3x + 2 = 7x + 2$

5. $x + 8 = 8(x + 1)$

6. $x + x + x = x$

7. $8(x - 3) = 6(3x - 4)$

8. $3(x - 2) = 4(3 - x)$

9. $3x + \dfrac{1}{2} = 8$

10. $\dfrac{x}{2} + 4 = \dfrac{1}{3}$

11. $\dfrac{5x}{2} - 4 = \dfrac{2x - 7}{6}$

12. $\dfrac{2x}{3} - 1 = \dfrac{x - 2}{2}$

13. $\dfrac{5x - 1}{9} = \dfrac{5(x - 1)}{6}$

14. $\dfrac{6x + 5}{2} = \dfrac{5(2 - x)}{3}$

15. $0.4x = 16$

16. $0.4x + 5 = 9$

17. $2x + 0.6 = 4$

18. $0.3x + 0.4 = 1$

19. $2(x - 7) = 5(x + 3) - x$

20. $3(x - 4) = 4 - 2(x + 2)$

21. $2[x - (4 + 2x) + 3] = 2x + 2$

22. $3[2x - (2 + 4x) - 1] = 3(x - 4)$

23. $x + \dfrac{1}{3} = 2\left(x - \dfrac{2}{3}\right) - 6x$

24. $\dfrac{3x}{4} - \dfrac{1}{3} = 1 - \dfrac{2}{3}\left(x - \dfrac{1}{6}\right)$

25. $x - 2\left(x - \dfrac{1}{5}\right) = 2\left(x - \dfrac{4}{5}\right) - 3$

26. $\left(\dfrac{3x}{4} - \dfrac{2}{5}\right)2 - x = x - 3\left(\dfrac{x}{2} - \dfrac{1}{5}\right)$

27. $\dfrac{1}{x + 2} - \dfrac{1}{2} = \dfrac{x}{x + 2}$ 28. $\dfrac{1}{4x} + \dfrac{2}{x} = \dfrac{5}{12}$

In problems 29–32, use a calculator to solve each equation. Round your answer to three decimal places.

29. $3.259x - 8.638 = -3.8(8.625x + 4.917)$

30. $3.319(14.1x - 5) = 9.95 - 4.6x$

31. $0.000316x + 9.18 = 2.1(3.1 - 0.0029x) - 4.68$

32. $3.814x = 2.916(4.2 - 0.06x) + 5.3$

In problems 33–36, solve each equation for y in terms of x.

33. $3x + 2y = 40$

34. $4x - 3y + 18 = 0$

35. $5(x - y - 1) = 6(y - x + 1)$

36. $4(x + y) = 3(x - 1) + 2(y - 1)$

APPLICATIONS

37. **Wildlife management** In wildlife management, the capture-mark-recapture technique is used to estimate the populations of fish or birds in an area or to measure the infestation of insects such as Japanese beetles. Suppose 100 individuals of the species being studied are caught, marked, and released, and one week later 100 more are caught. To estimate the total number of individuals, the following relationship is used:

$$\frac{\text{Total marked found in 2nd capture}}{\text{Total in 2nd capture}} = \frac{\text{Total number marked}}{\text{Total population}}.$$

(a) If in the second capture of 100, it is found that 3 are marked, what is the total population?

(b) Suppose that 1000 beetles are captured, marked, and released. Suppose further that in the second capture of 1000 it is found that 63 are marked. What is the population estimate?

38. **Simple interest** Joe bought a radio for $99.95. He paid $19.95 down and agreed to pay the balance plus a finance charge of $2.00 in three months. What rate of simple interest did he pay? (Use the formula: interest = principal × rate × time.)

39. **Fahrenheit-Celsius** The equation $F = \frac{9}{5}C + 32$ describes the relation between temperature readings in Fahrenheit and Celsius. What Celsius temperature is equivalent to $-40°F$?

40. **Height-weight relationships** It has been noted that for adults over 5 feet tall in the northeast United Sates, their weight is related to their height according to

$$3w + 110 = 11(h - 20)$$

where w is measured in pounds and h is measured in inches. Use the formula above to answer the following.

(a) Find the weight of an adult whose height is 5 feet, 6 inches.

(b) Find the height of adults weighing 160 pounds.

41. **Cigarette use** Data from the National Institute on Drug Abuse indicates that the percentage p of high school seniors who haves tried cigarettes can be modeled (that is, predicted with some accuracy) by the equation

$$p = 75.4509 - 0.706948t,$$

where t is the number of years past 1975.

(a) What percentage does this model predict for 1998?

(b) This model ceases to be effective when p reaches zero (and then becomes negative), and perhaps before that. Find the year when $p = 0$.

42. **Heat and humidity index** One way to measure people's discomfort during extremes of heat and humidity is with the apparent temperature A, which is given by

$$A = 2.70 + 0.885t - 78.7h + 1.20th,*$$

where t is the Fahrenheit temperature and h is the relative humidity (expressed as a decimal). On June 15,

---

* Bosch, W., and C. G. Cobb, "Temperature-Humidity Indices," UMAP Unit 691, *The UMAP Journal*, 10(3), Fall 1989, 237–256.

1994, in Washington, D.C., the temperature reached 101°F; for this reading the apparent temperature formula becomes

$$A = 92.085 + 42.5h.$$

If the local Washington, D.C., news reported that this 101°F reading was equivalent to an apparent temperature of 111°F, what was the relative humidity (as a decimal)? What was the percent relative humidity?

43. **Sales projections** Data in an article in the *Atlanta Journal* on July 14, 1993, indicated that the growth of sales, S, (in millions of dollars) of the Chick-fil-A's restaurant chain can be modeled by

$$S = 241.33 + 29t,$$

where t is the number of years past 1988. In what year does this model project Chick-fil-A's sales will reach $1 billion?

44. **Investing** In Federal Signal Corporation's publication of March 7, 1994, *Notice of Annual Meeting,* data was given that compared the cumulative total return on a $100 investment made in 1988. If t is the number of years past 1988, then this cumulative total return can be modeled by R_{FS} for Federal Signal Corporation and by R_{SP} for the S & P 400 Index as follows.

$$R_{FS} = 85.714t + 88.381$$
$$R_{SP} = 17.1714t + 104.238$$

Assuming these models are valid, find the year in which the cumulative total return of
(a) Federal Signal is predicted to reach $850,
(b) S & P 400 Index is predicted to reach $300.

45. **Organizational impact** The Beaver-Castle Girl Scout Council of Beaver, Pennsylvania, determined the number of youths it had served by adding its active registrations and the number of dropouts. If the drop-out rate was one-third of the total served, find the total number of youths served when there were 6000 youths actively registered.

46. **Population** A city has a population P at the beginning of a year. The birth rate during the year was 10 per thousand and the death rate was 12 per thousand (of the year's original population). During the year, 360 people moved into the city and 190 moved away. If the population at the end of the year is 30,110, what was the population at the beginning of the year?

47. **Insect behavior** The number of times n per minute that a cricket chirps is related to the Fahrenheit temperature T. This relationship can be approximated by

$$7n - 12T = 52.$$

(a) Solve this equation for T.
(b) If you count 28 chirps in 15 seconds, find the approximate temperature.

48. **Seawater pressure** In seawater, the pressure p is related to the depth d according to

$$33p - 18d = 495,$$

where d is in feet and p is in pounds per square inch.
(a) Solve this equation for p in terms of d.
(b) The *Titanic* was discovered at a depth of 12,460 ft. Find the pressure at this depth.

49. **Investments** The total amount of a simple interest investment is given by

$$A = P + Prt.$$

What principal P must be invested for $t = 5$ years at the simple interest rate $r = 10\%$ so that the amount A grows to $6,000?

50. **Sales tax** The total price of a new car (including 6% sales tax) is $10,033. How much of this is tax?

51. **Markups** A retailer wants a 30% markup on the selling price of an item that costs him $214.90. What selling price should he charge?

52. **Markups** A toaster costs the Ace Department Store $22.74. If the store marks the toaster up by 40% of the selling price, what is the selling price?

53. **Markups** A room air conditioner costs the wholesaler $154.98. If the wholesaler's markup is 10% of the wholesale selling price and if the retailer's markup is 30% of the retail selling price, for what price does the retailer sell the air conditioner?

54. **Markups** An electric mixer retails for $48.54, which includes a markup of 40% for the retailer and a markup of 20% for the wholesaler. If these markups are based on selling price,
(a) what did the mixer cost the retailer?
(b) what did the mixer cost the wholesaler?

55. **Markups** A dinette set costs a wholesaler $189 and her markup is 10% of her selling price. If the retailer's markup is 30% of his selling price for the item, what is the retailer's selling price?

56. ***Markups*** A wholesaler wants to establish a list price for an item so that his selling price gives a trade discount of 10% of the list price. If he makes a markup of 20% of this selling price, and if his cost is $160, what list price should he establish?

57. ***Course grades*** To earn an A in a course, a student must get at least a 90 average on four tests and a final exam, with the final exam weighted twice that of any one test. If the four test scores are 93, 69, 89, and 97, what is the lowest score the student can earn on the final exam and still get an A in the course?

58. ***Course grades*** Suppose a professor counts the final exam as being equal to each of the other tests in her course, and she will also change the lowest test score to match the final exam score if the final exam score is higher. If a student's four test scores are 83, 67, 52, and 90, what is the lowest score the student can earn on the final exam and still obtain at least an 80 average for the course?

59. ***Profit and loss*** An antique collector sold two pieces of furniture for $480 each. For one of them, this represented a 20% loss; for the other, it was a 20% profit (based on the cost of each item). How much did he make or lose on the transaction?

60. ***Salary increases*** A woman making $2000 per month has her salary reduced by 10% due to sluggish sales. One year later, after a dramatic improvement in sales, she is given a 20% raise over her reduced salary. Find her salary after the raise. What percent change is this from the $2000 per month?

61. ***Profit*** A car dealer purchases 20 new automobiles for $8000 each. If he sells 16 of them at a profit of 20%, for how much must he sell the remaining 4 to obtain an average profit of 18%?

62. ***Investment mix*** A woman has $90,000 to invest. She has chosen one relatively safe investment fund that has an annual yield of 10% and another riskier one that has a 15% annual yield. How much should she invest in each fund if she would like to earn $10,000 per year from her investments?

1.2   GRAPHING LINEAR EQUATIONS

Objectives

❏ To graph linear equations by plotting points
❏ To graph linear equations by using intercepts

APPLICATION PREVIEW

The dollar volume of transactions at automatic teller machines (ATMs) has increased as the number of machines has increased. This relationship can be modeled by

$$y = 0.1369x - 5.091255,*$$

where y is billions of dollars of transactions and x is the number of terminals (in thousands). To get a better feel for the relation between the number of machines and the dollar value of transactions, we could graph the equation that relates x and y. In this section, we'll look at how to graph equations with two variables.

Equations may have more than one variable. The equation $y = 3x - 2$, for example, has two variables. There are an infinite number of solutions to this equation because any pair of values that satisfies the equation is a solution. For example, $x = 1$, $y = 1$ is a solution, as is $x = 4$, $y = 10$. We can find

* Equation developed with data from the Electronic Funds Transfer Association published in *Statistical Abstract of the United States,* 1993, p. 517.

other solutions by substituting values for x and finding the value of y that satisfies the equation. Thus, if we let $x = 5$, then letting $y = 3(5) - 2 = 13$ will satisfy the equation.

We denote an **ordered pair** of real numbers as (a, b), where a and b are real numbers. The number a is called the **first component** or **first coordinate** of the ordered pair, and b is called the **second component** or **second coordinate.** The ordered pairs (a, b) and (c, d) are said to be equal if and only if $a = c$ and $b = d$. We may write a **solution** to an equation in two variables x and y as an *ordered pair* of numbers (a, b), where the first number a is the x-value of the solution and the second number b is the y-value.

Just as we can use a number line to graph the real numbers (see Section 0.2), we can use a rectangular (or Cartesian) coordinate plane to graph ordered pairs of real numbers. The coordinate plane is especially useful, since it lets us geometrically represent solutions to equations in two variables.

We construct the coordinate system by drawing two real number lines perpendicular to each other so that they intersect at their origins. (See Figure 1.2.) The point of intersection is called the **origin** of the system, and the two lines are called the **coordinate axes.** We call the horizontal axis the x-axis (or the first coordinate axis), and the vertical axis the y-axis (or the second coordinate axis). The scales on the two axes are frequently (but not always) the same. The axes divide the plane into four parts, called **quadrants.**

Any point in the plane can be represented by an ordered pair of real numbers. The point on the x-axis representing the real number a is denoted by the ordered pair $(a, 0)$, and the point on the y-axis representing the real number b is denoted by the ordered pair $(0, b)$. Any point P that is a units from the y-axis (right or left, depending on whether a is positive or negative) and b units from the x-axis (up or down, depending on whether b is positive or negative) is denoted by the ordered pair (a, b).

Note that the line drawn through the point denoted by (a, b) and perpendicular to the x-axis intersects the x-axis at $(a, 0)$ and that the line drawn through (a, b) and perpendicular to the y-axis intersects the y-axis at $(0, b)$. The values a and b in the ordered pair associated with the point P are called the **rectangular coordinates** of the point. The first coordinate of a point is called the **x-coordinate** (or **abscissa**), and the second coordinate is called the **y-coordinate** (or **ordinate**). To plot a point $P(a, b)$ means to locate the point with x-coordinate a and y-coordinate b.

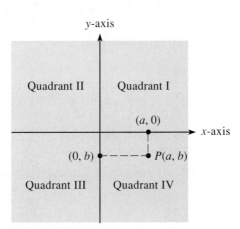

Figure I.2

EXAMPLE 1

Plot the points $A(5, 2)$, $B(-4, 1)$, $C(-4, -4)$, and $D(1, -5)$.

Solution

See Figure 1.3.

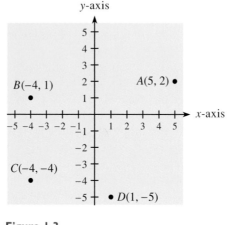

Figure 1.3

The **graph** of an equation is the picture that is drawn by plotting the points whose coordinates (x, y) satisfy the equation. Since there are an infinite number of such points, we cannot hope to plot them all. But we can plot enough points to determine the general outline of the graph. Then we usually connect the points with a smooth curve. This is called the **point-plotting method** of sketching a graph.

One very special equation is the **linear equation.** A linear equation can be written in the form $ax + by = c$ where a, b, and c are constants, and a and b are not both zero.

The graph of a linear equation may be drawn by plotting the points whose coordinates (x, y) satisfy the equation $ax + by = c$. Because the graph of a linear equation is always a straight line, only two points are required to determine its graph (two points determine a straight line). A third point may be plotted as a check on the graph.

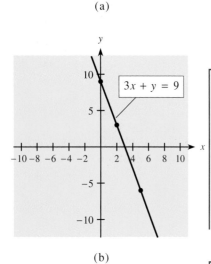

(a)

(b)

Figure 1.4

EXAMPLE 2

Graph the equation $3x + y = 9$.

Solution

1. Plot any points that satisfy the equation.
   (a) If $x = 0$, $y = 9$, so $(0, 9)$ is one point.
   (b) If $x = 5$, $y = -6$, so $(5, -6)$ is another point.

2. Plot a third point as a check: if $x = 2$, $y = 3$, so $(2, 3)$ should be in line with the other two points. (See Figure 1.4 (a).)

3. Draw a straight line through the points. (See Figure 1.4(b).)

EXAMPLE 3

Graph (a) $x = 2$ and   (b) $y = 3$.

Solution

(a) We may think of this equation as $x + 0y = 2$, which indicates that $x = 2$ for all values of y. That is, y can have any value and x will still be 2. (See Figure 1.5.)

(b) Similarly $y = 3$ is the same as $0x + y = 3$. Hence, x may have any value and y will still be 3. (See Figure 1.5.)

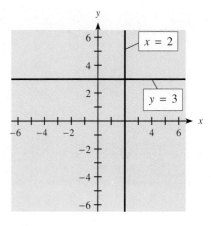

The points where $y = 0$ are points at which a graph intersects the x-axis, and are called the **x-intercepts** of the graph; **y-intercepts** are points where the graph intersects the y-axis.

Figure 1.5

Intercepts

a. To find the **y-intercepts(s)** of the graph of an equation, set $x = 0$ in the equation and solve for y.

b. To find the **x-intercept(s),** set $y = 0$ and solve for x.

For the graph in Figure 1.4 (a), we see that the y-intercept is $(0, 9)$. If we set $y = 0$ in $3x + y = 9$ (Figure 1.4 (b)), and solve for x, we see that the x-intercept is $(3, 0)$.

EXAMPLE 4

A business property is purchased for \$122,970 and depreciated over a period of 10 years. Its value y is related to the number of months of service x by the equation

$$4099x + 4y = 491{,}880.$$

Find the x-intercept and the y-intercept and use them to sketch the graph of the equation.

Solution

$$x\text{-intercept: } y = 0 \quad \text{gives} \quad 4099x = 491{,}880$$
$$x = 120.$$

Thus $(120, 0)$ is the x-intercept.

$$y\text{-intercept: } x = 0 \quad \text{gives} \quad 4y = 491{,}880$$
$$y = 122{,}970.$$

Thus (0, 122,970) is the y-intercept. The graph is shown in Figure 1.6. Notice that the units on the x- and y-axes are different and that the y-intercept corresponds to the value of the property 0 months after purchase. That is, the

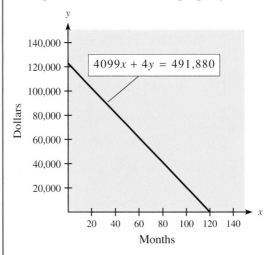

y-intercept gives the purchase price. The x-intercept corresponds to the number of months that have passed before the value is 0; that is, the property is fully depreciated after 120 months, or 10 years. Notice that only positive values for x and y make sense in this application, so only the Quadrant I portion of the graph is shown.

Figure 1.6

Despite the ease of using intercepts to graph linear equations, this method is not always the best. For example, vertical lines, horizontal lines, or lines that pass through the origin may have a single intercept, and if a line has both intercepts very close to the origin, using the intercepts may lead to an inaccurate graph.

CHECKPOINT

1. Graph $4x - y = 8$.
2. Find the intercepts of $2y = -(13x + 52)$.

EXAMPLE 5

As mentioned in the Application Preview, the dollar volume of transactions at ATMs can be modeled by

$$y = 0.1369x - 5.091255,$$

where y is billions of dollars of transactions and x is the number of terminals (in thousands).
(a) Find the point that corresponds to $x = 100$.
(b) Write a sentence that explains the meaning of the point found in (a).
(c) Graph the equation.

Solution

(a) If $x = 100$, then $y = 0.1369\,(100) - 5.091255$

$$y = 13.69 - 5.091255 = 8.598745.$$

Thus, the point is $(100, 8.598745)$.

(b) The point in (a) means that when there were 100 thousand ATM terminals, there were (approximately) \$8.598745 billion in ATM transactions.

(c) From (a) we have one point on the graph. Another is $x = 200$, $y = 0.1369\,(200) - 5.091255 = 22.288745$ or $(200, 22.288745)$. A third (check) point is $(150, 15.443745)$. The decimal values make it difficult to exactly locate these points. Figure 1.7 shows the points and the graph. Notice that part of the graph has points with negative y-coordinates. Certainly this portion cannot be part of the model.

$$y = 0.1369x - 5.091255$$

Figure 1.7

CHECKPOINT SOLUTIONS

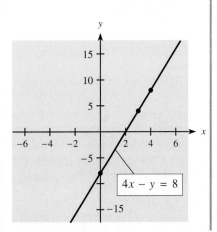

$$4x - y = 8$$

Figure 1.8

1. If $x = 0$, then $y = -8$, so $(0, -8)$ is on the graph. If $x = 3$, then $y = 4$, so $(3, 4)$ is on the graph. A third point $x = 4$, $y = 8$ or $(4, 8)$ is also plotted as a check. The graph is shown in Figure 1.8.

2. For $2y = -(13x + 52)$

x-intercept $(y = 0)$:

$$0 = -(13x + 52)$$
$$0 = 13x + 52$$
$$-52 = 13x$$
$$-4 = x$$

The x-intercept is $(-4, 0)$.

y-intercept $(x = 0)$:

$$2y = -(52)$$
$$y = -26$$

The y-intercept is $(0, -26)$.

EXERCISE 1.2

Graph the following linear equations.

1. $y = 6x - 2$ 2. $y = 2x + 8$ 3. $y = \frac{x}{3} + \frac{2}{3}$

4. $y = \frac{4x}{5} - \frac{1}{2}$ 5. $2x - 3y = 12$ 6. $3x + 2y = 0$

7. $6x + 5y = 9$ 8. $4x + 5y = 8$ 9. $y = 3$

10. $x = -5$ 11. $x = 0$ 12. $y = 0$

13. $5y + 2x = 3(y - 1)$

14. $3x - 4(x - 3) = 2(y + x)$

In problems 15–18, match each of the following equations with one of the graphs labeled (a)–(e) shown below these equations.

15. $3x + 4y = 12$ 16. $3x - 4y = 12$
17. $4x - 3y = 12$ 18. $4x + 3y = 12$

(a)

(b)

(c)

(d)

(e)

Graph the following linear equations.

19. $0.6x - 0.3y = 1.2$ 20. $0.1x - 0.3y = 0.9$

21. $\frac{x}{4} - \frac{y}{6} = 1$ 22. $\frac{y}{5} - \frac{x}{4} = 1$ 23. $x + \frac{y}{2} = 2$

24. $\frac{x}{3} - \frac{y}{5} = 1$ 25. $\frac{x}{2} - \frac{y}{4} = 1$ 26. $\frac{x}{2} + \frac{y}{5} = 1$

APPLICATIONS

27. **Fish population and pollution** The fish population y in a certain river is related to the tons of pollutants x according to the following:

$$y = 100{,}000 - 1500x.$$

(a) Sketch the graph of this equation.
(b) Suppose that 17.5 tons of pollutants is deemed critical. How many fish would be in the stream at the critical level of pollution?
(c) The points (0, 100,000) and (50, 25,000) lie on this line. Write a sentence that explains the meaning of each point.

28. **Chemical reactions** The activity of particles in a chemical reaction (such as digestion) or in the diffusion of a given solution is related to the concentration of the particles according to the formula $A = KC$, where A is the activity, C is the concentration, and K is the fraction of the concentration that is effective in determining the diffusion or the chemical reaction.
(a) Sketch the graph of the equation if $K = 1$.
(b) Sketch the graph of the equation if $K = 0.7$.

29. **Depreciation** A $360,000 building is depreciated by its owner. The value y of the building after x months of use is $y = 360{,}000 - 1500x$.
(a) Graph this equation for $x \geq 0$.
(b) How long before the building is completely depreciated (its value is zero)?
(c) The point (60, 270,000) lies on the graph. Explain its meaning.

30. **Engineering materials** A study by a construction company showed that the strength s of concrete, measured by the percent of design strength attained, depends on the length of time t (in days) since it was poured. The value can be approximated using $s = 1.8t + 27$.
(a) Graph this equation for $t \geq 0$.
(b) How long before $s = 100$?
(c) What is the significance of your answer to part (b)?
(d) The point (10, 55) satisfies the equation. Explain its meaning.

31. **Tax burden** Using data from the Internal Revenue Service, the per capita tax burden T (in hundreds of dollars) can be modeled by

$$T = 20.37 + 1.834t,$$

where t is the number of years past 1980.
(a) Find the point (t, T) that corresponds to $t = 20$.
(b) Write a sentence that explains the meaning of the point in part (a).
(c) Graph the equation for per capita tax burden.

32. **Median salary, gender differences** Data from selected years from 1973 to 1991 show that the median annual salary for females, y, is related to that of males, x, and can be modeled by

$$y = 0.838x - 1387.4.$$

(Source: U.S. Equal Employment Opportunity Commission)
(a) Find the point that corresponds to $x = 28,000$.
(b) Write a sentence that explains the meaning of the point found in (a).
(c) Graph the equation.

33. **Investing** The data below compares the cumulative total return on a $100 investment made in 1988, one in Federal Signal Corporation (FS) and one that represents the Standard and Poors 400 Index (S&P).

	1988	1989	1990	1991	1992	1993
FS	$100	$170	$248	$357	$402	$539
S & P	$100	$129	$128	$168	$171	$187

SOURCE: Federal Signal Corporation's *Notice of Annual Meeting*, March 7, 1994

If t is the number of years past 1988 (so $t = 0$ in 1988), then this cumulative return can be modeled by R_{FS} for Federal Signal Corporation and by R_{SP} for the S & P 400 Index as follows.

$$R_{FS} = 85.714t + 88.381$$
$$R_{SP} = 17.1714t + 104.238$$

(a) Graph $R_{FS} = 85.714t + 88.381$ and the points (t, FS) from the data in the table.
(b) Graph $R_{SP} = 17.1714t + 104.238$ and the points (t, SP) from the data in the table.
(c) What are the values of R_{FS} and R_{SP} when $t = 0$? Explain why they are different from the values in the tables.

(d) What is the danger of predicting future investment returns from these equations?

34. **Temperature-humidity models** Two models for measuring the effects of high temperature and humidity are the Summer Simmer Index and the Apparent Temperature.* For an outside temperature of 100°F, these indices relate the relative humidity, H (expressed as a decimal), to the perceived temperature as follows.

Summer Simmer: $S = 141.1 - 45.78 (1 - H)$

Apparent Temperature: $A = 91.2 + 41.3H$

(a) For each index, find the point that corresponds to a relative humidity of 40%.
(b) For each point in (a), write a sentence that explains its meaning.
(c) Graph both equations for $0 \le H \le 1$.

35. **Distance to a thunderstorm** The distance d (in miles) to a thunderstorm is given by

$$d = \frac{t}{4.8},$$

where t is the number of seconds that elapse between seeing the lightning and hearing the thunder.
(a) Graph this equation for $0 \le t \le 20$.
(b) The point $(9.6, 2)$ satisfies the equation. Explain its meaning.

36. **Gas laws** The volume V (in cubic centimeters) of a certain gas depends on its temperature T (in °C) according to

$$V - 120 = \frac{3}{8}(T - 50).$$

Graph this equation.

37. **Body-heat loss** Body-heat loss due to convection depends on a number of factors. If H_c is body-heat loss due to convection, A_c is the exposed surface area of the body, $T_s - T_a$ is skin temperature minus air temperature, and K_c is the convection coefficient (determined by air velocity and so on), then we have

$$H_c = K_c A_c (T_s - T_a).$$

When $K_c = 1$, $A_c = 1$, and $T_s = 90$, the equation is

$$H_c = 90 - T_a.$$

Sketch the graph.

* Bosch, W., and C. G. Cobb, "Temperature-Humidity Indices," UMAP Unit 691, *The UMAP Journal*, 10(3), Fall 1989, 237–256.

38. **Blood flow** The rate of oxygen consumption of the blood, x, measures the cardiac output (blood flow through the lungs), y, by

$$y = \frac{x}{A - V} = \left(\frac{1}{A - V}\right)x,$$

where A is the arterial concentration of oxygen and V is the venous concentration of oxygen. Sketch the graph for the above if $1/(A - V) = \frac{3}{4}$.

39. **Stimulus thresholds** In psychology, the study of threshold levels of stimulus is of interest. If s is the amount of stimulus, then Δs is the required change in s so that the subject notices a change. Research has shown that

$$\Delta s = ks \quad (k = \text{constant}).$$

Sketch graphs of this equation ($s > 0$) for the values $k = \frac{1}{2}, k = 1$, and $k = 2$. Label the horizontal axis s and the vertical axis Δs.

40. **Temperature scales** The equation $C = \frac{5}{9}(F - 32)$ describes the relationship between temperature readings in degrees Celsius and degrees Fahrenheit.
    (a) Graph this equation with C on the vertical axis.
    (b) The point $(68, 20)$ lies on the graph; explain its meaning.

▣ **1.3** GRAPHS AND GRAPHING UTILITIES

Objectives

❑ To use a graphing utility to graph equations in the standard viewing window
❑ To use a graphing utility and a specified range to graph equations

APPLICATION PREVIEW

Suppose that for a certain city the cost C of obtaining drinking water with p percent impurities (by volume) is given by

$$C = \frac{120,000}{p} - 1200,$$

where $0 < p \le 100$. A graph of this equation would illustrate how the benefit of pure water is related to the cost of obtaining it. Because this equation is not linear, graphing it would require many more than two points. Such a graph could be efficiently obtained with a graphing utility.

In Section 1.2, we saw that a linear equation could easily be graphed by plotting points because only two points are required. When we want the graph of an equation that is not linear, we can still use point plotting, but we must have enough points to sketch an accurate graph.

Figure 1.9(a) shows three points on the graph of $y = (x - 1)^3$; these points lie on a straight line. However, when we plot additional points on the graph of $y = (x - 1)^3$, we see that the graph is not a straight line (see Figures 1.9(b) and (c)). These figures show that plotting only a few points may give an inaccurate impression of the shape of the graph. On the other hand, plotting many points may provide an accurate graph, but the process can be tedious.

Some computer software and all graphics calculators have **graphing utilities** (also called *graphics utilities*) that can be used to generate an accurate graph. All graphing utilities use the point-plotting method to quickly plot scores of points and thereby graph an equation. The computer monitor or calculator screen consists of a fine grid that looks like a piece of graph paper.

(a)   (b)   (c)

Figure 1.9

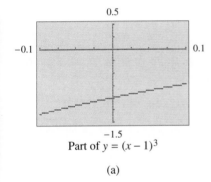

Part of $y = (x - 1)^3$

(a)

In this grid, each tiny area is called a **pixel.** Essentially, a pixel corresponds to a point on a piece of graph paper; and as a graphing utility plots points, the pixels corresponding to the points are lighted and then connected to reveal the graph.

With a graphing utility, as with a graph plotted by hand, a graph's appearance is determined by the portion we are viewing. For example, our graph of $y = (x - 1)^3$ in Figure 1.9 showed the graph in the region between $x = -3$ and $x = 5$ and between $y = -10$ and $y = 10$. If we had used the region between $x = -0.1$ and $x = 0.1$ and between $y = -1.5$ and $y = 0.5$, the graph would have appeared more like a line (see Figure 1.10(a)). Other regions for x and y might contain no points on the graph of $y = (x - 1)^3$ (see Figure 1.10(b)).

The region determined by the chosen values of x and y is called the **viewing window** or **viewing rectangle.** Many graphing utilities have a **standard viewing window** that lies between -10 and 10 on both the x-axis and the y-axis. For the graph of $y = (x - 1)^3$, the standard viewing window gives a graph that shows all the features (turns, bends, zeros, and trends) and hence provides an accurate graph.

The following guidelines may be used for most graphing utilities.

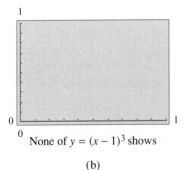

None of $y = (x - 1)^3$ shows

(b)

Figure 1.10

GUIDELINES FOR USING A GRAPHING UTILITY

To graph an equation in the variables x and y:

1. Solve the equation for y in terms of x.
2. Enter the equation into the graphing utility. Use parentheses as needed to assure correctness.
3. Determine a viewing window.
4. Activate the graphing utility.

Figure 1.11

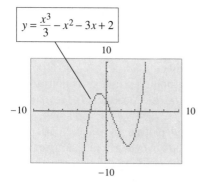

Figure 1.12

EXAMPLE 1

Graph $2x - 3y = 12$ with a graphing utility.

Solution

First notice that this is a linear equation and that the graph is therefore a line. Begin by solving $2x - 3y = 12$ for y.

$$2x - 3y = 12$$
$$-3y = -2x + 12$$
$$y = \frac{-2x}{-3} + \frac{12}{-3}$$
$$y = \frac{2x}{3} - 4$$

Enter this last equation, and use the standard viewing window. The resulting graph is shown in Figure 1.11.

EXAMPLE 2

Use a graphing utility and the standard viewing window to graph

$$y = \frac{x^3}{3} - x^2 - 3x + 2.$$

Solution

Because the equation is already solved for y in terms of x, we enter it into the graphing utility. The graph is shown in Figure 1.12.

CHECKPOINT

Use a graphing utility and the standard viewing window to graph

1. $x^2 + 4y = 0$,

2. $y = \frac{4(x + 1)^2}{x^2 + 1}.$

For some graphs, the standard viewing window is inappropriate because it gives a view of the graph that is misleading, incomplete, or perhaps even blank. Figure 1.13 shows three equations and the graph of each as it appears with the standard viewing window. Figure 1.14 shows the same equations and the graph of each with a different viewing window. In each case, notice how the standard window is inappropriate.

A viewing window is determined by the following four values.

x-min: the smallest x-value in the window
x-max: the largest x-value in the window
y-min: the smallest y-value in the window
y-max: the largest y-value in the window

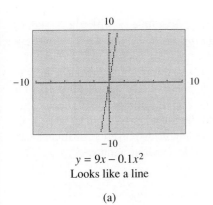

$y = 9x - 0.1x^2$
Looks like a line

(a)

$y = 0.1x^3 - 1.5x^2 + 8$
Looks like a cup

(b)

$y = 100 - 4x$
What happened?

(c)

Figure 1.13

On many graphing utilities, the standard viewing window automatically sets these four values as follows.

x-min: -10 y-min: -10
x-max: 10 y-max: 10

Specifying these four values is called setting the **range** for a graphing utility. Any nonstandard viewing window must be set manually. For example, the graph shown in Figure 1.14(b) was generated by setting the range as follows.

x-min: -5 y-min: -50
x-max: 25 y-max: 20

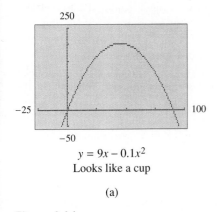

$y = 9x - 0.1x^2$
Looks like a cup

(a)

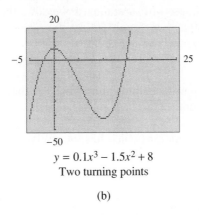

$y = 0.1x^3 - 1.5x^2 + 8$
Two turning points

(b)

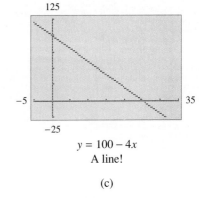

$y = 100 - 4x$
A line!

(c)

Figure 1.14

EXAMPLE 3
Use a graphing utility with the indicated range to graph $y = x^4 - 8x^2 - 10$.

x-min: -5 y-min: -40
x-max: 5 y-max: 10

Solution
The equation can be entered directly. Set the desired range and activate the graphing utility. The resulting graph is shown in Figure 1.15.

For comparison, graph the equation of Example 3 with the standard viewing window. How is it similar to the graph in Figure 1.15? How is it different?

Most graphing utilities have a TRACE capability that allows the user to trace along a graph. Tracing highlights a point on the graph and simultaneously displays its coordinates. Even if the graph is not visible in the viewing window (because you're at a point beyond the *x*- and/or *y*-range), the coordinates of the trace points are still displayed. These displayed coordinates can be used to help determine a more appropriate viewing window.

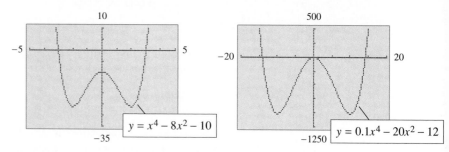

Figure 1.15   **Figure 1.16**

For example, if we graph $y = 0.1x^4 - 20x^2 - 12$ with the standard window, none of the graph appears in the window. When we trace along the graph with a standard window, the *y*-values range from -12 to -1012. Hence, we could adjust the *y*-range to account for these *y*-values and also expand the *x*-range. If we use *x*-min $= -20$, *x*-max $= 20$, *y*-min $= -1250$, *y*-max $= 500$, we obtain the graph shown in Figure 1.16.

We can use a so-called "friendly window" on a graphics calculator by choosing *x*-min and *x*-max so that *x*-max minus *x*-min is divisible by the number of pixels on the screen.

For example, if we set *x*-min $= -4.7$ and *x*-max $= 4.7$ on a T1-82 calculator (which has 94 pixels), each step of TRACE will change *x* by

$$\frac{4.7 - (-4.7)}{94} = 0.1.$$

We now return to the Application Preview problem.

EXAMPLE 4

For a certain city, the cost *C* of obtaining drinking water with *p* percent impurities (by volume) is given by

$$C = \frac{120,000}{p} - 1200.$$

Because p is the percentage of impurities, we know $0 \leq p \leq 100$. Use the restriction on p and a graphing utility to obtain an accurate graph of the equation.

Solution

To use a graphing utility we identify C with y and p with x. Thus we enter

$$y = \frac{120{,}000}{x} - 1200.$$

We will need an x-range on the graphing utility that corresponds to the p-range in the problem. Because $0 \leq p \leq 100$, we set the x-range to *include* these values, with the realization that only the portion of the graph above these values applies to our model. With this x-range we can determine a y-, or C-, range from the equation or by tracing. Figure 1.17(a) shows the graph for x from -25 to 100 and for y from $-25{,}000$ to $125{,}000$. We see that when p is near 0, the C-coordinates of the points are very large; indicating that water free of impurities is very costly. However, Figure 1.17(a) does not accurately show what happens for large p-values. Figure 1.17(b) shows another view of the equation with the C-range (that is, y-range) from -500 to 3500. We see the p-intercept is $p = 100$, indicating that water containing 100 percent impurities costs nothing ($C = 0$).

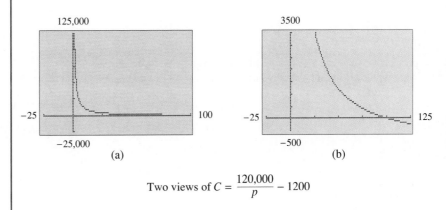

(a)   (b)

Two views of $C = \dfrac{120{,}000}{p} - 1200$

Figure 1.17

**CHECKPOINT
SOLUTIONS**

1. First solve $x^2 + 4y = 0$ for y.

$$4y = -x^2$$

$$y = \frac{-x^2}{4}$$

The graph is shown in Figure 1.18 (a).

2. The equation $y = \dfrac{4(x + 1)^2}{x^2 + 1}$ can be entered directly, but some care with parentheses is needed.

$$y = (4(x + 1)^2)/(x^2 + 1).$$

The graph is shown in Figure 1.18 (b)

$$y = \dfrac{-x^2}{4}$$

$$y = \dfrac{4(x + 1)^2}{x^2 + 1}$$

Figure 1.18   (a)   (b)

EXERCISE 1.3

 In problems 1–16, use a graphing utility with the standard viewing window to graph each equation. Identify any linear equations.

1. $5x + y = 10$
2. $3x - y = 9$
3. $-x + 4y = 4$
4. $x + 2y = -2$
5. $\dfrac{x}{8} - \dfrac{y}{4} = 1$
6. $\dfrac{x}{7} + \dfrac{y}{6} = 1$
7. $y = 4 - x^2$
8. $y = x(x + 2)$
9. $y = x^2 - 6x + 5$
10. $y = 6 + x - x^2$
11. $y = x^3 - 3x$
12. $y = x^3 - 6x^2$
13. $y = \dfrac{1}{3}x^3 - x^2 - 3x + 2$
14. $y = \dfrac{1}{4}x^4 - \dfrac{1}{3}x^3 - 3x^2 + 8$
15. $y = \dfrac{12x}{x^2 + 1}$
16. $y = \dfrac{8}{x^2 + 1}$

In problems 17–20, use a graphing utility with the specified range to graph each equation.

17. $y = x^3 - 12x - 1$
    x-min $= -5$, x-max $= 5$; y-min $= -20$, y-max $= 20$

18. $y = x^4 - 4x^3 + 6$
    x-min $= -4$, x-max $= 6$; y-min $= -40$, y-max $= 10$

19. $y = (x + 1)^2(x - 9)^2$
    x-min $= -5$, x-max $= 15$; y-min $= -100$,
    y-max $= 650$

20. $y = 8 + 16x + 3x^2 - \dfrac{1}{3}x^3$
    x-min $= -10$, x-max $= 15$; y-min $= -20$,
    y-max $= 160$

In problems 21–24, graph each equation with a graphing utility
(a) and the specified range.
(b) and the standard viewing window (for comparison).

21. $y = 0.01x^3 + 0.3x^2 - 72x + 150$
    x-min $= -100$, x-max $= 80$; y-min $= -2000$,
    y-max $= 4000$

22. $y = \dfrac{-0.01x^3 + 0.15x^2 + 60x + 700}{100}$
    x-min $= -100$, x-max $= 100$; y-min $= -50$,
    y-max $= 50$

23. $y = \dfrac{x + 15}{x^2 + 400}$

    x-min $= -200$, x-max $= 200$; y-min $= -0.02$,
    y-max $= 0.06$

24. $y = \dfrac{x - 80}{x^2 + 1700}$

    x-min $= -200$, x-max $= 400$; y-min $= -0.06$,
    y-max $= 0.01$

In problems 25 and 26,
(a) write a sentence that describes a plan to determine an appropriate viewing window for each graph,
(b) determine a window that would show the graph, and
(c) use a graphing utility (with your window) to graph each equation.

25. $y = 0.001x - 0.03$
26. $y = 50,000 - 100x$

In problems 27–30, a standard viewing window graph of an equation is shown. Experiment with the viewing window to obtain an accurate graph, adjusting the ranges where neccessary.

27. $y = -0.15(x - 10.2)^2 + 10$

28. $y = x^2 - x - 42$

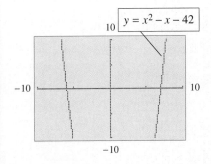

29. $y = \dfrac{x^3 + 19x^2 - 62x - 840}{20}$

30. $y = \dfrac{-x^3 + 33x^2 + 120x}{20}$

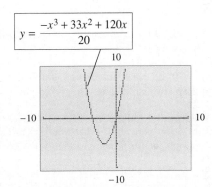

APPLICATIONS

31. **Advertising impact** An advertising agency has found that when it promotes a new product in a city of 350,000 people, the rate of change R of the number of people x who are aware of the product is given by

    $$R = 28,000 - 0.08x.$$

    Use the intercepts to determine a window, then use a graphing utility to graph the equation for $x \geq 0$.

32. **Learning rate** In a study using 50 foreign language vocabulary words, the learning rate L (in words per minute) was found to depend on the number of words already learned, x, according to the equation

    $$L = 20 - 0.4x.$$

    Use the intercepts to determine a window, then use a graphing utility to graph the equation for $x \geq 0$.

33. **Earnings and gender** With U.S. Equal Employment Opportunity Commission data found in the *Statistical Abstract of the U.S., 1993*, the model that relates the median annual salary (in thousands of dollars) of females, F, and males, M, in the United States was found to be

$$F = 0.838M - 1.364.$$

   (a) Use a graphing utility to graph this equation for the range M-min $= 0$, M-max $= 35$; F-min $= 0$, F-max $= 30$.
   (b) The point $(50, 40.536)$ lies on the graph of this equation. Explain its meaning.

34. **Earnings and minorities** With U.S. Equal Employment Opportunity Commission data found in the *Statistical Abstract of the U.S., 1993*, the model that relates the median annual salary (in thousands of dollars) of minorities, M, and whites, W, was found to be

$$M = 0.959W - 1.226.$$

   (a) Use a graphing utility to graph this equation for the range W-min $= 0$, W-max $= 30$; M-min $= 0$, M-max $= 30$.
   (b) The point $(40, 37.134)$ lies on the graph of this equation. Explain its meaning.

35. **Consumer expenditure** Suppose that the consumer expenditure E (in dollars) depends on the market price p per unit (in dollars) according to

$$E = 10,000p - 100p^2.$$

   (a) Graph this equation with a graphing utility and the range p-min $= -50$, p-max; $= 150$, E-min $= -50,000$, E-max $= 300,000$.
   (b) Because E represents consumer expenditure, we know $E \geq 0$. For what p-values is $E \geq 0$?

36. **Rectilinear motion** The height above ground (in feet) of a ball thrown vertically into the air is given by

$$S = 112t - 16t^2,$$

   where t is the time in seconds since the ball was thrown.
   (a) Graph this equation with a graphing utility and the range t-min $= -2$, t-max $= 10$, S-min $= -20$, S-max $= 250$.
   (b) Estimate the time at which the ball is at its highest point, and estimate the height of the ball at that time.

37. **Sales revenue** Sales revenue R (in billions of dollars) for Scott Paper Company can be modeled by

$$R = -0.031t^2 + 0.776t + 0.179,$$

where t is the number of years past 1980 (Source: Scott Paper Company, *1993 Annual Report*).
   (a) Graph this equation with a graphing utility and the range t-min $= 0$, t-max $= 15$; R-min $= 0$, R-max $= 8$, which corresponds to the years from 1980 to 1995.
   (b) If this model is valid until 2020, adjust the range to obtain a graph that extends to 2020.
   (c) Write a sentence that compares revenue from 1980 to 1995 with that from 1980 to 2020.

38. **Drug usage** The percentage p of high school seniors who have tried hallucinogens can be modeled by

$$p = 0.0241904(t + 70)^2 - 4.47459(t + 70) + 216.074,$$

   where t is the number of years past 1970 (Source: National Institute on Drug Abuse).
   (a) Graph this equation with a graphing utility and the range t-min $= 0$, t-max $= 25$; p-min $= 0$, p-max $= 20$.
   (b) Write a sentence that describes the trends in the percentage of high school students who have tried hallucinogens.

39. **Cost-benefit** The Millcreek watershed area was heavily strip-mined for coal during the late 1960s. Because of the resulting pollution, the streams can't support fish. Suppose the cost C of obtaining stream water that contains p percent of the current pollution levels is given by

$$C = \frac{285,000}{p} - 2850.$$

Because p is the percentage of current levels, $0 \leq p \leq 100$.
   (a) Use the restriction on p and determine a range for C so that a graphing utility can be used to obtain an accurate graph. Then graph the equation.
   (b) Describe what happens to the cost as p takes values near 0.
   (c) The point $(1, 282,150)$ lies on the graph of this equation. Explain its meaning.
   (d) Explain the meaning of the p-intercept.

40. **Cost-benefit** Suppose the cost C of removing p percent of the particulate pollution from the exhaust gases at an industrial site is gven by

$$C = \frac{8100p}{100 - p}.$$

Because p is the percentage of particulate pollution, we know $0 \leq p \leq 100$.

(a) Use the restriction on p and experiment with a C range to obtain an accurate graph of the equation with a graphing utility.

(b) Describe what happens to C as p gets close to 100.

(c) The point (98, 396,900) lies on the graph of this equation. Explain the meaning of the coordinates.

(d) Explain the meaning of the p-intercept.

41. ***Earning and gender*** The following table shows the 1986 median weekly earnings of men and women in various occupations.

	Earnings	
Occupation	Men (x)	Women (y)
Management	$666	$465
Technical sales	472	305
Service	299	208
Crafts, repair	446	302
Operators, fabricators	352	238
Fish, farm, forestry	234	201

SOURCE: U.S. Bureau of Labor Statistics

(a) Plot the data in the table, using a graphing utility.

(b) Graph

$$y = 0.630526x + 27.0386 \quad \text{and}$$

$$y = 0.914811x + 20.4118,$$

and decide which one fits the data better.

(c) Use the equation you chose in part (b) to see how well it predicts the 1986 median weekly earnings for a woman in a job where a man earned $472 (technical sales).

42. ***Training and productivity*** Suppose that a company purchased a new software package to handle their word processing, payroll, inventory, and accounting needs. After a few employees had received some training with the new system, their work with it was monitored and the following data were gathered.

Hours of Training (x)	Number of Errors (y)	Hours of Training (x)	Number of Errors (y)
1	8	12	0
2	7	7	3
10	2	6	3
8	2	4	4
6	8	4	6

(a) Plot the data in the table, using a graphing utility.

(b) Graph

$$y = -0.68868x + 8.4321 \quad \text{and}$$

$$y = -0.40116x + 6.3494,$$

and decide which one fits the data better.

(c) Use the equation you chose in part (b) to determine the number of hours of training needed for the predicted number of errors to equal 0.

1.4 FUNCTIONS

Objectives

❑ To determine if a relation is a function
❑ To state the domains and ranges of certain funtions
❑ To use function notation

APPLICATION PREVIEW

Rhea Lumber sells one grade of 3/4-inch plywood for $18.95 for a 4-by-8-foot sheet. Thus, their income I from the sale of this item is a function of (or depends on) the number of sheets they sell and can be expressed by the equation

$$I = 18.95x,$$

where x represents the number of sheets sold. For Rhea Lumber, the phrase *function of* means "depends on" and that relationship can be expressed by an equation. Understanding the mathematical meaning of the phrase *function of* and gaining the ability to interpret and apply such relationships is the goal of this section.

An equation containing two variables expresses a **relation** between those two variables. For example, the equation $R = 35x$ expresses a relation between the two variables x and R, and the equation $y = 4x - 3$ expresses a relation between the two variables x and y.

In addition to defining a relation by an equation or rule of correspondence, we may also define it as any set of **ordered pairs** of real numbers (a, b). For example, the solutions to $y = 4x - 3$ are pairs of numbers (one for x and one for y). We write the pairs (x, y) so that the first number is the x-value and the second is the y-value, and these ordered pairs define the relation between x and y. Some relations may be defined by a table, a graph, or an equation.

Relation	A **relation** is defined by a set of ordered pairs or by a rule that determines how the ordered pairs are found. It may also be defined by a table, a graph, or an equation.

For example, the set of ordered pairs

$$\{(1, 3), (2, 6), (3, 9), (4, 12)\}$$

expresses a relation between the set of first components, $\{1, 2, 3, 4\}$, and the set of second components, $\{3, 6, 9, 12\}$. The set of first components is called the **domain** of the relation, and the set of second components is called the **range** of the relation. Figure 1.19 uses arrows to indicate how the inputs from the domain (the first components) are associated with the outputs in the range (the second components). Table 1.1 and Figure 1.20 are also examples of relations.

An equation frequently expresses how the second component (the output) is obtained from the first component (the input). For example, the equation

$$y = 4x - 3$$

expresses how the output y results from the input x. This equation expresses a special relation between x and y, because each value of x that is substituted into the equation results in only one value for y. If each value of x put into an equation results in one value of y, we say that the equation expresses y as a **function** of x.

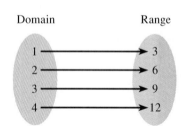

Domain   Range

Figure 1.19

Definition of a Function	A **function** is a relation between two sets such that to each element of the domain (input) there corresponds exactly one element of the range (output).

When a function is defined, the variable that represents the numbers in the domain (input) is called the **independent variable** of the function, and

Table 1.1 Single filer's tax table

Taxable Income (Domain)		Tax Due (Range)
at least	less than	
20,000	20,050	3,004
20,050	20,100	3,011
20,100	20,150	3,019
20,150	20,200	3,026
20,200	20,250	3,034
20,250	20,300	3,041
20,300	20,350	3,049
20,350	20,400	3,056
20,400	20,450	3,064
20,450	20,500	3,071
20,500	20,550	3,079
20,550	20,600	3,086
20,600	20,650	3,094
20,650	20,700	3,101
20,700	20,750	3,109
20,750	20,800	3,116
20,800	20,850	3,124
20,850	20,900	3,131
20,900	20,950	3,139
20,950	21,000	3,146

SOURCE: Internal Revenue Service,
1993 Form 1040, Instructions

Figure 1.20

Domain: time (days of the year)
Range: Dow Jones Average values
SOURCE: *Wall Street Journal*, January 24, 1994.

the variable that represents the numbers in the range (output) is called the **dependent variable** (because its values depend on the values of the independent variable). When we say "the equation $y = 4x - 3$ defines y as a function of x," we are saying that the equation defines a function with independent variable x and dependent variable y.

The equation

$$y = 4x^2$$

defines y as a function of x, because only one value of y will result from each value of x that is substituted into the equation. Thus x is the independent variable and y is the dependent variable.

We can also apply this idea to a relation defined by a table. For example, Table 1.1 defines tax due (the dependent variable) as a function of the taxable income of a single taxpayer (the independent variable).

EXAMPLE 1

Does $y^2 = 2x$ express y as a function of x?

Solution

No, because for some values of x there is more than one value for y. In fact, there are two y-values for each $x > 0$. For example, if $x = 8$, then $y = +4$ or $y = -4$, two different y-values for the same x-value. The equation $y^2 = 2x$ expresses a relation between x and y, but y is not a function of x.

We can determine whether a relation is a function by inspecting its graph. If the relation is a function, no two points on the graph will have the same first coordinate (component). Thus no two points of the graph will lie on the same vertical line.

Vertical Line Test If no vertical line exists that intersects the graph of a relation in more than one point, then the graph is the graph of a function.

Performing this test on the graph of $y = 4x^2$ (Figure 1.21(a)), we easily see that this equation describes a function. The vertical line test indicates, however, that the equation $y^2 = 2x$ (which is graphed in Figure 1.21(b)) is not a function. For example, a vertical line at $x = 2$ intersects the curve at $(2, 2)$ and $(2, -2)$.

We can use functional notation to indicate that y is a function of x. The function is denoted by f, and we write $y = f(x)$. This is read "y is a function of x" or "y equals f of x." For specific values of x, $f(x)$ represents the values of the function (that is, outputs, or y-values) at those x-values. Thus if

$$f(x) = 3x^2 + 2x + 1,$$

then $$f(2) = 3(2)^2 + 2(2) + 1 = 17$$

and $$f(-3) = 3(-3)^2 + 2(-3) + 1 = 22.$$

Figure 1.22 represents this functional notation as (a) an operator on x and as (b) a y-coordinate for a given x-value.

Letters other than f may also be used to denote functions. For example, $y = g(x)$ or $y = h(x)$ may be used.

(a)

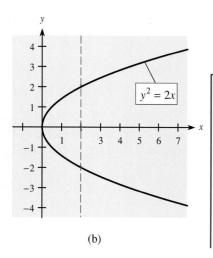

(b)

Figure 1.21

EXAMPLE 2

If $y = f(x) = 2x^3 - 3x^2 + 1$, find the following.
(a) $f(0)$ (b) $f(3)$ (c) $f(-1)$

Solution
(a) $f(0) = 2(0)^3 - 3(0)^2 + 1 = 1$
    Thus $y = 1$ when $x = 0$.
(b) $f(3) = 2(3)^3 - 3(3)^2 + 1 = 2(27) - 3(9) + 1 = 28$
    Thus $y = 28$ when $x = 3$.
(c) $f(-1) = 2(-1)^3 - 3(-1)^2 + 1 = 2(-1) - 3(1) + 1 = -4$
    Thus $y = -4$ when $x = -1$.

EXAMPLE 3

If $g(x) = 4x^2 - 3x + 1$, find the following.
(a) $g(a)$ (b) $g(-a)$ (c) $g(b)$ (d) $g(a + b)$

input x

f

output $f(x)$

(a)

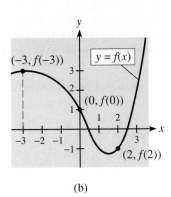

(b)

Figure 1.22

Solution
(a) $g(a) = 4(a)^2 - 3(a) + 1 = 4a^2 - 3a + 1$
(b) $g(-a) = 4(-a)^2 - 3(-a) + 1 = 4a^2 + 3a + 1$
(c) $g(b) = 4(b)^2 - 3(b) + 1 = 4b^2 - 3b + 1$
(d) $g(a + b) = 4(a + b)^2 - 3(a + b) + 1$

EXAMPLE 4
Given $f(x) = 3x - 1$, Find the following.

(a) $f(4)$ (b) $f(\lambda)$ (c) $\dfrac{f(x + h) - f(x)}{h}$, if $h \neq 0$

Solution
(a) $f(4) = 3(4) - 1 = 11$ (b) $f(\lambda) = 3(\lambda) - 1 = 3\lambda - 1$

(c) $\dfrac{f(x + h) - f(x)}{h} = \dfrac{[3(x + h) - 1] - (3x - 1)}{h}$

$$= \frac{3x + 3h - 1 - 3x + 1}{h} = \frac{3h}{h} = 3$$

EXAMPLE 5
We have seen in Table 1.1, on page 91, that a single taxpayer's federal tax T is a function of taxable income I. We can indicate this functional relationship by writing $T = f(I)$. Using this notation and Table 1.1, evaluate the following and then describe what each result means.
(a) $f(20{,}400)$ (b) $f(20{,}878)$

Solution
(a) $T = f(I)$, so $f(20{,}400)$ represents the tax due on an income of $20,400. From Table 1.1, we find $f(20{,}400) = 3{,}064$. That is, a single taxpayer with a taxable income of $20,400 owes $3,064 in federal income tax.
(b) Because 20,878 is between 20,850 and 20,900, Table 1.1 gives $f(20{,}878) = 3{,}131$. That is, a single taxpayer with a taxable income of $20,878 owes $3,131 in federal income tax.

We will limit our discussion in this text to **real functions,** which are functions whose domains and ranges contain only real numbers. If the domain and range of a function are not specified, it is assumed that the domain consists of all real numbers that result in real numbers in the range, and that the range is a subset of the real numbers.

In general, if the domain of a function is unspecified, it will include all real numbers except:

1. values that result in a denominator of 0, and

2. values that result in an even root of a negative number.

EXAMPLE 6

Find the domain of each of the following functions; find the range for the functions in (a) and (b).

(a) $y = 4x^2$ (b) $y = \sqrt{4 - x}$ (c) $y = 1 + \dfrac{1}{x - 2}$

Solution

(a) There are no restrictions on the numbers substituted for x, so the domain consists of all real numbers. Because the square of any real number is nonnegative, $4x^2$ must be nonnegative. Thus the range is $y \geq 0$. If we plot points or use a graphing utility, we will get the graph shown in Figure 1.23(a), which illustrates our conclusions about the domain and range.

(b) We note the restriction that $4 - x$ cannot be negative. Thus the domain consists of only numbers less than or equal to 4. That is, the domain is the set of real numbers satisfying $x \leq 4$. Since $\sqrt{4 - x}$ is always nonnegative, the range is all $y \geq 0$. Figure 1.23(b) shows the graph of $y = \sqrt{4 - x}$. Notice that the graph is located only where $x \leq 4$ and on or above the x-axis (where $y \geq 0$).

(c) $y = 1 + \dfrac{1}{x - 2}$ is undefined at $x = 2$ since $\dfrac{1}{0}$ is undefined. Hence, the domain consists of all real numbers except 2. Figure 1.23(c) shows the graph of $y = 1 + \dfrac{1}{x - 2}$. The break where $x = 2$ indicates that $x = 2$ is not part of the domain.

(a)

(b)

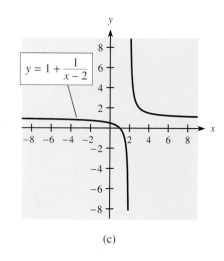

(c)

Figure 1.23

CHECKPOINT

1. If $y = f(x)$, the independent variable is _____ and the dependent variable is _____.

2. If $(1, 3)$ is on the graph of $y = f(x)$, then $f(1) = ?$

3. If $f(x) = 1 - x^3$, find $f(-2)$.

4. If $f(x) = 2x^2$, find $f(x + h)$.

5. If $f(x) = \dfrac{1}{x + 1}$, what is the domain of $f(x)$?

We can form new functions by performing algebraic operations with two or more functions. We define new functions that are the sum, difference, product, and quotient of two functions as follows.

Operations with Functions

Let f and g be functions of x, and define the following.

Sum $\qquad\qquad (f + g)(x) = f(x) + g(x)$

Difference $\qquad (f - g)(x) = f(x) - g(x)$

Product $\qquad\; (f \cdot g)(x) = f(x) \cdot g(x)$

Quotient $\qquad \left(\dfrac{f}{g}\right)(x) = \dfrac{f(x)}{g(x)} \quad$ if $\;g(x) \neq 0$

EXAMPLE 7

If $f(x) = 3x + 2$ and $g(x) = x^2 - 3$, find the following functions.

(a) $(f + g)(x)$ (b) $(f - g)(x)$

(c) $(f \cdot g)(x)$ (d) $\left(\dfrac{f}{g}\right)(x)$

Solution

(a) $(f + g)(x) = f(x) + g(x) = (3x + 2) + (x^2 - 3) = x^2 + 3x - 1$

(b) $(f - g)(x) = f(x) - g(x) = (3x + 2) - (x^2 - 3) = -x^2 + 3x + 5$

(c) $(f \cdot g)(x) = f(x) \cdot g(x) = (3x + 2)(x^2 - 3) = 3x^3 + 2x^2 - 9x - 6$

(d) $\left(\dfrac{f}{g}\right)(x) = \dfrac{f(x)}{g(x)} = \dfrac{3x + 2}{x^2 - 3}, \quad$ if $x^2 - 3 \neq 0$

We now consider a new way to combine two functions. Just as we can substitute a number for the independent variable in a function, we can substitute a second function for the variable. This creates a new function, called a **composite function.**

Composite Functions Let f and g be functions. Then the **composite functions** g of f (denoted $g \circ f$) and f of g (denoted $f \circ g$) are defined as follows:

$$(g \circ f)(x) = g(f(x))$$
$$(f \circ g)(x) = f(g(x))$$

Note that the domain of $g \circ f$ is the subset of the domain of f for which $g \circ f$ is defined. Similarly, the domain of $f \circ g$ is the subset of the domain of g for which $f \circ g$ is defined.

EXAMPLE 8

If $f(x) = 2x^3 + 1$ and $g(x) = x^2$, find

(a) $(g \circ f)(x)$ (b) $(f \circ g)(x)$.

Solution

(a) $(g \circ f)(x) = g(f(x))$
$$= g(2x^3 + 1)$$
$$= (2x^3 + 1)^2 = 4x^6 + 4x^3 + 1$$

(b) $(f \circ g)(x) = f(g(x))$
$$= f(x^2)$$
$$= 2(x^2)^3 + 1$$
$$= 2x^6 + 1$$

Figure 1.24 illustrates both composite functions found in Example 8.

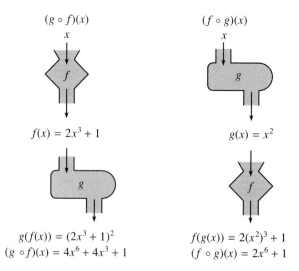

Figure 1.24

CHECKPOINT

6. If $f(x) = 1 - 2x$ and $g(x) = 3x^2$, find the following.
   (a) $(f \circ g)(x)$ (b) $(g \circ f)(x)$ (c) $(f \circ f)(x) = f(f(x))$

EXAMPLE 9

Suppose that a woman's ability to buy a house by borrowing y dollars is related to her ability to make a monthly payment x, according to

$$y = 100x,$$

and that her ability to make monthly payment x is related to her monthly income z by

$$x = 0.20z.$$

(a) Write a function in terms of her monthly income that describes her ability to buy a house by borrowing y dollars.
(b) If her monthly income is $3000, how much could she borrow in order to buy a house?

Solution

(a) Her ability to buy a house by borrowing y dollars is

$$y = f(x), \text{ where } x = g(z).$$

Thus

$$y = f(g(z)) = f(0.20z) = 100(0.20z)$$

or

$$y = 20z$$

gives y as a function of z, her monthly income.

(b) If her monthly income is $3000, then the amount she could borrow to buy a house is

$$y = f(g(3000)) = 20(3000) = 60,000 \text{ (dollars)}.$$

CHECKPOINT
SOLUTIONS

1. Independent variable is x; dependent variable is y.
2. $f(1) = 3$
3. $f(-2) = 1 - (-2)^3 = 1 - (-8) = 9$
4. $f(x + h) = 2(x + h)^2$
   $\qquad\qquad = 2(x^2 + 2xh + h^2)$

5. The domain is all real numbers *except* $x = -1$, because $f(x)$ is undefined when $x = -1$.

6. (a) $(f \circ g)(x) = f(g(x)) = f(3x^2) = 1 - 2(3x^2) = 1 - 6x^2$

   (b) $(g \circ f)(x) = g(f(x)) = g(1 - 2x) = 3(1 - 2x)^2$

   (c) $(f \circ f)(x) = f(f(x)) = f(1 - 2x) = 1 - 2(1 - 2x) = 4x - 1$

EXERCISE 1.4

1. If $y = 3x^3$, is y a function of x?

2. If $y = 6x^2$, is y a function of x?

3. If $y^2 = 3x$, is y a function of x?

4. If $y^2 = 10x^2$, is y a function of x?

5. If $p = \pm\sqrt{2q}$, is p a function of q?

6. If $R = \sqrt[3]{x}$, is R a function of x?

7. If $R(x) = 8x - 10$, find the following.
   (a) $R(0)$ (b) $R(2)$ (c) $R(-3)$ (d) $R(1.6)$

8. If $h(x) = 3x^2 - 2x$, find the following.
   (a) $h(3)$ (b) $h(-3)$
   (c) $h(2)$ (d) $h(\frac{1}{6})$

9. If $C(x) = 4x^2 - 3$, find the following.
   (a) $C(0)$ (b) $C(-1)$
   (c) $C(-2)$ (d) $C(-\frac{3}{2})$

10. If $R(x) = 100x - x^3$, find the following.
    (a) $R(1)$ (b) $R(10)$ (c) $R(2)$ (d) $R(-10)$

11. If $f(x) = x^3 - 4/x$, find the following.
    (a) $f(-\frac{1}{2})$ (b) $f(2)$ (c) $f(-2)$

12. If $C(x) = (x^2 - 1)/x$, find the following.
    (a) $C(1)$ (b) $C(\frac{1}{2})$ (c) $C(-2)$

13. Let $f(x) = 1 + x + x^2$.
    (a) Is $f(2 + 1) = f(2) + f(1)$?
    (b) Find $f(x + h)$.
    (c) Does $f(x + h) = f(x) + f(h)$?
    (d) Does $f(x + h) = f(x) + h$?

14. Let $f(x) = 3x^2 - 6x$.
    (a) Is $f(3+2) = f(3) + 2$?
    (b) Find $f(x + h)$.
    (c) Does $f(x + h) = f(x) + h$?
    (d) Does $f(x + h) = f(x) + f(h)$?

15. If $f(x) = x - 2x^2$, find the following.
    (a) $f(x + h)$ (b) $f(x + h) - f(x)$
    (c) $\dfrac{f(x + h) - f(x)}{h}$

16. If $f(x) = 2x^2 - x + 3$, find the following.
    (a) $f(x + h)$ (b) $f(x + h) - f(x)$
    (c) $\dfrac{f(x + h) - f(x)}{h}$

17. Do either of the graphs in Figure 1.25 represent y as a function of x? Explain your choices.

18. Do either of the graphs in Figure 1.26 represent y as a function of x? Explain your choices.

19. If $y = f(x)$ in Figure 1.25(a), find the following.
    (a) $f(9)$ (b) $f(5)$

20. Suppose $y = g(x)$ in Figure 1.26(b).
    (a) Find $g(0)$.
    (b) How many x-values in the domain of this function satisfy $g(x) = 0$?

21. The graph of $y = x^2 - 4x$ is shown in Figure 1.27.
    (a) If the coordinates of the point P on the graph are (a, b), how are a and b related?
    (b) What are the coordinates of the point Q? Do they satisfy the equation?
    (c) What are the coordinates of R? Do they satisfy the equation?
    (d) What are the x-values of the points on the graph whose y-coordinates are 0? Are these x-values solutions to the equation $x^2 - 4x = 0$?

22. The graph of $y = 2x^2$ is shown in Figure 1.28.
    (a) If the point P, with coordinates (a, b), is on the graph, how are a and b related?
    (b) Does the point $(1, 1)$ lie on the graph? Do the coordinates satisfy the equation?
    (c) What are the coordinates of point R? Do they satisfy the equation?
    (d) What is the x-value of the point whose y-coordinate is 0? Does this value of x satisfy the equation $0 = 2x^2$?

Figure 1.25   (a)   (b)

Figure 1.26   (a)   (b)

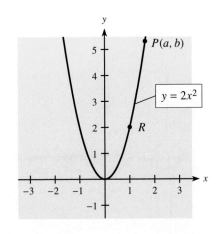

Figure 1.27   **Figure 1.28**

For $f(x)$ and $g(x)$ given in problems 23–26, find

(a) $(f + g)(x)$
(b) $(f - g)(x)$
(c) $(f \cdot g)(x)$
(d) $(f/g)(x)$

23. $f(x) = 3x \qquad g(x) = x^3$
24. $f(x) = \sqrt{x} \qquad g(x) = 1/x$
25. $f(x) = \sqrt{2x} \qquad g(x) = x^2$
26. $f(x) = (x - 1)^2 \qquad g(x) = 1 - 2x$

For $f(x)$ and $g(x)$ given in problems 27–30, find

(a) $(f \circ g)(x)$
(b) $(g \circ f)(x)$
(c) $f(f(x))$
(d) $f^2(x) = (f \cdot f)(x)$

27. $f(x) = (x - 1)^3 \qquad g(x) = 1 - 2x$
28. $f(x) = 3x \qquad g(x) = x^3 - 1$
29. $f(x) = 2\sqrt{x} \qquad g(x) = x^4 + 5$
30. $f(x) = \dfrac{1}{x^3} \qquad g(x) = 4x + 1$

State the domain and range of each of the following functions in problems 31–34. Check your conclusions by graphing each function with a graphing utility.

31. $y = x^2 + 4$ \qquad 32. $y = x^2 - 1$
33. $y = \sqrt{x} + 4$ \qquad 34. $y = \sqrt{x^2 + 1}$

 In problems 35–38, a function and its graph are given. In each problem find the domain.

35. $f(x) = \dfrac{\sqrt{x - 1}}{x - 2}$

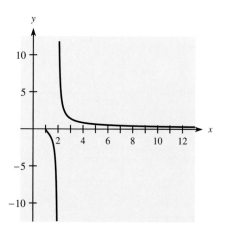

36. $f(x) = \dfrac{x + 1}{\sqrt{x + 3}}$

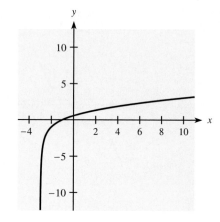

37. $f(x) = 4 + \sqrt{49 - x^2}$

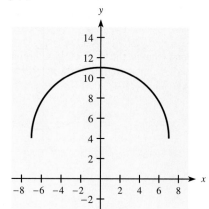

38. $f(x) = -2 - \sqrt{9 - x^2}$

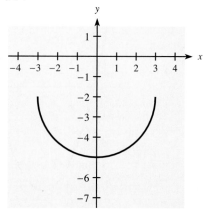

APPLICATIONS

39. Mortgage A couple seeking to buy a home decides that a monthly payment of $800 fits their budget. Their bank's interest rate is 7.5%. The amount they can borrow, A, is a function of the time t, in years, it will take to repay the debt. If we denote this function by $A = f(t)$, then the following table defines the function.

t	A
5	40,000
10	69,000
15	89,000
20	103,000
25	113,000
30	120,000

SOURCE: Comprehensive Mortgage Payment Tables, Publication No. 492, Financial Publishing Co., Boston

(a) Find $f(20)$ and write a sentence that explains its meaning.
(b) Does $f(5 + 5) = f(5) + f(5)$?
(c) If $f(t) = 115,000$, approximate t. Why would someone buying a home need to answer such a question?

40. Debt refinancing When a debt is refinanced, sometimes the term of the loan (that is, the time it takes to repay the debt) is shortened. Suppose the current interest rate is 7%, and the current debt is $100,000. The monthly payment R of the refinanced debt is a function of the term of the loan, t, in years. If we represent this function by $R = f(t)$, then the following table defines the function.

t	R
5	1980.12
10	1161.09
12	1028.39
15	898.83
20	775.30
25	706.78

SOURCE: Comprehensive Mortgage Payment Tables, Publication No. 492, Financial Publishing Co., Boston

(a) Find $f(10)$ and write a sentence that explains its meaning.
(b) Approximate t if $f(t) = \$1000$. Why would someone who is refinancing need to determine such a t?

41. Social Security benefits funding Social Security benefits paid to eligible beneficiaries are funded by individuals who are currently employed. The graph defines a function that gives the number of workers, n, supporting each retiree as a function of time t (given by calendar year). If we denote this function by $n = f(t)$,
(a) Find $f(1950)$ and explain its meaning.
(b) Find $f(1990)$.

(c) If data regarding workers per Social Security beneficiary was graphed in the year 2010, what parts of this graph *must* be the same and what parts *might* be the same? Explain.
(d) Find the domain and range of $n = f(t)$ if the function is defined by the graph.

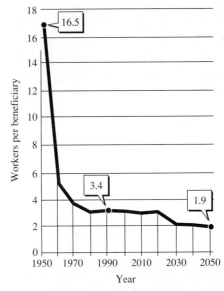

SOURCE: Social Security Administration

42. Dow Jones average If t represents the number of hours after 9:30 A.M. on Thursday, March 31, 1994, then the graph defines the Dow Jones industrial average D as a function of time t. If we represent this function by $D = f(t)$, use the graph to complete the following.

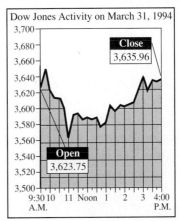

From *The Derrick*, Oil City, Pennsylvania, April 1, 1994
SOURCE: Telerate Systems, Inc.

(a) Find $f(0)$ and $f(6.5)$ and write a sentence that explains what each means.
(b) Find the domain and range for $D = f(t)$ as defined by the graph.
(c) How many t-values satisfy $f(t) = 3600$? Estimate one such t-value.

43. **Windchill factor** Dr. Paul Siple conducted studies testing the effect of wind on the formation of ice at various temperatures and developed the concept of the windchill factor, which we hear reported during winter weather reports. If the air temperature is $-5°F$, then the windchill, WC, is a function of the wind speed, s (in mph), and is given by

$$WC = f(s) = 45.694 + 1.75s - 29.26\sqrt{s}.$$

   (a) Based on the formula for $f(s)$ and the physical context of the problem, what is the domain of $f(s)$?
   (b) Find $f(10)$ and write a sentence that explains its meaning.
   (c) This formula does not work for $s = 0$. What should $f(0)$ equal?

44. **Marijuana use** For the years from 1975 to 1991, the percentage p of high school seniors who have tried marijuana can be considered as a function of time t according to

$$p = f(t) = -0.228596t^2 + 2.7783t + 49.813783,$$

   where t is the number of years past 1975 (Source: National Institute on Drug Abuse).
   (a) Find $f(15)$ and write a sentence that explains its meaning.
   (b) Find $f(25)$ and explain its meaning.
   (c) For $f(t)$ outside what range *must* this model no longer be accurate?

45. **Body-heat loss** The description of body-heat loss due to convection involves a coefficient of convection K_c, which depends on wind speed v according to the equation

$$K_c = 4\sqrt{4v + 1}.$$

   (a) Is K_c a function of v?
   (b) What is the domain?
   (c) What restrictions do nature and common sense put on v?

46. **Efficiency of a muscle** The efficiency E of a muscle performing a maximal contraction is related to the time t that the muscle is contracted according to

$$E = \frac{1 - 0.24t}{2 + t}.$$

   (a) Is E a function of t?
   (b) What is the domain?
   (c) What restrictions do nature and common sense put on the domain?

47. **Fahrenheit-Celsius** The equation

$$C = \frac{5}{9}F - \frac{160}{9}$$

   gives the relation between temperature readings in Celsius and Fahrenheit.
   (a) Is C a function of F?
   (b) What is the domain?
   (c) If we consider this equation as relating temperatures of water in its liquid state, what are the domain and range?
   (d) What is C when $F = 40°$?

48. **Pressure of a gas** The pressure P of a certain gas is related to volume V according to

$$P = \frac{100}{V}.$$

   (a) Is 0 in the domain of this function?
   (b) What is $P(100)$?   (c) What is $P(50)$?
   (d) As volume decreases, what happens to pressure?

49. **Population of organisms** The population size y of a certain organism at time t is given by

$$y = f(t) = 4t^2 + 2t.$$

   What is
   (a) $f(1)$?   (b) $f(2)$?   (c) $f(3)$?

50. **Drug reaction** The reaction R to an injection of a drug is related to the dosage x according to

$$R(x) = x^2\left(500 - \frac{x}{3}\right),$$

   where 1000 mg is the maximum dosage. What is $R(100)$?

51. **Nerve response** The number of action potentials produced by a nerve t seconds after a stimulus may be described by

$$N(t) = 25t + \frac{4}{t^2 + 2} - 2.$$

   Find the following and explain what each means.
   (a) $N(2)$ (b) $N(10)$

52. **Cost** The total cost of producing a product is given by

$$C = 300x + 0.1x^2 + 1200,$$

   where x represents the number of units produced. What is the
   (a) total cost of producing 10 units?
   (b) meaning of $C(100)$?   (c) value of $C(100)$?

53. **Cost-benefit** Suppose that the cost C (in dollars) of removing p percent of the particulate pollution from the smokestacks of an industrial plant is given by

$$C(p) = \frac{7300p}{100 - p}.$$

(a) Find the domain of this function. Recall that p represents the percentage of pollution that is removed. In parts (b)–(e), find the functional values and explain what each means.

(b) $C(45)$ (c) $C(90)$

(d) $C(99)$ (e) $C(99.6)$

54. **Test reliability** If a test having reliability r is lengthened by a factor n ($n \geq 1$), the reliability R of the new test is given by

$$R(n) = \frac{nr}{1 + (n - 1)r} \qquad 0 < r \leq 1.$$

If the reliability is $r = 0.6$, the equation becomes

$$R(n) = \frac{0.6n}{0.4 + 0.6n}.$$

(a) Find $R(1)$.

(b) Find $R(2)$; that is, find R when the test length is doubled.

(c) What percentage improvement is there in the reliability when the test length is doubled?

55. **Area** If 100 feet of fence is to be used to fence in a rectangular yard, then the resulting area of the fenced yard is given by

$$A = x(50 - x),$$

where x is the width of the rectangle.

(a) Is A a function of x?

(b) If $A = A(x)$, find $A(2)$ and $A(30)$.

(c) What restrictions must be placed on x (the domain) so that the problem makes physical sense?

56. **Postal restrictions** If a box with square cross section is to be sent by the postal service, there are restrictions on its size such that its volume is given by $V = x^2(108 - 4x)$, where x is the length of each side of the cross section (in inches).

(a) Is V a function of x?

(b) If $V = V(x)$, find $V(10)$ and $V(20)$.

(c) What restrictions must be placed on x (the domain) so that the problem makes physical sense?

57. **Profit** Suppose that the profit from the production and sale of x units of a product is given by

$$P(x) = 180x - \frac{x^2}{100} - 200.$$

In addition, suppose that for a certain month the number of units produced on day t of the month is

$$x = q(t) = 1000 + 10t.$$

(a) Find $(P \circ q)(t)$ to express the profit as a function of the day of the month.

(b) On the fifteenth day of the month, find the number of units produced, and find the profit.

58. **Fish species growth** For many species of fish, the weight W is a function of the length L that can be expressed by

$$W = W(L) = kL^3 \qquad k = \text{constant}.$$

Suppose that for a particular species $k = 0.02$, and that for this species the length (in cm) is a function of the number of years t that the fish has been alive that is given by

$$L = L(t) = 50 - \frac{(t - 20)^2}{10} \qquad 0 \leq t \leq 20.$$

Find $(W \circ L)(t)$ in order to express W as a function of the age t of the fish.

59. **Fencing a lot** A farmer wishes to fence the perimeter of a rectangular lot with an area of 1600 square feet. If the lot is x feet long, express the amount L of fence needed as a function of x.

60. **Cost** A shipping crate has a square base with sides of length x feet, and it is half as tall as it is wide. If the material for the bottom and sides of the box costs $2.00 per square foot and the material for the top costs $1.50 per square foot, express the total cost of material for the box as function of x.

61. **Revenue** An agency charges $10 per person for a trip to a concert if 30 people travel in a group. But for each person above the 30, the charge will be reduced by $0.20. If x represents the number of people above the 30, write the agency's revenue R as a function of x.

62. **Revenue** A company handles an apartment building with 50 units. Experience has shown that if the rent for each of the units is $360 per month, all of the units will be filled, but one unit will become vacant for each $10 increase in the monthly rate. If x represents the number of $10 increases, write the revenue R from the building as a function of x.

1.5 SPECIAL FUNCTIONS AND THEIR GRAPHS

Objectives

❑ To graph and apply basic functions
❑ To graph and apply polynomial functions
❑ To graph and apply rational functions
❑ To graph and apply piecewise defined functions

APPLICATION PREVIEW

The average cost per item for a product is calculated by dividing the total cost by the number of items. Hence, if the total cost function for x units of a product is

$$C(x) = 900 + 3x + x^2,$$

and if we denote the average cost function by $\overline{C}(x)$, then

$$\overline{C}(x) = \frac{C(x)}{x} = \frac{900 + 3x + x^2}{x}.$$

This average cost function is a special kind of function, called a **rational function.** In this section, we discuss this and other types of functions and answer questions such as "What is the minimum average cost?"

Linear equations that can be written in the form $y = ax + b$ express y as a function of x, so they are called **linear functions.** As we have seen in Section 1.2, graphs of these types of equations (and thus of linear functions) are lines. The special linear function

$$y = f(x) = x$$

is called the **identity function** (Figure 1.29(a)), and a linear function defined by

$$y = f(x) = C, \qquad C \text{ a constant}$$

is called a **constant function.** Figure 1.29(b) shows the graph of the constant function $y = f(x) = 2$.

Figure 1.29

(a)

(b)

(c)

Figure 1.29 (continued)

(d)

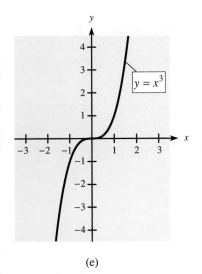

(e)

Some additional basic functions are $y = 1/x$, $y = x^2$, and $y = x^3$. (See Figures 1.29 (c), (d), and (e).) These basic functions (or variations of them) are the starting points for learning about other types of functions or classes of functions. Linear functions, $y = x^2$, and $y = x^3$ are part of a more general class of functions called **polynomial functions,** and $y = 1/x$ belongs to a class called **rational functions.**

POLYNOMIAL AND RATIONAL FUNCTIONS

A **polynomial function of degree n** has the form

$$y = a_n x^n + a_{n-1} x^{n-1} + \cdots + a_1 x + a_0,$$

where $a_n \neq 0$.

A **linear function** is a polynomial function of degree 1, and a polynomial function of degree 2 is called a **quadratic function** (studied in more detail in Chapter 4).

Accurate graphing of a polynomial function of degree greater than 2 may require the methods of calculus; we will investigate these methods in Chapter 10. For now, we will observe some characteristics of the graphs of polynomial functions of degrees 2, 3, and 4; these are summarized in Table 1.2. Using this information and a point-plotting or a graphing utility yields the graphs of these functions.

Table 1.2 Graphs of some polynomials

	Degree 2	Degree 3	Degree 4
Turning points	1	0 or 2	1 or 3
x-intercepts	0, 1, or 2	1, 2, or 3	0, 1, 2, 3, or 4
Possible shapes	∪ or ∩ (a) (b)	(a) (b) (c) (d)	(a) (b) (c) (d) (e) (f)

EXAMPLE 1

Graph $y = x^3 - 16x$.

Solution

This function is a third-degree polynomial function, so it has one of the four shapes shown in the "Degree 3" column in Table 1.2. By graphing the function with a large x-range and y-range, we obtain the graph shown in

Figure 1.30

(a)

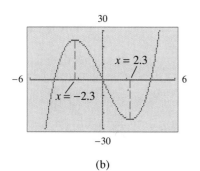

(b)

Two views of $y = x^3 - 16x$

Figure 1.30(a). We see that the graph has a shape like (b) in Table 1.2, so we know that a smaller viewing window (using a smaller x-range and y-range) will provide a more detailed view of all the graph's turning points (See Figure 1.30(b)). Figure 1.30(b) indicates that the turning points are at $x = -2.3$ and $x = 2.3$ (to the nearest tenth).

EXAMPLE 2
Graph $y = x^4 - 2x^2$.

Solution
This is a polynomial function of degree 4. Thus it has one or three turning points and has one of the six shapes in the "Degree 4" column of Table 1.2. We begin by graphing this function with a large viewing window to obtain the graph in Figure 1.31(a). We see that the graph appears to fit either shape (a) or (f) from Table 1.2. Figure 1.31(b) shows the same graph with a smaller viewing window, near $x = 0$. We now see that the graph has shape (f) from Table 1.2, with turning points at $x = -1$, $x = 0$, and $x = 1$.

(a)

(b)

Figure 1.31   Two views of $y = x^4 - 2x^2$

CHECKPOINT

1. All constant functions (such as $f(x) = 8$) have graphs that are _____.

2. Which of the following are polynomial functions?

   (a) $f(x) = x^3 - x + 4$ (b) $f(x) = \dfrac{x + 1}{4x}$ (c) $f(x) = 1 + \sqrt{x}$

   (d) $g(x) = \dfrac{1 + \sqrt{x}}{1 + x + \sqrt{x}}$ (e) $h(x) = 5$

3. A third-degree polynomial can have at most _____ turning points.

   A **rational function** is a function that can be written as the ratio of two polynomials. The graph of the rational function

$$y = \frac{1}{x}$$

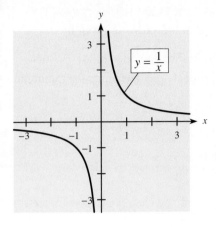

Figure 1.32

is shown in Figure 1.32. Because division by 0 is not possible, $x = 0$ is not in the domain of the function.

The graph of $y = 1/x$ approaches the y-axis but does not touch it. In this case, we call the y-axis a **vertical asymptote.**

On the graphs of polynomial functions, the turning points are usually the features of greatest interest. However, on the graphs of rational functions, vertical asymptotes frequently are the most interesting features. (Rational functions may or may not have turning points.)

In general, when a rational function has its denominator equal to 0 at a value $x = a$, then we should include $x = a$ in our viewing rectangle to see what shape the graph is taking near $x = a$. When $|y|$ becomes large as x gets close to a, we say $x = a$ is a **vertical asymptote.**

Rational functions sometimes have **horizontal asymptotes** as well as vertical asymptotes. Whenever the values of y approach some finite number b as $|x|$ becomes very large, we say that there is a **horizontal asymptote** at $y = b$.

Notice that the graph of $y = 1/x$ appears to get close to the x-axis as $|x|$ becomes large. Testing values of x for which $|x|$ is large, we see that y is close to 0. Thus, we say $y = 0$ (or the x-axis) is a **horizontal asymptote** for the graph of $y = 1/x$. The graph in the following example has both a vertical and a horizontal asymptote.

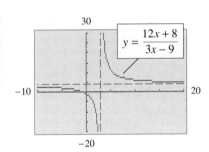

Figure 1.33

EXAMPLE 3

Sketch the graph of

$$y = \frac{12x + 8}{3x - 9}.$$

Determine any asymptotes.

Solution

Because $3x - 9 = 0$ when $x = 3$, it follows that $x = 3$ is not in the domain of the function. Thus, to search for a vertical asymptote, our window should include $x = 3$ and large values for y. Figure 1.33 shows a vertical asymptote at $x = 3$. Also, it looks like the graph has a horizontal asymptote. To verify this, we calculate y as $|x|$ becomes larger.

x	$y = f(x)$
$-10,000$	3.998
-1000	3.98
1000	4.01
10,000	4.001

This calculation and the graph indicate that $y = 4$ is a horizontal asymptote.

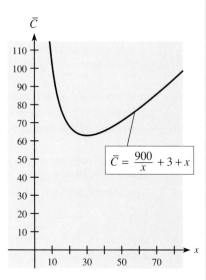

Figure 1.34

EXAMPLE 4

Suppose the total cost function for x units of a product is given by

$$C(x) = 900 + 3x + x^2.$$

Graph the average cost function for this product, and determine the minimum average cost.

Solution

The average cost per unit is

$$\overline{C}(x) = \frac{C(x)}{x} = \frac{900 + 3x + x^2}{x}$$

or

$$\overline{C}(x) = \frac{900}{x} + 3 + x.$$

Because x represents the number of units produced, $x \geq 0$. Because $x = 0$ cannot be in the domain of the function, we choose a window with $x \geq 0$. Figure 1.34 shows the graph of $\overline{C}(x)$. The graph appears to have a minimum near $x = 30$. By plotting points or using MINIMUM or TRACE on a utility, we can verify that the minimum point occurs at $x = 30$ and $\overline{C} = 63$. Thus, the minimum average cost is $63 per unit when 30 units are produced.

CHECKPOINT

4. Given $f(x) = \dfrac{3x}{x - 4}$, decide whether the following are true or false.

    (a) $f(x)$ has a vertical asymptote at $x = 4$.
    (b) $f(x)$ has a horizontal asymptote at $x = 3$.

Figure 1.35

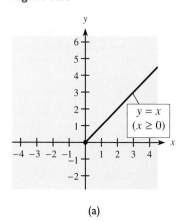

(a)

PIECEWISE DEFINED FUNCTIONS

Another special function comes from the definition of $|x|$. The **absolute value function** can be written as

$$f(x) = |x| \quad \text{or} \quad f(x) = \begin{cases} x & \text{if } x \geq 0 \\ -x & \text{if } x < 0 \end{cases}.$$

Notice that restrictions on the domain of the absolute value function specify different formulas for different parts of the domain. To graph $f(x) = |x|$, we graph the portion of the line $y = x$ for $x \geq 0$ (see Figure 1.35(a)) and the portion of the line $y = -x$ for $x < 0$ (see Figure 1.35(b)). When we put these pieces on the same graph (Figure 1.35(c)), they give us the graph of $f(x) = |x|$. Since the absolute value function is defined by two equations, we say it is a **piecewise defined function**.

Figure 1.35 (continued)

(b)

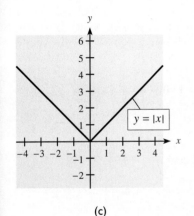

(c)

It is possible for the selling price S of a product to be defined as a piecewise function of the cost C of the product. For example, the selling price might be defined by two different equations on two different intervals, as follows:

$$S = f(C) = \begin{cases} 3C & \text{if } 0 \le C \le 20 \\ 1.5C + 30 & \text{if } C > 20 \end{cases}.$$

Written in this way, the value of S depends on the value of C, so C is the **independent variable** and S is the **dependent variable.** The set of possible values of C (the domain) is the set of nonnegative real numbers (because cost cannot be negative), and the range is the set of nonnegative real numbers (since all values of C will result in nonnegative values for S).

Note that the selling price of a product that costs \$15 would be $f(15) = 3(15) = 45$ (dollars) and the selling price of a product that costs \$25 would be $f(25) = 1.5(25) + 30 = 67.50$ (dollars).

EXAMPLE 5

Graph the selling price function

$$S = f(C) = \begin{cases} 3C & \text{if } 0 \le C \le 20 \\ 1.5C + 30 & \text{if } C > 20 \end{cases}.$$

Solution

Each of the two pieces of the graph of this function is a line and is easily graphed. It remains only to graph each in the proper interval. The graph is shown in Figure 1.36.

EXAMPLE 6

The 1993 monthly charge in dollars for x kilowatt hours (kWh) of electricity used by a residential consumer of Excelsior Electric Membership Corporation during the months of November through June is given by the function

$$C(x) = \begin{cases} 10 + .094x & \text{if } 0 \le x \le 100 \\ 19.40 + .075(x - 100) & \text{if } 100 < x \le 500. \\ 49.40 + .05(x - 500) & \text{if } x > 500 \end{cases}$$

(a) What is the monthly charge if 1100 kWh of electricity are consumed in a month?
(b) What is the monthly charge if 450 kWh are consumed in a month?
(c) Is there a charge if no electricity is used?

Solution

(a) We need to find $C(1100)$. Because $1100 > 500$, we use it in the third formula line.

$$C(1100) = 49.40 + 0.05(1100 - 500) = \$79.40$$

Figure 1.36

(b) We evaluate $C(450)$ by using the second formula line for $C(x)$.

$$C(450) = 19.40 + 0.075(450 - 100) = \$45.65$$

(c) If no electricity is used, the charge is $C(0)$; it is found by using the first formula line.

$$C(0) = 10 + 0.094(0) = \$10$$

There is a $10 charge even when no electricity is used. This could be thought of as a service charge or line charge for the privilege and convenience of having electrical service available.

CHECKPOINT

5. If $f(x) = \begin{cases} 0 & \text{if } x \leq 0 \\ 12x & \text{if } 0 < x < 10, \\ 10x + 20 & \text{if } x \geq 10 \end{cases}$ find the following.

(a) $f(-5)$ (b) $f(5)$ (c) $f(20)$

CHECKPOINT SOLUTIONS

1. horizontal lines

2. Polynomial functions are (a) and (e); also (e) is a constant function; function (b) is a quotient of polynomials (a rational function).

3. two

4. (a) True.
   (b) False. The horizontal asymptote is $y = 3$.

5. (a) $f(-5) = 0$ In fact, for any negative value of $x, f(x) = 0$.
   (b) $f(5) = 12(5) = 60$
   (c) $f(20) = 10(20) + 20 = 220$

EXERCISE 1.5

In problems 1–18, match each of the functions with one of the graphs labeled (a)–(r) shown following these functions. Recognizing special features of certain types of functions and plotting points for the functions will be helpful.

1. $y = x^3$

2. $y = x^2$

3. $f(x) = -3$

4. $y = x$

5. $y = x^3 - x$

6. $y = (x - 3)^2(x + 1)$

7. $y = 16x^2 - x^4$

8. $y = (x + 1)(x - 3)(x - 1)$

9. $y = \dfrac{1}{x}$

10. $y = \dfrac{1 - 2x}{x}$

11. $y = \dfrac{x - 3}{x + 1}$

12. $y = \dfrac{1 - 3x}{2x + 5}$

13. $y = |x|$

14. $y = |x - 2|$

15. $f(x) = \begin{cases} x^2 & \text{if } x \leq 2 \\ 4 & \text{if } x > 2 \end{cases}$ 16. $f(x) = \begin{cases} 2 & \text{if } x < 0 \\ x^3 & \text{if } x \geq 0 \end{cases}$

17. $y = \begin{cases} -x & \text{if } x < -1 \\ 1 & \text{if } -1 \leq x \leq 1 \\ x & \text{if } x > 1 \end{cases}$

18. $y = \begin{cases} x & \text{if } x \leq 0 \\ 7x/2 & \text{if } 0 < x < 2 \\ 7 & \text{if } x \geq 2 \end{cases}$

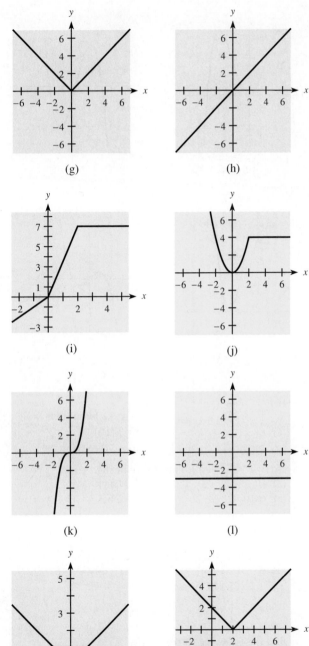

(g)

(h)

(a)

(b)

(i)

(j)

(c)

(d)

(k)

(l)

(e)

(f)

(m)

(n)

(o)

(p)

(q)

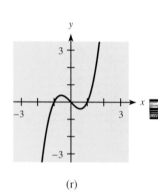

(r)

24. If $f(x) = \begin{cases} 1 - 4x & \text{if } x < -1 \\ -x^2 & \text{if } x \geq -1 \end{cases}$,

find the following, if they exist.

(a) $f(-3)$ (b) $f(-1.1)$ (c) $f(-1)$ (d) $f(10)$

25. If $k(x) = \begin{cases} 2 & \text{if } x < 0 \\ x + 4 & \text{if } 0 \leq x < 1, \\ 1 - x & \text{if } x \geq 1 \end{cases}$

find the following, if they exist.

(a) $k(-5)$ (b) $k(0)$ (c) $k(1)$ (d) $k(-0.001)$

26. If $g(x) = \begin{cases} 0.5x + 4 & \text{if } x < 0 \\ 4 - x & \text{if } 0 \leq x < 4, \\ 0 & \text{if } x > 4 \end{cases}$

find the following, if they exist.

(a) $g(-4)$ (b) $g(1)$ (c) $g(7)$ (d) $g(3.9)$

 In each of problems 27 and 28, use a graphing utility to graph the given function, $y = f(x)$, with $y = af(x)$ using $a = -1/10, 1/10, -2$ and 2. Based on these graphs, describe how the graph of $y = f(x)$ is altered when (a) $0 < |a| < 1$, (b) $|a| > 1$, and (c) $a < 0$.

27. $f(x) = x^2$

28. $f(x) = x^3$

In problems 29–32, use a graphing utility to graph the functions described using $a = -1, -3, 2,$ and 4. Based on these graphs, describe how the graph of $y = f(x)$ is altered for (a) $a < 0$ and (b) $a > 0$.

29. Graph $f(x) = |x|$ and $y = f(x - a) = |x - a|$.

30. Graph $f(x) = \dfrac{1}{x}$ and $y = f(x - a) = \dfrac{1}{x - a}$.

31. Graph $f(x) = \dfrac{1}{x}$ and $y = f(x) + a$.

32. Graph $f(x) = x^2$ and $y = f(x) + a$.

In problems 33–42, (a) graph each function (with a graphing utility if one is available); (b) classify each function as a polynomial function, a rational function, or a piecewise defined function; (c) identify any asymptotes; and (d) use the graphs to locate turning points.

33. $f(x) = -2(x - 4)^2$

34. $f(x) = -x^2 + 2x + 8$

35. $f(x) = (x - 3)^2(x + 1)$

36. $f(x) = (x + 1)(x - 3)(x - 1)$

19. If $F(x) = \dfrac{|x|}{x}$, find the following.

(a) $F(-\frac{1}{3})$ (b) $F(10)$

(b) $F(0.001)$ (d) Is $F(0)$ defined?

20. If $H(x) = |x - 1|$, find the following.
(a) $H(-1)$ (b) $H(1)$
(c) $H(0)$ (d) Does $H(-x) = H(x)$?

21. If $f(x) = \begin{cases} |x| + 1 & \text{if } x < 1 \\ 3 - 1/x & \text{if } x > 1 \end{cases}$,

find the following, if they exist.
(a) $f(-6)$ (b) $f(1)$ (c) $f(100)$ (d) $f(0.9)$

22. If $k(x) = \begin{cases} 2 & \text{if } x < 0 \\ x + 4 & \text{if } 0 < x < 1, \\ |1 - x| & \text{if } x \geq 1 \end{cases}$

find the following, if they exist.
(a) $k(-0.1)$ (b) $k(0.1)$ (c) $k(0.9)$ (d) $k(1.1)$

23. If $f(x) = \begin{cases} x^2 + 1 & \text{if } x < 1 \\ (x - 1)/x & \text{if } x \geq 1 \end{cases}$,

find the following, if they exist.
(a) $f(-6)$ (b) $f(1)$ (c) $f(100)$ (d) $f(0.9)$

37. $f(x) = 1.6 x^2 - 0.1x^4$ 38. $f(x) = \dfrac{x^4 - 4x^3}{3}$

39. $f(x) = \dfrac{2x + 4}{x + 1}$ 40. $f(x) = \dfrac{1 - 3x}{2x + 5}$

41. $f(x) = \begin{cases} -x & \text{if } x < 0 \\ 5x & \text{if } x \ge 0 \end{cases}$

42. $f(x) = \begin{cases} 2x - 1 & \text{if } x < 1 \\ -x & \text{if } x \ge 1 \end{cases}$

APPLICATIONS

43. **Fixed costs** Fixed costs FC are business costs that remain constant regardless of the number of units produced. Some items that might contribute to fixed costs are rent and utilities.

$$FC = 2000$$

is an equation indicating that a business has fixed costs of $2000. Graph $FC = 2000$ by putting x (the number of units produced) on the horizontal axis. (Note that FC does not mean the product of F and C.)

44. **Investments** If $1000 is invested at 10% simple interest for t years, then the future value A of the investment is given by

$$A = f(t) = 1000 + 100t.$$

(a) Find $f(5)$.
(b) Graph this function.
(c) Find how long it takes for this investment to double.

45. **Water usage costs** The monthly charge for water in a small town is given by

$$f(x) = \begin{cases} 18 & \text{if } 0 \le x \le 20 \\ 18 + 0.1(x - 20) & \text{if } x > 20 \end{cases}$$

where x is hundreds of gallons and $f(x)$ is in dollars. Find the monthly charge for each of the following usages:
(a) 30 gallons, (b) 3000 gallons, and (c) 4000 gallons.
(d) Graph the function for $0 \le x \le 100$.

46. **Commercial electrical usage** The monthly charge (in dollars) for x kilowatt hours (kWh) of electricity used by a commercial customer is given by the following function.

$$C(x) = \begin{cases} 7.52 + 0.1079x & \text{if } 0 \le x \le 5 \\ 19.22 + 0.1079x & \text{if } 5 < x \le 750 \\ 20.795 + 0.1058x & \text{if } 750 < x \le 1500 \\ 131.345 + 0.0321x & \text{if } x > 1500 \end{cases}$$

Find the monthly charges for the following usages.
(a) 5 kWh (b) 6 kWh (c) 3000 kWh

47. **Residential power costs** Excluding fuel adjustment costs and taxes, Georgia Power Company charges its residential customers during the months of June through September according to the following function.

$$C(x) = \begin{cases} 7.50 + 0.04783x & \text{if } 0 \le x \le 650 \\ 38.59 + 0.07948(x - 650) & \text{if } 650 < x \le 1000 \\ 66.41 + 0.08184(x - 1000) & \text{if } x > 1000 \end{cases}$$

where $C(x)$ is the charge in dollars for x kilowatt hours (kWh) of electricity.
(a) Find $C(780)$ and write a sentence that explains its meaning.
(b) Find the charge for using 1280 kWh in a month.

48. **First-class postage** The U.S. Postal Service fee for first-class postage P is a function of the weight w (in ounces) of the letter or package and is given by

$$P(w) = \begin{cases} 0.32 & 0 < w \le 1 \\ 0.55 & 1 < w \le 2 \\ 0.78 & 2 < w \le 3 \\ 1.01 & 3 < w \le 4 \\ \vdots & \vdots \end{cases}$$

(a) Find $P(2.3)$ and write a sentence that explains its meaning.
(b) Find the postage for a letter that weighs exactly 1 oz. and for one that weighs 1.01 oz.
(c) Find the postage for a letter that weighs 7.3 oz.

49. **Federal income tax** The 1993 U.S. federal income tax owed by a married couple filing jointly can be found from the following schedule.

Schedule Y-1—Use if your filing status is Married filing jointly or Qualifying widow(er)

If the amount on Form 1040, line 37, is: Over—	But not over—	Enter on Form 1040, line 38	of the amount over—
$0	$36,900	......15%	$0
36,900	89,150	$5,535.00 + 28%	36,900
89,150	140,000	20,165.00 + 31%	89,150
140,000	250,000	35,928.50 + 36%	140,000
250,000		75,528.50 + 39.6%	250,000

SOURCE: Internal Revenue Service, 1993 Form 1040, Instructions

From this schedule, the tax due T (entered on Form 1040, line 38) is a function of income x (the amount on Form 1040, line 37) as follows.

$$T(x) = \begin{cases} 0.15x & \text{if } 0 \le x \le 36{,}900 \\ 5535.00 + 0.28(x - 36{,}900) \\ \quad \text{if } 36{,}900 < x \le 89{,}150 \\ 20{,}165.00 + 0.31(x - 89{,}150) \\ \quad \text{if } 89{,}150 < x \le 140{,}000 \\ 35{,}928.50 + 0.36(x - 140{,}000) \\ \quad \text{if } 140{,}000 < x \le 250{,}000 \\ 75{,}528.50 + 0.396(x - 250{,}000) \\ \quad \text{if } 250{,}000 < x \end{cases}$$

(a) Find $T(58{,}676)$ and write a sentence that explains its meaning.

(b) When taxable income passes \$36,900, the tax rate changes from 15% to 28%. Does a couple whose taxable income is \$36,901 pay 13% more tax than a couple whose taxable income is \$36,899? Explain.

50. **Calories and temperature** The temperature of 1 g of H_2O as it changes from ice at $-10°C$ to steam at $110°C$ is given by

$$T = f(x) = \begin{cases} x - 10 & \text{if } 0 \le x < 10 \\ 0 & \text{if } 10 \le x < 100 \\ x - 100 & \text{if } 100 \le x < 200 \\ 100 & \text{if } 200 \le x < 700 \\ x - 600 & \text{if } 700 \le x \le 710 \end{cases}$$

where x represents calories of heat. Find the temperature for each of the following values of x.

(a) 5   (b) 50   (c) 150   (d) 500   (e) 705

(f) For $10 < x < 100$, the value of T is 0. Over this interval, what is happening to the 1 g of substance?

51. **Windchill** The windchill expresses the combined effects of low temperatures and wind speeds as a single temperature reading. When the outside temperature is 0°F, the windchill, WC, is a function of the wind speed s (in mph) and is given by the following.*

$$WC = f(s) = \begin{cases} 0 & \text{if } 0 \le s \le 4 \\ 91.4 - 7.46738(5.81 + 3.7\sqrt{s} - 0.25s) \\ \quad \text{if } 4 < s \le 45 \\ -55 & \text{if } s > 45 \end{cases}$$

(a) Find $f(15)$ and write a sentence that explains its meaning.

(b) Find the windchill when the wind speed is 48 mph.

52. **Photosynthesis** The amount of photosynthesis that takes place in a certain plant depends on the intensity of light x according to the equation

$$f(x) = 145x^2 - 30x^3.$$

Graph this function for values of x from 0 to 4.

53. **Parcel restrictions** If a box with a square base is to be sent by UPS, the volume of the box is restricted by

$$V = 108x^2 - 4x^3,$$

where x is the length of one side of the base in inches.

(a) Graph this function for $0 \le x \le 27$.

(b) Interpret the x-intercepts.

(c) Interpret any turning points.

(d) Does the domain extend beyond $0 \le x \le 27$? Explain.

(e) Estimate the maximum volume, in cubic inches, that can be sent by UPS.

54. **Container design** A rectangular box with a square base is to be formed from a square piece of cardboard with 12-in. sides. If a square piece with side x is cut from each corner of the cardboard and the sides are folded up to form an open box, the volume of the box is

$$V = (12 - 2x)^2 x.$$

(a) What domain makes sense for this problem?

(b) Sketch the graph of this function for values of x from $x = 0$ to $x = 6$.

(c) Interpret the x-intercepts.

(d) Interpret any turning points.

55. **Intensity of light** The relationship between the intensity C of light (in candlepower) and the distance x from the source of light (in meters) is given by

$$C = \frac{k}{x^2}.$$

If $k = 5000$, graph this function.

---

*Bosch, W., and L. G. Cobb, "Windchill," UMAP Unit 658, *The UMAP Journal*, 5(4), Winter 1984, 477–492.

56. **Demand** The demand function for a product is given by

$$p = \frac{200}{2 + 0.1x},$$

where x is the number of units and p is the price. Graph this demand function for $0 \leq x \leq 250$, with x on the horizontal axis.

57. **Minimum costs** A printer has a contract to run 10,000 posters for a fire company benefit. He can use any number of printing plates from 1 to 10 to run the posters, with the cost of printing given by

$$C(x) = 3x + \frac{48}{x},$$

where x is the number of plates he uses.
   (a) Graph this function for $1 \leq x \leq 10$.
   (b) Identify and interpret any turning points for $1 \leq x \leq 10$.
   (c) Does the domain extend beyond $1 \leq x \leq 10$? Explain.

58. **Production costs** A manufacturer estimates that the cost of a production run for a product is

$$C(x) = 30(x - 1) + \frac{3000}{x + 10}$$

where x is the number of machines used.

   (a) Graph this total cost function for values of $x \geq 0$.
   (b) Interpret any turning points.
   (c) Interpret the y-intercept.

59. **Mob behavior** In studying lynchings between 1899 and 1946, psychologist Brian Mullin found that the size of a lynch mob relative to the number of victims predicted the level of brutality. He developed a formula for the other-total ratio (y) that predicts the level of self-attentiveness of people in a crowd of size x with 1 victim.

$$y = \frac{1}{x + 1}$$

The lower the value of y, the more likely an individual is to be influenced by "mob psychology." Graph this function; use positive integers as its domain.

60. **Gravitational force** A person weighing 150 pounds on earth has weight given by

$$w = w(d) = \frac{2,400,000,000}{(4000 + d)^2}$$

d miles above the earth's surface.
   (a) Find $w(0)$ and interpret this.
   (b) How much would this person weigh while flying in an airplane at 29,000 feet?
   (c) An astronaut orbits the earth at an average of 80 miles above the surface. If he weighs 150 pounds on earth, how much does he weigh while in orbit?

1.6 APPLICATIONS OF FUNCTIONS IN BUSINESS AND ECONOMICS

Objectives

❏ To formulate and evaluate total cost, total revenue, and profit functions
❏ To evaluate and graph supply and demand functions
❏ To formulate and evaluate consumption functions

APPLICATION PREVIEW

We will discuss numerous applications of mathematics in this text, but in this section we will introduce three important applications, which will be expanded and used in different circumstances throughout the text as increased mathematical skills permit. These applications are total cost, total revenue, and profit as functions of quantity sold (theory of the firm); supply and demand as functions of price (market analysis); and national consumption (Keynesian analysis).

TOTAL COST, TOTAL REVENUE, AND PROFIT FUNCTIONS

Suppose you are the manager of a shoelace company that operates according to the following conditions:

1. It makes only shoelaces, and sells them for $.50 per pair.

2. Ten people are employed, and are paid $.10 for each pair of shoelaces they make. Note that they are paid *only* for what they produce.

3. Raw materials cost $.24 for each pair of shoelaces made.

4. Costs for equipment, plant, insurance, the manager's salary, and fringe benefits (these are called **fixed costs**) amount to $2000 per month.

If you were the manager of this shoelace factory, you would be interested in how many pairs of shoelaces you would have to make to break even. This means you would have to determine what your **total revenue** and your **total cost** would be at different levels of production. The information given is sufficient to formulate equations for the total revenue and the total cost, and from these equations we can formulate an equation that will determine the **profit** at different levels of production. From this equation you can determine when profit is 0; that is, when you break even. Let's discuss this problem in more detail.

The **total revenue function** states how much money is brought into the firm each month by the sale of its product. The total revenue for a month is the amount taken in before any expenses are paid. If x pairs of shoelaces are sold in a month, the total revenue for the shoelace factory is $0.50x$. Thus the total revenue function is

$$R = R(x) = 0.50x \quad \text{(in dollars).}$$

Here x is the independent variable and R is the dependent variable. The graph of $R(x)$ for one month is shown in Figure 1.37. Note that the domain of the function is $\{x: x \geq 0\}$ because a negative quantity of shoelaces is meaningless, even though the equation makes sense when $x < 0$.

We can also write the equation for the total cost function, which measures how many dollars the firm must pay to produce and sell its product. The total cost is comprised of two parts, fixed costs and variable costs. **Fixed costs** are those that remain constant regardless of the number of units produced. They include depreciation or rent on buildings, interest on investments, and so on. The fixed costs for the shoelace factory are given to be $2000 per month. Thus the cost for one month will always be at least $2000, even if no shoelaces are produced. The fixed costs are denoted by FC on the graph of Figure 1.38.

Variable costs are those directly related to the production of a commodity. The variable costs for producing x pairs of shoelaces are $0.10x$ for labor

Figure 1.37

Figure 1.38

and $0.24x$ for raw materials, for a total of $0.34x$. The **total cost function** is the sum of the fixed and variable costs, or

$$C = C(x) = 2000 + 0.34x \quad \text{(in dollars)}.$$

The total cost is denoted by *TC* on the graph of Figure 1.38. Note that the total costs equals the fixed cost when production is zero.

EXAMPLE 1

Find the total revenue and total cost if the shoelace factory produces and sells
(a) 10,000 pairs of shoelaces in a month.
(b) 15,000 pairs of shoelaces in a month.

Solution
(a) The total revenue is $R(10{,}000) = 0.50(10{,}000) = 5000$ (in dollars).
   The total cost is $C(10{,}000) = 2000 + 0.34(10{,}000) = 5400$ (dollars).

(b) The total revenue is $R(15{,}000) = 0.50(15{,}000) = 7500$ (dollars).
   The total cost is $C(15{,}000) = 2000 + 0.34(15{,}000) = 7100$ (dollars).

We see that the profit or loss for a month can be found by subtracting the total cost from the total revenue. In the previous example, producing and selling 10,000 pairs of shoelaces results in a loss of \$400 (that is, \$5000 − 5400) while producing and selling 15,000 pairs results in a profit of \$7500 − 7100 = \$400. So the firm will break even at some point between 10,000 and 15,000 pairs of shoelaces.

If x units of a commodity are produced and sold, the profit is found by subtracting the total cost from the total revenue.*

Profit (or loss) = total revenue − total cost

We can define the **profit function** as follows:

$$P(x) = R(x) - C(x).$$

Thus for the total revenue function $R(x) = 0.50x$ and the total cost function $C(x) = 2000 + 0.34x$ in the shoelace example, the profit function is given by the equation $P(x) = 0.50x - (2000 + 0.34x)$, or $P(x) = 0.16x - 2000$.

Figure 1.39 is the graph of $P(x)$ for positive values of x. Note that the profit function is negative for some values of x (a loss is incurred) and positive for other values (a profit is made). For example, $P(10,000) = 1600 - 2000 = -400$, so a loss of $400 will result from producing and selling 10,000 pairs of shoelaces. Also, $P(15,000) = 400$, so a profit of $400 will result from selling 15,000 pairs of shoelaces.

The point at which the profit function is zero is called the **break-even point.** Setting $P = 0$ and solving for x gives the break-even point.

$0 = 0.16x - 2000$

$x = 12{,}500$ units

Note that this x-value corresponds to the x-intercept in Figure 1.39.

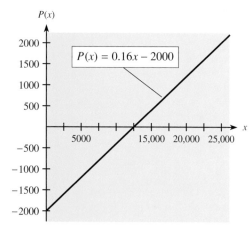

Figure 1.39

EXAMPLE 2

Suppose the selling price of a commodity is $.40 per item. If fixed costs are $200 and the variable costs amount to $.20 per item, find
(a) the total revenue function.
(b) the total cost function.
(c) the profit function.

---

* The symbols generally used in economics for total cost, total revenue, and profit are *TC*, *TR*, and π, respectively. We do not use these symbols to avoid confusion, especially with the use of π as a variable.

Solution
(a) The total revenue function is $R(x) = 0.40x$.
(b) The total cost function is $C(x) = 200 + 0.20x$.
(c) The profit function is

$$P(x) = 0.40x - (200 + 0.20x)$$

or

$$P(x) = 0.20x - 200.$$

CHECKPOINT

1. Suppose when a company produces its product, fixed costs are $12,500 and variable costs are $75 per item.
   (a) Write the total cost function.
   (b) Are fixed costs equal to $C(0)$?

2. Suppose the company in problem 1 sells its products for $175 per item.
   (a) Write the total revenue function.
   (b) Find $R(100)$ and give its meaning.

3. (a) Give the formula for profit in terms of revenue and cost.
   (b) Find the profit function for the company in problems 1 and 2.

4. (a) How do you find the break-even point ?
   (b) Find the break-even point for the profit function in problem 3(b).

SUPPLY AND DEMAND FUNCTIONS

It is a fact that consumers want more goods than they actually purchase. The goods that they actually buy are determined by (a) the price of the good desired, (b) the amount of money they have to spend on the item, (c) the strength of their tastes and preferences, (d) the prices of competing goods, and (e) their expectations of the future and its effect on this good.

If we concentrate on the relationship between the price of the good and the amount of the good a consumer is willing to purchase in a fixed period of time, the resulting function is called a **demand function.** Economists define demand for a commodity by using a schedule, or table, that shows the various amounts of the commodity that consumers are willing and able to purchase at specific prices in a set of possible prices during a specified period of time. The **law of demand** states that as price increases, the corresponding quantity demanded will fall or, as price decreases, the quantity demanded will increase.

Table 1.3 is a demand schedule that shows the relationship between the price of a particular brand of shirt and the number of shirts one person is willing and able to buy in a given period of time (the quantity demanded). This function (or schedule) is for only one person and for a fixed period (one month). Different people will have different demand schedules based on the other four factors mentioned. Usually we will be talking about the demand of

Table 1.3 One person's demand schedule for shirts for one month

q, Quantity Demanded	p, Price of One Shirt
0	24
1	21
2	18
3	15
4	12
5	9

Figure 1.40

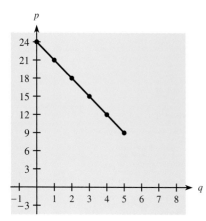

Figure 1.41

a group of consumers. Note also that the demand schedule is for a specified period of time. The relationship from Table 1.3 is for a period of one month. Figure 1.40 shows its graph.

The relationship between the price of a commodity and the quantity demanded is functional; namely, quantity demanded is a function of price, which is denoted $q = f(p)$. However, economists traditionally have graphed the demand function on a two-dimensional graph with the quantity demanded on the horizontal axis and the price on the vertical axis, as we have done in Figure 1.40. Throughout the text, we will follow this tradition. Equations relating price and quantity demanded can be solved for either p or q, and we will have occasion to use the equations in both forms.

Although we only have a series of points as the graph of the demand function in Figure 1.40, it is common practice to draw a smooth curve joining the points and to call the curve the **demand curve.** In this case the smooth curve would be a straight line (see Figure 1.41) with the equation $p = 24 - 3q$. The graph implies that, for example, the consumer will buy 0.5 shirts per month if the cost is $22.50 per shirt. This can be interpreted as meaning he or she will buy 1 shirt every two months. Though some points on the graph, such as $(\sqrt{2}, 24 - 3\sqrt{2})$, will not have much meaning in a business sense, the advantage of dealing with continuous demand curves far outweighs any possible disadvantages. Thus we assume that demand functions will have continuous curves.

Except in very sophisticated economic models, negative price* and quantity have no meaning, so the graphs of demand curves will be in the first quadrant. The demand curve will normally slope downward toward the right because the quantity demanded will be larger as the price decreases. The demand curve may be a vertical line, however, if price is not a factor in demand. For example, a diabetic person must continue to buy insulin regardless of its cost. The demand curve could also be a horizontal line if the price remains constant regardless of the demand.

Just as a consumer's willingness to buy is related to price, so is a manufacturer's willingness to supply goods. Economists define **supply** as a schedule that shows the various amounts of a commodity that a producer is willing and able to produce and make available for sale at specific prices during a specified time.

Supply shows the quantity producers will offer at various prices, so quantity supplied is a function of price. As with demand, price is placed on the vertical axis on the graph of the **supply curve.** The horizontal axis now represents quantity supplied rather than quantity demanded.

The **law of supply** states that as price rises, the corresponding quantity supplied for sale will also rise. Because the producer will supply more goods if the price increases, the supply curve will normally rise as it moves to the right (see Figure 1.42). Of course, certain special circumstances may result in horizontal lines representing supply curves. For example, a consumer may

---

* Negative price values would result if buyers were paid to remove goods from the market.

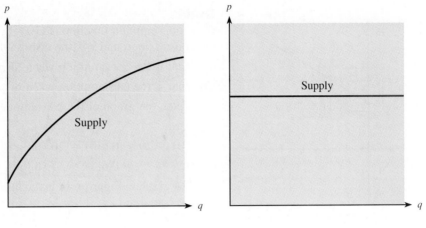

Figure 1.42   **Figure 1.43**

be supplied with the right to make an unlimited number of local phone calls per month for a fixed fee, resulting in a horizontal supply curve (see Figure 1.43). As with the demand curve, the supply curves will have meaning for us only in the first quadrant.

CHECKPOINT

5. (a) Does a typical supply function rise or fall?
   (b) Does a typical demand function rise or fall?

THE CONSUMPTION FUNCTION

The consumption function is one of the basic ingredients in a larger discussion of how an economy can have persistent high unemployment or persistent high inflation. This study is often called Keynesian analysis after its founder, John Maynard Keynes. During the Great Depression of the 1930s, the United States suffered through many years of high unemployment that threatened the collapse of capitalism in this country. The Keynesian theory was used by the Roosevelt administration in one of the most dramatic periods of our history.

A student versed in the language of functions could write a consumption function as

$$C = f(y) = 0.8y + 6,$$

where y is disposable income (in billions of dollars) and C is consumption (in billions of dollars). This statement says that consumption is a function, and expresses exactly what the functional relationship is. **Consumption functions** are frequently assumed to be linear over short periods of time. Over longer periods of time, consumption functions will be nonlinear.

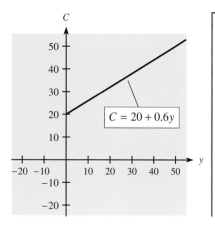

Figure 1.44

EXAMPLE 3

If the consumption function is given by $C = f(y) = 20 + 0.6y$, where C is the consumption and y is the disposable income (in billions of dollars),

(a) what is the consumption when disposable income is 0?

(b) what is the consumption when disposable income is $5 billion?

(c) sketch the graph of the consumption function.

Solution

(a) $f(0) = 20 + 0.6(0) = 20$ (billion dollars)

(b) $f(5) = 20 + 0.6(5) = 23$ (billion dollars)

(c) The graph in Figure 1.44 is the line passing through the points satisfying the equation $C = 20 + 0.6y$. Note that y cannot be negative.

CHECKPOINT SOLUTIONS

1. (a) $C(x) = 75x + 12,500$
   (b) $C(0) = 12,500 =$ Fixed costs. In fact, fixed costs are defined to be $C(0)$.

2. (a) $R(x) = 175x$
   (b) $R(100) = 175(100) = \$17,500$, which means that revenue is $17,500 when 100 units are sold.

3. (a) Profit = Revenue − Cost or $P(x) = R(x) − C(x)$
   (b) $P(x) = 175x − (75x + 12,500)$
   $$= 175x − 75x − 12,500 = 100x − 12,500.$$

4. (a) The break-even point occurs when $P(x) = 0$.
   (b) $100x − 12,500 = 0$
   $$100x = 12,500$$
   $$x = 125 \text{ units}$$

5. (a) rise   (b) fall

EXERCISE 1.6

TOTAL COST, TOTAL REVENUE, AND PROFIT FUNCTIONS

1. For a certain period, a calculator manufacturer has fixed costs of $3400 and variable costs of $17 for each item produced.
   (a) Write the equation that represents total cost.
   (b) What will it cost to produce 200 units during the given period?

2. A stereo receiver manufacturer has a fixed cost of $1650 and a variable cost of $105 for each item produced during a given month.
   (a) Write the equation that represents the total cost.
   (b) What will it cost to produce 215 items during the month?

3. Production of radios has a fixed cost of $1850 and a variable cost of $43 for each radio produced.
   (a) Write the equation that represents the total cost.

(b) What is the cost if no units are produced? What is this called?

(c) What is the cost if 37 units are produced?

4. A computer manufacturer has a fixed cost of $3300 and a variable cost of $85 for each item produced.
   (a) Write the equation that represents total cost.
   (b) What is the cost if no units are produced? What is this called?
   (c) What is the cost if 300 units are produced?

5. A calculator is sold for $34 per unit.
   (a) Write the total revenue function as an equation.
   (b) What will the revenue be if 300 units are sold?

6. A stereo receiver is sold for $215 per item.
   (a) Write the equation that represents the revenue function.
   (b) What will the revenue be from the sale of 50 items?

7. A radio is sold for $80 per unit.
   (a) Write the equation that represents the revenue function.
   (b) What will the revenue be from the sale of 75 radios?

8. A computer is sold for $385 per item.
   (a) Write the equation that represents the revenue function.
   (b) What will the revenue be from the sale of 130 items?

9. Suppose a calculator manufacturer has the total cost function $C(x) = 17x + 3400$ and the total revenue function $R(x) = 34x$.
   (a) What is the equation of the profit function for the calculator?
   (b) What is the profit on 300 units?
   (c) How many units must be sold to ensure no money will be lost during the period?

10. Suppose a stereo receiver manufacturer has the total cost function $C(x) = 105x + 1650$ and the total revenue function $R(x) = 215x$.
    (a) What is the equation of the profit function for this commodity?
    (b) What is the profit on 50 items?
    (c) How many items must be sold in a month to avoid losing money?

11. Suppose a radio manufacturer has the total cost function $C(x) = 43x + 1850$ and the total revenue function $R(x) = 80x$.
    (a) What is the equation of the profit function for this commodity?
    (b) What is the profit on 30 units? Interpret your result.
    (c) How many radios must be sold to avoid losing money?

12. Suppose a computer manufacturer has the total cost function $C(x) = 85x + 3300$ and the total revenue function $R(x) = 385x$.
    (a) What is the equation of the profit function for this commodity?
    (b) What is the profit on 351 items?
    (c) How many items must be sold to avoid losing money?

13. Figure 1.45 is the graph of a profit function for a product. Use the graph to answer the following.
    (a) How many units must be produced to break even? (Two values give the break-even points.)
    (b) How many units will give maximum profit?
    (c) What is the maximum possible profit?

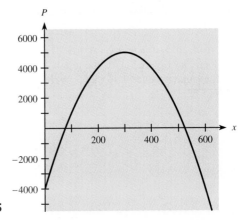

Figure 1.45

14. Refer to Figure 1.45.
    (a) How much money is lost if no units are produced?
    (b) Is it possible to lose more money than $4000? How?

SUPPLY AND DEMAND FUNCTIONS

15. Plot the points in the demand schedule below.

Price	Quantity
$52	0
48	1
44	3
40	6
36	12

16. Do the points in the demand schedule in problem 15 lie on a straight line?

17. Plot the points in the supply schedule below.

Price	Quantity
$ 8	0
16	2
24	4
40	8
48	10

18. Do the points in the supply schedule in problem 17 lie on a straight line?
19. As the price of a commodity increases, what happens to demand?
20. As the price of a commodity increases, what happens to supply?

Figure 1.46 is the graph of the demand function for a product, and Figure 1.47 is the graph of the supply function for the same product. Use these graphs to answer questions 21 and 22.

Figure 1.46

Figure 1.47

21. (a) How many units q are *demanded* when the price p is $100?
    (b) How many units q are *supplied* when the price p is $100?
    (c) Will there be a market surplus (more supplied) or shortage (more demanded) when $p = $100?

22. (a) How many units q are demanded when the price p is $200?
    (b) How many units q are supplied when the price p is $200?
    (c) Will there be a market surplus or shortage when the price p is $200?

23. If the demand for a pair of shoes is given by $2p + 5q = 200$ and the supply function for it is $p - 2q = 10$, compare the quantity demanded and supplied when the price is $60. Will there be a surplus or shortfall at this price?

24. If the demand function and supply function for Z-brand phones are $p + 2q = 100$ and $35p - 20q = 350$, respectively, compare the quantity demanded and the quantity supplied when $p = 14$. Are there surplus phones or not enough to meet demand?

NATIONAL CONSUMPTION

25. If $C = f(y) = 3 + 0.7y$ (in billions of dollars),
    (a) what is the consumption when disposable income is 0?
    (b) what is the consumption when disposable income is $6 billion?
    (c) sketch the graph of the consumption function.

26. If $C = 4 + 0.65y$ (in billions of dollars),
    (a) what is the consumption when disposable income is 0?
    (b) what is the consumption when disposable income is $18 billion?
    (c) sketch the graph of the consumption function.

27. If $C = 5.6 + 0.6y$ (in billions of dollars),
    (a) what is the consumption when disposable income is 0?
    (b) what is the consumption when disposable income is $15 billion?
    (c) sketch the graph of the consumption function.

28. If $C = 9 + 0.5y$ (in billions of dollars),
    (a) what is the consumption when disposable income is 0?
    (b) what is the consumption when disposable income is $15 billion?
    (c) sketch the graph of the consumption function.

29. If the consumption function is given by

    $$C(y) = 0.47y + 16 \qquad \text{(in billions of dollars),}$$

    (a) what is the consumption when there is no disposable income?
    (b) what is the consumption when disposable income is $1 billion?
    (c) use the difference in the answers to parts (a) and (b) to find what percent of the $1 billion of disposable income was spent.

30. If the consumption function is given by

    $$C(y) = 0.82y + 36 \qquad \text{(in billions of dollars),}$$

    (a) what is the consumption when there is no disposable income?
    (b) what is the consumption when disposable income is $1 billion?
    (c) use the difference in the answers to parts (a) and (b) to find what percent of the $1 billion of disposable income was spent.

1.7 SLOPE OF A LINE; WRITING EQUATIONS OF LINES

Objectives

❑ To find the slope of a line from its graph and from its equation
❑ To graph a line, given its slope and y-intercept or its slope and one point on the line
❑ To write the equation of a line, given information about its graph
❑ To find marginal cost, revenue, and profit, given linear total cost, total revenue, and profit functions
❑ To write the equations of linear total cost, total revenue, and profit functions, by using information given about the functions.

APPLICATION PREVIEW

U.S. Census Bureau data from 15 different years showed that the average price p (in dollars) of a color television set depended on the number of sets sold N (in thousands). The data indicated that when N increased by 1000 units, there was a decrease of $10.40 in the average price per set. The data also showed that when 6485 (thousand) sets were sold one year, the average price per set was $504.39. (Source: *Statistical Abstract of the United States*, Washington, D.C.) By using skills developed in this section, we could express p as a linear function of N and use that equation to better understand the relationship between p and N.

The cost of producing one additional unit at any level of production is called **marginal cost.** If the cost function is linear, the marginal cost will be the same at all levels of production. For example,

$$C = 2x + 10$$

is the cost function for a good whose fixed cost is $10 and whose variable cost is $2 per unit. Because producing one more unit adds $2 to the cost, the marginal cost is $2. Looking at the graph of this function (Figure 1.48), we see that the C-value changes $+2$ units (from 10 to 12) while the x-value changes $+1$ unit (from 0 to 1). The ratio of this change in C to the corresponding change in x is called the **slope** of the line. The slope measures the steepness and orientation (rise or fall) of the line and represents the rate of change in the total cost function.

We could also use points on the graph to find the slope of this line. To do this, we divide the change in C by the change in x for any two points on the line. For example, the points $(1, 12)$ and $(3, 16)$ lie on the line that

Figure 1.48

represents the graph of $C = 2x + 10$. The C-value changes $+4$ units (from 12 to 16) while the x-value changes $+2$ units (from 1 to 3). Thus the slope, symbolized by m, is

$$m = \frac{\text{change in } C}{\text{change in } x} = \frac{4}{2} = 2.$$

The slope of any nonvertical line can be found by using any two points on the line as follows.

Slope of a Line   If a nonvertical line passes through the points $P_1(x_1, y_1)$ and $P_2(x_2, y_2)$, its **slope** m is found by using either

$$m = \frac{y_2 - y_1}{x_2 - x_1}$$

or

$$m = \frac{y_1 - y_2}{x_1 - x_2}.$$

Note that for a given line, the slope is the same regardless of which two points are used in the calculation; this is because corresponding sides of similar triangles are in proportion.

We may also write the slope by using the notation

$$m = \frac{\Delta y}{\Delta x} \quad (\Delta y = y_2 - y_1 \quad \text{and} \quad \Delta x = x_2 - x_1),$$

where Δy means "change in y" and Δx means "change in x."

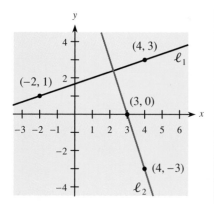

Figure 1.49

EXAMPLE 1

Find the slope of:
(a) line ℓ_1, passing through $(-2, 1)$ and $(4, 3)$.
(b) line ℓ_2, passing through $(3, 0)$ and $(4, -3)$.

Solution

(a) $m = \dfrac{3 - 1}{4 - (-2)} = \dfrac{2}{6} = \dfrac{1}{3}$ or $m = \dfrac{1 - 3}{-2 - 4} = \dfrac{-2}{-6} = \dfrac{1}{3}$

This means that a point 3 units to the right and 1 unit up from any point on the line is also on the line. Line ℓ_1 is shown in Figure 1.49.

(b) $m = \dfrac{0 - (-3)}{3 - 4} = \dfrac{3}{-1} = -3$

This means that a point 1 unit to the right and 3 units down from any point on the line is also on the line. Line ℓ_2 is also shown in Figure 1.49.

From the previous discussion we see that the slope describes the direction of a line as follows:

Orientation of a line and its slope

1. The slope is *positive* if the line slopes upward toward the right.

$$m = \frac{\Delta y}{\Delta x} > 0$$

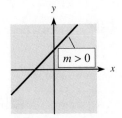

2. The slope is *negative* if the line slopes downward toward the right.

$$m = \frac{\Delta y}{\Delta x} < 0$$

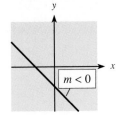

Two additional facts should be stated:

3. The slope of a *horizontal line* is 0, because the y-values are the same for any two points on the line $(y_2 - y_1 = 0)$.

$$m = \frac{\Delta y}{\Delta x} = 0 \quad \text{since } \Delta y = 0$$

4. The slope of a *vertical line* is *undefined,* because $x_2 - x_1$ would equal 0, and division by 0 is undefined. (*Note:* Here y is not a function of x.)

$$m = \frac{\Delta y}{\Delta x} \text{ is undefined since } \Delta x = 0.$$

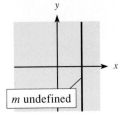

Clearly, two distinct nonvertical lines that have the same slope are parallel and, conversely, two parallel lines have the same slope.

Parallel Lines Two distinct nonvertical lines are *parallel* if and only if their slopes are *equal*.

Because the slope of a vertical line is undefined, we cannot use slope in discussing parallel and perpendicular relations involving vertical lines. Two vertical lines are parallel, and any horizontal line is perpendicular to any vertical line.

Note that the lines ℓ_1 and ℓ_2 of Figure 1.49 on page 126 appear to be perpendicular and that the slope of ℓ_1, $\frac{1}{3}$, is the negative reciprocal of the slope of ℓ_2, -3. In fact, any two nonvertical lines that are perpedicular have slopes that are negative reciprocals of each other.

Slopes of Perpendicular Lines   A line ℓ_1 with slope m, where $m \neq 0$, is *perpendicular* to line ℓ_2 if and only if the slope of ℓ_2 is $-1/m$. (The slopes are *negative* reciprocals.)

CHECKPOINT

1. Find the slope of the line through $(4, 6)$ and $(28, -6)$.

2. If a line has slope $m = 0$, then the line is _____. If a line has slope m undefined, then the line is _____.

3. Suppose that line 1 has slope $m_1 = 3$ and line 2 has slope m_2.
   (a) If line 1 is perpendicular to line 2, find m_2.
   (b) If line 1 is parallel to line 2, find m_2.

If the slope of a line is m, then the slope between a fixed point (x_1, y_1) and any other point (x, y) on the line is also m. That is,

$$m = \frac{y - y_1}{x - x_1}.$$

Solving for $y - y_1$ gives the point-slope form of the equation of a line.

Point-Slope Form   The equation of the line passing through the point (x_1, y_1) and with slope m can be written in the **point-slope form**

$$y - y_1 = m(x - x_1).$$

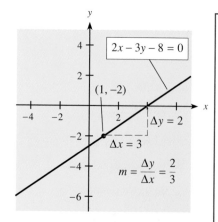

Figure 1.50

EXAMPLE 2

Write equations for the lines passing through $(1, -2)$ and having
(a) slope $\frac{2}{3}$.   (b) undefined slope.   (c) point $(2, 3)$ also on it.

Solution

(a) Here $m = \frac{2}{3}$, $x_1 = 1$, and $y_1 = -2$. An equation of the line is

$$y - (-2) = \frac{2}{3}(x - 1).$$

This equation may be written as $y = \frac{2}{3}x - \frac{8}{3}$ or in the **general form** as $2x - 3y - 8 = 0$. Figure 1.50 shows the graph of this line; the point $(1, -2)$ and the slope are highlighted.

(b) Because m is undefined, we cannot use the point-slope form. This line is vertical, so every point on it has x-coordinate 1. Thus, the equation is $x = 1$.

(c) First find

$$m = \frac{3 - (-2)}{2 - 1} = 5.$$

Using $m = 5$ and the point $(1, -2)$ (the other point could also be used) gives

$$y - (-2) = 5(x - 1) \quad \text{or} \quad y = 5x - 7.$$

The graph of $x = 1$ (from Example 2(b)) is a vertical line, as shown in Figure 1.51(a); the graph of $y = 1$ has slope 0, and its graph is a horizontal line, as shown in Figure 1.51(b). Note that $y = 1$ is a constant function, but $x = 1$ is not a function.

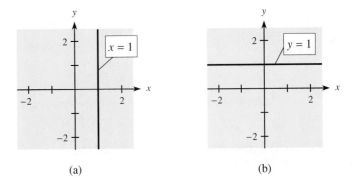

Figure 1.51   (a)   (b)

EXAMPLE 3

The U.S. Census Bureau data presented in the Application Preview indicates that the average price of color television sets can be expressed as a linear function of the number of sets sold N (in thousands). The relevant facts were that as N increased by 1000, p dropped by \$10.40, and when 6485 (thousand) sets were sold, the average price per set was \$504.39. Write the equation of the line determined by this information.

Solution

The slope is given by

$$m = \frac{\text{change in } p}{\text{change in } N} = \frac{-10.40}{1000} = -0.0104.$$

A point on the line is $(N_1, p_1) = (6485, 504.39)$. We use the point-slope form adapted to the variables N and p.

$$p - p_1 = m(N - N_1)$$
$$p - 504.39 = -0.0104(N - 6485)$$
$$p - 504.39 = -0.0104N + 67.444$$
$$p = -0.0104N + 571.834$$

If we think of N as representing quantity demanded, then
this is a typical demand function. The graph in Figure 1.52
has the characteristic shape
for a linear demand function
and shows that lower prices
result in more television sets
demanded.

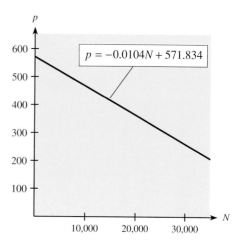

$$p = -0.0104N + 571.834$$

Figure 1.52

The point-slope form, with the intercept point $(0, b)$, can be used to derive
the following special form for the equation of a line.

Slope-Intercept Form

The **slope-intercept form** of the equation of a line with slope m and
y-intercept b is

$$y = mx + b.$$

If we consider the total cost function

$$C = 2x + 10,$$

we see that the equation is in slope-intercept form. We noted earlier that the
slope of the graph of this function is 2. As Figure 1.48 on page 125 shows,
the graph crosses the C-axis at 10.

When a linear equation does not appear in slope-intercept form, it can be
put into slope-intercept form by solving the equation for y.

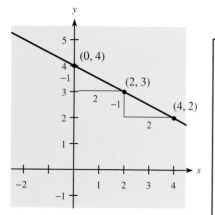

Figure 1.53

EXAMPLE 4

(a) Find the slope and y-intercept of the line whose equation is $x + 2y = 8$.
(b) Use this information to graph the equation.

Solution

(a) To put the equation in slope-intercept form, we must solve it for y:

$$2y = -x + 8 \quad \text{or} \quad y = -\frac{1}{2}x + 4.$$

Thus the slope is $-\frac{1}{2}$ and the y-intercept is 4.

(b) First we plot the y-intercept, $(0, 4)$. The point $(2, 3)$ is on the line because
it is 2 units to the right of and 1 unit below $(0, 4)$. A third point (for a
check) is plotted at $(4, 2)$. The graph is shown in Figure 1.53.

It is also possible to graph a straight line if we have its slope and any point that it passes through; we simply plot the point that is given, and then use the slope to plot other points.

EXAMPLE 5

Write the equation of the line with slope $\frac{1}{2}$ and y-intercept 3.

Solution

Substituting $m = \frac{1}{2}$ and $b = 3$ in the equation $y = mx + b$ gives $y = \frac{1}{2}x + 3$.

The following summarizes the forms of equations of lines.

Forms of Linear Equations		
General form:	$ax + by + c = 0$
Point-slope form:	$y - y_1 = m(x - x_1)$
Slope-intercept form:	$y = mx + b$
Vertical line:	$x = a$
Horizontal line:	$y = b$

CHECKPOINT

4. Write the point-slope form of the equation of the line that has slope $-\frac{3}{4}$ and passes through $(4, -6)$.

5. What are the slope and y-intercept of the graph of $x = -4y + 1$?

For a linear total cost function the **slope** represents the *rate of change in total cost* with respect to the number of units produced, or the **marginal cost.** If we are given a linear total revenue function, then the slope of the line represents the *rate of change in the total revenue.* This rate of change is called the **marginal revenue.** The slope of a linear profit function is the **marginal profit.**

EXAMPLE 6

If the total cost function is given by $C(x) = 300 + 2x$ (dollars), and if the total revenue function is given by $R(x) = 9x$ (dollars), find
(a) the marginal cost, (b) the marginal revenue, (c) the marginal profit.

Solution

(a) $C(x) = 300 + 2x$ is in slope-intercept form with $m = 2$ and $b = 300$. Thus the marginal cost is $2, so the production of each additional unit costs $2 at all levels of production. Note that $b = 300$ indicates that the fixed costs (when $x = 0$) are $300.

(b) $R(x) = 9x$ is also in slope-intercept form with $m = 9$ and $b = 0$. The slope represents the marginal revenue, which is $9. That is, the sale of each additional unit produces an additional $9 in revenue at all levels of production. Note that $b = 0$ indicates that if nothing is sold ($x = 0$), no revenue results.

(c) The profit function is

$$P(x) = R(x) - C(x)$$
$$P(x) = 9x - (300 + 2x)$$
$$P(x) = 7x - 300.$$

The profit function slope is equal to 7, so marginal profit is $7. That is, the sale of each extra unit results in $7 additional profit at all levels of production. Also note that $b = -300$ indicates that when no items have been sold ($x = 0$), the profit is $-$300$; that is, there is a loss of $300.

We have seen that the slope of a linear total cost function is the marginal cost, and the C-intercept (where C lies along the vertical axis of the graph) represents the fixed cost. Thus, if we know that a linear cost function has a marginal cost of $30 and a fixed cost of $1000, we can write the equation that defines the function:

$$C = 30x + 1000.$$

Note that when total cost functions are linear, the marginal cost is the same as the variable cost. This is not the case, however, if the functions are not linear, as we shall see later.

CHECKPOINT

6. If $C(x)$, $R(x)$, and $P(x)$ are linear functions, how are their marginals found?

EXAMPLE 7

Suppose the profit function for a good is linear, and the marginal profit is $5. If the profit is $200 when 125 units are sold, write the equation of the profit function.

Solution

The marginal profit gives us the slope of the line representing the profit function. Using this slope and the point (125, 200), we have the equation

$$P - 200 = 5(x - 125),$$

or

$$P = 5x - 425.$$

CHECKPOINT SOLUTIONS

1. $m = \dfrac{y_2 - y_1}{x_2 - x_1} = \dfrac{-6 - (6)}{28 - (4)} = \dfrac{-12}{24} = -\dfrac{1}{2}$ or

   $m = \dfrac{y_1 - y_2}{x_1 - x_2} = \dfrac{6 - (-6)}{4 - (28)} = \dfrac{12}{-24} = -\dfrac{1}{2}$

2. If $m = 0$, then the line is horizontal. If m is undefined, then the line is vertical.

3. (a) $m_2 = \dfrac{-1}{m_1} = -\dfrac{1}{3}$ (b) $m_2 = m_1 = 3$

4. Use $y - y_1 = m(x - x_1)$. Hence, $y - (-6) = \dfrac{-3}{4}(x - 4)$, or

   $3x + 4y = 12$.

5. If $x = -4y + 1$, then the slope-intercept form is $y = mx + b$ or $y = -\frac{1}{4}x + \frac{1}{4}$. Hence, $m = -\frac{1}{4}$ and the y-intercept is $(0, \frac{1}{4})$.

6. For linear functions, the marginals are the slopes.

EXERCISE 1.7

In problems 1–6, find the slopes of the lines passing through the given pair of points.

1. (1, 2) and (0, 0)
2. (−1, 2) and (2, −3)
3. (2, 1) and (3, −4)
4. (−1, −2) and (−2, −3)
5. (3, 2) and (−1, 2)
6. (−4, 2) and (−4, −2)

Find the slopes and y-intercepts of the lines whose equations are given in problems 7–18.

7. $y = \frac{7}{3}x - \frac{1}{4}$
8. $y = \frac{4}{3}x + \frac{1}{2}$
9. $y = 3$
10. $y = -2$
11. $x = -8$
12. $x = -\frac{1}{2}$
13. $x = 4y$
14. $x = 3y + 6$
15. $2x + 3y = 6$
16. $3x - 2y = 18$
17. $2.61x + 3.14y = 91.7$
18. $1.31x - 3.82y = 0.0213$

Write the equation and sketch the graph of each line with the given slope and y-intercept.

19. Slope $\frac{1}{2}$ and y-intercept 3
20. Slope 2 and y-intercept 3
21. Slope −2 and y-intercept $\frac{1}{2}$
22. Solve $\frac{2}{3}$ and y-intercept −1

In problems 23–28, write the equation and graph the line that passes through the given point and has the slope indicated.

23. (2, 0) with slope $\frac{1}{2}$
24. (1, 1) with slope $-\frac{1}{3}$
25. (−1, 3) with slope −2
26. (3, −1) with slope 1
27. (−1, 1) with undefined slope
28. (1, 1) with 0 slope

In problems 29–32, write the equations of the lines passing through the given pairs of points.

29. (3, 2) and (−1, −6)
30. (−4, 2) and (2, 4)
31. (7, 3) and (−6, 2)
32. (10, 2) and (5, 7)
33. Write the equation of the x-axis.
34. Write the equation of the y-axis.
35. Write the equation of the line through (−2, −7) that is parallel to $3x + 5y = 11$.
36. Write the equation of the line through (6, −4) that is parallel to $4x - 5y = 6$.
37. Write the equation of the line through (3, 1) that is perpendicular to $5x - 6y = 4$.
38. Write the equation of the line through (−2, −8) that is perpendicular to $x = 4y + 3$.
39. Write the equation of the line that is perpendicular to and has the same x-intercept as $5x - 8y = 80$.

40. Do the equations $4x - y = 2$ and $8x - 2y = 4$ represent parallel lines or the same line?

41. (a) Use a graphing utility to graph $y = mx$ for $m = \frac{1}{2}$, $-\frac{1}{2}$, 1, -1, 5, and -5.
    (b) What is the x-intercept of each graph? What is the y-intercept?
    (c) Write a sentence that describes the orientation of each line when
    (i) $m > 0$ (ii) $m < 0$
    (iii) $|m| \approx 0$ (iv) $|m|$ is large.

42. Consider $f(x) = 2x$ and $g(x) = -x$, and let $b = -2$, 1, and 5.
    (a) Use a graphing utility to simultaneously graph $f(x)$ and $f(x) + b$ for the given b-values.
    (b) Use a graphing utility to simultaneously graph $g(x)$ and $g(x) + b$ for the given b-values.
    (c) Write a sentence that explains how varying the value of b changes each graph. In particular, describe how the slope is affected and what differences there are when $b > 0$ and when $b < 0$.

APPLICATIONS

43. **ATM transactions** The dollar volume of transactions at automatic teller machines (ATMs) has increased as the number of machines has increased. This relationship can be modeled by

$$y = 0.1369x - 5.091255,$$

where y is billions of dollars of transactions and x is the number of terminals (in thousands). (Source: Equation developed with data from the Electronic Funds Transfer Association, published in *Statistical Abstract of the United States*, 1993, p. 517.)
    (a) Find the slope and y-intercept of this model.
    (b) What interpretation could be given to the y-intercept? Does this interpretation mean that there must be restrictions on x and y for the model to make sense?
    (c) What interpretation could be given to the slope?

44. **Cigarette use** For selected years from 1975 to 1991, the percentage p of high school seniors who have tried cigarettes can be modeled by

$$p = 75.4509 - 0.706948t,$$

where t is the number of years past 1975 (Source: Equation developed with data from the National Institute on Drug Abuse).
    (a) Find the slope and p-intercept of this model.
    (b) Write a sentence that interprets the meaning of the slope.
    (c) Write a sentence that interprets the meaning of the p-intercept.

45. **Residential electric costs** An electric utility company determines the monthly bill for a residential customer by adding an energy charge of 8.38 cents per kilowatt hour (kWh) to its base charge of $4.95 per month. Write an equation for the monthly charge y in terms of x, the number of kWh used.

46. **Residential heating costs** Residential customers who heat their homes with natural gas have their monthly bill calculated by adding a base service charge of $5.19 per month and an energy charge of 51.91 cents per hundred cubic feet (CCF). Write an equation for the monthly charge y in terms of x, the number of CCF used.

47. **Gender differences in median salaries** Data from selected years from 1973 to 1991 show that for each $1000 increase in the median annual salary of U.S. males, the median annual salary of females increased by about $838. Also, in 1985, the median annual salary for males was $22,300 and for females was $17,300 (Source: U.S. Equal Employment Opportunity Commission).
    (a) Let m represent the median salary for males and f the median salary for females, and write the equation of the line that gives f as a linear function of m.
    (b) When the median salary for males reaches $30,000, what does the equation in (a) predict for the median salary for females?

48. **Retirement plans** The retirement plan of Public Service Company of Colorado as described on page 21 of their *Notice of Annual Meeting of Shareholders*, May 11, 1994, is based on the following formula: "1.5% of average final compensation multiplied by years of credited service." Let p represent annual retirement pension, y years of service, and c average final compensation.
    (a) For someone with average final compensation of $80,000, write the linear equation that gives p in terms of y.
    (b) For someone intending to retire after 30 years, write the linear equation that gives p in terms of c.

49. ***Pollution effects*** It has been estimated that a certain stream can support 85,000 fish if it is pollution-free. It has further been estimated that for each ton of pollutants in the stream, 1700 fewer fish can be supported. Write the equation that gives the population of fish p in terms of the tons of pollutants x.

50. ***Age-sleep relationships*** Each day, a young person should sleep 8 hours plus $\frac{1}{4}$ hour for each year that the person is under 18 years of age. Assuming the relation is linear, write the equation relating hours of sleep y and age x.

51. ***Cost*** A linear cost function is $C(x) = 5x + 250$.
    (a) What are the slope and C-intercept?
    (b) What is the marginal cost, and what does it mean?
    (c) What are the fixed costs?
    (d) How are your answers to (a) and to (b) and (c) related?
    (e) What is the cost of producing *one more* item if 50 are currently being produced? What is it if 100 are currently being produced?

52. ***Cost*** A linear cost function is $C(x) = 21.75x + 4890$.
    (a) What are the slope and C-intercept?
    (b) What is the marginal cost, and what does it mean?
    (c) What are the fixed costs?
    (d) How are your answers to (a) and to (b) and (c) related?
    (e) What is the cost of producing *one more* item if 50 are currently being produced? What is it if 100 are currently being produced?

53. ***Revenue*** A linear revenue function is $R = 27x$.
    (a) What is the slope?
    (b) What is the marginal revenue, and what does it mean?
    (c) What is the revenue received from selling *one more* item if 50 are currently being sold? If 100 are being sold?

54. ***Revenue*** A linear revenue function is $R = 38.95x$.
    (a) What is the slope?
    (b) What is the marginal revenue, and what does it mean?
    (c) What is the revenue received from selling *one more* item if 50 are currently being sold? If 100 are being sold?

55. ***Profit*** Let $C(x) = 5x + 250$ and $R(x) = 27x$.
    (a) Write the profit function $P(x)$.
    (b) What is the slope of the profit function?

(c) What is the marginal profit?
(d) Interpret the marginal profit. What does this tell the manager of the firm with this profit function, if the objective is to make the most profit?

56. ***Profit*** Given $C(x) = 21.95x + 1400$ and $R(x) = 10x$, find the profit function.
    (a) What is marginal profit, and what does it mean?
    (b) What should a firm with these cost, revenue, and profit functions do? (Hint: Graph the profit function and see where it goes.)

57. ***Cost*** Suppose the total cost function for a radio is linear, that the marginal cost is \$27, and that the fixed costs amount to \$3000. Write the equation of this cost function and then graph it.

58. ***Revenue*** Suppose the total revenue function for a radio is linear, with marginal revenue \$30. Write the total revenue function; recall that the revenue from 0 units is \$0.

59. ***Demand*** Suppose a certain outlet chain selling appliances has found that for one brand of stereo system, the monthly demand is 240 when the price is \$900. However, when the price is \$850, the monthly demand is 315. Assuming that the demand function for this system is linear, write the equation for the demand function. Use p for price and q for quantity.

60. ***Demand*** Suppose a certain home improvements outlet knows that the monthly demand for framing studs is 2500 when the price is \$1.00 each, but the demand is 3500 when the price is \$0.90 each. Assuming that the demand function is linear, write its equation. Use p for price and q for quantity.

61. ***Supply*** Suppose the manufacturer of a board game will supply 10,000 games if the wholesale price is \$1.50 each but will supply only 5000 if the price is \$1.00 each. Assuming that the supply function is linear, write its equation. Use p for price and q for quantity.

62. ***Supply*** Suppose a mining company will supply 100,000 tons of ore per month if the price is \$30 per ton but will supply 80,000 tons per month if the price is \$25 per ton. Assuming that the supply function is linear, write its equation.

63. ***Insulation R-values*** The R-value of insulation is a measure of its ability to resist heat transfer. For fiberglass insulation, $3\frac{1}{2}$ inches is rated at R-11 and 6 inches is rated at R-19. Assuming this relationship is linear, write

the equation that gives the R-value of fiberglass insulation as a function of its thickness t (in inches).

64. **Fahrenheit-Celsius** Write the equation of the linear relationship between temperature in Celsius (C) and Fahrenheit (F) if water freezes at 0°C and 32°F and boils at 100°C and 212°F.

65. **Profit** A company charting its profits notices the relationship between the number of units sold, x, and the profit, P, is linear. If 200 units sold results in $3100 profit and 250 units sold results in $6000 profit, write the profit function for this company. Find the marginal profit.

66. **Depreciation** Suppose the cost of a business property is $960,000 and a company wants to use a straight line depreciation schedule for a period of 240 months. If y is the value of this property after x months, then the company's depreciation schedule will be the equation of a line through (0, 960,000) and (240, 0). Write the equation of this depreciation schedule.

67. **Cholesterol and coronary heart disease risk** The Seven Countries Study, conducted by Ancel Keys, was a long-term study of the relationship of cholesterol to coronary heart disease (CHD) mortality in men. The relationship was approximated by the line shown in the figure below. The line passes through the points (200, 25) and (250, 49), meaning that there were 25 CHD deaths per 1000 among men with 200 mg/dl of cholesterol and 49 CHD deaths among men with 250 mg/dl of cholesterol. Using x to represent the cholesterol and y to represent CHD deaths, write the equation that represents this relationship.

Church attendance The following table shows the percentages of adults and teens who attended church each year from 1980 to 1989. Assume that the percentage of adults who attended church during these years can be modeled with the

function $a(x)$ and the percentage of teenagers who behaved similarly can be modeled with the function $t(x)$ where

$$a(x) = 29.73 + 0.13x \qquad t(x) = 24.33 + 0.33x,$$

and x denotes the number of years past 1900. The functions a and t and the data in the table are all graphed in the figure below. Use this information in problems 68 and 69.

Year	Percentage of Adults	Percentage of Teens
1980	41	50
1981	41	54
1982	41	50
1983	40	53
1984	40	52
1985	42	52
1986	40	54
1987	40	52
1988	42	51
1989	43	57

SOURCE: *World Almanac*, 1991

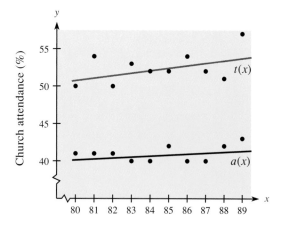

68. (a) Which equation appears to represent more accurately the percentage attending church?
    (b) For what year can you guarantee that $a(x)$ can no longer be used to model the percentage of adults attending church?

69. (a) Assuming that a and t are valid models, during what year will 50% of adults attend church?
    (b) In what year will 60% of teenagers attend church?

1.8 SOLUTIONS OF SYSTEMS OF LINEAR EQUATIONS

Objectives

- ❑ To solve systems of linear equations in two variables by graphing
- ❑ To solve systems of linear equations by substitution
- ❑ To solve systems of linear equations by elimination
- ❑ To solve systems of three linear equations in three variables

APPLICATION PREVIEW

Suppose a person has $200,000 invested, part at 9% and part at 8%, and that the yearly income from the two investments is $17,200. If x represents the amount invested at 9% and y represents the amount invested at 8%, then to find how much is invested at each rate we must find the values of x and y that satisfy both

$$x + y = 200{,}000 \quad \text{and} \quad 0.09x + 0.08y = 17{,}200.$$

The terminology associated with problems like this and the skills needed to determine x and y are the topics of this section.

In the previous sections we graphed linear equations in two variables, and observed that the graphs are straight lines. Each point on the graph represents an ordered pair of values (x, y) that satisfies the equation, so a point of intersection of two (or more) lines represents a solution to both (or all) the equations of those lines.

The equations are referred to as a **system of equations,** and the ordered pairs (x, y) that satisfy all the equations are the **solutions** (or **simultaneous solutions**) to the system. We can use graphing to find the solution to a system of equations.

EXAMPLE I

Use graphing to find the solution of the system

$$\begin{cases} 4x + 3y = 11 \\ 2x - 5y = -1 \end{cases}$$

Solution

The graphs of the two equations intersect (meet) at the point (2, 1). (See Figure 1.54.) The solution of the system is $x = 2$, $y = 1$. Note that these values satisfy both equations.

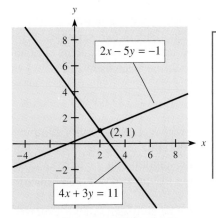

Two distinct nonparallel lines: one solution

Figure 1.54

If the graphs of two equations are parallel lines, they have no point in common and thus the system has no solution. Such a system of equations is called **inconsistent.** For example,

$$\begin{cases} 4x + 3y = 4 \\ 8x + 6y = 18 \end{cases}$$

is an **inconsistent system** (see Figure 1.55).

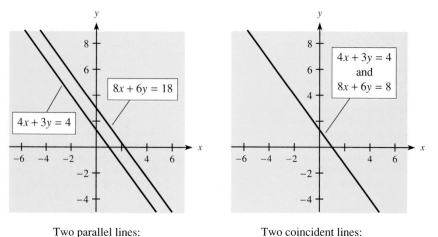

Two parallel lines:
no solution

Figure 1.55

Two coincident lines:
infinitely many solutions

Figure 1.56

It is also possible that two equations describe the same line. When this happens the equations are equivalent, and values that satisfy either equation are solutions to the system. For example,

$$\begin{cases} 4x + 3y = 4 \\ 8x + 6y = 8 \end{cases}$$

is called a **dependent system** because all points satisfying one equation also satisfy the other (see Figure 1.56).

Figures 1.54, 1.55 and 1.56 represent the three possibilities that can occur when solving a system of two linear equations in two variables.

Of course, graphing utilities also can be used to solve systems of equations, either by tracing or by using a special routine (such as INTERSECT or SOLVER). These capabilities are extremely helpful for solving a system when the point of intersection is in the viewing window. (Try to rework Example 1 with a graphing utility.)

However, graphing, whether by hand or with a graphing utility, is limited by two factors: (1) it may be difficult to determine a viewing window that contains the point of intersection and (2) the solution may be only approximate. With *some* systems of equations, the only practical method may be graphical approximation.

However, systems of *linear* equations can be consistently and accurately solved with algebraic methods. Computer algebra systems, some software packages (such as some spreadsheets), and some graphics calculators have the capability to solve systems of linear equations algebraically. Algebraic methods are based on the fact that equivalent systems result when any of the following operations are performed.

Equivalent Systems   Equivalent systems result when:

1. One expression is replaced by an equivalent expression.
2. Two equations are interchanged.
3. A multiple of one equation is added to another equation.
4. An equation is multiplied by a nonzero constant.

The **substitution method** is based on operation (1).

Substitution Method for Solving Systems

Procedure	Example
To solve a system of two equations in two variables by substitution:	Solve the system containing $2x + 3y = 4$ and $x - 2y = 3$.
1. Solve one of the equations for one of the variables in terms of the other.	1. Solving $x - 2y = 3$ for x gives $x = 2y + 3$.
2. Substitute this expression into the other equation to give one equation in one unknown.	2. Replacing x by $2y + 3$ in $2x + 3y = 4$ gives $2(2y + 3) + 3y = 4$.
3. Solve this linear equation for the unknown.	3. $4y + 6 + 3y = 4$ $$7y = -2 \Rightarrow y = -\frac{2}{7}$$
4. Substitute this solution into the equation in step 1 or into one of the original equations to solve for the other variable.	4. $x = 2\left(-\frac{2}{7}\right) + 3 \Rightarrow x = \frac{17}{7}$
5. Check the solution by substituting for x and y in both original equations.	5. $2\left(\frac{17}{7}\right) + 3\left(-\frac{2}{7}\right) = 4 \checkmark$ $$\frac{17}{7} - 2\left(-\frac{2}{7}\right) = 3 \checkmark$$

EXAMPLE 2
Solve the system

$$\begin{cases} 4x + 5y = 18 & (1) \\ 3x - 9y = -12 & (2) \end{cases}.$$

Solution

1. $x = \dfrac{9y - 12}{3} = 3y - 4$ Solving for x in equation (2)

2. $4(3y - 4) + 5y = 18$ Substituting for x in equation (1)

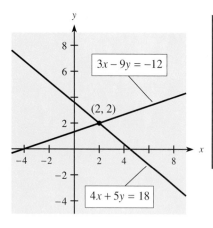

Figure 1.57

3. $12y - 16 + 5y = 18$ Solving for y
$$17y = 34$$
$$y = 2$$

4. $x = 3(2) - 4$ Using $y = 2$ to find x
$$x = 2$$

5. $4(2) + 5(2) = 18$ and $3(2) - 9(2) = -12$ Checking

Thus, the solution is $x = 2$, $y = 2$. This means that when the two equations are graphed simultaneously, their point of intersection is $(2, 2)$. See Figure 1.57.

We can also eliminate one of the variables in a system by the **elimination method,** which uses addition or subtraction of equations.

Elimination Method for Solving Systems

Procedure	Example
To solve a system of two equations in two variables by the elimination method:	Solve the system $$\begin{cases} 2x - 5y = 4 & (1) \\ x + 2y = 3 & (2) \end{cases}$$
1. If necessary, multiply one or both equations by a nonzero number that will make the coefficients of one of the variables identical, except perhaps for signs.	1. Multiply (2) by -2. $$2x - 5y = 4$$ $$-2x - 4y = -6$$
2. Add or subtract the equations to eliminate one of the variables.	2. Adding gives $$0x - 9y = -2.$$
3. Solve for the variable in the resulting equation.	3. $y = \dfrac{2}{9}$
4. Substitute the solution into one of the original equations and solve for the other variable.	4. $2x - 5\left(\dfrac{2}{9}\right) = 4$ $$2x = 4 + \dfrac{10}{9} = \dfrac{36}{9} + \dfrac{10}{9}$$ $$x = \dfrac{23}{9}$$
5. Check the solutions in both original equations.	5. $2\left(\dfrac{23}{9}\right) - 5\left(\dfrac{2}{9}\right) = 4 \ \checkmark$ $$\dfrac{23}{9} + 2\left(\dfrac{2}{9}\right) = 3 \quad \checkmark$$

We now return to the Application Preview problem.

EXAMPLE 3

A person has $200,000 invested, part at 9% and part at 8%. If the total yearly income from the two investments is $17,200, how much is invested at 9%, and how much at 8%?

Solution

If x represents the amount invested at 9% and y represents the amount invested at 8%, then $x + y$ is the total investment,

$$x + y = 200{,}000 \qquad (1),$$

and $0.09x + 0.08y$ is the total income earned.

$$0.09x + 0.08y = 17{,}200 \qquad (2)$$

We solve these equations as follows:

$$
\begin{array}{ll}
-8x - 8y = -1{,}600{,}000 & (3) \quad \text{Multiplying equation (1) by } -8 \\
\underline{9x + 8y = 1{,}720{,}000} & (4) \quad \text{Multiplying equation (2) by 100} \\
x = 120{,}000 & \quad \text{Adding (3) and (4)}
\end{array}
$$

Thus $120,000 is invested at 9%, and $80,000 is invested at 8%.

EXAMPLE 4

Solve the systems:

(a) $\begin{cases} 4x + 3y = 4 \\ 8x + 6y = 18 \end{cases}$ (b) $\begin{cases} 4x + 3y = 4 \\ 8x + 6y = 8 \end{cases}$

Solution

(a) $\begin{cases} 4x + 3y = 4 \\ 8x + 6y = 18 \end{cases}$

$$
\begin{array}{ll}
\text{Multiply by } -2 \text{ to get:} & -8x - 6y = -8 \\
\text{Leave as is, giving:} & 8x + 6y = 18 \\
\text{Add the equations to get:} & 0x + 0y = 10 \\
& 0 = 10
\end{array}
$$

The system is solved when $0 = 10$. This is impossible, so there are no solutions to the system. The equations are inconsistent. Their graphs are parallel lines; see Figure 1.55 on page 138.

(b) $\begin{cases} 4x + 3y = 4 \\ 8x + 6y = 8 \end{cases}$

$$
\begin{array}{ll}
\text{Multiply by } -2 \text{ to get:} & -8x - 6y = -8 \\
\text{Leave as is, giving:} & 8x + 6y = 8 \\
\text{Add the equations to get:} & 0x + 0y = 0 \\
& 0 = 0
\end{array}
$$

This is an identity, so the two equations share infinitely many solutions. The equations are dependent. Their graphs coincide, and each point on this line represents a solution to the system; see Figure 1.56 on page 138.

EXAMPLE 5

A nurse has two solutions that contain different concentrations of a certain medication. One is a 12.5% concentration and the other is a 5% concentration. How many cubic centimeters of each should she mix to obtain 20 cc of an 8% concentration?

Solution

Let x equal the number of cubic centimeters of the 12.5% solution, and let y equal the number of cubic centimeters of the 5% solution. The total amount of substance is

$$x + y = 20,$$

and the total amount of medication is

$$0.125x + 0.05y = (0.08)(20) = 1.6.$$

Solving this pair of equations simultaneously gives

$$
\begin{array}{rcr}
50x + 50y = & 1000 \\
-125x - 50y = & -1600 \\
\hline
-75x \qquad\quad = & -600
\end{array}
$$

$$x = 8$$
$$8 + y = 20, \text{ so } y = 12.$$

Thus 8 cubic centimeters of a 12.5% concentration and 12 cubic centimeters of a 5% concentration yield 20 cubic centimeters of an 8% concentration.

CHECKPOINT

1. Solve by substitution: $\begin{cases} 3x - 4y = -24 \\ x + y = -1 \end{cases}$

2. Solve by elimination: $\begin{cases} 2x + 3y = 5 \\ 3x + 5y = -25 \end{cases}$

3. In problems (1) and (2), the solution method is given.
   (a) In each case, explain why you think that method is appropriate.
   (b) In each case, would the other method work as well? Explain.

THREE EQUATIONS IN THREE VARIABLES

If a, b, c, and d represent constants, then

$$ax + by + cz = d$$

is a first-degree (linear) equation in three variables. When equations of this form are graphed in a three-dimensional coordinate system, their graphs are planes. Two different planes may intersect in a line (like two walls) or may not intersect at all (like a floor and ceiling). Three different planes may intersect in a single point (as when two walls meet the ceiling), in a line (as

Ceiling

Wall

Wall

Figure 1.58   Unique solution

Infinitely many
solutions

Figure 1.59

in a paddle wheel), or may not have a common intersection. (See Figures 1.58, 1.59, and 1.60.) Thus, three linear equations in three variables may have a unique solution, infinitely many solutions, or no solution. For example, the solution to the system

$$\begin{cases} 3x + 2y + z = 6 \\ x - y - z = 0 \\ x + y - z = 4 \end{cases}$$

is $x = 1$, $y = 2$, $z = -1$, since these three values satisfy all three equations, and these are the only values that satisfy them. In this section, we will only discuss systems of three linear equations in three variables that have unique solutions. Additional systems will be discussed in Section 2.3.

We can solve three equations in three variables using a systematic procedure, called the **left-to-right elimination method.**

No solution

Figure 1.60

Left-to-Right Elimination Method

Procedure	Example
To solve a system of three equations in three variables by the left-to-right elimination method:	Solve: $\begin{cases} 2x + 4y + 5z = 4 \\ x - 2y - 3z = 5 \\ x + 3y + 4z = 1 \end{cases}$
1. If necessary, interchange two equations or use multiplication to make the coefficient of the first variable in equation (1) a factor of the other first variable coefficients.	1. Interchange the first two equations: $$\begin{aligned} x - 2y - 3z &= 5 \quad (1) \\ 2x + 4y + 5z &= 4 \quad (2) \\ x + 3y + 4z &= 1 \quad (3) \end{aligned}$$
2. Add multiples of the first equation to each of the following equations so that the coefficients of the first variable in the second and third equations become zero.	2. Add $(-2) \times$ equation (1) to equation (2) and add $(-1) \times$ equation (1) to equation (3): $$\begin{aligned} x - 2y - 3z &= 5 \quad (1) \\ 0x + 8y + 11z &= -6 \quad (2) \\ 0x + 5y + 7z &= -4 \quad (3) \end{aligned}$$
3. Add a multiple of the second equation to the third equation so that the coefficient of the second variable in the third equation becomes zero.	3. Add $\left(-\frac{5}{8}\right) \times$ equation (2) equation (3): $$\begin{aligned} x - 2y - 3z &= -5 \quad (1) \\ 8y + 11z &= -6 \quad (2) \\ 0y + \tfrac{1}{8}z &= -\tfrac{2}{8} \quad (3) \end{aligned}$$
4. Solve the third equation and *back substitute* from the bottom to find the remaining variables.	4. $z = -2$ from equation (3) $y = \frac{1}{8}(-6 - 11z) = 2$ from equation (2) $x = 5 + 2y + 3z = 3$ from equation (1) so $x = 3$, $y = 2$, $z = -2$.

EXAMPLE 6

Solve: $\begin{cases} x + 2y + 3z = 6 & (1) \\ 2x + 3y + 2z = 6 & (2) \\ -x + y + z = 4 & (3) \end{cases}$

Solution

Using equation (1) to eliminate x from the other equations gives the equivalent system:

$\begin{cases} x + 2y + 3z = -6 & (1) \\ -y - 4z = -6 & (2) \qquad (-2) \times \text{equation (1) added to equation (2)} \\ 3y + 4z = 10 & (3) \qquad \text{Equation (1) added to equation (3)} \end{cases}$

Using equation (2) to eliminate y from equation (3) gives

$\begin{cases} x + 2y + 3z = 6 & (1) \\ -y - 4z = -6 & (2) \\ -8z = -8 & (3) \qquad (3) \times \text{equation (2) added to equation (3)} \end{cases}$

Solving for each **lead variable** gives

$$x = 6 - 2y - 3z$$
$$y = 6 - 4z$$
$$z = 1$$

and using **back substitution** from the bottom gives

$$z = 1,$$
$$y = 6 - 4 = 2,$$
$$x = 6 - 4 - 3 = -1.$$

Hence, the solution is $x = -1$, $y = 2$, $z = 1$.

Although other methods for solving systems of equations in three variables may be useful, the left-to-right elimination method is important because it is systematic and can easily be extended to larger systems and to systems solved with matrices (see Section 2.3).

CHECKPOINT

4. Use left-to-right elimination to solve.

$$\begin{cases} x - y - z = 0 & (1) \\ y - 2z = -18 & (2) \\ x + y + z = 6 & (3) \end{cases}$$

CHECKPOINT SOLUTIONS

1. $x + y = -1$ means $x = -y - 1$, so $3x - 4y = -24$ becomes

$$3(-y - 1) - 4y = -24$$
$$-3y - 3 - 4y = -24$$
$$-7y = -21$$
$$y = 3$$

Hence, $x = -3 - 1 = -4$ and the solution is $x = -4$, $y = 3$.

2. Multiply the first equation by (-3) and the second by (2). Add the resulting equations to eliminate the x-variable and solve for y.

$$\begin{aligned} -6x - 9y &= -15 \\ 6x + 10y &= -50 \\ \hline y &= -65 \end{aligned}$$

Hence, $2x + 3(-65) = 5$ or $2x - 195 = 5$, so $2x = 200$. Thus, $x = 100$ and $y = -65$.

3. (a) Substitution works well in problem (1) because it is easy to solve for x (or for y). Substitution would not work well in problem (2) because solving for x (or for y) would introduce fractions.

   (b) Elimination would work well in problem (1) by multiplying the first equation by 4. As stated in part (a), substitution wouldn't be as easy in problem (2).

4.
$$\begin{cases} x - y - z = 0 & (1) \\ \phantom{x -{}} y - 2z = -18 & (2) \\ x + y + z = 6 & (3) \end{cases}$$

Add $(-1) \times$ equation 1 to equation 3.

$$\begin{cases} x - y - z = 0 & (1) \\ \phantom{x -{}} y - 2z = -18 & (2) \\ \phantom{x -{}} 2y + 2z = 6 & (3) \end{cases}$$

Add $(-2) \times$ equation 2 to equation 3.

$$\begin{cases} x - y - z = 0 \\ \phantom{x -{}} y - 2z = -18 \\ \phantom{x -{}} 6z = 42 \end{cases}$$

Solve the equations for x, y, and z.

$$x = y + z$$
$$y = 2z - 18$$
$$z = 7$$

Back substitution gives $y = 14 - 18 = -4$ and $x = -4 + 7 = 3$. Thus, the solution is $x = 3$, $y = -4$, $z = 7$.

EXERCISE 1.8

In problems 1–4, solve the systems of equations by using graphical methods.

1. $\begin{cases} 4x - 2y = 4 \\ x - 2y = -2 \end{cases}$ 2. $\begin{cases} x - y = -2 \\ 2x + y = -1 \end{cases}$

3. $\begin{cases} 3x - y = 10 \\ 6x - 2y = 5 \end{cases}$ 4. $\begin{cases} 2x - y = 3 \\ 4x - 2y = 6 \end{cases}$

In problems 5–22, solve the systems of equations.

5. $\begin{cases} 4x - y = 3 \\ 2x + 3y = 19 \end{cases}$ 6. $\begin{cases} 5x - 3y = 9 \\ x + 2y = 7 \end{cases}$

7. $\begin{cases} 2x - y = 2 \\ 3x + 4y = 6 \end{cases}$ 8. $\begin{cases} x - y = 4 \\ 3x - 2y = 5 \end{cases}$

9. $\begin{cases} 3x - 2y = 6 \\ 4y = 8 \end{cases}$ 10. $\begin{cases} 5x - 2y = 4 \\ 2x - 3y = 5 \end{cases}$

11. $\begin{cases} -4x + 3y = -5 \\ 3x - 2y = 4 \end{cases}$ 12. $\begin{cases} x + 2y = 3 \\ 3x + 6y = 6 \end{cases}$

13. $\begin{cases} \frac{5}{2}x - \frac{7}{2}y = -1 \\ 8x + 3y = 11 \end{cases}$ 14. $\begin{cases} 3m + 2n = 2 \\ m - 2n = 1 \end{cases}$

15. $\begin{cases} 0.3u - 0.2v = 0.5 \\ 0.9u - 0.6v = 0.1 \end{cases}$ 16. $\begin{cases} x - \frac{1}{2}y = 1 \\ \frac{2}{3}x - \frac{1}{3}y = 1 \end{cases}$

17. $\begin{cases} 0.2x - 0.3y = 4 \\ 2.3x - y = 1.2 \end{cases}$ 18. $\begin{cases} 0.5x + y = 3 \\ 0.3x + 0.2y = 6 \end{cases}$

19. $\begin{cases} 4x + 6y = 4 \\ 2x + 3y = 2 \end{cases}$ 20. $\begin{cases} \dfrac{x}{4} + \dfrac{3y}{4} = 12 \\ \dfrac{y}{2} - \dfrac{x}{3} = -4 \end{cases}$

21. $\begin{cases} \dfrac{(x + y)}{4} = 2 \\ \dfrac{(y - 1)}{x} = 6 \end{cases}$ 22. $\begin{cases} 3x + 4y = 8 \\ \dfrac{(16 - 6x)}{y} = 8 \end{cases}$

Use the left-to-right elimination method to solve the systems in problems 23–28.

23. $\begin{cases} x + 2y + z = 2 \\ -y + 3z = 8 \\ 2z = 10 \end{cases}$ 24. $\begin{cases} x - 2y + 2z = -10 \\ y + 4z = -10 \\ -3z = 9 \end{cases}$

25. $\begin{cases} x - y - 8z = 0 \\ y + 4z = 8 \\ 3y + 14z = 22 \end{cases}$ 26. $\begin{cases} x + 3y - 8z = 20 \\ y - 3z = 11 \\ 2y + 7z = -4 \end{cases}$

27. $\begin{cases} x + 4y - 2z = 9 \\ x + 5y + 2z = -2 \\ x + 4y - 28z = 22 \end{cases}$ 28. $\begin{cases} x - 3y - z = 0 \\ x - 2y + z = 8 \\ 2x - 6y + z = 6 \end{cases}$

APPLICATIONS

29. **Investment yields** One safe investment pays 10% per year and a more risky investment pays 18% per year. A woman who has $145,600 to invest would like to have an income of $20,000 per year from her investments. How much should she invest at each rate?

30. **Loans** A bank loaned $118,500 to a company for the development of two products. If the loan for product A was for $34,500 more than that for product B, how much was loaned for each product?

31. **Rental income** A woman has $23,500 invested in two rental properties. One earns 10% on the investment and the other yields 12%. Her total income from them is $2550. How much is her income from each property?

32. **Loans** Mr. Jackson borrowed money from his bank and on his life insurance to start a business. His interest rate on the bank loan is 10% and his rate on the insurance loan is 12%. If the total amount borrowed is $10,000 and his total yearly interest payment is $1090, how much did he borrow from the bank?

33. **Nutrition** Each ounce of substance A supplies 5% of the required nutrition a patient needs. Substance B supplies 12% of the required nutrition per ounce. If digestive restrictions require that the ratio of substance A to substance B be 3/5, how many ounces of each should be in the diet to provide 100% of the required nutrition?

34. **Nutrition** A glass of skim milk supplies 0.1 mg of iron and 8.5 g of protein. A quarter pound of lean red meat provides 3.4 mg of iron and 22 g of protein. If a person on a special diet is to have 7.15 mg of iron and 73.75 g of protein, how many glasses of skim milk and how many quarter-pound servings of meat would provide this?

35. **Bacterial growth** Bacteria of species A and species B are kept in a single test tube, where they are fed two nutrients. Each day the test tube is supplied with 10,600 units of the first nutrient and 19,650 units of the second nutrient. Each bacteria of species A requires 2 units of the first nutrient and 3 units of the second, and each bacteria of species B requires 1 unit of the first nutrient and 4 units of the second. What populations of each species can coexist in the test tube so that all the nutrients are consumed each day?

36. **Botany** A biologist has a 40% solution and a 10% solution of the same plant nutrient. How many cubic centimeters of each solution should be mixed to obtain 25 cc of a 28% solution?

37. *Medications* A nurse has two solutions that contain different concentrations of a certain medication. One is a 20% concentration and the other is a 5% concentration. How many cubic centimeters of each should he mix to obtain 10 cc of a 15.5% solution?

38. *Medications* Medication A is given every 4 hours and medication B is given twice each day. The total intake of the two medications is restricted to 50.6 mg per day, for a certain patient. If the ratio of the dosage of A to the dosage of B is 5 to 8, find the dosage for each administration of each medication.

39. *Pricing* A concert promoter needs to take in $380,000 on the sale of 16,000 tickets. If the promoter charges $20 for some tickets and $30 for others, how many of each type must be sold to take in the $380,000?

40. *Pricing* A nut wholesaler sells a mix of peanuts and cashews. He charges $2.80 per pound for peanuts and $5.30 per pound for cashews. If the mix is to sell for $3.30 per pound, how many pounds each of peanuts and cashews should be used to make 100 pounds of the mix?

41. *Nutrient solutions* How many cubic centimeters of a 20% solution of a nutrient must be added to 100 cc of a 2% solution of the same nutrient to make a 10% solution of the nutrient?

42. *Mixtures* How many gallons of washer fluid that is 13.5% antifreeze must a manufacturer add to 200 gallons of washer fluid that is 11% antifreeze to have washer fluid that is 13% antifreeze?

Application problems 43–46 require systems of equations in three variables.

43. *Nutrition* Each ounce of substance A supplies 5% of the required nutrition a patient needs. Substance B supplies 15% of the required nutrition per ounce, and substance C supplies 12% of required nutrition per ounce. If digestive restrictions require that substances A and C be given in equal amounts, and the amount of substance B is one-fifth of either of these other amounts, find the number of ounces of each substance that should be in the meal to provide 100% of the nutrition.

44. *Dietary requirements* A glass of skim milk supplies 0.1 mg of iron, 8.5 g of protein, and 1 g of carbohydrates. A quarter pound of lean red meat provides 3.4 mg of iron, 22 g of protein, and 20 g of carbohydrates. Two slices of whole grain bread supply 2.2 mg of iron, 10 g of protein, and 12 g of carbohydrates. If a person on a special diet must have 10.5 mg of iron, 94.5 g of protein, and 61 g of carbohydrates, how many glasses of skim milk, how many quarter-pound servings of meat, and how many two-slice servings of whole grain bread will supply this?

45. *Social services* A social agency is charged with providing services to three types of clients, A, B, and C. A total of 500 clients are to be served, with $150,000 available for counseling and $100,000 available for emergency food and shelter. Type A clients require an average of $200 for counseling and $300 for emergencies, Type B clients require an average of $500 for counseling and $200 for emergencies, and Type C clients require an average of $300 for counseling and $100 for emergencies. How many of each type client can be served?

46. *Social services* If funding for counseling is cut to $135,000 and funding for emergency food and shelter is cut to $90,000, only 450 clients can be served. How many of each type can be served in this case? (See problem 45.)

1.9 BUSINESS APPLICATIONS OF LINEAR EQUATIONS IN TWO VARIABLES

Objectives

❑ To solve problems involving break-even analysis
❑ To solve problems involving market equilibrium
❑ To solve problems involving consumption functions

APPLICATIONS PREVIEW

The three applications introduced in Section 1.6 are extended here in terms of new techniques and vocabulary. In particular, for the theory of the firm we study break-even analysis. In market analysis, we introduce market equilibrium and use simultaneous solution techniques to find it. Finally, we extend marginality to the national consumption function.

BREAK-EVEN ANALYSIS

As we stated in Section 1.6, the profit that a firm makes on its product is the difference between its total revenue (the total amount it receives from sales) and its total cost (both fixed and variable).

$$\text{Profit (or loss)} = \text{total revenue} - \text{total cost}$$

We can solve the equations for total revenue and total cost simultaneously to find the point where cost and revenue are equal. This point is called the **break-even point.** Figure 1.61 is called a break-even chart. The quantity produced is measured along the horizontal axis; the vertical axis represents the dollar value of the costs or revenue. We will use x to represent the quantity produced and y to represent the dollar value of revenue *and* cost. The line C represents the sum of the fixed and variable costs. The line R represents the total revenue. Note that the total revenue is zero when no units are produced, and it increases more rapidly than the total cost line. (If it didn't, the firm would never make a profit.) The point where the total revenue line crosses the total cost line is called the break-even point. To the left of this point the total revenue is less than the total cost (resulting in a loss), and to the right of this point the total revenue is greater than the total cost (resulting in a profit).

EXAMPLE 1

A manufacturer sells a product for $10 per unit. The manufacturer's fixed costs are $1200 per month, and the variable costs are $2.50 per unit. How many units must be produced each month to break even?

Solution

The total revenue for x units of the product is $10x$, so the equation for total revenue is $R = 10x$. The fixed costs are $1200, so the total cost for x units is $2.50x + 1200$. Thus the equation for total cost is $C = 2.50x + 1200$. We

Figure 1.61

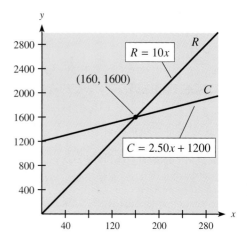

Figure 1.62

find the break-even point by solving the two equations simultaneously ($R = C$ at the break-even point).

By substitution,

$$10x = 2.50x + 1200$$

$$7.5x = 1200$$

$$x = 160.$$

Thus the manufacturer will break even if 160 units are produced per month. The manufacturer will make a profit if more than 160 units are produced. (See Figure 1.62.)

Using the fact that the profit function is found by subtracting the total cost function from the total revenue function, we can form the profit function for the previous example. The profit function is given by

$$P(x) = 10x - (2.50x + 1200) \quad \text{or} \quad P(x) = 7.50x - 1200.$$

We can find the point where the profit is zero (the break-even point) by setting $P(x) = 0$ and solving for x.

$$0 = 7.50x - 1200$$

$$1200 = 7.50x$$

$$x = 160$$

Note that this is the same break-even point that was found by solving the total revenue and total cost equations simultaneously.

EXAMPLE 2

Suppose a firm requires a 15% return on its investment (fixed costs) to break even. (a) If the good sells for $1, the variable cost is $0.50 per unit, and the fixed cost is $20,000, find the quantity the firm must produce to break even. (b) If the capacity of the firm's plant is 40,000 units, should it start production?

Solution

The firm's minimum desired profit is 15% of its fixed costs, so the desired profit is $0.15(20,000) = 3000$. If this desired profit is added to fixed costs, the break-even quantity is the quantity that will provide the firm with its 15% return. Thus we seek the quantity that makes $P = 3000$, where

$$P = R - C$$

$$3000 = 1x - (0.50x + 20,000)$$

$$3000 = 0.50x - 20,000$$

$$0.50x = 23,000$$

$$x = 46,000.$$

(a) The break-even quantity is 46,000.
(b) No, the plant does not have the capacity to produce enough units to reach the desired profit.

CHECKPOINT

1. Identify two ways that break-even points can be found.
2. If $C(x) = 17x + 8500$ and $R(x) = 37x$ are total cost and total revenue functions for a company, find the break-even point.

MARKET EQUILIBRIUM

We discussed demand functions and supply functions in Section 1.6. In many circumstances it is reasonable to assume that demand and supply functions are linear functions. Because demand will normally increase as price decreases, the slope of a demand curve (line) will be negative. On the other hand, a manufacturer will be willing to supply more goods as the price increases, so the supply curve will have a positive slope.

EXAMPLE 3

Suppose that the demand per month for a commodity is 24 if the price is $15 for each unit, 27 if the price is $13, 30 if the price is $11, and 33 if the price is $9. Assuming the demand curve is linear, write its equation. Graph the demand curve.

Solution

Using two of the points on the demand curve, say $(24, 15)$ and $(33, 9)$, we can find the slope of the line.

$$m = \frac{15 - 9}{24 - 33} = \frac{6}{-9} = -\frac{2}{3}$$

We can then use one of the points, say $(24, 15)$, to write the equation of the demand curve as $p - 15 = -\frac{2}{3}(q - 24)$ or $p = -\frac{2}{3}q + 31$. The graph of the demand function is shown in Figure 1.63.

Figure 1.63

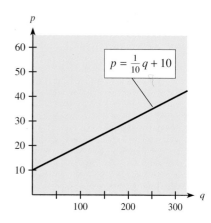

Figure 1.64

EXAMPLE 4

A manufacturer will supply 200 electric motors per month when the price is $30 and 300 per month when the price is $40. Assuming the supply curve is linear, write its equation and sketch its graph.

Solution

The slope of the line connecting the points (200, 30) and (300, 40) is

$$m = \frac{40 - 30}{300 - 200} = \frac{10}{100} = \frac{1}{10}.$$

The equation is

$$p - 30 = \tfrac{1}{10}(q - 200) \quad \text{or} \quad p = \tfrac{1}{10}q + 10.$$

The graph is shown in Figure 1.64.

Producers of a commodity like high prices and consumers like low prices, but an agreement is made between these two opposing forces in something called a "market" (thus this area is called *market analysis*). The effects on prices of gasoline and natural gas shortages are painfully apparent to all consumers, and the effects of farm surpluses have led many small farmers to stop producing.

Consider the demand and supply schedules (Table 1.4) for a commodity over a period of one month. We can see from these schedules that when the price is $15, the producers are willing to supply many more units than consumers are willing to purchase. The $15 could not be the prevailing market price, for the producers would be producing many more than consumers are willing to buy at that price. This price would create a *surplus* of 36 units of this commodity each month. Eventually this surplus would force the price down below $15. If the price drops to $13, the demand is still less than the supply, so there will still be a surplus on the market, and this will force the price even lower. So the market prices of $15 and $13 will be unstable because they are too high. The price that will put the market in balance will have to be lower than $13.

On the other hand, if the price is $9, consumers will be willing to buy 33 units per month but producers will only supply 15 units. Thus the market

Table 1.4

Demand Schedule		Supply Schedule	
Price Per Unit	*Quantity Demanded*	*Price Per Unit*	*Quantity Supplied*
$15	24	$15	60
13	27	13	45
11	30	11	30
9	33	9	15

will experience a *shortage.* Competition among consumers for the small number of items will force the price higher than the $9 price.

If the price is $11, consumers will be willing to purchase the same number of units (30) that the producers are willing to supply. So the market will be in balance if the price is $11.

When the quantity of a commodity demanded is equal to the amount supplied, **market equilibrium** is said to occur. If the demand and supply curves are graphed on the same coordinate system, with the same units, the point at which they intersect represents the **equilibrium point.** The price at that point is the **equilibrium price,** and the quantity at that point is the **equilibrium quantity.**

Graphing the curves associated with the schedules of Table 1.4 on the same graph, we see that the two lines intersect at (30, 11). This corresponds to the solution we have already found. (See Figure 1.65.)

In general, the equilibrium price and the equilibrium quantity must be positive for the market equilibrium to have meaning.

We can find the market equilibrium by graphing the demand and supply functions on the same coordinate system, but solving the two equations simultaneously by algebraic methods will generally provide more accurate results.

EXAMPLE 5

Find the equilibrium point for the following supply and demand functions:

$$\text{Demand:} \quad p = -3q + 26$$
$$\text{Supply:} \quad p = 4q - 9$$

Solution
Solving simultaneously by substitution gives

$$-3q + 26 = 4q - 9$$
$$35 = 7q$$

so
$$q = 5, \quad p = 11.$$

The equilibrium point is (5, 11).

EXAMPLE 6

A group of wholesalers will buy 50 dryers per month if the price is $200 and 30 per month if the price is $300. The manufacturer is willing to supply 20 if the price is $210 and 30 if the price is $230. Assuming the resulting supply and demand functions are linear, find the equilibrium point for the market.

Figure 1.65

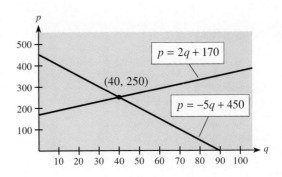

Figure 1.66

Solution

Demand function:

$$m = \frac{300 - 200}{30 - 50} = -5$$

$$p - 200 = -5(q - 50)$$

$$p = -5q + 450$$

Supply function:

$$m = \frac{230 - 210}{30 - 20} = 2$$

$$p - 230 = 2(q - 30)$$

$$p = 2q + 170$$

Solving simultaneously gives (by substitution)

$$-5q + 450 = 2q + 170$$

$$280 = 7q$$

$$q = 40$$

$$p = 250.$$

The equilibrium point is (40, 250). See Figure 1.66 for the graph of the functions.

If a supplier is taxed $K per unit sold, then the tax is passed on to the consumer by adding $K to the selling price of the product. If the original supply function is $p = f(q)$, then passing the tax on gives a new supply function, $p = f(q) + K$. Because the value of the product is not changed by

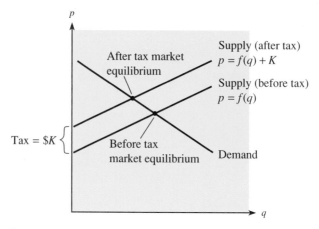

Figure 1.67

the tax, the demand function is unchanged. Figure 1.67 shows the effect that this has on market equilibrium.

Notice that the new market equilibrium point is the point of intersection of the original demand function and the new (after taxes) supply function.

EXAMPLE 7

In Example 6, the supply and demand functions for dryers were given as follows.

$$\text{Supply:} \quad p = 2q + 170$$
$$\text{Demand:} \quad p = -5q + 450$$

The equilibrium point was $q = 40, p = \$250$. If the wholesaler is taxed \$14 per unit sold, what is the new equilibrium point?

Solution

The \$14 tax per unit is passed on by the wholesaler, so the new supply function is

$$p = 2q + 170 + 14,$$

and the demand function is unchanged. Thus we solve the system

$$\begin{cases} p = 2q + 184 \\ p = -5q + 450. \end{cases}$$
$$2q + 184 = -5q + 450$$
$$7q = 266$$
$$q = 38$$
$$p = 2(38) + 184 = 260$$

The new equilibrium point is $q = 38, p = \$260$.

CHECKPOINT

3. (a) What do we call the point of intersection of a supply function and a demand function?
   (b) What algebraic technique is used to find the point named in part (a)?

4. If the supply and demand functions for a product are given, respectively, by

$$7p - 3q = 30 \quad \text{and} \quad 3p + q = 470,$$

find the market equilibrium point.

NATIONAL CONSUMPTION

As we stated in Section 1.6, the national consumption function is frequently considered to be linear over short periods of time. Thus we may write the general form of the consumption function as $C = my + b$, where y is the disposable consumption and the value of C for a given value of y is the **consumption.**

The value of b will always be greater than zero, for there is always some amount of consumption necessary to maintain life, even though disposable income (money) is zero. Mathematically, we say $b > 0$. Consumption will increase whenever income increases, but at a slower rate. Thus the slope m of the line representing the consumption function is greater than 0 but less than 1; that is, $0 < m < 1$.

We call m the **marginal propensity to consume** because it indicates the proportion of our income increases that we are willing to spend.

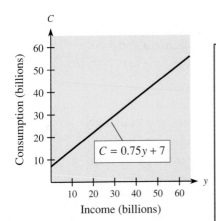

Figure 1.68

EXAMPLE 8

Suppose that national consumption is $7 billion when national disposable income is zero, and that at each level of income above zero, consumption is 0.75 of disposable income.
(a) Write this consumption function as an equation.
(b) What is the marginal propensity to consume?
(c) Graph the consumption function.

Solution

(a) The line representing the consumption function has $m = 0.75$ and $b = 7$, so the equation is $C = 0.75y + 7$ (in billions of dollars).
(b) The marginal propensity to consume is $m = 0.75$.
(c) See Figure 1.68.

CHECKPOINT

5. If the national consumption function is linear, how is the marginal propensity to consume found?

CHECKPOINT
SOLUTIONS

1. Break-even occurs at values of x where revenue equals cost ($R(x) = C(x)$) or at x-values where profit is zero ($P(x) = 0$).

2. $17x + 8500 = 37x$

   $8500 = 20x$ so $x = 425$

   The break-even point occurs when 425 units are sold.

3. (a) Market equilibrium point   (b) Solving simultaneously

4. $7p - 3q = 30$
   $\underline{9p + 3q = 1410}$
   $16p \qquad = 1440$ so $p = 90$

   and $3(90) + q = 470$

   $q = 200$

   Thus, market equilibrium occurs when $p = \$90$ and $q = 200$.

5. For a linear function, the marginal propensity to consume is the slope.

EXERCISE 1.9

BREAK-EVEN ANALYSIS

1. Figure 1.69 shows graphs of the total cost function and the total revenue function for a commodity.

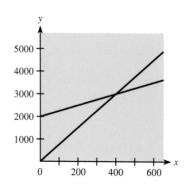

Figure 1.69

(a) Label each function correctly.
(b) Determine the fixed costs.
(c) Locate the break-even point and determine the number of units sold to break even.
(d) Estimate the marginal cost and marginal revenue.

2. A manufacturer of shower-surrounds has a revenue function of

$$R(x) = 81.50x$$

and a cost function of

$$C(x) = 63x + 1850.$$

Find the number of units that must be sold to break even.

3. A jewelry maker incurs costs for a necklace according to

$$C(x) = 35x + 1650.$$

If the revenue function for the necklaces is

$$R(x) = 85x,$$

how many necklaces must be sold to break even?

4. A small business recaps and sells tires. If a set of four tires has the revenue function

$$R(x) = 89x$$

and the cost function

$$C(x) = 1400 + 75x,$$

find the number of sets of recaps that must be sold to break even.

5. A manufacturer sells belts for $12 per unit. The fixed costs are $1600 per month, and the variable costs are $8 per unit.
   (a) Write the equations of the revenue and cost functions.
   (b) Find the break-even point.

6. A manufacturer sells watches for $50 per unit. The fixed costs related to this product are $10,000 per month, and the variable costs are $30 per unit.
   (a) Write the equations of the revenue and cost functions.
   (b) How many watches must be sold to break even?

7. (a) Write the profit function for problem 5.
   (b) Set profit equal to zero and solve for x. Compare this x-value with the break-even point from 5(b).

8. (a) Write the profit function for problem 6.
   (b) Set profit equal to zero and solve for x. Compare this x-value with the break-even point from 6(b).

9. A manufacturer of box fans has fixed costs of $45,000, and the marginal cost is $7.50. If the fans sell for $20 each, find the break-even point.

10. A company sells table lamps for $59. The fixed costs are $7876, and the marginal cost is $37. How many lamps must be sold to break even?

11. A company manufactures and sells bookcases. The selling price is $54.90 per bookcase. The total cost function is linear, and costs amount to $50,000 for 2000 bookcases and $32,120 for 800 bookcases.
   (a) Write the equation for revenue.
   (b) Write the equation for total costs.
   (c) Find the break-even point.

12. A company distributes college logo sweatshirts and sells them for $50 each. The total cost function is linear, and the total cost for 100 sweatshirts is $4360, while the total cost for 250 sweatshirts is $7060.
   (a) Write the equation for the revenue function.
   (b) Write the equation for the total cost function.
   (c) Find the break-even point.

13. Suppose a firm will not begin production of a product unless it receives a 20% return on its investment for fixed costs. If the fixed costs are $50,000, the variable costs are $2 per unit, and the product sells for $5 per unit, how many units must be produced to break even (that is, make the required return)?

14. Suppose a firm requires a 10% return on its investment for fixed costs to break even. If the product sells for $2 per unit, the variable costs are $0.50 per unit, and the fixed costs are $30,000, how many units must be produced to break even?

MARKET EQUILIBRIUM

Complete problems 15–17 by using Figure 1.70, which shows a supply function and a demand function.

Figure 1.70

15. (a) Label each function as "demand" or "supply."
   (b) Label the equilibrium point and determine the price and quantity at which market equilibrium occurs.

16. (a) If the price is $30, what quantity is demanded?
   (b) If the price is $30, what quantity is supplied?
   (c) Is there a surplus or shortage when the price is $30? How many units is this surplus or shortage?

17. (a) If the price is $20, what quantity is supplied?
   (b) If the price is $20, what quantity is demanded?
   (c) Is there a surplus or shortage when the price is $20? How many units is this surplus or shortage?

18. Will a price above the equilibrium price result in a market surplus or shortage?

19. Will a price below the equilibrium price result in a market surplus or shortage?

20. Suppose the demand per month for a product is 5 if the price is $310, 10 if the price is $300, and 15 if the price is $290. Assuming that the demand curve is linear, graph the curve and write its equation.

21. Suppose the demand per month for a commodity is 24 if the price is $16, 20 if the price is $18, 16 if the price is $20, and 12 if the price is $22. Assuming that the demand curve is linear, graph the curve and write its equation.

22. A manufacturer will supply a product according to the price it receives, as follows:

5 if the price is $42	15 if the price is $122
10 if the price is $82	

   Assuming that the supply curve is linear, graph it and write its equation.

23. A distributor will supply a commodity according to the price it receives, as follows:

8 if the price is $14	20 if the price is $18
14 if the price is $16	26 if the price is $20

   Assuming that the supply curve is linear, graph the curve and write its equation.

24. Find the market equilibrium point for these demand and supply functions (found in problems 20 and 22).

$$\text{Demand:} \quad p = -2q + 320$$
$$\text{Supply:} \quad p = 8q + 2$$

25. Find the market equilibrium point for these demand and supply functions (found in problems 21 and 23).

$$\text{Demand:} \quad 2p = -q + 56$$
$$\text{Supply:} \quad 3p - q = 34$$

26. Find the equilibrium point for the following supply and demand functions:

$$\text{Demand:} \quad p = 480 - 3q$$
$$\text{Supply:} \quad p = 17q + 80$$

27. Find the equilibrium point for the following supply and demand functions:

$$\text{Demand:} \quad p = -4q + 220$$
$$\text{Supply:} \quad p = 15q + 30$$

28. Retailers will buy 45 Z-brand phones from a wholesaler if the price is $10 each, and will buy 20 if the price is $60. The wholesaler will supply 35 phones at $30 each and 70 at $50 each. Assuming the supply and demand functions are linear, find the market equilibrium point.

31. A group of retailers will buy 80 televisions from a wholesaler if the price is $350 and 120 if the price is $300. The wholesaler is willing to supply 60 if the price is $280 and 140 if the price is $370. Assuming the resulting supply and demand functions are linear, find the equilibrium point for the market.

30. A shoe store owner will buy 10 pairs of a certain shoe if the price is $75 per pair, and 30 pairs if the price is $25. The supplier of the shoes is willing to provide 35 pairs if the price is $80 per pair, and 5 pairs if the price is $20. Assuming the supply and demand functions for the shoes are linear, find the market equilibrium point.

Problems 31–38 involve market equilibrium after taxation. Use Figure 1.71 to answer problems 31 and 32.

Figure 1.71

31. (a) What is the amount of the tax?
    (b) What is the original equilibrium price and quantity?
    (c) What is the new equilibrium price and quantity?
    (d) Does the supplier suffer from the tax even though it is passed on?

32. (a) If the tax is doubled, how many units will be sold?
    (b) Can a government lose money by increasing taxes?

33. If a $38 tax is placed on each unit of the product of problem 27, what are the new equilibrium price and quantity?

34. If a $56 tax is placed on each unit of the product of problem 26, what is the new equilibrium point?

35. Suppose that a certain product has the following demand and supply functions.

$$\text{Demand:} \quad p = -0.05q + 65$$
$$\text{Supply:} \quad p = 0.05q + 10$$

If a $5 tax per item is levied on the supplier and this tax is passed on to the consumer, find the market equilibrium point after the tax.

36. Suppose that a certain product has the following demand and supply functions.

$$\text{Demand:} \quad p = -8q + 2800$$
$$\text{Supply:} \quad p = 3q + 35$$

If a $15 tax per item is levied on the supplier, who passes it on to the consumer as a price increase, find the market equilibrium point after the tax.

37. Suppose that in a certain market the demand function for a product is given by $60p + q = 2100$ and the supply function is given by $120p - q = 540$. Then a tax of $0.50 per item is levied on the supplier, who passes it on to the consumer as a price increase. Find the equilibrium price and quantity after the tax is levied.

38. Suppose that in a certain market the demand function for a product is given by $10p + q = 2300$ and the supply function is given by $45p - q = 360$. If the government levies a tax of $2 per item on the supplier, who passes the tax on to the consumer as a price increase, find the equilibrium price and quantity after the tax is levied.

NATIONAL CONSUMPTION

39. Suppose that national consumption is $8 billion when national disposable income is zero, and that at each level of income above zero, consumption is 0.6 of disposable income.
    (a) Write this consumption function as an equation.
    (b) What is the marginal propensity to consume?

40. Suppose that national consumption is $10 billion when national disposable income is zero, and that at each level of income above zero, consumption is 0.7 of disposable income.
    (a) Write this consumption function as an equation.
    (b) What is the marginal propensity to consume?
41. Graph the equation from problem 39.
42. Graph the equation from problem 40.
43. If the national consumption function is given by the equation

    $$C = 0.65y + 5 \quad \text{(billions of dollars)},$$

    (a) what is the marginal propensity to consume?
    (b) what is the national consumption when disposable income is zero?
44. If the national consumption function is given by the equation

    $$C = 0.55y + 9 \quad \text{(billions of dollars)},$$

    (a) what is the marginal propensity to consume?
    (b) what is the national consumption when disposable income is zero?

45. Graph the consumption function is problem 43.
46. Graph the consumption function in problem 44.
47. What is the relation between the marginal propensity in problem 43 and the slope of the graph in problem 45?
48. If the national consumption is $7 billion when national disposable income is zero, and if the marginal propensity to consume is 0.8, write the linear consumption function as an equation.
49. Suppose that national consumption is $15.6 billion when income is $10 billion and $24 billion when income is $20 billion. Assume that the consumption function is linear.
    (a) Find the marginal propensity to consume.
    (b) Write the equation for the consumption function.
50. Suppose that national consumption is $18 billion when income is $20 billion and $50 billion when income is $60 billion. Assume that the consumption function is linear.
    (a) Find the marginal propensity to consume.
    (b) Write the equation for the consumption function.

KEY TERMS
& FORMULAS

Section	Key Terms	Formula
1.1	Equation, members; variable; solution Identities; conditional equations Properties of equality Linear equation in one variable	

1.2	Coordinate system
	  Ordered pair, origin, x-axis,
	  y-axis
	Graph
	Linear equation in two variables
	Intercepts
		
		$Ax + By = C$
		x-intercept: where $y = 0$
		y-intercept: where $x = 0$

1.3	Graphing utilities Standard viewing window Range	

1.4	Relation
	Function
	  Vertical line test
	  Domain, range	

Section	**Key Terms**	**Formula**
	Operations with functions	
	Sum	$(f + g)(x) = f(x) + g(x)$
	Difference	$(f - g)(x) = f(x) - g(x)$
	Product	$(f \cdot g)(x) = f(x) \cdot g(x)$
	Quotient	$\left(\dfrac{f}{g}\right)(x) = \dfrac{f(x)}{g(x)}, \quad g(x) \neq 0$
	Composite	$(f \circ g)(x) = f(g(x))$ and $(g \circ f)(x) = g(f(x))$
1.5	Linear functions	$f(x) = ax + b$
	Basic functions	$f(x) = C, \quad C = \text{constant}$
		$f(x) = x$ (identity)
		$f(x) = x^2, \quad f(x) = x^3, \quad f(x) = 1/x$
	Polynomial function	$f(x) = a_n x^n + a_{n-1} x^{n-1} + \cdots + a_1 x + a_0$
	Rational function	$f(x) = p(x)/q(x)$, where $p(x)$ and $q(x)$ are polynomials
	Vertical asymptote	
	Horizontal asymptote	
	Absolute value function	$f(x) = |x| = \begin{cases} x & \text{if } x \geq 0 \\ -x & \text{if } x < 0 \end{cases}$
	Piecewise defined functions	
1.6	Cost and revenue functions	$C(x)$ and $R(x)$
	Profit functions	$P(x) = R(x) - C(x)$
	Supply and demand functions	
	Consumption function	
1.7	Slope of a line	$m = \dfrac{y_2 - y_1}{x_2 - x_1}$
	Parallel lines	$m_1 = m_2$
	Perpendicular lines	$m_2 = -1/m_1$
	Point-slope form	$y - y_1 = m(x - x_1)$
	Slope-intercept form	$y = mx + b$
	Vertical line	$x = a$
	Horizontal line	$y = b$
	Marginal profit	Slope of linear profit function
	Marginal cost	Slope of linear cost function
	Marginal revenue	Slope of linear revenue function
1.8	Systems of linear equations	
	Solutions	
	Equivalent systems	
	Substitution method	
	Elimination method	

Section	Key Terms	Formula
	Left-to-right elimination method	
	Lead variable	
	Back substitution	
1.9	Break-even point	$C(x) = R(x)$ or $P(x) = 0$
	Market equilibrium	
	National consumption	
	Marginal propensity to consume	Slope of the consumption function

REVIEW EXERCISES

1.1

Solve the equations in problems 1–8.

1. $x + 7 = 14$

2. $3x - 8 = 23$

3. $2x - 8 = 3x + 5$

4. $\dfrac{6x + 3}{6} = \dfrac{5(x - 2)}{9}$

5. $2x + \dfrac{1}{2} = \dfrac{x}{2} + \dfrac{1}{3}$

6. $0.6x + 4 = x - 0.02$

7. $\dfrac{3}{x} - 2 = 3 + \dfrac{1}{x}$

8. $2.81\,(3.62x - 4.02) = \dfrac{0.061x - 3.82}{412}$

1.2

In problems 9–14, sketch the graph of each equation.

9. $y = 3x$

10. $y = 5$

11. $x = -2$

12. $2x - 3y = 6$

13. $5x + 3y = 6$

14. $2(y + 1) = \frac{1}{2}x + \frac{1}{2}$

1.3

In problems 15 and 16, graph each equation with a graphing utility and the standard viewing window.

15. $x^2 + y - 2x - 3 = 0$

16. $y = \dfrac{x^3 - 27x + 54}{15}$

In problems 17 and 18, use a graphing utility and

(a) graph each equation in the viewing window given.

(b) graph each equation in the standard viewing window.

(c) explain how the two views differ and why.

17. $y = (x + 6)(x - 3)(x - 15)$ with x-min $= -15$, x-max $= 25$; y-min $= -700$, y-max $= 500$

18. $y = x^2 - x - 42$ with x-min $= -15$, x-max $= 15$; y-min $= -50$, y-max $= 50$

1.4

19. If $p = 3q^3$, is p a function of q?

20. If $y^2 = 9x$, is y a function of x?

21. If $R = \sqrt[3]{x + 4}$, is R a function of x?

22. What are the domain and range of the function $y = \sqrt{9 - x}$? Check with a graphing utility.

23. What is the domain of $y = \dfrac{\sqrt{x + 3}}{x}$? Check with a graphing utility.

24. If $f(x) = x^2 + 4x + 5$, find the following.
    (a) $f(-3)$ (b) $f(4)$ (c) $f(\frac{1}{2})$

25. If $g(x) = x^2 + 1/x$, find the following.
    (a) $g(-1)$ (b) $g(\frac{1}{2})$ (c) $g(0.1)$

26. If $f(x) = 9x - x^2$, find the following.
    (a) $f(-4 + 1)$ (b) $f(x + 1)$ (c) $f(x + h)$

27. If $f(x) = 9x - x^2$, find $\dfrac{f(x + h) - f(x)}{h}$.

28. Does the graph in Figure 1.72 represent y as a function of x?

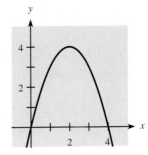

Figure 1.72

29. Does the graph in Figure 1.73 represent y as a function of x?

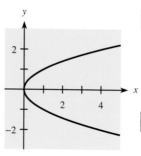

Figure 1.73

30. For the function f graphed in Figure 1.72, what is $f(2)$?
31. For the function f graphed in Figure 1.72, for what values of x does $f(x) = 0$?
32. For the relation whose graph is in Figure 1.73, when $x = 1$, what values does y assume?
33. The following table defines y as a function of x, denoted $y = f(x)$.

x	-2	-1	0	1	3	4
y	8	2	-3	4	2	7

Use the table to complete the following.
(a) Find $f(4)$.
(b) Find all x-values for which $f(x) = 2$.
(c) Graph $y = f(x)$.

1.5

34. Sketch a graph of each of the following basic functions.

(a) $f(x) = x^2$ (b) $f(x) = \dfrac{1}{x}$ (c) $f(x) = 3$

35. If $f(x) = \begin{cases} -x^2 & \text{if } x \le 0 \\ \dfrac{1}{x} & \text{if } x > 0 \end{cases}$, find the following.

(a) $f(0)$ (b) $f(0.0001)$
(c) $f(-5)$ (d) $f(10)$

36. If $f(x) = \begin{cases} x & \text{if } x \le 1 \\ 3x - 2 & \text{if } x > 1 \end{cases}$, find the following.

(a) $f(-2)$ (b) $f(0)$
(c) $f(1)$ (d) $f(2)$

In problems 37 and 38, graph each function.

37. $f(x) = \begin{cases} x & \text{if } x \le 1 \\ 3x - 2 & \text{if } x > 1 \end{cases}$ (See problem 36.)

38. $f(x) = \begin{cases} 1 & \text{if } x < 0 \\ 1 - x & \text{if } x \ge 0 \end{cases}$

 In problems 39 and 40, use a graphing utility to graph each function. Find any turning points.

39. $y = x^3 + 3x^2 - 9x$

40. $y = \dfrac{x^3 - 9x}{x^2 + 1}$

In problems 41 and 42, use a graphing utility to graph each function. Find and identify any asymptotes.

41. $y = \dfrac{1}{x - 2}$ 42. $y = \dfrac{2x - 1}{x + 3}$

1.7

In problems 43–45, find the slope of the line passing through each pair of points.
43. $(-1, 2)$ and $(3, -1)$
44. $(2, -1)$ and $(-1, -4)$
45. $(-3.8, -7.16)$ and $(-3.8, 1.16)$

In problems 46 and 47 find the slope and y-intercept of each line.
46. $2x + 5y = 10$
47. $x = -\frac{3}{4}y + \frac{3}{2}$

In problems 48–55, write the equation of each line described.
48. Slope 4 and y-intercept 2
49. Slope $-\frac{1}{2}$ and y-intercept 3
50. Through $(-2, 1)$ with slope $\frac{2}{5}$
51. Through $(6, -11)$ with slope -1
52. Through $(-2, 7)$ and $(6, -4)$
53. Through $(-1, 8)$ and $(-1, -1)$
54. Through $(1, 6)$ and parallel to $y = 4x - 6$
55. Through $(-1, 2)$ and perpendicular to $3x + 4y = 12$

1.8

In problems 56–63, solve each system of equations.

56. $\begin{cases} 4x - 2y = 6 \\ 3x + 3y = 9 \end{cases}$ 57. $\begin{cases} 2x + y = 19 \\ x - 2y = 12 \end{cases}$

58. $\begin{cases} 3x + 2y = 5 \\ 2x - 3y = 12 \end{cases}$ 59. $\begin{cases} 6x + 3y = 1 \\ y = -2x + 1 \end{cases}$

60. $\begin{cases} 4x - 3y = 253 \\ 13x + 2y = -12 \end{cases}$ 61. $\begin{cases} 0.6x + y = 2.2 \\ 0.3x - 0.1y = 0.5 \end{cases}$

62. $\begin{cases} x + 2y + 3z = 5 \\ y + 11z = 21 \\ 5y + 9z = 13 \end{cases}$ 63. $\begin{cases} x + y - z = 12 \\ 2y - 3z = -7 \\ 3x + 3y - 7z = 0 \end{cases}$

APPLICATIONS

1.1

64. *Course grades* In a certain course, grades are based on three tests worth 100 points each, three quizzes worth 50 points each, and a final exam worth 200 points. A student has test grades of 91, 82, and 88, and quiz grades of 50, 42, and 42. What is the lowest percent the student can get on the final and still earn an A (90% or more of the total points) in the course?

65. *Cost analysis* The owner of a small construction business needs a new truck. He can buy a diesel truck for $18,000 and it will cost him $0.16 per mile to operate it. He can buy a gas engine truck for $16,000 and it will cost him $0.21 per mile to operate. Find the number of miles he must drive before the costs are equal. If he normally keeps a truck for 5 years, which is the better buy?

1.4

66. *Heart disease risk* The Multiple Risk Factor Intervention Trial (MRFIT) used data from 356,222 men aged 35 to 57 to investigate the relationship between serum cholesterol and coronary heart disease (CHD) risk. Figure 1.74 shows the graph of the relationship of CHD risk and cholesterol, where a risk of 1 is assigned to 200 mg/dl of serum cholesterol and where the CHD risk is 4 times as high when serum cholesterol is 300 mg/dl.

Serum cholesterol (mg/dl)   **Figure 1.74**

(a) Does this graph indicate that the CHD risk is a function of the serum cholesterol?
(b) Is the relationship a linear function?
(c) If CHD risk is a function *f* of serum cholesterol, what is *f*(300)?

67. *Mortgage loans* When a couple purchases a home, one of the first questions they face deals with the relationship between the amount borrowed and the monthly pay

ment. In particular, if a bank offers 25-year loans at 7% interest, then the data in the following table would be available from that bank.

Amount Borrowed	Monthly Payment
$40,000	$282.72
50,000	353.39
60,000	424.07
70,000	494.75
80,000	565.44
90,000	636.11
100,000	706.78

(SOURCE: *Comprehensive Mortgage Payment Tables,* Publication No. 492, Financial Publishing Co., Boston)

If we assume that the monthly payment *P* is a function of the amount borrowed *A* (in thousands), and denoted by $P = f(A)$, answer the following.
(a) Find $f(80)$.
(b) Write a sentence that explains the meaning of $f(70) = 494.75$.
(c) Explain how to find $f(120)$, and then find it.

1.5

68. *Municipal water costs* The Borough Municipal Authority of Beaver, Pennsylvania, used the following function to determine charges for water.

$$C(x) = \begin{cases} 1.557x & 0 \le x \le 100 \\ 155.70 + 1.04(x - 100) & 100 < x \le 1000 \\ 1091.70 + 0.689(x - 1000) & x > 1000 \end{cases}$$

where $C(x)$ is the cost in dollars for *x* thousand gallons of water.
(a) Find the monthly charge for 12,000 gallons of water.
(b) Find the monthly charge for 825,000 gallons of water.

69. *Photosynthesis* The amount *y* of photosynthesis that takes place in a certain plant depends on the intensity *x* of the light present, according to

$$y = 120x^2 - 20x^3 \quad \text{for } x \ge 0.$$

(a) Graph this function with a graphing utility. (Use *y*-min = -100 and *y*-max = 700.)

(b) The model is valid only when $f(x) \ge 0$ (that is, on or above the *x*-axis). For what *x*-values is this true?

1.6

70. *Cost* Suppose that a commodity has fixed costs of $500 and variable costs of $2 per unit.
(a) Find the total cost function for this commodity.
(b) What is the cost of producing 100 units?

71. *Revenue* Suppose that the commodity of problem 70 sells for $4 per unit.
(a) Find the total revenue function for the commodity.
(b) Find the revenue if 100 units are sold.

72. *Profit*
(a) Form the total profit function for the commodity in problems 70 and 71.
(b) Find the profit if 100 units are produced and sold.
(c) Find the number of units required before no money is lost.

73. *Supply and demand* A certain product has supply and demand functions $p = 4q + 5$ and $p = -2q + 81$, respectively.
(a) If the price is $53, how many units are supplied and how many are demanded?
(b) Does this give a shortfall or a surplus?
(c) Is the price likely to increase from $53 or decrease from it?

74. *National consumption* Suppose the national consumption is described by $C = 0.8y + 2$, in billions of dollars.
(a) What is consumption when disposable income is 0?
(b) What is consumption when disposable income is $5 billion?

75. *National consumption* Graph the consumption function in problem 74.

1.8

76. *Investment mix* A retired couple has $150,000 to invest and wants to earn $15,000 per year in interest. The safer investment yields 9.5%, but they can supplement by investing some of their money at 11%. How much should they invest at each rate to earn $15,000 per year?

77. *Botony* A botanist has a 20% solution and a 70% solution of an insecticide. How much of each must be used to make 4.0 liters of a 35% solution?

1.7, 1.9

78. *Market analysis* Of the equations $p + 6q = 420$ and $p = 6q + 60$, one is the supply function for a product and one is the demand function for that product.
(a) Graph these equations on the same set of axes.
(b) Label the supply function and the demand function.
(c) Find the market equilibrium point.

79. *National consumption* A national consumption function is given by $C = 4 + 0.6y$ (billions of dollars).
(a) What is the consumption when disposable income is zero?
(b) What is the marginal propensity to consume?
(c) Graph the consumption function.

80. *Cost, revenue, and profit* The total costs and total revenues of a certain product are given by the following.
$$C(x) = 38.80x + 4500$$
$$R(x) = 61.30x$$
(a) Find the marginal cost.
(b) Find the marginal revenue.
(c) Find the marginal profit.
(d) Find the number of units required to break even.

81. *Cost, revenue and profit* A certain commodity has the following costs for a period:

Fixed costs: $1500
Variable costs: $22 per unit

If the commodity is sold for $52 per unit,
(a) what is the total cost function?
(b) what is the total revenue function?
(c) what is the profit function?
(d) what is the marginal cost?
(e) what is the marginal revenue?
(f) what is the marginal profit?
(g) what is the break-even point?

82. *Market analysis* The supply function and the demand function for a product are linear and determined by the tables below. Find the price that will give market equilibrium.

Supply function		Demand function	
Price	Quantity	Price	Quantity
100	200	200	200
200	400	100	400
300	600	0	600

83. *Market analysis* Suppose that for a certain product the supply and demand functions prior to any taxation are as follows.

Supply: $p = q/10 + 8$
Demand: $10p + q = 1500$

If a tax of $2 per item is levied on the supplier and is passed on to the consumer as a price increase, find the market equilibrium after the tax is levied.

ONSULTANT'S CORNER

Hospital Administration

Southwest Hospital has an operating room used only for eye surgery. The annual cost of rent, heat, and electricity for the operating room and its equipment is $180,000, and the annual salaries for the people who staff this room total $270,000.

Each surgery performed requires the use of $380 worth of medical supplies and drugs. To promote goodwill, every patient receives a bouquet of flowers the day after surgery. In addition, one quarter of the patients require dark glasses, which the hospital provides free of charge. It costs the hospital $15 for each bouquet of flowers and $20 for each pair of glasses.

The hospital receives a payment of $1000 for each eye operation performed.

1. Identify the revenue per case and the annual fixed and variable costs for running the operating room.
2. How many eye operations must the hospital perform each year in order to break even?
3. Southwest Hospital currently averages 70 eye operations per month. One of the nurses has just learned about a machine that would reduce by $50 per patient the amount of medical supplies needed. It can be leased for $50,000 annually. Keeping in mind the financial costs/benefits, advise the hospital on whether or not they should lease this machine.
4. An advertising agency has proposed to the hospital's president that she spend $10,000 per month on television and radio advertising to persuade people that Southwest Hospital is the best place to have any eye surgery performed. The advertising firm estimates that such publicity would increase business by 40 operations per month. If they are correct and if this increase is not big enough to affect fixed costs, what impact would this advertising have on the hospital's profits?
5. In case the advertising agency is being overly optimistic, how many extra operations per month are needed to cover the cost of the proposed ads?
6. If the ad campaign is approved and subsequently meets its projections, should the hospital review its decision about the machine discussed in question 3?

Warm-up

Prerequisite problem type	For section	Answer	Section for review
Evaluate: (a) $4 - (-2)$ (b) $-4(\frac{2}{3}) + 3$ (c) $(-5)(4) + (8)(1) + (2)(5)$ (d) Multiply each of the numbers $0, -3, -3,$ $-2, 2$ by $-\dfrac{1}{3}$.	2.1 2.2 2.3 2.4 2.5	(a) 6 (b) $\dfrac{1}{3}$ (c) -2 (d) $0, 1, 1, \dfrac{2}{3}, -\dfrac{2}{3}$	0.2 Signed numbers
Write the coefficients of x, y, and z and the constant term in each equation: (a) $2x + 5y + 4z = 4$ (b) $x - 3y - 2z = 5$ (c) $3x + y = 4$	2.3 2.4 2.5	*Coefficients* *Const.* *term* $\begin{array}{cccc} x & y & z & \\ \hline 2 & 5 & 4 & 4 \\ 1 & -3 & -2 & 5 \\ 3 & 1 & 0 & 4 \end{array}$	0.5 Algebraic expressions
Solve (a) $x_1 - 2x_3 = 2$ for x_1 (b) $x_2 + x_3 = -3$ for x_2	2.3 2.4 2.5	(a) $x_1 = 2 + 2x_3$ (b) $x_2 = -3 - x_3$	1.1 Linear equations
Solve the system: $\begin{cases} x - 0.25y = 11{,}750 \\ -0.20x + y = 10{,}000 \end{cases}$	2.3 2.4 2.5	$x = 15{,}000$ $y = 13{,}000$	1.8 Systems of linear equations

MATRICES

A business may collect and store or analyze various types of data as a regular part of its record-keeping procedures. The data may be presented in tabular form. For example, a building contractor who builds four different styles of houses may catalog the number of units of certain materials needed to build each style in a table like Table 2.1.

Table 2.1

	House Style			
Required Units	*Ranch*	*Colonial*	*Split Level*	*Cape Cod*
Wood	20	25	22	18
Siding	27	40	31	25
Roofing	16	16	19	16

If we write the numbers from Table 2.1 in the rectangular array

$$A = \begin{bmatrix} 20 & 25 & 22 & 18 \\ 27 & 40 & 31 & 25 \\ 16 & 16 & 19 & 16 \end{bmatrix},$$

we say A is a **matrix** (plural: matrices) representing Table 2.1. In addition to storing data in a matrix, we can analyze data and make business decisions by defining the operations of addition, subtraction, and multiplication for matrices. Matrices are also useful for analyzing many types of problems and for making comparisons of data. Matrices can be used to solve systems of linear equations, as we will see in this chapter; and they are useful in linear programming, which is discussed in Chapter 3. Matrix operations, discussed in this chapter, are the basis for computer spreadsheets, which are used extensively in education, research, and business.

2.1 MATRICES

Objectives

❏ To organize and interpret data
   stored in matrices
❏ To add and subtract matrices

APPLICATION PREVIEW

Their organizational format makes matrices useful for solving an assort-
ment of problems, including the following. Suppose the purchase prices
and delivery charges (in dollars per unit) for wood, siding, and roofing
used in construction are given by Table 2.2. If the supplier decides to raise
the purchase prices by $.60 per unit and the delivery charges by $.05 per
unit, the matrix that describes the new unit charges will be the sum of the
charges matrix, formed from Table 2.2, and the matrix representing the
increases.

As we noted in the introduction to this chapter, matrices can be used
to store data and to perform operations with the data. The matrix

$$C = \begin{bmatrix} 6 & 4 & 2 \\ 1 & 1 & 0.5 \end{bmatrix}$$

represents the original unit charges; it contains the heart of the information
from Table 2.2, without the labels. The matrix representing the increases
is given by

$$H = \begin{bmatrix} 0.60 & 0.60 & 0.60 \\ 0.05 & 0.05 & 0.05 \end{bmatrix}.$$

We will discuss how to find the sum of these two and other pairs of
matrices and how to find the difference of two matrices.

Table 2.2

	Wood	Siding	Roofing
Purchase	6	4	2
Delivery	1	1	0.5

In general, we define a **matrix** as any rectangular array of numbers. In
matrix A (representing Table 2.1), the *rows* correspond to the types of building materials and the *columns* correspond to the types of houses. The rows of
a matrix are numbered from the top to the bottom and the columns are
numbered from left to right.

	Column 1	Column 2	Column 3	Column 4
Row 1	20	25	22	18
Row 2	27	40	31	25
Row 3	16	16	19	16

Matrices are classified by the number of rows and columns they have.
The matrix A has three rows and four columns, so we say this is a 3×4 (read
"three by four") matrix.

The matrix

$$A = \begin{bmatrix} a_{11} & a_{12} & a_{13} & \cdots & a_{1n} \\ a_{21} & a_{22} & a_{23} & \cdots & a_{2n} \\ & \vdots & & & \vdots \\ a_{m1} & a_{m2} & a_{m3} & \cdots & a_{mn} \end{bmatrix}$$

has m rows and n columns, so it is an $m \times n$ matrix. When we designate A as an $m \times n$ matrix, we are indicating the size of the matrix. Two matrices are said to have the same **order** (be the same size) if they have the same number of rows and the same number of columns.

The numbers in a matrix are called its **entries** or **elements.** Note that the subscripts on an entry in matrix A correspond respectively to the row and column in which the entry is located. Thus a_{23} represents the entry in the second row and the third column, and we refer to it as the "two-three entry." In matrix B below, the entry denoted by b_{23} is 1.

Some matrices take special names because of their size. If the number of rows equals the number of columns, we say the matrix is a **square matrix.** Matrix B is an example of a 3×3 square matrix.

$$B = \begin{bmatrix} 4 & 3 & 0 \\ 0 & 0 & 1 \\ -4 & 0 & 0 \end{bmatrix}$$

EXAMPLE 1

In a study of labor assignment rules for workers who are not perfectly interchangeable, Paul M. Bobrowski and Paul Sungchil Park used Figure 2.1 to describe the tardiness in completing a job (in hours/job) at six different locations for the job priorities as follows.

(a) smallest ratio of time available until the due date to processing time remaining (the critical ratio, CR)
(b) first come, first served (FCFS)
(c) shortest processing time (SPT)
(d) earliest due date (EDD)

A partial table of values from the figure is shown below. Use the figure to complete the table and to create a 4×6 matrix describing the information.

Figure 2.1

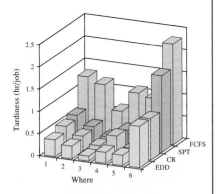

SOURCE: Adapted from Bobrowki, P. M., and P. S. Park, "An evaluation of labor assignment rules," *Journal of Operations Management*, Vol. 11, Sept., 1993

	1	*2*	*3*	*4*	*5*	*6*
EDD			0.1			
CR	0.5	0.4			0.2	1.0
SPT			0.2	0.4	0.4	1.5
FCFS		1.1	0.5	1.1	1.0	

Solution

Reading the data from the figure (to the nearest tenth) and completing the table gives

	1	2	3	4	5	6
EDD	0.4	0.3	0.1	0.2	0.2	1.0
CR	0.5	0.4	0.2	0.3	0.2	1.0
SPT	0.5	0.4	0.2	0.4	0.4	1.5
FCFS	1.2	1.1	0.5	1.1	1.0	2.3

Writing this data in a matrix gives

$$T = \begin{bmatrix} 0.4 & 0.3 & 0.1 & 0.2 & 0.2 & 1.0 \\ 0.5 & 0.4 & 0.2 & 0.3 & 0.2 & 1.0 \\ 0.5 & 0.4 & 0.2 & 0.4 & 0.4 & 1.5 \\ 1.2 & 1.1 & 0.5 & 1.1 & 1.0 & 2.3 \end{bmatrix}.$$

A matrix with one row, such as $\begin{bmatrix} 9 & 5 \end{bmatrix}$ or $\begin{bmatrix} 3 & 2 & 1 & 6 \end{bmatrix}$, is called a **row matrix,** and a matrix with one column, such as

$$\begin{bmatrix} 1 \\ 3 \\ 5 \end{bmatrix},$$

is called a **column matrix.** Row and column matrices are also called **vectors.**

Any matrix in which *every* entry is zero is called a **zero matrix;** examples are

$$\begin{bmatrix} 0 & 0 \\ 0 & 0 \end{bmatrix}, \quad \begin{bmatrix} 0 & 0 \\ 0 & 0 \\ 0 & 0 \end{bmatrix}, \quad \begin{bmatrix} 0 & 0 & 0 & 0 \\ 0 & 0 & 0 & 0 \\ 0 & 0 & 0 & 0 \end{bmatrix}.$$

We define two matrices to be **equal** if they are the same order and if each entry in one equals the corresponding entry in the other.

EXAMPLE 2

Are $A = \begin{bmatrix} 1 & 3 & 2 \\ 4 & 1 & 2 \end{bmatrix}$ and $B = \begin{bmatrix} 1 & 4 \\ 2 & 1 \\ 3 & 0 \end{bmatrix}$ the same order matrices?

Solution

No. A is a 2×3 matrix and B is a 3×2 matrix.

CHECKPOINT

1. (a) Do matrices A and B have the same order?
   (b) Does matrix A equal matrix B?

$$A = \begin{bmatrix} 1 & 2 & 3 \\ 4 & 5 & 6 \\ 7 & 8 & 9 \end{bmatrix} \quad \text{and} \quad B = \begin{bmatrix} 1 & 4 & 7 \\ 2 & 5 & 8 \\ 3 & 6 & 9 \end{bmatrix}$$

EXAMPLE 3
(a) Which element of

$$A = \begin{bmatrix} 1 & 2 & 3 \\ 4 & 5 & 6 \\ 7 & 8 & 9 \end{bmatrix}$$

is represented by a_{32}?
(b) Is A a square matrix?

Solution
(a) a_{32} represents the element in row 3 and column 2; that is, 8.
(b) Yes, it is a 3×3 (square) matrix.

If two matrices have the same number of rows and columns, we can add the matrices by adding their corresponding entries. More formally,

Sum of Two Matrices If matrix A and matrix B are the same order and have elements a_{ij} and b_{ij}, respectively, then their **sum** $A + B$ is a matrix C whose elements are $c_{ij} = a_{ij} + b_{ij}$ for all i and j.

That is,

$$\begin{bmatrix} a_{11} & a_{12} & \cdots & a_{1n} \\ a_{21} & a_{22} & \cdots & a_{2n} \\ \vdots & & & \vdots \\ a_{m1} & a_{m2} & \cdots & a_{mn} \end{bmatrix} + \begin{bmatrix} b_{11} & b_{12} & \cdots & b_{1n} \\ b_{21} & b_{22} & \cdots & b_{2n} \\ \vdots & & & \vdots \\ b_{m1} & b_{m2} & \cdots & b_{mn} \end{bmatrix}$$

$$= \begin{bmatrix} a_{11} + b_{11} & a_{12} + b_{12} & \cdots & a_{1n} + b_{1n} \\ a_{21} + b_{21} & a_{22} + b_{22} & \cdots & a_{2n} + b_{2n} \\ \vdots & & & \vdots \\ a_{m1} + b_{m1} & a_{m2} + b_{m2} & \cdots & a_{mn} + b_{mn} \end{bmatrix}.$$

EXAMPLE 4

Find the sum of A and B if

$$A = \begin{bmatrix} 1 & 2 & 3 \\ 4 & -1 & -2 \end{bmatrix} \quad \text{and} \quad B = \begin{bmatrix} -1 & 2 & -3 \\ -2 & 0 & 1 \end{bmatrix}.$$

Solution

$$A + B = \begin{bmatrix} 1 + (-1) & 2 + 2 & 3 + (-3) \\ 4 + (-2) & -1 + 0 & -2 + 1 \end{bmatrix} = \begin{bmatrix} 0 & 4 & 0 \\ 2 & -1 & -1 \end{bmatrix}.$$

The matrix $-B$ is called the **negative** of the matrix B, and each element of $-B$ is the negative of the corresponding element of B. For example, if

$$B = \begin{bmatrix} -1 & 2 & -3 \\ -2 & 0 & 1 \end{bmatrix}, \quad \text{then} \quad -B = \begin{bmatrix} 1 & -2 & 3 \\ 2 & 0 & -1 \end{bmatrix}.$$

Using the negative, we can define the difference $A - B$ (when A and B have the same order) by $A - B = A + (-B)$, or by subtracting corresponding elements.

EXAMPLE 5

For matrices A and B of Example 4, find $A - B$.

Solution

$A - B$ can be found by subtracting corresponding elements.

$$A - B = \begin{bmatrix} 1 - (-1) & 2 - 2 & 3 - (-3) \\ 4 - (-2) & -1 - 0 & -2 - 1 \end{bmatrix} = \begin{bmatrix} 2 & 0 & 6 \\ 6 & -1 & -3 \end{bmatrix}$$

CHECKPOINT

2. If $A = \begin{bmatrix} 1 & 2 & 0 \\ 3 & 1 & 2 \end{bmatrix}$, $B = \begin{bmatrix} 3 & 6 & 1 \\ 0 & 2 & 1 \end{bmatrix}$, and $C = \begin{bmatrix} 1 & 0 & 2 \\ 0 & 1 & 1 \end{bmatrix}$, find $A + B - C$.

3. (a) What matrix D must be added to matrix A so that their sum is matrix Z?

$$A = \begin{bmatrix} 1 & -2 & 3 \\ -5 & 1 & 2 \end{bmatrix} \quad \text{and} \quad Z = \begin{bmatrix} 0 & 0 & 0 \\ 0 & 0 & 0 \end{bmatrix}$$

(b) Does $D = -A$?

Graphing Utilities

Graphics calculators and computers have the capability to perform a number of operations on matrices. Computer spreadsheets are also very useful in performing operations with matrices.

EXAMPLE 6

Use a graphics calculator or computer to compute $A + B$ and $A - B$ for

$$A = \begin{bmatrix} -1 & 3 & 2 \\ 1 & 0 & 2 \\ 0 & 1 & -1 \\ -4 & 2 & 3 \end{bmatrix} \quad \text{and} \quad B = \begin{bmatrix} -3 & 2 & 1 \\ 0 & -2 & 2 \\ -1 & 6 & 8 \\ 3 & 2 & 4 \end{bmatrix}.$$

Solution

On most graphics calculators, you can use MATRIX and EDIT to define a matrix. To define matrix A, we select MATRIX and EDIT, select the name, A, set the dimensions of A as 4×3 (the number of rows and columns), and enter the elements of A. Notice that the address (row and column) of each entry is used when entering it. After quitting the edit mode, using the key for matrix A displays the elements of A as they appear above. We repeat the process to define matrix B. We find the sum and difference of the two matrices by using the regular sum and difference keys.

$$A + B = \begin{bmatrix} -4 & 5 & 3 \\ 1 & -2 & 4 \\ -1 & 7 & 7 \\ -1 & 4 & 7 \end{bmatrix}, \quad A - B = \begin{bmatrix} 2 & 1 & 1 \\ 1 & 2 & 0 \\ 1 & -5 & -9 \\ -7 & 0 & -1 \end{bmatrix}$$

Note that if matrix A and matrix B have different orders, they cannot be added or subtracted, and a graphics calculator will give an error statement.

We now return to the Application Preview.

EXAMPLE 7

Suppose the purchase prices and delivery charges (in dollars per unit) for wood, siding, and roofing used in construction are given by Table 2.3. If the supplier decides to raise the purchase prices by $.60 per unit and the delivery charges by $.05 per unit, write the matrix that describes the new unit charges.

Solution

The matrix representing the original unit charges is

$$C = \begin{bmatrix} 6 & 4 & 2 \\ 1 & 1 & 0.5 \end{bmatrix},$$

and the matrix representing the increases is given by

$$H = \begin{bmatrix} 0.60 & 0.60 & 0.60 \\ 0.05 & 0.05 & 0.05 \end{bmatrix}.$$

Table 2.3

	Wood	Siding	Roofing
Purchase	6	4	2
Delivery	1	1	0.5

The new unit charge for each item is its former charge plus the increase, so the new unit charge matrix is given by

$$C + H = \begin{bmatrix} 6 + 0.60 & 4 + 0.60 & 2 & + 0.60 \\ 1 + 0.05 & 1 + 0.05 & 0.5 + 0.05 \end{bmatrix} = \begin{bmatrix} 6.60 & 4.60 & 2.60 \\ 1.05 & 1.05 & 0.55 \end{bmatrix}.$$

Note that the sum of C and H in Example 7 could be found by adding the matrices in either order. That is, $C + H = H + C$. This is known as the **commutative law of addition** for matrices. We will see in the next section that multiplication of matrices is *not* commutative.

EXAMPLE 8

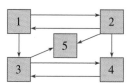

Suppose that in a government agency, paperwork is constantly flowing among offices according to the diagram.

(a) Construct matrix A with elements

$$a_{ij} = \begin{cases} 1 & \text{if paperwork flows directly from } i \text{ to } j. \\ 0 & \text{if paperwork does not directly flow from } i \text{ to } j. \end{cases}$$

(b) Construct matrix B with elements

$$b_{ij} = \begin{cases} 1 & \text{if paperwork can flow from } i \text{ to } j \text{ through at most 1} \\ & \text{intermediary, with } i \neq j. \\ 0 & \text{if this is not true.} \end{cases}$$

(c) The person in office i has the most power to influence others if the sum of the elements of row i in the matrix $A + B$ is the largest. What is the office number of this person?

Solution

$$\text{(a) } A = \begin{bmatrix} 0 & 1 & 1 & 0 & 0 \\ 1 & 0 & 0 & 1 & 1 \\ 0 & 0 & 0 & 1 & 1 \\ 0 & 0 & 1 & 0 & 0 \\ 0 & 0 & 0 & 0 & 0 \end{bmatrix} \qquad \text{(b) } B = \begin{bmatrix} 0 & 1 & 1 & 1 & 1 \\ 1 & 0 & 1 & 1 & 1 \\ 0 & 0 & 0 & 1 & 1 \\ 0 & 0 & 1 & 0 & 1 \\ 0 & 0 & 0 & 0 & 0 \end{bmatrix}$$

$$\text{(c) } A + B = \begin{bmatrix} 0 & 2 & 2 & 1 & 1 \\ 2 & 0 & 1 & 2 & 2 \\ 0 & 0 & 0 & 2 & 2 \\ 0 & 0 & 2 & 0 & 1 \\ 0 & 0 & 0 & 0 & 0 \end{bmatrix}$$

Because the sum of the elements in row 2 is the largest (7), the person in office 2 has the most power to influence others. Because row 5 has the smallest sum (0), the person in office 5 has the least power.

Graphics calculators, computer algebra systems, and computer spreadsheets have the capability to perform a number of complex operations on matrices. Many applications in business and industry call for repeated manipulations or operations with very large tables (matrices). Once the data is entered in a graphics calculator, computer program, or spreadsheet, it can be stored and then operated on or combined with other matrices.

The ability to store data in arrays (matrices), and the ability to add and subtract (and to perform other operations) with matrices is at the heart of the development of computer spreadsheets such as *Lotus 1-2-3, Quattro Pro,* and *Excel.* Table 2.4 shows the financial results (in millions of dollars) from the *1993 Annual Report of the Horsham Corporation* in a spreadsheet.

Table 2.4

A	B	C	D	E
1		1993	1992	1991
2		Financial	Results	($million)
3	Earnings	41.5	34.1	18.4
4	from Barrick Dilution gains	2.4	0.9	16.7
5	Earnings from Clark	2.8	(3.0)	23.0
6	R&M Holdings parent	(11.6)	0	0
7	Horsham parent	5.0	4.2	3.3
8	Net earnings	40.1	36.2	61.4
9	Operating cash flow	62.3	31.8	91.6
10	Capital expenditures	100.9	81.1	90.7

One of the easiest and most frequently performed operations on a computer spreadsheet is adding columns, which is accomplished in the same manner as adding two (or more) $n \times 1$ column matrices. Similarly, adding all the rows of a spreadsheet is performed in the same manner as adding 2 (or more) $1 \times n$ row matrices.

EXAMPLE 9

Use the financial results in Table 2.4 to answer the following.
(a) Write the earnings in each of 1993 and 1992 as matrices, and find the gain (or loss) from 1992 to 1993 by using a matrix operation.
(b) Determine which row matrix is the sum of the row matrices above it on the spreadsheet.

Solution

(a) Writing the gains or losses (indicated by inclusion in parentheses in the annual report) for 1993 as column matrix A and for 1992 as B, we have:

$$A = \begin{bmatrix} 41.5 \\ 2.4 \\ 2.8 \\ -11.6 \\ 5.0 \\ 40.1 \\ 62.3 \\ 100.9 \end{bmatrix} \quad \text{and} \quad B = \begin{bmatrix} 34.1 \\ 0.9 \\ -3.0 \\ 0 \\ 4.2 \\ 36.2 \\ 31.8 \\ 81.1 \end{bmatrix}.$$

The gain (or loss) in 1993 is given by $A - B$.

$$A - B = \begin{bmatrix} 7.4 \\ 1.5 \\ 5.8 \\ -11.6 \\ 0.8 \\ 3.9 \\ 30.5 \\ 19.8 \end{bmatrix}$$

This result indicates that gains were made in all categories but one.

(b) The row matrix that gives net earnings (row 8 of the spreadsheet) is the sum of the row matrices above it on the spreadsheet.

CHECKPOINT SOLUTIONS

1. (a) Yes, A and B have the same order (size).

   (b) No. They have the same elements, but they are not equal because their corresponding elements are not equal. The rows and columns of A and B are interchanged. They are **transposes** of each other.

2. $\begin{bmatrix} 3 & 8 & -1 \\ 3 & 2 & 2 \end{bmatrix}$ 3. (a) $\begin{bmatrix} -1 & 2 & -3 \\ 5 & -1 & -2 \end{bmatrix}$ (b) Yes

EXERCISE 2.1

Use the following matrices for problems 1–10.

$$A = \begin{bmatrix} 1 & 0 & 2 \\ 3 & 2 & 1 \\ 4 & 0 & 3 \end{bmatrix} \quad B = \begin{bmatrix} 1 & 1 & 3 & 0 \\ 4 & 2 & 1 & 1 \\ 3 & 2 & 0 & 1 \end{bmatrix}$$

$$C = \begin{bmatrix} 5 & 3 \\ 1 & 2 \end{bmatrix} \quad D = \begin{bmatrix} 4 & 2 \\ 3 & 5 \end{bmatrix} \quad E = \begin{bmatrix} 1 & 0 & 4 \\ 5 & 1 & 0 \end{bmatrix}$$

$$F = \begin{bmatrix} 1 & 2 & 3 \\ -1 & 0 & 1 \\ 2 & -3 & -4 \end{bmatrix} \quad Z = \begin{bmatrix} 0 & 0 & 0 \\ 0 & 0 & 0 \\ 0 & 0 & 0 \end{bmatrix}$$

1. How many rows does matrix B have?
2. What is the order of matrix E?
3. Write the negative of matrix F.
4. Write a zero matrix that is the same order as D.
5. Which of the matrices $A, B, C, D, E, F,$ and Z are square?
6. Which pairs of matrices, if any, are equal?
7. Which pairs of matrices, if any, are the same order?
8. Write the matrix that is the negative of matrix F.
9. What is the element a_{23}?
10. What is element b_{24}?

Use the matrices M and N in problems 11–14.

$$M = \begin{bmatrix} 1 & 0 & 2 \\ 3 & 2 & 1 \\ 4 & 0 & 3 \end{bmatrix} \quad \text{and} \quad N = \begin{bmatrix} 1 & 1 & 3 & 0 \\ 4 & 2 & 1 & 1 \\ 3 & 2 & 0 & 1 \end{bmatrix}$$

11. Can the sum of M and N be found?
12. What is the negative of N?
13. What is the sum of matrix M and its negative?
14. If matrix M has element $m_{3j} = 0$, what is j?

Use the matrices below for problems 15–26 to compute the indicated sums and differences, if possible.

$$A = \begin{bmatrix} 1 & 0 & 2 \\ 3 & 2 & 1 \\ 4 & 0 & 3 \end{bmatrix} \quad B = \begin{bmatrix} 1 & 1 & 3 & 0 \\ 4 & 2 & 1 & 1 \\ 3 & 2 & 0 & 1 \end{bmatrix}$$

$$C = \begin{bmatrix} 5 & 3 \\ 1 & 2 \end{bmatrix} \quad D = \begin{bmatrix} 4 & 2 \\ 3 & 5 \end{bmatrix}$$

$$E = \begin{bmatrix} 1 & 0 & 4 \\ 5 & 1 & 0 \end{bmatrix} \quad F = \begin{bmatrix} 1 & 2 & 3 \\ -1 & 0 & 1 \\ 2 & -3 & -4 \end{bmatrix}$$

$$Z = \begin{bmatrix} 0 & 0 & 0 \\ 0 & 0 & 0 \\ 0 & 0 & 0 \end{bmatrix}$$

15. $C + D$
16. $A + F$
17. $A - F$
18. $C - D$
19. $A + B$
20. $D - C$
21. $D - E$
22. $B + F$
23. $A + Z$
24. $A - Z$
25. $Z + C$
26. $Z + F$

In problems 27–30 find $x, y, z,$ and w.

27. $\begin{bmatrix} x & 1 & 0 \\ 0 & y & z \\ w & 2 & 1 \end{bmatrix} = \begin{bmatrix} 3 & 1 & 0 \\ 0 & 2 & 3 \\ 4 & 2 & 1 \end{bmatrix}$

28. $\begin{bmatrix} 0 & x & 1 \\ 3 & y & y \\ z & 0 & 2 \end{bmatrix} = \begin{bmatrix} 0 & 4 & 1 \\ 3 & 1 & y \\ 1 & 0 & w \end{bmatrix}$

29. $\begin{bmatrix} x & 3 & (2x - 1) \\ y & 4 & 4y \end{bmatrix} = \begin{bmatrix} 2x - 4 & z & 7 \\ 1 & (w + 1) & (3y + 1) \end{bmatrix}$

30. $\begin{bmatrix} x & y & (x + 3) \\ z & 4 & 4y \end{bmatrix} = \begin{bmatrix} (2x - 1) & -1 & w \\ x & (5 + y) & -4 \end{bmatrix}$

31. Solve for $x, y,$ and z if

$$\begin{bmatrix} x & y \\ y & z \end{bmatrix} + \begin{bmatrix} 2x & -y \\ 3y & -4z \end{bmatrix} = \begin{bmatrix} 6 & 0 \\ 8 & 9 \end{bmatrix}.$$

32. Find $x, y, z,$ and w if

$$\begin{bmatrix} x & 4 \\ 4y & w \end{bmatrix} - \begin{bmatrix} 4x & 2z \\ -3 & -2w \end{bmatrix} = \begin{bmatrix} 12 & 8 \\ y & 6 \end{bmatrix}.$$

APPLICATIONS

33. *Sales* Let matrix A represent the sales (in thousands of dollars) for the Walbash Company in 1994 in various cities and let matrix B represent the sales (in thousands of dollars) for the same company in 1995 in the same cities.

$$A = \begin{array}{c} \\ \\ \end{array} \begin{array}{ccc} \text{Chicago} & \text{Atlanta} & \text{Memphis} \\ \begin{bmatrix} 450 & 280 & 850 \\ 400 & 350 & 150 \end{bmatrix} \end{array} \begin{array}{l} \text{Wholesale} \\ \text{Retail} \end{array}$$

$$B = \begin{array}{c} \\ \\ \end{array} \begin{array}{ccc} \text{Chicago} & \text{Atlanta} & \text{Memphis} \\ \begin{bmatrix} 375 & 300 & 710 \\ 410 & 300 & 200 \end{bmatrix} \end{array} \begin{array}{l} \text{Wholesale} \\ \text{Retail} \end{array}$$

(a) Write the matrix that represents the total sales by type and city for both years.
(b) Write the matrix that represents the change in sales by type and city for both years.

34. *Opinion polls* A poll of 3320 people revealed that of the respondents that are registered Democrats, 843 approved the president's job performance, 426 did not, and 751 had no opinion. Of the registered Republicans, 257 approved of the president's job performance, 451 did not, and 92 had no opinion. Of those registered as Independents, 135 approved, 127 did not approve, and 38 had no opinion. Of the remaining respondents, who were not registered, 92 approved, 64 did not approve, and 44 had no opinion. Represent this data in a 3×4 matrix.

35. *Management* Management is attempting to identify the most active person in labor's efforts to unionize. The following diagram shows how influence flows from one employee to another among the four most active employees.

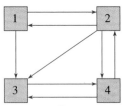

(a) Construct matrix A with elements
$$a_{ij} = \begin{cases} 1 & \text{if } i \text{ influences} \\ & j \text{ directly.} \\ 0 & \text{otherwise.} \end{cases}$$

(b) Construct matrix B with elements
$$b_{ij} = \begin{cases} 1 & \text{if } i \text{ influences } j \text{ through at most 1} \\ & \text{person, with } i \neq j. \\ 0 & \text{otherwise.} \end{cases}$$

(c) The person i is most active in influencing others if the sum of the elements in row i of the matrix $A + B$ is the largest. Who is the most active person?

36. *Ranking* In order to rank the five members of its chess team for play against another school, the coach draws the following diagram. An arrow from 1 to 2 means player 1 has defeated player 2.

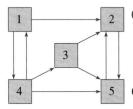

(a) Construct matrix A with elements
$$a_{ij} = \begin{cases} 1 & \text{if } i \text{ defeated } j. \\ 0 & \text{otherwise.} \end{cases}$$

(b) Construct matrix B with elements
$$b_{ij} = \begin{cases} 1 & \text{if } i \text{ defeated someone who defeated} \\ & j \text{ and } i \neq j. \\ 0 & \text{otherwise.} \end{cases}$$

(c) The player i is the top-ranked player if the sum of row i in the matrix $A + B$ is the largest. What is the number of this player?

37. *Nutrition* In a computer simulation, three groups of baby laboratory animals were used. Group I had an enriched diet, group II (control) had the regular diet, and group III had a deficient diet. At the beginning of the simulation the animals were weighed and measured according to their group, with group I averaging 140 g with a length of 5.5 cm. Group II and group III weighed 151 g and 141 g, with lengths of 5.7 cm and 5.5 cm, respectively.

(a) Make a 3 × 2 matrix that displays this information.

(b) At the end of two weeks, the same measurements were made, with the following results: 12.5 cm,

250 g; 11.8 cm, 215 g; 9.8 cm, 190 g. Make a matrix that displays this information. Make sure the matrix is the same size as your answer in part (a).

38. *Nutrition* Calculate the changes in weight and length from problems 37 (a) and 37 (b) by using matrix subtraction.

39. *Life expectancy* The table below gives the years of life expected at birth for blacks and whites born in the U.S. in the years 1920, 1940, 1960, and 1980.

	Whites		Blacks	
	Males	*Females*	*Males*	*Females*
1920	54.4	55.6	45.5	45.2
1940	62.1	66.6	51.5	54.9
1960	67.4	74.1	61.1	67.4
1980	70.7	78.1	65.3	73.6

SOURCE: National Center for Health Statistics

Make a matrix A containing the information for whites and a matrix B for blacks. Use these to find how many more years whites in each category are expected to live than blacks.

40. *Imports* (a) From the data in the table below make a matrix A that gives the value (in millions of dollars) of imports for the various country groupings in the years 1983–1985.

(b) Make a matrix B that gives the value of exports (in millions of dollars) for the same groupings in the same years.

(c) Find the trade balance for each country grouping in each year by finding $A - B$.

	Imports		
Countries	*83*	*84*	*85*
Developed	122,822	135,884	134,018
Developing	72,342	74,421	72,673
Communist	5,085	7,214	7,091
Other	289	369	365

	Exports		
	83	*84*	*85*
Developed	152,117	200,714	223,314
Developing	102,266	119,790	116,161
Communist	3,604	5,221	5,801
Other	1	1	0

SOURCE: U.S. Bureau of Census

Production and inventories Operating from two plants, the Book Equipment Company (BEC) produces bookcases and filing cabinets. Matrix A summarizes its production for a week, with row 1 representing the number of bookcases and row 2 representing the number of filing cabinets. Matrix B gives the production for the second week, and matrix C that of the third and fourth weeks combined.

$$A = \begin{bmatrix} 50 & 30 \\ 36 & 44 \end{bmatrix} \qquad B = \begin{bmatrix} 30 & 45 \\ 22 & 62 \end{bmatrix} \qquad C = \begin{bmatrix} 96 & 52 \\ 81 & 37 \end{bmatrix}$$

If column 1 in each matrix represents production from plant 1 and column 2 represents production from plant 2, answer the questions in problems 41–42.

41. (a) How many bookcases were produced in the first week?

   (b) How many filing cabinets were produced in the first week?

   (c) How many units were produced by plant 1 in the first week?

   (d) How many units were produced by plant 2 in the first week?

   (e) If production during the second week is given by matrix B, write a matrix that describes production for the first two weeks.

   (f) If production over the next two-week period is given by matrix C, describe production for the four-week period.

42. (a) If matrix D

$$D = \begin{bmatrix} 40 & 26 \\ 29 & 42 \end{bmatrix}$$

   describes the shipments made during the first week, write the matrix that describes the units added to the plants' inventories.

   (b) If D also describes the shipments during the second week, describe the change in inventory. What happened at plant 1?

43. ***Management*** The figure that follows depicts the mean (average) flow time for a job when the critical ratio rule is used to dispatch workers at several machines required to complete the job. In the figure, WHN 1, WHN 2, and WHN 3 represent three different rules for determining when a worker should be transferred to another machine and WHERE represents six rules for determining the machine to which the worker is transferred. Construct a 3×6 matrix A to represent this data, with entries rounded to the nearest integer.

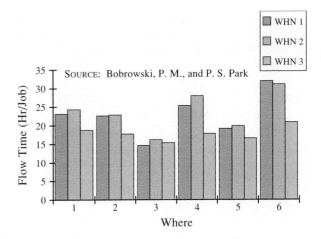

44. ***Management*** The figure below graphically depicts the mean flow time under the same conditions as those discussed in problem 43 except that it considers that the work force is not a homogeneous entity at all times. Construct a 3×6 matrix B to represent this data, with entries rounded to the nearest integer.

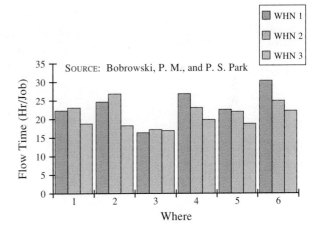

45. ***Management*** Under which of the six rules is flow time the same for matrix A (problem 43) and matrix B (problem 44)?

46. ***Management*** In an evaluation of labor assignment rules when workers are not perfectly interchangeable, Paul M. Bobrowski and Paul Sungchil Park created a dynamic job shop with 9 work centers and 9 workers, each numbered 1–9. The efficiency of each worker is specified in the labor efficiency matrix given below, which represents the degree of worker cross-training (Source: Bobrowski, P. M., and P. S. Park, "An evaluation of labor

assignment rules," *Journal of Operations Management,* Vol. 11, Sept. 1993).

(a) For what work center(s) is worker 7 least efficient?

(b) For what work center is worker 1 most efficient?

Work Centers

$$\begin{bmatrix} 1.00 & .95 & .95 & .95 & .95 & .85 & .85 & .85 & .85 \\ .85 & 1.00 & .95 & .95 & .95 & .95 & .85 & .85 & .85 \\ .85 & .85 & 1.00 & .95 & .95 & .95 & .95 & .85 & .85 \\ .85 & .85 & .85 & 1.00 & .95 & .95 & .95 & .95 & .85 \\ .85 & .85 & .85 & .85 & 1.00 & .95 & .95 & .95 & .95 \\ .95 & .85 & .85 & .85 & .85 & 1.00 & .95 & .95 & .95 \\ .95 & .95 & .85 & .85 & .85 & .85 & 1.00 & .95 & .95 \\ .95 & .95 & .95 & .85 & .85 & .85 & .85 & 1.00 & .95 \\ .95 & .95 & .95 & .95 & .85 & .85 & .85 & .85 & 1.00 \end{bmatrix} \begin{matrix} \\ W \\ o \\ r \\ k \\ e \\ r \\ s \\ \\ \end{matrix}$$

47. Suppose that for the workers and the work centers described in the matrix of problem 46, changes in rules cause efficiency to decrease by .01 at work centers 2 and 5, and to decrease by .02 at work center 7. Write the matrix that describes the new efficiencies.

48. For the data of problem 46, use a computer spreadsheet to find the average efficiency for each worker over the first four work centers. Which worker is the least efficient? Where does this worker perform best?

49. For the data of problem 46 use a computer spreadsheet to find the average efficiency for each work center over the first five workers. Which work center is the most efficient? Which work center should be studied for improvement?

2.2 MULTIPLICATION OF MATRICES

Objectives

❏ To multiply a matrix by a scalar (real number)

❏ To multiply two matrices

APPLICATION PREVIEW

Pentico Industries is a manufacturing company that has two divisions, located at Clarion and Brooks, each of which needs different amounts (units) of production materials as described by the following table.

	Steel	Wood	Plastic
Clarion	20	30	8
Brooks	22	25	15

These raw materials are supplied by Western Supply and Coastal Supply, with prices given in dollars per unit in the following table.

	Western	Coastal
Steel	300	290
Wood	100	90
Plastic	145	180

If one supplier must be chosen to provide all materials for either or both divisions of Pentico, company officials can decide which supplier to choose by constructing matrices from these tables and using **matrix multiplication.**

In this section we will discuss **scalar multiplication** of a matrix and finding the **product of two matrices.**

Consider the matrix

$$A = \begin{bmatrix} 3 & 2 \\ 1 & 1 \\ 2 & 0 \end{bmatrix}.$$

Since $2A$ is $A + A$, we see that

$$2A = \begin{bmatrix} 3 & 2 \\ 1 & 1 \\ 2 & 0 \end{bmatrix} + \begin{bmatrix} 3 & 2 \\ 1 & 1 \\ 2 & 0 \end{bmatrix} = \begin{bmatrix} 6 & 4 \\ 2 & 2 \\ 4 & 0 \end{bmatrix}.$$

Note that $2A$ could have been found by multiplying each entry of A by 2. In the same manner,

$$3A = A + A + A = \begin{bmatrix} 3 & 2 \\ 1 & 1 \\ 2 & 0 \end{bmatrix} + \begin{bmatrix} 3 & 2 \\ 1 & 1 \\ 2 & 0 \end{bmatrix} + \begin{bmatrix} 3 & 2 \\ 1 & 1 \\ 2 & 0 \end{bmatrix}$$

$$= \begin{bmatrix} 9 & 6 \\ 3 & 3 \\ 6 & 0 \end{bmatrix} = \begin{bmatrix} 3(3) & 3(2) \\ 3(1) & 3(1) \\ 3(2) & 3(0) \end{bmatrix}.$$

We can define **scalar multiplication** as follows.

Scalar Multiplication   Multiplying a matrix by a real number (called a *scalar*) results in a matrix in which each entry of the original matrix is multiplied by the real number.

Thus, if

$$A = \begin{bmatrix} a_{11} & a_{12} \\ a_{21} & a_{22} \end{bmatrix}, \quad \text{then} \quad cA = \begin{bmatrix} ca_{11} & ca_{12} \\ ca_{21} & ca_{22} \end{bmatrix}.$$

EXAMPLE 1

If

$$A = \begin{bmatrix} 4 & 1 & 4 & 0 \\ 2 & -7 & 3 & 6 \\ 0 & 0 & 2 & 5 \end{bmatrix},$$

find $5A$ and $-2A$.

Solution

$$5A = \begin{bmatrix} 5 \cdot 4 & 5 \cdot 1 & 5 \cdot 4 & 5 \cdot 0 \\ 5 \cdot 2 & 5(-7) & 5 \cdot 3 & 5 \cdot 6 \\ 5 \cdot 0 & 5 \cdot 0 & 5 \cdot 2 & 5 \cdot 5 \end{bmatrix} = \begin{bmatrix} 20 & 5 & 20 & 0 \\ 10 & -35 & 15 & 30 \\ 0 & 0 & 10 & 25 \end{bmatrix}$$

$$-2A = \begin{bmatrix} -2 \cdot 4 & -2 \cdot 1 & -2 \cdot 4 & -2 \cdot 0 \\ -2 \cdot 2 & -2(-7) & -2 \cdot 3 & -2 \cdot 6 \\ -2 \cdot 0 & -2 \cdot 0 & -2 \cdot 2 & -2 \cdot 5 \end{bmatrix}$$

$$= \begin{bmatrix} -8 & -2 & -8 & 0 \\ -4 & 14 & -6 & -12 \\ 0 & 0 & -4 & -10 \end{bmatrix}$$

CHECKPOINT

1. (a) Find $3D - D$ if $D = \begin{bmatrix} 1 & 3 & 2 & 10 \\ 20 & 5 & 6 & 3 \end{bmatrix}$.

(b) Does $3D - D = 2D$?

EXAMPLE 2

Suppose the purchase prices and delivery costs (per unit) for wood, siding, and roofing used in construction are given by Table 2.5. Then the table of unit costs may be represented by the matrix

$$C = \begin{bmatrix} 6 & 4 & 2 \\ 1 & 1 & 0.5 \end{bmatrix}.$$

If the supplier announces a 10% increase on both purchase and delivery of these items, find the new unit cost matrix.

Table 2.5

	Wood	Siding	Roofing
Purchase	6	4	2
Delivery	1	1	0.5

Solution

A 10% increase means that the new unit costs are the former costs plus 0.10 times the former cost. That is, the new costs are 1.10 times the former, so the new unit cost matrix is given by

$$1.10C = 1.10\begin{bmatrix} 6 & 4 & 2 \\ 1 & 1 & 0.5 \end{bmatrix}$$

$$= \begin{bmatrix} 6.60 & 4.40 & 2.20 \\ 1.10 & 1.10 & 0.55 \end{bmatrix}.$$

Suppose one store of a firm has 30 washers, 20 dryers, and 10 dishwashers in its inventory. If the value of each washer is $300, each dryer is $250, and each dishwasher is \$350, then the value of this inventory is

$$30 \cdot 300 + 20 \cdot 250 + 10 \cdot 350 = \$17,500.$$

If we write the value of each of the appliances in the *row matrix*

$$A = \begin{bmatrix} 300 & 250 & 350 \end{bmatrix}$$

and the number of each of the appliances in the *column matrix*

$$B = \begin{bmatrix} 30 \\ 20 \\ 10 \end{bmatrix},$$

then the value of the inventory may be represented by

$$AB = \begin{bmatrix} 300 & 250 & 350 \end{bmatrix} \begin{bmatrix} 30 \\ 20 \\ 10 \end{bmatrix}$$
$$= \begin{bmatrix} 300 \cdot 30 + 250 \cdot 20 + 350 \cdot 10 \end{bmatrix}$$
$$= \begin{bmatrix} \$17,500 \end{bmatrix}.$$

This useful way of operating with a row matrix and a column matrix is called the **product** of A times B.

In general, we can multiply the row matrix

$$A = \begin{bmatrix} a_1 & a_2 & a_3 & \cdots & a_n \end{bmatrix}$$

times the column matrix

$$B = \begin{bmatrix} b_1 \\ b_2 \\ b_3 \\ \cdot \\ \cdot \\ \cdot \\ b_n \end{bmatrix}$$

if A and B have the same number of elements. The product A times B is

$$AB = \begin{bmatrix} a_1 & a_2 & a_3 & \cdots & a_n \end{bmatrix} \begin{bmatrix} b_1 \\ b_2 \\ b_3 \\ \cdot \\ \cdot \\ \cdot \\ b_n \end{bmatrix} = \begin{bmatrix} a_1 b_1 + a_2 b_2 + a_3 b_3 + \cdots + a_n b_n \end{bmatrix}.$$

Suppose the firm has a second store, with 40 washers, 25 dryers, and 5 dishwashers. We can use matrix C to represent the inventories of the two stores.

$$C = \begin{array}{cc} \text{Store I} & \text{Store II} \\ \begin{bmatrix} 30 & 40 \\ 20 & 25 \\ 10 & 5 \end{bmatrix} & \begin{array}{l} \text{Washers} \\ \text{Dryers} \\ \text{Dishwashers} \end{array} \end{array}$$

Suppose the values of the appliances are \$300, \$250, and \$350, respectively. We have already seen that the value of the inventory of store I is \$17,500. The value of the inventory of store II is

$$300 \cdot 40 + 250 \cdot 25 + 350 \cdot 5 = \$20,000.$$

Because the value of the inventory of store I can be found by multiplying the row matrix A times the first column of the matrix C, and the value of the inventory of store II can be found by multiplying matrix A times the second column of matrix C, we can write this result as

$$AC = \begin{bmatrix} 300 & 250 & 350 \end{bmatrix} \begin{bmatrix} 30 & 40 \\ 20 & 25 \\ 10 & 5 \end{bmatrix} = \begin{bmatrix} 17{,}500 & 20{,}000 \end{bmatrix}.$$

The matrix we have represented as AC is called the **product** of the row matrix A and the 3×2 matrix C. We should note that matrix A is a 1×3 matrix, matrix C is a 3×2 matrix, and their product is a 1×2 matrix. In general, we can multiply an $m \times n$ matrix times an $n \times p$ matrix and the product will be an $m \times p$ matrix, as follows.

Product of Two Matrices Given an $m \times n$ matrix A and an $n \times p$ matrix B, the **matrix product** AB is an $m \times p$ matrix C, with the ij entry of C given by the formula

$$c_{ij} = a_{i1}b_{1j} + a_{i2}b_{2j} + \cdots + a_{in}b_{nj},$$

which is illustrated in Figure 2.2.

Figure 2.2

$$\begin{bmatrix} a_{11} & a_{12} & \cdots & a_{1n} \\ a_{21} & a_{22} & \cdots & a_{2n} \\ & & \vdots & \\ a_{i1} & a_{i2} & \cdots & a_{in} \\ & & \vdots & \\ a_{m1} & a_{m2} & \cdots & a_{mn} \end{bmatrix} \begin{bmatrix} b_{11} & b_{12} & \cdots & b_{1j} & \cdots & b_{1p} \\ b_{21} & b_{22} & \cdots & b_{2j} & \cdots & b_{2p} \\ & \vdots & & & & \\ b_{n1} & b_{n2} & \cdots & b_{nj} & \cdots & b_{np} \end{bmatrix} = \begin{bmatrix} c_{11} & c_{12} & \cdots & c_{1j} & \cdots & c_{1p} \\ c_{21} & c_{22} & \cdots & c_{2j} & \cdots & c_{2p} \\ & & \vdots & & & \\ c_{i1} & c_{i2} & \cdots & c_{ij} & \cdots & c_{ip} \\ & & \vdots & & & \\ c_{m1} & c_{m2} & \cdots & c_{mj} & \cdots & c_{mp} \end{bmatrix}$$

EXAMPLE 3

Find AB if

$$A = \begin{bmatrix} 3 & 4 \\ 2 & 5 \\ 6 & 10 \end{bmatrix} \quad \text{and} \quad B = \begin{bmatrix} a & b & c & d \\ e & f & g & h \end{bmatrix}.$$

Solution

A is a 3×2 matrix and B is a 2×4 matrix, so the number of columns of A equals the number of rows of B. Thus we can find the product AB, which is a 3×4 matrix, as follows.

$$AB = \begin{bmatrix} 3 & 4 \\ 2 & 5 \\ 6 & 10 \end{bmatrix} \begin{bmatrix} a & b & c & d \\ e & f & g & h \end{bmatrix}$$

$$= \begin{bmatrix} 3a + 4e & 3b + 4f & 3c + 4g & 3d + 4h \\ 2a + 5e & 2b + 5f & 2c + 5g & 2d + 5h \\ 6a + 10e & 6b + 10f & 6c + 10g & 6d + 10h \end{bmatrix}$$

This example shows that if $AB = C$, element c_{32} is found by multiplying each entry of A's *third* row by the corresponding entry of B's *second* column, and then adding these products.

EXAMPLE 4

Find AB and BA if

$$A = \begin{bmatrix} 3 & 2 \\ 1 & 0 \end{bmatrix} \quad \text{and} \quad B = \begin{bmatrix} 1 & 2 \\ 3 & 1 \end{bmatrix}.$$

Solution

Both A and B are 2×2 matrices, so both AB and BA are defined.

$$AB = \begin{bmatrix} 3 & 2 \\ 1 & 0 \end{bmatrix} \begin{bmatrix} 1 & 2 \\ 3 & 1 \end{bmatrix} = \begin{bmatrix} 3 \cdot 1 + 2 \cdot 3 & 3 \cdot 2 + 2 \cdot 1 \\ 1 \cdot 1 + 0 \cdot 3 & 1 \cdot 2 + 0 \cdot 1 \end{bmatrix} = \begin{bmatrix} 9 & 8 \\ 1 & 2 \end{bmatrix}$$

$$BA = \begin{bmatrix} 1 & 2 \\ 3 & 1 \end{bmatrix} \begin{bmatrix} 3 & 2 \\ 1 & 0 \end{bmatrix} = \begin{bmatrix} 1 \cdot 3 + 2 \cdot 1 & 1 \cdot 2 + 2 \cdot 0 \\ 3 \cdot 3 + 1 \cdot 1 & 3 \cdot 2 + 1 \cdot 0 \end{bmatrix} = \begin{bmatrix} 5 & 2 \\ 10 & 6 \end{bmatrix}$$

Note that for these matrices the two products AB and BA are the same size matrices, but they are not equal. That is $BA \neq AB$. Thus we say that *matrix multiplication is not commutative.*

CHECKPOINT

2. What is element c_{23} if $C = AB$ with

$$A = \begin{bmatrix} 1 & 2 & 0 \\ 2 & 1 & 3 \end{bmatrix} \quad \text{and} \quad B = \begin{bmatrix} 1 & 0 & 2 \\ 3 & 1 & 1 \\ -2 & 1 & 0 \end{bmatrix}?$$

3. Find the product AB.

EXAMPLE 5

Pentico Industries must choose a supplier for the raw materials that it uses in its two manufacturing divisions at Clarion and Brooks. Recall from the Application Preview on page 180 that each division uses different unit amounts of steel, wood, and plastic as shown in the table below.

	Steel	Wood	Plastic
Clarion	20	30	8
Brooks	22	25	15

The two supply companies being considered, Western and Coastal, can each supply all of these materials, but at different prices per unit, as described in the following table.

	Western	Coastal
Steel	300	290
Wood	100	90
Plastic	145	180

Use matrix multiplication to decide which supplier should be chosen to supply each of the following divisions.
(a) the Clarion division   (b) the Brooks division   (c) both divisions

Solution

The different amounts of products needed can be placed in the matrix A.

$$A = \begin{array}{c} \\ \end{array}\begin{array}{ccc} \text{Steel} & \text{Wood} & \text{Plastic} \end{array} \\ A = \begin{bmatrix} 20 & 30 & 8 \\ 22 & 25 & 15 \end{bmatrix} \begin{array}{c} \text{Clarion} \\ \text{Brooks} \end{array}$$

The prices charged by the suppliers are given in the matrix B.

$$\begin{array}{cc} \text{Western} & \text{Coastal} \end{array} \\ B = \begin{bmatrix} 300 & 290 \\ 100 & 90 \\ 145 & 180 \end{bmatrix} \begin{array}{c} \text{Steel} \\ \text{Wood} \\ \text{Plastic} \end{array}$$

The price from each supplier for each division is found from the product *AB*.

$$AB = \begin{bmatrix} 20 & 30 & 8 \\ 22 & 25 & 15 \end{bmatrix} \begin{bmatrix} 300 & 290 \\ 100 & 90 \\ 145 & 180 \end{bmatrix}$$

$$= \begin{bmatrix} 20 \cdot 300 + 30 \cdot 100 + 8 \cdot 145 & 20 \cdot 290 + 30 \cdot 90 + 8 \cdot 180 \\ 22 \cdot 300 + 25 \cdot 100 + 15 \cdot 145 & 22 \cdot 290 + 25 \cdot 90 + 15 \cdot 180 \end{bmatrix}$$

$$= \begin{bmatrix} 10,160 & 9940 \\ 11,275 & 11,330 \end{bmatrix}$$

(a) The price for the Clarion division for supplies from Western is in the first row, first column and is \$10,160; the price for Clarion for supplies from Coastal is in the first row, second column and is \$9940. Thus the best price for Clarion is from Coastal.

(b) The price for the Brooks division for supplies from Western is in the second row, first column and is \$11,275; the price for Brooks for supplies from Coastal is in the second row, second column and is \$11,330. Thus the best price for Brooks is from Western.

(c) The sum of the first column gives \$21,435, the price for supplies from Western for both divisions; the sum of the second column gives \$21,270, the price for supplies from Coastal for both divisions. Thus the best price is from Coastal if one supplier is used for both divisions.

An $n \times n$ (square) matrix (where n is any natural number) that has 1s down its diagonal and 0s everywhere else is called an **identity matrix.** The matrix

$$I = \begin{bmatrix} 1 & 0 & 0 \\ 0 & 1 & 0 \\ 0 & 0 & 1 \end{bmatrix}$$

is a 3×3 identity matrix. The matrix I is called an identity matrix because for any 3×3 matrix A, $AI = IA = A$. That is, if I is multiplied by a 3×3 matrix A, the product matrix is A. Note that when one of two square matrices being multiplied is an identity matrix, the product *is commutative.*

EXAMPLE 6

(a) Write the 2×2 identity matrix.

(b) Given $A = \begin{bmatrix} 4 & -7 \\ 13 & 2 \end{bmatrix}$, show $AI = IA = A$.

Solution

(a) We denote the 2×2 identity matrix by

$$I = \begin{bmatrix} 1 & 0 \\ 0 & 1 \end{bmatrix}.$$

(b) $AI = \begin{bmatrix} 4 & -7 \\ 13 & 2 \end{bmatrix} \begin{bmatrix} 1 & 0 \\ 0 & 1 \end{bmatrix} = \begin{bmatrix} 4+0 & 0-7 \\ 13+0 & 0+2 \end{bmatrix} = \begin{bmatrix} 4 & -7 \\ 13 & 2 \end{bmatrix}$

$IA = \begin{bmatrix} 1 & 0 \\ 0 & 1 \end{bmatrix} \begin{bmatrix} 4 & -7 \\ 13 & 2 \end{bmatrix} = \begin{bmatrix} 4 & -7 \\ 13 & 2 \end{bmatrix}$

Therefore, $AI = IA = A$.

EXAMPLE 7

Messages can be encoded through the use of a code and an encoding matrix. For example, given the code

a	b	c	d	e	f	g	h	i	j	k	l
1	2	3	4	5	6	7	8	9	10	11	12

m	n	o	p	q	r	s	t	u	v	w	x
13	14	15	16	17	18	19	20	21	22	23	24

y	z	blank
25	26	27

and the encoding matrix

$$A = \begin{bmatrix} 3 & 5 \\ 4 & 6 \end{bmatrix},$$

we can encode a message by separating it into number pairs (since A is a 2×2 matrix) and then multiplying each pair by A. Use this code and matrix to encode the message "good job."

Solution

The phrase "good job" is written

g	o	o	d	blank	j	o	b
7	15	15	4	27	10	15	2

and is encoded by writing each pair of numbers as a column matrix and multiplying each matrix by matrix A. Performing this multiplication gives the encoded message.

$$\begin{bmatrix} 3 & 5 \\ 4 & 6 \end{bmatrix} \begin{bmatrix} 7 \\ 15 \end{bmatrix} = \begin{bmatrix} 96 \\ 118 \end{bmatrix}$$

$$\begin{bmatrix} 3 & 5 \\ 4 & 6 \end{bmatrix} \begin{bmatrix} 15 \\ 4 \end{bmatrix} = \begin{bmatrix} 65 \\ 84 \end{bmatrix}$$

$$\begin{bmatrix} 3 & 5 \\ 4 & 6 \end{bmatrix} \begin{bmatrix} 27 \\ 10 \end{bmatrix} = \begin{bmatrix} 131 \\ 168 \end{bmatrix}$$

$$\begin{bmatrix} 3 & 5 \\ 4 & 6 \end{bmatrix} \begin{bmatrix} 15 \\ 2 \end{bmatrix} = \begin{bmatrix} 55 \\ 72 \end{bmatrix}$$

Thus the code sent is 96, 118, 65, 84, 131, 168, 55, 72.

We will discuss methods of decoding messages in Section 2.4.

CHECKPOINT

4. (a) Compute AB if $A = \begin{bmatrix} 2 & 4 \\ 3 & 6 \end{bmatrix}$ and $B = \begin{bmatrix} 2 & -6 \\ -1 & 3 \end{bmatrix}$.

   (b) Can the product of two matrices be a zero matrix even if neither matrix is a zero matrix?

EXAMPLE 8

Use a graphics calculator and the matrices A, B, and C below to perform the operations (a) $1.7A$, (b) $3.6B + 2.3C$, (c) AB, and (d) BA.

$$A = \begin{bmatrix} 2.1 & 3.6 & 4.2 & 5.4 & 2.2 \\ 3 & 0.1 & 3.4 & 5.6 & 2 \\ 3.4 & 4.1 & 3.3 & 1.2 & 2.1 \\ 8.1 & 4.5 & 2.4 & 1.5 & 3.8 \end{bmatrix}, \qquad B = \begin{bmatrix} 1.2 & 2.2 & 5.5 \\ 3.4 & 5 & 7.1 \\ 1.1 & 1.3 & 2.4 \\ 5.5 & 4.2 & 2.8 \\ 2.7 & 0.4 & 0.7 \end{bmatrix},$$

$$C = \begin{bmatrix} 0.5 & 3.5 & 2.6 \\ 2.1 & 7 & 0.4 \\ 5.6 & 3.6 & 7.8 \\ 2.4 & 4.5 & 3 \\ 0.9 & 1.9 & 9.7 \end{bmatrix}$$

Solution

We begin by entering matrices A, B, and C. The regular operation keys can then be used to perform the required operations.

(a) $1.7A$ is found by using the $\times$ key.

$$1.7A = \begin{bmatrix} 3.57 & 6.12 & 7.14 & 9.18 & 3.74 \\ 5.1 & 0.17 & 5.78 & 9.52 & 3.4 \\ 5.78 & 6.97 & 5.61 & 2.04 & 3.57 \\ 13.77 & 7.65 & 4.08 & 2.55 & 6.46 \end{bmatrix}$$

(b) The regular $\times$ and $+$ keys are used.

$$3.6B + 2.3C = \begin{bmatrix} 5.47 & 15.97 & 25.78 \\ 17.07 & 34.10 & 26.48 \\ 16.84 & 12.96 & 26.58 \\ 25.32 & 25.47 & 16.98 \\ 11.79 & 5.81 & 24.83 \end{bmatrix}$$

(c) Using the matrices A and B with the $\times$ key gives the product:

$$AB = \begin{bmatrix} 55.02 & 51.64 & 63.85 \\ 43.88 & 35.84 & 42.45 \\ 33.92 & 38.15 & 60.56 \\ 46.17 & 51.26 & 89.12 \end{bmatrix}.$$

(d) BA is not possible. A calculator gives an error message.

As with graphics calculators and software programs, modern computer spreadsheets, such as *Lotus 1-2-3,* can be used to multiply two matrices.

EXAMPLE 9

Suppose a bank has three main sources of income—business loans, auto loans, and home mortgages—and that it draws funds from these sources for venture capital that is used to provide start-up funds for new businesses. Suppose the income from these sources for each of three years is given in Table 2.6, and the bank uses 45% of its income from business loans, 20% of its income from auto loans, and 30% of its income from home mortgages to get its venture capital funds. Write a matrix product that gives the venture capital for these years and find the available venture capital in each of the three years.

Table 2.6

	A	B	C	D
1	Income	from	Loans	
2		Business	Auto	Home
3	1993	63,300	20,024	51,820
4	1994	48,305	15,817	63,722
5	1995	55,110	18,621	64,105

Solution

The matrix that describes the sources of income for the three years is

$$\begin{bmatrix} 63,300 & 20,024 & 51,820 \\ 48,305 & 15,817 & 63,722 \\ 55,110 & 18,621 & 64,105 \end{bmatrix},$$

and the matrix that describes the percent of each that is used for venture capital is

$$\begin{bmatrix} 0.45 \\ 0.20 \\ 0.30 \end{bmatrix}.$$

The product of these matrices gives the venture capital for the three years.

$$\begin{bmatrix} 63{,}300 & 20{,}024 & 51{,}820 \\ 48{,}305 & 15{,}817 & 63{,}722 \\ 55{,}110 & 18{,}621 & 64{,}105 \end{bmatrix}\begin{bmatrix} 0.45 \\ 0.20 \\ 0.30 \end{bmatrix} = \begin{bmatrix} 48{,}035.80 \\ 44{,}017.25 \\ 47{,}755.20 \end{bmatrix}$$

Thus the available venture capital for each of the three years is as follows.

1993	$48,035.80
1994	$44,017.25
1995	$47,755.20

CHECKPOINT SOLUTIONS

1. (a) $3D - D = \begin{bmatrix} 3 & 9 & 6 & 30 \\ 60 & 15 & 18 & 9 \end{bmatrix} - \begin{bmatrix} 1 & 3 & 2 & 10 \\ 20 & 5 & 6 & 3 \end{bmatrix}$

$$= \begin{bmatrix} 2 & 6 & 4 & 20 \\ 40 & 10 & 12 & 6 \end{bmatrix}$$

(b) Yes

2. $2 \cdot 2 + 1 \cdot 1 + 3 \cdot 0 = 5$ 3. $\begin{bmatrix} 7 & 2 & 4 \\ -1 & 4 & 5 \end{bmatrix}$

4. (a) $AB = \begin{bmatrix} 0 & 0 \\ 0 & 0 \end{bmatrix}$ (b) Yes. See (a)

EXERCISE 2.2

In problems 1–4, multiply the matrices.

1. $\begin{bmatrix} 1 & 2 & 3 \end{bmatrix}\begin{bmatrix} 4 \\ 5 \\ 6 \end{bmatrix}$

2. $\begin{bmatrix} 2 & 0 & 3 \end{bmatrix}\begin{bmatrix} 0 \\ 1 \\ -3 \end{bmatrix}$

3. $\begin{bmatrix} 1 & 2 \end{bmatrix}\begin{bmatrix} 3 & 5 \\ 4 & 6 \end{bmatrix}$

4. $\begin{bmatrix} 3 & 0 \end{bmatrix}\begin{bmatrix} 1 & 2 \\ 4 & 5 \end{bmatrix}$

In problems 5–28, use matrices A through F to perform the indicated operations when possible.

$$A = \begin{bmatrix} 1 & 0 & 2 \\ 3 & 2 & 1 \\ 4 & 0 & 3 \end{bmatrix} \quad B = \begin{bmatrix} 1 & 1 & 3 & 0 \\ 4 & 2 & 1 & 1 \\ 3 & 2 & 0 & 1 \end{bmatrix}$$

$$C = \begin{bmatrix} 5 & 3 \\ 1 & 2 \end{bmatrix} \quad D = \begin{bmatrix} 4 & 2 \\ 3 & 5 \end{bmatrix}$$

$$E = \begin{bmatrix} 1 & 0 & 4 \\ 5 & 1 & 0 \end{bmatrix} \quad F = \begin{bmatrix} 1 & 0 & -1 & 3 \\ 2 & -1 & 3 & -4 \end{bmatrix}$$

5. $3A$
6. $4D$
7. $4C + 2D$
8. $5C - 3D$
9. $2A - 3B$
10. $3B + 4F$
11. CD
12. DE
13. DC
14. CF
15. DF
16. FB
17. AB
18. EC
19. BA
20. CE
21. EB
22. BE
23. EA
24. AE
25. A^2
26. A^3
27. C^3
28. F^2
29. Are $(\frac{1}{4}A)B$ and $\frac{1}{4}(AB)$ equal?
30. Are $(CD)E$ and $C(DE)$ equal?
31. Does $CD = DC$? (see problems 11 and 13)
32. If the product CF is an $m \times p$ matrix, what are m and p?

In problems 33–42, use the matrices below. Perform the indicated operations.

$$A = \begin{bmatrix} 2 & 5 & 4 \\ 1 & 4 & 3 \\ 1 & -3 & -2 \end{bmatrix} \quad B = \begin{bmatrix} -1 & 2 & 1 \\ -5 & 8 & 2 \\ -7 & 11 & -3 \end{bmatrix}$$

$$I = \begin{bmatrix} 1 & 0 & 0 \\ 0 & 1 & 0 \\ 0 & 0 & 1 \end{bmatrix} \quad C = \begin{bmatrix} 3 & 0 & 4 \\ 1 & 7 & -1 \\ 3 & 0 & 4 \end{bmatrix}$$

$$D = \begin{bmatrix} 4 & 4 & -8 \\ -1 & -1 & 2 \\ -3 & -3 & 6 \end{bmatrix} \quad F = \begin{bmatrix} 0 & 1 & 2 \\ 0 & 0 & -4 \\ 0 & 0 & 0 \end{bmatrix}$$

$$Z = \begin{bmatrix} 0 & 0 & 0 \\ 0 & 0 & 0 \\ 0 & 0 & 0 \end{bmatrix}$$

33. AB 34. BA 35. CD 36. DC

37. F^3 38. AZ 39. AI 40. IA

41. ZCI 42. IFZ

43. Is it true for matrices (as it is for real numbers) that AB equals a zero matrix if and only if either A or B equals a zero matrix? (Refer to problems 33–42.)

44. Is it true for matrices (as it is for real numbers) that multiplication by a zero matrix gives a result of a zero matrix?

45. Find AB and BA if

$$A = \begin{bmatrix} a & b \\ c & d \end{bmatrix} \quad \text{and} \quad B = \frac{1}{ad - bc}\begin{bmatrix} d & -b \\ -c & a \end{bmatrix}.$$

46. For B (in problem 45) to exist, what restriction must $ad - bc$ satisfy?

In each of problems 47–50, multiply the matrices on the left side of the equation to create a new matrix equation. Write a system of equations that is equivalent to the matrix equation.

47. $\begin{bmatrix} 1 & 2 \\ 3 & 4 \end{bmatrix}\begin{bmatrix} x \\ y \end{bmatrix} = \begin{bmatrix} 0 \\ 2 \end{bmatrix}$ 48. $\begin{bmatrix} 3 & 1 \\ 2 & -2 \end{bmatrix}\begin{bmatrix} x \\ y \end{bmatrix} = \begin{bmatrix} 4 \\ 8 \end{bmatrix}$

49. $\begin{bmatrix} 1 & 0 & 2 \\ 3 & 1 & 0 \\ 1 & 2 & 1 \end{bmatrix}\begin{bmatrix} x \\ y \\ z \end{bmatrix} = \begin{bmatrix} 0 \\ 7 \\ 3 \end{bmatrix}$

50. $\begin{bmatrix} 1 & 1 & 2 \\ 4 & 0 & 1 \\ 2 & 1 & 1 \end{bmatrix}\begin{bmatrix} x \\ y \\ z \end{bmatrix} = \begin{bmatrix} 5 \\ 5 \\ 5 \end{bmatrix}$

In each of problems 51–54, determine whether the given values of x, y, and z are the solution to the given matrix equation by substituting the given values into the matrix equation and by using matrix multiplication.

51. $\begin{bmatrix} 1 & 2 & 1 \\ 3 & 4 & -2 \\ 2 & 0 & -1 \end{bmatrix}\begin{bmatrix} x \\ y \\ z \end{bmatrix} = \begin{bmatrix} 2 \\ -2 \\ 2 \end{bmatrix}$ $x = 2, y = -1, z = 2$

52. $\begin{bmatrix} 3 & 1 & 0 \\ 2 & -2 & 1 \\ 1 & 1 & 2 \end{bmatrix}\begin{bmatrix} x \\ y \\ z \end{bmatrix} = \begin{bmatrix} 4 \\ 9 \\ 2 \end{bmatrix}$ $x = 2, y = -2, z = 1$

53. $\begin{bmatrix} 1 & 1 & 2 \\ 4 & 0 & 1 \\ 2 & 1 & 1 \end{bmatrix}\begin{bmatrix} x \\ y \\ z \end{bmatrix} = \begin{bmatrix} 5 \\ 5 \\ 5 \end{bmatrix}$ $x = 1, y = 2, z = 1$

54. $\begin{bmatrix} 1 & 0 & 2 \\ 3 & 1 & 0 \\ 1 & 2 & 1 \end{bmatrix}\begin{bmatrix} x \\ y \\ z \end{bmatrix} = \begin{bmatrix} 0 \\ 7 \\ 3 \end{bmatrix}$ $x = 2, y = 1, z = 1$

In problems 55 and 56, use technology to find the product AB of the following matrices.

55. $A = \begin{bmatrix} 0.1 & 0.0 & 0.1 & 0.1 & 0.2 \\ 0.1 & 0.2 & -0.1 & 0.1 & -0.1 \\ 0.1 & -0.1 & 0.1 & -0.2 & 0.1 \\ 0.1 & 0.2 & 0.1 & 0.1 & 0.1 \\ 0.1 & 0.0 & 0.0 & 0.0 & 0.0 \end{bmatrix}$,

$B = \begin{bmatrix} 0 & 0 & 0 & 0 & 10 \\ 0 & 5 & \frac{10}{3} & \frac{5}{3} & -10 \\ -10 & -15 & -\frac{20}{3} & \frac{35}{3} & 20 \\ 0 & -5 & -\frac{20}{3} & \frac{5}{3} & 10 \\ 10 & 10 & \frac{20}{3} & -\frac{20}{3} & -20 \end{bmatrix}$

56. $A = \frac{1}{12}\begin{bmatrix} 3 & 15 & -9 & -15 & 6 \\ 3 & -9 & 3 & 9 & -6 \\ -1 & -17 & 3 & 23 & -4 \\ 0 & 12 & 0 & -12 & 0 \\ -1 & -5 & 3 & 5 & 2 \end{bmatrix}$

$B = \begin{bmatrix} 2 & 3 & 0 & 1 & 3 \\ 1 & 0 & 2 & 3 & 1 \\ 0 & 1 & 0 & 2 & 3 \\ 1 & 0 & 2 & 2 & 1 \\ 1 & 0 & 0 & 0 & 3 \end{bmatrix}$

APPLICATIONS

57. **Debt payment** Ace, Baker, and Champ are being purchased by ALCO, Inc., and their outstanding debt must be paid by the purchaser. The matrix below gives the amount of debt and the terms for the companies being purchased.

$$
\begin{array}{c} \\ \text{Ace} \\ \text{Baker} \\ \text{Champ} \end{array}
\begin{array}{cc} \text{Due in 30 days} & \text{Due in 60 days} \end{array}
\begin{bmatrix} \$40,000 & \$60,000 \\ \$25,000 & \$15,000 \\ \$35,000 & \$58,000 \end{bmatrix}
$$

(a) If ALCO pays 35% of the amount owed on each debt, write the matrix that gives the remaining debt.
(b) Suppose ALCO, Inc., decides to pay 80% of all debts due in 30 days and to increase the debts due in 60 days by 20%. Write a matrix that gives the debts after these transactions are made.

58. **Expense accounts** A sales associate's expense account for the first week of a certain month has the daily expenses (in dollars) shown in matrix A.

$$
A = \begin{array}{c} \\ \text{Meals} \quad \text{Lodging} \quad \text{Travel} \quad \text{Other} \end{array}
$$

$$
A = \begin{bmatrix} 22 & 40 & 100 & 5 \\ 20 & 40 & 20 & 0 \\ 28 & 70 & 45 & 0 \\ 15 & 70 & 20 & 10 \\ 20 & 0 & 100 & 5 \end{bmatrix}
\begin{array}{l} \text{Monday} \\ \text{Tuesday} \\ \text{Wednesday} \\ \text{Thursday} \\ \text{Friday} \end{array}
$$

(a) The associate finds that expenses for the second week are 5% more (in each category) than for the first week. Find the expenses matrix for the second week (note that this is $1.05A$).
(b) Find the associate's expenses matrix for the third week if expenses for that week are 10% more (in each category) than they were in the first week.

59. **Debt payment** When a firm buys another company, the company frequently has some outstanding debt that the purchaser must pay. Consider three companies, A, B, and C, that are purchased by Maxx Industries. The table below gives the amount of the debts of these companies, classified by the number of days remaining until the debt must be paid.

Company	30 Days	60 Days	More than 60 Days
A	$25,000	$26,000	$12,000
B	15,000	52,000	5,000
C	8,000	20,000	120,000

(a) Suppose Maxx Industries pays 20% of the amount owed on each loan. Write the matrix that would give the remaining debt in each category.
(b) What payment plan did Maxx Industries use if the outstanding debt is given by the following matrix?

$$
\begin{bmatrix} 25,000 & 26,000 & 12,000 \\ 15,000 & 52,000 & 5,000 \\ 8,000 & 20,000 & 120,000 \end{bmatrix}
$$

$$
-0.50 \begin{bmatrix} 0 & 26,000 & 0 \\ 0 & 52,000 & 0 \\ 0 & 20,000 & 0 \end{bmatrix} - \begin{bmatrix} 25,000 & 0 & 0 \\ 15,000 & 0 & 0 \\ 8,000 & 0 & 0 \end{bmatrix}
$$

60. **Revenue** A clothing manufacturer has factories in Atlanta, Chicago, and New York. Sales (in thousands) during the first quarter are summarized in the matrix below.

$$
\begin{array}{c} \\ \text{Coats} \\ \text{Shirts} \\ \text{Pants} \\ \text{Ties} \end{array}
\begin{array}{ccc} \text{Atl.} & \text{Chi.} & \text{NY} \end{array}
\begin{bmatrix} 40 & 63 & 18 \\ 85 & 56 & 42 \\ 6 & 18 & 8 \\ 7 & 10 & 8 \end{bmatrix}
$$

During this period the selling price of a coat is $100, of a shirt $10, of a pair of pants $25, and of a tie $5. Use matrix multiplication to find the total revenue received by each factory.

Encoding messages Multiplication by a matrix can be used to encode messages (and multiplication by its inverse can be used to decode messages). Given the code

a	b	c	d	e	f	g	h	i	j
1	2	3	4	5	6	7	8	9	10

k	l	m	n	o	p	q	r	s	t
11	12	13	14	15	16	17	18	19	20

u	v	w	x	y	z	blank
21	22	23	24	25	26	27

and the code matrix

$$
A = \begin{bmatrix} 5 & 9 \\ 6 & 11 \end{bmatrix},
$$

complete problems 61 and 62.

61. Use matrix A to encode the message "The die is cast."
62. Use matrix A to encode the message "To be or not to be."
63. **Area and population** Matrix A below gives the fraction of the world's area and the fraction of its population for five continents. Matrix B gives the earth's area (in

square miles) and its population. Find the area and population of each given continent by finding AB.

$$A = \begin{bmatrix} 0.162 & 0.051 \\ 0.119 & 0.084 \\ 0.066 & 0.120 \\ 0.298 & 0.588 \\ 0.202 & 0.151 \end{bmatrix} \begin{matrix} \text{North America} \\ \text{South America} \\ \text{Europe} \\ \text{Asia} \\ \text{Africa} \end{matrix}$$

with column headers: Fraction of Area, Fraction of Population

$$B = \begin{bmatrix} 57,850,000 & 0 \\ 0 & 5,423,000,000 \end{bmatrix} \begin{matrix} \text{Area} \\ \text{Population} \end{matrix}$$

with column headers: Area, Population

(*Source*: Rand McNally and Co., *World Almanac* 1993)

64. **Advertising** A business plans to use three methods of advertising: cable TV, radio, and newspaper. The cost per ad (in thousands of dollars) is given by matrix C.

$$C = \begin{bmatrix} 20 \\ 9 \\ 6 \end{bmatrix} \begin{matrix} \text{TV} \\ \text{Radio} \\ \text{Newspaper} \end{matrix}$$

Suppose the business has three target populations for its products, female teenagers, single women 20 to 35, and married women 35 to 60. Matrix T shows the number of ads per month directed at each of these groups.

$$T = \begin{bmatrix} 50 & 30 & 5 \\ 40 & 60 & 30 \\ 45 & 40 & 60 \end{bmatrix} \begin{matrix} \text{Female teens} \\ \text{Single women} \\ \text{Women 35–60} \end{matrix}$$

with column headers: TV, Radio, Papers

Find the matrix that gives the cost of each type of ad for each group of people.

65. **Ecology** In an ecological model, matrix A gives the fraction of several types of plants consumed by the herbivores in the ecosystem. Similarly, matrix B gives the fraction of each type of herbivore that is consumed by the carnivores in the system. Each row of matrix A refers to a type of plant and each column refers to a kind of herbivore. Similarly, each row of matrix B is a kind of herbivore and each column is a kind of carnivore.

$$A = \begin{bmatrix} 0.2 & 0.1 & 0.4 \\ 0.3 & 0.3 & 0.1 \\ 0.2 & 0.1 & 0.1 \\ 0.1 & 0.2 & 0.5 \\ 0.4 & 0.4 & 0 \end{bmatrix} \text{Plants}$$

with column header: Herbivores

$$B = \begin{bmatrix} 0.1 & 0.1 & 0.2 \\ 0.2 & 0 & 0.4 \\ 0.1 & 0.5 & 0.1 \end{bmatrix} \text{Herbivores}$$

with column header: Carnivores

The matrix AB gives the fraction of each plant that is consumed by each carnivore.
(a) Find AB.
(b) What fraction of plant type 1 (row 1) is consumed by each carnivore?
(c) Identity the plant type that each carnivore consumes the most.

66. **Nutrition** In problem 37 of Exercise 2.1, we found the weights and measures of the laboratory animals to be given by matrix A, where column 1 gives the measures.

$$A = \begin{bmatrix} 12.5 & 250 \\ 11.8 & 215 \\ 9.8 & 190 \end{bmatrix}$$

If the increase in both weight and length over the next two weeks is 20% for group I, 7% for group II, and 0% for group III, then the increases in the measures during the two weeks can be found by computing GA, where

$$G = \begin{bmatrix} 0.20 & 0 & 0 \\ 0 & 0.07 & 0 \\ 0 & 0 & 0 \end{bmatrix}.$$

(a) What are the increases in respective weights and measures at the end of these two weeks?
(b) Find the matrix that gives the new weights and measures at the end of this period by computing

$$(I + G)A,$$

where I is the 3×3 identity matrix.

67. **Manufacturing** Suppose products A and B are made from plastic, steel, and glass, with the number of units of each raw material required for each product given by the table below.

	Plastic	Steel	Glass
Product A	3	1	0.50
Product B	4	0.50	2

Because of transportation costs to the firm's two plants, X and Y, the unit costs for some of the raw materials are different. The table below gives the unit costs for each of the raw materials at the two plants.

	Plant X	Plant Y
Plastic	10	9
Steel	22	26
Glass	14	14

Using the information just given, find the total cost of producing each of the products at each of the factories.

68. *Manufacturing* Two departments of a firm, A and B, need differing amounts of the same products. The following table gives the amounts of the products needed by the departments.

	Steel	Plastic	Wood
Department A	30	20	10
Department B	20	10	20

These three products are supplied by two suppliers, Ace and Kink, with the unit prices given in the following table.

	Ace	Kink
Steel	300	280
Plastic	150	100
Wood	150	200

(a) Use matrix multiplication to find how much these two orders will cost at the two suppliers. The result should be a 2×2 matrix.

(b) From which supplier should each department make its purchase?

69. *Manufacturing* A furniture manufacturer produces four styles of chairs, and has orders for 1000 of style A, 4000 of style B, 2000 of style C, and 1000 of style D. The following table lists the number of units of raw material the manufacturer needs for each style. Suppose wood costs $5 per unit, nylon costs $3 per unit, velvet costs $4 per unit, and springs cost $4 per unit.

	Wood	Nylon	Velvet	Springs
Style A	10	5	0	0
Style B	5	0	20	10
Style C	5	20	0	10
Style D	5	10	10	10

(a) Form a 1×4 matrix containing the number of units of each style ordered.

(b) Use matrix multiplication to determine the total number of units of each raw material needed.

(c) Write a 4×1 matrix representing the costs for the raw materials.

(d) Use matrix multiplication to find the 1×1 matrix that represents the total investment to fill the orders.

(e) Use matrix multiplication to determine the cost for materials for each style of chair.

70. *Accounting* The annual budget of the Magnum Company has the following expenses in thousands of dollars for selected departments.

	Mfg.	Office	Sales	Shp.	Act.	Mgt.
Supplies	0.7	8.5	10.2	1.1	5.6	3.6
Phone	0.5	0.2	6.1	1.3	0.2	1.0
Transp.	2.2	0.4	8.8	1.2	1.2	4.8
Salaries	251.8	63.4	81.6	35.2	54.3	144.2
Utilities	30.0	1.0	1.0	1.0	1.0	1.0
Materials	788.9	0	0	0	0	0

Mfg. = Manufacturing; Shp. = Shipping;
Act. = Accounting; Mgt. = Management

Find the budget matrix for the following changes in budget.
(a) a 5% decrease
(b) an 8% increase

71. *Accounting* Consider the original budget matrix of problem 70. Assume there is a 20% increase in manufacturing, a 3% increase in office, a 5% increase in sales, a 20% increase in shipping, a 5% increase in accounting, and a 3% decrease in management. Find the new budget matrix by multiplying the following matrix times the original matrix from problem 70.

$$\begin{bmatrix} 1.20 & 0 & 0 & 0 & 0 & 0 \\ 0 & 1.03 & 0 & 0 & 0 & 0 \\ 0 & 0 & 1.05 & 0 & 0 & 0 \\ 0 & 0 & 0 & 1.20 & 0 & 0 \\ 0 & 0 & 0 & 0 & 1.05 & 0 \\ 0 & 0 & 0 & 0 & 0 & 0.97 \end{bmatrix}$$

72. **Management** In an evaluation of labor assignment rules when workers are not perfectly interchangeable, Paul M. Bobrowski and Paul Sungchil Park created a dynamic job shop with 9 work centers and 9 workers. The efficiency of each worker is specified in the labor efficiency matrix E given below, which represents the degree of worker cross-training. If a worker slowdown causes a 10% decrease in efficiency at all work centers, write the matrix that describes the efficiencies for the 9 workers at the 9 work centers. (Use a graphics calculator, a computer program, or a spreadsheet.)

$$E = \begin{bmatrix} 1.00 & .95 & .95 & .95 & .95 & .85 & .85 & .85 & .85 \\ .85 & 1.00 & .95 & .95 & .95 & .95 & .85 & .85 & .85 \\ .85 & .85 & 1.00 & .95 & .95 & .95 & .95 & .85 & .85 \\ .85 & .85 & .85 & 1.00 & .95 & .95 & .95 & .95 & .85 \\ .85 & .85 & .85 & .85 & 1.00 & .95 & .95 & .95 & .95 \\ .95 & .85 & .85 & .85 & .85 & 1.00 & .95 & .95 & .95 \\ .95 & .95 & .85 & .85 & .85 & .85 & 1.00 & .95 & .95 \\ .95 & .95 & .95 & .85 & .85 & .85 & .85 & 1.00 & .95 \\ .95 & .95 & .95 & .95 & .85 & .85 & .85 & .85 & 1.00 \end{bmatrix}$$

2.3 GAUSS-JORDAN ELIMINATION: SOLVING SYSTEMS OF EQUATIONS

Objectives

❑ To use matrices to solve systems of linear equations with unique solutions
❑ To use matrices to solve systems of linear equations with nonunique solutions

APPLICATION PREVIEW

The Walters Manufacturing Company makes three types of metal storage sheds. The company has three departments: stamping, painting, and packaging. The following table gives the number of hours each division requires for each shed.

Departments	Shed		
	Type I	Type II	Type III
Stamping	2	3	4
Painting	1	2	1
Packaging	1	1	2

By using the information in the table and by solving a system of equations, we can determine how many of each type of shed can be produced if the stamping department has 3200 hours available, the painting department has 1700 hours, and the packaging department has 1300 hours. The set of equations used to solve this problem is as follows.

$$\begin{cases} 2x_1 + 3x_2 + 4x_3 = 3200 \\ x_1 + 2x_2 + x_3 = 1700 \\ x_1 + x_2 + 2x_3 = 1300 \end{cases}$$

In this section, we will solve such systems by using matrices and the Gauss-Jordan Elimination Method.

SYSTEMS WITH UNIQUE SOLUTIONS

In solving systems with the left-to-right elimination method used in Example 6 of Section 1.8, we operated on the coefficients of the variables x, y, and z and on the constants. If we keep the coefficients of the variables x, y, and z in distinctive columns, we do not need to write the equations. In solving a system of linear equations with matrices, we first write the coefficients and constants from the system (on the left) in the **augmented matrix** (on the right).

$$\begin{cases} a_1x + b_1y + c_1z = d_1 \\ a_2x + b_2y + c_2z = d_2 \\ a_3x + b_3y + c_3z = d_3 \end{cases} \qquad \left[\begin{array}{ccc|c} a_1 & b_1 & c_1 & d_1 \\ a_2 & b_2 & c_2 & d_2 \\ a_3 & b_3 & c_3 & d_3 \end{array} \right]$$

In the augmented matrix, the numbers on the left side of the solid line form the **coefficient matrix,** with each column containing the coefficients of a variable (0 represents any missing variable). The column on the right side of the line (called the *augment*) contains the constants. Each row of the matrix gives the corresponding coefficients of an equation.

The augmented matrix associated with the system

$$\begin{cases} x + 2y + 3z = 6 \\ x \quad\;\; - z = 0 \\ x - y - z = -4 \end{cases} \text{ is } A = \left[\begin{array}{ccc|c} 1 & 2 & 3 & 6 \\ 1 & 0 & -1 & 0 \\ 1 & -1 & -1 & -4 \end{array} \right].$$

CHECKPOINT

1. (a) Write the augmented matrix for the system of linear equations below.

$$\begin{cases} 3x + 2y - z = 3 \\ x - y + 2z = 4 \\ 2x + 3y - z = 3 \end{cases}$$

   (b) Write the coefficient matrix for the system.

We can use matrices to solve systems of linear equations by performing the same operations on the rows of a matrix to reduce it as we do on equations in a linear system. The three different operations we can use to reduce the matrix are called **elementary row operations** and are similar to the operations with equations that result in equivalent systems. These operations are:

1. Interchange two rows.

2. Add a multiple of one row to another row.

3. Multiply a row by any nonzero constant.

When a new matrix results from one or more of these elementary row operations being performed on a matrix, the new matrix is said to be **equivalent** to the original because both these matrices represent equivalent systems of equations.

The process that we use to solve a system of equations using matrices (called the **elimination method**) is not always the fastest way, but it is a systematic procedure similar to the left-to-right elimination method discussed in Section 1.8. This method uses these row operations to attempt to reduce the coefficient matrix to an identity matrix. Example 1 illustrates this process, called the **Gauss-Jordan Elimination Method.**

EXAMPLE I

Use matrices to solve the system

$$\begin{cases} 2x + 5y + 4z = 4 \\ x + 4y + 3z = 1 \\ x - 3y - 2z = 5. \end{cases}$$

Solution

The augmented matrix for this system is

$$\left[\begin{array}{ccc|c} 2 & 5 & 4 & 4 \\ 1 & 4 & 3 & 1 \\ 1 & -3 & -2 & 5 \end{array}\right].$$

Note that this system has 3 equations in 3 variables, but the reduction procedure and row operations apply regardless of the size of the system.

The following display shows a series of step-by-step goals for reducing a 3×3 matrix and shows a series of row operations that reduces this matrix.

Gauss-Jordan Elimination Method

Goal (for Each Step)	Row Operation	Equivalent Matrix
1. Get a 1 in row 1, col. 1.	Interchange row 1 and row 2.	$\left[\begin{array}{ccc\|c} 1 & 4 & 3 & 1 \\ 2 & 5 & 4 & 4 \\ 1 & -3 & -2 & 5 \end{array}\right]$
2. Use row 1 *only* to get zeros in other entries of col. 1.	Add -2 times row 1 to row 2; put the result in row 2. Add -1 times row 1 to row 3; put the result in row 3.	$\left[\begin{array}{ccc\|c} 1 & 4 & 3 & 1 \\ 0 & -3 & -2 & 2 \\ 0 & -7 & -5 & 4 \end{array}\right]$
3. Use rows below row 1 to get a 1 in row 2, col. 2.	Multiply row 2 by $-\frac{1}{3}$; put the result in row 2.	$\left[\begin{array}{ccc\|c} 1 & 4 & 3 & 1 \\ 0 & 1 & \frac{2}{3} & -\frac{2}{3} \\ 0 & -7 & -5 & 4 \end{array}\right]$
4. Use row 2 *only* to get zeros as the other entries in col. 2.	Add -4 times row 2 to row 1; put the result in row 1. Add 7 times row 2 to row 3; put the result in row 3.	$\left[\begin{array}{ccc\|c} 1 & 0 & \frac{1}{3} & \frac{11}{3} \\ 0 & 1 & \frac{2}{3} & -\frac{2}{3} \\ 0 & 0 & -\frac{1}{3} & -\frac{2}{3} \end{array}\right]$

| 5. Use row below row 2 to get a 1 in row 3, col. 3. | Multiply row 3 by -3; put the result in row 3. | $\begin{bmatrix} 1 & 0 & \frac{1}{3} & \bigm| & \frac{11}{3} \\ 0 & 1 & \frac{2}{3} & \bigm| & -\frac{2}{3} \\ 0 & 0 & 1 & \bigm| & 2 \end{bmatrix}$ |
| 6. Use row 3 *only* to get zeros as the other entries in col. 3. | Add $-\frac{1}{3}$ times row 3 to row 1; put the result in row 1. Add $-\frac{2}{3}$ times row 3 to row 2; put the result in row 2. | $\begin{bmatrix} 1 & 0 & 0 & \bigm| & 3 \\ 0 & 1 & 0 & \bigm| & -2 \\ 0 & 0 & 1 & \bigm| & 2 \end{bmatrix}$ |

The system of equations corresponding to this final augmented matrix follows.

$$x + 0y + 0z = 3$$
$$0x + y + 0z = -2$$
$$0x + 0y + z = 2$$

Thus the solution is $x = 3$, $y = -2$, and $z = 2$.

This Gauss-Jordan Elimination Method (outlined in Example 1) may be used with systems of any size.

EXAMPLE 2

Solve the system

$$x_1 + x_2 + x_3 + 2x_4 = 6$$
$$x_1 + 2x_2 \qquad + x_4 = -2$$
$$x_1 + x_2 + 3x_3 - 2x_4 = 12$$
$$x_1 + x_2 - 4x_3 + 5x_4 = -16.$$

Solution

First note there is a variable missing in the second equation. When we form the augmented matrix, we must insert a zero in this place so that each column represents the coefficients of one variable.

To solve this system, we must reduce the augmented matrix.

$$\begin{bmatrix} 1 & 1 & 1 & 2 & \bigm| & 6 \\ 1 & 2 & 0 & 1 & \bigm| & -2 \\ 1 & 1 & 3 & -2 & \bigm| & 12 \\ 1 & 1 & -4 & 5 & \bigm| & -16 \end{bmatrix}$$

The reduction process follows. The entry in row 1, column 1 is 1. To get zeros in the first column, add -1 times row 1 to row 2, and put the result in row 2; add -1 times row 1 to row 3 and put the result in row 3; and add -1 times row 1 to row 4 and put the result in row 4.

$$\begin{bmatrix} 1 & 1 & 1 & 2 & | & 6 \\ 1 & 2 & 0 & 1 & | & -2 \\ 1 & 1 & 3 & -2 & | & 12 \\ 1 & 1 & -4 & 5 & | & -16 \end{bmatrix} \xrightarrow[\substack{(-1)R_1 + R_2 \to R_2 \\ (-1)R_1 + R_3 \to R_3 \\ \\ (-1)R_1 + R_4 \to R_4}]{} \begin{bmatrix} 1 & 1 & 1 & 2 & | & 6 \\ 0 & 1 & -1 & -1 & | & -8 \\ 0 & 0 & 2 & -4 & | & 6 \\ 0 & 0 & -5 & 3 & | & -22 \end{bmatrix}$$

The entry in row 2, column 2 is 1. To get zeros in the second column, add -1 times row 2 to row 1, and put the result in row 1.

$$\xrightarrow[]{(-1)R_2 + R_1 \to R_1} \begin{bmatrix} 1 & 0 & 2 & 3 & | & 14 \\ 0 & 1 & -1 & -1 & | & -8 \\ 0 & 0 & 2 & -4 & | & 6 \\ 0 & 0 & -5 & 3 & | & -22 \end{bmatrix}$$

The entry in row 3, column 3 is 2. To get a 1 in this position, multiply row 3 by $\frac{1}{2}$.

$$\xrightarrow[\phantom{\frac{1}{2}R_3 \to R_3}]{\frac{1}{2}R_3 \to R_3} \begin{bmatrix} 1 & 0 & 2 & 3 & | & 14 \\ 0 & 1 & -1 & -1 & | & -8 \\ 0 & 0 & 1 & -2 & | & 3 \\ 0 & 0 & -5 & 3 & | & -22 \end{bmatrix}$$

Using row 3, add -2 times row 3 to row 1 and put the result in row 1; add row 3 to row 2 and put the result in row 2; and add 5 times row 3 to row 4 and put the result in row 4.

$$\xrightarrow[\substack{5R_3 + R_4 \to R_4}]{\substack{(-2)R_3 + R_1 \to R_1 \\ R_3 + R_2 \to R_2 \\ \\ }} \begin{bmatrix} 1 & 0 & 0 & 7 & | & 8 \\ 0 & 1 & 0 & -3 & | & -5 \\ 0 & 0 & 1 & -2 & | & 3 \\ 0 & 0 & 0 & -7 & | & -7 \end{bmatrix}$$

Multiplying $-\frac{1}{7}$ times row 4 and putting the result in row 4 gives a 1 in row 4, column 4.

$$\xrightarrow[\phantom{-\frac{1}{7}R_4 \to R_4}]{-\frac{1}{7}R_4 \to R_4} \begin{bmatrix} 1 & 0 & 0 & 7 & | & 8 \\ 0 & 1 & 0 & -3 & | & -5 \\ 0 & 0 & 1 & -2 & | & 3 \\ 0 & 0 & 0 & 1 & | & 1 \end{bmatrix}$$

Adding appropriate multiples of row 4 to the other rows gives the following:

$$\xrightarrow[\substack{2R_4 + R_3 \to R_3}]{\substack{(-7)R_4 + R_1 \to R_1 \\ 3R_4 + R_2 \to R_2 \\ \\ }} \begin{bmatrix} 1 & 0 & 0 & 0 & | & 1 \\ 0 & 1 & 0 & 0 & | & -2 \\ 0 & 0 & 1 & 0 & | & 5 \\ 0 & 0 & 0 & 1 & | & 1 \end{bmatrix}.$$

This corresponds to the system

$$
\begin{aligned}
x_1 &= 1 \\
x_2 &= -2 \\
x_3 &= 5 \\
x_4 &= 1.
\end{aligned}
$$

So the solution is $x_1 = 1$, $x_2 = -2$, $x_3 = 5$, and $x_4 = 1$.

CHECKPOINT

2. Solve the system.

$$
\begin{cases}
3x + 2y - z = 3 \\
x - y + 2z = 4 \\
2x + 3y - z = 3
\end{cases}
$$

EXAMPLE 3

As mentioned in the Application Preview, the Walters Manufacturing Company needs to know how to best use the time available within its three manufacturing departments in the construction and packaging of the 3 types of metal storage sheds. Each one must be stamped, painted, and packaged. Table 2.7 shows the number of hours required for the processing of each type of shed. Using the information in the table, determine how many of each type of shed can be produced if the stamping department has 3200 hours available, the painting department has 1700 hours, and the packaging department has 1300 hours.

Table 2.7

| | Shed | | |
Departments	Type I	Type II	Type III
Stamping	2	3	4
Painting	1	2	1
Packaging	1	1	2

Solution

If we let x_1 be the number of type I sheds, x_2 be the number of type II sheds, and x_3 be the number of type III sheds, the equation

$$2x_1 + 3x_2 + 4x_3 = 3200$$

represents the hours used by the stamping department. Similarly,

$$1x_1 + 2x_2 + 1x_3 = 1700$$

and

$$1x_1 + 1x_2 + 2x_3 = 1300$$

represent the hours used by the painting department and the packaging department, respectively.

The augmented matrix for this system of equations is

$$\begin{bmatrix} 2 & 3 & 4 & 3200 \\ 1 & 2 & 1 & 1700 \\ 1 & 1 & 2 & 1300 \end{bmatrix}.$$

Reducing this augmented matrix proceeds as follows.

$$\begin{bmatrix} 2 & 3 & 4 & 3200 \\ 1 & 2 & 1 & 1700 \\ 1 & 1 & 2 & 1300 \end{bmatrix} \xrightarrow{\text{Switch } R_1 \text{ and } R_3} \begin{bmatrix} 1 & 1 & 2 & 1300 \\ 1 & 2 & 1 & 1700 \\ 2 & 3 & 4 & 3200 \end{bmatrix}$$

$$\xrightarrow[\,(-2)R_1 + R_3 \to R_3\,]{(-1)R_1 + R_2 \to R_2} \begin{bmatrix} 1 & 1 & 2 & 1300 \\ 0 & 1 & -1 & 400 \\ 0 & 1 & 0 & 600 \end{bmatrix}$$

$$\xrightarrow[\,(-1)R_2 + R_3 \to R_3\,]{(-1)R_2 + R_1 \to R_1} \begin{bmatrix} 1 & 0 & 3 & 900 \\ 0 & 1 & -1 & 400 \\ 0 & 0 & 1 & 200 \end{bmatrix}$$

$$\xrightarrow[\,R_3 + R_2 \to R_2\,]{(-3)R_3 + R_1 \to R_1} \begin{bmatrix} 1 & 0 & 0 & 300 \\ 0 & 1 & 0 & 600 \\ 0 & 0 & 1 & 200 \end{bmatrix}$$

Thus the solution to the system is

$$x_1 = 300$$
$$x_2 = 600$$
$$x_3 = 200.$$

Thus the company should make 300 type I, 600 type II, and 300 type III sheds.

SYSTEMS WITH NONUNIQUE SOLUTIONS

While all the systems considered so far had unique solutions, it is also possible for a system of linear equations to have an infinite number of solutions or no solution at all. Although coefficient matrices for systems with an infinite number of solutions or no solution will not reduce to identity matrices, row operations can be used to obtain a reduced form from which the solutions, if they exist, can be determined.

A matrix is said to be in **reduced form** when it is in the following form:

1. The first nonzero element in each row is 1.
2. Every column containing a first nonzero element for some row has zeros everywhere else.
3. The first nonzero element of each row is to the right of the first nonzero element of every row above it.
4. All rows containing zeros are grouped together below the rows containing nonzero entries.

The following matrices are in reduced form because they satisfy these conditions:

$$\left[\begin{array}{cc|c} 1 & 0 & 4 \\ 0 & 1 & 2 \\ 0 & 0 & 0 \end{array}\right] \quad \left[\begin{array}{ccc|c} 1 & 0 & 0 & 0 \\ 0 & 1 & 0 & 3 \\ 0 & 0 & 1 & 1 \\ 0 & 0 & 0 & 0 \end{array}\right] \quad \left[\begin{array}{ccc|c} 1 & 4 & 0 & 1 \\ 0 & 0 & 1 & 2 \\ 0 & 0 & 0 & 0 \end{array}\right].$$

The following matrices are *not* in reduced form:

$$\left[\begin{array}{cc|c} 1 & \textcircled{1} & 0 \\ 0 & 1 & 3 \\ 0 & 0 & 0 \end{array}\right] \quad \left[\begin{array}{cc|c} 1 & 0 & 1 \\ 0 & \textcircled{2} & 1 \\ 0 & 0 & 0 \end{array}\right] \quad \left[\begin{array}{cc|c} 0 & 1 & 0 \\ 1 & 0 & 0 \\ 0 & 0 & 1 \end{array}\right].$$

In the first two matrices the circled element must be changed to obtain a reduced form. Can you see what row operations would transform each of these matrices into reduced form? The third matrix does not satisfy point 3 above.

We can solve a system of linear equations by using row operations on the augmented matrix until the coefficient matrix is transformed to an equivalent matrix in reduced form.

EXAMPLE 4
Solve the system

$$\begin{cases} x_1 + x_2 - x_3 + x_4 = 3 \\ x_2 + x_3 + x_4 = 1 \\ x_1 \qquad - 2x_3 + x_4 = 6 \\ 2x_1 - x_2 - 5x_3 - 3x_4 = -5. \end{cases}$$

Solution
To solve this system we must reduce the augmented matrix

$$\left[\begin{array}{cccc|c} 1 & 1 & -1 & 1 & 3 \\ 0 & 1 & 1 & 1 & 1 \\ 1 & 0 & -2 & 1 & 6 \\ 2 & -1 & -5 & -3 & -5 \end{array}\right].$$

Attempting to reduce the coefficient matrix to an identity matrix requires the following.

$$\begin{bmatrix} 1 & 1 & -1 & 1 & | & 3 \\ 0 & 1 & 1 & 1 & | & 1 \\ 1 & 0 & -2 & 1 & | & 6 \\ 2 & -1 & -5 & -3 & | & -5 \end{bmatrix} \xrightarrow[\begin{subarray}{c} (-1)R_1 + R_3 \to R_3 \\ (-2)R_1 + R_4 \to R_4 \end{subarray}]{} \begin{bmatrix} 1 & 1 & -1 & 1 & | & 3 \\ 0 & 1 & 1 & 1 & | & 1 \\ 0 & -1 & -1 & 0 & | & 3 \\ 0 & -3 & -3 & -5 & | & -11 \end{bmatrix}$$

$$\xrightarrow[\begin{subarray}{c} (-1)R_2 + R_1 \to R_1 \\ R_2 + R_3 \to R_3 \\ 3R_2 + R_4 \to R_4 \end{subarray}]{} \begin{bmatrix} 1 & 0 & -2 & 0 & | & 2 \\ 0 & 1 & 1 & 1 & | & 1 \\ 0 & 0 & 0 & 1 & | & 4 \\ 0 & 0 & 0 & -2 & | & -8 \end{bmatrix}$$

The entry in row 3, column 3 cannot be made 1 using rows *below* row 2. Moving to column 4, the entry in row 3, column 4 is a 1. Using row 3 gives the following.

$$\xrightarrow[\begin{subarray}{c} (-1)R_3 + R_2 \to R_2 \\ 2R_3 + R_4 \to R_4 \end{subarray}]{} \begin{bmatrix} 1 & 0 & -2 & 0 & | & 2 \\ 0 & 1 & 1 & 0 & | & -3 \\ 0 & 0 & 0 & 1 & | & 4 \\ 0 & 0 & 0 & 0 & | & 0 \end{bmatrix}$$

The augmented matrix is now reduced; this matrix corresponds to the system

$$\begin{cases} x_1 - 2x_3 = 2 \\ x_2 + x_3 = -3 \\ x_4 = 4 \\ 0 = 0. \end{cases}$$

If we solve each of the equations for the leading variable (the variable corresponding to the first 1 in each row of the reduced form of a matrix), we obtain

$$x_1 = 2 + 2x_3$$
$$x_2 = -3 - x_3$$
$$x_4 = 4.$$

In this form we have the values of x_1 and x_2 dependent on the value of x_3, so we can get many different solutions to the system by specifying different values of x_3. For example, if $x_3 = 1$, then $x_1 = 4$, $x_2 = -4$, $x_3 = 1$, and $x_4 = 4$ is a solution to the system; if we let $x_3 = -2$, then $x_1 = -2, x_2 = -1$, $x_3 = -2$, and $x_4 = 4$ is a solution. The **general solution** to the system is given by

$$x_1 = 2 + 2x_3$$
$$x_2 = -3 - x_3$$
$$x_3 = x_3$$
$$x_4 = 4,$$

where x_3 is arbitrary (any real number).

We have seen two different possibilities for solutions to systems of linear equations. In Examples 1–3, there was only one solution, whereas in Example 4 we saw that there were many solutions. A third possibility exists, namely that the system will have no solution. Recall that these possibilities also existed for two equations in two unknowns, as discussed in Chapter 1.

EXAMPLE 5

Solve the system

$$\begin{cases} x + 2y - z = 3 \\ 3x + y = 4 \\ 2x - y + z = 2. \end{cases}$$

Solution

The augmented matrix is

$$\left[\begin{array}{ccc|c} 1 & 2 & -1 & 3 \\ 3 & 1 & 0 & 4 \\ 2 & -1 & 1 & 2 \end{array}\right].$$

This can be reduced as follows.

$$\left[\begin{array}{ccc|c} 1 & 2 & -1 & 3 \\ 3 & 1 & 0 & 4 \\ 2 & -1 & 1 & 2 \end{array}\right] \begin{array}{c} (-3)R_1 + R_2 \to R_2 \\ \xrightarrow{\hspace{1cm}} \\ (-2)R_1 + R_3 \to R_3 \end{array} \left[\begin{array}{ccc|c} 1 & 2 & -1 & 3 \\ 0 & -5 & 3 & -5 \\ 0 & -5 & 3 & -4 \end{array}\right]$$

$$\begin{array}{c} (-\frac{1}{5})R_2 \to R_2 \\ \xrightarrow{\hspace{1cm}} \end{array} \left[\begin{array}{ccc|c} 1 & 2 & -1 & 3 \\ 0 & 1 & -\frac{3}{5} & 1 \\ 0 & -5 & 3 & -4 \end{array}\right]$$

$$\begin{array}{c} (-2)R_2 + R_1 \to R_1 \\ \xrightarrow{\hspace{1cm}} \\ (5)R_2 + R_3 \to R_3 \end{array} \left[\begin{array}{ccc|c} 1 & 0 & \frac{1}{5} & 1 \\ 0 & 1 & -\frac{3}{5} & 1 \\ 0 & 0 & 0 & 1 \end{array}\right]$$

The system of equations corresponding to the reduced matrix is

$$\begin{cases} x + \frac{1}{5}z = 1 \\ y - \frac{3}{5}z = 1 \\ 0 = 1. \end{cases}$$

Thus we see that this system has $0 = 1$ as an equation. This is clearly an impossibility, so there is no solution.

Now that we have seen examples of each of the different solution possibilities, let us consider what we have learned. In all cases the solution procedure began by setting up and then reducing the augmented matrix. Once the matrix is reduced, we can write the corresponding reduced system of equations. The solutions, if they exist, are easily found from this reduced system.

Graphing Utilities

In Section 2.4, we will see how to use technology (1) to determine whether a system of linear equations has a unique solution, and (2) to find the unique solution. We will also see that some technology gives an error display if the solution to a system doesn't exist or if the solution is not unique. Whether or not a unique solution exists, we can perform the row operations necessary to reduce an augmented matrix to find any solution that exists. We can perform these row operations with a graphics calculator, and because computer spreadsheets are electronically stored matrices, we can solve systems of linear equations by instructing the spreadsheet to perform these row operations. As expected, the value of using a graphics calculator, a spreadsheet, or a computer algebra system to solve systems of linear equations increases as the size of the systems gets larger.

EXAMPLE 6

One of the matrices used in the matrix solution of the system

$$\begin{cases} 2x_1 + 3x_2 - x_3 + 3x_4 + x_5 = 20 \\ 3x_1 + x_2 - 4x_3 + 3x_4 - x_5 = 0 \\ x_1 + x_2 + 3x_3 - 4x_4 + 2x_5 = 0 \\ x_1 - 2x_2 - 3x_3 + 2x_4 - 2x_5 = -6 \\ 2x_1 \qquad\qquad + 4x_4 - 5x_5 = -7 \end{cases}$$

is the matrix

$$\begin{bmatrix} 1 & -2 & 3 & -4 & 2 & \bigm| & 0 \\ 0 & 7 & -13 & 15 & -7 & \bigm| & 0 \\ 0 & 7 & -7 & 11 & -3 & \bigm| & 20 \\ 0 & 4 & -6 & 6 & -4 & \bigm| & -6 \\ 2 & 0 & 0 & 4 & -5 & \bigm| & -7 \end{bmatrix}.$$

What is the next row operation in this solution?

Solution

The next step in the solution is to add (-2) times row 1 to row 5 and place the result in row 5, so that all elements of column 1 below the first row are zero. Performing this operation by hand or with technology gives the matrix

$$\begin{bmatrix} 1 & -2 & 3 & -4 & 2 & \bigm| & 0 \\ 0 & 7 & -13 & 15 & -7 & \bigm| & 0 \\ 0 & 7 & -7 & 11 & -3 & \bigm| & 20 \\ 0 & 4 & -6 & 6 & -4 & \bigm| & -6 \\ 0 & 4 & -6 & 12 & -9 & \bigm| & -7 \end{bmatrix}.$$

CHECKPOINT SOLUTIONS

1. (a) $\begin{bmatrix} 3 & 2 & -1 & | & 3 \\ 1 & -1 & 2 & | & 4 \\ 2 & 3 & -1 & | & 3 \end{bmatrix}$ (b) $\begin{bmatrix} 3 & 2 & -1 \\ 1 & -1 & 2 \\ 2 & 3 & -1 \end{bmatrix}$

2. $\begin{bmatrix} 3 & 2 & -1 & | & 3 \\ 1 & -1 & 2 & | & 4 \\ 2 & 3 & -1 & | & 3 \end{bmatrix} \longrightarrow \begin{bmatrix} 1 & 0 & 0 & | & 1 \\ 0 & 1 & 0 & | & 1 \\ 0 & 0 & 1 & | & 2 \end{bmatrix}$

Thus, $x = 1$, $y = 1$, $z = 2$

EXERCISE 2.3

UNIQUE SOLUTIONS

In problems 1 and 2, use the indicated row operation to change matrix A, where

$$A = \begin{bmatrix} -1 & 2 & 1 & | & 7 \\ 3 & 1 & 2 & | & 0 \\ 4 & 2 & 2 & | & 1 \end{bmatrix}.$$

1. Add 3 times row 1 to row 2 of matrix A and place the result in row 2 to get 0 in row 2, column 1.
2. Add 4 times row 1 to row 3 of matrix A and place the result in row 3 to get 0 in row 3, column 1.

In problems 3–6, write the augmented matrix associated with each system of linear equations.

3. $\begin{cases} 3x + 2y + 4z = 0 \\ 2x - y + 2z = 0 \\ x - 2y - 4z = 0 \end{cases}$ 4. $\begin{cases} x - 3y + 4z = 0 \\ 2x + 2y + z = 1 \\ 3x - 4y + 2z = 9 \end{cases}$

5. $\begin{cases} x - 3y + 4z = 2 \\ 2x + 2z = 1 \\ x + 2y + z = 1 \end{cases}$ 6. $\begin{cases} x + 2y + 2z = 3 \\ x - 2y = 4 \\ y - z = 1 \end{cases}$

In each of problems 7–10, the given matrix is an augmented matrix used in the solution of a system of linear equations. What is the solution of the system?

7. $\begin{bmatrix} 1 & 0 & 0 & | & 2 \\ 0 & 1 & 0 & | & \frac{1}{2} \\ 0 & 0 & 1 & | & -5 \end{bmatrix}$ 8. $\begin{bmatrix} 1 & 0 & 0 & | & -8 \\ 0 & 1 & 0 & | & 1 \\ 0 & 0 & 1 & | & \frac{1}{3} \end{bmatrix}$

9. $\begin{bmatrix} 1 & 0 & 0 & | & 18 \\ 0 & 1 & 0 & | & 10 \\ 0 & 0 & 1 & | & 0 \end{bmatrix}$ 10. $\begin{bmatrix} 1 & 0 & 0 & | & -1 \\ 0 & 1 & 0 & | & 7 \\ 0 & 0 & 1 & | & 0 \end{bmatrix}$

In problems 11–14, the given matrix is an augmented matrix representing a system of linear equations. Use the Gauss-Jordan Elimination Method to find the solution to the system.

11. $\begin{bmatrix} 1 & 1 & 2 & | & -1 \\ 0 & 3 & 1 & | & 7 \\ 0 & -2 & 4 & | & 0 \end{bmatrix}$ 12. $\begin{bmatrix} 1 & 5 & 2 & | & 6 \\ 0 & -2 & 3 & | & 9 \\ 0 & 1 & 3 & | & 0 \end{bmatrix}$

13. $\begin{bmatrix} 1 & 2 & 5 & | & -4 \\ 2 & -2 & 4 & | & -2 \\ 0 & 1 & -3 & | & 7 \end{bmatrix}$ 14. $\begin{bmatrix} 1 & 1 & 1 & | & 3 \\ 3 & -2 & 4 & | & 5 \\ 1 & 2 & 1 & | & 4 \end{bmatrix}$

In problems 15–26, use row operations on augmented matrices to solve the given system of linear equations.

15. $\begin{cases} 7x - 2y = -1 \\ 3x + 6y = 11 \end{cases}$ 16. $\begin{cases} 6x + 7y = 55 \\ 5x - 37y = 3 \end{cases}$

17. $\begin{cases} x + y - z = 0 \\ x + 2y + 3z = -5 \\ 2x - y - 13z = 17 \end{cases}$

18. $\begin{cases} x + 2y - z = 3 \\ 2x + 5y - 2z = 7 \\ -x + y + 5z = -12 \end{cases}$

19. $\begin{cases} 2x - 6y - 12z = 6 \\ 3x - 10y - 20z = 5 \\ 2x \qquad - 17z = -4 \end{cases}$

20. $\begin{cases} -3x + 6y - 9z = 3 \\ x - y - 2z = 0 \\ 5x + 5y - 7z = 63 \end{cases}$

21. $\begin{cases} x - 3y + 3z = 7 \\ x + 2y - z = -2 \\ 3x + 2y + 4z = 5 \end{cases}$

22. $\begin{cases} 3x + 2y - 4z = -12 \\ 3x - 3y + 2z = -5 \\ 4x + 6y + z = 0 \end{cases}$

23. $\begin{cases} x_1 + 3x_2 + 2x_3 + 2x_4 = 3 \\ x_1 + x_2 + 3x_3 \qquad = 4 \\ 2x_1 \qquad + 2x_3 - 3x_4 = 4 \\ x_1 - 3x_2 \qquad = 1 \end{cases}$

24. $\begin{cases} x_1 + x_2 + x_3 + 2x_4 = 1 \\ x_1 \qquad - 3x_3 \qquad = 2 \\ x_1 - 3x_2 \qquad + x_4 = -2 \\ x_2 - 4x_3 + x_4 = 0 \end{cases}$

25. $\begin{cases} x - 2y + 3z + w = -2 \\ x - 3y + z - w = -7 \\ x - y \qquad = -2 \\ x \qquad + z + w = 2 \end{cases}$

26. $\begin{cases} x + 4y - 6z - 3w = 3 \\ x - y \qquad + 2w = -5 \\ x \qquad + z + w = 1 \\ y + z + w = 0 \end{cases}$

NONUNIQUE SOLUTIONS

In each of problems 27–30, a system of linear equations and a reduced matrix for the system are given. Use the reduced matrix to find the general solution of the system, if one exists.

27. $\begin{cases} x + 2y + 3z = 1 \\ 2x - y = 3 \\ x + 2y + 3z = 2 \end{cases}$　$\begin{bmatrix} 1 & 0 & \frac{3}{5} & | & 0 \\ 0 & 1 & \frac{6}{5} & | & 0 \\ 0 & 0 & 0 & | & 1 \end{bmatrix}$

28. $\begin{cases} 2x + 3y + 4z = 2 \\ x + 2y + 2z = 1 \\ x + y + 2z = 2 \end{cases}$　$\begin{bmatrix} 1 & 0 & 2 & | & 0 \\ 0 & 1 & 0 & | & 0 \\ 0 & 0 & 0 & | & -1 \end{bmatrix}$

29. $\begin{cases} 2x + y - z = 7 \\ x - y - z = 4 \\ 3x + 3y - z = 10 \end{cases}$　$\begin{bmatrix} 1 & 0 & -\frac{2}{3} & | & \frac{11}{3} \\ 0 & 1 & \frac{1}{3} & | & -\frac{1}{3} \\ 0 & 0 & 0 & | & 0 \end{bmatrix}$

30. $\begin{cases} x - y + z = 3 \\ 3x \qquad + 2z = 7 \\ x - 4y + 2z = 5 \end{cases}$　$\begin{bmatrix} 1 & 0 & \frac{2}{3} & | & \frac{7}{3} \\ 0 & 1 & -\frac{1}{3} & | & -\frac{2}{3} \\ 0 & 0 & 0 & | & 0 \end{bmatrix}$

The systems of equations in problems 31–42 may have unique solutions, an infinite number of solutions, or no solution. Use matrices to find the general solution to each system, if a solution exists.

31. $\begin{cases} x + y + z = 0 \\ 2x - y - z = 0 \\ -x + 2y + 2z = 0 \end{cases}$

32. $\begin{cases} 2x - y + 3z = 0 \\ x + 2y + 2z = 0 \\ x - 3y + z = 0 \end{cases}$

33. $\begin{cases} 3x + 2y + z = 0 \\ x + y + 2z = 1 \\ 2x + y - z = -1 \end{cases}$

34. $\begin{cases} x + 3y + 2z = 2 \\ 2x - y - 2z = 1 \\ 3x + 2y = 3 \end{cases}$

35. $\begin{cases} 2x + 2y + z = 2 \\ x - 2y + 2z = 1 \\ -x + 2y - 2z = -1 \end{cases}$

36. $\begin{cases} 2x + y - z = 2 \\ x - y + 2z = 3 \\ x + y - z = 1 \end{cases}$

37. $\begin{cases} 2x - 5y + z = -9 \\ x + 4y - 6z = 2 \\ 3x - 4y - 2z = -10 \end{cases}$

38. $\begin{cases} 2x - y + z = 2 \\ 3x + y - 6z = -7 \\ x - y + 2z = 3 \end{cases}$

39. $\begin{cases} x + y + z = 3 \\ x - y + z = 4 \\ x + y - z = 5 \end{cases}$

40. $\begin{cases} 3x + 2y + z = 3 \\ x - y - z = 2 \\ 2x + y - 2z = 1 \end{cases}$

41. $\begin{cases} -0.6x_1 + 0.1x_2 + 0.3x_3 = 0 \\ 0.4x_1 - 0.7x_2 + 0.2x_3 = 0 \\ 0.2x_1 + 0.6x_2 - 0.5x_3 = 0 \end{cases}$

42. $\begin{cases} 0.1x_1 - 0.1x_2 - 0.3x_3 = 0 \\ 0.2x_1 + 0.3x_2 + 0.1x_3 = -0.3 \\ 0.3x_1 + 0.7x_2 + 0.5x_3 = -0.6 \end{cases}$

43. If the matrix below is a step in the matrix solution of a system of linear equations,
    (a) What row operation is used in the next step?
    (b) What is the resulting matrix?

$$\begin{bmatrix} 1 & 0 & 0 & -1 & -3 & | & -18 \\ 0 & 1 & -1.5 & 1.5 & -1 & | & -1.5 \\ 0 & 0 & 3.5 & 0.5 & 4 & | & 32.5 \\ 0 & 0 & -2.5 & 4.5 & 0 & | & 10.5 \\ 0 & 4 & -6 & 12 & -9 & | & -7 \end{bmatrix}$$

44. The matrix in problem 43 is a step in the solution of the system of linear equations below. Use technology to complete the solution of this system.

$$\begin{cases} 2x_1 + 3x_2 - x_3 + 3x_4 + x_5 = 22 \\ 3x_1 + x_2 - 4x_3 + 3x_4 - x_5 = 0 \\ x_1 + x_2 + 3x_3 - 4x_4 + 2x_5 = 6 \\ x_1 + 2x_2 - 3x_3 + 2x_4 - 2x_5 = -6 \\ 2x_1 + 4x_4 - 5x_5 = -7 \end{cases}$$

In problems 45–48, solve the system if a solution exists.

45. $$\begin{cases} x_1 + 2x_2 - x_3 + x_4 = 3 \\ x_1 + 3x_2 + 4x_3 + x_4 = -2 \\ 2x_1 + 5x_2 + 2x_3 + 2x_4 = 1 \\ 2x_1 + 3x_2 - 6x_3 + 2x_4 = 3 \end{cases}$$

46. $$\begin{cases} x_1 + 2x_2 + 3x_3 + 4x_4 = 2 \\ x_1 + 2x_2 + 4x_3 + 3x_4 = 2 \\ 2x_1 + 4x_2 + 6x_3 + 8x_4 = 4 \\ x_1 + 4x_2 + 3x_3 = 4 \end{cases}$$

47. $$\begin{cases} x_1 + 3x_2 + 4x_3 - x_4 = 1 \\ x_1 - x_2 - 2x_3 + x_4 = 2 \\ x_1 - 2x_2 + 3x_3 + x_4 = 0 \\ 2x_1 + 2x_2 + 2x_3 = 3 \end{cases}$$

48. $$\begin{cases} 2x_1 - x_2 + x_3 = 6 \\ x_1 + x_2 + x_3 + 2x_4 = 0 \\ 2x_1 + x_2 - x_3 + x_4 = -3 \\ x_1 + 2x_2 - 2x_3 + x_4 = -9 \end{cases}$$

49. Solve $\begin{cases} a_1x + b_1y = c_1 \\ a_2x + b_2y = c_2 \end{cases}$ for x. This gives a formula for solving two equations in two variables for x.

50. Solve $\begin{cases} a_1x + b_1y = c_1 \\ a_2x + b_2y = c_2 \end{cases}$ for y. This gives a formula for solving two equations in two variables for y.

APPLICATIONS

Application problems 51–60 involve systems with unique solutions.

51. **Nutrition** A psychologist studying the effects of nutrition on the behavior of laboratory rats is feeding one group a combination of three foods, I, II, and III. Each of these foods contains three additives, A, B, and C, that are being used in the study. Each additive is a certain percentage of each of the foods as follows.

	Foods		
	I	II	III
Additive A	10%	30%	60%
Additive B	0	4	5
Additive C	2	2	12

If the diet requires 53 g per day of A, 4.5 g per day of B, and 8.6 per day of C, find the number of grams per day of each food that must be used.

52. **Investment** A young couple with a $120,000 inheritance wants to invest all this money. They plan to diversify their investments, choosing some of each of the following types of stock.

Type	Cost/share	Expected total yield/share
Growth	$ 30	$ 4.60
Blue-chip	100	11.00
Utility	50	5.00

Their strategy is to have the total investment in growth stocks equal to the sum of the other investments, and their goal is a 13% return on their investment. How many shares of each type of stock should they purchase?

53. **Transportation** An airline company has three types of aircraft that carry three types of cargo. The payload of each type is summarized in the table below.

Units carried	Plane type		
	Passenger	Transport	Jumbo
First-class mail	100	100	100
Passengers	150	20	350
Air freight	20	65	35

Suppose that on a given day the airline must move 1100 units of first-class mail, 460 units of air freight, and 1930 passengers. How many aircraft of each type should be scheduled?

54. **Investment** A brokerage house offers 3 stock portfolios. Portfolio I consists of 2 blocks of common stock and 1 municipal bond. Portfolio II consists of 4 blocks of common stock, 2 municipal bonds, and 3 blocks of preferred stock. Portfolio III consists of 7 blocks of common stock, 3 municipal bonds, and 3 blocks of preferred stock. A

customer wants 21 blocks of common stock, 10 municipal bonds, and 9 blocks of preferred stock. How many of
each portfolio should be offered?

55. *Investment* Suppose that portfolios I and II in problem
54 are unchanged and portfolio III consists of 2 blocks
of common stock, 2 municipal bonds, and 3 blocks of
preferred stock. A customer wants 12 blocks of common
stock, 6 municipal bonds, and 6 preferred stock. How
many of each portfolio should be offered?

56. *Nutrition* Each ounce of substance A supplies 5% of the
required nutrition a patient needs. Substance B supplies
15% of the required nutrition per ounce and substance C
supplies 12% of required nutrition per ounce. If digestive restrictions require that substances A and C be
given in equal amounts and that the amount of substance B be one-fifth of these other amounts, find the
number of ounces of each substance that should be in the
meal to provide 100% nutrition.

57. *Nutrition* A glass of skim milk supplies 0.1 mg of iron,
8.5 g of protein, and 1 g of carbohydrates. A quarter
pound of lean red meat provides 3.4 mg of iron, 22 g of
protein, and 20 g of carbohydrates. Two slices of whole-
grain bread supply 2.2 mg of iron, 10 g of protein, and
12 g of carbohydrates. If a person on a special diet must
have 12.1 mg of iron, 97 g of protein, and 70 g of
carbohydrates, how many glasses of skim milk, how
many quarter-pound servings of meat, and how many
two-slice servings of whole-grain bread will supply this?

58. *Loans* A bank loaned $118,500 to a company for the
development of two products. If the loan for product A
was for $34,500 more than for product B, how much was
loaned for each product?

59. *Investment* A man has $23,500 invested in two rental
properties. One earns 15% on the investment and the
other yields 16%. His total income from the two properties is $3,625. How much is his income from each
property?

60. *Transportation* The King Trucking Company has an
order for three products to be delivered to a destination.
The table below gives the particulars for the products.

	Type I	Type II	Type III
Unit volume			
(cu ft)	10	8	20
Unit weight (lb)	10	20	40
Unit value ($)	100	20	200

If the carrier can carry 6,000 cu ft, 11,000 lbs, and is
insured for $36,900, how many units of each type can be
carried?

Application problems 61–68 involve systems with nonunique solutions.

61. *Nutrition* A botanist can purchase plant food of 4 different types, I, II, III, and IV. Each food comes in the
same size bag and the following table summarizes the
number of grams of each of three nutrients that each bag
contains.

| | Foods | | | |
	I	II	III	IV
Nutrient A	5	5	10	5
Nutrient B	10	5	30	10
Nutrient C	5	15	10	25

The botanist wants to use a food that has these nutrients
in a different proportion and determines that he will
need a total of 10,000 g of A, 20,000 g of B, and 20,000
of C. Find the number of bags of each type of food that
should be ordered.

62. *Traffic flow* In the analysis of traffic flow, a certain city
estimates the following situation for an area of its downtown district. Figure 2.3 indicates the traffic flow, with
the arrows indicating the flow of traffic. If x_1 represents
the number of cars traveling between intersections A
and B, x_2 represents the number of cars traveling between B and C, x_3 the number between C and D, and x_4
the number between D and A, we can formulate equa-

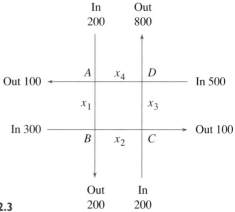

Figure 2.3

tions based on the principle that the number of vehicles entering an intersection equals the number leaving it. That is, for intersection A we obtain

$$200 + x_4 = 100 + x_1.$$

(a) Formulate equations for the traffic at B, C, and D.
(b) Solve the system of these four equations.

63. **Investment** An investment club has set a goal of earning 15% of the money they invest in stocks. The members are considering purchasing four possible stocks, with their cost per share (in dollars) and their earnings per share (in dollars) summarized in the following table.

	Oil	Bank	Computer	Retail
Cost/share	100	40	30	20
Yield/share	10.00	3.60	6.00	2.40

If they have $102,300 to invest, how many shares of each stock should they buy to meet their goal?

64. **Investment** A trust account manager has $200,000 to be invested. The investment choices have current yields of 8%, 7%, and 10%. Suppose that the investment goal is to earn interest of $16,000, and risk factors make it prudent to invest some money in all three investments. What amount should be invested at each rate?

65. **Nutrition** Three different bacteria are cultured in one dish and feed on three nutrients. Each individual of species I consumes one unit of each of the first and second nutrients and two units of the third nutrient. Each individual of species II consumes two units of the first nutrient and two units of the third nutrient. Species III con-

sumes two units of the first nutrient, three units of the second nutrient, and five units of the third nutrient. If the culture is given 5100 units of the first nutrient, 6900 units of the second nutrient, and 12,000 units of the third nutrient, how many of each species can be supported so all of the nutrients are consumed?

66. **Irrigation** An irrigation system allows water to flow in the pattern shown in Figure 2.4. Water flows into the system at A and exits at $B, C, D,$ and E with the amounts shown. Using the fact that at each point the water entering equals the water leaving, formulate an equation for water flow at each of the five points and solve the system.

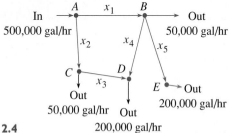

Figure 2.4

67. **Investment** A brokerage house offers 3 stock portfolios. Portfolio I consists of 2 blocks of common stock and 1 municipal bond. Portfolio II consists of 4 blocks of common stock, 2 municipal bonds, and 3 blocks of preferred stock. Portfolio III consists of 2 blocks of common stock, 1 municipal bond, and 3 blocks of preferred stock. A customer wants 16 blocks of common stock, 8 municipal bonds, and 6 blocks of preferred stock. If the numbers of each portfolio offered must be integers, find all possible offerings.

2.4 INVERSE OF A SQUARE MATRIX

Objectives

❑ To find the inverse of a square matrix
❑ To use inverse matrices to solve systems of linear equations
❑ To find determinants of certain matrices

APPLICATION PREVIEW

During World War II, the military commanders sent instructions that used simple substitution codes. These codes were easily broken, so they were further coded by using a coding matrix and matrix multiplication. The resulting coded messages, when received, could be unscrambled with a decoding matrix that is the **inverse matrix** of the coding matrix. In Section 2.2, we encoded messages by using the following code:

a	b	c	d	e	f	g	h	i	j	k	l
1	2	3	4	5	6	7	8	9	10	11	12

m	n	o	p	q	r	s	t	u	v	w	x
13	14	15	16	17	18	19	20	21	22	23	24

y	z	blank
25	26	27

and the encoding matrix

$$A = \begin{bmatrix} 3 & 5 \\ 4 & 6 \end{bmatrix},$$

which converts the numbers to the coded message (in pairs of numbers). To decode any message encoded by A, we must find the **inverse** of matrix A, denoted by A^{-1}, and multiply A^{-1} by the coded message (in pairs as 2×1 matrices). In this section, we will discuss how to find and apply the inverse of a given square matrix.

If the product of A and B is the identity matrix, I, we say that B is the inverse of A (and A is the inverse of B). The matrix B is called the **inverse matrix** of A, denoted A^{-1}.

Inverse Matrices   Two square matrices, A and B, are called **inverses** of each other if

$$AB = I \quad \text{and} \quad BA = I.$$

In this case, $B = A^{-1}$ and $A = B^{-1}$.

EXAMPLE I

Is B the inverse of A if $A = \begin{bmatrix} 1 & 2 \\ 1 & 1 \end{bmatrix}$ and $B = \begin{bmatrix} -1 & 2 \\ 1 & -1 \end{bmatrix}$?

Solution

$$AB = \begin{bmatrix} 1 & 2 \\ 1 & 1 \end{bmatrix} \cdot \begin{bmatrix} -1 & 2 \\ 1 & -1 \end{bmatrix} = \begin{bmatrix} 1 & 0 \\ 0 & 1 \end{bmatrix} = I$$

$$BA = \begin{bmatrix} -1 & 2 \\ 1 & -1 \end{bmatrix} \cdot \begin{bmatrix} 1 & 2 \\ 1 & 1 \end{bmatrix} = \begin{bmatrix} 1 & 0 \\ 0 & 1 \end{bmatrix} = I$$

Thus B is the inverse of A, or $B = A^{-1}$. We also say A is the inverse of B, or $A = B^{-1}$. Thus $(A^{-1})^{-1} = A$.

CHECKPOINT

1. Are the matrices A and B inverses if

$$A = \begin{bmatrix} 1 & 1 & -1 \\ 1 & -1 & 3 \\ 4 & 2 & 1 \end{bmatrix} \quad \text{and} \quad B = \begin{bmatrix} 3.5 & 1.5 & -1 \\ -5.5 & -2.5 & 2 \\ -3 & -1 & 1 \end{bmatrix}?$$

We have seen how to use elementary row operations on augmented matrices to solve systems of linear equations. We can also find the inverse of a matrix using elementary row operations.

If the inverse exists for a square matrix A, we find A^{-1} as follows.

Finding the Inverse of a Square Matrix

Procedure

To find the inverse of the square matrix A:

1. Form the augmented matrix $[A \,|\, I]$ where A is the $n \times n$ matrix and I is the $n \times n$ identity matrix.

2. Perform elementary operations on $[A \,|\, I]$ until we have an augmented matrix of the form $[I \,|\, B]$; that is, until the matrix A on the left is transformed into the identity matrix.

3. The matrix B (on the right) is the inverse of matrix A.

Example

Find the inverse of matrix $A = \begin{bmatrix} 1 & 2 \\ 1 & 1 \end{bmatrix}$.

1. $\left[\begin{array}{cc|cc} 1 & 2 & 1 & 0 \\ 1 & 1 & 0 & 1 \end{array}\right]$

2. $\left[\begin{array}{cc|cc} 1 & 2 & 1 & 0 \\ 1 & 1 & 0 & 1 \end{array}\right]$

$$\xrightarrow{-R_1 + R_2 \to R_2} \left[\begin{array}{cc|cc} 1 & 2 & 1 & 0 \\ 0 & -1 & -1 & 1 \end{array}\right]$$

$$\xrightarrow[-R_2 \to R_2]{2R_2 + R_1 \to R_1} \left[\begin{array}{cc|cc} 1 & 0 & -1 & 2 \\ 0 & 1 & 1 & -1 \end{array}\right]$$

3. The inverse of A is $B = \begin{bmatrix} -1 & 2 \\ 1 & -1 \end{bmatrix}$.

We have verified this in Example 1.

Note that if A has no inverse, the reduction process on $[A \,|\, I]$ will yield a row of zeros in the left half.

EXAMPLE 2

Find the inverse of matrix A, below.

$$A = \begin{bmatrix} 2 & 5 & 4 \\ 1 & 4 & 3 \\ 1 & -3 & -2 \end{bmatrix}$$

Solution

To find the inverse of matrix A, we reduce A in $[A\,|\,I]$ (the matrix A augmented with the identity matrix I).

$$\begin{bmatrix} 2 & 5 & 4 & | & 1 & 0 & 0 \\ 1 & 4 & 3 & | & 0 & 1 & 0 \\ 1 & -3 & -2 & | & 0 & 0 & 1 \end{bmatrix}$$

When the left side becomes an identity matrix, the inverse is in the augment.

$$\xrightarrow{\text{Switch } R_1 \text{ and } R_2} \begin{bmatrix} 1 & 4 & 3 & | & 0 & 1 & 0 \\ 2 & 5 & 4 & | & 1 & 0 & 0 \\ 1 & -3 & -2 & | & 0 & 0 & 1 \end{bmatrix}$$

$$\begin{array}{c} -2R_1 + R_2 \to R_2 \\ \xrightarrow{\hspace{2cm}} \\ -R_1 + R_3 \to R_3 \end{array} \begin{bmatrix} 1 & 4 & 3 & | & 0 & 1 & 0 \\ 0 & -3 & -2 & | & 1 & -2 & 0 \\ 0 & -7 & -5 & | & 0 & -1 & 1 \end{bmatrix}$$

$$\xrightarrow{-\frac{1}{3}R_2 \to R_2} \begin{bmatrix} 1 & 4 & 3 & | & 0 & 1 & 0 \\ 0 & 1 & \frac{2}{3} & | & -\frac{1}{3} & \frac{2}{3} & 0 \\ 0 & -7 & -5 & | & 0 & -1 & 1 \end{bmatrix}$$

$$\begin{array}{c} -4R_2 + R_1 \to R_1 \\ \xrightarrow{\hspace{2cm}} \\ 7R_2 + R_3 \to R_3 \end{array} \begin{bmatrix} 1 & 0 & \frac{1}{3} & | & \frac{4}{3} & -\frac{5}{3} & 0 \\ 0 & 1 & \frac{2}{3} & | & -\frac{1}{3} & \frac{2}{3} & 0 \\ 0 & 0 & -\frac{1}{3} & | & -\frac{7}{3} & \frac{11}{3} & 1 \end{bmatrix}$$

$$\begin{array}{c} R_3 + R_1 \to R_1 \\ 2R_3 + R_2 \to R_2 \\ \xrightarrow{\hspace{2cm}} \\ -3R_2 \to R_3 \end{array} \begin{bmatrix} 1 & 0 & 0 & | & -1 & 2 & 1 \\ 0 & 1 & 0 & | & -5 & 8 & 2 \\ 0 & 0 & 1 & | & 7 & -11 & -3 \end{bmatrix}$$

The inverse we seek is

$$\begin{bmatrix} -1 & 2 & 1 \\ -5 & 8 & 2 \\ 7 & -11 & -3 \end{bmatrix}.$$

CHECKPOINT

2. Find the inverse of the matrix

$$A = \begin{bmatrix} 1 & 2 & 1 \\ 0 & 1 & 1 \\ 1 & 2 & 2 \end{bmatrix}.$$

The technique of reducing $[A \mid I]$ can be used to find A^{-1} or to find that A^{-1} doesn't exist for any square matrix A. However, a special formula can be used to find the inverse of a 2×2 matrix, if the inverse exists.

Inverse of a 2 × 2 Matrix   If $A = \begin{bmatrix} a & b \\ c & d \end{bmatrix}$, then $A^{-1} = \dfrac{1}{ad - bc} \begin{bmatrix} d & -b \\ -c & a \end{bmatrix}$ provided $ad - bc \neq 0$.

If $ad - bc = 0$, then A^{-1} does not exist.

This result can be verified by direct calculation or by reducing $[A \mid I]$.

EXAMPLE 3

Find the inverse, if it exists, for each of the following.

(a) $A = \begin{bmatrix} 3 & 7 \\ 2 & 5 \end{bmatrix}$ (b) $B = \begin{bmatrix} 2 & -1 \\ -4 & 2 \end{bmatrix}$

Solution

(a) For A, $ad - bc = (3)(5) - (2)(7) = 1 \neq 0$, so A^{-1} exists.

$$A^{-1} = \frac{1}{1}\begin{bmatrix} 5 & -7 \\ -2 & 3 \end{bmatrix} = \begin{bmatrix} 5 & -7 \\ -2 & 3 \end{bmatrix}$$

(b) For B, $ad - bc = (2)(2) - (-4)(-1) = 4 - 4 = 0$, so B^{-1} does not exist.

EXAMPLE 4

In the Application Preview we recalled how to encode

a	b	c	d	e	f	g	h	i	j	k	l
1	2	3	4	5	6	7	8	9	10	11	12

m	n	o	p	q	r	s	t	u	v	w	x
13	14	15	16	17	18	19	20	21	22	23	24

y	z	blank
25	26	27

with the encoding matrix

$$A = \begin{bmatrix} 3 & 5 \\ 4 & 6 \end{bmatrix},$$

which converts the numbers to the coded message (in pairs of numbers).

To decode any message encoded by A, we must find the inverse of A, and multiply A^{-1} by the coded message (in pairs as 2×1 matrices). Use the inverse of A to decode the coded message 96, 118, 65, 84, 131, 168, 55, 72.

Solution

We first find the inverse of matrix A.

$$A^{-1} = \frac{1}{18-20}\begin{bmatrix} 6 & -5 \\ -4 & 3 \end{bmatrix} = -\frac{1}{2}\begin{bmatrix} 6 & -5 \\ -4 & 3 \end{bmatrix} = \begin{bmatrix} -3 & 2.5 \\ 2 & -1.5 \end{bmatrix}$$

Multiplying A^{-1} by the numbers, in pairs, gives

$$\begin{bmatrix} -3 & 2.5 \\ 2 & -1.5 \end{bmatrix}\begin{bmatrix} 96 \\ 118 \end{bmatrix} = \begin{bmatrix} 7 \\ 15 \end{bmatrix}$$

$$\begin{bmatrix} -3 & 2.5 \\ 2 & -1.5 \end{bmatrix}\begin{bmatrix} 65 \\ 84 \end{bmatrix} = \begin{bmatrix} 15 \\ 4 \end{bmatrix}$$

$$\begin{bmatrix} -3 & 2.5 \\ 2 & -1.5 \end{bmatrix}\begin{bmatrix} 131 \\ 168 \end{bmatrix} = \begin{bmatrix} 27 \\ 10 \end{bmatrix}$$

$$\begin{bmatrix} -3 & 2.5 \\ 2 & -1.5 \end{bmatrix}\begin{bmatrix} 55 \\ 72 \end{bmatrix} = \begin{bmatrix} 15 \\ 2 \end{bmatrix}.$$

The result is 7, 15, 15, 4, 27, 10, 15, 2, which is the message "good job."

 ## Graphing Utilities

Computer software programs and many types of calculators can be used to find the inverse of a matrix. Because of their ease of usage, it is logical to use graphing utilities to find inverses, especially for large matrices.

 EXAMPLE 5

Find the inverse of the following matrix.

$$A = \begin{bmatrix} 0.1 & 0 & 0.1 & 0.1 & 0.2 \\ 0.1 & 0.2 & -0.1 & 0.1 & -0.1 \\ 0.1 & -0.1 & 0.1 & -0.2 & 0.1 \\ 0.1 & 0.2 & 0.1 & 0.1 & 0.1 \\ 0.1 & 0 & 0 & 0 & 0 \end{bmatrix}$$

Solution

Entering this matrix and using an inverse key or command gives A^{-1} (to 9 decimal places).

$$A^{-1} = \begin{bmatrix} 0 & 0 & 0 & 0 & 10 \\ 0 & 5 & 3.333333333 & 1.666666667 & -10 \\ -10 & -15 & -6.666666667 & 11.666666667 & 20 \\ 0 & -5 & -6.666666667 & 1.666666667 & 10 \\ 10 & 10 & 6.666666667 & -6.666666667 & -20 \end{bmatrix}$$

Writing the fractional equivalents of the decimals gives the exact inverse matrix.

$$A^{-1} = \begin{bmatrix} 0 & 0 & 0 & 0 & 10 \\ 0 & 5 & \frac{10}{3} & \frac{5}{3} & -10 \\ -10 & -15 & -\frac{20}{3} & \frac{35}{3} & 20 \\ 0 & -5 & -\frac{20}{3} & \frac{5}{3} & 10 \\ 10 & 10 & \frac{20}{3} & -\frac{20}{3} & -20 \end{bmatrix}$$

The product of A and A^{-1} was found to be $I = \begin{bmatrix} 1 & 0 & 0 & 0 & 0 \\ 0 & 1 & 0 & 0 & 0 \\ 0 & 0 & 1 & 0 & 0 \\ 0 & 0 & 0 & 1 & 0 \\ 0 & 0 & 0 & 0 & 1 \end{bmatrix}$ in problem 55 of Exercise 2.2.

MATRIX EQUATIONS

In Example 1 of Section 2.3, we solved the system

$$\begin{cases} 2x + 5y + 4z = 4 \\ x + 4y + 3z = 1 \\ x - 3y - 2z = 5. \end{cases}$$

This system can be written as the following mat

$$\begin{bmatrix} 2x + 5y + 4z \\ x + 4y + 3z \\ x - 3y - 2z \end{bmatrix} =$$

The matrix equation is often written as f

$$\begin{bmatrix} 2 & 5 & 4 \\ 1 & 4 & 3 \\ 1 & -3 & - \end{bmatrix}$$

quivalent to the first
ritten in this second
both sides of the equation

We can see that this second form of
if we carry out the multiplication
form, we could also solve the syste
by the inverse of

$$3$$

$$-3 \quad -2$$

EXAMPLE 6

Solve the system

$$\begin{cases} 2x + 5y + 4z = 4 \\ x + 4y + 3z = 1 \\ x - 3y - 2z = 5 \end{cases}$$

by using inverse matrices.

Solution

To solve the system we multiply both sides of the associated matrix equation by the inverse, which we found in Example 2.

$$\begin{bmatrix} -1 & 2 & 1 \\ -5 & 8 & 2 \\ 7 & -11 & -3 \end{bmatrix} \begin{bmatrix} 2 & 5 & 4 \\ 1 & 4 & 3 \\ 1 & -3 & -2 \end{bmatrix} \begin{bmatrix} x \\ y \\ z \end{bmatrix} = \begin{bmatrix} -1 & 2 & 1 \\ -5 & 8 & 2 \\ 7 & -11 & -3 \end{bmatrix} \begin{bmatrix} 4 \\ 1 \\ 5 \end{bmatrix}$$

Notice that we must be careful to multiply both sides *from the left* since matrix multiplication is not commutative. If we carry out the multiplications, we obtain

$$\begin{bmatrix} 1 & 0 & 0 \\ 0 & 1 & 0 \\ 0 & 0 & 1 \end{bmatrix} \begin{bmatrix} x \\ y \\ z \end{bmatrix} = \begin{bmatrix} 3 \\ -2 \\ 2 \end{bmatrix}, \quad \text{or} \quad \begin{bmatrix} x \\ y \\ z \end{bmatrix} = \begin{bmatrix} 3 \\ -2 \\ 2 \end{bmatrix},$$

which yields $x = 3$, $y = -2$, $z = 2$, the same solution as that found in Example 1 of Section 2.3.

Thus inverse matrices can be used to solve systems of equations. Unfortunately, this method will work only if the solution to the system is unique. In fact, a system $AX = C$ has a unique solution if and only if A^{-1} exists.

ECKPOINT

3. Use inverse matrices to solve the system.

$$\begin{cases} 7x - 5y = 12 \\ 2x - 3y = 6 \end{cases}$$

Graphing Utilities

Because technology can so easily find the inverse of a matrix, it is easy to use graphing utilities to solve a system of linear equations by using the methods introduced in Example 6.

EXAMPLE 7

Use the inverse of the coefficient matrix to solve the following system of linear equations.

$$\begin{cases} 2x_1 + 3x_2 - x_3 + 3x_4 + x_5 = 22 \\ 3x_1 + x_2 - 4x_3 + 3x_4 - x_5 = 0 \\ x_1 - 2x_2 + 3x_3 - 4x_4 + 2x_5 = 0 \\ x_1 + 2x_2 - 3x_3 + 2x_4 - 2x_5 = -6 \\ 2x_1 + 4x_4 - 5x_5 = -7 \end{cases}$$

Solution

The coefficient matrix for this system is

$$A = \begin{bmatrix} 2 & 3 & -1 & 3 & 1 \\ 3 & 1 & -4 & 3 & -1 \\ 1 & -2 & 3 & -4 & 2 \\ 1 & 2 & -3 & 2 & -2 \\ 2 & 0 & 0 & 4 & -5 \end{bmatrix}$$

and the matrix equation for this system is

$$AX = \begin{bmatrix} 2 & 3 & -1 & 3 & 1 \\ 3 & 1 & -4 & 3 & -1 \\ 1 & -2 & 3 & -4 & 2 \\ 1 & 2 & -3 & 2 & -2 \\ 2 & 0 & 0 & 4 & -5 \end{bmatrix} \begin{bmatrix} x_1 \\ x_2 \\ x_3 \\ x_4 \\ x_5 \end{bmatrix} = \begin{bmatrix} 22 \\ 0 \\ 0 \\ -6 \\ -7 \end{bmatrix}.$$

If we use technology to find the inverse of the coefficient matrix and then multiply both sides of this matrix equation by the inverse, we get the solution. The inverse of the coefficient matrix has elements that are approximated by decimals, so the inverse may be difficult to read on a small screen. However, we need not record the inverse. If we multiply the coefficient matrix A by its inverse, we get

$$A^{-1}AX = IX = X,$$

so multiplying both sides by A^{-1}, we get the solution, X.

$$X = \begin{bmatrix} x_1 \\ x_2 \\ x_3 \\ x_4 \\ x_5 \end{bmatrix} = A^{-1} \begin{bmatrix} 22 \\ 0 \\ 0 \\ -6 \\ -7 \end{bmatrix}$$

Thus we can multiply the inverse we found times the column matrix to obtain

$$\begin{bmatrix} x_1 \\ x_2 \\ x_3 \\ x_4 \\ x_5 \end{bmatrix} = \begin{bmatrix} 1 \\ 2 \\ 3 \\ 4 \\ 5 \end{bmatrix}.$$

Thus the solution is $x_1 = 1$, $x_2 = 2$, $x_3 = 3$, $x_4 = 4$, $x_5 = 5$.

Like graphics calculators and computer software programs, computer spreadsheets also can be used to find the inverse of a matrix, if it exists.

EXAMPLE 8

Suppose a bank draws its venture capital funds from three main sources of income—business loans, auto loans, and home mortgages. The income from these sources for each of three years is given in Table 2.8, and the venture capital for these years is given in Table 2.9. If the bank uses a fixed percent of its income from each of the business loans (x), auto loans (y), and home mortgages (z) to get its venture capital funds, find the percent of income used for each of these sources of income.

Table 2.8

	A	B	C	D
1	Income	from	Loans	
2		Business	Auto	Home
3	1993	68,210	23,324	57,234
4	1994	43,455	19,335	66,345
5	1995	58,672	15,654	69,334

Table 2.9

Venture	Capital
1993	53,448.74
1994	43,447.50
1995	50,582.88

Solution

The matrix equation describing this problem is $AX = V$, or

$$\begin{bmatrix} 68,210 & 23,324 & 57,234 \\ 43,455 & 19,335 & 66,345 \\ 58,672 & 15,654 & 69,334 \end{bmatrix} \begin{bmatrix} x \\ y \\ z \end{bmatrix} = \begin{bmatrix} 53,448.74 \\ 43,447.50 \\ 50,582.88 \end{bmatrix}.$$

Finding the inverse of matrix of A and multiplying it times both sides of the matrix equation (on the left) gives the solution

$$\begin{bmatrix} x \\ y \\ z \end{bmatrix} = \begin{bmatrix} 0.47 \\ 0.23 \\ 0.28 \end{bmatrix}.$$

Thus 47%, 23%, and 28% come from business, auto, and home loans, respectively.

DETERMINANTS

The value $ad - bc$ used in finding the inverse of the 2×2 matrix

$$A = \begin{bmatrix} a & b \\ c & d \end{bmatrix}$$

is used so frequently that it is given a special name, the **determinant** of A (denoted *det A* or $|A|$).

Determinant of a 2 × 2 Matrix	$\det\begin{bmatrix} a & b \\ c & d \end{bmatrix} = \begin{vmatrix} a & b \\ c & d \end{vmatrix} = ad - bc$

EXAMPLE 9

Find det A if $A = \begin{bmatrix} 2 & 4 \\ 3 & -4 \end{bmatrix}$.

Solution

$$\det A = (2)(-4) - (3)(4) = -8 - 12 = -20$$

Formulas exist to find determinants of square matrices of orders larger than 2×2, but calculators and computers make it easy to find determinants. We have already seen that the inverse of a 2×2 matrix A does not exist if det $A = 0$. In general, the inverse of an $n \times n$ matrix B does not exist if det $B = 0$. This also means that if the coefficient matrix of a system of linear equations has a determinant equal to 0, the system does not have a unique solution.

EXAMPLE 10

(a) Use a calculator to find the determinant of

$$A = \begin{bmatrix} 2 & 3 & 0 & 1 & 3 \\ 1 & 0 & 2 & 3 & 1 \\ 0 & 1 & 0 & 2 & 3 \\ 1 & 0 & 2 & 2 & 1 \\ 1 & 0 & 0 & 0 & 3 \end{bmatrix}.$$

(b) Does the following system have a unique solution?

$$\begin{cases} 2x_1 + 3x_2 & + x_4 + 3x_5 = 16 \\ x_1 & + 2x_3 + 3x_4 + x_5 = 2 \\ x_2 & + 2x_4 + 3x_5 = 9 \\ x_1 & + 2x_3 + 2x_4 + x_5 = 3 \\ x_1 & + 3x_5 = 10 \end{cases}$$

Solution

(a) Entering matrix A and using the determinant key gives det $A = 24$.

(b) Because A is the coefficient matrix for the system and det $A = 24$, the system has a unique solution. Using the inverse matrix gives the solution $x_1 = 1, x_2 = 2, x_3 = 0.5, x_4 = -1, x_5 = 3$.

CHECKPOINT SOLUTIONS

1. Yes

2.
$$\left[\begin{array}{ccc|ccc} 1 & 2 & 1 & 1 & 0 & 0 \\ 0 & 1 & 1 & 0 & 1 & 0 \\ 1 & 2 & 2 & 0 & 0 & 1 \end{array}\right] \rightarrow \left[\begin{array}{ccc|ccc} 1 & 2 & 1 & 1 & 0 & 0 \\ 0 & 1 & 1 & 0 & 1 & 0 \\ 0 & 0 & 1 & -1 & 0 & 1 \end{array}\right]$$

$$\rightarrow \left[\begin{array}{ccc|ccc} 1 & 0 & -1 & 1 & -2 & 0 \\ 0 & 1 & 1 & 0 & 1 & 0 \\ 0 & 0 & 1 & -1 & 0 & 1 \end{array}\right] \rightarrow \left[\begin{array}{ccc|ccc} 1 & 0 & 0 & 0 & -2 & 1 \\ 0 & 1 & 0 & 1 & 1 & -1 \\ 0 & 0 & 1 & -1 & 0 & 1 \end{array}\right]$$

The inverse is $A^{-1} = \begin{bmatrix} 0 & -2 & 1 \\ 1 & 1 & -1 \\ -1 & 0 & 1 \end{bmatrix}$.

3. This system can be written

$$\begin{bmatrix} 7 & -5 \\ 2 & -3 \end{bmatrix}\begin{bmatrix} x \\ y \end{bmatrix} = \begin{bmatrix} 12 \\ 6 \end{bmatrix}$$

and for the coefficient matrix A, $ad - bc = -21 - (-10) = -11$. Thus, A^{-1} exists.

$$A^{-1} = -\frac{1}{11}\begin{bmatrix} -3 & 5 \\ -2 & 7 \end{bmatrix} = \begin{bmatrix} \frac{3}{11} & -\frac{5}{11} \\ \frac{2}{11} & -\frac{7}{11} \end{bmatrix},$$

so multiplying *from the left* gives the solution.

$$\begin{bmatrix} \frac{3}{11} & -\frac{5}{11} \\ \frac{2}{11} & -\frac{7}{11} \end{bmatrix}\begin{bmatrix} 7 & -5 \\ 2 & -3 \end{bmatrix}\begin{bmatrix} x \\ y \end{bmatrix} = \begin{bmatrix} \frac{3}{11} & -\frac{5}{11} \\ \frac{2}{11} & -\frac{7}{11} \end{bmatrix}\begin{bmatrix} 12 \\ 6 \end{bmatrix}$$

$$\begin{bmatrix} 1 & 0 \\ 0 & 1 \end{bmatrix}\begin{bmatrix} x \\ y \end{bmatrix} = \begin{bmatrix} \frac{6}{11} \\ -\frac{18}{11} \end{bmatrix}$$

So $x = \frac{6}{11}, y = -\frac{18}{11}$.

EXERCISE 2.4

1. If A is a 3×3 matrix and B is its inverse, what does the product AB equal?

2. If $C = \begin{bmatrix} 2 & -4 & 12 \\ 0 & 6 & -12 \\ 1 & -2 & 3 \end{bmatrix}$ and

   $D = \dfrac{1}{6}\begin{bmatrix} 1 & 2 & 4 \\ 2 & 1 & -4 \\ 1 & 0 & -2 \end{bmatrix}$, are C and D inverse matrices?

3. If $A = \begin{bmatrix} 1 & 2 & 1 \\ 0 & 0 & 3 \\ 1 & 0 & 1 \end{bmatrix}$ and $B = \begin{bmatrix} 0 & -\frac{1}{3} & 1 \\ \frac{1}{2} & 0 & -\frac{1}{2} \\ 0 & \frac{1}{3} & 0 \end{bmatrix}$, does $B = A^{-1}$?

4. If $D = \begin{bmatrix} 0 & 2 & -6 \\ -3 & 0 & 3 \\ 0 & -2 & 0 \end{bmatrix}$ and $C = -\dfrac{1}{3}\begin{bmatrix} 2 & 4 & 2 \\ 0 & 0 & 6 \\ 2 & 0 & 2 \end{bmatrix}$, does $C = D^{-1}$?

5. What is the inverse of $A = \begin{bmatrix} 3 & 0 & 0 \\ 0 & 3 & 0 \\ 0 & 0 & 3 \end{bmatrix}$?

6. What is the inverse of $B = \begin{bmatrix} 4 & 0 & 0 \\ 0 & 3 & 0 \\ 0 & 0 & 3 \end{bmatrix}$?

In problems 7–14, find the inverse matrix for each matrix that has an inverse.

7. $\begin{bmatrix} 4 & 7 \\ 1 & 2 \end{bmatrix}$ 8. $\begin{bmatrix} 4 & 5 \\ 7 & 9 \end{bmatrix}$ 9. $\begin{bmatrix} 2 & -1 \\ -4 & 2 \end{bmatrix}$

10. $\begin{bmatrix} 3 & 12 \\ 1 & 4 \end{bmatrix}$ 11. $\begin{bmatrix} 2 & 2 \\ 4 & 5 \end{bmatrix}$ 12. $\begin{bmatrix} 6 & -4 \\ -1 & 1 \end{bmatrix}$

13. $\begin{bmatrix} 4 & 7 \\ 2 & 1 \end{bmatrix}$ 14. $\begin{bmatrix} 1 & 2 \\ 3 & 0 \end{bmatrix}$

In problems 15–22, find the inverse matrix for each matrix that has an inverse.

15. $\begin{bmatrix} 0 & 1 & 0 \\ 1 & 1 & 0 \\ 0 & 1 & 1 \end{bmatrix}$ 16. $\begin{bmatrix} 0 & 2 & 1 \\ 3 & 0 & 1 \\ 1 & 1 & 1 \end{bmatrix}$

17. $\begin{bmatrix} 3 & 1 & 2 \\ 1 & 2 & 3 \\ 1 & 1 & 1 \end{bmatrix}$ 18. $\begin{bmatrix} 1 & 2 & 3 \\ -1 & 5 & 6 \\ -1 & 3 & 3 \end{bmatrix}$

19. $\begin{bmatrix} 1 & 3 & 5 \\ -1 & -1 & 2 \\ 1 & 5 & 12 \end{bmatrix}$ 20. $\begin{bmatrix} 1 & -1 & 4 \\ -1 & 0 & -2 \\ -1 & -3 & 4 \end{bmatrix}$

21. $\begin{bmatrix} 1 & 2 & 4 \\ 1 & -1 & -3 \\ 2 & 1 & 1 \end{bmatrix}$ 22. $\begin{bmatrix} 3 & 4 & -1 \\ 4 & 2 & 2 \\ 2 & 6 & -4 \end{bmatrix}$

23. Find the inverse of

$C = \begin{bmatrix} 0 & -1 & 0 & 1 & 0 \\ 1 & 1.5 & -1 & -1.5 & -0.5 \\ 0.5 & 1.75 & -1 & -1.25 & -0.25 \\ -0.5 & -1 & 1 & 1 & 0 \\ 0 & 0.5 & 0 & -0.5 & 0.5 \end{bmatrix}$

24. Find the inverse of $B = \begin{bmatrix} 2 & 3 & 0 & 1 & 3 \\ 1 & 0 & 2 & 3 & 1 \\ 0 & 1 & 0 & 2 & 3 \\ 1 & 0 & 2 & 2 & 1 \\ 1 & 0 & 0 & 0 & 3 \end{bmatrix}$.

In problems 25 and 26, use inverse matrices to find the solution to the given matrix equation.

25. $\begin{bmatrix} 6 & 3 & -8 \\ 1 & 0 & -3 \\ 20 & -12 & 2 \end{bmatrix}\begin{bmatrix} x \\ y \\ z \end{bmatrix} = \begin{bmatrix} 1 \\ -2 \\ 10 \end{bmatrix}$

26. $\begin{bmatrix} 3 & -1 & 0 \\ 2 & -1 & 1 \\ -1 & 1 & -2 \end{bmatrix}\begin{bmatrix} x \\ y \\ z \end{bmatrix} = \begin{bmatrix} 11 \\ 9 \\ -7 \end{bmatrix}$

27. Use the inverse found in problem 15 to solve

$\begin{bmatrix} 0 & 1 & 0 \\ 1 & 1 & 0 \\ 0 & 1 & 1 \end{bmatrix}\begin{bmatrix} x \\ y \\ z \end{bmatrix} = \begin{bmatrix} 1 \\ 2 \\ 3 \end{bmatrix}$.

28. Use the inverse found in problem 16 to solve

$\begin{bmatrix} 0 & 2 & 1 \\ 3 & 0 & 1 \\ 1 & 1 & 1 \end{bmatrix}\begin{bmatrix} x \\ y \\ z \end{bmatrix} = \begin{bmatrix} 1 \\ 0 \\ 4 \end{bmatrix}$.

In problems 29–38, use inverse matrices to solve each system of linear equations.

29. $\begin{cases} x + 2y = 4 \\ 3x + 4y = 10 \end{cases}$ 30. $\begin{cases} 3x - 4y = 11 \\ 2x + 3y = -4 \end{cases}$

31. $\begin{cases} 2x + y = 4 \\ 3x + y = 5 \end{cases}$ 32. $\begin{cases} 5x - 2y = 6 \\ 3x + 3y = 12 \end{cases}$

33. $\begin{cases} x + y + z = 3 \\ 2x + y + z = 4 \\ 2x + 2y + z = 5 \end{cases}$

34. $\begin{cases} 2x - y - 2z = 2 \\ 3x - y + z = -3 \\ x + y - z = 7 \end{cases}$

35. $\begin{cases} x + y + 2z = 8 \\ 2x + y + z = 7 \\ 2x + 2y + z = 10 \end{cases}$

36. $\begin{cases} x - 2y + z = 0 \\ 2x + y - 2z = 2 \\ 3x + 2y - 3z = 2 \end{cases}$

37. $\begin{cases} \quad\ -x_2 \qquad\quad + x_4 \qquad\qquad = 0.7 \\ x_1 + \quad x_2 - x_3 - 1.5x_4 - 0.5x_5 = -1.6 \\ 0.5x_1 + 1.75x_2 - x_3 - 1.25x_4 - 0.25x_5 = 1.275 \\ -0.5x_1 \quad - x_2 + x_3 \quad + x_4 \qquad\quad = 1.15 \\ \qquad\ \ 0.5x_2 \qquad\quad - 0.5x_4 + 0.5x_5 = -0.15 \end{cases}$

38. $\begin{cases} 2x_1 + 3x_2 \qquad + x_4 + 3x_5 = 0 \\ x_1 \qquad + 2x_3 + 3x_4 + x_5 = 27 \\ \quad\ x_2 \qquad + 2x_4 + 3x_5 = -15 \\ x_1 \qquad + 2x_3 + 2x_4 + x_5 = 24 \\ x_1 \qquad\qquad\qquad + 3x_5 = -24 \end{cases}$

In problems 39–42, evaluate each determinant.

39. $\begin{vmatrix} 1 & 2 \\ 3 & 4 \end{vmatrix}$

40. $\begin{vmatrix} 4 & 5 \\ -1 & -2 \end{vmatrix}$

41. $\begin{vmatrix} 3 & -1 \\ 2 & 4 \end{vmatrix}$

42. $\begin{vmatrix} -1 & 2 \\ 4 & -8 \end{vmatrix}$

 In problems 43–46, use technology to find the determinant.

43. $\begin{vmatrix} 3 & 2 & 1 \\ -1 & 0 & 2 \\ 0 & 1 & 1 \end{vmatrix}$

44. $\begin{vmatrix} 1 & 2 & 2 \\ 3 & 0 & 1 \\ -1 & 0 & -2 \end{vmatrix}$

45. $\begin{vmatrix} 0 & 1 & 2 \\ 3 & 1 & 1 \\ 4 & -1 & 3 \end{vmatrix}$

46. $\begin{vmatrix} 3 & 1 & 4 & 2 \\ -1 & 2 & 0 & 5 \\ 4 & 3 & 0 & -1 \\ 0 & 4 & 2 & 1 \end{vmatrix}$

In problems 47–50, use determinants to decide whether each matrix has an inverse.

47. $\begin{vmatrix} 2 & 3 \\ -1 & 4 \end{vmatrix}$

48. $\begin{vmatrix} 3 & 3 \\ 1 & -1 \end{vmatrix}$

49. $\begin{vmatrix} 1 & 3 & -2 \\ 2 & -1 & 5 \\ 3 & 2 & 3 \end{vmatrix}$

50. $\begin{vmatrix} 4 & 1 & 0 \\ 0 & 5 & 0 \\ 0 & 0 & -1 \end{vmatrix}$

APPLICATIONS

Decoding We have encoded messages by assigning the letters of the alphabet the numbers 1–26, with a blank assigned the number 27, and by using an encoding matrix A that converts the numbers to the coded message (in pairs of numbers). To decode any message encoded by A, we must find the inverse of A, denoted by A^{-1}, and multiply A^{-1} times the coded message. In problems 51–54 use this code and the given encoding matrix A to decode the given message.

51. The code matrix is

$$A = \begin{bmatrix} 5 & 9 \\ 6 & 11 \end{bmatrix}.$$

Use A^{-1} to decode the message

49, 59, 133, 161, 270, 327, 313, 381.

52. If the code matrix is

$$A = \begin{bmatrix} 3 & 5 \\ 1 & 2 \end{bmatrix},$$

find A^{-1} and use it to decode

157, 59, 73, 29, 147, 58, 63, 24, 119, 44.

53. If the code matrix is

$$A = \begin{bmatrix} 0 & 2 & 1 \\ 3 & 0 & 1 \\ 1 & 1 & 1 \end{bmatrix},$$

find A^{-1} and use it to decode

47, 22, 34, 28, 87, 46, 63, 66, 55, 56, 44, 43, 17, 14, 15.

(Since A and A^{-1} are 3×3 matrices, use triples of numbers.)

54. The code matrix is

$$A = \begin{bmatrix} 1 & 2 & 1 \\ 5 & 8 & 2 \\ 7 & 11 & 3 \end{bmatrix}.$$

Using A^{-1}, decode 49, 165, 231, 49, 154, 220, 39, 162, 226.

55. **Competition** A product is made by only two competing companies. If Company X retains two-thirds of its customers and loses one-third to Company Y each year, and Company Y retains three-quarters of its customers and loses one-quarter to Company X each year, then we can represent the fraction each company had last year by

$$\begin{bmatrix} x_0 \\ y_0 \end{bmatrix},$$

where x_0 is the fraction Company X had and y_0 is the fraction Company Y had. Then the fraction each will have this year can be represented by

$$\begin{bmatrix} x \\ y \end{bmatrix} = \begin{bmatrix} \frac{2}{3} & \frac{1}{4} \\ \frac{1}{3} & \frac{3}{4} \end{bmatrix} \begin{bmatrix} x_0 \\ y_0 \end{bmatrix}.$$

If Company X has 1900 customers and Company Y has 1700 customers this year, how many customers did each have last year?

56. **Politics** In a two-party system, it is observed that 70% of all Republicans vote for Republicans with the remainder voting for Democrats, while 60% of all Democrats vote for Democrats with the remainder voting for Republicans. (See the diagram below.)

The percent of each party that will win in this election is

$$\begin{bmatrix} r \\ d \end{bmatrix} = \begin{bmatrix} 0.70 & 0.40 \\ 0.30 & 0.60 \end{bmatrix} \begin{bmatrix} r_0 \\ d_0 \end{bmatrix},$$

where $\begin{bmatrix} r_0 \\ d_0 \end{bmatrix}$ represents the percents of Republicans and Democrats, respectively, that won in the last election. If 52% of those who win in this election are Republicans and 48% are Democrats, what were the percentages that won in the last election?

In problems 57–61, set up each system of equations and then solve it by using inverse matrices.

57. **Medication** Medication A is given every 4 hours and medication B is given twice a day. For patient I, the total intake of the two medications is restricted to 50.6 mg per day, and for patient II the total intake is restricted to 92 mg per day. The ratio of dosage of A to the dosage of B is always 5 to 8. Find the dosage for each administration of each medication for both patient I and patient II.

58. **Transportation** The Ace Freight Company has an order for two products to be delivered to two stores of a company. The matrix below gives information regarding the two products.

Product	I	II
Unit volume (cu ft)	20	30
Unit weight (lb)	100	400

If truck A can carry 2350 cu ft and 23,000 lbs, and truck B can carry 2500 cu ft and 24,500 lbs, how many of each product can be carried by each truck?

59. **Investment** One safe investment pays 10% per year and a more risky investment pays 18% per year. A woman has $145,600 to invest and would like to have an income of $20,000 per year from her investments. How much should she invest at each rate?

60. **Nutrition** A biologist has a 40% solution and a 10% solution of the same plant nutrient. How many cubic centimeters of each solution should be mixed to obtain 25 cc of a 28% solution?

61. **Medication** A nurse has two solutions that contain different concentrations of a certain medication. One is a 20% concentration and the other is a 5% concentration. How many cubic centimeters of each should he mix to obtain 10 cc of a 15.5% solution?

62. **Bee ancestry** Because a female bee comes from a fertilized egg and a male bee comes from an unfertilized egg, the number $(n_t + 2)$ of ancestors of a male bee $t + 2$ generations before the present generation is the sum of the number of ancestors t and $t + 1$ generations (n_t and n_{t+1}) before the present. If the number of ancestors of a male bee in a given generation t and in the previous generation is given by

$$N = \begin{bmatrix} n_t \\ n_{t+1} \end{bmatrix},$$

then there is a matrix M such that the number of ancestors in the two generations preceding generation t is given by

$$MN = \begin{bmatrix} 0 & 1 \\ 1 & 1 \end{bmatrix} \begin{bmatrix} n_t \\ n_{t+1} \end{bmatrix} = \begin{bmatrix} n_{t+1} \\ n_t + n_{t+1} \end{bmatrix}.$$

For a given male bee, the number of ancestors 5 and 6 generations back is given by

$$\begin{bmatrix} 8 \\ 13 \end{bmatrix}.$$

Find the number of ancestors 4 and 5 generations back by multiplying both sides of

$$MN = \begin{bmatrix} 8 \\ 13 \end{bmatrix}$$

by the inverse of M.

63. **Attendance** Suppose that during the first two weeks of a controversial art exhibit, the daily number of visitors to

a gallery turns out to be the sum of the number of visitors on the previous three days. We represent the numbers of visitors on three successive days by

$$N = \begin{bmatrix} n_t \\ n_{t+1} \\ n_{t+2} \end{bmatrix}.$$

(a) Find the matrix M so that

$$MN = \begin{bmatrix} n_{t+1} \\ n_{t+2} \\ n_t + n_{t+1} + n_{t+2} \end{bmatrix}.$$

(b) If the number of visitors on three successive days is given by

$$\begin{bmatrix} 191 \\ 346 \\ 645 \end{bmatrix},$$

use M^{-1} to find the number of visitors one day earlier (before the 191-visitor day).

64. **Bacteria Growth**　The population of a colony of bacteria grows so that the population size at any hour t is the sum of the populations of the three previous hours. Suppose that the matrix

$$N = \begin{bmatrix} n_t \\ n_{t+1} \\ n_{t+2} \end{bmatrix}.$$

describes the population size for three successive hours.

(a) Write matrix M so that MN gives the population size for three successive hours beginning 1 hour later, that is, so that

$$MN = \begin{bmatrix} n_{t+1} \\ n_{t+2} \\ n_t + n_{t+1} + n_{t+2} \end{bmatrix}.$$

(b) If the population at the end of three successive 1-hour periods was 200 at the end of the first hour, 370 at the end of the second hour, and 600 at the end of the third hour, what was the population 1 hour before it was 200? Use M^{-1}.

2.5　APPLICATIONS OF MATRICES: LEONTIEF INPUT-OUTPUT MODELS

Objectives

- To interpret Leontief technology matrices
- To use Leontief models to solve input-output problems

APPLICATION PREVIEW

Suppose we consider a simple economy as being based on three commodities: agricultural products, manufactured goods, and fuels. Suppose further that production of 10 units of agricultural products requires 5 units of agricultural products, 2 units of manufactured goods, and 1 unit of fuels; production of 10 units of manufactured goods requires 1 unit of agricultural products, 5 units of manufactured products, and 3 units of fuels; and production of 10 units of fuels requires 1 unit of agricultural products, 3 units of manufactured goods, and 4 units of fuels.

Table 2.10 summarizes this information in terms of production of one unit. The first column represents the units of agricultural products, manufactured goods, and fuels, respectively, that are needed to produce one unit of agricultural products. Column 2 represents the units required to produce one unit of manufactured goods, and column 3 represents the units required to produce one unit of fuels.

Table 2.10

| | *Outputs* | | |
Inputs	Agricultural Products	Manufactured Goods	Fuels
Agricultural products	0.5	0.1	0.1
Manufactured goods	0.2	0.5	0.3
Fuels	0.1	0.3	0.4

From Table 2.10 we can form matrix A, called a **technology matrix** or a **Leontief matrix.**

$$A = \begin{bmatrix} 0.5 & 0.1 & 0.1 \\ 0.2 & 0.5 & 0.3 \\ 0.1 & 0.3 & 0.4 \end{bmatrix}$$

The Leontief Input-Output model, named for Wassily Leontief, is an interesting application of matrices. Just as an airplane model approximates the real object, a *mathematical model* attempts to approximate a real-world situation. Models that are carefully constructed from past information may often provide remarkably accurate forecasts. The model Leontief developed was useful in predicting the effects on the economy of price changes or shifts in government spending.*

Leontief's work divided the economy into 500 sectors and in a subsequent article in *Scientific American* (October 1951), this was further reduced to a more manageable 42 departments of production. We can examine the workings of input-output analysis with a very simplified view of the economy.

In this section, we will interpret Leontief technology matrices for simple economies and solve input-output problems for these economies by using Leontief models.

In Table 2.10, we note that the sum of the units of agricultural products, manufactured goods, and fuels required to produce 1 unit of fuels (column 3) does not add up to 1 unit. This is because not all commodities or industries are represented in the model. In particular, it is customary to omit labor from models of this type.

EXAMPLE 1

Use Table 2.10 to answer the following questions.

(a) How many units of agricultural products and of fuels were required to produce 100 units of manufactured goods?

(b) Production of which commodity is least dependent on the other two?

(c) If fuel costs rise, which two industries will be most affected?

---

* Leontief's work dealt with a massive analysis of the American economy. He was awarded a Nobel Prize in economics in 1973 for his study.

Solution

(a) Referring to column 2, manufactured goods, we see that one unit requires 0.1 unit of agricultural products and 0.3 unit of fuels. Thus 100 units of manufactured goods require 10 units of agricultural products and 30 units of fuels.

(b) Looking down the columns, we see that one unit of agricultural products requires 0.3 unit of the other two commodities; one unit of manufactured goods requires 0.4 unit of the other two; and one unit of fuels requires 0.4 unit of the other two. Thus production of agricultural products is least dependent on the others.

(c) A rise in the cost of fuels would most affect those industries using the larger amount of fuels. One unit of agricultural products requires 0.1 unit of fuels, while a unit of manufactured goods requires 0.3 unit, and fuels requires 0.4 of its own units. Thus manufacturing and the fuel industry would be most affected by a cost increase in fuels.

CHECKPOINT

The following technology matrix for a simple economy describes the relationship of certain industries to each other in the production of one unit of product.

$$
\begin{array}{c}
\text{Agriculture} \\
\text{Manufacturing} \\
\text{Fuels} \\
\text{Utilities}
\end{array}
\begin{array}{cccc}
\text{Ag} & \text{Mfg} & \text{Fuels} & \text{Util}
\end{array}
\begin{bmatrix}
0.32 & 0.06 & 0.09 & 0.06 \\
0.11 & 0.41 & 0.19 & 0.31 \\
0.31 & 0.14 & 0.18 & 0.20 \\
0.18 & 0.26 & 0.21 & 0.25
\end{bmatrix}
$$

1. What industry is most dependent on its own goods for its operation?

2. How many units of fuels are required to produce 100 units of utilities?

For a simplified model of the economy, such as that described with Table 2.10, not all information is contained within the technology matrix. In particular, each industry has a gross production. The **gross production matrix** for the economy can be represented by the column matrix

$$
X = \begin{bmatrix} x_1 \\ x_2 \\ x_3 \end{bmatrix},
$$

where x_1 is the gross production of agricultural products, x_2 is the gross production of manufactured goods, and x_3 is the gross production of fuels.

Those units of gross production not used by these industries are called **final demands** or **surpluses,** and may be considered as being available for consumers, the government, or export. If we place these surpluses in a column matrix D, then the surplus can be represented by the equation

$$
X - AX = D, \qquad \text{or} \qquad (I - A)X = D,
$$

where I is an identity matrix. This matrix equation is called the **technological equation** for an **open Leontief model.**

Open Leontief Model   The technological equation for an open Leontief model is

$$(I - A)X = D.$$

EXAMPLE 2

Use matrix A from Table 2.10 as the technology matrix. If we wish to have a surplus of 85 units of agricultural products, 65 units of manufactured goods, and 0 units of fuels, what should the gross outputs be?

Solution

Let X be the matrix as indicated above. Then the technological equation is $(I - A)X = D$.

$$I - A = \begin{bmatrix} 1 & 0 & 0 \\ 0 & 1 & 0 \\ 0 & 0 & 1 \end{bmatrix} - \begin{bmatrix} 0.5 & 0.1 & 0.1 \\ 0.2 & 0.5 & 0.3 \\ 0.1 & 0.3 & 0.4 \end{bmatrix} = \begin{bmatrix} 0.5 & -0.1 & -0.1 \\ -0.2 & 0.5 & -0.3 \\ -0.1 & -0.3 & 0.6 \end{bmatrix}$$

Hence we must solve the matrix equation

$$\begin{bmatrix} 0.5 & -0.1 & -0.1 \\ -0.2 & 0.5 & -0.3 \\ -0.1 & -0.3 & 0.6 \end{bmatrix} \begin{bmatrix} x_1 \\ x_2 \\ x_3 \end{bmatrix} = \begin{bmatrix} 85 \\ 65 \\ 0 \end{bmatrix}.$$

The augmented matrix is

$$\begin{bmatrix} 0.5 & -0.1 & -0.1 & | & 85 \\ -0.2 & 0.5 & -0.3 & | & 65 \\ -0.1 & -0.3 & 0.6 & | & 0 \end{bmatrix}.$$

If we reduce by the methods of Section 2.3, we obtain

$$\begin{bmatrix} 1 & 0 & 0 & | & 300 \\ 0 & 1 & 0 & | & 400 \\ 0 & 0 & 1 & | & 250 \end{bmatrix},$$

so the gross outputs for the industries are

Agriculture:	$x_1 = 300$
Manufacturing:	$x_2 = 400$
Fuels:	$x_3 = 250.$

In Example 2 we solved the system of equations by using row operations on the augmented matrix. It is also possible to use the inverse of $I - A$ to solve the technology equation.

EXAMPLE 3

A primitive economy with a lumber industry and a power industry has the following technological matrix.

$$A = \begin{matrix} \text{Lumber} & \text{Power} \\ \begin{bmatrix} 0.1 & 0.2 \\ 0.2 & 0.4 \end{bmatrix} & \begin{matrix} \text{Lumber} \\ \text{Power} \end{matrix} \end{matrix}$$

If surpluses of 30 units of lumber and 70 units of power are desired, find the gross production of each industry.

Solution

Let x be the gross production of lumber and y be the gross production of power. Then

$$I - A = \begin{bmatrix} 0.9 & -0.2 \\ -0.2 & 0.6 \end{bmatrix}, \quad X = \begin{bmatrix} x \\ y \end{bmatrix}, \quad \text{and} \quad D = \begin{bmatrix} 30 \\ 70 \end{bmatrix}.$$

We must solve the matrix equation $(I - A)X = D$ and can do this by multiplying both sides, from the left, by $(I - A)^{-1}$ as follows:

$$(I - A)X = D$$
$$(I - A)^{-1} (I - A)X = (I - A)^{-1}D$$
$$IX = (I - A)^{-1}D$$
$$X = (I - A)^{-1}D.$$

We can find $(I - A)^{-1}$ with the formula from Section 2.4:

$$(I - A)^{-1} = \frac{1}{0.54 - 0.04} \begin{bmatrix} 0.6 & 0.2 \\ 0.2 & 0.9 \end{bmatrix} = 2 \begin{bmatrix} 0.6 & 0.2 \\ 0.2 & 0.9 \end{bmatrix} = \begin{bmatrix} 1.2 & 0.4 \\ 0.4 & 1.8 \end{bmatrix},$$

and then we can find the gross production of each industry:

$$X = (I - A)^{-1}D = \begin{bmatrix} 1.2 & 0.4 \\ 0.4 & 1.8 \end{bmatrix} \begin{bmatrix} 30 \\ 70 \end{bmatrix} = \begin{bmatrix} 64 \\ 138 \end{bmatrix}.$$

Hence the gross productions are

$$\text{Lumber:} \quad x = 64$$
$$\text{Power:} \quad y = 138.$$

EXAMPLE 4

The economy of a developing nation is based on agricultural products, steel, and coal. An output of one ton of agricultural products requires an input of 0.1 tons of agricultural products, 0.02 tons of steel, and 0.05 tons of coal. An output of one ton of steel requires an input of 0.01 tons of agricultural products, 0.13 tons of steel, and 0.18 tons of coal. An output of one ton of coal requires an input of 0.01 tons of agricultural products, 0.20 tons of steel, and 0.05 tons of coal. Write the technological matrix for this economy. Find the necessary gross productions to provide surpluses of 2350 tons of agricultural products, 4551 tons of steel, and 911 tons of coal.

Solution

The technology matrix for the economy is

$$
\begin{array}{ccc}
\text{Ag} & \text{Steel} & \text{Coal}
\end{array}
$$
$$
\begin{bmatrix}
0.1 & 0.01 & 0.01 \\
0.02 & 0.13 & 0.20 \\
0.05 & 0.18 & 0.05
\end{bmatrix}
\begin{array}{l}
\text{Ag} \\
\text{Steel} \\
\text{Coal}
\end{array}
$$

The technological equation for this economy is $(I - A)X = D$, or

$$
\left(
\begin{bmatrix}
1 & 0 & 0 \\
0 & 1 & 0 \\
0 & 0 & 1
\end{bmatrix}
-
\begin{bmatrix}
0.1 & 0.01 & 0.01 \\
0.02 & 0.13 & 0.20 \\
0.05 & 0.18 & 0.05
\end{bmatrix}
\right) X =
\begin{bmatrix}
2350 \\
4551 \\
911
\end{bmatrix}.
$$

The first step in solving this equation is to find the inverse of $I - A$ (this is easy with a graphing calculator). Multiplying both sides of this equation by $(I - A)^{-1}$ gives

$$
X =
\left(
\begin{bmatrix}
1 & 0 & 0 \\
0 & 1 & 0 \\
0 & 0 & 1
\end{bmatrix}
-
\begin{bmatrix}
0.1 & 0.01 & 0.01 \\
0.02 & 0.13 & 0.20 \\
0.05 & 0.18 & 0.05
\end{bmatrix}
\right)^{-1}
\begin{bmatrix}
2350 \\
4551 \\
911
\end{bmatrix},
$$

so the necessary gross productions of the industries are

$$
X =
\begin{bmatrix}
2700 \\
5800 \\
2200
\end{bmatrix}
\begin{array}{l}
\text{Ag} \\
\text{Steel} \\
\text{Coal}
\end{array}
$$

A few comments are in order. First, the surplus matrix must contain positive or zero entries. If this is not the case, then the gross production of one or more industries is insufficient to supply the needs of the others. That is, the economy cannot use more of a certain commodity than it produces.

Second, the model we have considered is referred to as an **open Leontief model** because not all inputs and outputs were incorporated within the technology matrix. In particular, since labor was omitted, some (or all) of the surpluses must be used to support this labor. Problems that fit this open Leontief model have unique solutions, so the matrix $(I - A)^{-1}$ can be used to solve the technological equation.

If a model is developed in which all inputs and outputs are used within the system, then such a model is called a **closed Leontief model.** In such a model, labor (and perhaps other factors) must be included. Labor is included by considering a new industry, households, which produces labor. When such a closed model is developed, all outputs are used within the system and the sum of the entries in each column equals 1. In this case, there is no surplus, so $D = 0$, and we have the following.

Closed Leontief Model The technological equation for a closed Leontief model is

$$(I - A)X = 0,$$

where 0 is a zero column matrix.

For closed models, the technological equation does *not* have a unique solution, so the matrix $(I - A)^{-1}$ does *not* exist and hence cannot be used to find the solution.

EXAMPLE 5

The following closed Leontief model with technology matrix A might describe the economy of the entire country, with x_1 equal to the government's budget, x_2 the value of industrial output (profit-making organizations), x_3 the budget of nonprofit organizations, and x_4 the household's budget.

$$A = \begin{array}{c} \\ \\ \end{array} \begin{array}{cccc} \text{G} & \text{I} & \text{N} & \text{H} \\ \begin{bmatrix} 0.4 & 0.2 & 0 & 0.3 \\ 0.2 & 0.2 & 0.2 & 0.1 \\ 0.2 & 0 & 0.2 & 0.1 \\ 0.2 & 0.6 & 0.6 & 0.5 \end{bmatrix} & \begin{array}{l} \text{Government} \\ \text{Industry} \\ \text{Nonprofit} \\ \text{Household} \end{array} \end{array}$$

Find the total budgets (or outputs) x_1, x_2, x_3, and x_4.

Solution

Since this is a closed Leontief model, we solve the technological equation $(I - A)X = 0$.

$$\left(\begin{bmatrix} 1 & 0 & 0 & 0 \\ 0 & 1 & 0 & 0 \\ 0 & 0 & 1 & 0 \\ 0 & 0 & 0 & 1 \end{bmatrix} - \begin{bmatrix} 0.4 & 0.2 & 0 & 0.3 \\ 0.2 & 0.2 & 0.2 & 0.1 \\ 0.2 & 0 & 0.2 & 0.1 \\ 0.2 & 0.6 & 0.6 & 0.5 \end{bmatrix} \right) X = \begin{bmatrix} 0 \\ 0 \\ 0 \\ 0 \end{bmatrix},$$

or

$$\begin{bmatrix} 0.6 & -0.2 & 0 & -0.3 \\ -0.2 & 0.8 & -0.2 & -0.1 \\ -0.2 & 0 & 0.8 & -0.1 \\ -0.2 & -0.6 & -0.6 & 0.5 \end{bmatrix} X = \begin{bmatrix} 0 \\ 0 \\ 0 \\ 0 \end{bmatrix}.$$

The augmented matrix for this system is

$$\begin{bmatrix} 0.6 & -0.2 & 0 & -0.3 & | & 0 \\ -0.2 & 0.8 & -0.2 & -0.1 & | & 0 \\ -0.2 & 0 & 0.8 & -0.1 & | & 0 \\ -0.2 & -0.6 & -0.6 & 0.5 & | & 0 \end{bmatrix}.$$

If we multiply each entry by -5 and rearrange the rows by moving row 3 to row 1, row 1 to row 2, and row 2 to row 3, we have the equivalent matrix

$$\left[\begin{array}{cccc|c} 1 & 0 & -4 & 0.5 & 0 \\ -3 & 1 & 0 & 1.5 & 0 \\ 1 & -4 & 1 & 0.5 & 0 \\ 1 & 3 & 3 & -2.5 & 0 \end{array}\right].$$

Then, if we reduce this matrix by the methods of Section 2.3, we obtain

$$\left[\begin{array}{cccc|c} 1 & 0 & 0 & -\frac{53}{86} & 0 \\ 0 & 1 & 0 & -\frac{15}{43} & 0 \\ 0 & 0 & 1 & -\frac{12}{43} & 0 \\ 0 & 0 & 0 & 0 & 0 \end{array}\right].$$

The system of equations that corresponds to this augmented matrix is

$$\begin{aligned} x_1 \quad & -\tfrac{53}{86}x_4 = 0 & \qquad & x_1 = \tfrac{53}{86}x_4 \\ & x_2 \quad -\tfrac{15}{43}x_4 = 0 & \text{or} \quad & x_2 = \tfrac{15}{43}x_4. \\ & x_3 - \tfrac{12}{43}x_4 = 0 & \qquad & x_3 = \tfrac{12}{43}x_4 \end{aligned}$$

Note that the economy satisfies the given equation if the government's budget is $\frac{53}{86}$ times the household's budget, if the value of industrial output is $\frac{15}{43}$ times the household's budget, and if the budget of nonprofit organizations is $\frac{12}{43}$ times the household's budget. The dependency here is expected. The fact that the system is closed suggests the dependency, but even more obvious is the fact that industrial output is limited by labor supply.

CHECKPOINT

3. Suppose a primitive economy has a wood products industry and a minerals industry, with technology matrix A (open Leontief model).

$$A = \begin{array}{cc} & \begin{array}{cc} \text{W} & \text{M} \end{array} \\ \begin{bmatrix} 0.12 & 0.11 \\ 0.09 & 0.15 \end{bmatrix} & \begin{array}{l} \text{Wood products} \\ \text{Minerals} \end{array} \end{array}$$

If surpluses of 1738 units of wood products and 1332 of minerals are desired, find the gross production of each industry.

4. For a closed Leontief model:
   (a) What is the sum of the column entries of the technology matrix?
   (b) State the technology equation.

We have examined two types of input-output models as they pertain to the economy as a whole. Since Leontief's original work with the economy, various other applications of input-output models have been developed. Consider the following parts-listing problem.

EXAMPLE 6

A storage shed consists of 4 walls and a roof. The walls and roof are made from stamped aluminum sheeting and are reinforced with 4 braces, each of which is held with 6 bolts. The final assembly joins the walls to each other and to the roof using a total of 20 more bolts. The parts listing for these sheds can be described by the following matrix.

$$
Q = \begin{array}{c} \\ \\ \\ \\ \\ \\ \\ \end{array}
\begin{bmatrix}
0 & 0 & 0 & 0 & 0 & 0 \\
4 & 0 & 0 & 0 & 0 & 0 \\
1 & 0 & 0 & 0 & 0 & 0 \\
0 & 1 & 1 & 0 & 0 & 0 \\
0 & 4 & 4 & 0 & 0 & 0 \\
20 & 24 & 24 & 0 & 0 & 0
\end{bmatrix}
\begin{array}{l}
\text{Sheds} \\
\text{Walls} \\
\text{Roofs} \\
\text{Sheets} \\
\text{Braces} \\
\text{Bolts}
\end{array}
$$

with column headers S　W　R　Sht　Br　Bo

This matrix is a form of input-output matrix, with the rows representing inputs that are used directly to produce the items that head the columns. We note that the entries here are quantities used rather than proportions used.

Thus to produce a shed requires 4 walls, 1 roof, and 20 bolts, and to produce a roof requires 1 aluminum sheet, 4 braces, and 24 bolts.

Suppose that an order is received for 4 completed sheds and spare parts including 2 walls, 1 roof, 8 braces, and 24 bolts. How many of each of the assembly items are required to fill the order?

Solution

If matrix D represents the order and matrix X represents the gross production, we have

$$
D = \begin{bmatrix}
4 \\ 2 \\ 1 \\ 0 \\ 8 \\ 24
\end{bmatrix}
\begin{array}{l}
\text{Sheds} \\ \text{Walls} \\ \text{Roofs} \\ \text{Sheets} \\ \text{Braces} \\ \text{Bolts}
\end{array}
\qquad
X = \begin{bmatrix}
x_1 \\ x_2 \\ x_3 \\ x_4 \\ x_5 \\ x_6
\end{bmatrix}
= \begin{bmatrix}
\text{total sheds required} \\
\text{total walls required} \\
\text{total roofs required} \\
\text{total sheets required} \\
\text{total braces required} \\
\text{total bolts required}
\end{bmatrix}.
$$

Then X must satisfy $X - QX = D$, which is the technological equation for an open Leontief model.

We can find X by solving the system $(I - Q)X = D$, which is represented by the augmented matrix

$$
\begin{bmatrix}
1 & 0 & 0 & 0 & 0 & 0 & | & 4 \\
-4 & 1 & 0 & 0 & 0 & 0 & | & 2 \\
-1 & 0 & 1 & 0 & 0 & 0 & | & 1 \\
0 & -1 & -1 & 1 & 0 & 0 & | & 0 \\
0 & -4 & -4 & 0 & 1 & 0 & | & 8 \\
-20 & -24 & -24 & 0 & 0 & 1 & | & 24
\end{bmatrix}.
$$

Although this matrix is quite large, it is easily reduced and yields the following:

$$\begin{bmatrix} 1 & 0 & 0 & 0 & 0 & 0 & | & 4 \\ 0 & 1 & 0 & 0 & 0 & 0 & | & 18 \\ 0 & 0 & 1 & 0 & 0 & 0 & | & 5 \\ 0 & 0 & 0 & 1 & 0 & 0 & | & 23 \\ 0 & 0 & 0 & 0 & 1 & 0 & | & 100 \\ 0 & 0 & 0 & 0 & 0 & 1 & | & 656 \end{bmatrix}.$$

This matrix tells us that the order will have 4 complete sheds. The 18 walls from the matrix include the 16 from the complete sheds and the 5 roofs include the 4 from the complete sheds. The 23 sheets, 100 braces, and 656 bolts are the total number of *primary assembly items* (bolts, braces, and sheets) required to fill the order.

CHECKPOINT SOLUTIONS

1. Manufacturing

2. Under the "Util" column and across from the "Fuels" row is 0.20, indicating that producing one unit of utilities requires 0.20 units of fuels. Thus producing 100 units of utilities requires 20 units of fuels.

3. Let x = the gross production of wood products
   y = the gross production of minerals.
   We must solve $(I - A)X = D$.

$$\left(\begin{bmatrix} 1 & 0 \\ 0 & 1 \end{bmatrix} - \begin{bmatrix} 0.12 & 0.11 \\ 0.09 & 0.15 \end{bmatrix} \right) \begin{bmatrix} x \\ y \end{bmatrix} = \begin{bmatrix} 1738 \\ 1332 \end{bmatrix}$$

$$\begin{bmatrix} 0.88 & -0.11 \\ -0.09 & 0.85 \end{bmatrix} \begin{bmatrix} x \\ y \end{bmatrix} = \begin{bmatrix} 1738 \\ 1332 \end{bmatrix}$$

$$\begin{bmatrix} 0.88 & -0.11 \\ -0.09 & 0.85 \end{bmatrix}^{-1} = \frac{1}{(0.88)(0.85) - (0.09)(0.11)} \begin{bmatrix} 0.85 & 0.11 \\ 0.09 & 0.88 \end{bmatrix}$$

Hence

$$\begin{bmatrix} x \\ y \end{bmatrix} = \frac{1}{0.7381} \begin{bmatrix} 0.85 & 0.11 \\ 0.09 & 0.88 \end{bmatrix} \begin{bmatrix} 1738 \\ 1332 \end{bmatrix} = \begin{bmatrix} 2200 \\ 1800 \end{bmatrix}.$$

4. (a) The sum of the column entries is 1.
   (b) $(I - A)X = 0$.

EXERCISE 2.5

1. The following technology matrix for a simple economy describes the relationship of certain industries to each other in the production of one unit of product.

$$
\begin{array}{cccc}
\text{A} & \text{M} & \text{F} & \text{U}
\end{array}
$$

$$
\begin{bmatrix}
0.36 & 0.03 & 0.10 & 0.04 \\
0.06 & 0.42 & 0.25 & 0.33 \\
0.18 & 0.15 & 0.10 & 0.41 \\
0.10 & 0.20 & 0.31 & 0.15
\end{bmatrix}
\begin{array}{l}
\text{Agriculture} \\
\text{Manufacturing} \\
\text{Fuels} \\
\text{Utilities}
\end{array}
$$

    (a) For each 100 units of manufactured products produced, how many units of fuels are required?

    (b) How many units of utilities are required to produce 40 units of agriculture products?

2. For the economy of problem 1, what industry is most dependent on utilities?

The following technology matrix describes the relationship of certain industries within the economy to each other.

(A&F, Agriculture and food; RM, Raw materials; M, Manufacturing; F, Fuels industry; U, Utilities; SI, Service industries)

$$
\begin{array}{cccccc}
\text{A\&F} & \text{RM} & \text{M} & \text{F} & \text{U} & \text{SI}
\end{array}
$$

$$
\begin{bmatrix}
0.410 & 0.008 & 0 & 0.002 & 0 & 0.006 \\
0.025 & 0.493 & 0.190 & 0.024 & 0.030 & 0.150 \\
0.015 & 0.006 & 0.082 & 0.009 & 0.001 & 0.116 \\
0.097 & 0.096 & 0.040 & 0.053 & 0.008 & 0.093 \\
0.028 & 0.129 & 0.039 & 0.058 & 0.138 & 0.409 \\
0.043 & 0.008 & 0.010 & 0.012 & 0.002 & 0.095
\end{bmatrix}
\begin{array}{l}
\text{A\&F} \\
\text{RM} \\
\text{M} \\
\text{F} \\
\text{U} \\
\text{SI}
\end{array}
$$

Use this matrix in problems 3–10.

3. For each 1000 units of raw materials produced, how many units of agricultural and food products were required?

4. For each 1000 units of raw materials produced, how many units of service were required?

5. How many units of fuels were required to produce 1000 units of manufactured goods?

6. How many units of fuels were required to produce 1000 units of power (utilities' goods)?

7. Which industry is most dependent on its own goods for its operations? Which industry is least dependent on its own goods?

8. Which industry is most dependent on the fuels industry?

9. Which three industries would be most affected by a rise in the cost of raw materials?

10. Which industry would be most affected by a rise in the cost of manufactured goods?

11. Given is the partially solved matrix for the gross production of each industry of an open Leontief model. Find the gross production for each industry.

$$
\begin{array}{ccc}
\text{U} & \text{A} & \text{M}
\end{array}
$$

$$
\left[\begin{array}{ccc|c}
1 & -2 & -1 & 80 \\
0 & 8 & -4 & 60 \\
0 & 0 & 10 & 50
\end{array}\right]
\begin{array}{l}
\text{Utilities} \\
\text{Agriculture} \\
\text{Manufacturing}
\end{array}
$$

12. Given is the partially solved matrix for the gross production of each industry of a closed Leontief model. Find the gross production for each industry.

$$
\begin{array}{ccc}
\text{U} & \text{A} & \text{H}
\end{array}
$$

$$
\left[\begin{array}{ccc|c}
3 & -10 & -7 & 0 \\
0 & 20 & -10 & 0 \\
0 & 0 & 0 & 0
\end{array}\right]
\begin{array}{l}
\text{Utilities} \\
\text{Agriculture} \\
\text{Households}
\end{array}
$$

13. Suppose a primitive economy consists of two industries, farm products and farm machinery. Suppose also that their technology matrix is

$$
\begin{array}{cc}
& \begin{array}{cc} \text{P} & \text{M} \end{array}
\end{array}
$$

$$
A = \begin{bmatrix} 0.5 & 0.1 \\ 0.1 & 0.3 \end{bmatrix}
\begin{array}{l}
\text{Products} \\
\text{Machinery}
\end{array}
$$

If surpluses of 96 units of farm products and 8 units of farm machinery are desired, find the gross production of each industry.

14. Suppose an economy has two industries, agriculture and minerals. Suppose further that the technology matrix for this economy is A.

$$
\begin{array}{cc}
& \begin{array}{cc} \text{A} & \text{M} \end{array}
\end{array}
$$

$$
A = \begin{bmatrix} 0.3 & 0.1 \\ 0.1 & 0.2 \end{bmatrix}
\begin{array}{l}
\text{Agriculture} \\
\text{Minerals}
\end{array}
$$

If a surplus of 60 agricultural units and 70 mineral units is desired, find the gross production for each industry.

15. Suppose the economy of an underdeveloped country has an agricultural industry and an oil industry, with technology matrix A.

$$
\begin{array}{cc}
& \begin{array}{cc} \text{A} & \text{O} \end{array}
\end{array}
$$

$$
A = \begin{bmatrix} 0.3 & 0.1 \\ 0.2 & 0.1 \end{bmatrix}
\begin{array}{l}
\text{Agricultural product} \\
\text{Oil product}
\end{array}
$$

If a surplus of 0 units of agricultural products and 610 units of oil products are desired, find the gross production of each industry.

16. Suppose a simple economy with only an agricultural industry and a steel industry has the following technology matrix.

$$A = \begin{bmatrix} 0.5 & 0.2 \\ 0.1 & 0.6 \end{bmatrix} \begin{matrix} \text{Agriculture} \\ \text{Steel} \end{matrix}$$

with column headers A S

If surpluses of 15 units of agricultural products and 33 units of steel are desired, find the gross production of each industry.

17. An economy is based on two industries: utilities and manufacturing. The technology matrix for these industries is A.

$$A = \begin{bmatrix} 0.3 & 0.15 \\ 0.3 & 0.4 \end{bmatrix} \begin{matrix} \text{Utility} \\ \text{Manufacturing} \end{matrix}$$

with column headers U M

If surpluses of 80 units of utility output and 180 units of manufacturing are desired, find the gross production of each industry.

18. A primitive economy has a mining industry and a fishing industry, with technology matrix A.

$$A = \begin{bmatrix} 0.25 & 0.05 \\ 0.05 & 0.40 \end{bmatrix} \begin{matrix} \text{Mining} \\ \text{Fishing} \end{matrix}$$

with column headers M F

If surpluses of 147 units of mining output and 26 units of fishing output are desired, find the gross production of each industry.

19. One sector of an economy consists of a mining industry and a manufacturing industry, with technology matrix A.

$$A = \begin{bmatrix} 0.2 & 0.4 \\ 0.3 & 0.3 \end{bmatrix} \begin{matrix} \text{Mining} \\ \text{Manufacturing} \end{matrix}$$

with column headers M Mfg

Surpluses of 36 units from mining and 278 units from manufacturing are desired. Find the gross production of each industry.

20. An underdeveloped country has an agricultural industry and a manufacturing industry, with technology matrix A.

$$A = \begin{bmatrix} 0.20 & 0.10 \\ 0.25 & 0.45 \end{bmatrix} \begin{matrix} \text{Agriculture} \\ \text{Manufacturing} \end{matrix}$$

with column headers A M

Surpluses of 8 units in agriculture and 620 units in manufacturing are desired. Find the gross production of each industry.

21. A simple economy has an electronics components industry and a computer industry, with technology matrix A.

$$A = \begin{bmatrix} 0.3 & 0.6 \\ 0.2 & 0.2 \end{bmatrix} \begin{matrix} \text{Electronics components} \\ \text{Computers} \end{matrix}$$

with column headers EC C

Surpluses of 648 units of electronics components and 16 computers are desired. Find the gross production of each industry.

22. An economy has an agricultural industry and a textile industry, with technology matrix A.

$$A = \begin{bmatrix} 0.4 & 0.1 \\ 0.1 & 0.2 \end{bmatrix} \begin{matrix} \text{Agriculture} \\ \text{Textiles} \end{matrix}$$

with column headers A T

Surpluses of 5 units in agriculture and 195 units of textiles are desired. Find the gross production of each industry.

23. A primitive economy consists of a fishing industry and an oil industry, with technology matrix A.

$$A = \begin{bmatrix} 0.30 & 0.04 \\ 0.35 & 0.10 \end{bmatrix} \begin{matrix} \text{Fishing} \\ \text{Oil} \end{matrix}$$

with column headers F O

Surpluses of 20 units from fishing and 1090 units from oil are desired. Find the gross production of each industry.

24. An economy has a manufacturing industry and a banking industry, with technology matrix A.

$$A = \begin{bmatrix} 0.5 & 0.3 \\ 0.2 & 0.3 \end{bmatrix} \begin{matrix} \text{Manufacturing} \\ \text{Banking} \end{matrix}$$

with column headers M B

If surpluses of 141 units from manufacturing and 106 units from banking are desired, find the gross production of each industry.

Interdepartment costs Within a company there is a (micro) economy that is monitored by the accounting procedures. In terms of the accounts, the various departments "produce" costs, some of which are internal and some of which are direct costs. Problems 25–28 show how an open Leontief model can be used to determine departmental costs.

25. Suppose the development department of a firm charges 10% of its total monthly costs to the promotional department, and the promotional department charges 5% of its total monthly costs to the development department. The

direct costs of the development department are $20,400, and the direct costs of the promotional department are $9,900. The solution to the matrix equation

$$\begin{bmatrix} 20,400 \\ 9,900 \end{bmatrix} + \begin{bmatrix} 0 & 0.05 \\ 0.1 & 0 \end{bmatrix} \begin{bmatrix} x \\ y \end{bmatrix} = \begin{bmatrix} x \\ y \end{bmatrix}$$

for the column matrix $\begin{bmatrix} x \\ y \end{bmatrix}$ gives the total costs for each department. Note that this equation can be written in the form

$$D + AX = X \quad \text{or} \quad X - AX = D,$$

which is the form for an open economy. Find the total costs for each department in this (micro) economy.

26. Suppose the shipping department of a firm charges 20% of its total monthly costs to the printing department, while the printing department charges 10% of its total monthly costs to the shipping department. If the direct costs of the shipping department are $16,500 and the direct costs of the printing department are $11,400, find the total costs for each department.

27. Suppose the engineering department of a firm charges 20% of its total monthly costs to the computer department, and the computer department charges 25% of its total monthly costs to the engineering department. If during a given month the direct costs were $11,750 for the engineering department and $10,000 for the computer department, what are the total costs of each department?

28. The sales department of an auto dealership charges 10% of its total monthly costs to the service department, and the service department charges 20% of its total monthly costs to the sales department. During a given month, the direct costs were $88,200 for sales and $49,000 for service. Find the total costs of each department.

29. Suppose an economy has the same technological matrix as that used in Example 2. If surpluses of 110 units of agricultural goods and 50 units each of manufactured goods and fuels are desired, find the gross production of each industry required to satisfy this.

30. Suppose that an economy has the same technological matrix as that used in Example 2. If surpluses of 180 units of agricultural goods, 90 units of manufactured goods, and 40 units of fuels are desired, find the gross production of each industry required to satisfy this.

31. Suppose that the economy of a small nation has an electronics industry, a steel industry, and an auto industry, with the following technology matrix.

$$A = \begin{bmatrix} 0.6 & 0.2 & 0.2 \\ 0.1 & 0.4 & 0.5 \\ 0.1 & 0.2 & 0.2 \end{bmatrix} \begin{matrix} \text{Electronics} \\ \text{Steel} \\ \text{Autos} \end{matrix}$$

with E, S, A column headers.

If the nation wishes to have surpluses of 100 units of electronics production, 200 units of steel production, and 200 automobiles, find the gross production of each industry.

32. Suppose an economy has the same technology matrix as that in problem 31. If surpluses of 540 units of electronics, 30 units of steel, and 140 autos are desired, find the gross production for each industry.

33. Suppose that for the primitive economy in problem 13 we form a closed model by including households. Suppose further that the resulting technology matrix is

$$A = \begin{bmatrix} 0.5 & 0.1 & 0.2 \\ 0.1 & 0.3 & 0 \\ 0.4 & 0.6 & 0.8 \end{bmatrix} \begin{matrix} \text{Products} \\ \text{Machinery} \\ \text{Households} \end{matrix}$$

with P, M, H column headers.

Find the gross productions for these industries.

34. In Example 5 we gave a closed Leontief model for an economy. In the following matrix we combined the non-profit organizations with industry (viewing nonprofit organizations as a service industry). Find the gross productions for the industries in this revised closed model.

$$A = \begin{bmatrix} 0.4 & 0.1 & 0.3 \\ 0.4 & 0.3 & 0.2 \\ 0.2 & 0.6 & 0.5 \end{bmatrix} \begin{matrix} \text{Government} \\ \text{Industry} \\ \text{Households} \end{matrix}$$

with G, I, H column headers.

35. Suppose the technology matrix for a closed model of a simple economy is given by matrix A. Find the gross productions for the industries.

$$A = \begin{bmatrix} 0.4 & 0.2 & 0.2 \\ 0.2 & 0.3 & 0.3 \\ 0.4 & 0.5 & 0.5 \end{bmatrix} \begin{matrix} \text{Government} \\ \text{Industry} \\ \text{Households} \end{matrix}$$

with G, I, H column headers.

36. Suppose the technology matrix for a closed model of an economy is given by matrix A. Find the gross productions for the industries.

$$A = \begin{bmatrix} 0.2 & 0.1 & 0.1 \\ 0.6 & 0.5 & 0.1 \\ 0.2 & 0.4 & 0.8 \end{bmatrix} \begin{matrix} \text{Shipping} \\ \text{Manufacturing} \\ \text{Households} \end{matrix}$$

with S, M, H column headers.

37. Suppose the technology matrix for a closed model of an economy is given by matrix A. Find the gross productions for the industries.

$$A = \begin{array}{c} \\ \\ \\ \end{array} \begin{array}{ccc} M & U & H \\ \begin{bmatrix} 0.5 & 0.4 & 0.3 \\ 0.4 & 0.5 & 0.3 \\ 0.1 & 0.1 & 0.4 \end{bmatrix} & \begin{array}{l} \text{Manufacturing} \\ \text{Utilities} \\ \text{Households} \end{array} \end{array}$$

38. Suppose the technology matrix for a closed model of an economy is given by matrix A. Find the gross productions for the industries.

$$A = \begin{array}{cccc} G & P & N & H \\ \begin{bmatrix} 0.3 & 0.2 & 0.1 & 0.05 \\ 0.2 & 0.3 & 0.1 & 0.1 \\ 0.2 & 0.1 & 0.2 & 0.1 \\ 0.3 & 0.4 & 0.6 & 0.75 \end{bmatrix} & \begin{array}{l} \text{Government} \\ \text{Profit sector} \\ \text{Nonprofit} \\ \text{Households} \end{array} \end{array}$$

39. For the storage shed in Example 6, find the number of each primary assembly item required to fill an order for 24 sheds, 12 braces, and 96 bolts.

40. Card tables are made by joining 4 legs and a top using 4 bolts. The legs are each made from a steel rod. The top has a frame made from 4 steel rods. A cover and four special clamps that brace the top and hold the legs are joined to the frame using a total of 8 bolts. The parts-listing matrix for the card table assembly is given by

$$A = \begin{array}{ccccccc} CT & L & T & R & Co & Cl & B \\ \begin{bmatrix} 0 & 0 & 0 & 0 & 0 & 0 & 0 \\ 4 & 0 & 0 & 0 & 0 & 0 & 0 \\ 1 & 0 & 0 & 0 & 0 & 0 & 0 \\ 0 & 1 & 4 & 0 & 0 & 0 & 0 \\ 0 & 0 & 1 & 0 & 0 & 0 & 0 \\ 0 & 0 & 4 & 0 & 0 & 0 & 0 \\ 4 & 0 & 8 & 0 & 0 & 0 & 0 \end{bmatrix} & \begin{array}{l} \text{Card table} \\ \text{Legs} \\ \text{Top} \\ \text{Rods} \\ \text{Cover} \\ \text{Clamps} \\ \text{Bolts} \end{array} \end{array}$$

If an order is received for 10 card tables, 4 legs, 1 top, 1 cover, 6 clamps, and 12 bolts, how many of each primary assembly item are required to fill the order?

41. A sawhorse is made from 2 pairs of legs and a top joined with 4 nails. The top is a 2×4 in. board 3 ft long. Each pair of legs is made from two 2×4 in. boards 3 ft long, a brace, and a special clamp, with the brace and the clamp joined using 8 nails. The parts-listing matrix for the sawhorse is given by

$$A = \begin{array}{cccccccc} SH & T & LP & 2 \times 4 & B & C & N \\ \begin{bmatrix} 0 & 0 & 0 & 0 & 0 & 0 & 0 \\ 1 & 0 & 0 & 0 & 0 & 0 & 0 \\ 2 & 0 & 0 & 0 & 0 & 0 & 0 \\ 0 & 1 & 2 & 0 & 0 & 0 & 0 \\ 0 & 0 & 1 & 0 & 0 & 0 & 0 \\ 0 & 0 & 1 & 0 & 0 & 0 & 0 \\ 4 & 0 & 8 & 0 & 0 & 0 & 0 \end{bmatrix} & \begin{array}{l} \text{Sawhorse} \\ \text{Top} \\ \text{Leg pair} \\ 2 \times 4\text{s} \\ \text{Brace} \\ \text{Clamp} \\ \text{Nails} \end{array} \end{array}$$

How many of each primary assembly item are required to fill an order for 10 sawhorses, 6 extra 2×4s, 6 extra clamps, and 100 nails?

42. A log carrier has a body made from a 4-ft length of reinforced material having a patch on each side and a dowel slid through each end to act as handles. The parts-listing matrix for the log carrier is given by

$$A = \begin{array}{cccccc} LC & B & H & C & P \\ \begin{bmatrix} 0 & 0 & 0 & 0 & 0 \\ 1 & 0 & 0 & 0 & 0 \\ 2 & 0 & 0 & 0 & 0 \\ 0 & 1 & 0 & 0 & 0 \\ 0 & 2 & 0 & 0 & 0 \end{bmatrix} & \begin{array}{l} \text{Log carrier} \\ \text{Body} \\ \text{Handles} \\ \text{Cloth} \\ \text{Patch} \end{array} \end{array}$$

How many of each primary assembly item are required to fill an order for 500 log carriers and 20 handles?

KEY TERMS & FORMULAS

Section	Key Terms	Formula
2.1	Matrices	
	Entries	
	Order	
	Square, row, column, and zero matrices	
	Sum of two matrices	
	Negative of a matrix	

Section	Key Terms	Formula
2.2	Scalar multiplication	
	Matrix product	
	Identity matrix	2×2 is $\begin{bmatrix} 1 & 0 \\ 0 & 1 \end{bmatrix}$
2.3	Augmented matrix	
	Coefficient matrix	
	Elementary row operations	
	Solving systems of equations	
	Gauss-Jordan Elimination Method	
	Nonunique solutions	
	Reduced form	
2.4	Inverse matrices	
	Determinants	$\begin{vmatrix} a & b \\ c & d \end{vmatrix} = ad - bc$
2.5	Technology matrix	
	Leontief Input-Output model	
	Gross production matrix	
	Final demands (surpluses)	
	Leontief model	
	Technology equation	
	Open	$(I - A)X = D$
	Closed	$(I - A)X = 0$

REVIEW EXERCISES

Use the matrices

$$A = \begin{bmatrix} 4 & 4 & 2 & -5 \\ 6 & 3 & -1 & 0 \\ 0 & 0 & -3 & 5 \end{bmatrix}, \quad B = \begin{bmatrix} 2 & -5 & -11 & 8 \\ 4 & 0 & 0 & 4 \\ -2 & -2 & 1 & 9 \end{bmatrix},$$

$$C = \begin{bmatrix} 4 & -2 \\ 5 & 0 \\ 6 & 0 \\ 1 & 3 \end{bmatrix}, \quad D = \begin{bmatrix} 3 & 5 \\ 1 & 2 \end{bmatrix}, \quad E = \begin{bmatrix} 1 & 1 \\ 1 & 1 \\ 4 & 6 \\ 0 & 5 \end{bmatrix},$$

$$F = \begin{bmatrix} -1 & 6 \\ 4 & 11 \end{bmatrix}, \quad G = \begin{bmatrix} 2 & -5 \\ -1 & 3 \end{bmatrix}, \quad \text{and} \quad I = \begin{bmatrix} 1 & 0 \\ 0 & 1 \end{bmatrix}$$

as needed to complete problems 1–25.

2.1

1. Find a_{12} in matrix A.
2. Find b_{23} in matrix B.
3. Which of the matrices, if any, are 3×4?

4. Which of the matrices, if any, are 2×4?
5. Which matrices are square?
6. Write the negative of matrix B.
7. If a matrix is added to its negative, what kind of matrix results?
8. Two matrices can be added if they have the same _____.

Again using the matrices above, perform the indicated operations in problems 9–25.

9. $A + B$	10. $C - E$	11. $D - I$

2.2

12. $3C$	13. $4I$	14. $-2F$
15. $4D - 3I$	16. $F + 2D$	17. $3A - 5B$
18. AC	19. CD	20. DF
21. FD	22. FI	23. IF
24. DG	25. $(DG)F$	

2.3

26. Given the following augmented matrix representing a system of linear equations, find the solution.

$$\left[\begin{array}{ccc|c} 1 & 1 & 2 & 5 \\ 4 & 0 & 1 & 5 \\ 2 & 1 & 1 & 5 \end{array}\right]$$

In problems 27–32, solve each system using matrices.

27. $\begin{cases} x - 2y = 4 \\ -3x + 10y = 24 \end{cases}$

28. $\begin{cases} x + y + z = 4 \\ 3x + 4y - z = -1 \\ 2x - y + 3z = 3 \end{cases}$

29. $\begin{cases} -x + y + z = 3 \\ 3x \quad - z = 1 \\ 2x - 3y - 4z = -2 \end{cases}$

30. $\begin{cases} x + y - 2z = 5 \\ 3x + 2y + 5z = 10 \\ -2x - 3y + 15z = 2 \end{cases}$

31. $\begin{cases} x - y = 3 \\ x + y + 4z = 1 \\ 2x - 3y - 2z = 7 \end{cases}$

32. $\begin{cases} x_1 + x_2 + x_3 + x_4 = 3 \\ x_1 - 2x_2 + x_3 - 4x_4 = -5 \\ x_1 \quad - x_3 + x_4 = 0 \\ x_2 + x_3 + x_4 = 2 \end{cases}$

2.4

33. Are D and G inverse matrices if

$$D = \begin{bmatrix} 3 & 5 \\ 1 & 2 \end{bmatrix} \quad \text{and} \quad G = \begin{bmatrix} 2 & -5 \\ -1 & 3 \end{bmatrix}?$$

Find the inverse of each of the following matrices.

34. $\begin{bmatrix} 7 & -1 \\ -10 & 2 \end{bmatrix}$
35. $\begin{bmatrix} 1 & 0 & 2 \\ 3 & 4 & -1 \\ 1 & 1 & 0 \end{bmatrix}$
36. $\begin{bmatrix} 3 & 3 & 2 \\ -1 & 4 & 2 \\ 2 & 5 & 3 \end{bmatrix}$

Solve the following systems of equations by using inverse matrices.

37. $\begin{cases} x \quad + 2z = 5 \\ 3x + 4y - z = 2 \\ x + y = -3 \end{cases}$
(See problem 35.)

38. $\begin{cases} 3x + 3y + 2z = 1 \\ -x + 4y + 2z = -10 \\ 2x + 5y + 3z = -6 \end{cases}$
(See problem 36.)

39. $\begin{cases} x + 3y + z = 0 \\ x + 4y + 3z = 2 \\ 2x - y - 11z = -12 \end{cases}$

40. Does the following matrix have an inverse?

$$\begin{bmatrix} 1 & 2 & 4 \\ 2 & 0 & -4 \\ 1 & 0 & -2 \end{bmatrix}$$

APPLICATIONS

The Burr Cabinet Company manufactures bookcases and filing cabinets at two plants, A and B. Matrix M gives the production for the two plants during June and matrix N gives the production for July. Use them in problems 41–43.

$$M = \begin{array}{c} \\ \end{array}\begin{array}{cc} A & B \\ \begin{bmatrix} 150 & 80 \\ 280 & 300 \end{bmatrix} & \begin{array}{l} \text{Bookcases} \\ \text{Files} \end{array} \end{array} \qquad N = \begin{array}{cc} A & B \\ \begin{bmatrix} 100 & 60 \\ 200 & 400 \end{bmatrix} \end{array}$$

$$S = \begin{bmatrix} 120 & 80 \\ 180 & 300 \end{bmatrix} \qquad P = \begin{bmatrix} 1000 & 800 \\ 600 & 1200 \end{bmatrix}$$

2.1

41. **Production** Write the matrix that represents total production at the two plants for the two months.

42. **Production** If matrix P represents the inventories at the plants at the beginning of June and matrix S represents shipments from the plants during June, write the matrix that represents the inventories at the end of June.

2.2

43. **Production** If the company sells its bookcases to wholesalers for $100 and its filing cabinets for $120, for which month was the value of production higher at (a) plant A? (b) plant B?

A small church choir is made up of men and women who wear choir robes in the sizes shown in matrix A.

$$A = \begin{array}{c} \\ \end{array}\begin{array}{cc} \text{Men} & \text{Women} \\ \begin{bmatrix} 1 & 14 \\ 12 & 10 \\ 8 & 3 \end{bmatrix} & \begin{array}{l} \text{Small} \\ \text{Medium} \\ \text{Large} \end{array} \end{array}$$

$$B = \begin{array}{c} \\ \end{array}\begin{array}{ccc} \text{S} & \text{M} & \text{L} \\ \begin{bmatrix} 25 & 40 & 45 \\ 10 & 10 & 10 \end{bmatrix} & & \begin{array}{l} \text{Robes} \\ \text{Hoods} \end{array} \end{array}$$

Matrix B contains the prices (in dollars) of new robes and hoods according to size. Use these matrices in problems 44–46.

44. **Cost** Find the product BA and label the rows and columns to show what each entry represents.

45. **Cost** To find a matrix that gives the cost of new robes and the cost of new hoods, find

$$BA\begin{bmatrix} 1 \\ 1 \end{bmatrix}.$$

46. **Cost** A matrix that gives the total cost of purchasing new robes and new hoods is given by

$$\begin{bmatrix} 1 & 1 \end{bmatrix} BA \begin{bmatrix} 1 \\ 1 \end{bmatrix}.$$

Find this total.

47. **Investment** Over the period of time 1984–1993, Chrysler's actual return (per dollar of stock) was 0.013469 per month and Ford's actual return was 0.013543 per month, while General Motor's actual return was 0.006504. Under certain assumptions, including that whatever underlying process generated the past set of observations will continue to hold in the present and near future, we can estimate the expected return for a stock portfolio containing these assets in some combination.

    Suppose we have a portfolio that has 20% of our total wealth in Chrysler, 30% in Ford, and 50% in GM stock. Use the following steps to estimate the portfolio's historical return, which can be used to estimate the future return.
(a) Write a 1 × 3 matrix that defines the decimal part of our total wealth in Chrysler, Ford, and GM.
(b) Write a 3 × 1 matrix that gives the monthly returns from each of these companies.
(c) Find the historical return of the portfolio by computing the product of these two matrices.
(d) What is the expected monthly return of the stock portfolio?

2.3

48. **Investment** A woman has $50,000 to invest. She has decided to invest all of it by purchasing some shares of stock in each of three companies: a fast-food chain that sells for $50 per share and has an expected growth of 11.5% per year, a software company that sells for $20 per share and has an expected growth of 15% per year, and a pharmaceutical company that sells for $80 per share and has an expected growth of 10% per year. She plans to buy twice as many shares of stock in the fast-food chain as in the pharmaceutical company. If her goal is 12% growth per year, how many shares of each stock should she buy?

49. **Nutrition** A biologist is growing three different types of slugs (types A, B, and C) in the same laboratory environment. Each day, the slugs are given a nutrient mix-

ture that contains three different ingredients (I, II and III). Each type A slug requires 1 unit of I, 3 units of II, and 1 unit of III per day. Each type B slug requires 1 unit of I, 4 units of II, and 2 units of III per day. Each type C slug requires 2 units of I, 10 units of II, and 6 units of III per day. If the daily mixture contains 2000 units of I, 8000 units of II, and 4000 units of III, find the number of slugs of each type that can be supported.

2.5

50. **Economy models** An economy has a shipping industry and an agricultural industry with technology matrix A.

$$A = \begin{bmatrix} 0.2 & 0.1 \\ 0.1 & 0.2 \end{bmatrix} \begin{array}{l} \text{Shipping} \\ \text{Agriculture} \end{array}$$

with S   A above the columns.

Surpluses of 4720 tons of shipping and 40 tons of agricultural output are desired. Find the gross production of each industry.

51. **Economy models** A simple economy has a shoe industry and a cattle industry with technology matrix A.

$$A = \begin{bmatrix} 0.1 & 0.1 \\ 0.2 & 0.05 \end{bmatrix} \begin{array}{l} \text{Shoes} \\ \text{Cattle} \end{array}$$

with S   C above the columns.

Surpluses of 850 units of shoes and 275 units of cattle production are desired. Find the gross production of each industry.

52. **Economy models** A look at the industrial sector of an economy can be simplified to include three industries: the mining industry, the manufacturing industry, and the fuels industry. The technology matrix for this sector of the economy is given by

$$A = \begin{bmatrix} 0.4 & 0.4 & 0.2 \\ 0.2 & 0.4 & 0.2 \\ 0.1 & 0.2 & 0.4 \end{bmatrix} \begin{array}{l} \text{Mining} \\ \text{Manufacturing} \\ \text{Fuels} \end{array}$$

with M   Mf   F above the columns.

Find the gross production of each industry if a surplus of 10 units of mined goods, 40 units of manufactured goods, and 140 units of fuels is desired.

53. **Economy models** Suppose a closed Leontief model for a nation's economy has the following technology matrix.

$$A = \begin{bmatrix} 0.4 & 0.2 & 0.2 & 0.2 \\ 0.2 & 0.4 & 0.1 & 0.2 \\ 0.2 & 0.1 & 0.3 & 0.1 \\ 0.2 & 0.3 & 0.4 & 0.5 \end{bmatrix} \begin{array}{l} \text{Government} \\ \text{Agriculture} \\ \text{Manufacturing} \\ \text{Households} \end{array}$$

with G   A   M   H above the columns.

Find the gross production of each industry.

CONSULTANT'S CORNER

Taxation

Federal income tax allows a deduction for any state income tax paid during the year. In addition, the state of Alabama allows a deduction from its state income tax for any federal income tax paid during the year. The federal corporate income tax rate is equivalent to a flat rate of 34% for taxable federal income between $335,000 and $10,000,000, and the Alabama rate is 5% of the taxable state income.*

Both the Alabama and federal taxable income for a corporation is $1,000,000 *before* either tax is paid. Because each tax is deductible on the other return, the taxable income will differ for the state and federal taxes. One procedure often used by tax accountants to find the tax due in this and similar situations is called *iteration* and is described by the first five steps below.[†]

1. Make an estimate of the federal taxes due by assuming no state tax is due, deduct this estimated federal tax due from the state taxable income, and calculate an estimate of the state taxes due based on this assumption.
2. Deduct the estimated state tax due as computed in (1) from the federal taxable income, and calculate a new estimate of the federal tax due under this assumption.
3. Deduct the new estimated federal income tax due as computed in (2) from the state taxable income, and calculate a new estimate of the state taxes due based on these calculations.
4. Repeat the process in (2) and (3) to get a better estimate of the taxes due to both governments.
5. Continue this process until the federal tax changes from one repetition to another by less than $1. What federal tax is due? What state tax is due?

To find the tax due each government directly, we can solve a matrix equation.

6. Create two linear equations that describe this taxation situation, convert the system to a matrix equation, and solve the system to see exactly what tax is due to each government.
7. What is the effective rate that this corporation pays to each government?

*Source: Federal Income Tax Tables, Alabama Income Tax Instructions.

[†] Performing these steps with an electronic spreadsheet is described by Kenneth H. Johnson in "A Simplified Calculation of Mutually Dependent Federal and State Tax Liabilities," in *The Journal of Taxation*, December 1994.

Warm-up

Prerequisite problem type	For section	Answer	Section for review
(a) Solve $3x - 2 = 7$. (b) Solve $2(x - 4) = \dfrac{x - 3}{3}$.	**3.1**	(a) $x = 3$ (b) $x = \frac{21}{5}$	1.1 Linear equations
Graph the equation $y = \frac{3}{2}x - 2$.	**3.1**		1.2 Graphing linear equations
Solve the systems: (a) $\begin{cases} x + 2y = 10 \\ 2x + y = 14 \end{cases}$ (b) $\begin{cases} x + 0.5y = 16 \\ x + y = 24 \end{cases}$	**3.2** **3.3**	(a) $x = 6, y = 2$ (b) $x = 8, y = 16$	1.8 Systems of linear equations
Write the system in an augmented matrix: $\begin{cases} x + 2y + s_1 = 10 \\ 2x + y + s_2 = 14 \\ -2x - 3y + f = 0 \end{cases}$	**3.4** **3.5** **3.6**	$\begin{bmatrix} 1 & 2 & 1 & 0 & 0 & \vert & 10 \\ 2 & 1 & 0 & 1 & 0 & \vert & 14 \\ -2 & -3 & 0 & 0 & 1 & \vert & 0 \end{bmatrix}$	2.3 Gauss-Jordan Elimination
Write a matrix equivalent to matrix A with the entry in row 1, column 2 equal to 1 and with all other entries in column 2 equal to 0. First multiply row 1 by $\frac{1}{2}$. $A = \begin{bmatrix} 1 & 2 & 1 & 0 & 0 & \vert & 10 \\ 2 & 1 & 0 & 1 & 0 & \vert & 14 \\ -2 & -3 & 0 & 0 & 1 & \vert & 0 \end{bmatrix}$	**3.4** **3.5** **3.6**	$\begin{bmatrix} \frac{1}{2} & 1 & \frac{1}{2} & 0 & 0 & \vert & 5 \\ \frac{3}{2} & 0 & -\frac{1}{2} & 1 & 0 & \vert & 9 \\ -\frac{1}{2} & 0 & \frac{3}{2} & 0 & 1 & \vert & 15 \end{bmatrix}$	2.3 Gauss-Jordan Elimination

244

CHAPTER

INEQUALITIES AND LINEAR PROGRAMMING

A firm frequently requires several components for the manufacture of the items it produces, and there are usually several stages for each item's assembly and final shipment. The company's costs and profits depend on the availability of the components (for example, labor and raw materials), the costs of these components, the unit profit for each product, and how many products are required. If the relationships among the various resources, the production requirements, the costs, and the profits are all linear, then these activities may be planned (or programmed) in the best possible (optimal) way by means of **linear programming.**

Because linear programming is useful in solving the problem of allocating limited resources among various activities in the best possible way, its impact has been tremendous. Although it is a relatively recent development, it is a standard tool for companies of all sizes, and its application has been responsible for saving many thousands of dollars. Numerous textbooks have been written on the subject; our intention here is to provide an introduction to the method.

Because the restrictions (constraints) on most business operations can often be expressed as linear inequalities, we begin this chapter by introducing methods for solving and graphing linear inequalities. We will show how graphs of the constraint inequalities can be used to solve linear programming problems. The simplex method provides a technique for converting a system of inequalities into a system of equations that can be used to solve linear programming problems.

3.1 LINEAR INEQUALITIES IN ONE VARIABLE

Objective

❏ To solve and graph linear inequalities in one variable

APPLICATION PREVIEW

The height H in inches and age A in years for boys between 4 and 16 years of age are related according to

$$H = 2.31A + 31.26.$$

To account for the normal variability among boys, normal height for a given age is ±5% of the height obtained in the equation.* We can express the range of normal height for a boy of a given age as an **inequality.** In this section we will solve inequalities involving one variable raised to the first power (**linear inequalities**).

An **inequality** is a statement that one quantity is greater than (or less than) another quantity. We have already encountered some very simple inequalities. For example, the number of items a firm produces and sells, x, must be a nonnegative quantity. Thus x is greater than or equal to zero, which is written $x \geq 0$. The inequality $3x - 2 > 2x + 1$ is a first-degree (linear) inequality that states that the left member is greater than the right member. Certain values of the variable will satisfy the inequality. These values form the solution set of the inequality. For example, 4 is in the solution set of $3x - 2 > 2x + 1$ because $3 \cdot 4 - 2 > 2 \cdot 4 + 1$. On the other hand, 2 is not in the solution set because $3 \cdot 2 - 2 \not> 2 \cdot 2 + 1$. *Solving* an inequality means finding its solution set, and two inequalities are *equivalent* if they have the same solution set. As with equations, we find the solutions to inequalities by finding equivalent inequalities from which the solutions can be easily seen. We use the following properties to reduce an inequality to a simple equivalent inequality.

Inequalities	
Properties	**Examples**
Substitution Property The inequality formed by substituting one expression for an equal expression is equivalent to the original inequality.	$5x - 4x < 6$ $x < 6$ The solution set is $\{x: x < 6\}$.

*Adapted from data from the National Center for Health Statistics.

Addition Property
The inequality formed by adding the same quantity to both sides of an inequality is equivalent to the original inequality.

$$2x - 4 > x + 6$$
$$2x - 4 + 4 > x + 6 + 4$$
$$2x > x + 10$$
$$2x + (-x) > x + 10 + (-x)$$
$$x > 10$$

Multiplication Property I
The inequality formed by multiplying both sides of an inequality by the same *positive* quantity is equivalent to the original inequality.

$$\tfrac{1}{2}x > 8 \qquad\qquad 3x < 6$$
$$\tfrac{1}{2}x(2) > 8(2) \qquad 3x(\tfrac{1}{3}) < 6(\tfrac{1}{3})$$
$$x > 16 \qquad\qquad x < 2$$

Multiplication Property II
The inequality formed by multiplying both sides of an inequality by the same *negative* number and reversing the sense of the inequality is equivalent to the original inequality.

$$-x < 6 \qquad\qquad -3x > -27$$
$$-x(-1) > 6(-1) \qquad -3x(-\tfrac{1}{3}) < -27(-\tfrac{1}{3})$$
$$x > -6 \qquad\qquad x < 9$$

We may graph the solution to inequalities in one unknown on the real number line. For example, the graph of $x < 2$ consists of all points to the left of 2 on the number line. The open circle on the graph in Figure 3.1 indicates that all points up to but *not* including 2 are in the solution set.

Figure 3.1

EXAMPLE I
Solve the inequality

$$\frac{s}{-2} < -4$$

and graph the solution set.

Solution

$$\frac{s}{-2} < -4$$

$$\frac{s}{-2}(-2) > -4(-2)$$

$$s > 8$$

The solution set is $\{s: s > 8\}$. The graph of the solution set is shown in Figure 3.2.

Figure 3.2

Some inequalities will require several operations to find their solution set. In this case, the order in which the operations are performed is the same as that used in solving linear equations.

EXAMPLE 2

Solve the inequality $2(x - 4) < \dfrac{x - 3}{3}$.

Solution

$$2(x - 4) < \frac{x - 3}{3}$$

$$6(x - 4) < x - 3 \qquad \text{Clear fractions}$$

$$6x - 24 < x - 3 \qquad \text{Remove parentheses}$$

$$5x < 21 \qquad \text{Perform additions and subtractions}$$

$$x < \frac{21}{5} \qquad \text{Multiply by } \frac{1}{5}$$

Now, if we want to check that this solution is a reasonable one, we can substitute the integer values around 21/5 into the original inequality. Note that $x = 4$ satisfies the inequality because

$$2[(4) - 4] < \frac{(4) - 3}{3},$$

but $x = 5$ does not:

$$2[(5) - 4] \not< \frac{(5) - 3}{3}.$$

Thus, $x < 21/5$ is a reasonable solution.

We may also solve inequalities of the form $a \leq b$. This means "a is less than b or $a = b$." The solution of $2x \leq 4$ is $x \leq 2$, since $x < 2$ is the solution of $2x < 4$ and $x = 2$ is the solution of $2x = 4$.

EXAMPLE 3

Solve the inequality $3x - 2 \leq 7$.

Solution

This inequality states that $3x - 2 = 7$ or that $3x - 2 < 7$. By solving in the usual manner, we get $3x \leq 9$, or $x \leq 3$. Then $x = 3$ is the solution to $3x - 2 = 7$ and $x < 3$ is the solution to $3x - 2 < 7$, so the solution set for $3x - 2 \leq 7$ is $\{x: x \leq 3\}$.

The graph of the solution set includes the point $x = 3$ and all points $x < 3$ (see Figure 3.3).

Figure 3.3

CHECKPOINT

Solve the following inequalities for *y*.
  1. $3y - 7 \leq 5 - y$ 2. $2y + 6 > 4y + 5$ 3. $4 - 3y \geq 4y + 5$

We may use the notation $-2 < x < 4$ to state that *x* satisfies the two conditions $-2 < x$ *and* $x < 4$. This **compound inequality** represents an **interval** on the number line. The graph of this compound inequality is called an **open interval** because neither *endpoint* (-2 or 4) is included in the graph (see Figure 3.4). We may denote the interval $-2 < x < 4$ by $(-2, 4)$.

Figure 3.4

The graph in Figure 3.5 of $-1 \leq x \leq 3$ is a **closed interval** because both endpoints are included. This closed interval may be denoted $[-1, 3]$.

Figure 3.5

The interval determined by $-2 < x \leq 4$ is denoted $(-2, 4]$ and is called **half-open** because one endpoint is excluded (see Figure 3.6).

Figure 3.6

In general, we denote the closed interval $a \leq x \leq b$ by $[a, b]$, the open interval $a < x < b$ by (a, b), and the half-open interval $a \leq x < b$ by $[a, b)$.

CHECKPOINT

Express each of the following inequalities in interval notation and name the type of interval.
  4. $3 \leq x \leq 6$ 5. $-6 \leq x < 4$

EXAMPLE 4

The Application Preview at the beginning of this section described how height and age are linearly related for boys between 4 and 16 years of age. That relation can be expressed as

$$H = 2.31A + 31.26,$$

where *H* is height in inches and *A* is age in years. To account for natural variation among individuals, normal is considered to be any measure falling within ±5% of the height obtained from the equation. Write as an inequality the range of normal height for a boy who is 9 years old.

Solution

The boy's height from the formula is $H = 2.31(9) + 31.26 = 52.05$ inches. For a 9-year-old boy to be considered of normal height, H would have to be within $\pm 5\%$ of 52.05 inches. That is, the boy's height H is considered normal if $H \geq 52.05 - (.05)(52.05)$ and $H \leq 52.05 + (.05)(52.05)$. We can write this range of normal height by the compound inequality

$$52.05 - (.05)(52.05) \leq H \leq 52.05 + (.05)(52.05)$$

or

$$49.45 \leq H \leq 54.65.$$

CHECKPOINT SOLUTIONS

1. $-7 \leq 5 - 4y$ 2. $6 > 2y + 5$ 3. $4 \geq 7y + 5$
   $y \leq 3$ $1 > 2y$ $-1 \geq 7y$
                             $y < 1/2$ $y \leq -1/7$

4. $[3,6]$; closed interval

5. $[-6, 4)$; half-open interval

EXERCISE 3.1

1. Write the inequality describing the interval $\left(-\frac{1}{2}, 3\right]$ on the x-axis.
2. Write the inequality describing the interval $[-3, 5)$ on the x-axis.

In problems 3–14, express each inequality or graph using interval notation and name the type of interval.

3. $1 < x \leq 3$ 4. $-4 \leq x \leq 3$

5.
   0   2   4   6   8   10

6.
   -2   0   2   4   6

7.
   -4   -2   0   2   4

8.
   -6   -4   -2   0   2

9. $-4 < x < 3$ 10. $-6 \leq x \leq -4$
11. $4 \leq x \leq 6$ 12. $-2 \leq x \leq -1$
13. $4 \leq x < 9$ 14. $5 < x \leq 8$

Solve the following inequalities.

15. $x + 2 < 4$ 16. $x - 3 > 5$
17. $x - 3 \geq 2x + 1$ 18. $2x - 1 \leq 3x + 4$
19. $3(x - 1) < 2x - 1$ 20. $2(x + 1) > x - 1$
21. $2(x - 1) - 3 > 4x + 1$ 22. $7x + 4 \leq 2(x - 1)$

23. $\dfrac{x}{3} > x - 1$ 24. $\dfrac{x - 3}{4} \geq 2x$

25. $\dfrac{3(x - 1)}{2} \leq x - 2$ 26. $\dfrac{x - 1}{2} + 1 > x + 1$

27. $\dfrac{5(2 - x)}{3} \geq \dfrac{6x + 5}{2}$ 28. $\dfrac{3(x - 1)}{4} \leq \dfrac{4 - x}{2}$

29. $2.9(2x - 4) \geq \dfrac{3.6x - 8.5}{5}$

30. $2.2(3.1x - 5) \leq \dfrac{3(4.6 - 6.1x)}{8}$

Graph the solutions to the following inequalities.

31. $2x + 1 < x - 3$ 32. $3x - 1 \geq 2x + 2$
33. $3(x - 1) < 2x$ 34. $3(x + 2) \geq 4x + 1$

35. $2x + \dfrac{1}{2} \ge x - 3$ 36. $5x - 2 > 3(x + 3)$

37. $\dfrac{3x}{4} - \dfrac{1}{6} < x - \dfrac{2(x-1)}{3}$ 38. $\dfrac{4x}{3} - 3 > \dfrac{1}{2} + \dfrac{5x}{12}$

39. $\dfrac{3x}{4} - \dfrac{1}{3} < 1 - \dfrac{2}{3}\left(x - \dfrac{1}{2}\right)$

40. $\dfrac{x}{2} - \dfrac{4x}{5} > \dfrac{3(x-1)}{10} - 2$

In problems 41–46, write an inequality that describes each interval or graph.

41. $(1, 4)$ 42. $[3, 7]$

43.

44.

45.

-22

-50 -30 -10 0 10 30 50

46.

-1.2

-4 -2 0 2 4 6

APPLICATIONS

47. **Profit** For a certain product, the revenue function is $R(x) = 40x$ and the cost function is $C(x) = 20x + 1600$. To obtain a profit, the revenue must be greater than the cost. For what values of x will there be a profit? Graph the solution.

48. **Car rental** Thrifty rents a compact car for $33 per day while Budget rents a similar car for $20 per day plus an intial fee of $78. For how many days would it be cheaper to rent from Budget? Graph the solution.

49. **Purchasing** Sean can spend at most $900 for a video camera and some video tapes. He plans to buy the camera for $695 and tapes for $5.75 each. Write an inequality that could be used to find the number of tapes (x) that he could buy. How many tapes could he buy?

50. **Taxes** In Sweetwater, Arizona, water bills are taxed based on the amount of the monthly bill in order to encourage conservation. If the bill is more than $0 but less than $60, the tax is 2% of the bill; if the bill is $60 but less than $80, the tax is 4% of the bill; and if the bill is $80 or more, the tax is 6% of the bill. Write the inequalities that represent the amount of tax owed.

51. **Income Taxes** The 1994 tax brackets for a single person claiming one personal exemption are given in the following table.

Taxable Income I	Tax T
$0–$22,750	15% I
$22,751–$55,100	28% $(I - 22{,}750) + 3412.50$
$55,101–$115,000	31% $(I - 55{,}100) + 12{,}470.50$
$115,001–$250,000	36% $(I - 115{,}000) + 31{,}039.50$
Over $250,000	39.6% $(I - 250{,}000) + 79{,}639.50$

Source: 1040 Forms and Instructions, IRS, 1994

(a) Write the income ranges in the table as inequalities.
(b) For each income bracket, write the inequality that represents the amount of tax owed.

52. **Health Statistics** From data adapted from the National Center for Health Statistics, the height H in inches and age A in years for boys between 4 and 16 years of age are related according to

$$H = 2.31A + 31.26.$$

To account for normal variability among boys, normal height for a given age is $\pm 5\%$ of the height obtained from the equation.
(a) Find the normal height range for a boy that is 10.5 years old, and write it as an inequality.
(b) Find the normal height range for a boy that is 5.75 years old, and write it as an inequality.

53. **Health Statistics** Normal body temperature is 98.6°F. What Celsius temperature C corresponds to $F \ge 98.6°$ if $F = \dfrac{9}{5}C + 32$?

54. **Sales** A saleswoman has monthly income I given by $I = 1000 + 0.062S$, where S is the monthly sales volume. How much must she sell to make at least $3500 in a month?

55. **Traffic Flow** The traffic flow problem illustrated in Figure 3.7 (on page 252) appeared as problem 62 of Section 2.3, where $x_1, x_2, x_3,$ and x_4 represent the number of cars on the indicated streets. The solution to that problem is described as follows.

$$x_1 = x_4 + 100$$
$$x_2 = x_4 + 200$$
$$x_3 = x_4 + 300$$

Since the total number of cars in the system is 1200 and each variable must be nonnegative we can develop the following inequalities.

$$x_4 + 100 \le 1200 \qquad (1)$$
$$x_4 + 200 \le 1200 \qquad (2)$$
$$x_4 + 300 \le 1200 \qquad (3)$$
$$x_4 \ge 0 \qquad (4)$$

(a) Solve each inequality for x_4.
(b) Determine an interval for x_4 that satisfies all these inequalities. (Hint: Find the intersection of the solution graphs.)
(c) Use the interval from (b) to develop an interval for each of the other variables.

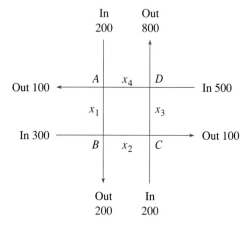

Figure 3.7

56. **Nutrition** Problem 61 of Section 2.3 described a botanist who could purchase 4 different plant foods with various nutrient values. If x_1 represents the number of bags of food I that were purchased, x_2 the number of bags of food II, and so on, then the number of bags of each food that the botanist can purchase and satisfy the requirements given in problem 61 of Section 2.3 can be described by the following.

$$x_1 = 160 - x_4$$
$$x_2 = 2x_4 - 220$$
$$x_3 = x_4 + 70$$

Since each variable must be nonnegative we can develop the following inequalities.

$$160 - x_4 \geq 0$$
$$2x_4 - 220 \geq 0$$
$$x_4 + 70 \geq 0$$
$$x_4 \geq 0$$

(a) Solve each inequality for x_4.
(b) Determine an interval for x_4 that solves all these inequalities.
(c) Develop an interval for each of the other variables.

3.2   LINEAR INEQUALITIES IN TWO VARIABLES

Objectives

❑ To graph linear inequalities in two variables
❑ To solve systems of linear inequalities in two variables

APPLICATION PREVIEW

A rental agency has a maximum of $240,000 to invest in the purchase of at most 13 cars of two different types, compact and full size. The cost per compact is $15,000 and the cost per full-size car is $24,000. The number of cars of each type is limited (constrained) by the budget available and the number of cars needed. These **constraints** can be expressed by a **system of inequalities** in two variables. Finding the values that satisfy all these constraints at the same time is called solving the system of inequalities.

Before we look at systems of inequalities, we will discuss solutions of one inequality in two variables, for example, $y < x$. The solutions to this inequality are the ordered pairs (x, y) that satisfy the inequality. Thus $(1, 0)$, $(3, 2)$, $(0, -1)$, and $(-2, -5)$ are the solutions to $y < x$, but $(3, 7)$, $(-4, -3)$ and $(2, 2)$ are not. The graph of $y < x$ consists of all points whose y-value is less than the x-value. Thus the graph consists of all points below the line $y = x$

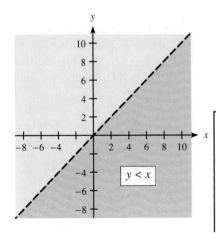

Figure 3.8

(the shaded area in Figure 3.8). Selecting any "test point" in the shaded region and using it to discover if it satisfies the inequality serves as a check.

Note that the points on the line $y = x$ do not satisfy the inequality, so they are not part of the graph. This is indicated by the dashed line in Figure 3.8.

EXAMPLE 1

Graph the inequality $3x + y < 3$.

Solution

If we solve the inequality for y, we get $y < 3 - 3x$. Thus the graph of this inequality will consist of all points *below* the line $y = 3 - 3x$. The graph is shown in Figure 3.9. Note that any point (such as $(0, 0)$) that is in the shaded region will satisfy the original inequality.

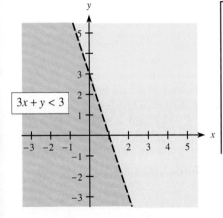

Figure 3.9

EXAMPLE 2

Graph the inequality $4x - 2y \leq 6$.

Solution

Solving the inequality for y gives $y \geq 2x - 3$. Because y is greater than or equal to $2x - 3$, the graph consists of all points *on* or *above* the graph of $y = 2x - 3$. The line representing $y = 2x - 3$ is solid in Figure 3.10 to indicate that it is included in the solution. The solution can be checked by picking a "test point" from the shaded region and verifying that its coordinates satisfy $y \geq 2x - 3$.

If we have two inequalities in two variables, we can find the solutions that satisfy both inequalities. We call the inequalities a **system of inequalities,** and the solution to the system can be found by finding the intersection of the solution sets of the two inequalities.

The solution set of the system of inequalities can be found by graphing the inequalities on the same set of axes and noting their points of intersection.

EXAMPLE 3

Graph the solution to the system

$$\begin{cases} 3x - 2y \geq 4 \\ x + y - 3 > 0. \end{cases}$$

Solution

The inequalities can be written in the form

$$y \leq \tfrac{3}{2}x - 2$$
$$y > -x + 3.$$

Figure 3.10

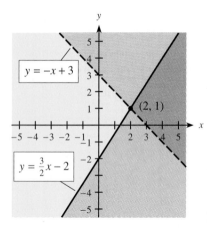

Figure 3.11

The graph of $y \le \frac{3}{2}x - 2$ consists of all points on or below the line $y = \frac{3}{2}x - 2$. The graph of $y > -x + 3$ consists of all points above the line $y = -x + 3$. The points that satisfy both of these inequalities lie in the intersection of the two solution regions, shown in Figure 3.11. This **solution region** is the graph of the solution to the system containing these two inequalities.

The point (2, 1) in Figure 3.11, where the two regions form a "corner," is found by solving the *equations* $y = \frac{3}{2}x - 2$ and $y = -x + 3$ simultaneously.

EXAMPLE 4

Graph the solution to the system

$$\begin{cases} x + 2y \le 10 \\ 2x + y \le 14 \\ \quad\quad x \ge 0 \\ \quad\quad y \ge 0. \end{cases}$$

Solution

We seek points in the first quadrant (on or above $y = 0$ and on or to the right of $x = 0$) that satisfy $x + 2y \le 10$ *and* $2x + y \le 14$. We can write these inequalities in their equivalent forms $y \le 5 - \frac{1}{2}x$ and $y \le 14 - 2x$. The points that satisfy these inequalities (in the first quadrant) are shown by the overlapping area in Figure 3.12. We can observe from the graph that the points (0, 0), (7, 0), and (0, 5) are corners of the solution region. The corner (6, 2) is found by solving the *equations* $y = 5 - \frac{1}{2}x$ and $y = 14 - 2x$ simultaneously as follows.

$$5 - \frac{1}{2}x = 14 - 2x$$

$$\frac{3}{2}x = 9$$

$$x = 6$$

$$y = 2$$

Figure 3.12

We will see that the corners of the solution region are important in solving linear programming problems.

We now return to the problem posed in the Application Preview.

EXAMPLE 5

A rental agency has a maximum of $240,000 to invest in the purchase of at most 13 cars of two different types, compact and full size. The cost per compact is $15,000 and the cost per full-size car is $24,000. Write the system of inequalities that describes the constraints, and graph the solution region for the system.

Solution

If x represents the number of compact cars and y represents the number of full-size cars, then the number of cars is limited by $x + y \leq 13$. The limitation on available funds is given by $15,000x + 24,000y \leq 240,000$. Because the number of cars must be positive, the system of inequalities is

$$x + y \leq 13$$
$$15,000x + 24,000y \leq 240,000$$
$$x \geq 0$$
$$y \geq 0$$

The solution region is shown in Figure 3.13. The solution region has corners at (0, 0), (13, 0), (8, 5), and (0, 10). The corner at (8, 5) is found by solving $x + y = 13$ and $15,000x + 24,000y = 240,000$ simultaneously.

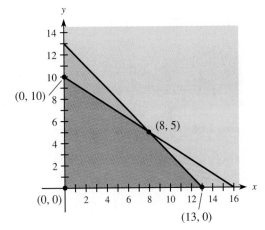

Figure 3.13

CHECKPOINT

1. Graph the region determined by the inequalities

$$2x + 3y \leq 12$$
$$4x + 2y \leq 16$$
$$x \geq 0$$
$$y \geq 0.$$

2. Determine the corners of the region.

Graphing Utilities

Many computer programs save considerable time and energy in determining regions that satisfy a system of inequalities. Graphics calculators can also be used in the graphical solution of a system of inequalities.

EXAMPLE 6

Use a graphing utility to find the following.
(a) Find the region determined by the inequalities below.

$$5x + 2y \le 54$$
$$2x + 4y \le 60$$
$$x \ge 0$$
$$y \ge 0$$

(b) Find the corners of this region.

Solution

(a) We write the inequalities above as equations, solved for y. Graphing these equations with a graphics calculator and using shading shows the region satisfying the inequalities (see Figure 3.14).

(b) Some utilities can determine the points of intersection of two equations with an INTERSECT command. This command can be used to find the intersection of the pair of equations from part (a). With other graphing utilities, using the command SOLVER or INTERSECT, or using TRACE along the lines that form the borders of this region will give the points where the boundaries intersect. These points, (0, 15), (6, 12), and (10.8, 0) can also be found algebraically. These points are the corners of the solution region. These corners will be important to us in the graphical solution to linear programming problems in the next section.

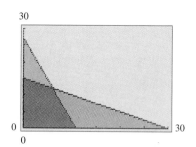

Figure 3.14

CHECKPOINT
SOLUTIONS

1.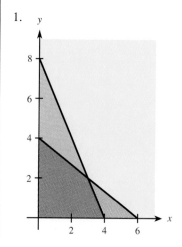

2. Corners occur at (0, 0) and where $x = 0$ or $y = 0$: $x = 0$ gives $y = 4$ and $y = 8$ so (0, 4) is a corner; $y = 0$ gives $x = 6$ and $x = 4$ so (4, 0) is a corner. A corner also occurs where $2x + 3y = 12$ and $4x + 2y = 16$ intersect. The point of intersection is $x = 3$, $y = 2$, so (3, 2) is a corner.

EXERCISE 3.2

In problems 1–10, graph each inequality.

1. $y > x$

2. $y < 3x$

3. $y \le 2x - 1$

4. $y \ge 4x - 5$

5. $\dfrac{x}{2} + \dfrac{y}{4} < 1$

6. $x - \dfrac{y}{3} < \dfrac{-2}{3}$

7. $2(x - y) \le y + 3$

8. $2(x - y) \ge x + 4$

9. $0.4x \ge 0.8$

10. $\dfrac{-y}{8} > \dfrac{1}{4}$

In problems 11–16, the graph of the solution to each system of inequalities is shown with that system. Find the corners of each solution region.

11. $\begin{cases} x + 4y \le 60 \\ 4x + 2y \le 100 \\ x \ge 0 \\ y \ge 0 \end{cases}$

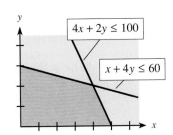

12. $\begin{cases} 4x + 3y \le 240 \\ 5x - y \le 110 \\ x \ge 0 \\ y \ge 0 \end{cases}$

13. $\begin{cases} -x + y \le 4 \\ x + 3y \le 20 \\ 5x + 2y \le 35 \\ x \ge 0 \\ y \ge 0 \end{cases}$

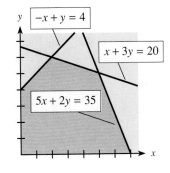

14. $\begin{cases} x + 2y \le 20 \\ 3x + 4y \le 48 \\ 3x + 2y \le 42 \\ x \ge 0 \\ y \ge 0 \end{cases}$

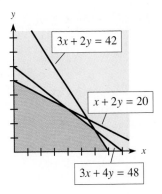

15. $\begin{cases} x + 3y \ge 6 \\ 2x + 4y \ge 10 \\ 3x + y \ge 5 \\ x \ge 0 \\ y \ge 0 \end{cases}$

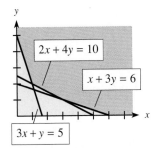

16. $\begin{cases} x + 4y \ge 10 \\ 4x + 2y \ge 10 \\ x + y \ge 4 \\ x \ge 0 \\ y \ge 0 \end{cases}$

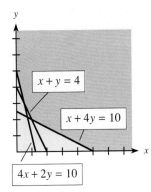

In problems 17–32 graph the solution to each system of inequalities.

17. $\begin{cases} y < 2x \\ y > x - 1 \end{cases}$

18. $\begin{cases} y > 3x - 4 \\ y < 2x + 3 \end{cases}$

19. $\begin{cases} 2x + y < 3 \\ x - 2y \ge -1 \end{cases}$

20. $\begin{cases} 3x + y > 4 \\ x - 2y < -1 \end{cases}$

21. $\begin{cases} y \ge 3 - 2x \\ y \ge \frac{1}{2}x + \frac{1}{2} \\ y \ge 3 \end{cases}$

22. $\begin{cases} y \le x + 1 \\ y \ge 2x - 1 \\ x \ge 0,\ y \ge 0 \end{cases}$

23. $\begin{cases} x + 5y \le 200 \\ 2x + 3y \le 134 \\ x \ge 0, y \ge 0 \end{cases}$

24. $\begin{cases} -x + y \le 2 \\ x + 2y \le 10 \\ 3x + y \le 15 \\ x \ge 0, y \ge 0 \end{cases}$

25. $\begin{cases} x + 2y \le 48 \\ x + y \le 30 \\ 2x + y \le 50 \\ x \ge 0, y \ge 0 \end{cases}$

26. $\begin{cases} 3x + y \le 9 \\ 3x + 2y \le 12 \\ x + 2y \le 8 \\ x \ge 0, y \ge 0 \end{cases}$

27. $\begin{cases} x + 2y \ge 19 \\ 3x + 2y \ge 29 \\ x \ge 0, y \ge 0 \end{cases}$

28. $\begin{cases} 4x + y \ge 12 \\ x + y \ge 9 \\ x + 3y \ge 15 \\ x \ge 0, y \ge 0 \end{cases}$

29. $\begin{cases} x + 3y \ge 3 \\ 2x + 3y \ge 5 \\ 2x + y \ge 3 \\ x \ge 0, y \ge 0 \end{cases}$

30. $\begin{cases} x + 2y \ge 10 \\ 2x + y \ge 11 \\ x + y \ge 9 \\ x \ge 0, y \ge 0 \end{cases}$

31. $\begin{cases} x + 3y \ge 6 \\ 3x + 4y \ge 13 \\ x \ge 0 \\ y \ge 0 \end{cases}$

32. $\begin{cases} x + 2y \le 8 \\ 5x + 4y \le 28 \\ x \ge 0 \\ y \ge 0 \end{cases}$

APPLICATIONS

33. **Management** The Wellbuilt Company produces two types of wood chippers, Economy and Deluxe. The Deluxe model requires 3 hours to assemble and 1/2 hour to paint, and the Economy model requires 2 hours to assemble and 1 hour to paint. The maximum number of assembly hours available is 24 per day and the maximum number of painting hours available is 8 per day.
    (a) Write the system of inequalities that describes the constraints on the production of these models.
    (b) Graph the solution to the system of inequalities and find the corners of the solution region.

34. **Learning environments** An experiment involving learning in animals requires placing white mice and rabbits into separate, controlled environments, Environment I and Environment II. The maximum amount of time available in Environment I is 500 minutes and the maximum amount of time available in Environment II is 600 minutes. The white mice must spend 10 minutes in Environment I and 25 minutes in Environment II, while the rabbits must spend 15 minutes in Environment I and 15 minutes in Environment II.

    (a) Write a system of inequalities that describes the constraints on the experiment.
    (b) Graph the solution to the system of inequalities and find the corners of the solution region.

35. **Production** A company manufactures two types of electric hedge-trimmers, one of which is cordless. The cord-type trimmer requires 2 hours to make, and the cordless model requires 4 hours. The company has only 800 work hours to use in manufacturing each day, and the packing department can package only 300 trimmers per day.
    (a) Write the inequalities that describe these constraints on production.
    (b) Graph the region determined by these constraint inequalities.

36. **Manufacturing** A firm manufactures bumper bolts and fender bolts for cars. One machine can produce 130 fender bolts per hour and another machine can produce 120 bumper bolts per hour. The combined number of fender bolts and bumper bolts the packaging department can handle is 230 per hour.
    (a) Write the inequalities that describe these constraints on production.
    (b) Graph the region determined by these constraint inequalities.

37. **Politics** A candidate wishes to use a combination of radio and television advertisements in his campaign. Research has shown that each 1-minute spot on television reaches 0.09 million people and each 1-minute spot on radio reaches 0.006 million. The candidate feels he must reach at least 2.16 million people, and he must buy a total of at least 80 minutes of advertisements.
    (a) Write the inequalities that describe his needs.
    (b) Graph the region determined by these constraint inequalities.

38. **Nutrition** In a hospital ward, the patients can be grouped into two general categories depending on their condition and the amount of solid foods they require in their diet. A combination of two diets is used for solid foods because they supply essential nutrients for recovery. The table below summarizes the patient group and minimum daily requirements.

	Diet A	Diet B	Daily Requirements
Group 1	4 oz per serving	1 oz per serving	26 oz
Group 2	2 oz per serving	1 oz per serving	18 oz

(a) Write the inequalities that describe these needs.
(b) Graph the region determined by these constraint inequalities.

39. ***Manufacturing*** A sausage company makes two different kinds of hot dogs, regular and all beef. Each pound of all-beef hot dogs requires 0.75 lb of beef and 0.2 lb of spices, and each pound of regular hot dogs requires 0.18 lb of beef, 0.3 lb of pork, and 0.2 lb of spices. Suppliers can deliver at most 1020 lb of beef, at most 600 lb of pork, and at least 500 lb of spices.
(a) Write the inequalities that describe these needs.
(b) Graph the region determined by these constraint inequalities.

40. ***Manufacturing*** A cereal manufacturer makes two different kinds of cereal, Senior Citizen's Feast and Kids Go. Each pound of Senior Citizen's Feast requires 0.6 lb of wheat and 0.2 lb of vitamin-enriched syrup, and each pound of Kids Go requires 0.4 lb of wheat, 0.2 lb of sugar, and 0.2 lb of vitamin-enriched syrup. Suppliers can deliver at most 2800 lb of wheat, at most 800 lb of sugar, and at least 1000 lb of the vitamin-enriched syrup.
(a) Write the inequalities that describe these needs.
(b) Graph the region determined by these constraint inequalities.

3.3 LINEAR PROGRAMMING: GRAPHICAL METHODS

Objective

❑ To use graphical methods to find the optimum value of a linear function subject to constraints

APPLICATION PREVIEW

Many practical problems in business and economics involve complex relationships between capital, raw material, manpower, and so forth. Consider the following example. Suppose that two chemical plants produce three types of fertilizer, low phosphorus (LP), medium phosphorus (MP), and high phosphorus (HP). The fertilizer is produced in a single production run, so the three types are produced in fixed proportions. The Macon plant produces 1 ton of LP, 2 tons of MP, and 3 tons of HP in a single operation, and it charges $600 for what is produced in one operation, while one operation of the Jonesboro plant produces 1 ton of LP, 5 tons of MP, and 1 ton of HP, and it charges $1000 for what it produces in one operation. If a customer needs a specific amount of each type of fertilizer, the minimum cost can be found by using a mathematical technique called **linear programming.** Linear programming can be used to solve problems such as this if the limits on the variables (called **constraints**) can be expressed as linear inequalities and if the function that is to be maximized or minimized (called the **objective function**) is a linear function.

Frequently linear programming problems involve many variables, but in this section we restrict our discussion to problems involving two variables. With two variables we can use graphical methods to help solve the problem. The constraints form a system of linear inequalities in two variables that we can solve by graphing. The solution to the system of inequalities determines a region, any point of which may yield the *optimum* (maximum or minimum) value for the objective function.* Hence any point in the region determined by the constraints is called a **feasible solution,** and the region itself is called the **feasible region.**

In a linear programming problem, we seek the feasible solution that maximizes (or minimizes) the objective function. For example, suppose we wish to maximize the function $C = 2x + y$ subject to the constraints $x \geq 0$, $y \geq 0$, $x + 3y \leq 6$, and $x + y \leq 4$. The solution set (feasible region) for the system of constraints is shown in Figure 3.15(a). *Any* point inside the shaded region or on its boundary is a feasible solution to the problem. To determine which point will maximize the objective function, we graph $C = 2x + y$, for different values of C, on the same graph containing the feasible region. The different values of C change the position of the line, *but the slope of the line does not change.* Letting $C = 0, 2, 3, 7$, and 8, we get the graphs shown in Figure 3.15(b). If any part of the line lies within the feasible region, we have feasible solutions.

Since we seek the values of x and y that will maximize $2x + y$, subject to the constraints, we keep trying larger values of C while keeping some part of the line intersecting the feasible region. It is clear from Figure 3.15(b) that any value of C larger than 8 will cause the line $C = 2x + y$ to "miss" the region, and so the maximum value for $2x + y$, subject to the constraints, is 8. The point where the line $8 = 2x + y$ intersects the feasible region is $(4, 0)$, so the function $2x + y$ is maximized when $x = 4$, $y = 0$. Note that the objective function was maximized at one of the "corners" (or vertices) of the feasible region.

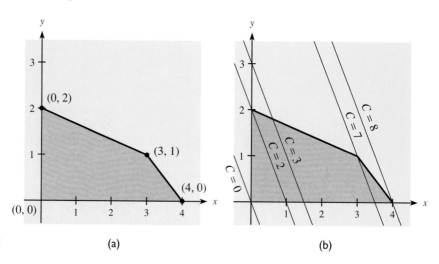

Figure 3.15   (a)   (b)

---

*The region determined by the constraints must be *convex* for the optimum to exist. A convex region is one such that for any two points in the region, the segment joining those points lies entirely within the region. We restrict our discussion to convex regions.

The feasible region in Figure 3.15 is an example of a closed and bounded region because it is entirely enclosed by, and includes, the lines associated with the constraints. When the feasible region for a linear programming problem is closed and bounded, the objective function has a maximum value and a minimum value. When the feasible region is not closed and bounded, the objective function may have a maximum only, a minimum only, or no solution. However, if a solution exists, we have the following.

If a linear programming problem has a solution, then the optimum (maximum or minimum) value of an objective function occurs at a corner of the feasible region determined by the constraints.

Thus, to find the maximum or minimum value of the objective function, we need only evaluate the function at each of the corners of the feasible region formed by the solution of the constraint inequalities. If the objective function has its optimum values at two corners, then it also has that optimum value at any point on the line (boundary) connecting those two corners.

EXAMPLE 1

Maximize $C = 2x + 3y$ subject to the constraints

$$\begin{cases} x + 2y \le 10 \\ 2x + y \le 14 \\ x \ge 0 \\ y \ge 0. \end{cases}$$

Solution

The feasible region determined by the constraint inequalities is shown in Figure 3.16 below. Note that the region is identical to that in Example 4 of Section 3.2—see Figure 3.12. The corners of this region were found in Example 4 to be $(0, 0)$, $(0, 5)$, $(6, 2)$ and $(7, 0)$. We can find the maximum value by evaluating $C = 2x + 3y$ at each of the corner points.

At $(0, 0)$, $2x + 3y = 0$

At $(0, 5)$, $2x + 3y = 15$

At $(6, 2)$, $2x + 3y = 18$

At $(7, 0)$, $2x + 3y = 14$

Thus we see that C is maximized when $x = 6$ and $y = 2$ and has a maximum value of $C = 18$. See Figure 3.16.

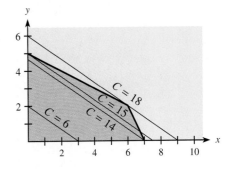

Figure 3.16

The steps involved in solving a linear programming problem are as follows.

Linear Programming (Graphical Method)

Procedure	**Example**
To find the optimal value of a function subject to constraints:	Example 1 (again)

1. Write the objective function and constraint inequalities from the problem.	1. Objective function: $C = 2x + 3y$
	   Constraints: $x + 2y \le 10$
	   $\qquad\qquad\quad 2x + y \le 14$
	   $\qquad\qquad\qquad\quad x \ge 0$
	   $\qquad\qquad\qquad\quad y \ge 0$

2. Graph the solution to the constraint system.	2. See Figure 3.16.
3. Find the corners of the resulting feasible region. This may require simultaneous solution of two boundary equations.	3. Corners are $(0, 0)$, $(0, 5)$, $(6, 2)$, $(7, 0)$.

4. Evaluate the objective function at each corner of the feasible region determined by the constraints.	4. At $(0, 0)$, $2x + 3y = 0$
	   At $(0, 5)$, $2x + 3y = 15$
	   At $(6, 2)$, $2x + 3y = 18$
	   At $(7, 0)$, $2x + 3y = 14$

5. If two corners give the optimum value for the objective function, all points on the boundary line joining these two corners also optimize the function.	5. The function is maximized at $x = 6$, $y = 2$. The maximum value is 18.

CHECKPOINT

1. The shaded region in the figure is determined by the following constraints.

$$\begin{cases} 2x + 3y \le 12 \\ 4x - 2y \le 8 \\ \qquad\quad x \ge 0 \\ \qquad\quad y \le 0 \end{cases}$$

Find the maximum value of the objective function $f = 4x + 3y$ on the region.

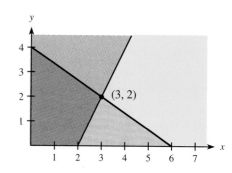

EXAMPLE 2

A small firm manufactures necklaces and bracelets. The combined number of necklaces and bracelets that it can handle per day is 24. Each bracelet takes 1 hour of labor to make and each necklace takes a half hour. The total number of hours of labor available per day is 16. If the profit on the bracelet is $2 and the profit on the necklace is $1, how many of each product should be produced daily to maximize profit?

Solution

Let x be the number of bracelets produced per day and y be the number of necklaces produced per day. Then the daily profit function is given by the equation $P = 2x + 1y$. Thus we seek to maximize $2x + y$, subject to the constraints.

The constraint on labor indicates that the total number of hours per day is 16, so if x bracelets and y necklaces are made, $1x + \frac{1}{2}y$ must be no more than 16. That is, $x + \frac{1}{2}y \le 16$. Since only 24 items can be handled each day, the second constraint is expressed as $x + y \le 24$. Since all quantities must be nonnegative, two additional constraints are $x \ge 0$ and $y \ge 0$.

Thus we seek to solve the following problem.

$$\text{Maximize} \quad P = 2x + y$$
$$\text{subject to} \quad \begin{cases} x + \frac{1}{2}y \le 16 \\ x + \ y \le 24 \\ x \ge 0, y \ge 0. \end{cases}$$

The set that satisfies the constraints is the shaded area in Figure 3.17.

The corners of the polygon are $(0, 0)$, $(0, 24)$, $(16, 0)$, and $(8, 16)$. All of these are obvious except for $(8, 16)$, which can be found by solving $x + \frac{1}{2}y = 16$ and $x + y = 24$ simultaneously. Testing the objective function at the corners gives

At $(0, 0)$, $2x + y = \ \ 0$

At $(0, 24)$, $2x + y = 24$

At $(16, 0)$, $2x + y = 32$

At $(8, 16)$, $2x + y = 32.$

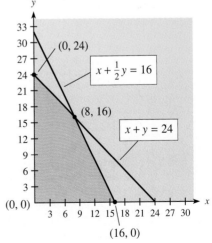

Thus the maximum value of the function occurs at *either* the point $x = 16$, $y = 0$ or the point $x = 8$, $y = 16$, which means that the profit will be maximized at any point on the line joining $(8, 16)$ and $(16, 0)$. For example, the point $(12, 8)$ is on this line, and $2x + y = 32$ at $x = 12$, $y = 8$.

Figure 3.17

Although the examples so far have sought to maximize an objective function, the same procedures apply when a minimum is sought.

EXAMPLE 3

Minimize $C = x + y$ subject to the constraints

$$\begin{cases} 3x + 2y \geq 12 \\ x + 3y \geq 11 \\ x \geq 0, y \geq 0. \end{cases}$$

Solution

The graph of the constraint system is shown in Figure 3.18. Notice that even though the feasible region is not an enclosed polygon, the corners still hold the key to the solution.

The corners $(0, 6)$ and $(11, 0)$ can be identified from the graph. The third corner, $(2, 3)$, can be found by solving the equations of the two lines

$$3x + 2y = 12 \quad \text{and}$$
$$x + 3y = 11$$

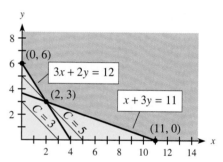

Figure 3.18

simultaneously, as follows.

$$\begin{aligned} 3x + 2y &= 12 \\ -3x - 9y &= -33 \\ \hline -7y &= -21 \\ y &= 3 \\ x &= 2 \end{aligned}$$

Examining the value of C at each corner point, we have

$$\begin{aligned} \text{At } (0, 6), \quad & C = x + y = 6 \\ \text{At } (11, 0), \quad & C = x + y = 11 \\ \text{At } (2, 3), \quad & C = x + y = 5. \end{aligned}$$

So C is minimized at $(2, 3)$ with minimum value $C = 5$. Notice that for any smaller value of C, the graph of $C = x + y$ misses the feasible region.

CHECKPOINT

2. Minimize the objective function $g = 3x + 4y$ subject to the following constraints.

$$\begin{aligned} x + 2y &\geq 12, & x &\geq 0 \\ 3x + 4y &\geq 30, & y &\geq 2 \end{aligned}$$

EXAMPLE 4

Let's return to the Application Preview example of the two chemical plants that produce three types of fertilizer, low phosphorus (LP), medium phosphorus (MP), and high phosphorus (HP). Recall that the fertilizer is produced in a single production run, so the three types are produced in fixed proportions. The Macon plant produces 1 ton of LP, 2 tons of MP, and 3 tons of HP in a single operation, and it charges $600 for what is produced in one operation, while one operation of the Jonesboro plant produces 1 ton of LP, 5 tons of MP, and 1 ton of HP, and it charges $1000 for what it produces in one operation. If a customer needs 100 tons of LP, 260 tons of MP, and 180 tons of HP, how many production runs should be ordered from each plant to minimize costs?

Solution

If x represents the number of operations requested from the Macon plant and y represents the number of operations requested from the Jonesboro plant, then we seek to minimize cost

$$C = 600x + 1000y.$$

The following table summarizes production capabilities and requirements.

	Macon Plant	Jonesboro Plant	Requirements
Units of LP	1	1	100
Units of MP	2	5	260
Units of HP	3	1	180

Using the number of operations requested and the fact that requirements must be met or exceeded, we can formulate the following constraints.

$$\begin{cases} x + y \geq 100 \\ 2x + 5y \geq 260 \\ 3x + y \geq 180 \\ x \geq 0, y \geq 0 \end{cases}$$

Graphing this system gives the feasible set shown in Figure 3.19. The corners are $(0, 180)$, $(40, 60)$, $(80, 20)$, and $(130, 0)$, where $(40, 60)$ is obtained by solving $x + y = 100$ and $3x + y = 180$ simultaneously, and where $(80, 20)$ is obtained by solving $x + y = 100$ and $2x + 5y = 260$ simultaneously.

Evaluating $C = 600x + 1000y$ at each corner we obtain

At $(0, 180)$, $C = 180,000$
At $(40, 60)$, $C = 84,000$
At $(80, 20)$, $C = 68,000$
At $(130, 0)$, $C = 78,000$.

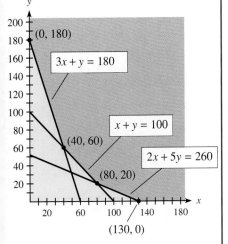

Figure 3.19

So to minimize costs the customer should place orders requiring 80 production runs from the Macon plant and 20 production runs from the Jonesboro plant.

EXAMPLE 5

Use a graphing utility to find the feasible region and to find the maximum value of $5x + 11y$ subject to the constraints

$$5x + 2y \leq 54$$
$$2x + 4y \leq 60$$
$$x \geq 0, y \geq 0.$$

Solution

We write the inequalities above as equations, solved for y. Graphing these equations with a graphics calculator and using shading shows the region satisfying the inequalities (see Figure 3.20). By using an INTERSECT command or by using TRACE along the lines that form the borders of this region, we see that the points where the boundaries intersect are $(0, 15)$, $(6, 12)$, and $(10.8, 0)$. These points can also be found algebraically. Testing the objective function at each of these corners gives the following values of f:

$$\text{At } (0, 15), f = 165$$
$$\text{At } (6, 12), f = 162$$
$$\text{At } (10.8, 0), f = 54$$

The maximum value of f is 165 at $x = 0$, $y = 15$.

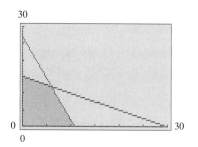

Figure 3.20

CHECKPOINT SOLUTIONS

1. The values of f at the corners are found as follows.

   At $(0, 0)$, $f = 0$
   At $(2, 0)$, $f = 8$
   At $(3, 2)$, $f = 12 + 6 = 18$
   At $(0, 4)$, $f = 12$

   The maximum value of f is 18 at $x = 3$, $y = 2$.

2. The graph of the feasible region is shown in Figure 3.21. The values of g at the corners are found as follows.

   At $(0, 7.5)$, $g = 30$
   At $(6, 3)$, $g = 18 + 12 = 30$
   At $(8, 2)$, $g = 24 + 8 = 32$

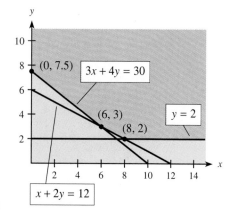

Figure 3.21

The minimum value of g is 30 at both (0, 7.5) and (6, 3). Thus any point on the border joining (0, 7.5) and (6, 3) will give the minimum value 30. For example, (2, 6) is on this border and gives the value $6 + 24 = 30$.

EXERCISE 3.3

1. The corners of a feasible region in a linear programming problem are (0, 0), (0, 8), (12, 0), and (6, 5). Which corner or corners give the maximum for the objective function $f = 6x + 3y$?
2. The corners of a feasible region in a linear programming problem are (0, 2), (3, 4), (5, 3), and (7, 0). Which corner or corners give the maximum for the objective function $f = 6x + 2y$?

In problems 3–8, use the given feasible region determined by the constraint inequalities to find the maximum and minimum of the given objective function (if they exist).

3. $C = 2x + 3y$

4. $f = 6x + 4y$

5. $C = 5x + 2y$

6. $C = 4x + 7y$

7. $f = 3x + 4y$

8. $f = 4x + 5y$

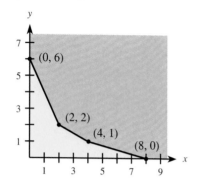

In each of problems 9–12, the graph of the feasible region is shown. Find the corners of each feasible region, and maximize or minimize the function as directed.

9. Maximize $f = 3x + 2y$

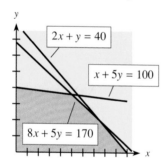

10. Maximize $f = 5x + 8y$

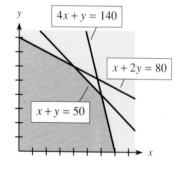

11. Minimize $g = 3x + 2y$

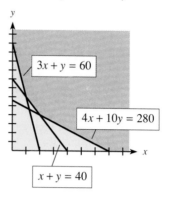

12. Minimize $g = x + 3y$

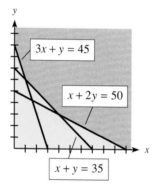

In problems 13–20, find the indicated maximum or minimum value of the objective function in the linear programming problem. Note that the feasible regions for these problems are the solution regions sketched in problems 23–30 of Exercise 3.2.

13. Maximize $f = 4x + 9y$ subject to

$$x + 5y \leq 200$$
$$2x + 3y \leq 134$$
$$x \geq 0, y \geq 0.$$

14. Maximize $f = 2x + y$ subject to

$$-x + y \leq 2$$
$$x + 2y \leq 10$$
$$3x + y \leq 15$$
$$x \geq 0, y \geq 0.$$

15. Maximize $f = 3x + 2y$ subject to

$$x + 2y \leq 48$$
$$x + y \leq 30$$
$$2x + y \leq 50$$
$$x \geq 0, y \geq 0.$$

16. Maximize $f = 7x + 10y$ subject to

$$3x + y \leq 9$$
$$3x + 2y \leq 12$$
$$x + 2y \leq 8$$
$$x \geq 0, y \geq 0.$$

17. Minimize $g = 9x + 10y$ subject to

$$x + 2y \geq 19$$
$$3x + 2y \geq 29$$
$$x \geq 0, y \geq 0.$$

18. Minimize $g = 5x + 2y$ subject to

$$4x + y \geq 12$$
$$x + y \geq 9$$
$$x + 3y \geq 15$$
$$x \geq 0, y \geq 0.$$

19. Minimize $g = 12x + 48y$ subject to

$$x + 3y \geq 3$$
$$2x + 3y \geq 5$$
$$2x + y \geq 3$$
$$x \geq 0, y \geq 0.$$

20. Minimize $g = 12x + 8y$ subject to

$$x + 2y \geq 10$$
$$2x + y \geq 11$$
$$x + y \geq 9$$
$$x \geq 0, y \geq 0.$$

In problems 21–36, solve the following linear programming problems.

21. Maximize $f = x + 3y$ subject to

$$x + 2y \leq 4$$
$$2x + y \leq 4$$
$$x \geq 0, y \geq 0.$$

22. Maximize $f = 3x + 2y$ subject to

$$2x + y \leq 8$$
$$2x + 3y \leq 12$$
$$x \geq 0, y \geq 0.$$

23. Maximize $f = 3x + 4y$ subject to

$$x + y \leq 6$$
$$2x + y \geq 10$$
$$y \leq 4$$
$$x \geq 0, y \geq 0.$$

24. Maximize $f = x + 3y$ subject to

$$x + 4y \leq 12$$
$$y \leq 2$$
$$x + y \leq 9$$
$$x \geq 0, y \geq 0.$$

25. Maximize $f = 2x + 6y$ subject to

$$x + y \leq 7$$
$$2x + y \leq 12$$
$$x + 3y \leq 15$$
$$x \geq 0, y \geq 0.$$

26. Maximize $f = 4x + 2y$ subject to

$$x + 2y \leq 20$$
$$x + y \leq 12$$
$$4x + y \leq 36$$
$$x \geq 0, y \geq 0.$$

27. Minimize $g = 7x + 6y$ subject to

$$5x + 2y \geq 16$$
$$3x + 7y \geq 27$$
$$x \geq 0, y \geq 0.$$

28. Minimize $g = 22x + 17y$ subject to

$$8x + 5y \geq 100$$
$$12x + 25y \geq 360$$
$$x \geq 0, y \geq 0.$$

29. Minimize $g = 3x + y$ subject to

$$4x + y \geq 11$$
$$3x + 2y \geq 12$$
$$x \geq 0, y \geq 0.$$

30. Minimize $g = 50x + 70y$ subject to

$$11x + 15y \geq 225$$
$$x + 3y \geq 27$$
$$x \geq 0, y \geq 0.$$

31. Minimize $f = x + 4y$ subject to

$$y \leq 30$$
$$3x + 2y \geq 75$$
$$-3x + 5y \geq 30$$
$$x \geq 0, y \geq 0.$$

32. Minimize $f = 2x + y$ subject to

$$x \leq 12$$
$$x + 2y \geq 20$$
$$-3x + 2y \leq 4$$
$$x \geq 0, y \geq 0.$$

33. Maximize $f = x + 2y$ subject to

$$x + y \geq 4$$
$$2x + y \leq 8$$
$$y \leq 4.$$

34. Maximize $f = 3x + 5y$ subject to

$$2x + 4y \geq 8$$
$$3x + y \leq 7$$
$$y \leq 4.$$

35. Minimize $g = 40x + 25y$ subject to

$$x + y \geq 100$$
$$-x + y \leq 20$$
$$-2x + 3y \geq 30$$
$$x \geq 0, y \geq 0.$$

36. Minimize $g = 3x + 8y$ subject to

$$4x - 5y \geq 50$$
$$-x + 2y \geq 4$$
$$x + y \leq 80$$
$$x \geq 0, y \geq 0.$$

APPLICATIONS

37. **Manufacturing** The Wellbuilt Company produces two types of wood chippers, Economy and Deluxe. The Deluxe model requires 3 hours to assemble and $\frac{1}{2}$ hour to paint, and the Economy model requires 2 hours to assemble and 1 hour to paint. The maximum number of assembly hours available is 24 per day and the maximum number of painting hours available is 8 per day. If the profit on the Deluxe model is $15 per unit and the profit on the Economy model is $12 per unit, how many units of each model will maximize profit? (See problem 33 of Exercise 3.2.)

38. **Learning environment** An experiment involving learning in animals requires placing white mice and rabbits into separate, controlled environments, Environment I and Environment II. The maximum amount of time available in Environment I is 500 minutes and the maximum amount of time available in Environment II is 600 minutes. The white mice must spend 10 minutes in Environment I and 25 minutes in Environment II, while the rabbits must spend 15 minutes in Environment I and 15 minutes in Environment II. Find the maximum possible number of animals that can be used in the experiment, and find the number of white mice and the number of rabbits that can be used. (See problem 34 of Exercise 3.2.)

39. **Manufacturing** A company manufactures two types of electric hedge-trimmers, one of which is cordless. The cord-type trimmer requires 2 hours to make, and the cordless model requires 4 hours. The company has only 800 work hours to use in manufacturing each day, and the packing department can package only 300 trimmers per day. If the company sells the cord-type model for $30 and the cordless model for $40, how many of each type should it produce per day maximize its sales? (See problem 35 of Exercise 3.2.)

40. **Manufacturing** A firm manufactures bumper bolts and fender bolts for cars. One machine can produce 130 fender bolts per hour and another machine can produce 120 bumper bolts per hour. The combined number of fender bolts and bumper bolts the packaging department can handle is 230 per hour. How many of each type of bolt should the firm produce hourly to maximize its sales if fender bolts sell for $1 and bumper bolts sell for $2? (See problem 36 of Exercise 3.2.)

41. **Nutrition** A privately owned lake contains two types of game fish, bass and trout. The owner provides two types of food, A and B, for these fish. Bass require 2 units of food A and 4 units of food B, and trout require 5 units of food A and 2 units of food B. If the owner has 800 units of each food, find the maximum number of fish that the lake can support.

42. **Manufacturing** A company makes two different products, Wigbats and Widgets, each of which is begun at one factory and finished at a second factory. The Atlanta

factory has 1000 labor hours available per day and the Waycross factory has 1200 labor hours available per day. The Atlanta factory works 10 hours on each Wigbat and 5 hours on each Widget while the Waycross factory works 6 hours on each Wigbat and 10 hours on each Widget. If only 100 Widgets can be sold, and if the profit on Wigbats is $50 and the profit on Widgets is $25, producing how many of each of these products will maximize profit?

43. **Management** A bank has two types of branches. A satellite branch employs 3 people, requires $100,000 to construct and open, and generates an average daily revenue of $10,000. A full-service branch employs 6 people, requires $140,000 to construct and open, and generates an average daily revenue of $18,000. The bank has up to $2.98 million available to open new branches, and has decided to limit the new branches to a maximum of 25. If the bank further decides to hire at most 120 new employees, how many branches of each type should the bank open in order to maximize the average daily revenues?

44. **Nutrition** In a zoo, there is a natural habitat containing several feeding areas. One of these areas serves as a feeding area for two species, I and II, and it is supplied each day with 120 pounds of food A, 110 pounds of food B, and 57 pounds of food C. Each individual of species I requires 5 lb of A, 5 lb of B, and 2 lb of C, and each individual of species II requires 6 lb of A, 4 lb of B, and 3 lb of C. Find the maximum number of these species that can be supported.

45. **Politics** A candidate wishes to use a combination of radio and television advertisements in his campaign. Research has shown that each 1-minute spot on television reaches 0.09 million people and each 1-minute spot on radio reaches 0.006 million. The candidate feels he must reach at least 2.16 million people, and he must buy a total of at least 80 minutes of advertisements. How many minutes of each medium should be used to minimize costs if television costs $500/minute and radio costs $100/minute? (See problem 37 of Exercise 3.2.)

46. **Nutrition** In a hospital ward, the patients can be grouped into two general categories depending on their condition and the amount of solid foods they require in their diet. A combination of two diets is used for solid foods because they supply essential nutrients for recovery, but each diet has an amount of a substance deemed detrimental. The table summarizes the patient group, minimum diet requirements, and the amount of the detrimental substance. How many servings from each

diet should be given each day in order to minimize the intake of this detrimental substance? (See problem 38 of Exercise 3.2.)

	Diet A	Diet B	Daily Reqs.
Group 1	4 oz per serving	1 oz per serving	26 oz
Group 2	2 oz per serving	1 oz per serving	18 oz
Detrimental substance	0.18 oz per serving	0.07 oz per serving	

47. **Manufacturing** Two factories produce three different types of kitchen appliances. The table below summarizes the production capacity, the number of each type of appliance ordered, and the daily operating costs for the factories. How many days should each factory operate to fill the orders at minimum cost?

	Factory 1	Factory 2	No. ordered
Appliance 1	80/day	20/day	1600
Appliance 2	10/day	10/day	500
Appliance 3	20/day	70/day	2000
Daily cost	$10,000	$20,000	

48. **Nutrition** In a laboratory experiment, two separate foods are given to experimental animals. Each food contains essential ingredients, A and B, for which the animals have a minimum requirement, and each food also has an ingredient C, which can be harmful to the animals. The table below summarizes this information.

	Food 1	Food 2	Reqs.
Ingredient A	10 units/g	3 units/g	49 units
Ingredient B	6 units/g	12 units/g	60 units
Ingredient C	3 units/g	1 unit/g	

How many grams of foods 1 and 2 should be given to the animals in order to satisfy the requirements for A and B while minimizing the amount of ingredient C ingested?

49. **Manufacturing** The Janie Gioffre Drapery Company makes three types of draperies at two different locations. At location I, they can make 10 pairs of deluxe drapes, 20 pairs of better drapes, and 13 pairs of standard drapes per day. At location II, they can make 20

pairs of deluxe, 50 pairs of better, and 6 pairs of standard per day. The company has orders for 2000 pairs of deluxe drapes, 4200 pairs of better drapes, and 1200 pairs of standard drapes. If the daily costs are $500 per day at location I and $800 per day at location II, how many days should Janie schedule at each location in order to fill the orders at minimum costs? Find the minimum costs.

50. *Nutrition* Two foods contain only proteins, carhohydrates, and fats. Food A costs $1 per pound, and contains 30% protein and 50% carbohydrates. Food B costs $1.50 per pound, and contains 20% protein and 75% carbohydrates. What combination of these two foods provides at least 1 pound of protein, $2\frac{1}{2}$ pounds of carbohydrates, and $\frac{1}{4}$ pound of fat at the lowest cost?

51. *Manufacturing* A sausage company makes two different kinds of hot dogs, regular and all beef. Each pound of all-beef hot dogs requires 0.75 lb of beef and 0.2 lb of spices, and each pound of regular hot dogs requires 0.18 lb of beef, 0.3 lb of pork, and 0.2 lb of spices. Suppliers can deliver at most 1020 lb of beef, at most 600 lb of pork, and at least 500 lb of spices. If the profit is $0.60 on each pound of all-beef hot dogs and $0.40 on each pound of regular hot dogs, how many pounds of each should be produced to obtain maximum profit?

What is the maximum profit? (See problem 39 of Exercise 3.2.)

52. *Manufacturing* A cereal manufacturer makes two different kinds of cereal, Senior Citizen's Feast and Kids Go. Each pound of Senior Citizen's Feast requires 0.6 lb of wheat and 0.2 lb of vitamin-enriched syrup, and each pound of Kids Go requires 0.4 lb of wheat, 0.2 lb of sugar, and 0.2 lb of vitamin-enriched syrup. Suppliers can deliver at most 2800 lb of wheat, at most 800 lb of sugar, and at least 1000 lb of the vitamin-enriched syrup. If the profit is $0.90 on each pound of Senior Citizen's Feast and $1.00 on each pound of Kids Go, find the number of pounds of each cereal that should be produced to obtain maximum profit. Find the maximum profit. (See problem 40 of Exercise 3.2.)

53. *Construction* A contractor builds two types of homes. The Carolina requires one lot, $160,000 capital, and 160 man-days of labor, whereas the Savannah requires one lot, $240,000 capital, and 160 man-days of labor. The contractor owns 300 lots and has $48,000,000 available capital and 43,200 man-days labor. The profit on the Carolina is $40,000 and the profit on the Savannah is $50,000. List the corner points of the feasible region and find how many of each type of home should be built to maximize profit. Find the maximum possible profit.

3.4 THE SIMPLEX METHOD: MAXIMIZATION

Objective

❏ To use the simplex method to maximize functions subject to constraints

APPLICATION PREVIEW

Society has become increasingly aware of the hazards of toxic spill accidents and of pollution from dump sites, chemical plants, and nuclear power plants. Because it is difficult or impossible to eliminate a given facility, a desirable goal is to place it in a location that will maximize the distance from the facility to population centers and to environmentally sensitive locations. Solving such problems usually requires the use of more variables than can be handled with graphical linear programming methods. Consider the following business application.

The Solar Technology Company manufactures three different types of hand calculators and classifies them as small, medium, and large according to their calculating capabilities. The three types have production requirements given by the following table.

	Small	*Medium*	*Large*
Electronic circuit components	5	7	10
Assembly time (in hours)	1	3	4
Cases	1	1	1

The firm has a monthly limit of 90,000 circuit components, 30,000 hours of labor, and 9000 cases. If the profit is \$6 for the small, \$13 for the medium, and \$20 for the large calculators, the number of each that should be produced to yield maximum profit and the maximum possible profit can be found by using the **simplex method** of linear programming.

The graphical method for solving linear programming problems is suitable only when there are three or fewer variables. If there are more than three variables, we could still find the corners of the convex region by simultaneously solving the equations of the boundaries. The objective function could then be evaluated at these corners. However, as the number of variables and the number of constraints increases, it becomes increasingly difficult to discover the corners and extremely time consuming to evaluate the function. Furthermore, some problems have no solution, and this cannot be determined by using the method of solving the equations simultaneously.

The method discussed in this section is called the **simplex method** for solving linear programming problems. Basically, this method gives a systematic way of moving from one feasible corner of the convex region to another one in such a way that the value of the objective function increases until an optimum value is reached or it is discovered that no solution exists.

The simplex method was first developed by George Dantzig in the late 1940s. One of the earliest applications of the method was to the scheduling problem that arose in connection with the Berlin airlift, begun in 1948. There the objective was to maximize the amount of goods delivered, subject to such constraints as the number of personnel, the number and size of available aircraft, and the number of runways.

In discussing the simplex method, we initially restrict ourselves to linear programming problems satisfying the following conditions.

1. The objective function is to be maximized.

2. All variables are nonnegative.

3. The constraints are of the form

$$a_1x_1 + a_2x_2 + \cdots + a_nx_n \leq b,$$

where $b > 0$.

These may seem to restrict unduly the types of problems, but in applied situations where the objective function is to be maximized, the constraints often satisfy conditions 2 and 3.

Before outlining a procedure for the simplex method, let us develop the procedure and investigate its rationale by seeing how the simplex method compares to the graphical method.

EXAMPLE I

Maximize $f = 2x + 3y$ subject to

$$x + 2y \leq 10$$
$$2x + y \leq 14.$$

Since we assume $x \geq 0$ and $y \geq 0$ in these problems, we shall no longer state these conditions.

Solution

The simplex method uses matrix methods on systems of equations, so we convert the constraint inequalities to equations by using **slack variables.** When we write $x + 2y \leq 10$, this means that there is some nonnegative number, say s_1, such that s_1 added to $x + 2y$ equals 10. This variable s_1 is called a slack variable because it changes with x and y so that the sum is always 10. Thus

$$x + 2y + s_1 = 10.$$

We can use another slack variable s_2 to write

$$2x + y \leq 14 \quad \text{as} \quad 2x + y + s_2 = 14.$$

If we write the objective function as an equation in the form

$$-2x - 3y + f = 0,$$

then we have a system of equations that describes the problem.

$$\begin{cases} x + 2y + s_1 && = 10 \\ 2x + y && + s_2 & = 14 \\ -2x - 3y &&& + f = 0 \end{cases}$$

We seek to maximize f in the last equation subject to the constraints in the first two equations.

We can place this system of equations in a matrix called a **simplex matrix** or **simplex tableau.** Note that the objective function is in the last row of this matrix with all variables on the left side, and that the first two rows correspond to the constraints.

$$A = \begin{array}{c} \begin{array}{ccccc} x & y & s_1 & s_2 & f \end{array} \\ \left[\begin{array}{ccccc|c} 1 & 2 & 1 & 0 & 0 & 10 \\ 2 & 1 & 0 & 1 & 0 & 14 \\ \hline -2 & -3 & 0 & 0 & 1 & 0 \end{array} \right] \begin{array}{l} \text{constraint 1} \\ \text{constraint 2} \\ \text{objective} \end{array} \end{array}$$

From the graph in Figure 3.22, we know that $(0, 0)$ is a feasible solution giving a value of 0 for f. Since the origin is always a feasible solution to the type of linear programming problems we are considering, the simplex method begins there and systematically moves to new corners while increasing the value of f.

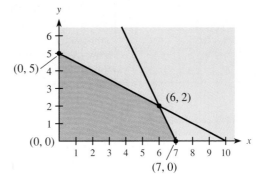

Figure 3.22

Notice that if $x = 0$ and $y = 0$, then from rows 1, 2, and 3 we can read the values of s_1, s_2, and f, respectively.

$$s_1 = 10, \qquad s_2 = 14, \qquad f = 0$$

Recall that the last row still corresponds to the objective function $f = 2x + 3y$. If we seek to improve (that is, increase the value of) $f = 2x + 3y$ by changing the value of only one of the variables, then since the coefficient of y is larger, the greatest improvement can be made by increasing y.

In general, we can find the variable that will improve f most by looking for the *most negative* value in the last row of the simplex matrix. The column containing this value is called the **pivot** column.

The simplex matrix A is shown again below with the y column indicated as the pivot column.

$$A = \begin{bmatrix} x & y & s_1 & s_2 & f & \\ 1 & ② & 1 & 0 & 0 & | & 10 \\ 2 & 1 & 0 & 1 & 0 & | & 14 \\ -2 & -3 & 0 & 0 & 1 & | & 0 \end{bmatrix}$$

— most negative entry, so pivot column

The amount by which y can be increased is limited by the constraining equations (rows 1 and 2 of the simplex matrix). As the graph in Figure 3.22 shows, the constraints limit y to no larger than $y = 5$. Note that when the positive coefficients in the y-column of the simplex matrix are divided into the constants in the augment, the *smallest* quotient is $10/2 = 5$. The 2 (circled) in row 1, column 2 of the matrix A is called the **pivot entry.** The quotient $y = 5$ is found by setting $x = 0$ and $s_1 = 0$ in the row 1 equation. The point $(0, 5)$ corresponds to a corner adjoining $(0, 0)$ on the constraint region.

We will see that if we continue this method, then the next values for x and y will correspond to another corner of the constraint region.

To continue the solution, note that we used row 1 to determine the largest value for y. We can "transform" the simplex matrix into an equivalent simplex matrix by eliminating y from the remaining rows. If we also multiply the first row by $\frac{1}{2}$, we obtain the equivalent matrix B. We say that the variable y is now **basic.**

$$B = \begin{bmatrix} \frac{1}{2} & 1 & \frac{1}{2} & 0 & 0 & 5 \\ \boxed{\frac{3}{2}} & 0 & -\frac{1}{2} & 1 & 0 & 9 \\ -\frac{1}{2} & 0 & \frac{3}{2} & 0 & 1 & 15 \end{bmatrix}$$

When we look at this transformed simplex matrix B, we notice that the bottom row represents $-\frac{1}{2}x + \frac{3}{2}s_1 + f = 15$, so $f = 15 + \frac{1}{2}x - \frac{3}{2}s_1$ depends on x and s_1 and $f = 15$ if $x = 0$ and $s_1 = 0$. Since x was held at 0 and s_1 must be nonnegative, the maximum value of f occurs at this point when $s_1 = 0$. Using $x = 0$ and $s_1 = 0$, rows 1, 2, and 3 allow us to read the values of y, s_2, and f, respectively.

$$y = 5, \qquad s_2 = 9, \qquad f = 15$$

This solution corresponds to the corner $(0, 5)$ in Figure 3.22. We also see that the value of f can be increased by increasing the value of x while holding the value s_1 at 0. This is clear from the equation

$$f = 15 + \frac{1}{2}x - \frac{3}{2}s_1$$

or by observing that the most negative number in the last row occurs in the x column of the new matrix B.

The amount of the increase for x is limited by the constraints. We can discover the limitations on x by dividing the coefficient of x (if positive) into the constant term in the augment of each row and choosing the smaller quotient.

$$5 \div \frac{1}{2} = (5)(2) = 10 \qquad \text{from row 1}$$

$$9 \div \frac{3}{2} = (9)\left(\frac{2}{3}\right) = 6 \qquad \text{from row 2}$$

The smaller quotient is 6, so the pivot entry is in row 2, column 1 of B (circled). Note that if we move to a corner adjoining $(0, 5)$ on the region in Figure 3.22, the largest x-value is 6 (at $(6, 2)$).

Let us again transform the simplex matrix B by eliminating x from all but the pivot row (the second row) in B and then multiplying the second row by $\frac{2}{3}$. The new simplex matrix is C.

$$x \quad y \quad s_1 \quad s_2 \quad f$$

$$C = \begin{bmatrix} 0 & 1 & \frac{2}{3} & -\frac{1}{3} & 0 & 2 \\ 1 & 0 & -\frac{1}{3} & \frac{2}{3} & 0 & 6 \\ 0 & 0 & \frac{4}{3} & \frac{1}{3} & 1 & 18 \end{bmatrix}$$

The last row of this new matrix C corresponds to

$$\tfrac{4}{3}s_1 + \tfrac{1}{3}s_2 + f = 18 \quad \text{or} \quad f = 18 - \tfrac{4}{3}s_1 - \tfrac{1}{3}s_2$$

so any increase in s_1 or s_2 will cause a *decrease* in f. Observing this, or the fact that all entries of the last row are nonnegative, tells us that we have the optimum solution when $s_1 = 0$ and $s_2 = 0$. Using this, rows 1, 2, and 3 give the values for y, x, and f, respectively.

$$y = 2, \qquad x = 6, \qquad f = 18$$

This corresponds to the corner $(6, 2)$ on the graph and to the maximum value of 18 at this corner (which was found in the graphical solution in Example 1 of Section 3.3).

As we review the simplex solution in Example 1, observe that each of the feasible solutions corresponded to a corner (x, y) of the feasible region, and that two of the four variables x, y, s_1, and s_2 were equal to 0 at each feasible solution.

At $(0, 0)$: x and y were equal to 0.

At $(0, 5)$: x and s_1 were equal to 0.

At $(6, 2)$: s_1 and s_2 were equal to 0.

The values that were set equal to zero at each feasible solution are called the **nonbasic variables** for that feasible solution, and the other variables are called the **basic variables** for that feasible solution. At each step in the simplex method, we can identify the basic variables directly from the simplex matrix by noting that their columns contain a single entry of 1 with all other entries zeros. (Check this.) The remaining columns identify the nonbasic variables. Thus the values of the basic variables and of the objective function can be read from the rows of the simplex matrix, because the nonbasic variables are set equal to 0 at each step.

The simplex method involves a series of decisions and operations using matrices. It involves three major tasks.

Simplex Method Tasks

Task A: Setting up the matrix for the simplex method.

Task B: Determining necessary operations and implementing those to reach a solution.

Task C: Reading the solution from the simplex matrix.

From Example 1, it should be clear that the pivot selection and row operations of Task B are designed to move the system to the corner that achieves the greatest increase in the function and at the same time is sensitive to the nonnegative limitations on the variables. The procedure that combines these three tasks to maximize a function is given below.

Linear Programming (Simplex Method)

Procedure	Example

To use the simplex method to solve linear programming problems:

Maximize $f = 2x + 3y$ subject to
$$x + 2y \leq 10$$
$$2x + y \leq 14.$$

1. Use slack variables to write the constraints as equations with positive constants on the right side.

1. $x + 2y \leq 10$ becomes $x + 2y + s_1 = 10$.
$2x + y \leq 14$ becomes $2x + y + s_2 = 14$.
Maximize f in $-2x - 3y + f = 0$.

2. Set up the simplex matrix. Put the objective function in the last row with all variables on the left and coefficient 1 for f.

2.
$$
\begin{array}{ccccc}
x & y & s_1 & s_2 & f \\
\end{array}
$$
$$
\left[\begin{array}{ccccc|c}
1 & 2 & 1 & 0 & 0 & 10 \\
2 & 1 & 0 & 1 & 0 & 14 \\
\hline
-2 & -3 & 0 & 0 & 1 & 0
\end{array}\right]
$$

3. Find the pivot:
   (a) The pivot column has the most negative number in the last row. (If a tie occurs, use either column.)

3. (a)
$$
\left[\begin{array}{c|c|ccc|c}
1 & 2 & 1 & 0 & 0 & 10 \\
2 & 1 & 0 & 1 & 0 & 14 \\
\hline
-2 & -3 & 0 & 0 & 1 & 0
\end{array}\right]
$$
↑——— most negative entry, so
pivot column

   (b) The positive coefficient in the pivot column that gives the smallest quotient when divided into the constant in the constraint rows of the augment is the pivot entry. If there is a tie, either coefficient may be chosen. If there are no positive coefficients in the pivot column, no solution exists and we stop.

   (b) Row 1 quotient: $\dfrac{10}{2} = 5$

   Row 2 quotient: $\dfrac{14}{1} = 14$

   Therefore, pivot row is row 1.

$$
\left[\begin{array}{c|c|ccc|c}
1 & ② & 1 & 0 & 0 & 10 \\
2 & 1 & 0 & 1 & 0 & 14 \\
\hline
-2 & -3 & 0 & 0 & 1 & 0
\end{array}\right]
$$

   Pivot entry is circled.

4. Use row operations with only the row containing the pivot entry to make the pivot entry a 1 and all other entries in the pivot column zeros. This makes the variable basic.

4. Row operations:
Multiply row 1 by $\frac{1}{2}$.
Add -1 times new row 1 to row 2.
Add 3 times new row 1 to row 3.

Result:
$$
\left[\begin{array}{ccccc|c}
\frac{1}{2} & 1 & \frac{1}{2} & 0 & 0 & 5 \\
\frac{3}{2} & 0 & -\frac{1}{2} & 1 & 0 & 9 \\
\hline
-\frac{1}{2} & 0 & \frac{3}{2} & 0 & 1 & 15
\end{array}\right]
$$
Indicators

5. The numerical entries in the last row are the indicators of what to do next.
   (a) If there is a negative indicator, return to step 3.
   (b) If all indicators are positive or zero, an optimum value has been obtained for the objective function.

5. $-\frac{1}{2}$ is negative, so we identify a new pivot column and reduce again.

$$\left[\begin{array}{ccccc|c} \frac{1}{2} & 1 & \frac{1}{2} & 0 & 0 & 5 \\ \frac{3}{2} & 0 & -\frac{1}{3} & 1 & 0 & 9 \\ \hline -\frac{1}{2} & 0 & \frac{2}{3} & 0 & 1 & 15 \end{array}\right] \begin{array}{l} 5 \div \frac{1}{2} = 10 \\ 9 \div \frac{3}{2} = 6* \end{array}$$

↑ Pivot column   *6 is smaller quotient.

The new pivot entry is $\frac{3}{2}$ (circled). Now reduce, using the new pivot row, to obtain:

$$\left[\begin{array}{ccccc|c} 0 & 1 & \frac{2}{3} & -\frac{1}{3} & 0 & 2 \\ 1 & 0 & -\frac{1}{3} & \frac{2}{3} & 0 & 6 \\ \hline 0 & 0 & \frac{1}{2} & \frac{1}{3} & 1 & 18 \end{array}\right]$$

Indicators all 0 or positive, so the solution is complete.

6. Read the values of the basic variables and the objective function from the rows of the matrix (after setting the nonbasic variables equal to 0).

6. f is maximized at 18 when $y = 2$, $x = 6$, $s_1 = 0$, and $s_2 = 0$.

As we look back over our work, we can observe the following:

1. Steps 1 and 2 complete Task A.
2. Steps 3–5 complete Task B.
3. Step 6 completes Task C.

EXAMPLE 2
Complete Tasks A, B, and C to find the maximum value of $f = 4x + 3y$ subject to

$$x + 2y \le 8$$
$$2x + y \le 10.$$

Solution
Task A: Introducing the slack variables gives the system of equations

$$\begin{cases} x + 2y + s_1 = 8 \\ 2x + y + s_2 = 10 \\ -4x - 3y + f = 0 \end{cases}$$

Writing this in a matrix with the objective function in the last row gives

$$A = \begin{array}{c} \\ \\ \end{array} \begin{array}{ccccc} x & y & s_1 & s_2 & f \\ \left[\begin{array}{ccccc|c} 1 & 2 & 1 & 0 & 0 & 8 \\ ② & 1 & 0 & 1 & 0 & 10 \\ \hline -4 & -3 & 0 & 0 & 1 & 0 \end{array}\right] \end{array} \begin{array}{l} 8/1 = 8 \\ 10/2 = 5* \end{array}$$

↑ Most negative   *smallest quotient

Task B: The most negative entry in the last row is -4 (indicated by the arrow), so the pivot column is column 1. The smallest quotient formed by dividing the positive coefficients from column 1 into the constants in the augment is 5 (indicated by the asterisk), so row 2 is the pivot row. Thus, the pivot entry is in row 2, column 1 (circled in matrix A above).

Next use row operations with row 2 only to make the pivot entry equal 1 and all other entries in column 1 equal zero. The row operations are:

(a) Multiply row 2 by $\frac{1}{2}$.

(b) Add -1 times the new row 2 to row 1.

(c) Add 4 times the new row 2 to row 3.

This gives

$$
\begin{array}{cccccc}
 & x & y & s_1 & s_2 & f \\
B = & \begin{bmatrix} 0 & \textcircled{$\frac{3}{2}$} & 1 & -\frac{1}{2} & 0 \\ 1 & \frac{1}{2} & 0 & \frac{1}{2} & 0 \\ 0 & -1 & 0 & 2 & 1 \end{bmatrix} & & & & \begin{bmatrix} 3 \\ 5 \\ 20 \end{bmatrix}
\end{array}
\qquad
\begin{array}{l}
3 \div \frac{3}{2} = 2* \\
5 \div \frac{1}{2} = 10
\end{array}
$$

Most negative   *smallest quotient

Since the last row contains a negative entry, we locate another pivot column and continue. The pivot column is indicated with an arrow, and the smallest quotient is indicated with an asterisk. The new pivot entry is circled.

The row operations using this pivot row are:

(a) Multiply row 1 by $\frac{2}{3}$.

(b) Add $-\frac{1}{2}$ times the new row 1 to row 2.

(c) Add the new row 1 to row 3.

This gives

$$
\begin{array}{cccccc}
 & x & y & s_1 & s_2 & f \\
C = & \begin{bmatrix} 0 & 1 & \frac{2}{3} & -\frac{1}{3} & 0 \\ 1 & 0 & -\frac{1}{3} & \frac{2}{3} & 0 \\ 0 & 0 & \frac{2}{3} & \frac{5}{3} & 1 \end{bmatrix} & & & & \begin{bmatrix} 2 \\ 4 \\ 22 \end{bmatrix}
\end{array}.
$$

Because there are no negative indicators in the last row, the solution is complete.

Task C: s_1 and s_2 are nonbasic variables, so they equal zero. f is maximized at 22 when $y = 2$, $x = 4$, $s_1 = 0$, and $s_2 = 0$.

CHECKPOINT

1. Write the following constraints as equations by using slack variables.

$$
\begin{aligned}
3x_1 + x_2 + x_3 &\le 9 \\
2x_1 + x_2 + 3x_3 &\le 8 \\
2x_1 + x_2 &\le 5
\end{aligned}
$$

2. Set up the simplex matrix to maximize $f = 2x_1 + 5x_2 + x_3$ subject to the constraints discussed in problem 1.

Using the simplex matrix from problem 2, answer the following.

3. Find the first pivot entry.

4. Use row operations to make the variable of the first pivot entry basic.

5. Find the second pivot entry and make a second variable basic.

6. Find the solution.

EXAMPLE 3

The Solar Technology Company manufactures three different types of hand calculators and classifies them as small, medium, and large according to their calculating capabilities. The three types have production requirements given by the following table.

	Small	Medium	Large
Electronic circuit components	5	7	10
Assembly time (in hours)	1	3	4
Cases	1	1	1

The firm has a monthly limit of 90,000 circuit components, 30,000 hours of labor, and 9000 cases. If the profit is $6 for the small, $13 for the medium, and $20 for the large calculators, how many of each should be produced to yield maximum profit? What is the maximum profit?

Solution

Let x_1 be the number of small calculators produced, x_2 the number of medium calculators produced, and x_3 the number of large calculators produced. Then the problem is to maximize the profit $f = 6x_1 + 13x_2 + 20x_3$ subject to

Inequalities	*Equations with slack variables*
$5x_1 + 7x_2 + 10x_3 \le 90,000$	$5x_1 + 7x_2 + 10x_3 + s_1 = 90,000$
$x_1 + 3x_2 + 4x_3 \le 30,000$	$x_1 + 3x_2 + 4x_3 + s_2 = 30,000$
$x_1 + x_2 + x_3 \le 9,000$	$x_1 + x_2 + x_3 + s_3 = 9,000$

The simplex matrix, with the pivot entry circled, follows.

Slack variables

$$
\begin{array}{ccccccc|c}
x_1 & x_2 & x_3 & s_1 & s_2 & s_3 & f & \\
5 & 7 & 10 & 1 & 0 & 0 & 0 & 90,000 \\
1 & 3 & ④ & 0 & 1 & 0 & 0 & 30,000 \\
1 & 1 & 1 & 0 & 0 & 1 & 0 & 9,000 \\
\hline
-6 & -13 & -20 & 0 & 0 & 0 & 1 & 0
\end{array}
$$

$90,000/10 = 9,000$
$30,000/4 = 7,500*$
$9,000/1 = 9,000$

Most negative ↑

*Smallest quotient

We now use row operations to change the pivot entry to 1 and create zeros elsewhere in the pivot column. The row operations are

1. Multiply row 2 by $\frac{1}{4}$ to convert the pivot entry to 1.
2. Using the new row 2,
   (a) add -10 times new row 2 to row 1.
   (b) add -1 times new row 2 to row 3.
   (c) add 20 times new row 2 to row 4.

The result is the second simplex matrix.

Slack variables

$$
\begin{array}{ccccccc}
x_1 & x_2 & x_3 & s_1 & s_2 & s_3 & f
\end{array}
$$

$$
\left[\begin{array}{ccccccc|c}
\frac{5}{2} & -\frac{1}{2} & 0 & 1 & -\frac{5}{2} & 0 & 0 & 15{,}000 \\
\frac{1}{4} & \frac{3}{4} & 1 & 0 & \frac{1}{4} & 0 & 0 & 7500 \\
\boxed{\frac{3}{4}} & \frac{1}{4} & 0 & 0 & -\frac{1}{4} & 1 & 0 & 1500 \\
-1 & 2 & 0 & 0 & 5 & 0 & 1 & 150{,}000
\end{array}\right]
\begin{array}{l}
15{,}000 \div \frac{5}{2} = 6000 \\
7500 \div \frac{1}{4} = 30{,}000 \\
1500 \div \frac{3}{4} = 2000* \\
\phantom{150{,}000}
\end{array}
$$

$\uparrow$
Most negative   *Smallest quotient

For this new simplex matrix, we check the last row indicators. Since there is an entry that is negative, we repeat the simplex process. Again the pivot entry is circled and we apply row operations similar to the above. What are the row operations this time? The resulting matrix is

$$
\begin{array}{ccccccc}
x_1 & x_2 & x_3 & s_1 & s_2 & s_3 & f
\end{array}
$$

$$
\left[\begin{array}{ccccccc|c}
0 & -\frac{4}{3} & 0 & 1 & -\frac{5}{3} & -\frac{10}{3} & 0 & 10{,}000 \\
0 & \frac{2}{3} & 1 & 0 & \frac{1}{3} & -\frac{1}{3} & 0 & 7000 \\
1 & \frac{1}{3} & 0 & 0 & -\frac{1}{3} & \frac{4}{3} & 0 & 2000 \\
\hline
0 & \frac{7}{3} & 0 & 0 & \frac{14}{3} & \frac{4}{3} & 1 & 152{,}000
\end{array}\right].
$$

Checking the last row, we see all the entries are 0 or positive, so the solution is complete. Setting $x_2 = 0$, $s_2 = 0$, and $s_3 = 0$ (the nonbasic variables), we see from the matrix that $x_1 = 2000$, $x_3 = 7000$, $s_1 = 10{,}000$, and $f = 152{,}000$.

Thus the number of calculators that should be produced is

$$x_1 = 2000 \text{ small calculators,}$$
$$x_2 = 0 \text{ medium calculators,}$$
$$x_3 = 7000 \text{ large calculators,}$$

in order to obtain a maximum profit of $152,000 for the month. Note that s_1 being nonzero means that some circuit components are not used in this optimal situation.

As our discussion indicates, the simplex method is quite complicated, even when the number of variables is relatively small (2 or 3). For this reason, computer software packages are often used to solve linear programming problems with the simplex method. It is useful to examine how the output from a typical simplex software package is similar to and different from the simplex matrices we've used.

EXAMPLE 4

In this example the calculator problem of maximizing $6x_1 + 13x_2 + 20x_3$ (Example 3) is solved again using a typical simplex software package. The constraints are the same:

$$5x_1 + 7x_2 + 10x_3 \leq 90,000 \quad \text{circuit components}$$
$$1x_1 + 3x_2 + 4x_3 \leq 30,000 \quad \text{assembly hours}$$
$$1x_1 + 1x_2 + 1x_3 \leq 9000 \quad \text{number of cases}$$

A computer output gives the following (partial) tableau for this problem.

C		6	13	20	0	0	0	
		x_1	x_2	x_3	s_1	s_2	s_3	RHS
6	x_1	1	.33	0	0	−.33	1.33	2000
20	x_3	0	.67	1	0	.33	−.33	7000
0	s_1	0	−1.33	0	1	−1.67	−3.33	10,000
Z		6	15.33	20	0	4.67	1.33	152,000
C − Z		0	−2.33	0	0	−4.67	−1.33	

The C (first) row gives the coefficients from the objective function, and the second row identifies the variables in the problem.

(a) The solution is optimal when the values in the $C - Z$ (last) row are all nonpositive entries, and the optimal solution is the value of Z in the RHS (last) column of the ouput. What is the optimal solution to this problem?

(b) Row 3 is headed by x_1 and (the objective function coefficient of x_1) 6. The value of x_1 that gives the optimal solution is in the last (RHS) column of this row. What value of x_1 gives the optimal solution? What value of x_3 gives the optimal solution?

Solution

(a) The given solution in Row Z of the RHS column is optimal because all the values in the $C - Z$ (last) row are nonpositive entries. Row 6 (headed by Z) gives the maximum value of the objective function. This agrees with the solution in Example 3, where it was obtained with the simplex method.

(b) Row 3 (headed by x_1) gives $x_1 = 2000$ in the last column, so $x_1 = 2000$ gives the optimal solution. Row 4 (headed by x_3) gives $x_3 = 7000$, and row 5 (headed by s_1) gives $s_1 = 10,000$, so $x_3 = 7000$ gives the optimal solution. The absence of a row headed by x_2 tells us that $x_2 = 0$ gives the optimal solution. Note that these values agree with those found in Example 3, even though the computer output has a different form than the simplex matrix. Thus the profit is maximized at $152,000 when 2000 small calculators, 0 medium calculators, and 7000 large calculators are produced.

Some modern computer spreadsheets have features that solve linear programming problems. For example, *Quattro Pro* has an Optimizer that uses a technique called *optimization analysis* to solve linear programming problems.

NONUNIQUE SOLUTIONS

Not all linear programming problems result in unique solutions. In fact, we have seen that the problem in Example 2 of Section 3.3 had an infinite number of solutions (on the line between $(8, 16)$ and $(16, 0)$). Let us solve this same problem using the simplex method.

The objective function to be maximized is

$$f = 2x + y$$

subject to the constraints

$$x + \tfrac{1}{2}y \le 16$$
$$x + \ y \le 24.$$

The simplex matrix (with the slack variables introduced) is

$$
\begin{array}{ccccc}
x & y & s_1 & s_2 & f \\
\end{array}
$$
$$
\left[
\begin{array}{ccccc|c}
\textcircled{1} & \tfrac{1}{2} & 1 & 0 & 0 & 16 \\
1 & 1 & 0 & 1 & 0 & 24 \\
\hline
-2 & -1 & 0 & 0 & 1 & 0 \\
\end{array}
\right].
$$

The most negative value in the last row is -2, so the x column is the pivot column. The smallest quotient occurs in row 1; the pivot entry (circled) is 1. Making x a basic variable gives the following transformed simplex matrix.

$$
\begin{array}{ccccc}
x & y & s_1 & s_2 & f \\
\end{array}
$$
$$
\left[
\begin{array}{ccccc|c}
1 & \tfrac{1}{2} & 1 & 0 & 0 & 16 \\
0 & \tfrac{1}{2} & -1 & 1 & 0 & 8 \\
\hline
0 & 0 & 2 & 0 & 1 & 32 \\
\end{array}
\right]
$$

This simplex matrix has no negative indicators, so the optimum value of f has been found ($f = 32$ when $x = 16$, $y = 0$). Note that one of the nonbasic variables (y) has a zero indicator. When this occurs, there frequently are multiple solutions to the linear programming problem. (We have seen that $x = 16$, $y = 0$ and $x = 8$, $y = 16$ are solutions in Section 3.3.)

When the simplex matrix for the optimum value of f has a nonbasic variable with a zero indicator in its column, there may be multiple solutions giving the *same* optimum value for f. We can discover if another solution exists by using the column of that nonbasic variable as the pivot column.

In the example above, if we use the y-column as the pivot column, then we could find the second solution. You will be asked to do this in problems 41 and 42 of Exercise 3.4.

It is also possible for a linear programming problem to have an unbounded solution (and thus no maximum value for f). This is identified by the following conditions in the simplex method.

If, after finding the pivot column, there are no positive coefficients in that column, no maximum solution exists.

EXAMPLE 5

Maximize $f = 2x_1 + x_2 + 2x_3$ subject to

$$x_1 - 3x_2 + x_3 \leq 3$$
$$x_1 - 6x_2 + 2x_3 \leq 6$$

if a maximum exists.

Solution

The simplex matrix for this problem (after introduction of slack variables) is

$$\begin{array}{ccccccc} & x_1 & x_2 & x_3 & s_1 & s_2 & f \\ \left[\begin{array}{cccccc|c} 1 & -3 & \boxed{1} & 1 & 0 & 0 & 3 \\ 1 & -6 & 2 & 0 & 1 & 0 & 6 \\ \hline -2 & -1 & \boxed{-2} & 0 & 0 & 1 & 0 \end{array}\right]. \end{array}$$

We see that the x_1 column and the x_3 column have the same most negative number, so we can use either column as the pivot column.

Using the x_3 column as the pivot column, we see that the smallest quotient is 3. Both coefficients give this quotient, so we can use either coefficient as the pivot entry. Using the element in row 1, we can make x_3 basic. The new simplex matrix is

$$\begin{array}{ccccccc} & x_1 & x_2 & x_3 & s_1 & s_2 & f \\ \left[\begin{array}{cccccc|c} 1 & -3 & 1 & 1 & 0 & 0 & 3 \\ -1 & 0 & 0 & -2 & 1 & 0 & 0 \\ \hline 0 & -7 & 0 & 2 & 0 & 1 & 6 \end{array}\right]. \end{array}$$

The most negative value in the last row of this matrix is -7, so the pivot column is the x_2 column. But there are no positive coefficients in the x_2 column, so no solution exists for this problem. That is, there is no maximum value for f.

Some difficulties can arise in using the simplex method to solve linear programming problems. In the solution process, it is possible for the same simplex matrix to recur before an optimum value for the objective function is obtained. These "degeneracies" are discussed in more advanced courses. They rarely occur in applied linear programming problems, and we shall not encounter them in this text.

CHECKPOINT SOLUTIONS

1. $3x_1 + x_2 + x_3 + s_1 \qquad\qquad = 9$
    $2x_1 + x_2 + 3x_3 \qquad + s_2 \qquad = 8$
    $2x_1 + x_2 \qquad\qquad\qquad + s_3 = 5$

2. $\begin{bmatrix} 3 & 1 & 1 & 1 & 0 & 0 & 0 & | & 9 \\ 2 & 1 & 3 & 0 & 1 & 0 & 0 & | & 8 \\ 2 & 1 & 0 & 0 & 0 & 1 & 0 & | & 5 \\ \hline -2 & -5 & -1 & 0 & 0 & 0 & 1 & | & 0 \end{bmatrix}$

3. The first pivot entry is the 1 in row 3, column 2.

4. $\begin{bmatrix} 1 & 0 & 1 & 1 & 0 & -1 & 0 & | & 4 \\ 0 & 0 & 3 & 0 & 1 & -1 & 0 & | & 3 \\ 2 & 1 & 0 & 0 & 0 & 1 & 0 & | & 5 \\ \hline 8 & 0 & -1 & 0 & 0 & 5 & 1 & | & 25 \end{bmatrix}$

5. The second pivot entry is the 3 in row 2, column 3.
    $\begin{bmatrix} 1 & 0 & 0 & 1 & -\frac{1}{3} & -\frac{2}{3} & 0 & | & 3 \\ 0 & 0 & 1 & 0 & \frac{1}{3} & -\frac{1}{3} & 0 & | & 1 \\ 2 & 1 & 0 & 0 & 0 & 1 & 0 & | & 5 \\ \hline 8 & 0 & 0 & 0 & \frac{1}{3} & \frac{14}{3} & 1 & | & 26 \end{bmatrix}$

6. All indicators are 0 or positive, so f is maximized at 26 when $x_1 = 0, x_2 = 5,$ and $x_3 = 1.$

EXERCISE 3.4

1. Convert the constraint inequalities $3x + 5y \leq 15$ and $3x + 6y \leq 20$ to equations containing slack variables.

2. Convert the constraint inequalities $2x + y \leq 4$ and $3x + 2y \leq 8$ to equations containing slack variables.

Task A: In problems 3–8 set up the simplex matrix used to solve each linear programming problem. Assume all variables are nonnegative.

3. Maximize $f = 2x + 4y$ subject to
$$2x + 5y \leq 30$$
$$x + 5y \leq 25.$$

4. Maximize $f = x + 5y$ subject to
$$3x + 7y \leq 20$$
$$7x + 4y \leq 44.$$

5. Maximize $f = 4x + 9y$ subject to
$$x + 5y \leq 200$$
$$2x + 3y \leq 134.$$

6. Maximize $f = x + 3y$ subject to
$$5x + y \leq 50$$
$$x + 2y \leq 50.$$

7. Maximize $f = 2x + 5y + 2z$ subject to
$$2x + 7y + 9z \leq 100$$
$$6x + 5y + z \leq 145$$
$$x + 2y + 7z \leq 90.$$

8. Maximize $f = 3x + 5y + 11z$ subject to
$$2x + 3y + 4z \leq 60$$
$$x + 4y + z \leq 48$$
$$5x + y + z \leq 48.$$

Task B: In problems 9–12, a simplex matrix is given. In each case identify the pivot element.

9. $\begin{bmatrix} 2 & 4 & 1 & 0 & 0 & | & 24 \\ 1 & 1 & 0 & 1 & 0 & | & 5 \\ -4 & -11 & 0 & 0 & 1 & | & 0 \end{bmatrix}$

10. $\begin{bmatrix} 3 & 1 & 1 & 1 & 0 & 0 & | & 100 \\ 4 & 3 & 6 & 0 & 1 & 0 & | & 250 \\ -5 & -3 & -4 & 0 & 0 & 1 & | & 0 \end{bmatrix}$

11. $\begin{bmatrix} 10 & 27 & 1 & 0 & 0 & 0 & | & 200 \\ 4 & 51 & 0 & 1 & 0 & 0 & | & 400 \\ 15 & 27 & 0 & 0 & 1 & 0 & | & 350 \\ -6 & -7 & 0 & 0 & 0 & 1 & | & 0 \end{bmatrix}$

12. $\begin{bmatrix} 5 & 5 & 7 & 1 & 0 & 0 & 0 & | & 12 \\ 4 & 0 & 6 & 0 & 1 & 0 & 0 & | & 48 \\ 0 & 1 & 1 & 0 & 0 & 1 & 0 & | & 8 \\ -1 & -3 & -1 & 0 & 0 & 0 & 1 & | & 0 \end{bmatrix}$

Task B: In problems 13–20, a simplex matrix is given. Examine the indicators and determine if the solution is complete. If it is not complete, find the next pivot element; if it is complete, indicate this.

13. $\begin{bmatrix} 2 & 0 & 1 & -\frac{3}{4} & 0 & | & 12 \\ 3 & 1 & 0 & \frac{1}{3} & 0 & | & 15 \\ -4 & 0 & 0 & 3 & 1 & | & 15 \end{bmatrix}$

14. $\begin{bmatrix} 0 & 4 & 1 & -\frac{1}{5} & 0 & | & 2 \\ 1 & 2 & 0 & 1 & 0 & | & 4 \\ 0 & 2 & 0 & 2 & 1 & | & 12 \end{bmatrix}$

15. $\begin{bmatrix} 4 & 1 & 0 & 0 & \frac{3}{4} & 4 & 0 & | & 14 \\ -2 & 0 & 0 & 1 & -\frac{5}{8} & -2 & 0 & | & 6 \\ 3 & 0 & 1 & 0 & 2 & 6 & 0 & | & 11 \\ 4 & 0 & 0 & 0 & 2 & \frac{1}{2} & 1 & | & 525 \end{bmatrix}$

16. $\begin{bmatrix} 2 & 1 & 1 & 1 & 0 & 0 & 0 & | & 12 \\ -2 & 0 & -1 & 0 & 0 & 1 & 0 & | & 5 \\ 4 & 0 & 2 & 0 & 1 & 0 & 0 & | & 6 \\ -2 & 0 & 5 & 0 & 0 & 0 & 1 & | & 30 \end{bmatrix}$

17. $\begin{bmatrix} 4 & 4 & 1 & 0 & 0 & 2 & 0 & | & 12 \\ 2 & 4 & 0 & 1 & 0 & -1 & 0 & | & 4 \\ -3 & -11 & 0 & 0 & 1 & -1 & 0 & | & 6 \\ -3 & -3 & 0 & 0 & 0 & 4 & 1 & | & 150 \end{bmatrix}$

18. $\begin{bmatrix} 0 & 0 & -3 & 4 & -4 & 0 & 0 & | & 12 \\ 0 & 1 & -4 & 2 & 5 & 0 & 0 & | & 100 \\ 1 & 0 & -1 & -6 & -3 & 1 & 0 & | & 40 \\ 0 & 0 & 6 & -1 & -3 & 0 & 1 & | & 380 \end{bmatrix}$

19. $\begin{bmatrix} 4 & -1 & 0 & 1 & -5 & 0 & 0 & | & 5 \\ 1 & 0 & 1 & 0 & 1 & 0 & 0 & | & 12 \\ 3 & -3 & 0 & 0 & -2 & 1 & 0 & | & 6 \\ -2 & -5 & 0 & 0 & 10 & 0 & 1 & | & 120 \end{bmatrix}$

20. $\begin{bmatrix} 0 & 20 & -6 & 1 & 0 & -3 & 0 & | & 4 \\ 0 & 5 & -6 & 0 & 1 & -2 & 0 & | & 2 \\ 1 & -1 & 0 & 0 & 0 & 4 & 0 & | & 6 \\ 0 & -9 & -12 & 0 & 0 & 10 & 1 & | & 20 \end{bmatrix}$

Task C: In problems 21–26, a simplex matrix is given. In each case the solution is complete, so identify the maximum value of f and a set of values of the variables that gives this maximum value.

21. $\begin{bmatrix} 1 & 0 & \frac{3}{4} & -\frac{3}{4} & 0 & | & 11 \\ 0 & 1 & \frac{5}{12} & \frac{11}{12} & 0 & | & 9 \\ 0 & 0 & 2 & 4 & 1 & | & 20 \end{bmatrix}$

22. $\begin{bmatrix} 0 & 4 & 1 & -\frac{1}{5} & 0 & \bigm| & 2 \\ 1 & 2 & 0 & 1 & 0 & \bigm| & 4 \\ \hline 0 & 2 & 0 & 2 & 1 & \bigm| & 12 \end{bmatrix}$

23. $\begin{bmatrix} 4 & 1 & 0 & 0 & \frac{3}{4} & 4 & 0 & \bigm| & 14 \\ -2 & 0 & 0 & 1 & -\frac{8}{5} & -2 & 0 & \bigm| & 6 \\ 3 & 0 & 1 & 0 & 2 & 6 & 0 & \bigm| & 11 \\ \hline 4 & 0 & 0 & 0 & 2 & \frac{1}{2} & 1 & \bigm| & 525 \end{bmatrix}$

24. $\begin{bmatrix} 0 & 1 & 0 & -3 & -1 & -4 & 0 & \bigm| & 12 \\ 1 & 0 & 0 & -2 & -1 & -5 & 0 & \bigm| & 16 \\ 0 & 0 & 1 & -1 & 2 & 1 & 0 & \bigm| & 22 \\ \hline 0 & 0 & 0 & 4 & 1 & 2 & 1 & \bigm| & 150 \end{bmatrix}$

25. $\begin{bmatrix} 1 & 0 & 3 & 0 & 6 & 0 & \bigm| & 50 \\ 0 & 0 & 4 & 1 & -4 & 0 & \bigm| & 6 \\ 0 & 1 & -2 & 0 & 2 & 0 & \bigm| & 10 \\ \hline 0 & 0 & 9 & 0 & 0 & 1 & \bigm| & 100 \end{bmatrix}$

26. $\begin{bmatrix} 0 & 1 & 2 & 1 & 0 & 0 & \bigm| & 20 \\ 1 & 0 & 6 & 1 & 0 & 0 & \bigm| & 50 \\ 0 & 0 & -1 & -2 & 1 & 0 & \bigm| & 10 \\ \hline 0 & 0 & 4 & 0 & 0 & 1 & \bigm| & 88 \end{bmatrix}$

In problems 27–34, use the simplex method to maximize the given functions. Assume $x \geq 0$, $y \geq 0$ in all problems.

27. Maximize $f = 3x + 10y$ subject to
$$14x + 7y \leq 35$$
$$5x + 5y \leq 50.$$

28. Maximize $f = 7x + 10y$ subject to
$$14x + 14y \leq 98$$
$$8x + 10y \leq 100.$$

29. Maximize $f = 2x + 3y$ subject to
$$x + 2y \leq 10$$
$$x + y \leq 7.$$

30. Maximize $f = 5x + 30y$ subject to
$$2x + 10y \leq 96$$
$$x + 10y \leq 90.$$

31. Maximize $f = x + 3y$ subject to
$$x + 2y \leq 12$$
$$y \leq 5$$
$$x + y \leq 9.$$

32. Maximize $f = 2x + 6y$ subject to
$$2x + 5y \leq 30$$
$$x + 5y \leq 25$$
$$x + y \leq 11.$$

33. Maximize $f = 2x + y$ subject to
$$-x + y \leq 2$$
$$x + 2y \leq 10$$
$$3x + y \leq 15.$$

34. Maximize $f = 3x + 2y$ subject to
$$x + 2y \leq 48$$
$$x + y \leq 30$$
$$2x + y \leq 50$$
$$x + 10y \leq 200.$$

In problems 35–40, assume $x \geq 0$, $y \geq 0$, and $z \geq 0$.

35. Maximize $f = 7x + 10y + 4z$ subject to
$$3x + 5y + 4z \leq 30$$
$$3x + 2y \leq 4$$
$$x + 2y \leq 8.$$

36. Maximize $f = 64x + 36y + 38z$ subject to
$$4x + 2y + 2z \leq 40$$
$$8x + 10y + 5z \leq 90$$
$$4x + 4y + 3z \leq 30.$$

37. Maximize $f = 20x + 12y + 12z$ subject to
$$x + z \leq 40$$
$$x + y \leq 30$$
$$y + z \leq 40.$$

38. Maximize $f = 10x + 5y + 4z$ subject to
$$2x + y + z \leq 50$$
$$x + 3y \leq 20$$
$$y + z \leq 15.$$

39. Maximize $f = 10x + 8y + 5z$ subject to
$$2x + y + z \leq 40$$
$$x + 2y \leq 10$$
$$y + 3z \leq 80.$$

40. Maximize $f = x + 3y + z$ subject to
$$x + 4y \leq 12$$
$$3x + 6y + 4z \leq 48$$
$$y + z \leq 8.$$

Problems 41–46 involve linear programming problems that have nonunique solutions. The simplex matrices shown in problems 41 and 42 indicate that an optimal solution has been found, but that a second solution is possible. Find the second solution.

41. $\begin{bmatrix} 1 & \frac{1}{2} & 1 & 0 & 0 & | & 16 \\ 0 & \frac{1}{2} & -1 & 1 & 0 & | & 8 \\ 0 & 0 & 2 & 0 & 1 & | & 32 \end{bmatrix}$

42. $\begin{bmatrix} 1 & 2 & 0 & 1 & 0 & | & 40 \\ 0 & 1 & 1 & 2 & 0 & | & 15 \\ 0 & 0 & 0 & 4 & 1 & | & 90 \end{bmatrix}$

In problems 43–46, use the simplex method to maximize each function (whenever possible) subject to the given constraints. If there is no solution, indicate this; if multiple solutions exist, find two of them.

43. Maximize $f = 3x + 2y$ subject to

$$x - 10y \le 10$$
$$-x + y \le 40.$$

44. Maximize $f = 10x + 15y$ subject to

$$-5x + y \le 24$$
$$2x - 3y \le 9.$$

45. Maximize $f = 3x + 12y$ subject to

$$2x + y \le 120$$
$$x + 4y \le 200.$$

46. Maximize $f = 8x + 6y$ subject to

$$4x + 3y \le 24$$
$$10x + 9y \le 198.$$

APPLICATIONS

47. **Manufacturing** Newjet Inc. manufactures two types of printers, an inkjet printer and a laser printer. The company can make a total of 60 printers per day and it has 120 man-hours per day available. It takes 1 man-hour to make an inkjet printer and 3 man-hours to make a laser printer. If the profit is $40 per inkjet printer and $60 per laser printer, (a) write the simplex matrix to maximize the daily profit, and (b) find the maximum profit and the number of each type of printer that will give the maximum profit.

48. **Construction** A contractor builds two types of homes. The Carolina requires one lot, $160,000 capital, and 160 man-days of labor, and the Savannah requires one lot, $240,000 capital, and 160 man-days of labor. The contractor owns 300 lots and has $48,000,000 available capital and 43,200 man-days labor. The profit on the Carolina is $40,000 and the profit on the Savannah is $50,000. (a) Write the simplex matrix to maximize the profit. (b) Use the simplex method to find how many of each type of home should be built to maximize profit, and find the maximum possible profit.

49. **Manufacturing** The Standard Steel Company produces two products, railroad car wheels and axles. Production requires processing in two departments: the smelting department (Department A) and the machining department (Department B). Department A has 50 hours available per week and Department B has 43 hours available per week. Manufacturing one axle requires 4 hours in Department A and 3 hours in Department B, while manufacturing one wheel requires 3 hours in Department A and 5 hours in Department B. The profit is $300 per axle and $300 per wheel.
    (a) Set up the simplex matrix used to maximize the profit for the company.
    (b) Find the maximum possible profit and the number of axles and wheels that will maximize the profit.

50. **Experimentation** An experiment involves placing the males and females of a laboratory animal species in two separate controlled environments. There is a limited time available in these environments, and the experimenter wishes to maximize the number of animals subject to the constraints described.

	Males	Females	Time Available
Environment A	20 min	25 min	800 min
Environment B	20 min	15 min	600 min

How many males and how many females will maximize the total number of animals?

51. **Shipping** A produce wholesaler has determined that it takes $\frac{1}{2}$ hour of labor to sort and pack a crate of tomatoes, and it takes $1\frac{1}{4}$ hours to sort and pack a crate of peaches. The crate of tomatoes weighs 60 pounds, and the crate of peaches weighs 50 pounds. The wholesaler has 2500 hours of labor available each week and can ship 120,000 pounds per week. If profits are $1 per crate of tomatoes and $2 per crate of peaches, how many crates of each should be sorted, packed, and shipped to maximize profits? What is the maximum profit?

52. **Manufacturing** A manufacturer has two resources available, R_1 and R_2. These resources can be used to produce two different products, A and B, in the following way.

> 2 units of R_1 and 5 units of R_2 give 1 unit of A
> 5 units of R_1 and 5 units of R_2 give 1 unit of B
> 30 units of R_1 and 25 units of R_2 are available

The profit for A is $2 per unit, and the profit for B is $4 per unit. How many units of products A and B should be produced to maximize profit?

53. **Advertising** Tire Corral has $6000 available per month for advertising. Newspaper ads cost $100 each and can occur a maximum of 21 times per month. Radio ads cost $300 each and can occur a maximum of 28 times per month at this price. Each newspaper ad reaches 6000 men over 20 years of age and each radio ad reaches 8000 of these men. The company wants to maximize the number of ad exposures to this group. How many of each ad should it purchase?

54. **Manufacturing** A bicycle manufacturer makes a ten-speed and a regular bicycle. The ten-speed requires 2 units of steel and 6 units of aluminum in its frame and 12 special components for the hub, sprocket, and gear assembly. The regular bicycle requires 5 units each of steel and aluminum for its frame and 5 of the special components. Shipments are such that steel is limited to 100 units per day, aluminum is limited to 120 units per day, and the special components are limited to 180 units per day. If the profit is $30 on each ten-speed and $20 on each regular bike, how many of each should be produced to yield maximum profit? What is the maximum profit?

55. **Advertising** The total advertising budget for a firm is $200,000. The following table gives the costs per ad package for each medium and the number of exposures per ad package (with all numbers in thousands).

	Med. 1	Med. 2	Med. 3
Cost/package	10	4	5
Exposures/ package	3100	2000	2400

If the maximum number of medium 1, medium 2, and medium 3 packages that can be purchased is 18, 10, and 12, respectively, how many of each ad package should be purchased to maximize the number of ad exposures?

56. **Advertising** The Laposata Pasta Company has $12,000 available for advertising. The following table gives the cost per ad and number of people exposed to its ads in three different media (with numbers in thousands).

Ad Packages	Newspaper	Radio	TV
Cost	2	2	4
Total audience	30	21	54
Working mothers	6	12	8

If the total available audience is 420,000, and if the company wishes to maximize the number of exposures to working mothers, how many ads of each type should it purchase?

57. **Manufacturing** A firm has decided to discontinue production of an unprofitable product. This will create excess capacity, and the firm is considering one or more of three possible new products, A, B, and C. The available weekly hours in the plant will be 477 hours in tool and die, 350 hours on the drill presses, and 150 hours on lathes. The hours of production required in each of these areas are as follows for each of the products.

	Tool and Die	Drill Press	Lathe
A	9	5	3
B	3	4	0
C	0.5	0	2

Furthermore, the sales department foresees no limitations on the sale of products A and C, but sees sales of only 20 or fewer per week for B. If the unit profits expected are $30 for A, $9 for B, and $15 for C, how many of each should be produced to maximize profits? What is the maximum profit?

58. **Diagnostics** A medical clinic performs three types of medical tests that use the same machines. Tests A, B, and C take 15 minutes, 30 minutes, and 1 hour, respectively, with respective profits of $30, $50, and $100. The clinic has 4 machines available. One person is qualified to do test A, two to do test B, and one to do test C. If the clinic has a rush of customers for these tests, how many of each type should it schedule in a 12-hour day to maximize its profit?

59. **Manufacturing** A manufacturer of blenders produces three different sizes: Regular, Special, and Kitchen Magic. The production of each type requires the materials and amounts given in the table on page 291. Suppose the manufacturer is opening a new plant, and anticipates weekly supplies of 90,000 electrical components, 12,000 units of aluminum, and 9,000 containers. The

unit profit is \$3 for the Regular, \$4 for the Special, and \$8 for the Kitchen Magic. A partial computer output for the problem is given below. Use the output to maximize $3x_1 + 4x_2 + 8x_3$ subject to the following constraints. How many of each type of blender should be produced to maximize profits? What is the maximum possible profit?

$5x_1 + 12x_2 + 24x_3 \leq 90{,}000$ electrical components

$1x_1 + 2x_2 + 4x_3 \leq 12{,}000$ aluminum

$1x_1 + 1x_2 + 1x_3 \leq 9{,}000$ containers

		Regular	Special	Kitchen Magic
Electrical	components	5	12	24
Units of	aluminum	1	2	4
Blending	containers	1	1	1

C		3	4	8	0	0	0	
		x_1	x_2	x_3	s_1	s_2	s_3	RHS
3	x_1	1	.67	0	0	−.33	1.33	8000
8	x_3	0	.33	1	0	.33	−.33	1000
0	s_1	0	.67	0	1	−6.33	1.33	26,000
Z		3	4.67	8	0	1.67	1.33	32,000
C − Z		0	−.67	0	0	−1.67	−1.33	

3.5 THE SIMPLEX METHOD: DUALITY AND MINIMIZATION

Objectives

❑ To formulate the dual for minimization problems
❑ To solve minimization problems using the simplex method on the dual

APPLICATION PREVIEW

A beef producer is considering two different types of feed. Each feed contains some or all of the necessary ingredients for fattening beef. Brand 1 feed costs 2 cents per pound and brand 2 costs 3 cents per pound. The producer would like to determine how much of each brand to buy in order to satisfy the nutritional requirements at minimum cost. Table 3.1 contains all the relevant data about nutrition and cost of each brand and the minimum requirements per unit of beef.

Table 3.1

	Brand 1	Brand 2	Minimum Requirement
Ingredient A	3/lb	5/lb	40/lb
Ingredient B	4/lb	3/lb	46/lb
Cost per pound	2¢	3¢	

In this section we will use the simplex method to solve minimization problems such as this one.

We have solved both maximization and minimization problems using graphical methods, but the simplex method, as discussed in Section 3.4, applies only to maximization problems. Now let us turn our attention to extending the use of the simplex method to solve minimization problems. Such problems might arise when a company seeks to minimize its production costs yet fill customer's orders or purchase items necessary for production.

As with maximization problems, we shall limit our discussion of minimization problems to only those in which the constraints satisfy the following conditions.

1. All variables are nonnegative.

2. The constraints are of the form

$$a_1 y_1 + a_2 y_2 + \cdots + a_n y_n \geq b,$$

where b is positive.

In this section, we will show how minimization problems of this type can be solved using the simplex method.

EXAMPLE I

Minimize $g = 11y_1 + 7y_2$ subject to

$$y_1 + 2y_2 \geq 10$$
$$3y_1 + y_2 \geq 15.$$

Solution

Since the simplex method specifically seeks to increase the objective function, it does not apply to this minimization problem. So, rather than solve this problem, let us solve a different but related problem called the **dual problem.**

In forming the dual problem, we write a matrix A that has a form similar to the simplex matrix, but without slack variables and with positive coefficients for the variables in the last row (and with the function to be minimized in the augment).

The matrix for the dual problem is formed by interchanging the rows and columns of matrix A. Matrix B is the result of this procedure, and we say that B is the **transpose** of A.

$$A = \begin{bmatrix} 1 & 2 & | & 10 \\ 3 & 1 & | & 15 \\ 11 & 7 & | & g \end{bmatrix} \qquad B = \begin{bmatrix} 1 & 3 & | & 11 \\ 2 & 1 & | & 7 \\ 10 & 15 & | & g \end{bmatrix}$$

Notice that row 1 of A became column 1 of B, row 2 of A became column 2 of B, and row 3 of A became column 3 of B.

In the same way that we formed matrix A, we can now "convert back" from matrix B to the maximization problem that is the dual of the original minimization problem. We state this dual problem, using different letters to emphasize that this is a different problem from the original.

Maximize $f = 10x_1 + 15x_2$ subject to
$$x_1 + 3x_2 \leq 11$$
$$2x_1 + x_2 \leq 7.$$

The simplex method does apply to this maximization problem, with the following simplex matrix.

$$
\begin{array}{ccccc}
x_1 & x_2 & s_1 & s_2 & f \\
\end{array}
$$
$$
\left[
\begin{array}{ccccc|c}
1 & 3 & 1 & 0 & 0 & 11 \\
2 & 1 & 0 & 1 & 0 & 7 \\
\hline
-10 & -15 & 0 & 0 & 1 & 0
\end{array}
\right]
$$

Successive matrices that occur using the simplex method on the maximization problem are shown below with each pivot entry circled.

Beside the maximization problem is the solution to the original minimization problem, obtained using graphical methods.

Comparison of a Minimization Problem and Its Dual

Maximization Problem

1.
$$
\begin{array}{ccccc}
x_1 & x_2 & s_1 & s_2 & f \\
\end{array}
$$
$$
\left[
\begin{array}{ccccc|c}
1 & ③ & 1 & 0 & 0 & 11 \\
2 & 1 & 0 & 1 & 0 & 7 \\
\hline
-10 & -15 & 0 & 0 & 1 & 0
\end{array}
\right]
$$

Minimization Problem

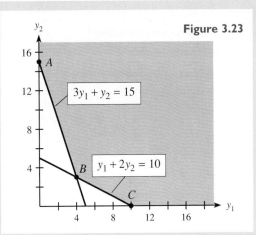

Figure 3.23

$3y_1 + y_2 = 15$

$y_1 + 2y_2 = 10$

2.
$$
\left[
\begin{array}{ccccc|c}
\frac{1}{3} & 1 & \frac{1}{3} & 0 & 0 & \frac{11}{3} \\
⑤⁄₃ & 0 & -\frac{1}{3} & 1 & 0 & \frac{10}{3} \\
\hline
-5 & 0 & 5 & 0 & 1 & 55
\end{array}
\right]
$$

Corners:
$A = (0, 15)$
$B = (4, 3)$ $\quad$ {obtained by solving simultaneously}
$C = (10, 0)$

3.
$$
\left[
\begin{array}{ccccc|c}
0 & 1 & \frac{2}{5} & -\frac{1}{5} & 0 & 3 \\
1 & 0 & -\frac{1}{5} & \frac{3}{5} & 0 & 2 \\
\hline
0 & 0 & 4 & 3 & 1 & 65
\end{array}
\right]
$$

$g = 11y_1 + 7y_2$;
at A, $g = 105$
at B, $g = 65$
at C, $g = 110$

Maximum $f = 65$ occurs at $x_1 = 2$, $x_2 = 3$.

Minimum $g = 65$ occurs at $y_1 = 4$, $y_2 = 3$.

In comparing the solutions to the previous two problems the first thing to notice is that the maximum value for f in the maximization problem and the minimum value for g in the minimization problem are the same. Furthermore, looking at the final simplex matrix, we can also find the values of y_1 and y_2 that give the minimum value for g by looking at the last row of the matrix.

$$
\begin{array}{ccccc}
x_1 & x_2 & s_1 & s_2 & f \\
\end{array}
$$
$$
\left[
\begin{array}{ccccc|c}
0 & 1 & \frac{2}{5} & -\frac{1}{5} & 0 & 3 \\
1 & 0 & -\frac{1}{5} & \frac{3}{5} & 0 & 2 \\
\hline
0 & 0 & 4 & 3 & 1 & 65 \\
\end{array}
\right]
$$

These values give the values for y_1
and y_2 in the minimization problem.

From this example we see that the given minimization problem and its dual maximization problem are very closely related. Furthermore, it appears that when problems enjoy this relationship, the simplex method might be used to solve them both. The question is whether the simplex method will always solve them both. The answer is yes. This fact was proved by John von Neumann, and it is summarized by the Principle of Duality.

Principle of Duality

1. When a minimization problem and its dual have a solution, the maximum value of the function to be maximized is the same value as the minimum value of the function to be minimized.

2. When the simplex method is used to solve the maximization problem, the values for the variables that solve the corresponding minimization problem are *the last entries in the columns corresponding to the slack variables.*

EXAMPLE 2

Given the problem

$$\text{minimize } 18y_1 + 12y_2 = g \text{ subject to}$$
$$2y_1 + y_2 \geq 8$$
$$6y_1 + 6y_2 \geq 36.$$

(a) State the dual of this minimization problem.
(b) Use the simplex method to solve the given problem.

Solution

(a) The dual of this minimization problem will be a maximization problem. To form the dual, write the matrix A (without slack variables and with positive coefficients for the variables in the last row) for the given problem.

$$A = \begin{bmatrix} 2 & 1 & 8 \\ 6 & 6 & 36 \\ \hline 18 & 12 & g \end{bmatrix}$$

Transpose matrix A to yield matrix B.

$$B = \begin{bmatrix} 2 & 6 & 18 \\ 1 & 6 & 12 \\ \hline 8 & 36 & g \end{bmatrix}$$

Write the dual maximization problem, renaming the variables and the function.

Dual problem

Maximize $f = 8x_1 + 36x_2$ subject to

$$2x_1 + 6x_2 \leq 18$$
$$x_1 + 6x_2 \leq 12.$$

(b) To solve the given minimization problem, we use the simplex method on its dual maximization problem. The complete simplex matrix for each step is given and each pivot entry is circled.

1. $\begin{bmatrix} 2 & 6 & 1 & 0 & 0 & 18 \\ 1 & ⑥ & 0 & 1 & 0 & 12 \\ \hline -8 & -36 & 0 & 0 & 1 & 0 \end{bmatrix}$

2. $\begin{bmatrix} ① & 0 & 1 & -1 & 0 & 6 \\ \frac{1}{6} & 1 & 0 & \frac{1}{6} & 0 & 2 \\ \hline -2 & 0 & 0 & 6 & 1 & 72 \end{bmatrix}$

3. $\begin{array}{ccccccc} x_1 & x_2 & s_1 & s_2 & f \end{array}$
$\begin{bmatrix} 1 & 0 & 1 & -1 & 0 & 6 \\ 0 & 1 & -\frac{1}{6} & \frac{2}{6} & 0 & 1 \\ \hline 0 & 0 & 2 & 4 & 1 & 84 \end{bmatrix}$

The last entries in the slack variable columns give the values of y_1 and y_2. Thus $y_1 = 2$, $y_2 = 4$, and $g = 84$.

CHECKPOINT

Perform the steps below to begin the process of finding the minimum value of $g = x + 4y$ subject to the following constraints.

$$2y_1 + 4y_2 \geq 18$$
$$y_1 + 5y_2 \geq 15$$

1. Form the matrix associated with the minimization problem.

2. Find the transpose of this matrix.

3. State the dual problem by using the matrix from problem 2.

4. Write the simplex matrix of the dual of this problem.

5. The dual of this minimization problem has the solution matrix below. What is the solution to the minimization problem?

$$\begin{bmatrix} 1 & 0 & \frac{5}{6} & -\frac{1}{6} & 0 & \Big| & \frac{1}{6} \\ 0 & 1 & -\frac{2}{3} & \frac{1}{3} & 0 & \Big| & \frac{2}{3} \\ 0 & 0 & 5 & 2 & 1 & \Big| & 13 \end{bmatrix}$$

We now return to the Application Preview problem.

EXAMPLE 3

A beef producer is considering two different types of feed. Each feed contains some or all of the necessary ingredients for fattening beef. Brand 1 feed costs 2 cents per pound and brand 2 costs 3 cents per pound. How much of each brand should the producer buy in order to satisfy the nutritional requirements at minimum cost? Table 3.2 contains all the relevant data about nutrition and cost of each brand and the minimum requirements per unit of beef.

Table 3.2

	Brand 1	Brand 2	Minimum Requirement
Ingredient A	3/lb	5/lb	40/lb
Ingredient B	4/lb	3/lb	46/lb
Cost per pound	2¢	3¢	

Solution

Let y_1 be the number of pounds of brand 1 and y_2 the number of pounds of brand 2. Then we can formulate the problem as follows:

$$\text{Minimize costs } C = 2y_1 + 3y_2 \text{ subject to}$$
$$3y_1 + 5y_2 \geq 40$$
$$4y_1 + 3y_2 \geq 46.$$

The original linear programming problem is usually called the **primal problem.** If we write this in an augmented matrix, without the slack variables, then we can transpose to find the dual problem.

Primal
$$\begin{bmatrix} 3 & 5 & \Big| & 40 \\ 4 & 3 & \Big| & 46 \\ 2 & 3 & \Big| & C \end{bmatrix}$$

Dual (with function renamed)
$$\begin{bmatrix} 3 & 4 & \Big| & 2 \\ 5 & 3 & \Big| & 3 \\ 40 & 46 & \Big| & f \end{bmatrix}$$

Thus the dual maximization problem is

$$\text{maximize } f = 40x_1 + 46x_2 \text{ subject to}$$
$$3x_1 + 4x_2 \leq 2$$
$$5x_1 + 3x_2 \leq 3.$$

The simplex matrix for this maximization problem, with slack variables included, is

1. $$\begin{bmatrix} 3 & ④ & 1 & 0 & 0 & 2 \\ 5 & 3 & 0 & 1 & 0 & 3 \\ \hline -40 & -46 & 0 & 0 & 1 & 0 \end{bmatrix}$$

Solving the problem gives

2. $$\begin{bmatrix} \frac{3}{4} & 1 & \frac{1}{4} & 0 & 0 & \frac{1}{2} \\ ⑪⁄₄ & 0 & -\frac{3}{4} & 1 & 0 & \frac{3}{2} \\ \hline -\frac{11}{2} & 0 & \frac{23}{2} & 0 & 1 & 23 \end{bmatrix}$$

3. $$\begin{bmatrix} 0 & 1 & \frac{5}{11} & -\frac{3}{11} & 0 & \frac{1}{11} \\ 1 & 0 & -\frac{3}{11} & \frac{4}{11} & 0 & \frac{6}{11} \\ \hline 0 & 0 & 10 & 2 & 1 & 26 \end{bmatrix}$$

The solution to this problem is

$$y_1 = 10 \text{ lb of brand 1}$$
$$y_2 = 2 \text{ lb of brand 2}$$
$$\text{Minimum cost} = 26 \text{ cents per unit of beef.}$$

In a more extensive study of linear programming, the duality relationship that we have used in this section as a means to solve minimization problems could be shown to have more properties than we have mentioned. For example, in applied problems where the objective is to minimize costs, it is possible to interpret the variables in the dual as "shadow prices" and interpret their values as marginal costs. Moreover, if the given problem is a profit maximization problem, then we can also form its dual, a minimization problem, by using matrix transposition, and we can think of the dual variables as representing "shadow prices" and interpret their values as marginal profits. A further discussion of shadow prices is given in the Consultant's Corner at the end of this chapter.

CHECKPOINT SOLUTIONS

1. $$\begin{bmatrix} 2 & 4 & 18 \\ 1 & 5 & 15 \\ \hline 1 & 4 & g \end{bmatrix}$$

2. $$\begin{bmatrix} 2 & 1 & 1 \\ 4 & 5 & 4 \\ \hline 18 & 15 & g \end{bmatrix}$$

3. Find the maximum value of $f = 18x_1 + 15x_2$ subject to the following constraints.

$$2x_1 + x_2 \leq 1$$
$$4x_1 + 5x_2 \leq 4$$

4. The simplex matrix for this dual problem is

$$\left[\begin{array}{ccccc|c} 2 & 1 & 1 & 0 & 0 & 1 \\ 4 & 5 & 0 & 1 & 0 & 4 \\ \hline -18 & -15 & 0 & 0 & 1 & 0 \end{array}\right].$$

5. The minimum value is 13 when $y_1 = 5$ and $y_2 = 2$.

EXERCISE 3.5

1. Write the transpose of the matrix associated with minimizing $g = 3x + y$ subject to the constraints $x \geq 0$, $y \geq 0$, $5x + 2y \geq 10$, and $x + 2y \geq 6$.

2. Write the transpose of the matrix associated with minimizing $g = 3x + 2y$ subject to the constraints $x \geq 0$, $y \geq 0$, $3x + y \geq 6$, and $3x + 4y \geq 12$.

3. Write the dual maximization problem associated with minimizing $g = 3x + y$ subject to the constraints $x \geq 0$, $y \geq 0$, $5x + 2y \geq 10$, and $x + 2y \geq 6$.

4. Write the dual maximization problem associated with minimizing $g = 3x + 2y$ subject to the constraints $x \geq 0$, $y \geq 0$, $3x + y \geq 3$, and $2x + 3y \geq 6$.

5. Write the simplex matrix for the dual maximization problem associated with minimizing $g = 3x + y$ subject to the constraints $x \geq 0$, $y \geq 0$, $5x + 2y \geq 10$, and $x + 2y \geq 6$.

6. Write the simplex matrix for the dual maximization problem associated with minimizing $g = 3x + 2y$ subject to the constraints $x \geq 0$, $y \geq 0$, $3x + y \geq 6$, and $3x + 4y \geq 12$.

7. Find the minimum value of $3x + y$ subject to the given constraints $x \geq 0$, $y \geq 0$, $5x + 2y \geq 10$, and $x + 2y \geq 6$.

8. Find the minimum value of $3x + 2y$ subject to the given constraints $x \geq 0$, $y \geq 0$, $3x + y \geq 6$, and $3x + 4y \geq 12$.

In problems 9–14, form the matrix associated with each problem and find the transpose of the matrix.

9. Minimize $g = 5y_1 + 2y_2$ subject to

$$y_1 + y_2 \geq 9$$
$$y_1 + 3y_2 \geq 15.$$

10. Minimize $g = 9y_1 + 10y_2$ subject to

$$y_1 + 2y_2 \geq 21$$
$$3y_1 + 2y_2 \geq 27.$$

11. Minimize $g = 3y_1 + y_2$ subject to

$$4y_1 + y_2 \geq 11$$
$$3y_1 + 2y_2 \geq 12$$
$$3y_1 + y_2 \geq 6.$$

12. Minimize $g = 2y_1 + 5y_2$ subject to

$$4y_1 + y_2 \geq 12$$
$$y_1 + y_2 \geq 9$$
$$y_1 + 3y_2 \geq 15.$$

13. Minimize $g = 12y_1 + 48y_2 + 8y_3$ subject to

$$y_1 + 3y_2 \geq 1$$
$$4y_1 + 6y_2 + y_3 \geq 3$$
$$4y_2 + y_3 \geq 1.$$

14. Minimize $g = 12y_1 + 8y_2 + 10y_3$ subject to

$$y_1 + 2y_3 \geq 10$$
$$y_1 + y_2 \geq 12$$
$$2y_1 + 2y_2 + y_3 \geq 8.$$

15–20. Write the dual problem for problems 9–14. Be sure to rename the variables and the function.

21–26. Solve the linear programming problem given in problems 9–14. Use the simplex method.

27. Minimize $g = 2x + 10y$ subject to

$$2x + y \geq 11$$
$$x + 3y \geq 11$$
$$x + 4y \geq 16.$$

28. Minimize $g = 8x + 4y$ subject to

$$3x - 2y \geq 6$$
$$2x + y \geq 11.$$

29. Minimize $g = 8x + 7y + 12z$ subject to

$$x + y + z \geq 3$$
$$y + 2z \geq 2$$
$$x \geq 2.$$

30. Minimize $g = 20x + 30y + 36z$ subject to

$$x + 2y + 3z \geq 48$$
$$2x + 2y + 3z \geq 70$$
$$2x + 3y + 4z \geq 96.$$

APPLICATIONS

31. **Manufacturing** The Video Star Company makes two different models of VCRs, which are assembled on two different assembly lines. Line 1 can assemble 30 units of the Star model and 40 units of the Prostar model per hour, and Line 2 can assemble 150 units of the Star model and 40 units of the Prostar model per hour. The company needs to produce at least 270 units of the Star model and 200 units of the Prostar model to fill an order. If it costs $200 per hour to run Line 1 and $400 per hour to run Line 2, how many hours should each line be run to fill the order at the minimum cost? What is the minimum cost?

32. **Nutrition** A pork producer is considering two types of feed that contain the necessary ingredients for the nutritional requirements for fattening hogs. Red Star Brand contains 9 units of ingredient A and 12 units of ingredient B, and Blue Chip Brand contains 15 units of ingredient A and 9 units of ingredient B. The nutritional requirement for the hogs is at least 120 units of ingredient A and at least 138 units of ingredient B.
    (a) If Red Star Brand costs 50 cents per pound and Blue Chip Brand costs 75 cents, how many pounds of each brand should be bought to satisfy the nutritional requirements at the minimum cost?
    (b) What is the minimum cost?

33. **Production** A small company produces two products, I and II, at 3 facilities, A, B, and C. They have orders for 2000 of product I and 1200 of product II. The production capacity and cost per week to operate each facility are summarized in the following table.

	A	B	C
I	200	200	400
II	100	200	100
Cost/week	$1,000	$3,000	$4,000

How many weeks should each facility operate to fill the company's orders at a minimum cost, and what is the minimum cost?

34. **Nutrition** In a hospital ward, the patients can be grouped into two general categories depending on their condition and the amount of solid foods they require in their diet. A combination of two diets is used for solid foods because they supply essential nutrients for recovery, but each diet has an amount of a substance deemed detrimental. The table below summarizes the patient group, diet requirements, and the amount of the detrimental substance. How many servings from each diet should be given each day in order to minimize the intake of this detrimental substance?

	Diet A	Diet B	Daily Reqs.
Group 1	4 oz per serving	1 oz per serving	26 oz
Group 2	2 oz per serving	1 oz per serving	18 oz
Detrimental substance	0.18 oz per serving	0.07 oz per serving	

35. **Production** Two factories produce three different types of kitchen appliances. The following table summarizes the production capacity, the number of each type of appliance ordered, and the daily operating costs for the factories. How many days should each factory operate to fill the orders at minimum cost?

	Factory 1	Factory 2	No. Ordered
Appliance 1	80/day	20/day	1600
Appliance 2	10/day	10/day	500
Appliance 3	50/day	20/day	1900
Daily cost	$10,000	$20,000	

36. **Nutrition** In a laboratory experiment, two separate foods are given to experimental animals. Each food contains essential ingredients, A and B, which have minimum requirements for the animals, and each food also

has an ingredient C, which can be harmful to the animals. The table below summarizes this information.

	Food 1	Food 2	Reqs.
Ingredient A	10 units/g	3 units/g	49 units
Ingredient B	6 units/g	12 units/g	60 units
Ingredient C	3 units/g	1 unit/g	

How many grams of foods 1 and 2 will satisfy the requirements for A and B, and minimize the amount of ingredient C that is ingested?

37. **Politics** A political candidate wishes to use a combination of radio and TV advertisements in her campaign. Research has shown that each 1-minute spot on TV reaches 0.9 million people and each 1 minute on radio reaches 0.6 million. The candidate feels she must reach 63 million people, and she must buy at least 90 minutes of advertisements. How many minutes of each medium should be used if TV costs $500 per minute and radio costs $100 per minute, and the candidate wishes to minimize costs?

38. **Production** The James MacGregor Mining Company owns three mines, I, II, and III. Three ores, A, B, and C, are mined at these mines. For each ore, the number of tons per week available from each mine and the number of tons per week required to fill orders are given in the following table.

		I	II	III	Required Tons/Week
Ore Types	A	10	10	10	90
	B	0	10	10	50
	C	10	0	10	60
Cost/day		$6000	$8000	$12,000	

Find the number of days the company should operate each mine so that orders are filled at minimum cost. Find the minimum cost.

39. **Nutrition** A hospital wishes to provide at least 24 units of nutrient A and 16 units of nutrient B in a meal, while minimizing the cost of the meal. If three types of food are available, with the nutritional value and costs (per ounce) given in the following table, how many ounces of each food should be served to minimize cost?

	Units of Nutrient A	Units of Nutrient B	Cost
Food I	2	1	1
Food II	2	5	5
Food III	2	1	2

40. **Scheduling** Each nurse works 8 consecutive hours at the Beaver Medical Center. The center has the following staffing requirements for each 4-hour work period.

Work Period	Nurses Needed
1 (7–11)	40
2 (11–3)	20
3 (3–7)	30
4 (7–11)	40
5 (11–3)	20
6 (3–7)	10

(a) If y_1 represents the number of nurses starting in period 1, y_2 the number starting in period 2, and so on, write the linear programming problem that will minimize the total number of nurses needed. (Note that the nurses who begin work in period 6, work periods 6 and 1 for their 8-hour shift.)

(b) Solve the problem in (a).

3.6 THE SIMPLEX METHOD WITH MIXED CONSTRAINTS

APPLICATION PREVIEW

The Laser Company manufactures two models of stereo systems, the Star and Allstar, at two plants, located in Ashville and in Cleveland. The maximum daily output at the Ashville plant is 900, and the profit there is

$200 per unit of the Allstar and $100 per unit of the Star. The maximum daily output at the Cleveland plant is 800, and the profit there is $210 per unit of the Allstar and $80 per unit of the Star. In addition, restrictions at the Ashville plant mean that the number of units of the Star model cannot exceed 100 more than the number of units of the Allstar model produced. If the company gets a rush order for 800 units of the Allstar model and 600 units of the Star model, finding the number of units of each model that should be produced at each location to fill the order and obtain the maximum profit is a **mixed constraint linear programming** problem. The term *mixed constraint* is used because some inequalities describing the constraints contain "$\leq$" and some contain "$\geq$" signs.

In Sections 3.4 and 3.5, we used the simplex method to solve two different types of linear programming problems. These types are summarized as follows.

	Maximization Problems	*Minimization Problems*
Function	Maximized	Minimized
Variables	Nonnegative	Nonnegative
Constraints	$a_1x_1 + a_2x_2 + \cdots + a_nx_n \leq b \ (b \geq 0)$	$c_1y_1 + c_2y_2 + \cdots + c_ny_n \geq d \ (d \geq 0)$
Simplex method	Applied directly	Applied to dual

In this section we will show how to apply the simplex method to mixed constraint linear programming problems that do not exactly fit either of these types.

We begin by considering maximization problems with constraints that have a form different from those noted in the table above. This can happen in one of the following ways:

1. If the constraint has the form $a_1x_1 + a_2x_2 + \cdots + a_nx_n \leq b$, where $b < 0$, such as with $x - 2y \leq -8$.

2. If the constraint has the form $a_1x_1 + a_2x_2 + \cdots + a_nx_n \geq b$, where $b \geq 0$, such as with $2x - 3y \geq 6$.

Notice in the latter case that multiplying both sides of the inequality by (-1) changes it as follows:

$$a_1x_1 + a_2x_2 + \cdots + a_nx_n \geq b$$

$$\text{becomes} \quad -a_1x_1 - a_2x_2 - \cdots - a_nx_n \leq -b.$$

That is,

$$2x - y \geq 6 \quad \text{becomes} \quad -2x + y \leq -6.$$

Thus we see that the two possibilities for mixed constraints are actually one, namely constraints that have the form $a_1x_1 + a_2x_2 + \cdots + a_nx_n \leq b$, where b is any constant. We call these **less than or equal to constraints** and denote them as **≤ constraints.**

Let's consider an example and see how we might proceed with a problem of this type.

EXAMPLE I

Maximize $f = x + 2y$ subject to

$$x + y \leq 13$$
$$2x - 3y \geq 6$$
$$x \geq 0, y \geq 0.$$

Solution

We begin by expressing all constraints as "≤" constraints. In this case, we multiply $2x - 3y \geq 6$ by (-1) to obtain $-2x + 3y \leq -6$. We can now use slack variables to write the inequalities as equations and form the simplex matrix.

$$
\begin{array}{c}
x + y + s_1 = 13 \\
-2x + 3y + s_2 = -6 \\
-x - 2y + f = 0
\end{array}
\quad \text{gives} \quad
\begin{array}{ccccc}
x & y & s_1 & s_2 & f \\
\end{array}
\left[
\begin{array}{ccccc|c}
1 & 1 & 1 & 0 & 0 & 13 \\
-2 & 3 & 0 & 1 & 0 & -6 \\
\hline
-1 & -2 & 0 & 0 & 1 & 0
\end{array}
\right]
$$

The -6 in the upper portion of the last column is a problem. If we compute values for the variables x, y, s_1, and s_2 that are associated with this matrix, we get $s_2 = -6$. This violates the condition of the simplex method that all variables are nonnegative. In order to use the simplex method, we must change the sign of any negative entry that appears in the upper portion of the last column (above the line separating the function from the constraints). There will always be another negative entry in the same row as this negative entry, but in a different column. We choose this column as the pivot column, because pivoting with it will give a positive entry in the last column.

$$
\begin{array}{ccccc}
x & y & s_1 & s_2 & f \\
\end{array}
\left[
\begin{array}{ccccc|c}
1 & 1 & 1 & 0 & 0 & 13 \\
-2 & 3 & 0 & 1 & 0 & -6 \\
\hline
-1 & -2 & 0 & 0 & 1 & 0
\end{array}
\right]
\begin{array}{l}
\leftarrow \text{ Negative in} \\
\quad \text{last column}
\end{array}
$$

Pivot column (negative entry in the same row as the negative entry in the last column)

Once we have identified the pivot column, we choose the pivot entry by forming *all* quotients and choosing the entry that gives the *smallest positive quotient*. Often this will mean pivoting by a negative number, which is the case in this problem.

$$\begin{array}{ccccc} x & y & s_1 & s_2 & f \end{array}$$

$$\left[\begin{array}{ccccc|c} 1 & 1 & 1 & 0 & 0 & 13 \\ \boxed{-2} & 3 & 0 & 1 & 0 & -6 \\ \hline -1 & -2 & 0 & 0 & 1 & 0 \end{array}\right] \quad \begin{array}{l} 13/1 = 13 \\ (-6)/(-2) = 3* \end{array}$$

Pivot entry circled   *smallest positive quotient

Pivoting with this entry gives the following matrix.

$$\begin{array}{ccccc} x & y & s_1 & s_2 & f \end{array}$$

$$\left[\begin{array}{ccccc|c} 0 & \boxed{\frac{5}{2}} & 1 & \frac{1}{2} & 0 & 10 \\ 1 & -\frac{3}{2} & 0 & -\frac{1}{2} & 0 & 3 \\ \hline 0 & -\frac{7}{2} & 0 & -\frac{1}{2} & 1 & 3 \end{array}\right] \left.\begin{array}{l} \\ \\ \end{array}\right\} \text{No negatives}$$

Most negative

This new matrix has no negatives in the upper portion of the last column, so we proceed with the simplex method as we have used it previously. The new pivot column is found from the most negative entry in the last row (indicated), and the pivot is found from the smallest positive quotient (circled). Using this pivot completes the solution.

$$\begin{array}{ccccc} x & y & s_1 & s_2 & f \end{array}$$

$$\left[\begin{array}{ccccc|c} 0 & 1 & \frac{2}{5} & \frac{1}{5} & 0 & 4 \\ 1 & 0 & \frac{3}{5} & -\frac{1}{5} & 0 & 9 \\ \hline 0 & 0 & \frac{7}{5} & \frac{1}{5} & 1 & 17 \end{array}\right]$$

We see that when $x = 9$ and $y = 4$, then the maximum value of $f = x + 2y$ is 17.

CHECKPOINT

1. Write the simplex matrix to maximize $f = 4x + y$ subject to the constraints

$$x + 2y \le 12$$
$$3x - 4y \ge 6$$
$$x \ge 0, y \ge 0.$$

2. Using the matrix from problem 1, find the maximum value of $f = 4x + y$ subject to the constraints.

If mixed constraints occur in a minimization problem, the simplex method does not apply to the dual problem. However, the same techniques used in Example 1 can be slightly modified and applied to minimization problems with mixed constraints. In particular, if our objective is to minimize f, then we alter the problem so as to maximize $-f$ and then proceed as in Example 1.

EXAMPLE 2

Minimize the function $f = 3x + 4y$ subject to the constraints

$$x + y \geq 20$$
$$x + 2y \geq 25$$
$$-5x + y \leq 4$$
$$x \geq 0, y \geq 0.$$

Solution

Because of the mixed constraints, we seek to maximize $-f = -3x - 4y$ subject to

$$-x - y \leq -20$$
$$-x - 2y \leq -25$$
$$-5x + y \leq 4.$$

The simplex matrix for this problem is

$$
\begin{array}{c}
\begin{array}{cccccc} x & y & s_1 & s_2 & s_3 & -f \end{array} \\
\left[
\begin{array}{cccccc|c}
-1 & -1 & 1 & 0 & 0 & 0 & -20 \\
-1 & -2 & 0 & 1 & 0 & 0 & -25 \\
-5 & 1 & 0 & 0 & 1 & 0 & 4 \\
\hline
3 & 4 & 0 & 0 & 0 & 1 & 0
\end{array}
\right]
\end{array} \quad \Leftarrow \text{Negatives}
$$

In this case, there are two negative entries in the upper portion of the last column. When this happens, we can start with either one; we choose -20 (row 1). In row 1 (to the left of -20), we find negative entries in columns 1 and 2. We can choose our pivot column for either of these columns; we choose column 1. Once this choice is made, the pivot is determined from the quotients.

$$
\begin{array}{c}
\begin{array}{cccccc} x & y & s_1 & s_2 & s_3 & -f \end{array} \\
\left[
\begin{array}{cccccc|c}
\boxed{-1} & -1 & 1 & 0 & 0 & 0 & -20 \\
-1 & -2 & 0 & 1 & 0 & 0 & -25 \\
-5 & 1 & 0 & 0 & 1 & 0 & 4 \\
\hline
3 & 4 & 0 & 0 & 0 & 1 & 0
\end{array}
\right]
\end{array}
\qquad
\begin{array}{l}
-20/(-1) = 20^* \\
-25/(-1) = 25 \\
4/(-5) = -4/5 \\
\\
\end{array}
$$

$\uparrow$
Pivot column (pivot entry circled)   *Smallest positive quotient

Pivoting with this entry gives the following matrix.

$$
\begin{array}{cccccc}
x & y & s_1 & s_2 & s_3 & -f \\
\end{array}
$$

$$
\begin{bmatrix}
1 & 1 & -1 & 0 & 0 & 0 & 20 \\
0 & -1 & -1 & 1 & 0 & 0 & -5 \\
0 & 6 & -5 & 0 & 1 & 0 & 104 \\
0 & 1 & 3 & 0 & 0 & 1 & -60
\end{bmatrix}
\begin{array}{l}
\\
\leftarrow \text{Negative} \\
\\
\\
\end{array}
$$

$$\uparrow$$
Pivot column

Now there is only one negative in the upper portion of the last column. Looking for negatives in this same row, we see that we can choose either column 2 or column 3 for our pivot column. We choose column 2, and the pivot is circled. Pivoting gives the following matrix.

$$
\begin{array}{cccccc}
x & y & s_1 & s_2 & s_3 & -f \\
\end{array}
$$

$$
\begin{bmatrix}
1 & 0 & -2 & 1 & 0 & 0 & 15 \\
0 & 1 & 1 & -1 & 0 & 0 & 5 \\
0 & 0 & -11 & 6 & 1 & 0 & 74 \\
0 & 0 & 2 & 1 & 0 & 1 & -65
\end{bmatrix}
$$

At this point we have all positives in the upper portion of the last column, so the simplex method can be applied. (The negative value in the lower portion is not a problem.) Looking at the indicators, we see that the solution is complete.

From the matrix we have $x = 15$, $y = 5$, and $-f = -65$. Thus the solution is $x = 15$ and $y = 5$, which gives the minimum value for the function $f = 3x + 4y = 65$.

A careful review of these examples allows us to summarize the key procedural steps for solving linear programming problems with mixed constraints.

SUMMARY: SIMPLEX METHOD FOR MIXED CONSTRAINTS

1. If the problem is to minimize f, then maximize $-f$.

2. a. Make all constraints "≤" constraints by multiplying both sides of any "≥" constraints by (-1).
   b. Use slack variables and form the simplex matrix.

3. In the simplex matrix, scan the *upper portion* of the last column for any negative entries.
   a. If there are no negative entries, apply the simplex method.
   b. If there are negative entries, go to Step 4.

4. When there is a negative value in the upper portion of the last column, proceed as follows:
   a. Select any negative entry in the same row and use its column as the pivot column.
   b. In the pivot column, compute all quotients for the entries above the line and determine the pivot from the *smallest positive quotient*.
   c. After completing the pivot operations, return to Step 3.

We now return to the Application Preview.

EXAMPLE 3

The Laser Company manufactures two models of stereo systems, the Star and Allstar, at two plants, located in Ashville and in Cleveland. The maximum daily output at the Ashville plant is 900, and the profit there is $200 per unit of the Allstar and $100 per unit of the Star. The maximum daily output at the Cleveland plant is 800, and the profit there is $210 per unit of the Allstar and $80 per unit of the Star. In addition, restrictions at the Ashville plant mean that the number of units of the Star model cannot exceed 100 more than the number of units of the Allstar model produced. If the company gets a rush order for 800 units of the Allstar model and 600 units of the Star model, how many units of each model should be produced at each location to fill the order and obtain the maximum profit?

Solution

The following table identifies our variables x and y and relates other important facts in the problem. In particular, we see that only variables x and y are required.

	Ashville Plant	Total Needed	Cleveland Plant
Allstar	x produced (profit = $200 each)	800	$800 - x$ produced (profit = $210 each)
Star	y produced (profit = $100 each)	600	$600 - y$ produced (profit = $80 each)
Capacity	900	—	800

We wish to maximize profit, and from the table we see that total profit is given by

$$P = 200x + 100y + 210(800 - x) + 80(600 - y)$$

or

$$P = 216{,}000 - 10x + 20y.$$

We can read capacity constraints from our table.

$$x + y \leq 900 \quad \text{(Ashville)}.$$

$$(800 - x) + (600 - y) \leq 800 \quad \text{or} \quad x + y \geq 600 \quad \text{(Cleveland)}.$$

An additional Ashville plant constraint from the statement of the problem is

$$y \leq x + 100 \quad \text{or} \quad -x + y \leq 100.$$

Thus our problem is to maximize

$$P = 216{,}000 - 10x + 20y$$

subject to

$$x + y \leq 900$$
$$x + y \geq 600$$
$$-x + y \leq 100$$
$$x \geq 0, y \geq 0.$$

We must express $x + y \geq 600$ as a "$\leq$" constraint; multiplying both sides by (-1) gives $-x - y \leq -600$. The simplex matrix is as follows.

$$
\begin{array}{cccccc}
x & y & s_1 & s_2 & s_3 & P \\
\end{array}
$$

$$
\left[
\begin{array}{cccccc|c}
1 & 1 & 1 & 0 & 0 & 0 & 900 \\
-1 & -1 & 0 & 1 & 0 & 0 & -600 \\
-1 & 1 & 0 & 0 & 1 & 0 & 100 \\
10 & -20 & 0 & 0 & 0 & 1 & 216{,}000 \\
\end{array}
\right]
\quad \leftarrow \text{Negative}
$$

↑
Pivot column

The negative in the upper portion of the last column is indicated. In row 2 we find negatives in both column 1 and column 2, so either of these may be our pivot column. Our choice is indicated, and the pivot entry is circled. The matrix that results from the pivot operations follows.

$$
\begin{array}{cccccc}
x & y & s_1 & s_2 & s_3 & P \\
\end{array}
$$

$$
\left[
\begin{array}{cccccc|c}
0 & 0 & 1 & 1 & 0 & 0 & 300 \\
1 & 1 & 0 & -1 & 0 & 0 & 600 \\
0 & 2 & 0 & -1 & 1 & 0 & 700 \\
0 & -30 & 0 & 10 & 0 & 1 & 210{,}000 \\
\end{array}
\right]
$$

↑
Pivot column

No entry in the upper portion of the last column is negative, so we can proceed with the simplex method. From the indicators we see that column 2 is the pivot column (indicated above), and the pivot entry is circled. The simplex matrix that results from pivoting follows.

$$
\begin{array}{cccccc}
x & y & s_1 & s_2 & s_3 & P \\
\end{array}
$$

$$
\left[\begin{array}{cccccc|c}
0 & 0 & 1 & \boxed{1} & 0 & 0 & 300 \\
1 & 0 & 0 & -\frac{1}{2} & -\frac{1}{2} & 0 & 250 \\
0 & 1 & 0 & -\frac{1}{2} & \frac{1}{2} & 0 & 350 \\
\hline
0 & 0 & 0 & -5 & 15 & 1 & 220{,}500
\end{array}\right]
$$

$$\underset{\text{Pivot column}}{\uparrow}$$

The new pivot column is indicated, and the pivot entry is circled. The pivot operation yields the following matrix.

$$
\begin{array}{cccccc}
x & y & s_1 & s_2 & s_3 & P \\
\end{array}
$$

$$
\left[\begin{array}{cccccc|c}
0 & 0 & 1 & 1 & 0 & 0 & 300 \\
1 & 0 & \frac{1}{2} & 0 & -\frac{1}{2} & 0 & 400 \\
0 & 1 & \frac{1}{2} & 0 & \frac{1}{2} & 0 & 500 \\
\hline
0 & 0 & 5 & 0 & 15 & 1 & 222{,}000
\end{array}\right]
$$

This matrix shows that the solution is complete. We see that $x = 400$ (so $800 - x = 400$), $y = 500$ (so $600 - y = 100$), and $P = \$222{,}000$.

Thus, the company should operate the Ashville plant at capacity and produce 400 units of the Allstar model and 500 units of the Star model. The remainder of the order, 400 units of the Allstar model and 100 units of the Star model, should be produced at the Cleveland plant.

CHECKPOINT SOLUTIONS

1.
$$
\left[\begin{array}{ccccc|c}
1 & 2 & 1 & 0 & 0 & 12 \\
-3 & 4 & 0 & 1 & 0 & -6 \\
-4 & -1 & 0 & 0 & 1 & 0
\end{array}\right]
$$

2.
$$
\left[\begin{array}{ccccc|c}
0 & \frac{10}{3} & 1 & \frac{1}{3} & 0 & 10 \\
1 & -\frac{4}{3} & 0 & -\frac{1}{3} & 0 & 2 \\
0 & -\frac{19}{3} & 0 & -\frac{4}{3} & 1 & 8
\end{array}\right]
\rightarrow
\left[\begin{array}{ccccc|c}
0 & 1 & \frac{3}{10} & \frac{1}{10} & 0 & 3 \\
1 & 0 & \frac{2}{5} & -\frac{1}{5} & 0 & 6 \\
0 & 0 & \frac{19}{10} & -\frac{7}{10} & 1 & 27
\end{array}\right]
$$

$$
\rightarrow
\left[\begin{array}{ccccc|c}
0 & 10 & 3 & 1 & 0 & 30 \\
1 & 2 & 1 & 0 & 0 & 12 \\
0 & 7 & 4 & 0 & 1 & 48
\end{array}\right]
$$

The maximum value of f is 48 at $x = 12$, $y = 0$, $s_1 = 0$, and $s_2 = 30$.

EXERCISE 3.6

In problems 1–4, express each inequality as a "$\leq$" constraint.

1. $3x - y \geq 5$

2. $4x - 3y \geq 6$

3. $y \geq 40 - 6x$

4. $x \geq 60 - 8y$

5. Write the simplex matrix used to maximize $f = 4x + 5y$ subject to the constraints $x \geq 0$, $y \geq 0$, $x + 2y \leq 6$, and $4x + 2y \geq 12$.

6. Write the simplex matrix used to maximize $f = 4x + y$ subject to the constraints $x \geq 0$, $y \geq 0$, $3x + 5y \leq 6$, and $x + y \geq 8$.

7. Write the simplex matrix used to minimize $g = 2x + 2y$ subject to the constraints $x \geq 0$, $y \geq 0$, $6x + 4y \leq 24$, and $5x + 2y \geq 16$.

8. Write the simplex matrix used to minimize $g = 10x + 2y$ subject to the constraints $x \geq 0$, $y \geq 0$, $2x + 3y \leq 16$, and $2x + y \geq 8$.

In problems 9–12, complete both of the following.
(a) State the given problem in a form from which the simplex matrix can be formed (that is, as a maximization problem with "$\leq$" constraints).
(b) Form the simplex matrix, and circle the first pivot.

9. Maximize $f = 2x + 3y$ subject to
$$7x + 4y \leq 28$$
$$-3x + y \geq 2$$
$$x \geq 0, y \geq 0.$$

10. Maximize $f = 5x + 11y$ subject to
$$x - 3y \geq 3$$
$$-x + y \leq 1$$
$$x \leq 10$$
$$x \geq 0, y \leq 0.$$

11. Minimize $g = 3x + 8y$ subject to
$$4x - 5y \leq 50$$
$$x + y \leq 80$$
$$-x + 2y \geq 4$$
$$x \geq 0, y \geq 0.$$

12. Minimize $g = 40x + 25y$ subject to
$$x + y \leq 100$$
$$-x + y \leq 20$$
$$-2x + 3y \geq 6$$
$$x \geq 0, y \geq 0.$$

In problems 13 and 14, a final simplex matrix for a minimization problem is given. In each case, find the solution.

13.

x	y	z	s_1	s_2	s_3	$-f$	
0	1	0	$\frac{7}{3}$	$\frac{1}{3}$	$-\frac{2}{3}$	0	8
0	0	1	$\frac{2}{3}$	$-\frac{1}{3}$	$\frac{8}{3}$	0	12
1	0	0	$-\frac{4}{3}$	$\frac{2}{3}$	$\frac{4}{3}$	0	6
0	0	0	4	$\frac{4}{3}$	2	1	-120

14.

x	y	z	s_1	s_2	s_3	$-f$	
1	0	0	$\frac{2}{5}$	$-\frac{2}{5}$	0	0	50
0	0	1	$\frac{1}{5}$	$\frac{1}{5}$	$\frac{11}{5}$	0	12
0	1	0	1	$-\frac{8}{5}$	$\frac{14}{5}$	0	30
0	0	0	3	1	4	1	-1200

In problems 15–34, use the simplex method to find the optimal solution.

15. Maximize $f = 4x + y$ subject to
$$5x + 2y \leq 84$$
$$-3x + 2y \leq 4$$
$$x \geq 0, y \geq 0.$$

16. Maximize $f = x + 2y$ subject to
$$-x + 2y \leq 60$$
$$-7x + 4y \geq 20$$
$$x \geq 0, y \geq 0.$$

17. Minimize $f = 2x + 3y$ subject to
$$x \geq 5$$
$$y \leq 13$$
$$-x + y \geq 2$$
$$x \geq 0, y \geq 0.$$

18. Minimize $f = 3x + 2y$ subject to
$$x \leq 20$$
$$y \leq 20$$
$$x + y \geq 21$$
$$x \geq 0, y \geq 0.$$

19. Minimize $f = 2x + y$ subject to
$$x \le 12$$
$$x + 2y \ge 20$$
$$-3x + 2y \le 4$$
$$x \ge 0, y \ge 0.$$

20. Minimize $f = x + 4y$ subject to
$$y \le 30$$
$$3x + 2y \ge 75$$
$$-3x + 5y \ge 30$$
$$x \ge 0, y \ge 0.$$

21. Maximize $f = 5x + 2y$ subject to
$$y \le 20$$
$$2x + y \le 32$$
$$-x + 2y \ge 4$$
$$x \ge 0, y \ge 0.$$

22. Maximize $f = 3x + 4y$ subject to
$$2x + 3y \le 90$$
$$x + y \le 35$$
$$-x + 2y \ge 10$$
$$x \ge 0, y \ge 0.$$

23. Maximize $f = 3x + 2y$ subject to
$$-x + y \le 10$$
$$x + y \ge 20$$
$$x + y \le 35$$
$$x \ge 0, y \ge 0.$$

24. Minimize $f = 4x + y$ subject to
$$-x + y \le 4$$
$$3x + y \ge 12$$
$$x + y \le 20$$
$$x \ge 0, y \ge 0.$$

25. Maximize $f = 2x + 5y$ subject to
$$-x + y \le 10$$
$$x + y \le 30$$
$$-2x + y \ge -24$$
$$x \ge 0, y \ge 0.$$

26. Maximize $f = 3x + 2y$ subject to
$$-x + 2y \le 20$$
$$-3x + 2y \le -36$$
$$x + y \le 22$$
$$x \ge 0, y \ge 0.$$

27. Minimize $f = x + 2y + 3z$ subject to
$$x + z \le 20$$
$$x + y \ge 30$$
$$y + z \le 20$$
$$x \ge 0, y \ge 0, z \ge 0.$$

28. Minimize $f = x + 2y + z$ subject to
$$x + 3y + 2z \ge 40$$
$$x + y + z \ge 30$$
$$x + y + z \le 100$$
$$x \ge 0, y \ge 0, z \ge 0.$$

29. Maximize $f = 2x - y + 4z$ subject to
$$x + y + z \le 8$$
$$x - y + z \ge 4$$
$$x + y - z \ge 2$$
$$x \ge 0, y \ge 0, z \ge 0.$$

30. Maximize $f = 5x + 2y + z$ subject to
$$2x + 3y + z \le 30$$
$$x - 2y + z \ge 20$$
$$2x + 5y + 2z \ge 25$$
$$x \ge 0, y \ge 0, z \ge 0.$$

31. Maximize $f = 2x + 3y - z$ subject to
$$3x + y + 2z \le 240$$
$$-2x - 2y + z \ge 10$$
$$x \ge 0, y \ge 0, z \ge 0.$$

32. Maximize $f = -x + 2y + 4z$ subject to
$$x + y + z \le 40$$
$$-x + y + z \ge 20$$
$$x + y - z \ge 10$$
$$x \ge 0, y \ge 0, z \ge 0.$$

33. Minimize $f = 10x + 30y - 35z$ subject to
$$x + y + z \le 250$$
$$x + y + 2z \ge 150$$
$$2x + y + z \le 180$$
$$x \ge 0, y \ge 0, z \ge 0.$$

34. Minimize $f = 2x + 3y - z$ subject to
$$3x + y + 2z \le 120$$
$$-2x - 2y + z \ge 5$$
$$x \ge 0, y \ge 0, z \ge 0.$$

APPLICATIONS

35. *Manufacturing* A company manufactures commercial heating system components and domestic furnaces at its factories in Monaca, PA, and Hamburg, NY. The Monaca plant can produce no more than 1000 units per day, and the number of commercial components cannot exceed 100 more than half the number of domestic furnaces. The Hamburg plant can produce no more than 850 units per day. The profit on each commercial component is $400 at the Monaca plant and $390 at the Hamburg plant. The profit on each domestic furnace is $200 at the Monaca plant and $215 at the Hamburg plant. If there is a rush order for 500 commercial components and 750 domestic furnaces, how many of each should be produced at each plant in order to maximize profits? Find the maximum profit.

36. *Manufacturing* A manufacturer makes products A and B at locations I and II. Location I can produce at most 1800 products, and the production of B can be at most 200 units less than the production of A. Location II can produce at most 1200 products. The profit on product A is $100 at location I and $90 at location II, and the profit on product B is $70 at location I and $75 at location II. If the manufacturer gets a rush order for 1500 units of product A and 1300 units of product B, how many units of each should be produced at each location so as to maximize profits? Find the maximum profit.

37. *Manufacturing* Refer to problem 35. Suppose that the cost of each commercial component is $380 at the Monaca plant and $400 at the Hamburg plant. The cost of each domestic furnace is $200 at the Monaca plant and $185 at the Hamburg plant. If the same rush order for 500 commercial components and 750 domestic furnaces is received, how many of each should be produced at each location to minimize cost? Find the minimum costs.

38. *Manufacturing* Refer to problem 36. Suppose the costs associated with product A are $50 at location I and $60 at location II, and the costs associated with product B are $40 at location I and $25 at location II. If the manufacturer gets the same rush order for 1500 units of A and 1300 units of B, how many units of each should be produced at each location in order to minimize costs? Find the minimum cost.

39. *Production* A sausage company makes two different kinds of hot dogs, regular and all beef. Each pound of all-beef hot dogs requires 0.75 lb of beef and 0.2 lb of spices, and each pound of regular hot dogs requires 0.18 lb of beef, 0.3 lb of pork, and 0.2 lb of spices.

Suppliers can deliver at most 1020 lb of beef, at most 600 lb of pork, and at least 500 lb of spices. If the profit is $0.60 on each pound of all-beef hot dogs and $0.40 on each pound of regular hot dogs, how many pounds of each should be produced to obtain maximum profit? What is the maximum profit?

40. *Manufacturing* A cereal manufacturer makes two different kinds of cereal, Senior Citizen's Feast and Kids Go. Each pound of Senior Citizen's Feast requires 0.6 lb of wheat and 0.2 lb of vitamin-enriched syrup, and each pound of Kids Go requires 0.4 lb of wheat, 0.2 lb of sugar, and 0.2 lb of vitamin-enriched syrup. Suppliers can deliver at most 2800 lb of wheat, at most 800 lb of sugar, and at least 1000 lb of the vitamin-enriched syrup. If the profit is $0.90 on each pound of Senior Citizen's Feast and $1.00 on each pound of Kids Go, find the number of pounds of each cereal that should be produced to obtain maximum profit. Find the maximum profit.

41. *Water purification* Three water purification facilities can handle at most 10 million gallons in a certain time period. Plant 1 leaves 20% of certain impurities, and costs $20,000 per million gallons. Plant 2 leaves 15% of these impurities and costs $30,000 per million gallons. Plant 3 leaves 10% impurities and costs $40,000 per million gallons. The desired level of impurities in the water from all three plants is at most 15%. If Plant 1 and Plant 3 combined must handle at least 6 million gallons, find the number of gallons each plant should handle so as to achieve the desired level of purity at minimum cost. Find the minimum cost.

42. *Mixture* A chemical storage tank has a capacity of 200 tons. Currently the tank contains 50 tons of mixture that has 10% of a certain active chemical and 1.8% of other inert ingredients. The owners of the tank want to replenish the supply in the tank and will purchase some combination of two available mixes. Mix 1 contains 70% of the active chemical and 3% of the inert ingredients; its cost is $100 per ton. Mix 2 contains 30% of the active chemical and 1% of the inert ingredients; its cost is $40 per ton. The desired final mixture should have at least 40% of the active chemical and at most 2% of the inert ingredients. How many tons of each mix should be purchased to obtain the desired final mixture at minimum cost? Find the minimum cost. Note that at least 40% of the active chemical means

70% (Mix 1) + 30% (Mix 2) + 10% (mix on hand)
$\geq$ 40% (Mix 1 + Mix 2 + mix on hand).

KEY TERMS

Section	Key Term
3.1	Linear inequalities
	Properties
	Solutions
	Graphs
	Intervals
	Open
	Closed
	Half-open
3.2	Linear inequalities in two variables
	Graphs
	Systems
	Solution region
3.3	Linear programming
	Objective function
	Constraints
	Optimum values
	Feasible region
	Graphical solution
3.4	Simplex method (maximization)
	Slack variables
	Simplex matrix
	Pivot entry
	Basic variable
	Nonunique solutions
3.5	Simplex method (minimization)
	Dual problem
	Transpose
	Primal problem
3.6	Simplex method with mixed constraints

REVIEW EXERCISES

3.1

1. Solve $3x - 9 \leq 4(3 - x)$ and graph the solution.
2. Solve $\frac{2}{3}x \leq x + 4$ and graph the solution.
3. Solve $5x + 1 \geq \frac{2}{3}(x - 6)$ and graph the solution.
4. Solve $\dfrac{4(x - 2)}{3} \geq 3x - \dfrac{1}{6}$ and graph the solution.

5. Determine if the following represent open, closed, or half-open intervals. Write each in interval notation.
   (a) $0 \leq x \leq 5$ (b) $3 \leq x < 7$
   (c) $-3 < x < 2$

6. Write the inequality represented by each of the following.
   (a) $(-1, 16)$ (b) $(-12, -8)$
   (c)

3.2

In problems 7–10, graph the solutions sets of each inequality or system of inequalities.

7. $2x + 3y \leq 12$

8. $4x + 5y \geq 100$

9. $\begin{cases} x + 2y \leq 20 \\ 3x + 10y \leq 80 \\ x \geq 0, y \geq 0 \end{cases}$

10. $\begin{cases} 3x + y \geq 4 \\ x + y \geq 2 \\ -x + y \leq 4 \\ x \quad\quad \leq 5 \end{cases}$

3.3

Solve the following linear programming problems using graphical methods. Restrict $x \geq 0$ and $y \geq 0$.

11. Maximize $f = 5x + 6y$ subject to

$$x + 3y \leq 24$$
$$4x + 3y \leq 42$$
$$2x + y \leq 20.$$

12. Maximize $f = 9x + 5y$ subject to the constraints in problem 11.

13. Maximize $f = x + 4y$ subject to

$$7x + 3y \leq 105$$
$$2x + 5y \leq 59$$
$$x + 7y \leq 70.$$

14. Maximize $f = 2x + y$ subject to the constraints in problem 13.

15. Minimize $g = 5x + 3y$ subject to

$$3x + y \geq 12$$
$$x + y \geq 6$$
$$x + 6y \geq 11.$$

16. Minimize $g = 3x + 5y$ subject to the constraints in problem 15.

17. Minimize $g = x + 5y$ subject to

$$8x + y \geq 85$$
$$x + y \geq 50$$
$$x + 4y \geq 80$$
$$x + 10y \geq 94.$$

18. Minimize $g = 7x + y$ subject to the constraints in problem 17.

3.4

Use the simplex method to solve the linear programming problems 19–32. Assume all variables are nonnegative.

19. Maximize $f = 7x + 12y$ subject to the conditions in problem 13.

20. Maximize $f = 3x + 4y$ subject to

$$x + 4y \leq 160$$
$$x + 2y \leq 100$$
$$4x + 3y \leq 300.$$

21. Maximize $f = 3x + 8y$ subject to the conditions in problem 20.

22. Maximize $f = 39x + 5y + 30z$ subject to

$$x + z \leq 7$$
$$3x + 5y \leq 30$$
$$3x + y \leq 18.$$

Problems 23 and 24 have nonunique solutions. If there is no solution, indicate this; if there are multiple solutions, find two different solutions. Use the simplex method.

23. Maximize $f = 4x + 4y$ subject to

$$x + 5y \leq 500$$
$$x + 2y \leq 230$$
$$x + y \leq 160.$$

24. Maximize $f = 2x + 5y$ subject to

$$-4x + y \leq 40$$
$$x - 7y \leq 70.$$

3.5

Form the dual and use the simplex method to solve the minimization problems 25–28.

25. Minimize $g = 7y_1 + 6y_2$ subject to

$$5y_1 + 2y_2 \geq 16$$
$$3y_1 + 7y_2 \geq 27.$$

26. Minimize $g = 3y_1 + 4y_2$ subject to

$$3y_1 + y_2 \geq 8$$
$$y_1 + y_2 \geq 6$$
$$2y_1 + 5y_2 \geq 18.$$

27. Minimize $g = 2y_1 + y_2$ subject to the conditions of problem 26.

28. Minimize $g = 12y_1 + 48y_2 + 8y_3$ subject to

$$y_1 + 3y_2 \geq 1$$
$$4y_1 + 6y_2 + y_3 \geq 3.$$
$$4y_2 + y_3 \geq 1.$$

3.6

In problems 29–32, use the simplex method.

29. Maximize $f = 3x + 5y$ subject to

$$x + y \geq 19$$
$$-x + y \geq 1$$
$$-x + 10y \leq 190$$
$$x \geq 0, y \geq 0.$$

30. Maximize $f = 4x + 6y$ subject to

$$2x + 5y \leq 37$$
$$5x - y \leq 34$$
$$-x + 2y \geq 4$$
$$x \geq 0, y \geq 0.$$

31. Minimize $f = 10x + 3y$ subject to

$$-x + 10y \geq 5$$
$$4x + y \geq 62$$
$$x + y \leq 50$$
$$x \geq 0, y \geq 0.$$

32. Minimize $f = 4x + 3y$ subject to

$$-x + y \geq 1$$
$$x + y \leq 45$$
$$10x + y \geq 45$$
$$x \geq 0, y \geq 0.$$

APPLICATIONS

3.3

33. **Manufacturing** A company manufactures two different size backyard swing sets. The larger requires 5 hours of labor to complete, the smaller requires 2 hours, and there are 700 hours of labor available each week. The packing department can pack at most 185 swing sets per week. If the profit is $100 on each larger set and $50 on each smaller set, how many of each should be produced to yield maximum profit? What is the maximum profit? Use graphical methods.

34. **Production** A company produces two different grades of steel, A and B, at two different factories, 1 and 2. The table below summarizes the production capabilities of

the factories, the cost per day, and the number of units of each grade of steel that is required to fill orders.

	Factory 1	Factory 2	Required
Grade A steel	1 unit	2 units	80 units
Grade B steel	3 units	2 units	140 units
Cost per day	$5000	$6000	

How many days should each factory operate in order to fill the orders at minimum cost? What is the minimum cost? Use graphical methods.

Use the simplex method to solve problems 35–40.

3.4

35. **Production** A small industry produces two items, I and II. It operates at capacity and makes a profit of $6 on each item I and $4 on each item II. The following table gives the hours required to produce each item and the hours available per day.

	I	II	Hours Available
Assembly	2 hours	1 hour	100
Packaging & inspection	1 hour	1 hour	60

Find the number of item that should be produced each day to maximize profits, and find the maximum daily profit.

3.5

36. **Nutrition** A nutritionist wants to find the least expensive combination of two foods that meet minimum daily vitamin requirements, which are 5 units of A and 30 units of B. Each ounce of food I provides 2 units of A and 1 unit of B, and each ounce of food II provides 10 units of A and 10 units of B. If food I costs 30 cents per ounce and food II costs 20 cents per ounce, find the number of ounces of each food that will provide the vitamins and minimize the cost.

37. **Nutrition** A laboratory wishes to purchase two different feeds, A and B, for their animals. The following table summarizes the nutritional content of the feeds, the required amounts of each ingredient, and the cost of each type of feed.

	Feed A	*Feed B*	*Reqs.*
Carbohydrates	1 unit/lb	4 units/lb	40 units
Protein	2 units/lb	1 unit/lb	80 units
Cost	14¢/lb	16¢/lb	

How many pounds of each type of feed should the laboratory buy in order to satisfy their needs at minimum cost?

38. **Production** A company makes three products, I, II, and III at three different factories. At factory A, they can make 10 units of each product per day. At factory B, they can make 20 units of II and 20 units of III per day. At factory C, they can make 20 units of I, 20 units of II, and 10 units of III per day. The company has orders for 200 units of I, 500 units of II, and 300 units of III. If the daily costs are $200 at A, $300 at B, and $500 at C, find the number of days that each factory should operate in order to fill their orders at minimum cost. Find the minimum cost.

3.6

39. **Profit** A company makes pancake mix and cake mix. Each pound of pancake mix uses 0.6 lb of flour and 0.1 lb of shortening. Each pound of cake mix uses 0.4 lb of flour, 0.1 lb of shortening, and 0.4 lb of sugar. Suppliers can deliver at most 6000 lb of flour, at least 500 lb of shortening, and at most 1200 lb of sugar. If the profit per pound is $0.35 for pancake mix and $0.25 for cake mix, how many pounds of each mix should be made to earn maximum profit? What is the maximum profit?

40. **Manufacturing** A company manufactures desks and computer tables at plants in Texas and Louisiana. At the Texas plant, production costs are $12 for each desk and $20 for each computer table, and the plant can produce at most 120 units per day. At the Louisiana plant, costs are $14 for each desk and $19 per computer table, and the plant can produce at most 150 units per day. The company gets a rush order for 130 desks and 130 computer tables at a time when the Texas plant is further limited by the fact that the number of computer tables it produces must be at least 10 more than the number of desks. How should production be scheduled at each location in order to fill the order at minimum cost? What is the minimum cost?

ONSULTANT'S CORNER

I. Transportation

The Kimble Firefighting Equipment Company has two assembly plants, Plant A in Atlanta and Plant B in Buffalo. It has two distribution centers, Center I in Pittsburgh and Center II in Savannah. The company can produce at most 800 alarm valves per week at Plant A and at most 1000 per week at Plant B. Center I must have at least 900 alarm valves per week and Center II must have at least 600 alarm valves per week. The costs per unit for transportation from the plants to the centers follow.

Plant A to Center I	$2
Plant A to Center II	3
Plant B to Center I	4
Plant B to Center II	1

What weekly shipping plan will meet the market demands and minimize the total cost of shipping the valves from the assembly plants to the distribution centers? What is the minimum transportation cost? Should the company eliminate the route with the most expensive unit shipment cost?

1. To assist the company in making these decisions, use linear programming with the following variables.

   y_1 = the number of units shipped from Plant A to Center I
   y_2 = the number of units shipped from Plant A to Center II
   y_3 = the number of units shipped from Plant B to Center I
   y_4 = the number of units shipped from Plant B to Center II

2. Form the constraint inequalities that give the limits for the plants and distribution centers, and form the total cost function that is to be minimized.
3. Solve the mixed constraint problem for the minimum transportation cost.
4. Determine whether any shipment route is unnecessary.

II. Sensitivity Analysis

A manufacturer of television sets makes one or more of four models of color sets: a 19-inch portable, a 23-inch table model, a 35-inch table model, and a 35-inch floor model. The assembly and testing time requirements for each model are shown in the table below together with the maximum amount of time available for assembly and testing. In addition to these constraints, the supplier of picture tubes indicated that it would supply no more than 180 picture tubes in the next month, and of these no more than 100 could be 35-inch picture tubes.

	19-Inch Portable	23-Inch Table	35-Inch Table	35-Inch Floor	Total Available
Assembly time (hours)	8	10	12	15	2000
Test time (hours)	2	2	4	5	500
Marginal profit (dollars)	40	60	80	100	

The computer output from a linear programming package analysis of this problem is given on page 318. Use it to answer the following questions.

1. What is the optimum production schedule for the television manufacturer for this month? (It is only profitable to bring a variable into the solution if its $C_j - Z_j$ entry in the output summary is greater than or equal to 0.)
2. By how much could marginal profit for 19-inch portable televisions increase before a new optimal solution should be sought? (The nonbasic variables will not affect the current optimal solution if the coefficients of these variables, C_j, are held in the range $-\infty < C_j \le Z_j$, called the *range of insignificance.*)
3. If the marginal profit for the 35-inch table model increases beyond $80, should a new optimal solution be sought?
4. Over what range of values can the marginal profit for the 23-inch table model vary without the optimal solution changing? This is the range of optimality for this model, with the lower and upper limits for the variables given in the table titled "Sensitivity Analysis of Objective Function Coefficients."
5. Over what range of values can the marginal profit for the 35-inch floor model vary without the optimal solution changing?
6. The $C_j - Z_j$ values tell how much the objective function changes as one unit of a variable is introduced into the solution, and the negative of the $C_j - Z_j$ values for a slack variable associated with a constraint tells how much the objective function will increase if one additional unit of the resource is made available. This can be considered as the price we would be willing to pay to obtain one additional unit of this resource, so we call it the **shadow price** of the resource. What is the marginal value of an additional hour of assembly time? Over what range of values is this marginal value valid?
7. What is the marginal value of an additional hour of test time? Over what range of values is this marginal value valid?
8. Management is considering the addition of a new 23-inch model that uses 10 hours of assembly and 3 hours of test time. If the marginal profit from this model would be $70, can the model be produced at a profit? What would the profit or loss per unit be?

```
0   MAX 40X1+60X2+80X3+100X4
    SUBJECT TO:
1   8X1+10X2+12X3+15X4<2000
2   2X1+2X2+4X3+5X4<500
3   X1+X2+X3+X4<180
4   X3+X4<100
```

OUTPUT SUMMARY:

C		40	60	80	100	0	0	0	0		
		X1	X2	X3	X4	S1	S2	S3	S4		RHS
60	X2	.5	1	0	0	.25	−.75	0	0	:	125
100	X4	.2	0	.8	1	−.1	.5	0	0	:	50
0	S3	.3	0	.2	0	−.15	.25	1	0	:	5
0	S4	−.2	0	.2	0	.1	−.5	0	1	:	50
Z		50	60	80	100	5	5	0	0	:	12500
C − Z		−10	0	0	0	−5	−5	0	0		

VARIABLE	QUANTITY	VARIABLE	SHADOW PRICE	
X2	125	X1	− 10	
X4	50	X2	0	
S3	5	X3	0	
S4	50	X4	0	
		S1	− 5	ASSEMBLY TIME
		S2	− 5	TEST TIME
OPTIMAL Z = 12500		S3	0	PICTURE TUBES
		S4	0	35" TUBES

SENSITIVITY ANALYSIS OF OBJECTIVE FUNCTION COEFFICIENTS

VARIABLE NAME	LOWER LIMIT	ORIGINAL VALUE	UPPER LIMIT
X2	40	60	66.67
X4	90	100	150
X1	NO LIMIT	40	50
X3	NO LIMIT	80	80

SENSITIVITY ANALYSIS OF RIGHT HAND SIDE RANGES

CONSTRAINT	LOWER LIMIT	ORIGINAL VALUE	UPPER LIMIT	
S1	1500	2000	2033.33	ASSEMBLY TIME
S2	480	500	600	TEST TIME
S3	175	180	NO LIMIT	PICTURE TUBES
S4	50	100	NO LIMIT	35" TUBES

NONLINEAR MODELS

Warm-up

Prerequisite problem type	For section	Answer	Section for review
Find $b^2 - 4ac$ if (a) $a = 1, b = 2, c = -2$ (b) $a = -2, b = 3, c = -1$ (c) $a = -3, b = -3, c = -2$	**4.1** **4.2** **4.3**	(a) 12 (b) 1 (c) -15	0.2 Signed numbers
(a) Factor $6x^2 - x - 2$. (b) Factor $6x^2 - 9x$.	**4.1** **4.2** **4.3**	(a) $(3x - 2)(2x + 1)$ (b) $3x(2x - 3)$	0.6 Factoring
Is $\dfrac{1 + \sqrt{-3}}{2}$ a real number?	**4.1** **4.2**	No	0.4 Radicals
(a) Find the y-intercept of $y = x^2 - 6x + 8$. (b) Find the x-intercepts of $y = 3x(2x - 3)$.	**4.2** **4.3**	(a) 8 (b) $0, \frac{3}{2}$	1.2 x and y-intercepts

CHAPTER

QUADRATIC FUNCTIONS

In previous chapters we studied linear equations and inequalities and their applications in detail, including their analysis by using matrices and linear programming. Even though linear equations are relatively easy to formulate and to use, they are not sufficient to solve all problems that arise in business, the social sciences, or the life sciences. Thus we now direct our attention to equations of second degree, called **quadratic equations.**

In this chapter, we solve quadratic equations in one variable; that is, equations of the form $ax^2 + bx + c = 0$. We also graph quadratic functions of the form $y = ax^2 + bx + c$, whose graphs are parabolas.

The use of these quadratic functions makes it possible to greatly expand our study of the special business functions we have been discussing in detail. In particular, we will see that the mathematical principles governing market equilibrium and break-even analysis still apply, but now the functions are quadratics. Furthermore, our analysis allows us to find maximum values for revenue and profit functions, if they are quadratic functions of the form $f(x) = ax^2 + bx + c$.

4.1 QUADRATIC EQUATIONS

Objectives

❏ To solve quadratic equations with factoring methods
❏ To solve quadratic equations with the quadratic formula

APPLICATION PREVIEW

With fewer workers, future income from payroll taxes used to fund Social Security benefits won't keep pace with scheduled benefits. The function

$$B = -1.056785714t^2 + 8.259285714t + 74.07142857$$

models the Social Security trust fund balance B, in billions of dollars, where t is the number of years past the year 2000 (Source: Social Security Administration). For planning purposes, it is important to know when the trust fund balance will be 0. That is, for what t-value does

$$0 = -1.056785714t^2 + 8.259285714t + 74.07142857?$$

In this section, we learn how to solve equations of this type.

FACTORING METHODS

It is important for us to know how to solve equations, such as the equation in the Application Preview, that arise as we study different types of functions and consider their graphs and applications. This was true for linear equations and is again true for quadratic equations.

A **quadratic equation** in one variable is an equation that can be written in the *general form*

$$ax^2 + bx + c = 0 \qquad (a \neq 0),$$

where a, b, and c represent constants. For example, the equations

$$3x^2 + 4x + 1 = 0 \qquad \text{and} \qquad 2x^2 + 1 = x^2 - x$$

are quadratic equations; the first of these is in general form, and the second may easily be rewritten in general form.

When we solve quadratic equations we will only be interested in real number solutions and will consider two methods of solution: factoring and the quadratic formula. We will discuss solving quadratic equations by factoring first. (For a review of factoring, see Section 0.6.)

Solution by factoring is based on the following property of the real numbers.

Zero Product Property   For real numbers a and b, $ab = 0$ if and only if $a = 0$ or $b = 0$ or both.

Hence, to solve by factoring, the equation must be written with zero on one side.

EXAMPLE 1

Solve $6x^2 + 3x = 4x + 2$.

Solution

$$6x^2 + 3x = 4x + 2$$

$$6x^2 - x - 2 = 0 \qquad \text{Proper form for factoring}$$

$$(3x - 2)(2x + 1) = 0 \qquad \text{Factored}$$

$$3x - 2 = 0 \quad \text{or} \quad 2x + 1 = 0 \qquad \text{Factors equal to zero}$$

$$3x = 2 \qquad\qquad 2x = -1$$

$$x = \frac{2}{3} \qquad\qquad x = -\frac{1}{2} \qquad \text{Solutions}$$

We now check that these values are, in fact, solutions to our original equation.

$$6\left(\frac{2}{3}\right)^2 + 3\left(\frac{2}{3}\right) \overset{?}{=} 4\left(\frac{2}{3}\right) + 2 \qquad 6\left(-\frac{1}{2}\right)^2 + 3\left(-\frac{1}{2}\right) \overset{?}{=} 4\left(-\frac{1}{2}\right) + 2$$

$$\frac{14}{3} = \frac{14}{3} \quad \checkmark \qquad\qquad\qquad\qquad 0 = 0 \quad \checkmark$$

EXAMPLE 2

Solve $6x^2 = 9x$.

Solution

$$6x^2 = 9x$$
$$6x^2 - 9x = 0$$
$$3x(2x - 3) = 0$$

$$3x = 0 \quad \text{or} \quad 2x - 3 = 0$$
$$x = 0 \qquad\qquad 2x = 3$$
$$x = \frac{3}{2}$$

CHECK: $6(0)^2 = 9(0)$ $\checkmark$ $\qquad 6\left(\frac{3}{2}\right)^2 = 9\left(\frac{3}{2}\right)$ $\checkmark$

Thus, the solutions are $x = 0$ and $x = \frac{3}{2}$.

Note that in the previous example (and always) it is incorrect to divide both sides of the equation by x, since this results in the loss of the solution $x = 0$.

EXAMPLE 3

Solve $(y - 3)(y + 2) = -4$ for y.

Solution

Note that the left side of the equation is factored, but the right member is not 0. Therefore we must multiply the factors before we can rewrite the equation in general form.

$$(y - 3)(y + 2) = -4$$
$$y^2 - y - 6 = -4$$
$$y^2 - y - 2 = 0$$
$$(y - 2)(y + 1) = 0$$

$$y - 2 = 0 \quad \text{or} \quad y + 1 = 0$$
$$y = 2 \qquad\qquad y = -1$$

CHECK: $(2 - 3)(2 + 2) = -4$ $\checkmark$ $\qquad (-1 - 3)(-1 + 2) = -4$ $\checkmark$

CHECKPOINT

1. The factoring method for solving a quadratic equation is based on the _____ product property. Hence, to solve a quadratic equation by factoring, one side of the equation must equal _____ .

2. Solve the following equations by factoring.

    (a) $x^2 - 19x = 20$　　　(b) $2x^2 = 6x$

THE QUADRATIC FORMULA

Factoring does not lend itself easily to solving quadratic equations like

$$x^2 - 5 = 0.$$

However, we can solve this equation by writing

$$x^2 = 5$$
$$x = \pm\sqrt{5}.$$

In general, we can solve quadratic equations of the form $x^2 = C$ (no x term) by taking the square root of both sides.

Square Root Property　　The solution of $x^2 = C$ is

$$x = \pm\sqrt{C}.$$

This property also can be used to solve equations like those in the following example.

EXAMPLE 4

Solve the following equations.

(a) $4x^2 = 5$　　　(b) $(3x - 4)^2 = 9$

Solution

We can use the square root property for both parts.

(a) $4x^2 = 5$ is equivalent to $x^2 = \frac{5}{4}$. Thus,

$$x = \pm\sqrt{\frac{5}{4}} = \pm\frac{\sqrt{5}}{\sqrt{4}} = \pm\frac{\sqrt{5}}{2}.$$

(b) $(3x - 4)^2 = 9$ is equivalent to $3x - 4 = \pm\sqrt{9}$. Thus

$3x - 4 = 3$	$3x - 4 = -3$
$3x = 7$	$3x = 1$
$x = \dfrac{7}{3}$	$x = \dfrac{1}{3}$

By using the fact that for variable x and constant k

$$(x + k)^2 = x^2 + 2kx + k^2,$$

we can solve the general quadratic equation

$$ax^2 + bx + c = 0 \qquad (a \neq 0)$$

by creating the square of a binomial on one side of the equation and then taking the square roots of both sides.* This procedure, which follows, results in a formula for solving any quadratic equation.

Completing the Square/Derivation of the Quadratic Formula

Procedure	Example
	Solve $ax^2 + bx + c = 0$.
1. Divide by a: then subtract $\dfrac{c}{a}$ from both sides.	1. $x^2 + \dfrac{bx}{a} = \dfrac{-c}{a}$
2. Take half the coefficient of x and square it.	2. The coefficient of x is b/a. $$\left[\frac{1}{2}\left(\frac{b}{a}\right)\right]^2 = \left[\frac{b}{2a}\right]^2 = \frac{b^2}{4a^2}$$
3. Add $\dfrac{b^2}{4a^2}$ to both sides.	3. $x^2 + \dfrac{bx}{a} + \dfrac{b^2}{4a^2} = \dfrac{-c}{a} + \dfrac{b^2}{4a^2}$
4. Factor; write as a square and simplify.	4. $\left(x + \dfrac{b}{2a}\right)^2 = \dfrac{b^2 - 4ac}{4a^2}$
5. Take the square roots of both sides.	5. $x + \dfrac{b}{2a} = \pm\sqrt{\dfrac{b^2 - 4ac}{4a^2}}$
6. Simplify.	6. $x + \dfrac{b}{2a} = \pm\dfrac{\sqrt{b^2 - 4ac}}{2|a|} = \pm\dfrac{\sqrt{b^2 - 4ac}}{2a}$
7. Solve for x.	7. $x = \dfrac{-b}{2a} \pm \dfrac{\sqrt{b^2 - 4ac}}{2a}$

Quadratic Formula   If $ax^2 + bx + c = 0$, where $a \neq 0$, then

$$x = \frac{-b + \sqrt{b^2 - 4ac}}{2a} \quad \text{or} \quad x = \frac{-b - \sqrt{b^2 - 4ac}}{2a}.$$

* This technique is called **completing the square** and can be used to solve any quadratic equation.

We may use the quadratic formula to solve all quadratic equations, but especially those in which factorization is difficult or impossible to see. The proper identification of values for a, b, and c to be substituted into the formula requires that the equation be in general form.

Example 5

Use the quadratic formula to solve $2x^2 - 3x - 6 = 0$ for x.

Solution

The equation is already in general form, with $a = 2$, $b = -3$, and $c = -6$. Hence,

$$x = \frac{-b \pm \sqrt{b^2 - 4ac}}{2a}$$

$$= \frac{-(-3) \pm \sqrt{(-3)^2 - 4(2)(-6)}}{2(2)}$$

$$= \frac{3 \pm \sqrt{9 + 48}}{4}$$

$$= \frac{3 \pm \sqrt{57}}{4}.$$

Thus, the solutions are

$$x = \frac{3 + \sqrt{57}}{4} \quad \text{and} \quad x = \frac{3 - \sqrt{57}}{4}.$$

Example 6

Using the quadratic formula, find all real solutions to $x^2 = x - 1$.

Solution

We must rewrite the equation in general form before we can determine the values of a, b, and c.

$$x^2 = x - 1$$

$$x^2 - x + 1 = 0 \qquad \text{Note: } a = 1, b = -1, \text{ and } c = 1.$$

$$x = \frac{-(-1) \pm \sqrt{(-1)^2 - 4(1)(1)}}{2(1)}$$

$$x = \frac{1 \pm \sqrt{1 - 4}}{2}$$

$$x = \frac{1 \pm \sqrt{-3}}{2}$$

Since $\sqrt{-3}$ is not a real number, the values of x are not real. Hence, there are no real solutions to the given equation.

The quadratic formula is especially useful when the coefficients of a quadratic equation are decimal values that make factorization impractical. This occurs in many applied problems, such as the one in the Application Preview, to which we now return.

EXAMPLE 7

Solve $0 = -1.056785714t^2 + 8.259285714t + 74.07142857$ from the Application Preview to find when the Social Security trust fund balance is zero. Recall that t is the number of years past the year 2000.

Solution

We use the quadratic formula with

$$a = -1.056785714, \quad b = 8.259285714, \quad c = 74.07142857.$$

Thus

$$
\begin{aligned}
t &= \frac{-b \pm \sqrt{b^2 - 4ac}}{2a} \\
&= \frac{-8.259285714 \pm \sqrt{(8.259285714)^2 - 4(-1.056785714)(74.07142857)}}{2(-1.056785714)} \\
&= \frac{-8.259285714 \pm \sqrt{381.3263106}}{-2.113571428} \\
&\approx \frac{-8.259285714 \pm 19.52757821}{-2.113571428}.
\end{aligned}
$$

We are only interested in the positive solution; hence

$$t = \frac{-8.259285714 - 19.52757821}{-2.113571428} \approx 13.14687393.$$

Therefore the trust fund balance is projected to reach zero sometime after 13 years, during the year 2013.

In Example 6, there were no real solutions to the quadratic equation because the radicand of the quadratic formula was negative. In general, when solving a quadratic equation, we can use the sign of the radicand in the quadratic formula, i.e., the sign of $b^2 - 4ac$, to determine how many real solutions there are. Thus, we refer to $b^2 - 4ac$ as the **quadratic discriminant.**

Quadratic Discriminant   Given $ax^2 + bx + c = 0$ and $a \neq 0$,

if $b^2 - 4ac > 0$, the equation has 2 distinct real solutions.
if $b^2 - 4ac = 0$, the equation has 1 real solution.
if $b^2 - 4ac < 0$, the equation has no real solutions.

CHECKPOINT

3. The statement of the Quadratic Formula says, if _____, then $x =$ _____ .

4. Solve $2x^2 - 5x = 9$ using the Quadratic Formula.

💻 *Graphing Utilities*

Many graphing utilities have commands (such as ROOT or SOLVER) or can be programmed to find the solutions of a quadratic equation. We can also use a graphing utility to determine (or approximate) solutions to a quadratic equation. For example, suppose we want to solve

$$x^2 - 10x - 24 = 0.$$

If we let $y = x^2 - 10x - 24$ and graph this function, then the zeros (or x-intercepts) of the graph correspond to x-values where $y = 0$; that is, solutions to

$$x^2 - 10x - 24 = 0.$$

These zeros can be located by observation or with TRACE and ZOOM. Figure 4.1 shows the graphical solution to this equation. The algebraic solution follows.

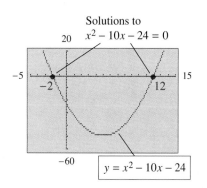

Figure 4.1

$$x^2 - 10x - 24 = 0$$
$$(x - 12)(x + 2) = 0$$

$x - 12 = 0$	$x + 2 = 0$
$x = 12$	$x = -2$

CHECKPOINT SOLUTIONS

1. Zero; zero

2. (a) $\quad\quad x^2 - 19x - 20 = 0$
   $$(x - 20)(x + 1) = 0$$

$x - 20 = 0$	$x + 1 = 0$
$x = 20$	$x = -1$

   (b) $\quad\quad 2x^2 - 6x = 0$
   $$2x(x - 3) = 0$$

$2x = 0$	$x - 3 = 0$
$x = 0$	$x = 3$

3. If $ax^2 + bx + c = 0$ with $a \neq 0$, then

$$x = \frac{-b \pm \sqrt{b^2 - 4ac}}{2a}.$$

4. $2x^2 - 5x - 9 = 0$ so $a = 2, b = -5, c = -9$

$$x = \frac{-b \pm \sqrt{b^2 - 4ac}}{2a} = \frac{5 \pm \sqrt{25 - 4(2)(-9)}}{4}$$

$$= \frac{5 \pm \sqrt{25 + 72}}{4}$$

$$= \frac{5 \pm \sqrt{97}}{4}$$

$$x = \frac{5 + \sqrt{97}}{4} \approx 3.712 \quad \text{or} \quad x = \frac{5 - \sqrt{97}}{4} \approx -1.212$$

EXERCISE 4.1

Write the following equations in general form.

1. $x^2 + 5x = 3$
2. $y^2 + 4y = 2$
3. $2x^2 + 3 = x^2 - 2x + 4$
4. $x^2 - 2x + 5 = 2 - 2x^2$
5. $(y + 1)(y + 2) = 4$
6. $(z - 1)(z - 3) = 1$

In problems 7–14, solve each equation by factoring.

7. $9 - 4x^2 = 0$
8. $25x^2 - 16 = 0$
9. $x = x^2$
10. $t^2 - 4t = 3t^2$
11. $0.05x^2 = 7x$
12. $0.8y^2 + 6.4y = 0$
13. $z^2 - z = z + 3$
14. $x^2 + 10x = 18x - 15$

In problems 15–20, solve each equation by factoring. You may check your solution graphically with a graphing utility.

15. $x^2 + 5x = 21 + x$
16. $x^2 + 17x = 8x - 14$
17. $4t^2 - 4t + 1 = 0$
18. $49z^2 + 14z + 1 = 0$
19. $(x - 1)(x + 5) = 7$
20. $(x - 3)(1 - x) = 1$

In problems 21–30, solve each equation by using the quadratic formula. Give real answers (a) exactly and (b) rounded to two decimal places.

21. $6x^2 = x + 1$
22. $3y^2 + 5y = 2$
23. $x^2 - 4x - 4 = 0$
24. $x^2 - 6x + 7 = 0$
25. $2w^2 + w + 1 = 0$
26. $z^2 + 2z + 4 = 0$
27. $16z^2 + 16z - 21 = 0$
28. $10y^2 - y - 65 = 0$
29. $5x^2 = 2x + 6$
30. $3x^2 = -6x - 2$

In problems 31–38, find the exact real solutions to each equation, if they exist.

31. $y^2 = 7$
32. $z^2 = 12$
33. $y^2 + 9 = 0$
34. $z^2 + 121 = 0$
35. $(x + 4)^2 = 25$
36. $(x + 1)^2 = 2$
37. $(3x - 2)^2 = 5$
38. $(2x - 6)^2 = 49$

In problems 39–48, (a) solve each equation by using a graphing utility. (b) Check your work by solving algebraically.

39. $49x^2 + 28x + 4 = 0$
40. $25x^2 + 120x + 144 = 0$
41. $21x + 70 - 7x^2 = 0$
42. $4x^2 + 12x + 8 = 0$
43. $3x^2 - 11x + 6 = 0$
44. $6x^2 + 5x - 6 = 0$
45. $300 - 2x - 0.01x^2 = 0$
46. $-9.6 + 2x - 0.1x^2 = 0$
47. $25.6x^2 - 16.1x - 1.1 = 0$
48. $6.8z^2 - 4.9z - 2.6 = 0$

APPLICATIONS

49. **Profit** If the profit from the sale of x units of a product is $P = 90x - 200 - x^2$, what level(s) of production yields a profit of \$1200?

50. **Profit** If the profit from the sale of x units of a product is $P = 16x - 0.1x^2 - 100$, what level(s) of production yields a profit of \$180?

51. **Profit** If the profit from the sale of x units of a product is $P = 6400x - 18x^2 - 400$, what level(s) of production yields a profit of \$61,800?

52. **Profit** If the profit from the sale of x units of a product is $P = 50x - 300 - 0.01x^2$, what level(s) of production yields a profit of \$250?

53. **Flow rates of water** The speed at which water travels in a pipe can be measured by directing the flow through an elbow and measuring the height it spurts out the top. If the elbow height is 10 cm, the equation relating the height of the water above the elbow (in centimeters) and its velocity v (in centimeters per second) is given by

$$v^2 = 1960(h + 10).$$

(a) Find v if $h = 2$.

(b) Find v if $h = 18$.

(c) The height h is a function of the velocity v. Solve the equation for h to express this function.

54. **Windchill** Weather forecasters frequently report wind-chill factors because a body exposed to wind loses heat due to convection. The amount of loss depends on many factors, but for a given situation there is a positive number called the coefficient of convection, K_c, which depends on the wind velocity v. The approximate relationship between K_c and v is given by

$$\frac{(K_c)^2}{64} - \frac{1}{4} = v.$$

(a) Find the coefficient of convection when the wind velocity is 10 mph.

(b) Find K_c when $v = 40$ mph.

(c) What is the change in K_c for the change in v from 10 mph to 40 mph?

55. **Velocity of blood** The velocity of a blood corpuscle in a vessel depends on how far the corpuscle is from the center of the vessel. Let R be the constant radius of the vessel, v_m the constant maximum velocity of the corpuscle, r the distance from the center to a particular blood corpuscle (variable), and v_r the velocity of that corpuscle. Then the velocity v_r is related to the distance r according to

$$v_r = v_m \left(1 - \frac{r^2}{R^2}\right).$$

(a) Find r when $v_r = \frac{1}{2}v_m$.

(b) Find r when $v_r = \frac{1}{4}v_m$.

(c) Find v_r when $r = R$.

56. **Ladder movement** A ladder 13 feet long is leaning against a wall with its foot 5 feet from the wall. If the foot of the ladder is pulled 4 feet farther from the wall, how much does the top of the ladder slide down the wall?

57. **Toyota Supra acceleration** The time t, in seconds, that it takes a 1993 Toyota Supra to accelerate to x mph can be modeled by

$$t = 0.01(0.0839x^2 + 2.6536x + 92.3571)$$

(Source: *Motor Trend,* March 1993, p. 7). How fast is the Supra going after 5.5 seconds? Give your answer to the nearest tenth.

58. **Social Security trust fund** Social Security benefits are paid from a trust fund. As mentioned in the Application

Preview, the trust fund balance, B, in billions of dollars t years past the year 2000 is projected to be modeled by

$$B = -1.056785714t^2 + 8.259285714t + 74.07142857.$$

(Source: Social Security Administration projections). In what year is the trust fund balance projected to be $1000 billion in the red; i.e. when $B = -1000$?

59. **AT&T total revenues** For the years 1985–1993, total revenues of AT&T can be modeled by

$$R = 0.253t^2 - 4.03t + 76.84,$$

where R is in billions of dollars and t is the number of years past 1980 (Source: *AT&T Annual Report,* 1993).

(a) For what t-values will total revenue be $70 billion?

(b) Use your answer to part (a) to determine what year (after 1993) total revenue will reach $70 billion.

60. **Depth of a fissure** A fissure in the earth appeared after an earthquake. To measure its vertical depth a stone was dropped into it, and the sound of the stone's impact was heard 3.9 seconds later. The distance (in feet) the stone fell is given by $s = 16t_1^2$, and the distance (in feet) the sound traveled is given by $s = 1090t_2$. In these equations, the distances traveled by the sound and the stone are the same, but their times are not. Using the fact that the total time is 3.9 seconds, find the depth of the fissure.

61. **Marijuana use** The percentage of high school seniors, p, who have tried marijuana can be modeled by

$$p = -0.228596t^2 + 2.7783t + 49.813783,$$

where t is the number of years past 1975 (Source: National Institute on Drug Abuse). In what year, after 1990, will the percentage predicted by the model reach 33% if this model remains valid?

62. **Projectile motion** Two projectiles are shot into the air over a lake. The paths of the projectiles are given by

(a) $y = -0.0013x^2 + x + 10$, and

(b) $y = -\dfrac{x^2}{81} + \dfrac{4}{3}x + 10$,

where y is the height and x is the horizontal distance traveled. Determine which projectile travels farther by substituting $y = 0$ in each equation and finding x.

63. **Percentage profit** The Ace Jewelry Store sold a necklace for $144. If the percentage of profit (based on cost) equals the cost of the necklace to the store, how much did the store pay for it? (Use $P = \left(\dfrac{C}{100}\right) \cdot C$ where P is profit and C is cost.)

4.2 QUADRATIC FUNCTIONS: PARABOLAS

Objectives

- ❏ To determine if a vertex is a maximum point or a minimum point
- ❏ To find the vertex of the graph of a quadratic function
- ❏ To find the zeros of a quadratic function
- ❏ To graph quadratic functions

APPLICATION PREVIEW

Because additional equipment, raw materials, and manpower may cause variable costs of some products to increase dramatically as more units are produced, total cost functions are not always linear functions. For example, suppose that the total cost of producing a product is given by the equation

$$C = C(x) = 300x + 0.1x^2 + 1200,$$

where x represents the number of units produced. That is, the cost of this product is represented by a **second-degree polynomial function,** or a **quadratic function.** We can find the cost of producing 10 units by evaluating

$$C(10) = 300(10) + 0.1(10)^2 + 1200 = 4210.$$

If the revenue function for this product is

$$R = R(x) = 600x,$$

then the profit is also a quadratic function

$$P = R - C = 600x - (300x + 0.1x^2 + 1200)$$
$$P = -0.1x^2 + 300x - 1200.$$

It is natural to ask how many units give maximum profit and what that profit is. In this section we will describe ways to find the maximum point, or minimum point, for a quadratic function.

As we noted in Chapter 1, the equation $y = f(x)$, where $f(x)$ is a polynomial in the variable x, is called a polynomial function. One special class of polynomial functions is the linear function (or first degree function) $y = mx + b$, studied in Chapter 1. Increasing the degree of a linear function by one gives a **second-degree polynomial function,** or **quadratic function.** The general equation of a quadratic function has the form

$$y = f(x) = ax^2 + bx + c,$$

where a, b, and c are real numbers and $a \neq 0$.

We begin our study of quadratic functions by reconsidering the quadratic profit function from our Application Preview.

Figure 4.2 (a)

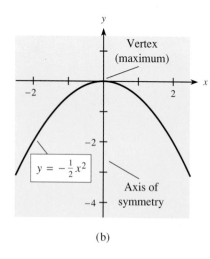

(b)

Figure 4.2

EXAMPLE 1

In the Application Preview, we discussed a product whose profit from the production and sale of x units is modeled by $P(x) = -0.1x^2 + 300x - 1200$. Find the profit if

(a) 1000 units are produced.
(b) 1500 units are produced.
(c) 2000 units are produced.

Solution

(a) $P(1000) = -0.1(1000)^2 + 300(1000) - 1200 = 198{,}800$
(b) $P(1500) = -0.1(1500)^2 + 300(1500) - 1200 = 223{,}800$
(c) $P(2000) = -0.1(2000)^2 + 300(2000) - 1200 = 198{,}800$

Note that the profit function of Example 1 indicates that as production and sales increase from 1000 to 2000 units, the profit first increases and then decreases. Thus the graphs of (this and any other) quadratic functions are not lines. However, the graph of a quadratic function

$$y = ax^2 + bx + c \qquad (a \neq 0)$$

does have a distinctive shape, called a **parabola.**

The basic function $y = x^2$ and a variation of it, $y = -\frac{1}{2}x^2$, are parabolas whose graphs are shown in Figure 4.2.

As these examples illustrate, the graph of $y = ax^2$ is a parabola that opens upward if $a > 0$ and downward if $a < 0$. The turning point, or **vertex,** is a **minimum point** if $a > 0$ and a **maximum point** if $a < 0$. The vertical line through the vertex of a parabola is called the **axis of symmetry** because one half of the graph is a reflection of the other half through this line.

The graph of $y = (x - 2)^2 - 1$ is the graph of $y = x^2$ shifted to a new location that is 2 units to the right and 1 unit down; its vertex is shifted from $(0, 0)$ to $(2, -1)$ and its axis of symmetry is shifted 2 units to the right. (See Figure 4.3.) The graph of $y = -\frac{1}{2}(x + 1)^2 + 2$ is the graph of $y = -\frac{1}{2}x^2$ shifted 1 unit to the left and 2 units upward, with its vertex at $(-1, 2)$. (See Figure 4.4.)

In general, if a quadratic function $y = f(x) = ax^2 + bx + c$ has the form

$$y = a(x - h)^2 + k,$$

then its graph is that of $y = ax^2$, with its vertex shifted to $(h, k) = (h, f(h))$. In addition, by completing the square on the x-terms, $f(x) = ax^2 + bx + c$ can be written in the form

$$f(x) = a\left(x + \frac{b}{2a}\right)^2 + \left(c - \frac{b^2}{4a}\right),$$

so we have the following.

Figure 4.3

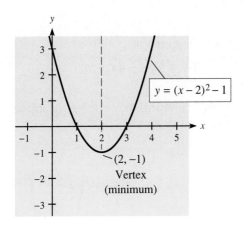

$y = (x - 2)^2 - 1$

$(2, -1)$
Vertex
(minimum)

Figure 4.4

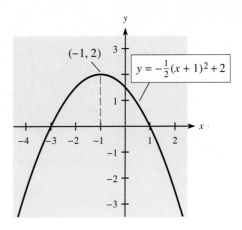

$(-1, 2)$

$y = -\frac{1}{2}(x + 1)^2 + 2$

Vertex of a Parabola   The quadratic function $y = f(x) = ax^2 + bx + c$ has its **vertex** at

$$\left(\frac{-b}{2a}, \ f\left(\frac{-b}{2a}\right) \right).$$

See Figure 4.5.

Figure 4.5

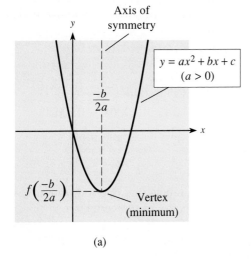

Axis of symmetry

$y = ax^2 + bx + c$
$(a > 0)$

$\frac{-b}{2a}$

$f\left(\frac{-b}{2a}\right)$

Vertex
(minimum)

(a)

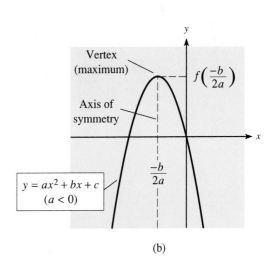

Vertex (maximum)

$f\left(\frac{-b}{2a}\right)$

Axis of symmetry

$\frac{-b}{2a}$

$y = ax^2 + bx + c$
$(a < 0)$

(b)

Knowing the location of the vertex of a parabola and whether the parabola opens upward or downward, we need very few other points to make a good sketch.

EXAMPLE 2

Find the vertex and sketch the graph of

$$f(x) = 2x^2 - 4x + 4.$$

Solution

Since the graph of $f(x)$ opens upward, the vertex is the minimum point. We can calculate its coordinates as follows:

$$x = \frac{-b}{2a} = \frac{-(-4)}{2(2)} = 1$$

$$y = f(1) = 2.$$

Thus the vertex is (1, 2). Using *x*-values on either side of the vertex to plot additional points allows us to sketch the graph accurately. (See Figure 4.6.)

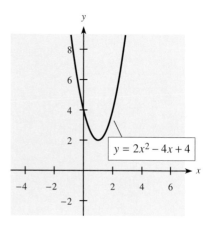

$y = 2x^2 - 4x + 4$

Figure 4.6

We can also use the coordinates of the vertex to find maximum or minimum values without a graph.

EXAMPLE 3

For the profit function in our Application Preview,

$$P(x) = -0.1x^2 + 300x - 1200,$$

find the number of units that give maximum profit and find the maximum profit.

Solution

The graph of $y = P(x)$ is a parabola that opens downward, so the vertex is a maximum point. The coordinates of the vertex are

$$x = \frac{-b}{2a} = \frac{-300}{2(-0.1)} = 1500$$

$$P = P(1500) = -0.1(1500)^2 + 300(1500) - 1200 = 223,800.$$

Therefore, the maximum profit is $223,800 when 1500 units are sold.

Graphing Utilities

We can also use a graphing utility to graph a quadratic function. Even with a graphing tool, recognizing that the graph is a parabola and locating the vertex are important. They help us to set the range (to include the vertex) and to know when we have a complete graph.

EXAMPLE 4

Suppose that the profit received from the production and sale of x units $(x \geq 0)$ of a product is given by

$$P(x) = -0.4x^2 + 40x - 360.$$

(a) Find the maximum profit.
(b) Sketch the graph with a graphing utility.

Solution

(a) Since the graph of $P(x)$ is a parabola that opens downward, its vertex is a maximum point. Thus maximum profit can be found by finding the vertex of the parabola.

$$x = \frac{-b}{2a} = \frac{-40}{2(-0.4)} = \frac{-40}{-0.8} = 50$$

$$P(50) = -0.4(50^2) + 40(50) - 360 = 640$$

Hence, $(50, 640)$ is the maximum point and the maximum profit is $640.

(b) Because $x \geq 0$ and 50 is the x-coordinate of the vertex (and $x = 50$ is the axis of symmetry), we set the x-range so that it includes the y-axis and is centered at $x = 50$. For the P-range (or y-range), we know that the maximum is $P = 640$. Also, $P(0) = -360$, so we include these values in the P-range. Figure 4.7 shows the graph.

$P(x) = -0.4x^2 + 40x - 360$

Figure 4.7

Notice in Figure 4.7 that the P-intercept is -360, corresponding to a profit of -360 (dollars) or a loss of $360. Notice also that the graph of this parabola has *two* x-intercepts.

As we noted in Chapter 1, the x-intercepts of the graph of a function $y = f(x)$ are the values of x for which $f(x) = 0$, called the **zeros** of the function. As we saw in Section 4.1, the zeros of the quadratic function $y = f(x) = ax^2 + bx + c$ are the solutions to the quadratic equation

$$ax^2 + bx + c = 0.$$

Therefore, the zeros are given by the quadratic formula as

$$x = \frac{-b + \sqrt{b^2 - 4ac}}{2a}$$

and

$$x = \frac{-b - \sqrt{b^2 - 4ac}}{2a}.$$

The information that is useful in graphing quadratic functions is summarized on the following page.

Graphs of Quadratic Functions

Form: $y = ax^2 + bx + c$

Graph: parabola (See Figure 4.5 on page 333.)

$a > 0$ parabola opens upward; vertex is a minimum point

$a < 0$ parabola opens downward; vertex is a maximum point

Coordinates of vertex: $x = \dfrac{-b}{2a}, \quad y = f\left(\dfrac{-b}{2a}\right)$

Axis of symmetry equation: $x = \dfrac{-b}{2a}$

x-intercepts or zeros (if real*):

$$x = \frac{-b + \sqrt{b^2 - 4ac}}{2a}, \quad x = \frac{-b - \sqrt{b^2 - 4ac}}{2a}$$

y-intercept: Let $x = 0$; then $y = c$.

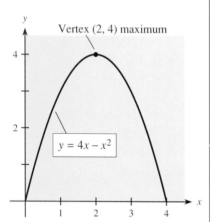

Vertex (2, 4) maximum

$y = 4x - x^2$

Figure 4.8

EXAMPLE 5

For the function $y = 4x - x^2$: determine if its vertex is a maximum point or minimum point and find the coordinates of this point; find the zeros, if any exist, and sketch the graph.

Solution

The proper form is $y = -x^2 + 4x + 0$, so $a = -1$. Thus, the parabola opens downward, and the vertex is the highest (maximum) point.

The vertex occurs at $x = \dfrac{-b}{2a} = \dfrac{-4}{2(-1)} = 2$.

The y-coordinate of the vertex is $f(2) = -(2)^2 + 4(2) = 4$.

The zeros for the parabola are solutions to

$$-x^2 + 4x = 0$$
$$x(-x + 4) = 0,$$

or $x = 0$ and $x = 4$.

Plotting these three points (or using a graphing utility) gives the graph (see Figure 4.8).

CHECKPOINT

1. Name the graph of a quadratic function.

2. (a) What is the x-coordinate of the vertex of $y = ax^2 + bx + c$?

   (b) For $y = 12x - \dfrac{1}{2}x^2$, what is the x-coordinate of the vertex? What is the y-coordinate of the vertex?

* If the zeros are not real, the graph does not cross the x-axis.

3. (a) How can you tell whether the vertex of $f(x) = ax^2 + bx + c$ is a maximum point or a minimum point?
   (b) In 2(b), is the vertex a maximum point or a minimum point?
4. The zeros of a function correspond to what feature of its graph?

EXAMPLE 6

Figure 4.9 shows the graph of two different quadratic functions. Use the figure to answer the following.
(a) Determine the vertex of each function.
(b) Determine the real solutions to $f_1(x) = 0$ and to $f_2(x) = 0$.
(c) One of the graphs in Figure 4.9 is the graph of $y = 7 + 6x - x^2$ and one is the graph of $y = x^2 - 6x + 10$. Determine which is which and why.

Solution
(a) For $y = f_1(x)$ the vertex is the maximum point, at $(3, 16)$.
    For $y = f_2(x)$ the vertex is the minimum point, at $(3, 1)$.
(b) For $y = f_1(x)$, the real solutions to $f_1(x) = 0$ are the zeros or x-intercepts, at $x = -1$ and $x = 7$. For $y = f_2(x)$, the graph has no x-intercepts, so $f_2(x) = 0$ has no real solutions.
(c) Because the graph of $y = f_1(x)$ opens downward, the coefficient of x^2 must be negative. Hence Figure 4.9(a) shows the graph of $y = f_1(x) = 7 + 6x - x^2$. Similarly, the coefficient of x^2 in $y = f_2(x)$ must be positive, so Figure 4.9(b) shows the graph of $y = f_2(x) = x^2 - 6x + 10$.

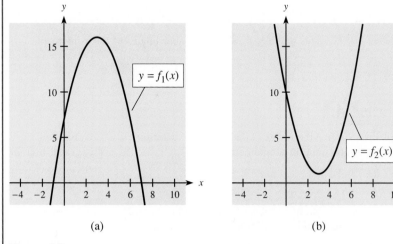

(a)   (b)

Figure 4.9

Thus far we have discussed quadratic functions of the form $y = ax^2 + bx + c$. More generally, we define any equation that has at least one second-degree term and no term with degree higher than 2 as a **quadratic equation.**

The general form of a quadratic equation in two unknowns is

$$Ax^2 + Bxy + Cy^2 + Dx + Ey + F = 0,$$

with A, B, and C not all 0. The graphs of some of these equations will not be parabolas, but all equations that can be written in the form $y = ax^2 + bx + c$, where $a \neq 0$, will have parabolas as their graphs.

Techniques exist that make graphing the other forms of quadratic equations fairly easy, but we are primarily interested in parabolas, so we will not discuss those techniques. We can graph quadratic equations in two variables that are not of the form $y = ax^2 + bx + c$ by plotting a number of points or by using techniques we will learn in later chapters.

CHECKPOINT SOLUTIONS

1. Parabola

2. (a) $x = \dfrac{-b}{2a}$

   (b) $x = \dfrac{-12}{2(-1/2)} = \dfrac{-12}{(-1)} = 12$

   $y = 12(12) - \dfrac{1}{2}(12)^2 = 144 - 72 = 72$

3. (a) Maximum point if $a < 0$; minimum point if $a > 0$.

   (b) (12, 72) is a maximum point because $a = -\dfrac{1}{2}$.

4. x-coordinates of the x-intercepts

EXERCISE 4.2

In problems 1–4, find the vertex of the graph and determine if it is a maximum point or a minimum point.

1. $y = \frac{1}{2}x^2 + x$ 2. $y = x^2 - 2x$

3. $y = 8 + 2x - x^2$ 4. $y = 6 - 4x - 2x^2$

For each function given in problems 5 and 6, find (a) where the function has its maximum and (b) the maximum value of the function.

5. $f(x) = 6x - x^2$ 6. $f(x) = 4 + 3x - x^2$

For each function given in problems 7 and 8, find (a) where the function has its minimum and (b) the minimum value of the function.

7. $f(x) = x^2 + 2x - 3$ 8. $f(x) = \frac{1}{2}x^2 + 2x.$

In problems 9–12, find the zeros of each function and sketch its graph. (See problems 1–4 for information.)

9. $y = \frac{1}{2}x^2 + x$ 10. $y = x^2 - 2x$

11. $y = 8 + 2x - x^2$ 12. $y = 6 - 4x - 2x^2$

In problems 13–20, determine whether each function's vertex is a maximum point or a minimum point and find the coordinates of this point. Find the zeros, if any exist, and sketch the graph of the function.

13. $y = x^2 - 4$ 14. $y = x^2 - 2$

15. $y = x - \frac{1}{4}x^2$ 16. $y = -2x^2 + 18x$

17. $y = x^2 + 4x + 4$ 18. $y = x^2 - 6x + 9$

19. $\frac{1}{2}x^2 + x - y - 3 = 0$ 20. $x^2 + x + 2y = 5$

In problems 21–26, graph each function with a graphing utility. Use the graph to find the vertex and zeros. Check your results algebraically.

21. $y = \frac{1}{2}x^2 - x - \frac{15}{2}$ 22. $y = 0.1(x^2 + 4x - 32)$

23. $y = -5x - x^2$ 24. $y = 8 - \frac{1}{2}x^2$

25. $y = \frac{1}{4}x^2 + 3x + 12$ 26. $y = x^2 - 2x + 5$

In problems 27–30, find the vertex and zeros and use them to determine a range for a graphing utility that includes these values; then graph the function with that range.

27. $y = 63 + 0.2x - 0.01x^2$

28. $y = 0.2x^2 + 16x + 140$

29. $y = 0.0001x^2 - 0.01$

30. $y = 0.01x - 0.001x^2$

In problems 31–33, use a graphing utility to graph the basic function $f(x) = x^2$ with the requested variations.

31. (a) Graph $f(x) = x^2$ and $y = af(x) = ax^2$ for $a = \pm 3$, ± 2, $\pm \frac{1}{2}$, and $\pm \frac{1}{4}$.
    (b) Describe the differences between these graphs when $a > 0$ and when $a < 0$.
    (c) Describe how $af(x)$ differs from $f(x)$ when $0 < a < 1$ and when $a > 1$.

32. (a) Graph $f(x) = x^2$ and $y = f(x - h) = (x - h)^2$ for $h = 3, 1, -2$, and -4.
    (b) Write a sentence that precisely describes how the graph of $y = (x - h)^2$ is related to the graph of $y = x^2$.

33. (a) Graph $f(x) = x^2$ and $y = f(x) + k = x^2 + k$ for $k = 4, 2, -1$, and -3.
    (b) Write a sentence that precisely describes how the graph of $y = x^2 + k$ is related to the graph of $y = x^2$.

34. How does the discriminant $b^2 - 4ac$ affect the graph of $y = ax^2 + bx + c$ if
    (a) $b^2 - 4ac = 0$?   (b) $b^2 - 4ac > 0$?
    (c) $b^2 - 4ac < 0$?

APPLICATIONS

35. **Profit** If the daily profit from the sale of a product is given by $P = 16x - 0.1x^2 - 100$,
    (a) what level of production maximizes profit?
    (b) what is the maximum possible profit?

36. **Profit** If the daily profit from the sale of x units of a product is $P = 80x - 0.4x^2 - 200$,
    (a) what level of production maximizes profit?
    (b) what is the maximum possible profit?

37. **Wind and pollution** The amount of particulate pollution x depends on the wind velocity v, among other things. If the relationship between x and v can be approximated by

$$x = 20 - 0.01v^2,$$

sketch the graph relating these quantities, with v on the horizontal axis.

38. **Velocity of blood** The velocity of a blood corpuscle in a vessel, v_r, depends on the distance of the corpuscle from the center of the vessel, r, according to

$$v_r = v_m\left(1 - \frac{r^2}{R^2}\right),$$

where v_m is the maximum velocity and R the radius of the vessel.

The blood pressure also affects the velocity of a corpuscle. If we make some simplifications in the formula, namely, let $v_m = 1$ and let $x^2 = r^2/R^2$, then we can observe the following.

Pressure	Equation
$p = 20$ mm	$v(x) = 8(1 - x^2)$
$p = 10$ mm	$v(x) = 3(1 - x^2)$
$p = 5$ mm	$v(x) = 1 - x^2$

Graph each of these equations on the same set of axes.

39. **Crop yield** The yield from a grove of orange trees is given by $Y = x(800 - x)$, where x is the number of orange trees per acre. How many trees will maximize the yield?

40. **Stimulus-response** One of the early results in psychology relating the magnitude of a stimulus x to the magnitude of a response y is expressed by the equation

$$y = kx^2,$$

where k is an experimental constant. Sketch this graph for $k = 1$, $k = 2$, and $k = 4$.

41. **Drug sensitivity** The sensitivity S to a drug is related to the dosage x by

$$S = 1000x - x^2.$$

Sketch the graph of this function and determine what dosage gives maximum sensitivity. Use the graph to determine the maximum sensitivity.

42. **Maximizing an enclosed area** If 100 feet of fence is used to fence in a rectangular yard, then the resulting

area is given by

$$A = x(50 - x),$$

where x is the width of the rectangle and $50 - x$ is the length. Graph this equation and determine the length and width that give maximum area.

43. **Photosynthesis** The rate of photosynthesis R for a certain plant depends on the intensity of light x according to

$$R = 270x - 90x^2.$$

Sketch the graph of this function, and determine the intensity that gives the maximum rate.

44. **Projectiles** A ball thrown vertically into the air has its height above ground given by

$$s = 112t - 16t^2.$$

Find the maximum height of the ball.

45. **Projectiles** Two projectiles are shot into the air from the same location. The paths of the projectiles are parabolas and given by
    (a) $y = -0.0013x^2 + x + 10$, and
    (b) $y = \dfrac{-x^2}{81} + \dfrac{4}{3}x + 10$.

Determine which projectile goes higher by locating the vertex of each parabola.

46. **Flow rates of water** The speed at which water travels in a pipe can be measured by directing the flow through an elbow and measuring the height it spurts out the top. If the elbow height is 10 cm, the equation relating the height h (in centimeters) of the water above the elbow and its velocity v (in centimeters per second) is given by

$$v^2 = 1960(h + 10).$$

Solve this equation for h and graph the result, using the velocity as the independent variable.

47. **U.S. trade deficit** The figure to the right shows the U.S. trade deficit and the 12-month moving average U.S. trade deficit.
    (a) What kind of function might be used to model the 12-month moving average?
    (b) If a function of the form $f(x) = ax^2 + bx + c$ were used to model the 12-month moving average, would $a > 0$ or $a < 0$? Why?

48. **Projectile motion** When a stone is thrown upward, it follows a parabolic path given by a form of the equation

$$y = ax^2 + bx + c.$$

If $y = 0$ represents ground level, find the equation of a stone thrown from ground level at $x = 0$ and landing on the ground 40 units away if the stone reaches a maxi-

mum height of 40 units. (*Hint:* Find the coordinates of the vertex of the parabola and two other points.)

49. **Health care** Many politicians are discussing national health insurance because health care costs are increasing so rapidly. Health care costs in the United States are given in the table below.

Year	Costs (in Millions)
1960	27.1
1965	41.6
1970	74.4
1975	132.9
1980	249.1
1985	420.1
1989	539.9

SOURCE: *World Almanac,* 1991

(a) Plot the points from the table. Use x to represent the number of years since 1960 and y to represent costs in millions of dollars.
(b) What type of function appears to be the best fit for these points?
(c) Graph different quadratic functions of the form $y = ax^2 + c$ until you find a curve that is a reasonable fit for the points.
(d) Use your "model" function to predict health care costs in 1990.
(e) Actual health care costs for 1990 were $666.2 million. Do health care costs seem to be increasing even faster than your model predicts? Does it appear that some measures are necessary to help slow the growth of health care costs?

U. S. Trade Deficit

In billions of dollars

SOURCE: *The Wall Street Journal,*
December 17, 1993

50. **Social Security** In 1995, America's 45 million Social Security recipients received a 2.6% cost-of-living increase, the second smallest increase in nearly 20 years, a reflection of lower inflation. The following data, which shows the percent increase for the previous 6 years, can be modeled by the function

$$p(t) = -0.3375t^2 + 7.3t - 34.3625,$$

where t is the number of years past 1980.

Year	Percent increase	Year	Percent increase
1989	4.0	1992	4.1
1990	4.9	1993	3.5
1991	5.1	1994	2.6

SOURCE: Social Security Administration

(a) Graph the data points with the model $p(t)$.
(b) From the graph, identify t-values where $p(t) < 0$, and hence where the model is not valid.
(c) From the graph, identify the maximum point on the graph of $p(t)$. From the data, in what year was the maximum percentage cost-of-living increase? What was the amount of the increase?

Union participation The following table shows the percentages of U.S. workers who belonged to unions for selected years from 1930 to 1989.

Year	Percentage	Year	Percentage
1930	11.6	1975	25.5
1940	26.9	1980	21.9
1950	31.5	1985	18
1960	31.4	1989	16.4
1970	27.3		

SOURCE: *World Almanac,* 1991

Assume that this data can be modeled with the function

$$u(x) = -39.3 + 2.38x - 0.02x^2,$$

where x is the number of years past 1900. Use this information in problems 51–53.

51. (a) Plot the values from the table.
    (b) Graph the function $u(x)$ on the same set of axes with the data points.

52. (a) For what year does the model $u(x)$ indicate a maximum percentage of workers belonged to unions?
    (b) Does your answer in (a) agree with the data in the table?

53. (a) For what years does the model $u(x)$ predict that 0% of U.S. workers will belong to unions?
    (b) When can you guarantee that $u(x)$ can no longer be used to model the percentage of U.S. workers who belong to unions?

4.3 BUSINESS APPLICATIONS OF QUADRATIC FUNCTIONS

Objectives

❏ To graph quadratic supply and demand functions
❏ To find market equilibrium by using quadratic supply and demand functions
❏ To find break-even points by using quadratic cost and revenue functions
❏ To maximize quadratic revenue and profit functions

APPLICATION PREVIEW

In this section, we revisit applications concerning market analysis (supply and demand) and theory of the firm (cost, revenue, and profit).

We extend our work to graphing supply and demand functions when they are quadratic. We also reconsider market equilibrium and find it by simultaneously solving supply and demand equations in situations when at least one of the equations is quadratic.

We consider graphs of cost, revenue, and profit functions when they are quadratic. We also revisit break-even points in a quadratic setting and notice that the notion of break-even points does not change even though the mathematics used to find them does. Also we use the vertex of quadratic functions to determine maximum and minimum points, such as for revenue and profit.

SUPPLY, DEMAND, AND MARKET EQUILIBRIUM

The first-quadrant parts of parabolas or other quadratic equations are frequently used to represent supply and demand functions. For example, the first-quadrant part of $p = q^2 + q + 2$ (Figure 4.10) may represent a supply curve, while the first-quadrant part of $q^2 + 2q + 6p - 23 = 0$ (Figure 4.11) may represent a demand curve.

When quadratic equations are used to represent supply or demand curves, we can solve their equations simultaneously to find the market equilibrium as we did with linear supply and demand functions. As in Section 1.8, we can solve two equations in two variables by eliminating one variable and obtaining an equation in the other variable. When the functions are quadratic, the substitution method of solution is perhaps the best, and the resulting equation in one unknown will usually be quadratic.

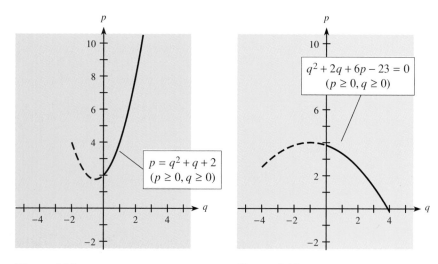

Figure 4.10

Figure 4.11

EXAMPLE 1

If the supply function for a commodity is given by $p = q^2 + 100$ and the demand function is given by $p = -20q + 2500$, find the point of market equilibrium.

Solution

At market equilibrium, both equations will have the same p-value. So substituting $q^2 + 100$ for p in $p = -20q + 2500$, we get

$$q^2 + 100 = -20q + 2500$$
$$q^2 + 20q - 2400 = 0$$
$$(q - 40)(q + 60) = 0$$
$$q = 40 \quad \text{or} \quad q = -60.$$

Figure 4.12

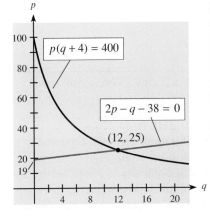

Figure 4.13

Since a negative quantity has no meaning, the equilibrium point occurs when 40 units are sold, at (40, 1700). The graphs of the functions are shown (in the first quadrant only) in Figure 4.12.

EXAMPLE 2

If the demand function for a commodity is given by $p(q + 4) = 400$ and the supply function is given by $2p - q - 38 = 0$, find the market equilibrium.

Solution

First note that the demand is defined by a quadratic equation, but its graph is not a parabola. This causes no problems for us in trying to solve the equations simultaneously; we proceed as before. Solving $2p - q - 38 = 0$ for p gives $p = \frac{1}{2}q + 19$. Substituting for p in $p(q + 4) = 400$ gives

$$(\tfrac{1}{2}q + 19)(q + 4) = 400$$
$$\tfrac{1}{2}q^2 + 21q - 324 = 0.$$

Multiplying both sides of the equation by 2 gives $q^2 + 42q - 648 = 0$. Factoring gives

$$(q - 12)(q + 54) = 0$$
$$q = 12 \quad \text{or} \quad q = -54.$$

Thus the market equilibrium occurs when 12 items are sold, at a price of $25 each. The graphs of the demand and supply functions are shown in Figure 4.13.

Graphing utilities also can be used to sketch these graphs. TRACE and ZOOM, or a special command such as INTERSECT, could be used to determine points of intersection that give market equilibrium.

CHECKPOINT

1. The point of intersection of the supply and demand functions is called
_____.

2. If the demand and supply functions for a product are

$$p + \frac{1}{10}q^2 = 1000 \quad \text{and} \quad p = \frac{1}{10}q + 10,$$

respectively, finding the market equilibrium point requires solution of what equation?

BREAK-EVEN POINTS AND MAXIMIZATION

In Chapter 1, we discussed linear total cost and total revenue functions. Many total revenue functions may be linear, but costs tend to increase sharply after a certain level of production. Thus functions other than linear functions, including quadratic functions, are used to predict the total costs of products.

For example, the monthly total cost curve for a commodity may be the parabola with equation $C(x) = 360 + 40x + 0.1x^2$. If the total revenue function is $R(x) = 60x$, we can find the break-even point by finding the quantity x that makes $C(x) = R(x)$. (See Figure 4.14.)

Setting $C(x) = R(x)$, we have

$$360 + 40x + 0.1x^2 = 60x$$

$$0.1x^2 - 20x + 360 = 0$$

$$x^2 - 200x + 3600 = 0$$

$$(x - 20)(x - 180) = 0$$

$$x = 20 \quad \text{or} \quad x = 180.$$

Figure 4.14

Thus $C(x) = R(x)$ at $x = 20$ and at $x = 180$. If 20 items are produced and sold, $C(x)$ and $R(x)$ are both $1200; if 180 items are sold, $C(x)$ and $R(x)$ are both $10,800. Thus, there are two break-even points.

In a monopoly market, the revenue of a company is restricted by the demand for the product. In this case, the relationship between the price p of the product and the number of units sold x is described by the demand function $p = f(x)$, and the total revenue function for the product is given by

$$R = px = [f(x)]x.$$

If, for example, the demand for a product is given by

$$p = 300 - x,$$

where x is the number of units sold and p is the price, then the revenue function for this product is

$$R = px = (300 - x)x$$

or

$$R = 300x - x^2,$$

a quadratic function.

EXAMPLE 3

Suppose that in a monopoly market the total cost per week of producing a high-tech product is given by $C = 3600 + 100x + 2x^2$. Suppose further that the weekly demand function for this product is $p = 500 - 2x$. Find the number of units that will give break-even for the product.

Solution
The total cost function is $C(x) = 3600 + 100x + 2x^2$, and the total revenue function is $R(x) = (500 - 2x)x = 500x - 2x^2$.

Setting $C(x) = R(x)$ and solving for x gives

$$3600 + 100x + 2x^2 = 500x - 2x^2$$
$$4x^2 - 400x + 3600 = 0$$
$$x^2 - 100x + 900 = 0$$
$$(x - 90)(x - 10) = 0$$
$$x = 90 \quad \text{or} \quad x = 10.$$

Does this mean the firm will break even at 10 units and at 90 units? Yes. Figure 4.15 shows the graphs of $C(x)$ and $R(x)$. From the graph we can observe that the firm makes a profit after $x = 10$ *until* $x = 90$. At $x = 90$, the profit is 0, and the firm loses money if more than 90 units are produced per week.

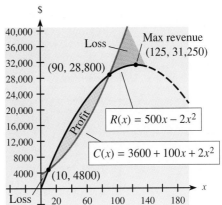

Figure 4.15

CHECKPOINT

3. The point of intersection of the revenue function and the cost function is called _____ .

4. If $C(x) = 120x + 15{,}000$ and $R(x) = 370x - x^2$, finding the break-even points requires solution of what equation?

Notice that for Example 3, the revenue function

$$R(x) = (500 - 2x)x = 500x - 2x^2$$

is a parabola that opens downward. Thus the vertex is the point at which revenue is maximum. We can locate this vertex by using the methods discussed in the previous section.

$$\text{Vertex}: x = \frac{-b}{2a} = \frac{-500}{2(-2)} = \frac{500}{4} = 125 \text{ (units)}$$

It is interesting to note that when $x = 125$, the firm achieves its maximum revenue of

$$R(125) = 500(125) - 2(125)^2 = 31{,}250 \text{ (dollars)},$$

but the costs when $x = 125$ are

$$C(125) = 3600 + 100(125) + 2(125)^2 = 47{,}350 \text{ (dollars)},$$

which results in a loss. This illustrates that maximizing revenue is not a good goal. We should seek to maximize profit.

EXAMPLE 4

If total costs are given by $C(x) = 3600 + 100x + 2x^2$ and total revenues are given by $R(x) = 500x - 2x^2$, form the profit function and find the maximum profit.

Solution

Note that these are the same cost and revenue functions that appear in the previous example.

$$\text{Profit} = \text{revenue} - \text{cost},$$

so we have

$$P(x) = (500x - 2x^2)$$
$$- (3600 + 100x + 2x^2),$$

or

$$P(x) = -3600 + 400x - 4x^2.$$

This profit function is a parabola that opens downward, so the vertex will be the maximum point.

$$\text{Vertex}: x = \frac{-b}{2a} = \frac{-400}{2(-4)}$$
$$= \frac{-400}{-8} = 50$$

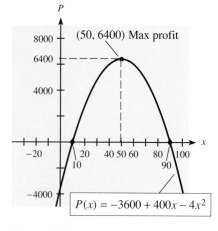

Figure 4.16

Furthermore, when $x = 50$, we have

$$P(50) = -3600 + 400(50) - 4(50)^2 = 6400 \text{ (dollars)}.$$

Thus, when 50 items are produced and sold, a maximum profit of \$6400 results (see Figure 4.16).

Comparing Figures 4.15 and 4.16, notice first that the maximum profit of $6400 at 50 units in Figure 4.16 corresponds to the vertical distance between revenue and cost at 50 units in Figure 4.15. In addition, the break-even points at $x = 10$ and $x = 90$ in Figure 4.15 correspond to x-intercepts (Profit $= 0$) in Figure 4.16.

It is important to note that the procedures for finding maximum revenue and profit in these examples depend on the fact that these functions are parabolas. Using methods discussed in Sections 4.1 and 4.2, we can use graphing utilities to locate maximum points, minimum points, and break-even points. For more general functions, procedures for finding maximum or minimum values are discussed in Chapter 10.

CHECKPOINT SOLUTIONS

1. The market equilibrium point or market equilibrium

2. Demand: $p = -\dfrac{1}{10}q^2 + 1000$; supply: $p = \dfrac{1}{10}q + 10$

   Solution of $-\dfrac{1}{10}q^2 + 1000 = \dfrac{1}{10}q + 10$

   $$-q^2 + 10{,}000 = q + 100$$
   $$0 = q^2 + q - 9900$$

3. The break-even point

4. Solution of $C(x) = R(x)$. That is, solution of $120x + 15{,}000 = 370x - x^2$, or $x^2 - 250x + 15{,}000 = 0$.

EXERCISE 4.3

SUPPLY, DEMAND, AND MARKET EQUILIBRIUM

1. Sketch the first-quadrant portions of the following on the same set of axes.
   (a) The supply function whose equation is
      $p = \frac{1}{4}q^2 + 10$
   (b) The demand function whose equation is
      $p = 86 - 6q - 3q^2$
   (c) Label the market equilibrium point.
   (d) Algebraically determine the equilibrium point for the supply and demand functions.

2. Sketch the first-quadrant portions of the following on the same set of axes.
   (a) The supply function whose equation is
      $p = q^2 + 8q + 16$
   (b) The demand function whose equation is
      $p = 216 - 2q$
   (c) Label the market equilibrium point.
   (d) Algebraically determine the equilibrium point for the supply and demand functions.

3. Sketch the first-quadrant portions of the following on the same set of axes.
   (a) The supply function whose equation is
      $p = 0.2q^2 + 0.4q + 1.8$

(b) The demand function whose equation is
$$p = 9 - 0.2q - 0.1q^2$$
(c) Label the market equilibrium point.
(d) Algebraically determine the market equilibrium point.

4. Sketch the first-quadrant portions of the following on the same set of axes.
   (a) The supply function whose equation is
   $$p = q^2 + 8q + 22$$
   (b) The demand function whose equation is
   $$p = 198 - 4q - \tfrac{1}{4}q^2$$
   (c) Label the market equilibrium point.
   (d) Algebraically determine the market equilibrium point for the supply and demand functions.

5. If the supply function for a commodity is $p = q^2 + 8q + 16$ and the demand function is $p = -3q^2 + 6q + 436$, find the equilibrium quantity and equilibrium price.

6. If the supply function for a commodity is $p = q^2 + 8q + 20$ and the demand function is $p = 100 - 4q - q^2$, find the equilibrium quantity and equilibrium price.

7. If the demand function for a commodity is given by the equation $p^2 + 4q = 1600$ and the supply function is given by the equation $300 - p^2 + 2q = 0$, find the equilibrium quantity and equilibrium price.

8. If the supply and demand functions for a commodity are given by $4p - q = 42$ and $(p + 2)q = 2100$, respectively, find the price that will result in market equilibrium.

9. If the supply and demand functions for a commodity are given by $q = p - 10$ and $q = 1200/p$, what is the equilibrium price and what is the corresponding number of units supplied and demanded?

10. If the supply and demand functions for a certain product are given by the equations $2p - q + 6 = 0$ and $(p + q)(q + 10) = 3696$, respectively, find the price and quantity that give market equilibrium.

11. The supply function for a product is $2p - q - 10 = 0$, while the demand function for the same product is $(p + 10)(q + 30) = 7200$. Find the market equilibrium point.

12. The supply and demand for a product are given by $2p - q = 50$ and $pq = 100 + 20q$, respectively. Find the market equilibrium point.

13. For the product in problem 11, if a $22 tax is placed on production of the item, then the supplier passes this tax on by adding $22 to his selling price. Find the new equilibrium point for this product when the tax is passed on. (The new supply function is given by $p = \tfrac{1}{2}q + 27$.)

14. For the product in problem 12, if a $12.50 tax is placed on production and passed through by the supplier, find the new equilibrium point.

BREAK-EVEN POINTS AND MAXIMIZATION

15. The total costs for a company are given by
$$C(x) = 2000 + 40x + x^2$$
and the total revenues are given by
$$R(x) = 130x.$$
Find the break-even points.

16. If a firm has the following cost and revenue functions,
$$C(x) = 3600 + 25x + \frac{1}{2}x^2; \quad R(x) = \left(175 - \frac{1}{2}x\right)x,$$
find the break-even points.

17. If a company has total costs given by $C(x) = 15{,}000 + 35x + 0.1x^2$ and total revenues given by $R(x) = 385x - 0.9x^2$, find the break-even points.

18. If total costs are $C(x) = 1600 + 1500x$ and total revenues are $R(x) = 1600x - x^2$, find the break-even points.

19. Given that $C(x) = 150 + x + 0.09x^2$ and $R(x) = 12.5x - 0.01x^2$, and given that production is restricted to fewer than 75 units, find the break-even points.

20. If the profit function for a firm is given by $P(x) = -1100 + 120x - x^2$ and limitations on space require that production is less than 100 units, find the break-even points.

21. Find the maximum revenue for the revenue function $R(x) = 385x - 0.9x^2$.

22. Find the maximum revenue for the revenue function $R(x) = 1600x - x^2$.

23. If, in a monopoly market, the demand for a product is $p = 175 - 0.50x$ and the revenue function is $R = px$, where x is the number of units sold, what price will maximize revenue?

24. If, in a monopoly market, the demand for a product is $p = 1600 - x$ and the revenue is $R = px$, where x is the number of units sold, what price will maximize revenue?

25. The profit function for a certain commodity is $P(x) = 110x - x^2 - 1000$. Find the level of production that yields maximum profit, and find the maximum profit.

26. The profit function for a firm making widgets is $P(x) = 88x - x^2 - 1200$. Find the number of units at which maximum profit is achieved and find the maximum profit.

27. (a) Form the profit function for the cost and revenue functions in problem 17, and find maximum profit.
    (b) Compare the level of production to maximize profit with the level to maximize revenue (see problem 21). Do they agree?
    (c) How do the break-even points compare with the zeros of $P(x)$?

28. (a) Form the profit function for the cost and revenue functions in problem 18, and find maximum profit.
    (b) Compare the level of production to maximize profit with the level to maximize revenue (see problem 22). Do they agree?
    (c) How do the break-even points compare with the zeros of $P(x)$?

29. Suppose a company has fixed costs of $28,000 and variable costs of $\frac{2}{5}x + 222$ dollars per unit, where x is the total number of units produced. Suppose further that the selling price of its product is $1250 - \frac{3}{5}x$ dollars per unit.
    (a) Find the break-even points.
    (b) Find the maximum revenue.
    (c) Form the profit function from the cost and revenue functions and find maximum profit.
    (d) What price will maximize the profit?

30. Suppose a company has fixed costs of $300 and variable costs of $\frac{3}{4}x + 1460$ dollars per unit, where x is the total number of units produced. Suppose further that the selling price of its product is $1500 - \frac{1}{4}x$ dollars per unit.
    (a) Find the break-even points.
    (b) Find the maximum revenue.
    (c) Form the profit function from the cost and revenue functions and find maximum profit.
    (d) What price will maximize the profit?

31. The following table gives the total revenues of AT&T for selected years.

Year	Total revenues (in billions)
1985	$63.13
1986	$69.906
1987	$60.53
1989	$61.1
1990	$62.191
1991	$63.089
1992	$64.904
1993	$67.156

Source: *AT&T Annual Report,* 1993

Suppose the data can be modeled by the equation

$$R(t) = 0.253t^2 - 4.03t + 76.84,$$

where t is the number of years past 1980.
(a) Use the model to find the year in which revenue was minimum and find the minimum predicted revenue.
(b) Check the result from (a) against the data in the table.
(c) Graph the model $R(t)$ and the data points from the table.
(d) Write a sentence to describe how well the data fit the model.

The data in the table give sales revenues and costs and expenses for Scott Paper Co. for various years, and is used in problems 32 and 33.

Year	Sales Revenue (in Billions)	Costs and Expenses (in Billions)
1983	$2.6155	$2.4105
1984	2.7474	2.4412
1985	2.934	2.6378
1986	3.3131	2.9447
1987	3.9769	3.5344
1988	4.5494	3.8171
1989	4.8949	4.2587
1990	5.1686	4.8769
1991	4.9593	4.9088
1992	5.0913	4.6771
1993	4.7489	4.9025

Source: Scott Paper Company, *1993 Annual Report*

32. Assume that sales revenues for Scott Paper can be modeled by

$$R(t) = -0.031t^2 + 0.776t + 0.179,$$

where t is the number of years past 1980.
   (a) Use the model to determine the year in which maximum revenue occurs and the maximum revenue it predicts.
   (b) Check the result from (a) against the data in the table.
   (c) Graph the model $R(t)$ and the data points from the table.
   (d) Write a sentence to describe how well the data fit the model.

33. Assume that costs and expenses for Scott Paper can be modeled by

$$C(t) = -0.012t^2 + 0.492t + 0.725,$$

where t is the number of years past 1980.
   (a) Use $R(t)$ as given in problem 32 and form the profit function (as a function of time).
   (b) Use the model (from part (a)) to find the year in which maximum profit occurs.
   (c) Graph the profit function from part (b) and the data points from the table.
   (d) Through the mid to late 1990s, does the model project increasing or decreasing profits? Does the data support this trend (as far as it goes)?
   (e) How might management respond to this kind of projection?

KEY TERMS
& FORMULAS

Section	Key Term	Formula
4.1	Quadratic equation	$ax^2 + bx + c = 0$
	Quadratic formula	$x = \dfrac{-b \pm \sqrt{b^2 - 4ac}}{2a}$
	Quadratic discriminant	$b^2 - 4ac$
4.2	Quadratic function; parabola	$y = f(x) = ax^2 + bx + c$
	Vertex of a parabola	$(-b/2a, \quad f(-b/2a))$ Maximum point if $a < 0$ Minimum point if $a > 0$
	Axis of symmetry	$x = -b/2a$
	Zeros of a quadratic function	$x = \dfrac{-b \pm \sqrt{b^2 - 4ac}}{2a}$
4.3	Supply function Demand function Market equilibrium Cost, revenue and profit functions Break-even points	$P(x) = R(x) - C(x)$ $R(x) = C(x)$ or $P(x) = 0$

REVIEW EXERCISES

4.1

In problems 1–10, find the real solutions to each quadratic equation.

1. $3x^2 + 10x = 5x$
2. $4x - 3x^2 = 0$
3. $x^2 + 5x + 6 = 0$
4. $11 - 10x - 2x^2 = 0$
5. $(x - 1)(x + 3) = -8$
6. $4x^2 = 3$
7. $20x^2 + 3x = 20 - 15x^2$
8. $8x^2 + 8x = 1 - 8x^2$
9. $7 = 2.07x - 0.02x^2$
10. $46.3x - 117 - 0.5x^2 = 0$

In problems 11–14, solve each equation by using a graphing utility to find the zeros of a function. Solve the equation algebraically to check your results.

11. $4z^2 + 25 = 0$
12. $z(z + 6) = 27$
13. $3x^2 - 18x - 48 = 0$
14. $3x^2 - 6x - 9 = 0$
15. Solve $x^2 + ax + b = 0$ for x.
16. Solve $xr^2 - 4ar - x^2c = 0$ for r.

In problems 17 and 18, approximate the real solutions to each quadratic equation to two decimal places.

17. $23.1 - 14.1x - 0.002x^2 = 0$
18. $1.03x^2 + 2.02x - 1.015 = 0$

4.2

For each function in problems 19–24, find the vertex and determine if it is a maximum or minimum point, find the zeros if they exist, and sketch the graph.

19. $y = \frac{1}{2}x^2 + 2x$
20. $y = 4 - \frac{1}{4}x^2$
21. $y = 6 + x - x^2$
22. $y = x^2 - 4x + 5$
23. $y = x^2 + 6x + 9$
24. $y = 12x - 9 - 4x^2$

In problems 25–30, use a graphing utility to graph each function. Use the vertex and zeros to determine an appropriate range. Be sure to label the maximum or minimum point.

25. $y = \frac{1}{3}x^2 - 3$
26. $y = \frac{1}{2}x^2 + 2$
27. $y = x^2 + 2x + 5$
28. $y = -10 + 7x - x^2$
29. $y = 20x - 0.1x^2$
30. $y = 50 - 1.5x + 0.01x^2$

In problems 31–34, a graph is given. Use the graph to
(a) locate the vertex,
(b) determine the zeros,
(c) match the graph with one of the equations A, B, C, or D.

A. $y = 7x - \frac{1}{2}x^2$
B. $y = \frac{1}{2}x^2 - x - 4$
C. $y = 8 - 2x - x^2$
D. $y = 49 - x^2$

31.

32.

33.

34.
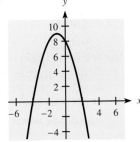

APPLICATIONS

4.2

35. **Maximum area** A rectangular lot is to be fenced in, then divided down the middle to create two identical fenced lots (see figure). If 1200 ft of fence is to be used, the area of each lot is given by

$$A = x\left(\frac{1200 - 3x}{4}\right).$$

(a) Find the x value that maximizes this area.
(b) Find the maximum area.

4.3

36. **Supply** Graph the first-quadrant portion of the supply function

$$p = 2q^2 + 4q + 6.$$

37. **Demand** Graph the first-quadrant portion of the demand function

$$p = 18 - 3q - q^2.$$

38. **Market equilibrium**

(a) Suppose the supply function for a product is $p = 0.1q^2 + 1$ and the demand function is $p = 85 - 0.2q - 0.1q^2$. Sketch the first-quadrant portion of the graph of each function. Use the same set of axes for both and label the market equilibrium point.

(b) Use algebraic methods to find the equilibrium price and quantity.

39. **Market equilibrium** The supply function for a product is given by $p = q^2 + 300$, and the demand is given by $p + q = 410$. Find the equilibrium quantity and price.

40. **Market equilibrium** If the demand function for a commodity is given by the equation $p^2 + 5q = 200$ and the supply function is given by $40 - p^2 + 3q = 0$, find the equilibrium quantity and price.

41. **Break-even points** If total costs for a product are given by $C(x) = 1760 + 8x + 0.6x^2$ and total revenues are given by $R(x) = 100x - 0.4x^2$, find the break-even points.

42. **Break-even points** If total costs for a commodity are given by $C(x) = 900 + 25x$ and total revenues are given by $R(x) = 100x - x^2$, find the break-even points.

43. **Maximizing revenue and profit** Find the maximum revenue and maximum profit for the functions described in problem 42.

44. **Break-even points and profit maximization** Given total profit $P(x) = 1.3x - 0.01x^2 - 30$, find maximum profit and the break-even points and sketch the graph.

45. **Maximum profit** Given $C(x) = 360 + 10 + 0.2x^2$ and $R(x) = 50x - 0.2x$, find the level of production that gives maximum profit and find the maximum profit.

46. **Break-even points and profit maximization** A certain company has fixed costs of $15,000 for its product and variable costs given by $140 + 0.04x$ dollars per unit, where x is the total number of units. The selling price of the product is given by $300 - 0.06x$ dollars per unit.

(a) Formulate the functions for total cost and total revenue.

(b) Find the break-even points.

(c) Find the level of sales that maximizes revenue.

(d) Form the profit function and find the level of production and sales that maximizes profit.

(e) Find the profit (or loss) at the production levels found in parts (c) and (d).

CONSULTANT'S CORNER

Maximizing Profit

The Digital Electronics Company has monopolistic control in the marketplace for its new product. Its average daily costs for the product are given by

$$\overline{C} = \frac{8000}{x} + 1000 + x,$$

where x is the number of units produced and sold, and the daily demand for its product is

$$x = 150 - \frac{p}{40},$$

where x is the number of units demanded and p is the price.

For the short run, management wants to know how many units to produce per day to maximize its daily profit. In addition, Digital management expects Congress to impose a new excise tax of $100 per unit on its product, so they want to know how many units will maximize daily profit after the tax is imposed and how daily profit will be affected by the tax.

To assist the company, find the following.

1. The daily total cost function.
2. The daily revenue function. (Recall that $R = R(x) = px$.)
3. The daily profit function.
4. The number of units, x, that maximizes the profit function and the maximum daily profit.
5. The new total cost function after the tax is added and the new profit function.
6. The number of units that maximizes the new profit function and the new maximum daily profit.
7. The difference in daily profit caused by the tax.

Warm-up

Prerequisite problem type	For section	Answer	Section for review
Write the following with positive exponents: (a) x^{-3} (b) $\dfrac{1}{x^{-2}}$ (c) $\sqrt{x}$	**5.1** **5.2** **5.3**	(a) $\dfrac{1}{x^3}$ (b) x^2 (c) $x^{1/2}$	0.3, 0.4 Exponents and radicals
Simplify: (a) 2^0 (b) x^0 $(x \neq 0)$ (c) $49^{1/2}$ (d) 10^{-2}	**5.1** **5.2** **5.3**	(a) 1 (b) 1 (c) 7 (d) $\frac{1}{100}$	0.3, 0.4 Exponents
Answer true or false. (a) $(\frac{1}{2})^x = 2^{-x}$ (b) $\sqrt{50} = 50^{1/2}$ (c) If $8 = 2^y$, then $y = 4$. (d) If $x^3 = 8$, then $x = 2$.	**5.1** **5.2**	(a) True (b) True (c) False; $y = 3$ (d) True	0.3, 0.4 Exponents and radicals
(a) If $f(x) = 2^{-2x}$, what is $f(-2)$? (b) If $f(x) = 2^{-2x}$, what is $f(1)$? (c) If $f(t) = (1 + 0.02)^t$, what is $f(0)$? (d) If $f(t) = 100(0.03)^{0.02t}$, what is $f(0)$?	**5.1** **5.2** **5.3**	(a) 16 (b) $\frac{1}{4}$ (c) 1 (d) 3	1.4 Functional notation

CHAPTER

EXPONENTIAL AND LOGARITHMIC FUNCTIONS

In this chapter we study exponential and logarithmic functions, which provide models for many applications that at first seem remote and unrelated. For example, a business manager uses these functions to study the growth of money or corporations or the decay of new sales volume. A social scientist uses these functions to study the growth of organizations or populations or the dating of fossilized remains. And a biologist uses these functions to study the growth of microorganisms in a laboratory culture, the spread of disease, the measurement of pH, or the decay of radioactive material.

In our study of exponential and logarithmic functions, we will examine their descriptions, their properties, their graphs, and the special inverse relationship between these two functions. We will see how these functions are applied to some of the concerns of social scientists, business managers, and life scientists. In these applications, the inverse relationship of the exponential and logarithmic functions is used to solve some of the equations that arise. We will assume that you have a calculator that computes powers of e and logarithms to the base e (denoted e^x and $\ln x$, respectively).

5.1 EXPONENTIAL FUNCTIONS

Objective

❏ To graph exponential functions

Table 5.1

Minutes Passed	Number of Organisms
0	1
1	2
2	4
3	8
4	16

APPLICATION PREVIEW

Suppose a culture of bacteria has the characteristic that each minute, every microorganism splits into two new organisms. We can describe the number of bacteria in the culture as a function of time. That is, if we begin the culture with one microorganism, we know that after one minute we will have two organisms, after two minutes, four, and so on. Table 5.1 gives a few of the values that describe this growth. If x represents the number of minutes that have passed and y represents the number of organisms, the points (x, y) lie on the graph of the function with equation

$$y = 2^x.$$

The equation $y = 2^x$ in the Application Preview is an example of a special group of functions called **exponential functions.** In general, we define these functions as follows:

Exponential Functions	If a is a positive real number and $a \neq 1$, then the function

$$f(x) = a^x$$

is an **exponential function.**

A table of some values satisfying $y = 2^x$ and the graph of this function are given in Figure 5.1. This function is said to model the growth of the number of organisms in the Application Preview discussion, even though some points on the graph do not correspond to a time and a number of organisms. For example, time x could not be negative, and the number of organisms y could not be fractional.

We have defined rational powers of x in terms of radicals in Chapter 0, so 2^x makes sense for any rational power x. It can also be shown that the laws of exponents apply for irrational numbers. We will assume that if we graphed $y = 2^x$ for irrational values of x, those points would lie on the curve in Figure 5.1. Thus in general, we can graph an exponential function by plotting easily calculated points, such as those in the table in Figure 5.1, and drawing a smooth curve through the points.

x	$y = 2^x$
-3	$2^{-3} = \frac{1}{8}$
-2	$2^{-2} = \frac{1}{4}$
-1	$2^{-1} = \frac{1}{2}$
0	$2^0 = 1$
1	$2^1 = 2$
2	$2^2 = 4$
3	$2^3 = 8$

EXAMPLE I
Graph $y = 10^x$.

Solution
A table of values and the graph are given in Figure 5.2.

x	y
-3	$10^{-3} = 1/1000$
-2	$10^{-2} = 1/100$
-1	$10^{-1} = 1/10$
0	$10^0 = 1$
1	$10^1 = 10$
2	$10^2 = 100$
3	$10^3 = 1000$

Figure 5.1

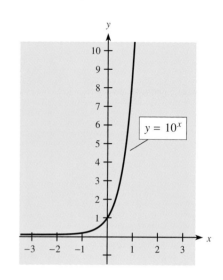

Figure 5.2

We also could have used a graphing utility to graph $y = 2^x$ and $y = 10^x$. Figure 5.3 shows graphs of $f(x) = 1.5^x$, $f(x) = 4^x$, and $f(x) = 20^x$ that were obtained using a graphing utility. Notice the range for each graph in the figure.

Figure 5.3

Notice that the three graphs in the figure and the graphs of $y = 2^x$ and $y = 10^x$ are similar. The graphs of $y = 2^x$ and $y = 10^x$ clearly approach, but do not touch, the negative x-axis. However, the graphs in Figure 5.3 appear as if they might eventually merge with the negative x-axis. But, by adjusting the range, we would see that these graphs also approach, but do not touch, the negative x-axis. That is, the negative x-axis is an asymptote for these functions.

In fact, the shapes of the graphs of functions of the form $y = f(x) = a^x$, with $a > 1$, are similar to those in Figures 5.1, 5.2, and 5.3. Exponential functions of this type are called **exponential growth functions** because they are used to model growth in diverse applications. Their graphs have the basic shape shown in the following box.

Graphs of Exponential Growth Functions

Function: $y = f(x) = a^x$ $(a > 1)$
y-intercept: $(0, 1)$
Domain: All reals
Range: All positive reals
Asymptote: x-axis (negative half)

Basic shape

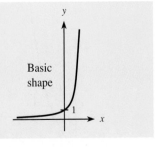

One exponential growth model concerns money invested at compound interest. Often we seek to evaluate rather than graph these functions.

EXAMPLE 2

If $10,000 is invested at 6%, compounded monthly, then the future value of the investment S after x years is given by

$$S = 10{,}000(1.005)^{12x}.$$

Find the future value of the investment after (a) 5 years and (b) 30 years.

Solution

These future values can be found with a calculator.

(a) $S = 10{,}000(1.005)^{12(5)}$
$= 10{,}000(1.005)^{60}$
$= \$13{,}488.50$ (nearest cent)

(b) $S = 10{,}000(1.005)^{12(30)}$
$= 10{,}000(1.005)^{360}$
$= \$60{,}225.75$ (nearest cent)

Notice that the amount after 30 years is significantly more than the amount after 5 years, a result consistent with exponential growth models.

CHECKPOINT

1. Can any value of x give a negative value for y if $y = a^x$ and $a > 1$?

2. If $a > 1$, what asymptote does the graph of $y = a^x$ approach?

A special function that occurs frequently in economics and biology is $y = e^x$, where e is a fixed irrational number (approximately 2.71828 . . .). We will see how e arises in Section 6.2 when we discuss interest that is compounded continuously, and we will formally define e in Section 9.2.

Since $e > 1$, the graph of $y = e^x$ will have the same basic shape as other growth exponentials. We can calculate the y-coordinate for points on the graph of this function with a calculator. A table of some values (with y-values rounded to two places) and the graph are shown in Figure 5.4.

Exponential functions with base e often arise in natural ways. As we will see in Section 6.2, the growth of money that is compounded continuously is given by $S = Pe^{jt}$, where P is the original principal, j the annual interest rate, and t the time in years. Certain populations (of insects, for example) grow exponentially, and the number of individuals can be closely approximated by the equation $y = P_0e^{ht}$, where P_0 is the original population size, h is a constant that depends on the type of population, and y is the population size at any instant t. These are both examples of exponential growth with base e.

In fact, many types of data can be modeled using an exponential growth function. Figure 5.5(a) shows a graph of amounts of carbon in the atmosphere due to emissions from fossil fuel burning. With curve fitting tools available with some computer software or on some graphing calculators, we can develop an equation that models, or approximates, this data. The equation model and its graph are shown in Figure 5.5(b).

x	$y = e^x$
-2	0.14
-1	0.37
0	1.00
1	2.72
2	7.39

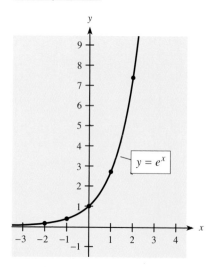

$y = e^x$

Figure 5.4

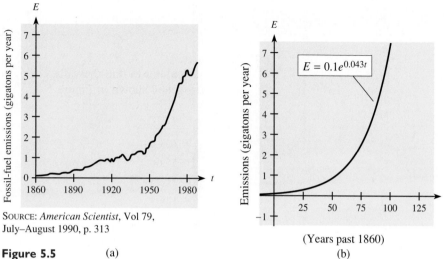

SOURCE: *American Scientist*, Vol 79,
July–August 1990, p. 313

Figure 5.5   (a)   (b)

    Exponentials whose bases are between 0 and 1, such as $y = (\frac{1}{2})^x$, have graphs different from those of the exponentials just discussed. Using the properties of exponents, we have

$$y = \left(\frac{1}{2}\right)^x = (2^{-1})^x = 2^{-x}.$$

This suggests that exponentials of the form $y = b^x$, where $0 < b < 1$, can be rewritten in the form $y = a^{-x}$, where $a > 1$.

EXAMPLE 3

Graph $y = 2^{-x}$.

Solution

A table of values and the graph are given in Figure 5.6.

x	$y = 2^{-x}$
-3	$2^3 = 8$
-2	$2^2 = 4$
-1	$2^1 = 2$
0	$2^0 = 1$
1	$2^{-1} = \frac{1}{2}$
2	$2^{-2} = \frac{1}{4}$
3	$2^{-3} = \frac{1}{8}$

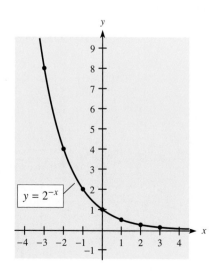

Figure 5.6

EXAMPLE 4

Graph $y = e^{-2x}$.

Solution

Using a calculator to find the values of powers of e (to two decimal places), we get the graph shown in Figure 5.7.

x	y
-3	$e^6 = 403.43$
-2	$e^4 = 54.60$
-1	$e^2 = 7.39$
0	$e^0 = 1.00$
1	$e^{-2} = 0.14$
2	$e^{-4} = 0.02$

Figure 5.7

We can also use a graphing utility to graph exponential functions like those in Examples 3 and 4. Additional examples of functions of this type would yield graphs with the same basic shape. **Exponential decay functions** have the form $y = a^{-x}$, where $a > 1$, model decay for various phenomena, and their graphs have the characteristics and shape shown in the following box.

Graphs of Exponential Decay Functions

Function: $y = f(x) = a^{-x}$ $(a > 1)$
y-intercept: $(0, 1)$
Domain: All reals
Range: All positive reals
Asymptote: x-axis (positive half)

Basic shape

One example of exponential decay is the decrease in the number of atoms of a radioactive element. The number of atoms y present at an instant in time t is given by

$$y = w_0 e^{-h(t-t_0)},$$

where w_0 is the number of atoms at time t_0 and h is a constant that depends on the element. There are numerous other applications of decay exponentials.

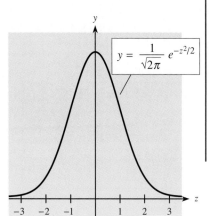

Standard normal probability curve

Figure 5.8

EXAMPLE 5

The purchasing power P of a fixed income of \$30,000 per year (such as a pension) after t years of 4% inflation can be modeled by

$$P = 30{,}000e^{-0.04t}.$$

Find the purchasing power after (a) 5 years and (b) 20 years.

Solution

We can use a calculator to answer both parts.

(a) $P = 30{,}000e^{-0.04(5)} = 30{,}000e^{-0.2} = \$24{,}562$ (nearest dollar)

(b) $P = 30{,}000e^{-0.04(20)} = 30{,}000e^{-0.8} = \$13{,}480$ (nearest dollar)

  Notice that the impact of inflation over time significantly erodes the purchasing power and provides some insight into the plight of elderly people who live on a fixed income.

  There are some important exponential functions that use base e, but whose graphs are different from those we have discussed. For example, the standard normal probability curve (often referred to as a bell-shaped curve) is the graph of an exponential function with base e (see Figure 5.8). Later in this chapter we will study other exponential functions that model growth, but whose graphs are also different from those discussed previously.

CHECKPOINT

3. True or false: The graph of $y = \left(\dfrac{1}{a}\right)^x$, with $a > 1$, is the same as the graph of $y = a^{-x}$.

4. True or false: The graph of $y = a^{-x}$, with $a > 1$, approaches the positive x-axis as an asymptote.

5. True or false: The graph of $y = 2^{-x}$ is the graph of $y = 2^x$ reflected about the y-axis.

 Graphing Utilities

Graphing utilities allow us to investigate variations of the growth and decay exponentials. For example, we can examine the similarities and differences between the graphs of $y = f(x) = a^x$ and $y = mf(x)$, $y = f(kx)$, $y = f(x + h)$, and $y = f(x) + C$ for constants m, k, h, and C. We also can explore these similarities and differences for $y = f(x) = a^{-x}$.

EXAMPLE 6

(a) Use a graphing utility to graph $y = f(x) = e^x$ and $y = mf(x) = me^x$ for $m = -6, -3, 2,$ and 10.

(b) Explain the effect that m has on the shape of the graph, the asymptote, and the y-intercept.

Solution
(a) The graphs are shown in Figure 5.9.
(b) From the graphs, we see that when $m > 0$, the shape of the graph is still that of a growth exponential. When $m < 0$, the shape is that of a growth exponential turned upside down, or reflected through the x-axis (making the range $y < 0$). In all cases, the asymptote is unchanged and is the negative x-axis. In each case, the y-intercept is $(0, m)$.

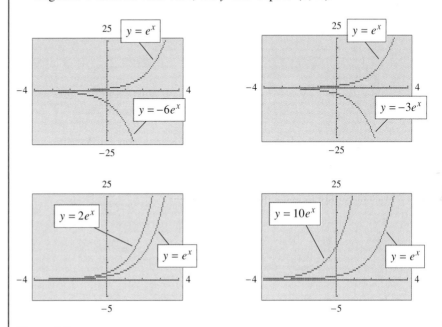

Figure 5.9

CHECKPOINT SOLUTIONS

1. No, all values of y are positive.
2. The left side of the x-axis ($y = 0$)
3. True   4. True   5. True

EXERCISE 5.1

In problems 1–10, use a calculator or table to evaluate each expression.

1. $10^{0.5}$
2. $10^{3.6}$
3. $5^{-2.7}$
4. $8^{-2.6}$
5. $3^{1/3}$
6. $2^{11/6}$
7. e^2
8. e^5
9. e^{-3}
10. $e^{-1.5}$

In problems 11–22, graph each function.

11. $y = 4^x$
12. $y = 8^x$
13. $y = 2(x^3)$
14. $y = 3(2^x)$
15. $y = 5^x$
16. $y = 3^{x-1}$
17. $y = e^x$
18. $y = 2e^x$

19. $y = 3^{-x}$

20. $y = 3^{-2x}$

21. $y = e^{-x}$

22. $y = \frac{1}{3}e^{x}$

In problems 23–26, use a graphing utility to make the graphs.

23. Given $f(x) = e^{-x}$. Graph $y = f(x)$ and $y = f(kx) = e^{-kx}$ for each k, where $k = 0.1, 0.5, 2,$ and 5. Explain the effect that different values of k have on the graph.

24. Given $f(x) = 2^{-x}$. Graph $y = f(x)$ and $y = mf(x) = m(2^{-x})$ for each m, where $m = -7, -2, 3,$ and 8. Explain the effect that different values of m have on the graph.

25. Given $f(x) = 4^{x}$. Graph $y = f(x)$ and $y = f(x) + C = 4^{x} + C$ for each C, where $C = -5, -2, 3,$ and 6. Explain the effect that C has on the graph.

26. Given $f(x) = 3^{x}$. Graph $y = f(x)$ and $y = f(x - h) = 3^{x-h}$ for each h, where $h = 4, 1, -2,$ and -5. Explain the effect that h has on the graph.

For problems 27 and 28, let $f(x) = c(1 + e^{-ax})$ with $a > 0$. Use a graphing utility to make the requested graphs.

27. (a) Fix $a = 1$ and graph $y = f(x) = c(1 + e^{-x})$ for $c = 10, 50,$ and 100.

(b) What effect does c have on the graphs?

28. (a) Fix $c = 50$ and graph $y = f(x) = 50(1 + e^{-ax})$ for $a = 0.1, 1,$ and 10.

(b) What effect does a have on the graphs?

For problems 29 and 30, let $f(x) = \dfrac{A}{1 + ce^{-x}}$. Use a graphing utility to make the requested graphs.

29. (a) Fix $A = 100$ and graph $y = f(x) = \dfrac{100}{1 + ce^{-x}}$ for $c = 0.25, 1, 9,$ and 49.

(b) What effect does c have on the graphs?

30. (a) Fix $c = 1$ and graph $y = f(x) = \dfrac{A}{1 + e^{-x}}$ for $A = 50, 100,$ and 150.

(b) What effect does A have on the graphs?

APPLICATIONS

31. **Social Security benefits** The graph in the figure shows the number of active workers who will be (or have been) supporting each Social Security beneficiary. What type of function might be an appropriate model for this situation?

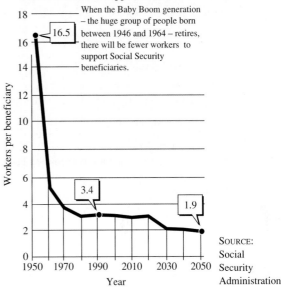

Fewer workers to support retirees

When the Baby Boom generation – the huge group of people born between 1946 and 1964 – retires, there will be fewer workers to support Social Security beneficiaries.

SOURCE: Social Security Administration

32. **Postal rates** The graph in the figure shows the history of first-class postage stamp rates from 1890 to 1995. If this data were modeled by a smooth curve, what type of function might be appropriate?

FIRST-CLASS POSTAGE RATES

SOURCE: *Oil City Derrick*, Oil City, PA, December 1, 1994

33. **Compound interest** If $1000 is invested for x years at 8%, compounded quarterly, the future value that will result is

$$S = 1000(1.02)^{4x}.$$

What amount will result in 8 years?

34. **Compound interest** If $3200 is invested for x years at 8%, compounded quarterly, the interest earned is

$$I = 3200(1.02)^{4x} - 3200.$$

What interest is earned in 5 years?

Population growth Use the following information to answer problems 35–38. World population can be considered as growing according to the equation

$$N = N_0(1 + r)^t,$$

where N_0 is the number of individuals at time $t = 0$, r is the yearly rate of growth, and t is the number of years.

35. Sketch the graph for $t = 0$ to $t = 10$ when the growth rate is 2% and N_0 is 4.1 billion.

36. Sketch the graph for $t = 0$ to $t = 10$ when the growth rate is 3% and N_0 is 4.1 billion.

37. Repeat problem 35 when the growth rate is 5%.

38. Repeat problem 36 when the growth rate is 7%.

39. **Chemical reactions** The number of molecules of a certain substance that have enough energy to activate a reaction is given by

$$y = 100,000e^{-1/x},$$

where y is the number of molecules and x is the (absolute) temperature of the substance. Plot the graph of this equation for $x > 0$.

40. **Newton's law of cooling** When a body is moved from one medium to another, its temperature T will change according to the equation

$$T = T_0 + ce^{kt},$$

where T_0 is the temperature of the new medium, c the temperature difference between the mediums (old minus new), t the time in the new medium, and k a constant. Given that $T_0 = 70$, $c = 23$, and $k = -0.2$, sketch the graph for $t \geq 0$.

41. **Compound interest** We will show in Chapter 6 that if P is invested for n years at 10% compounded continuously, the future value of the investment is given by

$$S = Pe^{0.1n}.$$

Use $P = 1000$ and graph this function for $0 \leq n \leq 20$.

42. **Compound interest** If $1000 is invested for x years at 10%, compounded continuously, the future value that results is

$$S = 1000e^{0.10x}.$$

What amount will result in 5 years?

43. **Compound interest** If $10,000 is invested for x years at 12%, compounded continuously, the interest earned is

$$I = 10,000e^{0.12x} - 10,000.$$

What interest is earned in $1\frac{1}{2}$ years?

44. **Bacterial growth** A single bacterium splits into two bacteria every half hour, so that the number of bacteria in a culture quadruples every hour. Thus the equation by which a colony of 10 bacteria multiplies in t hours is given by

$$y = 10(4^t).$$

Graph this equation for $0 \leq t \leq 8$.

45. **Drug in the bloodstream** The concentration y of a certain drug in the bloodstream at any time t is given by the equation

$$y = 100(1 - e^{-0.462t}).$$

Graph this equation for $0 \leq t \leq 10$. Write a sentence that interprets the graph.

46. **Product reliability** A statistical study shows that the fraction of television sets of a certain brand that are still in service after x years is given by $f(x) = e^{-0.15x}$. Graph this equation for $0 \leq x \leq 10$. Write a sentence that interprets the graph.

47. **Atmospheric pressure** The atmospheric pressure P (in lb/sq in.) is given by

$$P = 14.7e^{-0.2x},$$

where x is the number of miles above sea level. Graph this equation for $0 \leq x \leq 20$.

48. **Advertising and sales** Suppose that sales are related to advertising expenditures according to one of the following two models, where S_1 and S_n are sales and x is advertising, all in millions of dollars.

$$S_1 = 30 + 20x - 0.4x^2$$
$$S_n = 24.58 + 325.18(1 - e^{-x/14})$$

(a) Graph both of these functions on the same set of axes. Use a graphing utility.

(b) Do these two functions give approximately the same sales per million dollars of advertising for $0 \leq x \leq 20$?

(c) How do these functions differ for $x > 20$? Which more realistically represents the relationship between sales and advertising expenditures after $20 million is spent on advertising? Why?

49. ***Revenues from wireless technology*** According to data in *USA Today* (March 1, 1994), the revenue R from wireless technology can be modeled by

$$R = 0.572e^{0.3860t},$$

where t is the number of years since 1985 and R is in billions of dollars. Graph this equation with a graphing utility to show the graph through the year 2010.

50. ***National health care expenditures*** Using data from the Congressional Budget Office (reported in *Newsweek*, October 4, 1993), the national health care expenditure H can be modeled by

$$H = 45e^{0.0898t},$$

where t is the number of years from 1960 and H is in billions of dollars. Graph this equation with a graphing utility to show the graph through the year 2010.

51. ***U.S. debt*** The following table shows the U.S. national debt for selected years from 1900 to 1989. Assume that the amount of the national debt (in billions of dollars) can be modeled by the function

$$d(x) = 1.53361e^{0.0853458x},$$

where x is the number of years past 1900. Graph this function and the data points. Use a graphing utility, and use the table to set the range.

Year	U.S. Debt (in Billions)	Year	U.S. Debt (in Billions)
1900	$1.2	1955	272.8
1910	1.1	1965	313.8
1920	24.2	1975	533.2
1930	16.1	1985	1823.1
1940	43.0	1987	2350.3
1945	258.7	1989	2857.4

SOURCE: *World Almanac*, 1991

52. ***Dow-Jones industrial average*** The following table shows the high values achieved by the Dow-Jones industrial average for the years from 1979 to 1989. Assume that the high h can be modeled with the function

$$h(x) = 821.958356e^{0.119314418x},$$

where x is the number of years past 1900. Graph this function and the data points. Use a graphing utility, and use the table to set the range.

Year	Dow-Jones High	Year	Dow-Jones High
1979	897	1985	1553
1980	1000	1986	1955
1981	1024	1987	2722
1982	1070	1988	2183
1983	1287	1989	2791
1984	1286		

SOURCE: *World Almanac*, 1991

5.2 LOGARITHMIC FUNCTIONS AND THEIR PROPERTIES

Objectives

❏ To convert equations for logarithmic functions from logarithmic to exponential form and vice versa
❏ To evaluate some special logarithms
❏ To graph logarithmic functions
❏ To use properties of logarithmic functions to simplify expressions involving logarithms
❏ To use the change of base formula

APPLICATION PREVIEW

When interest is compounded continuously, if P dollars are invested at an annual rate j (as a decimal), then the future value of the investment after t years is given by

$$S = Pe^{jt}.$$

A common question with investments such as this is how long before the investment doubles, that is, when does $S = 2P$. To answer this question, and hence to develop a "doubling time" formula, requires that we solve for t, and this requires the use of **logarithmic functions**.

LOGARITHMIC FUNCTIONS AND GRAPHS

Before the development and easy availability of calculators and computers, certain arithmetic computations, such as $(1.37)^{13}$ and $\sqrt[16]{3.09}$, were difficult to perform. The computations could be performed relatively easily using **logarithms,** which were developed in the seventeenth century by John Napier, or by using a slide rule, which is based on logarithms. The use of logarithms as a computing technique has all but disappeared today, but the study of **logarithmic functions** is still very important because of the many applications of these functions.

For example, let us again consider the culture of bacteria described at the beginning of Section 5.1. If we know that the culture is begun with one microorganism and that each minute every microorganism present splits into two new ones, then we can find the number of minutes it takes until there are 1024 organisms by solving

$$1024 = 2^y.$$

The solution of this equation may be written in the form

$$y = \log_2 1024,$$

which is read "*y* equals the logarithm of 1024 to the base 2."

In general, we may express the equation $x = a^y (a > 0, a \neq 1)$ in the form $y = f(x)$ by defining a **logarithmic function.**

Logarithmic Function   For $a > 0$ and $a \neq 1$, the **logarithmic function**

$$y = \log_a x \qquad \text{(logarithmic form)}$$

has domain $x > 0$, and is defined by

$$a^y = x \qquad \text{(exponential form)}.$$

The *a* is called the **base** in both $\log_a x = y$ and $a^y = x$, and *y* is the *logarithm* in $\log_a x = y$ and the *exponent* in $a^y = x$. Thus **a logarithm is an exponent.**

Table 5.2 shows some logarithmic equations and their equivalent exponential forms.

Table 5.2

Logarithmic Form	*Exponential Form*
$\log_{10} 100 = 2$	$10^2 = 100$
$\log_{10} 0.1 = -1$	$10^{-1} = 0.1$
$\log_2 x = y$	$2^y = x$
$\log_a 1 = 0 \quad (a > 0)$	$a^0 = 1$
$\log_a a = 1 \quad (a > 0)$	$a^1 = a$

EXAMPLE 1

(a) Write $64 = 4^3$ in logarithmic form.

(b) Write $\log_4 \left(\frac{1}{64}\right) = -3$ in exponential form.

(c) If $4 = \log_2 x$, find *x*.

Solution

(a) $64 = 4^3$ is equivalent to $3 = \log_4 64$.

(b) $\log_4 \left(\frac{1}{64}\right) = -3$ is equivalent to $4^{-3} = \frac{1}{64}$.

(c) If $4 = \log_2 x$, then $2^4 = x$ and $x = 16$.

EXAMPLE 2

Evaluate:

(a) $\log_2 8$ (b) $\log_3 9$ (c) $\log_5(\frac{1}{25})$

Solution

(a) If $y = \log_2 8$, then $8 = 2^y$. Since $2^3 = 8$, $\log_2 8 = 3$.

(b) If $y = \log_3 9$, then $9 = 3^y$. Since $3^2 = 9$, $\log_3 9 = 2$.

(c) If $y = \log_5(\frac{1}{25})$, then $\frac{1}{25} = 5^y$. Since $5^{-2} = \frac{1}{25}$, $\log_5(\frac{1}{25}) = -2$.

EXAMPLE 3

In the Application Preview we noted that a doubling time formula for a continuously compounded investment can be found by solving for t in $S = Pe^{jt}$ when $S = 2P$. That is, we must solve $2P = Pe^{jt}$, or equivalently $2 = e^{jt}$. Express $2 = e^{jt}$ in logarithmic form and then solve for t to find the doubling time formula.

Solution

In logarithmic form $2 = e^{jt}$ is equivalent to $\log_e 2 = jt$. Solving for t gives the doubling time formula

$$t = \frac{\log_e 2}{j}.$$

We can use the exponential form of a logarithmic equation to sketch its graph.

EXAMPLE 4

Graph $y = \log_2 x$.

Solution

We may graph $y = \log_2 x$ by graphing $x = 2^y$. The table of values (found by substituting values for y and calculating x) and the graph are shown in Figure 5.10.

$x = 2^y$	y
$\frac{1}{8}$	-3
$\frac{1}{4}$	-2
$\frac{1}{2}$	-1
1	0
2	1
4	2
8	3

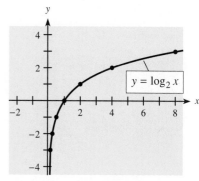

Figure 5.10

From the definition of logarithms, we see that every logarithm has a base. Most applications of logarithms involve logarithms to the base 10 (called **common logarithms**) or logarithms to the base e (called **natural logarithms**). In fact, logarithms to the base 10 and to the base e are the only ones that have function keys on scientific calculators. Thus it is important to be familiar with their names and designations.

Common and Natural Logarithms	Common logarithms: $\log x$ means $\log_{10} x$. Natural logarithms: $\ln x$ means $\log_e x$.

Values of the natural logarithm function are usually found with a scientific calculator. In light of this fact and the new notation for natural logarithms, we can rewrite the doubling time formula from Example 3 as

$$t = \frac{\log_e 2}{j} = \frac{\ln 2}{j} \approx \frac{0.693}{j}.$$

Thus, the doubling time for a continuously compounded investment at 6.5% is

$$t \approx \frac{0.693}{0.065} \approx 10.66 \text{ years.}$$

EXAMPLE 5

Suppose that after a company introduces a new product, the number of months m before their market share is s percent can be modeled by

$$m = 20 \ln\left(\frac{40}{40 - s}\right).$$

When will their product have a 35% share of the market?

Solution

A 35% market share means $s = 35$. Hence

$$m = 20 \ln\left(\frac{40}{40 - s}\right)$$

$$= 20 \ln\left(\frac{40}{40 - 35}\right) = 20 \ln\left(\frac{40}{5}\right) = 20 \ln(8) \approx 41.6.$$

Thus, after about 41.6 months, their market share will be 35%.

EXAMPLE 6

Graph $y = \ln x$.

Solution

We can graph $y = \ln x$ by writing the equation in its exponential form, $x = e^y$, and then finding x for integer values of y. We can also graph by evaluating $y = \ln x$ for $x > 0$ (including some values $0 < x < 1$) with a calculator. The graph is shown in Figure 5.11.

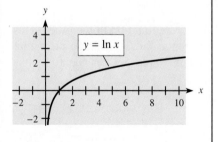

y	$x = e^y$
-3	0.050
-2	0.135
-1	0.368
0	1.000
1	2.718
2	7.389
3	20.086

x	$y = \ln x$
0.05	-3.000
0.10	-2.303
0.50	-0.693
1	0.000
2	0.693
3	1.099
5	1.609
10	2.303

Figure 5.11

Note that because $\ln x$ is a standard function on a graphing utility, such a utility could be used to obtain its graph. Note also that the graphs of $y = \log_2 x$ (Figure 5.10) and $y = \ln x$ (Figure 5.11) are very similar. The shapes of graphs of equations of the form $y = \log_a x$ with $a > 1$ are similar to these two graphs.

Graphs of Logarithmic Functions

Equation: $y = \log_a x$ $(a > 1)$
x-intercept: $(1, 0)$
Domain: All positive reals
Range: All reals
Asymptote: y-axis (negative half)

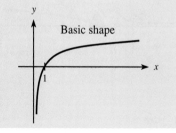

The basic shape of a logarithmic function is important for two reasons. First, when we graph a logarithmic function, we know that the graph should have this shape. Second, when data points have this basic shape, they suggest a logarithmic model.

CHECKPOINT

1. What asymptote does the graph of $y = \log_a x$ approach when $a > 1$?

2. For $a > 1$, does the equation $y = \log_a x$ represent the same function as the equation $x = a^y$?

3. For what values of x is $y = \log_a x$, $a > 0$, $a \neq 1$, defined?

The definition of the logarithmic function and the previous examples suggest a special relationship between the logarithmic function $y = \log_a x$

and the exponential function $y = a^x (a > 0, a \neq 1)$. Since we can write $y = \log_a x$ in exponential form as $x = a^y$, we see that the connection between

$$y = \log_a x \qquad \text{and} \qquad y = a^x$$

is that x and y have been interchanged from one function to the other. This is true for the functional description and hence for the ordered pairs that satisfy these functions. This is illustrated in Table 5.3 for the functions $y = \log_{10} x$ and $y = 10^x$.

Table 5.3

$y = \log_{10} x$		$y = 10^x$	
Coordinates	*Justification*	*Coordinates*	*Justification*
$(1000, 3)$	$3 = \log_{10} 1000$	$(3, 1000)$	$1000 = 10^3$
$(100, 2)$	$2 = \log_{10} 100$	$(2, 100)$	$100 = 10^2$
$(\frac{1}{10}, -1)$	$-1 = \log_{10} \frac{1}{10}$	$(-1, \frac{1}{10})$	$\frac{1}{10} = 10^{-1}$

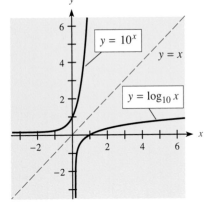

Figure 5.12

In general we say that $y = f(x)$ and $y = g(x)$ are **inverse functions** if, whenever the pair (a, b) satisfies $y = f(x)$, the pair (b, a) satisfies $y = g(x)$. Furthermore, because the values of the x and y coordinates are interchanged for inverse functions, their graphs are reflections of each other about the line $y = x$.

Thus for $a > 0$ and $a \neq 1$, the logarithmic function $y = \log_a x$ (also written $x = a^y$) and the exponential function $y = a^x$ are inverse functions.

The logarithmic function $y = \log x$ is the inverse of the exponential function $y = 10^x$. Thus the graphs $y = \log_{10} x$ and $y = 10^x$ are reflections of each other about the line $y = x$. Some values of x and y for these functions are given in Table 5.3, and their graphs are shown in Figure 5.12.

LOGARITHM PROPERTIES

As we stated previously, if $y = \log_a x$, then $x = a^y$. This means that the logarithm y is an exponent. For example, $\log_3 81 = 4$ because $3^4 = 81$. In this case the logarithm, 4, was the exponent to which we had to raise the base to obtain 81. In general, if $y = \log_a x$, then y is the exponent to which the base a must be raised to obtain x.

Since logarithms are exponents, the properties of logarithms can be derived from the properties of exponents. (The properties of exponents are discussed in Chapter 0.) The following properties of logarithms are useful in simplifying expressions containing logarithms.

Logarithm Property 1   If $a > 0$, $a \neq 1$, then $\log_a a^x = x$, for any real number x.

evaluating $y = \ln x$ for $x > 0$ (including some values $0 < x < 1$) with a calculator. The graph is shown in Figure 5.11.

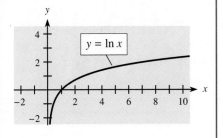

y	$x = e^y$
-3	0.050
-2	0.135
-1	0.368
0	1.000
1	2.718
2	7.389
3	20.086

x	$y = \ln x$
0.05	-3.000
0.10	-2.303
0.50	-0.693
1	0.000
2	0.693
3	1.099
5	1.609
10	2.303

Figure 5.11

Note that because $\ln x$ is a standard function on a graphing utility, such a utility could be used to obtain its graph. Note also that the graphs of $y = \log_2 x$ (Figure 5.10) and $y = \ln x$ (Figure 5.11) are very similar. The shapes of graphs of equations of the form $y = \log_a x$ with $a > 1$ are similar to these two graphs.

Graphs of Logarithmic Functions

Equation: $y = \log_a x$ $(a > 1)$
x-intercept: $(1, 0)$
Domain: All positive reals
Range: All reals
Asymptote: y-axis (negative half)

Basic shape

The basic shape of a logarithmic function is important for two reasons. First, when we graph a logarithmic function, we know that the graph should have this shape. Second, when data points have this basic shape, they suggest a logarithmic model.

CHECKPOINT

1. What asymptote does the graph of $y = \log_a x$ approach when $a > 1$?

2. For $a > 1$, does the equation $y = \log_a x$ represent the same function as the equation $x = a^y$?

3. For what values of x is $y = \log_a x$, $a > 0$, $a \neq 1$, defined?

The definition of the logarithmic function and the previous examples suggest a special relationship between the logarithmic function $y = \log_a x$

and the exponential function $y = a^x (a > 0, a \neq 1)$. Since we can write $y = \log_a x$ in exponential form as $x = a^y$, we see that the connection between

$$y = \log_a x \qquad \text{and} \qquad y = a^x$$

is that x and y have been interchanged from one function to the other. This is true for the functional description and hence for the ordered pairs that satisfy these functions. This is illustrated in Table 5.3 for the functions $y = \log_{10} x$ and $y = 10^x$.

Table 5.3

$y = \log_{10} x$		$y = 10^x$	
Coordinates	Justification	Coordinates	Justification
$(1000, 3)$	$3 = \log_{10} 1000$	$(3, 1000)$	$1000 = 10^3$
$(100, 2)$	$2 = \log_{10} 100$	$(2, 100)$	$100 = 10^2$
$(\frac{1}{10}, -1)$	$-1 = \log_{10} \frac{1}{10}$	$(-1, \frac{1}{10})$	$\frac{1}{10} = 10^{-1}$

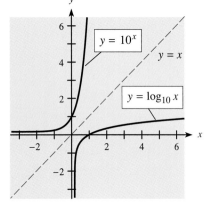

Figure 5.12

In general we say that $y = f(x)$ and $y = g(x)$ are **inverse functions** if, whenever the pair (a, b) satisfies $y = f(x)$, the pair (b, a) satisfies $y = g(x)$. Furthermore, because the values of the x and y coordinates are interchanged for inverse functions, their graphs are reflections of each other about the line $y = x$.

Thus for $a > 0$ and $a \neq 1$, the logarithmic function $y = \log_a x$ (also written $x = a^y$) and the exponential function $y = a^x$ are inverse functions.

The logarithmic function $y = \log x$ is the inverse of the exponential function $y = 10^x$. Thus the graphs $y = \log_{10} x$ and $y = 10^x$ are reflections of each other about the line $y = x$. Some values of x and y for these functions are given in Table 5.3, and their graphs are shown in Figure 5.12.

LOGARITHM PROPERTIES

As we stated previously, if $y = \log_a x$, then $x = a^y$. This means that the logarithm y is an exponent. For example, $\log_3 81 = 4$ because $3^4 = 81$. In this case the logarithm, 4, was the exponent to which we had to raise the base to obtain 81. In general, if $y = \log_a x$, then y is the exponent to which the base a must be raised to obtain x.

Since logarithms are exponents, the properties of logarithms can be derived from the properties of exponents. (The properties of exponents are discussed in Chapter 0.) The following properties of logarithms are useful in simplifying expressions containing logarithms.

Logarithm Property 1   If $a > 0$, $a \neq 1$, then $\log_a a^x = x$, for any real number x.

To prove this result, note that the exponential form of $y = \log_a a^x$ is $a^y = a^x$, so $y = x$. That is, $\log_a a^x = x$.

EXAMPLE 7

Use Property I to simplify each of the following.

(a) $\log_4 4^3$ (b) $\log_e e^x$

Solution

(a) $\log_4 4^3 = 3$

CHECK: The exponential form is $4^3 = 4^3$.

(b) $\log_e e^x = x$

CHECK: The exponential form is $e^x = e^x$.

We note that two special cases of Property I are used frequently; these are when $x = 1$ and $x = 0$.

Special Cases of Logarithm Property I

Since $a^1 = a$, we have $\log_a a = 1$.

Since $a^0 = 1$, we have $\log_a 1 = 0$.

The logarithmic form of $y = a^{\log_a x}$ is $\log_a y = \log_a x$, so $y = x$. This means that $a^{\log_a x} = x$ and proves Property II.

Logarithm Property II

If $a > 0$, $a \neq 1$, then $a^{\log_a x} = x$, for any positive real number x.

EXAMPLE 8

Use Property II to simplify each of the following.

(a) $2^{\log_2 4}$ (b) $e^{\ln x}$

Solution

(a) $2^{\log_2 4} = 4$

CHECK: The logarithmic form is $\log_2 4 = \log_2 4$.

(b) $e^{\ln x} = x$

CHECK: The logarithmic form is $\log_e x = \ln x$.

If $u = \log_a M$ and $v = \log_a N$, then the exponential forms are $a^u = M$ and $a^v = N$. Thus

$$\log_a(MN) = \log_a(a^u \cdot a^v) = \log_a(a^{u+v}) = u + v = \log_a M + \log_a N,$$

and Property III is established.

Logarithm Property III   If $a > 0$, $a \neq 1$, and M and N are positive real numbers, then

$$\log_a(MN) = \log_a M + \log_a N.$$

EXAMPLE 9
(a) Find $\log_2(4 \cdot 16)$, if $\log_2 4 = 2$ and $\log_2 16 = 4$.
(b) Find $\log_{10}(4 \cdot 5)$ if $\log_{10} 4 = 0.6021$ and $\log_{10} 5 = 0.6990$ (to four decimal places).

Solution
(a) $\log_2(4 \cdot 16) = \log_2 4 + \log_2 16 = 2 + 4 = 6$
CHECK: $\log_2(4 \cdot 16) = \log_2(64) = 6$, because $2^6 = 64$.
(b) $\log_{10}(4 \cdot 5) = \log_{10} 4 + \log_{10} 5 = 0.6021 + 0.6990 = 1.3011$
CHECK: Using a calculator, we can see that $10^{1.3011} = 20 = 4 \cdot 5$.

Logarithm Property IV   If $a > 0$, $a \neq 1$, and M and N are positive real numbers, then

$$\log_a(M/N) = \log_a M - \log_a N.$$

The proof of this property is left to the student in problem 45 of the exercises.

EXAMPLE 10
(a) Evaluate $\log_3(\frac{9}{27})$.
(b) Find $\log_{10}(\frac{16}{5})$, if $\log_{10} 16 = 1.2041$ and $\log_{10} 5 = 0.6990$ (to 4 decimal places).

Solution
(a) $\log_3(\frac{9}{27}) = \log_3 9 - \log_3 27 = 2 - 3 = -1$
CHECK: $\log_3(\frac{1}{3}) = -1$ because $3^{-1} = \frac{1}{3}$.
(b) $\log_{10}(\frac{16}{5}) = \log_{10} 16 - \log_{10} 5 = 1.2041 - 0.6990 = 0.5051$
CHECK: Using a calculator, we can see that $10^{0.5051} = 3.2 = \frac{16}{5}$.

Logarithm Property V   If $a > 0$, $a \neq 1$, M is a positive real number, and N is any real number, then

$$\log_a(M^N) = N \log_a M.$$

The proof of this property is left to the student in problem 46 of the exercises.

EXAMPLE 11
(a) Simplify $\log_3(9^2)$.
(b) Simplify $\ln 8^{-4}$, if $\ln 8 = 2.0794$ (to 4 decimal places).

Solution
(a) $\log_3(9^2) = 2 \log_3 9 = 2 \cdot 2 = 4$
CHECK: $\log_3 81 = 4$ because $3^4 = 81$.
(b) $\ln 8^{-4} = -4 \ln 8 = -4(2.0794) = -8.3176$
CHECK: Using a calculator, we can see that $8^{-4} \approx 0.000244$ and $\ln (0.000244) \approx -8.3176$.

CHECKPOINT

4. Simplify:
   (a) $6^{\log_6 x}$
   (b) $\log_7 7^3$
   (c) $\ln \left(\dfrac{1}{e^2} \right)$
   (d) $\log 1$

CHANGE OF BASE

By using a calculator we can directly evaluate only those logarithms with base 10 or base e. Also logarithms with base 10 or base e are the only ones that are standard functions on a graphing utility. Thus, if we had a way to express a logarithmic function with any base in terms of a logarithm with base 10 or base e, we would be able to evaluate the original logarithmic function with a scientific calculator and graph it with a graphing utility.

In general, if we use the properties of logarithms, we can write

$$y = \log_b x \quad \text{in the form} \quad b^y = x.$$

If we take the base a logarithms of both sides, we have

$$\log_a b^y = \log_a x$$
$$y \log_a b = \log_a x$$
$$y = \frac{\log_a x}{\log_a b}.$$

This gives us the **change of base formula** from base b to base a.

Change of Base Formula   If $a \neq 1, b \neq 1, a > 0, b > 0$, then

$$\log_b x = \frac{\log_a x}{\log_a b}.$$

For the purposes of calculation, we usually convert logarithms to base e or base 10.

| **Change of Base Calculation Formula** | $\log_b x = \dfrac{\ln x}{\ln b}$ or $\log_b x = \dfrac{\log x}{\log b}$ |

EXAMPLE 12

Evaluate $\log_7 15$ by using the change of base calculation formula.

Solution

(a) $\log_7 15 = \dfrac{\ln 15}{\ln 7} = \dfrac{2.70805}{1.94591} = 1.39166$ (approximately)

(b) $\log_7 15 = \dfrac{\log 15}{\log 7} = \dfrac{1.17609}{0.84510} = 1.39166$ (approximately)

EXAMPLE 13

Use the change of base calculation formula to rewrite each of the following logarithms in base e. Then graph each with a graphing utility.

(a) $y = \log_2 x$ (See Example 4 and Figure 5.10.)

(b) $y = \log_7 x$

Solution

(a) $y = \log_2 x = \dfrac{\ln x}{\ln 2}$. See Figure 5.13 for the graph. Notice that it is exactly the same as the graph in Figure 5.10.

(b) $y = \log_7 x = \dfrac{\ln x}{\ln 7}$. See Figure 5.14 for the graph.

Figure 5.13

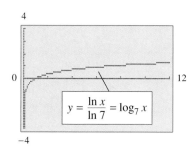

Figure 5.14

Natural logarithms, $y = \ln x$ (and the inverse exponential with base e), have many practical applications, some of which are considered in Section 5.3. Common logarithms, $y = \log x$, were widely used for computation before computers and calculators became popular. They also have several applications to scaling variables, where the purpose is to reduce the scale of variation when a natural physical variable covers a wide range.

EXAMPLE 14

The Richter scale is used to measure the intensity of an earthquake. The magnitude on the Richter scale of an earthquake of intensity I is given by

$$R = \log(I/I_0),$$

where I_0 is a certain minimum intensity used for comparison.
(a) Find R if I is 3,160,000 times as great as I_0.
(b) The 1964 Alaskan earthquake measured 8.5 on the Richter scale. Find the intensity of the 1964 Alaskan earthquake.

Solution
(a) With $I = 3{,}160{,}000 I_0$, then $I/I_0 = 3{,}160{,}000$. Hence

$$R = \log(3{,}160{,}000)$$
$$= 6.5 \quad \text{(approximated to one decimal place).}$$

(b) For $R = 8.5$, it follows that

$$8.5 = \log(I/I_0).$$

Rewriting this in exponential form gives

$$10^{8.5} = I/I_0,$$

and from a calculator we obtain

$$I/I_0 = 316{,}000{,}000 \qquad \text{(approximately).}$$

Thus the intensity is 316,000,000 times I_0.
Notice that a Richter scale measurement that is 2 units larger means that the intensity is $10^2 = 100$ *times* greater.

CHECKPOINT SOLUTIONS

1. The (negative) y-axis
2. Yes
3. $x > 0$ is the domain of $\log_a x$.
4. (a) x, by Property II
   (b) 3, by Property I
   (c) $\ln\left(\dfrac{1}{e^2}\right) = \ln(e^{-2}) = -2$, by Property I
   (d) 0, because $\log_a 1 = 0$ for any base a.

EXERCISE 5.2

In problems 1–4, write each equation in exponential form.

1. $4 = \log_2 16$ 2. $4 = \log_3 81$

3. $\frac{1}{2} = \log_4 2$ 4. $-2 = \log_3(\frac{1}{9})$

In problems 5–8, solve for x by writing the equations in exponential form.

5. $\log_2 x = 3$ 6. $\log_4 x = -2$

7. $\log_8 x = -\frac{1}{3}$ 8. $\log_{25} x = \frac{1}{2}$

In problems 9–12, write each equation in logarithmic form.

9. $2^5 = 32$ 10. $5^3 = 125$

11. $4^{-1} = \frac{1}{4}$ 12. $9^{1/2} = 3$

In problems 13–18, graph the functions.

13. $y = \log_3 x$ 14. $y = \log_4 x$

15. $y = \ln x$ 16. $y = \log_9 x$

17. $y = \log_2(-x)$ 18. $y = \ln(-x)$

In problems 19–26, use properties of logarithms or a definition to simplify each expression.

19. (a) $\log_3 27$ (b) $\log_5(\frac{1}{5})$

20. (a) $\log_4 16$ (b) $\log_9 3$

21. (a) $\log_7(7^{2/3})$ (b) $\log_{13}(13^5)$

22. (a) $\log_{11}(11^4)$ (b) $\log_8(8^{-1/4})$

23. (a) $\ln(e^x)$ (b) $\log 100$

24. (a) $\log(10^x)$ (b) $\ln \sqrt{e}$

25. (a) $3^{\log_3 x}$ (b) $e^{\ln 3}$

26. (a) $10^{\log 2}$ (b) $4^{\log_4 5}$

In problems 27 and 28, evaluate each logarithm by using properties of logarithms and the following facts.

$$\log_a x = 3.1 \qquad \log_a y = 1.8 \qquad \log_a z = -2.7$$

27. (a) $\log_a(xy)$ (b) $\log_a\left(\frac{x}{z}\right)$ (c) $\log_a(x^4)$ (d) $\log_a \sqrt{y}$

28. (a) $\log_a(yz)$ (b) $\log_a\left(\frac{z}{y}\right)$ (c) $\log_a(y^6)$ (d) $\log_a \sqrt[3]{z}$

Write each expression in problems 29–32 as the sum or difference of two logarithmic functions containing no exponents.

29. $\log\left(\dfrac{x}{x+1}\right)$ 30. $\ln[(x+1)(4x+5)]$

31. $\log_7(x\sqrt[3]{x+4})$ 32. $\log_5\left(\dfrac{x^2}{\sqrt{x+4}}\right)$

Use the properties of logarithms to write each expression in problems 33–36 as a single logarithm.

33. $\ln x - \ln y$

34. $\log_3(x+1) + \log_3(x-1)$

35. $\log_5(x+1) + \frac{1}{2}\log_5 x$

36. $\log(2x+1) - \frac{1}{3}\log(x+1)$

37. (a) Given $f(x) = \ln x$. Use a graphing utility to graph $f(x) = \ln x$ with $f(x-c) = \ln(x-c)$ for $c = -4$, -2, 1, and 5.

    (b) For each c-value, identify the vertical asymptote and the domain of $y = \ln(x-c)$.

    (c) For each c-value, find the x-intercept of $y = \ln(x-c)$.

    (d) Write a sentence that precisely explains how the graphs of $y = f(x)$ and $y = f(x-c)$ are related.

38. Given $f(x) = \ln x$. Use a graphing utility to graph $f(ax) = \ln(ax)$ for $a = -2, -1, -0.5, 0.2,$ and 3 with $f(x) = \ln x$. Explain the differences between the graphs

    (a) when $a > 0$ and when $a < 0$;

    (b) when $|a| > 1$ and when $0 < |a| < 1$.

In problems 39 and 40, use the change of base calculation formula to evaluate each logarithm.

39. (a) $\log_2 17$ (b) $\log_5(0.78)$

40. (a) $\log_3 12$ (b) $\log_8(0.15)$

 In problems 41–44, use the change of base calculation formula to rewrite each logarithm, then use a graphing utility to graph the function.

41. $y = \log_5 x$ 42. $y = \log_6 x$

43. $y = \log_{13} x$ 44. $y = \log_{16} x$

45. Prove logarithm Property IV.

46. Prove logarithm Property V.

APPLICATIONS

47. **Doubling time** If money is invested at 6% compounded monthly, then the doubling time t, in months, can be found from $2 = (1.005)^t$.

    (a) Rewrite this equation in logarithmic form.

    (b) Use your answer to part (a) and the change of base calculation formula to find the doubling time.

48. **Demand** If the demand function for a product is given by $p = 100/\ln(q+1)$,

    (a) what will be the price if 19 units are demanded?

    (b) how many units, to the nearest unit, will be demanded if the price is $29.40? (*Hint:* Find p for $q = 28$, 29, and 30.)

49. **Supply** If the supply function for a product is given by $p = 100 \ln(q + 1)$,
    (a) at what price will 9 items be supplied?
    (b) how many units, to the nearest unit, will be supplied if the price is \$346.57?
    (*Hint:* Find p for q between 30 and 35.)

Richter scale Use the formula $R = \log(I/I_0)$ in problems 50–55.

50. Find the Richter scale reading of an earthquake with intensity 20,000 times I_0.

51. Find the Richter scale reading of an earthquake with intensity 800,000 times I_0.

52. In May 1983, an earthquake measuring 7.7 on the Richter scale occurred in Japan. This was the first major earthquake in Japan since 1948, when one registered 7.3 on the Richter scale. How many times more severe was the 1983 shock?

53. The strongest earthquake ever to strike Japan occurred in 1933 and measured 8.9 on the Richter scale. How many times more severe was this 1933 quake than the one in May 1983? (See problem 52.)

54. The San Francisco earthquake of 1906 measured 8.25 on the Richter scale, and the San Francisco earthquake of 1989 measured 7.1. How much more intense was the 1906 quake?

55. The largest earthquakes ever recorded measured 8.9 on the Richter scale; the San Francisco earthquake of 1906 measured 8.25. Calculate the ratio of their intensities. (These readings correspond to devastating quakes.)

Decibel readings Problems 56–59. The loudness of sound (in decibels) perceived by the human ear depends upon intensity levels according to

$$L = 10 \log(I/I_0),$$

where I_0 is the threshold of hearing for the average human ear.

56. Find the loudness when I is 10,000 times I_0. This is the intensity level of the average voice.

57. A sound that causes pain has intensity about 10^{14} times I_0. Find the decibel reading for this threshold.

58. Graph the equation for loudness. Use I/I_0 as the independent variable.

59. A relatively quiet room has a background noise level of about $L_1 = 32$ decibels, and a heated argument has a decibel level of about $L_2 = 66$. Find the ratio I_2/I_1 of the associated intensities.

pH levels Problems 60–64. Chemists use the pH (hydrogen potential) of a solution to measure its acidity or basicity. The pH is given by the formula

$$pH = -\log[H^+],$$

where $[H^+]$ is the concentration of hydrogen ions in moles per liter.

60. Most common solutions have a pH range between 1 and 14. What values of H^+ are associated with these extremes?

61. Find the approximate pH of each of the following.
    (a) blood: $[H^+] = 3.98 \times 10^{-8}$
       $= 0.0000000398$
    (b) beer: $[H^+] = 6.31 \times 10^{-5} = 0.0000631$
    (c) vinegar: $[H^+] = 6.3 \times 10^{-3} = 0.0063$

62. Sometimes pH is defined as the logarithm of the reciprocal of the concentration of hydrogen ions. Write an equation that represents this sentence, and explain how it and the equation given above can both represent pH.

63. Find the approximate hydrogen ion concentration $[H^+]$ for each of the following.
    (a) apples: pH = 3.0
    (b) eggs: pH = 7.79
    (c) water (neutral): pH = 7.0

64. A solution is considered acidic if $[H^+] > 10^{-7}$ or basic if $[H^+] < 10^{-7}$. What pH values correspond to acids and what values correspond to bases?

65. **Expected life span** The following table shows the expected life span at birth of people born in certain years in the United States. Assume that for these years, the expected life span at birth can be modeled with the function $l(x) = 11.7 + 14.1 \ln x$, where x is the number of years past 1900. Graph this function and the data points.

Year	Life Span (in Years)
1920	54.1
1930	59.7
1940	62.9
1950	68.2
1960	69.7
1970	70.8
1975	72.6
1980	73.7
1987	75.0
1988	74.9
1989	75.2

Source: *World Almanac*, 1991

66. **Violent crimes** The following data represent the number of violent crimes (murders, rapes, and armed robberies) per 100,000 people for the years 1987 to 1992.

Year	Violent Crimes (per 100,000)
1987	610
1988	637
1989	663
1990	732
1991	758
1992	765

SOURCE: FBI Crime Report

Letting $t = 0$ in 1980, the above data can be modeled by

$$y = -22.9 + 321 \ln t.$$

Assuming this model is accurate, graph this function and the data points.

67. **Deforestation** One of the major causes for rain forest deforestation is agricultural and residential development. The data below reflect the number of hectares destroyed in particular years.

Year (t)	Millions of Hectares (y)
1960	2.21
1970	3.79
1980	4.92
1988	5.77

If $t = 0$ in 1950, this data can be modeled by

$$y = -3.91435 + 2.62196 \ln t.$$

(a) Graph this function and the data points from the table.
(b) Project the number of hectares that will be destroyed in the year 2000 because of agricultural and residential development.

5.3 SOLUTION OF EXPONENTIAL EQUATIONS; APPLICATIONS OF EXPONENTIAL AND LOGARITHMIC FUNCTIONS

Objectives

❏ To solve exponential growth or decay equations when sufficient data are known
❏ To solve exponential equations representing demand, supply, total revenue, or total cost when sufficient data are known

APPLICATION PREVIEW

In this section, we discuss applications of exponential and logarithmic functions to the solution of problems involving exponential growth and decay and supply and demand. With exponential decay, we consider various applications including decay of radioactive materials and decay of sales following an advertising campaign. Exponential growth focuses on population growth and growth that can be modeled by Gompertz curves. Finally, we consider applications of demand functions that can be modeled by exponential functions.

GROWTH AND DECAY

Exponential decay models are those that can be described by a function of the following form.

Decay Models $f(x) = ma^{-x}$ with $a > 1$ and $m > 0$

We have already mentioned that radioactive decay has this form, as do the demand curves discussed later. Another example of this phenomenon occurs in the life sciences when the valves to the aorta are closed, and blood flows into the heart. During this period of time, the blood pressure in the aorta falls exponentially. This is illustrated in the following example.

EXAMPLE 1

Medical research has shown that over short periods of time when the valves to the aorta of a normal adult close, the pressure in the aorta is a function of time and can be modeled by the equation

$$P = 95e^{-0.491t},$$

where t is in seconds.
(a) What is the aortic pressure when the valves are first closed ($t = 0$)?
(b) What is the aortic pressure after 0.1 seconds?
(c) How long will it be before the pressure reaches 80?

Solution
(a) If $t = 0$, the pressure is given by

$$P = 95e^0 = 95.$$

(b) If $t = 0.1$, the pressure is given by

$$P = 95e^{-0.491(.01)}$$
$$= 95e^{-0.0491}$$
$$= 95(0.952)$$
$$= 90.44.$$

(c) Setting $P = 80$ and solving for t will give us the length of time before the pressure reaches 80.

$$80 = 95e^{-0.491t}$$

To solve this equation for t, we must rewrite it so that t is not in an exponent. We first isolate the exponential containing t by dividing both sides by 95.

$$\frac{80}{95} = e^{-0.491t}$$

Rewriting this equation in logarithmic form gives

$$\ln\left(\frac{80}{95}\right) = -0.491t.$$

Using a calculator, we have $-0.172 = -0.491t$. Thus,

$$\frac{-0.172}{-0.491} = t$$

$$t = 0.35 \text{ seconds} \qquad \text{(approximately)}.$$

Note that in Example 1 we could have solved

$$\frac{80}{95} = e^{-0.491t}$$

for t by taking the logarithm, base e, of both sides and then using properties of logarithms.

$$\ln\left(\frac{80}{95}\right) = \ln e^{-0.491t}$$

By logarithm Property I, we have

$$\ln\left(\frac{80}{95}\right) = -0.491t.$$

Note that taking the natural logarithm of both sides gives the same result as writing $80/95 = e^{-0.491t}$ in logarithmic form.

The advantage of taking the natural logarithm (or the common logarithm) of both sides of an exponential equation to solve the equation becomes apparent when the base of the exponential is not e (or 10).

To solve an exponential equation, we must be able to rewrite the equation so that the variable is not in an exponent. For equations involving a single exponential, we solve for that exponential. (That is, we isolate the exponential expression on one side of the equation.)

$$Na^X = M \quad \text{yields} \quad a^X = \frac{M}{N}.$$

Then taking the logarithm (base 10 or base e) of both sides of the equation, using logarithm Property V, and using algebraic methods, we can solve the equation. Part (b) of the following example illustrates this solution method.

EXAMPLE 2

A company finds that its daily sales begin to fall after the end of an advertising campaign, and the decline is such that the number of sales is $S = 2000(2^{-0.1x})$, where x is the number of days after the campaign's end.
(a) How many sales will be made 10 days after the end of the campaign?
(b) If they do not want sales to drop below 500 per day, when should they start a new campaign?

Solution

(a) If $x = 10$, sales are given by $S = 2000(2^{-1}) = 1000$.
(b) Setting $S = 500$ and solving for x will give us the number of days after the end of the campaign when sales will reach 500.

$$500 = 2000(2^{-0.1x})$$

$$\frac{500}{2000} = 2^{-0.1x} \qquad \text{Isolate the exponential.}$$

$$0.25 = 2^{-0.1x}$$

Since the base of this exponential is 2 rather than e or 10, we choose to take the logarithm, base 10, of both sides of the equation instead of rewriting in logarithmic form.

$$\log(0.25) = \log(2^{-0.1x})$$

$$\log(0.25) = (-0.1x)(\log 2) \qquad \text{Property V}$$

$$\frac{\log(0.25)}{\log 2} = -0.1x$$

$$\frac{-0.6021}{0.3021} = -0.1x$$

$$1.993 = -0.1x$$

$$x = 20 \qquad \text{(approximately)}$$

Thus sales will be 500 on the twentieth day after the end of campaign. If a new campaign isn't begun on or before the twenty-first day, sales will drop below 500. (See Figure 5.15.)

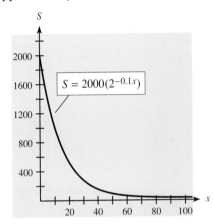

Figure 5.15

$S = 2000(2^{-0.1x})$

Example 2 is an example of *sales decay* and is typical of many exponential decay models. In particular, once some action is completed, such as an advertising campaign, its effect on sales volume diminishes, or decays, with time.

In business, economics, biology, and the social sciences, the growth of money, bacteria, or population is frequently of interest. **Exponential growth models** are those that can be described by a function of the following form.

Growth Models $f(x) = ma^x$ with $a > 1$ and $m > 0$

As we mentioned in Section 5.1, the curve that models the growth of some populations is given by $y = P_0 e^{ht}$, where P_0 is the population size at a particular time t, h is a constant that depends on the population involved, and y is the total population at time t. This function may be used to model population growth for humans, insects, or bacteria.

EXAMPLE 3
The population of a certain city was 30,000 in 1970 and 40,500 in 1980. If the formula $y = P_0 e^{ht}$ applies to the growth of the city's population, what should the population be in the year 2000?

Solution
We can first use the data from 1970 and 1980 to find the value of h in the formula. Letting $P_0 = 30,000$ and $t = 10$, we get

$$40,500 = 30,000 e^{h(10)}$$
$$1.35 = e^{10h}. \qquad \text{Isolate the exponential.}$$

Taking the natural logarithm of both sides gives

$$\ln 1.35 = \ln e^{10h} = 10h(\ln e)$$
$$\ln 1.35 = 10h, \quad \text{since} \quad \ln e = 1$$
$$0.3001 = 10h$$
$$h = 0.0300 \qquad \text{(approximately).}$$

Thus the formula for this population is $y = P_0 e^{0.03t}$. To predict the population for the year 2000, we use $P_0 = 40,500$ (for 1980) and $t = 20$. This gives

$$y = 40,500 e^{0.03(20)}$$
$$= 40,500 e^{0.6}$$
$$= 40,500(1.8221)$$
$$= 73,795 \qquad \text{(approximately).}$$

One family of curves that has been used to describe human growth and development, the growth of organisms in a limited environment, and the growth of many types of organizations is the family of **Gompertz curves.** These curves are graphs of equations of the form

$$N = Ca^{R^t},$$

where t represents the time, R $(0 < R < 1)$ is a constant depending on the population, a represents the proportion of initial growth, C is the maximum possible number of individuals, and N is the number of individuals at a given time t.

For example, the equation $N = 100(0.03)^{0.2^t}$ could be used to predict the size of a deer herd introduced on a small island. Here the maximum number of deer C would be 100, the proportion of the initial growth a is 0.03, and R is 0.2. For this example, t represents time, measured in decades. The graph of this equation is given in Figure 5.16.

Figure 5.16

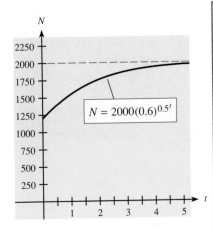

Figure 5.17

EXAMPLE 4
A hospital administrator predicts that the growth in the number of hospital employees will follow the Gompertz equation

$$N = 2000(0.6)^{0.5^t},$$

where t represents the number of years after the opening of a new facility.
(a) What is the number of employees when the facility opens ($t = 0$)?
(b) How many employees are predicted after one year of operation ($t = 1$)?
(c) Graph the curve.
(d) What is the maximum value for N that the curve will approach?

Solution
(a) If $t = 0$, $N = 2000(0.6)^1 = 1200$
(b) If $t = 1$, $N = 2000(0.6)^{0.5} = 2000\sqrt{0.6} = 1549$ (approximately)
(c) The graph is shown in Figure 5.17.
(d) From the graph we can see that as larger values of t are substituted in the function, the values of N approach, but never reach, 2000. We say that the line $N = 2000$ (dashed) is an **asymptote** for this curve, and that 2000 is the maximum possible value.

EXAMPLE 5
The Gompertz equation

$$N = 100(0.03)^{0.2^t}$$

predicts the size of a deer herd on a small island t decades from now. During what year will the deer population reach or exceed 70?

Solution
We solve the equation with $N = 70$.

$$70 = 100(0.03)^{0.2^t}$$

$$0.7 = 0.03^{0.2^t} \qquad \text{Isolate the exponential.}$$

$$\ln 0.7 = \ln 0.03^{0.2^t} = 0.2^t \ln 0.03$$

$$\frac{\ln 0.7}{\ln 0.03} = 0.2^t \qquad \text{Again, isolate the exponential.}$$

$$\ln\left(\frac{\ln 0.7}{\ln 0.03}\right) = t \ln 0.2$$

$$\ln(0.10172) = t(-1.6094)$$

$$\frac{-2.2855}{-1.6094} = t$$

$$t = 1.42 \text{ decades} \qquad \text{(approximately)}$$

The population will exceed 70 in just over 14 years, or during the 15th year.

CHECKPOINT

1. Suppose the sales of a product, in dollars, are given by $S = 1000e^{-0.07x}$, where x is the number of days after the end of an advertising campaign.
   (a) What are sales 2 days after the end of the campaign?
   (b) How long before sales are $300?

2. Suppose the number of employees at a new regional hospital is predicted by the Gompertz curve

$$N = 3500(0.1)^{0.5^t},$$

   where t is the number of years after the hospital opens.
   (a) How many employees did the hospital have when it opened?
   (b) What is the expected upper limit on the number of employees?

ECONOMIC AND MANAGEMENT APPLICATIONS

In Section 4.3, we discussed quadratic cost, revenue, demand, and supply functions. But cost, revenue, demand, and supply may also be modeled by exponential or logarithmic equations. For example, suppose the demand for a product is given by $p = 30(3^{-q/2})$, where q is the number of units demanded at a price of p dollars per unit. Then the graph of the demand curve is as given in Figure 5.18.

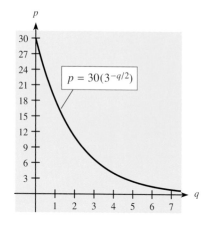

Figure 5.18

EXAMPLE 6

If the demand function for a certain commodity is given by $p = 30(3^{-q/2})$,
(a) at what price per unit will the demand equal 4 units?
(b) how many units, to the nearest unit, will be demanded if the price is $17.32?

Solution

(a) If $q = 4$, then $p = 30(3^{-4/2})$
$$= 30(0.1111)$$
$$= 3.33 \text{ dollars} \qquad \text{(approximately)}$$

(b) If $p = 17.32$, then $17.32 = 30(3^{-q/2})$
$$0.5773 = 3^{-q/2} \qquad \text{Isolate the exponential.}$$
$$\ln 0.5773 = \ln 3^{-q/2}$$
$$\ln 0.5773 = -\frac{q}{2} \ln 3$$
$$\frac{-2 \ln 0.5773}{\ln 3} = q$$
$$1 = q \qquad \text{(approximately)}$$

The number of units demanded would be 1, to the nearest unit.

CHECKPOINT

3. Suppose the monthly demand for a product is given by $p = 400e^{-0.003x}$, where p is the price in dollars and x is the number of units. How many units will be demanded when the price is $100?

EXAMPLE 7

If the demand function for a commodity is given by $p = 100e^{-x/10}$, where p is the price per unit when x units are sold,
(a) what is the total revenue for the commodity?
(b) what would be the total revenue if 30 units are demanded and supplied?

Solution

(a) The total revenue can be computed by multiplying the quantity sold times the price per unit. Since the demand function gives the price per unit when x units are sold, the total revenue for x units is $x \cdot p = x(100e^{-x/10})$. Thus the total revenue function is $R(x) = 100xe^{-x/10}$.
(b) If 30 units are sold, the total revenue is

$$R(30) = 100(30)e^{-30/10} = 100(30)(0.0498)$$
$$= 149.40 \text{ (dollars) (approximately)}.$$

CHECKPOINT
SOLUTIONS

1. (a) $S = 1000e^{-0.07(2)} = 1000e^{-0.14} \approx \$869.36$ nearest cent

   (b) $300 = 1000e^{-0.07x}$

   $0.3 = e^{-0.07x}$

   $\ln(0.3) = -0.07x$

   $\dfrac{\ln(0.3)}{-0.07} = x \approx 17$ nearest day

2. (a) $3500(0.1) = 350$
   (b) 3500

3. $100 = 400e^{-0.003x}$

   $0.25 = e^{-0.003x}$

   $\ln(0.25) = -0.003x$

   $\dfrac{\ln(0.25)}{-0.003} = x \approx 462$ nearest unit

EXERCISE 5.3

GROWTH AND DECAY

1. **Sales decay** The sales decay for a product is given by $S = 50,000e^{-0.8x}$, where S is the monthly sales and x is the number of months that have passed since the end of a promotional campaign.
   (a) What will be the sales 4 months after the end of the campaign?
   (b) How many months after the end of the campaign will sales drop below 1000, if no new campaign is initiated?

2. **Sales decay** The sales of a product decline after the end of an advertising campaign, with the sales decay given by $S = 100,000e^{-0.5x}$, where S represents the weekly sales and x represents the number of weeks since the end of the campaign.
   (a) What will be the sales for the tenth week after the end of the campaign?
   (b) During what week after the end of the campaign will sales drop below 400?

3. **Radioactive decay** Radioactive carbon-14 can be used to determine the age of fossils. Carbon-14 decays according to the equation
$$y = y_0 e^{-0.00012378t},$$
   where y is the amount of carbon-14 at time t, in years, and y_0 is the original amount.
   (a) How much of the original amount would be left after 3000 years?
   (b) If a fossil is found to have $\frac{1}{100}$ the original amount of carbon-14, how old is the fossil?

4. **Product reliability** A statistical study shows that the fraction of television sets of a certain brand that are still in service after x years is given by $f(x) = e^{-0.15x}$.
   (a) What fraction of the sets are still in service after 5 years?
   (b) After how many years will the fraction still in service be 1/10?

5. **Radioactive half-life** An initial amount of 100 g of the radioactive isotope thorium-234 decays according to
$$Q(t) = 100e^{-0.02828t},$$
   where t is in years. How long before half the initial amount has disintegrated? This time is called the half-life of this isotope.

6. **Radioactive half-life** A breeder reactor converts stable uranium-238 into the isotope plutonium-239. The decay of this isotope is given by
$$A(t) = A_0 e^{-0.00002876t},$$
   where $A(t)$ is the amount of the isotope at time t, in years, and A_0 is the original amount.
   (a) If $A_0 = 500$ lb, how much will be left after a human lifetime (use $t = 70$ years)?
   (b) Find the half-life of this isotope.

7. **Population growth** If the population of a certain county was 100,000 in 1980 and 110,517 in 1990 and if the formula $y = P_0 e^{ht}$ applies to the growth of the county's population, estimate the population of the county in 2005.

8. **Population growth** The population of a certain city grows according to the formula $y = P_0 e^{0.03t}$. If the population was 250,000 in 1980, estimate the population in the year 2100.

9. **Sales growth** The president of a company predicts that sales will increase after she assumes office, and that the number of monthly sales will follow the curve given by $N = 3000(0.2)^{0.6t}$, where t represents the months since she assumed office.
   (a) What will be the sales when she assumes office?
   (b) What will be the sales after three months?
   (c) What is the expected upper limit on sales?
   (d) Graph the curve.

10. **Organizational growth** Because of a new market opening, the number of employees of a firm is expected to increase according to the equation $N = 1400(0.5)^{0.3t}$, where t represents the number of years after the new market opens.
    (a) What is the level of employment when the new market opens?
    (b) How many employees should be working at the end of two years?
    (c) What is the expected upper limit on the number of employees?
    (d) Graph the curve.

11. **Organizational growth** Suppose that the equation $N = 500(0.02)^{0.7t}$ represents the number of employees working t years after a company begins operations.
    (a) How many employees are there when the company opens (at $t = 0$)?
    (b) After how many years will at least 100 employees be working?

12. **Sales growth** A firm predicts that sales will increase during a promotional campaign, and that the number of daily sales will be given by $N = 200(0.01)^{0.8^t}$, where t represents the number of days after the campaign begins. How many days after the beginning of the campaign would the firm expect to sell at least 60 units per day?

Gompertz curves describe situations in which growth is limited. There are other equations that describe this phenomenon under different assumptions. Two examples are

(a) $y = c(1 - e^{-ax})$, $a > 0$

(b) $y = \dfrac{A}{1 + ce^{-ax}}$, $a > 0$ \quad (logistic curve).

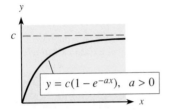

$$y = c(1 - e^{-ax}), \quad a > 0$$

$$y = \frac{A}{1 + ce^{-ax}}, \quad a > 0$$

These equations have many applications. In general, both (a) and (b) can be used to describe learning, sales of new products, and population growth, and (b) can be used to describe the spread of epidemics. Problems 13–16 illustrate applications.

13. **Drugs in the bloodstream** The concentration y of a certain drug after t hours in the bloodstream (with $0 \le t \le 15$) is given by the equation

$$y = 100(1 - e^{-0.462t}).$$

(a) What is y after one hour ($t = 1$)?
(b) How long does it take for y to reach 50?

14. **Population growth** Suppose that the size y of a deer herd t years after being introduced onto an island is given by

$$y = 2500 - 2490e^{-0.1t}.$$

(a) Find the population when the herd was introduced (at $t = 0$).
(b) How long before the herd numbers 1500?

15. **Spread of disease** On a college campus of 10,000 students, a single student returned to campus infected by a disease. The spread of the disease through the student body is given by

$$y = \frac{10{,}000}{1 + 9999e^{-0.99t}},$$

where y is the total number infected at time t (in days).
(a) How many are infected after 4 days?
(b) If the school will shut down if 50% of the students are ill, during what day will it close?

16. **Spread of a rumor** The number of people $N(t)$ in a community who are reached by a particular rumor at time t is given by the equation

$$N(t) = \frac{50{,}500}{1 + 100e^{-0.7t}}.$$

(a) Find $N(0)$.
(b) What is the upper limit on the number of people affected?
(c) How long before 75% of the upper limit is reached?

ECONOMIC AND MANAGEMENT APPLICATIONS

17. **Demand** If the demand function for a certain commodity is given by $p = 100e^{-q/2}$,
(a) at what price per unit will the quantity demanded equal 6 units?
(b) if the price is $1.83 per unit, how many units will be demanded, to the nearest unit?

18. **Demand** If the demand function for a product is given by $p = 3000e^{-q/3}$,
(a) at what price per unit will the quantity demanded equal 6 units?
(b) if the price is $149.40 per unit, how many units will be demanded, to the nearest unit?

19. **Supply** If the supply function for a product is given by $p = 100e^q/(q + 1)$, where q represents the number of hundreds of units, what will be the price when the producers are willing to supply 300 units?

20. **Supply** If the supply function for a product is given by $p = 200(2^q)$, where q represents the number of hundreds of units, what will be the price when the producers are willing to supply 500 units?

21. **Total cost** If the total cost function for a product is $C(x) = e^{0.1x} + 400$, where x is the number of items produced, what is the total cost of producing 30 units?

22. **Total cost** If the total cost for a product is given by $C(x) = 400 \ln(x + 10) + 100$, what is the total cost of producing 100 units?

23. **Total revenue** If the demand function for a product is given by $p = 200e^{-0.02x}$, where p is the price per unit when x units are demanded, then what is the total revenue when 100 units are demanded and supplied?

24. **Total revenue** If the demand function for a product is given by $p = 4000/\ln(x + 10)$, where p is the price per unit when x units are demanded, then what is the total revenue when 40 units are demanded and supplied?

25. **Compound interest** If $8500 is invested at 11.5% compounded continuously, then the future value A at any time t (in years) is given by

$$A = 8500e^{0.115t}.$$

(a) What is the amount after 18 months?
(b) How long before the investment doubles?

26. **Compound interest** If $1000 is invested at 10% compounded continuously, then the future value A at any time t (in years) is given by $A = 1000e^{0.1t}$.
(a) What is the amount after 1 year?
(b) How long before the investment doubles?

27. **Compound interest** If $5000 is invested at 9% per year compounded monthly, then the future value A at any time t (in months) is given by $A = 5000(1.0075)^t$.
(a) What is the amount after 1 year?
(b) How long before the investment doubles?

28. **Compound interest** If $10,000 is invested at 1% per month, then the future value A at any time t (in months) is given by $A = 10,000(1.01)^t$.
(a) What is the amount after 1 year?
(b) How long before the investment doubles?

29. **Market share** Suppose that the market share y (as a percent) that a company expects t months after a new product is introduced is given by $y = 40 - 40e^{-0.05t}$.
(a) What is the market share after the first month (to the nearest percent)?
(b) How long (to the nearest month) before the market share is 25%?

30. **Advertising** An advertising agency has found that when it promotes a new product in a certain market of 350,000, then the number of people x who are aware of the product t days after initiating the ad campaign is

given by

$$x = 350,000(1 - e^{-0.077t}).$$

(a) How many people (to the nearest thousand) are aware after 1 week?
(b) How long (to the nearest day) before 300,000 are aware of the new product?

ADDITIONAL APPLICATIONS

31. **Health care** For selected years from 1960 to 1992, the national health care expenditures H, in billions of dollars, can be modeled by

$$H = 45e^{0.08988t},$$

where t is the number of years past 1960. (Source: Congressional Budget Office data as reported in *Newsweek*, October 4, 1993.)
(a) If this model remains accurate, how long before national health care expenditures reach $2.5 trillion (that is, $2500 billion)?
(b) How do you think a national health care plan would alter this model?

32. **U.S. debt** For selected years from 1900 to 1989, the national debt d, in billions of dollars, can be modeled by

$$d = 1.53361\, e^{0.0853458t},$$

where t is the number of years past 1900. (Source: *World Almanac*, 1991.) How long before the debt is predicted to reach $5 trillion (that is, $5000 billion)?

33. **Inflation** The purchasing power P (in dollars) of an annual amount of A dollars after t years of 5% inflation is given by

$$P = Ae^{-0.05t}.$$

(Source: *Viewpoints*, VALIC, Summer 1993.)
(a) How long before a pension of $20,000 per year has a purchasing power of $8000?
(b) How much pension A would be needed so that the purchasing power P is $20,000 after 15 years?

34. **Mutual funds** For selected years from 1940 to 1990, the number of mutual funds N, excluding money market funds, can be modeled by

$$N = 3.6992e^{0.662056t},$$

where t is the number of decades past 1900. (Source: Investment Company Institute data, published in *Investment Digest*, VALIC, Summer 1992.)

(a) Use the model to estimate the number of mutual funds in 1970.
(b) Use the model to estimate the year when the number of mutual funds will reach 5000.

35. **Pollution** Pollution levels in Lake Erie have been modeled* by the equation

$$x = 0.05 + 0.18e^{-0.38t},$$

where x is the volume of pollutants (in cubic kilometers) and t is the time (in years).
(a) Find the initial pollution levels; that is, find x when $t = 0$.
(b) How long before x is 30% of that initial level?

36. **Fish length** Suppose that the length x (in centimeters) of an individual of a certain species of fish is given by

$$x = 50 - 40e^{-0.05t},$$

where t is its age in months.
(a) Find the length after 1 year.
(b) How long (to the nearest month) until the length is 45 cm?

37. **Chemical reaction** When two chemicals, A and B, react to form another chemical C (such as in the digestive process), this is a special case of the **law of mass action,** which is fundamental to studying chemical reaction rates. Suppose that chemical C is formed from A and B according to

$$x = \frac{120[1 - (0.6)^{3t}]}{4 - (0.6)^{3t}},$$

where x is the number of pounds of C formed in t minutes.
(a) How much of C is present when the reaction begins?
(b) How much of C is formed in 4 minutes?
(c) How long does it take to form 10 lb of C?

38. **Newton's law of cooling** A certain object at 90°F that is allowed to cool in a room where the temperature is 20°F has its temperature T given as a function of time t (in minutes) according to

$$T = 20 + 70e^{-0.056t}.$$

(a) Find the temperature of the object 30 seconds after it is in the room.
(b) How long does it take before the temperature of the object is 20.5°F?

39. **Ventilation** The ventilation system in a building operates when the concentration of carbon dioxide (CO_2) reaches a certain level. Suppose that when the ventilation system operates, the cubic feet of CO_2, x, in an 8000 cu ft room depends on time t (in minutes) according to

$$x = 4.8 + 11.2e^{-t/4}.$$

(a) Find the initial amount of CO_2 in the room, and find the concentration of CO_2 (as a percent) at this time.
(b) How long does it take to have a concentration of 0.07% CO_2?
(c) The steady-state or equilibrium concentration is what would result if the ventilation system were left on indefinitely (as t becomes larger and larger). Determine the steady-state concentration.

KEY TERMS & FORMULAS

Section	Key Terms	Formula
5.1	Exponential functions	
	Growth functions	$f(x) = a^x \quad (a > 1)$
	Decay functions	$f(x) = a^{-x} \quad (a > 1)$
	e	$e \approx 2.71828$
5.2	Logarithmic function	$y = \log_a x$, defined by $x = a^y$
	Common logarithm	$\log x = \log_{10} x$
	Natural logarithm	$\ln x = \log_e x$
	Inverse functions	

*Adapted from R. H. Rainey, *Science* 155 (1967), pp. 1242–1243.

Logarithmic Properties I–V	$\log_a a^x = x$; $a^{\log_a x} = x$; $\log_a(MN) = \log_a M + \log_a N$; $\log_a(M/N) = \log_a M - \log_a N$; $\log_a(M^N) = N(\log_a M)$
Change of base formula	$\log_b x = \dfrac{\log_a x}{\log_a b}$
Calculation formula	$\log_b x = \dfrac{\ln x}{\ln b}$ or $\log_b x = \dfrac{\log x}{\log b}$

5.3   Exponential growth and decay
Decay models $f(x) = ma^{-x}$ $(a > 1, m > 0)$
Growth models $f(x) = ma^x$ $(a > 1, m > 0)$
Gompertz curves $N = Ca^{R^t}$

REVIEW EXERCISES

5.1, 5.2

1. Write each statement in logarithmic form.
   (a) $2^x = y$ (b) $3^y = 2x$
2. Write each statement in exponential form.
   (a) $\log_7(\frac{1}{49}) = -2$ (b) $\log_4 x = -1$

Graph the following functions.

3. $y = e^x$ 4. $y = e^{-x}$
5. $y = \log_2 x$ 6. $y = 2^x$
7. $y = 4^x$ 8. $y = \ln x$
9. $y = \log_4 x$ 10. $y = 3^{-2x}$
11. $y = \log x$ 12. $y = \log(x + 2)$
13. $y = \ln(x - 3)$ 14. $y = 10(2^{-x})$

In problems 15–22 evaluate each logarithm without using a calculator.

15. $\log_5 1$ 16. $\log_8 64$
17. $\log_{25} 5$ 18. $\log_3(\frac{1}{3})$
19. $\log_3 3^8$ 20. $\ln e$
21. $e^{\ln 5}$ 22. $10^{\log 3.15}$

If $\log_a x = 1.2$ and $\log_a y = 3.9$, find each of the following by using the properties of logarithms.

23. $\log_a\left(\dfrac{x}{y}\right)$ 24. $\log_a \sqrt{x}$
25. $\log_a(xy)$ 26. $\log_a(y^4)$

In problems 27 and 28, use the properties of logarithms to write each expression as the sum or difference of two logarithmic functions containing no exponents.

27. $\log(yz)$ 28. $\ln \sqrt{\dfrac{x + 1}{x}}$

29. Is it true that $\ln x + \ln y = \ln(x + y)$ for all values of x?
30. If $f(x) = \ln x$, find $f(e^{-2})$.
31. If $f(x) = 2^x + \log(7x - 4)$, find $f(2)$.
32. If $f(x) = e^x + \ln(x + 1)$, find $f(0)$.
33. If $f(x) = \ln(3e^x - 5)$, find $f(\ln 2)$.

In problems 34 and 35, use the change of base calculation formula to evaluate each logarithm.

34. $\log_9 2158$ 35. $\log_{12}(0.0195)$

In problems 36 and 37, rewrite each logarithm by using the change of base calculation formula, then graph the function with a graphing utility.

36. $y = \log_{\sqrt{3}} x$ 37. $f(x) = \log_{11}(2x - 5)$

APPLICATIONS

5.1, 5.2

38. **Stock price history** The graph in the figure on page 391 shows the share price history of the stock of Federal Signal Corporation from 1978 to 1993. If the share price were modeled as a function of time, which of the following would give the best model: a growth exponential, a decay exponential, or a logarithm? Justify your choice.

SOURCE: Federal Signal Corporation, 1993
Annual Report

39. **HIV infections** The figure shows the number of world
HIV infections since AIDS was discovered in 1981.
(a) Suppose t is the number of years since 1980 and y is
the number of world HIV infections (in hundreds of
thousands). Graph each of the following functions to
decide which one models the low estimate and which
one models the high estimate. Use a graphing utility.

$$y = 0.099t^{1.969} \quad \text{and} \quad y = 0.08356(1.454)^t$$

(b) Which of the two functions in part (a) is an exponential function? Explain your choice.

40. **Mutual funds** The figure shows the history of the
growth in the number of mutual funds. If the number of
mutual funds was modeled by a function, do you think
the best model would be linear, exponential, logarithmic, or quadratic? Explain.

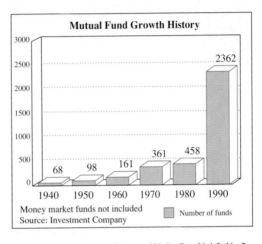

Published in *Investment Digest* of Valic Co., Vol 5, No 2,
Summer, 1992

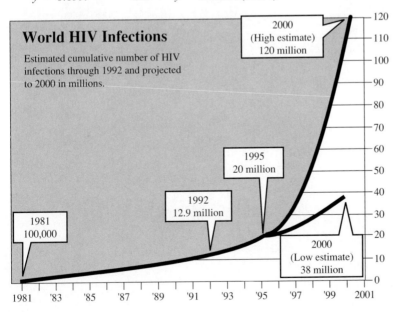

SOURCE: World Health Organization and the Global AIDS Policy Coalition, as reported
in the *El Paso Times*, December 6, 1992.

41. ***Stellar magnitude*** The stellar magnitude M of a star is related to its brightness B as seen from earth according to

$$M = -\frac{5}{2}\log{(B/B_0)},$$

where B_0 is a standard level of brightness (the brightness of the star Vega).
   (a) Find the magnitude of Venus if its brightness is 36.3 times B_0.
   (b) Find the brightness of the North Star if its magnitude is 2.1 (as a multiple of B_0).
   (c) If the faintest stars have magnitude 6, find their brightness (as a multiple of B_0).
   (d) Is a star with magnitude -1.0 brighter than a star with magnitude $+1.0$?

5.3

42. ***Advertising and sales*** Because of a new advertising campaign, a company predicts that sales will increase, and that the yearly sales will be given by the equation

$$N = 10{,}000(0.3)^{0.5^t},$$

where t represents the number of years after the start of the campaign.
   (a) What are the sales when the campaign begins?
   (b) What are the predicted sales for the third year?
   (c) What are the maximum predicted sales?

43. ***Sales decay*** The sales decay for a product is given by $S = 50{,}000e^{-0.1x}$, where S is the weekly sales and x is the number of weeks that have passed since the end of an advertising campaign.
   (a) What will sales be 6 weeks after the end of the campaign?
   (b) How many weeks will pass before sales drop below 15,000?

44. ***Sales decay*** The sales decay for a product is given by $S = 50{,}000e^{-0.6x}$, where S is the monthly sales and x is the number of months that have passed since the end of an advertising campaign. What will sales be 6 months after the end of a campaign?

45. ***Compound interest*** If \$1000 is invested at 12%, compounded monthly, the future value A at any time t (in years) is given by

$$A = 1000(1.01)^{12t}.$$

How long will it take for the amount to double?

46. ***Compound interest*** If \$5000 is invested at 13.5%, compounded continuously, then the future value A at any time t (in years) is given by $A = 5000e^{0.135t}$.
   (a) What is the amount after 9 months?
   (b) How long before the investment doubles?

ONSULTANT'S CORNER

Agricultural Business Management

A commercial vegetable and fruit grower carefully observes the relationship between the amount of fertilizer used on a certain variety of squash and the extra revenue made from the sales of the resulting squash crop. These observations are recorded in the table below.

Amount of fertilizer (lbs/acre)	0	250	500	750	1000	1250	1500	1750	2000
Extra revenue earned (dollars/acre)	0	96	145	172	185	192	196	198	199

The fertilizer costs 14¢ per pound. What would you advise the grower is the most profitable amount of fertilizer to use?

Check your advice by answering the following questions.

1. Is each pound of fertilizer equally effective?
2. Estimate the maximum amount of extra revenue that can be earned per acre. (Round to the nearest dollar.)
3. Graph the extra revenue made from improved squash sales as a function of the amount of fertilizer used. What does the graph tell you about the growth of the extra revenue as a function of the amount of fertilizer?
4. Which of the following equations best models the graph you obtained in question 3?

   (a) $N = Ca^{Rx}$ (b) $y = \dfrac{A}{1 + ce^{-ax}}$ (c) $y = c(1 - e^{-ax})$

   Determine a specific equation that fits the observed data. That is, find the value of the constants in the model you have chosen.
5. Graph the total cost of the fertilizer as a function of the amount of fertilizer used on the graph you created for question 3. Determine an equation that models this relationship. For what amount of fertilizer is the extra revenue earned from squash sales offset by the cost of the fertilizer?
6. Use the graphs or the equations you have created to estimate the maximum extra profit that can be made and the amount of fertilizer per acre that would give that extra profit.

Warm-up

Prerequisite problem type	For section	Answer	Section for review
What is $\dfrac{(-1)^n}{2n}$, if (a) $n = 1$ (b) $n = 2$ (c) $n = 3$	**6.2**	(a) $-\frac{1}{2}$ (b) $\frac{1}{4}$ (c) $-\frac{1}{6}$	0.2 Signed numbers
(a) What is $(-2)^6$? (b) What is $\dfrac{\frac{1}{4}\left[1 - \left(\frac{1}{2}\right)^6\right]}{1 - \frac{1}{2}}$?	**6.2**	(a) 64 (b) $\frac{63}{128}$	0.2 Signed numbers
If $f(x) = \dfrac{1}{2x}$, what is (a) $f(2)$? (b) $f(4)$?	**6.1**	(a) $\frac{1}{4}$ (b) $\frac{1}{8}$	1.4 Functional notation
Evaluate $e^{(.08)(20)}$	**6.2**	4.95303	5.1 Exponential functions
Evaluate $\dfrac{1 - (1.05)^{-16}}{0.05}$	**6.3** **6.4**	10.83777	0.3 Integral exponents

CHAPTER 6

MATHEMATICS OF FINANCE

Regardless of whether or not your career is in business, understanding how interest is computed on investments and loans is important to you as a consumer. The proliferation of personal finance and money management software for home computers attests to this importance. In this chapter we'll explain the mathematics that is the basis for such software packages.

We begin by studying simple interest, sometimes used on short-term investments or loans. Most investments (for example, savings accounts or bonds) pay compound interest, where the interest is calculated over short periods of time and added to the principal. Most loans are not simple interest loans because they are repaid by making partial payments of both principal and interest during the period of the loan; one such method of repaying a loan is called *amortization*.

The goal of this chapter is to provide some understanding of the methods used to determine the interest and the future value (principal plus interest) resulting from savings plans and the methods used in repayment of debts. We will also study sequences, and pay special attention to arithmetic and geometric sequences because of their applications in mathematics of finance. From these we will develop formulas for computing compound interest, the future and present values of annuities, amortization of debts, and the amount paid into a sinking fund to discharge a debt.

6.1 SIMPLE INTEREST; SEQUENCES

Objectives

❏ To compute simple interest
❏ To write a specified number of terms of a sequence
❏ To find specified terms of arithmetic sequences
❏ To find sums of specified numbers of terms of arithmetic sequences

APPLICATION PREVIEW

In this section we begin our study of the mathematics of finance by considering **simple interest.** Simple interest forms the basis for all calculations involving interest that is paid on an investment or that is due on a loan. Also, we can evaluate short-term investments (such as in stocks or real estate) by calculating their simple interest rate. For example, on April 29, 1994, Mary Spaulding purchased 100 shares of Exxon stock for

$6125.00. After 6 months, the stock had risen in value by $138.00 and had paid dividends totaling $144.14. One way to evaluate this investment and compare it to a bank savings plan is to find the simple interest rate that this gain represents.

SIMPLE INTEREST

If a sum of money (called the **principal**) P is invested for a time period t (frequently years) at an interest rate r per period, the **simple interest** is given by the following formula.

Simple Interest

$$I = Prt$$
I = interest (in dollars)
P = principal (in dollars)
r = annual interest rate (written as a decimal)
t = time (in years)*

Simple interest is paid on investments involving time certificates issued by banks and on certain types of bonds, such as U.S. government series H bonds and municipal bonds. The interest for a given period is paid to the investor and the principal remains the same.

EXAMPLE 1

If $4000 is invested for 2 years at an annual interest rate of 9%, how much interest will be received at the end of the 2-year period?

Solution
The interest is

$$I = \$4000(0.09)(2) = \$720.$$

The **future amount of an investment** or its **future value** at the end of an interest period is the sum of the principal and the interest. Thus in Example 1, the future value is

$$S = \$4000 + \$720 = \$4720.$$

Interest on loans may also be calculated using simple interest.

---

*Periods of time other than years can be used, as can other monetary systems.

EXAMPLE 2

If $4000 is borrowed for 39 weeks at an annual interest rate of 15%, how much interest is due at the end of the 39 weeks?

Solution

The principal is $P = 4000$ (dollars).
The annual interest rate is $r = 15\% = 0.15$.
The time in years is $t = \frac{39}{52} = \frac{3}{4}$
The interest due is $I = 4000(0.15)(\frac{3}{4})$
$$= 450 \text{ (dollars).}$$

The amount of money that would have to be repaid at the end of six months in the loan of Example 2 is the principal plus the interest, or

$$S = \$4000 + \$450 = \$4450.$$

This is called the **future amount of the loan** or its **future value.**

Future Value

If we use the letter S to denote the future value of an investment or a loan, then we have the following:

Future value of investment or loan: $S = P + I$

where P is the principal (in dollars) and I is the interest (in dollars).

The principal P of a loan is also called the **face value** or the **present value** of the loan.

EXAMPLE 3

If $2000 is invested for 5 months at a simple interest rate of 1% per month, what is the future value of the investment at the end of the 5-month period?

Solution

The interest for the 5-month period is $I = \$2000(0.01)(5) = \$100$. Thus, the future value of the investment for the period is $S = \$2000 + \$100 = \$2100$.

We now return to the Application Preview problem.

EXAMPLE 4

Find the simple interest rate Mary Spaulding earned on her Exxon stock investment. Recall that she bought the stock for $6125.00, and after 6 months the value of her shares had risen by $138.00 and dividends totaling $144.14 had been paid.

Solution

We can view the principal as \$6125.00, the time as 1/2 year, and the interest earned as the total of all gains (that is, interest $= I = 138.00 + 144.14 = 282.14$). Hence

$$I = Prt$$
$$282.14 = (6125)r(0.5)$$
$$282.14 = 3062.5r$$
$$\frac{282.14}{3062.5} = r \approx 0.092 = 9.2\%$$

Thus Mary's return was equivalent to an annual simple interest rate of about 9.2%.

CHECKPOINT

1. What is the simple interest formula?

2. If \$8000 is invested at 6% simple interest for 9 months, find the future value of the investment.

SEQUENCES

Let us look at the future values of the investment in Example 3, at the end of each of the 5 months it was invested.

Month	Interest $(I = Prt)$	Future Value of the Investment
1	(\$2000)(0.01)(1) = \$20	\$2000 + \$20 = \$2020
2	(\$2000)(0.01)(1) = \$20	\$2020 + \$20 = \$2040
3	(\$2000)(0.01)(1) = \$20	\$2040 + \$20 = \$2060
4	(\$2000)(0.01)(1) = \$20	\$2060 + \$20 = \$2080
5	(\$2000)(0.01)(1) = \$20	\$2080 + \$20 = \$2100

These future values are outputs that result when the inputs are positive integers that correspond to the number of months of the investment. Outputs (such as these future values) that arise uniquely from positive integer inputs define a function whose domain is the positive integers. Such a function is called a **sequence function.** Because the domain of a sequence function is the positive integers, the outputs form an ordered list, called a **sequence.**

Sequence   The set of functional values $a_1, a_2, a_3, \ldots$ of a sequence function is called a **sequence.** The values $a_1, a_2, a_3, \ldots$ are called **terms** of the sequence, with a_1 the first term, a_2 the second term, and so on.

Because calculations involving interest often result from using positive integer inputs, sequences are the basis for most of the financial formulas derived in this chapter.

EXAMPLE 5

Write the first four terms of the sequence whose nth term is $a_n = (-1)^n/(2n)$.

Solution

The first four terms of the sequence are as follows:

$$a_1 = \frac{(-1)^1}{2(1)} = -\frac{1}{2} \qquad a_2 = \frac{(-1)^2}{2(2)} = \frac{1}{4}$$

$$a_3 = \frac{(-1)^3}{2(3)} = -\frac{1}{6} \qquad a_4 = \frac{(-1)^4}{2(4)} = \frac{1}{8}$$

We usually write these terms in the form $-\frac{1}{2}, \frac{1}{4}, -\frac{1}{6}, \frac{1}{8}$.

EXAMPLE 6

If the nth term of a sequence is

$$a_n = \frac{n-2}{n(n+1)},$$

write the first four terms and the eighth term of the sequence.

Solution

The first four terms and the eighth term of the sequence are as follows:

$$a_1 = \frac{1-2}{1(1+1)} = -\frac{1}{2} \qquad a_2 = \frac{2-2}{2(2+1)} = 0$$

$$a_3 = \frac{3-2}{3(3+1)} = \frac{1}{12} \qquad a_4 = \frac{4-2}{4(4+1)} = \frac{1}{10}$$

$$a_8 = \frac{8-2}{8(8+1)} = \frac{1}{12}$$

ARITHMETIC SEQUENCES

Earlier in this section, we described the sequence

$$2020, 2040, 2060, 2080, 2100, 2120, 2140, \ldots$$

This sequence can also be described in the following way:

$$a_1 = 2020, \qquad a_n = a_{n-1} + 20 \quad \text{for } n > 1.$$

This sequence is an example of a special kind of sequence called an **arithmetic sequence.** In such a sequence, each term after the first can be found by adding a constant to the preceding term. Thus, we have the following definition.

Arithmetic Sequence   A sequence is called an **arithmetic sequence** (progression) if there exists a number d, called the **common difference,** such that

$$a_n = a_{n-1} + d \qquad \text{for } n > 1.$$

EXAMPLE 7

Write the next three terms of the following arithmetic sequences:
(a) $1, 3, 5, \ldots$ (b) $9, 6, 3, \ldots$ (c) $\frac{1}{2}, \frac{5}{6}, \frac{7}{6}, \ldots$

Solution
(a) The common difference is 2, so the next three terms are 7, 9, 11.
(b) The common difference is -3, so the next three terms are 0, -3, -6.
(c) The common difference is $\frac{1}{3}$, so the next three terms are $\frac{3}{2}, \frac{11}{6}, \frac{13}{6}$.

Since each term after the first in an arithmetic sequence is obtained by adding d to the preceding term, the second term is $a_1 + d$, the third is $(a_1 + d) + d = a_1 + 2d, \ldots$ and the nth term is $a_1 + (n - 1)d$. Thus we have the following formula.

The **n th term of an arithmetic sequence** (progression) is given by

$$a_n = a_1 + (n - 1)d,$$

where a_1 is the first term and d is the common difference between successive terms.

EXAMPLE 8

Find the eleventh term of the arithmetic sequence with first term 3 and common difference -2.

Solution
The eleventh term is $a_{11} = 3 + (11 - 1)(-2) = -17$.

EXAMPLE 9

If the first term of an arithmetic sequence is 4 and the ninth term is 20, find the fifth term.

Solution
Substituting the values $a_1 = 4$, $a_n = 20$, and $n = 9$ in $a_n = a_1 + (n - 1)d$ gives $20 = 4 + (9 - 1)d$. Solving this equation gives $d = 2$. Therefore, the fifth term is $a_5 = 4 + (5 - 1)2 = 12$.

Consider the arithmetic sequence with first term a_1, common difference d, and nth term a_n. The first n terms of an arithmetic sequence can be written from a_1 to a_n as

$$a_1, a_1 + d, a_1 + 2d, a_1 + 3d, \ldots, a_1 + (n - 1)d$$

or backward from a_n to a_1 as

$$a_n, a_n - d, a_n - 2d, a_n - 3d, \ldots, a_n - (n-1)d.$$

If we let s_n represent the sum of the first n terms of the sequence described above, then we have

$$s_n = a_1 + (a_1 + d) + (a_1 + 2d) + \cdots + [a_1 + (n-1)d], \quad (1)$$

or alternately,

$$s_n = a_n + (a_n - d) + (a_n - 2d) + \cdots + [a_n - (n-1)d]. \quad (2)$$

If we add equations (1) and (2) together, term by term, we obtain

$$2s_n = (a_1 + a_n) + (a_1 + a_n) + (a_1 + a_n) + \cdots + (a_1 + a_n),$$

in which $(a_1 + a_n)$ appears as a term n times. Thus,

$$2s_n = n(a_1 + a_n)$$

$$s_n = \frac{n}{2}(a_1 + a_n).$$

We may state this result as follows.

The **sum of the first n terms of an arithmetic sequence** is given by the formula

$$s_n = \frac{n}{2}(a_1 + a_n),$$

where a_1 is the first term of the sequence and a_n is the nth term.

EXAMPLE 10

Find the sum of the first 10 terms of the arithmetic sequence with first term 2 and common difference 4.

Solution

We are given the values $n = 10$, $a_1 = 2$, and $d = 4$. Thus, the tenth term is $a_{10} = 2 + (10 - 1)4 = 38$, and the sum is

$$s_{10} = \frac{10}{2}(2 + 38) = 200.$$

EXAMPLE 11

Find the sum of the first 7 terms of the arithmetic sequence $\frac{1}{4}, \frac{7}{12}, \frac{11}{12} \ldots$.

Solution

The first term is $\frac{1}{4}$ and the common difference is $\frac{1}{3}$. Therefore, the seventh term is $a_7 = \frac{1}{4} + (7 - 1)(\frac{1}{3}) = \frac{9}{4}$. The sum is $s_7 = \frac{7}{2}(\frac{1}{4} + \frac{9}{4}) = \frac{35}{4} = 8\frac{3}{4}$.

CHECKPOINT

3. Of the following sequences, identify the arithmetic sequences.
    (a) 1, 4, 9, 16, . . .   (b) 1, 4, 7, 10, . . .   (c) 1, 2, 4, 8, . . .
4. Given the arithmetic sequence $-10, -6, -2, . . .$, find
    (a) the 51st term,   (b) the sum of the first 51 terms.

Finally, we should note that many graphics calculators, some graphics software packages, and spreadsheets have the capability of defining sequence functions and then operating on them by finding additional terms, graphing the terms, or summing a fixed number of terms.

CHECKPOINT SOLUTIONS

1. $I = Prt$ 2. $S = P + I = 8000 + 8000(0.06)(\frac{9}{12}) = \$8360$
3. Only (b) is an arithmetic sequence; the common difference is 3.
4. (a) The common difference is $d = 4$. The 51st term $(n = 51)$ is
       $a_n = a_1 + (n - 1)d = -10 + 50(4) = -10 + 200 = 190.$
    (b) For $a_1 = -10$ and $a_n = 190$, the sum is $s_n = \frac{51}{2}(-10 + 190) = 4590.$

EXERCISE 6.1

SIMPLE INTEREST

1. If $10,000 is invested for 6 years at an annual simple interest rate of 16%,
    (a) how much interest will be earned?
    (b) what is the future value of the investment at the end of the 6 years?
2. If $800 is invested for 5 years at an annual simple interest rate of 14%,
    (a) how much interest will be earned?
    (b) what is the future value of the investment at the end of the 5 years?
3. If $1000 is invested for 3 months at an annual simple interest rate of 12%,
    (a) how much interest will be earned?
    (b) what is the future value of the investment after 3 months?
4. If $1800 is invested for 9 months at an annual simple interest rate of 15%,
    (a) how much interest will be earned?
    (b) what is the future value of the investment after 9 months?
5. If you borrow $800 for 6 months at 16%, simple interest, how much must you repay at the end of the 6 months?

6. If you borrow $1600 for 2 years at 14%, simple interest, how much must you repay at the end of the 2 years?
7. If you loan $3500 to a friend for 15 months at 8% simple interest, find the future value of the loan.
8. Mrs. Gonzalez loaned $2500 to her son Luis for 7 months at 9% simple interest. What is the future value of this loan?
9. A couple bought some stock for $30 per share that pays an annual dividend of $0.90 per share. After one year the price of the stock was $33. Find the simple interest rate they earned on their investment.
10. Jenny Reed bought SSX stock for $16 per share. If the annual dividend was $1.50 per share and after one year SSX was selling for $35 per share, find the simple interest rate of growth of her money.
11. (a) To buy a treasury bill (T-bill) that matures to $10,000 in 6 months, you must pay $9685.23. What rate does this earn?
    (b) If the bank charges a fee of $40 to buy a T-bill, what is the actual interest rate you earn?
12. Janie Christopher loaned $6000 to a friend for 90 days at 12%. After 30 days, she sold the note to a third party for $6000. What interest rate did the third party receive? Use 360 days in a year.

13. A firm buys 12 file cabinets at $140 each, with the bill due in 90 days. How much must the firm deposit now to have enough to pay the bill if money is worth 12% per year? Use 360 days in a year.

14. A student has a savings account earning 9% simple interest. She must pay $1500 for first semester tuition by September 1 and $1500 for second semester on January 1. How much must she earn in the summer (by September 1) in order to pay the first semester bill on time and still have the remainder of her summer earnings grow to $1500 between September 1 and January 1?

15. A retailer owes a wholesaler $500,000 due in 45 days. If the bill is 15 days late, there is a 1% penalty charge. The retailer can get a 45-day certificate paying 6% or a 60-day certificate paying 7%. Is it better to take the 45-day certificate and pay on time or the 60-day certificate and pay late with the penalty?

16. An investor owns several apartment buildings. The taxes on these buildings total $30,000 per year and are due before April 1. The late fee is 1/2% per month up to 6 months, at which time the buildings are seized by the authorities and sold for back taxes. If the investor has $30,000 available on March 31, will he save money by paying the taxes at that time or by investing the money at 8% and paying the taxes and the penalty on September 30?

SEQUENCES

17. Write the first ten terms of the sequence defined by $a_n = 3n$.

18. Find the first six terms of the sequence defined by $a_n = 4n$.

19. Write the first eight terms of the sequence defined by $a_n = n/3$.

20. Write the first seven terms of the sequence defined by $a_n = 2/n$.

21. Write the first six terms of the sequence whose nth term is $(-1)^n/4$.

22. Write the first five terms of the sequence whose nth term is $(-1)^n/3$.

23. Write the first six terms of the sequence whose nth term is $(-1)^n/(2n)$.

24. Write the first five terms of the sequence whose nth term is
$$a_n = \frac{(-1)^n}{n + 1}.$$

25. Write the first four terms and the tenth term of the sequence whose nth term is
$$a_n = \frac{n - 4}{n(n + 2)}.$$

26. Write the sixth term of the sequence whose nth term is
$$a_n = \frac{n(n - 1)}{n + 3}.$$

ARITHMETIC SEQUENCES

In problems 27–30, (a) identify d and a_1 and (b) write the next 3 terms.

27. 2, 5, 8, . . .

28. 3, 9, 15, . . .

29. $3, \frac{9}{2}, 6, \ldots$

30. 2, 2.75, 3.5, . . .

31. Find the 28th term of the arithmetic sequence with first term -3 and common difference 4.

32. Find the 31st term of the arithmetic sequence with first term 7 and common difference 4.

33. Find the 83rd term of the arithmetic sequence with first term 6 and common difference $-\frac{1}{2}$.

34. Find the 66th term of the arithmetic sequence with first term $\frac{1}{2}$ and common difference $-\frac{1}{3}$.

35. Find the 100th term of the arithmetic sequence with first term 5 and eighth term 19.

36. Find the 73rd term of the arithmetic sequence with first term 20 and tenth term 47.

37. Find the sum of the first 38 terms of the arithmetic sequence with first term 2 and common difference 3.

38. Find the sum of the first 56 terms of the arithmetic sequence with first term 6 and common difference 4.

39. Find the sum of the first 70 terms of the arithmetic sequence with first term 10 and common difference $\frac{1}{2}$.

40. Find the sum of the first 80 terms of the arithmetic sequence with first term 12 and common difference -3.

41. Find the sum of the first 50 terms of the arithmetic sequence 2, 4, 6, ...

42. Find the sum of the first 60 terms of the arithmetic sequence 6, 9, 12, ...

43. Find the sum of the first 150 terms of the arithmetic sequence $6, \frac{9}{2}, 3, \ldots$.

44. Find the sum of the first 200 terms of the arithmetic sequence 12, 9, 6, ...

APPLICATIONS

45. **Bee reproduction** A female bee hatches from a fertilized egg while a male bee hatches from an unfertilized egg. Thus a female bee has a male parent and a female

parent while a male bee only has a female parent. Therefore the number of ancestors of a male bee follows the *Fibonacci sequence*

$$1, 2, 3, 5, 8, 13, \ldots$$

Observe the pattern and write three more terms of the sequence.

46. **Salaries** Suppose you are offered a job with a relatively low starting salary but with a $1500 raise for each of the next 7 years. How much more than your starting salary would you be making in the eighth year?

47. **Profit** A new firm loses $2000 in its first month, but its profit increases by $400 in each succeeding month for the next year. What is its profit in the twelfth month?

48. **Pay raises** If you make $27,000 and get $1800 raises each year, in how many years will your salary double?

49. **Salaries** Suppose you are offered two identical jobs: one paying a starting salary of $20,000 with yearly raises of $1000 and one paying a starting salary of $18,000 with yearly raises of $1200. Which job will be paying you more for your tenth year on the job?

50. **Profit** A new firm loses $2000 in its first month, but its profit increases by $400 in each succeding month for the next year. What is its profit for the year?

51. **Pay raises** If you are an employee, would you rather be given a raise of $1000 at the end of each year (Plan I) or $300 at the end of each 6-month period (Plan II)? Consider the following table for an employee whose base salary is $20,000 per year (or $10,000 per 6-month period), and answer parts (a)–(g).

Period (in Months)	Salary Received per 6-Month Period	
	Plan I	Plan II
0–6	$10,000	$10,000
6–12	10,000	10,300
12–18	10,500	10,600
18–24	10,500	10,900
24–30	11,000	11,200
30–36	11,000	11,500

(a) Find the sum of the raises for Plan I for the first 3 years.
(b) Find the sum of the raises for Plan II for the first 3 years.
(c) Which plan is better, and by how much?
(d) Find the sum of the raises in Plan I for 5 years.
(e) Find the sum of the raises in Plan II for 5 years.
(f) Which plan is better, and by how much?
(g) Do you want Plan I or Plan II?

52. **Pay raises** As an employee, would you prefer being given a $1200 raise each year for 5 years or a $200 raise each quarter for 5 years?

6.2 COMPOUND INTEREST; GEOMETRIC SEQUENCES

Objectives

❑ To find the future value of a compound interest investment and the amount of interest earned when interest is compounded at regular intervals
❑ To find future value and interest earned on money where interest is compounded continuously
❑ To find the effective annual interest rate of money invested at compound interest
❑ To find specified terms of geometric sequences
❑ To find the sums of specified numbers of terms of geometric sequences

APPLICATION PREVIEW

A compound interest investment is one in which interest is paid into an account at regular intervals throughout the duration of the investment. After an interest payment is made, that interest, as well as the principal, earns interest. Because of the seeming complexity of computing interest on the interest, interest on the interest on the interest, and so on, we need to develop formulas that allow us to find the **future value** of such an investment. It seems reasonable that this future value depends on the duration of the investment and on the interest rate, but it also depends on how the periodic interest payments are made (that is, on how the compounding is done).

  Because the compounding method affects the future value, another application of this section develops formulas for the **effective annual rate** of an investment. By calculating the effective annual rates of investments with different compounding schemes, we can meaningfully compare those investments.

Finally, we should note that the methods we use with compound interest in finding periodic interest rates and numbers of interest periods are important for all subsequent financial applications of this chapter, including such things as savings plans (annuities) and debt repayment with regular payments (amortization).

COMPOUND INTEREST

In Section 6.1 we discussed simple interest. A second method of paying interest is the **compound interest** method, where the interest for each period is added to the principal before interest is calculated for the next period. With this method the principal grows as the interest is added to it. This method is used in investments such as savings accounts and U.S. government series E bonds.

An understanding of compound interest is important not only for people planning careers with financial institutions, but also for anyone planning to invest money. To see how compound interest is computed, we will first calculate the *future value* that will result if $20,000 is invested for three years at 10%, *compounded annually* (each year).

The principal for the first year is $20,000.

The interest at the end of the first year is

$$I = \$20{,}000(0.10)(1) = \$2000.$$

The future value at the end of the first year is

$$S = P + I = \$20{,}000 + \$2000 = \$22{,}000.$$

Thus the principal for the second year is $22,000.

The interest at the end of the second year is

$$I = \$22{,}000(0.10)(1) = \$2200.$$

The future value at the end of the second year is

$$S = \$22{,}000 + \$2200 = \$24{,}200.$$

Thus the principal for the third year is $24,200.

The interest at the end of the third year is

$$I = \$24{,}200(0.10)(1) = \$2420.$$

The future value at the end of the third year is

$$S = \$24{,}200 + \$2420 = \$26{,}620.$$

We see that calculation of the future value of a compound interest investment is very tedious if we proceed in this manner. But notice that the future value for each year can be found by multiplying the amount from the preceding year by 1.10, or $1 + 0.10$. That is, the future values at the end of the first, second, and third years are, respectively,

$$\$20,000(1.10) = \$22,000,$$

$$[\$20,000(1.10)](1.10) = \$20,000(1.10)^2 = \$24,200, \qquad \text{and}$$

$$[\$20,000(1.10)^2](1.10) = \$20,000(1.10)^3 = \$26,620.$$

This suggests that if we maintained this investment for n years, the future value at the end of this time would be $\$20,000(1.10)^n$. Thus, we have the following general formula.

If $\$P$ is invested at an interest rate of i per year, compounded annually, the future value S at the end of the nth year is

$$S = P(1 + i)^n.$$

EXAMPLE 1

If $3000 is invested for 4 years at 9%, compounded annually, how much interest is earned?

Solution

The future value is

$$S = \$3000(1 + 0.09)^4$$

$$= \$3000(1.4115816)$$

$$= \$4234.7448$$

$$= \$4234.74, \text{ to the nearest cent.}$$

Since $3000 of this amount was the original investment, the interest earned is $4234.74 − $3000 = $1234.74.

Some accounts have the interest compounded semiannually, quarterly, monthly, or daily. Unless specifically stated otherwise, a stated interest rate is the rate per year, and is called the **nominal annual rate.** The interest rate *per period* is the nominal rate divided by the number of interest periods per year. The interest periods are also called *conversion periods*. Thus, if $100 is invested for five years at 6% compounded semiannually (twice a year), it has been invested for $n = 10$ periods (5 years × 2 periods per year) at $i = 3\%$ per period (6% per year ÷ 2 per year). The future value of investments of this type is found using the following formula.

Future Value (Periodic Compounding)	If P is invested for t years at a nominal interest rate, j, compounded m times per year, then the total number of compounding periods is $$n = mt,$$ the interest rate per compounding period is $$i = \frac{j}{m},$$ and the future value is $$S = P(1 + i)^n = P\left(1 + \frac{j}{m}\right)^{mt}.$$

Because compound interest is used so frequently in business, tables are available that give the future value of $1 invested for different periods and at different rates. Table II in the appendix is a sample of such a table. Note that the left-hand column of this table lists values for the total number of compounding periods, n, and the top row lists values for the interest rate per period, i. The future value can also be calculated from the formula using a calculator. Further, if we want to track the future value for several periods, the "fill down" or "fill across" capabilities of a spreadsheet are ideal.

EXAMPLE 2

If $8000 is invested for 6 years at 8%, compounded quarterly, find (a) the future value, (b) the compound interest.

Solution

(a) Since $P = \$8000$, $t = 6$ years, $j = 0.08$, and $m = 4$ compounding periods per year, the future value is given by

$$S = P\left(1 + \frac{j}{m}\right)^{mt}$$

$$= \$8000\left(1 + \frac{0.08}{4}\right)^{4 \cdot 6} = \$8000(1 + 0.02)^{24}$$

$$= \$8000(1.608437) = \$12{,}867.50.$$

To use Table II in the appendix, we must find the total number of compounding periods ($n = (4)(6) = 24$) and the interest rate per period ($i = 0.08/4 = 0.02$). From Table II we see that the future value of $1 invested for 24 periods at 2% per period is $1.608437. Thus the future value is

$$S = \$8000(1.608437) = \$12{,}867.50.$$

(b) The compound interest is given by $12,867.50 − $8000 = $4867.50.

EXAMPLE 3

If \$500 is invested for 3 years 8%, compounded quarterly (four times a year), what interest is earned?

Solution

Since $P = \$500$, $t = 3$ years, $j = 0.08$, and $m = 4$ compounding periods per year, the total number of compounding periods is

$$n = (4)(3) = 12,$$

the interest rate per period is

$$i = \frac{0.08}{4} = 0.02,$$

and the future value is

$$
\begin{aligned}
S &= P(1 + i)^n \\
&= \$500(1 + 0.02)^{12} \\
&= \$500(1.2682418) \\
&= \$634.1209 \\
&= \$634.12.
\end{aligned}
$$

Thus the interest is $\$634.12 - \$500 = \$134.12$.

If we invest a sum of money, say \$100, the higher the interest rate, the greater the future value. Figure 6.1 shows a graphical comparison of the future value when \$100 is invested at 5%, 8%, and 10%, all compounded annually over a period of 30 years. Notice that higher interest rates yield consistently higher future values and have dramatically higher future values after 15–20 years. Notice also that these graphs of the future values are growth exponentials.

Figure 6.1

Suppose $1 is invested for one year at a nominal rate of 8%, compounded semiannually. Then the future value will be $1(1.04)^2 = 1.0816, and the interest for the year will be $1.0816 - $1 = 0.0816. Note that this amount of interest represents a *simple* interest rate of 8.16%. Since over one year the nominal rate of 8%, compounded semiannually, is equivalent to a simple interest rate of 8.16%, we say that the effective annual rate in this case is 8.16%. We can find the effective annual rate with the following formula.

Effective Annual Rate If j represents the nominal rate and m the number of compounding periods per year, then the effective annual rate, r, is given by

$$r = \left(1 + \frac{j}{m}\right)^m - 1.$$

Thus, although we cannot directly compare two nominal rates having different compounding periods, we can compare their equivalent simple interest rates, i.e., their corresponding effective annual rates.

To prove that this formula gives the true (effective) interest rate, consider the following: If I, S, and P represent interest, future value, and principal, respectively, for a 1-year period, then

$$I = S - P$$

$$I = P\left(1 + \frac{j}{m}\right)^{mt} - P$$

$$\frac{I}{P} = \left(1 + \frac{j}{m}\right)^{mt} - 1.$$

But if the time is 1 year, then $mt = m(1)$ and $I = Prt = Pr(1)$. Thus, the true yearly rate is

$$r = \frac{I}{P} = \left(1 + \frac{j}{m}\right)^m - 1.$$

EXAMPLE 4

Find the effective annual rate equivalent to the nominal rate of 6% compounded quarterly.

Solution

Since the number of periods per year is $m = 4$ and the nominal rate is $j = 0.06$, the rate per period is $j/m = 0.06/4 = 0.015$. Thus,

$$r = (1 + 0.015)^4 - 1 = 1.061364 - 1$$
$$= 0.061364$$
$$= 6.14\%.$$

It is reasonable to assume that the more frequently the interest is compounded, the larger the future value will become. In order to determine the interest that results from *continuous* compounding (compounding every instant), consider an investment of $1 for one year at a 100% interest rate. If the interest is compounded m times per year, the future value is given by

$$S = \left(1 + \frac{1}{m}\right)^m.$$

Table 6.1 shows the future value that results as the number of compounding periods increases.

Table 6.1

Compounded	Number of Periods per Year	Future Value
Annually	1	$(1 + \frac{1}{1})^1 = 2$
Monthly	12	$(1 + \frac{1}{12})^{12} = 2.1630 \ldots$
Daily	360 (business year)	$(1 + \frac{1}{360})^{360} = 2.7145 \ldots$
Hourly	8640	$(1 + \frac{1}{8640})^{8640} = 2.71812 \ldots$
Each minute	518,400	$(1 + \frac{1}{518400})^{518400} = 2.71827 \ldots$

Table 6.1 shows that as the number of periods per year increases, the future value increases, although not very rapidly. In fact, no matter how often the interest is compounded, the future value will never exceed $2.72. We say that as the number of periods increases, the future value approaches a **limit,** which is the number e;

$$e = 2.7182818. \ldots$$

We have discussed the number e and the function $y = e^x$ in Chapter 5. The discussion here shows one way the number we call e may be derived. We will define e more formally later.

Future Value (Continuous Compounding)

In general, if P is invested for t years at a nominal rate, j, compounded continuously, then the future value is given by the exponential function

$$S = Pe^{jt}.$$

The Internal Revenue Service charges interest with continuous compounding on the unpaid balance of taxes due or paid late.

EXAMPLE 5
Find the future value if $1000 is invested for 20 years at 8%, compounded continuously.

Solution

The future value is

$$S = \$1000e^{(0.08)(20)} = \$1000e^{1.6}$$
$$= \$1000(4.95303) \quad (\text{since } e^{1.6} = 4.95303)$$
$$= \$4953.03.$$

EXAMPLE 6

(a) How much more will you earn if you invest $1000 for 5 years at 8% compounded continuously instead of 8% compounded quarterly?
(b) What is the effective annual rate of 8% compounded continuously?

Solution

(a) If the interest is compounded continuously, the future value at the end of the 5 years is

$$S = \$1000e^{(0.08)(5)} = \$1000e^{0.4} = \$1000(1.49182)$$
$$= \$1491.82.$$

If the interest is compounded quarterly, the future value at the end of the 5 years is

$$S = \$1000(1.02)^{20} = \$1000(1.485947)$$
$$= \$1485.95, \text{ to the nearest cent.}$$

Thus the extra interest earned by compounding continuously is

$$\$1491.82 - \$1485.95 = \$5.87.$$

(b) For periodic compounding, the effective annual rate was found from the interest earned on an investment of $1 for 1 year. Using that same idea for continuous compounding, we get

$$r = e^j - 1 = e^{0.08} - 1 \approx 0.0833.$$

Thus, the effective annual rate is 8.33%.

We have seen that if P dollars is invested for t years at a nominal interest rate, j, compounded m times per year, then the future value S that results is

$$S = P\left(1 + \frac{j}{m}\right)^{mt}.$$

We also saw that if P dollars is invested for t years at a nominal rate, j, compounded continuously, the future value is given by

$$S = Pe^{jt}.$$

From our development of this continuous compounding formula, we saw that when the number of compounding periods is very large, there is little difference between periodic and continuous compounding. To better understand the effect of compounding periods on an investment, consider the following example.

EXAMPLE 7

If $100 is invested at 10%, then the function

$$f(t) = 100\left(1 + \frac{0.10}{m}\right)^{mt}$$

gives the future value of the investment for t years, if the investment is compounded m times per year.

(a) Graph $f(t)$ for $m = 1$ (annual compounding), $m = 4$ (quarterly compounding), and $m = 365$ (daily compounding).

(b) Which graph lies above the others? What does this mean?

Solution

(a) The graphs are shown in Figure 6.2

(b) The graph for $m = 365$ (daily compounding) lies above the others. This indicates that more compounding periods gives a greater future value for a given interest rate. In addition, if $f(t)$ with $m = 365$ is graphed with $g(t) = 100e^{0.10t}$ (the function for continuous compounding), it is difficult to see any difference between the graphs. However, the graph for continuous compounding does lie above the one for daily compounding. This suggests, and it is indeed a fact, that at a given interest rate, more frequent compounding earns more interest (with continuous compounding earning the most interest).

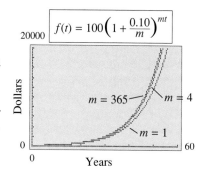

Figure 6.2

CHECKPOINT

1. Which future value formula below is used for interest that is compounded periodically and which is used for interest that is compounded continuously?
   (a) $S = P(1 + i)^n$ (b) $S = Pe^{it}$

2. If $5000 is invested at 6% compounded quarterly for 5 years, find
   (a) the number of compounding periods per year, m;
   (b) the number of compounding periods for the investment, n;
   (c) the interest rate for each compounding period, i;
   (d) the future value of the investment;
   (e) the effective annual rate.

3. If $5000 is invested at 6% compounded continuously for 5 years, find the future value of the investment.

GEOMETRIC SEQUENCES

If P is invested at an interest rate of i per year, compounded at the end of each year, the future value at the end of each succeeding year is

$$P(1 + i), P(1 + i)^2, P(1 + i)^3, \ldots, P(1 + i)^n, \ldots.$$

The future values for each of the succeeding years form a sequence in which each term (after the first) is found by multiplying the previous term by the same number. Such a sequence is called a **geometric sequence.**

A sequence is called a **geometric sequence** (progression) if there exists a number r, called the **common ratio,** such that

$$a_n = ra_{n-1} \qquad \text{for } n > 1.$$

Geometric sequences form the foundation for other applications involving compound interest.

EXAMPLE 8
Write the next three terms of the following geometric sequences:
(a) $1, 3, 9, \ldots$ (b) $4, 2, 1, \ldots$ (c) $3, -6, 12, \ldots$

Solution
(a) The common ratio is 3, so the next three terms are 27, 81, 243.
(b) The common ratio is $\frac{1}{2}$, so the next three terms are $\frac{1}{2}, \frac{1}{4}, \frac{1}{8}$.
(c) The common ratio is -2, so the next three terms are $-24, 48, -96$.

Because each term after the first in a geometric sequence is obtained by multiplying the previous term by r, the second term is $a_1 r$, the third is $a_1 r^2, \ldots$ and the nth term is $a_1 r^{n-1}$. Thus we have the following formula.

The **nth term of a geometric sequence** (progression) is given by

$$a_n = a_1 r^{n-1},$$

where a_1 is the first term of the sequence and r is the common ratio.

EXAMPLE 9
Find the seventh term of the geometric sequence with first term 5 and common ratio -2.

Solution
The seventh term is $a_7 = 5(-2)^{7-1} = 5(64) = 320$.

EXAMPLE 10

A ball is dropped from a height of 125 feet. If it rebounds $\frac{3}{5}$ of the height from which it falls every time it hits the ground, how high will it bounce after it strikes the ground for the fifth time?

Solution

The first rebound is $\frac{3}{5}(125) = 75$ feet; the second rebound is $\frac{3}{5}(75) = 45$. The heights of the rebounds form a geometric sequence with first term 75 and common ratio $\frac{3}{5}$. Thus the fifth term is

$$a_5 = 75(\tfrac{3}{5})^4 = 75(\tfrac{81}{625}) = \tfrac{243}{25} = 9\tfrac{18}{25} \text{ feet.}$$

Next we develop a formula for the sum of a geometric sequence, a formula that is important in our study of annuities in Section 6.3. The sum of the first n terms of a geometric sequence is

$$s_n = a_1 + a_1 r + a_1 r^2 + \cdots + a_1 r^{n-1}. \tag{1}$$

If we multiply equation (1) by r, we have

$$rs_n = a_1 r + a_1 r^2 + a_1 r^3 + \cdots + a_1 r^n. \tag{2}$$

Subtracting equation (2) from equation (1), we obtain

$$s_n - rs_n =$$
$$a_1 + (a_1 r - a_1 r) + (a_1 r^2 - a_1 r^2) + \cdots + (a_1 r^{n-1} - a_1 r^{n-1}) - a_1 r^n.$$

Thus

$$s_n(1 - r) = a_1 - a_1 r^n$$

$$s_n = \frac{a_1 - a_1 r^n}{1 - r}, \qquad \text{if } r \neq 1.$$

This gives the following.

> The **sum of the first n terms of the geometric sequence** with first term a_1 and common ratio r is
>
> $$s_n = \frac{a_1(1 - r^n)}{1 - r}, \qquad \text{provided } r \neq 1.$$

EXAMPLE 11

Find the sum of the first 5 terms of the geometric progression with first term 4 and common ratio -3.

Solution

We are given that $n = 5$, $a_1 = 4$, and $r = -3$. Thus,

$$s_5 = \frac{4[1 - (-3)^5]}{1 - (-3)} = \frac{4[1 - (-243)]}{4} = 244.$$

EXAMPLE 12

Find the sum of the first six terms of the geometric sequence $\frac{1}{4}, \frac{1}{8}, \frac{1}{16}, \ldots$

Solution

We know that $n = 6$, $a_1 = \frac{1}{4}$, and $r = \frac{1}{2}$. Thus,

$$s_6 = \frac{\frac{1}{4}[1 - (\frac{1}{2})^6]}{1 - \frac{1}{2}} = \frac{\frac{1}{4}(1 - \frac{1}{64})}{\frac{1}{2}} = \frac{1 - \frac{1}{64}}{2} = \frac{64 - 1}{128} = \frac{63}{128}.$$

CHECKPOINT

4. Identify any geometric sequences among the following.
   (a) 1, 4, 9, 16, ...   (b) 1, 4, 7, 10, ...   (c) 1, 4, 16, 64, ...
5. (a) Find the 40th term of the geometric sequence 8, 6, $\frac{9}{2}$, ...
   (b) Find the sum of the first 20 terms of the geometric sequence 2, 6, 18, ...

CHECKPOINT SOLUTIONS

1. (a) Periodic compounding   (b) Continuous compounding

2. (a) $m = 4$ (b) $n = m \cdot t = 4 \cdot 5 = 20$ (c) $i = \dfrac{j}{m} = \dfrac{0.06}{4} = 0.015$

   (d) $S = P(1 + i)^n = \$5000(1 + 0.015)^{20} \approx \$6734.28$ (nearest cent)

   (e) $r = (1 + 0.015)^4 - 1 \approx 0.0614$ or   6.14%

3. $S = Pe^{jt} = \$5000e^{(0.06)(5)} = \$5000e^{0.3} \approx \$6749.29$ (nearest cent)

4. Only sequence (c) is geometric with common ratio $r = 4$.

5. (a) $a_{40} = a_1 r^{40-1} = 8(\frac{3}{4})^{39} \approx 0.0001$

   (b) $s_{20} = \dfrac{a_1(1 - r^{20})}{1 - r} = \dfrac{2(1 - 3^{20})}{1 - 3} = -(1 - 3^{20})$
   $= 3^{20} - 1 = 3{,}486{,}784{,}400$

EXERCISE 6.2

COMPOUND INTEREST

For each investment situation in problems 1–6, identify the (a) annual interest rate, (b) length of the investment in years, (c) periodic interest rate, and (d) number of periods of the investment.

1. 6% compounded semiannually for 6 years
2. 9% compounded semiannually for 10 years
3. 8% compounded quarterly for 7 years
4. 12% compounded monthly for 3 years
5. 9% compounded monthly for 5 years
6. 10% compounded quarterly for 8 years

7. Find the future value if $8000 is invested for 10 years at 12%, compounded annually.
8. Find the interest that will be earned if $5000 is invested for 3 years at 10%, compounded annually.
9. Find the interest that will be earned if $10,000 is invested for 3 years at 9%, compounded annually.
10. What is the future value if $8600 is invested for 8 years at 10%, compounded semiannually?
11. What is the future value if $3200 is invested for 5 years at 8%, compounded quarterly?
12. What interest will be earned if $6300 is invested for 3 years at 12%, compounded monthly?

w

13. Find the interest that will be earned if $8600 is invested for 6 years at 10%, compounded semiannually.

14. Find the future value if $3500 is invested for 6 years at 8%, compounded quarterly.

15. Find the future value if $5100 is invested for 4 years at 9%, compounded continuously.

16. Find the interest that will result if $8000 is invested at 7%, compounded continuously, for 8 years.

17. What is the compound interest if $410 is invested for 10 years at 8%, compounded continuously?

18. If $8000 is invested at 8.5%, compounded continuously, find the future value after $4\frac{1}{2}$ years.

19. Which investment will earn more money, a $1000 investment for 5 years at 8%, compounded annually, or a $1000 investment for 5 years, compounded continuously at 7%?

20. How much more interest will be earned if $5000 is invested for 6 years at 7%, compounded continuously, instead of 7%, compounded quarterly?

21. What is the effective annual rate equivalent to a nominal rate of 8%, compounded quarterly?

22. If money is invested at 9%, compounded monthly, what is the equivalent effective annual rate?

23. If money is invested at 6% compounded continuously, what is the effective annual rate?

24. What is the effective annual rate equivalent to a nominal rate of 10%, compounded continuously?

In problems 25 and 26, rank each interest rate and compounding scheme in order from highest yield to lowest yield.

25. 8% compounded quarterly, 8% compounded monthly, 8% compounded annually

26. 6% compounded continuously, 6% compounded semi-annually, 6% compounded monthly

27. The figure below shows a graph of the future value of $100 at 8% compounded annually along with the graph of $100 at 8% compounded continuously. Which is which? Explain.

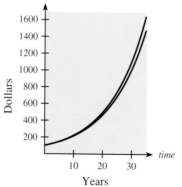

28. The figure below shows a graph of the future value of $350 at 6% compounded monthly along with the graph of $350 at 6% compounded annually. Which is which? Explain.

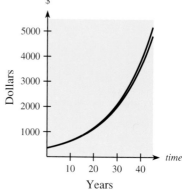

29. A shareholder who purchased 100 shares of Federal Signal Corporation's stock for $1438 at the beginning of 1979 and reinvested the dividends would have had an investment worth $50,620 at the end of 1993 (Source: Federal Signal Corporation, *1993 Annual Report*). What interest rate, compounded annually, did this investment earn?

30. An investment of $10,000 in Templeton World Fund at the beginning of 1978 would have been worth $93,575 in mid-1993. (Source: advertisement in *Newsweek*, November 8, 1993). What interest rate, compounded annually, did this investment earn?

31. How long (in years) would $700 have to be invested at 11.9%, compounded continuously, to earn $300 interest?

32. How long (in years) would $600 have to be invested at 8%, compounded continuously, to amount to $970?

33. At what nominal rate, compounded quarterly, would $20,000 have to be invested to amount to $26,425.82 in 7 years?

34. At what nominal rate, compounded annually, would $10,000 have to be invested to amount to $14,071 in 7 years?

35. For her first birthday, Polly's grandparents invested $1000 in an 18-year certificate for her that pays 8% compounded annually. How much will the certificate be worth on Polly's 19th birthday?

36. To help their son buy a car on his 16th birthday, a boy's parents invest $1500 on his tenth birthday. If the investment pays 9%, compounded continuously, how much is available on his 16th birthday?

37. (a) A forty-year-old man no longer qualifies for additional IRAs. If his present IRAs have a balance of

$12,860 and if he expects them to earn interest at 7.5%, compounded annually, how much does he expect to have when he retires at age 62?

(b) How much more money would the man have if his investments earned 8.5%, compounded annually?

38. (a) The purchase of Alaska cost the U.S. $7 million in 1869. If this money had been placed in a savings account paying 6%, compounded annually, how much money would be available from this investment in 1999?

(b) If the $7 million earned 7%, compounded annually since 1869, how much would be available in 1999?

(c) Do you think either amount would purchase Alaska in 1999? Explain in light of the value of Alaska's resources or perhaps the price per acre of land.

39. If a couple needs $15,000 as a down payment for a home and if they invest the $10,000 they have at 8%, compounded quarterly, how long will it take for the money to grow into $15,000?

40. If the couple in problem 39 invests the $10,000 at 10%, compounded continuously, how long will it take to grow into $15,000?

GEOMETRIC SEQUENCES

For each geometric sequence given in problems 41 and 42 write the next 3 terms.

41. (a) 3, 6, 12, . . .   (b) 81, 54, 36, . . .
42. (a) 4, 12, 36, . . .   (b) 32, 40, 50, . . .

43. Find the 13th term of the geometric sequence with first term 10 and common ratio 2.

44. Find the 11th term of the geometric sequence with first term 6 and common ratio 3.

45. Find the 16th term of the geometric sequence with first term 4 and common ratio $\frac{3}{2}$.

46. Find the 20th term of the geometric sequence with first term 3 and common ratio -2.

47. Find the sum of the first 17 terms of the geometric sequence with first term 6 and common ratio 3.

48. Find the sum of the first 14 terms of the geometric sequence with first term 3 and common ratio 4.

49. Find the sum of the first 21 terms of the geometric sequence with first term 4 and common ratio $-\frac{1}{2}$.

50. Find the sum of the first 25 terms of the geometric sequence with first term 9 and common ratio $\frac{1}{3}$.

51. Find the sum of the first 35 terms of the geometric sequence 1, 3, 9, . . .

52. Find the sum of the first 14 terms of the geometric sequence 16, 64, 256, . . .

53. Find the sum of the first 18 terms of the geometric sequence 6, 4, $\frac{8}{3}$, . . .

54. Find the sum of the first 31 terms of the geometric sequence 9, -6, 4, . . .

APPLICATIONS

55. **Inflation** A house worth $160,000 twenty years ago has increased in value by 4% each year because of inflation. What is its worth today?

56. **Inflation** If inflation causes the cost of automobiles to increase by 10% each year, what should a car that cost $5000 six years ago cost today?

57. **Population growth** Suppose a country has a population of 20 million and it projects a growth rate of 2% per year for the next 20 years. What will the population of this country be in 10 years?

58. **Spread of AIDS** Suppose a country is so devastated by the AIDS epidemic that its population decreases by 0.5% each year for a four-year period. If the population was originally 10 million, what is the population at the end of the four-year period?

59. **Population growth** If the rate of growth of a population continues at 2%, in how many years will the population double?

60. **Population** If a population of 8 million begins to increase at a rate of 0.1% each month, in how many months will it be 10 million?

61. **Ball rebounding** A ball is dropped from a height of 128 feet. If it rebounds $\frac{3}{4}$ of the height from which it falls every time it hits the ground, how high will it bounce after it strikes the ground for the fourth time?

62. **Water pumping** A pump removes $\frac{1}{3}$ of the water in a container with every stroke. What amount of water is still in a container after 5 strokes if it originally contained 81 cm^3?

63. **Depreciation** A machine is valued at $10,000. If the depreciation at the end of each year is 20% of its value at the beginning of the year, find its value at the end of 4 years.

64. **Profit** Suppose a new business makes a $1000 profit in its first month and has its profit increase by 10% each month for the next 2 years. How much profit will the business earn in its twelfth month?

65. **Bacterial growth** The size of a certain bacteria culture doubles each hour. If the number of bacteria present initially is 5000, how many would be present at the end of 6 hours?

66. ***Bacterial growth*** If a bacteria culture increases by 20% every hour and 2000 are present initially, how many will be present at the end of 10 hours?

67. ***Profit*** If changing market conditions cause a company earning $8,000,000 in 1992 to project a loss of 2% of its profit in each of the next 5 years, what profit does it project in 1997?

68. ***Profit*** Suppose a new business makes a $1000 profit in its first month and has its profit increase by 10% each month for the next 2 years. How much profit will it earn in its first year?

69. ***Loans*** An interest-free loan of $12,000 requires monthly payments of 10% of the outstanding balance. What is the outstanding balance after 18 payments?

70. ***Loans*** If Sherri must repay a $9000 interest-free loan by making monthly payments of 15% of the unpaid balance, what is the unpaid balance after one year?

71. ***Chain letters*** Suppose you receive a chain letter with six names on it, and to keep the chain unbroken you are to mail a dime to the person whose name is at the top, cross out the top name, add your name to the bottom, and mail it to five friends. If your friends mail out five letters each, and no one breaks the chain, you will eventually receive dimes. How many dimes would you receive? (This is a geometric sequence with first term 5.)

72. ***Chain letters*** Mailing chain letters through the mail was declared illegal because most people would receive nothing while a comparative few would profit. If the chain letter in problem 71 were to go through 12 unbroken progressions,
   (a) how many people would receive money?
   (b) how much money would these people receive as a group?

73. ***Chain letters*** How many letters would be mailed if the chain letter in problem 71 went through 12 unbroken progressions?

6.3 ANNUITIES

Objectives

- ❏ To compute the future value of an ordinary annuity
- ❏ To compute the future value of an annuity due
- ❏ To compute the present value of an ordinary annuity
- ❏ To compute the present value of a deferred annuity

APPLICATION PREVIEW

An **annuity** is a financial plan characterized by regular payments. We can view an annuity as a savings plan in which the regular payments are contributions to the account, and then we can ask what the total value of the account will become. Also, we can view an annuity as a payment plan (such as for retirement) in which regular payments are made from an account, often to an individual, and we can ask what amount is needed to generate payments of a certain size and duration.

In this section we consider three types of annuities: ordinary annuities, annuities due, and deferred annuities. For ordinary annuities and annuities due, we develop a formula for each type of annuity that allows us to find its **future value;** that is, a formula for the value of the account after regular deposits have been made over a period of time. With ordinary annuities and deferred annuities, we develop formulas that allow us to determine the **present value** of the annuity; that is, the single sum of money needed to generate a specified number of regular payments (such as for a retirement supplement). Note that it is regular payments that distinguish annuities from the future values discussed in Section 6.2.

ORDINARY ANNUITIES

In Section 6.2 we learned how to compute the future value and amount of interest if a fixed sum of money was deposited in an account that pays interest that is compounded periodically or continuously. But not many people are in a position to deposit a large sum of money at one time in an account. Most people save (or invest) money by depositing relatively small amounts at different times. If a depositor makes equal deposits at regular intervals, he or she is contributing to an **annuity.** The payments (deposits) may be made weekly, monthly, quarterly, yearly, or for any other period of time. The sum of all payments plus all interest earned is called the **future amount** or **future value of the annuity.**

Annuities may be classified into two categories—annuities certain and contingent annuities. An **annuity certain** is one in which the payments begin and end on fixed dates. In a **contingent annuity** the payments are related to events that cannot be paced regularly, so the payments are not regular. We will deal with annuities certain in this text, and we will deal first with annuities in which the payments are made at the end of each of the equal payment intervals. This type of annuity certain is called an **ordinary annuity.** The ordinary annuities we will consider have payment intervals that coincide with the compounding period of the interest.

Suppose you invested $100 at the end of each year for five years in an account that paid interest at 10%, compounded annually. How much money would you have in the account at the end of the five years?

Because you are making payments at the end of each period (year), this annuity is called an **ordinary annuity.**

To find the future value of your annuity at the end of the five years, we compute the future value of each payment separately, and add the amounts. (See Figure 6.3.) The $100 invested *at the end* of the first year will draw interest for 4 years, so it will amount to $100(1.10)^4$. The $100 invested at the end of the second year will draw interest for 3 years, so it will amount to $100(1.10)^3$. Similarly, the $100 invested at the end of the third year will amount to $100(1.10)^2$, and the $100 invested at the end of the fourth year will amount to $100(1.10)$. The $100 invested at the end of the fifth year will draw no interest, so it will amount to $100.

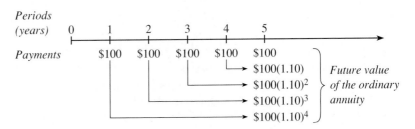

Figure 6.3

Thus the future value of the annuity is given by

$$S = 100 + 100(1.10) + 100(1.10)^2 + 100(1.10)^3 + 100(1.10)^4.$$

The terms of this sum are the first five values of the geometric sequence having $a_1 = 100$ and $r = 1.10$. Thus,

$$S = \frac{100[1 - (1.10)^5]}{1 - 1.10} = \frac{100(-0.61051)}{-0.10}$$

$$= 610.51.$$

Thus the investment ($100 at the end of each year for 5 years, at 10%, compounded annually) would return $610.51.

Because every such annuity will take the same form, we can state that if a periodic payment R is made for n periods at an interest rate i *per period,* the **future amount of the annuity,** or its **future value,** will be given by

$$S = R \cdot \frac{1 - (1 + i)^n}{1 - (1 + i)}.$$

This simplifies to the following.

Future Value of an Ordinary Annuity   If $R is deposited at the end of each period for n periods in an annuity that earns interest at a rate of i per period, the future value of the annuity will be

$$S = R \cdot s_{\overline{n}|i} = R \cdot \left[\frac{(1 + i)^n - 1}{i}\right],$$

where $s_{\overline{n}|i}$ is read "s, n angle i" and represents the future value of an ordinary annuity of $1 per period for n periods with an interest rate of i per period.

The value of $s_{\overline{n}|i}$ can be computed directly with a calculator or found in a table like Table III in the appendix.

EXAMPLE 1
Ruth Gearhart deposits $500 at the end of each year in a savings account that pays 6%, compounded annually. How much money will she have in the account at the end of 10 years?

Solution
Her savings form an ordinary annuity with $R = 500, $i = 6\%$, and $n = 10$. Thus, we have

$$S = \$500 \cdot s_{\overline{10}|6\%} = \$500 \cdot \frac{(1 + 0.06)^{10} - 1}{0.06}.$$

Using Table III in the appendix, we see that $s_{\overline{10}|6\%}$, the future value of an ordinary annuity of $1 per period for 10 periods with an interest rate of 6% per period, is $13.180795. Therefore,

$$S = \$500(13.180795) = \$6590.40.$$

EXAMPLE 2

Richard Lloyd deposits $100 at the end of each month in an account that pays 12%, compounded monthly. How much money will he have in his account in a year and a half?

Solution

The number of periods is $n = 18$, and the rate *per period* is $i = 1\%$. At the end of 18 months the future value of the annuity will be

$$S = \$100 \cdot s_{\overline{18}|1\%} = \$100(19.614748)$$
$$= \$1961.47.$$

Figure 6.4 shows a comparison (each graphed as a smooth curve) of the future values of two annuities that are invested at 6% compounded monthly. One annuity has monthly payments of $100 and one has payments of $125. Notice the impact of a slightly larger contribution on the future value of the annuity after 5 years, 10 years, 15 years, and 20 years. Of course, a slightly higher interest rate also can have a substantial impact over time; try a graphical comparison yourself.

A spreadsheet also can be used to compare investments, such as those with a given interest rate and different periodic contributions (as in Figure 6.4) or those with different interest rates and the same periodic contributions.

Figure 6.4

ANNUITIES DUE

Deposits in saving accounts, rent payments, and insurance premiums are examples of **annuities due.** An annuity due differs from an ordinary annuity in that the periodic payments are made at the *beginning* of the period with an annuity due. The *term* of an annuity due is from the first payment to the end of one period after the last payment. Thus an annuity due draws interest for one period more than the ordinary annuity.

We can find the future value of an annuity due by treating each payment as if it were made at the *end* of the preceding period in an ordinary annuity. Thus we calculate $s_{\overline{n}|i}$ for one additional period. But increasing the number of periods also adds one more payment than should be paid. To compensate for this, we subtract the amount of one payment (see Figure 6.5).

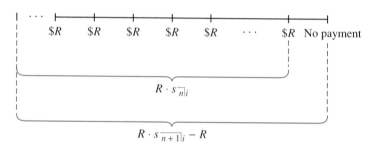

Figure 6.5

Thus the future value of an annuity due with n payments can be found by using a calculator or Table III in the appendix to find the future value of an ordinary annuity for $n + 1$ periods and by subtracting one annuity payment.

Of course, calculating $s_{\overline{n}|i}$ for one additional period gives $s_{\overline{n+1}|i}$ and gives one more payment than should be paid. Thus the formula for the future value of an annuity due is as follows.

Future Value of an Annuity Due	$$S_{\text{due}} = Rs_{\overline{n+1}|i} - R$$ $$= R\frac{(1+i)^{n+1} - 1}{i} - R$$

EXAMPLE 3

Find the future value of an annuity due of $1 for 3 years at 8%, compounded quarterly.

Solution

The number of quarterly periods in 3 years is $n = 12$ and the interest rate per period is $i = 2\%$. The future value of the annuity due is

$$S_{\text{due}} = Rs_{\overline{n+1}|i} - R = \$1s_{\overline{13}|2\%} - \$1.$$

$s\overline{_{13}}_{2\%}$ can be found using Table III in the appendix or $\dfrac{(1.02)^{13} - 1}{0.02}$ to obtain

$$\$14.680332 - \$1 = \$13.680332.$$

EXAMPLE 4

What is the future value of a savings account if $50 is deposited at the beginning of every 6 months for 5 years and the account pays interest at the rate of 6%, compounded semiannually?

Solution

This is an annuity due with 10 periods at 3% per period. The future value is

$$S_{\text{due}} = Rs\overline{_{n+1}}_{i} - R$$
$$= \$50\, s\overline{_{11}}_{3\%} - \$50$$
$$= \$50(12.807796) - \$50$$
$$= \$590.39$$

CHECKPOINT

Answer parts (a)–(d) below for each of the following problems.
(a) What is the total number of payments (or periods)?
(b) What is the interest rate per period?
(c) What formula is used to find the future value of the annuity?
(d) Find the future value.

1. Suppose $500 is deposited at the end of every quarter for 6 years into an account that pays 8% compounded quarterly.

2. Suppose $100 is deposited at the beginning of each month for 3 years into an account that pays 6% compounded monthly.

PRESENT VALUE OF AN ANNUITY

We have been discussing how contributing to an annuity program will result in a sum of money, and we have called that sum the future value of the annuity. Just as the term *annuity* is used to describe an account into which a person makes equal periodic payments (deposits), this term is also used to describe an account from which a person receives equal periodic payments (withdrawals); that is, if you invest a lump sum of money in an account today, so that at regular intervals you will receive a fixed sum of money, you have established an annuity. We call this lump sum the **present value** of the annuity.

Many people who are retiring purchase an annuity to supplement their income. This annuity pays them a fixed sum of money at regular intervals (usually each month). The single sum of money required to purchase an annuity that will provide these payments at regular intevals is the **present value** of the annuity. For example, you could receive \$1000 at the end of each year for 16 years if you invested a lump sum of \$10,837.77 in an annuity that paid interest at 5%, compounded annually. This lump sum is called the present value of the annuity. But how do we determine that the present value of this annuity (of \$1000 per year) is \$10,837.77?

Suppose we wish to invest a lump sum of money (which we will denote A_n) in an annuity for n periods at an interest rate of i per period, so that we will receive (withdraw) a payment of \$$R$ at the end of each period.

To find a formula for A_n, we must first note that the value today of a series of payments is the same whether we call these payments withdrawals or deposits. This means that A_n also represents the present value of an annuity in which we deposit \$$R$ at the end of each period for n periods with an interest rate of i per period.

Thus, we can express the future value of A_n in two ways. It is the future value of a compound interest investment that would result if the lump sum of \$$A_n$ were invested and left in an account for n periods at an interest rate of i per period:

$$S = A_n(1 + i)^n;$$

and it is also the future amount that would result if \$$R$ is deposited at the end of each period for n periods in an annuity that earns interest at a rate of i per period: $S = R \cdot s_{\overline{n}|i}$ (see Figure 6.6).

Figure 6.6

Equating these two expressions for S, we have

$$A_n(1 + i)^n = R \cdot s_{\overline{n}|i}$$

or

$$A_n(1 + i)^n = R \cdot \frac{(1 + i)^n - 1}{i}.$$

Multiplying both sides of this equation by $(1 + i)^{-n}$, we obtain a formula for the present value:

$$A_n = R \cdot \frac{1 - (1 + i)^{-n}}{i}.$$

Thus, we have the following.

Present Value of an Ordinary Annuity   If a payment of $R is made at the end of each period for n periods, into (or out of) an annuity that earns interest at a rate of i per period, the present value of the annuity is

$$A_n = R \cdot a_{\overline{n}|i} = R \cdot \left[\frac{1 - (1 + i)^{-n}}{i} \right],$$

where $a_{\overline{n}|i}$ represents the present value of an ordinary annuity of $1 per period for n periods, with an interest rate of i per period.

The values of $a_{\overline{n}|i}$ can be computed directly with a calculator or found in a table like Table IV in the appendix.

The present value of the annuity we discussed earlier (paying $1000 at the end of each year for 16 years) is

$$A_n = \$1000 \cdot a_{\overline{16}|5\%}$$
$$= \$1000 \cdot \frac{1 - (1.05)^{-16}}{0.05}$$
$$= \$1000 \cdot 10.83777$$
$$= \$10{,}837.77.$$

EXAMPLE 5

What lump sum would have to be invested at 6%, compounded annually, to provide an ordinary annuity of $1000 per year for 10 years?

Solution
We are given that $i = 6\%$, $R = \$1000$, and $n = 10$. Thus, the lump sum, or present value, is

$$A_n = \$1000 \cdot a_{\overline{10}|6\%} = \$1000 \cdot \frac{1 - (1.06)^{-10}}{.06}.$$

Using Table IV in the appendix or a calculator, we see that $a_{\overline{10}|6\%}$ is 7.360087. Therefore,

$$A_n = \$1000(7.360087)$$
$$= \$7360.087$$
$$= \$7360.09, \text{ to the nearest cent.}$$

EXAMPLE 6

What is the present value of an annuity of $1500 payable at the end of each 6-month period for 2 years if money is worth 8%, compounded semi-annually?

Solution

We are given that $R = \$1500$ and $i = 8\%/2 = 4\%$. Since a payment is made twice a year for 2 years, the number of periods is $n = (2)(2) = 4$. Thus,

$$A_n = \$1500 \cdot a_{\overline{4}|4\%}$$
$$= \$1500(3.69895)$$
$$= \$5444.84, \text{ to the nearest cent.}$$

Figure 6.7 shows a graph that compares the present value of an annuity of \$1000 per year at an interest rate of 6% compounded annually to the same annuity at an interest rate of 10% compounded annually. Notice the impact that the higher interest rate has on the present value needed to establish such an annuity for a longer period of time. Specifically, at 6% interest, a present value of about \$12,500 is needed to generate 25 years of \$1000 payments, but at an interest rate of 10%, the necessary present value is about \$10,000.

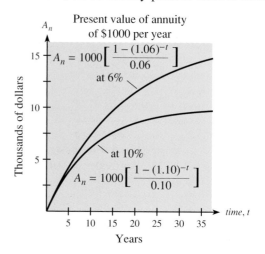

Figure 6.7

Our graphical comparisons in Figures 6.4 and 6.7 emphasize the truth of the saying "time is money." We have seen that, given enough time, relatively small differences in contributions or in interest rates result in substantial differences in amounts (present or future).

DEFERRED ANNUITIES

A **deferred annuity** is one in which the first payment is not made at the beginning or end of the first period, but at some later date. An annuity that is deferred for k periods and then has payments of $\$R$ per period at the end of each of the next n periods can be illustrated by Figure 6.8.

We now consider how to find the present value of such a deferred annuity when the interest rate is i per period. If we first assume that payments are made for the $k + n$ periods, then the annuity is an ordinary annuity with present value $Ra_{\overline{n+k}|i}$. However, for an annuity that is deferred for k periods,

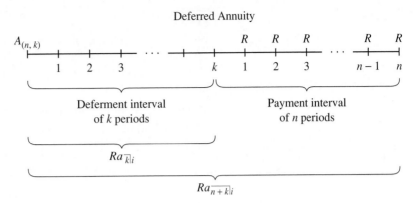

Figure 6.8

$Ra_{\overline{n+k}|i}$ must be reduced by an amount equal to the present value of the first k payments, $Ra_{\overline{k}|i}$, since these payments are not made. Thus, the present value of the deferred annuity is found by subtracting the present value for the period of deferment.

Present Value of a Deferred Annuity   The present value of a deferred annuity of $R per period for n periods deferred for k periods with interest rate i per period is given by

$$A_{(n,k)} = Ra_{\overline{n+k}|i} - Ra_{\overline{k}|i} = R(a_{\overline{n+k}|i} - a_{\overline{k}|i})$$
$$= R\left[\frac{1 - (1+i)^{-(n+k)}}{i} - \frac{1 - (1+i)^{-k}}{i}\right].$$

The formula for $A_{(n,k)}$ can be written in the form

$$A_{(n,k)} = R\left[\frac{1 - (1+i)^{-n}}{i}\right](1+i)^{-k};$$

this may be easier to use with a calculator.

EXAMPLE 7

A deferred annuity is purchased that will pay $10,000 per year for 15 years after being deferred for 5 years. If money is worth 6% per year, what is the present value of this annuity?

Solution

We have $R = \$10,000$, $n = 15$, $k = 5$, and $i = .06$. The present value is

$$A_{(15,5)} = \$10,000(a_{\overline{20}|6\%} - a_{\overline{5}|6\%})$$
$$= \$10,000(11.469921 - 4.212364)$$
$$= \$72,575.57.$$

As Figure 6.8 shows, the deferment interval ends **one** period before the first payment.

EXAMPLE 8

Find the present value of an annuity of $1000 every month for 2 years if the first payment is to be made one year after the annuity is established, and money is worth 12%, compounded monthly.

Solution

The first payment is made at the end of 12 months, so the payment is deferred 11 months; that is, $k = 11$. Using $n = 24$, $i = .01$, and $R = \$1000$, we have

$$A_{(24,11)} = \$1000\,(a_{\overline{35}|.01} - a_{\overline{11}|.01})$$
$$= \$1000(28.408580 - 10.367628)$$
$$= \$14,040.95.$$

CHECKPOINT

Answer parts (a)–(d) below for each of the following problems.
(a) What is the total number of periods?
(b) What is the interest rate per period?
(c) What formula is used to find the present value of the annuity?
(d) Find the present value.

3. Suppose an annuity pays $2000 at the end of each 3-month period for $3\frac{1}{2}$ years and money is worth 4% compounded quarterly.

4. Suppose an annuity will pay $5000 at the end of each 6-month period for 5 years with the first payment deferred 10 years. Suppose also that money is worth 6% compounded semiannually.

CHECKPOINT SOLUTIONS

1. (a) $n = 24$ periods
   (b) $i = 2\% = 0.02$ per period
   (c) $S = R\left[\dfrac{(1 + i)^n - 1}{i}\right]$ for an ordinary annuity
   (d) $S = 500\left[\dfrac{(1 + 0.02)^{24} - 1}{0.02}\right] = \$15,210.93$ (nearest cent)

2. (a) $n = 36$ periods
   (b) $i = 0.5\% = 0.005$ per period
   (c) $S_{due} = R\left[\dfrac{(1 + i)^{n+1} - 1}{i}\right] - R$ for an annuity due
   (d) $S_{due} = 100\left[\dfrac{(1 + 0.005)^{37} - 1}{0.005}\right] - 100 = \$4053.28$ (nearest cent)

3. (a) $n = 14$ periods

 (b) $i = 1\% = 0.01$ per period

 (c) $A_n = R\left[\dfrac{1 - (1 + i)^{-n}}{i}\right]$ for an ordinary annuity

 (d) $A_n = 2000\left[\dfrac{1 - (1.01)^{-14}}{0.01}\right] = \$26,007.41$ (nearest cent)

4. (a) $n = 10$ periods and $k = 20$ periods

 (b) $i = 3\% = 0.03$ per period

 (c) $A_{(n,k)} = R\left[\dfrac{1 - (1 + i)^{-(n+k)}}{i} - \dfrac{1 - (1 + i)^{-k}}{i}\right]$ for deferred annuities

 (d) $A_{(n,k)} = 5000\left[\dfrac{1 - (1.03)^{-30}}{0.03} - \dfrac{1 - (1.03)^{-20}}{0.03}\right]$

 $= \$23,614.83$ (nearest cent)

EXERCISE 6.3

ORDINARY ANNUITIES

For each set of investment conditions in problems 1–4, find (a) the periodic interest rate as a decimal and (b) the number of periods.

1. 7% compounded semiannually for 11 years
2. $5\frac{1}{4}\%$ compounded annually for 17 years
3. 3% compounded monthly for 5 years
4. 10% compounded quarterly for $7\frac{1}{2}$ years
5. The figure below shows a graph that compares the future value, at 8% compounded annually, of an annuity of $1000 per year and one of $1120 per year.
 (a) Decide which graph corresponds to which annuity.
 (b) Verify your conclusion to part (a) by finding the value of each annuity and the difference between them at $t = 25$ years.

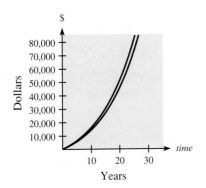

6. The figure below shows a graph that compares the future value, at 9% compounded monthly, of an annuity of $50 per month and one of $60 per month.
 (a) Decide which graph corresponds to which annuity.
 (b) Use the graph to estimate (to the nearest 10 months) how long before the larger annuity is $10,000 more than the smaller one.

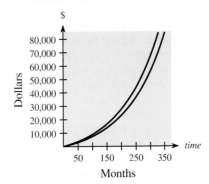

7. Find the future value of an annuity of $1 paid at the end of each year for 10 years, if it earns interest at 8%, compounded annually.
8. Find the future value of an ordinary annuity of $1 paid every 6 months for 3 years, if the interest rate is 6%, compounded semiannually.
9. Find the future value of an annuity of $1300 paid at the end of each year for 5 years, if interest is earned at a rate of 6%, compounded annually.

10. Find the future value of an annuity of $50 paid at the end of each year for 10 years, if it earns 9%, compounded annually.

11. Find the future value of an ordinary annuity of $80 paid quarterly for 3 years, if the interest rate is 8%, compounded quarterly.

12. Find the future value of an ordinary annuity of $300 paid quarterly for 5 years, if the interest rate is 12%, compounded quarterly.

13. Find the future value of an ordinary annuity of $40 paid at the end of each 6-month period for 10 years, if the interest rate is 6%, compounded semiannually.

14. Find the future value of an ordinary annuity of $75 paid every 6 months for 7 years, if the interest rate is 8%, compounded semiannually.

15. Mr. Gordon plans to invest $300 at the end of each year for 12 years in a savings account. If the account pays 9%, compounded annually, how much will he have at the end of the 12 years?

16. Sam deposits $500 at the end of every 6 months into an account that pays 8%, compounded semiannually. How much will he have at the end of 8 years?

17. Parents agree to invest $500 for their son (at 10%, compounded semiannually) on the December 31 or June 30 following each semester that he makes the Dean's list during his four years in college. If he makes the Dean's list in each of the eight semesters, how much money will his parents have to give him when he graduates?

18. Jake Werkheiser qualifies to invest $2000 in an IRA each April 15 for the next ten years. If he makes these investments, and if the certificates pay 12%, compounded annually, how much will he have at the end of the ten years?

19. (a) Patty Stacey deposited $2000 at the end of each of the 5 years she qualified for an IRA. If she leaves the money that has accumulated in the IRA account for 25 additional years, how much is in her account at the end of the 30-year period? Assume an interest rate of 9%, compounded annually.
    (b) Suppose that Patty's husband delayed starting an IRA for the first 10 years he worked, but he made $2000 deposits at the end of each of the next 15 years. If the interest rate is 9%, compounded annually, and if he leaves the money in his account for 5 additional years, how much will be in his account at the end of the 30-year period?
    (c) Does Patty or her husband have more IRA money?

20. A pair of twins graduate from college together and start their careers.
    (a) Twin 1 invests $2000 at the end of each of seven years in an account that pays 10% compounded annually. After the initial 7 years, no additional contributions are made, but the investment continues to earn 10% compounded annually. How much will Twin 1 have after 37 more years, at age 65?
    (b) Twin 2 invests no money for 7 years, but then contributes $2000 per year to an account that pays 10% compounded annually for a period of 37 years, to age 65. How much will Twin 2 have at age 65?
    (c) How much did each twin contribute? Which twin do you think made the wiser choice?

ANNUITIES DUE

21. Find the future value of an annuity due of $1 payable at the beginning of each year for 5 years if the interest rate is 7%, compounded annually.

22. Find the future value of an annuity due of $1 payable at the beginning of each 6-month period for 7 years if the interest rate is 8%, compounded semiannually.

23. Find the future value of an annuity due of $100 each year for $2\frac{1}{2}$ years at 12%, compounded quarterly.

24. Find the future value of an annuity due of $1500 each month for 3 years if the interest rate is 12%, compounded monthly.

25. Find the future value of an annuity due of $200 paid at the beginning of each 6-month period for 8 years if the interest rate is 6%, compounded semiannually.

26. For 3 years $400 is placed in a savings account at the beginning of each 6-month period. If the account pays interest at 10%, compounded semiannually, how much will be in the account at the end of the 3 years?

27. Jane Adele deposits $500 in an account at the beginning of each 3-month period for 9 years. If the account pays interest at the rate of 8%, compounded quarterly, how much will she have in her account?

28. A house is rented for $900 per quarter, with each quarter's rent payable in advance. If money is worth 8%, compounded quarterly, and the rent is deposited in an account, what is the future value of the rent for one year?

PRESENT VALUE OF AN ANNUITY

29. Find the present value of an annuity of $100 paid at the end of each year for 17 years if the interest rate is 7%, compounded annually.

30. Find the present value of an annuity of $800 paid at the end of each year for 15 years if the interest rate is 12%, compounded annually.

31. Find the present value of an annuity of $100 paid at the end of each 6-month period for 8 years if the interest rate is 8%, compounded semiannually.

32. Find the present value of an annuity that pays $300 at the end of each 6-month period for 6 years if the interest rate is 6%, compounded semiannually.

33. What lump sum would have to be invested at 8%, compounded quarterly, to provide an annuity that pays $200 at the end of each quarter for 3 years?

34. A woman buying a home paid $10,000 down and agreed to make 18 quarterly payments of $500 beginning in 3 months. If money is worth 12%, compounded quarterly, how much should the house cost if she paid for it in cash?

35. Joe Burkhart wants to purchase an annuity that will give him $1000 at the end of each 6-month period for 9 years. If money is worth 10%, compounded semiannually, how much will he have to pay for the annuity?

36. Is it more economical to buy an automobile for $3900 cash or to pay $800 down and $400 each quarter for 2 years, if money is worth 8%, compounded quarterly?

37. Suppose a state lottery prize of $5 million is to be paid in 20 payments of $250,000 each at the end of the next 20 years. If money is worth 10%, compounded annually, what is the present value of the prize?

38. Suppose Becky has her choice of $10,000 per month for life or a single prize of $1.5 million. She is 35 years old and her life expectancy is 40 more years. (a) Find the present value of the annuity if money is worth 7.2%, compounded monthly. (b) If she takes the $1.5 million, spends $700,000 of it, and invests the remainder at 7.2%, what monthly annuity will she receive?

39. The figure below shows a graph that compares the present values of two annuities of $1000 annually, one at 8% compounded annually and one at 10% compounded annually.
    (a) Determine which graph corresponds to the 8% rate and which to the 10% rate.
    (b) Use the graph to estimate the difference between the present values of these annuities for 25 years.
    (c) Write a sentence that explains this difference.

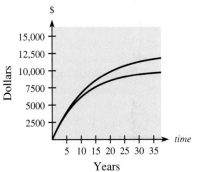

40. The figure below shows a graph that compares the present values of two annuities of $800 quarterly, one at 6% compounded quarterly and one at 9% compounded quarterly.
    (a) Determine which graph corresponds to the 6% rate and which to the 9% rate.
    (b) Use the graph to estimate the difference between the present values of these annuities for 25 years (100 quarters).
    (c) Write a sentence that explains this difference.

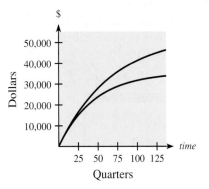

DEFERRED ANNUITIES

41. Find the present value of an annuity of $2000 per year at the end of each of 8 years after being deferred for 6 years, if money is worth 7%, compounded annually.

42. Find the present value of an annuity of $2000, at the end of each quarter for 5 years after being deferred for 3 years, if money is worth 8%, compounded quarterly.

43. The terms of a single parent's will indicate that a child will receive an ordinary annuity of $16,000 per year from age 18 to age 24 (so the child can attend college) with the balance of the estate going to a niece. If the parent dies on the child's 15th birthday, how much money must be removed from the estate to purchase the annuity? (Assume an interest rate of 6%, compounded annually).

44. On his 48th birthday, a man wants to set aside enough money to provide an income of $300 per month at the end of each month from his 60th birthday to his 65th birthday. If he earns 6%, compounded monthly, on his money, how much will this supplemental retirement plan cost him on his 48th birthday?

45. The semiannual tuition payment at a major university is expected to be $10,000 for the four years beginning in the year 2016. What lump sum payment should the university accept on January 1, 1998, in lieu of tuition payments beginning on July 1, 2016? Assume money is

worth 7%, compunded semiannually, and tuition is paid on July 1 and January 1 of the four years.

46. A grateful alumnus wishes to provide a scholarship of $2000 per year for 5 years to his alma mater, with the first scholarship awarded on his 60th birthday. If money is worth 6%, compounded annually, how much money must he donate on his 50th birthday?

If the present value of a deferred annuity $A_{(n,k)}$ is known, we can use the formula

$$A_{(n,k)} = R \left[a_{\overline{n+k}|i} - a_{\overline{k}|i} \right]$$

to find the size of the deferred periodic payment, R. Use this formula in problems 47 and 48.

47. Danny Metzger's parents invested $1600 when he was born. This money is to be used for Danny's college education and is to be withdrawn in four equal annual payments beginning when Danny is age 18. Find the amount that will be available each year, if money is worth 6%, compounded annually.

48. Carol Goldsmith received a trust fund inheritance of $10,000 on her 30th birthday. She plans to use the money to supplement her income with 20 quarterly payments beginning on her 60th birthday. If money is worth 7.6%, compounded quarterly, how much will each quarterly payment be?

6.4 LOANS AND AMORTIZATION; SINKING FUNDS

Objectives

❑ To find the regular payments required to amortize a debt
❑ To find the amount that must be invested periodically in a sinking fund to discharge a debt

APPLICATIONS PREVIEW

This section considers two applications that deal with the typical ways that consumers or businesses discharge debts: regular payments made on a loan (**amortization**) or regular payments to a fund that is used to pay a debt when it is due (**sinking fund**). Most debts incurred by consumers or businesses, such as for automobiles, equipment, or real estate, are amortized with regular payments. Sinking funds sometimes are established by a business that anticipates a major expenditure (such as for modernization) or by a municipal authority or school district.

In this section we study how each type of debt repayment is related to the future value or present value of an annuity and how payments are determined. With amortization, we study how payments are allocated (to pay the interest due and discharge the debt) and how payoff amounts are determined when the debt is paid in full before the final payment is due.

LOANS AND AMORTIZATION

Just as we invest money to earn interest, banks and lending institutions loan money and collect interest for its use. While your aunt may loan you money with the understanding that you will repay the full amount of the money plus simple interest at the end of a year, financial institutions generally expect you to make partial payments on a regular basis (frequently monthly). Most consumer loans (for automobiles, appliances, televisions, and the like) are classed as **installment loans.**

Federal law now requires that the full cost of any loan and the **true annual percentage rate (APR)** be disclosed with the loan. The APR gives the interest rate as though the interest is always charged on the unpaid balance.

Prior to enactment of the law that requires "truth in lending," banks offered installment loans whose stated interest rates were different from the true APR. However, since the federal regulations have gone into effect, banks have phased out these types of loans and treat most consumer loans in the same manner as home mortgages and long-term commercial loans. That is, loans now are usually paid off by a series of partial payments with interest charged on the unpaid balance at the end of each payment period. In this case, the stated interest rate is the same as the APR.

A loan of this type can be made in one of two ways. One repayment plan applies an equal amount to the principal each payment period plus the interest for the period, which is the interest on the unpaid balance. For example, a loan of $120,000 for 10 years at 12% could be repaid in 120 payments of $1000 plus 1% of the remaining balance of the loan. When this payment method is used, the payments will decrease as the unpaid balance decreases.

This type of loan can also be repaid by making all payments (including principal and interest) of equal size. This process of repaying the loan is called **amortization.**

When a bank makes a loan of this type, it is purchasing an ordinary annuity that pays a fixed return each payment period. The lump sum the bank gives to the borrower (i.e., the principal of the loan) is the present value of the annuity, and each payment the bank receives from the borrower is a payment from the annuity. To find the size of these periodic payments, we solve the formula for the present value of an ordinary annuity, $A_n = R \cdot a_{\overline{n}|i}$, for **R**. Thus, we have the following.

Amortization Formula

If the debt of $\$A_n$, with interest at a rate of i per period, is amortized by n equal, periodic payments (each payment being made at the end of a period), the size of each payment is

$$R = A_n \cdot \frac{1}{a_{\overline{n}|i}} = A_n \cdot \left[\frac{i}{1 - (1 + i)^{-n}} \right],$$

where $1/a_{\overline{n}|i}$ represents the periodic payment necessary to amortize $1 over n periods, at an interest rate of i per period.

Note that in the formula above, we can divide by $a_{\overline{n}|i}$ or multiply by $1/a_{\overline{n}|i}$. Values for $a_{\overline{n}|i}$ can be found in Table IV in the appendix, and values for $1/a_{\overline{n}|i}$ can be found in Table V.

EXAMPLE 1

A debt of $1000 with interest at 16%, compounded quarterly, is to be amortized by 20 quarterly payments (all the same size) over the next 5 years. What will the size of these payments be?

Solution

The 20 equal payments form an annuity whose present value is $1000. Therefore, $A_n = \$1000$ and $n = 20$. Since we are given that the nominal interest rate is 16% and the number of compounding periods per year is $m = 4$, we know that the rate per period $i = j/m = 16\%/4 = 4\%$. Thus, using Table V or a calculator, we have

$$R = \frac{\$1000}{a_{\overline{20}|4\%}} = \$1000\left[\frac{.04}{1 - (1.04)^{-20}}\right]$$

$$= \$1000[0.073582]$$

$$= \$73.58, \text{ to the nearest cent.}$$

EXAMPLE 2

A man buys a house for $200,000. He makes a $50,000 down payment and agrees to amortize the rest of the debt with quarterly payments over the next 10 years. If interest on the debt is 12%, compounded quarterly, find

(a) the size of the quarterly payments,
(b) the total amount of the payments,
(c) the total amount of interest paid.

Solution

(a) We know that $A_n = \$200,000 - \$50,000 = \$150,000$, $n = (4)(10) = 40$, and $i = 12\%/4 = 3\%$. Thus, using Table V or a calculator, we have

$$R = \$150,000 \cdot \frac{1}{a_{\overline{40}|3\%}} = \$150,000 \cdot 0.043262 = \$6489.30.$$

The quarterly payment is $6489.30.

(b) The man made 40 payments of $6489.30, so his payments totaled

$$(40)(6489.30) = \$259,572$$

plus the $50,000 down payment, or $309,572.

(c) Of the $309,572 paid, $200,000 was for payment of the house, and the remainder, $109,572, was interest.

We can construct an amortization schedule that summarizes all the information regarding the amortization of a loan.

For example, a loan of $10,000 with interest at 10% could be repaid in 5 equal annual payments of size

$$R = \$10,000\left(\frac{1}{a_{\overline{5}|10\%}}\right) = \$10,000(0.263797) = \$2,637.97.$$

Each time this $2637.97 payment is made, some is used to pay the interest on the unpaid balance and some is used to reduce the principal. For the first payment, the unpaid balance is $10,000, so the interest payment is 10% of $10,000, or $1000. The remaining $1,637.97 is applied to the principal. Hence, after this first payment, the unpaid balance is $10,000 − $1637.97 = $8,362.03.

For the second payment of $2,637.97, the amount used for interest is 10% of $8,362.03 or $836.20; the remainder, $1801.77, is used to reduce the principal.

This information for these two payments and for the remaining payments is summarized in the amortization schedule below, which gives the payment, the interest on the unpaid balance, the reduction of the balance, and the remining unpaid balance for each payment period.

Period	Payment	Interest	Balance Reduction	Unpaid Balance
				10000.00
1	2637.97	1000.00	1637.97	8362.03
2	2637.97	836.20	1801.77	6560.26
3	2637.97	656.03	1981.94	4578.32
4	2637.97	457.83	2180.14	2398.18
5	2638.00	239.82	2398.18	0.00
Total	13189.88	3189.88	10000.00	

Note that the last payment was increased by 3¢ so that the balance was reduced to $0 at the end of the five years. Such an adjustment is normal in amortizing a loan.

The amortization schedule above involves only a few payments. For loans involving more payments, a spreadsheet program is an excellent tool to develop an amortization schedule. The spreadsheet, which is fast and accurate, requires the user to understand the interrelationships among the columns. The spreadsheet output on page 436 shows the amortization schedule for the first 24 monthly payments on a $100,000 loan amortized for 30 years at 6% compounded monthly. It is interesting to observe how little the unpaid balance changes during these first two years, and consequently how much of each payment is devoted to paying interest due rather than to reducing the debt. In fact, the column labeled "Total Interest" shows a running total of how much has been paid toward interest. We see that after 12 payments of $599.55, a total of $7194.60 has been paid, and $5966.59 of this has been interest payments.

Many people who borrow money, such as for a car or a home, do not pay on the loan for its entire term, but pay off the loan early by making a final lump sum payment. The unpaid balance found in the amortization schedule is the "payoff amount" of the loan and represents the lump sum payment that must be made to complete payment on the loan. When the number of payments is large, we may wish to find the unpaid balance of a loan without constructing an amortization schedule.

	A	B	C	D	E	F
1	Payment Number	Payment Amount	Interest	Balance Reduction	Unpaid Balance	Total Interest
2	0	0	0	0	$100,000.00	$0.00
3	1	$599.55	$500.00	$99.55	$99,900.45	$500.00
4	2	$599.55	$499.50	$100.05	$99,800.40	$999.50
5	3	$599.55	$499.00	$100.55	$99,699.85	$1,498.50
6	4	$599.55	$498.50	$101.05	$99,598.80	$1,997.00
7	5	$599.55	$497.99	$101.56	$99,497.25	$2,495.00
8	6	$599.55	$497.49	$102.06	$99,395.18	$2,992.48
9	7	$599.55	$496.98	$102.57	$99,292.61	$3,489.46
10	8	$599.55	$496.46	$103.09	$99,189.52	$3,985.92
11	9	$599.55	$495.95	$103.60	$99,085.92	$4,481.87
12	10	$599.55	$495.43	$104.12	$98,981.80	$4,977.30
13	11	$599.55	$494.91	$104.64	$98,877.16	$5,472.21
14	12	$599.55	$494.39	$105.16	$98,771.99	$5,966.59
15	13	$599.55	$493.86	$105.69	$98,666.30	$6,460.45
16	14	$599.55	$493.33	$106.22	$98,560.09	$6,953.79
17	15	$599.55	$492.80	$106.75	$98,453.34	$7,446.59
18	16	$599.55	$492.27	$107.28	$98,346.05	$7,938.85
19	17	$599.55	$491.73	$107.82	$98,238.23	$8,430.58
20	18	$599.55	$491.19	$108.36	$98,129.87	$8,921.77
21	19	$599.55	$490.65	$108.90	$98,020.97	$9,412.42
22	20	$599.55	$490.10	$109.45	$97,911.53	$9,902.53
23	21	$599.55	$489.56	$109.99	$97,801.54	$10,392.09
24	22	$599.55	$489.01	$110.54	$97,690.99	$10,881.09
25	23	$599.55	$488.45	$111.10	$97,579.90	$11,369.55
26	24	$599.55	$487.90	$111.65	$97,468.25	$11,857.45

EXAMPLE 3

In Example 2, find the unpaid balance immediately after the fifteenth payment.

Solution

The unpaid balance after the fifteenth payment is the present value of the annuity with 40 − 15 = 25 payments to be made. So the principal is the payment times $a\,\overline{_{25}}_{|3\%}$ (from Table IV or a calculator):

$$A_n = \$6489.30 \cdot a\,\overline{_{25}}_{|3\%} = \$6489.30 \cdot 17.413148$$
$$= \$112,999.14.$$

The unpaid balance is $112,999.14. Thus, he could pay off the loan after the fifteenth payment by making a lump sum payment of $112,999.14.

The curve in Figure 6.9 shows the unpaid balance of a $100,000 loan at 6% compounded monthly for 25 years (with monthly payments of $644.30). In the figure, the straight line represents how the unpaid balance would decrease if each payment diminished the debt by the same amount. Notice how the curve decreases slowly at first and more rapidly as the unpaid balance nears zero. The reason for this behavior is that when the unpaid balance is large, much of each payment is devoted to interest. Similarly, when the debt decreases, more of each payment goes toward the principal.

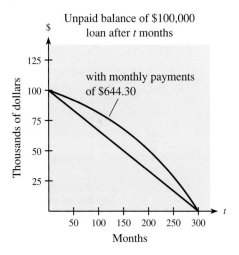

Figure 6.9

CHECKPOINT

1. A debt of $25,000 is to be amortized with equal quarterly payments over 6 years, and money is worth 7% compounded quarterly.
   (a) Find the total number of payments.
   (b) Find the interest rate per period.
   (c) Write the formula used to find the size of each payment.
   (d) Find the amount of each payment.

2. Write the formula that would be used to find the unpaid balance after the 10th payment for the loan in Checkpoint 1.

SINKING FUNDS

Some borrowers, such as municipalities, may have a debt that must be paid in a single large sum on a specified date. If they make periodic deposits that will produce that sum on the specified date, we say they have established a **sinking fund.**

If the deposits are all the same size and are made regularly, they form an ordinary annuity whose future value (on a specified date) is the amount of the principal of the debt. To find the size of these periodic deposits, we solve the equation for the future value of an ordinary annuity, $S = R \cdot s_{\overline{n}|i}$, for R. Thus, we have the following.

Sinking Fund Formula   Suppose n equal, periodic payments are deposited into a sinking fund earning interest at a rate of i per period so that at the end of n periods, the fund contains enough money to pay off a debt of $\$S$. If each deposit is made at the end of a period, the size of each payment is

$$R = S \cdot \frac{1}{s_{\overline{n}|i}} = S \cdot \left[\frac{i}{(1 + i)^n - 1} \right],$$

where $1/s_{\overline{n}|i}$ represents the periodic payment into a sinking fund that would amount to $\$1$ over n periods, at an interest rate of i per period.

Note that in the formula above, we can divide S by $s_{\overline{n}|i}$ or multiply S by $1/s_{\overline{n}|i}$. Values for $s_{\overline{n}|i}$ can be found in Table III in the appendix and values for $1/s_{\overline{n}|i}$ can be found in Table VI.

EXAMPLE 4

How much will have to be invested at the end of each year at 8%, compounded annually, to pay off a debt of $60,000 in 5 years?

Solution

We are given that $i = 8\%$, $S = \$60,000$, and $n = 5$. Using Table III, we have

$$R = \frac{\$60,000}{s_{\overline{5}|8\%}} = \frac{\$60,000}{5.866601}$$

$$= \$10227.39, \text{ to the nearest cent.}$$

Note that an annuity of $10,227.39 for 5 years at 8% compounded annually will amount to $10,227.39 \cdot 5.866601 = \$60,000.01$.

EXAMPLE 5

A company establishes a sinking fund to discharge a debt of $100,000 due in 5 years by making equal semiannual deposits, the first due in six months. If the deposits are placed in an account that pays 6%, compounded semiannually, what is the size of the deposits?

Solution

We know that $S = \$100,000$, $n = (2)(5) = 10$, and $i = 6\%/2 = 3\%$. Using Table VI, we have

$$R = \$100,000 \cdot \frac{1}{s\,\overline{_{10|3\%}}} = \$100,000 \cdot 0.087231 = \$8723.10.$$

The semiannual deposit is $8723.10.

Notice that a sinking fund is really a savings plan used to discharge a future debt, but it can also be used to provide a lump sum at a future date for any purpose, such as a college education.

CHECKPOINT

3. A sinking fund of $100,000 is to be established with equal payments at the end of each half-year for 15 years, and money is worth 10% compounded semiannually.
   (a) Find the number of payments.
   (b) Find the interest rate per period.
   (c) Write the formula used to find the payment size.
   (d) Find the amount of each payment.

CHECKPOINT SOLUTIONS

1. (a) $n = 6(4) = 24$ payments
   (b) $i = 7\%/4 = 1.75\% = 0.0175$ per period
   (c) $R = A_n \left[\dfrac{i}{1 - (1 + i)^{-n}} \right]$
   (d) $R = 25{,}000 \left[\dfrac{0.0175}{1 - (1.0175)^{-24}} \right] = \$1284.64$ (nearest cent)

2. $A_n = R \left[\dfrac{1 - (1 + i)^{-n}}{i} \right] = 1284.64 \left[\dfrac{1 - (1.0175)^{-14}}{0.0175} \right]$
   because after 10 payments, 14 payments remain.

3. (a) $n = 2(15) = 30$ payments
   (b) $i = 10\%/2 = 5\% = 0.05$ per period
   (c) $R = S \left[\dfrac{i}{(1 + i)^n - 1} \right]$
   (d) $R = 100{,}000 \left[\dfrac{0.05}{(1.05)^{30} - 1} \right] = \$1505.14$ (nearest cent)

EXERCISE 6.4

LOANS AND AMORTIZATION

1. Answer parts (a) and (b) with respect to loans for 10 years and 25 years. Assume the same interest rate.
   (a) Which loan results in more of each payment being directed toward principal? Explain.
   (b) Which loan results in a lower periodic payment? Explain.

2. When a debt is amortized, which interest rate is better for the borrower? Explain.
   (a) 10%   (b) 6%.

3. A debt of $10,000 is to be amortized by equal payments at the end of each year for 5 years. If the interest charged is 8%, compounded annually, find the periodic payment.

4. A debt of $15,000 is to be amortized by equal payments at the end of each year for 10 years. If the interest charged is 9%, compounded annually, find the periodic payment.

5. A debt of $8000 is to be amortized with 8 equal semiannual payments. If the interest rate is 12%, compounded semiannually, what is the size of each payment?

6. A loan of $10,000 is to be amortized with 10 equal quarterly payments. If the interest rate is 6%, compounded quarterly, what is the periodic payment?

7. Develop an amortization schedule for a loan of $100,000 to be repaid in 3 annual payments of equal size. The interest rate is 9%.

8. Develop an amortization schedule for a loan of $30,000 to be repaid in 5 annual payments of equal size. The interest rate is 7%.

9. Develop an amortization schedule for a loan of $20,000 with interest at 12%, compounded quarterly, if it is to be repaid by 4 quarterly payments of equal size.

10. Develop an amortization schedule for a loan of $50,000 with interest at 10%, if it is to be repaid in $2\frac{1}{2}$ years by making equal semiannual payments.

11. A debt of $10,000 is to be amortized by equal payments of $2504.56 at the end of each year for 5 years. Find the outstanding principal (unpaid balance) immediately after the third payment. Interest is 8%, compounded annually.

12. A debt of $15,000 is to be amortized by equal payments of $2337.30 at the end of each year for 10 years. Find the unpaid balance immediately after the eighth payment. Interest is 9%, compounded annually.

13. A debt of $8000 is to be amortized with 8 equal semiannual payments of $1288.29. If the interest rate is 12%, compounded semiannually, find the unpaid balance immediately after the fifth payment.

14. A $10,000 loan is to be amortized for 10 years with quarterly payments of $334.27. If the interest rate is 6%, compounded quarterly, what is the unpaid balance immediately after the sixth payment?

Problems 15–18 describe a debt to be amortized. In each problem, find:
(a) the size of each payment,
(b) the total amount paid over the life of the loan, and
(c) the total interest paid over the life of the loan.

15. A man buys a house for $75,000. He makes a $25,000 down payment and amortizes the rest of the debt with semiannual payments over the next 10 years. The interest rate on the debt is 12%, compounded semiannually.

16. Sean Lee purchases $20,000 worth of supplies for his restaurant by making a $3000 down payment and amortizing the rest with quarterly payments over the next 5 years. The interest rate on the debt is 16%, compounded quarterly.

17. John Fare purchased $10,000 worth of equipment by making a $2000 down payment and promising to pay the remainder of the cost in semiannual payments over the next 4 years. The interest rate on the debt is 10%, compounded semiannually.

18. A woman buys an apartment house for $250,000 by making a down payment of $100,000 and amortizing the rest of the debt with semiannual payments over the next 10 years. The interest rate on the debt is 12%, compounded semiannually.

19. A man buys a car for $12,000. If the interest rate on the loan is 12%, compounded monthly, and he wants to make monthly payments of $300 for 36 months, how much must he put down?

20. A woman buys a car for $15,000. If the interest rate on the loan is 12%, compounded monthly, and she wants to make monthly payments of $500 for 3 years, how much must she have for a down payment?

21. A couple purchasing a home wants their payment to be $500 per month. If they have $15,000 available for a down payment and the mortgage rate on a 25-year loan is 12%, compounded monthly, how much can they spend on a house?

22. An investor interested in purchasing an apartment building determines that she can make payments of $3000 per month. If a loan is available at 15%, compounded monthly for 20 years, how much can she afford to pay for the building?

23. Suppose the man of problem 15 receives an inheritance and wishes to pay off his loan after he has made 8 payments. Find the payoff amount.

24. What is the payoff amount for Sean Lee's loan (see problem 16) immediately after the eleventh payment?

25. A couple that borrows $90,000 for 30 years at 7.2%, compounded monthly, must make monthly payments of $610.91.
    (a) Find their unpaid balance after 1 year.
    (b) During that first year, how much interest did the couple pay?

26. A company that purchases a piece of equipment by borrowing $250,000 for 10 years at 6%, compounded monthly, has monthly payments of $2775.51.
    (a) Find the unpaid balance on this loan after 1 year.
    (b) During that first year, how much interest did the company pay?

A debt of $100,000 is amortized at 6%, compounded monthly over 25 years with 300 monthly payments of $644.30 each. Figure 6.10 shows two graphs: one shows the total amount paid (in monthly payments) as a function of time (in months), and the other shows the amount paid toward the principal of the debt as a function of time. Use this figure to complete problems 27–30.

Figure 6.10

27. Correctly label each graph.

28. Draw a vertical segment whose length represents the outstanding principal of the debt (or the payoff amount of the loan) after 150 months.

29. Draw a vertical segment whose length represents the total amount of interest paid on the debt after 250 months.

30. Each year, if you made an extra payment devoted entirely toward payment of principal, would the total amount paid over the life of the loan be greater than, the same as, or less than the total amount that would be paid if you just made scheduled monthly payments? Explain.

SINKING FUNDS

31. When you establish a sinking fund, which interest rate is better? Explain.
    (a) 10%   (b) 6%

32. If you set up a sinking fund, which interest rate is better? Explain.
    (a) 12% compounded monthly
    (b) 12% compounded annually

33. What amount will have to be invested at the end of each year for 10 years to form a sinking fund of $100,000 if the interest rate is 10%, compounded annually?

34. What amount will have to be invested at the end of each year for 5 years to form a sinking fund of $25,000 if the interest rate is 12%, compounded annually?

35. How much will have to be invested at the end of each year at 10%, compounded annually, to pay off a debt of $50,000 in 8 years?

36. How much will have to be invested at the end of each year at 12%, compounded annually, to pay off a debt of $30,000 in 6 years?

37. A company establishes a sinking fund to discharge a debt of $75,000 due in 8 years by making equal semiannual deposits, the first due in 6 months. If their investment pays 12%, compounded semiannually, what is the size of the deposits?

38. A sinking fund is established to discharge a debt of $80,000 in 10 years. If deposits are made at the end of each 6-month period and interest is paid at the rate of 8%, compounded semiannually, what is the size of the deposits?

39. A sinking fund is established by a working couple so that they will have $20,000 to pay for part of their daughter's education when she enters college. If they make deposits at the end of each three-month period for 10 years, and if interest is paid at 12%, compounded quarterly, what size deposits must they make?

40. The Weidmans want to save $20,000 in two years for a down payment on a house. If they make monthly deposits into an account paying 12%, compounded monthly, what size payments are required to meet their goal?

41. A company can borrow $100,000 for ten years to purchase new equipment. They can either amortize the loan at an annual rate of 12% or they can pay interest on the loan at 11% for each of the ten years and set up a sinking fund at 10% per year, compounded annually. Which method of repayment costs less per year?

42. A company borrows $200,000 to be repaid in equal payments at the end of each semiannual period for 10 years. Find the total semiannual cost if:
(a) the loan is amortized at an interest rate of 12%.
(b) the interest on the loan is paid at 12% in each of the 20 semiannual periods, and a sinking fund paying 10%, compounded semiannually, is set up to discharge the debt.

Key Terms & Formulas

Section	Key Term	Formula
6.1	Simple interest	$I = Prt$
	Future value of a simple interest investment or loan	$S = P + I$
	Sequence function	
	Arithmetic sequence	$a_n = a_{n-1} + d \quad (n > 1)$
	Common difference	d
	nth term of	$a_n = a_1 + (n-1)d$
	Sum of first n terms	$s_n = \dfrac{n}{2}(a_1 + a_n)$
6.2	Future value of compound interest investment	
	Periodic compounding	$S = P(1 + i)^n = P\left(1 + \dfrac{j}{m}\right)^{mt}$
	Continuous compounding	$S = Pe^{jt}$
	Effective annual rate	$r = \left(1 + \dfrac{j}{m}\right)^m - 1$
	Geometric sequence	$a_n = ra_{n-1} \quad (n > 1)$
	Common ratio	r
	nth term of	$a_n = a_1 r^{n-1}$
	Sum of first n terms	$s_n = \dfrac{a_1(1 - r^n)}{1 - r} \quad (\text{if } r \neq 1)$

6.3   Annuity
                 Future value of

                     Ordinary annuity $S = Rs_{\overline{n}|i} = R\left[\dfrac{(1 + i)^n - 1}{i}\right]$

                     Annuity due $S_{\text{due}} = Rs_{\overline{n+1}|i} - R$

                                                        $= R\left[\dfrac{(1 + i)^{n+1} - 1}{i}\right] - R$

                 Present value of

                     Ordinary annuity $A_n = R \cdot a_{\overline{n}|i} = R\left[\dfrac{1 - (1 + i)^{-n}}{i}\right]$

                     Deferred annuity $A_{(n,\,k)} = R(a_{\overline{n+k}|i} - a_{\overline{k}|i})$

                                                        $= R\left[\dfrac{1 - (1 + i)^{-(n+k)}}{i} - \dfrac{1 - (1 + i)^{-k}}{i}\right]$

6.4   True annual percentage
                 rate (APR)

                     Amortization $R = A_n\left(\dfrac{1}{a_{\overline{n}|i}}\right) = A_n\left[\dfrac{i}{1 - (1 + i)^{-n}}\right]$

                     Sinking fund $R = S\left(\dfrac{1}{s_{\overline{n}|i}}\right) = S\left[\dfrac{i}{(1 + i)^n - 1}\right]$

REVIEW EXERCISES

6.1

1. Find the first 4 terms of the sequence whose nth term is

$$a_n = \frac{1}{n^2}.$$

2. Find the tenth term of the arithmetic sequence with first term -2 and common difference 3.

3. Find the sixth term of the arithmetic sequence with third term 10 and eighth term 25.

4. Find the sum of the first 6 terms of the arithmetic sequence $\frac{1}{3}, \frac{1}{2}, \frac{2}{3}, \ldots$.

6.2

5. Find the fourth term of the geometric sequence with first term 64 and eighth term $\frac{1}{2}$.

6. Find the sum of the first 6 terms of the geometric sequence $\frac{1}{9}, \frac{1}{3}, 1, \ldots$.

APPLICATIONS

6.1

7. *Simple interest* If $8000 is borrowed at 12% simple interest for 3 years, what is the future value of the loan at the end of the 3 years?

8. **Simple interest** Mary Toy borrowed $2000 from her parents and repaid them $2100 after 9 months. What simple interest rate did she pay?

9. **Contest payments** Suppose the winner of a contest receives $10 on the first day of the month, $20 on the second day, $30 on the third day, and so on for a 30-day month. What is the total won?

10. **Salaries** Suppose you are offered two identical jobs: one paying a starting salary of $20,000 with yearly raises of $1000 and one paying a starting salary of $18,000 with yearly raises of $1200. Which job will pay more money over a 10-year period?

6.2

11. **Investments** If an investment is made at 8% compounded quarterly for 10 years, find
    (a) the number of periods and
    (b) the interest rate per period.

12. **Future value** Write the formula for the future value of a compound interest investment, if interest is compounded
    (a) periodically
    (b) continuously.

13. **Future value** If $8000 is invested at 12%, compounded annually, for 3 years, what is the future value at the end of the 3 years?

14. **Interest** If $1000 is invested for 4 years at 8%, compounded quarterly, how much interest will be earned?

15. **Future value** What is the future value if $1000 is invested for 6 years at 8%, compounded semiannually?

16. **Future value** What is the future value if $1000 is invested for 6 years at 8%, compounded continuously?

17. **Future value** If $8000 is invested at 12% compounded continuously for 3 years, what is the future value at the end of the 3 years?

18. **Interest** What interest is earned in problem 17?

19. **Continuous compounding** If $8000 is invested at 7% compounded continuously, how long before it grows to $22,000?

20. **Investments** If $15,000 is invested at 6% compounded quarterly, how long before it grows to $25,000?

21. **Compound rate** If an initial investment of $2500 grows to $38,000 in 18 years, what annual interest rate, compounded annually, did this investment earn?

22. **Continuous compounding** If an initial investment of $35,000 grows to $257,000 in 15 years, what continuously compounded annual interest rate was earned?

23. **Effective annual rate** Find the effective annual rate equivalent to a nominal rate of 7.2% compounded monthly.

24. **Compound interest** Which compounding method earns more money?
    (a) semiannually   (b) monthly

25. **Chess legend** Legend has it that when the king of Persia wanted to reward the inventor of chess with whatever he desired, the inventor asked for one grain of wheat on the first square of the chessboard, with the number of grains doubled on each square thereafter for the remaining 63 squares. Find the number of grains on the 64th square. (The sum of all the grains of wheat on all the squares would cover Alaska more than 3 inches deep with wheat!)

26. **Chess legend** If, in problem 25, the king granted the inventor's wish for *half* a chessboard, how many grains would the inventor receive?

6.3

27. **Ordinary annuity** Find the future value of an ordinary annuity of $800 paid at the end of every 6-month period for 10 years, if it earns interest at 12%, compounded semiannually.

28. **Annuity due** Find the future value of an annuity due of $800 paid at the beginning of every 6-month period for 10 years, if it earns interest at 12%, compounded semiannually.

29. **Annuity** If $100 is deposited at the end of each quarterly period for 4 years in an account that pays 8% interest, compounded quarterly, how much money will accrue in the account?

30. **Present value** What lump sum would have to be invested at 9%, compounded annually, to provide an annuity of $10,000 per year for 10 years?

31. **Present value** Find the present value of an annuity of $6500 paid semiannually for 4 years after being deferred for 10 years, if money is worth 8%, compounded semiannually.

32. **Annuity** A couple wishes to set up an annuity that would provide 6 monthly payments of $3000 while they take an extended cruise. How much of a $15,000 inher-

itance must be set aside if they plan to leave in 5 years and want the first payment before they leave? Assume money is worth 7.8%, compounded monthly.

6.4

33. **Amortization** A debt of $1000 with interest at 12%, compounded monthly, is amortized by 12 monthly payments (of equal size). What is the size of each payment?

34. **Amortization** A couple borrowed $92,000 to buy a condominium. If their loan was for 25 years and money is worth 6%, compounded monthly, find
    (a) their monthly payment size,
    (b) the total amount they would pay over 25 years, and
    (c) the total interest they would pay over 25 years.

35. **Payoff amount** A debt of $8000 is amortized with eight semiannual payments of $1288.29 each. If money is worth 12%, compounded semiannually, find the unpaid balance after 5 payments have been made.

36. **Cash value** A woman paid $10,000 down for a house and agreed to pay 18 quarterly payments of $1500 each. If money is worth 4%, compounded quarterly, how much would the house cost if she paid cash?

37. **Sinking fund** How much would have to be invested at the end of each year at 6%, compounded annually, to pay off a debt of $80,000 in 10 years?

38. **Divorce settlement** A divorce settlement of $40,000 is paid in 40 monthly payments of $1000 each. What is the present value of this settlement if money is worth 12%, compounded monthly?

Profit Reinvestment

T. C. Hardware Store wants to construct a new building at a second location. If they could begin immediately, the cost would be $250,000. At this time, however, the owners do not have the required 20% down payment, so they plan to invest $1000 per month of their profits until they have the necessary amount. They can invest their money in an annuity account that pays 6% compounded monthly, but they are concerned about the 3% average inflation rate in the construction industry. They would like you to give them some projections about how 3% inflation will affect the time required to accrue the down payment and the building's eventual cost. They would also like to know how their projected profits, after the new building is complete, will affect their mortgage loan payoff schedule.

Specifically, the owners would like you to prepare a report that answers the following questions.

1. If the 3% annual inflation is accurate, how long will it take to get the down payment? *Hint:* Assume inflation is compounded monthly. Then the required down payment will be 20% of

$$250{,}000\left(1 + \frac{0.03}{12}\right)^{n},$$

where n is in months. Thus you must find n so that this amount equals the future value of the owners' annuity. The solution to the resulting equation is difficult to obtain by algebraic methods but can be found with a graphing utility.

2. If the 3% annual inflation rate is accurate (compounded monthly), what will T. C. Hardware's projected construction costs be (to the nearest hundred dollars) when they have their down payment?

3. If T. C. Hardware borrows 80% of their construction costs and amortizes that amount at 7.8% compounded monthly for 15 years, what will their monthly payment be?

4. Finally, the owners believe that two years after this new building is begun (that is, after two years of payments on the loan) their profits will increase enough so that they'll be able to make double mortgage payments each month until the loan is paid. Under these assumptions, how long will it take T. C. Hardware to pay off the loan?

PART 4

PROBABILISTIC MODELS

Warm-up

Prerequisite problem type	For section	Answer	Section for review
(a) The sum of x, $\frac{1}{4}$, and $\frac{2}{3}$ is 1. What is x? (b) If the sum of x plus $\frac{1}{3}$ plus $\frac{1}{3}$ is 1, what is x?	**7.1**	(a) $x = \frac{1}{12}$ (b) $x = \frac{1}{3}$	1.1 Linear equations
If 12 numbers are weighted so that each of the 12 numbers has an equal weight and the sum of the weights is 1, what weight should be assigned to each number?	**7.1**	$\frac{1}{12}$	1.1 Linear equations
(a) List the set of integers that satisfies $2 \le x < 6$. (b) List the set of integers that satisfies $4 \le s \le 12$.	**7.1**	(a) $\{2, 3, 4, 5\}$ (b) $\{4, 5, 6, 7, 8, 9,$ $10, 11, 12\}$	3.1 Inequalities
For the sets $E = \{1, 2, 3, 4, 5\}$ and $F = \{2, 4, 6, 8\}$ find (a) $E \cup F$ (b) $E \cap F$	**7.1** **7.2**	 (a) $\{1, 2, 3, 4, 5, 6, 8\}$ (b) $\{2, 4\}$	0.1 Sets
Find the product: $[.9 \quad .1]\begin{bmatrix} .8 & .2 \\ .3 & .7 \end{bmatrix}$	**7.7**	$[.75 \quad .25]$	2.2 Matrix multiplication
Solve the system using matrices. $\begin{cases} .5V_1 + .4V_2 + .3V_3 = V_1 \\ .4V_1 + .5V_2 + .3V_3 = V_2 \\ .1V_1 + .1V_2 + .4V_3 = V_3 \end{cases}$	**7.7**	$V_1 = 3V_3$ $V_2 = 3V_3$	2.3 Gauss-Jordan elimination

INTRODUCTION TO PROBABILITY

An economist cannot predict exactly how the gross national product will change, a sociologist cannot determine exactly how group behavior is affected by climatic conditions, and a psychologist cannot determine the exact effect of environment on behavior. Since the behavior of the economy and of people is subject to chance, its prediction involves probability. Answers to questions about the behavioral and life sciences, business, and economics are frequently found by conducting experiments with samples from the larger population and by using statistics (discussed in Chapter 8). The experiments are designed to compare what actually happens with what probability theory predicted would happen. For example, if a rat runs a T-maze, can it learn which way to turn to get food? If the food is on the right and the rat turns right many more times than probability indicates that it should, then we have evidence that it has learned. The conclusions drawn from experiments are usually given in terms of probability because they are extended to larger populations. For example, substances that cause cancer in experimental animals may not be proven to cause cancer in humans, but will be removed from the market because the probability is high that they will.

We will solve a number of problems involving games in this chapter because they lend themselves easily to probability. (In fact, probability theory really started as the result of gambling problems.) We will also see that the theory of probability can be useful in many applications in the management, social, and life sciences.

7.1 PROBABILITY

Objectives

- ❏ To compute the probability of the occurrence of an event
- ❏ To construct a sample space for a probability experiment
- ❏ To compute the odds that an event will occur
- ❏ To compute the empirical probability that an event will occur

APPLICATION PREVIEW

Before the General Standard Company can place a three-year warranty on all of the faucets it produces, it needs to find the **probability** that each faucet it produces will function properly (not leak) for three years. In this section we will use **sample spaces** to find the probability that an event will occur by using assumptions or knowledge of the entire **population** of data under consideration. When knowledge of the entire population is not available, we can compute the **empirical probability** that an event will occur by using data that is collected about the probability experiment. This method can be used to compute the probability that the General Standard faucets will work properly if information about a large **sample** of the faucets is gathered.

If we toss a coin, it can land in one of two ways, heads or tails. If the coin is a "fair" coin, these two possible **outcomes** have an equal chance of occurring, and we say the outcomes of this **probability experiment** are **equally likely.** If we seek the probability that an experiment has a certain result, that result is called an **event.**

Suppose an experiment can have a total of n equally likely outcomes, and that k of these outcomes would be considered successes. Then the probability of achieving a success in the experiment is k/n. That is, the probability of a success in an experiment is the number of ways the experiment can result in a success divided by the total number of possible outcomes.

Probability of a Single Event   If an event E can happen in k ways out of a total of n equally likely possibilities, the probability of the occurrence of the event is denoted by

$$\Pr(E) = \frac{\text{number of successes}}{\text{number of possible outcomes}} = \frac{k}{n}.$$

EXAMPLE I
If we draw a ball from a bag containing 4 white balls and 6 black balls, what is the probability of
(a) getting a white ball?
(b) getting a black ball?
(c) not getting a white ball?

Solution
(a) A white ball can occur (be drawn) in 4 ways out of a total of 10 equally likely possibilities. Thus the probability of drawing a white ball is

$$\Pr(W) = \frac{4}{10} = \frac{2}{5}.$$

(b) We have 6 chances to succeed at drawing a black ball out of a total of 10 possible outcomes. Thus the probability of getting a black ball is

$$\Pr(B) = \frac{6}{10} = \frac{3}{5}.$$

(c) The probability of not getting a white ball is $\Pr(B)$, since not getting a white ball is the same as getting a black ball. Thus $\Pr(\text{not } W) = \Pr(B) = 3/5$.

Suppose a card is selected at random* from a box containing cards numbered 1 through 12. To find the probability that the card drawn has an even number on it, it is necessary (at least mentally) to list all the possible outcomes of the experiment (selecting a number at random). A set that contains all the possible outcomes of an experiment is called a **sample space.** For the experiment of drawing a numbered card, a sample space is

$$S = \{1, 2, 3, 4, 5, 6, 7, 8, 9, 10, 11, 12\}.$$

Each element of the sample space is called a **sample point,** and an **event** is a subset of the sample space. Each sample point in the sample space is assigned a nonnegative **probability measure** or **weight** such that the sum of the weights in the sample space is 1. The probability of an event is the sum of the weights of the sample points in the event's **subspace** (subset of S).

For example, the experiment "drawing a number at random from the numbers 1 through 12" has the sample space S given above. Because each element of S is equally likely to occur, we assign a probability weight of $1/12$ to each element. The event "drawing an even number" in this experiment has the subspace

$$E = \{2, 4, 6, 8, 10, 12\}.$$

Because each of the 6 elements in the subspace has weight $1/12$, the probability of event E is

$$\Pr(\text{even}) = \Pr(E) = \frac{1}{12} + \frac{1}{12} + \frac{1}{12} + \frac{1}{12} + \frac{1}{12} + \frac{1}{12} = \frac{6}{12} = \frac{1}{2}.$$

Note that using the sample space S for this experiment gives the same result as using the formula $\Pr(E) = k/n$.

Some experiments can have an infinite number of outcomes, but we will concern ourselves only with experiments having finite sample spaces. We can usually construct more than one sample space for an experiment. The sample space in which each sample point is equally likely is called an **equiprobable sample space.** Suppose the number of elements in a sample space is $n(S) = n$, with each element representing an equally likely outcome of a probability experiment, then the weight (probability) assigned to each element is

$$\frac{1}{n(S)} = \frac{1}{n}.$$

---

*Selecting a card at random means every card has an equal chance of being selected.

If an event E that is a subset of an equiprobable sample space contains $n(E) = k$ elements of S, then we can restate the probability of an event E in terms of the number of elements of E.

Probability of an Event If an event E can occur in $n(E) = k$ ways out of $n(S) = n$ equally likely ways, then

$$Pr(E) = \frac{n(E)}{n(S)} = \frac{k}{n}.$$

EXAMPLE 2

If a number is to be selected at random from the integers 1 through 12, what is the probability that it is
(a) divisible by 3?
(b) even and divisible by 3?
(c) even or divisible by 3?

Solution

The set $S = \{1, 2, 3, 4, 5, 6, 7, 8, 9, 10, 11, 12\}$ is an equiprobable sample space for this experiment (as mentioned previously).
(a) The set $E = \{3, 6, 9, 12\}$ contains the numbers that are divisible by 3. Thus

$$Pr(\text{divisible by 3}) = \frac{n(E)}{n(S)} = \frac{4}{12} = \frac{1}{3}.$$

(b) The numbers 1 through 12 that are even *and* divisible by 3 are 6 and 12, so

$$Pr(\text{even and divisible by 3}) = \frac{2}{12} = \frac{1}{6}.$$

(c) The numbers that are even *or* divisible by 3 are 2, 3, 4, 6, 8, 9, 10, 12, so

$$Pr(\text{even or divisible by 3}) = \frac{8}{12} = \frac{2}{3}.$$

CHECKPOINT

1. If a coin is drawn from a box containing 4 gold, 10 silver, and 16 copper coins, all the same size, what is the probability of
   (a) getting a gold coin?
   (b) getting a copper coin?
   (c) not getting a copper coin?

If an event E is certain to occur, E contains all of the elements of the sample space, S. Hence the sum of the probability weights of E is the same as that of S, so

$$\Pr(E) = 1 \qquad \text{if } E \text{ is certain to occur.}$$

If event E is impossible, $E = \emptyset$, so

$$P(E) = 0 \qquad \text{if } E \text{ is impossible.}$$

Because all probability weights of elements of a sample space are non-negative,

$$0 \leq \Pr(E) \leq 1 \qquad \text{for any event } E.$$

EXAMPLE 3

Suppose a coin is tossed 3 times.
(a) Construct an equiprobable sample space for the experiment.
(b) Find the probability of obtaining 0 heads.
(c) Find the probability of obtaining 2 heads.

Solution

(a) Perhaps the most obvious way to record the possibilities for this experiment is to list the number of heads that could result: $\{0, 1, 2, 3\}$. But the probability of obtaining 0 heads is different from the probability of obtaining 2 heads, so this sample space is not equiprobable. A sample space in which each outcome is equally likely is {HHH, HHT, HTH, THH, HTT, THT, TTH, TTT}, where HHT indicates the first two tosses were heads and the third was a tail.

(b) Because there are 8 equally likely possible outcomes, $n(S) = 8$. Only one of the eight possible outcomes, $E = \{TTT\}$, gives 0 heads, so $n(E) = 1$. Thus

$$\Pr(0 \text{ heads}) = \frac{n(E)}{n(S)} = \frac{1}{8}.$$

(c) The event "two heads" is $F = \{HHT, HTH, THH\}$, so

$$\Pr(2 \text{ heads}) = \frac{n(F)}{n(S)} = \frac{3}{8}.$$

To find the probability of obtaining a given sum when a pair of dice is rolled, we need to determine how many outcomes are possible. If we distinguish between the two dice we are rolling, and we record all the possible outcomes for each die, we see that there are 36 possibilities, each of which is equally likely (see Table 7.1).

This list of possible outcomes for finding the sum of two dice is an equiprobable sample space for the experiment. Because each *element* of this sample space has the same chance of occurring, this is an equiprobable sample space. Because the 36 elements in the sample space are equally likely, we can find the probability that a given sum results by determining the

Table 7.1

First die \ Second die	1	2	3	4	5	6
1	(1, 1)	(1, 2)	(1, 3)	(1, 4)	(1, 5)	(1, 6)
2	(2, 1)	(2, 2)	(2, 3)	(2, 4)	(2, 5)	(2, 6)
3	(3, 1)	(3, 2)	(3, 3)	(3, 4)	(3, 5)	(3, 6)
4	(4, 1)	(4, 2)	(4, 3)	(4, 4)	(4, 5)	(4, 6)
5	(5, 1)	(5, 2)	(5, 3)	(5, 4)	(5, 5)	(5, 6)
6	(6, 1)	(6, 2)	(6, 3)	(6, 4)	(6, 5)	(6, 6)

number of ways that sum can occur and dividing that number by 36. Thus the probability that a sum 6 will occur is 5/36, and the probability that a 9 will occur is $4/36 = 1/9$. (See the 4 ways the sum 9 can occur in Table 7.1.) Note that we could have made a sample space for this experiment (finding the sum of a pair of dice) by listing the sums: {2, 3, 4, 5, 6, 7, 8, 9, 10, 11, 12}. However, these outcomes are not equally likely; the probability of obtaining a 2 is different from the probability of obtaining a 6. We can see that the probabilities of events can be found more easily if an equiprobable sample space is used to determine the possibilities.

EXAMPLE 4

Use Table 7.1 to find the following probabilities if a pair of distinguishable dice is rolled:
(a) Pr(sum is 5)   (b) Pr(sum is 2)   (c) Pr(sum is 8)

Solution
(a) If E is the event "sum of 5" then $E = \{(4, 1), (3, 2), (2, 3), (1, 4)\}$.

Therefore Pr(sum is 5) $= \dfrac{n(E)}{n(S)} = \dfrac{4}{36} = \dfrac{1}{9}$.

(b) The sum 2 results only from {(1, 1)}. Thus Pr(sum is 2) $= 1/36$.
(c) The sample points that give a sum of 8 are in the subspace
$F = \{(6, 2), (5, 3), (4, 4), (3, 5), (2, 6)\}$.

Thus Pr(sum is 8) $= \dfrac{n(F)}{n(S)} = \dfrac{5}{36}$.

CHECKPOINT

2. A ball is to be drawn from a bag containing balls numbered 1, 2, 3, 4, and 5. To find the probability that a ball with an even number is drawn, we can use the sample space $S_1 = \{\text{even, odd}\}$ or $S_2 = \{1, 2, 3, 4, 5\}$.
   (a) Which of these sample spaces is an equiprobable sample space?
   (b) What is the probability of drawing a ball that is even-numbered?

3. A coin is tossed and a die is rolled.
   (a) Write an equiprobable sample space listing the possible outcomes for this experiment.

(b) Find the probability of getting a head on the coin and an even number on the die.

(c) Find the probability of getting a tail on the coin and a number divisible by 3 on the die.

If we know the probability that an event E will occur, we can compute the **odds** in favor of E occurring and the odds against E occurring.

Odds   If the probability that event E occurs is $\Pr(E) \neq 1$, then the odds that E will occur are

$$\text{Odds in favor of } E = \frac{\Pr(E)}{1 - \Pr(E)}$$

and the odds that E will not occur, if $\Pr(E) \neq 0$, are

$$\text{Odds against } E = \frac{1 - \Pr(E)}{\Pr(E)}.$$

The odds $\dfrac{a}{b}$ are stated as "a to b" or "$a : b$."

EXAMPLE 5

If the probability of drawing a queen from a deck of playing cards is $1/13$, what are the odds

(a) in favor of drawing a queen?   (b) against drawing a queen?

Solution

(a) $\dfrac{1/13}{1 - 1/13} = \dfrac{1/13}{12/13} = \dfrac{1}{12}$

The odds in favor of drawing a queen are 1 to 12, which we write as $1 : 12$.

(b) $\dfrac{12/13}{1/13} = \dfrac{12}{1}$

The odds against drawing a queen are 12 to 1, denoted $12 : 1$.

Odds and Probability   If the odds in favor of an event E occurring are $a : b$, the probability that E will occur is

$$\Pr(E) = \frac{a}{a + b}.$$

If the odds against E are $b : a$, then

$$\Pr(E) = \frac{a}{a + b}.$$

EXAMPLE 6
If the odds against dying in an industrial accident are $3992:3$, what is the probability of dying in an industrial accident?

Solution
The probability is

$$\frac{3}{3 + 3992} = \frac{3}{3995}.$$

Up to this point, the probabilities assigned to events were determined *theoretically,* either by assumption (a fair coin has probability $1/2$ of obtaining a head in one toss) or by knowing the entire population under consideration (if 1000 people are in a room and 400 of them are males, the probability of picking a person at random and getting a male is $400/1000 = 2/5$). We can also determine probabilities **empirically,** where the assignment of the probability of an event is frequently derived from our experiences or from data that has been gathered. Probabilities developed in this way are called **empirical probabilities.** An empirical probability is formed by conducting an experiment or observing a situation a number of times, then noting how many times a certain event occurs.

Empirical Probability of an Event

$$\text{Pr(Event)} = \frac{\text{Number of observed occurrences of the event}}{\text{Total number of trials or observations}}$$

In some cases empirical probability is the only way to assign a probability to an event. For example, life insurance premiums are based on the probability that an individual will live to a certain age. These probabilities are developed empirically and organized into mortality tables.

In other cases, empirical probabilities may have a theoretical framework. For example, in tossing a coin,

$$\text{Pr(Head)} = \text{Pr(Tail)} = \frac{1}{2}$$

gives the theoretical probability. If a coin was actually tossed 1000 times, it might come up heads 517 times and tails 483 times, giving empirical probabilities for these 1000 tosses of

$$\text{Pr(Head)} = \frac{517}{1000} = .517$$

$$\text{Pr(Tail)} = \frac{483}{1000} = .483.$$

On the other hand, if the coin was tossed only 4 times, it might come up heads 3 times and tails once, giving empirical probabilities vastly different from the theoretical ones. In general, empirical probabilities can differ from theoretical probabilities, but they are likely to reflect the theoretical probabilities when the number of observations is large. Moreover, when this is not the case, there may be reason to suspect that the theoretical model does not fit the event (for example, that a coin may be biased).

EXAMPLE 7

Experience has shown that 800 of the 500,000 microcomputer diskettes manufactured by a firm are defective. Based on this evidence, what is the probability that a diskette picked at random from those manufactured by this firm will be defective?

Solution

The 800 defective diskettes occurred in a very large number of diskettes manufactured, so the ratio of the number of defective diskettes to the total number of diskettes manufactured provides the *empirical* probability of a diskette picked at random being defective.

$$\text{Pr(defective)} = \frac{800}{500,000} = \frac{1}{625}$$

EXAMPLE 8

As mentioned in the Application Preview, the General Standard Company can use information about a large sample of its faucets to compute the empirical probability that a faucet chosen at random will function properly (not leak) for three years. Of a sample of 12,316 faucets produced in 1992, 137 were found to fail within three years. Find the empirical probability that any faucet the company produces will work properly for three years.

Solution

If 137 of these faucets fail to work properly, then $12,316 - 137 = 12,179$ will work properly. Thus the probability that any faucet selected at random will function properly is

$$\frac{12,179}{12,316} \approx 0.989.$$

CHECKPOINT SOLUTIONS

1. (a) The number of coins is 30 and the number of gold coins is 4, so
       Pr(gold) = 4/30 = 2/15.
   (b) The number of copper coins is 16, so Pr(copper) = 16/30 = 8/15.
   (c) The number of coins that are not copper is 14, so
       Pr(not copper) = 14/30 = 7/15.

2. (a) S_1 = {even, odd} is not an equiprobable sample space for this experiment because there are not equal numbers of even and odd numbers.
        S_2 = {1, 2, 3, 4, 5} is an equiprobable sample space.
    (b) Pr(even) = 2/5

3. (a) {H1, H2, H3, H4, H5, H6, T1, T2, T3, T4, T5, T6}
    (b) Pr(head and even number) = 3/12 = 1/4
    (c) Pr(tail and number divisible by 3) = 2/12 = 1/6

EXERCISE 7.1

1. If you draw one card at random from a deck of 12 cards numbered 1 through 12, inclusive, what is the probability that the number you draw is divisible by 4?
2. An ordinary die is tossed. What is the probability of getting a 3 or a 4?
3. One ball is drawn at random from a bag containing 4 red balls and 6 white balls. What is the probability that the ball is
    (a) red?   (b) green?   (c) red or white?
4. One ball is drawn at random from a bag containing 4 red balls and 6 white balls. What is the probability that the ball is
    (a) white?   (b) white or red?   (c) red and white?
5. A die is rolled. What is the probability that
    (a) a 4 will result?
    (b) a 7 will result?
    (c) an odd number will result?
6. A die is rolled. Find the probability of getting a number greater than 0.
7. An urn contains three red balls numbered 1, 2, 3, four white balls numbered 4, 5, 6, 7, and three black balls numbered 8, 9, 10. A ball is drawn from the urn. What is the probability that
    (a) it is red?
    (b) it is odd-numbered?
    (c) it is red and odd-numbered?
    (d) it is red or odd-numbered?
    (e) it is not black?
8. An urn contains three red balls numbered 1, 2, 3, four white balls numbered 4, 5, 6, 7, and three black balls numbered 8, 9, 10. A ball is drawn from the urn. What is the probability that the ball is
    (a) white?   (b) white and odd?
    (c) white or even?   (d) black or white?
    (e) black and white?

9. From a deck of 52 ordinary playing cards, one card is drawn. Find the probability that it is
    (a) a queen.   (b) a red card.   (c) a spade.
10. From a deck of 52 ordinary playing cards, one card is drawn. Find the probability that it is
    (a) a red king.
    (b) a king or a black card.
11. Suppose a fair coin is tossed two times. Construct an equiprobable sample space for the experiment, and determine each of the following probabilities:
    (a) Pr(0 heads)   (b) Pr(1 head)   (c) Pr(2 heads)
12. Suppose a fair coin is tossed four times. Construct an equiprobable sample space for the experiment, and determine each of the following probabilities:
    (a) Pr(2 heads)   (b) Pr(3 heads)   (c) Pr(4 heads)
13. Use Table 7.1 to determine the following probabilities if a distinguishable pair of dice is rolled.
    (a) Pr(sum is 4)
    (b) Pr(sum is 10)
    (c) Pr(sum is 12)
14. (a) When a pair of distinguishable dice is rolled, what sum is most likely to occur?
    (b) When a pair of distinguishable dice is rolled, what is $Pr(4 \leq S \leq 8)$, where S represents the sum rolled?
15. If a pair of dice, one green and one red, is rolled, what is
    (a) $Pr(4 \leq S \leq 7)$, where S is the sum rolled?
    (b) $Pr(8 \leq S \leq 12)$, where S is the sum rolled?
16. If a green die and a red die are rolled, find
    (a) $Pr(2 \leq S \leq 6)$, where S is the sum rolled on the two dice.
    (b) $Pr(4 \leq S)$, where S is the sum rolled on the two dice.
17. Suppose a die is tossed 1200 times and a 6 comes up 431 times.
    (a) Find the empirical probability for a 6 to occur.

(b) Based on a comparison of the empirical probability and the theoretical probability, do you think the die is fair or biased?

18. Suppose that a coin is tossed 3000 times and 1800 heads result. What is the empirical probability that a head will occur with this coin? Is there evidence that the coin is a fair coin?

19. If the probability that an event will occur is 2/5, what are the odds
   (a) in favor of the event occurring?
   (b) against the event occurring?

20. If the probability that an event will occur is 7/15, what are the odds
   (a) that the event will occur?
   (b) that the event will not occur?

21. If the probability that an event E will not occur is 8/11, what are the odds
   (a) that E will occur?
   (b) that E will not occur?

22. If the probability that an event E will not occur is 4/19, what are the odds
   (a) that E will not occur?
   (b) that E will occur?

23. If the odds that a particular horse will win a race are $1:20$, what is the probability
   (a) that the horse will win the race?
   (b) that the horse will lose the race?

24. If the odds that an event E will occur are $3:19$, what is the probability
   (a) that E will occur?
   (b) that E will not occur?

25. If the odds that an event E will not occur are $23:57$, what is the probability
   (a) that event E will occur?
   (b) that event E will not occur?

26. If the odds that an event F will not occur are $21:23$, what is the probability
   (a) that event F will occur?
   (b) that event F will not occur?

APPLICATIONS

27. **Sales promotion** In a sales promotion, a clothing store gives its customers a chance to draw a ticket from a box that contains a discount on their next purchase. The box contains 3000 tickets giving a 10% discount, 500 giving a 30% discount, 100 giving a 50% discount, and 1 giving

a 100% discount. What is the probability that a given customer will randomly draw a ticket giving
   (a) a 100% discount?
   (b) a 50% discount?
   (c) a discount of less than 50%?

28. **Management** A dry cleaning firm has 12 employees: 7 women and 5 men. Three of the women and five of the men are 40 years old or older. The remainder are over 20 years of age and under 40. If a person is chosen at random from this firm, what is the probability that the person is
   (a) a woman?
   (b) under 40 years of age?
   (c) 20 years old?

29. **Blood types** Human blood is classified by blood type, which indicates the presence or absence of the antigens A, B, and Rh, as follows.

A present	Type A
B present	Type B
Both A and B present	Type AB
Neither A nor B present	Type O

   Each of these types is combined with a $+$ or a $-$ sign to indicate whether the Rh antigen is present or not. Write a sample space containing all possible blood types.

30. **Genetics** Construct an equiprobable sample space that gives the possible combinations of male and female children in a family with three children.

31. **Drug use** Forty-six percent of all drug use occurs in cities (Source: Partnership for a Drugfree America). If an instance of drug use is chosen at random, what is the probability that the use occurs in a city?

32. **Breast cancer** According to the American Cancer Society, 199 of 200 mammograms turn out to be normal. What is the probability that the mammogram of a woman chosen at random will be normal?

33. **Blood type** Four percent of the population of the United States has type AB blood (Source: *Encyclopedia Americana*, 1993 ed.). What is the probability that a U.S. resident chosen at random
   (a) will have type AB blood?
   (b) will not have type AB blood?

34. **Blood type** Fifteen percent of the population of the United States has a negative Rh (Rhesus) factor in their blood, and the remainder have a positive Rh factor (Source: American Medical Association, *Encyclopedia*

of Medicine, 1989). What is the probability that a resident of the U.S. chosen at random
(a) will have a negative Rh factor?
(b) will have a positive Rh factor?

35. **Minorities** Among the residents of Los Angeles, 62.7% are classified as minorities (Source: *Upclose Census Digest,* City and Town Edition, 1992). If an L.A. resident is chosen at random, what is the probability that he or she is
(a) a member of a minority?
(b) not a member of a minority?

36. **Minorities** Among the population of Miami, 62.5% are of Hispanic descent (Source: *Upclose Census Digest,* City and Town Edition, 1992). What is the probability that a Miami resident chosen at random is
(a) of Hispanic descent?
(b) not of Hispanic descent?

37. **Quality control** A car rental firm has 350 cars. Seventy of the cars have only defective windshield wipers and 25 have only defective tail lights. Two hundred of the cars have no defects; the remainder have other defects. What is the probability that a car chosen at random
(a) has defective windshield wipers?
(b) has defective tail lights?
(c) does not have defective tail lights?

38. **Inventory** Forty percent of a company's total output consists of baseballs, 30% consists of softballs, and 10% consists of tennis balls. Its only remaining product is handballs. If they placed balls in a box in the same proportion as the company's output and selected a ball at random from the box, what is the probability that
(a) the ball is a baseball?
(b) the ball is a tennis ball?
(c) the ball is not a softball?
(d) the ball is a handball?

39. **Management** Because of a firm's growth, it is necessary to transfer one of its employees to one of its branch stores. Three of the nine employees are women and each of the nine employees is equally qualified for the transfer. If the person to be transferred is chosen at random, what is the probability that the transferred person is a woman?

40. **Lactose intolerance** Lactose intolerance affects about 75% of African, Asian, and Native Americans, (Source: Jean Carper, "Eat Smart," *USA Weekend,* Sept. 4, 1994). If a person is selected from this group of people, what is the probability that the person will have lactose intolerance?

41. **Lactose intolerance** Lactose intolerance affects about 20% of non-Hispanic white Americans (Source: Jean Carper, "Eat Smart," *USA Weekend,* Sept. 4, 1994). If a person is selected from this group of people, what is the probability that the person will be lactose intolerant?

42. **Quality control** A supply of 500 television picture tubes has 6 defective tubes. What is the probability that a tube picked at random from the supply is not defective?

43. **Education** A professor assigns 6 homework projects and then during the next class rolls a die to determine which one of the 6 projects to grade. If a student has completed 2 of the assignments, what is the probability that the professor will grade a project that the student has completed?

44. **Marketing** Meow Mix offers a free coin in each bag of cat food. If each 10,000-bag shipment has one gold coin, 2 silver dollars, 5 quarters, 20 dimes, 50 nickels, and 9922 pennies, what is the probability that a bag of Meow Mix cat food selected at random will have a penny in it?

45. **Minorities** The NAACP claims that in Volusia County, Florida, police stop dark-skinned drivers on I-95 who are not violating the law much more frequently than they stop white drivers, and then they conduct searches of the cars for drugs or for large sums of money that could be related to drug sales (Source: *American Journal,* August 30, 1994). If police made routine stops of randomly selected motorists who are not violating a law, and if 22% of the drivers on I-95 passing through this county are dark-skinned, what is the probability that the police stop a driver who is not violating a law but who is dark-skinned? If 39% of the motorists stopped for routine checks are dark-skinned people, does the NAACP claim seem reasonable?

46. **Quality control** A computer store offered used 64K computers free to local middle schools. Of the 54 machines available, four have defective memories, six have defective keys, and the remainder have no defects. If a teacher picks one at random, what is the probability that she will select a defective computer?

47. **Football** Before the 1994 football season, a *USA Today* oddsmaker gave the San Francisco 49ers' odds against winning Super Bowl XXIX at 3 : 1, saying they were the favorite to win (Source: "Sheridan's Odds," *USA Today,* Sept. 2, 1994). If the odds were accurate, what probability would be assigned to the 49ers winning this Super Bowl?

48. **Football** Before the 1994 football season, a *USA Today* oddsmaker gave the Cincinnati Bengals' odds against winning Super Bowl XXIX at 1 million : 1, saying they were the least likely to win (Source: "Sheridan's Odds," *USA Today,* Sept. 2, 1994). If the odds were accurate, what probability would be assigned to the Bengals winning this Super Bowl?

49. **Genetics** A newly married couple plans to have three children. Assuming the probability of a girl being born equals that of a boy, what is the probability that exactly two of the three children born will be girls? (Hint: Construct the sample space.)

50. **Management** A frustrated store manager is asked to make four different yes-no decisions that have no relation to each other. Because he is impatient to leave work, he flips a coin for each decision. If the correct decision in each case was yes, what is the probability that
    (a) all of his decisions were correct?
    (b) none of his decisions was correct?
    (c) half of his decisions were correct?

51. **Genetics** A couple plans to have two children and wants to know the probability of having one boy and one girl.
    (a) Does {0, 1, 2} represent an equiprobable sample space for finding the probability of exactly one boy in two children?
    (b) Construct an equiprobable sample space for this problem.
    (c) What is the probability of getting one boy and one girl in two children?

52. **Genetics** What is the probability that a couple planning to have two children will have
    (a) at least one child of each sex?
    (b) two children of the same sex?

53. **Testing** A quiz consists of 6 multiple-choice questions with 5 possible answers for each question. If a student guesses on the first question, what is the probability he will get the correct answer?

54. **Testing** In taking a quiz with 6 multiple-choice questions, a student knows that two of the possible five answers to the first question are not correct. What is the probability of her getting the correct answer to the first question?

55. **Quality control** On the average, 6 articles out of each 250 produced by a certain machine are defective. What is the probability that an article chosen at random is defective?

56. **Baseball** If a baseball player has 1150 at bats and 320 hits, what is the probability he will get a hit his next time at bat?

57. **Traffic safety** In a certain town, citizens' groups have identified three intersections as potentially dangerous. They collected the following data.

	Intersections		
	A	B	C
Number of vehicles/day	25,500	3,890	8,580
Number of vehicles in accidents/year	182	51	118

At each intersection, find the probability that a vehicle entering the intersection will be involved in an accident. Which intersection is the most dangerous?

58. **Traffic safety** A traffic survey showed that 5680 cars entered an intersection of the main street of a city and that 1460 of them turned onto the intersecting street at this intersection.
    (a) What is the empirical probability a car will turn onto the intersecting street?
    (b) What is the probability a car will not turn onto the intersecting street?

59. **Elections** Prior to an election, a poll of 1200 citizens is taken. Suppose 557 plan to vote for A, 533 plan to vote for B, and the remainder indicate they will not vote.
    (a) If the poll is accurate, find the probability that a citizen chosen at random will vote for A.
    (b) If the poll is accurate, what is the probability that a person picked at random will not vote?

60. **Elections** Suppose a county has 800,000 registered voters, 480,000 of whom are Democrats and 290,000 of whom are Republicans. If a registered voter is chosen at random, what is the probability that
    (a) he or she is a Democrat?
    (b) he or she is neither a Democrat nor a Republican?

61. **Births** Suppose that at a certain hospital the births on a given day consisted of 8 girls and 2 boys. At the same hospital, suppose the births in a given year consisted of 1220 girls and 1194 boys.
    (a) Find the empirical probabilities for the birth of each sex on the given day.
    (b) Find the empirical probabilities for the given year.
    (c) What are the theoretical probabilities, and which empirical probabilities do you think more accurately reflect reality?

7.2 Unions and Intersections of Events; One-Trial Experiments

Objectives

❏ To find the probability of the intersection of two events
❏ To find the probability of the union of two events
❏ To find the probability of the complement of an event

APPLICATION PREVIEW

Suppose that the directors of the Chatham County School Board consist of 8 Democrats, 3 of whom are female, and 4 Republicans, 2 of whom are female. If one director is chosen at random to discuss an issue on television, we can find the probability that the person will be a Democrat or female by finding the probability of the union of the two events "a Democrat is chosen" and "a female is chosen." In this section we will discuss methods of finding the probability of the intersection and union of events by using sample spaces and formulas.

As we stated in Section 7.1, events determine subsets of the sample space for a probability experiment. The intersection, union, and complement of events are defined as follows.

If E and F are two events in a sample space S, then the **intersection of E and F** is

$$E \cap F = \{a: a \in E \quad \text{and} \quad a \in F\},$$

the **union of E and F** is

$$E \cup F = \{a: a \in E \quad \text{or} \quad a \in F\},$$

and the **complement of E** is

$$E' = \{a: a \in S \quad \text{and} \quad a \notin E\}.$$

Using these definitions, we can write the probability that certain events occur as follows.

$$\Pr(E \text{ and } F \text{ both occur}) = \Pr(E \cap F)$$
$$\Pr(E \text{ or } F \text{ occurs}) = \Pr(E \cup F)$$
$$\Pr(E \text{ does } \textbf{not} \text{ occur}) = \Pr(E')$$

EXAMPLE 1

A card is drawn from a box containing 15 cards numbered 1 to 15. What is the probability that the card is

(a) not even?

(b) even and divisible by 3?

(c) even or divisible by 3?

Solution

The sample space S contains the 15 numbers.

$$S = \{1, 2, 3, 4, 5, 6, 7, 8, 9, 10, 11, 12, 13, 14, 15\}$$

If we let E represent "even-numbered" and D represent "number divisible by 3," we have

$$E = \{2, 4, 6, 8, 10, 12, 14\} \text{ and } D = \{3, 6, 9, 12, 15\}.$$

These sets are shown in the Venn diagram in Figure 7.1.

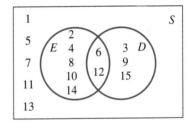

Figure 7.1

(a) The event "not even," or E', contains all elements of S not in E, so $E' = \{1, 3, 5, 7, 9, 11, 13, 15\}$, and the probability that the card is not even is

$$\Pr(\text{not } E) = \Pr(E') = \frac{n(E')}{n(S)} = \frac{8}{15}.$$

(b) The event "even and divisible by 3" is $E \cap D = \{6, 12\}$, so

$$\Pr(\text{even and divisible by 3}) = \Pr(E \cap D) = \frac{n(E \cap D)}{n(S)} = \frac{2}{15}.$$

(c) The event "even or divisible by 3" is $E \cup D = \{2, 3, 4, 6, 8, 9, 10, 12, 14, 15\}$, so

$$\Pr(\text{even or divisible by 3}) = \Pr(E \cup D) = \frac{n(E \cup D)}{n(S)} = \frac{10}{15} = \frac{2}{3}.$$

CHECKPOINT

1. If a ball is drawn from a bag containing 4 red balls numbered 1, 2, 3, 4 and 3 white balls numbered 5, 6, 7, what is the probability that the ball is
   (a) red and even?
   (b) white and even?

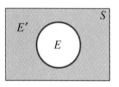

Figure 7.2

Because the complement of an event contains all the elements of *S except* for those elements in *E* (see Figure 7.2), the number of elements in *E'* is $n(S) - n(E)$, so

$$\Pr(E') = \frac{n(S) - n(E)}{n(S)} = 1 - \frac{n(E)}{n(S)} = 1 - \Pr(E).$$

$$\Pr(\text{not } E) = \Pr(E') = 1 - \Pr(E)$$

EXAMPLE 2

A dry cleaning firm has 12 employees: 7 women and 5 men. Three of the women and five of the men are 40 years old or older. If a person is chosen at random from this firm, what is the probability that the person is under 40 years of age?

Solution

Eight people are 40 or older, so

$$\Pr(\text{under } 40) = 1 - \Pr(40+) = 1 - \frac{8}{12} = \frac{4}{12} = \frac{1}{3}.$$

The number of elements in the set $E \cup F$ is the number of elements in *E* plus the number of elements in *F*, *minus* the number of elements in $E \cap F$ (because this number was counted twice in $n(E) + n(F)$). Thus

$$n(E \cup F) = n(E) + n(F) - n(E \cap F).$$

Thus,

$$\Pr(E \cup F) = \frac{n(E \cup F)}{n(S)} = \frac{n(E)}{n(S)} + \frac{n(F)}{n(S)} - \frac{n(E \cap F)}{n(S)}$$
$$= \Pr(E) + \Pr(F) - \Pr(E \cap F).$$

If *E* and *F* are any two events, then the probability that one event or the other will occur, denoted $\Pr(E \text{ or } F)$, is given by

$$\Pr(E \text{ or } F) = \Pr(E \cup F) = \Pr(E) + \Pr(F) - \Pr(E \cap F).$$

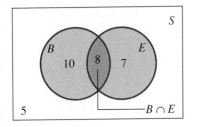

Figure 7.3

Suppose that of 30 people interviewed for a position, 18 had a business degree, 15 had previous experience, and 8 of those with experience also had a business degree. The Venn diagram in Figure 7.3 illustrates this situation.

Because 8 of the 30 people have both a business degree and experience,

$$\Pr(B \cap E) = \frac{8}{30}.$$

We also see that

$$\Pr(B) = \frac{18}{30} \quad \text{and} \quad \Pr(E) = \frac{15}{30}.$$

Thus, the probability of choosing a person at random who has a business degree or experience is

$$\Pr(B \cup E) = \Pr(B) + \Pr(E) - \Pr(B \cap E)$$
$$= \frac{18}{30} + \frac{15}{30} - \frac{8}{30} = \frac{25}{30} = \frac{5}{6}.$$

EXAMPLE 3

An integer between 1 and 6 inclusive is chosen at random. What is the probability that it is even or divisible by 3?

Solution

Since Pr(even) = 3/6 = 1/2, Pr(divisible by 3) = 2/6 = 1/3, and Pr(even *and* divisible by 3) = Pr(6) = 1/6,

$$\Pr(\text{even or divisible by 3}) = \frac{1}{2} + \frac{1}{3} - \frac{1}{6} = \frac{5}{6} - \frac{1}{6} = \frac{4}{6} = \frac{2}{3}.$$

We now consider the election problem mentioned in the Application Preview.

EXAMPLE 4

The directors of the Chatham County School Board consist of 8 Democrats, 3 of whom are female, and 4 Republicans, 2 of whom are female. If a director is chosen at random to discuss an issue on television, what is the probability that the person will be a Democrat or female?

Solution

There are 12 directors, of whom 8 are Democrats and 5 are female. But 3 of the directors are Democrats *and* females and cannot be counted twice. Thus

$$\Pr(D \text{ or } f) = \Pr(D \cup f) = \Pr(D) + \Pr(f) - \Pr(D \text{ and } f)$$
$$= \frac{8}{12} + \frac{5}{12} - \frac{3}{12} = \frac{10}{12}.$$

CHECKPOINT

2. If a ball is drawn from a bag containing 4 red balls numbered 1, 2, 3, 4 and 3 white balls numbered 5, 6, 7, what is the probability that the ball is
   (a) red or even?   (b) white or even?

We say that events E and F are **mutually exclusive** if and only if $E \cap F = \emptyset$. Thus

$$\Pr(E \cup F) = \Pr(E) + \Pr(F) - 0 = \Pr(E) + \Pr(F)$$

if and only if E and F are mutually exclusive.

If E and F are **mutually exclusive,** then $\Pr(E \cap F) = 0$, and

$$\Pr(E \text{ or } F) = \Pr(E \cup F) = \Pr(E) + \Pr(F).$$

Mutually exclusive

EXAMPLE 5
Find the probability of obtaining a 6 or a 4 in one roll of a die.

Solution
Rolling a 6 and rolling a 4 on one roll of a die are mutually exclusive events. Let E be the event "rolling a 6" and F be the event "rolling a 4."

$$\Pr(E) = \Pr(\text{rolling } 6) = \frac{1}{6} \quad \text{and} \quad \Pr(F) = \Pr(\text{rolling } 4) = \frac{1}{6}.$$

Then

$$\Pr(E \cup F) = \Pr(E) + \Pr(F) = \frac{1}{6} + \frac{1}{6} = \frac{1}{3}.$$

EXAMPLE 6
Sacco and Rosen are among 3 candidates running for a public office and polls indicate that the probability that Sacco will win is 1/3 and the probability that Rosen will win is 1/2. Only one candidate can win.
(a) What is the probability that Sacco or Rosen will win?
(b) What is the probability that neither Sacco nor Rosen will win?

Solution

(a) The two events are mutually exclusive, so

$$Pr(\text{Sacco or Rosen}) = Pr(\text{Sacco} \cup \text{Rosen})$$

$$= Pr(\text{Sacco}) + Pr(\text{Rosen}) = \frac{1}{3} + \frac{1}{2} = \frac{5}{6}.$$

(b) The probability that neither will win is $1 - 5/6 = 1/6$.

The formula for the probability of the union of mutually exclusive events can be extended beyond two events.

$$Pr(E_1 \cup E_2 \cup \ldots \cup E_n) = Pr(E_1) + Pr(E_2) + \cdots + Pr(E_n)$$

for mutually exclusive events $E_1, E_2, \ldots, E_n$.

EXAMPLE 7

Find the probability of tossing 2 distinct dice and obtaining a sum greater than 9.

Solution

Since the event "a sum greater than 9" consists of the mutually exclusive events "sum = 10," "sum = 11," and "sum = 12," and these probabilities can be found in Table 7.1 on page 454,

$$Pr(10 \text{ or } 11 \text{ or } 12) = Pr(10 \cup 11 \cup 12)$$

$$= Pr(10) + Pr(11) + Pr(12)$$

$$= \frac{3}{36} + \frac{2}{36} + \frac{1}{36} = \frac{6}{36} = \frac{1}{6}.$$

CHECKPOINT
SOLUTIONS

1. (a) The sample space contains 7 balls: $S = \{R1, R2, R3, R4, W5, W6, W7\}$. The probability that a ball drawn at random is red and even is found by using the subspace that describes that event. Thus $Pr(\text{red and even}) = Pr(\text{red} \cap \text{even}) = 2/7$.

   (b) $F = \{W6\}$ describes the event "white and even," so $Pr(\text{white and even}) = Pr(\text{white} \cap \text{even}) = 1/7$.

2. (a) $Pr(\text{red or even})$
   $$= Pr(\text{red} \cup \text{even}) = Pr(\text{red}) + Pr(\text{even}) - Pr(\text{red} \cap \text{even})$$
   $$= \frac{4}{7} + \frac{3}{7} - \frac{2}{7} = \frac{5}{7}.$$

   (b) $Pr(\text{white or even})$
   $$= Pr(\text{white} \cup \text{even}) = Pr(\text{white}) + Pr(\text{even}) - Pr(\text{white} \cap \text{even})$$
   $$= \frac{3}{7} + \frac{3}{7} - \frac{1}{7} = \frac{5}{7}.$$

EXERCISE 7.2

1. An ordinary die is tossed. What is the probability of getting a 3 or an odd number?

2. A card is drawn at random from a deck of 52 playing cards. Find the probability that it is either a club or a king.

3. If you draw one card at random from a deck of 12 cards numbered 1 through 12, inclusive, what is the probability that the number you draw is divisible by 3 and even?

4. If you draw one card at random from the deck of 12 cards numbered 1–12, what is the probability that the number you draw is even and divisible by 4?

5. If the probability that event E will occur is 3/5, what is the probability that E will not occur?

6. If the probability that an event will occur is 7/9, what is the probability that the event will not occur?

7. A bag contains 5 red balls numbered 1, 2, 3, 4, 5 and 9 white balls numbered 6, 7, 8, 9, 10, 11, 12, 13, 14. If a ball is drawn, what is the probability that
   (a) the ball is red and even-numbered?
   (b) the ball is red or even-numbered?

8. A bag contains 5 red balls numbered 1, 2, 3, 4, 5 and 9 white balls numbered 6, 7, 8, 9, 10, 11, 12, 13, 14. If a ball is drawn, what is the probability that the ball is
   (a) white and odd-numbered?
   (b) white or odd-numbered?
   (c) white or even-numbered?

9. If you draw one card from a deck of 12 cards, numbered 1 through 12 inclusive, what is the probability you will get an odd number or a number divisible by 4?

10. Suppose a die is biased (unfair) so that each odd-numbered face has probability 1/4 of resulting, and each even one has probability 1/12 of resulting. Find the probability of getting a number greater than 3.

11. If you draw one card from a deck of 12 cards numbered 1 through 12 inclusive, what is the probability that the card will be odd or divisible by 3?

12. A card is drawn from a deck of 52. What is the probability that it will be an ace, king, or jack?

13. A bag contains 4 white, 7 black, and 6 green balls. What is the probability that a ball drawn at random from the bag is white or green?

14. In a game where only one player can win, the probability that Jack will win is 1/5 and that Bill will win is 1/4. Find the probability that one of them will win.

15. A cube has 2 faces painted red, 2 painted white, and 2 painted blue. What is the probability of getting a red face or a white face in one roll?

16. A cube has 3 faces painted white, 2 faces painted red, and 1 face painted blue. What is the probability that a roll will result in a red or blue face?

17. A ball is drawn from a bag containing 13 red balls numbered 1–13 and 5 white balls numbered 14–18. What is the probability that
    (a) the ball is not even-numbered?
    (b) the ball is red and even-numbered?
    (c) the ball is red or even-numbered?
    (d) the ball is neither red nor even numbered?

18. A ball is drawn from a bag containing 13 red balls numbered 1–13 and 5 white balls numbered 14–18. What is the probability that
    (a) the ball is not red?
    (b) the ball is white and odd-numbered?
    (c) the ball is white or odd?

APPLICATIONS

19. **Drug use** Forty-six percent of marijuana use among youth occurs in the inner cities (Source: Partnership for a Drugfree America). If an instance of such marijuana use is chosen at random, what is the probability that the use does not occur in an inner city?

20. **Breast cancer** According to the American Cancer Society, 199 of 200 mammograms turn out to be normal. What is the probability that the mammogram of a woman chosen at random will show an abnormality?

21. **Car maintenance** A car rental firm has 425 cars. Sixty-three of these cars have defective turn signals and 32 have defective tires. What is the probability that one of these cars selected at random
    (a) does not have defective turn signals?
    (b) has no defects?

22. **Testing** An unprepared student must take a 7-question multiple-choice test that has 5 possible answers per question. If the student guesses on the first question, what is the probability that she will answer that question incorrectly?

23. **Linguistics** Of 100 students, 24 can speak French, 18 can speak German, and 8 can speak both French and German. If a student is picked at random, what is the probability that he or she can speak French or German?

24. **Management** A company employs 65 people. Eight of the 30 men and 21 of the 35 women work in the business office. What is the probability that an employee picked at random is a woman or works in the business office?

25. **Site selection** A firm is considering three possible locations for a new factory. The probability that site A will

be selected is 1/3 and the probability that site B will be selected is 1/5. If only one location will be chosen,

(a) what is the probability that site A or site B will be chosen?

(b) what is the probability that neither site A nor site B will be chosen?

26. **Photography** Rob Lee knows his camera will take a good picture unless the flash is defective or the batteries are dead. The probability of having a defective flash is 0.05, the probability of the batteries being dead is 0.3, and the probability that both these problems occur is 0.01. What is the probability that the picture will be good?

27. **Cognitive complexity** The cognitive complexity of a structure was studied by Scott* using a technique in which a person was asked to specify a number of objects and group them into as many groupings as he or she found meaningful. If a person groups 12 objects into three groups in such a way that

7 objects are in group A,
7 objects are in group B,
8 objects are in group C,
3 objects are in both group A and group B,
5 objects are in both group B and group C,
4 objects are in both group A and group C,
2 objects are in all three groups,

(a) what is the probability an object chosen at random has been placed into group A or group B?

(b) what is the probability an object chosen at random has been placed into group B or group C?

28. **Politics** A group of 100 people contains 60 Democrats and 35 Republicans. If there are 60 women and if 40 of the Democrats are women, what is the probability that a person selected at random is a Democrat or a woman?

29. **Education** A mathematics class has 16 engineering majors, 12 science majors, and 4 liberal arts majors.

(a) What is the probability that a student selected at random will be a science or liberal arts major?

(b) What is the probability that a student selected at random will be an engineering or science major?

(c) 5 of the engineering students, 6 of the science majors, and 2 of the liberal arts majors are female. What is the probability that a student selected at random is an engineering major or is a female?

30. **Scheduling** Maria has ordered a washer and dryer from two different companies. Both the washer and the dryer

*W. Scott, "Cognitive Complexity and Cognitive Flexibility," *Sociometry* 25 (1962), pp. 405–414.

are to be delivered Thursday. The probability that the washer will be delivered in the morning is .6 and the probability that the dryer will be delivered in the morning is .8. If the probability that either the washer or dryer is delivered in the morning is .9, what is the probability that both will be delivered in the morning?

Military spending Suppose the table below summarizes the opinions of various groups on the issue of increased military spending. Use this table to calculate probabilities in problems 31–34.

| Opinion | Whites | | Non-whites | | |
	Reps.	Dems.	Reps.	Dems.	Total
Favor	300	100	25	10	435
Oppose	100	250	25	190	565
Total	400	350	50	200	1000

31. Find the probability that an individual chosen at random is a Republican or favors increased military spending.

32. Find the probability that an individual chosen at random is a Democrat or opposes increased military spending.

33. Find the probability that an individual chosen at random is white or opposes increased military spending.

34. Find the probability that an individual chosen at random is non-white or favors increased military spending.

35. **Drinking age** A survey questioned 1000 people regarding raising the legal drinking age from 18 to 21. Of the 560 favoring raising the age, 390 were female. Of the 440 opposition responses, 160 were female. If a person is selected at random,

(a) what is the probability that the person is a woman or favors raising the age?

(b) what is the probability the person is a male or favors raising the age?

(c) what is the probability the person is male or opposes raising the age?

36. **Job bids** Three construction companies have bid for a job. Max knows that the two companies he is competing with have probabilities 1/3 and 1/6, respectively, of getting the job. What is the probability Max will get the job?

37. **Management** A company employs 65 people. Eight of the 30 men and 21 of the 35 women work in the business office. What is the probability that an employee picked at random is a man or works in the business office?

38. **Television** If the probability that a wife watches a certain television show is .55, that her husband watches it is

.45, and that both watch the show is .30, what is the probability that the husband or the wife watches this show?

39. **Cancer testing** A long-term study has revealed that the prostate specific antigens (PSA) test for prostate cancer in men is very effective. The study shows that 87% of the men for which the test is positive actually have prostate cancer (Source: *Journal of the American Medical Association,* January 1995). If a man selected at random tests positively for prostate cancer with this test,

what is the probability that he does not have prostate cancer?

40. **Cancer testing** The study described in problem 39 also revealed that 91% of the men for whom the PSA test was negative do not have prostate cancer (Source: *Journal of the American Medical Association,* January 1995). If a man selected at random tests negatively for prostate cancer with this test, what is the probability that he does have prostate cancer?

7.3 CONDITIONAL PROBABILITY; THE PRODUCT RULE

Objectives

❑ To solve probability problems involving conditional probability
❑ To compute the probability that two or more independent events will occur
❑ To compute the probability that two or more dependent events will occur

APPLICATION PREVIEW

Until recently, AT&T calling cards used each subscriber's home phone number plus a 4-digit personal identification number (PIN). Because criminals observe users of public phones to record and sell PINs for fraudulent use, AT&T has discontinued using home phone numbers and is now using 14-digit PINs. To see how this reduces the chance of someone obtaining a PIN, we can find the probability that someone can guess a PIN in these two cases if they have observed 3 of the digits and know the person's home phone number. To see which PIN would be safer, we can calculate the probability that someone can guess your PIN if you have a 4-digit PIN and if you have a 14-digit PIN by using the **Product Rule.** In this section we will use the Product Rule to find the probability that two events will occur, and we will solve problems involving **conditional probability.**

Each probability that we have computed has been relative to the sample space for the experiment. Sometimes information is given that reduces the sample space needed to solve a stated problem. For example, if a die is rolled, the probability that a 5 occurs, given that an odd number is rolled, is denoted by

$$\text{Pr}(5 \text{ rolled} \mid \text{odd number rolled}).$$

The knowledge that an odd number occurs reduces the sample space from the numbers on a die, $S = \{1, 2, 3, 4, 5, 6\}$, to the sample space for odd numbers on a die, $S_1 = \{1, 3\ 5\}$, so the event $E =$ "a 5 occurs" has probability

$$\text{Pr}(5 \mid \text{odd number}) = \frac{n(5 \text{ and odd number rolled})}{n(\text{odd number rolled})} = \frac{n(E \cap S_1)}{n(S_1)} = \frac{1}{3}.$$

To find $\text{Pr}(A \mid B)$, we seek the probability that A occurs, given that B occurs. Figure 7.4 shows the original sample space S with the reduced sample space

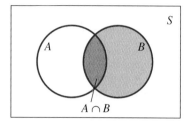

Figure 7.4

B shaded. Because all elements must be contained in *B*, we evaluate $\Pr(A \mid B)$ by dividing the number of elements in $A \cap B$ by the number of elements in *B*.

The probability that *A* occurs, given that *B* occurs, is denoted by $\Pr(A \mid B)$ and is given by

$$\Pr(A \mid B) = \frac{n(A \cap B)}{n(B)}.$$

EXAMPLE I

Suppose the table below summarizes the opinions of various groups on the issue of increased military spending. What is the probability that a person selected at random from this group of people

(a) is non-white?
(b) is non-white and favors increased military spending?
(c) favors increased military spending, given that the person is non-white?

	Whites		Non-Whites		
Opinion	Republicans	Democrats	Republicans	Democrats	Totals
Favor	300	100	25	10	435
Oppose	100	250	25	190	565
Total	400	350	50	200	1000

Solution

(a) Of the 1000 people in the group, $50 + 200 = 250$ are non-white, so

$$\Pr(\text{non-white}) = 250/1000 = 1/4.$$

(b) There are 1000 people in the group and $25 + 10 = 35$ who are both non-white and favor increased spending, so

$$\Pr(\text{non-white and favor}) = 35/1000 = 7/200.$$

(c) There are $50 + 200 = 250$ non-white and $25 + 10 = 35$ people who are both non-white and favor increased spending, so

$$\Pr(\text{favor} \mid \text{non-white}) = \frac{n(\text{favor and non-white})}{n(\text{non-white})} = \frac{35}{250} = \frac{7}{50}.$$

The probability found in Example 1 is called **conditional probability.** The conditional probability $\Pr(A \mid B)$ can be found by using the number of elements in $A \cap B$ and in *B* or by using the probability of $A \cap B$ and *B*. To find the formula by using probabilities, we divide both the numerator and denominator of the right side of

$$\Pr(A \mid B) = \frac{n(A \cap B)}{n(B)}$$

by $n(S)$, the number of elements in the sample space, which gives

$$Pr(A \mid B) = \frac{\dfrac{n(A \cap B)}{n(S)}}{\dfrac{n(B)}{n(S)}} = \frac{Pr(A \cap B)}{Pr(B)}.$$

Conditional Probability   Let A and B be events in the sample space S with $Pr(B) > 0$. The conditional probability that event A occurs, given that event B occurs, is given by

$$Pr(A \mid B) = \frac{Pr(A \cap B)}{Pr(B)}.$$

EXAMPLE 2

A red die and a green die are rolled. What is the probability the sum rolled on the dice is 6, given that the sum is less than 7?

Solution

Using the conditional probability formula and the sample space in Table 7.1 on page 454 we have

$$Pr(\text{sum is 6} \mid \text{sum} < 7) = \frac{Pr(\text{sum is 6} \cap \text{sum} < 7)}{Pr(\text{sum} < 7)}.$$

Because

$$Pr(\text{sum is 6} \cap \text{sum} < 7) = Pr(\text{sum is 6}) = \frac{5}{36},$$

the probability is

$$Pr(\text{sum is 6} \mid \text{sum} < 7) = \frac{5/36}{1/36 + 2/36 + 3/36 + 4/36 + 5/36}$$

$$= \frac{5/36}{15/36} = \frac{1}{3}.$$

CHECKPOINT

1. Suppose that one ball is drawn from a bag containing 4 red balls numbered 1, 2, 3, 4 and 3 white balls numbered 5, 6, 7. What is the probability that the ball is white, given that it is an even-numbered ball?

We can solve the formula

$$Pr(A \mid B) = \frac{Pr(A \cap B)}{Pr(B)}$$

for $\Pr(A \cap B)$ by multiplying both sides of the equation by $\Pr(B)$. This gives

$$\Pr(A \cap B) = \Pr(B) \cdot \Pr(A \mid B).$$

Similarly, we can solve

$$\Pr(B \mid A) = \frac{\Pr(B \cap A)}{\Pr(A)} = \frac{\Pr(A \cap B)}{\Pr(A)}$$

for $\Pr(A \cap B)$, getting

$$\Pr(A \cap B) = \Pr(A) \cdot \Pr(B \mid A).$$

Product Rule   If A and B are probability events, then $\Pr(A \text{ and } B)$ can be found by one or the other of these two formulas.

$$\Pr(A \text{ and } B) = \Pr(A \cap B) = \Pr(A) \cdot \Pr(B \mid A)$$

or

$$\Pr(A \text{ and } B) = \Pr(A \cap B) = \Pr(B) \cdot \Pr(A \mid B).$$

EXAMPLE 3

Suppose that from a deck of 52 cards, two cards are drawn in succession, without replacement. Find the probability that both cards are kings.

Solution

We can find the probability that two cards (drawn without replacement) are both kings as follows:

$$\Pr(\text{1st card is a king}) = \frac{4}{52}$$

$$\Pr(\text{2d card is a king} \mid \text{1st card is a king}) = \frac{3}{51}$$

$$\Pr(\text{2 kings}) = \Pr(\text{1st king}) \cdot \Pr(\text{2d king} \mid \text{1st king})$$

$$= \frac{4}{52} \cdot \frac{3}{51} = \frac{1}{221}.$$

The probability

$$\Pr(\text{2d king} \mid \text{1st king})$$

is the probability that the second king will be drawn given that the first king has already been drawn. It is 3/51 because one king has been removed from the deck.

EXAMPLE 4

A box contains 3 black balls, 2 red balls, and 5 white balls. One ball is drawn, is *not* replaced, and a second ball is drawn. Find the probability that the first ball is red and the second is black.

Solution

Let E_1 be "red ball first" and E_2 be "black ball second." Then

$$\Pr(E_1 \cap E_2) = \Pr(E_1) \cdot \Pr(E_2 \mid E_1) = \frac{2}{10} \cdot \frac{3}{9} = \frac{1}{15}.$$

CHECKPOINT

2. If two balls are drawn, without replacement, from the box described in Example 4
   (a) find the probability that both balls are white.
   (b) find the probability that the first ball is white and the second is red.

EXAMPLE 5

Polk's Department Store has observed that 80% of its charge accounts have men's names on them and that 16% of the accounts with men's names on them have been delinquent at least once, while 5% of the accounts with women's names on them have been delinquent at least once. What is the probability that an account selected at random
(a) is in a man's name and is delinquent?
(b) is in a woman's name and is delinquent?

Solution

(a) Pr(man's account) = 0.80
    Pr(delinquent | man's) = 0.16
    Pr(man's and delinquent) = Pr(man's ∩ delinquent)
                             = Pr(man's) · Pr(delinquent | man's)
                             = (0.80)(0.16) = 0.128
(b) Pr(woman's account) = 0.20
    Pr(delinquent | woman's) = 0.05
    Pr(woman's and delinquent) = (0.20)(0.05) = 0.01

EXAMPLE 6

All products on an assembly line must pass two inspections. It has been determined that the probability that the first inspector will miss a defective item is 0.09. If a defective item gets past the first inspector, the probability that the second inspector will not detect it is 0.01. What is the probability that a defective item will not be rejected by either inspector? (All good items pass both inspections.)

Solution

The second inspector only inspects items passed by the first inspector. The probability a defective item will pass both inspections is

Pr(pass both) = Pr(pass first ∩ pass second)

= Pr(pass first) · Pr(pass second | passed first)

= 0.09 · 0.01 = 0.0009.

If a coin tossed twice, the probability that both results will be heads if the first toss gives a head is

$$\text{Pr}(2\text{H} \mid 1\text{st is H}) = \frac{\text{Pr}(1\text{st is H} \cap 2\text{H})}{\text{Pr}(1\text{st is H})} = \frac{\text{Pr}(\text{Both H})}{\text{Pr}(1\text{st is H})}.$$

The sample space for this experiment is $S = \{\text{HH, HT, TH, TT}\}$, so

$$\text{Pr}(2\text{H} \mid 1\text{st H}) = \frac{1/4}{2/4} = \frac{1}{2}.$$

We can see that the probability that the second toss of a coin is head, given that the first toss gave a head, was 1/2, which is the same as the probability that the second toss in a head, regardless of what happened on the first toss. This is because the results of the coin tosses are **independent events.** We can define *independent events* as follows.

Independent Events   The events A and B are independent if and only if

$$\text{Pr}(A \mid B) = \text{Pr}(A) \qquad \text{and} \qquad \text{Pr}(B \mid A) = \text{Pr}(B).$$

This means that the occurrence or nonoccurrence of one event does not affect the other.

If A and B are independent events, then this definition can be used to simplify the product rule:

$$\text{Pr}(A \cap B) = \text{Pr}(A) \cdot \text{Pr}(B \mid A) = \text{Pr}(A) \cdot \text{Pr}(B).$$

Thus, we have the following.

Product Rule for Independent Events   If A and B are independent events, then

$$\text{Pr}(A \text{ and } B) = \text{Pr}(A \cap B) = \text{Pr}(A) \cdot \text{Pr}(B).$$

EXAMPLE 7
A die is rolled and a coin is tossed. Find the probability of getting a 4 on the die and a head on the coin.

Solution
Let E_1 be "4 on the die" and E_2 be "head on the coin." The events are independent because what occurs on the die does not affect what happens to the coin.

$$\text{Pr}(E_1) = \text{Pr}(4 \text{ on die}) = \frac{1}{6}, \quad \text{and} \quad \text{Pr}(E_2) = \text{Pr}(\text{head on coin}) = \frac{1}{2}.$$

Then

$$\text{Pr}(E_1 \cap E_2) = \text{Pr}(E_1) \cdot \text{Pr}(E_2) = \frac{1}{6} \cdot \frac{1}{2} = \frac{1}{12}.$$

The product rule can be expanded to include more than two events, as the next example shows.

EXAMPLE 8

A bag contains 3 red marbles, 4 white marbles, and 3 black marbles. Find the probability of getting a red marble on the first draw, a black marble on the second draw, and a white marble on the third draw if (a) the marbles are drawn with replacement, (b) the marbles are drawn without replacement.

Solution

(a) The marbles are replaced after each draw, so the events are independent. Let E_1 be "red on first," E_2 be "black on second," and E_3 be "white on third." Then

$$\Pr(E_1 \cap E_2 \cap E_3) = \Pr(E_1) \cdot \Pr(E_2) \cdot \Pr(E_3)$$
$$= \frac{3}{10} \cdot \frac{3}{10} \cdot \frac{4}{10} = \frac{36}{1000} = \frac{9}{250}.$$

(b) The marbles are not replaced, so the events are dependent.

$$\Pr(E_1 \cap E_2 \cap E_3) = \Pr(E_1) \cdot \Pr(E_2 \mid E_1) \cdot \Pr(E_3 \mid E_1 \text{ and } E_2)$$
$$= \frac{3}{10} \cdot \frac{3}{9} \cdot \frac{4}{8} = \frac{1}{20}.$$

We now turn to the problem posed in the Application Preview.

EXAMPLE 9

Suppose that a person knows your home phone number and has observed the first 3 digits of your PIN. Calculate the probability that the person can guess your PIN if
(a) you have a 4-digit PIN.
(b) you have a 14-digit PIN.
With which PIN would you feel safer?

Solution

(a) Because the person knows the first 3 digits, he or she must only guess the fourth digit, so the probability is

$$1 \cdot 1 \cdot 1 \cdot \frac{1}{10} = \frac{1}{10}.$$

(b) The person knows the first 3 digits, so the person must guess the remaining 11 digits. The probability he or she can do this is

$$1 \cdot 1 \cdot 1 \cdot \frac{1}{10} \cdot \frac{1}{10} \cdot \frac{1}{10} \cdot \frac{1}{10} \cdot \frac{1}{10} \cdot \frac{1}{10} \cdot \frac{1}{10} \cdot \frac{1}{10} \cdot \frac{1}{10} \cdot \frac{1}{10} \cdot \frac{1}{10}$$
$$= \frac{1}{100,000,000,000}.$$

Clearly the 14-digit PIN is much safer than the 4-digit PIN. However, some professional criminals are able to steal even the 14-digit PIN if it is not guarded carefully.

EXAMPLE 10

A coin is tossed 4 times. Find the probability that
(a) no heads occur.   (b) at least one head is obtained.

Solution

(a) each toss is independent, and on each toss, the probability that a tail occurs is 1/2. Thus the probability of 4 tails in 4 tosses is

$$\text{Pr(TTTT)} = \frac{1}{2} \cdot \frac{1}{2} \cdot \frac{1}{2} \cdot \frac{1}{2} = \frac{1}{16}.$$

(b) The probability that at least 1 head will occur can be found by computing Pr(1 head or 2 heads or 3 heads or 4 heads).
The equiprobable sample space for this probability experiment is

$$S = \{\text{HHHH, HHHT, HHTH, HTHH, THHH, HHTT, HTHT, HTTH,}$$
$$\text{THTH, TTHH, THHT, HTTT, THTT, TTHT, TTTH, TTTT}\}$$

It contains 16 elements, and 15 of these satisfy the conditions of the problem. The only element that does not satisfy the conditions of the problem is the element TTTT, which gives 0 heads. Thus the probability that at least one head occurs is

$$\text{Pr(at least one head occurs)} = \text{Pr(not 0 heads)}$$
$$= 1 - \text{Pr(0 heads)} = 1 - \frac{1}{16} = \frac{15}{16}.$$

In general, we can calculate the probability of at least one success in n trials as follows:

Pr(at least 1 success in n trials) = 1 − Pr(0 successes)

CHECKPOINT
SOLUTIONS

1. $\text{Pr(white | even numbered)} = \dfrac{\text{Pr(white and even)}}{\text{Pr(even)}} = \dfrac{1/7}{3/7} = \dfrac{1}{3}$

2. (a) $\text{Pr(white on both draws)} = \text{Pr}(W_1 \cap W_2) = \text{Pr}(W_1) \cdot \text{Pr}(W_2 \mid W_1)$
$$= \frac{5}{10} \cdot \frac{4}{9} = \frac{2}{9}$$

   (b) $\text{Pr(white on first and red on second)} = \text{Pr}(W_1 \cap R_2) = \text{Pr}(W_1) \cdot \text{Pr}(R_2 \mid W_1)$
$$= \frac{5}{10} \cdot \frac{2}{9} = \frac{1}{9}$$

EXERCISE 7.3

1. A card is drawn from a deck of 52 playing cards. Given that it is a red card, what is the probability that
   (a) it is a heart?   (b) it is a king?

2. A card is drawn from a deck of 52 playing cards. Given that it is an ace, king, queen, or jack, what is the probability that it is a jack?

3. A die has been "loaded" so that the probability of rolling any even number is 2/9 and any odd number is 1/9. What is the probability of
   (a) rolling a 6, given that a number divisible by 3 is rolled?
   (b) rolling a 3, given that a number divisible by 3 is rolled?

4. A die has been "loaded" so that the probability of rolling any even number is 2/9 and any odd number is 1/9. What is the probability of
   (a) rolling a 5?
   (b) rolling a 5, given that an even number is rolled?
   (c) rolling a 5, given that an odd number is rolled?

5. A bag contains 9 red balls numbered 1, 2, 3, 4, 5, 6, 7, 8, 9 and 6 white balls numbered 10, 11, 12, 13, 14, 15. One ball is drawn from the bag. What is the probability that the ball is even-numbered, given that the ball is red?

6. If one ball is drawn from the bag of problem 5, what is the probability that the ball is white, given that the ball is even-numbered?

7. A bag contains 4 red balls and 6 white balls. Two balls are drawn without replacement.
   (a) What is the probability that the second ball is white, given that the first ball is red?
   (b) What is the probability that the second ball is red, given that the first ball is white?
   (c) Answer part (a) if the first ball is replaced before the second is drawn.

8. A fair die is rolled. Find the probability that the result is a 4, given that the result is even.

9. A fair coin is tossed 3 times. Find the probability of
   (a) throwing 3 heads, given that the first toss is a head.
   (b) throwing 3 heads, given that the first two tosses result in heads.

10. A fair coin is tossed 14 times. What is the probability of tossing 14 heads, given that the first 13 tosses are heads?

11. A die is thrown twice. What is the probability that a 3 will result the first time and a 6 will result the second time?

12. A die is rolled and a coin is tossed. What is the probability that the die results in an even number and the coin results in a head?

13. (a) A coin is to tossed three times. What is the probability of getting a head on all three tosses?
    (b) A coin is tossed three times. What is the probability of getting at least one tail? (Hint: Use the result from part (a)).

14. The probability that Sam will win in a certain game whenever he plays is 2/5. If he plays two games, what is the probability that he will win just the first game?

15. (a) A box contains 3 red balls, 2 white balls, and 5 black balls. Two balls are drawn at random from the box (with replacement of the first before the second is drawn); what is the probability of getting a red ball on the first draw and a white ball on the second draw?
    (b) Answer the question if the first ball is not replaced before the second is drawn.

16. (a) One card is drawn at random from a deck of 52 cards. The first card is replaced, and a second card is drawn. Find the probability that both cards are hearts.
    (b) Answer the question if the first card is not replaced before the second card is drawn.

17. Two balls are drawn from a bag containing 3 white balls and 2 red balls. If the first ball is replaced before the second is drawn, what is the probability that
    (a) both balls are red?
    (b) both balls are white?
    (c) the first ball is red and the second is white?
    (d) one of the balls is black?

18. Answer problem 17 if the first ball is not replaced before the second is drawn.

19. A bag contains 9 nickels, 4 dimes, and 5 quarters. If you draw 3 coins at random from the bag, without replacement, what is the probability that you will get a nickel, a quarter, a nickel, in that order?

20. A bag contains 6 red balls and 8 green balls. If two balls are drawn together, find the probability that
    (a) both are red.
    (b) both are green.

21. A red ball and 4 white balls are in a box. If two balls are drawn, without replacement, what is the probability
    (a) of getting a red ball on the first draw and a white ball on the second?
    (b) of getting 2 white balls?
    (c) of getting 2 red balls?

22. From a deck of 52 playing cards two cards are drawn, one after the other without replacement. What is the probability that
    (a) the first will be a king and the second will be a jack?
    (b) the first will be a king and the second will be a jack of the same suit?

23. One card is drawn at random from a deck of 52 cards. The first card is not replaced, and a second card is drawn. Find the probability that
    (a) both cards are spades.
    (b) the first card is a heart and the second is a club.

24. Two cards are drawn from a deck of 52 cards. What is the probability both are aces,
    (a) if the first card was replaced before the second was drawn?
    (b) if the cards were drawn without replacement?

25. Two balls are drawn, without replacement, from a bag containing 13 red balls numbered 1–13 and 5 white balls numbered 14–18. What is the probability that
    (a) the second ball is red, given that the first ball is white?
    (b) both balls are even-numbered?
    (c) the first ball is red and even-numbered and the second ball is even?

26. A red die and a green die are rolled. What is the probability that a 6 results on the green die and that a 4 results on the red die?

APPLICATIONS

Universal health care The following table gives the result of a survey of 1000 people regarding the funding of universal health care by employers (employer mandate). Use the table to answer problems 27–30.

	Favor	Oppose	No opinion	Total
Democrat	310	150	60	520
Republican	125	345	10	480
Total	435	495	70	1000

27. What is the probability that a person selected at random from this group will favor universal health care with an employer mandate, given that the person is a Democrat?

28. What is the probability that a person selected at random from this group will oppose universal health care with an employer mandate, if the person is a Republican?

29. If the group surveyed represents the people of the United States, what is the probability that a citizen selected at random will favor universal health care with an employer mandate, given that the person is a Republican?

30. If the group surveyed represents the people of the United States, what is the probability that a citizen selected at random will oppose universal health care with an employer mandate, given that the person is a Democrat?

Military spending Suppose the table below summarizes the opinion of various groups on the issue of increased military spending. Use this table to calculate the empirical probabilities in problems 31–38.

Opinion	Whites		Non-whites		
	Reps.	Dems.	Reps.	Dems.	Total
Favor	300	100	25	10	435
Oppose	100	250	25	190	565
Total	400	350	50	200	1000

31. Given that a randomly selected individual is non-white, find the probability that he or she opposes increased military spending.

32. Given that a randomly selected individual is a Democrat, find the probability that he or she opposes increased military spending.

33. Given that a randomly selected individual is in favor of increased military spending, find the probability that he or she is a Republican.

34. Given that a randomly selected individual is opposed to increased military spending, find the probability that he or she is a Democrat.

35. Find the probability that a person who favors increased military spending is non-white.

36. Find the probability that an individual is white and opposes increased military spending.

37. Find the probability that an individual is a white Republican opposed to increased military spending.

38. Find the probability that an individual is a Democrat and opposes increased military spending.

39. ***Car maintenance*** NACO Body Shops has found that 14% of the cars on the road need to be painted and that 3% need major body work and painting. What is the probability that a car selected at random needs major body work if it is known that it needs to be painted?

40. ***Lottery*** A name is drawn at random from the 96 entrants in a golf tournament. The person whose name is selected draws a ball from a bag containing four balls,

one of which is embossed with the Peggos Company logo. If the person draws the ball with the logo, he or she wins a set of Ping irons. If you are one of the entrants in the tournament, what is the probability that you will win the Ping irons?

41. **Blood types** In the pretrial hearing of the O. J. Simpson case, the prosecution stated that Mr. Simpson's blood markers include type A blood, which 33.7% of the population has; blood SD subtype 1, which 79.6% of the population has; and PGM 2+2−, which 1.6% of the population has. If these blood markers are independent, what is the probability that a person selected at random will have the same blood markers as O. J. Simpson?

42. **Education** Fifty-one percent of the U.S. population is female and 18.6 percent of the female population has a college degree (Source: *Upclose Census Digest,* City and Town Edition, 1992). If a U.S. resident is chosen at random, what is the probability that the person is a female with a college degree?

43. **Quality control** Each computer component that the Peggos Company produces is tested twice before it is shipped. There is a 0.7 probability that a defective component will be so identified by the first test and a 0.8 probability that it will be identified as being defective by the second test. What is the probability that a defective component will not be identified as defective before it is shipped?

44. **Lactose intolerance** Lactose intolerance affects about 20% of non-Hispanic white Americans, and 75.6% of the residents of the United States are non-Hispanic whites (Source: Jean Carper, "Eat Smart," *USA Weekend,* Sept. 4, 1994). If a United States resident is selected at random, what is the probability that the person will be a non-Hispanic white and have lactose intolerance?

45. **Lactose intolerance** Lactose intolerance affects about 50% of Hispanic Americans, and 9% of the residents of the United States are Hispanic (Source: Jean Carper, "Eat Smart," *USA Weekend,* Sept. 4, 1994). If a United States resident is selcted at random, what is the probability that the person will be Hispanic and have lactose intolerance?

46. **Quality control** If 3% of all light bulbs a company manufactures are defective, the probability of any one bulb being defective is 0.03. What is the probability that three bulbs drawn independently from the company's stock will be defective?

47. **Quality control** To test its shotgun shells, a company fires 5 of them. What is the probability that all 5 will fire properly if 5% of its shells are actually defective?

48. **Quality control** One machine produces 30% of a product for a company. If 10% of the products from this machine are defective and the other machines produce no defective items, what is the probability that an article produced by this company is defective?

49. **Advertising** A company estimates that 30% of the country has seen its commercial and that if a person sees its commercial there is a 20% probability that the person will buy its product. What is the probability that a person chosen at random in the country will have seen the commercial and bought its product?

50. **Employment** Ronald Lee has been told by a company that the probability that he will be offered a job in the quality control department is 0.6 and the probability that he will be asked to be foreman of the department, if he is offered the job, is 0.1. What is the probability that he will be offered the job and asked to be foreman?

51. **Birth control** Suppose a birth control pill is 99% effective in preventing pregnancy.
    (a) What is the probability that none of 100 women using the pill will become pregnant?
    (b) What is the probability that at least one woman per 100 users will become pregnant?

52. **Combat** If a fighter pilot has a 4% chance of being shot down on each mission during a war, is it certain he will be shot down in 25 missions?

53. **Maintenance** Twenty-three percent of the cars owned by a car rental firm have some defect. What is the probability that of 3 cars selected at random
    (a) none have defects?
    (b) at least one has defects?

54. **Testing** An unprepared student must take a 7-question multiple-choice test that has 5 possible answers per question. If the student guesses on every question, what is the probability that
    (a) she will answer every question correctly?
    (b) she will answer every question incorrectly?
    (c) she will answer at least one question correctly?

55. **Testing** An unprepared student must take a 7-question multiple-choice test that has 5 possible answers per question. If the student can eliminate two of the possible answers on the first three questions, and guesses on every question, what is the probability that

(a) she will answer every question correctly?

(b) she will answer every question incorrectly?

(c) she will answer at least one question correctly?

56. *Testing* What is the probability that a student who guesses on every question of a 10-question true-or-false test will get at least one answer correct?

57. *Birth dates* Assuming there are 365 different birthdays, find the probability that two people chosen at random will have

(a) different birthdays.

(b) the same birthday.

58. *Birth dates* Assuming there are 365 different birthdays, find the probability that of three people chosen at random,

(a) no two will have the same birthday.

(b) at least two will have the same birthday.

59. *Birth dates* Assuming there are 365 different birthdays, find the probability that of 20 people chosen randomly,

(a) no two will have the same birthday.

(b) at least two will have the same birthday.

60. *Crime* In an actual case,* probability was used to convict a couple of mugging an elderly woman. Shortly after the mugging, a young, white woman with blonde hair worn in a ponytail was seen running from the scene of the crime and entering a yellow car that was driven away by a black man with a beard. A couple matching this description was arrested for the crime. A prosecuting attorney argued that the couple arrested had to be the couple at the scene of the crime because the probability of a second couple matching the description was very small. He estimated the probabilities of six events as follows:

> Probability of black-white couple: 1/1000
> Probability of black man: 1/3
> Probability of bearded man: 1/10
> Probability of blonde woman: 1/4
> Probability of hair in ponytail: 1/10
> Probability of yellow car: 1/10

He multiplied these probabilities and concluded that the probability that another couple would have these characteristics is 1/12,000,000. Based on this circumstantial evidence, the couple was convicted and sent to prison. The conviction was overturned by the state supreme court because the prosecutor made an incorrect assumption. What error do you think he made?

7.4 PROBABILITY TREES AND BAYES' FORMULA

Objectives

❏ To use probability trees to solve problems

❏ To use Bayes' formula to solve probability problems

APPLICATION PREVIEW

A security system is manufactured with a "fail-safe" provision so that it functions properly if any two or more of its three main components, A, B, and C, are functioning properly. The probabilities that components A, B, and C are functioning properly are 0.95, 0.90, and 0.92, respectively. There are four different ways in which at least 2 components will function properly, so there are four mutually exclusive possible outcomes in which the system will function properly. We can find the probability that the system functions properly by using a **probability tree.** Probability trees are especially useful in dealing with probability problems that involve two or more stages.

*Time, January 8, 1965, p. 42, and April 26, 1968, p. 41.

PROBABILITY TREES

When a probability experiment involves more than one trial, it is frequently helpful to represent the possible outcomes in a tree diagram. We can think of the possible outcomes of the first trial as the beginning branches of the tree, as in Figure 7.5(a), where the first trial has three possible outcomes. The possible outcomes on the second draw can be treated as branches that eminate from each of the outcomes of the first trial, as in Figure 7.5(b), where there are two possible outcomes of the second trial. We know from the previous section that the probability that event E_1 occurs on the first trial and F_2 occurs on the second trial is

$$Pr(E_1 \cap F_2) = Pr(E_1) \cdot Pr(F_2 \mid E_1).$$

If we put these probabilities on the branches of the tree, then the "event E_1 occurs on trial 1 and event F_2 occurs on trial 2" is represented by a "path" through the tree, and the probability that this occurs is the product of the probabilities on the branches.

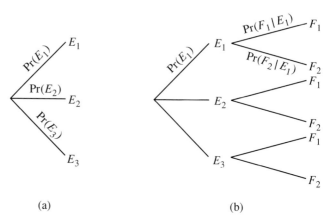

Figure 7.5   (a)   (b)

To illustrate how a probability tree is constructed, consider the following example.

EXAMPLE 1

A bag contains 5 red balls, 4 blue balls, and 3 white balls. Two balls are drawn, one after the other, without replacement. Draw a tree representing the experiment.

Solution

The first draw could result in a red ball (with probability 5/12), a blue ball (with probability 4/12), or a white ball (with probability 3/12). We can represent the results of the first draw by the tree shown in Figure 7.6.

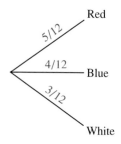

Figure 7.6

The probabilities for red, blue, and white balls occurring on the second draw depend on the result of the first draw. We can represent all the possibilities by adding a second stage to the tree we have drawn, with the conditional probabilities noted (see Figure 7.7).

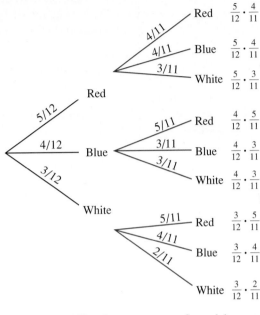

Figure 7.7   First draw   Second draw

The first stage of the experiment in Example 1 has three possible outcomes, so the tree has three branches in the first stage. The second stage has three possible outcomes, so the tree has three branches in its second stage for each first outcome. The total number of possible outcomes in the compound experiment is the same as the total number of terminating branches, so each path through the tree represents a sample point for the experiment. For example, drawing a red ball and then drawing a blue ball is one possible outcome, and it is represented by one path through the tree, namely the branch with red on the first draw and blue on the second draw.

Probability Tree   The probability attached to *each branch* of the tree is the *conditional probability* that the specified event will occur, given that the events on the preceding branches have occurred. The probability of a **sample point** for the experiment described by a tree is the **product** of the probabilities attached to the branches **along the path** representing the sample point.

For example, the probability of drawing a red ball first and then a blue ball in the experiment described by Figure 7.7 is the product of 5/12 and 4/11, or 5/33.

EXAMPLE 2

In the experiment described by Example 1 (and Figure 7.7), find
(a) Pr(blue on first draw and white on second draw).
(b) Pr(white on both draws).
(c) Pr(drawing a blue ball and a white ball).
(d) Pr(second ball is red).

Solution

(a) By multiplying the probabilities along the path that represents blue on the first draw and white on the second draw, we obtain

$$\text{Pr(blue on first draw and white on second draw)} = \frac{4}{12} \cdot \frac{3}{11} = \frac{1}{11}.$$

(b) Multiplying the probabilities along the path that represents white on both draws gives

$$\text{Pr(white on both draws)} = \frac{3}{12} \cdot \frac{2}{11} = \frac{1}{22}.$$

(c) This result can occur by drawing a blue ball first, then a white ball *or* by drawing a white ball, then a blue ball. Thus either of two paths leads to this result. Because the results represented by these paths are mutually exclusive, the probability is found by *adding* the probabilities from the two paths. That is, letting B represent a blue ball and W represent a white ball,

$$\begin{aligned} \text{Pr}(B \text{ and } W) &= \text{Pr}(B \cap W) \\ &= \text{Pr}(B \text{ first, then } W) + \text{Pr} (W \text{ first, then } B) \\ &= \frac{4}{12} \cdot \frac{3}{11} + \frac{3}{12} \cdot \frac{4}{11} = \frac{2}{11}. \end{aligned}$$

(d) $$\begin{aligned} \text{Pr(2d ball is red)} &= \text{Pr}(R \text{ then } R, \text{ or } B \text{ then } R, \text{ or } W \text{ then } R) \\ &= \text{Pr}(R \text{ then } R) + \text{Pr}(B \text{ then } R) + \text{Pr}(W \text{ then } R) \\ &= \frac{5}{12} \cdot \frac{4}{11} + \frac{4}{12} \cdot \frac{5}{11} + \frac{3}{12} \cdot \frac{5}{11} = \frac{5}{12} \end{aligned}$$

CHECKPOINT

1. Urn I contains 3 gold coins, Urn II contains 1 gold coin and 2 silver coins, and Urn III contains 1 gold coin and 1 silver coin. If an urn is selected at random and a coin is drawn from the urn, what is the probability that a gold coin will be drawn?

EXAMPLE 3

As we mentioned in the Application Preview, a security system is manufactured with a "fail-safe" provision so that it functions properly if any two or more of its three main components, A, B, and C, are functioning properly.

The probabilities that components A, B, and C are functioning properly are 0.95, 0.90, and 0.92, respectively. What is the probability that the system functions properly?

Solution

The probability tree in Figure 7.8 shows all of the possibilities for the components of the system. Each stage of the tree corresponds to a component, and G indicates that the component is good; N indicates it is not good. There are four paths through the probability tree that give at least 2 good (functioning) components, so there are four mutually exclusive possible outcomes. The probability that the system functions properly is

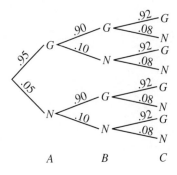

Figure 7.8

$$\Pr(\text{at least 2 } G) = \Pr(GGG \cup GGN \cup GNG \cup NGG)$$
$$= (0.95)(0.90)(0.92) + (0.95)(0.90)(0.08)$$
$$+ (0.95)(0.10)(0.92) + (0.05)(0.90)(0.92)$$
$$= 0.9838$$

BAYES' FORMULA

Using a probability tree permits us to answer more difficult questions regarding conditional probability. For example, in the experiment described by Example 1, we can find the probability that the first ball is red, given that the second ball is blue. To find this probability we recall the formula for conditional probability, which is

$$\Pr(E_2 \mid E_1) = \frac{\Pr(E_1 \cap E_2)}{\Pr(E_1)}.$$

So

$$\Pr(\text{1st is red} \mid \text{2d is blue}) = \frac{\Pr(\text{1st is red and 2d is blue})}{\Pr(\text{2d is blue})}.$$

One path in Figure 7.7 (red-blue) describes "1st is red and 2d is blue" and three paths (red-blue, blue-blue, white-blue) describe "2d is blue," so

$$\Pr(\text{1st is red} \mid \text{2d is blue}) = \frac{\dfrac{5}{12} \cdot \dfrac{4}{11}}{\dfrac{5}{12} \cdot \dfrac{4}{11} + \dfrac{4}{12} \cdot \dfrac{3}{11} + \dfrac{3}{12} \cdot \dfrac{4}{11}}$$
$$= \frac{5/33}{5/33 + 1/11 + 1/11} = \frac{5}{11}.$$

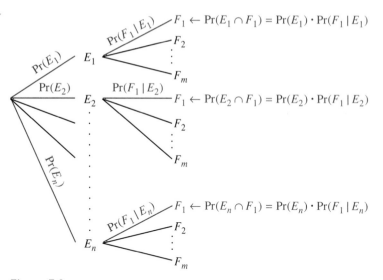

$F_1 \leftarrow \Pr(E_1 \cap F_1) = \Pr(E_1) \cdot \Pr(F_1 \mid E_1)$

$F_1 \leftarrow \Pr(E_2 \cap F_1) = \Pr(E_2) \cdot \Pr(F_1 \mid E_2)$

$F_1 \leftarrow \Pr(E_n \cap F_1) = \Pr(E_n) \cdot \Pr(F_1 \mid E_n)$

Figure 7.9

The preceding example illustrates a special type of problem, which we call a **Bayes problem.** In this type of problem, we know the result of the second stage of a two-stage experiment and we wish to find the probability of a specified result in the first stage. In the preceding example we knew the second ball drawn was blue and we sought the probability that the first ball was red.

Suppose there are n possible outcomes in the first stage of the experiment, denoted E_1, $E_2, \ldots, E_n$ and m possible outcomes in the second stage, denoted $F_1, F_2, \ldots, F_m$ (see Figure 7.9). Then the probability that event E_1 occurs in the first stage, given that F_1 occured in the second stage, is

$$\Pr(E_1 \mid F_1) = \frac{\Pr(E_1 \cap F_1)}{\Pr(F_1)}. \qquad (1)$$

By looking at the probability tree in Figure 7.9 we see that

$$\Pr(E_1 \cap F_1) = \Pr(E_1) \cdot \Pr(F_1 \mid E_1)$$

and that

$$\Pr(F_1) = \Pr(E_1 \cap F_1) + \Pr(E_2 \cap F_1) + \cdots + \Pr(E_n \cap F_1).$$

Thus using these facts, equation (1) becomes

$$\Pr(E_1 \mid F_1) = \frac{\Pr(E_1) \cdot \Pr(F_1 \mid E_1)}{\Pr(E_1 \cap F_1) + \Pr(E_2 \cap F_1) + \cdots + \Pr(E_n \cap F_1)}. \qquad (2)$$

Using the fact that for any E_i, $\Pr(E_i$ and $F_1) = \Pr(E_i) \cdot \Pr(F_1 \mid E_i)$, we can rewrite equation (2) in a form called **Bayes' formula.**

Bayes' Formula   If a probability experiment has n possible outcomes in its first stage given by $E_1, E_2, \ldots, E_n$, and if F_1 is an event in the second stage, then the probability that even E_1 occurs in the first stage, given that F_1 has occurred in the second stage is

$$\Pr(E_1 \mid F_1) = \frac{\Pr(E_1) \cdot \Pr(F_1 \mid E_1)}{\Pr(E_1) \cdot \Pr(F_1 \mid E_1) + \Pr(E_2) \cdot \Pr(F_1 \mid E_2) + \cdots + \Pr(E_n) \cdot \Pr(F_1 \mid E_n)}.$$

Note that Bayes problems can be solved either by this formula or by the use of a probability tree. Of the two methods, many students find using the tree easier because it is less abstract than the formula.

Bayes' Formula and Trees

$$\Pr(E_1 \mid F_1) = \frac{\text{Product of branch probabilities on path leading to } F_1 \text{ through } E_1}{\text{Sum of all branch products on paths leading to } F_1}$$

The following example illustrates both methods of solving a Bayes problem.

EXAMPLE 4

Suppose a test for diagnosing a certain serious disease is successful in detecting the disease in 95% of all persons infected, but that it incorrectly diagnoses 4% of all healthy people as having the serious disease. Suppose also that it incorrectly diagnoses 12% of all people having another minor disease as having the serious disease. If it is known that 2% of the population has the serious disease, 90% of the population is healthy, and 8% has the minor disease, find the probability that a person selected at random has the serious disease if the test indicates that he or she does. Use H to represent healthy, M to represent having the minor disease, and D to represent having the serious disease.

Solution
The tree that represents the health condition of a person chosen at random and the results of the test on that person is shown in Figure 7.10. (A test that indicates that a person has the disease is called a *positive* test.)

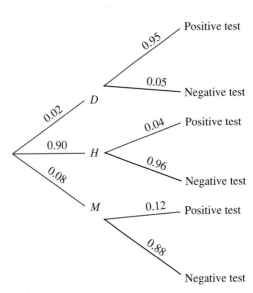

Figure 7.10

We seek $\Pr(D \mid \text{pos.test})$. Using the tree and the conditional probability formula gives

$$\Pr(D \mid \text{pos. test}) = \frac{\Pr(D \cap \text{pos. test})}{\Pr(\text{pos. test})}$$

$$= \frac{(.02)(.95)}{(.90)(.04) + (.02)(.95) + (.08)(.12)}$$

$$= \frac{.0190}{.0360 + .0190 + .0096}$$

$$= 0.2941.$$

Using Bayes' formula directly gives the same result.

$\Pr(D \mid \text{pos. test})$

$$= \frac{\Pr(D) \cdot \Pr(\text{pos. test} \mid D)}{\Pr(H) \cdot \Pr(\text{pos. test} \mid H) + \Pr(D) \cdot \Pr(\text{pos. test} \mid D) + \Pr(M) \cdot \Pr(\text{pos. test} \mid M)}$$

$$= \frac{(.02)(.95)}{(.90)(.04) + (.02)(.95) + (.08)(.12)} = 0.2941$$

CHECKPOINT

2. Use the information (and tree) from Example 4 to find the probability that a person chosen at random is healthy if the test is negative (indicates that the person does not have the disease).

CHECKPOINT SOLUTIONS

1. The probability tree for this probability experiment is given below. From the tree we have

   $\Pr(\text{gold}) = \Pr[(\text{I and } G) \text{ or } (\text{II and } G) \text{ or } (\text{III and } G)]$

   $$= \frac{1}{3} \cdot \frac{3}{3} + \frac{1}{3} \cdot \frac{1}{3} + \frac{1}{3} \cdot \frac{1}{2} = \frac{11}{18}.$$

2. The probability that a person is healthy, given that the test is negative, is

   $$\Pr(H \mid \text{neg.}) = \frac{\Pr(H \cap \text{neg.})}{\Pr(\text{neg.})}$$

   $$= \frac{(.90)(.96)}{(.02)(.05) + (.90)(.96) + (.08)(.88)}$$

   $$\approx 0.9237.$$

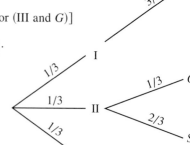

EXERCISE 7.4

In problems 1–14, use probability trees to find the probabilities of the indicated outcomes.

1. Two cards are drawn, without replacement, from a regular deck of 52 cards. What is the probability that
   (a) both cards are the same suit?
   (b) both cards are the same suit, given that the first card is the king of hearts?

2. A bag contains 5 coins, 4 of which are fair and 1 that has a head on each side. If a coin is selected from the box and tossed twice, what is the probability of obtaining 2 heads?

3. Three balls are drawn, without replacement, from a bag containing 4 red balls and 5 white balls. Find the probability that
   (a) three white balls are drawn.
   (b) two white balls and one red ball are drawn.
   (c) the third ball drawn is red.

4. Two cards are drawn, without replacement, from a regular deck of 52 cards. What is the probability that both cards are aces, given that they are the same color?

5. An urn contains 4 red, 5 white, and 6 black balls. One ball is drawn from the urn, replaced, and a second ball is drawn.
   (a) What is the probability that both balls are white?
   (b) What is the probability that one ball is white and one is red?
   (c) What is the probability that at least one ball is black?

6. An urn contains 4 red, 5 white, and 6 black balls. One ball is drawn from the urn, not replaced, and a second ball is drawn.
   (a) What is the probability that both balls are white?
   (b) What is the probability that one ball is white and one is red?
   (c) What is the probability that at least one ball is black?

7. A bag contains 4 white balls and 7 red balls. If 2 balls are drawn at random from the bag, with the first ball not replaced before the second ball is drawn, what is the probability that
   (a) the first ball is red and the second is white?
   (b) one ball is red and the other is white?

8. An urn contains 4 red, 5 white, and 6 black balls. Three balls are drawn, without replacement, from the urn.
   (a) What is the probability that all three balls are red?
   (b) What is the probability that exactly two balls are red?

(c) What is the probability that at least two balls are red?

9. A bag contains 4 white balls and 6 red balls. Three balls are drawn, without replacement, from the bag.
   (a) What is the probability that all three balls are white?
   (b) What is the probability that exactly one ball is white?
   (c) What is the probability that at least one ball is white?

10. A bag contains 5 coins, of which 4 are fair; the remaining coin has a head on both sides. If a coin is selected at random from the bag and tossed three times, what is the probability that heads will occur exactly twice?

11. A bag contains 5 coins, of which 4 are fair; the remaining coin has a head on both sides. If a coin is selected at random from the bag and tossed three times, what is the probability that heads will occur at least twice?

12. A bag contains 4 white balls and 7 red balls. If 2 balls are drawn at random from the bag, with the first ball replaced before the second ball is drawn, what is the probability that
    (a) the first ball is red and the second is white?
    (b) one ball is red and the other is white?

13. A bag contains 4 white balls and 6 red balls. What is the probability that if 2 balls are drawn (with replacement)
    (a) the first ball is red and the second is white?
    (b) both balls are red?
    (c) one ball is red and one is white?
    (d) the first ball is red or the second is white?

14. Two balls are drawn, without replacement, from a bag containing 13 red balls numbered 1–13 and 5 white balls numbered 14–18. What is the probability that
    (a) the second ball is even, given that the first ball is even?
    (b) the first ball is red and the second ball is even-numbered?
    (c) the first ball is even and the second is white?

In problems 15–18, use (a) a probability tree and (b) Bayes' formula to find the probabilities. Urns I and II each have 5 red balls, 3 white balls, and 2 green balls. Urn III has 1 red ball, 1 white ball, and 8 green balls. Use this information in problems 15 and 16.

15. An urn is selected at random and a ball is drawn. If the ball is green, find the probability that urn III was selected.

16. An urn is selected at random and a ball is drawn. If the ball is red, what is the probability that urn III was selected?

17. There are 3 urns containing coins. Urn I contains 3 gold coins, urn II contains 1 gold coin and 1 silver coin, and

urn III contains 2 silver coins. An urn is selected and a coin is drawn from the urn. If the selected coin is gold, what is the probability the urn selected was urn I?

18. In problem 17, what is the probability that the urn selected was urn III if the coin selected was silver?

APPLICATIONS

19. **Lactose intolerance** Lactose intolerance affects about 20% of non-Hispanic white Americans, 50% of Hispanic Americans and 75% of African, Asian, and Native Americans (Source: Jean Carper, "Eat Smart," *USA Weekend,* Sept. 4, 1994). Seventy-six percent of United States residents are non-Hispanic whites, 9% of them are Hispanic, and 15% are African, Asian, or Native American. If a person is selected from this group of people, what is the probability that the person will have lactose intolerance?

20. **Genetics** What is the probability that a couple will have at least two sons if a they plan to have 3 children and if the probability of having a son equals the probability of having a daughter?

21. **Marksmanship** Suppose that a marksman hits the bullseye 15,000 times in 50,000 shots. If the next 4 shots are independent, find the probability that
    (a) the next 4 shots hit the bullseye.
    (b) two of the next 4 shots hit the bullseye.

22. **Majors** In a random survey of students concerning student activities, 30 engineering majors, 25 business majors, 20 science majors, and 15 liberal arts majors were selected. If two students are selected at random, what is the probability of getting
    (a) two science majors?
    (b) a science major and an engineering major?

23. **Quality control** Suppose a box contains 3 defective transistors and 12 good transistors. If 2 transistors are drawn from the box without replacement, what is the probability that
    (a) the first transistor is good and the second transistor is defective?
    (b) the first transistor is defective and the second one is good?
    (c) one of the transistors drawn is good and one of them is defective?

24. **Education** The probability that an individual without a college education earns more than $20,000 is 0.2, whereas the probability that a person with a B.S. or higher degree will earn more than $20,000 is 0.6. The probability that a person chosen at random has a B.S. degree is 0.3. What is the probability a person has at least a B.S. degree if it is known that he or she earns more than $20,000?

25. **Alcoholism** A small town has 8000 adult males and 6000 adult females. A sociologist conducted a survey and found that 40% of the males and 30% of the females drink heavily. An adult is selected at random from the town.
    (a) What is the probability the person is a male?
    (b) What is the probability the person drinks heavily?
    (c) What is the probability the person is a male or drinks heavily?
    (d) What is the probability the person is a male, if it is known that the person drinks heavily?

26. **TV violence** One hundred boys and one hundred girls were asked if they had ever been frightened by a television program. Thirty of the boys and sixty of the girls said they had been frightened. If one of these children is selected at random,
    (a) what is the probability that he or she has been frightened?
    (b) what is the probability the child is a girl, given that he or she has been frightened?

27. **Drinking age** A survey questioned 1000 people regarding raising the legal drinking age from 18 to 21. Of the 560 favoring raising the age, 390 were female. Of the 440 opposition responses, 160 were female. If a person selected at random from this group is a man, what is the probability that the person favors raising the drinking age?

28. **Politics** A candidate for public office knows that he has a 60% chance of being elected to the office if he is nominated by his political party, and that his party has a 50% chance of winning with another nominee. If the probability he will be nominated by his party is 0.25, and his party wins the election, what is the probability that he is the winning candidate?

29. **Pregnancy test** A self-administered pregnancy test detects 85% of those who are pregnant, but does not detect pregnancy in 15%. It is 90% accurate in indicating women who are not pregnant, but indicates 10% of this group as being pregnant. Suppose it is known that 1% of the women in a neighborhood are pregnant. If a woman is chosen at random from those living in this neighborhood and if the test indicates she is pregnant, what is the probability she really is?

30. **Drinking age** A survey questioned 1000 people regarding raising the legal drinking age from 18 to 21. Of the 560 favoring raising the age, 390 were female. Of the 440 opposition responses, 160 were female.
    (a) What is the probability that a person selected at random is a female?
    (b) What is the probability that a person selected at random favors raising the age if the person is a woman?

31. **Survey** In a SAPS (Student Activity Participation Study) survey, 30 engineering majors, 25 business majors, 20 science majors, 15 liberal arts majors, and 10 human development majors were selected. Ten of the

engineering, 12 of the science, 6 of the human development, 13 of the business, and 8 of the liberal arts majors selected for the study were female.
    (a) What is the probability of selecting a female if one person from this group is selected randomly?
    (b) What is the probability that a student selected randomly from this group is a science major, given that she is female?

32. **Survey** If a student is selected at random from the group in problem 31, what is the probability the student is
    (a) an engineering major, given that she is female?
    (b) an engineering major, given that he is male?

7.5 COUNTING; PERMUTATIONS AND COMBINATIONS

Objectives

❑ To use the Fundamental Counting Principle and permutations to solve counting problems
❑ To use combinations to solve counting problems

APPLICATION PREVIEW

During a national television advertising campaign, Little Caesar's Pizza stated that for $9.95, you could get 2 medium-sized pizzas, each with any of 0 to 5 toppings chosen from 11 that are available. The commercial asked the question "How many different pairs of pizzas can you get?" Answering this question uses the computation of **combinations,** which are discussed in this section. Other counting techniques, including **permutations,** are also discussed in this section.

Suppose that you decide to dine at a restaurant that offers 3 appetizers, 8 entrees, and 6 desserts. How many different meals can you have? We can answer questions of this type using the **Fundamental Counting Principle.**

Fundamental Counting Principle

If there are $n(A)$ ways in which an event A can occur and if there are $n(B)$ ways in which a second event B can occur after the first event has occurred, then the two events can occur in $n(A) \cdot n(B)$ ways.

The Fundamental Counting Principle can be extended to any number of events as long as they are independent. Thus the total number of possible meals at the restaurant mentioned above is

$$n(\text{appetizers}) \cdot n(\text{entrees}) \cdot n(\text{desserts}) = 3 \cdot 8 \cdot 6 = 144.$$

EXAMPLE 1

The state of West Virginia has auto license plates that begin with a digit followed by two letters and three more digits. If no other plates are made for automobiles in this state, how many automobiles could be licensed in West Virginia?

Solution

The following numbers represent the number of possible digits or letters in each of the 6 spaces on the license plates.

$$\underline{10} \quad \underline{26} \quad \underline{26} \quad \underline{10} \quad \underline{10} \quad \underline{10}$$

The product of these numbers gives the number of license plates that can be made satisfying the given conditions.

$$\underline{10} \cdot \underline{26} \cdot \underline{26} \cdot \underline{10} \cdot \underline{10} \cdot \underline{10} = 6{,}760{,}000$$

Thus 6,760,000 automobiles can be licensed with these plates.

CHECKPOINT

1. If a state permits either a letter or a nonzero digit to be used in each of six places on its license plates, how many different plates can it issue?

EXAMPLE 2

Suppose a person planning a banquet cannot decide how to seat 6 honored guests at the head table. In how many arrangements can they be seated in the six chairs on one side of a table?

Solution

If we think of the six chairs as spaces, we can determine how many ways each space can be filled. The planner could place any one of the 6 people in the first space (say the first chair on the left). One of the 5 remaining persons could then be placed in the second space, one of the 4 remaining persons in the third space, and so on, as follows:

$$\underline{6} \ \underline{5} \ \underline{4} \ \underline{3} \ \underline{2} \ \underline{1}$$

By the Fundamental Counting Principle, the total number of arrangements that can be made is the product of these numbers; that is, the total number of possible arrangements is

$$6 \cdot 5 \cdot 4 \cdot 3 \cdot 2 \cdot 1 = 720.$$

Because special products such as $6 \cdot 5 \cdot 4 \cdot 3 \cdot 2 \cdot 1$ frequently occur in counting theory, we use special notation to denote them. We write 6! (read "6 factorial") to denote $6 \cdot 5 \cdot 4 \cdot 3 \cdot 2 \cdot 1$. Likewise $4! = 4 \cdot 3 \cdot 2 \cdot 1$.

For any positive integer n, we define **n-factorial,** written $n!$, as

$$n! = n(n - 1)(n - 2) \cdots 3 \cdot 2 \cdot 1.$$

We define $0! = 1$.

EXAMPLE 3

Suppose the planner in Example 2 knows that 8 people feel they should be at the head table, but only 6 spaces are available. How many arrangements can be made that place 6 of the 8 people at the head table?

Solution

There are again 6 spaces to fill, but any of 8 people can be placed in the first space, any of 7 people in the second space, and so on. Thus the total number of possible arrangements is

$$\underline{8} \cdot \underline{7} \cdot \underline{6} \cdot \underline{5} \cdot \underline{4} \cdot \underline{3} = 20,160.$$

The number of possible arrangements in Example 3 is called the number of **permutations** of 8 things taken 6 at a time, and is denoted $_8P_6$. Note that $_8P_6$ gives the first 6 factors of $8!$, so we can use factorial notation to write the product.

$$_8P_6 = \frac{8!}{2!} = \frac{8!}{(8 - 6)!}$$

Permutations The number of possible distinct arrangements of r objects chosen from a set of n objects is called the number of **permutations** of n objects taken r at a time, and equals

$$_nP_r = \frac{n!}{(n - r)!}.$$

Most scientific and graphics calculators have keys that can be used to calculate permutations.

EXAMPLE 4

In how many ways can a president, vice president, secretary, and treasurer be selected from an organization with 20 members?

Solution

We seek the number of orders (arrangements) in which 4 people can be selected from a group of 20. This number is

$$_{20}P_4 = \frac{20!}{(20 - 4)!} = 116,280.$$

CHECKPOINT

2. In a 5-question matching test, how many different answer sheets are possible if no answer can be used twice and there are
   (a) 5 answers available for matching?
   (b) 7 answers available for matching?

When we talk about arrangements of people at a table or arrangements of digits on a license plate, order is important. We now consider counting problems in which order is not important. Suppose you are the president of a company and you want to pick 2 secretaries to work for you. If 5 people are qualified, how many different pairs of people can you select? If you select 2 people from the 5 in a specific order, the number of *arrangements* of the 5 people taken 2 at a time would be

$$_5P_2 = 20.$$

But $_5P_2$ gives the number of ways you an select *and* order the pairs, and we seek only the number of ways you can select them. If the names of the people are A, B, C, D, and E, then two of the arrangements, AB and BA, would represent only one pair. Because each of the pairs of people can be arranged in $2! = 2$ ways, the number of pairs that can be selected without regard to order is

$$\frac{_5P_2}{2!} = \frac{20}{2} = 10.$$

To find the number of ways you can select 3 people from the 5 without regard to order, we would first find $_5P_3$, then divide by the number of ways the 3 people could be ordered ($3!$). Thus 3 people can be selected from 5 (without regard for order) in $_5P_3/3! = 10$ ways. We say that the number of combinations of 5 things taken 3 at a time is 10.

Combinations The number of ways r objects can be chosen from a set of n objects without regard to the order of selection is called the number of **combinations** of n objects taken r at a time, and equals

$$_nC_r = \frac{_nP_r}{r!}.$$

Note that the fundamental difference between permutations and combinations is that permutations are used when order is a factor in the selection, and combinations are used when it is not. Because $_nP_r = n!/(n-r)!$, we have

$$_nC_r = \frac{_nP_r}{r!} = \frac{\dfrac{n!}{(n-r)!}}{r!} = \frac{n!}{r!(n-r)!}.$$

Combination Formula The number of combinations of n objects taken r at a time is also denoted by $\binom{n}{r}$; that is,

$$\binom{n}{r} = {}_nC_r = \frac{n!}{r!(n-r)!}.$$

Most scientific and graphics calculators have keys that can be used to calculate combinations.

EXAMPLE 5
An auto dealer is offering any of 6 special options at one price on a specially equipped car being sold. If the dealer will sell as many of these options as you want, and you decide to buy the car with 4 options, how many different choices of specially equipped cars do you have?

Solution
The order in which you choose the options is not relevant, so we seek the number of combinations of 6 things taken 4 at a time. Thus the number of possible choices is

$$_6C_4 = \frac{6!}{4!(6-4)!} = \frac{6!}{4!2!} = 15.$$

EXAMPLE 6
Suppose you are the junior class president, and you want to pick 4 people to serve on a committee. If 10 people are willing to serve, in how many different ways can you pick your committee?

Solution
Since the order in which the committee members are selected is not important, there are

$$_{10}C_4 = \frac{10!}{4!6!} = 210$$

ways in which the committee could be picked.

EXAMPLE 7
Six men and eight women have volunteered to serve on a committee. How many different committees can be formed containing three men and three women?

Solution
Three men can be selected from six in $\binom{6}{3}$ ways, and three women can be selected from eight in $\binom{8}{3}$ ways. Therefore, three men *and* three women can be selected in

$$\binom{6}{3}\binom{8}{3} = \frac{6!}{3!3!} \cdot \frac{8!}{5!3!} = 1120 \text{ ways.}$$

CHECKPOINT

Determine whether each of the following counting problems can be solved by using permutations or combinations, and answer each question with a calculator.

3. Find the number of ways that 8 members of a space shuttle crew can be selected from 20 available astronauts.

4. If the command structure on a space shuttle flight is determined by the order in which astronauts are selected for a flight, how many different command structures are possible if 8 astronauts are selected from the 20 that are available?

5. If 14 men and 6 women are available for a space shuttle flight, how many crews are possible that have 5 men and 3 women astronauts?

We now consider the television commercial mentioned in the Application Preview.

EXAMPLE 8

During a national television advertising campaign, Little Caesar's Pizza stated that for $9.95, you could get 2 medium-sized pizzas, each with any of 0 to 5 toppings chosen from 11 that are available. The commercial asked the question "How many different pairs of pizzas can you get?" Answer the question, if the first pizza has a thin crust and the second has a thick crust.

Solution

Because you can get 0, 1, 2, 3, 4, or 5 toppings from the 11 choices, the possible choices for the first pizza are

$$_{11}C_0 + {}_{11}C_1 + {}_{11}C_2 + {}_{11}C_3 + {}_{11}C_4 + {}_{11}C_5$$
$$= 1 + 11 + 55 + 165 + 330 + 462 = 1024.$$

Because the same number of choices are available for the second pizza, the number of choices for two pizzas are

$$1024 \cdot 1024 = 1,048,576.$$

CHECKPOINT SOLUTIONS

1. Any of 35 items (26 letters or 9 digits) can be used in any of the six places, so the number of different license plates is

$$\underline{35} \cdot \underline{35} \cdot \underline{35} \cdot \underline{35} \cdot \underline{35} \cdot \underline{35} = 35^6 = 1,838,265,625.$$

2. (a) There are 5 answers available for the first question, 4 for the second, 3 for the third, 2 for the fourth, and 1 for the fifth, so there are

$$5 \cdot 4 \cdot 3 \cdot 2 \cdot 1 = 120$$

possible answer sheets. Note that this is the number of permutations of 5 things taken 5 at a time,

$$_5P_5 = 120.$$

(b) There are 7 answers available for the first question, 6 for the second, 5 for the third, 4 for the fourth, and 3 for the fifth, so there are

$$7 \cdot 6 \cdot 5 \cdot 4 \cdot 3 = 2520$$

possible answer sheets. Note that this is the number of permutations of 7 things taken 5 at a time, $_7P_5 = 2520$.

3. The members of the crew are to be selected without regard to the order of selection, so we seek the number of combinations of 8 people selected from 20. There are

$$_{20}C_8 = 125,970$$

possible crews.

4. The order in which the members are selected is important, so we seek the number of permutations of 8 people selected from 20. There are

$$_{20}P_8 = 5,079,110,400$$

possible command structures.

5. As in 3, the order of selection is not relevant, so we seek the number of combinations of 5 men out of 14 *and* 3 women out of 6. There are

$$_{14}C_5 \cdot {_6}C_3 = 2002 \cdot 20 = 40,040$$

possible crews that have 5 men and 3 women.

EXERCISE 7.5

PERMUTATIONS

1. Compute $_6P_4$.
2. Compute $_8P_5$.
3. Compute $_{10}P_6$.
4. Compute $_7P_4$.
5. Compute $_5P_5$.
6. Compute $_4P_1$.
7. Compute $_5P_0$.
8. Compute $_nP_n$.
9. How many four-digit numbers can be formed from the digits 1, 3, 5, 7, 8, and 9
    (a) if each digit may be used once in each number?
    (b) if each digit may be used repeatedly in each number?
10. How many four-digit numbers can be formed from the digits 1, 3, 5, 7, 8, and 9
    (a) if the numbers must be even and digits are not used repeatedly?

(b) if the numbers are less than 3000 and digits are not repeated?
(Hint: In both parts, begin with the digit where there is a restriction on the choices.)

11. Compute $_nP_0$.
12. Compute $_nP_1$.
13. Find $\dfrac{(n + 1)!}{n!}$.
14. Find $\dfrac{(2n + 2)!}{(2n)!}$.
15. *If* $(n + 1)! = 17n!$, find n.
16. If $(n + 1)! = 30(n - 1)!$, find n.

COMBINATIONS

17. Compute $\binom{6}{2}$.

18. Compute $\binom{8}{6}$.

19. Compute $_7C_5$.

20. Compute $_5C_2$.

21. Compute $_4C_4$.

22. Compute $_3C_1$.

23. Compute $\binom{5}{0}$.

24. Compute $\binom{n}{0}$.

25. Compute $\binom{n}{n}$.

26. Compute $\binom{n}{1}$.

27. Verify $_{13}C_{10} = {}_{12}C_9 + {}_{12}C_{10}$.

28. Verify $_{20}C_{15} = {}_{19}C_{14} + {}_{19}C_{15}$.

29. If $_nC_6 = {}_nC_4$, find n.   30. If $_nC_8 = {}_nC_7$, find n.

APPLICATIONS

31. **License plates** If a state wants each of its license plates to contain 7 different digits, how many different license plates could it make?

32. **License plates** If a state wants each of it license plates to contain any 7 digits, how many different license plates are possible?

33. **Signaling** A sailboat owner received 6 different signal flags with his new sailboat. If the order in which the flags are arranged on the mast determines the signal being sent, how many 3-flag signals can be sent?

34. **Racing** Eight horses are entered in a race. In how many ways can the horses finish?

35. **Management** Four candidates for manager of a department store are ranked according to the weighted average of several criteria. How many different ranking are possible if no two candidates receive the same rank?

36. **Testing** An examination consists of 12 questions. If 10 questions must be answered, find the number of different orders in which a student can answer the questions.

37. **Molecules** Biologists have identified 4 kinds of small molecules: adenine, cytosine, quanine, and thymine, which link together to form larger molecules in genes. How many 3-molecule chains can be formed if the order of linking is important and each small molecule can occur more than once in a chain?

38. **Families** A member of a family of four is asked to rank all members of the family in the order of their power in the family. How many possible rankings are there?

39. **Management** A department store manager wants to display 6 brands of a product along one shelf of an aisle. In how many ways can he arrange the brands?

40. **Call letters** The call letters for radio stations begin with K or W, followed by 3 additional letters. How many sets of call letters having 4 letters are possible?

41. **Testing** How many ways can a 10-question true-false test be answered?

42. **Testing** How many ways can a 10-question multiple-choice test be answered if each question has 4 possible answers?

43. **Testing** In a 10-part matching question, how many ways can the matches be made if no two parts are used twice and if there are 10 possible matches?

44. **Testing** Answer problem 43 if there are 10 parts and 15 possible matches.

45. **Committees** In how many ways can a committee of 5 be selected from 10 people willing to serve?

46. **Astronauts** If 8 people are qualified for the next flight to a space station, how many different groups of 3 people can be chosen for the flight?

47. **Sales** A traveling salesperson has 30 products to sell, but has room in her sample case for only 20 of the products. If all the products are the same size, in how many ways can she select 20 different products for her case?

48. **Gambling** A poker hand consists of 5 cards dealt from a deck of 52 cards. How many different poker hands are possible?

49. **Psychology** A psychological study of outstanding salespeople is to be made. If 5 salespeople are needed and 12 are qualified, in how many ways can the 5 be selected?

50. **Banking** A company wishes to use the services of 3 different banks in the city. If 15 banks are available, in how many ways can it choose the 3 banks?

51. **Politics** To determine voters' feelings regarding an issue, a candidate asked a sample of people to pick 4 words out of 10 that they felt best describe the issue. How many different groupings of 4 words are possible?

52. **Sociology** A sociologist wants to pick 3 fifth-grade students from each of four schools. If each of the schools has 50 fifth-grade students, how many different groups of 12 students can he select? (Set up in combination symbols; do not work out.)

53. **Committees** In how many ways can a committee consisting of 6 men and 6 women be selected from a group consisting of 20 men and 22 women?

54. **Bridge** In how many ways can a hand consisting of 6 spades, 4 hearts, 2 clubs, and 1 diamond be selected from a deck of 52 cards?

7.6 PERMUTATIONS, COMBINATIONS, AND PROBABILITY

Objective

❑ To use counting techniques to solve probability problems

APPLICATION PREVIEW

Suppose that of the 20 prospective jurors for a trial, 12 favor the death penalty and 8 do not, and that 12 are chosen at random from these 20. To find the probability that 7 of the 12 will favor the death penalty, we can use combinations to find the number of ways to choose 12 jurors from 20 prospective jurors, and to find the number of ways to choose 7 from the 12 who favor the death penalty and 5 from the 8 who oppose it. We frequently use permutations, combinations, or other counting principles to determine how many ways an event can occur, $n(E)$, and the total number of outcomes, $n(S)$.

As we saw in Section 7.1, if an event E can happen in $n(E)$ ways out of a total of $n(S)$ equally likely possibilities, the probability of the occurrence of the event is

$$\Pr(E) = \frac{n(E)}{n(S)}.$$

EXAMPLE 1

A psychologist claims she can teach four-year-old children to spell three-letter words very quickly. To test her, one of her students was given cards with the letters A, C, D, K, and T on them and told to spell CAT. What is the probability the child will spell CAT by chance?

Solution

The number of different three-letter "words" that can be formed using these 5 cards is $n(S) = {}_5P_3 = 5!/(5 - 3)! = 60$. Since only one of those 60 arrangements will give CAT, $n(E) = 1$, and the probability that the child will be successful by guessing is $1/60$.

EXAMPLE 2

Suppose that license plates for a state each contain six digits. If a license plate from this state is selected at random, what is the probability that it will have six different digits?

Solution

The number of ways that six different digits can be selected from 0, 1, 2, 3, 4, 5, 6, 7, 8, 9 is $n(E) = n(\text{all digits different}) = {}_{10}P_6$.

The total number of license plates that can be made with six digits will include those with digits repeated, so the number is

$$n(S) = 10^6.$$

Thus the probability that a plate chosen at random will have six different digits is

$$\Pr(E) = \frac{n(E)}{n(S)} = \frac{_{10}P_6}{10^6} = \frac{10 \cdot 9 \cdot 8 \cdot 7 \cdot 6 \cdot 5}{1,000,000} = 0.1512.$$

CHECKPOINT

1. If three wires (red, black, and white) are randomly attached to a 3-way switch (which has 3 poles to which wires can be attached), what is the probability that the wires will be attached at random in the one order that makes the switch work properly?

EXAMPLE 3

A box of 24 shotgun shells contains 4 shells that will not fire. If 8 shells are selected from the box, what is the probability
(a) that all 8 shells will be good?
(b) that 6 shells will be good and 2 will not fire?

Solution

(a) The 8 shells can be selected from 24 in $n(S) = {}_{24}C_8 = \binom{24}{8}$ ways, and 8 good shells can be selected from the 20 good ones in $n(E) = {}_{20}C_8 = \binom{20}{8}$ ways. Thus the probability that all 8 shells will be good is

$$\Pr(8 \text{ good}) = \frac{n(E)}{n(S)} = \frac{\dfrac{20!}{12!8!}}{\dfrac{24!}{16!8!}} = 0.17.$$

(b) Six good shells *and* two defective shells can be selected in $n(E) = {}_{20}C_6 \cdot {}_4C_2 = \binom{20}{6} \cdot \binom{4}{2}$ ways, so

$$\Pr(6 \text{ good and 2 defective}) = \frac{n(E)}{n(S)} = \frac{\dfrac{20!}{14!6!} \cdot \dfrac{4!}{2!2!}}{\dfrac{24!}{16!8!}} = 0.316.$$

EXAMPLE 4

A manufacturing process for computer chips is such that 5 out of 100 chips are defective. If 10 chips are chosen at random from a box containing 100 newly manufactured chips, what is the probability that none of the chips will be defective?

Solution

The number of ways that any 10 chips can be chosen from 100 is

$$n(S) = {}_{100}C_{10} = \binom{100}{10} = \frac{100!}{10!90!}.$$

The number of ways that 10 chips can be chosen with none of them being defective is the same as the number of ways 10 good chips can be chosen from the 95 good chips in the box. Thus,

$$n(E) = {}_{95}C_{10} = \binom{95}{10} = \frac{95!}{10!85!}.$$

Thus the probability that none of the chips chosen will be defective is

$$\Pr(E) = \frac{n(E)}{n(S)} = \frac{\dfrac{95!}{10!85!}}{\dfrac{100!}{10!90!}} = 0.58375.$$

We now consider the problem discussed in the Application Preview.

EXAMPLE 5

Suppose that of the 20 prospective jurors for a trial, 12 favor the death penalty and 8 do not. If 12 jurors are chosen at random from these 20, what is the probability that 7 of the jurors will favor the death penalty?

Solution

The number of ways that 7 jurors of the 12 will favor the death penalty is the number of ways that 7 favor and 5 do not favor, or

$$n(E) = {}_{12}C_7 \cdot {}_8C_5 = \frac{12!}{7!5!} \cdot \frac{8!}{5!3!}.$$

The total number of ways that 12 jurors can be selected from 20 people is

$$n(S) = {}_{20}C_{12} = \frac{20!}{12!8!}.$$

Thus the probability we seek is

$$\Pr(E) = \frac{n(E)}{n(S)} = \frac{\dfrac{12!}{7!5!} \cdot \dfrac{8!}{5!3!}}{\dfrac{20!}{12!8!}} = 0.3521.$$

CHECKPOINT

2. If 5 people are chosen out of a group that contains 4 men and 6 women, what is the probability that 3 men and 2 women will be chosen?

EXAMPLE 6

Pennsylvania auto license plates have 3 letters followed by 3 numbers. If Janie's initials are J. J. R. and she lives at 125 Spring Street, what is the probability that the license plate she draws at random will
(a) have her initials and house numbers on it?
(b) have her initials on it?

Solution

(a) The total number of possible license plates with 3 letters followed by 3 numbers is

$$26 \cdot 26 \cdot 26 \cdot 10 \cdot 10 \cdot 10 = 17,576,000.$$

Of all these plates, only one will have Janie's initials and house number, so the probability that she will get this plate at random is $1/17,576,000$.

(b) The total number of possible license plates is 17,576,000, and the number with Janie's initials and any numbers on it is $1 \cdot 1 \cdot 1 \cdot 10 \cdot 10 \cdot 10 = 1000$, so the probability that she will get a plate with her initials is $1000/17,576,000 = 1/17,576$.

EXAMPLE 7

Beginning in 1990, General Motors began to use a theft deterrent key on some of its cars. The key has six parts, with three patterns for each part, plus an electronic chip containing a code from 1 to 15. What is the probability that one of these keys selected at random will start a GM car requiring a key of this type?

Solution

Each of the six parts has three patterns and the chip can have any of 15 codes, so there are

$$3 \cdot 3 \cdot 3 \cdot 3 \cdot 3 \cdot 3 \cdot 15 = 10,935$$

possible keys. Because there is only one key that will start the car, the probability that a key selected at random will start a given car is $1/10,935$.

CHECKPOINT
SOLUTIONS

1. To find the total number of orders in which the wires can be attached, we note that the first pole can have any of the 3 wires attached to it, the second pole can have any of 2, and the third pole can have only the remaining wire, so there are $3 \cdot 2 \cdot 1 = 6$ possible different orders. Only one of these orders permits the switch to work properly, so the probability is $1/6$.

2. Five people can be chosen out of 10 people in $_{10}C_5 = 252$ ways. Three men and 2 women can be chosen in $_4C_3 \cdot _6C_2 = 4 \cdot 15 = 60$ ways. Thus the probability is $60/252 = 5/21$.

EXERCISE 7.6

APPLICATIONS

1. *Education* If a child is given cards with A, C, D, G, O, and T on them, what is the probability he or she could spell DOG by guessing the correct arrangement of 3 cards from the six?

2. *Racing* Eight horses in a race wear numbers 1, 2, 3, 4, 5, 6, 7, and 8. What is the probability that the first three horses to finish the race will be numbered 1, 2, and 3, respectively?

3. *Politics* A senator asks his constituents to rank 5 issues in order of importance.
   (a) How many rankings are possible?
   (b) What is the probability that one reply chosen at random will rank the issues in alphabetical order?

4. *Photography* Two men and a woman are lined up to have their picture taken. If they are arranged at random, what is the probability that
   (a) the woman will be on the left in the picture?
   (b) the woman will be in the middle in the picture?

5. *Licenses* Suppose that all license plates in a state have three letters and three digits. If a plate is chosen at random, what is the probability that all three letters and all three numbers on the plate will be different?

6. *Spelling*
   (a) In how many different orders can the letters R, A, N, D, O, M be written?
   (b) If the letters R, A, N, D, O, M are placed in a random order, what is the probability they will spell RANDOM?

7. *ATMs*
   (a) An automatic teller machine requires that each customer enter a four-digit personal identification number (PIN) when he or she inserts a bank card. If a person finds a bank card and guesses at a PIN to use the card fraudulantly, what is the probability that the person will succeed in one attempt?
   (b) If the person knows that the PIN will not have any digit repeated, what is the probability the person will succeed in guessing in one attempt?

8. *Keys* Keys for older General Motors cars had six parts, with three patterns for each part.
   (a) How many different key designs are possible for these cars?
   (b) If you find an older GM key and own an older GM car, what is the probability it will fit your trunk?

9. *Telephones* If the first digit of a seven-digit telephone number cannot be a 0 or a 1, what is the probability that a number chosen at random will have all seven digits the same?

10. *Licenses* If a license plate has seven different digits, what is the probability that it contains a zero?

11. *Testing* A quiz consists of 10 multiple-choice questions with 5 possible choices for each question. If a student guesses the answer to each question, what is the probability that
    (a) she gets all questions correct?
    (b) she gets no questions correct?
    (c) she gets at least one question correct?

12. *Testing* Suppose that the student in problem 11 was certain of the first 5 questions, and on each of the last 5 she could narrow the answer to one of three choices. Find the probability that
    (a) she scores 50%.
    (b) she scores 100%.
    (c) she passes the test (at least 60%).

13. *Testing* In a 10-question matching test with 10 possible answers to match and no answers used more than once, what is the probability of guessing and getting every answer correct?

14. *Testing* In a 10-question matching test with 15 possible answers to match and no answers used more than once, what is the probability of guessing and getting every answer correct?

15. *Quality control* A box of 12 transistors has 3 defective ones. If 2 transistors are drawn from the box together, what is the probability
    (a) that both transistors are defective?
    (b) that neither transistor is defective?
    (c) that one transistor is defective?

16. *Quality control* Suppose that 6 transistors are drawn at random from a box containing 18 good transistors and 2 defective transistors. What is the probability that
    (a) all 6 of the transistors are good?
    (b) exactly 4 of the transistors are good?
    (c) exactly 2 of the transistors are good?

17. *Quality control* A retailer purchases 100 of a new brand of digital tape deck, of which 2 are defective. The purchase agreement says that if he tests 5 chosen at random and finds 1 or more defective, he receives all the tape decks free of charge. What is the probability that he will not have to pay for the tape decks?

18. *Quality control* A box containing 500 central processing units (CPUs) for microcomputers has 20 defective

CPUs. If 10 CPUs are selected together from the box, what is the probability that
(a) all 10 are defective?   (b) none are defective?
(c) half of them are defective?

19. **Banking** To see if a bank has enough minority construction company loans, a social agency selects 30 loans to construction companies at random and finds that 2 of them are loans to minority construction companies. If the bank's claim that 10 of every 100 of its loans to construction companies are minority loans is true, what is the probability that 2 loans out of 30 are minority loans? Leave your answer with combination symbols.

20. **Diversity** A high school principal must select 12 girls at random from the freshman class to serve as hostesses at the junior-senior prom. There are 200 freshman girls, including 20 from minorities, and the principal would like at least one minority girl to have this honor. If he selects the girls at random, what is the probability that
(a) he will select exactly one minority girl?
(b) he will select no minority girls?
(c) he will select at least one minority girl?

21. **Diversity** Suppose that an employer plans to hire four people from a group of nine equally qualified people, of whom three are minority candidates. If the employer does not know which candidates are minority candidates and she selects her employees at random, what is the probability that
(a) no minority candidates are hired?
(b) all three minority candidates are hired?
(c) one minority candidate is hired?

22. **Lottery** Four men and three women are semifinalists in a lottery. From this group, three finalists are to be selected by a drawing. What is the probability that all three finalists will be men?

23. **Management** Suppose that a little league basketball coach knows the best 6 players to use on his team, but pressure from parents to give everyone a chance to start in a game causes him to pick the starting team by choosing 5 players at random from the 10 team members. What is the probability that this will give him a team with 5 of his 6 best players?

24. **Sales** A car dealer has 12 different cars he would like to display, but he only has room to display 5.
(a) In how many ways can he pick 5 cars to display?
(b) Suppose 8 of the cars are the same color, with the remaining 4 having distinct colors. If the dealer tells a salesperson to display any 5 cars, what is the probability that all 5 cars will be the same color?

25. **Diversity** Suppose that two openings on an appellate court bench are to be filled from current municipal court judges. If the municipal court judges consist of 23 men and 4 women, find the probability
(a) that both appointees are men.
(b) that one man and one woman are appointed.
(c) that at least one woman is appointed.

26. **Politics** Suppose that four of the eight students running for class officers (president, vice president, secretary, and treasurer) have grade-point averages (GPAs) above 3.0. If the officers are selected at random, what is the probability that all four officers will be students with GPAs above 3.0?

27. **Management** Suppose that an indecisive company owner has selected the three top officers of his company at random but claims that they earned their jobs because of ability. What is
(a) the probability that the most able person is at the top?
(b) the probability that the top three officers are ranked according to their ability?

28. **Quality control** Suppose that 10 computer chips are drawn from a box containing 12 good chips and 4 defective chips. What is the probability that
(a) exactly 4 of the chips are defective?
(b) all 10 of the chips are good?
(c) exactly 8 of the chips are good?

29. **Lotteries** In New York State there is a state lottery game called Pick 10. The state chooses 20 numbers from 80 numbers, and each player chooses 10 numbers from the same 80 numbers. If all 10 of a player's numbers are among the 20 numbers that the state picked, then that player is a "big winner." Find the probability of being a big winner.

30. **Diversity** A task force studying sex discrimination wishes to establish a subcommittee. If the subcommittee is to consist of 11 members chosen from a group of 23 men and 19 women, find the probability that the subcommittee consists of
(a) 6 men and 5 women.
(b) all women.

31. **Insurance** An insurance company receives 25 claims in a certain month. Company policy mandates that a random selection of 5 of these claims be thoroughly investigated. If 2 of the 25 claims are fraudulent, find the probability that
(a) both fraudulent claims are thoroughly investigated.
(b) neither fraudulent claim is thoroughly investigated.

32. ***Evaluation*** Employees of a firm receive annual reviews. In a certain department, 4 employees received excellent ratings, 15 received good ratings, and 1 received a marginal rating. If 3 employees in this department are randomly selected to complete a form for an internal study of the firm, find the probability that
 (a) all 3 selected were rated excellent.
 (b) one from each category was selected.

33. ***Poker*** What is the probability of being dealt a poker hand of 5 cards containing
 (a) 5 spades?   (b) 5 cards of the same suit?

34. ***Bridge*** A bridge deck contains 13 spades, 13 hearts, 13 diamonds, and 13 clubs. A bridge hand contains 13 cards selected at random from the 52-card deck. Find the probability that a hand contains
 (a) 13 hearts.
 (b) exactly 8 spades.

35. ***Bridge*** A bridge deck contains 13 spades, 13 hearts, 13 diamonds, and 13 clubs. A bridge hand contains 13 cards selected at random from the 52-card deck. Find the probability that a hand contains 4 spades, 4 hearts, 4 diamonds, and 1 club.

36. ***Bridge*** Find the probability of being dealt a hand containing 1 spade, 6 hearts, 4 diamonds, and 2 clubs. Assume that a normal deck of 52 cards is used.

7.7 MARKOV CHAINS

Objectives

- To use transition matrices to find specific stages of Markov chains
- To write transition matrices for Markov chain problems
- To find steady-state vectors for certain Markov chain problems

APPLICATION PREVIEW

Suppose a department store has determined that a woman who used a credit card last month will use it again this month with probability 0.8. That is,

 Pr(woman will use a charge card this month | she used it last month)
  = 0.8

and

 Pr(woman will not use card this month | she used it last month) = 0.2.

Furthermore, suppose the store has determined that the probability that a woman who did not use a credit card last month will not use it this month is 0.7. That is,

Pr(woman will not use card this month | she did not use it last month)
  = 0.7

and

Pr(woman will use card this month | she did not use it last month)
  = 0.3.

To find the probability that the woman will use the credit card in some later month, we use **Markov chains.** Markov chains are named for the famous Russian mathematician A. A. Markov, who developed this theory in 1906. They are useful in the analysis of price movements, consumer behavior, laboratory animal behavior, and many other processes in business, the social sciences, and the life sciences.

The Markov chain experiment described in the Application Preview has two states: using a credit card and not using a credit card. The probability of moving from one state to another is called the **transition** probability; these transition probabilities are the conditional probabilities stated in the preview.

Transition probabilities can be organized into a **transition matrix** P, illustrated below.

$$P = \begin{array}{c} \\ \\ \text{Card used} \\ \text{Card not used} \end{array} \begin{array}{cc} \text{\small Next month:} \\ \begin{array}{cc} \text{Uses} & \text{Card is} \\ \text{card} & \text{not used} \end{array} \\ \begin{bmatrix} .8 & .2 \\ .3 & .7 \end{bmatrix} \end{array}$$

Given month

In general, a transition matrix is characterized by the following:

1. The ij-entry in the matrix is: Pr(moving from State i to State j). Note that this is really the probability of moving to State j given that you are in State i.
2. The entries are nonnegative.
3. The sum of the entries in each row equals 1.
4. The matrix is square.

Now suppose the department store data shows that during the first month a credit card is received, the probabilities for each state are as follows:

$$\text{Pr(a woman used a charge card)} = .9$$
$$\text{Pr(a woman did not use a charge card)} = .1.$$

These probabilities can be placed in the row matrix

$$A = [.9 \quad .1],$$

called the **initial-probability** vector for the problem. This vector gives the probabilities for each state at the outset of the trials. In general, a probability vector is defined as follows.

Probability Vector	For a vector to be a probability vector, the sum of its entries must be 1 and each of its entries must be nonnegative.

Notice that of the 90% of women who used the credit card in their initial month, 0.8 of them will use it again, and of the 10% who didn't use the card, 0.3 of them will use it next month. Thus

$$\text{Pr(will use 2d month)} = \text{Pr(used 1st month)} \cdot \text{Pr(will use 2d} \mid \text{used 1st)}$$
$$+ \text{Pr(not used 1st month)} \cdot \text{Pr(will use 2d} \mid \text{not used 1st)}$$
$$= (.9)(.8) + (.1)(.3)$$
$$= 0.72 + 0.03 = 0.75.$$

7.7 Markov Chains   507

Note that this is one element of the product matrix AP. Similarly, the second element of the product matrix is the probability that a customer will not use the charge card in the second month.

$$AP = [.9 \quad .1] \begin{bmatrix} .8 & .2 \\ .3 & .7 \end{bmatrix} = [.75 \quad .25]$$

$$\begin{array}{ccc} \text{initial} & \text{transition} & \text{second} \\ \text{month} & \text{matrix} & \text{month} \end{array}$$

Thus

Pr(a woman will use a charge card in 2d month) = 0.75

and

Pr(a woman will not use a charge card in 2d month) = 0.25.

To find the probabilities for each state for the third month, we multiply the second month probability vector times the transition matrix.

$$[.75 \quad .25] \begin{bmatrix} .8 & .2 \\ .3 & .7 \end{bmatrix} = [.675 \quad .325]$$

$$\begin{array}{ccc} \text{second} & \text{transition} & \text{third} \\ \text{month} & \text{matrix} & \text{month} \end{array}$$

Continuing in this manner, we can find the probabilities for later months.

EXAMPLE I

Suppose that in a certain city the Democratic, Republican, and Consumer parties always nominate candidates for mayor. The probability of winning in any election depends on the party in power and is given by the following transisiton matrix P.

$$\begin{array}{c} \text{Party in office now} \end{array} \quad \begin{array}{c} \text{Party in office next term} \\ \text{D} \quad \text{R} \quad \text{C} \end{array}$$

$$P = \begin{array}{c} \text{Democrat} \\ \text{Republican} \\ \text{Consumer} \end{array} \begin{bmatrix} .5 & .4 & .1 \\ .4 & .5 & .1 \\ .3 & .3 & .4 \end{bmatrix}$$

Suppose the probability of winning the next election is given by $[.4 \quad .4 \quad .2]$; that is, Pr(D) = .4, Pr(R) = .4, Pr(C) = .2. Find the probability of each party winning the *fourth* election (the third election after this one).

Solution

The probability of each party winning the second election is found using the probabilities given for the first.

$$[.4 \quad .4 \quad .2] \begin{bmatrix} .5 & .4 & .1 \\ .4 & .5 & .1 \\ .3 & .3 & .4 \end{bmatrix} = [0.42 \quad 0.42 \quad 0.16]$$

Using the probabilities for the second election, we can find the probabilites for the third.

$$[.42 \quad .42 \quad .16] \begin{bmatrix} .5 & .4 & .1 \\ .4 & .5 & .1 \\ .3 & .3 & .4 \end{bmatrix} = [0.426 \quad 0.426 \quad 0.148]$$

Using the probabilities for the third election, we can find the probabilities for the fourth election.

$$[.426 \quad .426 \quad .148] \begin{bmatrix} .5 & .4 & .1 \\ .4 & .5 & .1 \\ .3 & .3 & .4 \end{bmatrix} = [0.4278 \quad 0.4278 \quad 0.1444]$$

We can perform matrix multiplication with a graphics calculator or we can use a computer spreadsheet to perform repeated multiplications. Continuing the multiplication process with the vectors in Example 1 eventually gives a vector with the probabilities

$$[0.42857 \quad 0.42857 \quad 0.14286]$$

to 5 decimal places, and repeated multiplications will not change the probabilities, so the system appears to be stabilizing. This vector is the vector $[V_1 \quad V_2 \quad V_3]$ such that

$$[V_1 \quad V_2 \quad V_3] \cdot P = [V_1 \quad V_2 \quad V_3].$$

The exact probability vector for which this is true in Example 1 is $[\frac{3}{7} \quad \frac{3}{7} \quad \frac{1}{7}]$, since

$$[\tfrac{3}{7} \quad \tfrac{3}{7} \quad \tfrac{1}{7}] \begin{bmatrix} .5 & .4 & .1 \\ .4 & .5 & .1 \\ .3 & .3 & .4 \end{bmatrix} = [\tfrac{3}{7} \quad \tfrac{3}{7} \quad \tfrac{1}{7}].$$

This vector is called the **fixed-probability vector** for the given transition matrix. Because this vector is where the process eventually stabilizes, it is also called the **steady-state vector.**

Rather than trying to guess this steady-state vector by repeated multiplication, we can solve the equation

$$[V_1 \quad V_2 \quad V_3] \begin{bmatrix} .5 & .4 & .1 \\ .4 & .5 & .1 \\ .3 & .3 & .4 \end{bmatrix} = [V_1 \quad V_2 \quad V_3]$$

for V_1, V_2, and V_3 as follows.

$$[V_1 \quad V_2 \quad V_3] \begin{bmatrix} .5 & .4 & .1 \\ .4 & .5 & .1 \\ .3 & .3 & .4 \end{bmatrix} = [V_1 \quad V_2 \quad V_3]$$

$$.5V_1 + .4V_2 + .3V_3 = V_1$$
$$.4V_1 + .5V_2 + .3V_3 = V_2$$
$$.1V_1 + .1V_2 + .4V_3 = V_3$$

Rewriting with zeros on the right and solving by Gauss-Jordan elimination gives

$$\begin{bmatrix} -.5 & .4 & .3 & | & 0 \\ .4 & -.5 & .3 & | & 0 \\ .1 & .1 & -.6 & | & 0 \end{bmatrix} \rightarrow \begin{bmatrix} 1 & 1 & -6 & | & 0 \\ 0 & -9 & 27 & | & 0 \\ 0 & 9 & -27 & | & 0 \end{bmatrix}$$

$$\rightarrow \begin{bmatrix} 1 & 1 & -6 & | & 0 \\ 0 & 1 & -3 & | & 0 \\ 0 & 0 & 0 & | & 0 \end{bmatrix}$$

$$\rightarrow \begin{bmatrix} 1 & 0 & -3 & | & 0 \\ 0 & 1 & -3 & | & 0 \\ 0 & 0 & 0 & | & 0 \end{bmatrix}$$

so

$$V_1 = 3V_3 \qquad V_2 = 3V_3.$$

Any vector satisfying these conditions is a solution to the system. Since $[V_1 \quad V_2 \quad V_3]$ is a probability vector, we have $V_1 + V_2 + V_3 = 1$. Substituting gives $3V_3 + 3V_3 + V_3 = 1$, so $V_3 = \frac{1}{7}$, and the steady-state vector is

$$\begin{bmatrix} \dfrac{3}{7} & \dfrac{3}{7} & \dfrac{1}{7} \end{bmatrix}.$$

Any transition matrix P such that some power of P, P^m, contains only positive entries is called a **regular transition matrix,** and it can be shown that each Markov chain using a regular transition matrix will have a steady-state vector.

Note that a regular transition matrix P may contain zeros; the only requirement is that some power of P contain all positive entries. In this text we will limit ourselves to regular transition matrices.

Furthermore, if a steady-state vector exists, then the effect of the initial probabilities diminishes as the number of steps in the process increases. This means that the probabilities for the elections would approach

$$\begin{bmatrix} \dfrac{3}{7} & \dfrac{3}{7} & \dfrac{1}{7} \end{bmatrix}$$

regardless of the probabilities for the inital election. For example, even if a Democrat were certain to be elected in the initial election (i.e., the initial vector is $[1 \quad 0 \quad 0]$), the probabilities for the fifth election are $[.4292 \quad .4291 \quad .1417]$, which approximates the steady-state vector. (See problem 17 of Exercise 7.7.)

CHECKPOINT

1. Use the transition matrix $\begin{bmatrix} .1 & .3 & .6 \\ .3 & .5 & .2 \\ .1 & .1 & .8 \end{bmatrix}$ and the initial probability vector $[.2 \quad .3 \quad .5]$ to

(a) find the probability vector for the third stage (two steps after the initial stage).

(b) find the steady-state vector associated with the matrix.

EXAMPLE 2

Suppose that in a certain city the probability that a woman with less than a high school education has a daughter with less than a high school education, a high school degree, or more than a high school education is .2, .6, and .2, respectively. The probability that a woman with a high school degree has a daughter with less than a high school education, a high school degree, or more than a high school education is .1, .5, and .4, respectively. The probability that a woman with more than a high school education has a daughter with less than a high school education, a high school degree, or education beyond high school is .1, .1, and .8, respectively. If the population of women in the city is now 60% with less than a high school education, 30% with a high school degree, and 10% with education beyond high school, answer these questions.

(a) What is the transition matrix for this information?

(b) What is the likely distribution of women according to educational level two generations from now?

(c) What is the steady-state vector for this information?

Solution

(a) The transition matrix is

	Daughter's educational level		
Mother's educational level	Less than H.S.	H.S. degree	More than H.S.
Less than H.S.	.2	.6	.2
H.S. degree	.1	.5	.4
More than H.S.	.1	.1	.8

(b) The likely distribution after one generation is

$$[.6 \quad .3 \quad .1] \begin{bmatrix} .2 & .6 & .2 \\ .1 & .5 & .4 \\ .1 & .1 & .8 \end{bmatrix} = [.16 \quad .52 \quad .32].$$

The likely distribution after two generations is

$$[.16 \quad .52 \quad .32] \begin{bmatrix} .2 & .6 & .2 \\ .1 & .5 & .4 \\ .1 & .1 & .8 \end{bmatrix} = [.116 \quad .388 \quad .496].$$

(c) The steady-state vector is found by solving

$$[V_1 \quad V_2 \quad V_3] \begin{bmatrix} .2 & .6 & .2 \\ .1 & .5 & .4 \\ .1 & .1 & .8 \end{bmatrix} = [V_1 \quad V_2 \quad V_3].$$

$$\begin{cases} .2V_1 + .1V_2 + .1V_3 = V_1 \\ .6V_1 + .5V_2 + .1V_3 = V_2 \\ .2V_1 + .4V_2 + .8V_3 = V_3 \end{cases} \quad \text{or} \quad \begin{cases} -.8V_1 + .1V_2 + .1V_3 = 0 \\ .6V_1 - .5V_2 + .1V_3 = 0 \\ .2V_1 + .4V_2 - .2V_3 = 0 \end{cases}$$

$$\begin{bmatrix} -.8 & .1 & .1 & | & 0 \\ .6 & -.5 & .1 & | & 0 \\ .2 & .4 & -.2 & | & 0 \end{bmatrix} \rightarrow \begin{bmatrix} 1 & 2 & -1 & | & 0 \\ 0 & -17 & 7 & | & 0 \\ 0 & 17 & -7 & | & 0 \end{bmatrix}$$

$$\rightarrow \begin{bmatrix} 1 & 0 & -\frac{3}{17} & | & 0 \\ 0 & 1 & -\frac{7}{17} & | & 0 \\ 0 & 0 & 0 & | & 0 \end{bmatrix}$$

Thus $V_1 = \frac{3}{17}V_3$ and $V_2 = \frac{7}{17}V_3$. The probability vector satisfying these conditions and $V_1 + V_2 + V_3 = 1$ must satisfy

$$\frac{3}{17}V_3 + \frac{7}{17}V_3 + V_3 = 1.$$

Thus

$$\frac{27}{17}V_3 = 1, \quad \text{so} \quad V_3 = \frac{17}{27},$$

and the steady-state vector is

$$\begin{bmatrix} \dfrac{3}{27} & \dfrac{7}{27} & \dfrac{17}{27} \end{bmatrix}.$$

CHECKPOINT SOLUTIONS

1. (a) The second-stage probability vector is

$$[.2 \quad .3 \quad .5] \begin{bmatrix} .1 & .3 & .6 \\ .3 & .5 & .2 \\ .1 & .1 & .8 \end{bmatrix} = [.16 \quad .26 \quad .58]$$

and the third stage is

$$[.16 \quad .26 \quad .58] \begin{bmatrix} .1 & .3 & .6 \\ .3 & .5 & .2 \\ .1 & .1 & .8 \end{bmatrix} = [.152 \quad .236 \quad .612].$$

(b) To find the steady-state vector, we solve

$$[V_1 \quad V_2 \quad V_3] \begin{bmatrix} .1 & .3 & .6 \\ .3 & .5 & .2 \\ .1 & .1 & .8 \end{bmatrix} = [V_1 \quad V_2 \quad V_3].$$

Multiplying and combining like terms gives the following system of equations.

$$\begin{cases} -.9V_1 + .3V_2 + .1V_3 = 0 \\ .3V_1 - .5V_2 + .1V_3 = 0 \\ .6V_1 + .2V_2 - .2V_3 = 0 \end{cases}$$

We solve this by using the augmented matrix

$$\begin{bmatrix} -.9 & .3 & .1 & | & 0 \\ .3 & -.5 & .1 & | & 0 \\ .6 & .2 & -.2 & | & 0 \end{bmatrix}.$$

The solution gives the steady-state vector $\begin{bmatrix} \dfrac{1}{7} & \dfrac{3}{14} & \dfrac{9}{14} \end{bmatrix}$.

EXERCISE 7.7

Which of the vectors in problems 1–4 cound *not* be probability vectors?

1. $\begin{bmatrix} \dfrac{1}{4} & \dfrac{3}{4} \end{bmatrix}$

2. $\begin{bmatrix} \dfrac{5}{6} & \dfrac{2}{3} \end{bmatrix}$

3. $\begin{bmatrix} \dfrac{1}{4} & \dfrac{3}{5} & \dfrac{1}{6} \end{bmatrix}$

4. $\begin{bmatrix} \dfrac{2}{3} & \dfrac{1}{12} & \dfrac{1}{4} \end{bmatrix}$

Which of the matrices in problems 5–8 could *not* be transition matrices? Explain.

5. $\begin{bmatrix} .3 & .5 & .2 \\ .1 & .6 & .3 \end{bmatrix}$

6. $\begin{bmatrix} .6 & .2 & .2 \\ .1 & .5 & .4 \\ .3 & .5 & .1 \end{bmatrix}$

7. $\begin{bmatrix} .1 & .3 & .6 \\ .5 & .2 & .3 \\ .1 & .7 & .2 \end{bmatrix}$

8. $\begin{bmatrix} .6 & .5 & -.1 \\ .2 & .6 & .2 \\ .3 & .3 & .4 \end{bmatrix}$

In problems 9–12, use the transition matrix and the initial-probability vector to find the probability vector for the third stage (two after the initial stage) of the Markov chain.

$$A = \begin{bmatrix} .1 & .9 \\ .3 & .7 \end{bmatrix} \quad B = \begin{bmatrix} .5 & .5 \\ .9 & .1 \end{bmatrix}$$

$$C = \begin{bmatrix} .5 & .3 & .2 \\ .3 & .5 & .2 \\ .1 & .1 & .8 \end{bmatrix} \quad D = \begin{bmatrix} .8 & .1 & .1 \\ .2 & .6 & .2 \\ .3 & .3 & .4 \end{bmatrix}$$

9. Transition matrix A, initial-probability vector [.2 .8].
10. Transition matrix B, initial-probability vector [.6 .4].
11. Transition matrix C, initial-probability vector [.1 .3 .6].
12. Transition matrix D, initial-probability vector [.9 0 .1].

In problems 13–16, find the steady-state vector associated with the given transition matrix from problems 9–12.

13. A 14. B
15. C 16. D

APPLICATIONS

Elections Use the following information for problems 17 and 18. In a certain city the Democratic, Republican, and Consumer parties always nominate candidates for mayor. The probability of winning in any election depends on the party in power and is given by the transition matrix below.

Party in office now	Party in office next term D	R	C
Democrat	.5	.4	.1
Republican	.4	.5	.1
Consumer	.3	.3	.4

17. Using the transition matrix and assuming the initial-probability vector is [1   0   0], find the probability vectors for the next four steps of the Markov chain. (This initial-probability vector indicates a Democrat is certain to win the initial election.)

18. Using the transition matrix above and assuming that a Republican is certain to win the initial election, find the probability vectors for the next four steps of the Markov chain.

Church attendance Use the following information for problems 19–22. The probability that daughters of a mother who attends church regularly will also attend church regularly is .8, while the probability that daughters of a mother who does not attend regularly will attend regularly is .3.

19. What is the transition matrix for this information?

20. If a woman attends church and has one daughter who in turn has one daughter, what is the probability that the granddaughter attends church regularly?

21. If a woman does not attend church, what is the probability that her granddaughter attends church regularly?

22. What is the steady-state vector for this information?

Car selection Use the following information for problems 23–28. A man owns an Audi, a Ford, and a VW. He drives every day and never drives the same car two days in a row. This is the probability that he drives the other cars the next day:

$$\text{Pr(Ford after Audi)} = .7, \quad \text{Pr(VW after Audi)} = .3$$
$$\text{Pr(Audi after Ford)} = .6, \quad \text{Pr(VW after Ford)} = .4$$
$$\text{Pr(Audi after VW)} = .8, \quad \text{Pr(Ford after VW)} = .2$$

23. Write the transition matrix for his selection of a car.

24. If he drove the Ford today, what are the probabilities for the cars that he will drive 2 days from today?

25. If he drove the Ford today, what are the probabilities for the cars that he will drive 4 days from today?

26. If he drove the Audi today, what are the probabilities for the cars that he will drive 4 days from now?

27. What is the steady-state vector for this problem?

28. Would the probabilities for which car he drives 100 days from now depends on whether he drove a Ford or Audi today?

29. *Population demographics* Suppose a government study estimated that the probability of successive generations of a rural family remaining in a rural area was .7, and the probability of successive generations of an urban family remaining in an urban area was .9. Assuming a Markov chain applies to these facts, find the steady-state vector.

30. *Psychology* A psychologist found that a laboratory mouse placed in a T-maze will go down the same branch with probability .8 if there was food there on the previous trial, and that it will go down the same branch with probability .6 if there was no food there on the previous trial. Find the steady-state vector for this experiment. If the mouse takes the branches with equal probability when no reward is found, does it appear that the mouse "learns" to choose the food path?

31. *Charitable contributions* The local community-service funding organization in a certain county has classified the population into those who did not give the previous year, those who gave less than their "fair share," and those who met or exceeded their "fair share." Suppose the organization developed the following transition matrix for these groups.

	This year		
Last year	N	L	F
No contribution	.7	.2	.1
Less than fair share	.1	.6	.3
Fair share	0	.1	.9

Find the steady-state vector for this county's contribution habits.

Advertising Use the following information for problems 32 and 33. A local business A has two competitors, B and C. Initially the probability that a customer patronizes A, B, or C is .2, .6, and .2, respectively. Suppose A initiates an advertising campaign to improve its business and finds the following transition matrix to describe the effect.

Last week's customers went to	*This week's customers go to*		
	A	B	C
A	.7	.2	.1
B	.4	.4	.2
C	.4	.4	.2

32. If A runs the advertising campaign for 3 weeks, find the probability of a customer patronizing each business.

33. Find the steady-state vector for this market, that is, the long-range share of the market that each business can expect if the transition matrix holds.

Genetics Use the following information for problems 34–36. For species that reproduce sexually, characteristics are determined by a gene from each parent. Suppose that for a certain trait there are two possible genes available from each parent, a dominant gene D and a recessive gene r. Then the different gene combinations (called *genotypes*) for the offspring are DD, Dr, rD, and rr, where Dr and rD produce the

same trait. Suppose further that these genotypes are
the states of a Markov chain with the transition matrix
below.

34. If the initial occurence of those genotypes in the popula-
    tion is .5 for DD, .4 for Dr, and .1 for rr, find the distri-
    bution (probability) of each type after the first and
    second generations.
35. Find the fixed-probability vector for this genotype.
36. To which generation(s) does the fixed-probability vector
    correspond? (See problems 34 and 35.) This is called
    the **Hardy-Weinberg model.**

KEY TERMS & FORMULAS

Section	Key Terms	Formula
7.1	Equally likely events	$\Pr(E) = k/n$
	Sample space	
	Certain event	$\Pr(E) = 1$
	Impossible event	$\Pr(E) = 0$
	Odds	
	Empirical probability	
7.2	Intersection of events	$\Pr(E \text{ and } F) = \Pr(E \cap F)$
	Union of events	$\Pr(E \text{ or } F) = \Pr(E \cup F)$
		$\quad = \Pr(E) + \Pr(F) - \Pr(E \cap F)$
	Complement	$\Pr(\text{not } E) = \Pr(E') = 1 - \Pr(E)$
7.3	Conditional probability	$\Pr(E \mid F) = \dfrac{\Pr(E \cap F)}{\Pr(F)}$
	Product Rule	$\Pr(E \cap F) = \Pr(F) \cdot \Pr(E \mid F)$
	Independent events	$\Pr(A \mid B) = \Pr(A) \quad \text{and}$
		$\Pr(A \cap B) = \Pr(A) \cdot \Pr(B)$
7.4	Probability tree	
	Bayes' formula	
7.5	Fundamental Counting Principle	$n(A) \cdot n(B)$
	n-factorial	$n! = n(n - 1) \cdots (3)(2)(1) \quad \text{and} \quad 0! = 1$
	Permutations	$_nP_r = \dfrac{n!}{(n - r)!}$
	Combinations	$_nC_r = \dfrac{n!}{r!(n - r)!}$
7.6	Probability using permutations and combinations	
7.7	Markov chains	
	Vectors	
	Initial-probability	
	Fixed-probability or steady-state	

REVIEW EXERCISES

7.1

1. A bag contains 6 red and 3 blue balls. If 1 ball is drawn at random, what is the probability that it is blue?
2. From a deck of 52 cards, 1 card is to be drawn. What is the probability that it will be a queen?
3. What is the probability that a letter chosen at random from the word "wooden" is an *o*?
4. Suppose that a fair coin is tossed two times. Construct an equiprobable sample space for the experiment and determine each of the following probabilities.
   (a) Pr(0 heads)   (b) Pr(1 head)   (c) Pr(2 heads)
5. Suppose that a fair coin is tossed three times. Construct an equiprobable sample space for the experiment and determine each of the following probabilities.
   (a) Pr(2 heads)   (b) Pr(3 heads)   (c) Pr(1 head)

7.2

6. A card is drawn at random from an ordinary deck of 52 playing cards. What is the probability that it is a queen or a jack?
7. A deck of 52 cards is shuffled. A card is drawn, then replaced, the pack is again shuffled, and a second card is drawn. What is the probability that each card drawn is an ace, king, queen, or jack?
8. If the probability of winning a game is 1/4 and there can be no ties, what is the probability of losing the game?
9. A card is drawn from a deck of 52 playing cards. What is the probability that it is an ace or a 10?
10. A card is drawn at random from a deck of playing cards. What is the probability that it is a king or a red card?
11. A bag contains 4 red balls numbered 1, 2, 3, 4 and 5 white balls numbered 5, 6, 7, 8, 9. A ball is drawn. What is the probability that the ball
    (a) is red and even-numbered?
    (b) is red or an even number?
    (c) is white or an odd number?

7.3

12. A bag contains 4 red balls and 3 black balls. Two balls are drawn at random from the bag without replacement. Find the probability that both balls are red.
13. A box contains 2 red balls and 3 black balls. Two balls are drawn from the box without replacement. Find the probability that the second ball is red, given that the first ball is black.

7.4

14. An urn contains 4 red and 6 white balls. One ball is drawn, is not replaced, and a second ball is drawn. What is the probability that one ball is white and one is red?
15. Urn I contains 3 red and 4 white balls and urn II contains 5 red and 2 white balls. An urn is selected and a ball drawn.
    (a) What is the probability that urn I is selected and a red ball is drawn?
    (b) What is the probability that a red ball is selected?
    (c) If a red ball is selected, what is the probability that urn I was selected?

7.5

16. Compute $_6P_2$.   17. Compute $_7C_3$.
18. How many 3-letter sets of initials are possible?

APPLICATIONS

7.1

19. **Senior citizens** In a certain city, 30,000 citizens out of 80,000 are over 50 years of age. What is the probability that a citizen selected at random will be 50 years old or younger?

7.2

20. **United Nations** Of 100 job applicants to the United Nations, 30 speak French, 40 speak German, and 12 speak both French and German. If an applicant is chosen at random, what is the probability the applicant speaks French or German?

Productivity Suppose that in a study of leadership style versus industrial productivity, the following data was obtained. Use this empirical data to answer problems 21–23.

| | Leadership style | | | |
Productivity	Demo-cratic	Authori-tarian	Laissez-faire	Total
Low	40	15	40	95
Medium	25	75	10	110
High	25	30	20	75
Total	90	120	70	280

21. Find the probability that a person chosen at random has a democratic style and medium productivity.
22. Find the probability that an individual chosen at random has high productivity or has an authoritarian style.

7.3

23. Find the probability that an individual chosen at random has an authoritarian style given that he or she has medium productivity.

7.5

24. **Management** A personnel director ranks 4 applicants for a job. How many rankings are possible?

25. **Management** An organization wants to select a president, vice president, secretary, and treasurer. If 8 people are willing to serve and each of them is eligible for any of the offices, in how many different ways can the offices be filled?

26. **Utilities** A utility company sends teams of 4 people each to perform repairs. If it has 12 qualified people, how many different ways can people be assigned to a team?

27. **Committees** An organization wants to select a committee of 4 members from a group of 8 eligible members. How many different committees are possible?

28. **Juries** A jury can be deadlocked if one person disagrees with the rest. There are 12 ways a jury can be deadlocked if one person disagrees because any one of the 12 jurors could disagree.
    (a) In how many ways can a jury be deadlocked if 2 people disagree?
    (b) If 3 people disagree?

29. **License plates** Because Atlanta is the site of the 1996 Olympics, the state of Georgia released vanity license plates containing the Olympic symbol, and a large number of these plates have been sold. Because of the symbol, each plate can contain at most 5 digits and/or letters. How many possible plates can be produced using from 1 to 5 digits or letters?

30. **Blood types** In a book describing the mass-murder of 18 people in Northern California, a policewoman was quoted as stating that there are blood groups O, A, B, and AB, positive and negative, blood types M, N, MN, and P, and 8 Rh blood types, so there are 288 unique groups. (Source: Joseph Harrington and Robert Berger, *Eye of Evil,* St. Martins Press, 1993.) Comment on this conclusion.

7.4

31. **Color blindness** Sixty men out of 1000 and 3 women out of 1000 are color blind. A person is picked at random from a group containing 10 men and 10 women.
    (a) What is the probability that the person is color blind?
    (b) What is the probability the person is a man if the person is color blind?

7.6

32. **Scheduling** A college registrar enters 4 new courses to the list of offerings for the spring semester. If he added the course names in random order at the end of the list, what is the probability that these 4 courses are listed in alphabetical order?

33. **Lottery** If a person "boxes" a 4-digit number with 4 different digits in Pennsylvania's Pick Four Lottery, what is the probability that the person will win?

34. **Lottery** Pennsylvania's Daily Number pays 500 to 1 to people who play the winning 3-digit number exactly as it is drawn. However, players can "box" a number so they can win $80 from a $1 bet if the 3 digits they pick come out in any order. What is the probability that a "boxed" number with 3 different digits will be a winner?

35. **Quality control** A supplier has 200 compact discs, of which 10% are known to be defective. If a music store purchases 10 of these discs, what is the probability that
    (a) none of the discs is defective?
    (b) 2 of the discs are defective?

36. **Stocks** Mr. Way must sell stocks from 3 of the 5 companies whose stock he owns so that he can send his children to college. If he chooses the companies at random, what is the probability that the 3 companies will be the 3 with the best future earnings?

37. **Quality control** A sample of 6 fuses is drawn from a lot containing 10 good fuses and 2 defective fuses. What is the probability that the number of defective fuses is
    (a) exactly 1   (b) at least 1

7.7

Income level Use the following information for problems 38 and 39. Suppose people in a certain community are classified as being low income, middle income, or high income. Suppose further that the probabilities for children being in a given state depends on which state their parents were in according to the Markov chain transition matrix below.

	Income level of children		
Income level of parents	L	M	H
Low	.15	.6	.25
Middle	.15	.35	.5
High	0	.2	.8

38. If the initial distribution probabilities of families in a certain population are $[.7 \quad .2 \quad .1]$, find the distribution for the next three generations.

39. Find the steady-state vector for the distribution of families by income level.

CONSULTANT'S CORNER

Competition in the Telecommunications Industry

Of the 100,000 households in a particular city having telephones, 85,000 use AT&T as their long-distance phone company, 10,000 use MCI, and 5,000 use US Sprint. A research firm has surveyed residents to determine the level of customer satisfaction and to learn what company each expects to be using in a year's time. The results are depicted below.

Although AT&T clearly dominates this city's long-distance service now, the situation could change in the future.

1. Calculate the percentage of the population that will be using each phone company in a year's time.
2. What will the long-term end result be if this pattern of change continues year after year?
3. You should have found in your answer to question 1 that the number of US Sprint customers increases in a year's time. Explain how this can happen when US Sprint loses 30% of its customers and gains only 5% of AT&T's and MCI's customers.
4. MCI management has been presented with two proposals for increasing their market share, with each proposal having the same cost. It is estimated that Plan A would increase their customer retention rate from 90% to 95% through an enhanced customer services department. Plan B, an advertising campaign targeting AT&T customers, is intended to increase the percentage of people switching from AT&T to MCI from 10% to 11%. If management has sufficient funds to implement one plan only, which one would you advise them to approve and why?
5. Suppose that because of budget constraints, MCI implements neither plan. A year later they face the same choice. What should they do then?

Warm-up

Prerequisite problem type	For section	Answer	Section for review
If a fair coin is tossed 3 times, find Pr(2 heads and 1 tail).	8.1	$\frac{3}{8}$	7.4 Probability trees
Evaluate: (a) $\binom{4}{3}$ (b) $\binom{5}{2}$ (c) $\binom{24}{0}$	8.3	(a) 4 (b) 10 (c) 1	7.5 Counting; permutations and combinations
Expand: (a) $(p + q)^2$ (b) $(p + q)^3$	8.3	(a) $p^2 + 2pq + q^2$ (b) $p^3 + 3p^2q +$ $3pq^2 + q^3$	0.5 Special products
Graph all real numbers satisfying (a) $100 \le x \le 115$, (b) $-1 \le z \le 2$.	8.5	(a) (b)	3.1 Linear inequalities
If 68% of the IQ scores of adults lie between 85 and 115, what is the probability that an adult chosen at random will have a IQ score between 85 and 115?	8.5	.68	7.1 Probability
Identify the graph of $y = -0.33x + 16.43$.	8.6	Straight line	1.2 Graphs of first-degree equations

CHAPTER 8

FURTHER TOPICS IN PROBABILITY; STATISTICS

In Chapter 7, we used sample spaces or probability trees to find the probability that the birth of three children resulted in 2 boys and 1 girl. But finding the probability that the birth of 6 children results in 4 boys and 2 girls is time consuming using either of these methods. There is a more efficient method of finding this probability and other probabilities of events occurring in **binomial experiments.**

In this chapter we continue our discussion of probability by considering binomial experiments, discrete probability distributions, binomial probability distributions, and normal probability distributions. We will also find the mean and standard deviation of discrete probability distributions and of normal probability distributions.

We will discuss how a set of data can be taken from a source of data for the purpose of learning about the source. This set of data is called a **sample,** and the source of the data is called a **population.** The branch of statistics that deals with the collection, organization, summarization, and presentation of this sample data is called **descriptive statistics.** Descriptive statistics is used by businesses to summarize data about advertising effectiveness, production costs, sales, wages, and profits. Behavioral scientists also collect statistical data about carefully selected samples in order to reach conclusions about general behavioral characteristics. The descriptive statistics methods in this chapter include graphical representation of data, measures of central tendency, and measures of dispersion.

In this chapter, we will also discuss **linear regression,** which is used to model linear functions throughout the text.

8.1 BINOMIAL PROBABILITY EXPERIMENTS

Objective

❑ To solve probability problems related to binomial experiments

APPLICATION PREVIEW

Suppose it has been determined with probability 0.01 that any child whose mother lives in a specific area near a chemical dump will be born with a birth defect. To find the probability that 2 of 10 children whose mothers live in the area will be born with birth defects, we could use a probability tree that lists all the possibilities and then select the paths that give 2 children with birth defects. However, it is easier to find the probability by using a formula developed in this section. This formula can be used when the probability experiment is a **binomial probability experiment.**

Suppose you have a coin that is biased so that when the coin is tossed the probability of getting a head is 2/3 and the probability of getting a tail is 1/3. What is the probability of tossing 2 heads in 3 tosses of this coin?

If the question were, What is the probability that the first two tosses will be heads and the third will be tails? the answer would be

$$\text{Pr}(\text{HHT}) = \frac{2}{3} \cdot \frac{2}{3} \cdot \frac{1}{3} = \frac{4}{27}.$$

However, the original question did not specify the order, so we must consider other orders that will give us 2 heads and 1 tail. We can use a tree to find all the possibilities (see Figure 8.1). We can see that there are 3 paths through the tree, which correspond to 2 heads and 1 tail in 3 tosses. Since the probability for each successful path is 4/27,

$$\text{Pr}(2\text{H and }1\text{T}) = 3 \cdot \frac{4}{27} = \frac{4}{9}.$$

In this problem we can find the probability of 2 heads in 3 tosses by finding the probability for any one path of the tree (like HHT) and then multiplying

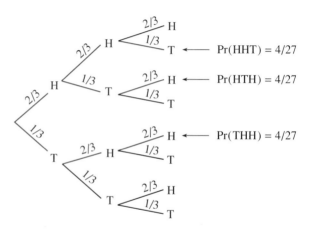

Figure 8.1

that probability by the number of paths that result in 2 heads. We can also determine how many ways 2 heads and 1 tail can result by considering the 3 blanks below:

$$\underline{\qquad} \qquad \underline{\qquad} \qquad \underline{\qquad}$$

The number of ways we can pick 2 blanks (for the 2 heads) from the 3 blanks will be equivalent to the number of paths that result in 2 heads. But the number of ways we can select 2 blanks from the 3 blanks is $_3C_2 = \binom{3}{2}$. Thus the probability of 2 heads resulting in 3 tosses is

$$Pr(2H \text{ and } 1T) = \binom{3}{2}\frac{2}{3} \cdot \frac{2}{3} \cdot \frac{1}{3} = \binom{3}{2}\left(\frac{2}{3}\right)^2\left(\frac{1}{3}\right)^1 = \frac{4}{9}.$$

The experiment discussed above (tossing a coin 3 times) is an example of a general class of experiments called **binomial probability experiments.** These experiments are also called **Bernoulli experiments,** or **Bernoulli trials,** after the eighteenth-century mathematician Jacob Bernoulli.

A **binomial probability experiment** satisfies the following properties:

1. There are n repeated trials of the experiment.
2. Each trial results in one of two outcomes, with one denoted success (S) and the other failure (F).
3. The trials are independent, with the probability of success, p, the same for every trial. The probability of failure is $q = 1 - p$ for every trial.
4. The probability of x successes in n trials is
$$Pr(x) = \binom{n}{x}p^x q^{n-x}.$$

$Pr(x)$ is called a **binomial probability.**

EXAMPLE 1
If a die is rolled 4 times and the number of times a 6 results is recorded,
(a) is the experiment a binomial experiment?
(b) what is the probability that three 6's will result?

Solution
(a) Yes. There are 4 trials, resulting in success (a 6) or failure (not a 6). The result of each roll is independent of previous rolls; the probability of success on each roll is $1/6$.
(b) There are $n = 4$ trials with $Pr(\text{rolling a } 6) = p = 1/6$ and $Pr(\text{not rolling a } 6) = q = 5/6$, and we seek the probability of $x = 3$ successes (because we want three 6's). Thus

$$Pr(\text{three } 6\text{'s}) = \binom{4}{3}\left(\frac{1}{6}\right)^3\left(\frac{5}{6}\right)^1 = \frac{4!}{1!3!}\left(\frac{1}{216}\right)\left(\frac{5}{6}\right) = \frac{5}{324}.$$

CHECKPOINT

Determine whether the following experiments satisfy the conditions for a binomial probability experiment.

1. A coin is tossed 8 times and the number of times a tail occurs is recorded.

2. A die is rolled twice and the sum that occurs is recorded.

3. A ball is drawn from a bag containing 4 white and 7 red balls, and then replaced. This is done a total of 6 times, and the number of times a red ball is drawn is recorded.

We now consider the problem mentioned in the Application Preview.

EXAMPLE 2

Suppose it has been found with probability 0.01 that any child whose mother lives in an area near a chemical dump will be born with a birth defect. If 10 children whose mothers live in the area are born in a given month,
(a) what is the probability that 2 of the 10 will be born with birth defects?
(b) what is the probability that at least 2 of them will have a birth defect?

Solution
(a) Note that we may consider a child having a birth defect as a success for this problem even though it is certainly not true in reality. Each of the 10 births is independent with probability of success $p = 0.01$ and probability of failure $q = 0.99$. Therefore the experiment is a binomial experiment, and

$$\Pr(2) = \binom{10}{2}(.01)^2(.99)^8 = .00415.$$

(b) The probability of at least 2 of the children having a birth defect is

$$1 - [\Pr(0) + \Pr(1)] = 1 - \left[\binom{10}{0}(.01)^0(.99)^{10} + \binom{10}{1}(.01)^1(.99)^9\right]$$

$$= 1 - [.90438 + .09135] = .00427.$$

We interpret this to mean that there is a 0.427% likelihood that at least 2 of these children will have a birth defect.

EXAMPLE 3

A manufacturer of motorcycle parts guarantees that a box of 24 parts will contain at most one defective part. If the records show that the manufacturer's machines produce 1% defective parts, what is the probability that a box of parts will satisfy the guarantee?

Solution
The problem is a binomial experiment problem with $n = 24$ and (considering getting a defective part as a success) $p = 0.01$. Then the probability the manufacturer will have no more than 1 defective part in a box is

$$Pr(x \le 1) = Pr(0) + Pr(1) = \binom{24}{0}(.01)^0(.99)^{24} + \binom{24}{1}(.01)^1(.99)^{23}$$

$$= 1(1)(.7857) + 24(.01)(.7936) = 0.9762.$$

We interpret this to mean that there is a 97.62% likelihood that there will be no more than 1 defective part in a box of 24 parts.

CHECKPOINT

4. A bag contains 5 red balls and 3 white balls. If 3 balls are drawn, with replacement, from the bag, what is the probability that
   (a) two balls will be red?
   (b) at least two balls will be red?

CHECKPOINT SOLUTIONS

1. A tail is a success on each trial and each of the conditions of a binomial probability distribution are satisfied.

2. Each trial in this experiment does not result in a success or failure, so this is not a binomial probability experiment.

3. A red ball is a success on each trial, and each condition of a binomial distribution is satisfied.

4. (a) $Pr(2 \text{ red}) = \binom{3}{2}\left(\frac{5}{8}\right)^2\left(\frac{3}{8}\right)$

   $= \dfrac{225}{512}$

   (b) $Pr(\text{at least 2 red}) = \binom{3}{2}\left(\frac{5}{8}\right)^2\left(\frac{3}{8}\right) + \binom{3}{3}\left(\frac{5}{8}\right)^3\left(\frac{3}{8}\right)^0$

   $= \dfrac{225}{512} + \dfrac{125}{512} = \dfrac{175}{256}$

EXERCISE 8.1

1. If the probability of success on each trial of an experiment is 0.3, what is the probability of 4 successes in 6 independent trials?

2. If the probability of success on each trial of an experiment is 0.4, what is the probability of 5 successes in 8 trials?

3. Suppose a fair coin is tossed 6 times. What is the probability that
   (a) 6 heads will occur? (b) 3 heads will occur?
   (c) 2 heads will occur?

4. If a fair die is rolled 3 times, what is the probability that

   (a) a 5 will result 2 times out of the 3 rolls?
   (b) an odd number will result 2 times?
   (c) a number divisible by 3 will occur 3 times?

5. Suppose a fair die is rolled 12 times. What is the probability that 5 "aces" (1's) will occur?

6. In problem 5, what is the probability that no aces will occur?

7. A bag contains 6 red balls and 4 black balls. If we draw 5 balls, with each one replaced before the next is drawn,
   (a) what is the probability that 2 balls drawn will be red?

(b) what is the probability that 5 black balls will be drawn?

(c) what is the probability that at least 3 black balls will be drawn?

8. A bag contains 6 red balls and 4 black balls. If we draw 5 balls, with each one replaced before the next is drawn, what is the probability that at least 2 balls drawn will be red?

9. Suppose a pair of dice is thrown 4 times. What is the probability that a sum of 9 occurs exactly 2 times?

10. If a pair of dice is thrown 4 times, find the probability that a sum of 7 occurs at least 3 times.

11. Suppose the probability that a marksman will hit a target each time he shoots is 0.85. If he fires 10 shots at a target, what is the probability he will hit it 8 times?

12. The probability that a sharpshooter will hit a target each time he shoots is 0.98. What is the probability that he will hit the target 10 times in 10 shots?

APPLICATIONS

13. **Genetics** A family has 4 children. If the probability that each child is a girl is 0.5, what is the probability that
(a) half of the children are girls?
(b) all of the children are girls?

14. **Testing** A multiple-choice test has 10 questions and 5 choices for each question. If a student is totally unprepared and guesses on each question, what is the probability that
(a) she will answer every question correctly?
(b) she will answer half of the questions correctly?

15. **Management** The manager of a store buys portable radios in lots of 12. Suppose that, on the average, 2 out of each group of 12 are defective. If the manager randomly selects 4 radios out of the group to test,
(a) what is the probability that he will find 2 defective radios?
(b) what is the probability that he will find no defective radios?

16. **Quality control** A hospital buys thermometers in lots of 1000. On the average, one out of 1000 is defective. If 10 are selected from one lot, what is the probability that none is defective?

17. **Genetics** If the probability that a certain couple will have a blue-eyed child is 1/4 and they have 4 children, what is the probability that
(a) 1 of their children has blue eyes?
(b) 2 of their children have blue eyes?
(c) none of their children has blue eyes?

18. **Genetics** If the probability that a certain couple will have a blue-eyed child is 1/2, and they have 4 children, what is the probability that
(a) none of the children has blue eyes?
(b) at least 1 child has blue eyes?

19. **Health care** Suppose that 10% of the patients having a certain disease die from it. If 5 patients have the disease, what is the probability that
(a) exactly 2 patients will die from it?
(b) no patients will die from it?
(c) no more than 2 patients will die from it?

20. **Employee benefits** In a certain school district, 3% of the faculty use none of their sick days in a school year. Find the probability that 5 faculty members selected at random used no sick days in a given year.

21. **Genetics** If the ratio of boys born to girls born is 105 to 100, and if 6 children are born in a certain hospital in a day, what is the probability that 4 of them are boys?

22. **Biology** It has been determined empirically that the probability that a given cell will survive for a given period of time is 0.4. Find the probability that 3 out of 6 of these cells will survive for this period of time.

23. **Insurance** If records indicate that 4 houses out of 1000 are expected to be damaged by fire in any year, what is the probability that a woman owning 10 houses will have fire damage in 2 of them in a year?

24. **Psychology** Suppose 4 rats are placed in a T-maze in which they must turn right or left. If each rat makes a choice by chance, what is the probability that 2 of the rats will turn to the right?

25. **Sports** A baseball player has a lifetime batting average of .300. If he comes to bat 5 times in a given game, what is the probability that
(a) he will get 3 hits?
(b) he will get more than 3 hits?

26. **Quality control** If the probability of an automobile part being defective is 0.02, find the probability of getting exactly 3 defective parts in a sample of 12.

27. **Quality control** A company produces shotgun shells in batches of 300. A sample of 10 is tested from each batch, and if more than one defect is found, the entire batch is tested. If 1% of the shells are actually defective,
(a) what is the probability of 0 defective shells in the sample?
(b) what is the probability of 1 defective shell?
(c) what is the probability of more than 1 defective shell?

28. **Quality control** A baseball pitching machine throws a bad pitch with probability 0.1. What is the probability that the machine will throw 5 or fewer bad pitches out of 20 pitches? (Set up only.)

29. **Testing** A quiz consists of 10 multiple-choice questions with 5 choices for each question. Suppose a student is sure of the first 5 answers and has each of the last 5 questions narrowed to 3 of the possible 5 choices. If the student guesses among the narrowed choices on the last 5 questions, find
    (a) the probability of passing the quiz (getting at least 60%).
    (b) the probability of getting at least a B (at least 80%).

30. **Demographics** In a certain community, 10% of the population is Jewish. A study shows that of 7 social service agencies, 4 have board presidents that are Jewish. Find the probability that this could happen by chance.

31. **Suicide rates** Suppose the probability of suicide among a certain age group is 0.003. If a randomly selected group of 100 Native Americans within this age group had no suicides, find the probability of this occurring by chance.

8.2 DISCRETE PROBABILITY DISTRIBUTIONS; DECISION MAKING

Objectives

❑ To identify random variables
❑ To verify that a table or formula describes a discrete probability distribution
❑ To compute the expected value of a discrete probability distribution
❑ To make decisions by using expected value
❑ To find the variance and standard deviation of a discrete probability distribution

APPLICATION PREVIEW

The T. J. Cooper Insurance Company insures 100,000 automobiles. Their records over a 5-year period indicate that during each 6-month period they will pay out the following amounts for accidents.

$100,000 with probability 0.0001
$50,000 with probability 0.001
$25,000 with probability 0.002
$5,000 with probability 0.008
$1,000 with probability 0.02

To determine the premium the company should charge each driver, it must determine the **expected value** of its payments for accidents per car for each 6-month period. In this section we will find the expected value of **discrete probability distributions.**

If a player rolls a die and receives $1 for each dot on the face he rolls, the amount of money won on one roll can be represented by the variable x. If the die is rolled once, the following table gives the possible outcomes of the experiment and their probabilities.

x	1	2	3	4	5	6
$\Pr(x)$	1/6	1/6	1/6	1/6	1/6	1/6

Note that we have used the variable x to denote the possible numerical outcomes of this experiment. Because x results from a probability experi-

Table 8.1

x	Pr(x)
0	1/16
1	1/4
2	3/8
3	1/4
4	1/16

ment, we call x a **random variable,** and because there are a finite number of possible values for x, we say that x is a **discrete random variable.** Whenever all possible values for a random variable can be listed (or counted), the random variable is a **discrete random variable.**

If x represents the possible number of heads that can occur in the toss of four coins, we can use the sample space {0, 1, 2, 3, 4} to list the possible numerical values for x. Looking at Table 8.1, we see the values that the random variable of the coin-tossing experiment can assume and how the probability is distributed over these values.

Discrete Probability Distribution	A table, graph, or formula that assigns to each value of a discrete random variable x a probability Pr(x) describes a **discrete probability distribution** if the following two conditions hold:

(i) $0 \le \text{Pr}(x) \le 1$,   for any value of x.

(ii) The sum of all the probabilities is 1. We use "Σ" to denote "the sum of" and write $\Sigma\text{Pr}(x) = 1$, where the sum is taken over all values of x.

EXAMPLE 1

Verify that the following table describes a discrete probability distribution.

x	Pr(x)
80	1/16
120	3/8
160	5/16
200	1/4

Solution

Note that for each value of x, $0 \le \text{Pr}(x) \le 1$, and

$$\sum\text{Pr}(x) = \frac{1}{16} + \frac{3}{8} + \frac{5}{16} + \frac{1}{4} = 1.$$

Thus the table describes a discrete probability distribution.

EXAMPLE 2

An experiment consists of selecting a ball from a bag containing 15 balls numbered 1 through 5. If the probability of selecting a ball with the number x on it is $\text{Pr}(x) = x/15$, where x is an integer and $1 \le x \le 5$, verify that $f(x) = \text{Pr}(x) = x/15$ describes a discrete probability distribution for the random variable x.

Solution
For each integer x $(1 \leq x \leq 5)$, $\Pr(x)$ satisfies $0 \leq \Pr(x) \leq 1$.

$$\sum \Pr(x) = \Pr(1) + \Pr(2) + \Pr(3) + \Pr(4) + \Pr(5)$$

$$= \frac{1}{15} + \frac{2}{15} + \frac{3}{15} + \frac{4}{15} + \frac{5}{15} = 1.$$

Hence $\Pr(x) = x/15$ describes a discrete probability distribution.

We can visualize the possible values of a discrete random variable and their associated probabilities by constructing a graph. This graph, called a **probability density histogram,** is designed so that centered over each value of the discrete random variable x along the horizontal axis is a bar having width equal to one unit and height (and thus, area) equal to $\Pr(x)$. Figure 8.2 gives the probability density histogram for the experiment described in Example 2.

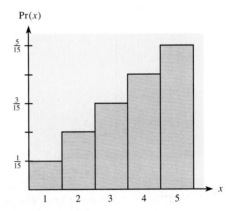

Figure 8.2

Many decisions in business and science are made on the basis of what the outcomes of specific decisions will be. One important property that is not a probability statement, but that can be developed using probability, is the **expected value,** or **mean,** of a distribution. Expected value is very useful in making decisions in business and the sciences.

Mean and Expected Value

If x is a discrete random variable with values $x_1, x_2, \ldots, x_n$, then the **mean, μ, of the distribution is the expected value of x,** denoted by $E(x)$, and is given by

$$\mu = E(x) = \sum x \Pr(x) = x_1 \Pr(x_1) + x_2 \Pr(x_2) + \cdots + x_n \Pr(x_n).$$

The mean μ is called a **measure of central tendency** because the probability density histogram can be balanced on a fulcrum located at $x = \mu$.

EXAMPLE 3

The probability distribution for the number of heads x that occur when a fair coin is tossed 4 times is given at the right. What is the expected value (expected number of heads) for this experiment?

x	Pr(x)
0	1/16
1	1/4
2	3/8
3	1/4
4	1/16

Solution

The expected value is

$$E(x) = 0 \cdot \frac{1}{16} + 1 \cdot \frac{1}{4} + 2 \cdot \frac{3}{8} + 3 \cdot \frac{1}{4} + 4 \cdot \frac{1}{16} = 2.$$

Note that an expected value of 2 heads in 4 tosses means that if 4 coins were tossed repeatedly, then the average (or mean) number of heads that would occur is expected to be 2.

CHECKPOINT

1. Consider a game in which you roll a die and receive $1 for each spot that occurs.
   (a) What are the expected winnings from this game?
   (b) If you paid $4 to play this game, how much would you lose, on average, each time you played the game?

2. An experiment has the following possible outcomes for the random variable x: 1, 2, 3, 4, 6, 7, 8, 9, 10. The probability that the value x occurs is $x/50$. Find the expected value of x for the experiment.

EXAMPLE 4

A fire company sells chances to win a new car, with each ticket costing $5. If the car is worth $8000 and the company sells 3000 tickets, what is the expected value for a person buying one ticket?

Solution

For a person buying one ticket, there are two possible outcomes from the drawing, winning or losing. The probability of winning is 1/3000, and the amount won would be $8000 − $5 (the cost of the ticket). The probability of losing is 2999/3000, and the amount lost is written as a winning of −$5. Thus the expected value is

$$7995\left(\frac{1}{3000}\right) + (-5)\left(\frac{2999}{3000}\right) = \frac{7995}{3000} - \frac{14{,}995}{3000} = -2.33.$$

Thus on the average, a person can expect to lose $2.33 on every ticket. It is not possible to lose exactly $2.33 on a ticket—the person will either win $8000 or lose $5; the expected value means that a person who buys large numbers of tickets can expect to lose an average of $2.33 on each ticket. That is, the fire company will make $2.33 on each ticket they sell if they sell all 3000.

We now return to the Application Preview problem.

EXAMPLE 5

The T. J. Cooper Insurance Company insures 100,000 cars. Their records over a 5-year period indicate that during each 6-month period they will pay out the following amounts for accidents:

$100,000 with probability 0.0001

$50,000 with probability 0.001

$25,000 with probability 0.002

$5,000 with probability 0.008

$1,000 with probability 0.02

How could the company use the expected payout per car for each 6-month period to help determine their rates?

Solution

The expected value of the company's payments is

$$\$100{,}000(.0001) + \$50{,}000(.001) + \$25{,}000(.002) + \$5000(.008)$$
$$+ \$1000(.02) = \$10 + \$50 + \$50 + \$40 + \$20 = \$170.$$

Thus the average premium the company would charge each driver per car for a 6-month period would be

$$\$170 + \text{operating costs} + \text{profit} + \text{reserve for bad years}.$$

Figure 8.3 shows two probability histograms that have the same mean (expected value). However, these histograms look very different. The reason for this difference is that the **dispersion** or **variation** of the values of the random variables differs.

Figure 8.3

(a)

(b)

$$\mu = E(x) = 4\left(\frac{2}{10}\right) + 5\left(\frac{6}{10}\right) + 6\left(\frac{2}{10}\right)$$

$$= \frac{8 + 30 + 12}{10} = 5$$

$$\mu = E(x) = 1\left(\frac{2}{10}\right) + 2\left(\frac{1}{10}\right) + 4\left(\frac{1}{10}\right) + 5\left(\frac{2}{10}\right) + 6\left(\frac{1}{10}\right) + 8\left(\frac{1}{10}\right) + 9\left(\frac{2}{10}\right)$$

$$= \frac{2 + 2 + 4 + 10 + 6 + 8 + 18}{10} = 5$$

Two useful measures of the variation of the values of the random variable about the mean are the **variance** of the distribution and the square root of the variance, called the **standard deviation.**

Variance and Standard Deviation   If x is a discrete random variable with values $x_1, x_2, \ldots, x_n$ and mean μ, then the variance of the distribution of x is

$$\sigma^2 = \sum (x - \mu)^2 \Pr(x).$$

The standard deviation is

$$\sigma = \sqrt{\sigma^2} = \sqrt{\sum (x - \mu)^2 \Pr(x)}.$$

EXAMPLE 6

Find the standard deviation of the distributions described in (a) Figure 8.3(a) and (b) Figure 8.3(b).

Solution

(a) $\sigma^2 = \sum (x - \mu)^2 \Pr(x)$

$$= (4 - 5)^2 \cdot \frac{2}{10} + (5 - 5)^2 \cdot \frac{6}{10} + (6 - 5)^2 \cdot \frac{2}{10}$$

$$= 1 \cdot \frac{2}{10} + 0 \cdot \frac{6}{10} + 1 \cdot \frac{2}{10}$$

$$= 0.4$$

$$\sigma = \sqrt{0.4} \approx 0.632$$

(b) $\sigma^2 = (1 - 5)^2 \cdot \frac{2}{10} + (2 - 5)^2 \cdot \frac{1}{10} + (4 - 5)^2 \cdot \frac{1}{10} + (5 - 5)^2 \cdot \frac{2}{10}$

$$+ (6 - 5)^2 \cdot \frac{1}{10} + (8 - 5)^2 \cdot \frac{1}{10} + (9 - 5)^2 \cdot \frac{2}{10}$$

$$= 16 \cdot \frac{2}{10} + 9 \cdot \frac{1}{10} + 1 \cdot \frac{1}{10} + 0 \cdot \frac{2}{10} + 1 \cdot \frac{1}{10}$$

$$+ 9 \cdot \frac{1}{10} + 16 \cdot \frac{2}{10}$$

$$= 8.4$$

$$\sigma = \sqrt{8.4} \approx 2.898$$

As we can see by referring to Figure 8.3 and to the standard deviations calculated in Example 6, smaller values of σ indicate that the values of x are clustered nearer μ. When σ is large, the values of x are more widely dispersed from μ.

CHECKPOINT

3. Find the standard deviation for the probability distribution in Example 3.

CHECKPOINT SOLUTIONS

1. (a) $E(x) = \$1\left(\frac{1}{6}\right) + \$2\left(\frac{1}{6}\right) + \$3\left(\frac{1}{6}\right) + \$4\left(\frac{1}{6}\right) + \$5\left(\frac{1}{6}\right) + \$6\left(\frac{1}{6}\right)$

   $= \$3.50$

   (b) $\$4 - \$3.50 = \$.50$. You would lose \$0.50, on average, from each play.

2. The expected value of x for this experiment is

$$E(x) = 1 \cdot \frac{1}{50} + 2 \cdot \frac{2}{50} + 3 \cdot \frac{3}{50} + 4 \cdot \frac{4}{50} + 6 \cdot \frac{6}{50}$$

$$+ 7 \cdot \frac{7}{50} + 8 \cdot \frac{8}{50} + 9 \cdot \frac{9}{50} + 10 \cdot \frac{10}{50} = 7.2.$$

3. $\sigma^2 = (0-2)^2\left(\frac{1}{16}\right) + (1-2)^2\left(\frac{1}{4}\right) + (2-2)^2\left(\frac{3}{8}\right) + (3-2)^2\left(\frac{1}{4}\right)$

   $+ (4-2)^2\left(\frac{1}{16}\right) = 1$

   $\sigma = \sqrt{1} = 1$

EXERCISE 8.2

Determine whether each table in problems 1–4 describes a discrete probability distribution.

Determine whether each formula in problems 5–8 describes a discrete probability distribution.

5. $\Pr(x) = \dfrac{x}{21}$; $x = 0, 1, 2, 3, 4, 5, 6$

6. $\Pr(x) = \dfrac{x}{10}$; $x = 1, 2, 3, 4$

7. $\Pr(x) = \dfrac{10 - x}{10}$; $x = 1, 2, 3, 4, 5$

8. $\Pr(x) = \dfrac{x + 1}{15}$; $x = 1, 2, 3, 4$

1.

x	$\Pr(x)$
1	$-1/5$
2	$1/4$
3	$5/10$
4	$9/20$

2.

x	$\Pr(x)$
0	$-4/5$
1	$2/5$
2	$3/5$
3	$4/5$

3.

x	$\Pr(x)$
-1	$1/4$
1	$3/8$
2	$1/4$
4	$1/8$

4.

x	$\Pr(x)$
0	$1/3$
4	$1/6$
8	$1/2$
12	0

Each of the following tables defines a discrete probability distribution. Find the expected value of each distribution.

9.

x	0	1	2	3
$\Pr(x)$	$1/8$	$1/4$	$1/4$	$3/8$

10.

x	1	2	3	4
$\Pr(x)$	1/10	1/5	3/10	2/5

11.

x	4	5	6	7
$\Pr(x)$	1/3	1/3	1/3	0

12.

x	1	2	3	4	5	6
$\Pr(x)$	1/12	1/4	1/12	1/4	1/12	1/4

13.

x	0	2	4
$\Pr(x)$	1/6	2/6	3/6

14.

x	3	4	5	6
$\Pr(x)$	3/10	1/5	3/10	1/5

15.

x	1	3	5
$\Pr(x)$	2/3	1/6	1/6

16.

x	1	2	3	4
$\Pr(x)$	1/15	4/15	1/5	7/15

In problems 17–22, find the mean, variance, and standard deviation for each probability distribution.

17.

x	$\Pr(x)$
0	1/4
1	1/4
2	1/8
3	3/8

18.

x	$\Pr(x)$
3	1/4
4	1/4
5	1/4
6	1/4

19.

x	$\Pr(x)$
0	1/5
2	1/4
4	1/10
6	3/20
8	3/10

20.

x	$\Pr(x)$
1	1/8
3	1/8
5	1/8
7	1/4
9	3/8

21. $\Pr(x) = \dfrac{x}{21}; \quad x = 0, 1, 2, 3, 4, 5, 6$

22. $\Pr(x) = \dfrac{x}{10}; \quad x = 2, 3, 5$

23. Suppose an experiment has five possible outcomes for x: 0, 1, 2, 3, 4. The probability that each of these outcomes occurs is $x/10$. What is the expected value of x for the experiment?

24. Suppose an experiment has six possible outcomes for x: 0, 1, 2, 3, 4, 5. The probability that each of these outcomes occurs is $x/15$. What is the expected value of x for the experiment?

25. Five slips of paper containing the numbers 0, 1, 2, 3, 4 are placed in a hat. If the experiment consists of drawing one number, and if the experiment is repeated a large number of times, what is the expected value of the number drawn?

26. An experiment consists of rolling a die once. If the experiment is repeated a large number of times, what is the expected value of the number rolled?

APPLICATIONS

27. **Animal relocation** In studying endangered species, scientists have found that when animals are relocated it takes x years without offspring before the first young are born, where x and the probability of x are given below. What is the expected number of years before the first young are born?

x	0	1	2	3	4
$\Pr(x)$	.04	.35	.38	.18	.05

28. **Raffle** A charity sells 1000 raffle tickets for \$1 each. There is one grand prize of \$250, two prizes of \$75 each, and five prizes of \$10 each. If x is the net winnings for someone who buys a ticket, then the following table gives the probability distribution for x. Find the expected winnings of someone who buys a ticket.

x	$\Pr(x)$
249	0.001
74	0.002
9	0.005
−1	0.992

29. **Campaigning** A candidate must decide if he should spend his time and money on TV commercials or making personal appearances. His staff determines that by using TV he can reach 100,000 people with probability 0.01, 50,000 people with probability 0.47, and 25,000 with probability 0.52; and that by making personal appearances he can reach 80,000 people with probability 0.02, 50,000 people with probability 0.47, and 20,000 with probability 0.51. In the tables, x represents the number of people reached by each choice. In each case,

find the expected value of x to decide which method will reach more people.

TV Commercials		Personal Appearances	
x	Pr(x)	x	Pr(x)
100,000	0.01	80,000	0.02
50,000	0.47	50,000	0.47
25,000	0.52	20,000	0.51

30. *Gambling* Suppose that a student is offered a chance to draw a card from an ordinary deck of 52 playing cards and win $10 if he or she draws an ace, $2 for a king, and $1 for a queen. If $2 must be paid to play the game, and if x is the net winnings (in dollars), use the table below to find the expected winnings each time the game is played.

x	Pr(x)
8	1/13
0	1/13
−1	1/13
−2	10/13

31. *Genetics* If the probability that a newborn child is a male is 1/2, what is the expected number of male children in a family having 4 children?

32. *Testing* A student is completely unprepared for a 4-choice multiple-choice test. If the test has 4 questions and all are guessed at, how many questions should he or she expect to get correct?

33. *Testing* Suppose the instructor of the student in problem 32 tries to account for guessing on the test by counting the number of questions the student answers correctly and then subtracting from this number 1/4 of the number of questions a student answers. If the student guesses on all 4 problems on this test, for how many problems should he or she expect to get credit?

34. *Gambling* Suppose you are invited to play the following game: You toss 3 coins and receive $1 if one head results, $4 if two heads result, and $9 if three heads result. How much should you pay to play if the game is fair?

35. *Lottery* A charity sells raffle tickets for $1 each. First prize is $500, second prize is $100, third prize is $10. If you bought one of the 1000 tickets sold, what are your expected winnings?

36. *Gambling* Suppose the coin used in problem 34 has been altered so that the probability of a head on any toss is 1/3. How much should you pay to play if the game is fair?

37. *Gambling* Suppose a student is offered a chance to draw a card from an ordinary deck of 52 playing cards and win $15 for an ace, $10 for a king, and $1 for a queen. If $4 must be paid to play the game, what is the expected winnings every time the game is played by the student?

38. *Revenue* The Rent-to-Own Company estimates that 35% of its rentals result in the sale of the product, with an average revenue of 100%, 56% of the rentals are returned in good condition, with an average revenue of 35% on these rentals, and the remainder of the rentals are stolen or returned in poor condition, giving a loss of 15% on these rentals. If these estimates are accurate, what is the expected percent of revenue for this company?

39. *Sales* A young man plans to sell umbrellas at the city's Easter Parade. He knows that he can sell 180 umbrellas at $5 each if it rains hard, he can sell 50 if it rains lightly, and he can sell 10 if it doesn't rain at all. Past records show it rains hard 25% of the time on Easter, rains lightly 20% of the time, and does not rain at all 55% of the time. If he can buy 0, 100, or 200 umbrellas at $2 each and return the unsold ones for $1 each, how many should he buy?

40. *Sales* If the young man in problem 39 learns from a highly reliable weather forecaster that the probability of a hard rain on Easter is 15%, of a light rain is 20%, and of no rain is 65%, should he still plan to sell umbrellas at the Easter Parade?

41. *Advertising* A candidate must decide if he should spend his advertising dollars on TV commercials or newspaper advertisements. His staff determines that by using TV he can reach 100,000 people with probability 0.01, reach 50,000 people with probability 0.47, and reach 25,000 with probability 0.52; and that by purchasing newspaper ads he can reach 70,000 people with probability 0.04, reach 50,000 people with probability 0.38, and reach 30,000 with probability 0.58. Which method will reach more people?

42. *Insurance* A car owner must decide if she should take out a $100 deductible collision policy in addition to her liability insurance policy. Records show that each year, in her area, 8% of the drivers have an accident that is their fault or for which no fault is assigned, and that the average cost of repairs for these types of accidents is

$1000. If the $100 deductible collision policy costs $100 per year, would she save money in the long run by buying the insurance or "taking the chance"? (Hint: Find the expected values if she has the policy and if she doesn't have the policy and compare them.)

43. ***Budgeting*** Suppose a youth has a part-time job in an ice cream shop. He receives $20 if he is called to work a full day, and $10 if he is called to work a half day. Over the past year he has been called to work a full day an average of 8 days per month and for a half day an average of 14 days per month. How much can be expect to earn *per day* during a 30-day month?

44. ***Testing*** On a multiple-choice test with each question having five possible choices, a correct answer earns 4 points, an incorrect answer loses 1 point, and a blank receives 0 points.
    (a) Find the expected point-value earned when a student has no idea of the correct answer but makes a guess on every question.
    (b) Find the expected point-value earned when a student has narrowed the choice by eliminating two of the possible choices on each question, but must guess among the remaining choices.

8.3 THE BINOMIAL DISTRIBUTION

Objectives

❑ To find the mean and standard deviation of a binomial distribution
❑ To graph a binomial distribution
❑ To expand a binomial to a power, using the binomial formula

APPLICATION PREVIEW

Opponents to the distribution of condoms in high schools and colleges object on moral grounds and on the grounds that only abstinence is safe. If condoms are 95% effective in preventing disease (such as HIV) and pregnancy, the probability of a failure is 5% each time a condom is used. Because the number of failures follows the **binomial probability distribution,** we can use the methods of this section to find the expected number of failures per thousand condom uses.

To begin our discussion of the binomial distribution, suppose a fair coin is tossed 6 times and x represents the number of heads. Then Table 8.2 gives all possible outcomes and their probabilities. This is an example of a special probability distribution, called the **binomial probability distribution.**

Table 8.2

x	$\Pr(x)$	x	$\Pr(x)$
0	$\binom{6}{0}\left(\frac{1}{2}\right)^0\left(\frac{1}{2}\right)^6 = \frac{1}{64}$	4	$\binom{6}{4}\left(\frac{1}{2}\right)^4\left(\frac{1}{2}\right)^2 = \frac{15}{64}$
1	$\binom{6}{1}\left(\frac{1}{2}\right)^1\left(\frac{1}{2}\right)^5 = \frac{6}{64}$	5	$\binom{6}{5}\left(\frac{1}{2}\right)^5\left(\frac{1}{2}\right)^1 = \frac{6}{64}$
2	$\binom{6}{2}\left(\frac{1}{2}\right)^2\left(\frac{1}{2}\right)^4 = \frac{15}{64}$	6	$\binom{6}{6}\left(\frac{1}{2}\right)^6\left(\frac{1}{2}\right)^0 = \frac{1}{64}$
3	$\binom{6}{3}\left(\frac{1}{2}\right)^3\left(\frac{1}{2}\right)^3 = \frac{20}{64}$		

If x is a variable that assumes the values $0, 1, 2, \ldots, r, \ldots, n$ with probabilities $\binom{n}{0}p^0q^n$, $\binom{n}{1}p^1q^{n-1}$, $\binom{n}{2}p^2q^{n-2}, \ldots, \binom{n}{r}p^rq^{n-r}, \ldots, \binom{n}{n}p^nq^0$, respectively, then x is called a **binomial variable.**

The values of x and their corresponding probabilities described above form the **binomial probability distribution.**

CHECKPOINT

1. Show that the values of x and the associated probabilities in Table 8.2 satisfy the conditions for a discrete probability distribution.

The expected value of the number of successes for the coin toss experiment above (expected value was discussed in Section 8.2) is given by

$$E(x) = 0 \cdot \frac{1}{64} + 1 \cdot \frac{6}{64} + 2 \cdot \frac{15}{64} + 3 \cdot \frac{20}{64} + 4 \cdot \frac{15}{64} + 5 \cdot \frac{6}{64} + 6 \cdot \frac{1}{64}$$

$$= \frac{192}{64} = 3.$$

This expected number of successes seems reasonable. If the probability of success is $1/2$, we would expect to succeed on half of the 6 trials. For any binomial distribution the expected number of successes is given by np, where n is the number of trials and p is the probability of success. Since the expected value of any probability distribution is defined as the mean of that distribution, we have the following.

Mean of a Binomial Distribution

The theoretical mean of any binomial distribution is

$$\mu = np,$$

where n is the number of trials in the corresponding binomial experiment and p is the probability of success on each trial.

A simple formula can also be developed for the standard deviation of a binomial distribution.

Standard Deviation of a Binomial Distribution

The standard deviation of a binomial distribution is

$$\sigma = \sqrt{npq},$$

where n is the number of trials, p is the probability of success on each trial, and $q = 1 - p$.

The standard deviation of the binomial distribution corresponding to the number of heads resulting when a coin is tossed 16 times is

$$\sigma = \sqrt{16 \cdot \frac{1}{2} \cdot \frac{1}{2}} = \sqrt{4} = 2.$$

EXAMPLE 1

A fair coin is tossed 4 times.

(a) Construct a table with each value of the binomial variable x, where x is the number of heads resulting, and the probability of each value of x.

(b) Graph the distribution.

(c) Find the mean of this binomial distribution.

(d) Find the standard deviation of this distribution.

Solution

(a)

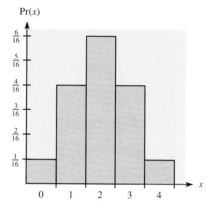

Figure 8.4

(b) The graph of the distribution is shown in Figure 8.4.

(c) The mean of this distribution is

$$\mu = np = 4 \cdot \frac{1}{2} = 2.$$

(d) The standard deviation of this binomial distribution is

$$\sigma = \sqrt{npq} = \sqrt{4 \cdot \frac{1}{2} \cdot \frac{1}{2}} = 1.$$

As mentioned in the previous section, the mean of a distribution is also the expected value of the distribution. We use this information to solve the problem posed in the Application Preview.

EXAMPLE 2

Latex condoms are reported to be 95% effective in preventing the transmission of disease (such as HIV) and pregnancy. Find the expected number of failures per thousand condom uses.

Solution

In this case, we seek the expected number of failures where the probability of failure in each case is $1 - .95 = .05$. We find the expected number of failures by using $E(x) = \mu = np$, where $n = 1000$ and $p = .05$.

$$E(x) = \mu = 1000(.05) = 50$$

CHECKPOINT

2. A cube is colored so 4 faces are green and 2 are red. The cube is tossed 9 times. If success is when a red face is up, find
   (a) the mean and
   (b) the standard deviation of the number of red faces that occur.

Table 8.3

x	$Pr(x)$
3	$\binom{3}{3} p^3 q^0 = p^3$
2	$\binom{3}{2} p^2 q = 3p^2 q$
1	$\binom{3}{1} pq^2 = 3pq^2$
0	$\binom{3}{0} p^0 q^3 = q^3$

The binomial probability distribution is closely related to the powers of a binomial. For example, if a binomial experiment has 3 trials with probability of success p on each trial, then the binomial probability distribution is given by Table 8.3 with the values of x written from 3 to 0. Compare this to the expansion of $(p + q)^3$, which we introduced in Chapter 0.

$$(p + q)^3 = p^3 + 3p^2 q + 3pq^2 + q^3$$

The formula that we can use to expand a binomial $(a + b)$ to any positive integer power n is

Binomial Formula

$$(a + b)^n = \binom{n}{n} a^n + \binom{n}{n-1} a^{n-1} b + \binom{n}{n-2} a^{n-2} b^2 + \cdots$$
$$+ \binom{n}{2} a^2 b^{n-2} + \binom{n}{1} ab^{n-1} + \binom{n}{0} b^n$$

EXAMPLE 3

Expand $(x + y)^4$.

Solution

$$(x + y)^4 = \binom{4}{4} x^4 + \binom{4}{3} x^3 y + \binom{4}{2} x^2 y^2 + \binom{4}{1} xy^3 + \binom{4}{0} y^4$$
$$= x^4 + 4x^3 y + 6x^2 y^2 + 4xy^3 + y^4$$

Note that the coefficients in this expansion are related to the probabilities in Example 1.

EXAMPLE 4
(a) If a die is rolled 5 times, what is the probability that exactly 2 faces with
    4 on them will result?
(b) What is the coefficient of p^2q^3 in $(p + q)^5$?

Solution

(a) If getting a 4 is a success, then $\text{Pr(success)} = \dfrac{1}{6}$ and we seek the probabil-
    ity of 2 successes in 5 trials. Thus,

$$\text{Pr(two 4's)} = \binom{5}{2}\left(\frac{1}{6}\right)^2\left(\frac{5}{6}\right)^3 = 10 \cdot \frac{1}{36} \cdot \frac{125}{216} = \frac{625}{3888}.$$

(b) By the Binomial Formula, the coefficient of p^2q^3 is $\dbinom{5}{2} = 10$.

CHECKPOINT SOLUTIONS

1. Each value of the random variable x has a probability, with $0 \le \text{Pr}(x) \le 1$, and
   with $\sum \text{Pr}(x) = 1$. Thus the conditions of a probability distribution are satisfied.

2. $\text{Pr(red face)} = \dfrac{1}{3} = p$, so $q = \dfrac{2}{3}$ and $n = 9$.

   (a) $\mu = np = 9\left(\dfrac{1}{3}\right) = 3$

   (b) $\sigma = \sqrt{npq} = \sqrt{9\left(\dfrac{1}{3}\right)\left(\dfrac{2}{3}\right)} = \sqrt{2}$

EXERCISE 8.3

1. A die is rolled 3 times. If success is rolling a 5,
   (a) construct the binomial distribution that describes
       this experiment.
   (b) find the mean of this distribution.
   (c) find the standard deviation of this distribution.
2. A die is rolled 3 times. If success is rolling a number
   divisible by 3,
   (a) construct the binomial distribution that describes
       this experiment.
   (b) write the mean of this distribution.
   (c) write the standard deviation of this distribution.

3. A variable x has a binomial distribution with probability
   of sucess 0.7 for each trial and a total of 60 trials.
   What is
   (a) the mean and
   (b) the standard deviation of the distribution?
4. A variable x has a binomial distribution with probability
   0.8 for each trial and a total of 40 trials. What is
   (a) the mean and
   (b) the standard deviation of the distribution?
5. Suppose a fair die is tossed 12 times. What is the mean
   and standard deviation of the number of 5's that occur?

6. Suppose a fair coin is tossed 8 times. What is the mean and standard deviation of the number of tails?

7. Suppose the probability that a certain coin will result in heads is 2/3 when it is tossed. If it is tossed 6 times, how many heads would we expect?

8. A coin is loaded so that the probability of tossing a head is 3/5. If it is tossed 50 times, what is
   (a) the mean and
   (b) the standard deviation of the number of heads that occur?

9. Suppose a pair of dice is thrown 12 times. How many times would we expect a sum of 7 to occur?

10. Suppose a pair of dice is thrown 20 times. How many times would we expect a sum of 6 to occur?

11. Suppose that a die is "loaded" so that the probability of getting each even number is 1/4 and the probability of getting each odd number is 1/12. If it is rolled 24 times, how many times would we expect to get a 3?

12. What is the coefficient of a^3b^5 in the expansion of $(a + b)^8$?

13. What is the coefficient of the term containing a^4b^2 in $(a + b)^6$?

14. (a) What is the coefficient of the term containing a^2b^4 in $(a + b)^6$?
    (b) Does $\binom{6}{4} = \binom{6}{2}$?

15. Expand $(a + b)^6$.

16. Expand $(x + y)^5$.

17. Expand $(x + h)^4$.

18. Write the general expression for $(x + h)^n$.

APPLICATION

19. **Health care** Suppose that 10% of the patients having a certain disease will die from it.
    (a) If 100 people have the disease, how many would we expect to die from it?
    (b) What is the standard deviation of the number of deaths that could occur?

20. **Cancer research** Suppose it has been determined that the probability that a rat injected with cancerous cells will live is 0.6. If 35 rats are injected, how many would be expected to die?

21. **Voting** A candidate claims 60% of the people in his district will vote for him.
    (a) If his district contains 100,000 voters, how many votes does he expect to get from his district?
    (b) What is the standard deviation of the number of his votes?

22. **Genetics** In a family with 2 children, the probability that both children will be boys is 1/4.
    (a) If 1200 families with two children are selected at random, how many of the families would we expect to have 2 boys?
    (b) What is the standard deviation for the number of families with 2 boys?

23. **Voting** If the number of votes the candidate in problem 21 gets is 2 standard deviations below what he expected, how many votes did he get (approximately)?

24. **Genetics** If the sample drawn in problem 22 had 315 families with two boys in it, how many standard deviations above the mean is the number of families?

25. **Testing** A multiple-choice test has 20 questions and 5 choices for each question. If a student is totally unprepared and guesses on each question, the mean is the number of questions she can expect to answer correctly.
    (a) How many questions can she expect to answer correctly?
    (b) What is the standard deviation of the distribution?

26. **Birth control** Suppose a birth control pill is 99% effective in preventing pregnancy. What number of women would be expected to get pregnant out of 100 women using this pill?

27. **Quality control** A certain calculator circuit board is manufactured in lots of 200. If 1% of the boards are defective, find the mean and standard deviation for the number of defects in each lot.

28. **CPA exam** Forty-eight percent of accountants taking exams to become a CPA fail the first time. If 1000 candidates take the exams for the first time, what is the expected number that will pass?

29. **Seed germination** A certain type of corn seed has an 85% germination rate. If 2000 seeds are planted, what is the expected number of seeds that would not germinate?

30. **Quality control** The probability that a manufacturer produces a defective medical thermometer is 0.001. What is the expected number of defective thermometers in a shipment of 3000?

8.4 DESCRIPTIVE STATISTICS

Objectives

❑ To set up frequency tables and construct frequency histograms for sets of data
❑ To find the mode of a set of scores (numbers)
❑ To find the median of a set of scores
❑ To find the mean of a set of scores
❑ To find the range of a set of data
❑ To find the variance and standard deviation

APPLICATION PREVIEW

In modern business, a vast amount of data is collected for use in making decisions about the production, distribution, and sale of merchandise. Businesses also collect and summarize data about advertising effectiveness, production costs, sales, wages, and profits. Behavioral scientists attempt to reach conclusions about general behavioral characteristics by studying the characteristics of a small sample of people. For example, election predictions are based on a careful sampling of the votes; correct predictions are frequently announced on television even though only 5% of the vote has been counted. Life scientists can use statistical methods with laboratory animals to detect substances that may be dangerous to humans.

For example, the table below gives the net income (or loss) in millions of dollars for the 11-year period 1983–1993 for the USF&G Corporation. This data can be summarized by using descriptive statistics. We will see in this section how to display this data with a histogram and how to summarize the data by finding the mean and standard deviation of the earnings over the 11-year period.

	1983	1984	1985	1986	1987	1988	1989	1990	1991	1992	1993
Net income	270	−64	−108	296	279	247	119	−569	−176	28	165

SOURCE: USF&G *1993 Annual Report*

A set of data taken from some source of data for the purpose of learning about the source is called a **sample;** the source of the data is called a **population.** The branch of statistics that deals with the collection, organization, summarization, and presentation of sample data is called **descriptive statistics.** The descriptive statistics methods discussed in this section include graphical representation of data, measures of central tendency, and measures of dispersion.

The first step in statistical work is the collection of data. Data are regarded as statistics data if the numbers collected have some relationship. The heights of all female first-year students are statistical data because a definite relationship exists among the collected numbers.

A *graph* frequently is used to provide a picture of the statistical data that have been gathered and to show relationships among various quantities. The advantage of the graph for showing data is that a person can quickly get an idea of the relationships that exist. The disadvantage is that the data can be shown only approximately on a graph. One type of graph that is used frequently in statistical work is the **frequency histogram.** A histogram is really a bar graph. It is constructed by putting the scores (numbers collected) along the horizontal axis and the frequency with which they occur along the vertical axis. A frequency table is usually set up to prepare the data for the histogram.

EXAMPLE 1

Construct a frequency histogram for the following scores: 38, 37, 36, 40, 35, 40, 38, 37, 36, 37, 39, 38.

Solution

We first construct the frequency table, and then use the table to construct the histogram shown in Figure 8.5.

Score	Frequency
35	1
36	2
37	3
38	3
39	1
40	2

Figure 8.5

Graphing calculators and computers can be used to plot the histogram of a set of data. The width of each bar of the histogram can be determined by the calculator or adjusted by the user when plotting histograms.

EXAMPLE 2

Construct a histogram from the following frequency table.

Interval	Frequency
0–5	0
6–10	2
11–15	5
16–20	1
21–25	3

Solution

The histogram is shown in Figure 8.6.

Figure 8.6

A set of data can be described by listing all the scores or by drawing a frequency histogram. But we often wish to discribe a set of scores by giving the *average score* or the typical score in the set. There are three types of measures that are called *averages,* or measures of central tendency. They are the **mode,** the **median,** and the **mean.**

To determine what value represents the most "typical" score in a set of scores, we use the **mode** of the scores.

> The **mode** of a set of scores is the score that occurs most frequently.

That is, the mode is the most popular score, the one that is most likely to occur. The mode can be readily determined from a frequency table or frequency histogram because it is determined according to the frequency of the scores.

EXAMPLE 3
Find the mode of the following scores: 10, 4, 3, 6, 4, 2, 3, 4, 5, 6, 8, 10, 2, 1, 4, 3.

Solution
The mode is 4, since it occurs more frequently than any other score.

EXAMPLE 4
Determine the mode of the scores shown in the histogram in Figure 8.7.

Solution
The most frequent score, as observed from the histogram, is 2.

Figure 8.7

EXAMPLE 5
Find the mode of the following scores: 4, 3, 2, 4, 6, 5, 5, 7, 6, 5, 7, 3, 1, 7, 2.

Solution
Both 5 and 7 occur three times. Thus they are both modes. This set of data is said to be *bimodal* because it has two modes.

Although the mode of a set of scores tells us the most frequent score, it does not always represent the typical performance, or central tendency, of the set of scores. For example, the scores 2, 3, 3, 3, 5, 8, 9, 10, 13 have a mode of 3. But 3 is not a good measure of central tendency for this set of scores, for it is nowhere near the middle of the distribution.

One value that gives a better measure of central tendency is the **median.**

> The **median** is the score or point above and below which an equal number of the scores lie.

It is the middle of the distribution of scores. If there is an odd number of scores, we find the median by ranking the scores from smallest to largest and picking the middle score. That score will be the median.

EXAMPLE 6
Find the median of the following scores: 2, 3, 16, 5, 15, 38, 18, 17, 12.

Solution
We rank the scores from smallest to largest: 2, 3, 5, 12, 15, 16, 17, 18, 38. The median is the middle score, which is 15. Note that there are 4 scores above and 4 below 15.

If the number of scores is even, there will be no middle score when the scores are ranked, so the *median* is the point that is the *average of the two middle scores.*

EXAMPLE 7
Find the median of the following scores: 3, 2, 6, 8, 12, 4, 3, 2, 1, 6.

Solution
We rank the scores from smallest to largest: 1, 2, 2, 3, 3, 4, 6, 6, 8, 12. The two scores that lie in the middle of the distribution are 3 and 4. Thus the median is a point that is midway between 3 and 4; that is, it is the arithmetic average of 3 and 4. The median is

$$\frac{3+4}{2} = \frac{7}{2} = 3.5.$$

In this case we say the median is a point because there is no *score* that is 3.5.

Some calculators have keys or functions for finding the median of a set of data. Such calculators, or computers, are valuable when the set of data is large.

The median is the most easily interpreted measure of central tendency, and it is the best indicator of central tendency when the set of scores contains a few extreme values. However, the most frequently used measure of central tendency is the **mean.**

The symbol $\bar{x}$ is used to represent the **mean** of a set of scores in a sample. Thus if n is the number of scores, the formula for the mean is

$$\bar{x} = \frac{\Sigma x}{n} = \frac{\text{sum of scores}}{\text{number of scores}}.$$

The mean is used most often as a measure of central tendency because it is more useful than the median in the general applications of statistics.

Scientific calculators and graphing calculators have keys or built-in functions that compute the mean of a set of data.

EXAMPLE 8

Find the mean of the following sample of numbers: 12, 8, 7, 10, 6, 14, 7, 6, 12, 9.

Solution

$$\bar{x} = \frac{\Sigma x}{10} = \frac{12 + 8 + 7 + 10 + 6 + 14 + 7 + 6 + 12 + 9}{10}$$

So $\bar{x} = 91/10 = 9.1$.

Note that the mean need not be one of the numbers (scores) given.

EXAMPLE 9

Find the mean of the following numbers: 2.1, 6.3, 7.1, 4.8, 3.2.

Solution

$$\bar{x} = \frac{\Sigma x}{5} = \frac{2.1 + 6.3 + 7.1 + 4.8 + 3.2}{5}$$

So $\bar{x} = 23.5/5 = 4.7$.

If we are given data in this frequency table,

x	1	2	3	4
f	3	1	4	2

we can find the mean $\bar{x}$ by using

$$\bar{x} = \frac{\Sigma x}{n},$$

where 1 is added 3 times, 2 is added once, 3 is added 4 times, and 4 is added twice. Or we can use $\Sigma(x \cdot f)$ and $n = \Sigma f$, so

$$\bar{x} = \frac{\Sigma(x \cdot f)}{\Sigma f},$$

where each value of x is used once.

The mean for the frequency table above is

$$\bar{x} = \frac{1 \cdot 3 + 2 \cdot 1 + 3 \cdot 4 + 4 \cdot 2}{3 + 1 + 4 + 2} = 2.5.$$

EXAMPLE 10

Find the mean of the following sample of test scores for a math class.

Scores	Class marks	Frequencies
40-49	44.5	2
50-59	54.5	0
60-69	64.5	6
70-79	74.5	12
80-89	84.5	8
90-99	94.5	2

Solution

For the purpose of finding the mean of interval data, we assume that all scores within an interval are represented by the class mark (midpoint of the interval). Thus

$$\bar{x} = \frac{(44.5)(2) + (54.5)(0) + (64.5)(6) + (74.5)(12) + (84.5)(8) + (94.5)(2)}{30}$$

$$= 74.5.$$

CHECKPOINT

Figure 8.8 shows the payroll of the Ace Cap Company.

1. Find the mean, median, and mode of the salaries.

2. Which of the three measures of central tendency gives the lowest value?

3. Which of these measures gives the highest value?

4. Which of these measures best represents the "average" salary for the company?

Payroll of Ace Cap Company

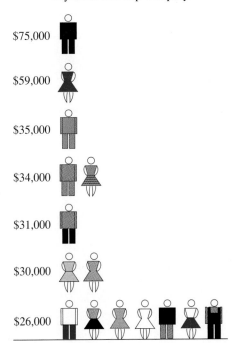

Figure 8.8

Although the mean of a set of data is useful in locating the center of the distribution of the data, it doesn't tell us as much about the distribution as we might think at first. For example, a basketball team with five 6-ft tall players is quite different from a team with one 6-ft, two 5-ft, and two 7-ft tall players. The difference in the distribution of the heights of these teams is not the mean height, but how the heights *vary* from the mean.

One measure of how a distribution varies is the **range** of the distribution.

The **range** of a set of numbers is the difference between the largest and smallest numbers in the set.

Consider the following set of heights of players on a basketball team, in inches: 69, 70, 75, 69, 73, 78, 74, 73, 78, 71. The range of this set of heights is $78 - 69 = 9$ inches. Note that this range is determined by only two numbers, and does not give any information about how the other heights vary.

If we calculated the deviation of each score from the mean, $(x - \bar{x})$, and then attempted to average the deviations, we would see that the deviations add to 0. In an effort to find a meaningful measure of dispersion about the mean, statisticians developed the **variance,** which squares the deviations (to make them all positive) and then averages the squared deviations.

Frequently we do not have all the data for a population and must use a sample of these data. To estimate the variance of a population from a sample, statisticians compensate for the fact that there is usually less variability in a sample than in the population itself by summing the squared deviations and dividing them by $n - 1$ rather than averaging them.

To get a measure comparable to the original deviations before they were squared, the **standard deviation,** which is the square root of the variance, was introduced. The formulas for the variance and standard deviation of sample data follow.

$$\text{Sample Variance} \quad s^2 = \frac{\Sigma(x - \bar{x})^2}{n - 1}$$

$$\text{Sample Standard Deviation} \quad s = \sqrt{\frac{\Sigma(x - \bar{x})^2}{n - 1}}$$

As with probability distributions, the standard deviation of a sample is a measure of the concentration of the scores about their mean. The smaller the standard deviation, the closer the scores lie to the mean. For example, the histograms in Figure 8.9 describe two sets of data with the same means and the same ranges. From looking at the histograms we see that more of the data are concentrated about the mean $\bar{x} = 4$ in Figure 8.9(b) than in Figure 8.9(a). We will see that the standard deviation is smaller for the data in Figure 8.9(b) than in Figure 8.9(a).

Figure 8.9

(a)

(b)

EXAMPLE 11
Figure 8.9(a) is the histogram for the sample data 1, 1, 1, 3, 3, 4, 4, 5, 6, 6, 7, 7. Figure 8.9(b) is the histogram for the data 1, 2, 3, 3, 4, 4, 4, 4, 5, 5, 6, 7. Both sets of data have a range of 6 and a mean of 4. Find the standard deviations of these samples.

Solution
For the sample data of Figure 8.9(a), we have

$$\Sigma(x - \bar{x})^2 = (-3)^2 + (-3)^2 + (-3)^2 + (-1)^2 + (-1)^2$$
$$+ 0^2 + 0^2 + 1^2 + 2^2 + 2^2 + 3^2 + 3^2$$
$$= 9 + 9 + 9 + 1 + 1 + 0 + 0 + 1$$
$$+ 4 + 4 + 9 + 9 = 56$$

$$s^2 = \frac{\Sigma(x - \bar{x})^2}{n - 1} = \frac{56}{11} = 5.0909$$

$$s = \sqrt{s^2} = \sqrt{5.0909} = 2.26.$$

For the sample data of Figure 8.9(b), we have

$$\Sigma(x - \bar{x})^2 = (-3)^2 + (-2)^2 + (-1)^2 + (-1)^2 + 0^2$$
$$+ 0^2 + 0^2 + 0^2 + 1^2 + 1^2 + 2^2 + 3^2$$
$$= 9 + 4 + 1 + 1 + 0 + 0 + 0 + 0$$
$$+ 1 + 1 + 4 + 9 = 30$$

$$s^2 = \frac{30}{11} = 2.7273$$

$$s = \sqrt{\frac{30}{11}} = 1.65.$$

Scientific calculators and graphing calculators have keys or built-in functions that compute the mean and standard deviation of a set of data. The ease of using calculators to find standard deviation makes it the preferred method.

EXAMPLE 12
The heights of the basketball team members mentioned earlier are 69, 70, 75, 69, 73, 78, 74, 73, 78, and 71 inches. Find the mean and standard deviation of this sample.

Solution
By using the formula or a calculator, the mean is 73 and the sample standard deviation is

$$s = 3.333.$$

We now return to the USF&G earnings discussion begun in the Application Preview.

EXAMPLE 13

The table below gives the net income (or loss) in millions of dollars for the 11-year period 1983–1993 for the USF&G Corporation.

(a) Construct a bar graph to represent earnings over this period.

(b) What is the mean earnings for this 11-year period?

(c) What is the standard deviation of the earnings over this period?

Year	Net Income
1983	270
1984	−64
1985	−108
1986	296
1987	279
1988	247
1989	119
1990	−569
1991	−176
1992	28
1993	165

SOURCE: USF&G *1993 Annual Report*

Solution

(a) The bar graph displaying the earnings over this period is shown in Figure 8.10.

(b) The mean earnings over this period is

$$\bar{x} = \frac{\Sigma x}{n} = 44.27$$

(c) The standard deviation of the earnings over this period is

$$s = 262.77.$$

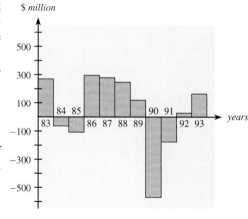

Figure 8.10

This shows that the net income fluctuates widely.

CHECKPOINT SOLUTIONS

1. We can see that the mode (most frequent) salary is $26,000. Also, the figure has the 15 employees arranged in order, so we see that the median (middle) salary is $30,000. The mean salary is

$$\bar{x} = \frac{\Sigma(x \cdot f)}{\Sigma f}$$

$$= \frac{75,000 + 59,000 + 35,000 + 2(34,000) + 31,000 + 2(30,000) + 7(26,000)}{15}$$

$$= \$34,000.$$

2. The mode gives the lowest salary for this company because most people receive the lowest salary.

3. The mean gives the highest measure of central tendency because a few high salaries are added to many low salaries.

4. The median is most representative of the "average" salary because it is not affected by the few very high salaries or by the large number of very low salaries.

EXERCISE 8.4

Construct a frequency histogram for the data given in the following frequency tables.

1.
Score	Frequency
12	2
13	3
14	4
15	3
16	1

2.
Score	Frequency
22	1
23	4
24	3
25	2
26	3

3.
x	3	4	5	6	7	8	9	10	11	12
$f(x)$	6	0	5	3	4	3	2	1	0	4

4.
x	5	6	8	9	11	12
$f(x)$	3	4	2	7	3	2

5.
Interval	Frequency
1–4	1
5–8	0
9–12	2
13–16	3
17–20	1

6.
Interval	Frequency
10–14	2
15–19	4
20–24	3
25–29	2
30–34	3

Construct a frequency histogram for the data in problems 7 and 8.

7. 3, 2, 5, 6, 3, 2, 6, 5, 4, 2, 1, 6
8. 5, 4, 6, 5, 4, 6, 3, 6, 5, 4, 1, 7

Find the modes of the following sets of scores.

9. 3, 4, 3, 2, 2, 3, 5, 7, 6, 2, 3
10. 5, 8, 10, 12, 5, 4, 6, 3, 5
11. 14, 17, 13, 16, 15, 12, 13, 12, 13

Find the medians of the following sets of scores.

12. 1, 3, 6, 7, 5
13. 2, 1, 3, 4, 2, 1, 2
14. 4, 7, 9, 18, 36, 14, 12
15. 1, 0, 2, 1, 1, 0, 2, 14, 37

Find the mode, median, and mean of the following scores.

16. 5, 7, 7, 7, 9, 12, 4
17. 3, 2, 1, 6, 8, 12, 14, 2
18. 5, 8, 7, 6, 1, 1, 31
19. 14, 17, 20, 31, 17, 42
20. 3.2, 3.2, 3.5, 3.7, 3.4
21. 2.8, 6.4, 5.3, 5.3, 6.8
22. 1.14, 2.28, 7.58, 6.32, 5.17

In problems 23–26, find the range of the set of numbers given.

23. 3, 5, 7, 8, 2, 11, 6, 5
24. 5, 1, 3, 1.4, 6.3, 8
25. −1, 2, 4, 3, 6, 11, −3, 4, 2
26. 2, 3, 3, 3, 3, 3, 3, 9

Find the mean, variance, and standard deviation of the following sets of sample data.

27. 5, 7, 1, 3, 0, 8, 6, 2
28. 7, 13, 5, 11, 8, 10, 9
29. 11, 12, 13, 14, 15, 16, 17
30. 4, 3, 6, 7, 8, 9, 12

In problems 31–34, find the mean and standard deviation of the sample data in the table.

31.
x	1	2	3	4	5
$f(x)$	3	1	4	2	1

32.
x	3	4	5	6	8
$f(x)$	1	3	4	1	3

33.
x	3	4	5	6	7	8	9	10	11	12
$f(x)$	6	0	5	3	4	3	2	1	0	4

34.
x	5	6	7	8	9	10	11	12	13	14	15
$f(x)$	3	2	1	0	2	5	3	1	2	4	6

APPLICATIONS

35. **Death penalty** The data in the table that follows give the total number of prisoners executed in the United States from 1950 to 1979 and the number of black prisoners executed during the same period. Construct a histogram

with x as the total number of executions (in groups of 10) and with $f(x)$ as the number of years in each group.

Year	Total	Blacks	Year	Total	Blacks
1950	82	42	1965	7	1
1951	105	47	1966	1	0
1952	83	47	1967	2	1
1953	62	31	1968	0	0
1954	81	42	1969	0	0
1955	76	32	1970	0	0
1956	65	43	1971	0	0
1957	65	31	1972	0	0
1958	49	28	1973	0	0
1959	49	33	1974	0	0
1960	56	35	1975	0	0
1961	42	22	1976	0	0
1962	47	19	1977	1	0
1963	21	8	1978	0	0
1964	15	7	1979	2	0

SOURCE: U.S. Department of Justice, Bureau of Statistics

36. **Salaries** Suppose a company has 10 employees, 1 earning $80,000, 1 earning $60,000, 2 earning $30,000, 1 earning $20,000, and 5 earning $16,000.
    (a) What is the mean salary for the company?
    (b) What is the median salary?
    (c) What is the mode of the salaries?

37. **Salaries** A survey revealed that of University of North Carolina alumni entering employment in 1984, those with a major in geography had the highest mean starting salary of all majors. Discuss this conclusion, using the fact that Michael Jordan was in this group and majored in geography.

Real estate Use the following information for problems 38–40. Suppose you live in a neighborhood with a few expensive homes and many modest homes.

38. If you wanted to impress people with the neighborhood where you lived, which measure would you give as the "average" property value?

39. Which "average" would you cite to the property tax committee if you wanted to convince them that property values aren't very high in your neighborhood?

40. What measure of central tendency would give the most representative "average" property value for your neighborhood? Explain.

41. **Salaries** S & S Printing has 15 employees and 5 job classifications. Position and wages for these employees are given in the table.

Job	Number	Weekly Salary
Supervisor	1	$800
Printers	8	320
Camera-Darkroom	3	500
Secretaries	2	360
Delivery	1	420

(a) What is the mean salary per job classification?
(b) What is the mean salary per person?

42. **Product testing** A taxi company tests a new brand of tires by putting a set of four on each of three taxis. The following table indicates the number of miles the tires lasted. What is the mean number of miles the tires lasted?

Number of Miles	Number of Tires
15,000	4
16,000	2
17,000	3
18,000	2
19,000	1

43. **Sales** A new car with a $19,000 list price can be bought for different prices from different dealers. In one city the car can be bought for $18,200 from 2 dealers, $18,000 from 1 dealer, $17,800 from 3 dealers, $17,600 from 2 dealers, and $17,500 from 2 dealers. What is the mean and standard deviation of this sample of car prices?

44. **Birth weights** The birth weights (in kilograms) of a sample of 160 children are given in the following table. Find the mean and standard deviation of the weights.

Weight (kg)	Frequency	Weight (kg)	Frequency
2.0	4	3.5	26
2.3	12	3.8	20
2.6	20	4.1	16
2.9	26	4.4	10
3.2	20	4.7	6

45. **Gold reserves** American Barrick Resources Corporation has North America's largest gold reserves, 30.6 million ounces. The table gives the operating statistics for Barrick for the years 1991–1993. Use an electronic spreadsheet or other technology to find the mean of each category over the 3-year period.

Year ended	1991	1992	1993
Gold production (thousand ounces)	789.8	1325.4	1632.0
Gold sales (thousand ounces)	787.7	1280.3	1633.7
Revenue per ounce ($)	438	422	409
Average spot price per ounce ($)	362	345	360
Operating costs per ounce ($)	205	164	170
Net income per ounce ($)	126	148	141

46. **Gold reserves** Use the gold production data for Barrick from the table in problem 45 to find the standard deviation of gold production for 1991–1993.

47. **Automotive testing** The acceleration time from 0 to 60 mph and the braking distance from 60 mph to a stop for five 1995 subcompact cars is given in Table 8.4. Use the data in the accompanying figure to find the mean and standard deviation of the braking distances of these cars.

Table 8.4

Acceleration and Braking – 1995 Subcompacts

	0-60 mph	Braking from 60 mph
Honda	8.8 sec	165 ft
Mazda	9.1 sec	158 ft
Nissan	11.0 sec	148 ft
Plymouth	8.4 sec	142 ft
Toyota	10.2 sec	186 ft

48. **Military presence** The figure shows the number of military troops in Asia for each of ten countries. Add the number of U.S. troops in Japan to the number of Japanese troops, and add the number of U.S. troops in South Korea to the number of South Korean troops. Then compute the mean number of troops in Asia per country. How many of the countries have more troops than the mean number per country?

Military Troops in Asia

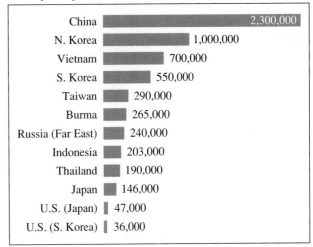

Country	Troops
China	2,300,000
N. Korea	1,000,000
Vietnam	700,000
S. Korea	550,000
Taiwan	290,000
Burma	265,000
Russia (Far East)	240,000
Indonesia	203,000
Thailand	190,000
Japan	146,000
U.S. (Japan)	47,000
U.S. (S. Korea)	36,000

SOURCES: Chicago Tribune, Japan Self-Defense Agency, "Defense of Japan, 1994"; Knight-Ridder Tribune/Ken Marshall and Rick Tuma; *Savannah Morning News*, Jan 5, 1995

49. **Automotive testing** Use Table 8.4 to find the mean and standard deviation of the acceleration time from 0 to 60 mph for the 5 cars.

8.5 NORMAL PROBABILITY DISTRIBUTION

Objectives

❑ To calculate the probability that a random variable following the standard normal distribution has values in a certain interval

❑ To convert normal distribution values to standard normal values (z-scores)

❑ To find the probability that normally distributed values lie in a certain interval

APPLICATION PREVIEW

Discrete probability distributions are important, but they do not apply to many kinds of measurements. For example, the weights of people, the heights of trees, and the IQ scores of college students cannot be measured with whole numbers because each of them can assume any one of an infinite number of values on a measuring scale, and the values cannot be listed or counted as is possible for a discrete random variable. The measurements mentioned above follow a special **continuous probability distribution** called the **normal distribution**.

The IQ scores of college students follow the normal distribution with a mean of 100 and a standard deviation of 15. We will see in this section that this information permits us to graph the distribution of the IQ scores and to find the probability that a person's IQ score lies within a given range of values.

The normal distribution is perhaps the most important probability distribution, because so many measurements occuring in nature follow this particular distribution.

The **normal distribution** has the following properties:

1. Its graph is a bell-shaped curve like that of Figure 8.11.* The graph is called the **normal curve.** It approaches but never touches the horizontal axis as it extends in both directions.

Figure 8.11

Normal distribution

2. The curve is symmetric about a vertical centerline which passes through the value that is the mean, the median, *and* the mode of the distribution. That is, the mean, median, and mode are the same for a normal distribution. (That is why they are all called average.)

3. A normal distribution is completely determined when its mean μ and its standard deviation σ are known.

4. Approximately 68% of all scores lie within one standard deviation of the mean. Approximately 95% of all scores lie within two standard deviations of the mean. More than 99.5% of all scores will lie within three standard deviations of the mean.

EXAMPLE I

If IQ scores follow a normal distribution with mean 100 and standard deviation 15, what percentage of the scores will be
(a) between 100 and 115?   (b) between 85 and 115?
(c) between 85 and 130?   (d) between 70 and 130?
(e) greater than 130?

*Percentages shown are approximate.

Solution
Figure 8.12 is a graph of the distribution, with approximate percentages shown. We can find the percentages from this graph.
(a) 34%   (b) 68%   (c) 81.5%   (d) 95%   (e) 2.5%

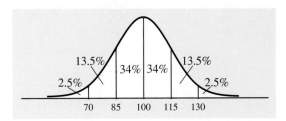

Figure 8.12

CHECKPOINT

1. If the mean of a normal distribution is 25 and the standard deviation is 3,
   (a) what percent of the scores will be between 22 and 28?
   (b) what is the probability that a score chosen at random will be between 22 and 28?

The total area under the normal curve is 1. The area under the curve from value x_1 to value x_2 represents the percentage of the scores that lie between x_1 and x_2. Thus the *area* under the curve from x_1 to x_2 *represents the probability* that a score chosen at random will lie between x_1 and x_2. Thus, in Example 1, the probability that a score chosen at random would lie between 85 and 115 is .68, and the probability that it would lie between 70 and 130 is .95.

We have seen that 34% of the scores lie between 100 and 115 in the normal distribution graph in Figure 8.12. As we have mentioned, this also means that the area under the curve from $x = 100$ to $x = 115$ is .34, and that the probability that a score chosen at random from this normal population lies between 100 and 115 is .34. We can write this as

$$\Pr(100 \le x \le 115) = .34.$$

Because the probability of obtaining a score from a normal distribution is always related to how many standard deviations the score is away from the mean, it is desirable to convert all scores from a normal distribution to **standard scores,** or **z-scores.** The z-score for any score x is found by determining how many standard deviations x is from the mean μ. This is done using the formula

$$z = \frac{x - \mu}{\sigma}.$$

If we convert all the scores from any normal distribution to z-scores, the distribution of z-scores will always be a normal distribution with mean 0 and standard deviation 1. This distribution is called the **standard normal**

distribution. Figure 8.13 shows the graph of the standard normal distribution, with approximate percentages shown. The total area under the curve is 1, with .5 on either side of the mean 0.

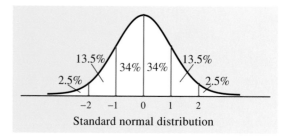

Figure 8.13

Standard normal distribution

By comparing Figure 8.13 with Figure 8.11, we see that each unit from 0 in the standard normal distribution corresponds to one standard deviation from the mean of the normal distribution.

We can use a table to more accurately determine the area under the standard normal curve between two z-scores. Table I in the appendix gives the area under the normal curve from $z = 0$ to $z = z_0$, for values of z_0 from 0 to 3. As with the normal curve, the area under the curve from $z = 0$ to $z = z_0$ is the probability that a z-score lies between 0 and z_0.

EXAMPLE 2

(a) Find the area under the standard normal curve from $z = 0$ to $z = 1.50$.
(b) Find $\Pr(0 \le z \le 1.50)$.

Solution

(a) Looking in the column headed by z in Table I, we see 1.50. Across from 1.50 in the column headed by A is .4332. Thus the area under the standard normal curve between $z = 0$ and $z = 1.50$ is $A_{1.50} = .4332$. (See Figure 8.14.)

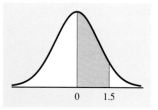

Figure 8.14

(b) The area under the curve from 0 to 1.50 equals the probability that z lies between 0 and 1.50. Thus

$$\Pr(0 \le z \le 1.50) = A_{1.50} = .4332.$$

The following facts, which are direct results of the symmetry of the normal curve about μ, are useful in calculating probabilities using Table I.

1. $\Pr(-z_0 \le z \le 0) = \Pr(0 \le z \le z_0) = A_{z_0}$
2. $\Pr(z \ge 0) = \Pr(z \le 0) = .5$

EXAMPLE 3

Find the following probabilities for the random variable z with standard normal distribution.

(a) $\Pr(-1 \leq z \leq 0)$ (b) $\Pr(-1 \leq z \leq 1.5)$ (c) $\Pr(1 \leq z \leq 1.5)$

(d) $\Pr(z > 2)$ (e) $\Pr(z < 1.35)$

Solution

(a) $\Pr(-1 \leq z \leq 0) = \Pr(0 \leq z \leq 1) = A_1 = .3413$

(b) We find $\Pr(-1 \leq z \leq 1.5)$ by using A_1 and $A_{1.5}$ as follows.

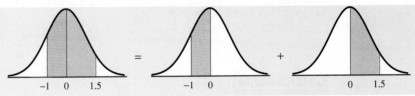

$$\begin{aligned} \Pr(-1 \leq z \leq 1.5) &= \Pr(-1 \leq z \leq 0) &&+ \Pr(0 \leq z \leq 1.5) \\ &= A_1 &&+ A_{1.5} \\ &= .3413 &&+ .4332 \\ &= .7745 \end{aligned}$$

(c) We find $\Pr(1 \leq z \leq 1.5)$ by using A_1 and $A_{1.5}$ as follows.

$$\begin{aligned} \Pr(1 \leq z \leq 1.5) &= \Pr(0 \leq z \leq 1.5) &&- \Pr(0 \leq z \leq 1) \\ &= A_{1.5} &&- A_1 \\ &= .4332 &&- .3413 \\ &= .0919 \end{aligned}$$

(d) We find $\Pr(z > 2)$ by using .5 and A_2 as follows.

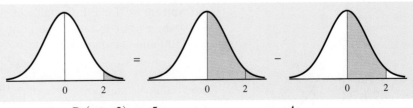

$$\begin{aligned} \Pr(z > 2) &= .5 &&- A_2 \\ &= .5 &&- .4772 \\ &= .0228 \end{aligned}$$

(e) We find $\Pr(z < 1.35)$ by using .5 and $A_{1.35}$ as follows.

$$\Pr(z < 1.35) = .5 \qquad + A_{1.35}$$
$$= .5 \qquad + .4115$$
$$= .9115$$

CHECKPOINT

2. For the standard normal distribution, find the following probabilities.
   (a) $\Pr(0 \le z \le 2.5)$
   (b) $\Pr(z > 2.5)$
   (c) $\Pr(z \le 2.5)$

If a population follows a normal distribution, but with mean and/or standard deviation different from 0 and 1, respectively, it is desirable to convert all scores from a normal distribution to **standard scores,** or **z-scores.** The z-score for any score x is found by determining how many standard deviations x is from the mean μ. This is done using the formula

$$z = \frac{x - \mu}{\sigma}.$$

Thus

$$\Pr(a \le x \le b) = \Pr(z_a \le z \le z_b),$$

where

$$z_a = \frac{a - \mu}{\sigma} \qquad \text{and} \qquad z_b = \frac{b - \mu}{\sigma},$$

and we can use Table I, the table of standard scores (z-scores) to find these probabilities. For example, the normal distribution of IQ scores has mean $\mu = 100$ and standard deviation $\sigma = 15$. Thus the z-score for $x = 115$ is

$$z = \frac{115 - 100}{15} = 1.$$

A z-score of 1 indicates that 115 is 1 standard deviation above the mean (see Figure 8.15), and

$$\Pr(100 \le x \le 115) = \Pr(0 \le z \le 1) = .3413.$$

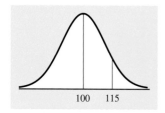

Figure 8.15

EXAMPLE 4

If the mean height of a population of students is $\mu = 68$ in. with standard deviation $\sigma = 3$ in., what is the probability that a person chosen at random from the population will be between
(a) 68 and 74 in. tall?   (b) 65 and 74 in. tall?

Solution

(a) To find $\Pr(68 \leq x \leq 74)$, we convert 68 and 74 to z-scores.

$$\text{For 68: } z = \frac{68 - 68}{3} = 0 \qquad \text{For 74: } z = \frac{74 - 68}{3} = 2$$

Thus

$$\Pr(68 \leq x \leq 74) = \Pr(0 \leq z \leq 2) = A_2 = .4772.$$

(b) The z-score for 65 is

$$z = \frac{65 - 68}{3} = -1,$$

and the z-score for 74 is 2. Thus

$$\Pr(65 \leq x \leq 74) = \Pr(-1 \leq z \leq 2)$$
$$= A_1 + A_2$$
$$= .3413 + .4772 = .8185.$$

CHECKPOINT

3. For the normal distribution with mean 70 and standard deviation 5, find the following probabilities.
   (a) $\Pr(70 \leq x \leq 77)$ (b) $\Pr(68 \leq x \leq 73)$ (c) $\Pr(x > 75)$

EXAMPLE 5

Recessionary pressures have forced an electrical supplies distributor to consider reducing its inventory. It has determined that, over the past year, the daily demand for 48-in. fluorescent light fixtures is normally distributed with a mean of 432 and a standard deviation of 86. If the distributor decides to reduce its daily inventory to 500 units, what percentage of the time will its inventory be insufficient to meet the demand? Will this cost the distributor business?

Solution

The distribution is normal, with $\mu = 432$ and $\sigma = 86$. The probability that more than 500 units will be demanded is

$$\Pr(x > 500) = \Pr\left(z > \frac{500 - 432}{86}\right) = \Pr(z > .79)$$
$$= .5 - A_{.79} = .5 - .2852 = .2148.$$

The distributor will have an insufficient inventory 21.48% of the time. This will probably result in a loss of customers, so the distributor should not decrease the inventory to 500 units per day.

EXAMPLE 6

The function

$$f(x) = \frac{1}{\sigma\sqrt{2\pi}}e^{-(x-\mu)^2/(2\sigma^2)}$$

gives the equation for a normal curve with μ and standard deviation σ.

(a) Graph the following normal curves to see how different means yield different distributions:

$$\mu = 0, \sigma = 1, y = \frac{1}{\sqrt{2\pi}}e^{-x^2/2}; \quad \text{and}$$

$$\mu = 3, \sigma = 1, y = \frac{1}{\sqrt{2\pi}}e^{-(x-3)^2/2}.$$

If the standard deviations are the same, how does increasing the size of the mean affect the graph of the distribution?

(b) To see how normal distributions differ for different standard deviations, graph the normal distributions with

$$\mu = 0, \sigma = 3; \quad \text{and}$$
$$\mu = 0, \sigma = 2.$$

What happens to the height of the curve as the standard deviation decreases? Do the data appear to be clustered nearer the mean when the standard deviation is smaller?

Solution

(a) The graphs are shown in Figure 8.16. The graphs are the same size but are centered over different means. Different means cause the graphs to shift to the right or left.

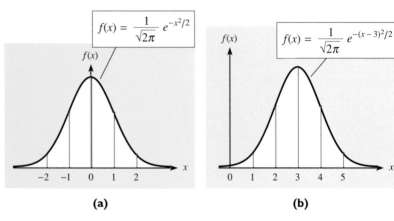

Figure 8.16

(a)   (b)

(b) The graphs are shown in Figure 8.17. The height of the graph will be larger if the standard deviation decreases. Yes, the data will be clustered closer to the mean when the standard deviation is smaller.

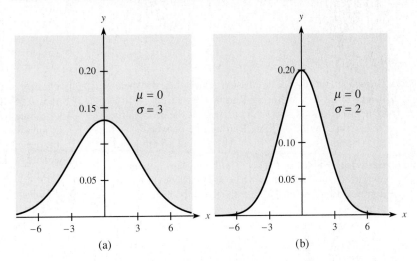

Figure 8.17

(a) (b)

CHECKPOINT SOLUTIONS

1. (a) These scores lie within one standard deviation of the mean, 25, so 68% lie in this range.
   (b) The percent of the scores in this range is 68%, so the probability that a score chosen at random from this normal disribution will be in this range is .68.

2. (a) $\Pr(0 \le z \le 2.5) = A_{2.5} = .4938$
   (b) $\Pr(z > 2.5) = .5 - A_{2.5} = .5 - .4938 = .0062$
   (c) $\Pr(z \le 2.5) = .5 + A_{2.5} = .5 + .4938 = .9938$

3. (a) $\Pr(70 \le x \le 77) = \Pr(0 \le z \le 1.4) = A_{1.4} = .4192$
   (b) $\Pr(68 \le x \le 73) = \Pr(-.4 \le z \le .6) = A_{.4} + A_{.6}$
      $= .1554 + .2258 = .3812$
   (c) $\Pr(x > 75) = \Pr(z > 1) = .5 - A_1 = .5 - .3413 = .1587$

EXERCISE 8.5

Use Table I in the appendix to find the probability that a z-score from the standard normal distribution will lie within each of the following intervals.

1. 0 and 1.8 2. 0 and 2.4 3. −1.8 and 0
4. −3 and 0 5. −1.5 and 2.1 6. −1.25 and 3
7. −1.9 and −1.1 8. −2.45 and −1.45
9. 2.1 and 3.0 10. 1.85 and 2.85
11. $z > 2$ 12. $z > 1.5$
13. $z < 1.2$ 14. $z > -1.6$

If a population of scores x is normally distributed with $\mu = 20$ and $\sigma = 5$, use the standard normal distribution to find the following.

15. $\Pr(20 \le x \le 22.5)$
16. $\Pr(20 \le x \le 21.25)$
17. $\Pr(13.75 \le x \le 20)$
18. $\Pr(18.5 \le x \le 20)$
19. $\Pr(12.25 \le x \le 21.25)$
20. $\Pr(18.5 \le x \le 27.75)$

21. $Pr(13.75 \leq x \leq 18.75)$
22. $Pr(21.25 \leq x \leq 25)$
23. $Pr(x < 18)$
24. $Pr(x < 14)$
25. $Pr(x > 17)$
26. $Pr(x < 24)$

APPLICATIONS

27. **Growth** If the Fish Commission states that the mean length of all fish in Spring Run is $\mu = 15$ cm, with a standard deviation of $\sigma = 4$ cm, what is the probability that a fish caught in Spring Run will be between
    (a) 15 and 19 cm long?   (b) 10 and 15 cm long?

28. **Butterfat** A quart of Parker's milk contains a mean of 39 grams of butterfat, with a standard deviation of 2 grams. Find the probability that a quart of this brand of milk chosen at random will contain between
    (a) 39 and 43 grams of butterfat.
    (b) 36 and 39 grams of butterfat.

29. **Growth** The heights of a certain species of plant are normally distributed with a mean $\mu = 20$ cm and $\sigma = 4$ cm. What is the probability that a plant chosen at random will be between 10 and 30 cm tall?

30. **Mating calls** The mean duration of the mating call of a population of tree toads is 189 msec, with a standard deviation of 32 msec. What proportion of these calls would be expected to last between
    (a) 157 and 221 msec?   (b) 205 and 253 msec?

31. **Growth** The mean weight of a group of boys is 160 lb, with a standard deviation of 15 lb. Find the probability that one of the boys picked at random from the group weighs
    (a) between 160 and 181 lb.
    (b) more than 190 lb.
    (c) between 181 and 190 lb.
    (d) between 130 and 181 lb.

32. **Testing** The Scholastic Aptitude Test (SAT) scores in mathematics at a certain high school are normally distributed with a mean of 500 and a standard deviation of 100. What is the probability that an individual chosen at random has a score
    (a) greater than 700?
    (b) less than 300?
    (c) between 550 and 600?

33. **Mileage** A certain model of automobile has its gas mileage (in miles per gallon, or mpg) normally distributed with a mean of 28 mpg and a standard deviation of 4 mpg. Find the probability that a car selected at random has gas mileage
    (a) less than 22 mpg.
    (b) greater than 30 mpg.
    (c) between 26 and 30 mpg.

34. **Quality control** A machine precision cuts tubing such that $\mu = 40.00$ inches and $\sigma = 0.03$ inches. If a piece of tubing is selected at random, find the probability that it measures
    (a) greater than 40.05 inches.
    (b) less than 40.045 inches.
    (c) between 39.95 and 40.05 inches.

35. **Blood pressure** Systolic blood pressure for a group of women is normally distributed with a mean of 120 and a standard deviation of 12. Find the probability that a woman selected at random has blood pressure
    (a) greater than 140.
    (b) less than 110.
    (c) between 110 and 130.

36. **Industrial waste** In a certain state, the daily amounts of industrial waste are normally distributed with a mean of 8000 tons and a standard deviation of 2000 tons. Find the probability that on a randomly selected day the amount of industrial waste is
    (a) greater than 11,000 tons.
    (b) less than 6000 tons.
    (c) between 9000 and 11,000 tons.

37. **Reaction time** Reaction time is normally distributed with a mean of 0.7 seconds and a standard deviation of 0.1 seconds. Find the probability that an individual selected at random has a reaction time
    (a) greater than 0.9 seconds.
    (b) less than 0.6 seconds.
    (c) between 0.6 and 0.9 seconds.

38. **Inventory** Suppose that the electrical supplies distributor in Example 5 wants to reduce its daily inventory, but wants to limit to 5% the number of days that it will have an insufficient inventory of 48-in. fluorescent light bulbs to meet daily demand. At what level should it maintain its daily inventory? (Recall that the daily demand over the past year had a mean of 432 and a standard deviation of 86; test values of 550, 575, and 600 units.)

8.6 LINEAR REGRESSION AND CORRELATION

Objectives

❑ To write the equation of the line that is the best fit for a set of data points
❑ To compute the correlation coefficient for a set of data points

APPLICATION PREVIEW

The data in Table 8.5 give the amount of federal tax paid per capita for selected years.

We can use **linear regression** to find the equation of the linear function that best fits this data, and we can use **correlation** to see how well the data fit the linear function we have found.

Table 8.5

Year	Federal Tax (per Capita)
1983	$2500
1984	2750
1985	3000
1986	3100
1987	3400
1988	3500
1989	3800
1990	4000
1991	4000
1992	4100
1993	4300
1994	4700 (estimated)

SOURCE: Internal Revenue Service

We have seen in Section 1.7 that it is possible to write the equation of a straight line if we have two points on the line. Business firms frequently like to treat demand functions as though they are linear, even when they are not exactly linear. They do this because linear functions are much easier to handle than other functions. If a firm has more than two points describing the demand for its product, it is likely that the points will not all lie on the same straight line. However, by using a technique called **linear regression** the firm can determine the "best line" that fits these points.

Suppose we have the points shown in Figure 8.18. We seek the line that is the "best fit" for the points. We can eyeball a line that appears to lie along the points; we now find the equation of the line that fits best.

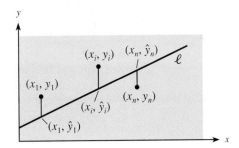

Figure 8.18

Corresponding to each point (x_1, y_1) that we we know, we can find a point $(x_1, \hat{y}_1)$ on the line. If the line we have is a good fit for the points, the differences between the y values of the given points and the corresponding $\hat{y}$ values on the line should be small. The line shown is the best line for the points only if the sum of the squares of these differences is a minimum.*

Because we can write the equation of a line as $y = mx + b$, we can find the equation of the line that best fits n given points by assuming the points lie

---

*Because linear regression involves minimizing the sum of squared numbers, it is also called the **least squares method.**

on or near a line with equation

$$\hat{y} = ax + b.$$

For each value of x, we get a value $\hat{y}$, which lies on the line and is an estimate of y. We seek the numbers a and b such that the sum of the squares of the differences,

$$\Sigma(y - \hat{y})^2,$$

is a minimum. If we can find a and b, then we have the equation of the line that describes the data. Using this equation, we can estimate unknown y values for other values of x.

Linear Regression Equation Using calculus, we can show that $\Sigma(y - \hat{y})^2$ is a *minimum, and thus* $\hat{y} = ax + b$ is the equation of the line that is the best fit, if

$$a = \frac{\Sigma x \cdot \Sigma y - n\Sigma xy}{(\Sigma x)^2 - n\Sigma x^2}$$

and

$$b = \frac{\Sigma y - a\Sigma x}{n}.$$

EXAMPLE 1

Accountants frequently use linear regression (least squares) to determine a linear equation for their total costs. If the following table gives the total cost for a product at different levels of production, find the best linear equation for the total cost function.

x (units)	100	200	300	400
C (total cost)	2400	4100	6000	8500

Solution

We seek the values a and b such that $C = ax + b$ is the equation that best fits the points. To find a and b, we first find

$$\Sigma x = 100 + 200 + 300 + 400 = 1000$$

$$\Sigma x^2 = 10{,}000 + 40{,}000 + 90{,}000 + 160{,}000 = 300{,}000$$

$$\Sigma y = 2400 + 4100 + 6000 + 8500 = 21{,}000$$

$$\Sigma xy = 240{,}000 + 820{,}000 + 1{,}800{,}000 + 3{,}400{,}000 = 6{,}260{,}000.$$

Then

$$a = \frac{1000(21,000) - 4(6,260,000)}{1000^2 - 4(300,000)} = \frac{21,000,000 - 25,040,000}{1,000,000 - 1,200,000}$$

$$= 20.2$$

$$b = \frac{21,000 - (20.2)(1000)}{4} = \frac{800}{4} = 200.$$

Thus the linear equation that best predicts total cost is

$$C = 20.2x + 200.$$

Figure 8.19 shows the points and the line.

This equation gives (approximate) values for total cost at different levels of production. For example, at $x = 500$, total cost should be approximately

$$C = 20.2(500) + 200 = 10,300,$$

and at $x = 250$, it is approximately

$$C = 20.2(250) + 200 = 5250.$$

Figure 8.19

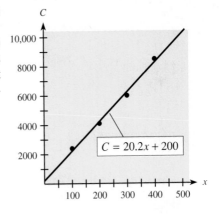

CHECKPOINT

1. Use linear regression to write the equation of the line that is the best fit for the points.

x	50	25	10	5
y	2	4	10	20

Linear regression gives us the equation of the line that is the best fit for our data, but it does not tell us how closely our data fit the line. One measure of goodness of fit is the correlation coefficient, denoted by r.

The **correlation coefficient** r is found using

$$r = \frac{n(\Sigma xy) - (\Sigma x)(\Sigma y)}{\sqrt{n(\Sigma x^2) - (\Sigma x)^2} \; \sqrt{n(\Sigma y^2) - (\Sigma y)^2}}$$

Although this formula looks very complicated, it is not difficult to use if we have already found the regression line. We have already calculated Σxy, Σx,

r close to +1

r close to −1

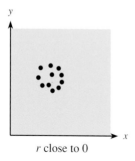

r close to 0

Figure 8.20

Σy and Σx^2 in finding the regression line for Example 1. Thus we need only compute Σy^2 and substitute these values (with $n = 4$) into the formula for r. Since the values are $\Sigma xy = 6{,}260{,}000$, $\Sigma x = 1000$, $\Sigma y = 21{,}000$, $\Sigma x^2 = 300{,}000$, and $\Sigma y^2 = 130{,}820{,}000$,

$$r = \frac{4(6{,}260{,}000) - (1000)(21{,}000)}{\sqrt{4(300{,}000) - (1000)^2}\;\sqrt{4(130{,}820{,}000) - (21{,}000)^2}}$$

$$= \frac{25{,}040{,}000 - 21{,}000{,}000}{\sqrt{1{,}200{,}000 - 1{,}000{,}000}\;\sqrt{523{,}280{,}000 - 441{,}000{,}000}}$$

$$= \frac{4{,}040{,}000}{\sqrt{200{,}000}\;\sqrt{82{,}280{,}000}} = \frac{4{,}040{,}000}{(447.21)(9070.83)}$$

$$= .9959.$$

The formula for the correlation coefficient, developed by Karl Pearson, is designed so that $-1 \le r \le 1$, with a value of r close to 1 meaning the two variables will increase or decrease together, and there is a strong mathematical relationship between them. It does not necessarily mean that one of the variables has a direct effect on the other. For example, there is a high correlation between the increase in the number of ministers in a certain area and the sale of liquor in that area. This does not mean the ministers are drinking the liquor; both increases reflect an increase in population in that area.

A correlation coefficient close to −1 indicates that there is a strong negative correlation; that is, one variable will tend to decrease as the other one increases. It is generally agreed that correlations between −0.2 and 0.2 indicate no significant relation between the variables. (See Figure 8.20.)

Graphing Utilities

Most scientific and graphing calculators can find the linear regression equation for a set of data. The data are entered and the calculator automatically calculates the slope, a, and the y-intercept, b, of the linear regression equation, $\hat{y} = ax + b$.* Graphing calculators (and others that do linear regression) also calculate the correlation coefficient, r.

We now find the linear equation that is the best fit for the data given in the Application Preview.

EXAMPLE 2

The data in Table 8.6 give the amount of federal tax paid per capita for selected years.
(a) Use a calculator (or other techonology) to find the linear regression equation and the correlation coefficient.

---

* Some graphing calculators give the regression equation in the form $\hat{y} = a + bx$.

(b) Use a graphing utility to graph the regression line and the data points on the same set of axes.

Table 8.6

Year	Federal Tax (per Capita)	Year	Federal Tax (per Capita)	Year	Federal Tax (per Capita)
1983	$2500	1987	3400	1991	4000
1984	2750	1988	3500	1992	4100
1985	3000	1989	3800	1993	4300
1986	3100	1990	4000	1994	4700 (estimated)

SOURCE: Internal Revenue Service

Solution

(a) Let x be the number of years past 1980 and y be the federal tax in dollars. Then the first line of data in the table corresponds to the point (3, 2500), and the last line corresponds to (14, 4700). We enter the data into the calculator and activate the linear regression routine to obtain the following result.

$$\hat{y} = ax + b \qquad a = 183.3916084$$
$$b = 2037.004662 \qquad r = 0.9904850045$$

Hence, the regression equation is (rounded)

$$\hat{y} = 183.39x + 2037.00$$

and the correlation coefficient is $r = 0.99$. This r-value indicates that the equation is an excellent fit to the data and also indicates that as x increases, so does y (the taxes go up each year).

(b) Figure 8.21 shows a graph of the data points and the linear regression equation and also shows how well the line fits the data.

$$\hat{y} = 183.39x + 2037.00$$

Figure 8.21

CHECKPOINT SOLUTION

1.
$$\Sigma x = 50 + 25 + 10 + 5 = 90$$
$$\Sigma x^2 = 2500 + 625 + 100 + 25 = 3250$$
$$\Sigma y = 2 + 4 + 10 + 20 = 36$$
$$\Sigma xy = 100 + 100 + 100 + 100 = 400$$

Then

$$a = \frac{\Sigma x \cdot \Sigma y - n\Sigma xy}{(\Sigma x)^2 - n\Sigma x^2} = \frac{90 \cdot 36 - 4 \cdot 400}{90^2 - 4 \cdot 3250} = \frac{1640}{-4900} = -.33,$$

and

$$b = \frac{\Sigma y - a\Sigma x}{n} = \frac{36 - (-.33)(90)}{4} = \frac{65.7}{4} = 16.43.$$

Thus the line that gives the best fit to these points is

$$\hat{y} = -.33x + 16.43.$$

EXERCISE 8.6

In problems 1–6, use the points given in the tables to write the equation of the line that is the best fit for the points.

1. | x | 3 | 4 | 5 | 6 |
    |---|---|---|---|---|
    | y | 15 | 22 | 28 | 32 |

2. | x | 0 | 1 | 2 | 3 |
    |---|---|---|---|---|
    | y | 6 | 8 | 12 | 15 |

3. | x | 10 | 20 | 30 | 40 |
    |---|---|---|---|---|
    | y | 2 | 6 | 5 | 6 |

4. | x | 60 | 70 | 80 | 90 |
    |---|---|---|---|---|
    | y | 10 | 20 | 31 | 38 |

5. | x | 1 | 2 | 3 | 4 |
    |---|---|---|---|---|
    | y | 1 | −4 | 8 | −12 |

6. | x | 2 | 3 | 4 | 5 |
    |---|---|---|---|---|
    | y | 8 | 25 | 50 | 100 |

7. Find the correlation coefficient for the data in problem 1.
8. Find the correlation coefficient for the data in problem 2.
9. Find the correlation coefficient for the data in problem 3.
10. Find the correlation coefficient for the data in problem 4.
11. Find the correlation coefficient for the data in problem 5.
12. Find the correlation coefficient for the data in problem 6.

APPLICATIONS

13. **Demand** (a) Write the linear equation (in the form $\hat{p} = aq + b$) that is the best fit for the following demand schedule.

q	10	20	30	40	50
p	300	250	220	200	180

    (b) Find the correlation coefficient relating quantity demanded and price.

14. **Supply** (a) Write the linear regression equation for the points in the following supply schedule.

q	10	20	30	40	50
p	100	110	115	125	150

    (b) Find the correlation coefficient relating quantity supplied and price.

15. **Earnings and gender** (a) The following table compares the weekly wages for men and women. Write the linear equation (in the form $\hat{y} = ax + b$) that is the best fit for the data.

Year	Men (x)	Women (y)
1920	$31.69	$17.71
1930	27.66	15.98
1940	30.64	17.43
1944	54.65	31.21

    (b) Find the correlation between the weekly wages for men and women.

16. **Temperature** The following table shows the temperature readings from two different thermometers placed in
water as the temperature of the water changes.

Thermometer 1 (x)	Thermometer 2 (y)
12	60
30.5	88
50	110
76.5	165
90	203

(a) Write the linear equations (in the form $\hat{y} = ax + b$)
that is the best fit for the data.
(b) How are the thermometers different?
(c) Find the correlation between the temperatures on
the two thermometers.

17. **Grades** (a) The following table gives the high school
and college grade point averages for seven students,
where the highest possible grade point in both cases is
4.0. Write the equation of the least squares line for the
data.

High school (x)	College (y)
2.50	1.80
3.00	2.50
3.25	3.50
3.50	3.00
3.60	2.20
3.75	3.20
4.00	3.50

(b) Find the correlation between high school and college grade point averages.

18. **Income and savings** (a) The following table compares
yearly income for families and their yearly savings.
Write the linear regression equation that is the best fit for
the data.

Income (x)	Savings (y)
20,000	1000
30,000	1600
40,000	2200
50,000	3000
60,000	3500

(b) Find the correlation between yearly income and
yearly savings.

19. **Average cost** (a) The following table gives the average
cost for a product at different levels of production. Write

the linear regression equation that predicts average cost
from production.

Units (x)	Average Cost ($\overline{C}$)
400	30
500	28
600	27
700	27
800	26

(b) Find the correlation between production units and
average cost.

20. **Cost** (a) The following table gives the cost for a product
at different levels of production. Find the regression line
relating cost to units produced.

Units (x)	Cost (C)
1000	50
2000	72
3000	86
4000	100
5000	120

(b) Find the correlation between number of units produced and cost.

21. **Earnings and gender** The following table shows the
1986 median weekly earnings of men and women in
various occupations.
(a) Write the linear regression equation that is the best
fit for the data.
(b) Use the equation found in (a) to see how well it
predicts the 1986 median weekly earnings for a
woman in a job where a man earned $472 (techni-
cal, sales).
(c) Find the correlation between men's and women's
earnings.

Occupation	Earnings	
	Men (x)	Women (y)
Management	$666	$465
Technical, sales	472	305
Service	299	208
Crafts, repair	446	302
Operators, fabricators	352	238
Fish, farm, forestry	234	201

SOURCE: U.S. Bureau of Labor Statistics

22. **Profit** The following table shows the 1988 sales and profit for several of the largest U.S. corporations.
    (a) Write the linear regression equation that is the best fit for the data.
    (b) Find the correlation between sales and profits.

Corporation	Sales (x) (in Billions)	Profit (y) (in Millions)
GM	$121.1	$4,856
IBM	59.7	5,806
Chrysler	35.5	1,050
Philip Morris	25.9	2,337
Boeing	17.0	614
Pepsico	13.0	762
Coca-Cola	8.3	1,044
Philips Petroleum	11.3	650
GE	49.4	3,386
Du Pont	32.5	2,190

SOURCE: *Fortune Magazine*

23. **Gestation** The following data show the gestation period (in days) and the average longevity (in years) of several North American mammals in captivity. Write the linear regression equation that best fits these data.

Animal	Gestation (x) (in Days)	Longevity (y) (in Years)
Grizzly bear	225	25
Beaver	122	5
Cow	284	15
Elk	250	15
Red fox	52	7
White-tailed deer	201	8
Horse	330	20
Moose	240	12
Sheep	154	12

SOURCE: *1990 World Almanac and Book of Facts*, by Pharos Books

24. **Training and productivity** Suppose that a company purchased a new software package to handle their word processing, payroll, inventory, and accounting needs. After a few employees had received some training with the new system, their work with it was monitored and the following data were gathered.

Hours of Training (x)	Number of Errors (y)	Hours of Training (x)	Number of Errors (y)
1	8	12	0
2	7	7	3
10	2	6	3
8	2	4	4
6	8	4	6

(a) Develop the linear regression equation for these data.
(b) Use the equation developed in part (a) to determine the number of hours of training needed for the predicted number of errors to equal 0.

25. **Supply and demand** The data below give the number of color television sets (in thousands) sold in 15 different years, along with the corresponding average price per set for the year. Use linear regression to find the best linear equation defining the demand function $q = f(p)$.

Price (p) (in Dollars)	Quantity Sold (q) (in Thousands)
471.56	9793
487.79	10236
487.32	9107
510.78	7700
504.39	6485
466.29	8411
449.81	10071
509.86	7908
524.96	6349
514.10	4822
515.43	5962
520.82	5981
525.01	5777
462.32	5892
560.09	2646

SOURCE: U.S. Bureau of the Census, *Statistical Abstract of the United States*, Washington, D.C.

26. **Research expenditures and profits** The data below give the money spent in 1990 on research and development (R & D) in 18 different aerospace industries, along with the corresponding profits for the year. Find the linear regression equation for the line that is the best fit for the data, with profit as a function of R & D.

R & D (in Millions)	Profit (in Millions)
$ 827	$1972
2.7	7.3
0.7	−15.5
36	82
390	−988
11.8	−10.3
9.9	32.8
450	430
212.4	48.8
620	402
155.7	49.8
2.6	−5.5
6.9	1.8
22.7	75.1
124.0	189.5
10.3	68.2
1021.4	1291.2
12.4	46.2

SOURCE: *Business Week*, Jan. 15, 1992

27. **Minority salary differences** The following data give the median annual salary in thousands of dollars for whites and minorities for selected years from 1973 to 1991.

Year	Median Annual Salary (Thousands)	
	White	Minority
1973	$ 8.8	$ 7.5
1975	10.2	8.8
1976	10.7	9.2
1977	11.3	9.7
1978	12.0	10.4
1979	12.8	10.9
1980	13.8	11.8
1981	16.1	13.5
1983	18.5	15.9
1984	19.6	17.4
1985	20.6	18.4
1986	21.5	19.6
1987	22.4	20.9
1989	24.1	22.1
1990	25.2	23.3
1991	26.4	23.8

SOURCE: U.S. Equal Employment Opportunity Commission Data; *Statistical Abstract of the U.S., 1993*

(a) If the median salary for whites is x and for minorities is y, find the linear regression equation and the correlation coefficient for these data.

(b) Graph the regression line and the data points on the same set of axes.

28. **ATM transactions** The following data give the number of automatic teller machines (ATMs) and the total dollar transactions with those ATMs for 1985 to 1992.

Year	ATM Terminals	Transactions (Millions)
1985	60,000	$3565
1986	64,000	3625
1987	68,000	4048
1988	72,492	4480
1889	75,632	5116
1990	80,156	5751
1991	83,545	6418
1992	87,330	7206

SOURCE: Electronic Transfer Association, Herndon, VA, unpublished data

(a) If x is the number of ATM terminals (in thousands) and y is millions of dollars of ATM transactions for a given year, find the linear regression equation and the correlation coefficient for these data. (Note that the first data point is (60, 3565).)

(b) Graph the regression line and the data points on the same set of axes.

29. **Tuition** The following table gives the annual tuition for universities in Georgia from 1985 to 1994.

Year	Tuition	Year	Tuition
1985	$377	1990	528
1986	424	1991	552
1987	460	1992	574
1988	487	1993	597
1989	506	1994	615

SOURCE: University System of Georgia

(a) Let $t = 0$ in 1985 and use linear regression to write a linear equation representing the annual tuition at universities in Georgia as a function of the number of years from 1985.

(b) Find the correlation coefficient r for this data and decide whether or not the equation in part (a) is reasonable.

(c) Use the equation to predict tuition in 1997.

KEY TERMS
& FORMULAS

Section	*Key Term*	*Formula*
8.1	Binomial probability experiments	$\Pr(x) = \dbinom{n}{x} p^x q^{n-x}$
8.2	Discrete probability distributions Random variable Probability density histogram	
	Expected value	$E(x) = \sum x \Pr(x)$
	Mean	$\mu = E(x) = \sum x \Pr(x)$
	Variance	$\sigma^2 = \sum (x - \mu)^2 \Pr(x)$
	Standard deviation	$\sigma = \sqrt{\sigma^2}$
8.3	Binomial probability distribution	
	Mean	$\mu = np$
	Standard deviation	$\sigma = \sqrt{npq}$
	Binomial formula	
8.4	Descriptive statistics Frequency histogram Mode Median	
	Mean	$\overline{x} = \dfrac{\sum x}{n}$
	Range	
	Sample variance	$s^2 = \dfrac{\sum (x - \overline{x})^2}{n - 1}$
	Sample standard deviation	$s = \sqrt{\dfrac{\sum (x - \overline{x})^2}{n - 1}}$
8.5	Normal distribution Standard normal distribution	
	z-scores	$z = \dfrac{x - \mu}{\sigma}$
8.6	Linear regression	$\hat{y} = ax + b$
		$a = \dfrac{\sum x \cdot \sum y - n \sum xy}{(\sum x)^2 - n \sum x^2}$
		$b = \dfrac{\sum y - a \sum x}{n}$
	Correlation coefficient	$r = \dfrac{n(\sum xy) - (\sum x)(\sum y)}{\sqrt{n(\sum x^2) - (\sum x)^2}\ \sqrt{n(\sum y^2) - (\sum y)^2}}$

REVIEW EXERCISES

8.1

1. Suppose 20% of the population are victims of crime. Out of 5 people selected at random, find the probability that 2 are crime victims.
2. If one person in 100,000 develops a certain disease, then calculate
   (a) Pr(exactly one person in 100,000 has the disease).
   (b) Pr(at least one person in 100,000 has the disease).
3. If a die is rolled 4 times, what is the probability that a number greater than 4 is rolled at least 2 times?

8.2

4. For the following probability distribution, find the expected value $E(x)$.

x	Pr(x)
1	0.2
2	0.3
3	0.4
4	0.1

Determine whether or not each function or table in problems 5–8 represents a discrete probability distribution.

5. $Pr(x) = x/15$, $x = 1, 2, 3, 4, 5$
6. $Pr(x) = x/55$, $x = 1, 2, 3, 4, 5, 6, 7, 8, 9$

7.
x	Pr(x)
3	1/6
4	1/3
5	1/3
6	1/6

8.
x	Pr(x)
2	1/2
3	1/4
4	−1/4
5	1/2

9. For the following probability distribution, find the expected value $E(x)$.

x	1	2	3	4
Pr(x)	.4	.3	.2	.1

10. For the discrete probability distribution described by $Pr(x) = x/16$, $x = 1, 2, 3, 4, 6$, find the
    (a) mean.   (b) variance.   (c) standard deviation.

11. For the probability distribution described by the table, find the
    (a) mean.
    (b) variance.
    (c) standard deviation.

x	Pr(x)
1	1/12
2	1/6
3	1/3
4	5/12

8.3

12. A coin has been altered so that the probability that a head will occur is 2/3. If the coin is tossed 6 times, give the mean and standard deviation of the distribution of the number of heads.
13. Refer to problem 1.
    (a) What is the expected number of people in this sample that are victims?
    (b) Find the probability that exactly the expected number of people in this sample are victims of crime.
14. Expand $(x + y)^5$.

8.4

Consider the data in the following frequency table, and use the data in problems 15–18.

Score	Frequency
1	4
2	6
3	8
4	3
5	5

15. Construct a frequency histogram for the data.
16. What is the mode of the data?
17. What is the mean of the data?
18. What is the median of the data?
19. Construct a frequency histogram for the following set of test scores. 14, 16, 15, 14, 17, 16, 12, 12, 13, 14

20. What is the median of the scores in problem 19?
21. What is the mode of the scores in problem 19?
22. What is the mean of the scores in problem 19?
23. Find the mean, variance, and standard deviation of the following data: 4, 3, 4, 6, 8, 0, 2.
24. Find the mean, variance, and standard deviation of the following data: 3, 2, 1, 5, 1, 4, 0, 2, 1, 1.

8.5

25. What is the area under the standard normal curve between $z = -1.6$ and $z = 1.9$?
26. If z is a standard normal score, find $Pr(-1 \le z \le -.5)$.
27. Find $Pr(1.23 \le z \le 2.55)$.
28. If a variable x is normally distributed with $\mu = 25$ and $\sigma = 5$, find $Pr(25 \le x \le 30)$.
29. If a variable x is normally distributed with $\mu = 25$ and $\sigma = 5$, find $Pr(20 \le x \le 30)$.
30. If a variable x is normally distributed with $\mu = 25$ and $\sigma = 5$, find $Pr(30 \le x \le 35)$.

APPLICATIONS

8.1

31. *Genetics* Suppose the probability that a certain couple will have a blond child is 1/4. If they have 6 children, what is the probability that 2 of them will be blond?

32. **Sampling** Suppose 70% of a population opposes a proposal and a sample of size 5 is drawn from the population. What is the probability the majority of the sample will favor the proposal?

8.2

33. **Lotteries** A state lottery pays $500 to anyone who selects the correct 3-digit number from a random drawing. If it costs $1 to play, then the probability distribution for the amount won is given in the table.

x	$Pr(x)$
−1	0.999
499	0.001

What are the expected winnings of a person who plays?

8.3

34. **Cancer testing** The probability that a man with prostate cancer will test positively with the Prostate Specific Antigen (PSA) test is 0.91. If 500 men with prostate cancer are tested, how many would we expect to test positively?

8.4

Farm families The distribution of farm families (as a percentage of the population) in a 50-county survey is given in the table. Use this data in problems 35–37.

Percentage	Number of Counties
10–19	5
20–29	16
30–39	25
40–49	3
50–59	1

35. Make a frequency histogram for these data.
36. Find the mean percentage of farm families in these 50 counties.
37. Find the standard deviation of the percentage of farm families in these 50 counties.

8.5

38. **Testing** Suppose the mean SAT score for students admitted to a university is 1000, with a standard deviation of 200. Suppose that a student is selected at random. Find the probability that the student's SAT score is
   (a) between 1000 and 1400.
   (b) between 1200 and 1400.
   (c) greater than 1400.

8.6

39. **Salary differences by gender** The following data give the median annual salary in thousands of dollars for males and females for selected years from 1973 to 1991.

	Median Annual Salary (thousands)	
Year	Males	Females
1973	$ 9.6	$ 7.0
1975	11.3	8.2
1976	11.8	8.6
1977	12.4	9.1
1978	13.3	9.7
1979	14.1	10.4
1980	15.2	11.4
1981	17.7	13.1
1983	20.1	15.3
1984	21.4	16.2
1985	22.3	17.3
1986	23.4	18.1
1987	24.2	18.9
1989	26.1	20.6
1990	27.3	21.8
1991	28.4	22.7

SOURCE: U.S. Equal Employment Opportunity Commission Data, *Statistical Abstract of the U.S.*, 1993

(a) If the median salary for males is x and for females is y, find the linear regression equation and the correlation coefficient for these data.
(b) Use a graphing utility to graph the regression line and the data points on the same set of axes.

40. **Taxes** The following data show U.S. national personal income and personal taxes for selected years.
   (a) Write the linear regression equation that best fits these data.
   (b) Use the equation found in part (a) to see how well it predicts the taxes for the income from 1980.

Year	Income (x) (in Billions)	Taxes (y) (in Billions)
1974	$1168.6	$170.2
1976	1391.2	196.8
1978	1732.7	258.7
1980	2165.3	336.5
1982	2584.6	404.1
1984	3108.7	440.2
1986	3526.2	512.9
1988	4064.5	586.6

SOURCE: Bureau of Economic Analysis, U.S. Commerce Dept.

Statistics in Medical Research; Hypothesis Testing

Bering Research Laboratories is involved in a national study organized by the National Center for Disease Control to determine if a new birth control pill affects the blood pressure of women who are using it. A random sample of 100 women using the pill was selected and the participants' blood pressures were measured. The mean blood pressure was $\bar{x} = 112.5$. If the blood pressure of all women is normally distributed with mean $\mu = 110$ and standard deviation $\sigma = 12$, is the mean of this sample sufficiently far from the mean of the population of all women to indicate that the birth control pill affects blood pressure?

To answer this question, researchers at Bering Labs first need to know how the means of samples drawn from a normal population are distributed. If all possible samples of size n are drawn from a normal population that has mean μ and standard deviation σ, then the distribution of the means of these samples will also be normally distributed with mean μ, but with standard deviation given by

$$\sigma_{\bar{x}} = \frac{\sigma}{\sqrt{n}}.$$

To conclude whether or not women using this birth control pill have blood pressure that is significantly different from that of all women in the population, researchers must decide whether or not the mean $\bar{x}$ of their sample is so far from the population mean μ that it is unlikely that the sample was chosen from the population.

To use a familiar analogy, the statistical proof that $\bar{x}$ does not come from the population of all women is like the trial of an accused criminal. In a criminal trial, we assume that the person who stands accused is innocent unless there is evidence "beyond all reasonable doubt" that the person is guilty. In this statistical test, the researchers will assume that the sample has been drawn from the population of women with mean $\mu = 110$ unless the blood pressures collected in the sample are so different from that of the population that it is not reasonable to make that assumption. In statistics, we can say that the assumption is unreasonable if the probability that the sample mean, $\bar{x} = 112.5$, could be drawn from this population, with mean 110, is less than .05. If this probability is less than .05, we say that the sample is drawn from a population with a mean different from 110, which strongly suggests that the women who have taken the birth control pill have, on average, different blood pressure from that of the women who are not using the pill. That is, there is statistically significant evidence that the pill affects blood pressure of women.

Because 95% of all normally distributed data points lie within 2 standard deviations of the mean, researchers at Bering Labs need to determine whether or not the sample mean $\bar{x} = 112.5$ is more than 2 standard deviations from $\mu = 110$. To determine how many standard deviations $\bar{x} = 112.5$ is from $\mu = 110$, they compute the z-score for $\bar{x}$ with the formula

$$z_{\bar{x}} = \frac{\bar{x} - \mu}{\sigma/\sqrt{n}}.$$

Prior to completion of the broad national study, an administrator at the National Center for Disease Control asks you to submit a preliminary report on the work being conducted at Bering Research Laboratories. Answer the following questions as the basis for your preliminary report.

1. Calculate the z-score, $z_{\bar{x}}$, associated with $\bar{x} = 112.5$, to see how many standard deviations $\bar{x} = 112.5$ is from $\mu = 110$.
2. Is the computed z-score more than 2 standard deviations from the population mean μ?
3. What conclusion would you expect the researchers at Bering Labs to reach about the effect of this birth control pill on the blood pressure of women using it?

P A R T 5

CALCULUS

Warm-up

Prerequisite problem type	For section	Answer	Section for review
If $f(x) = \dfrac{x^2 - x - 6}{x + 2}$, then find (a) $f(-3)$ (b) $f(-2.5)$ (c) $f(-2.1)$ (d) $f(-2)$	**9.1–9.9**	(a) -6 (b) -5.5 (c) -5.1 (d) undefined	1.4 Functional notation
Factor: (a) $x^2 - x - 6$ (b) $x^2 - 4$ (c) $x^2 + 3x + 2$	**9.1 9.7**	(a) $(x + 2)(x - 3)$ (b) $(x - 2)(x + 2)$ (c) $(x + 1)(x + 2)$	0.6 Factoring
Write as a power: (a) $\sqrt{t}$ (b) $\dfrac{1}{x}$ (c) $\dfrac{1}{\sqrt[3]{x^2 + 1}}$	**9.4–9.8**	(a) $t^{1/2}$ (b) x^{-1} (c) $(x^2 + 1)^{-1/3}$	0.3, 0.4 Exponents and radicals
Simplify: (a) $\dfrac{4(x + h)^2 - 4x^2}{h}$, if $h \neq 0$ (b) $(2x^3 + 3x + 1)(2x) + (x^2 + 4)(6x^2 + 3)$ (c) $\dfrac{x(3x^2) - x^3(1)}{x^2}$, if $x \neq 0$	**9.3 9.5 9.7**	(a) $8x + 4h$ (b) $10x^4 + 33x^2$ $+ 2x + 12$ (c) $2x$	0.5 Simplifying algebraic expressions
Simplify: (a) $\dfrac{x^2 - x - 6}{x + 2}$ if $x \neq -2$ (b) $\dfrac{x^2 - 4}{x - 2}$ if $x \neq 2$	**9.1 9.7**	(a) $x - 3$ (b) $x + 2$	0.7 Simplifying fractions
If $f(x) = 3x^2 + 2x$, find $\dfrac{f(x + h) - f(x)}{h}$.	**9.3**	$6x + 3h + 2$	1.4 Functional notation
Find the slope of the line passing through $(1, 2)$ and $(2, 4)$.	**9.3**	2	1.7 Slopes
Write the equation of the line passing through $(1, 5)$ with slope 8.	**9.3 9.4 9.6**	$y = 8x - 3$	1.7 Point-slope equation of a line

DERIVATIVES

If a firm receives $30,000 in revenue during a 30-day month, its average revenue per day is $30,000/30 = $1000. This does not necessarily mean that the actual revenue was $1000 on any one day, just that the average is $1000 per day. Similarly, if a person drives a car 50 miles in an hour's time, the car's average velocity is 50 miles per hour, but the driver could have gotten a speeding ticket for traveling 70 miles per hour on this trip. When we say a car is moving at a velocity of 50 miles per hour, we are talking about the velocity of the car at an instant in time (the instantaneous velocity). We can use the average velocity to find the instantaneous velocity, as follows.

If a car travels in a straight line from position y_1 at time x_1 and arrives at position y_2 at time x_2, then it has traveled the distance $y_2 - y_1$ in the elapsed time $x_2 - x_1$. If we represent the distance traveled by Δy and the elapsed time by Δx, the average velocity is given by

$$V_{av} = \frac{\Delta y}{\Delta x}.$$

The smaller the time interval, the more nearly the average velocity will be to the instantaneous velocity. For example, knowing that a car traveled 50 miles in an hour does not tell us much about its instantaneous velocity at any time during that hour. But knowing that it traveled 1 mile in 1 minute, or 50 feet in one second, tells us much more about the velocity at a given time. Continuing to decrease the length of the time interval (Δx) will get us closer and closer to the instantaneous velocity.

Some police departments have equipment that measures how fast a car is traveling by measuring how much time elapses while the car travels between two sensors placed 60 inches apart on the road. This is not the instantaneous velocity of the car, but it is an excellent approximation.

We define the **instantaneous velocity** to be the *limit* of $\Delta y/\Delta x$ as Δx approaches 0. We write this as

$$V = \lim_{\Delta x \to 0} \frac{\Delta y}{\Delta x}.$$

Thus we may think of velocity as the instantaneous rate of change of distance with respect to time.

577

This chapter is concerned with *limits* and *rates of change*. We will see that the *derivative* of a function can be used to determine the rate of change of the dependent variable with respect to the independent variable. In this chapter the derivative will be used to find the marginal profit, marginal cost, and marginal revenue, given the respective profit, total cost, and total revenue functions, and we will find other rates of change, such as rates of change of populations and velocity. We will also use the derivative to determine the slope of a tangent to a curve at a point on the curve. In Chapter 10, more applications of the derivative will be discussed. For example, we will use differentiation to minimize average cost, maximize total revenue, maximize profit, and find the maximum dosage for certain medications.

9.1 LIMITS

Objectives

❏ To use graphs and numerical tables to find limits of functions, when they exist
❏ To find limits of polynomial functions
❏ To find limits of rational functions

APPLICATION PREVIEW

Although everyone recognizes the value of eliminating any and all particulate pollution from smokestack emissions of factories, company owners are concerned about the cost of removing this pollution. Suppose that USA Steel has shown that the cost C of removing p percent of the particulate pollution from the emissions at one of its plants is

$$C = C(p) = \frac{7300p}{100 - p}.$$

To investigate the cost of removing as much of the pollution as possible, we can evaluate the **limit** as p (the percent) approaches 100 from values less than 100. Using a limit is important in this case because we cannot evaluate this function at $p = 100$.

We have used the notation $f(c)$ to indicate the value of a function $f(x)$ at $x = c$. If we need to discuss a value that $f(x)$ approaches as x approaches c, we use the idea of a **limit**. For example, if

$$f(x) = \frac{x^2 - x - 6}{x + 2},$$

then we know that -2 is not in the domain of $f(x)$ so that $f(-2)$ does not exist. Figure 9.1 shows the graph of $y = f(x)$ with an open circle where $x = -2$. Even though $f(-2)$ is not defined, the figure shows that as x approaches -2 from either side of -2, the graph approaches the open circle at $(-2, -5)$ and the values of $f(x)$ approach -5. Thus -5 is the limit of $f(x)$ as x approaches -2, and we write

$$\lim_{x \to -2} f(x) = -5, \quad \text{or} \quad f(x) \to -5 \quad \text{as} \quad x \to -2.$$

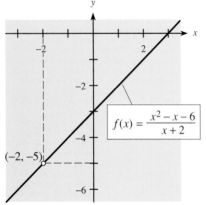

Figure 9.1

Table 9.1

Left of −2	
x	$f(x) = \dfrac{x^2 - x - 6}{x + 2}$
−3.000	−6.000
−2.500	−5.500
−2.100	−5.100
−2.010	−5.010
−2.001	−5.001

Right of −2	
x	$f(x) = \dfrac{x^2 - x - 6}{x + 2}$
−1.000	−4.000
−1.500	−4.500
−1.900	−4.900
−1.990	−4.990
−1.999	−4.999

This conclusion is fairly obvious from the graph, but not so obvious from the equation for $f(x)$.

We can use the values near $x = -2$ in Table 9.1 to help verify that $f(x) \to -5$ as $x \to -2$. Notice that to the left of -2, the values of x increase from -3.000 to -2.001 in small increments, and in the corresponding column for $f(x)$, the values of the function $f(x)$ increase from -6.000 to -5.001. To the right of -2, the values of x decrease from -1.000 to -1.999 while the corresponding values of $f(x)$ decrease from -4.000 to -4.999. Hence, Table 9.1 and Figure 9.1 indicate that the value of $f(x)$ approaches -5 as x approaches -2 from both sides of $x = -2$.

From our discussion of the graph in Figure 9.1 and Table 9.1, we see that as x approaches -2 from either side of -2, the limit of the function is the value L that the function approaches. This limit L is not necessarily the value of the function at $x = -2$.

EXAMPLE I

Figure 9.2 shows three functions for which the limit exists as x approaches 2. Use this figure to find the following.

(a) $\lim\limits_{x \to 2} f(x)$ and $f(2)$ (if it exists)

(b) $\lim\limits_{x \to 2} g(x)$ and $g(2)$ (if it exists)

(c) $\lim\limits_{x \to 2} h(x)$ and $h(2)$ (if it exists)

(a)

(b)

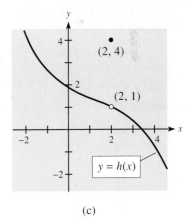

(c)

Figure 9.2

Solution

(a) From the graph in Figure 9.2(a), we see that as x approaches 2 from both the left and the right, the graph approaches the point $(2, 3)$. Thus $f(x)$ approaches the single value 3. That is,

$$\lim_{x \to 2} f(x) = 3.$$

The value of $f(2)$ is the y-coordinate of the point on the graph at $x = 2$. Thus $f(2) = 3$.

(b) Figure 9.2(b) shows that as x approaches 2 from both the left and the right, the graph approaches the open circle at $(2, -1)$. Thus

$$\lim_{x \to 2} g(x) = -1.$$

The figure also shows that at $x = 2$ there is no point on the graph. Thus $g(2)$ is undefined.

(c) Figure 9.2(c) shows that

$$\lim_{x \to 2} h(x) = 1.$$

The figure also shows that at $x = 2$ there is a point on the graph at $(2, 4)$. Thus $h(2) = 4$, and we see that $\lim_{x \to 2} h(x) \neq h(2)$.

As Example 1 shows, the limit of the function as x approaches c may or may not be the same as the value of the function at $x = c$. This leads to our intuitive definition of *limit*.

Limit   Let $f(x)$ be a function defined on an open interval containing c, except perhaps at c. Then

$$\lim_{x \to c} f(x) = L$$

is read "the limit of $f(x)$ as x approaches c equals L." The number L exists if we can make values of $f(x)$ as close to L as we desire by choosing values of x sufficiently close to c. When the values of $f(x)$ do not approach a single finite value L as x approaches c, we say the limit does not exist.

As the definition states, a limit as $x \to c$ can exist only if the function approaches a single finite value as x approaches c from both the left and right of c. In the next example, we consider some cases where a limit does not exist.

Figure 9.3

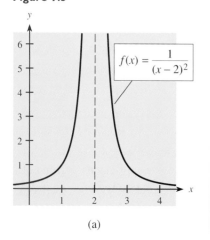

$$f(x) = \frac{1}{(x-2)^2}$$

(a)

EXAMPLE 2

Using the functions graphed in Figure 9.3, determine why the limit as $x \to 2$ does not exist for

(a) $f(x)$ (b) $g(x)$ (c) $h(x)$.

Solution

(a) As $x \to 2$ from the left side and the right side of $x = 2$, $f(x)$ increases without bound, which we denote by saying that $f(x)$ approaches $+\infty$ as $x \to 2$. In this case, $\lim_{x \to 2} f(x)$ does not exist (briefly noted by DNE) because $f(x)$ does not approach a finite value as $x \to 2$. The graph has a vertical asymptote at $x = 2$. We say that this is an **infinite limit.**

Figure 9.3 (continued)

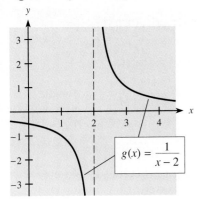

$$g(x) = \frac{1}{x - 2}$$

(b)

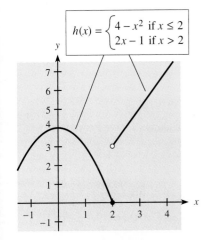

$$h(x) = \begin{cases} 4 - x^2 & \text{if } x \le 2 \\ 2x - 1 & \text{if } x > 2 \end{cases}$$

(c)

(b) As $x \to 2$ from the left, $g(x)$ approaches $-\infty$, and as $x \to 2$ from the right, $g(x)$ approaches $+\infty$, so $g(x)$ does not approach a finite value as $x \to 2$. Therefore, the limit does not exist. The graph of $y = g(x)$ has a vertical asymptote at $x = 2$.

(c) As $x \to 2$ from the left (while $x < 2$, denoted by $x \to 2^-$), the graph approaches the point at $(2, 0)$, so $h(x)$ approaches the value 0, but as $x \to 2$ from the right (while $x > 2$, denoted by $x \to 2^+$), the graph approaches the open circle at $(2, 3)$, so $h(x)$ approaches the value 3. Because $h(x)$ approaches two different numbers as x approaches 2, the limit does not exist.

Examples 1 and 2 illustrate the following two important facts regarding limits.

1. The limit of a function as x approaches c is independent of the value of the function at c. When $\lim\limits_{x \to c} f(x)$ exists, the value of the function at c may be (i) the same as the limit, (ii) undefined, or (iii) defined but different from the limit (see Figure 9.2 and Example 1).

2. The limit is said to exist only if the following conditions are satisfied:
   (a) The limit L is a finite value (real number).
   (b) The limit as x approaches c from the left equals the limit as x approaches c from the right. That is, we must have

   $$\lim_{x \to c^-} f(x) = \lim_{x \to c^+} f(x).$$

   Figure 9.3 and Example 2 illustrate cases where $\lim\limits_{x \to c} f(x)$ does not exist.

CHECKPOINT

1. Can $\lim\limits_{x \to c} f(x)$ exist if $f(c)$ is undefined?
2. Does $\lim\limits_{x \to c} f(x)$ exist if $f(c) = 0$?
3. Does $f(c) = 1$ if $\lim\limits_{x \to c} f(x) = 1$?
4. If $\lim\limits_{x \to c^-} f(x) = 0$, does $\lim\limits_{x \to c} f(x)$ exist?

Fact 1 regarding limits tells us that the value of the limit of a function as $x \to c$ will not always be the same as the value of the function at $x = c$. However, there are many functions for which the limit and the functional value agree (see Figure 9.2(a)), and for these functions we can easily evaluate limits. The following properties of limits allow us to identify certain classes or types of functions for which $\lim\limits_{x \to c} f(x)$ equals $f(c)$.

PROPERTIES OF LIMITS

If k is a constant, $\lim\limits_{x \to c} f(x) = L$ and $\lim\limits_{x \to c} g(x) = M$, then the following are true.

I. $\lim\limits_{x \to c} k = k$

II. $\lim\limits_{x \to c} x = c$

III. $\lim\limits_{x \to c} [f(x) \pm g(x)] = L \pm M$

IV. $\lim\limits_{x \to c} [f(x) \cdot g(x)] = LM$

V. $\lim\limits_{x \to c} \dfrac{f(x)}{g(x)} = \dfrac{L}{M}$ if $M \neq 0$

VI. $\lim\limits_{x \to c} \sqrt[n]{f(x)} = \sqrt[n]{\lim\limits_{x \to c} f(x)} = \sqrt[n]{L}$

Note that if f is a polynomial function, then Properties I–IV imply that $\lim\limits_{x \to c} f(x)$ can be found by evaluating $f(c)$. Moreover, if h is a rational function whose denominator is not zero at $x = c$, then Property V implies that $\lim\limits_{x \to c} h(x)$ can be found by evaluating $h(c)$.

Function	Definition	Limit
Polynomial Function	The function $$f(x) = a_n x^n + a_{n-1} x^{n-1} + \cdots + a_1 x + a_0,$$ where n is a positive integer, is called a **polynomial function.**	$\lim\limits_{x \to c} f(x) = f(c)$ for all values c (by Properties I–IV)
Rational Function	The function $$h(x) = \frac{f(x)}{g(x)},$$ where both $f(x)$ and $g(x)$ are polynomial functions, is called a **rational function.**	$\lim\limits_{x \to c} h(x) = \lim\limits_{x \to c} \dfrac{f(x)}{g(x)} = \dfrac{f(c)}{g(c)}$ when $g(c) \neq 0$ (by Property V)

EXAMPLE 3

Find the following limits, if they exist.

(a) $\lim\limits_{x \to -1} (x^3 - 2x)$

(b) $\lim\limits_{x \to 4} \dfrac{x^2 - 4x}{x - 2}$

Solution

(a) Note that $f(x) = x^3 - 2x$ is a polynomial, so

$$\lim_{x \to -1} f(x) = f(-1) = (-1)^3 - 2(-1) = 1.$$

Figure 9.4(a) shows the graph of $f(x) = x^3 - 2x$.

(b) Note that this limit has the form

$$\lim_{x \to c} \frac{f(x)}{g(x)},$$

where $f(x)$ and $g(x)$ are polynomials and $g(c) \neq 0$. Therefore, we have

$$\lim_{x \to 4} \frac{x^2 - 4x}{x - 2} = \frac{4^2 - 4(4)}{4 - 2} = \frac{0}{2} = 0.$$

Figure 9.4(b) shows the graph of $g(x) = \dfrac{x^2 - 4x}{x - 2}$.

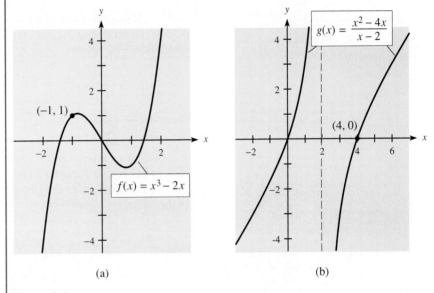

(a)   (b)

Figure 9.4

We have seen that we can use Property V to find the limit of a rational function $f(x)/g(x)$ as long as the denominator is *not* zero. If the limit of the denominator of $f(x)/g(x)$ *is* zero, then there are two possible cases.

I.   Both $\lim_{x \to c} g(x) = 0$ and $\lim_{x \to c} f(x) = 0$, or

II. $\lim_{x \to c} g(x) = 0$ and $\lim_{x \to c} f(x) \neq 0$.

In case I we say that $f(x)/g(x)$ has the form $0/0$ at $x = c$. We call this the **0/0 indeterminate form;** the limit cannot be evaluated until $x - c$ is factored from both $f(x)$ and $g(x)$ and the fraction is reduced. Example 4 will illustrate this case.

In case II, the limit has the form $a/0$, where a is a constant, $a \neq 0$. This expression is undefined, and the limit does not exist. Example 5 will illustrate this case.

Figure 9.5

(a)

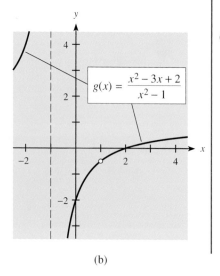

(b)

EXAMPLE 4

Evaluate the following limits, if they exist.

(a) $\lim\limits_{x \to 2} \dfrac{x^2 - 4}{x - 2}$ (b) $\lim\limits_{x \to 1} \dfrac{x^2 - 3x + 2}{x^2 - 1}$

Solution

(a) We cannot find the limit by using Property V because the denominator is zero at $x = 2$. Since the numerator is also zero at $x = 2$, the expression

$$\frac{x^2 - 4}{x - 2}$$

has the 0/0 indeterminate form at $x = 2$. Thus we can factor $x - 2$ from both the numerator and the denominator and reduce the fraction. (We can divide by $x - 2$ because $x - 2 \neq 0$ while $x \to 2$.)

$$\lim_{x \to 2} \frac{x^2 - 4}{x - 2} = \lim_{x \to 2} \frac{(x - 2)(x + 2)}{x - 2} = \lim_{x \to 2}(x + 2) = 4$$

Figure 9.5(a) shows the graph of $f(x) = (x^2 - 4)/(x - 2)$. Note the open circle at (2, 4).

(b) By substituting 1 for x in $(x^2 - 3x + 2)/(x^2 - 1)$ we see that the expression has the 0/0 indeterminate form at $x = 1$, so $x - 1$ is a factor of both the numerator and the denominator. (We can then reduce the fraction because $x - 1 \neq 0$ while $x \to 1$.)

$$\lim_{x \to 1} \frac{x^2 - 3x + 2}{x^2 - 1} = \lim_{x \to 1} \frac{(x - 1)(x - 2)}{(x - 1)(x + 1)}$$

$$= \lim_{x \to 1} \frac{x - 2}{x + 1}$$

$$= \frac{1 - 2}{1 + 1} = \frac{-1}{2} \quad \text{(by Property V)}$$

Figure 9.5(b) shows the graph of $g(x) = (x^2 - 3x + 2)/(x^2 - 1)$. Note the open circle at $(1, -\frac{1}{2})$.

Notice that although both problems in Example 4 had the 0/0 indeterminate form, they had different answers.

EXAMPLE 5

Find $\lim\limits_{x \to 1} \dfrac{x^2 + 3x + 2}{x - 1}$, if it exists.

Solution

Substituting 1 for x in the function results in $6/0$, so this limit has the form $a/0$, with $a \neq 0$, and is like case II discussed previously. Hence the limit does not exist. Because the numerator is not zero when $x = 1$, we know that $x - 1$ is *not* a factor of the numerator, and we cannot divide numerator and denominator as we did in Example 4. Table 9.2 verifies that this limit does not exist,

Table 9.2

Left of $x = 1$		Right of $x = 1$	
x	$\dfrac{x^2 + 3x + 2}{x - 1}$	x	$\dfrac{x^2 + 3x + 2}{x - 1}$
0	-2	2	12
0.5	-7.5	1.5	17.5
0.7	-15.3	1.2	35.2
0.9	-55.1	1.1	65.1
0.99	-595.01	1.01	605.01
0.999	$-5,995.001$	1.001	6,005.001
0.9999	$-59,999.0001$	1.001	60,005.0001

$$\lim_{x \to 1^-} \frac{x^2 + 3x + 2}{x - 1} = -\infty \qquad \lim_{x \to 1^+} \frac{x^2 + 3x + 2}{x - 1} = +\infty$$

because the values of the expression are unbounded near $x = 1$.

The left-hand and right-hand limits are not finite, so they do not exist. Thus $\lim\limits_{x \to 1} \dfrac{x^2 + 3x + 2}{x - 1}$ does not exist.

The results of Examples 4 and 5 can be summarized as follows.

Rational Functions: Evaluating Limits of the Form $\lim\limits_{x \to c} \dfrac{f(x)}{g(x)}$, where $\lim\limits_{x \to c} g(x) = 0$

Type I. If $\lim\limits_{x \to c} f(x) = 0$ and $\lim\limits_{x \to c} g(x) = 0$, then the fractional expression has the **0/0 indeterminate form** at $x = c$. We can factor $x - c$ from $f(x)$ and $g(x)$, reduce the fraction, and then find the limit of the resulting expression, if it exists.

Type II. If $\lim\limits_{x \to c} f(x) \neq 0$ and $\lim\limits_{x \to c} g(x) = 0$, then $\lim\limits_{x \to c} \dfrac{f(x)}{g(x)}$ does not exist.

In this case, the values of $f(x)/g(x)$ are unbounded near $x = c$.

In Example 5, even though the left-hand and right-hand limits do not exist (see Table 9.2), knowledge that the functional values are unbounded (that is, that they become infinite) is helpful in graphing. The graph is shown in Figure 9.6. We see that $x = 1$ is a vertical asymptote.

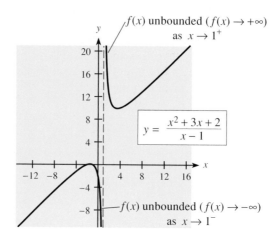

$f(x)$ unbounded ($f(x) \to +\infty$) as $x \to 1^+$

$$y = \frac{x^2 + 3x + 2}{x - 1}$$

$f(x)$ unbounded ($f(x) \to -\infty$) as $x \to 1^-$

Figure 9.6

CHECKPOINT

Assume that f, g, and h are polynomials.

5. Does $\lim_{x \to c} f(x) = f(c)$?

6. Does $\lim_{x \to c} \dfrac{g(x)}{h(x)} = \dfrac{g(c)}{h(c)}$?

7. If $g(c) = 0$ and $h(c) = 0$, can we be certain that

    (a) $\lim_{x \to c} \dfrac{g(x)}{h(x)} = 0$?

    (b) $\lim_{x \to c} \dfrac{g(x)}{h(x)}$ exists?

8. If $g(c) \neq 0$ and $h(c) = 0$, what can be said about

    $\lim_{x \to c} \dfrac{g(x)}{h(x)}$ and $\lim_{x \to c} \dfrac{h(x)}{g(x)}$?

As we noted in Section 1.5, many applications are modeled by piecewise defined functions. To see how we evaluate a limit involving a piecewise defined function, consider the following example.

EXAMPLE 6

Find $\lim_{x \to 1^-} f(x)$, $\lim_{x \to 1^+} f(x)$, and $\lim_{x \to 1} f(x)$, if they exist for

$$f(x) = \begin{cases} x^2 + 1 & \text{for } x \leq 1 \\ x + 2 & \text{for } x > 1 \end{cases}.$$

Solution

We select values of x that approach 1 from the left and right and evaluate f at these values (see Table 9.3). For values of x less than 1, we use $f(x) = x^2 + 1$ to find the value of the function (see the left side of the table). For values of x that are greater than 1, we use $f(x) = x + 2$ to find the value of f (see the right side of the table).

Table 9.3

	Left of 1		Right of 1
x	$f(x) = x^2 + 1$	x	$f(x) = x + 2$
0.1	1.01	1.2	3.2
0.9	1.81	1.01	3.01
0.99	1.98	1.001	3.001
0.999	1.998	1.0001	3.0001
0.9999	1.9998	1.00001	3.00001

Figure 9.7

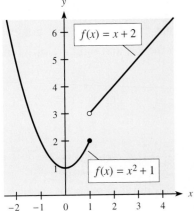

In this case, we observe that $f(x)$ appears to be approaching 2 as x approaches 1 from the left, while $f(x)$ appears to be approaching 3 as x approaches 1 from the right. Figure 9.7 shows the graph of $f(x)$, with $\lim_{x \to 1^-} f(x) = 2$ and $\lim_{x \to 1^+} f(x) = 3$, as we determined from Table 9.3. We show this result algebraically as follows. Because $f(x)$ is defined by $x^2 + 1$ when $x < 1$,

$$\lim_{x \to 1^-} f(x) = \lim_{x \to 1^-} (x^2 + 1) = 2.$$

Because $f(x)$ is defined by $x + 2$ when $x > 1$,

$$\lim_{x \to 1^+} f(x) = \lim_{x \to 1^+} (x + 2) = 3.$$

Since

$$2 = \lim_{x \to 1^-} f(x) \neq \lim_{x \to 1^+} f(x) = 3,$$

$\lim_{x \to 1} f(x)$ does not exist.

Graphing Utilities

To investigate the properties of a function, it is frequently useful to graph the function with a graphing utility.

Figure 9.8

(a)

EXAMPLE 7

Graph the following four functions and answer the questions below. Justify your answers.

$$f(x) = \frac{(x + 1)^2}{x + 2}, \qquad g(x) = \frac{2x}{x + 1},$$

$$h(x) = \frac{2\sqrt{(x + 1)^2}}{x + 1}, \qquad k(x) = \frac{x^3 + 1}{x + 1}$$

(a) Which of these functions exist at $x = -1$?

(b) For which of these functions does the limit exist as x approaches -1 and for which functions does the limit not exist as x approaches -1?

Figure 9.8 (continued)

(b)

(c)

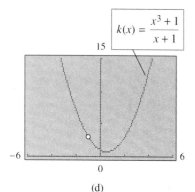

(d)

Solution

The graphs are shown in Figure 9.8.

(a) Only function f is defined at $x = -1$. None of the graphs of the other functions have a point at $x = -1$.

(b) $\lim_{x \to -1} f(x)$ appears to be 0 from the graph. This can be easily verified algebraically.

The vertical asymptote appears to exist at $x = -1$ on the graph of $y = g(x)$, which indicates that $\lim_{x \to -1} g(x)$ does not "appear" to exist.

Evaluating this limit algebraically verifies this.

$$\lim_{x \to -1} g(x) = \lim_{x \to -1} \frac{2x}{x + 1} = \text{DNE}$$

The "jump" in the graph of $y = h(x)$ indicates that $\lim_{x \to -1} h(x)$ does not exist because the limits as x approaches -1 from the left and right are different. Evaluating these limits gives

$$\lim_{x \to -1^-} \frac{2\sqrt{(x + 1)^2}}{x + 1} = \lim_{x \to -1^-} \frac{2[-(x + 1)]}{x + 1} = \lim_{x \to -1^-} (-2) = -2,$$

and

$$\lim_{x \to -1^+} \frac{2\sqrt{(x + 1)^2}}{x + 1} = \lim_{x \to -1^+} \frac{2[+(x + 1)]}{x + 1} = \lim_{x \to -1^-} (2) = 2.$$

Because the left and right limits are not equal, $\lim_{x \to 1} h(x)$ does not exist.

$\lim_{x \to -1} k(x)$ appears to be near 3 from the graph. Evaluating gives

$$\lim_{x \to -1} k(x) = \lim_{x \to -1} \frac{x^3 + 1}{x + 1} = \lim_{x \to -1} \frac{(x + 1)(x^2 - x + 1)}{x + 1}$$

$$= \lim_{x \to -1} (x^2 - x + 1) = 3.$$

EXAMPLE 8

As mentioned in the Application Preview, USA Steel has shown that the cost C of removing p percent of the particulate pollution from the smokestack emissions at one of its plants is

$$C = C(p) = \frac{7300p}{100 - p}.$$

To investigate the cost of removing as much of the pollution as possible, find:

(a) the cost of removing 50% of the pollution,
(b) the cost of removing 90% of the pollution,
(c) the cost of removing 99% of the pollution, and
(d) the cost of removing 100% of the pollution.

Solution

(a) The cost of removing 50% of the pollution is $7300, because

$$C(50) = \frac{7300(50)}{100 - 50} = \frac{365,000}{50} = 7300.$$

(b) The cost of removing 90% of the pollution is $65,700 because

$$C(90) = \frac{7300(90)}{100 - 90} = \frac{657,000}{10} = 65,700.$$

(c) The cost of removing 99% of the pollution is $722,700 because

$$C(99) = \frac{7300(99)}{100 - 99} = \frac{722,700}{1} = 722,700.$$

(d) The cost of removing 100% of the pollution is undefined because the denominator of the function is 0 when $p = 100$. To see what the cost approaches as p approaches 100 from values smaller than 100, we evaluate $\lim\limits_{x \to 100^-} \dfrac{7300p}{100 - p}$. Because $\lim\limits_{x \to 100^-} \dfrac{7300}{100 - p} = +\infty$, the cost of removing 100% of the pollution approaches infinity. (That is, it is impossible to remove 100% of the pollution.)

CHECKPOINT SOLUTIONS

1. Yes. For example, Figure 9.1 on page 578 and Table 9.1 show that this is possible for $g(x) = \dfrac{x^2 - x - 6}{x + 2}$. Remember that $\lim\limits_{x \to c} f(x)$ does not depend on $f(c)$.

2. Not necessarily. Figure 9.3(c) on page 581 shows the graph of $y = h(x)$ with $h(2) = 0$, but $\lim\limits_{x \to 2} h(x)$ does not exist.

3. Not necessarily. Figure 9.2(c) on page 579 shows the graph of $y = h(x)$ with $\lim\limits_{x \to 2} h(x) = 1$ but $h(2) = 4$.

4. Not necessarily. For example, Figure 9.3(c) on page 581 shows the graph of $y = h(x)$ with $\lim\limits_{x \to 2^-} h(x) = 0$, but with $\lim\limits_{x \to 2^+} h(x) = 2$, so the limit doesn't exist. Recall that if $\lim\limits_{x \to c^-} f(x) = \lim\limits_{x \to c^+} f(x) = L$, then $\lim\limits_{x \to c} f(x) = L$.

5. Yes, Properties I–IV yield this result.

6. Not necessarily. If $h(c) \neq 0$, then this is true. Otherwise it is not true.

7. For both (a) and (b), $g(x)/h(x)$ has the $0/0$ indeterminate form at $x = c$. In this case we can make no general conclusion about the limit. It is possible for the

limit to exist (and be zero or nonzero) or not to exist. Consider the following 0/0 indeterminate forms.

(i) $\lim\limits_{x \to 0} \dfrac{x^2}{x} = \lim\limits_{x \to 0} x = 0$ (ii) $\lim\limits_{x \to 0} \dfrac{x(x + 1)}{x} = \lim\limits_{x \to 0}(x + 1) = 1$

(iii) $\lim\limits_{x \to 0} \dfrac{x}{x^2} = \lim\limits_{x \to 0} \dfrac{1}{x}$, which does not exist

8. $\lim\limits_{x \to c} \dfrac{g(x)}{h(x)}$ does not exist and $\lim\limits_{x \to c} \dfrac{h(x)}{g(x)} = 0$

EXERCISE 9.1

In problems 1–8, a graph of $y = f(x)$ is shown and a c-value is given. For each problem, use the graph to find (a) $\lim\limits_{x \to c} f(x)$ and (b) $f(c)$, whenever they exist.

1. $c = 2$

2. $c = 2$

3. $c = 4$

4. $c = 6$

5. $c = 20$

6. $c = -10$

7. $c = -8$

8. $c = -2$

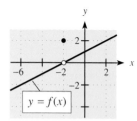

In problems 9–12, use the graph of $y = f(x)$ and the given c-value to find the following, whenever they exist.

(a) $\lim\limits_{x \to c^-} f(x)$ (b) $\lim\limits_{x \to c^+} f(x)$

(c) $\lim\limits_{x \to c} f(x)$ (d) $f(c)$

9. $c = -10$

10. $c = 2$

11. $c = -4\frac{1}{2}$

12. $c = 2$

In problems 13–18, complete each table and predict the limit, if it exists.

13. $f(x) = \dfrac{x^2 - 4x + 4}{x^2 - 2x}$

$\lim\limits_{x \to 2} f(x) = ?$

x	$f(x)$
1.9	
1.99	
1.999	
$\downarrow$	$\downarrow$
2	?
$\uparrow$	$\uparrow$
2.001	
2.01	
2.1	

14. $f(x) = \dfrac{2x - 10}{x^2 - 25}$

$\lim\limits_{x \to 5} f(x) = ?$

x	$f(x)$
4.9	
4.99	
4.999	
$\downarrow$	$\downarrow$
5	?
$\uparrow$	$\uparrow$
5.001	
5.01	
5.1	

15. $f(x) = \dfrac{1 - x^2}{x - 1}$

$\lim\limits_{x \to 1} f(x) = ?$

x	$f(x)$
0.9	
0.99	
0.999	
$\downarrow$	$\downarrow$
1	?
$\uparrow$	$\uparrow$
1.001	
1.01	
1.1	

16. $f(x) = \dfrac{2x + 1}{\frac{1}{4} - x^2}$

$\lim\limits_{x \to -0.5} f(x) = ?$

x	$f(x)$
-0.51	
-0.501	
-0.5001	
$\downarrow$	$\downarrow$
-0.5	?
$\uparrow$	$\uparrow$
-0.4999	
-0.499	
-0.49	

17. $f(x) = \begin{cases} 5x - 1 & \text{for } x < 1 \\ 8 - 2x - x^2 & \text{for } x \geq 1 \end{cases}$

$\lim\limits_{x \to 1} f(x) = ?$

x	$f(x)$
0.9	
0.99	
0.999	
$\downarrow$	$\downarrow$
1	?
$\uparrow$	$\uparrow$
1.001	
1.01	
1.1	

18. $f(x) = \begin{cases} 4 - x^2 & \text{for } x \leq -2 \\ x^2 + 2x & \text{for } x > -2 \end{cases}$

$\lim\limits_{x \to -2} f(x) = ?$

x	$f(x)$
-2.1	
-2.01	
-2.001	
$\downarrow$	$\downarrow$
-2	?
$\uparrow$	$\uparrow$
-1.999	
-1.99	

In problems 19–40, use properties of limits and algebraic methods to find the limits, if they exist.

19. $\lim\limits_{x \to 2}(x - 3)$

20. $\lim\limits_{x \to 2}(3x - 4)$

21. $\lim\limits_{x \to -1}(4x^3 - 2x^2 + 2)$

22. $\lim\limits_{x \to 3}(2x^3 - 12x^2 + 5x + 3)$

23. $\lim\limits_{x \to 4} \dfrac{4x^2 - 12x - 16}{x + 4}$

24. $\lim\limits_{x \to -1} \dfrac{x^3 - 3x + 1}{x - 1}$

25. $\lim\limits_{x \to 1/2} \dfrac{4x - 2}{4x^2 + 1}$

26. $\lim\limits_{x \to -1/3} \dfrac{1 - 3x}{9x^2 + 1}$

27. $\lim\limits_{x\to 3} \dfrac{x^2 - 9}{x - 3}$

28. $\lim\limits_{x\to -4} \dfrac{x^2 - 16}{x + 4}$

29. $\lim\limits_{x\to -2} \dfrac{x^2 + 4x + 4}{x + 2}$

30. $\lim\limits_{x\to 1} \dfrac{x^2 + x - 2}{x - 1}$

31. $\lim\limits_{x\to 3} f(x)$, where $f(x) = \begin{cases} 10 - 2x & \text{for } x < 3 \\ x^2 - x & \text{for } x \geq 3 \end{cases}$

32. $\lim\limits_{x\to 5} f(x)$, where $f(x) = \begin{cases} 7x - 10 & \text{for } x < 5 \\ 25 & \text{for } x \geq 5 \end{cases}$

33. $\lim\limits_{x\to -1} f(x)$, where $f(x) = \begin{cases} x^2 + \dfrac{4}{x} & \text{for } x \leq -1 \\ 3x^3 - x - 1 & \text{for } x > -1 \end{cases}$

34. $\lim\limits_{x\to 2} f(x)$, where $f(x) = \begin{cases} \dfrac{x^3 - 4}{x - 3} & \text{for } x \leq 2 \\ \dfrac{3 - x^2}{x} & \text{for } x > 2 \end{cases}$

35. $\lim\limits_{x\to 2} \dfrac{x^2 + 6x + 9}{x - 2}$

36. $\lim\limits_{x\to 5} \dfrac{x^2 - 6x + 8}{x - 5}$

37. $\lim\limits_{x\to -1} \dfrac{x^2 + 5x + 6}{x + 1}$

38. $\lim\limits_{x\to 3} \dfrac{x^2 + 2x - 3}{x - 3}$

39. $\lim\limits_{h\to 0} \dfrac{(x + h)^3 - x^3}{h}$

40. $\lim\limits_{h\to 0} \dfrac{2(x + h)^2 - 2x^2}{h}$

In problems 41–50, evaluate each limit by the method of your choice—algebraic, numeric, or graphic. Then check your work by evaluating with another method.

41. $\lim\limits_{x\to 2} \dfrac{x^3 - 4x}{2x^2 - x^3}$

42. $\lim\limits_{x\to 1/2} \dfrac{x^2 - \dfrac{1}{4}}{2x - 1}$

43. $\lim\limits_{x\to 10} \dfrac{x^2 - 19x + 90}{3x^2 - 30x}$

44. $\lim\limits_{x\to -3} \dfrac{x^4 + 3x^3}{2x^4 - 18x^2}$

45. $\lim\limits_{x\to -1} \dfrac{x^3 - x}{x^2 + 2x + 1}$

46. $\lim\limits_{x\to 5} \dfrac{x^2 - 7x + 10}{x^2 - 10x + 25}$

47. $\lim\limits_{x\to 4} f(x)$, where $f(x) = \begin{cases} 12 - \frac{3}{4}x & \text{for } x \leq 4 \\ x^2 - 7 & \text{for } x > 4 \end{cases}$

48. $\lim\limits_{x\to 7} f(x)$, where $f(x) = \begin{cases} 2 + x - x^2 & \text{for } x \leq 7 \\ 23 - 9x & \text{for } x > 7 \end{cases}$

49. $\lim\limits_{x\to -3} f(x)$, where $f(x) = \begin{cases} \dfrac{x + 1}{x - 7} & \text{for } x < -3 \\ \dfrac{x^2 - x}{18 - 10x} & \text{for } x \geq -3 \end{cases}$

50. $\lim\limits_{x\to 6} f(x)$, where $f(x) = \begin{cases} 3x + 5 & \text{for } x < 6 \\ x^2 - x - 9 & \text{for } x \geq 6 \end{cases}$

51. Use values 0.1, 0.01, 0.001, 0.0001, and 0.00001 with your calculator to approximate

$$\lim\limits_{a\to 0} (1 + a)^{1/a}$$

to 3 decimal places. This limit equals the special number e that was discussed in Sections 5.1 and 6.2.

52. If $\lim\limits_{x\to 2}[f(x) + g(x)] = 5$ and $\lim\limits_{x\to 2} g(x) = 11$, find

(a) $\lim\limits_{x\to 2} f(x)$.

(b) $\lim\limits_{x\to 2} \{[f(x)]^2 - [g(x)]^2\}$.

(c) $\lim\limits_{x\to 2} \dfrac{3g(x)}{f(x) - g(x)}$.

53. If $\lim\limits_{x\to 3} f(x) = 4$ and $\lim\limits_{x\to 3} g(x) = -2$, find

(a) $\lim\limits_{x\to 3}[f(x) + g(x)]$. (b) $\lim\limits_{x\to 3}[f(x) - g(x)]$.

(c) $\lim\limits_{x\to 3}[f(x) \cdot g(x)]$. (d) $\lim\limits_{x\to 3} \dfrac{g(x)}{f(x)}$.

54. (a) If $\lim\limits_{x\to 2^+} f(x) = 5$, $\lim\limits_{x\to 2^-} f(x) = 5$, and $f(2) = 0$, find $\lim\limits_{x\to 2} f(x)$, if it exists. Explain your conclusions.

(b) If $\lim\limits_{x\to 0^+} f(x) = 3$, $\lim\limits_{x\to 0^-} f(x) = 0$, and $f(0) = 0$, find $\lim\limits_{x\to 0} f(x)$, if it exists. Explain your conclusions.

APPLICATIONS

55. **Revenue** The total revenue for a product is given by

$$R(x) = 1600x - x^2,$$

where x is the number of units sold. What is $\lim\limits_{x\to 100} R(x)$?

56. **Profit** If the profit function for a product is given by

$$P(x) = 92x - x^2 - 1760,$$

find $\lim\limits_{x\to 40} P(x)$.

57. **Sales and training** The average monthly sales volume (in thousands of dollars) for a firm depends on the number of hours x of training of its sales staff, according to

$$S(x) = \dfrac{4}{x} + 30 + \dfrac{x}{4}, \quad 4 \leq x \leq 100.$$

(a) Find $\lim\limits_{x\to 4^+} S(x)$. (b) Find $\lim\limits_{x\to 100^-} S(x)$.

58. **Sales and training** During the first four months of employment, the monthly sales S (in thousands of dollars) for a new salesperson depends on the number of hours x of training, as follows:

$$S = S(x) = \frac{9}{x} + 10 + \frac{x}{4}, \quad \text{with } x \geq 4.$$

(a) Find $\lim\limits_{x \to 4^+} S(x)$.   (b) Find $\lim\limits_{x \to 10} S(x)$.

59. **Advertising and sales** Suppose that the daily sales S (in dollars) t days after the end of an advertising campaign is

$$S = S(t) = 400 + \frac{2400}{t + 1}.$$

(a) Find $S(0)$.   (b) Find $\lim\limits_{t \to 7} S(t)$.

(c) Find $\lim\limits_{t \to 14} S(t)$.

60. **Advertising and sales** Sales y (in thousands of dollars) are related to advertising expenses x (in thousands of dollars) according to

$$y = y(x) = \frac{200x}{x + 10}, \quad x \geq 0.$$

(a) Find $\lim\limits_{x \to 10} y(x)$.   (b) Find $\lim\limits_{x \to 0^+} y(x)$.

61. **Productivity** During an 8-hour shift, the rate of change of productivity (in units per hour) of children's phonographs assembled after t hours on the job is

$$r(t) = \frac{128(t^2 + 6t)}{(t^2 + 6t + 18)^2}, \quad 0 \leq t \leq 8.$$

(a) Find $\lim\limits_{x \to 4} r(t)$.   (b) Find $\lim\limits_{x \to 8^-} r(t)$.

(c) Is the rate of productivity higher near the lunch break (at $t = 4$) or near quitting time (at $t = 8$)?

62. **Revenue** If the revenue for a product is $R(x) = 100x - 0.1x^2$, and the average revenue per unit is

$$\bar{R}(x) = \frac{R(x)}{x}, \quad x > 0,$$

find (a) $\lim\limits_{x \to 100} \dfrac{R(x)}{x}$ and (b) $\lim\limits_{x \to 0^+} \dfrac{R(x)}{x}$.

63. **Cost-benefit** Suppose that the cost C of obtaining water that contains p percent impurities is given by

$$C(p) = \frac{120,000}{p} - 1200.$$

(a) Find $\lim\limits_{p \to 100^-} C(p)$, if it exists. Interpret this result.

(b) Find $\lim\limits_{p \to 0^+} C(p)$, if it exists.

(c) Is complete purity possible? Explain.

64. **Cost-benefit** Suppose that the cost C of removing p percent of the particulate pollution from the smokestacks of an industrial plant is given by

$$C(p) = \frac{730,000}{100 - p} - 7300.$$

(a) Find $\lim\limits_{p \to 80} C(p)$

(b) Find $\lim\limits_{p \to 100^-} C(p)$, if it exists.

(c) Can 100% of the particulate pollution be removed? Explain.

65. **Federal income tax** Use the following tax rate schedule for single taxpayers and create a table of values that could be used to find the following limits, if they exist. Let x represent the amount on Form 1040, line 37, and $T(x)$ represent the tax due (entered on Form 1040, line 38).

(a) $\lim\limits_{x \to 22,100^-} T(x)$ (b) $\lim\limits_{x \to 22,100^+} T(x)$

(c) $\lim\limits_{x \to 22,100} T(x)$

Schedule X — Use if your filing status is Single

If the amount on Form 1040, line 37, is: over—	But not over—	Enter on Form 1040, line 38	of the amount over—
$0	$22,100	- - - - - 15%	$0
22,100	53,500	$3,315 + 28%	22,100
53,500	115,500	12,107 + 31%	53,500
115,000	250,000	31,172 + 36%	115,000
250,000		- - - - - 79,772 + 39.6%	250,000

SOURCE: Internal Revenue Service, 1993 Form 1040 Instructions

66. **Parking costs** The Ace Parking Garage charges $2.00 for parking for 2 hours or less, and 50 cents for each extra hour or part of an hour after the 2-hour minimum. The parking charges for the first 5 hours could be written as a function of the time as follows.

$$f(t) = \begin{cases} \$2.00 & \text{if } 0 < t \leq 2 \\ \$2.50 & \text{if } 2 < t \leq 3 \\ \$3.00 & \text{if } 3 < t \leq 4 \\ \$3.50 & \text{if } 4 < t \leq 5 \end{cases}$$

(a) Find $\lim\limits_{t \to 1} f(t)$, if it exists.

(b) Find $\lim\limits_{t \to 2} f(t)$, if it exists.

67. **Municipal water rates** The Beaver, Pennsylvania, Bor-ough Municipal Authority has the following rates per 1000 gallons of water used.

Usage (x)	Cost per 1000 Gallons (C(x))
First 100,000 gallons	1.557
Next 900,000 gallons	1.040
Over 1,000,000 gallons	0.689

Write a function $C = C(x)$ that models the charges, and find $\lim\limits_{x \to 1000} C(x)$ (that is, as usage approaches 1,000,000 gallons).

68. **Telephone charges** A direct-dial call from Savannah, Georgia, to Atlanta, Georgia, costs $0.28 for the first minute and $0.24 for each additional minute or part of a minute. If $C = C(t)$ is the charge for a call lasting t minutes, create a table of charges for calls lasting close to one minute and use it to find the following limits, if they exist.

(a) $\lim\limits_{t \to 1^-} C(t)$ (b) $\lim\limits_{t \to 1^+} C(t)$ (c) $\lim\limits_{t \to 1} C(t)$

Dow-Jones average The graph in the figure shows the Dow-Jones Industrial Average (DJIA) at 5-minute intervals for Monday, January 17, 1994. Use the graph for problems 69 and 70, with t as the time of day and $D(t)$ as the DJIA at time t.

69. Find $\lim\limits_{t \to 9:30 \text{ AM}^+} D(t)$, if it exists. Explain what this limit corresponds to.

70. Find $\lim\limits_{t \to 4:00 \text{ PM}^-} D(t)$, if it exists. Explain what this limit corresponds to.

DJIA at 5-minute intervals

Friday's close

9:30 10 11 12 1 2 3 4
am   pm

SOURCE: *Wall Street Journal*, January 18, 1994

Farm Workers The percent of U.S. workers in farm occupa-tions during certain years is shown in the table.

Year	Percent	Year	Percent
1820	71.8	1950	11.6
1850	63.7	1960	6.1
1870	53	1970	3.6
1900	37.5	1980	2.7
1920	27	1985	2.8
1930	21.2	1990	2.4
1940	17.4		

SOURCE: *The World Almanac and Book of Facts*, 1993

Assume that the percentage of U.S. workers in farm occupa-tions can be modeled with the function

$$f(t) = 1000 \cdot \frac{-8.0912t + 1558.9}{1.09816t^2 - 122.183t + 21472.6},$$

where t is the number of years past 1800. (A graph of $f(t)$ along with the data in the table for the years corresponding to 1800 to 1990 is shown in the figure.) Use the table and equation in problems 71 and 72.

71. (a) Find $\lim\limits_{t \to 200} f(t)$, if it exists.
    (b) What does this limit predict?
    (c) Is the equation accurate as $t \to 200$? Explain.

72. (a) Find $\lim\limits_{t \to 100} f(t)$, if it exists.
    (b) What does this limit predict?
    (c) Is the equation accurate as $t \to 100$? Explain.

9.2 CONTINUOUS FUNCTIONS; LIMITS AT INFINITY

Objectives

- To determine whether a function is continuous or discontinuous
- To determine where a function is discontinuous
- To find limits at infinity

APPLICATION PREVIEW

Suppose that a friend of yours and her husband have a taxable income of $89,150, and she tells you that she doesn't want to make any more money since that would put them in a higher tax bracket. She makes this statement because the tax rate schedule for married taxpayers filing a joint return (shown in the table) appears to have a jump in taxes for taxable income at $89,150.

Schedule Y-I — Use if your filing status is Married filing jointly or Qualifying widow(er)

If the amount on Form 1040, line 37, is: Over—	But not over—	Enter on Form 1040, line 38	of the amount over—
$0	$36,900	- - - - - - - 15%	$0
36,900	89,150	$5,535.00 + 28%	36,900
89,150	140,000	20,165.00 + 31%	89,150
140,000	250,000	35,928.50 + 36%	140,000
250,000	- - - - -	75,528.50 + 39.6%	250,000

SOURCE: Internal Revenue Service, 1993 Form 1040 Instructions

To see whether the couple's taxes would jump to some higher level, we will write the function that gives income tax for married taxpayers as a function of taxable income and show that the function is **continuous.** That is, we will see that the tax paid does not jump at $89,150 even though the tax on income above $89,150 is collected at a higher rate. In this section, we will show how to determine whether a function is continuous and we will investigate some different types of discontinuous functions.

We have found that $f(c)$ is the same as the limit as $x \to c$ for any polynomial function $f(x)$ and any real number c. Any function for which this special property holds is called a **continuous function.** The graphs of such functions can be drawn without lifting the pencil from the paper, while graphs of others may have holes, vertical asymptotes, or jumps that make it impossible to draw them without lifting the pencil. Even though a function may not be continuous everywhere, it will likely have some points where the limit of the function as $x \to c$ is the same as $f(c)$. In general, we define continuity of a function at the value $x = c$ as follows.

Continuity at a Point

The function f is **continuous at $x = c$** if *all* of the following conditions are satisfied.

1. $f(c)$ exists   2. $\lim_{x \to c} f(x)$ exists   3. $\lim_{x \to c} f(x) = f(c)$

If one or more of the conditions above do not hold, we say the function is **discontinuous at $x = c$.** Figure 9.9 shows graphs of some functions that are discontinuous at $x = 2$.

Figure 9.9

(a) $f(x) = \dfrac{1}{x - 2}$

(i) $\lim_{x \to 2} f(x)$ and (ii) $f(2)$ do not exist.

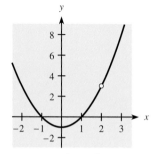

(b) $f(x) = \dfrac{x^3 - 2x^2 - x + 2}{x - 2}$

$f(2)$ does not exist.

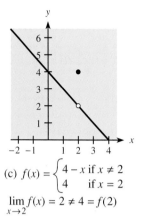

(c) $f(x) = \begin{cases} 4 - x & \text{if } x \neq 2 \\ 4 & \text{if } x = 2 \end{cases}$

$\lim_{x \to 2} f(x) = 2 \neq 4 = f(2)$

(d) $f(x) = \begin{cases} 1/(x - 2)^2 & \text{if } x \neq 2 \\ 0 & \text{if } x = 2 \end{cases}$

$\lim_{x \to 2} f(x)$ does not exist.

In Section 9.1 we saw that if f is a polynomial function, then $\lim_{x \to c} f(x) = f(c)$ for every real number c, and also that $\lim_{x \to c} h(x) = h(c)$ if $h(x) = \dfrac{f(x)}{g(x)}$ is a rational function and $g(c) \neq 0$. Thus, by definition, we have the following.

Every polynomial function is continuous for all real numbers.

Every rational function is continuous at all values of x except those that make the denominator 0.

EXAMPLE 1

For what values of x, if any, is the function $h(x) = \dfrac{3x + 2}{4x - 6}$ continuous?

Solution
This is a rational function, so it is continuous for all values of x except for those that make the denominator, $4x - 6$, equal to 0. Because $4x - 6 = 0$ at $x = 3/2$, $h(x)$ is continuous for all real numbers except $x = 3/2$.

EXAMPLE 2

For what values of x, if any, is the function f discontinuous if

$$f(x) = \frac{x^2 - x - 2}{x^2 - 4}?$$

Solution
This is a rational function, so it is continuous everywhere except where the denominator is 0. To find the zeros of the denominator, we factor $x^2 - 4$.

$$f(x) = \frac{x^2 - x - 2}{x^2 - 4} = \frac{x^2 - x - 2}{(x - 2)(x + 2)}$$

Because the denominator is 0 for $x = 2$ and for $x = -2$, $f(2)$ and $f(-2)$ do not exist (recall that division by 0 is undefined). Thus, the function is discontinuous at $x = 2$ and $x = -2$. The graph of this function (see Figure 9.10) shows a hole at $x = 2$ and a vertical asymptote at $x = -2$.

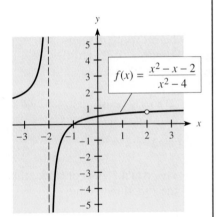

$$f(x) = \frac{x^2 - x - 2}{x^2 - 4}$$

Figure 9.10

CHECKPOINT

1. Find any x-values where the following functions are discontinuous.

(a) $f(x) = x^3 - 3x + 1$ (b) $g(x) = \dfrac{x^3 - 1}{(x - 1)(x + 2)}$

If the pieces of a piecewise defined function are polynomials, the only values of x where the function might be discontinuous are those at which the definition of the function changes.

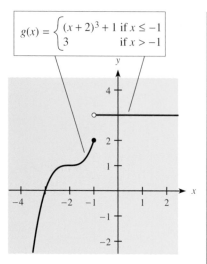

$$g(x) = \begin{cases} (x+2)^3 + 1 & \text{if } x \le -1 \\ 3 & \text{if } x > -1 \end{cases}$$

Figure 9.11

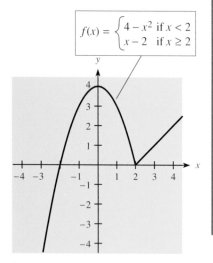

$$f(x) = \begin{cases} 4 - x^2 & \text{if } x < 2 \\ x - 2 & \text{if } x \ge 2 \end{cases}$$

Figure 9.12

EXAMPLE 3

Determine the values of x, if any, for which the following functions are discontinuous.

(a) $g(x) = \begin{cases} (x + 2)^3 + 1 & \text{if } x \le -1 \\ 3 & \text{if } x > -1 \end{cases}$ (b) $f(x) = \begin{cases} 4 - x^2 & \text{if } x < 2 \\ x - 2 & \text{if } x \ge 2 \end{cases}$

Solution

(a) $g(x)$ is a piecewise defined function in which each part is a polynomial. Thus, to see if a discontinuity exists, we need only check the value of x for which the definition of the function changes, that is, at $x = -1$. Since $x = -1$ satisfies $x \le -1$, $g(-1) = (-1 + 2)^3 + 1 = 2$. Evaluating the left- and right-hand limits gives

$$\lim_{x \to -1^-} g(x) = \lim_{x \to -1^-} [(x + 2)^3 + 1] = (-1 + 2)^3 + 1 = 2$$

and

$$\lim_{x \to -1^+} g(x) = \lim_{x \to -1^+} 3 = 3.$$

Because the left- and right-hand limits differ, $\lim_{x \to -1} g(x)$ does not exist, so $g(x)$ is discontinuous at $x = -1$. This result is confirmed by examining the graph of g, shown in Figure 9.11.

(b) As with $g(x)$, $f(x)$ is continuous everywhere except perhaps at $x = 2$, where the definition of $f(x)$ changes. Since $x = 2$ satisfies $x \ge 2$, $f(2) = 2 - 2 = 0$. The left- and right-hand limits are

$$\lim_{x \to 2^-} f(x) = \lim_{x \to 2^-} (4 - x^2) = 4 - 2^2 = 0$$

and

$$\lim_{x \to 2^+} f(x) = \lim_{x \to 2^+} (x - 2) = 2 - 2 = 0.$$

Since the right- and left-hand limits are equal, we conclude that $\lim_{x \to 2} f(x) = 0$. Since the limit is equal to the functional value

$$\lim_{x \to 2} f(x) = f(2),$$

we conclude that f is continuous at $x = 2$ and thus f is continuous for all values of x. This result is confirmed by the graph of f, shown in Figure 9.12.

We now consider the problem posed in the Application Preview.

EXAMPLE 4

The tax rate schedule for married taxpayers filing a joint return (shown in the table) appears to have a jump in taxes for taxable income at $89,150.

Schedule Y-I — Use if your filing status is Married filing jointly or Qualifying widow(er)

If the amount on Form 1040, line 37, is: Over—	But not over—	Enter on Form 1040, line 38	of the amount over—
$0	$36,900	– – – – – 15%	$0
36,900	89,150	$5,535.00 + 28%	36,900
89,150	140,000	20,165.00 + 31%	89,150
140,000	250,000	35,928.50 + 36%	140,000
250,000	– – – – –	75,528.50 + 39.6%	250,000

SOURCE: Internal Revenue Service, 1993 Form 1040 Instructions

(a) Use the table and write the function that gives income tax for married taxpayers as a function of taxable income, and graph the function.

(b) Is the function in (a) continuous at $x = 89{,}150$?

(c) A married friend of yours and her husband have a taxable income of $89,150, and she tells you that she doesn't want to make any more money since that would put her in a higher tax bracket. What would you tell her to do if she is offered a raise?

Solution

(a) The function that gives the tax due for married taxpayers is

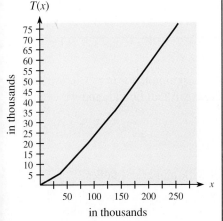

T(x)

in thousands

in thousands

Figure 9.13

$$T(x) = \begin{cases} 0.15x & \text{if } 0 \le x \le 36{,}900 \\ 5{,}535 + 0.28(x - 36{,}900) & \text{if } 36{,}900 < x \le 89{,}150 \\ 20{,}165 + 0.31(x - 89{,}150) & \text{if } 89{,}150 < x \le 140{,}000 \\ 35{,}928.50 + 0.36(x - 140{,}000) & \text{if } 140{,}000 < x \le 250{,}000 \\ 75{,}528.50 + 0.396(x - 250{,}000) & \text{if } x > 250{,}000 \end{cases}$$

The graph of this function is shown in Figure 9.13.

(b) This function is continuous at $x = 89{,}150$, because

  (i) $T(89{,}150) = 20{,}165$, so $T(89{,}150)$ exists.

  (ii) Because the function is piecewise defined near 89,150, we evaluate $\lim\limits_{x \to 89{,}150} T(x)$ by evaluating $\lim\limits_{x \to 89{,}150^-} T(x)$ and $\lim\limits_{x \to 89{,}150^+} T(x)$.

$$\lim_{x \to 89{,}150^-} T(x) = \lim_{x \to 89{,}150^-} [5535 + 0.28(x - 36{,}900)] = 20{,}165$$

$$\lim_{x \to 89{,}150^+} T(x) = \lim_{x \to 89{,}150^+} [20{,}165 + 0.31(x - 89{,}150)] = 20{,}165$$

Because these limits are the same, $\lim\limits_{x \to 89{,}150} T(x) = 20{,}165$, and so the limit exists.

  (iii) Because $T(89{,}150) = \lim\limits_{x \to 89{,}150} T(x) = 20{,}165$, the function is continuous at 89,150.

(c) If your friend earned more than \$89,150, she and her husband would pay taxes at a higher rate on the money earned *above* the \$89,150, but it would not increase the tax rate on any income *up to* \$89,150. Thus she should take any raise that's offered.

CHECKPOINT

2. If $f(x)$ and $g(x)$ are polynomials, $h(x) = \begin{cases} f(x) & \text{if } x \le a \\ g(x) & \text{if } x > a \end{cases}$ is continuous everywhere except perhaps at _____ .

We noted earlier (in Chapter 1) that the graph of $y = 1/x$ has a vertical asymptote at $x = 0$ (shown in Figure 9.14(a)). By graphing $y = 1/x$ with a large x-range or by using TRACE to let x get very large, we can see that $y = 1/x$ never becomes negative for positive x-values regardless of how large the x-value is. Although no value of x makes $1/x$ equal to 0, it is easy to see that $1/x$ approaches 0 as x gets very large. This is denoted by

$$\lim_{x \to +\infty} \frac{1}{x} = 0.$$

(a)

We say that $y = 0$ (the x-axis) is a horizontal asymptote for $y = 1/x$. We also see that $y = 1/x$ approaches 0 as x decreases without bound, and we denote this by

$$\lim_{x \to -\infty} \frac{1}{x} = 0.$$

These limits can also be established with numerical tables.

(b)

Figure 9.14

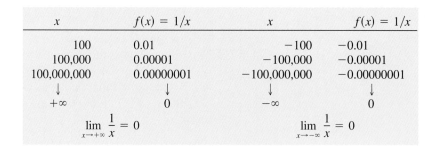

x	$f(x) = 1/x$	x	$f(x) = 1/x$
100	0.01	-100	-0.01
100,000	0.00001	$-100,000$	-0.00001
100,000,000	0.00000001	$-100,000,000$	-0.00000001
$\downarrow$	$\downarrow$	$\downarrow$	$\downarrow$
$+\infty$	0	$-\infty$	0
$\displaystyle\lim_{x \to +\infty} \frac{1}{x} = 0$		$\displaystyle\lim_{x \to -\infty} \frac{1}{x} = 0$	

It is also true that $\displaystyle\lim_{x \to +\infty} \frac{c}{x^p} = 0$ and $\displaystyle\lim_{x \to -\infty} \frac{c}{x^p} = 0$ for any positive p and any constant c (see Figure 9.14(b)) for the graph of $y = 2/x^2$. By using graphs and/or tables of values, we can conclude the following.

Limits at Infinity If c is any constant, then

1. $\lim\limits_{x \to +\infty} c = c$ and $\lim\limits_{x \to -\infty} c = c$.

2. $\lim\limits_{x \to +\infty} \dfrac{c}{x^p} = 0$, where $p > 0$.

3. $\lim\limits_{x \to -\infty} \dfrac{c}{x^n} = 0$, where $n > 0$ is any integer.

In order to use these properties for finding the limits of rational functions as x approaches $+\infty$ or $-\infty$, we first divide each term of the numerator and denominator by the highest power of x present and then determine the limit of the resulting expression.

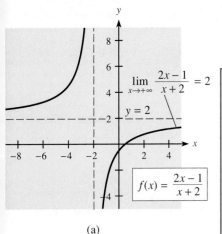

$$\lim\limits_{x \to +\infty} \frac{2x-1}{x+2} = 2$$

$$y = 2$$

$$f(x) = \frac{2x-1}{x+2}$$

(a)

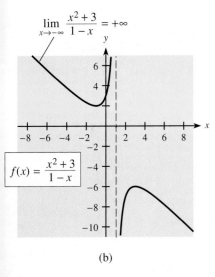

$$\lim\limits_{x \to -\infty} \frac{x^2+3}{1-x} = +\infty$$

$$f(x) = \frac{x^2+3}{1-x}$$

(b)

Figure 9.15

EXAMPLE 5

Find each of the following limits, if they exist.

(a) $\lim\limits_{x \to +\infty} \dfrac{2x-1}{x+2}$ 　　　　　(b) $\lim\limits_{x \to -\infty} \dfrac{x^2+3}{1-x}$

Solution

(a) The highest power of x present is x^1, so we divide each term in the numerator and denominator by x and then use the properties for limits at infinity.

$$\lim\limits_{x \to +\infty} \frac{2x-1}{x+2} = \lim\limits_{x \to +\infty} \frac{\dfrac{2x}{x} - \dfrac{1}{x}}{\dfrac{x}{x} + \dfrac{2}{x}} = \lim\limits_{x \to +\infty} \frac{2 - \dfrac{1}{x}}{1 + \dfrac{2}{x}}$$

$$= \frac{2-0}{1+0} = 2 \quad \text{(by Properties 1 and 2)}$$

Figure 9.15(a) shows the graph of this function with the y-coordinates of the graph approaching 2 as x approaches $+\infty$ and as x approaches $-\infty$. That is, $y = 2$ is a horizontal asymptote. Note also that there is a discontinuity (vertical asymptote) where $x = -2$.

(b) We divide each term in the numerator and denominator by x^2 and then use the properties.

$$\lim\limits_{x \to -\infty} \frac{x^2+3}{1-x} = \lim\limits_{x \to -\infty} \frac{\dfrac{x^2}{x^2} + \dfrac{3}{x^2}}{\dfrac{1}{x^2} - \dfrac{x}{x^2}} = \lim\limits_{x \to -\infty} \frac{1 + \dfrac{3}{x^2}}{\dfrac{1}{x^2} - \dfrac{1}{x}} = +\infty$$

This limit is $+\infty$ because the numerator approaches 1 and the denominator approaches 0 through positive values. Thus

$$\lim_{x \to -\infty} \frac{x^2 + 3}{1 - x} \quad \text{does not exist.}$$

The graph of this function, shown in Figure 9.15(b), has y-coordinates that increase without bound as x approaches $-\infty$ and that decrease without out bound as x approaches $+\infty$. (There is no horizontal asymptote.) Note also that there is a vertical asymptote at $x = 1$.

CHECKPOINT

3. Evaluate $\displaystyle\lim_{x \to +\infty} \frac{x^2 - 4}{2x^2 - 7}$.

Graphing Utilities

Using a graphing utility can be very helpful in locating the possible points of discontinuity of a function or in verifying conclusions about the discontinuity of a function.

Figure 9.16

(a)

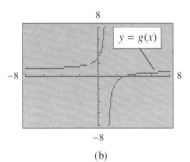

(b)

EXAMPLE 6

Graph the following four functions and answer the questions below. Justify your answers. (On some graphics calculators and computer programs, using the dot mode rather than the continuous mode may make it easier to locate points of discontinuity.)

$$f(x) = \frac{x^2 + 1}{x + 1}, \qquad g(x) = \frac{x^2 - 2x - 3}{x^2 - 1},$$

$$h(x) = \frac{|x + 1|}{x + 1}, \qquad k(x) = \begin{cases} \dfrac{-x^2}{2} - 2x & \text{if } x \le -1 \\[2mm] \dfrac{x}{2} + 2 & \text{if } x > -1 \end{cases}$$

(a) Which functions are discontinuous at $x = -1$?
(b) What happens at $x = -1$ for the discontinuous functions?
(c) Which of these functions is discontinuous at an x-value other than $x = -1$?

Solution

The graphs are shown in Figure 9.16.

(a) From the graphs, $f(x)$, $g(x)$, and $h(x)$ appear to be discontinuous at $x = -1$. Analytically, we see that the denominators of $f(x)$, $g(x)$, and $h(x)$ are 0 at $x = -1$; $f(-1)$, $g(-1)$, and $h(-1)$ do not exist, so these

Figure 9.16 (continued)

(c)

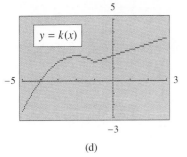

(d)

functions are not continuous at $x = -1$. The function k appears to be continuous. It is defined at $x = -1$, with
$$k(-1) = -(-1)^2/2 - 2(-1) = 3/2.$$
To find the limit of $k(x)$ as x approaches -1, we must find the left and right limits as x approaches -1.

$$\lim_{x \to -1^-} k(x) = \lim_{x \to -1^-} \frac{-(x)^2}{2} - 2(x) = \frac{-(-1)^2}{2} - 2(-1) = \frac{3}{2}$$

$$\lim_{x \to -1^+} k(x) = \lim_{x \to -1^+} \frac{x}{2} + 2 = \frac{-1}{2} + 2 = \frac{3}{2}$$

Thus $\lim_{x \to -1} k(x) = 3/2$, and because $k(-1) = 3/2$, the function $k(x)$ is continuous at $x = -1$.

(b) The graph of $y = f(x)$ has a vertical asymptote at $x = -1$.
The graph of $y = g(x)$ has a hole (missing point) at $x = -1$.
The graph of $y = h(x)$ has a jump at $x = -1$.

(c) The graph of $y = g(x)$ has a vertical asymptote at $x = 1$, so the function $g(x)$ is discontinuous at $x = 1$. Analytically, we see that the denominator of $g(x)$ is 0 at $x = 1$.

SUMMARY

The following information is useful in discussing continuity of functions.

A. A polynomial function is continuous everywhere.

B. A rational function is a function of the form $\dfrac{f(x)}{g(x)}$, where $f(x)$ and $g(x)$ are polynomials.

    1. If $g(x) \neq 0$ at any value of x, the function is continuous everywhere.
    2. If $g(c) = 0$, the function is discontinuous at $x = c$.
        (a) If $g(c) = 0$ and $f(c) \neq 0$, then there is a vertical asymptote at $x = c$.
        (b) If $g(c) = 0$ and $\lim_{x \to c} \dfrac{f(x)}{g(x)} = L$, then the graph has an open circle at $x = c$.

C. A piecewise defined function *may* have a discontinuity at any x-value where the function changes its formula. One-sided limits must be used to see if the limit exists.

The following steps are useful when evaluating limits at infinity for a rational function $f(x) = p(x)/q(x)$.

    1. Divide both $p(x)$ and $q(x)$ by the highest power of x found in either polynomial.

    2. Use the properties of limits at infinity to complete the evaluation.

CHECKPOINT
SOLUTIONS

1. (a) This is a polynomial function, so it is continuous at all values of x (discontinuous at none).
   (b) This is a rational function. It is discontinuous at $x = 1$ and $x = -2$ because these values make its denominator 0.

2. $x = a$.

3. $\lim\limits_{x \to +\infty} \dfrac{x^2 - 4}{2x^2 - 7} = \lim\limits_{x \to +\infty} \dfrac{1 - \dfrac{4}{x^2}}{2 - \dfrac{7}{x^2}} = \dfrac{1 - 0}{2 - 0} = \dfrac{1}{2}$

EXERCISE 9.2

Problems 1 and 2 refer to the figure below. For each given x-value, use the figure to determine whether the function is continuous or discontinuous at that x-value. If the function is discontinuous, state which of the three conditions that define continuity are not satisfied.

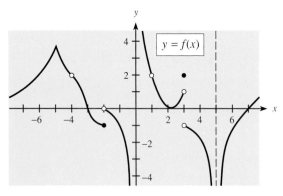

1. (a) $x = -5$ (b) $x = 1$ (c) $x = 3$ (d) $x = 0$
2. (a) $x = 2$ (b) $x = -4$ (c) $x = -2$ (d) $x = 5$

In problems 3–14, determine whether each function is continuous or discontinuous at the given x-value. Examine the three conditions in the definition of continuity.

3. $f(x) = x^2 - 5x$, $x = 0$

4. $f(x) = 3x - 5x^3$, $x = 2$

5. $f(x) = \dfrac{x^2 - 4}{x - 2}$, $x = -2$

6. $y = \dfrac{x^2 - 9}{x + 3}$, $x = 3$

7. $y = \dfrac{x^2 - 9}{x + 3}$, $x = -3$

8. $f(x) = \dfrac{x^2 - 4}{x - 2}$, $x = 2$

9. $y = \dfrac{x^2 + 5x - 6}{x + 1}$, $x = -1$

10. $y = \dfrac{x^2 - 2x - 3}{x - 1}$, $x = 1$

11. $f(x) = \begin{cases} 2 & \text{if } x \le 0 \\ x + 2 & \text{if } x > 0 \end{cases}$ $x = 0$

12. $f(x) = \begin{cases} x - 3 & \text{if } x \le 2 \\ 4x - 7 & \text{if } x > 2 \end{cases}$ $x = 2$

13. $f(x) = \begin{cases} x^2 + 1 & \text{if } x \le 1 \\ 2x^2 - 1 & \text{if } x > 1 \end{cases}$ $x = 1$

14. $f(x) = \begin{cases} x^2 - x & \text{if } x \le 2 \\ 8 - 3x & \text{if } x > 2 \end{cases}$ $x = 2$

In problems 15–26, determine whether or not the given function is continuous. If it is not, identify where it is discontinuous and which condition fails to hold. You can verify your conclusions by graphing each function with a graphing utility, if one is available.

15. $f(x) = 4x^2 - 1$

16. $y = 5x^2 - 2x$

17. $g(x) = \dfrac{x^2 + 4x + 4}{x + 2}$

18. $y = \dfrac{x^2 - 5x + 4}{x + 4}$

19. $g(x) = \dfrac{4x^2 + 3x + 2}{x + 2}$

20. $y = \dfrac{4x^2 + 4x + 1}{x + 1/2}$

21. $y = \dfrac{x}{x^2 + 1}$

22. $y = \dfrac{2x - 1}{x^2 + 3}$

23. $f(x) = \begin{cases} 3 & \text{if } x \le 1 \\ x^2 + 2 & \text{if } x > 1 \end{cases}$

24. $f(x) = \begin{cases} x^3 + 1 & \text{if } x \leq 1 \\ 2 & \text{if } x > 1 \end{cases}$

25. $f(x) = \begin{cases} x - 4 & \text{if } x \leq 3 \\ x^2 - 8 & \text{if } x > 3 \end{cases}$

26. $f(x) = \begin{cases} x^2 + 4 & \text{if } x \neq 1 \\ 5 & \text{if } x = 1 \end{cases}$

Problems 27–30 each contain a function and its graph. For each problem, answer (a) and (b).

(a) Use the graph to determine, as well as you can, each of the following.
   (i) vertical asymptotes   (ii) $\lim\limits_{x \to +\infty} f(x)$
   (iii) $\lim\limits_{x \to -\infty} f(x)$

(b) Check your conclusions in part (a) by using the functions to analytically determine items (i)–(iii).

27. $f(x) = \dfrac{8}{x + 2}$

28. $f(x) = \dfrac{x - 3}{x - 2}$

29. $f(x) = \dfrac{2(x + 1)^3(x + 5)}{(x - 3)^2(x + 2)^2}$

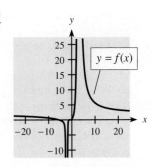

30. $f(x) = \dfrac{4x^2}{x^2 - 4x + 4}$

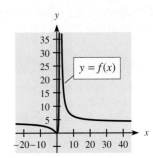

Use analytic methods to evaluate the limits in problems 31–40. You can verify your conclusions by graphing the functions with a graphing utility, if one is available.

31. $\lim\limits_{x \to +\infty} \dfrac{3}{x + 1}$

32. $\lim\limits_{x \to -\infty} \dfrac{4}{x^2 - 2x}$

33. $\lim\limits_{x \to +\infty} \dfrac{x^3 - 1}{x^3 + 4}$

34. $\lim\limits_{x \to -\infty} \dfrac{3x^2 + 2}{x^2 - 4}$

35. $\lim\limits_{x \to -\infty} \dfrac{5x^3 - 4x}{3x^3 - 2}$

36. $\lim\limits_{x \to +\infty} \dfrac{4x^2 + 5x}{x^2 - 4x}$

37. $\lim\limits_{x \to -\infty} \dfrac{4x + 3}{x^2 - 1}$

38. $\lim\limits_{x \to -\infty} \dfrac{5x^2 + 4x}{2x^3 - 1}$

39. $\lim\limits_{x \to +\infty} \dfrac{3x^2 + 5x}{6x + 1}$

40. $\lim\limits_{x \to -\infty} \dfrac{5x^3 - 8}{4x^2 + 5x}$

APPLICATIONS

41. **Sales volume** Suppose that the weekly sales volume (in thousands of units) for a product is given by

$$y = \dfrac{32}{(p + 8)^{2/5}}$$

where p is the price in dollars per unit. Is this function continuous

(a) for all values of p?
(b) at $p = 24$?
(c) for all $p \geq 0$?
(d) What is the domain for this application?

42. **Worker productivity** Suppose that the average number of minutes M that it takes a new employee to assemble one unit of a product is given by

$$M = \frac{40 + 30t}{2t + 1},$$

where t is the number of days on the job. Is this function continuous

(a) for all values of t?
(b) at $t = 14$?
(c) for all $t \geq 0$?
(d) What is the domain for this application?

43. **Demand** Suppose that the demand for a product is defined by the equation

$$p = \frac{200{,}000}{(q + 1)^2},$$

where p is the price and q is the quantity demanded.

(a) Is this function discontinuous at any value of q? What value?
(b) Because q represents quantity, we know that $q \geq 0$. Is this function continuous for $q \geq 0$?

44. **Advertising and sales** The sales volume y (in thousands of dollars) is related to advertising expenditures x (in thousands of dollars) according to

$$y = \frac{200x}{x + 10}.$$

(a) Does this function have any points of discontinuity?
(b) Since advertising expenditures x must be nonnegative, is this function continuous for these values of x?

45. **Annuities** If an annuity makes an infinite series of equal payments at the end of the interest periods, it is called a **perpetuity.** If a lump-sum investment of A_n is needed to result in n periodic payments of R when the interest rate per period is i, then

$$A_n = R\left[\frac{1 - (1 + i)^{-n}}{i}\right].$$

(a) Evaluate $\lim_{n \to \infty} A_n$ to find a formula for the lump-sum payment for a perpetuity.

(b) Find the lump-sum investment needed to make payments of $100 per month in perpetuity if interest is 12%, compounded monthly.

46. **Response to adrenalin** Experimental evidence suggests that the response y of the body to the concentration x of injected adrenalin is given by

$$y = \frac{x}{a + bx},$$

where a and b are experimental constants.

(a) Is this function continuous for all x?
(b) Based on your conclusion in (a) and the fact that in reality $x \geq 0$ and $y \geq 0$, must a and b be both positive, be both negative, or have opposite signs?

47. **Cost-benefit** Suppose that the cost C of removing p percent of the impurities from the waste water in a manufacturing process is given by

$$C(p) = \frac{9800p}{101 - p}.$$

Is this function continuous for all those p-values for which the problem makes sense?

48. **Cost-benefit** Suppose that the cost C of removing p percent of the particulate pollution from the exhaust gases at an industrial site is given by

$$C(p) = \frac{8100p}{100 - p}.$$

Describe any discontinuities for $C(p)$. Explain what this discontinuity means.

49. **Cost-benefit** The percentage p of particulate pollution that can be removed from the smokestacks of an industrial plant by spending C dollars is given by

$$p = \frac{100C}{7300 + C}.$$

Find the percentage of the pollution that could be removed if spending C was allowed to increase without bound. Can 100% of the pollution be removed? Explain.

50. **Cost-benefit** The percentage p of impurities that can be removed from the waste water of a manufacturing process at a cost of C dollars is given by

$$p = \frac{100C}{8100 + C}.$$

Find the percentage of the impurities that could be removed if cost were no object (that is, if cost were allowed to increase without bound). Can 100% of the impurities be removed? Explain.

51. ***Federal income tax*** The tax owed by a married couple filing jointly and their tax rates can be found in the following tax rate schedule.

Schedule Y-I — Use if your filing status is Married filing jointly or Qualifying widow(er)

If the amount on Form 1040, line 37, is: over—	But not over—	Enter on Form 1040, line 38	of the amount over—
$0	$36,900	- - - - - 15%	$0
36,900	89,150	$5,535.00 + 28%	36,900
89,150	140,000	20,165.00 + 31%	89,150
140,000	250,000	35,928.50 + 36%	140,000
250,000	- - - - -	75,528.50 + 39.6%	250,000

SOURCE: Internal Revenue Service, 1993 Form 1040 Instructions

From this schedule, the tax rate $R(x)$ is a function of income x (the amount on Form 1040, line 37) as follows.

$$R(x) = \begin{cases} 0.15 & \text{if} \quad 0 \le x \le 36,900 \\ 0.28 & \text{if} \quad 36,900 < x \le 89,150 \\ 0.31 & \text{if} \quad 89,150 < x \le 140,000 \\ 0.36 & \text{if} \quad 140,000 < x \le 250,000 \\ 0.396 & \text{if} \quad 250,000 < x \end{cases}$$

Identify any discontinuities in $R(x)$.

52. ***Calories and temperature*** Suppose that the number of calories of heat required to raise 1 gram of water (or ice) from $-40°C$ to $x°C$ is given by

$$f(x) = \begin{cases} \frac{1}{2}x + 20, & \text{if } -40 \le x < 0 \\ x + 100, & \text{if } 0 \le x \end{cases}.$$

(a) What can be said about the continuity of the function $f(x)$?
(b) What accounts for the behavior of the function at 0°C?

53. ***Electrical usage costs*** The monthly charge in dollars for x kilowatt hours (kWh) of electricity used by a residential consumer of Excelsior Electric Membership Corporation from November through June is given by the function

$$C(x) = \begin{cases} 10 + .094x & \text{if} \quad 0 \le x \le 100 \\ 19.40 + .075(x - 100) & \text{if } 100 < x \le 500. \\ 49.40 + .05(x - 500) & \text{if} \quad x > 500 \end{cases}$$

(a) What is the monthly charge if 1100 kWh of electricity are consumed in a month?
(b) Find $\lim_{x \to 100} C(x)$ and $\lim_{x \to 500} C(x)$, if the limits exist.
(c) Is C continuous at $x = 100$ and at $x = 500$?

54. ***Postage costs*** First-class postage is 32 cents for the first ounce or part of an ounce that a letter weighs and is an additional 23 cents for each additional ounce or part of an ounce above one ounce. Use the table or graph of the postage function, $f(x)$, to determine the following.
(a) $\lim_{x \to 2.5} f(x)$ (b) $f(2.5)$
(c) Is $f(x)$ continuous at 2.5?
(d) $\lim_{x \to 4} f(x)$ (e) $f(4)$
(f) Is $f(x)$ continuous at 4?

Weight x	Postage f(x)
$0 < x \le 1$	$0.32
$1 < x \le 2$	0.55
$2 < x \le 3$	0.78
$3 < x \le 4$	1.01
$4 < x \le 5$	1.24

Postage Function

Public debt of the United States The interest paid on the public debt of the United States of America as a percentage of federal expenditures for selected years is shown in the following table.

Year	Interest Paid as a Percentage of Federal Expenditures	Point Coordinates if t = 0 in 1900
1930	0	(30, 0)
1940	10.5	(40, 10.5)
1950	13.4	(50, 13.4)
1955	9.4	(55, 9.4)
1960	10.0	(60, 10.0)
1965	9.6	(65, 9.6)
1970	9.9	(70, 9.9)
1975	9.8	(75, 9.8)
1980	12.7	(80, 12.7)
1985	18.9	(85, 18.9)
1990	21.1	(90, 21.1)

SOURCE: Bureau of Public Debt, Department of the Treasury

Assume that the percentage of federal expenditures devoted to payment of interest can be modeled with the function

$$d(t) = 0.000593t^3 - 0.1056t^2 + 6.098t - 103.37,$$

where t is the number of years past 1900. Use the table and the function to answer problems 55 and 56.

55. (a) Use $d(t)$ to predict the percentage of federal expenditures devoted to payment of interest in 1995.

   (b) Calculate $\lim_{t \to +\infty} d(t)$.

   (c) Can $d(t)$ be used to predict the percentage of federal expenditures devoted to payment of interest on the public debt for large values of t? Explain.

56. (a) Graph $d(t)$ and the points in the table.

   (b) For what years can you guarantee that $d(t)$ cannot be used to predict the percentage of federal expenditures devoted to payment of interest on the public debt? Explain.

57. **Postal rates** The following graphic shows the history of postal rates from 1890 to 1995. Use the figure to answer the following.

   (a) If $P(t)$ represents first-class postage in year t, is $P(t)$ continuous?

   (b) Identify the longest period of years after World War I when $P(t)$ was continuous.

   (c) If $P(t)$ were modeled by a continuous curve, do you think the best model would be linear, exponential, or logarithmic? Explain.

SOURCE: *Oil City Derrick*, Oil City, PA, December 1, 1994.

9.3 THE DERIVATIVE: RATES OF CHANGE; TANGENT TO A CURVE

Objectives

- ❏ To define the derivative as a rate of change
- ❏ To use the definition of derivative to find derivatives of functions
- ❏ To use derivatives to find slopes of tangents to curves

APPLICATION PREVIEW

Suppose an oil company's revenue (in thousand of dollars) is given by

$$R = 100x - x^2, \, x \geq 0,$$

where x is the number of thousands of barrels of oil sold per day. Then we can use the **derivative** of this function to find the marginal revenue when 20,000 barrels are sold.

In Chapter 1, we studied linear revenue functions and defined the marginal revenue for a product as the rate of change of the revenue function. For linear revenue functions, this rate is also the slope of the line that is the graph of the revenue function. In this section, we will define **marginal revenue** as the rate of change of the revenue function, even when the revenue function is not linear. We will discuss the relationship between the marginal revenue at a given point and the slope of the line tangent to the revenue function at that point. We will see how the derivative of the revenue function can be used to find both the slope of this tangent line and the marginal revenue.

Figure 9.17

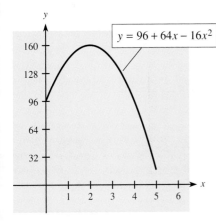

We will begin our study of rates of change (that is, derivatives) by investigating a common rate of change, velocity.

Suppose a ball is thrown straight upward at 64 feet per second from a spot 96 feet above ground level. The equation that describes the height y of the ball after x seconds is

$$y = f(x) = 96 + 64x - 16x^2.$$

Figure 9.17 shows the graph of this function for $0 \leq x \leq 5$. The average velocity of the ball over a given time interval is the change in the height divided by the length of time that has passed. Table 9.4 shows some average velocities over time intervals beginning at $x = 1$.

Table 9.4 Average velocities

Time			Height			
Beginning	*Ending*	*Change*	*Beginning*	*Ending*	*Change*	*Average Velocity*
1	2	1	144	160	16	16/1 = 16
1	1.5	0.5	144	156	12	12/0.5 = 24
1	1.1	0.1	144	147.04	3.04	3.04/0.1 = 30.4
1	1.01	0.01	144	144.3184	0.3184	0.3184/0.01 = 31.84

In Table 9.4, the smaller the time interval, the more closely the average velocity approximates the instantaneous velocity at $x = 1$. Thus the instantaneous velocity at $x = 1$ is closer to 31.84 ft/s than to 30.4 ft/s.

If we represent the change in time by h, then the average velocity from $x = 1$ to $x = 1 + h$ approaches the instantaneous velocity at $x = 1$ as h approaches 0. This is illustrated in the following example.

EXAMPLE 1

Suppose that a ball is thrown straight upward so that its height $f(x)$ (in feet) is given by the equation

$$f(x) = 96 + 64x - 16x^2,$$

where x is time (in seconds).
(a) Find the average velocity from $x = 1$ to $x = 1 + h$.
(b) Find the instantaneous velocity at $x = 1$.

Solution

(a) Let h represent the change in x (time) from 1 to $1 + h$. Then the corresponding change in $f(x)$ (height) is

$$
\begin{aligned}
f(1 + h) - f(1) &= [96 + 64(1 + h) - 16(1 + h)^2] - [96 + 64 - 16]\\
&= 96 + 64 + 64h - 16(1 + 2h + h^2) - 144\\
&= 16 + 64h - 16 - 32h - 16h^2\\
&= 32h - 16h^2.
\end{aligned}
$$

The average velocity V_{av} is the change in height divided by the change in time.

$$
\begin{aligned}
V_{av} &= \frac{f(1 + h) - f(1)}{h}\\[2mm]
&= \frac{32h - 16h^2}{h}\\[2mm]
&= 32 - 16h
\end{aligned}
$$

(b) The instantaneous velocity V is the limit of the average velocity as h approaches 0.

$$
\begin{aligned}
V = \lim_{h \to 0} V_{av} &= \lim_{h \to 0} (32 - 16h)\\
&= 32 \text{ ft/s}
\end{aligned}
$$

Note that average velocity is found over a time interval. Instantaneous velocity is usually called **velocity,** and it can be found at any time x, as follows.

Velocity   Suppose that an object moving in a straight line has its position y at time x given by $y = f(x)$. Then the **velocity** of the object at time x is

$$V = \lim_{h \to 0} \frac{f(x + h) - f(x)}{h},$$

provided this limit exists.

The instantaneous rate of change of any function (commonly called *rate of change*) can be found in the same way we find velocity. The function that gives this instantaneous rate of change of a function *f* is called the **derivative** of *f*.

Derivative If *f* is a function defined by $y = f(x)$, then the **derivative** of $f(x)$ at any value *x*, denoted $f'(x)$, is

$$f'(x) = \lim_{h \to 0} \frac{f(x + h) - f(x)}{h},$$

if this limit exists. If $f'(c)$ exists, we say *f* is **differentiable** at *c*.

The following procedure illustrates how to find the derivative of a function $y = f(x)$ at any value *x*.

Derivative Using the Definition

Procedure	Example
To find the derivative of $y = f(x)$ at any value *x*:	Find the derivative of $f(x) = 4x^2$.
1. Let *h* represent the change in *x* from *x* to $x + h$.	1. The change in *x* from *x* to $x + h$ is *h*.
2. The corresponding change in $y = f(x)$ is $$f(x + h) - f(x).$$	2. The change in $f(x)$ is $$\begin{aligned} f(x + h) - f(x) &= 4(x + h)^2 - 4x^2 \\ &= 4(x^2 + 2xh + h^2) - 4x^2 \\ &= 4x^2 + 8xh + 4h^2 - 4x^2 \\ &= 8xh + 4h^2. \end{aligned}$$
3. Form the difference quotient $\dfrac{f(x + h) - f(x)}{h}$ and simplify.	3. $\dfrac{f(x + h) - f(x)}{h} = \dfrac{8xh + 4h^2}{h}$ $\qquad\qquad = 8x + 4h$
4. Find $\displaystyle\lim_{h \to 0} \dfrac{f(x + h) - f(x)}{h}$ to determine $f'(x)$, the derivative of $f(x)$.	4. $f'(x) = \displaystyle\lim_{h \to 0} \dfrac{f(x + h) - f(x)}{h}$ $f'(x) = \displaystyle\lim_{h \to 0} (8x + 4h) = 8x$

Note that in the previous example, we could have found the derivative of the function $f(x) = 4x^2$ at a particular value of *x*, say $x = 3$, by evaluating the derivative formula at that value:

$$f'(x) = 8x \quad \text{so} \quad f'(3) = 8(3) = 24.$$

In addition to $f'(x)$, the derivative at any point x may be denoted by

$$\frac{dy}{dx}, \quad y', \quad \frac{d}{dx}f(x), \quad D_xy, \quad \text{or} \quad D_xf(x).$$

We can, of course, use variables other than x and y to represent functions and their derivatives. For example, we can represent the derivative of the function defined by $p = 2q^2 - 1$ by dp/dq.

CHECKPOINT

1. For the function $y = f(x) = x^2 - x + 1$, find

(a) $f(x + h) - f(x)$

(b) $\dfrac{f(x + h) - f(x)}{h}$

(c) $f'(x) = \lim\limits_{h \to 0} \dfrac{f(x + h) - f(x)}{h}$

(d) $f'(2)$

In Section 1.7 we defined the **marginal revenue** for a product as the rate of change of the total revenue function for the product. If the total revenue function for a product is not linear, we define the marginal revenue for the product as the instantaneous rate of change, or the derivative, of the revenue function.

Marginal Revenue   Suppose that the total revenue function for a product is given by $R = R(x)$, where x is the number of units sold. Then the **marginal revenue** at x units is

$$\overline{MR} = R'(x) = \lim\limits_{h \to 0} \frac{R(x + h) - R(x)}{h},$$

provided the limit exists.

Note that the marginal revenue (derivative of the revenue function) can be found by using the steps in the Procedure/Example above. These steps can also be combined, as they are in Example 2, which is the Application Preview problem.

EXAMPLE 2

Suppose that an oil company's revenue (in thousands of dollars) is given by the equation

$$R = R(x) = 100x - x^2, x \geq 0,$$

where x is the number of thousands of barrels of oil sold each day.
(a) Find the function that gives the marginal revenue at any value of x.
(b) Find the marginal revenue when 20,000 barrels are sold (that is, at $x = 20$).

Solution

(a) The marginal revenue function is found by evaluating

$$R'(x) = \lim_{h \to 0} \frac{R(x + h) - R(x)}{h}$$

$$= \lim_{h \to 0} \frac{[100(x + h) - (x + h)^2] - (100x - x^2)}{h}$$

$$= \lim_{h \to 0} \frac{100x + 100h - (x^2 + 2xh + h^2) - 100x + x^2}{h}$$

$$= \lim_{h \to 0} \frac{100h - 2xh - h^2}{h} = \lim_{h \to 0}(100 - 2x - h) = 100 - 2x.$$

(b) The marginal revenue function found in part (a) gives the marginal revenue at *any* value of *x*. To find the marginal revenue when 20 units are sold, we evaluate $R'(20)$.

$$R'(20) = 100 - 2(20) = 60$$

The marginal revenue at $x = 20$ is $60,000.

(a)

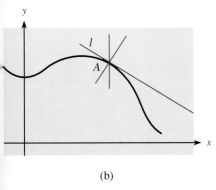

(b)

Figure 9.18

   As mentioned earlier, the rate of change of revenue (the marginal revenue) for a linear revenue function is given by the slope of the line. In fact, the slope of the revenue curve gives us the marginal revenue even if the revenue function is not linear. We will show that the slope of the graph of a function at any point is the same as the derivative at that point. In order to show this, we must define the slope of a curve at a point on the curve. We will define the slope of a curve at a point as the slope of the line tangent to the curve at the point.

   In geometry, a **tangent** to a circle is defined as a line that has one point in common with the circle. (See Figure 9.18(a).) This definition does not apply to all curves, as Figure 9.18(b) shows. Many lines can be drawn through the point *A* that touch the curve only at *A*. One of the lines, line *l*, looks like it is tangent to the curve.

   We can use **secant lines** (lines intersecting the curve at two points) to determine the tangent to a curve at a point. In Figure 9.19, we have a set of secant lines s_1, s_2, s_3, and s_4 that pass through a point *A* on the curve and points Q_1, Q_2, Q_3, and Q_4 on the curve near *A*. The line *l* represents the tangent line to the curve at point *A*. We can get a secant line as close as we wish to the tangent line *l* by choosing a "second point" *Q* sufficiently close to point *A*.

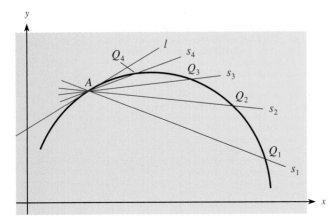

Figure 9.19

Figure 9.20

As we choose points on the curve closer and closer to A, the limiting position of the secant lines that pass through A is the **tangent line** to the curve at point A, and the slopes of those secant lines approach the slope of the tangent line at A. Thus we can find the slope of the tangent line by finding the slope of a secant line and taking the limit of this slope as the "second point" Q approaches A. To find the slope of the tangent to the graph of $y = f(x)$ at A $(x_1, f(x_1))$, we first draw a secant line from point A to a second point Q $(x_1 + h, f(x_1 + h))$ on the curve (see Figure 9.20).

The slope of this secant line is

$$m_{AQ} = \frac{f(x_1 + h) - f(x_1)}{h}.$$

As Q approaches A, we see that the difference between the x-coordinates of these two points decreases, so h approaches 0. Thus the slope of the tangent is given by the following.

Slope of the Tangent   The **slope of the tangent** to the graph of $y = f(x)$ at point $A(x_1, f(x_1))$ is

$$m = \lim_{h \to 0} \frac{f(x_1 + h) - f(x_1)}{h}$$

if this limit exists. That is, $m = f'(x_1)$.

EXAMPLE 3
Find the slope of $y = f(x) = x^2$ at the point A $(2, 4)$.

Solution
The formula for the slope of the tangent to $y = f(x)$ at $(2, 4)$ is

$$m = f'(2) = \lim_{h \to 0} \frac{f(2 + h) - f(2)}{h}.$$

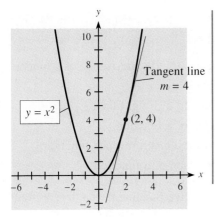

Figure 9.21

So for $f(x) = x^2$, we have

$$m = f'(2) = \lim_{h \to 0} \frac{(2 + h)^2 - 2^2}{h}.$$

Taking the limit immediately would result in both the numerator and the denominator approaching 0. To avoid this, we simplify the fraction before taking the limit.

$$m = \lim_{h \to 0} \frac{4 + 4h + h^2 - 4}{h} = \lim_{h \to 0} \frac{4h + h^2}{h} = \lim_{h \to 0}(4 + h) = 4$$

Thus the slope of the tangent to $y = x^2$ at $(2, 4)$ is 4 (see Figure 9.21).

The statement "the slope of the tangent to the curve at $(2, 4)$ is 4" is frequently simplified to the statement "the slope of the curve at $(2, 4)$ is 4." Knowledge that the slope is a positive number on an interval tells us that the function is increasing on that interval, which means that a point moving along the graph of the function rises as it moves to the right on that interval. If the derivative (and thus the slope) is negative on an interval, the curve is decreasing on the interval; that is, a point moving along the graph falls as it moves to the right on that interval.

EXAMPLE 4
Given $y = f(x) = 3x^2 + 2x$, find
(a) the derivative of $f(x)$ at any point $(x, f(x))$.
(b) the slope of the curve at $(1, 5)$.
(c) the equation of the line tangent to $y = 3x^2 + 2x$ at $(1, 5)$.

Solution
(a) The derivative of $f(x)$ at any value x is denoted by $f'(x)$, and is

$$y' = f'(x) = \lim_{h \to 0} \frac{f(x + h) - f(x)}{h}$$

$$= \lim_{h \to 0} \frac{[3(x + h)^2 + 2(x + h)] - (3x^2 + 2x)}{h}$$

$$= \lim_{h \to 0} \frac{3(x^2 + 2xh + h^2) + 2x + 2h - 3x^2 - 2x}{h}$$

$$= \lim_{h \to 0} \frac{6xh + 3h^2 + 2h}{h}$$

$$= \lim_{h \to 0}(6x + 3h + 2)$$

$$= 6x + 2.$$

(b) The derivative is $f'(x) = 6x + 2$, so the slope of the tangent to the curve at $(1, 5)$ is $f'(1) = 6(1) + 2 = 8$.

(c) The equation of the tangent line uses the given point $(1, 5)$ and the slope $m = 8$. It is $y - 5 = 8(x - 1)$, or $y = 8x - 3$.

The discussion in this section indicates that the derivative of a function can be used to accomplish the following.

1. Find the velocity of an object moving in a straight line.
2. Find the instantaneous rate of change of a function.
3. Find the marginal revenue function for a given revenue function.
4. Find the slope of the tangent to the graph of a function.

So far we have talked about how the derivative is defined, what it represents, and how to find it. However, there are functions for which derivatives do not exist at every value of x. Figure 9.22 shows some common cases where $f'(c)$ does not exist but where $f'(x)$ exists for all other values of x.

Figure 9.22

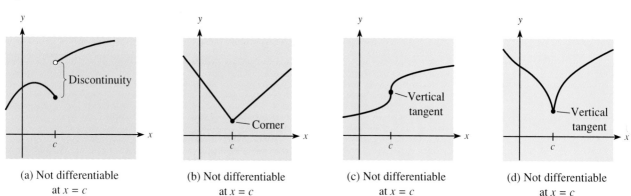

(a) Not differentiable at $x = c$

(b) Not differentiable at $x = c$

(c) Not differentiable at $x = c$

(d) Not differentiable at $x = c$

From Figure 9.22 we see that a function may be continuous at $x = c$ even though $f'(c)$ does not exist. Thus continuity does not imply differentiability at a point. However, differentiability does imply continuity.

Differentiability Implies Continuity   If a function f is differentiable at $x = c$, then f is continuous at $x = c$.

EXAMPLE 5

The monthly charge for water in a small town is given by

$$y = f(x) = \begin{cases} 18 & \text{if } 0 \le x \le 20 \\ 0.1x + 16 & \text{if } x > 20 \end{cases}.$$

(a) Is this function continuous at $x = 20$?
(b) Is this function differentiable at $x = 20$?

Solution

(a) We must check the three properties for continuity.

1. $f(x) = 18$ for $x \le 20$, so $f(20) = 18$

2. $\displaystyle\lim_{x\to 20^-} f(x) = \lim_{x\to 20^-} 18 = 18$

   $\displaystyle\lim_{x\to 20^+} f(x) = \lim_{x\to 20^+} (0.1x + 16) = 18$ $\Big\} \Rightarrow \displaystyle\lim_{x\to 20} f(x) = 18$

3. $\displaystyle\lim_{x\to 20} f(x) = f(20)$

Thus $f(x)$ is continuous at $x = 20$.

(b) Because the function is defined differently on either side of $x = 20$, we need to test to see if $f'(20)$ exists by evaluating both

(i) $\displaystyle\lim_{h\to 0^-} \frac{f(20 + h) - f(20)}{h}$ and (ii) $\displaystyle\lim_{h\to 0^+} \frac{f(20 + h) - f(20)}{h}$

and determining whether they are equal.

(i) $\displaystyle\lim_{h\to 0^-} \frac{f(20 + h) - f(20)}{h} = \lim_{h\to 0^-} \frac{18 - 18}{h}$

$\displaystyle = \lim_{h\to 0^-} 0 = 0$

(ii) $\displaystyle\lim_{h\to 0^+} \frac{f(20 + h) - f(20)}{h} = \lim_{h\to 0^+} \frac{[0.1(20 + h) + 16] - 18}{h}$

$\displaystyle = \lim_{h\to 0^+} \frac{0.1h}{h}$

$\displaystyle = \lim_{h\to 0^+} 0.1 = 0.1$

Because these limits are not equal, the derivative does not exist.

CHECKPOINT

2. Which of the following are given by $f'(c)$?
   (a) The slope of the tangent when $x = c$.
   (b) The y-coordinate of the point where $x = c$.
   (c) The instantaneous rate of change of $f(x)$ at $x = c$.
   (d) The marginal revenue at $x = c$, if $f(x)$ is the revenue function.

3. Must a graph that has no discontinuity, corner, or cusp at $x = c$ be differentiable at $x = c$?

EXAMPLE 6

If the point (a, b) lies on the graph of $y = x^2$, then the equation of the secant line to $y = x^2$ from $(1, 1)$ to (a, b) has the equation

$$y - 1 = \frac{b - 1}{a - 1}(x - 1), \quad \text{or} \quad y = \frac{b - 1}{a - 1}(x - 1) + 1.$$

Figure 9.23

(a)

(b)

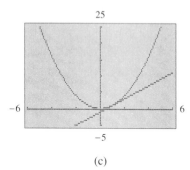

(c)

(a) Write the equation of the secant line from (1, 1) to (5, 25) and graph $y = x^2$ and this secant line.

(b) Write the equation of the secant line from (1, 1) to (3, 9) and graph $y = x^2$ and this secant line.

(c) Write the equation of the secant line from (1, 1) to (1.01, 1.0201) and graph $y = x^2$ and this secant line.

(d) Which secant line appears as if it might be closest to the tangent line at (1, 1)?

(e) Express the slope of the secant line from (1, 1) to (a, b), in terms of a and find the limit of this slope as $a \to 1$. Is this limit the slope of the tangent line to $y = x^2$ at (1, 1)?

Solution

(a) The equation of the secant line from (1, 1) to $a = 5$, $b = 25$ is

$$y = \frac{25 - 1}{5 - 1}(x - 1) + 1 \quad \text{or} \quad y = 6x - 5.$$

The graph of $y = x^2$ and the secant line is shown in Figure 9.23(a).

(b) The equation of the secant line from (1, 1) to $a = 3$, $b = 9$ is

$$y = \frac{9 - 1}{3 - 1}(x - 1) + 1 \quad \text{or} \quad y = 4x - 3.$$

The graph of $y = x^2$ and the secant line is shown in Figure 9.23(b).

(c) The equation of the secant line from (1, 1) to $a = 1.01$, $b = 1.0201$ is

$$y = \frac{1.0201 - 1}{1.01 - 1}(x - 1) + 1 \quad \text{or} \quad y = 2.01x - 1.01.$$

The graph of $y = x^2$ and the secant line is shown in Figure 9.23(c).

(d) The secant line from (1, 1) to (1.01, 1.0201) is closest to the tangent line at (1, 1).

(e) The slope of the secant line from (1, 1) to (a, b) is

$$\frac{b - 1}{a - 1} = \frac{a^2 - 1}{a - 1}.$$

The limit of this slope as a approaches 1, the x-value of the point (1, 1), is

$$\lim_{a \to 1} \frac{b - 1}{a - 1} = \lim_{a \to 1} \frac{a^2 - 1}{a - 1} = \lim_{a \to 1}(a + 1) = 2.$$

This limit, 2, is the slope of the tangent line at (1, 1). That is, the derivative of $y = x^2$ at (1, 1) is 2.

CHECKPOINT SOLUTIONS

1. (a) $f(x + h) - f(x) = [(x + h)^2 - (x + h) + 1] - (x^2 - x + 1)$
   $$= x^2 + 2xh + h^2 - x - h + 1 - x^2 + x - 1$$
   $$= 2xh + h^2 - h$$

   (b) $\dfrac{f(x + h) - f(x)}{h} = \dfrac{2xh + h^2 - h}{h}$
   $$= 2x + h - 1$$

   (c) $f'(x) = \lim\limits_{h \to 0} \dfrac{f(x + h) - f(x)}{h} = \lim\limits_{h \to 0}(2x + h - 1)$
   $$= 2x - 1$$

   (d) $f'(x) = 2x - 1$, so $f'(2) = 3$.

2. Parts (a), (c), and (d) are given by $f'(c)$. The y-coordinate where $x = c$ is given by $f(c)$.

3. No. Figure 9.22(c) shows such an example.

EXERCISE 9.3

1. In the Procedure/Example table in this section we were given $f(x) = 4x^2$ and found $f'(x) = 8x$. Find the
   (a) instantaneous rate of change of $f(x)$ at $x = 4$.
   (b) slope of the tangent to the graph of $y = f(x)$ at $x = 4$.
   (c) point on the graph of $y = f(x)$ at $x = 4$.

2. In Example 4 of this section we were given $f(x) = 3x^2 + 2x$ and found $f'(x) = 6x + 2$. Find the
   (a) instantaneous rate of change of $f(x)$ at $x = 6$.
   (b) slope of the tangent to the graph of $y = f(x)$ at $x = 6$.
   (c) point on the graph of $y = f(x)$ at $x = 6$.

3. Let $f(x) = 2x^2 - x$.
   (a) Use the Procedure/Example in this section to verify that $f'(x) = 4x - 1$.
   (b) Find the instantaneous rate of change of $f(x)$ at $x = -1$.
   (c) Find the slope of the tangent to the graph of $y = f(x)$ at $x = -1$.
   (d) Find the point on the graph of $y = f(x)$ at $x = -1$.

4. Let $f(x) = 9 - \dfrac{1}{2}x^2$.
   (a) Use the Procedure/Example in this section to verify that $f'(x) = -x$.

(b) Find the instantaneous rate of change of $f(x)$ at $x = 2$.
(c) Find the slope of the tangent to the graph of $y = f(x)$ at $x = 2$.
(d) Find the point on the graph of $y = f(x)$ at $x = 2$.

In problems 5–8, the tangent line to the graph of $f(x)$ at $x = 1$ is shown. On the tangent line, P is the point of tangency and A is another point on the line.
(a) Find the coordinates of the points P and A.
(b) Use the coordinates of P and A to find the slope of the tangent line.
(c) Find $f'(1)$.
(d) Find the instantaneous rate of change of $f(x)$ at P.

5.

6.

7.

8.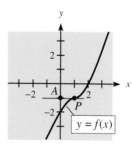

In problems 9–12, use the definition to find the derivative of each function.

9. $f(x) = 1 - 6x$

10. $f(x) = 4 - 5x$

11. $f(x) = 4x^2 - 2x + 1$

12. $f(x) = 16x^2 - 4x + 2$

In problems 13–16, find (a) the instantaneous rate of change of the function at any point and (b) the instantaneous rate of change at the given value.

13. (a) $p(q) = q^2 + 4q + 1$
    (b) $q = 5$

14. (a) $p(q) = 2q^2 - 4q + 5$
    (b) $q = 2$

15. (a) $C(x) = 14x^2 + 5x + 60$
    (b) $x = 4$

16. (a) $C(x) = 500 + 40x + 0.5x^2$
    (b) $x = 5$

In problems 17 and 18, a point (a, b) on the graph of $y = f(x)$ is given and the equation of the line tangent to the graph of $f(x)$ at (a, b) is given. In each case, find $f'(a)$ and $f(a)$.

17. $(-3, -9)$; $5x - 2y = 3$

18. $(-1, 6)$; $x + 10y = 59$

19. If the instantaneous rate of change of $f(x)$ at $(1, -1)$ is 3, write the equation of the line tangent to the graph of $f(x)$ at $x = 1$.

20. If the instantaneous rate of change of $g(x)$ at $(-1, -2)$ is $1/2$, write the equation of the line tangent to the graph of $g(x)$ at $x = -1$.

Since the derivative of a function represents both the slope of the tangent to the curve and the instantaneous rate of change of the function, it is possible to use information about one to gain information about the other. In problems 21 and 22, use the graph of the function $y = f(x)$ given in Figure 9.24.

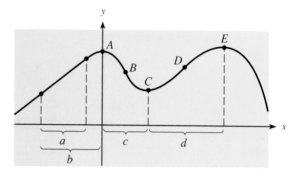

Figure 9.24

21. (a) Over what interval(s) (*a*) through (*d*) is the rate of change of $f(x)$ positive?
    (b) Over what interval(s) (*a*) through (*d*) is the rate of change of $f(x)$ negative?
    (c) At what point(s) *A* through *E* is the rate of change of $f(x)$ equal to zero?

22. (a) At what point(s) *A* through *E* does the rate of change of $f(x)$ change from positive to negative?
    (b) At what point(s) *A* through *E* does the rate of change of $f(x)$ change from negative to positive?

23. Given the graph of $y = f(x)$ in Figure 9.25, determine for which *x*-values *A*, *B*, *C*, *D*, or *E* the function is
    (a) continuous.
    (b) differentiable.

24. Given the graph of $y = f(x)$ in Figure 9.25, determine for which *x*-values *F*, *G*, *H*, *I*, or *J* the function is
    (a) continuous.
    (b) differentiable.

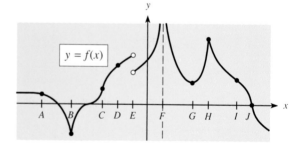

Figure 9.25

In problems 25–28, (a) find the slope of the tangent to the graph of $f(x)$ at any point, (b) find the slope of the tangent at the given x-value, (c) write the equation of the line tangent to the graph of $f(x)$ at the given point, and (d) graph both $f(x)$ and its tangent line (use a graphing utility if one is available).

25. (a) $f(x) = x^2 + x$
    (b) $x = 2$
    (c) $(2, 6)$

26. (a) $f(x) = x^2 + 3x$
    (b) $x = -1$
    (c) $(-1, -2)$

27. (a) $f(x) = x^3 + 3$
    (b) $x = 1$
    (c) $(1, 4)$

28. (a) $f(x) = 5x^3 + 2$
    (b) $x = -1$
    (c) $(-1, -3)$

APPLICATIONS

29. **Average velocity** When a ball is dropped from a height of 256 feet, its position (height above the ground) after x seconds is given by

$$S(x) = 256 - 16x^2.$$

    (a) What is the average velocity in the first 2 seconds of the fall?
    (b) What does the negative average velocity in part (a) mean?

30. **Average velocity** If an object is thrown upward at 64 ft/s from a height of 20 feet, its height S after x seconds is given by

$$S(x) = 20 + 64x - 16x^2.$$

What is the average velocity in the
    (a) first 2 seconds after it is thrown?
    (b) next 2 seconds?

31. **Demand** If the demand for a product is given by

$$D(p) = \frac{1000}{\sqrt{p}} - 1,$$

what is the average rate of change of demand if p increases from
    (a) 1 to 25?
    (b) 25 to 100?

32. **Revenue** If the total revenue function for a blender is

$$R(x) = 36x - 0.01x^2,$$

where x is the number of units sold, what is the average rate of change in revenue $R(x)$ as x increases from 10 to 20 units?

33. **Speed limit enforcement** If 0.05 second elapses while a car on a road travels over two sensors that are 5 feet

apart, what is the average velocity (in miles per hour) of the car as it travels between the sensors (88 feet per second is equivalent to 60 miles per hour)?

34. **Speed limit enforcement** One speed-check system used by local police measures the elasped time between two marks 0.1 mile apart. If 5.85 seconds elapse while a car on the highway travels between the two marks, find the average velocity of the car (in miles per hour) as it travels between the two marks.

35. **Marginal revenue** If the revenue function for a stereo system is

$$R(x) = 300x - x^2,$$

where x denotes the number of units sold, what is
    (a) the function that gives marginal revenue?
    (b) the marginal revenue if 50 units are sold and what does it mean?
    (c) the marginal revenue if 200 units are sold and what does it mean?
    (d) the marginal revenue if 150 units are sold and what does it mean?
    (e) As the number of units sold passes through 150, what happens to revenue?

36. **Marginal revenue** If the total revenue function for a blender is

$$R(x) = 36x - 0.01x^2,$$

where x is the number of units sold, what
    (a) function gives the marginal revenue?
    (b) is the marginal revenue when 600 units are sold and what does it mean?
    (c) is the marginal revenue when 2000 units are sold and what does it mean?
    (d) is the marginal revenue when 1800 units are sold and what does it mean?

37. **Labor force and output** The monthly output at the Olek Carpet Mill is

$$Q(x) = 15{,}000 + 2x^2 \text{ units}, \quad (40 \le x \le 60),$$

where x is the number of workers employed at the mill. If there are currently 50 workers, find the instantaneous rate of change of monthly output with respect to the number of workers. (That is, find $Q'(50)$.)

38. **Consumer expenditure** Suppose that the demand x for a product is

$$x = 10{,}000 - 100p,$$

where p dollars is the price per unit. Then the consumer expenditure for the product is

$$E(p) = px = p(10,000 - 100p)$$
$$= 10,000p - 100p^2.$$

What is the instantaneous rate of change of consumer expenditure with respect to price at
(a) any price p?   (b) $p = 5$?   (c) $p = 20$?

39. **Profit** Suppose that the profit function for the monthly sales of a car by a dealership is

$$P(x) = 500x - x^2 - 100,$$

where x is the number of cars sold. What is the instantaneous rate of change of profit when
(a) 200 cars are sold? Explain its meaning.
(b) 300 cars are sold? Explain its meaning.

40. **Profit** If the total revenue function for a toy is

$$R(x) = 2x,$$

and the total cost function is

$$C(x) = 100 + 0.2x^2 + x,$$

what is the instantaneous rate of change of profit if 10 units are produced and sold? Explain its meaning.

41. **Heat index** The highest recorded temperature on record in the state of Alaska was 100°F, which occurred on June 27, 1915, at Fort Yukon. The *heat index* is the apparent temperature of the air at a given temperature and humidity level. If x denotes the relative humidity (in percent), then the heat index (in degrees Fahrenheit) for an air temperature of 100°F can be approximated with the function

$$f(x) = 0.009x^2 + 0.139x + 91.875.$$

(a) At what rate is the heat index changing when the humidity is 50%?
(b) Write a sentence that explains the meaning of your answer in part (a).

42. **Receptivity** In learning theory, receptivity is defined as the ability of students to understand a complex concept. Receptivity is highest when the topic is introduced, and tends to decrease as time passes in a lecture. Suppose that the receptivity of a group of students in a mathematics class is given by

$$g(t) = -0.2t^2 + 3.1t + 32,$$

where t is minutes after the lecture begins.
(a) At what rate is receptivity changing 10 minutes after the lecture begins?

(b) Write a sentence that explains the meaning of your answer in part (a).

43. **HIV infections** The figure shows the number of world HIV infections since AIDS was discovered in 1981.
(a) Find the average rate of change in world HIV infections from 1981 to 1992.
(b) Find the average rates of change in world HIV infections from 1995 to 2000 using first the low estimate and then the high estimate.

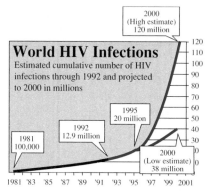

Source: World Health Organization and the Global AIDS Policy Coalition, as reported in the *El Paso Times*, December 6, 1992.

44. **Mutual funds** The figure shows the history of the growth of mutual funds.
(a) Find the average rate of change in the number of funds from 1950 to 1960.
(b) Find the average rate of change in the number of funds from 1980 to 1990.

Published in *Investment Digest* of Valic Co., Vol. 5, No. 2, Summer, 1992.

9.4 DERIVATIVE FORMULAS

Objectives

- ❏ To find derivatives of powers of x
- ❏ To find derivatives of constant functions
- ❏ To find derivatives of functions involving constant coefficients
- ❏ To find derivatives of sums and differences of functions

APPLICATION PREVIEW

The killer bees bred in South America have entered the United States in spite of efforts to halt their spread. The first bees were recorded entering California in 1985 and were quickly destroyed. However, the first American was killed by the bees on July 15, 1993. Suppose that the bees enter a county in Texas and that the bee population in that county grows over a six-week period, with the number of bees given by the equation

$$P(t) = 2t^2 + 10t + 1,$$

where t is the number of weeks since the first bee is discovered. We can find the rate of growth of the bee population two weeks after the first bee is discovered by using the derivative $P'(t)$ of the growth function.

As we discussed in Section 9.3, the derivative of a function can be used to find the rate of change of the function. In this section we will develop formulas that will make it easier to find certain derivatives.

We can use the definition of derivative to show the following:

If $f(x) = x^2$, then $f'(x) = 2x$.
If $f(x) = x^3$, then $f'(x) = 3x^2$.
If $f(x) = x^4$, then $f'(x) = 4x^3$.
If $f(x) = x^5$, then $f'(x) = 5x^4$.

Do you recognize a pattern that could be used to find the derivative of $f(x) = x^6$? What is the derivative of $f(x) = x^n$? If you said that the derivative of $f(x) = x^6$ is $f'(x) = 6x^5$ and the derivative of $f(x) = x^n$ is $f'(x) = nx^{n-1}$, you're right. We can use the definition of derivative to show this. If n is a positive integer, then

$$f'(x) = \lim_{h \to 0} \frac{f(x + h) - f(x)}{h}$$

$$= \lim_{h \to 0} \frac{(x + h)^n - x^n}{h}.$$

Since we are assuming that n is a positive integer, we can use the binomial formula to expand $(x + h)^n$. You may recall from Section 8.3 that this formula is stated as follows:

$$(a + b)^n = a^n + na^{n-1}b + \frac{n(n - 1)}{1 \cdot 2}a^{n-2}b^2 + \cdots + b^n.$$

Thus replacing a with x and b with h gives the following.

$$f'(x) = \lim_{h \to 0} \frac{\left[x^n + nx^{n-1}h + \frac{n(n-1)}{1 \cdot 2}x^{n-2}h^2 + \cdots + h^n \right] - x^n}{h}$$

$$= \lim_{h \to 0} \left[nx^{n-1} + \frac{n(n-1)}{1 \cdot 2}x^{n-2}h + \cdots + h^{n-1} \right]$$

Now, each term after nx^{n-1} contains h as a factor, so all terms except nx^{n-1} will approach 0 as $h \to 0$. Thus

$$f'(x) = nx^{n-1}.$$

Even though we proved this derivative rule only for the case when n is a positive integer, the rule applies for any real number n.

Powers of x Rule   If $f(x) = x^n$, where n is a real number, then $f'(x) = nx^{n-1}$.

EXAMPLE 1

Find the derivatives of the following functions.

(a) $g(x) = x^6$ (b) $f(x) = x^{-2}$

(c) $y = x^4$ (d) $y = x^{1/3}$

Solution

(a) If $g(x) = x^6$, then $g'(x) = 6x^{6-1} = 6x^5$.

(b) The Powers of x Rule applies for all real values. Thus for $f(x) = x^{-2}$, we have

$$f'(x) = -2x^{-2-1} = -2x^{-3} = \frac{-2}{x^3}.$$

(c) If $y = x^4$, then $dy/dx = 4x^{4-1} = 4x^3$.

(d) The Powers of x Rule applies to $y = x^{1/3}$.

$$\frac{dy}{dx} = \frac{1}{3}x^{1/3-1} = \frac{1}{3}x^{-2/3} = \frac{1}{3x^{2/3}}.$$

In Example 1 we took the derivative with respect to x of *both sides* of each equation. We denote the operation "take the derivative with respect to x" by $\frac{d}{dx}$. Thus for $y = x^4$ (in part (c)),

$$\frac{d}{dx}(y) = \frac{d}{dx}(x^4) \quad \text{gives} \quad \frac{dy}{dx} = 4x^3.$$

Similarly, for $f(x) = x^{-2}$ (in part (b)),

$$\frac{d}{dx}[f(x)] = \frac{d}{dx}(x^{-2}) \quad \text{gives} \quad f'(x) = -2x^{-3}.$$

The differentiation rules are stated and proved for the independent variable x, but they also apply to other independent variables. The following examples illustrate differentiation with variables other than x.

EXAMPLE 2
Find the derivatives of the following functions.

(a) $u(s) = s^8$ (b) $p = q^{2/3}$ (c) $C(t) = \sqrt{t}$ (d) $s = \dfrac{1}{\sqrt{t}}$

Solution
(a) If $u(s) = s^8$, then $u'(s) = 8s^{8-1} = 8s^7$.
(b) If $p = q^{2/3}$, then

$$\frac{dp}{dq} = \frac{2}{3}q^{2/3-1} = \frac{2}{3}q^{-1/3} = \frac{2}{3q^{1/3}}.$$

(c) Writing $\sqrt{t}$ in its equivalent form, $t^{1/2}$, permits us to use the derivative formula.

$$C'(t) = \frac{1}{2}t^{1/2-1} = \frac{1}{2}t^{-1/2}$$

Writing the derivative in radical form gives

$$C'(t) = \frac{1}{2} \cdot \frac{1}{t^{1/2}} = \frac{1}{2\sqrt{t}}.$$

(d) Writing $1/\sqrt{t}$ as a power of t gives

$$s = \frac{1}{t^{1/2}} = t^{-1/2}, \quad \text{so} \quad \frac{ds}{dt} = -\frac{1}{2}t^{-1/2-1} = -\frac{1}{2}t^{-3/2}.$$

Writing the derivative in a form similar to that of the original function gives

$$\frac{ds}{dt} = -\frac{1}{2} \cdot \frac{1}{t^{3/2}} = -\frac{1}{2\sqrt{t^3}}.$$

EXAMPLE 3
Find the slope of the tangent to the curve $y = x^3$ at $x = 1$.

Solution
Finding the slope of the tangent line to $y = x^3$ at $x = 1$ involves two steps.

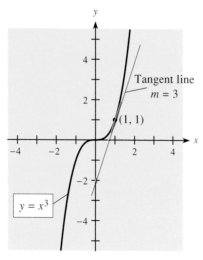

Figure 9.26

1. Find the derivative of $y = x^3$.

$$y' = 3x^2$$

2. Evaluate the derivative at $x = 1$.

$$m_{\text{tan}} = y'|_{x=1} = y'(1) = 3(1)^2 = 3$$

The graph of $y = x^3$ and the tangent line to the graph at $x = 1$, $y = 1^3 = 1$ are shown in Figure 9.26.

A function of the form $y = f(x) = c$, where c is a constant, is called a **constant function.** We can show that the derivative of a constant function is 0, as follows.

$$f'(x) = \lim_{h \to 0} \frac{f(x + h) - f(x)}{h} = \lim_{h \to 0} \frac{c - c}{h} = \lim_{h \to 0} 0 = 0$$

We can state this rule formally.

Constant Function Rule   If $f(x) = c$, where c is a constant, then $f'(x) = 0$.

EXAMPLE 4

Find the derivative of the function defined by $y = 4$.

Solution

Because 4 is a constant, $\dfrac{dy}{dx} = 0$.

Recall that the function defined by $y = 4$ has a horizontal line as its graph. Thus the slope of the line (and the derivative of the function) is 0.

We now can take derivatives of constant functions and powers of x. But we do not yet have a rule for taking derivatives of functions of the form $f(x) = 4x^5$ or $g(t) = \frac{1}{2}t^2$. The following rule provides a method for handling functions of this type.

Coefficient Rule   If $f(x) = c \cdot u(x)$, where c is a constant and $u(x)$ is a differentiable function of x, then $f'(x) = c \cdot u'(x)$.

The above formula says that the derivative of a constant times a function is the constant times the derivative of the function.

We can use the fact that

$$\lim_{h \to 0} c \cdot g(h) = c \cdot \lim_{h \to 0} g(h),$$

which was discussed in Section 9.1, to verify the coefficient rule. If $f(x) = c \cdot u(x)$, then

$$f'(x) = \lim_{h \to 0} \frac{f(x + h) - f(x)}{h} = \lim_{h \to 0} \frac{c \cdot u(x + h) - c \cdot u(x)}{h}$$

$$= \lim_{h \to 0} c \cdot \left[\frac{u(x + h) - u(x)}{h} \right] = c \cdot \lim_{h \to 0} \frac{u(x + h) - u(x)}{h},$$

so $f'(x) = c \cdot u'(x)$.

EXAMPLE 5

Find the derivatives of the following functions.

(a) $f(x) = 4x^5$ (b) $f(t) = \frac{1}{2}t^2$ (c) $p = \dfrac{5}{\sqrt{q}}$

Solution

(a) $f'(x) = 4(5x^4) = 20x^4$ (b) $g'(t) = \frac{1}{2}(2t) = t$

(c) $p = \dfrac{5}{\sqrt{q}} = 5q^{-1/2}$, so

$$\frac{dp}{dq} = 5\left(-\frac{1}{2}q^{-3/2} \right) = -\frac{5}{2\sqrt{q^3}}.$$

In Example 4 of Section 9.3 we found the derivative of $f(x) = 3x^2 + 2x$ to be $f'(x) = 6x + 2$. This result, along with the results of several of the derivatives calculated in Exercise 9.3, suggest that we can find the derivative of a function by finding the derivatives of its terms and combining them. The following rules state this formally.

Sum Rule   If $f(x) = u(x) + v(x)$, where u and v are differentiable functions of x, then $f'(x) = u'(x) + v'(x)$.

We can prove this rule as follows. If $f(x) = u(x) + v(x)$, then

$$f'(x) = \lim_{h \to 0} \frac{f(x + h) - f(x)}{h}$$

$$= \lim_{h \to 0} \frac{[u(x + h) + v(x + h)] - [u(x) + v(x)]}{h}$$

$$= \lim_{h \to 0} \left[\frac{u(x + h) - u(x)}{h} + \frac{v(x + h) - v(x)}{h} \right]$$

$$= \lim_{h \to 0} \frac{u(x + h) - u(x)}{h} + \lim_{h \to 0} \frac{v(x + h) - v(x)}{h}$$

$$= u'(x) + v'(x).$$

Difference Rule   If $f(x) = u(x) - v(x)$, where u and v are differentiable functions of x, then
$f'(x) = u'(x) - v'(x)$.

EXAMPLE 6
Find the derivatives of the following functions.

(a) $y = x^2 + 3$ (b) $p = q^2 - 4q$ (c) $y = 3x + 5$

Solution
(a) $y' = 2 \cdot x + 0 = 2x$
(b) $dp/dq = 2 \cdot q - 4 \cdot 1 = 2q - 4$
(c) $y' = 3 \cdot 1 + 0 = 3$

In Example 6(c) we saw that the derivative of $y = 3x + 5$ is 3. Because the slope of a line is the same at all points on the line, it is reasonable that the derivative of a linear equation is a constant. In particular, the slope of the graph of the equation $y = mx + b$ is m at all points on its graph because the derivative of $y = mx + b$ is $y' = f'(x) = m$.

The rules regarding the derivatives of sums and differences of two functions also apply if more than two functions are involved. For example, the derivative of $f(x) = 4x^3 - 2x^2 + 5x - 3$ is $f'(x) = 12x^2 - 4x + 5$. We may think of the functions that are added and subtracted as terms of the function f. Then it would be correct to say that we may take the derivative of a function term by term.

EXAMPLE 7
Find the derivatives of the following functions.

(a) $y = 3x^3 - 4x^2$

(b) $p = \frac{1}{3}q^3 + 2q^2 - 3$

(c) $u(x) = 5x^4 + x^{1/3}$

Solution
(a) $y' = 3(3x^2) - 4(2x) = 9x^2 - 8x$

(b) $\dfrac{dp}{dq} = \frac{1}{3}(3q^2) + 2(2q) - 0 = q^2 + 4q$

(c) $u'(x) = 5(4x^3) + \frac{1}{3}x^{-2/3} = 20x^3 + \dfrac{1}{3x^{2/3}}$

EXAMPLE 8
Find the derivatives of the following functions.

(a) $y = 4x^3 + \sqrt{x}$ (b) $s = 5t^6 - \dfrac{1}{t^2}$

Solution

(a) We may write the equation as

$$y = 4x^3 + x^{1/2},$$

so

$$y' = 4(3x^2) + \frac{1}{2}x^{-1/2} = 12x^2 + \frac{1}{2x^{1/2}}$$

$$= 12x^2 + \frac{1}{2\sqrt{x}}.$$

(b) We may write $s = 5t^6 - 1/t^2$ as

$$s = 5t^6 - t^{-2},$$

so

$$\frac{ds}{dt} = 5(6t^5) - (-2t^{-3}) = 30t^5 + 2t^{-3}$$

$$= 30t^5 + \frac{2}{t^3}.$$

EXAMPLE 9

Find the slope of the tangent to $f(x) = \frac{1}{2}x^2 + 5x$ at each of the following.

(a) $x = 2$ (b) $x = -5$

Solution

The derivative of $f(x) = \frac{1}{2}x^2 + 5x$ is $f'(x) = x + 5$.

(a) At $x = 2$, the slope is $f'(2) = 2 + 5 = 7$.

(b) At $x = -5$, the slope is $f'(-5) = -5 + 5 = 0$. Thus when $x = -5$, the tangent to the curve is a horizontal line.

CHECKPOINT

1. True or false: The derivative of a constant times a function is equal to the constant times the derivative of the function.

2. True or false: The derivative of the sum of two functions is equal to the sum of the derivatives of the two functions.

3. True or false: The derivative of the difference of two functions is equal to the difference of the derivatives of the two functions.

4. Does the Coefficient Rule apply to $f(x) = x^n/c$, where c is a constant? Explain.

The marginal revenue $R'(x)$ is used to estimate the change in revenue caused by the sale of one additional unit.

EXAMPLE 10

Suppose that a manufacturer of a product knows that because of the demand for this product, his revenue is given by

$$R(x) = 1500x - 0.02x^2, \qquad 0 \le x \le 1000,$$

where x is the number of units sold and $R(x)$ is in dollars.

(a) Find the marginal revenue at $x = 500$.
(b) Find the change in revenue caused by the increase in sales from 500 to 501 units.
(c) Find the difference between the marginal revenue found in part (a) and the change in revenue found in part (b).

Solution

(a) The marginal revenue for any value of x is

$$R'(x) = 1500 - 0.04x.$$

The marginal revenue at $x = 500$ is

$$R'(500) = 1500 - 20 = 1480 \text{ (dollars).}$$

(b) The revenue at $x = 500$ is $R(500) = 745{,}000$, and the revenue at $x = 501$ is $R(501) = 746{,}479.98$, so the change in revenue is

$$R(501) - R(500) = 746{,}479.98 - 745{,}000 = 1479.98 \text{ (dollars).}$$

(c) The difference is $1480 - 1479.98 = 0.02$. Thus we see that the marginal revenue at $x = 500$ is a good estimate of the revenue from the 501st unit.

EXAMPLE 11

Suppose that the killer bees mentioned in the Application Preview enter a county in Texas and that the bee population in that county grows over a six-week period, with the number of bees given by the equation

$$P(t) = 2t^2 + 10t + 1,$$

where t is the number of weeks since the first bee is discovered. Find the rate of growth of the bee population two weeks after the first bee is discovered.

Solution

The rate of growth of the bees is given by $P'(t) = 4t + 10$, so the rate of growth two weeks after the first bee is discovered is $P'(2) = 18$ bees per week.

EXAMPLE 12

(a) Graph $f(x) = x^3 - 3x + 3$ and its derivative $f'(x)$ on the same set of axes so that all values of x that make $f'(x) = 0$ are in the x-range.
(b) Investigate the graph of $y = f(x)$ near values of x where $f'(x) = 0$. Does the graph of $y = f(x)$ appear to turn at values where $f'(x) = 0$?
(c) Compare the interval of x values where $f'(x) < 0$ with the interval where the graph of $y = f(x)$ is decreasing from left to right.
(d) What is the relationship between the intervals where $f'(x) > 0$ and where the graph of $y = f(x)$ is increasing from left to right?

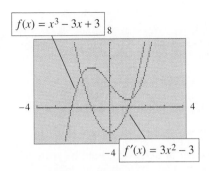

$f(x) = x^3 - 3x + 3$

$f'(x) = 3x^2 - 3$

Figure 9.27

Solution

(a) The graphs of $f(x) = x^3 - 3x + 3$ and $f'(x) = 3x^2 - 3$ are shown in Figure 9.27.

(b) The values where $f'(x) = 0$ are the x-intercepts, $x = -1$ and $x = 1$. The graph of $y = x^3 - 3x + 3$ appears to turn at both these values.

(c) $f'(x) < 0$ where the graph of $y = f'(x)$ is below the x-axis, for $-1 < x < 1$. The graph of $y = f(x)$ appears to be decreasing on this interval.

(d) They appear to be the same intervals.

CHECKPOINT SOLUTIONS

1. True, by the Coefficient Rule.

2. True, by the Sum Rule.

3. True, by the Difference Rule.

4. Yes, $f(x) = x^n/c = (1/c)x^n$, so the coefficient is $(1/c)$.

EXERCISE 9.4

Find the derivatives of the following functions.

1. $y = 4$ 2. $f(s) = 6$ 3. $y = x$

4. $s = t^2$ 5. $f(x) = 2x^3 - x^5$ 6. $f(x) = 3x^4 - x^9$

7. $y = 6x^4 - 5x^2 + x - 2$

8. $y = 3x^5 - 5x^3 - 8x + 8$

9. $g(x) = 10x^9 - 5x^5 + 7x^3 + 5x - 6$

10. $h(x) = 12x^{20} + 8x^{10} - 2x^7 + 17x - 9$

In problems 11–14, at the indicated points, find
(a) the slope of the tangent to the curve, and
(b) the instantaneous rate of change of the function.

11. $y = 4x^2 + 3x, \quad x = 2$

12. $C(x) = 3x^2 - 5, \quad (3, 22)$

13. $P(x) = x^2 - 4x, \quad (2, -4)$

14. $R(x) = 16x + x^2, \quad x = 1$

In problems 15–24, find the derivative of each function.

15. $y = x^{-5} + x^{-8} - 3$

16. $y = x^{-1} - x^{-2} + 13$

17. $y = 3x^{11/3} - 2x^{7/4} - x^{1/2} + 8$

18. $y = 5x^{8/5} - 3x^{5/6} + x^{1/3} + 5$

19. $f(x) = 5x^{-4/5} + 2x^{-4/3}$

20. $f(x) = 6x^{-8/3} - x^{-2/3}$

21. $g(x) = \dfrac{3}{x^4} + \dfrac{2}{x^5} + 5\sqrt[3]{x}$

22. $h(x) = \dfrac{7}{x^7} - \dfrac{3}{x^3} + 8\sqrt{x}$

23. $h(x) = \dfrac{10}{x^3} - \dfrac{10}{\sqrt[5]{x^2}} + x^2 + 1$

24. $g(x) = \dfrac{5}{x^{10}} + \dfrac{4}{\sqrt[4]{x^3}} + x^5 - 4$

In problems 25–30, write the equations of the tangent lines to the curves at the indicated points.

25. $y = x^3 - 3x^2 + 5$ at $x = 1$

26. $y = x^4 - 4x^3 - 2$ at $x = 2$

27. $f(x) = 4x^2 - \dfrac{1}{x}$ at $x = -\dfrac{1}{2}$

28. $f(x) = \dfrac{x^3}{3} - \dfrac{3}{x^3}$ at $x = -1$

29. $y = 5 - 2\sqrt{x}$ at $x = 4$

30. $y = 1 + 3x^{2/3}$ at $x = -8$

The tangent line to a curve at a point closely approximates the curve near the point. In fact, for x-values close enough to the point of tangency, the function and its tangent line are virtually indistinguishable. Problems 31 and 32 explore this relationship. Use each given function and the indicated point to complete the following.

(a) Write the equation of the tangent line to the curve at the point.

(b) Use a graphing utility to graph both the function and its tangent line. Be sure your graph shows the point of tangency.

(c) Repeatedly zoom on the point of tangency until the function and the tangent line cannot be distinguished. Identify the x- and y-ranges in this window.

31. $f(x) = 3x^2 + 2x$, $(1, 5)$

32. $f(x) = 4x - x^2$, $(5, -5)$

For each function in problems 33–36, do the following.

(a) Find $f'(x)$.

(b) Graph both $f(x)$ and $f'(x)$ with a graphing utility.

(c) Use the graph of $f'(x)$ to identify x-values where $f'(x) = 0$, $f'(x) > 0$, and $f'(x) < 0$.

(d) Use the graph of $f(x)$ to identify x-values where $f(x)$ has a maximum or minimum point, where the graph of $f(x)$ is rising, and where the graph of $f(x)$ is falling.

33. $f(x) = 8 - 2x - x^2$

34. $f(x) = x^2 + 4x - 12$

35. $f(x) = x^3 - 12x - 5$

36. $f(x) = 7 - 3x^2 - \dfrac{x^3}{3}$

37. (a) Find the derivatives of $y = x^3$, $u(x) = x$, and $v(x) = x^2$.

(b) Does $\dfrac{d}{dx}(x \cdot x^2) = \dfrac{d}{dx}(x) \cdot \dfrac{d}{dx}(x^2)$?

(c) We know that

$$\frac{d}{dx}[u(x) + v(x)] = \frac{d}{dx}[u(x)] + \frac{d}{dx}[v(x)].$$

Does $\dfrac{d}{dx}[u(x) \cdot v(x)] = \dfrac{d}{dx}[u(x)] \cdot \dfrac{d}{dx}[v(x)]$ for all differentiable functions $u(x)$ and $v(x)$?

38. If $u(x)$ and $v(x)$ are differentiable functions of x, prove the Difference Rule:
If $f(x) = u(x) - v(x)$, then $f'(x) = u'(x) - v'(x)$.

APPLICATIONS

39. **Revenue** Suppose that a wholesaler expects that his monthly revenue for small television sets will be

$$R(x) = 100x - 0.1x^2, \quad 0 \le x \le 800,$$

where x is the number of units sold. Find his marginal revenue and interpret it when the quantity sold is
(a) $x = 300$.   (b) $x = 600$.

40. **Revenue** The total revenue for a commodity is described by the function

$$R = 300x - 0.02x^2.$$

(a) What is the marginal revenue when 40 units are sold?

(b) Interpret your answer to part (a).

41. **Workers and output** The weekly output of a certain product is

$$Q(x) = 200x + 6x^2,$$

where x is the number of workers on the assembly line. If there are presently 60 workers on the line,

(a) find $Q'(x)$ and estimate the change in the weekly output caused by the addition of one worker.

(b) calculate $Q(61) - Q(60)$ to see the actual change in the weekly output.

42. **Capital investment and output** The monthly output of a certain product is

$$Q(x) = 800x^{5/2},$$

where x is the capital investment in millions of dollars. Find dQ/dx, which can be used to estimate the effect on the output if an additional capital investment of $1 million is made.

43. **Demand** The demand q for a product depends on the price p (in dollars) according to

$$q = \frac{1000}{\sqrt{p}} - 1, \quad \text{for } p > 0.$$

Find and explain the meaning of the instantaneous rate of change of demand with respect to price when the price is
(a) $25.   (b) $100.

44. **Demand** Suppose that the demand for a product depends on the price p according to

$$D(p) = \frac{50,000}{p^2} - \frac{1}{2}, \qquad p > 0,$$

where p is in dollars. Find and explain the meaning of the instantaneous rate of change of demand with respect to price when

(a) $p = 50$.   (b) $p = 100$.

45. **Cost and average cost** Suppose that the total cost function for the production of x units of a product is given by

$$C(x) = 4000 + 55x + 0.1x^2.$$

Then the average cost of producing x items is

$$\overline{C(x)} = \frac{\text{total cost}}{x} = \frac{4000}{x} + 55 + 0.1x.$$

(a) Find the instantaneous rate of change of average cost with respect to the number of units produced, at any level of production.

(b) Find the level of production at which this rate of change equals zero.

46. **Cost and average cost** Suppose that the total cost function for a certain commodity is given by

$$C(x) = 40,500 + 190x + 0.2x^2,$$

where x is the number of units produced.

(a) Find the instantaneous rate of change of the average cost

$$\overline{C} = \frac{40,500}{x} + 190 + 0.2x$$

for any level of production.

(b) Find the level of production where this rate of change equals zero.

47. **Cost-benefit** Suppose that for a certain city the cost C of obtaining drinking water that contains p percent impurities (by volume) is given by

$$C = \frac{120,000}{p} - 1200.$$

(a) Find the rate of change of cost with respect to p when impurities account for 1% (by volume).

(b) Write a sentence that explains the meaning of your answer in part (a).

48. **Cost-benefit** Suppose that the cost C of processing the exhaust gases at an industrial site so that only p percent of the particulate pollution escapes is given by

$$C(p) = \frac{8100(100 - p)}{p}.$$

(a) Find the rate of change of cost C with respect to the percent of particulate pollution that escapes when $p = 2$ (percent).

(b) Write a sentence that interprets your answer in part (a).

49. **Windchill** One form of the formula that meteorologists use to calculate windchill temperature (WC) is

$$WC = 48.064 + 0.474t - 0.020ts - 1.85s$$
$$+ 0.304t\sqrt{s} - 27.74\sqrt{s},$$

where s is the wind speed in mph and t is the actual air temperature in degrees Fahrenheit. If temperature is constant at $15°$,

(a) express windchill WC as a function of wind speed s.

(b) find the rate of change of windchill with respect to wind speed when the wind speed is 25 mph.

(c) interpret your answer to part (b).

50. **Allometric relationships—crabs** For fiddler crabs, data gathered by Thompson* show that the allometric relationship between the weight C of the claw and the weight W of the body is given by

$$C = 0.11W^{1.54}.$$

Find the function that gives the rate of change of claw weight with respect to body weight.

51. **Union participation** The following table shows the percentage of U.S. workers who belonged to unions for selected years from 1930 to 1989. Assume that these data can be modeled with the function

$$u(x) = -39.3 + 2.38x - 0.02x^2,$$

where x is the number of years past 1900. The function and data points are graphed in the figure on p. 634.

graphed in the figure on p. 634.

Year	Percentage
1930	11.6
1940	26.9
1950	31.5
1960	31.4
1970	27.3
1975	25.5
1980	21.9
1985	18
1989	16.4

SOURCE: *World Almanac*, 1991

*d'Arcy Thompson, *On Growth and Form* (Cambridge: Cambridge University Press, 1961).

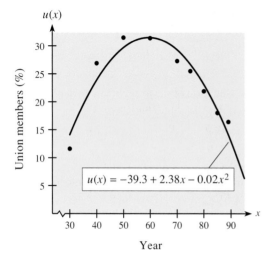

$$u(x) = -39.3 + 2.38x - 0.02x^2$$

Year

(a) For the period from 1980 to 1985, use the data points to find the average rate of change of the percentage of U.S. workers who belonged to unions.

(b) Find the instantaneous rate of change of the modeling function $u(x)$ for the year 1980.

52. **Population below poverty level** The table below shows the number of millions of people in the United States who lived below the poverty level for selected years between 1960 and 1990.

Year	Persons Living Below the Poverty Level (in Millions)
1960	39.9
1965	33.2
1970	25.4
1975	25.9
1980	29.3
1986	32.4
1989	31.5
1990	33.6

SOURCE: *The World Almanac and Book of Facts,* 1993

Because these data approximately fit the graph of the function

$$p(t) = -0.0020595t^3 + 0.508851t^2 - 41.3634t + 1135.4,$$

where p is the number of millions of people and t represents the number of years since 1900, this function mod-

els the number of people below the poverty level. A graph of $p(t)$ along with the data from the table is shown in the figure.

(a) Find the rate of growth in the number of persons living below the poverty level in 1980.

(b) Interpret your answer from part (a).

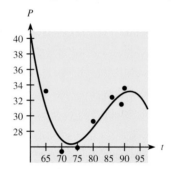

53. **Inflation rate** The annual change in the Consumer Price Index (CPI) during certain years is shown in the table below. Assume that the annual change in the Consumer Price Index can be modeled with the function

$$f(t) = -0.0068567913t^4 + 2.32783445t^3 - 295.9642445t^2 + 16701.32789t - 352921,$$

where t is the number of years past 1900. A graph of $f(t)$ along with the data in the table is shown in the figure.

Year	Consumer Price Indexes Annual Percent Change
1977	6.5
1978	7.6
1979	11.3
1980	13.5
1981	10.3
1982	6.2
1983	3.2
1984	4.3
1985	3.6
1986	1.9
1987	3.6
1988	4.1
1989	4.8
1990	5.4
1991	4.2

SOURCE: *The World Almanac and Book of Facts,* 1993

(a) Find the function that models the instantaneous rate of change of the Consumer Price Index.
(b) Use the model found in part (a) to find the instantaneous rate of change in 1981 and in 1988.
(c) Interpret the two rates of change from part (b).

$f(t)$

54. **Consumer debt** The percentage of disposable income spent on consumer debt during certain years is shown in the table below.

Year	Consumer Debt as a Percentage of Disposable Income	Year	Consumer Debt as a Percentage of Disposable Income
1970	18.3	1986	21.4
1975	17.9	1987	21.2
1980	18.2	1988	21.0
1982	16.9	1989	20.9
1983	17.7	1990	20.0
1984	19.1	1991	18.9
1985	20.8		

SOURCE: Federal Reserve System

Assume that consumer debt as a percentage of disposable income can be modeled with the function

$$d(t) = -0.000739658t^4 + 0.235441t^3 - 28.0073t^2 + 1475.75t - 29{,}045.2,$$

where t is the number of years past 1900. A graph of $d(t)$ along with the data in the table is shown in the figure.

d

(a) Find the function that models the instantaneous rate of change of consumer debt as a percentage of disposable income.
(b) Use the model found in part (a) to find the instantaneous rate of change in 1985 and in 1991.
(c) Interpret the two rates of change found in part (b).

55. **Drug use** The percentage of high school seniors who have tried certain drugs is shown in the table below for selected years.

Year of Graduation	Marijuana	Hallucinogens
1975	47.3	16.3
1978	59.2	14.3
1980	60.3	13.3
1983	57	11.9
1984	54.9	10.7
1985	54.2	10.3
1986	50.9	9.7
1987	50.2	10.3
1988	47.2	8.9
1989	43.7	9.4
1990	40.7	9.4
1991	36.7	9.6

SOURCE: National Institute on Drug Abuse

Assume that the percentage of high school seniors who have tried marijuana can be modeled with the function

$$f_m(t) = -0.228596t^2 + 37.0677t - 1444.41,$$

where t is the number of years past 1900.
(a) Find the function that models the instantaneous rate of change with respect to time of the percentage of seniors who have tried marijuana.
(b) Use the model in part (a) to find the instantaneous rate of change in 1975 and in 1985.
(c) Graph both the data from the table and the function $f_m(t)$.
(d) In what years can you guarantee that $f_m(t)$ cannot be used to predict the percentage of seniors who tried marijuana? Explain.

9.5 PRODUCT AND QUOTIENT RULES

Objectives

☐ To use the Product Rule to find the derivative of certain functions
☐ To use the Quotient Rule to find the derivative of certain functions

APPLICATION PREVIEW

When medicine is administered, reaction (measured in change of blood pressure or temperature) can be modeled by

$$R = m^2 \left(\frac{c}{2} - \frac{m}{3} \right),$$

where c is a positive constant and m is the amount of medicine absorbed into the blood.[*] The rate of change of R with respect to m is the sensitivity of the body to medicine. To find an expression for sensitivity as a function of m, we calculate dR/dm. We can find this derivative with the **Product Rule** for derivatives.

We have simple formulas for finding the derivatives of the sums and differences of functions. But we are not so lucky with products. The derivative of a product is *not* the product of the derivatives. To see this, we consider the function $f(x) = x \cdot x$. Since this function is $f(x) = x^2$, its derivative is $f'(x) = 2x$. But the product of the derivatives of x and x would give $1 \cdot 1 = 1 \neq 2x$. Thus we need a different formula to find the derivative of a product. This formula is given by the Product Rule.

Product Rule   If $f(x) = u(x) \cdot v(x)$, where u and v are differentiable functions of x, then

$$f'(x) = u(x) \cdot v'(x) + v(x) \cdot u'(x).$$

Thus the derivative of a product of two functions is the first function times the derivative of the second plus the second function times the derivative of the first.

We can prove the Product Rule as follows. If $f(x) = u(x) \cdot v(x)$, then

$$\lim_{h \to 0} \frac{f(x+h) - f(x)}{h} = \lim_{h \to 0} \frac{u(x+h) \cdot v(x+h) - u(x) \cdot v(x)}{h}.$$

Subtracting and adding $u(x+h) \cdot v(x)$ in the numerator gives

$$f'(x) = \lim_{h \to 0} \frac{u(x+h) \cdot v(x+h) - u(x+h) \cdot v(x) + u(x+h) \cdot v(x) - u(x) \cdot v(x)}{h}$$

$$= \lim_{h \to 0} \left(u(x+h) \left[\frac{v(x+h) - v(x)}{h} \right] + v(x) \left[\frac{u(x+h) - u(x)}{h} \right] \right).$$

[*] Source: Thrall, R. M., et al, *Some Mathematical Models in Biology,* U.S. Department of Commerce, 1967.

Properties III and IV of limits give

$$f'(x) = \lim_{h \to 0} u(x + h) \cdot \lim_{h \to 0} \frac{v(x + h) - v(x)}{h} + \lim_{h \to 0} v(x) \cdot \lim_{h \to 0} \frac{u(x + h) - u(x)}{h}$$

Now, because u is differentiable and hence continuous, it follows that $\lim_{h \to 0} u(x + h) = u(x)$. So we have $f'(x) = u(x) \cdot v'(x) + v(x) \cdot u'(x)$.

EXAMPLE 1

Use the Product Rule to find the derivative of $f(x) = x^2 \cdot x$.

Solution

Using the formula with $u(x) = x^2$, $v(x) = x$, we have

$$\begin{aligned} f'(x) &= u(x) \cdot v'(x) + v(x) \cdot u'(x) \\ &= x^2 \cdot 1 + x(2x) \\ &= x^2 + 2x^2 \\ &= 3x^2. \end{aligned}$$

Note that we could have found the same result by multiplying the factors and finding the derivative of $f(x) = x^3$. But we will soon see how valuable the Product Rule is.

EXAMPLE 2

Find dy/dx if $y = (2x^3 + 3x + 1)(x^2 + 4)$.

Solution

Using the Product Rule with $u(x) = 2x^3 + 3x + 1$ and $v(x) = x^2 + 4$. we have

$$\begin{aligned} \frac{dy}{dx} &= (2x^3 + 3x + 1)(2x) + (x^2 + 4)(6x^2 + 3) \\ &= 4x^4 + 6x^2 + 2x + 6x^4 + 3x^2 + 24x^2 + 12 \\ &= 10x^4 + 33x^2 + 2x + 12. \end{aligned}$$

We could, of course, avoid using the Product Rule by multiplying the two factors before taking the derivative. But multiplying the factors first may involve more work than using the Product Rule.

EXAMPLE 3

Given $f(x) = (4x^3 + 5x^2 - 6x + 5)(x^3 - 4x^2 + 1)$, find the slope of the tangent to the graph of $y = f(x)$ at $x = 1$.

Solution

$$f'(x) = (4x^3 + 5x^2 - 6x + 5)(3x^2 - 8x)$$
$$+ (x^3 - 4x^2 + 1)(12x^2 + 10x - 6).$$

If we substitute $x = 1$ into $f'(x)$, we find that the slope of the curve at $x = 1$ is $f'(1) = 8(-5) + (-2)(16) = -72$.

The rule for finding the derivative of a function that is the quotient of two functions requires a new formula.

Quotient Rule   If $f(x) = u(x)/v(x)$, where u and v are differentiable functions of x, with $v(x) \neq 0$, then

$$f'(x) = \frac{v(x) \cdot u'(x) - u(x) \cdot v'(x)}{[v(x)]^2}.$$

The preceding formula says that the derivative of a quotient is the denominator times the derivative of the numerator minus the numerator times the derivative of the denominator, all divided by the square of the denominator.

To see that this rule is reasonable, consider the function $f(x) = x^3/x$, $x \neq 0$. Using the Quotient Rule, with $u(x) = x^3$ and $v(x) = x$, we get

$$f'(x) = \frac{x(3x^2) - x^3(1)}{x^2} = \frac{3x^3 - x^3}{x^2} = \frac{2x^3}{x^2} = 2x.$$

Since $f(x) = x^3/x = x^2$ if $x \neq 0$, we see that $f'(x) = 2x$ is the correct derivative. The proof of the Quotient Rule is left for the student in problem 39 of Exercise 9.5.

EXAMPLE 4

If $f(x) = \dfrac{x^2 - 4x}{x + 5}$, find $f'(x)$.

Solution

Using the Quotient Rule with $u(x) = x^2 - 4x$ and $v(x) = x + 5$, we get

$$f'(x) = \frac{(x + 5)(2x - 4) - (x^2 - 4x)(1)}{(x + 5)^2}$$

$$= \frac{2x^2 + 6x - 20 - x^2 + 4x}{(x + 5)^2}$$

$$= \frac{x^2 + 10x - 20}{(x + 5)^2}.$$

EXAMPLE 5

If $f(x) = \dfrac{x^3 - 3x^2 + 2}{x^2 - 4}$, find $f'(x)$.

Solution

Using the Quotient Rule with $u(x) = x^3 - 3x^2 + 2$ and $v(x) = x^2 - 4$, we get

$$f'(x) = \frac{(x^2 - 4)(3x^2 - 6x) - (x^3 - 3x^2 + 2)(2x)}{(x^2 - 4)^2}$$

$$= \frac{(3x^4 - 6x^3 - 12x^2 + 24x) - (2x^4 - 6x^3 + 4x)}{(x^2 - 4)^2}$$

$$= \frac{x^4 - 12x^2 + 20x}{(x^2 - 4)^2}.$$

EXAMPLE 6

Use the Quotient Rule to find the derivative of $y = 1/x^3$.

Solution

Letting $u(x) = 1$ and $v(x) = x^3$, we get

$$y' = \frac{x^3(0) - 1(3x^2)}{(x^3)^2}$$

$$= -\frac{3x^2}{x^6}$$

$$= -\frac{3}{x^4}.$$

Note that we could have found the derivative more easily by writing

$$y = 1/x^3 = x^{-3},$$

so

$$y' = -3x^{-4} = -\frac{3}{x^4}.$$

Recall that we proved the Powers of x Rule for positive integer powers and assumed that it was true for all real number powers. In problem 40 of Exercise 9.5, you will be asked to use the Quotient Rule to show that the Powers of x Rule applies to negative integers.

It is not necessary to use the Quotient Rule when the denominator of the function in question contains only a constant. For example, the function $y = (x^3 - 3x)/3$ can be written $y = \frac{1}{3}(x^3 - 3x)$, so the derivative is $y' = \frac{1}{3}(3x^2 - 3) = x^2 - 1$.

CHECKPOINT

1. True or false: The derivative of the product of two functions is equal to the product of the derivatives of the two functions.

2. True or false: The derivative of the quotient of two functions is equal to the quotient of the derivatives of the two functions.

3. If $y = \frac{4}{3}(x^2 + 3x - 4)$, does finding y' require the Product Rule? Explain.

4. If $y = f(x)/c$, where c is a constant, does finding y' require the Quotient Rule? Explain.

EXAMPLE 7

Suppose that the revenue function for a product is given by

$$R(x) = 10x + \frac{100x}{3x + 5},$$

where x is the number of units sold and R is in dollars.
(a) Find the marginal revenue function.
(b) Find the marginal revenue when $x = 15$.

Solution

(a) We must use the Quotient Rule to find the marginal revenue (the derivative).

$$\overline{MR} = R'(x) = 10 + \frac{(3x + 5)(100) - 100x(3)}{(3x + 5)^2}$$

$$= 10 + \frac{300x + 500 - 300x}{(3x + 5)^2} = 10 + \frac{500}{(3x + 5)^2}$$

(b) The marginal revenue when $x = 15$ is $R'(15)$.

$$R'(15) = 10 + \frac{500}{[(3)(15) + 5]^2} = 10 + \frac{500}{(50)^2}$$

$$= 10 + \frac{500}{2500} = 10.20$$

Recall that $R'(15)$ estimates the revenue from the sale of the 16th item.

We now consider the problem posed in the Application Preview.

EXAMPLE 8

When medicine is administered, reaction (measured in change of blood pressure or temperature) can be modeled by

$$R = m^2 \left(\frac{c}{2} - \frac{m}{3} \right),$$

where c is a positive constant and m is the amount of medicine absorbed into the blood. The rate of change of R with respect to m is the sensitivity of the body to medicine. Find an expression for sensitivity s as a function of m.

Solution

The sensitivity is the rate of change of R with respect to m, or the derivative:

$$\frac{dR}{dm} = m^2 \left(0 - \frac{1}{3}\right) + \left(\frac{c}{2} - \frac{1}{3}m\right)(2m)$$

$$= -\frac{1}{3}m^2 + mc - \frac{2}{3}m^2 = mc - m^2,$$

so sensitivity is given by

$$s = \frac{dR}{dm} = mc - m^2.$$

EXAMPLE 9

(a) Graph $f(x) = \dfrac{8x}{x^2 + 4}$ and its derivative on the same set of axes over an interval that contains the x-values where $f'(x) = 0$

(b) Determine the values of x where $f'(x) = 0$ and the intervals where $f'(x) > 0$ and where $f'(x) < 0$.

(c) What is the relation between the value of the derivative, $f'(x)$, and the direction of the graph of $y = f(x)$?

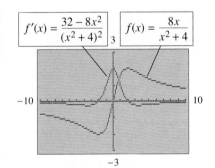

Figure 9.28

Solution

We find the derivative of $f(x)$ by using the Quotient Rule.

$$f'(x) = \frac{(x^2 + 4)(8) - (8x)(2x)}{(x^2 + 4)^2}$$

$$= \frac{32 - 8x^2}{(x^2 + 4)^2}$$

(a) The graphs of $f(x) = \dfrac{8x}{x^2 + 4}$ and $f'(x) = \dfrac{32 - 8x^2}{(x^2 + 4)^2}$ are shown in Figure 9.28.

(b) $f'(x) = 0$ at $x = -2$ and $x = 2$. In the interval $-2 < x < 2$, the graph of $y = f'(x)$ is above the x-axis, so $f'(x)$ is positive for $-2 < x < 2$. In the intervals $x < -2$ and $x > 2$, the graph of $y = f'(x)$ is below the x-axis, so $f'(x)$ is negative there.

(c) The graph of $y = f(x)$ appears to turn at $x = -2$, and $x = 2$, where $f'(x) = 0$. The graph of $y = f(x)$ is increasing where $f'(x) > 0$ and decreasing where $f'(x) < 0$.

CHECKPOINT
SOLUTIONS

1. False. The derivative of a product is equal to the first function times the derivative of the second plus the second function times the derivative of the first. That is,

$$\frac{d}{dx}(fg) = f \cdot \frac{dg}{dx} + g \cdot \frac{df}{dx}.$$

2. False. The derivative of a quotient is equal to the denominator times the derivative of the numerator minus the numerator times the derivative of the denominator, all divided by the square of the denominator. That is,

$$\frac{d}{dx}\left(\frac{f}{g}\right) = \frac{g \cdot f' - f \cdot g'}{g^2}.$$

3. No, y' can be found with the Coefficient Rule:

$$y' = \frac{4}{3}(2x + 3).$$

4. No, y' can be found with the Coefficient Rule:

$$y' = \left(\frac{1}{c}\right)f'(x).$$

EXERCISE 9.5

1. Find y' if $y = (x + 3)(x^2 - 2x)$.

2. Find $f'(x)$ if $f(x) = (3x - 1)(x^3 + 1)$.

3. Find $\dfrac{dp}{dq}$ if $p = (3q - 1)(q^2 + 2)$.

4. Find $\dfrac{ds}{dt}$ if $s = (t^4 + 1)(t^3 - 1)$.

5. If $f(x) = (x^{12} + 3x^4 + 4)(4x^3 - 1)$, find $f'(x)$.

6. If $y = (3x^7 + 4)(8x^6 - 6x^4 - 9)$, find $\dfrac{dy}{dx}$.

In problems 7–10, find the derivative, but do not simplify your answer.

7. $y = (7x^6 - 5x^4 + 2x^2 - 1)(4x^9 + 3x^7 - 5x^2 + 3x)$

8. $y = (9x^9 - 7x^7 - 6x)(3x^5 - 4x^4 + 3x^3 - 8)$

9. $y = (x^2 + x + 1)(\sqrt[3]{x} - 2\sqrt{x} + 5)$

10. $y = (\sqrt[5]{x} - 2\sqrt[4]{x} + 1)(x^3 - 5x - 7)$

In problems 11–14, at each indicated point find
(a) the slope of the tangent line, and
(b) the instantaneous rate of change of the function.

11. $y = (x^2 + 1)(x^3 - 4x)$ at $(-2, 0)$

12. $y = (x^3 - 3)(x^2 - 4x + 1)$ at $(2, -15)$

13. $y = \left(4\sqrt{x} + \dfrac{3}{x}\right)\left(3\sqrt[3]{x} - \dfrac{5}{x^2} - 25\right)$ at $x = 1$

14. $y = (3\sqrt[4]{x^5} + \sqrt[5]{x^4} - 1)\left(\dfrac{2}{x^3} - \dfrac{1}{\sqrt{x}}\right)$ at $x = 1$

In problems 15–26, find the indicated derivatives.

15. y' for $y = \dfrac{x}{x^2 - 1}$

16. $f'(x)$ for $f(x) = \dfrac{x^2}{x - 3}$

17. $\dfrac{dp}{dq}$ for $p = \dfrac{q^2 + 1}{q - 2}$

18. $C'(x)$ for $C(x) = \dfrac{x^2 + 1}{x^2 - 1}$

19. $\dfrac{dy}{dx}$ for $y = \dfrac{1 - 2x^2}{x^4 - 2x^2 + 5}$

20. $\dfrac{ds}{dt}$ for $s = \dfrac{t^3 - 4}{t^3 - 2t^2 - t - 5}$

21. $\dfrac{dz}{dx}$ for $z = x^2 + \dfrac{x^2}{1 - x - 2x^2}$

22. $\dfrac{dy}{dx}$ for $y = 200x - \dfrac{100x}{3x + 1}$

23. $\dfrac{dp}{dq}$ for $p = \dfrac{3\sqrt[3]{q}}{1 - q}$

24. $\dfrac{dy}{dx}$ for $y = \dfrac{2\sqrt{x} - 1}{1 - 4\sqrt{x^3}}$

25. y' for $y = \dfrac{x(x^2 + 4)}{x - 2}$

26. $f'(x)$ for $f(x) = \dfrac{(x + 1)(x - 2)}{x^2 + 1}$

In problems 27 and 28, at the indicated point for each function, find
(a) the slope of the tangent line, and
(b) the instantaneous rate of change of the function.

27. $y = \dfrac{x^2 + 1}{x + 3}$ at $(2, 1)$.

28. $y = \dfrac{x^2 - 4x}{x^2 + 2x}$ at $\left(2, -\dfrac{1}{2}\right)$.

In problems 29–32, write the equation of the tangent line to the graph of the function at the indicated point.

29. $y = (9x^2 - 6x + 1)(1 + 2x)$ at $x = 1$

30. $y = (4x^2 + 4x + 1)(7 - 2x)$ at $x = 0$

31. $y = \dfrac{3x^4 - 2x - 1}{4 - x^2}$ at $x = 1$

32. $y = \dfrac{x^2 - 4x}{2x - x^3}$ at $x = 2$

In problems 33–36,
(a) find points where the slope of the tangent to the function is 0, and
(b) use a graphing utility to graph the function and identify the points found in part (a).

33. $f(x) = (x^2 + 4x + 4)(x - 7)$

34. $f(x) = (x^2 - 14x + 49)(2x + 1)$

35. $y = \dfrac{x^2}{x - 2}$

36. $y = \dfrac{x^2 - 7}{4 - x}$

 In problems 37 and 38,
(a) find $f'(x)$,
(b) graph both $f(x)$ and $f'(x)$ with a graphing utility,
(c) identify the x-values where $f'(x) = 0$, $f'(x) > 0$, and $f'(x) < 0$, and
(d) identify x-values where $f(x)$ has a maximum point or a minimum point, where $f(x)$ is increasing, and where $f(x)$ is decreasing.

37. $f(x) = \dfrac{10x^2}{x^2 + 1}$

38. $f(x) = \dfrac{8 - x^2}{x^2 + 4}$

39. Prove the Quotient Rule for differentiation. (Hint: Add $[-u(x) \cdot v(x) + u(x) \cdot v(x)]$ to the expanded numerator and use steps similar to those used to prove the Product Rule).

40. Use the Quotient Rule to show that the Powers of x Rule applies to negative integer powers. That is, show that $(d/dx)x^n = nx^{n-1}$ when $n = -k$, $k > 0$, by finding the derivative of $f(x) = 1/(x^k)$.

APPLICATIONS

41. **Cost-benefit** If the cost C of removing p percent of the particulate pollution from the exhaust gases at an industrial site is given by

$$C(p) = \dfrac{8100p}{100 - p},$$

find the rate of change of C with respect to p.

42. **Cost-benefit** If the cost C of removing p percent of the impurities from the waste water in a manufacturing process is given by

$$C(p) = \dfrac{9800p}{101 - p},$$

find the rate of change of C with respect to p.

43. **Revenue** Suppose the revenue function for a product is given by

$$R(x) = \dfrac{60x^2 + 74x}{2x + 2}.$$

Find the marginal revenue when 49 units are sold. Interpret your result.

44. **Revenue** The revenue from the sale of x units of a product is given by

$$R(x) = \dfrac{3000}{2x + 2} + 80x - 1500.$$

Find the marginal revenue when 149 units are sold. Interpret your result.

45. **Revenue** A travel agency will plan a group tour for groups of size 25 or larger. If the group contains exactly 25 people, the cost is $300 per person. If each person's cost is reduced by $10 for each additional person above the 25, then the revenue is given by

$$R(x) = (25 + x)(300 - 10x),$$

where x is the number of additional people above 25. Find the marginal revenue if the group contains 30 people. Interpret your result.

46. **Revenue** McRobert's TV Shop sells 200 sets per month at a price of $400 per unit. Market research indicates that they can sell one additional set for each $1 they reduce the price, and in this case the total revenue is

$$R(x) = (200 + x)(400 - x),$$

where x is the number of additional sets beyond the 200. If the shop sells a total of 250 sets, find the marginal revenue. Interpret your result.

47. **Response to a drug** The reaction R to an injection of a drug is related to the dosage x according to

$$R(x) = x^2\left(500 - \frac{x}{3}\right),$$

where 1000 mg is the maximum dosage. If the rate of reaction with respect to the dosage defines the sensitivity to the drug, find the sensitivity.

48. **Nerve response** The number of action potentials produced by a nerve, t seconds after a stimulus is given by

$$N(t) = 25t + \frac{4}{t^2 + 2} - 2.$$

Find the rate at which the action potentials are produced by the nerve.

49. **Test reliability** If a test having reliability r is lengthened by a factor n, the reliability of the new test is given by

$$R = \frac{nr}{1 + (n - 1)r}, \quad 0 < r \le 1.$$

Find the rate at which R changes with respect to n.

50. **Advertising and sales** The sales of a product s (in thousands of dollars) are related to advertising expenses (in thousands of dollars) by

$$s = \frac{200x}{x + 10}.$$

Find and interpret the meaning of the rate of change of sales with respect to advertising expenses when
(a) $x = 10$ (b) $x = 20$.

51. **Candidate recognition** Suppose that the proportion P of voters who recognize a candidate's name t months after the start of the campaign is given by

$$P(t) = \frac{13t}{t^2 + 100} + 0.18.$$

(a) Find the rate of change of P when $t = 6$ and explain its meaning.
(b) Find the rate of change of P when $t = 12$ and explain its meaning.
(c) One month prior to the election, is it better for $P'(t)$ to be positive or negative?

52. **Endangered species population** It is determined that a wildlife refuge can support a group of up to 120 of a certain endangered species. If 75 are introduced onto the refuge and their population after t years is given by

$$p(t) = 75\left(1 + \frac{4t}{t^2 + 16}\right),$$

find the rate of population growth after t years. Find the rate after each of the first seven years.

53. **Windchill** According to the National Climatic Data Center, during 1991, the lowest temperature recorded in Indianapolis, Indiana, was 0°F. If x is the wind speed in miles per hour and $x \ge 5$, then the windchill (in degrees Fahrenheit) for an air temperature of 0°F can be approximated by the function

$$f(x) = \frac{289.173 - 58.5731x}{x + 1}.$$

(a) At what rate is the windchill changing when the wind speed is 20 mph?
(b) Explain the meaning of your answer to part (a).

54. **Response to injected adrenalin** Experimental evidence has shown that the concentration of injected adrenaline x is related to the response y of a muscle according to the equation

$$y = \frac{x}{a + bx},$$

where a and b are constants. Find the rate of change of response with respect to the concentration.

55. **Social Security** America's 45 million Social Security recipients got a 2.6% cost-of-living increase in 1994. That was the second-smallest increase in nearly 20 years, a reflection of the low inflation rate. The following data shows the percent increases for selected years. (We will assume that $t = 0$ in 1985.)

Year	Percent Increase, Cost of Living
1989	4.0
1990	4.9
1991	5.1
1992	4.1
1993	3.5
1994	2.6

SOURCE: Social Security Administration

Assume this data can be modeled by the equation

$$C(t) = (t + 5)(-0.218t + 3.57) - 20.13.$$

(a) Find the rate of change of the cost-of-living increase as predicted by the model for 1990.
(b) Interpret your answer to part (a).
(c) Graph both the data and the model using a graphing utility.
(d) From the graph in part (c), identify t-values for which the model must be invalid.

56. **Drug use** The percentage of high school seniors who have tried hallucinogens is shown in the table below for selected years.

Year of Graduation	Hallucinogens
1975	16.3
1978	14.3
1980	13.3
1983	11.9
1984	10.7
1985	10.3
1986	9.7
1987	10.3
1988	8.9
1989	9.4
1990	9.4
1991	9.6

SOURCE: National Institute on Drug Abuse

The data from the table indicate that the percentage of high school seniors who have tried hallucinogens can be modeled with the function

$$f_h(t) = 216.074 - (t + 70)(2.782 - 0.0242t),$$

where t is the number of years past 1970.
(a) Find the function that models the instantaneous rate of change with respect to time of the percentage of seniors who have tried hallucinogens.
(b) Use the model in part (a) to find the instantaneous rate of change in 1978 and in 1988.
(c) Interpret each rate of change found in part (b).

57. **Farm workers** The percent of U.S. workers in farm occupations during certain years is shown in the table.

Year	Percent of All Workers in Farm Occupations
1820	71.8
1850	63.7
1870	53
1900	37.5
1920	27
1930	21.2
1940	17.4
1950	11.6
1960	6.1
1970	3.6
1980	2.7
1985	2.8
1990	2.4

SOURCE: *The World Almanac and Book of Facts*, 1993

Assume that the percentage of U.S. workers in farm occupations can be modeled with the function

$$f(t) = 1000 \cdot \frac{-8.0912t + 1558.9}{1.09816t^2 - 122.183t + 21472.6},$$

where t is the number of years past 1800.
(a) Find the function that models the instantaneous rate of change of the percentage of U.S. workers in farm occupations.
(b) Use the model in part (a) to find the instantaneous rate of change in 1870 and in 1970.
(c) Interpret each of the rates of change in part (b).

9.6 THE CHAIN RULE AND POWER RULE

Objectives

❏ To use the Chain Rule to differentiate functions
❏ To use the Power Rule to differentiate functions

APPLICATION PREVIEW

The demand x for a product is given by

$$x = \frac{98}{\sqrt{2p + 1}} - 1,$$

where p is the price per unit. To find how fast demand is changing when price is \$24, we take the derivative of x with respect to p. If we write this function with a power rather than a radical, it has the form

$$x = 98(2p + 1)^{-1/2} - 1.$$

The formulas learned so far cannot be used to find this derivative. Rather than use the limit definition of derivative, we use a new formula, the **Power Rule,** to find this derivative. In this section we will discuss the **Chain Rule** and the Power Rule, which is one of the results of the Chain Rule, and we will use these formulas to solve applied problems.

Recall from Section 1.4 that if f and g are functions, then the composite functions g of f (denoted $g \circ f$) and f of g (denoted $f \circ g$) are defined as follows:

$$(g \circ f)(x) = g(f(x)) \quad \text{and} \quad (f \circ g)(x) = f(g(x)).$$

EXAMPLE I

If $f(x) = 3x^2$ and $g(x) = 2x - 1$, find $F(x) = f(g(x))$.

Solution

Substituting $g(x) = 2x - 1$ for x in $f(x)$ gives

$$f(g(x)) = f(2x - 1) = 3(2x - 1)^2.$$

Thus $F(x) = 3(2x - 1)^2$.

We could find the derivative of the function $F(x) = 3(2x - 1)^2$ by multiplying out the expression $3(2x - 1)^2$. Then

$$F(x) = 3(4x^2 - 4x + 1) = 12x^2 - 12x + 3,$$

so $F'(x) = 24x - 12$. But we can also use a very powerful rule, called the **Chain Rule,** to find derivatives of functions of this type. If we write the composite function $y = f(g(x))$ in the form $y = f(u)$, where $u = g(x)$, we state the Chain Rule as follows.

Chain Rule If f and g are differentiable functions with $y = f(u)$ and $u = g(x)$, then y is a differentiable function of x, and

$$\frac{dy}{dx} = \frac{d}{du} f(u) \cdot \frac{d}{dx} g(x)$$

or, written another way,

$$\frac{dy}{dx} = \frac{dy}{du} \cdot \frac{du}{dx}.$$

Note that dy/du represents the derivative of $y = f(u)$ *with respect to u* and du/dx represents the derivative of $u = g(x)$ *with respect to x.* For example, if $y = 3(2x - 1)^2$, we may write $y = f(u) = 3u^2$, where $u = 2x - 1$. Then the derivative is

$$\frac{dy}{dx} = \frac{dy}{du} \cdot \frac{du}{dx} = 6u \cdot 2 = 12u.$$

To write this derivative in terms of x, we substitute $2x - 1$ for u. Thus

$$\frac{dy}{dx} = 12(2x - 1) = 24x - 12.$$

Note that we get the same result by using the Chain Rule as we did by multiplying out $F(x) = 3(2x - 1)^2$. The Chain Rule is important because it is not always possible to rewrite the function as a polynomial. Consider the following example.

EXAMPLE 2

If $y = \sqrt{x^2 - 1}$, find $\dfrac{dy}{dx}$.

Solution
If we write this function as $y = f(u) = \sqrt{u}$, when $u = x^2 - 1$, we can find the derivative.

$$\frac{dy}{dx} = \frac{dy}{du} \cdot \frac{du}{dx} = \frac{1}{2} \cdot u^{-1/2} \cdot 2x = u^{-1/2} \cdot x = \frac{1}{\sqrt{u}} \cdot x = \frac{x}{\sqrt{u}}$$

To write this derivative in terms of x alone, we substitute $x^2 - 1$ for u. Then

$$\frac{dy}{dx} = \frac{x}{\sqrt{x^2 - 1}}.$$

Note that we could not find the derivative of a function like that of Example 2 by the methods learned previously.

EXAMPLE 3

If $y = \dfrac{1}{(x^2 + 3x + 1)^2}$, find $\dfrac{dy}{dx}$.

Solution

If we let $u = x^2 + 3x + 1$, we can write $y = f(u) = \dfrac{1}{u^2}$, or $y = u^{-2}$. Then

$$\frac{dy}{dx} = \frac{dy}{du} \cdot \frac{du}{dx} = -2u^{-3}(2x + 3) = \frac{-4x - 6}{u^3}.$$

Substituting for u gives

$$\frac{dy}{dx} = \frac{-4x - 6}{(x^2 + 3x + 1)^3}.$$

EXAMPLE 4

The relationship between the length L (in meters) and weight W (in kilograms) of a species of fish in the Pacific Ocean is given by $W = 10.375L^3$.

The rate of growth in length is given by $\dfrac{dL}{dt} = 0.36 - 0.18L$, where t is measured in years.

(a) Determine a formula for the rate of growth in weight $\dfrac{dW}{dt}$ in terms of L.

(b) If a fish weighs 30 kilograms, then approximate its rate of growth in weight using the formula found in (a).

Solution

(a) The rate of change uses the Chain Rule, as follows.

$$\frac{dW}{dt} = \frac{dW}{dL} \cdot \frac{dL}{dt} = 31.125L^2(0.36 - 0.18L) = 11.205L^2 - 5.6025L^3$$

(b) From $W = 10.375L^3$, we have

$$L = \sqrt[3]{\frac{W}{10.375}}.$$

If $W = 30$ kilograms, $L = \sqrt[3]{\dfrac{30}{10.375}} = 1.4247$ meters, so the rate of growth in weight is

$$\frac{dW}{dt} = 11.205(1.4247)^2 - 5.6025(1.4247)^3 = 6.530 \text{ kilograms/year.}$$

The Chain Rule is very useful and will be extremely important with functions that we will study later, but a special case of the Chain Rule, called the **Power Rule,** is useful for the algebraic functions we have studied so far.

Power Rule   If $y = u^n$, where u is a differentiable function of x, then

$$\frac{dy}{dx} = nu^{n-1} \cdot \frac{du}{dx}.$$

EXAMPLE 5

If $y = (x^2 - 4x)^6$, find $\dfrac{dy}{dx}$.

Solution
The right side of the equation is in the form u^n, with $u = x^2 - 4x$. Thus by the Power Rule,

$$\frac{dy}{dx} = nu^{n-1} \cdot \frac{du}{dx} = 6u^5(2x - 4).$$

Substituting for u gives

$$\frac{dy}{dx} = 6(x^2 - 4x)^5(2x - 4)$$

$$= (12x - 24)(x^2 - 4x)^5.$$

EXAMPLE 6

If $y = 3\sqrt[3]{x^2 - 3x + 1}$, find y'.

Solution
Because $y = 3(x^2 - 3x + 1)^{1/3}$, we can make use of the Power Rule with $u = x^2 - 3x + 1$.

$$y' = 3\left(nu^{n-1} \frac{du}{dx}\right) = 3\left[\frac{1}{3}u^{-2/3}(2x - 3)\right]$$

$$= (x^2 - 3x + 1)^{-2/3}(2x - 3)$$

$$= \frac{2x - 3}{(x^2 - 3x + 1)^{2/3}}$$

EXAMPLE 7

If $p = \dfrac{4}{3q^2 + 1}$, find $\dfrac{dp}{dq}$.

Solution
We can use the Power Rule to find dp/dq if we write the equation in the form

$$p = 4(3q^2 + 1)^{-1}.$$

Then

$$\frac{dp}{dq} = 4[-1(3q^2 + 1)^{-2}(6q)] = \frac{-24q}{(3q^2 + 1)^2}.$$

The derivative of the function in Example 7 can also be found by using the Quotient Rule, but the Power Rule provides a more efficient method.

EXAMPLE 8

Find the derivative of $g(x) = \dfrac{1}{\sqrt{(x^2 + 1)^3}}$.

Solution
Writing $g(x)$ as a power gives

$$g(x) = (x^2 + 1)^{-3/2}.$$

Then

$$g'(x) = -\frac{3}{2}(x^2 + 1)^{-5/2}(2x) = -3x \cdot \frac{1}{(x^2 + 1)^{5/2}}$$

$$= \frac{-3x}{\sqrt{(x^2 + 1)^5}}.$$

CHECKPOINT

1. (a) If $f(x) = (3x^4 + 1)^{10}$, does $f'(x) = 10(3x^4 + 1)^9$?
    (b) If $f(x) = (2x + 1)^5$, does $f'(x) = 10(2x + 1)^4$?
    (c) If $f(x) = \dfrac{[u(x)]^n}{c}$, where c is a constant, does $f'(x) = \dfrac{n[u(x)]^{n-1} \cdot u'(x)}{c}$?

2. (a) If $f(x) = \dfrac{12}{2x^2 - 1}$, find $f'(x)$ by using the Power Rule (not the Quotient Rule).
    (b) If $f(x) = \dfrac{\sqrt{x^3 - 1}}{3}$, find $f'(x)$ by using the Power Rule (not the Quotient Rule).

EXAMPLE 9

The demand x for a product is given by

$$x = 98(2p + 1)^{-1/2} - 1,$$

where p is the price per unit. Find the rate of change of the demand with respect to price when $p = 24$.

Solution

The rate of change of demand with respect to price is

$$\frac{dx}{dp} = 98\left[-\frac{1}{2}(2p + 1)^{-3/2}(2)\right] = -98(2p + 1)^{-3/2}.$$

When $p = 24$, the rate of change is

$$\frac{dx}{dp}\bigg|_{p\,=\,24} = -98(48 + 1)^{-3/2} = -98 \cdot \frac{1}{49^{3/2}}$$

$$= -98 \cdot \frac{1}{343}$$

$$= -\frac{2}{7}.$$

CHECKPOINT SOLUTIONS

1. (a) No, $f'(x) = 10(3x^4 + 1)^9(12x^3)$. (b) Yes (c) Yes

2. (a) $f(x) = 12(2x^2 - 1)^{-1}$

$$f'(x) = -12(2x^2 - 1)^{-2}(4x) = \frac{-48x}{(2x^2 - 1)^2}$$

   (b) $f(x) = \frac{1}{3}(x^3 - 1)^{1/2}$

$$f'(x) = \frac{1}{6}(x^3 - 1)^{-1/2}(3x^2) = \frac{x^2}{2\sqrt{x^3 - 1}}$$

EXERCISE 9.6

Differentiate the functions in problems 1–20.

1. $y = (x^2 + 1)^3$
2. $p = (q^2 + 4q)^4$
3. $y = (4x^2 - x + 1)^4$
4. $r = (s^2 + 5s)^{10}$
5. $f(x) = \dfrac{1}{(x^2 + 2)^3}$
6. $g(x) = \dfrac{1}{4x^3 + 1}$
7. $g(x) = (x^2 + 4x)^{-2}$
8. $p = (q^3 + 1)^{-5}$
9. $c(x) = (x^2 + 3x + 4)^{-3}$
10. $y = (x^2 - 8x)^{2/3}$
11. $g(x) = \dfrac{1}{(2x^3 + 3x + 5)^{3/4}}$
12. $y = \dfrac{1}{(3x^3 + 4x + 1)^{3/2}}$

13. $y = \sqrt{x^2 + 4x + 5}$
14. $y = \sqrt{x^2 + 3x}$
15. $s = 4\sqrt{3x - x^2}$
16. $y = 3\sqrt[3]{(x - 1)^2}$
17. $y = \dfrac{8(x^2 - 3)^5}{5}$
18. $y = \dfrac{5\sqrt{1 - x^3}}{6}$
19. $y = \dfrac{(3x + 1)^5 - 3x}{7}$
20. $y = \dfrac{\sqrt{2x - 1} - \sqrt{x}}{2}$

At the indicated point, for each function in problems 21–24, find

(a) the slope of the tangent line, and
(b) the instantaneous rate of change of the function.

21. $y = (x^3 + 2x)^4$ at $x = 2$

22. $y = \sqrt{5x^2 + 2x}$ at $x = 1$

23. $y = \sqrt{x^3 + 1}$ at $(2, 3)$

24. $y = (4x^3 - 5x + 1)^3$ at $(1, 0)$

In problems 25–28, write the equation of the line tangent to the graph of each function at the indicated point.

25. $y = (x^2 - 3x + 3)^3$ at $(1, 1)$

26. $y = (x^2 + 1)^3$ at $(2, 125)$

27. $y = \sqrt{3x^2 - 2}$ at $x = 3$

28. $y = \left(\dfrac{1}{x^3 - x}\right)^3$ at $x = 2$

 In problems 29 and 30, for each function
(a) find x-values for which the slope of the tangent is 0,
(b) find points (x, y) where the slope of the tangent is 0, and
(c) use a graphing utility to graph the function and locate the points found in part (b).

29. $f(x) = (x^2 - 4)^3 + 12$

30. $f(x) = 10 - (x^2 - 2x - 8)^2$

 In problems 31 and 32, for each function $f(x)$
(a) find $f'(x)$,
(b) graph both $f(x)$ and $f'(x)$ with a graphing utility,
(c) determine x-values where $f'(x) = 0, f'(x) > 0,$
$f'(x) < 0,$ and
(d) determine x-values for which $f(x)$ has a maximum or minimum point, where the graph is increasing, and where it is decreasing.

31. $f(x) = 5 - 3(1 - x^2)^{4/3}$

32. $f(x) = 3 + \dfrac{1}{16}(x^2 - 4x)^4$

In problems 33 and 34, find the derivative of each function.

33. (a) $y = \dfrac{2x^3}{3}$ (b) $y = \dfrac{2}{3x^3}$

    (c) $y = \dfrac{(2x)^3}{3}$ (d) $y = \dfrac{2}{(3x)^3}$

34. (a) $y = \dfrac{3}{(5x)^5}$ (b) $y = \dfrac{3x^5}{5}$

    (c) $y = \dfrac{3}{5x^5}$ (b) $y = \dfrac{(3x)^5}{5}$

APPLICATIONS

35. **Ballistics** Ballistics experts are able to identify the weapon that fired a certain bullet by studying the markings on the bullet. Tests are conducted by firing into a bale of paper. If the distance s, in inches, that the bullet travels into the paper is given by

$$s = 27 - (3 - 10t)^3$$

for $0 \le t \le 0.3$ seconds, find the velocity of the bullet one-tenth of a second after it hits the paper.

36. **Population of microorganisms** Suppose that the population of a certain microorganism at time t (in minutes) is given by

$$P = 1000 - 1000(t + 10)^{-1}.$$

Find the rate of change of population.

37. **Revenue** The revenue from the sale of x units of a product is

$$R = 1500x + 3000(2x + 3)^{-1} - 1000,$$

where x is the number of units sold. Find the marginal revenue when 100 units are sold. Interpret your result.

38. **Revenue** The revenue from the sale of x units of a product is

$$R = 15(3x + 1)^{-1} + 50x - 15.$$

Find the marginal revenue when 40 units are sold. Interpret your result.

39. **Pricing and sales** Suppose that the weekly sales volume y (in thousands of units sold) depends on the price per unit of the product according to

$$y = 32(3p + 1)^{-2/5}, \quad p > 0,$$

where p is in dollars.
(a) What is the rate of change in sales volume when the price is $21?
(b) Interpret your answer in part (a).

40. **Pricing and sales** A chain of auto service stations has found that their monthly sales volume y (in thousands of dollars) is related to the price p (in dollars) of an oil change according to

$$y = \dfrac{90}{\sqrt{p + 5}}, \quad p > 10.$$

(a) What is the rate of change of sales volume when the price is \$20?

(b) Interpret your answer in part (a).

41. **Demand** Suppose that the demand for a product is described by

$$p = \frac{200,000}{(q + 1)^2}.$$

(a) What is the rate of change of price with respect to the quantity demanded when $q = 49$?

(b) Interpret your answer in part (a).

Stimulus-response The relation between the magnitude of a sensation y and the magnitude of the stimulus x is given by

$$y = k(x - x_0)^n,$$

where k is a constant, x_0 is the threshold of effective stimulus, and n depends on the type of stimulus. Find the rate of change of sensation with respect to the amount of stimulus for each of problems 42–44.

42. For the stimulus of visual brightness $y = k(x - x_0)^{1/3}$

43. For the stimulus of warmth $y = k(x - x_0)^{8/5}$

44. For the stimulus of electrical stimulation
$y = k(x - x_0)^{7/2}$

45. **Demand** If the demand for a product is described by the equation

$$p = \frac{100}{\sqrt{2q + 1}},$$

find the rate of change of p with respect to q.

46. **Advertising and sales** The daily sales S (in thousands of dollars) attributed to an advertising campaign are given by

$$S = 1 + \frac{3}{t + 3} - \frac{18}{(t + 3)^2},$$

where t is the number of weeks the campaign runs. What is the rate of change of sales at

(a) $t = 8$? (b) $t = 10$?

(c) Should the campaign be continued after the 10th week? Explain.

47. **Body-heat loss** The description of body-heat loss due to convection involves a coefficient of convection, K_c, which depends on wind velocity according to the following equation.

$$K_c = 4\sqrt{4v + 1}$$

Find the rate of change of the coefficient with respect to the wind velocity.

48. **Typing speed** The typing speed (in words per minute) of a secretarial student is

$$S = 10\sqrt{0.8x + 4}, \quad 0 \le x \le 100,$$

where x is the number of hours of training he has had. What is the rate at which his speed is changing and what does this rate mean when he has had

(a) 15 hours of training?

(b) 40 hours of training?

49. **Investments** If an IRA is a variable-rate investment for 20 years at rate r percent per year, compounded monthly, then the future value S that accumulates from an initial investment of \$1000 is

$$S = 1000\left[1 + \frac{0.01r}{12}\right]^{240}.$$

What is the rate of change of S with respect to r and what does it tell us if the interest rate is

(a) 6%? (b) 12%?

50. **Concentration of body substances** The concentration C of a substance in the body depends on the quantity of the substance Q and the volume V through which it is distributed. For a static substance this is given by

$$C = \frac{Q}{V}.$$

For a situation like that in the kidneys, where the fluids are moving, the concentration is the ratio of the rate of change of quantity with respect to time and the rate of change of volume with respect to time.

(a) Formulate the equation for concentration of a moving substance.

(b) Show this is equal to the rate of change of quantity with respect to volume.

51. **Public debt of the United States** The interest paid on the public debt of the United States of America as a percentage of federal expenditures for selected years is shown in the table on p. 654.

Assume that the percentage of federal expenditures devoted to payment of interest can be modeled with the function

$$d(t) = 0.593(0.1t + 3)^3 - 10.56(0.1t + 3)^2$$
$$+ 60.98(0.1t + 3) - 103.37,$$

Year	Interest Paid as a Percentage of Federal Expenditures
1930	0
1940	10.5
1950	13.4
1955	9.4
1960	10.0
1965	9.6
1970	9.9
1975	9.8
1980	12.7
1985	18.9
1990	21.1

SOURCE: Bureau of Public Debt, Department of the Treasury

where t is the number of years past 1930. Use this model to determine and interpret the instantaneous rate of change of the percentage of federal expenditures devoted to payment of interest on the public debt in 1960 and in 1990.

52. **Union membership** The table shows the percentage of United States workers who belonged to unions for selected years from 1930 to 1990.

Year	Union Membership as a Percentage of the Labor Force
1930	11.6
1935	13.2
1940	26.9
1945	35.5
1950	31.5
1955	33.2
1960	31.4
1965	28.4
1970	27.3
1975	25.5
1980	21.9
1985	18.0
1990	16.1

SOURCE: Bureau of Labor Statistics, Department of Labor

Assume that the percentage of the labor force that belongs to unions can be modeled with the function

$$f(t) = 0.485(0.1t + 2)^3 - 10.9(0.1t + 2)^2 + 75.2(0.1t + 2) - 131.47,$$

where t is the number of years past 1920. Use the model to find and interpret the instantaneous rate of change of union membership in 1940 and in 1990.

53. **Persons living below the poverty level** The table below shows the number of millions of people in the United States who lived below the poverty level for selected years between 1960 and 1990.

Year	Persons Living Below the Poverty Level (in Millions)
1960	39.9
1965	33.2
1970	25.4
1975	25.9
1980	29.3
1986	32.4
1989	31.5
1990	33.6

SOURCE: *The World Almanac and Book of Facts*, 1993

Assume the number of persons below the poverty level can be modeled by

$$p(t) = -0.0020595(t + 60)^3 + 0.508851(t + 60)^2 - 41.3634(t + 60) + 1135.4,$$

where t is the number of years past 1960. Use the model to find and interpret the instantaneous rate of change of the number of persons below the poverty level in 1970 and in 1990.

54. **Inflation Rate** The annual change in the Consumer Price Index (CPI) during certain years is shown in the table below. Assume that the annual change in the Consumer Price Index can be modeled with the function

$$f(t) = -0.00685679(t + 75)^4 + 2.32783445(t + 75)^3 - 295.9642445(t + 75)^2 + 16701.32789(t + 75) - 352921,$$

where t is the number of years past 1975. A graph of $f(t)$ along with the data in the table is shown in the figure.

Year	Consumer Price Index Annual Percent Change	Year	Consumer Price Index Annual Percent Change
1977	6.5	1985	3.6
1978	7.6	1986	1.9
1979	11.3	1987	3.6
1980	13.5	1988	4.1
1981	10.3	1989	4.8
1982	6.2	1990	5.4
1983	3.2	1991	4.2
1984	4.3		

SOURCE: *The World Almanac and Book of Facts*, 1993

Use the model to find and interpret the instantaneous rate of change of the Consumer Price Index in 1979 and in 1990.

9.7 USING DERIVATIVE FORMULAS

Objective

☐ To use derivative formulas separately and in combination with each other

APPLICATION PREVIEW

Suppose the weekly revenue function for a product is given by

$$R(x) = \frac{36,000,000x}{(2x + 500)^2},$$

where x is the number of units sold. We can find marginal revenue by finding the derivative of the revenue function. This revenue function contains both a quotient and a power, so its derivative is found by using both the Quotient Rule and the Power Rule. But before we do this, we must first decide the order in which to apply these formulas.

We have used the Power Rule to find the derivative of functions like

$$y = (x^3 - 3x^2 + x + 1)^5,$$

but we have not found the derivative of functions like

$$y = [(x^2 + 1)(x^3 + x + 1)]^5.$$

This function is different because the function u (which is raised to the fifth power) is the product of two functions $(x^2 + 1)$ and $(x^3 + x + 1)$. The equation is of the form $y = u^5$, where $u = (x^2 + 1)(x^3 + x + 1)$. This means that the Product Rule should be used to find du/dx. Then

$$\frac{dy}{dx} = 5u^4 \cdot \frac{du}{dx}$$

$$= 5[(x^2 + 1)(x^3 + x + 1)]^4[(x^2 + 1)(3x^2 + 1) + (x^3 + x + 1)(2x)]$$

$$= 5[(x^2 + 1)(x^3 + x + 1)]^4(5x^4 + 6x^2 + 2x + 1)$$
$$= (25x^4 + 30x^2 + 10x + 5)[(x^2 + 1)(x^3 + x + 1)]^4.$$

A different type of problem involving the Power Rule and the Product Rule is finding the derivative of $y = (x^2 + 1)^5(x^3 + x + 1)$. We may think of y as the *product* of two functions, one of which is a power. Thus the fundamental formula we should use is the Product Rule. The two functions are $u(x) = (x^2 + 1)^5$ and $v(x) = x^3 + x + 1$. The Product Rule gives

$$\frac{dy}{dx} = u(x) \cdot v'(x) + v(x) \cdot u'(x)$$
$$= (x^2 + 1)^5(3x^2 + 1) + (x^3 + x + 1)[5(x^2 + 1)^4 2x].$$

Note that the Power Rule was used to find $u'(x)$, since $u(x) = (x^2 + 1)^5$. We can simplify dy/dx by factoring $(x^2 + 1)^4$ from both terms:

$$\frac{dy}{dx} = (x^2 + 1)^4[(x^2 + 1)(3x^2 + 1) + (x^3 + x + 1) \cdot 5 \cdot 2x]$$
$$= (x^2 + 1)^4(13x^4 + 14x^2 + 10x + 1).$$

EXAMPLE 1

If $y = \left(\dfrac{x^2}{x - 1}\right)^2$, find y'.

Solution

We again have an equation of the form $y = u^n$, but this time u is a quotient. Thus we will need the Quotient Rule to find du/dx.

$$y' = nu^{n-1} \cdot \frac{du}{dx} = 2u \frac{(x - 1) \cdot 2x - x^2 \cdot 1}{(x - 1)^2}$$

Substituting for u and simplifying gives

$$y' = 2 \cdot \frac{x^2}{x - 1} \cdot \frac{2x^2 - 2x - x^2}{(x - 1)^2} = \frac{2x^2(x^2 - 2x)}{(x - 1)^3}$$
$$= \frac{2x^4 - 4x^3}{(x - 1)^3}.$$

EXAMPLE 2

Find $f'(x)$ if $f(x) = \dfrac{(x - 1)^2}{(x^2 + 1)^3}$.

Solution

This function is the quotient of two functions, $(x - 1)^2$ and $(x^2 + 1)^3$, so we must use the Quotient Rule to find the derivative of $f(x)$, but taking the derivatives of $(x - 1)^2$ and $(x^2 + 1)^3$ will require the Power Rule.

$$f'(x) = \frac{[v(x) \cdot u'(x) - u(x) \cdot v'(x)]}{[v(x)]^2}$$

$$= \frac{(x^2 + 1)^3[2(x - 1)(1)] - (x - 1)^2[3(x^2 + 1)^2 \, 2x]}{[(x^2 + 1)^3]^2}$$

$$= \frac{2(x^2 + 1)^3(x - 1) - 6x(x - 1)^2(x^2 + 1)^2}{(x^2 + 1)^6}$$

We see that 2, $(x^2 + 1)^2$, and $(x - 1)$ are all factors in both terms of the numerator, so we can factor them from both terms and reduce the fraction.

$$f'(x) = \frac{2(x^2 + 1)^2(x - 1)[(x^2 + 1) - 3x(x - 1)]}{(x^2 + 1)^6}$$

$$= \frac{2(x - 1)(-2x^2 + 3x + 1)}{(x^2 + 1)^4}$$

EXAMPLE 3
Find $f'(x)$ if $f(x) = (x^2 - 1)\sqrt{3 - x^2}$.

Solution
The function is the product of two functions, $x^2 - 1$ and $\sqrt{3 - x^2}$. Therefore we will use the Product Rule to find the derivative of $f(x)$, but the derivative of $\sqrt{3 - x^2} = (3 - x^2)^{1/2}$ will require the Power Rule.

$$f'(x) = u(x) \cdot v'(x) + v(x) \cdot u'(x)$$

$$= (x^2 - 1)\left[\frac{1}{2}(3 - x^2)^{-1/2}(-2x)\right] + (3 - x^2)^{1/2}(2x)$$

$$= (x^2 - 1)[-x(3 - x^2)^{-1/2}] + (3 - x^2)^{1/2}(2x)$$

$$= \frac{-x^3 + x}{(3 - x^2)^{1/2}} + 2x(3 - x^2)^{1/2}$$

We can combine these terms over the common denominator $(3 - x^2)^{1/2}$ as follows.

$$f'(x) = \frac{-x^3 + x}{(3 - x^2)^{1/2}} + \frac{2x(3 - x^2)^1}{(3 - x^2)^{1/2}} = \frac{-x^3 + x + 6x - 2x^3}{(3 - x^2)^{1/2}}$$

$$= \frac{-3x^3 + 7x}{(3 - x^2)^{1/2}}$$

We should note that in Example 3 we could have written $f'(x)$ in the form

$$f'(x) = (-x^3 + x)(3 - x^2)^{-1/2} + 2x(3 - x^2)^{1/2}.$$

Now, the factor $(3 - x^2)$, to different powers, is contained in both terms of the expression. Thus we can factor $(3 - x^2)^{-1/2}$ from both terms. (We choose the $-1/2$ power because it is the smaller of the two powers.) Dividing

$(3 - x^2)^{-1/2}$ into the first term gives $(-x^3 + x)$, and dividing it into the second term gives $2x(3 - x^2)^1$. Why? Thus we have

$$f'(x) = (3 - x^2)^{-1/2}[(-x^3 + x) + 2x(3 - x^2)]$$
$$= \frac{-3x^2 + 7x}{(3 - x^2)^{1/2}},$$

which agrees with our previous answer.

CHECKPOINT

1. If a function has the form $y = [u(x)]^n \cdot v(x)$, where n is a constant, we begin to find the derivative by using the _____ Rule and then use the _____ Rule to find the derivative of $[u(x)]^n$.

2. If a function has the form $y = [u(x)/v(x)]^n$, where n is a constant, we begin to find the derivative by using the _____ Rule and then use the _____ Rule.

We now return to the Application Preview problem.

EXAMPLE 4

Suppose that the weekly revenue function for a product is given by

$$R(x) = \frac{36,000,000x}{(2x + 500)^2},$$

where x is the number of units sold.
(a) Find the marginal revenue function.
(b) Find the marginal revenue when 50 units are sold.

Solution
(a) $\overline{MR} = R'(x)$

$$= \frac{(2x + 500)^2(36,000,000) - 36,000,000x[2(2x + 500)^1(2)]}{(2x + 500)^4}$$

$$= \frac{36,000,000(2x + 500)(2x + 500 - 4x)}{(2x + 500)^4}$$

$$= \frac{36,000,000(500 - 2x)}{(2x + 500)^3}$$

(b) $\overline{MR}(50) = R'(50) = \dfrac{36,000,000(500 - 100)}{(100 + 500)^3}$

$$= \frac{36,000,000(400)}{(600)^3}$$

$$= \frac{200}{3} = 66.67$$

The marginal revenue is \$66.67 when 50 units are sold. That is, the predicted revenue from the sale of the 51st unit is approximately \$66.67.

It may be helpful to review the formulas needed to find the derivatives of various types of functions. Table 9.5 presents examples of different types of functions and the formulas needed to find their derivatives.

Table 9.5   Derivative formulas summary

Example	Formula
$f(x) = 14$	If $f(x) = c$, then $f'(x) = 0$.
$y = x^4$	If $f(x) = x^n$, then $f'(x) = nx^{n-1}$.
$g(x) = 5x^3$	If $g(x) = cf(x)$, then $g'(x) = cf'(x)$.
$y = 3x^2 + 4x$	If $f(x) = u(x) + v(x)$, then $f'(x) = u'(x) + v'(x)$.
$y = (x^2 - 2)(x + 4)$	If $f(x) = u(x) \cdot v(x)$, then $f'(x) = u(x) \cdot v'(x) + v(x) \cdot u'(x)$.
$f(x) = \dfrac{x^3}{x^2 + 1}$	If $f(x) = \dfrac{u(x)}{v(x)}$, then $f'(x) = \dfrac{v(x) \cdot u'(x) - u(x) \cdot v'(x)}{[v(x)]^2}$.
$y = (x^3 - 4x)^{10}$	If $y = u^n$ and $u = g(x)$, then $\dfrac{dy}{dx} = nu^{n-1} \cdot \dfrac{du}{dx}$.
$y = \left(\dfrac{x - 1}{x^2 + 3}\right)^3$	Power Rule, then Quotient Rule to find $\dfrac{du}{dx}$, where $u = \dfrac{x - 1}{x^2 + 3}$
$y = (x + 1)\sqrt{x^3 + 1}$	Product Rule, then Power Rule to find $v'(x)$, where $v(x) = \sqrt{x^3 + 1}$
$y = \dfrac{(x^2 - 3)^4}{x + 1}$	Quotient Rule, then Power Rule to find the derivative of the numerator

CHECKPOINT SOLUTIONS

1. Product, Power
2. Power, Quotient

EXERCISE 9.7

Find the derivatives of the following functions. Simplify and express the answer using positive exponents only.

1. $f(x) = \pi^4$

2. $f(x) = \dfrac{1}{4}$

3. $g(x) = \dfrac{4}{x^4}$

4. $y = \dfrac{x^4}{4}$

5. $g(x) = 5x^3 + \dfrac{4}{x}$

6. $y = 3x^2 + 4\sqrt{x}$

7. $y = (x^2 - 2)(x + 4)$

8. $y = (x^3 - 5x^2 + 1)(x^3 - 3)$

9. $f(x) = \dfrac{x^3 + 1}{x^2}$

10. $y = \dfrac{1 + x^2 - x^4}{1 + x^4}$

11. $y = \dfrac{(x^3 - 4x)^{10}}{10}$ 12. $y = \dfrac{5}{2}(3x^4 - 6x^2 + 2)^5$

13. $y = \dfrac{5}{3}x^3(4x^5 - 5)^3$ 14. $y = 3x^4(2x^5 + 1)^7$

15. $y = (x - 1)^2(x^2 + 1)$

16. $f(x) = (5x^3 + 1)(x^4 + 5x)^2$

17. $y = \dfrac{(x^2 - 4)^3}{x^2 + 1}$ 18. $y = \dfrac{(x^2 - 3)^4}{x}$

19. $p = [(q + 1)(q^3 - 3)]^3$

20. $y = [(4 - x^2)(x^2 + 5x)]^4$

21. $R(x) = [x^2(x^2 + 3x)]^4$ 22. $c(x) = [x^3(x^2 + 1)]^{-3}$

23. $y = \left(\dfrac{2x - 1}{x^2 + x}\right)^4$ 24. $y = \left(\dfrac{5 - x^2}{x^4}\right)^3$

25. $g(x) = (8x^4 + 3)^2(x^3 - 4x)^3$

26. $y = (3x^3 - 4x)^3(4x^2 - 8)^2$

27. $f(x) = \dfrac{\sqrt[3]{x^2 + 5}}{4 - x^2}$ 28. $g(x) = \dfrac{\sqrt[3]{2x - 1}}{2x + 1}$

29. $y = x^2\sqrt[4]{4x - 3}$ 30. $y = 3x^3\sqrt{x^4 + 3}$

31. $c(x) = 2x\sqrt{x^3 + 1}$ 32. $R(x) = x\sqrt[3]{3x^3 + 2}$

In problems 33 and 34, find the derivative of each function.

33. (a) $F_1(x) = \dfrac{3(x^4 + 1)^5}{5}$ (b) $F_2(x) = \dfrac{3}{5(x^4 + 1)^5}$

     (c) $F_3(x) = \dfrac{(3x^4 + 1)^5}{5}$ (d) $F_4(x) = \dfrac{3}{(5x^4 + 1)^5}$

34. (a) $G_1(x) = \dfrac{2(x^3 - 5)^3}{3}$ (b) $G_2(x) = \dfrac{(2x^3 - 5)^3}{3}$

     (c) $G_3(x) = \dfrac{2}{3(x^3 - 5)^3}$ (d) $G_4(x) = \dfrac{2}{(3x^3 - 5)^3}$

APPLICATIONS

35. **Physical output** The total physical output P of workers is a function of the number of workers, x. The function $P = f(x)$ is called the physical productivity function. Suppose that the physical productivity of x construction workers is given by

$$P = 10(3x + 1)^3 - 10.$$

Find the marginal physical productivity, dP/dx.

36. **Revenue** Suppose that the revenue function for a certain product is given by

$$R(x) = 15(2x + 1)^{-1} + 30x - 15,$$

where x is in thousands of units and R is in thousands of dollars.
(a) Find the marginal revenue when 2000 units are sold.
(b) How is revenue changing when 2000 units are sold?

37. **Revenue** Suppose that the revenue function for a computer is given by

$$R(x) = 60,000x + 40,000(10 + x)^{-1} - 4000.$$

(a) Find the marginal revenue when 10 units are sold.
(b) How is revenue changing when 10 units are sold?

38. **Production** Suppose that the production of x items of a new line of products is given by

$$x = 200[(t + 10) - 400(t + 40)^{-1}],$$

where t is the number of weeks the line has been in production. Find the rate of production.

39. **National consumption** If the national consumption function is given by

$$C(y) = 2(y + 1)^{1/2} + 0.4y + 4,$$

find the marginal propensity to consume, dC/dy.

40. **Demand** Suppose that the demand function for an appliance is given by

$$p = \dfrac{400(q + 1)}{(q + 2)^2}.$$

Find the rate of change of price with respect to the number of appliances.

41. **Volume** When squares of side x are cut from the corners of a 12-inch-square piece of cardboard, an open-top box can be formed by folding up the sides. The volume of this box is given by

$$V = x(12 - 2x)^2.$$

Find the rate of change of volume with respect to the size of the squares.

42. **Advertising and sales** Suppose that sales (in thousands of dollars) are directly related to an advertising campaign according to

$$S = 1 + \dfrac{3t - 9}{(t + 3)^2},$$

where t is the number of weeks of the campaign.
(a) Find the rate of change of sales after 3 weeks.
(b) Interpret the result in part (a).

43. **Advertising and sales** An inferior product with an extensive advertising campaign does well when it is released, but sales decline as people discontinue use of the product. If the sales S after t weeks are given by

$$S(t) = \frac{200t}{(t+1)^2}, \quad t \ge 0,$$

what is the rate of change of sales when $t = 9$? Interpret your result.

44. **Advertising and sales** An excellent film with a very small advertising budget must depend largely on word-of-mouth advertising. If attendance at the film after t weeks is given by

$$A = \frac{100t}{(t+10)^2},$$

what is the rate of change in attendance and what does it mean when
(a) $t = 10$?   (b) $t = 20$?

45. **Farm workers** The percent of U.S. workers in farm occupations during certain years is shown in the table.

Year	Percent	Year	Percent
1820	71.8	1950	11.6
1850	63.7	1960	6.1
1870	53	1970	3.6
1900	37.5	1980	2.7
1920	27	1985	2.8
1930	21.2	1990	2.4
1940	17.4		

SOURCE: *The World Almanac and Book of Facts*, 1993

Assume that the percentage of U.S. workers in farm occupations can be modeled with the function

$$f(t) = \frac{1000(-8.0912(t+20) + 1558.9)}{1.09816(t+20)^2 - 122.183(t+20) + 21472.6},$$

where t is the number of years past 1820.
(a) Find the function that models the instantaneous rate of change of the percent of U.S. workers in farm occupations.
(b) Use the result of part (a) to find the instantaneous rate of change in 1850 and in 1950.
(c) Interpret the two rates of change in part (b).

9.8 HIGHER-ORDER DERIVATIVES

Objective

❑ To find second derivatives and higher derivatives of certain functions

APPLICATION PREVIEW

Suppose a particle travels according to the equation $s = 100t - 16t^2 + 200$, where s is the distance and t is the time. Then $\dfrac{ds}{dt}$ is the velocity, and the acceleration of the particle is the rate of change of the velocity of the particle; so acceleration is $\dfrac{dv}{dt}$. That is, acceleration is the derivative of the derivative of the distance function. This is called the second derivative. In this section we will discuss second and higher-order derivatives.

Because the derivative of a function is itself a function, we can take a derivative of the derivative. The derivative of a first derivative is called a **second derivative.** We can find the second derivative of a function f by differentiating it twice. If f' represents the first derivative of a function, then f'' represents the second derivative of that function.

EXAMPLE 1

If $f(x) = 3x^3 - 4x^2 + 5$, find $f''(x)$.

Solution

The first derivative is $f'(x) = 9x^2 - 8x$.
The second derivative is $f''(x) = 18x - 8$.

EXAMPLE 2

Find the second derivative of $y = x^4 - 3x^2 + x^{-2}$.

Solution

The first derivative is $y' = 4x^3 - 6x - 2x^{-3}$.
The second derivative, which we may denote by y'', is

$$y'' = 12x^2 - 6 + 6x^{-4}.$$

It is also common to use $\dfrac{d^2y}{dx^2}$ and $\dfrac{d^2}{dx^2}f(x)$ to denote the second derivative of a function.

EXAMPLE 3

If $y = \sqrt{2x - 1}$, find d^2y/dx^2.

Solution

The first derivative is

$$\frac{dy}{dx} = \frac{1}{2}(2x - 1)^{-1/2}(2) = (2x - 1)^{-1/2}.$$

The second derivative is

$$\frac{d^2y}{dx^2} = -\frac{1}{2}(2x - 1)^{-3/2}(2) = -(2x - 1)^{-3/2}$$

$$= \frac{-1}{(2x - 1)^{3/2}} = \frac{-1}{\sqrt{(2x - 1)^3}}.$$

We can also find third, fourth, fifth, and higher derivatives, continuing indefinitely. The third, fourth, and fifth derivatives of a function f are denoted by f''', $f^{(4)}$, and $f^{(5)}$, respectively. Other notations for the third and fourth derivatives include:

$$y''' = \frac{d^3y}{dx^3} = \frac{d^3f(x)}{dx^3} \qquad\qquad y^{(4)} = \frac{d^4y}{dx^4} = \frac{d^4f(x)}{dx^4}$$

EXAMPLE 4

Find the first four derivatives of $f(x) = 4x^3 + 5x^2 + 3$.

Solution

$$f'(x) = 12x^2 + 10x, \quad f''(x) = 24x + 10, \quad f'''(x) = 24, \quad f^{(4)}(x) = 0$$

Just as the first derivative, $f'(x)$, can be used to determine the rate of change of a function $f(x)$, the second derivative, $f''(x)$, can be used to determine the rate of change of $f'(x)$.

EXAMPLE 5

If $f(x) = 3x^4 + 6x^3 - 3x^2 + 4$,
(a) how fast is $f(x)$ changing at $(1, 10)$?
(b) how fast is $f'(x)$ changing at $(1, 10)$?
(c) is $f'(x)$ increasing or decreasing at $(1, 10)$?

Solution
(a) Since $f'(x) = 12x^3 + 18x^2 - 6x$, we have

$$f'(1) = 12 + 18 - 6 = 24.$$

Thus the rate of change of $f(x)$ at $(1, 10)$ is 24.
(b) Since $f''(x) = 36x^2 + 36x - 6$, we have

$$f''(1) = 66.$$

Thus the rate of change of $f'(x)$ at $(1, 10)$ is 66.
(c) Since $f''(1) = 66 > 0$, $f'(x)$ is increasing at $(1, 10)$.

EXAMPLE 6

Suppose that a particle travels according to the equation

$$s = 100t - 16t^2 + 200,$$

where s is the distance and t is the time. Then ds/dt is the velocity, and $d^2s/dt^2 = dv/dt$ is the acceleration of the particle. Find the acceleration.

Solution
The velocity is $v = ds/dt = 100 - 32t$, and the acceleration is

$$\frac{dv}{dt} = \frac{d^2s}{dt^2} = -32.$$

CHECKPOINT

Suppose that the vertical distance a particle travels is given by

$$s = 4x^3 - 12x^2 + 6,$$

where s is in feet and x is in seconds.
1. Find the function that describes the velocity of this particle.

2. Find the function that describes the acceleration of this particle.

3. Is the acceleration always positive?

4. When does the *velocity* of this particle increase?

EXAMPLE 7

(a) Given $f(x) = x^4 - 12x^2 + 2$, graph $f(x)$ and its second derivative on the same set of axes over an interval that contains all x-values where $f''(x) = 0$.

(b) When the graph of $y = f(x)$ is opening downward, is $f''(x) > 0$, $f''(x) < 0$, or $f''(x) = 0$?

(c) When the graph of $y = f(x)$ is opening upward, is $f''(x) > 0$, $f''(x) < 0$, or $f''(x) = 0$?

Solution

(a) $f'(x) = 4x^3 - 24x$ and $f''(x) = 12x^2 - 24 = 12(x^2 - 2)$. Because $f''(x) = 0$ at $x = -\sqrt{2}$ and at $x = \sqrt{2}$, we use an x-range that contains $x = -\sqrt{2}$ and $\sqrt{2}$. The graph of $f(x) = x^4 - 12x^2 + 2$ and its second derivative, $f''(x) = 12x^2 - 24$, are shown in Figure 9.29.

(b) The graph of $y = f(x)$ appears to be opening downward on the same interval for which $f''(x) < 0$.

(c) The graph appears to be opening upward on the same intervals for which $f''(x) > 0$.

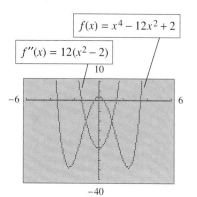

$f(x) = x^4 - 12x^2 + 2$

$f''(x) = 12(x^2 - 2)$

Figure 9.29

CHECKPOINT SOLUTIONS

1. The velocity is described by $s'(x) = 12x^2 - 24x$.

2. The acceleration is described by $s''(x) = 24x - 24$.

3. No, the acceleration is positive when $s''(x) > 0$, that is, when $24x - 24 > 0$. It is zero when $24x - 24 = 0$ and negative when $24x - 24 < 0$. Thus acceleration is negative when $x < 1$ second, zero when $x = 1$ second, and positive when $x > 1$ second.

4. The velocity increases when the acceleration is positive. Thus the velocity is increasing after 1 second.

EXERCISE 9.8

In problems 1–8, find the second derivative.

1. $f(x) = 4x^3 - 15x^2 + 3x + 2$

2. $f(x) = 2x^{10} - 18x^5 - 12x^3 + 4$

3. $y = 10x^3 - x^2 + 14x + 3$

4. $y = 6x^5 - 3x^4 + 12x^2$

5. $g(x) = x^3 - \dfrac{1}{x}$

6. $h(x) = x^2 - \dfrac{1}{x^2}$

7. $y = x^3 - \sqrt{x}$

8. $y = 3x^2 - \sqrt[3]{x^2}$

In problems 9–16, find the third derivative.

9. $y = x^5 - 16x^3 + 12$

10. $y = 6x^3 - 12x^2 + 6x$

11. $f(x) = 2x^9 - 6x^6$

12. $f(x) = 3x^5 - x^6$

13. $y = 1/x$

14. $y = 1/x^2$

15. $y = \sqrt{x}$

16. $y = \sqrt[3]{x}$

In problems 17–30, find the indicated derivative.

17. If $y = x^5 - x^{1/2}$, find $\dfrac{d^2y}{dx^2}$.

18. If $y = x^4 + x^{1/3}$, find $\dfrac{d^2y}{dx^2}$.

19. If $f(x) = \sqrt{x + 1}$, find $f'''(x)$.

20. If $f(x) = \sqrt{x - 5}$, find $f'''(x)$.

21. Find $\dfrac{d^4y}{dx^4}$ if $y = 4x^3 - 16x$.

22. Find $y^{(4)}$ if $y = x^6 - 15x^3$.

23. Find $f^{(4)}(x)$ if $f(x) = \sqrt{x}$.

24. Find $f^{(4)}(x)$ if $f(x) = 1/x$.

25. Find $y^{(4)}$ if $y' = \sqrt{x - 1}$.

26. Find $y^{(5)}$ if $\dfrac{d^2y}{dx^2} = \sqrt[3]{x + 2}$.

27. Find $\dfrac{d^3y}{dx^3}$ if $\dfrac{dy}{dx} = \sqrt{x^2 + 4}$.

28. Find $\dfrac{d^4y}{dx^4}$ if $y'' = \sqrt{4x^2 - 2}$.

29. Find $f^{(6)}(x)$ if $f^{(4)}(x) = x(x + 1)^{-1}$.

30. Find $f^{(3)}(x)$ if $f'(x) = \dfrac{x^2}{x^2 + 1}$.

31. If $f(x) = 16x^2 - x^3$, what is the rate of change of $f'(x)$ at $(1, 15)$?

32. If $y = 36x^2 - 6x^3 + x$, what is the rate of change of y' at $(1, 31)$?

In problems 33–36, do the following for each function $f(x)$.
(a) Find $f'(x)$ and $f''(x)$.
(b) Graph $f(x)$, $f'(x)$, and $f''(x)$ with a graphing utility.
(c) Identify x-values where $f''(x) = 0$, $f''(x) > 0$, and $f''(x) < 0$.
(d) Identify x-values where $f'(x)$ has a maximum point or a minimum point, where $f'(x)$ is increasing, and where $f'(x)$ is decreasing.
(e) When $f(x)$ has a maximum point, is $f''(x) > 0$ or $f''(x) < 0$?
(f) When $f(x)$ has a minimum point, is $f''(x) > 0$ or $f''(x) < 0$?

33. $f(x) = x^3 - 3x^2 + 5$ 34. $f(x) = 2 + 3x - x^3$

35. $f(x) = -\dfrac{1}{3}x^3 - x^2 + 3x + 7$

36. $f(x) = \dfrac{1}{3}x^3 - \dfrac{3x^2}{2} - 4x + 10$

APPLICATIONS

37. **Acceleration** If a particle travels as a function of time according to the formula

$$s = 100 + 10t + 0.01t^3,$$

find the acceleration of the particle when $t = 2$.

38. **Acceleration** If the formula describing the distance an object travels as a function of time is

$$s = 100 + 160t - 16t^2,$$

what is the acceleration of the object when $t = 4$?

39. **Revenue** If the revenue from sales of a product can be described by

$$R(x) = 100x - 0.01x^2,$$

find the instantaneous rate of change of the marginal revenue.

40. **Revenue** Suppose that the revenue from the sale of a product is given by

$$R = 70x + 0.5x^2 - 0.001x^3,$$

where x is the number of units sold. How fast is the marginal revenue $\overline{MR}$ changing when $x = 100$?

41. **Sensitivity** When medicine is administered, reaction (measured in change of blood pressure or temperature) can be modeled by

$$R = m^2\left(\frac{c}{2} - \frac{m}{3}\right),$$

where c is a positive constant and m is the amount of medicine absorbed into the blood (Source: Thrall, R. M., et. al., *Some Mathematical Models in Biology*, U.S. Department of Commerce, 1967). The sensitivity to the medication is defined to be the rate of change of reaction R with respect to the amount of medicine m absorbed in the blood.
(a) Find the sensitivity.
(b) Find the instantaneous rate of change of sensitivity with respect to the amount of medicine absorbed in the blood.
(c) Which order derivative of reaction gives the rate of change of sensitivity?

42. **Photosynthesis** The amount of photosynthesis that takes place in a certain plant depends on the intensity of light x according to the equation

$$f(x) = 145x^2 - 30x^3.$$

(a) Find the rate of change of photosynthesis with respect to the intensity.

(b) What is the rate of change when $x = 1$? when $x = 3$?

(c) How fast is the rate found in (a) changing when $x = 1$? when $x = 3$?

43. **Revenue** The revenue (in thousands of dollars) from the sale of x units of a product is

$$R = 15x + 30(4x + 1)^{-1} - 30,$$

where x is the number of units sold.

(a) At what rate is the marginal revenue $\overline{MR}$ changing when the number of units being sold is 25?

(b) Interpret your result in (a).

44. **Advertising and sales** The sales of a product S (in thousands of dollars) are given by

$$S = \frac{600x}{x + 40},$$

where x is the advertising expenditure (in thousands of dollars).

(a) Find the rate of change of sales with respect to advertising expenditure.

(b) Use the second derivative to find how this rate is changing at $x = 20$.

(c) Interpret your result in (b).

45. **Advertising and sales** The daily sales S (in thousands of dollars) that are attributed to an advertising compaign are given by

$$S = 1 + \frac{3}{t + 3} - \frac{18}{(t + 3)^2},$$

where t is the number of weeks the campaign runs.

(a) Find the rate of change of sales at any time t.

(b) Use the second derivative to find how this rate is changing at $t = 15$.

(c) Interpret your result in (b).

46. **Advertising and sales** A product with a large advertising budget has its sales S (in millions of dollars) given by

$$S = \frac{500}{t + 2} - \frac{1000}{(t + 2)^2},$$

where t is the number of months the product has been on the market.

(a) Find the rate of change of sales at any time t.

(b) What is the rate of change of sales at $t = 2$?

(c) Use the second derivative to find how this rate is changing at $t = 2$.

(d) Interpret your results from (b) and (c).

47. **Persons living below the poverty level** The table below shows the number of millions of people in the United States who lived below the poverty level for selected years between 1960 and 1990.

Year	Persons Living Below the Poverty Level (in Millions)
1960	39.9
1965	33.2
1970	25.4
1975	25.9
1980	29.3
1986	32.4
1989	31.5
1990	33.6

SOURCE: *The World Almanac and Book of Facts,* 1993

Assume that the number of millions of people in the United States who lived below the poverty level can be modeled by the function

$$p(t) = -0.002t^3 + 0.138t^2 - 2.54t + 40.61,$$

where t is the number of years past 1960.

(a) Find the function that models the instantaneous rate of change of $p(t)$.

(b) Use the second derivative to find how this rate is changing in 1970 and in 1990.

(c) Interpret the meaning of $p'(10)$ and $p''(10)$. Write a sentence that explains each.

48. **Consumer debt** The percentage of disposable income spent on consumer debt during certain years is shown in the table below.

Year	Consumer Debt as a Percentage of Disposable Income
1970	18.3
1975	17.9
1980	18.2
1982	16.9
1983	17.7
1984	19.1
1985	20.8
1986	21.4
1987	21.2
1988	21.0

Year	Consumer Debt as a Percentage of Disposable Income
1989	20.9
1990	20.0
1991	18.9

SOURCE: Federal Reserve System

Assume that consumer debt as a percentage of disposable income can be modeled with the function

$$d(t) = -0.0007396584t^4 + 0.0283367028t^3$$
$$- 0.3106532484t^2 + 0.8945254071t$$
$$+ 18.28059641,$$

where t is the number of years past 1970.

(a) Find the function that models the instantaneous rate of change of the percentage of disposable income spent on consumer debt.

(b) Use the second derivative to determine how this rate is changing in 1985 and in 1990.

(c) Write a sentence that explains the meaning of $d'(15)$ and another that explains the meaning of $d''(15)$.

49. **Consumer Price Index** The annual change in the Consumer Price Index (CPI) during certain years is shown in the table below. Assume that the annual change in the Consumer Price Index can be modeled with the function

$$f(t) = -0.0068568(t + 70)^4 + 2.32783(t + 70)^3$$
$$- 295.964(t + 70)^2 + 16701.327(t + 70)$$
$$- 352921,$$

where t is the number of years past 1970.

Year	Consumer Price Index Annual Percent Change
1977	6.5
1978	7.6
1979	11.3
1980	13.5
1981	10.3
1982	6.2
1983	3.2
1984	4.3
1985	3.6
1986	1.9
1987	3.6
1988	4.1

Year	Consumer Price Index Annual Percent Change
1989	4.8
1990	5.4
1991	4.2

SOURCE: *The World Almanac and Book of Facts*, 1993

(a) Find the function that models the instantaneous rate of change of the annual change in the CPI.

(b) Use the second derivative to determine how this rate is changing in 1980 and in 1991.

(c) Write sentences that explain the meanings of $f'(10)$ and $f''(10)$.

50. **Union membership** The table shows the percentage of United States workers who belonged to unions for selected years from 1930 to 1990.

Year	Union Membership as a Percentage of the Labor Force
1930	11.6
1935	13.2
1940	26.9
1945	35.5
1950	31.5
1955	33.2
1960	31.4
1965	28.4
1970	27.3
1975	25.5
1980	21.9
1985	18.0
1990	16.1

SOURCE: Bureau of Labor Statistics, Department of Labor

Assume that the percentage of the labor force that belonged to unions can be modeled with the function

$$f(t) = 0.485(0.1t + 2)^3 - 10.9(0.1t + 2)^2$$
$$+ 75.2(0.1t + 2) - 131.47,$$

where t is the number of years past 1920.

(a) Find the function that models the instantaneous rate of change of the percentage of the U.S. labor force that belonged to unions.

(b) Use the second derivative to determine how this rate was changing in 1970 and in 1990.

(c) Write sentences that explain the meanings of $f'(50)$ and $f''(50)$.

9.9 APPLICATIONS OF DERIVATIVES IN BUSINESS AND ECONOMICS

Objectives

- ❑ To find the marginal cost and marginal revenue at different levels of production
- ❑ To find the marginal profit function, given information about cost and revenue

APPLICATION PREVIEW

In Chapter 1, we defined marginal cost as the rate of change of the total cost function. For a linear total cost function, the marginal cost was defined as the slope of the function's graph. For any total cost function defined by an equation, we can find the instantaneous rate of change of cost (the marginal cost) at any level of production by finding the derivative of the function.

In the same way, we can find the marginal revenue from the total revenue function, and we can use total revenue and total cost functions to find marginal profit.

We begin this section by defining the marginal cost function.

Marginal Cost   If $C = C(x)$ is a total cost function for a commodity, then its derivative, $\overline{MC} = C'(x)$, is the **marginal cost function.**

The linear cost function with equation

$$C(x) = 300 + 6x \qquad \text{(in dollars)}$$

has marginal cost $6 because its slope is 6. Taking the derivative of $C(x)$ gives

$$\overline{MC} = C'(x) = 6,$$

which verifies that the marginal cost is $6 at all levels of production.
    The cost function

$$C(x) = 1000 + 6x + x^2$$

has derivative

$$C'(x) = 6 + 2x.$$

Thus the *marginal cost* at $x = 10$ (when 10 units are produced) is

$$C'(10) = 6 + 2(10) = 26,$$

and the marginal cost at 40 units is

$$C'(40) = 6 + 2(40) = 86.$$

Note that when a cost function is linear, the marginal cost gives the amount by which cost would change if production were increased by one unit. When a cost function is *not* linear (as with $C(x) = 1000 + 6x + x^2$), the marginal cost is used to estimate the amount by which cost would change if production were increased by one unit. Thus at 10 units, the marginal cost is \$26, so costs would increase by approximately \$26 if one more unit were produced. (Note that $C(11) - C(10) = \$27$ gives the actual increase in costs.) Also, at 40 units, the marginal cost is \$86, so costs would increase by approximately \$86 if one more unit were produced.

As noted previously, when the derivative of a function is positive (and thus the slope of the tangent to the curve is positive), the function is increasing, and the value of the derivative gives us a measure of how fast it is increasing. As we saw, the marginal cost for

$$C(x) = 1000 + 6x + x^2$$

is 86 at $x = 40$ and 26 at $x = 10$. This tells us that the cost is increasing faster at $x = 40$ than it is at $x = 10$.

Since producing more units can never reduce the total cost of production, the following properties are valid:

1. The total cost can never be negative. If there are fixed costs, the cost of producing 0 units is positive; otherwise, the cost of producing 0 units is 0.

2. The total cost function is always increasing; the more units produced, the higher the total cost. Thus the marginal cost is always positive.

3. There may be limitations on the units produced, by factors such as plant space.

The graphs of many marginal cost functions tend to be U-shaped; they eventually will rise, even though there may be an initial interval where they decrease.

Figure 9.30

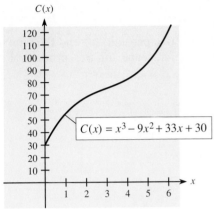

$C(x) = x^3 - 9x^2 + 33x + 30$

(a) Total cost function

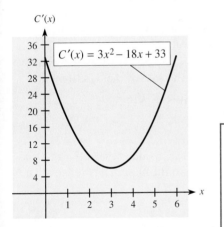

$C'(x) = 3x^2 - 18x + 33$

(b) Marginal cost function

EXAMPLE 1
If the total cost function for a commodity is $C(x) = x^3 - 9x^2 + 33x + 30$, find the marginal cost.

Solution
The marginal cost is $\overline{MC} = C'(x) = 3x^2 - 18x + 33$.

The graph of the total cost function is shown in Figure 9.30(a), and the graph of the marginal cost function is shown in Figure 9.30(b).

As we have seen in Section 9.3, the instantaneous rate of change (the derivative) of the revenue function is the marginal revenue.

Marginal Revenue   If $R = R(x)$ is the total revenue function for a commodity, then the **marginal revenue function** is $\overline{MR} = R'(x)$.

If the demand function for a product in a monopoly market is $p = f(x)$, then the total revenue from the sale of x units is

$$R(x) = px = f(x) \cdot x.$$

EXAMPLE 2

If the demand for a product in a monopoly market is given by

$$p = 16 - 0.02x,$$

where x is the number of units and p is the price per unit, (a) find the total revenue function, and (b) find the marginal revenue for this product at $x = 40$.

Solution

(a) The total revenue function is

$$R(x) = px = (16 - 0.02x)x$$
$$= 16x - 0.02x^2.$$

(b) The marginal revenue function is

$$\overline{MR} = R'(x) = 16 - 0.04x.$$

At $x = 40$, $R'(40) = 16 - 1.6 = 14.40$. Thus the 41st item sold will increase the total revenue by approximately \$14.40.

The marginal revenue is an approximation of the revenue gained from the sale of one additional unit. We have used marginal revenue in Example 2 to find that the revenue from the sale of the 41st item will be approximately \$14.40. The actual increase in revenue from the sale of the 41st item is

$$R(41) - R(40) = 622.38 - 608 = \$14.38.$$

EXAMPLE 3

Use the graphs in Figure 9.31 to determine the x-value where the revenue function has its maximum. What is happening to the marginal revenue at and near this x-value?

Solution

Figure 9.31(a) shows that the total revenue function has a maximum value at $x = 400$. After that, the total revenue function decreases. This means that the total revenue will be reduced each time a unit is sold if more than 400 are produced and sold. The graph of the marginal revenue function in Figure 9.31(b) shows that the marginal revenue is positive to the left of 400. This indicates that the rate at which the total revenue is changing is positive until 400 units are sold; thus the total revenue is increasing. Then, at 400 units, the rate of change is 0. After 400 units are sold, the marginal revenue is negative,

(a) Total revenue function

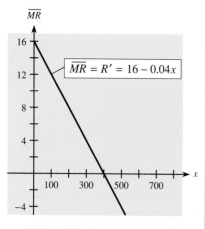

(b) Marginal revenue function

Figure 9.31

which indicates that the total revenue is now decreasing. It is clear from looking at either graph that 400 units should be produced and sold to maximize the total revenue function $R(x)$. That is, the *total revenue* function has its maximum at $x = 400$.

CHECKPOINT

The total cost function for the commodity considered in Example 1 is $C(x) = x^3 - 9x^2 + 33x + 30$, and the marginal cost is $C'(x) = 3x^2 - 18x + 33$.

1. What is the marginal cost if $x = 10$ units are produced?

2. Use marginal cost to estimate the cost of producing the 11th unit.

3. Calculate $C(11) - C(10)$ to find the actual cost of producing the 11th unit.

4. True or false: For products having linear cost functions, the actual cost of producing the $(x + 1)$st unit is equal to the marginal cost at x.

As with marginal cost and marginal revenue, the derivative of a profit function for a commodity will give us the marginal profit function for the commodity.

Marginal Profit   If $P = P(x)$ is the profit function for a commodity, then the **marginal profit function** is $\overline{MP} = P'(x)$.

EXAMPLE 4

If the total profit, in thousands of dollars, for a product is given by $P(x) = 20\sqrt{x + 1} - 2x$, what is the marginal profit at a production level of 15 units?

Solution

The marginal profit function is

$$\overline{MP} = P'(x) = 20 \cdot \frac{1}{2}(x + 1)^{-1/2} - 2 = \frac{10}{\sqrt{x + 1}} - 2.$$

If 15 units are produced, the marginal profit is

$$P'(15) = \frac{10}{\sqrt{15 + 1}} - 2 = \frac{1}{2}.$$

This means that the profit from the sale of the 16th unit is approximately $\frac{1}{2}$(thousand dollars), or $500.

In a **competitive market,** each firm is so small that its actions in the market cannot affect the price of the product. The price of the product is determined in the market by the intersection of the market demand curve (from all consumers) and the market supply curve (from all firms that supply this product). The firm can sell as little or as much as it desires at the given market price, which it cannot change.

Therefore, a firm in a competitive market has a total revenue function given by $R(x) = px$, where p is the market equilibrium price for the product and x is the quantity sold.

EXAMPLE 5

A firm in a competitive market must sell its product for $200 per unit. The cost per unit (per month) is $80 + x$, where x represents the number of units sold per month. Find the marginal profit.

Solution

If the cost per unit is $80 + x$, then the total cost of x units is given by $C(x) = (80 + x)x = 80x + x^2$. The revenue per unit is $200, so the total revenue is given by $R(x) = 200x$. Thus the profit function is

$$P(x) = R(x) - C(x) = 200x - (80x + x^2) \quad \text{or} \quad P(x) = 120x - x^2.$$

The marginal profit is $P'(x) = 120 - 2x$.

The marginal profit in Example 5 is not always positive, so producing and selling a certain number of items will maximize profit. Note that the marginal profit will be negative (that is, profit will decrease) if more than 60 items per month are produced. We will discuss methods of maximizing total revenue and profit, and for minimizing average cost, in Chapter 10.

CHECKPOINT

If the total profit function for a product is $P(x) = 20\sqrt{x + 1} - 2x$, then the marginal profit is

$$P'(x) = \frac{10}{\sqrt{x + 1}} - 2 \quad \text{and} \quad P''(x) = \frac{-5}{\sqrt{(x + 1)^3}}.$$

5. Is $P''(x) < 0$ for all values of $x \geq 0$?

6. Is the marginal profit decreasing for all $x \geq 0$?

EXAMPLE 6

In Example 4, we found that the profit (in thousands of dollars) for a company's products is given by $P(x) = 20\sqrt{x + 1} - 2x$ and its marginal profit is given by $P'(x) = \frac{10}{\sqrt{x + 1}} - 2$.

(a) Use the graphs of $P(x)$ and $P'(x)$ to determine the relationship between the two functions.

(b) When is the marginal profit 0? What is happening to profit at this level of production?

Solution

(a) By comparing the graphs of the two functions (shown in Figure 9.32) we see that for $x > 0$, profit $P(x)$ is increasing over the interval where the marginal profit $P'(x)$ is positive, and the profit is decreasing over the interval where the marginal profit $P'(x)$ is negative.

(b) By using SOLVER, INTERSECT, or TRACE, or by using algebra, we see that $P'(x) = 0$ when $x = 24$. This level of production ($x = 24$) is where profit is maximized, at 52 (thousand dollars).

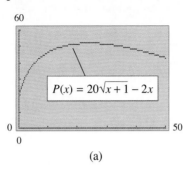

$P(x) = 20\sqrt{x + 1} - 2x$

(a)

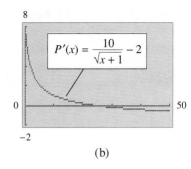

$P'(x) = \dfrac{10}{\sqrt{x + 1}} - 2$

(b)

Figure 9.32

CHECKPOINT SOLUTIONS

1. $C'(10) = 153$

2. Since $C'(10) = 153$, it will cost approximately $153 to produce the 11th unit.

3. $C(11) - C(10) = 635 - 460 = 175$

4. True

5. Yes

6. Yes, because $P''(x) < 0$ for $x \geq 0$.

EXERCISE 9.9

MARGINAL COST, REVENUE, AND PROFIT

Find the marginal cost functions related to the cost functions in problems 1–8.

1. $C(x) = 40 + 8x$

2. $C(x) = 200 + 16x$

3. $C(x) = 500 + 13x + x^2$

4. $C(x) = 300 + 10x + \frac{1}{100}x^2$

5. $C = x^3 - 6x^2 + 24x + 10$

6. $C = x^3 - 12x^2 + 63x + 15$

7. $C = 400 + 27x + x^3$

8. $C(x) = 50 + 48x + x^3$

9. Suppose that the cost function for a commodity is

$$C(x) = 40 + x^2.$$

(a) Find the marginal cost at $x = 5$ and tell what this predicts about the cost of producing one additional unit.

(b) Calculate $C(6) - C(5)$ to find the actual cost of producing one additional unit.

10. Suppose that the cost function for a commodity is

$$C(x) = 300 + 6x + \tfrac{1}{20}x^2.$$

(a) Find the marginal cost at $x = 8$ and tell what this predicts about the cost of producing one additional unit.

(b) Calculate $C(9) - C(8)$ to find the actual cost of producing one additional unit.

11. If the cost function for a commodity is

$$C(x) = x^3 - 4x^2 + 30x + 20,$$

find the marginal cost at $x = 4$ and tell what this predicts about the cost of producing one additional unit.

12. If the cost function for a commodity is

$$C(x) = \tfrac{1}{90}x^3 + 4x^2 + 4x + 10,$$

find the marginal cost at $x = 3$ and tell what this predicts about the cost of producing one additional unit.

13. If the cost function for a commodity is

$$C(x) = 300 + 4x + x^2,$$

graph the marginal cost function.

14. If the cost function for a commodity is

$$C(x) = x^3 - 12x^2 + 63x + 15,$$

graph the marginal cost function.

15. (a) If the total revenue function for a product is $R(x) = 4x$, what is the marginal revenue function?

(b) What does this marginal revenue function tell us?

16. If the total revenue function for a product is $R(x) = 32x$, what is the marginal revenue for the product? What does this mean?

17. Suppose that the total revenue function for a commodity is $R = 36x - 0.01x^2$.

(a) Find $R(100)$ and tell what it represents.

(b) Find the marginal revenue function.

(c) Find the marginal revenue at $x = 100$, and tell what it predicts about the sale of the next unit.

(d) Find $R(101) - R(100)$ and explain what this value represents.

18. Suppose that the total revenue function for a commodity is $R(x) = 25x - 0.05x^2$.

(a) Find $R(50)$ and tell what it represents.

(b) Find the marginal revenue function.

(c) Find the marginal revenue at $x = 50$, and tell what it predicts about the sale of the next unit.

(d) Find $R(51) - R(50)$ and explain what it represents.

19. (a) Graph the marginal revenue function from problem 17.

(b) At what value of x will total revenue be maximized?

(c) What is the maximum revenue?

20. (a) Graph the marginal revenue function from problem 18.

(b) Determine the number of units that must be sold to maximize total revenue.

(c) What is the maximum revenue?

21. If the total profit function is $P(x) = 5x - 25$, find the marginal profit.

22. If the total profit function is $P(x) = 16x - 32$, find the marginal profit.

23. Suppose that the total revenue function for a product is $R(x) = 32x$ and that the total cost function is $C(x) = 200 + 2x + x^2$.

(a) Find the profit from the production and sale of 20 units.

(b) Find the marginal profit function.

(c) Find $\overline{MP}$ at $x = 20$ and explain what it predicts.

(d) Find $P(21) - P(20)$ and explain what it represents.

24. Suppose that the total revenue function is given by

$$R(x) = 46x$$

and that the total cost function is given by

$$C(x) = 100 + 30x + \tfrac{1}{10}x^2.$$

(a) Find $P(100)$.

(b) Find the marginal profit function.

(c) Find $\overline{MP}$ at $x = 100$ and explain what it predicts.

(d) Find $P(101) - P(100)$ and explain what it represents.

25. (a) Graph the marginal profit function for the profit function $P(x) = 30x - x^2 - 200$.

(b) What level of production and sales will give a 0 marginal profit?

(c) At what level of production and sales will profit be at a maximum?

(d) What is the maximum profit?

26. (a) Graph the marginal profit function for the profit function $P(x) = 16x - 0.1x^2 - 100$.
    (b) What level of production and sales will give a 0 marginal profit?
    (c) At what level of production and sales will profit be at a maximum?
    (d) What is the maximum profit?

27. The price of a product in a competitive market is $300. If the cost per unit of producing the product is $160 + x$, where x is the number of units produced per month, how many units should the firm produce and sell to maximize its profit?

28. The cost per unit of producing a product is $60 + 2x$, where x represents the number of units produced per week. If the equilibrium price determined by a competitive market is $220, how many units should the firm produce and sell each week to maximize its profit?

29. If the daily cost per unit of producing a product by the Ace Company is $10 + 2x$, and if the price on the competitive market is $50, what is the maximum daily profit the Ace Company can expect on this product?

30. The Mary Ellen Candy Company produces chocolate Easter bunnies at a cost per unit of $0.10 + 0.01x$, where x is the number produced. If the price on the competitive market for a bunny this size is $2.50, how many should the company produce to maximize its profit?

31. The following table gives the total revenues of AT&T for selected years.

Year	Total Revenues (in Billions)
1985	$63.13
1986	$69.906
1987	$60.53
1989	$61.1
1990	$62.191
1991	$63.089
1992	$64.904
1993	$67.156

SOURCE: *AT&T Annual Report*, 1993

Suppose the data can be modeled by the equation

$$R(t) = 0.253t^2 - 4.03t + 76.84,$$

where t is the number of years past 1980.
(a) Find $R(7)$ from the data and compare it to $R(7)$ as found from the model. What does $R(7)$ represent?

(b) Find the function that gives the instantaneous rate of change of revenue.
(c) Find the instantaneous rate of change of revenue in 1992.
(d) Interpret your result to part (c).

The data in the table give sales revenues and costs and expenses for Scott Paper Company for various years. Use this data for problems 32–34.

Year	Sales Revenue (in Billions)	Costs and Expenses (in Billions)
1983	$2.6155	$2.4105
1984	2.7474	2.4412
1985	2.934	2.6378
1986	3.3131	2.9447
1987	3.9769	3.5344
1988	4.5494	3.8171
1989	4.8949	4.2587
1990	5.1686	4.8769
1991	4.9593	4.9088
1992	5.0913	4.6771
1993	4.7489	4.9025

SOURCE: Scott Paper Company, 1993 *Annual Report*

32. Assume that sales revenues for Scott Paper can be modeled by
$$R(t) = -0.031t^2 + 0.776t + 0.179,$$
where t is the number of years past 1980.
(a) Use this model to find the instantaneous rate of change of revenue in 1990.
(b) Interpret your answer to part (a).

33. Assume that costs and expenses for Scott Paper Company can be modeled by
$$C(t) = -0.012t^2 + 0.492t + 0.725,$$
where t is the number of years past 1980.
(a) Use this model to find the instantaneous rate of change of costs and expenses in 1990.
(b) Interpret your answer to part (a) and check this interpretation against the data in the table.

34. Let income from operations, $I(t)$, be revenues minus costs.
(a) Find $I(t)$.
(b) Find the instataneous rate of change of income in 1990.
(c) Interpret your result in part (b).
(d) Would the board of directors be interested in altering this model? Explain.

KEY TERMS
& FORMULAS

Section	Key Term	Formula
9.1	Limit, infinite limit	
	0/0 indeterminate form	
9.2	Continuous function	
	Vertical asymptote	
	Horizontal asymptote	
	Limit at infinity	
9.3	Average velocity	
	Velocity	
	Instantaneous rate of change	
	Derivative	$f'(x) = \lim\limits_{h \to 0} \dfrac{f(x + h) - f(x)}{h}$
	Marginal revenue	$\overline{MR} = R'(x)$
	Tangent line	
	Secant line	
	Slope of a curve	
	Differentiability and continuity	
9.4	Powers of x Rule	$\dfrac{d(x^n)}{dx} = nx^{n-1}$
	Constant Function Rule	$\dfrac{d(c)}{dx} = 0$ for constant c
	Coefficient Rule	$\dfrac{d}{dx}[c \cdot f(x)] = c \cdot f'(x)$
	Sum Rule	$\dfrac{d}{dx}[u + v] = \dfrac{du}{dx} + \dfrac{dv}{dx}$
	Difference Rule	$\dfrac{d}{dx}[u - v] = \dfrac{du}{dx} - \dfrac{dv}{dx}$
9.5	Product Rule	$\dfrac{d}{dx}[uv] = uv' + vu'$
	Quotient Rule	$\dfrac{d}{dx}\left(\dfrac{u}{v}\right) = \dfrac{vu' - uv'}{v^2}$
9.6	Chain Rule	$\dfrac{dy}{dx} = \dfrac{dy}{du} \cdot \dfrac{du}{dx}$
	Power Rule	$\dfrac{d}{dx}(u^n) = nu^{n-1}\dfrac{du}{dx}$

Section	Key Term	Formula
9.8	Second derivative; third derivative; higher-order derivatives	
9.9	Marginal cost function	$\overline{MC} = C'(x)$
	Marginal revenue function	$\overline{MR} = R'(x)$
	Marginal profit function	$\overline{MP} = P'(x)$

REVIEW EXERCISES

9.1

In problems 1–6, use the graph of $y = f(x)$ in Figure 9.33 to find the functional values and limits, if they exist.

1. (a) $f(-2)$ (b) $\lim\limits_{x \to -2} f(x)$

2. (a) $f(-1)$ (b) $\lim\limits_{x \to -1} f(x)$

3. (a) $f(4)$ (b) $\lim\limits_{x \to 4^-} f(x)$

4. (a) $\lim\limits_{x \to 4^+} f(x)$ (b) $\lim\limits_{x \to 4} f(x)$

5. (a) $f(1)$ (b) $\lim\limits_{x \to 1} f(x)$

6. (a) $f(2)$ (b) $\lim\limits_{x \to 2} f(x)$

Figure 9.33

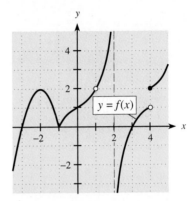

In problems 7–20, find each limit, if it exists.

7. $\lim\limits_{x \to 4} (3x^2 + x + 3)$

8. $\lim\limits_{x \to 4} \dfrac{x^2 - 16}{x + 4}$

9. $\lim\limits_{x \to -1} \dfrac{x^2 - 1}{x + 1}$

10. $\lim\limits_{x \to 3} \dfrac{x^2 - 9}{x - 3}$

11. $\lim\limits_{x \to 2} \dfrac{4x^3 - 8x^2}{4x^3 - 16x}$

12. $\lim\limits_{x \to -\frac{1}{2}} \dfrac{x^2 - \frac{1}{4}}{6x^2 + x - 1}$

13. $\lim\limits_{x \to 3} \dfrac{x^2 - 16}{x - 3}$

14. $\lim\limits_{x \to -3} \dfrac{x^2 - 9}{x - 3}$

15. $\lim\limits_{x \to 1} \dfrac{x^2 - 9}{x - 3}$

16. $\lim\limits_{x \to 2} \dfrac{x^2 - 8}{x - 2}$

17. $\lim\limits_{x \to 1} f(x)$ where $f(x) = \begin{cases} 4 - x^2 & \text{if } x < 1 \\ 4 & \text{if } x = 1 \\ 2x + 1 & \text{if } x > 1 \end{cases}$

18. $\lim\limits_{x \to -2} f(x)$ where $f(x) = \begin{cases} x^3 - x & \text{if } x < -2 \\ 2 - x^2 & \text{if } x \geq -2 \end{cases}$

19. $\lim\limits_{h \to 0} \dfrac{3(x + h)^2 - 3x^2}{h}$

20. $\lim\limits_{h \to 0} \dfrac{[(x + h) - 2(x + h)^2] - (x - 2x^2)}{h}$

9.2

Use the graph of $y = f(x)$ in Figure 9.33 to answer the questions in problems 21 and 22.

21. Is $f(x)$ continuous at
    (a) $x = -1$?   (b) $x = 1$?

22. Is $f(x)$ continuous at
    (a) $x = -2$?   (b) $x = 2$?

In problems 23–28, suppose that

$$f(x) = \begin{cases} x^2 + 1 & \text{if } x \leq 0 \\ x & \text{if } 0 < x < 1. \\ 2x^2 - 1 & \text{if } x \geq 1 \end{cases}$$

23. What is $\lim\limits_{x \to -1} f(x)$?

24. What is $\lim_{x \to 0} f(x)$, if it exists?

25. What is $\lim_{x \to 1} f(x)$, if it exists?

26. Is $f(x)$ continuous at $x = 0$?
27. Is $f(x)$ continuous at $x = 1$?
28. Is $f(x)$ continuous at $x = -1$?

Which of the functions in problems 29–32 are continuous?

29. $y = \dfrac{x^2 + 25}{x - 5}$

30. $y = \dfrac{x^2 - 3x + 2}{x - 2}$

31. $f(x) = \begin{cases} x + 2 & \text{if } x \leq 2 \\ 5x - 6 & \text{if } x > 2 \end{cases}$

32. $y = \begin{cases} x^4 - 3 & \text{if } x \leq 1 \\ 2x - 3 & \text{if } x > 1 \end{cases}$

In problems 33 and 34, use the graphs to find (a) the points of discontinuity, (b) $\lim_{x \to +\infty} f(x)$, and (c) $\lim_{x \to -\infty} f(x)$.

33.

34.

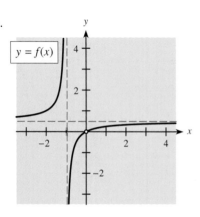

In problems 35 and 36 evaluate the limits, if they exist.

35. $\lim_{x \to -\infty} \dfrac{2x^2}{1 - x^2}$

36. $\lim_{x \to +\infty} \dfrac{3x^{2/3}}{x + 1}$

9.3

In problems 37 and 38, decide whether the statements are true or false.

37. $\lim_{h \to 0} \dfrac{f(x + h) - f(x)}{h}$ gives the formula for the slope of the tangent and the instantaneous rate of change of $f(x)$ at any value of x.

38. $\lim_{h \to 0} \dfrac{f(c + h) - f(c)}{h}$ gives the equation of the tangent line to $f(x)$ at $x = c$.

39. Use the definition of *derivative* to find $f'(x)$ for $f(x) = 3x^2 + 2x - 1$.

40. Use the definition of *derivative* to find $f'(x)$ if $f(x) = x - x^2$.

Use the graph of $y = f(x)$ in Figure 9.33 on page 677 to answer the questions in problems 41 and 42.

41. Is $f(x)$ differentiable at
    (a) $x = -1$?   (b) $x = 1$?
42. Is $f(x)$ differentiable at
    (a) $x = -2$?   (b) $x = 2$?

9.4

43. If $c = 4x^5 - 6x^3$, find c'.
44. If $f(x) = 4x^2 - 1$, find $f'(x)$.
45. If $p = 3q + \sqrt{7}$, find dp/dq.
46. If $y = \sqrt{x}$, find y'.
47. If $f(z) = \sqrt[3]{2^4}$, find $f'(z)$.
48. If $v(x) = 4/\sqrt[3]{x}$, find $v'(x)$.
49. If $y = \dfrac{1}{x} - \dfrac{1}{\sqrt{x}}$ find y'.

50. If $f(x) = \dfrac{3}{2x^2} - \sqrt[3]{x} + 4^5$, find $f'(x)$.

51. Write the equation of the line tangent to the graph of $y = 3x^5 - 6$ at $x = 1$.
52. Write the equation of the line tangent to the curve $y = 3x^3 - 2x$ at the point $(2, 20)$.

In problems 53 and 54, (a) find all *x*-values where the slope of the tangent equals zero, (b) find points (x, y) where the slope of the tangent equals zero, and (c) use a graphing utility to graph the function and label the points found in part (b).

53. $f(x) = x^3 - 3x^2 + 1$

54. $f(x) = x^6 - 6x^4 + 8$

9.5

55. If $f(x) = (3x - 1)(x^2 - 4)$, find $f'(x)$.

56. Find y' if $y = (x^2 + 1)(3x^3 + 1)$.

57. If $p = \dfrac{2q - 1}{q^2}$, find $\dfrac{dp}{dq}$.

58. Find $\dfrac{ds}{dt}$ if $s = \dfrac{\sqrt{t}}{(3t + 1)}$.

59. Find $\dfrac{dy}{dx}$ for $y = \sqrt{x}(3x + 2)$.

60. Find $\dfrac{dC}{dx}$ for $C = \dfrac{5x^4 - 2x^2 + 1}{x^3 + 1}$.

9.6

61. If $y = (x^3 - 4x^2)^3$, find y'.

62. If $y = (5x^6 + 6x^4 + 5)^6$, find y'.

63. If $y = (2x^4 - 5)^9$, find $\dfrac{dy}{dx}$.

64. Find $g'(x)$ if $g(x) = \dfrac{1}{\sqrt{x^3 - 4x}}$.

9.7

65. Find $f'(x)$ if $f(x) = x^2(2x^4 + 5)^8$.

66. Find S' if $S = \dfrac{(3x + 1)^2}{x^2 - 4}$.

67. Find $\dfrac{dy}{dx}$ if $y = [(3x + 1)(2x^3 - 1)]^{12}$.

68. Find y' if $y = \left(\dfrac{x + 1}{1 - x^2}\right)^3$.

69. Find y' if $y = x\sqrt{x^2 - 4}$.

70. Find $\dfrac{dy}{dx}$ if $y = \dfrac{x}{\sqrt[3]{3x - 1}}$.

9.8

In problems 71 and 72, find the second derivatives.

71. $y = \sqrt{x} - x^2$

72. $y = x^4 - \dfrac{1}{x}$

In problems 73 and 74, find the fifth derivatives.

73. $y = (2x + 1)^4$

74. $y = \dfrac{(1 - x)^6}{24}$

75. If $\dfrac{dy}{dx} = \sqrt{x^2 - 4}$, and $\dfrac{d^3y}{dx^3}$.

76. If $\dfrac{d^2y}{dx^2} = \dfrac{x}{x^2 + 1}$, find $\dfrac{d^4y}{dx^4}$.

APPLICATIONS

9.4

77. **Demand** Suppose that the demand *x* for a product is given by $x = (100/p) - 1$, where *p* is the price per unit of the product. Find and interpret the rate of change of demand with respect to the price if the price is
   (a) \$10.   (b) \$20.

9.6

78. **Demand** The demand *q* for a product at price *p* is given by
$$q = 10{,}000 - 50\sqrt{0.02p^2 + 500}.$$
Find the rate of change of demand with respect to price.

79. **Supply** The number of units *x* of a product that is supplied at price *p* is given by
$$x = \sqrt{p - 1}, \quad p \geq 1.$$
If the price *p* is \$10, what is the rate of change of the supply with respect to the price and what does it tell us?

9.9

80. **Cost** If the cost function for a particular good is $C(x) = 3x^2 + 6x + 600$, what is the
   (a) marginal cost function?
   (b) marginal cost if 30 units are produced?
   (c) interpretation of your answer in (b)?

81. **Cost** If the total cost function for a commodity is $C(x) = 400 + 5x + x^3$, what is the marginal cost when 4 units are produced and what does it mean?

82. **Revenue** If the total revenue function for a commodity is $R = 40x - 0.02x^2$, with x representing the number of units,
    (a) find the marginal revenue function.
    (b) at what level of production will marginal revenue be 0?

83. **Profit** If the total revenue function is given by $R(x) = 60x$ and the total cost function is given by $C = 200 + 10x + 0.1x^2$, what is the marginal profit at $x = 10$? What does the marginal profit at $x = 10$ predict?

84. **Revenue** If the total revenue function for a commodity is given by $R = 80x - 0.04x^2$,
    (a) find the marginal revenue function.
    (b) what is the marginal revenue at $x = 100$?
    (c) interpret your answer in (b).

85. **Revenue** If the revenue function for a product is

$$R(x) = \frac{60x^2}{2x + 1},$$

find the marginal revenue.

86. **Profit** A firm has monthly costs given by

$$C = 45,000 + 100x + x^3,$$

where x is the number of units produced per month. The firm can sell its product in a competitive market for $4600 per unit. Find the marginal profit.

87. **Profit** A small business has weekly costs of

$$C = 100 + 30x + \frac{x^2}{10},$$

where x is the number of units produced each week. The competitive market price for this business's product is $46 per unit. Find the marginal profit.

CONSULTANT'S CORNER

Marginal Return to Sales

A tire manufacturer studying the effectiveness of television advertising and other promotions on sales of its GRIPPER-brand tires attempted to fit data it had gathered to the equation

$$S = a_0 + a_1x + a_2x^2 + b_1y,$$

where S is sales in millions of dollars, x is millions of dollars spent on television advertising, y is millions of dollars spent on other promotions, and a_0, a_1, a_2, and b_1 are constants. The data, gathered in two different regions of the country where expenditures for other promotions were kept constant (at B_1 and B_2), resulted in the following quadratic equations relating TV advertising and sales.

$$\text{Region 1:} \quad S_1 = 30 + 20x - 0.4x^2 + B_1$$
$$\text{Region 2:} \quad S_2 = 20 + 36x - 0.6x^2 + B_2$$

The company wants to know how to make the best use of its advertising dollars in the regions, and whether the current allocation could be improved. Advise management about current advertising effectiveness, allocation of additional expenditures, and reallocation of current advertising expenditures by answering the following questions.

1. In the analysis of sales and advertising, **marginal return to sales** is usually used, and it is given by dS_1/dx for Region 1 and dS_2/dx for Region 2.
   (a) Find $\dfrac{dS_1}{dx}$ and $\dfrac{dS_2}{dx}$.
   (b) If \$10 million is being spent on TV advertising in each region, what is the marginal return to sales in each region?
2. Which region would benefit more from each additional dollar of advertising expenditure, if \$10 million is currently being spent in each region?
3. If any additional money is made available for advertising, in which region should it be spent?
4. How could money already being spent be reallocated to produce more sales?

Warm-up

Prerequisite problem type	For section	Answer	Section for review
Write $\frac{1}{3}(x^2 - 1)^{-2/3}(2x)$ with positive exponents.	10.2	$\dfrac{2x}{3(x^2 - 1)^{2/3}}$	0.3, 0.4 Exponents and radicals
Factor: (a) $x^3 - x^2 - 6x$ (b) $8000 - 80x - 3x^2$	10.1 10.2 10.3	(a) $x(x - 3)(x + 2)$ (b) $(40 - x)(200 + 3x)$	0.6 Factoring
(a) For what values of x is $\dfrac{2}{3\sqrt[3]{x + 2}}$ undefined? (b) For what values of x is $\frac{1}{3}(x^2 - 1)^{-2/3}(2x)$ undefined?	10.1 10.2 10.5	(a) $x = -2$ (b) $x = -1, x = 1$	1.4 Domains of functions
If $f(x) = \frac{1}{3}x^3 - x^2 - 3x + 2$, and $f'(x) = x^2 - 2x - 3$, (a) find $f(-1)$. (b) find $f'(-2)$.	10.1	 (a) $\frac{11}{3}$ (b) 5	1.4 Functional notation
(a) Solve $0 = x^2 - 2x - 3$. (b) If $f'(x) = 3x^2 - 3$, what values of x make $f'(x) = 0$?	10.1– 10.5	(a) $x = -1, x = 3$ (b) $x = -1, x = 1$	4.1 Solving quadratic equations
Does $\displaystyle\lim_{x \to -2} \frac{2x - 4}{3x + 6}$ exist?	10.5	No; unbounded	9.1 Limits
(a) Find $f''(x)$ if $f(x) = x^3 - 4x^2 + 3$. (b) Find $P''(x)$ if $P(x) = 48x - 1.2x^2$.	10.2	(a) $f''(x) = 6x - 8$ (b) $P''(x) = -2.4$	9.8 Higher-order derivatives
Find the derivatives: (a) $y = \frac{1}{3}x^3 - x^2 - 3x + 2$ (b) $f = x + 2\left(\dfrac{80,000}{x}\right)$ (c) $p(t) = 1 + \dfrac{4t}{t^2 + 16}$ (d) $y = (x + 2)^{2/3}$ (e) $y = \sqrt[3]{x^2 - 1}$	10.1 10.2 10.3 10.4 10.5	 (a) $y' = x^2 - 2x - 3$ (b) $f' = 1 - \dfrac{160,000}{x^2}$ (c) $p'(t) = \dfrac{64 - 4t^2}{(t^2 + 16)^2}$ (d) $y' = \dfrac{2}{3(x + 2)^{1/3}}$ (e) $y' = \dfrac{2x}{3(x^2 - 1)^{2/3}}$	9.4, 9.5, 9.6 Derivatives

CHAPTER 10

APPLICATIONS OF DERIVATIVES

In this chapter we will discuss applications of the derivative. In particular, we will consider methods of determining when a function has a "turning point" on its graph, so that we can determine when the graph of the function reaches its highest or lowest point within a particular interval. These points are called **relative maxima** and **relative minima,** respectively, and are useful in sketching the graph of a function whose equation is given and that has a maximum or minimum value within a particular interval. The endpoints of a given interval and the relative maxima and minima within the interval can be used to solve many types of applied problems. For example, we can use these points in determining the level of production that maximizes revenue or profit for a product and in minimizing the average cost of producing a product.

In addition to using the first derivative to help graph a function, we can use the second derivative to determine where the graph will be concave up or concave down and where it will change from concave up to concave down or vice versa. Points where this change occurs are called **points of inflection.** The second derivative uses concavity to determine where the graph of a function has a relative maximum or relative minimum. Knowledge of this information can be used to sketch the graph of a function or to determine the appropriate viewing window to use when graphing the function with a graphing utility. Because horizontal and vertical asymptotes are not always apparent when using a graphing utility, we will discuss methods of finding and accounting for asymptotes when graphing functions.

10.1 RELATIVE MAXIMA AND MINIMA; CURVE SKETCHING

Objectives

❑ To find relative maxima and minima and horizontal points of inflection of functions
❑ To sketch graphs of functions by using information about maxima, minima, and horizontal points of inflection

APPLICATION PREVIEW

When a company initiates an advertising campaign, there is typically a surge in weekly sales. As the effect of the campaign lessens, sales attributable to it usually decrease. For example, suppose a company models its weekly sales during an advertising campaign by

$$S = \frac{100t}{t^2 + 100},$$

where t is the number of weeks since the beginning of the campaign. The company would like to accurately determine when the revenue function is increasing, when it is decreasing, and when sales revenue is maximized.

In this section we will use the derivative of a function to decide whether the function is increasing or decreasing on an interval and to find where the function has relative maximum points and relative minimum points. We will use the information about derivatives of functions to graph the functions and to solve applied problems.

Except for very simple graphs (lines and parabolas, for example), plotting points to sketch a graph may be tedious. Even when using a graphing utility, special features of a graph of a function may be difficult to locate accurately. In addition to intercepts and asymptotes, we can use the first derivative as an aid in graphing. The first derivative identifies the "turning points" of the graph, which help in determining the general shape of the graph and in choosing a viewing window that includes the interesting points of the graph if a graphing utility is used.

In Figure 10.1(a) we see that the graph of $y = \frac{1}{3}x^3 - x^2 - 3x + 2$ has two "turning points," at $(-1, \frac{11}{3})$ and $(3, -7)$. The curve has a relative maximum at $(-1, \frac{11}{3})$ because this point is higher than any other point "near" it on the curve; the curve has a relative minimum at $(3, -7)$ because this point is lower than any other point "near" it on the curve. A formal definition follows.

Relative Maxima and Minima

The point $(x_1, f(x_1))$ is a **relative maximum point** for the function f if there is an interval around x_1 on which $f(x_1) \geq f(x)$ for all x in the interval. In this case, we say the relative maximum *occurs* at $x = x_1$ and the relative maximum *is* $f(x_1)$.

The point $(x_2, f(x_2))$ is a **relative minimum point** for the function f if there is an interval around x_2 on which $f(x_2) \leq f(x)$ for all x in the interval. In this case, we say the relative minimum *occurs* at $x = x_2$ and the relative minimum *is* $f(x_2)$.

In order to determine whether a turning point of a function is a maximum point or a minimum point, it is frequently helpful to know what the graph of the function does in intervals on either side of the turning point. We say a function is **increasing** on an interval if the functional values increase as the x-values increase (that is, if the graph rises as we move from left to right on the interval). Similarly, a function is **decreasing** on an interval if the func-

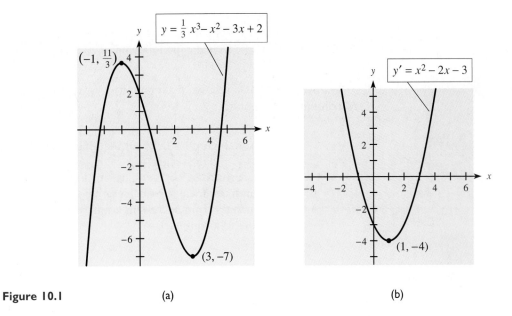

Figure 10.1   (a)   (b)

tional values decrease as the *x*-values increase (that is, if the graph falls as we move from left to right on the interval).

We have seen that if the slope of a line is positive, then the linear function is increasing and its graph is rising. Similarly, if $f(x)$ is differentiable over an interval and if each tangent line to the curve over that interval has positive slope, then the curve is rising over the interval and the function is increasing. Because the derivative of the function gives the slope of the tangent to the curve, we see that if $f'(x) > 0$ on an interval, then $f(x)$ is increasing on that interval. A similar conclusion can be reached when the derivative is negative on the interval.

Increasing and Decreasing Functions

If f is a function that is differentiable on an interval (a, b), then

if $f'(x) > 0$ for all x in (a, b), f is increasing on (a, b).
if $f'(x) < 0$ for all x in (a, b), f is decreasing on (a, b).

Figure 10.1 (a) shows the graph of a function and Figure 10.1 (b) shows the graph of its derivative. The figures show that the graph of $y = f(x)$ is increasing for the same *x*-values that the graph of $y' = f'(x)$ is above the *x*-axis (when $f'(x) > 0$). Similarly, the graph of $y = f(x)$ is decreasing for the same *x*-values $(-1 < x < 3)$ that the graph of $y' = f'(x)$ is below the *x*-axis (when $f'(x) < 0$).

The derivative $f'(x)$ can change signs only at values of x where $f'(x) = 0$ or $f'(x)$ is undefined. We call these values of x **critical values.** The point corresponding to a critical value for x is a **critical point.*** Because a curve changes from increasing to decreasing at a relative maximum (see Figure 10.1(a)), we have the following fact.

Relative Maximum	If f has a relative maximum at $x = x_0$, then $f'(x_0) = 0$ or $f'(x_0)$ is undefined.

From Figure 10.2, we see that this function has two relative maxima, one at $x = x_1$ and the second at $x = x_3$. At $x = x_1$ the derivative is 0, and at $x = x_3$ the derivative does not exist. In Figure 10.2, we also see that a relative minimum occurs at $x = x_2$ and $f'(x_2) = 0$. As Figure 10.2 shows, the function changes from decreasing to increasing at a relative minimum. Thus we have the following fact.

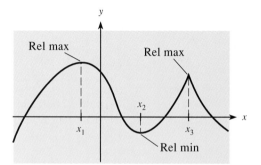

Figure 10.2

Relative Minimum	If f has a relative minimum at $x = x_0$, then $f'(x_0) = 0$ or $f'(x_0)$ is undefined.

Thus we can find relative maxima and minima for a curve by finding values of x for which the function has critical points. The behavior of the derivative to the left and right of (and near) these points will tell us if they are relative maxima or relative minima, or neither one.

Because the critical values are the only values where the graph can have turning points, the derivative cannot change sign anywhere except at a critical value. Thus in an interval between two critical values, the sign of the derivative at any value in the interval will be the sign of the derivative at all values in the interval.

Using the critical values of $f(x)$ and the sign of $f'(x)$ between those critical values, we can create a **sign diagram for $f'(x)$.** The sign diagram for

---

* There may be some critical values where $f'(x)$ and $f(x)$ are undefined. Critical points do not occur at these values, but studying the derivative on either side of such values may be of interest.

Figure 10.4

$f'(x_1) = 0$

$f'(x) > 0$ $f'(x) < 0$

Critical
point

Relative maximum

(a)

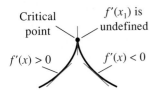

Critical $f'(x_1)$ is
point   undefined

$f'(x) > 0$ $f'(x) < 0$

Relative maximum

(b)

Figure 10.5

Critical
point

$f'(x) < 0$ $f'(x) > 0$

$f'(x_2) = 0$

Relative minimum

(a)

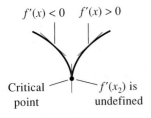

$f'(x) < 0$ $f'(x) > 0$

Critical $f'(x_2)$ is
point   undefined

Relative minimum

(b)

the graph in Figure 10.2 is shown in Figure 10.3. This sign diagram was created from the graph of f, but it is also possible to predict the shape of a graph from a sign diagram.

Direction of graph of $f(x)$:

Signs and values of $f'(x)$: $+ + + \; 0 \; - - - \; 0 \; + + + \; * \; - - -$

x-axis with critical values:

$x_1 \qquad x_2 \qquad x_3$

* means $f'(x_3)$ is undefined.

Figure 10.3

Suppose that the point (x_1, y_1) is a critical point. If $f'(x)$ is positive to the left of, and near, this critical point, and if $f'(x)$ is negative to the right of, and near, this critical point, then the curve is increasing to the left of the point and decreasing to the right. This means that a relative maximum occurs at the point. If we draw tangent lines to the curve on the left and right of this critical point, they would fit on the curve in one of the two ways shown in Figure 10.4.

Similarly, suppose that the point (x_2, y_2) is a critical point, $f'(x)$ is negative to the left of and near (x_2, y_2), and $f'(x)$ is positive to the right of and near this critical point. Then the curve is decreasing to the left of the point and increasing to the right, and a relative minimum occurs at the point. If we draw tangent lines to the curve to the left of and to the right of this critical point, they would fit on the curve in one of the two ways shown in Figure 10.5.

EXAMPLE 1

Show that the graph of $f(x) = 3x^2 - 2x^3$ has a relative maximum at $x = 1$.

Solution

We need to show that:

$f'(1) = 0$ or $f'(1)$ is undefined,
$f'(x) > 0$ to the left of and near $x = 1$, and
$f'(x) < 0$ to the right of and near $x = 1$.

Because $f'(x) = 6x - 6x^2 = 6x(x - 1)$ is 0 only at $x = 0$ and at $x = 1$, we can test (evaluate) the derivative at any value in the interval $(0, 1)$ to see what the curve is doing to the left of $x = 1$, and we can test the derivative at any value to the right of $x = 1$ to see what the curve is doing for $x > 1$.

$$f'\left(\frac{1}{2}\right) = 3 - \frac{3}{2} = \frac{3}{2} > 0 \Rightarrow \text{increasing to left of } x = 1$$

$$f'(1) = 0 \Rightarrow \text{horizontal tangent at } x = 1$$

$$f'\left(\frac{3}{2}\right) = 9 - \frac{27}{2} = -\frac{9}{2} < 0 \Rightarrow \text{decreasing to right of } x = 1$$

A partial sign diagram for $f'(x)$ is given at the right.
The sign diagram shows that the graph of the function has a relative maximum at $x = 1$.

The preceding discussion suggests the following procedure for finding relative maxima and minima of a function.

First-Derivative Test

Procedure	**Example**
To find relative maxima and minima of a function:	Find the relative maxima and minima of $f(x) = \frac{1}{3}x^3 - x^2 - 3x + 2$.
1. Find the first derivative of the function.	1. $f'(x) = x^2 - 2x - 3$
2. Set the derivative equal to 0, and solve for values of x that satisfy $f'(x) = 0$. These are called **critical values.** Values that make $f'(x)$ undefined are also critical values.	2. $0 = x^2 - 2x - 3 = (x + 1)(x - 3)$ has solutions $x = -1$, $x = 3$. No values of x make $x^2 - 2x - 3$ undefined. Critical values are -1 and 3.
3. Substitute the critical values into the *original function* to find the **critical points.**	3. $f(-1) = \frac{11}{3}$ $f(3) = -7$ The critical points are $(-1, \frac{11}{3})$ and $(3, -7)$.
4. Evaluate $f'(x)$ at some value of x to the left and right of each critical point: (a) If $f'(x) > 0$ to the left and $f'(x) < 0$ to the right of the critical value, the critical point is a relative maximum point. (b) If $f'(x) < 0$ to the left and $f'(x) > 0$ to the right of the critical value, the critical point is a relative minimum point.	4. $f'(-2) = 5 > 0$ and $f'(0) = -3 < 0$ Thus $(-1, 11/3)$ is a relative maximum point. $f'(2) = -3 < 0$ and $f'(4) = 5 > 0$ Thus $(3, -7)$ is a relative minimum point.
5. Develop a sign diagram and use the information and selected points to sketch the graph.	5. The sign diagram for $f'(x)$ is

The information from this sign diagram is shown in Figure 10.6(a). Plotting additional points gives the graph of the function, which is shown in Figure 10.6(b).

Figure 10.6

(a)

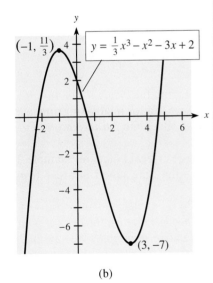

(b)

Because the critical values are the only *x*-values where the graph can have turning points, we can test to the left and right of each critical value by testing to the left of the smallest critical value, then testing a value *between* each critical value, and then testing to the right of the largest critical value. The following example illustrates this procedure.

EXAMPLE 2

Find the relative maxima and minima of $f(x) = \frac{1}{4}x^4 - \frac{1}{3}x^3 - 3x^2 + 8$, and sketch its graph.

Solution

1. $f'(x) = x^3 - x^2 - 6x$

2. Setting $f'(x) = 0$ gives $0 = x^3 - x^2 - 6x$. Solving for *x* gives
   $0 = x(x - 3)(x + 2)$.

$x = 0$	$x - 3 = 0$	$x + 2 = 0$
	$x = 3$	$x = -2$

   Thus the critical value are $x = 0$, $x = 3$, and $x = -2$.

3. Substituting the critical values into the original function gives the critical points:

   $$f(-2) = \tfrac{8}{3}, \qquad \text{so } (-2, \tfrac{8}{3}) \text{ is a critical point.}$$
   $$f(0) = 8, \qquad \text{so } (0, 8) \text{ is a critical point.}$$
   $$f(3) = -\tfrac{31}{4}, \qquad \text{so } (3, -\tfrac{31}{4}) \text{ is a critical point.}$$

4. Testing $f'(x)$ to the left of the smallest critical value, then between the critical values, then to the right of the largest critical value gives

   $f'(-3) = -18 < 0$
   $f'(-2) = 0$
   $f'(-1) = 4 > 0$ $\Big\}$ ⇒ $(-2, \tfrac{8}{3})$ is a relative minimum.

   $f'(0) = 0$
   $f'(1) = -6 < 0$ $\Big\}$ ⇒ $(0, 8)$ is a relative maximum.

   $f'(3) = 0$
   $f'(4) = 24 > 0$ $\Big\}$ ⇒ $(3, -\tfrac{31}{4})$ is a relative minimum.

5. The sign diagram for $f'(x)$ is

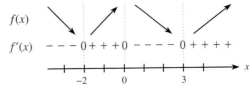

Figure 10.7(a) shows the graph of the function near the critical points, and Figure 10.7(b) shows the graph of the function.

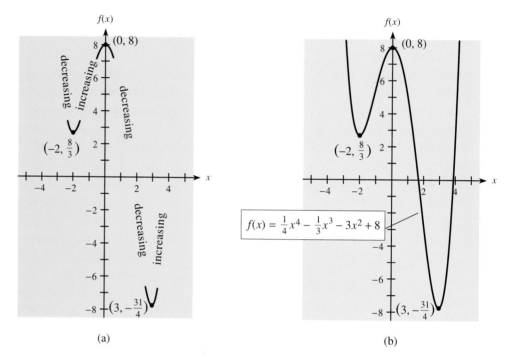

Figure 10.7   (a)   (b)

Note that we substitute the critical values into the *original function* $f(x)$ to find the y-values of the critical points, but we test for relative maxima and minima by substituting values near the critical values into the *derivative of the function,* $f'(x)$.

Only four values were needed to test three critical points in Example 2. This method will work *only* if the critical values are tested in order from smallest to largest.

If the first derivative of f is 0 at x_0 but does not change from positive to negative or negative to positive as x passes through x_0, then the critical point at x_0 is neither a relative maximum nor a relative minimum. We say that f has a **horizontal point of inflection** at x_0.

EXAMPLE 3

Find the relative maxima, relative minima, and horizontal points of inflection of $h(x) = \frac{1}{4}x^4 - \frac{2}{3}x^3 - 2x^2 + 8x + 4$, and sketch its graph.

Solution

1. $h'(x) = x^3 - 2x^2 - 4x + 8$

2. $0 = x^3 - 2x^2 - 4x + 8$ or $0 = x^2(x - 2) - 4(x - 2)$. Therefore, we have $0 = (x - 2)(x^2 - 4)$. Thus $x = -2$ and $x = 2$ are solutions.

3. The critical points are $(-2, -\frac{32}{3})$ and $(2, \frac{32}{3})$.

4. $h'(-3) = -25 < 0$
   $h'(-2) = 0$
   $h'(0) = 8 > 0$
   $h'(2) = 0$
   $h'(3) = 5 > 0$

   $(-2, -\frac{32}{2})$ is a relative minimum point.

   $(2, \frac{32}{3})$ is a horizontal point of inflection since $h'(2) = 0$ and $h'(x)$ does not change sign around $x = 2$.

5. The sign diagram for $h'(x)$ is

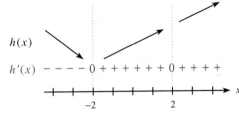

$$h(x)$$

$$h'(x) \quad - - - - 0 + + + + + + 0 + + +$$

Figure 10.8(a) shows the graph of the function near the critical points, and Figure 10.8(b) shows the graph of the function.

Figure 10.8

(a)

(b)

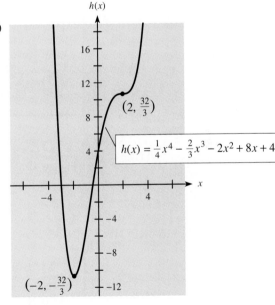

$$h(x) = \frac{1}{4}x^4 - \frac{2}{3}x^3 - 2x^2 + 8x + 4$$

EXAMPLE 4

Find the relative maxima and minima (if any) of the graph of $y = (x + 2)^{2/3}$.

Solution

1. $y' = f'(x) = \frac{2}{3}(x + 2)^{-1/3} = \dfrac{2}{3\sqrt[3]{x + 2}}$

2. $0 = \dfrac{2}{3\sqrt[3]{x+2}}$ has no solutions; $f'(x)$ is undefined at $x = -2$.

3. $f(-2) = 0$, so the critical point is $(-2, 0)$.

4. $\left.\begin{array}{l} f'(-3) = -\frac{2}{3} < 0 \\ f'(-2) \text{ undefined} \\ f'(-1) = \frac{2}{3} > 0 \end{array}\right\} \Rightarrow$ a relative minimum occurs at $(-2, 0)$.

5. The sign diagram for $f'(x)$ is

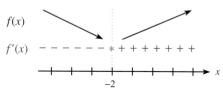

* means $f'(-2)$ is undefined.

Figure 10.9(a) shows the graph of the function near the critical point, and Figure 10.9(b) shows the graph.

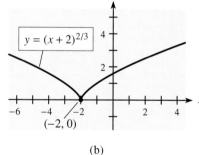

Figure 10.9   (a)   (b)

CHECKPOINT

1. The x-values of critical points are found where $f'(x)$ is _____ or _____ .

2. Decide whether the following are true or false.
   (a) If $f'(1) = 7$, then $f(x)$ is increasing at $x = 1$.
   (b) If $f'(-2) = 0$, then a relative maximum or a relative minimum occurs at $x = -2$.
   (c) If $f'(-3) = 0$ and $f'(x)$ changes from positive on the left to negative on the right of $x = -3$, then a relative minimum occurs at $x = -3$.

3. If $f(x) = 7 + 3x - x^3$, then $f'(x) = 3 - 3x^2$. Use these to decide whether the following are true or false.
   (a) The only critical value is $x = 1$.
   (b) The critical points are $(1, 0)$ and $(-1, 0)$.

4. If $f'(x)$ has the following partial sign diagram, make a "stick-figure" sketch of $f(x)$ and label where any maxima and minima occur. Assume that $f(x)$ is defined for all real numbers.

$f'(x)$ $+ + + 0 - - - * + + + + + + 0 + + +$

$* $ means $f'(0)$ is undefined.

Let us now return to the advertising and sales revenue discussion that we began in the Application Preview.

EXAMPLE 5

The weekly sales S of a product during an advertising campaign are given by

$$S = \frac{100t}{t^2 + 100}, \qquad 0 \le t \le 20,$$

where t is the number of weeks since the beginning of the campaign and S is in thousands of dollars.
(a) Over what interval are sales increasing? decreasing?
(b) What is the maximum weekly sales?
(c) Sketch the graph for $0 \le t \le 20$.

Solution
(a) To find where S is increasing, we first find $S'(t)$.

$$S'(t) = \frac{(t^2 + 100)100 - (100t)2t}{(t^2 + 100)^2}$$

$$= \frac{10{,}000 - 100t^2}{(t^2 + 100)^2}$$

We see that $S'(t) = 0$ when $10{,}000 - 100t^2 = 0$, or

$$100(100 - t^2) = 0$$
$$(10 + t)(10 - t) = 0$$
$$t = -10 \text{ or } t = 10.$$

Because $S'(t)$ is never undefined ($t^2 + 100 \ne 0$ for any real t) and because $0 \le t \le 20$, our only critical value is $t = 10$. Testing $S'(t)$ to the left and right of $t = 10$ gives

left of $t = 10$: $S'(9) = \dfrac{10{,}000 - 8100}{(81 + 100)^2} > 0,$

so S is increasing on the interval $[0, 10)$, and

right of $t = 10$: $S'(11) = \dfrac{10{,}000 - 12{,}100}{(121 + 100)^2} < 0,$

so S is decreasing on the interval $(10, 20]$.

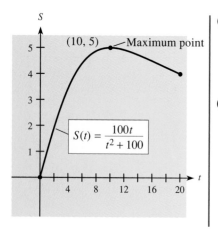

$$S(t) = \frac{100t}{t^2 + 100}$$

Figure 10.10

Figure 10.11

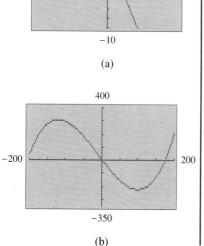

(a)

(b)

(b) Because S is increasing to the left of $t = 10$ and S is decreasing to the right of $t = 10$, the maximum value of S occurs at $t = 10$ and is

$$S = S(10) = \frac{100(10)}{10^2 + 100} = \frac{1000}{200} = 5 \text{ (thousand dollars).}$$

(c) The sign graph for S' on $[0, 20]$ is

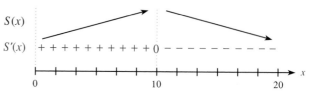

Plotting some additional points gives the graph, as shown in Figure 10.10.

Graphing Utilities

With a graphing utility, choosing an appropriate window is the key to understanding the graph of a function. We saw that the derivative can be used to determine the critical values of a function and hence the interesting points of its graph. Therefore, the derivative can be used to determine the viewing window that provides an accurate representation of the graph. The derivative can also be used to discover graphical behavior that might be overlooked in a graph with a standard window.

EXAMPLE 6

Find the critical points for $f(x) = 0.0001x^3 + 0.003x^2 - 3.6x + 5$. Use them to determine an appropriate viewing window, then sketch the graph.

Solution

Suppose that we first graph this function using a standard viewing window. The graph of this function, for $-10 \leq x \leq 10$, is shown in Figure 10.11(a). The graph looks like a line in this window, but the function is not linear. We could explore the function by tracing or zooming, but using the critical points is more helpful in graphing the function. We begin by finding $f'(x)$.

$$f'(x) = 0.0003x^2 + 0.006x - 3.6$$

Solve $f'(x) = 0$ to find critical values.

$$0 = 0.0003x^2 + 0.006x - 3.6$$
$$0 = 0.0003(x^2 + 20x - 12000)$$
$$0 = 0.0003(x + 120)(x - 100)$$
$$x = -120 \quad \text{or} \quad x = 100$$

To find critical points, we use these x-values in the original function.

$f(-120) = 0.0001(-120)^3 + 0.003(-120)^2 - 3.6(-120) + 5 = 307.4$,

so $(-120, 307.4)$ is a critical point.

$f(100) = 0.0001(100)^3 + 0.003(100)^2 - 3.6(100) + 5 = -225$,

so $(100, -225)$ is a critical point.

From these critical points we see that not all features of this graph are apparent unless the x-range includes the region from $x = -120$ to $x = 100$, and the y-range includes the region from $y = -225$ to $y = 307.4$. While there are many suitable viewing windows, an x-range containing values outside those mentioned above will be best. A graph of the function with x-range -200 to 200 and y-range from -350 to 400 produces the graph of $y = f(x)$ shown in Figure 10.11(b).

From the figure, we see that $(-120, 307.4)$ is a relative maximum and $(100, -225)$ is a relative minimum. Just as we have used the derivative of a function to learn about the function, we can use the graph of $y = f(x)$ in Figure 10.11 to learn about $f'(x)$. The following table shows the relationship between $f(x)$ and $f'(x)$.

Interval	$x < -120$	$-120 < x < 100$	$x > 100$
$f(x)$	increasing	decreasing	increasing
$f'(x)$	positive	negative	positive

The following example shows how the graph of a derivative of a function can be used to obtain information about the graph of the function.

EXAMPLE 7

Use the graph of

$$f'(x) = \frac{1}{3}(x^4 - 5x^3 + x^2 + 21x - 18)$$

to determine the following.
(a) Find critical values for f.
(b) Describe the intervals where f is increasing and where it is decreasing.
(c) Locate the relative maxima, relative minima, and horizontal points of inflection.

Solution

From the graph of f', shown in Figure 10.12, we can observe that

$f' = 0$ at its x-intercepts ($x = -2, 1, 3$),

$f' > 0$ when the graph is above the x-axis ($x < -2, x > 1, x \neq 3$), and

$f' < 0$ when the graph is below the x-axis ($-2 < x < 1$).

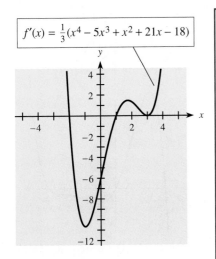

$f'(x) = \frac{1}{3}(x^4 - 5x^3 + x^2 + 21x - 18)$

Figure 10.12

Thus, using the information about f' from Figure 10.12, we can create the following sign diagram for the function f.

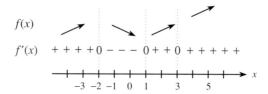

From this sign diagram, we can answer (a), (b), and (c).

(a) Critical values of f occur at $x = -2, 1, 3$.

(b) f is increasing for $x < -2$ and $x > 1$ (except when $x = 3$).
    f is decreasing for $-2 < x < 1$.

(c) f has a relative maximum at $x = -2$.
    f has a relative minimum at $x = 1$.
    f has a horizontal point of inflection at $x = 3$.

Putting together this information, we can sketch a graph that could be the graph of $y = f(x)$, shown in Figure 10.13(a). Figure 10.13(b) shows the graph of

$$y = \frac{x^5}{15} - \frac{5x^4}{12} + \frac{x^3}{9} + \frac{7x^2}{2} - 6x,$$

which is one function that has $f'(x)$ as its derivative. Notice that both graphs in Figure 10.13 satisfy the conditions described in the sign diagram above.

Figure 10.13

(a)

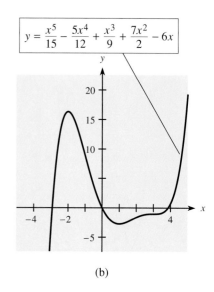

(b)

CHECKPOINT
SOLUTIONS

1. $f'(x) = 0$ or $f'(x)$ is undefined.

2. (a) True, $f(x)$ is increasing when $f'(x) > 0$.
   (b) False. There may be a horizontal point of inflection at $x = -2$ (see Figure 10.8 on page 691).
   (c) False. A relative maximum occurs at $x = -3$.

3. (a) False. Critical values are solutions to $3 - 3x^2 = 0$ or $x = 1$ and $x = -1$.
   (b) False, y-coordinates of critical points come from $f(x) = 7 + 3x - x^3$. Thus critical points are $(1, 9)$ and $(-1, 5)$.

4.

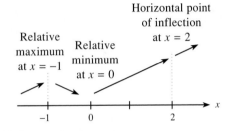

Relative maximum at $x = -1$

Relative minimum at $x = 0$

Horizontal point of inflection at $x = 2$

EXERCISE 10.1

For each function and graph in problems 1–4,

(a) find $\dfrac{dy}{dx}$,

(b) use $\dfrac{dy}{dx}$ to find the critical values,

(c) find the critical points, and

(d) use the graph to classify each critical point as a relative maximum, relative minimum, or horizontal point of inflection.

1. $y = x^3 - 3x + 4$ 2. $y = x - \frac{1}{3}x^3$

3. $y = x^3 + 3x^2 + 3x - 2$ 4. $y = x^3 - 6x^2 + 12x + 1$

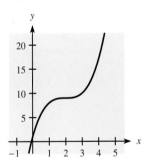

For each function in problems 5–10,

(a) find $\dfrac{dy}{dx}$,

(b) find critical values,   (c) find critical points,

(d) find intervals of x-values within which the function is increasing and decreasing, and

(e) classify the critical points as relative maxima, relative minima, or horizontal points of inflection. In each case, you may check your conclusions with a graphing utility.

5. $y = \frac{1}{2}x^2 - x$ 6. $y = x^2 + 4x$

7. $y = \dfrac{x^3}{3} + \dfrac{x^2}{2} - 2x + 1$ 8. $y = \dfrac{x^4}{4} - \dfrac{x^3}{3} - 2$

9. $y = x^{2/3}$ 10. $y = -(x - 3)^{2/3}$

In problems 11–18, both a function and its derivative are given. Use them to find critical values, critical points, intervals where the function is increasing and decreasing, relative maxima, relative minima, and horizontal points of inflection; sketch the graph of each function.

11. $y = (x^2 - 2x)^2$

     $\dfrac{dy}{dx} = 4x(x - 1)(x - 2)$

12. $y = x(x - 1)^2$

     $\dfrac{dy}{dx} = (x - 1)(3x - 1)$

13. $f(x) = x(x - 2)^3$

     $f'(x) = 2(2x - 1)(x - 2)^2$

14. $f(x) = (x^2 - 4)^2$

     $f'(x) = 4x(x + 2)(x - 2)$

15. $y = \dfrac{x^3(x - 5)^2}{27}$

     $\dfrac{dy}{dx} = \dfrac{5x^2(x - 3)(x - 5)}{27}$

16. $y = \dfrac{x^2(x - 5)^3}{27}$

     $\dfrac{dy}{dx} = \dfrac{5x(x - 2)(x - 5)^2}{27}$

17. $f(x) = x^{2/3}(x - 5)$

     $f'(x) = \dfrac{5(x - 2)}{3x^{1/3}}$

18. $f(x) = x - 3x^{2/3}$

     $f'(x) = \dfrac{x^{1/3} - 2}{x^{1/3}}$

For each function in problems 19–26, find the relative maxima, relative minima, horizontal points of inflection, and sketch the graph. You may check your graph with a graphing utility.

19. $y = \frac{1}{3}x^3 - x^2 + x + 1$

20. $y = \frac{1}{4}x^4 - \frac{2}{3}x^3 + \frac{1}{2}x^2 - 2$

21. $y = \frac{1}{3}x^3 + x^2 - 24x + 20$

22. $C(x) = x^3 - \frac{3}{2}x^2 - 18x + 5$

23. $y = x^4 - 4x^3$ 24. $y = x^3(x - 1)$

25. $y = 3x^5 - 5x^3 + 1$ 26. $y = \frac{1}{6}x^6 - 3x^4 + 7$

For each function and graph in problems 27–30,
(a) use the graph to identify x-values for which $y' > 0$, $y' < 0$, $y' = 0$, and y' does not exist, and
(b) use the derivative to check your conclusions.

27. $y = 6 - x - x^2$ 28. $y = \frac{1}{2}x^2 - 4x + 1$

   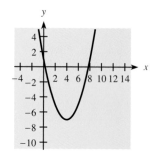

29. $y = 6 + x^3 - \frac{1}{15}x^5$ 30. $y = 0.002x^5 + 0.04x^4 + 0.2x^3 + 8$

In problems 31–34, use the derivative to locate critical points and determine an appropriate viewing window. Use a graphing utility to sketch a complete graph.

31. $f(x) = 0.001x^3 - 0.225x^2 + 15x - 12$

32. $f(x) = 0.0001x^3 - 0.0015x^2 - 1.68x + 8$

33. $f(x) = 0.00025x^4 + 0.04x^3 + 1.8x^2 - 10$

34. $f(x) = 0.00025x^4 - 0.06x^3 + 4.05x^2 - 15$

In problems 35–38,
(a) use a graphing utility to graph each function in the standard window,
(b) use the derivative to precisely locate and classify critical points and discover any hidden behavior,
(c) adjust the viewing window, if necessary, to show all features of the graph.

35. $y = 0.6\left(\dfrac{10x^3}{3} - \dfrac{19x^2}{2} + 9x - \dfrac{20}{3}\right)$

36. $y = \dfrac{10x^3}{3} + \dfrac{39x^2}{2} + 38x + 20$

37. $y = 7.5x^4 - x^3 + 2$ 38. $y = 2 - x^3 - 7.5x^4$

In each of problems 39–42, a graph of $f'(x)$ is given. Use the graph to determine the critical values of $f(x)$, where $f(x)$ is increasing, where it is decreasing, and where it has relative maxima, relative minima, and horizontal points inflection. In each case sketch a possible graph for $f(x)$ that passes through $(0, 0)$.

39. $f'(x) = x^2 - x - 2$ 40. $f'(x) = 4x - x^2$

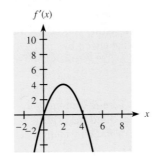

41. $f'(x) = x^3 - 3x^2$ 42. $f'(x) = x(x - 2)^2$

In problems 43 and 44, two graphs are given. One is the graph of f and the other is the graph of f'. Decide which is which and explain your reasoning.

43.

44.

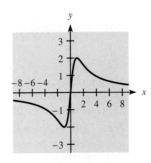

APPLICATIONS

45. **Advertising and sales** Suppose that the daily sales (in dollars) t days after the end of an advertising campaign are given by

$$S = 1000 + \frac{400}{t + 1}, \quad t \geq 0.$$

Does S increase for all $t \geq 0$, decrease for all $t \geq 0$, or change direction at some point in time?

46. **Pricing and sales** Suppose that a chain of auto service stations, Quick-Oil, Inc., has found that their monthly sales volume y (in thousands of dollars) is related to the price p (in dollars) of an oil change by

$$y = \frac{90}{\sqrt{p + 5}}, \quad p > 10.$$

Is y increasing or decreasing for all values of $p > 10$?

47. **Productivity** A time study showed that, on average, the productivity of a worker after t hours on the job can be modeled by

$$P(t) = 27t + 6t^2 - t^3, \quad 0 \leq t \leq 8,$$

where P is the number of units produced per hour.
(a) Find the critical values for this function.
(b) Which critical value makes sense in this model?
(c) For what values of t is P increasing?
(d) Graph the function for $0 \leq t \leq 8$.

48. **Production** Analysis of daily output of a factory shows that, on average, the number of units per hour y produced after t hours of production is

$$y = 70t + \tfrac{1}{2}t^2 - t^3, \quad 0 \leq t \leq 8.$$

(a) Find the critical values for this function.
(b) Which critical values make sense in this problem?
(c) For what values of t, for $0 \leq t \leq 8$, is y increasing?
(d) Graph this function.

49. **Production costs** Suppose that the total cost of producing a shipment of a certain product is

$$C = 5000x + \frac{125,000}{x}, \qquad x > 0,$$

where x is the number of machines used in production.
(a) Find the critical values for this function.
(b) Over what interval does the total cost decrease?
(c) Over what interval does the total cost increase?

50. **Total costs** Suppose the total costs of a mining operation depend on the number of machines used, and total
costs are given by

$$C(x) = 2900x + \frac{1,278,900}{x}, \qquad x > 0,$$

where x is the number of machines used.
(a) Find the critical values for $C(x)$ that lie in the
domain of the problem.
(b) Over what interval in the domain do total costs
decrease?
(c) Over what interval in the domain do total costs
increase?
(d) How many machines give minimum total costs?
(e) What are the minimum total costs?

51. **Marginal revenue** Suppose the weekly marginal revenue function for selling x units of a product is given by
the graph in the figure.

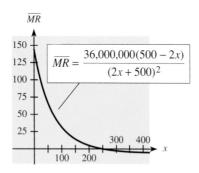

$$\overline{MR} = \frac{36,000,000(500 - 2x)}{(2x + 500)^2}$$

(a) At each of $x = 150$, $x = 250$, and $x = 350$, what is
happening to revenue?
(b) Over what interval is revenue increasing?
(c) How many units must be sold to maximize revenue?

52. **Earnings** Suppose that the rate of change $f'(x)$ of the
average annual earnings of new car salespersons is
shown in the figure.
(a) If a, b, and c represent certain years, what is happening to $f(x)$, the average annual earnings of the
salespersons, at a, b, and c?

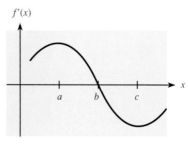

(b) Over what interval (involving a, b, or c) is there an
increase in $f(x)$, the average annual earnings of the
salespersons?

53. **Revenue** A recently released film has its weekly revenue given by

$$R(t) = \frac{50t}{t^2 + 36}, \qquad t \geq 0,$$

where R is in millions of dollars and t is in weeks.
(a) Find the critical values.
(b) For how many weeks will weekly revenue increase?

54. **Medication** Suppose that the concentration C of a medication in the bloodstream t hours after an injection is
given by

$$C(t) = \frac{0.2t}{t^2 + 1}.$$

(a) Determine the number of hours before C attains its
maximum.
(b) Find the maximum concentration.

55. **Candidate recognition** Suppose that the proportion P
of voters who recognize a candidate's name t months
after the start of the campaign is given by

$$P(t) = \frac{13t}{t^2 + 100} + 0.18.$$

(a) How many months after the start of the campaign is
recognition at its maximum?
(b) To have greatest recognition on November 1, when
should a campaign be launched?

56. **Medication** The number of milligrams x of a medication in the bloodstream t hours after taking a dosage can
be modeled by

$$x(t) = \frac{2000t}{t^2 + 16}.$$

(a) For what t-values is x increasing?
(b) Find the t-value at which x is maximum.
(c) Find the maximum value for x.

Poverty in the United States The following table shows the number of people in the United States who lived below the poverty level for selected years between 1960 and 1988. Assume that these data can be modeled with the function $p(x) = 1184.01 - 43.163x + 0.5296x^2 - 0.00214x^3$, where x is the number of years past 1900. Use these data and model for problems 57 and 58.

Year	Number of People Below the Poverty Level (in Millions)
1960	39.9
1965	33.2
1970	25.4
1973	24.5
1975	25.9
1980	29.3
1986	32.4
1987	32.5
1988	31.7

SOURCE: *World Almanac*, 1991

57. (a) For which year does the model $p(x)$ indicate that a minimum number of people lived below the poverty level?
    (b) Does your answer to (a) agree with the data in the table?
58. (a) For which intervals of years does the model $p(x)$ predict that the number of people who lived below the poverty level was decreasing?
    (b) Does your answer to (a) agree with the data in the table?

Public debt The interest paid on the public debt of the United States as a percentage of federal expenditures for selected years is shown in the following table.

Assume that the percentage of federal expenditures devoted to payment of interest on the public debt can be modeled with the function

$$d(t) = 0.000593t^3 - 0.0522t^2 + 1.365t + 0.56,$$

where t is the number of years past 1930. Use these data and model for problems 59 and 60.

Year	Interest Paid as a Percentage of Federal Expenditures
1930	0
1940	10.5
1950	13.4
1955	9.4
1960	10.0
1965	9.6
1970	9.9
1975	9.8
1980	12.7
1985	18.9
1990	21.1

SOURCE: Bureau of Public Debt, Department of the Treasury

59. (a) Use the model $d(t)$ to find critical values and critical points.
    (b) Classify these as either relative maxima or relative minima.
    (c) Do the relative extrema provided by the model accurately reflect the data? Explain.
60. (a) Use the model $d(t)$ to find when interest paid is increasing in the interval $0 \le t \le 60$.
    (b) Do the conclusions from (a) agree entirely or in part with the data in the table? Explain.
    (c) Graph both the data and the model to check your conclusions in (a) and (b).

10.2 CONCAVITY; POINTS OF INFLECTION

Objectives

❏ To find points of inflection of graphs of functions
❏ To use the second-derivative test to graph functions

APPLICATION PREVIEW

Suppose that in 1996 a retailer wishes to sell his store, and uses the graph in Figure 10.14 to show how profits have increased since he opened the store and the potential for profit in the future. Can we conclude that profits will continue to grow, or should we be concerned about future earnings?

Notice that although profits are still increasing in 1995, they are increasing more slowly than in previous years. Indeed, they appear to have been growing at a decreasing rate since about 1990. This means that 1990 was the year that the rate of change of profits was maximum, and since then the rate of change of profits has been diminishing. For this reason, 1990 is called the *point of diminishing returns*. We say that this profit curve is **concave down** over the interval from 1990 to 1995. Based on Figure 10.14, it would be unwise to expect a large increase in profit after 1995.

Figure 10.14

Figure 10.15

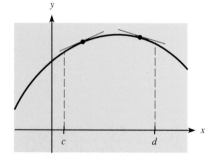

Concave up

(a)

Concave down

(b)

Just as we used the first derivative to determine if a curve was increasing or decreasing on a given interval, we can use the second derivative to determine if the curve is concave up or concave down on an interval.

A curve is said to be **concave up** on an interval $[a, b]$ if at each point on the interval the curve is above its tangent at the point (Figure 10.15 (a)). If the curve is below all its tangents on a given interval, it is **concave down** on the interval (Figure 10.15(b)).

Looking at Figure 10.15(a), we see that the *slopes* of the tangent lines increase over the interval where the graph is concave up. Because $f'(x)$ gives the slopes of those tangents, it follows that $f'(x)$ is increasing over the interval where $f(x)$ is concave up. However, if we know that $f'(x)$ is increasing, then its derivative, $f''(x)$, must be positive. That is, the second derivative is positive if the curve is concave up. Conversely, it can be shown that the graph of a function is concave up if the second derivative is positive.

Similarly, if the second derivative of a function is negative over an interval, the slopes of the tangents to the graph decrease over that interval. This happens when the tangent lines are above the graph, as in Figure 10.15(b), so the graph must be concave down on this interval.

Thus we see that the second derivative can be used to determine the concavity of a curve.

Concave Up and Concave Down	Assume that the first and second derivatives of function f exist. If $f''(x) > 0$ on an interval I, the graph of f is **concave up** on the interval. If $f''(x) < 0$ on an interval I, then the graph of f is **concave down** on I.

EXAMPLE 1

Is the graph of $f(x) = x^3 - 4x^2 + 3$ concave up or down at the point
(a) $(1, 0)$? (b) $(2, -5)$?

Solution
(a) We must find $f''(x)$ before we can answer this question.

$$f'(x) = 3x^2 - 8x \qquad f''(x) = 6x - 8$$

Then $f''(1) = 6(1) - 8 = -2$, so the graph is concave down at $(1, 0)$.
(b) Since $f''(2) = 6(2) - 8 = 4$, the graph is concave up at $(2, -5)$. The graph of $f(x) = x^3 - 4x^2 + 3$ is shown in Figure 10.16(a).

Looking at the graph of $y = x^3 - 4x^2 + 3$ (Figure 10.16(a)), we see that the curve is concave down on the left and concave up on the right. Thus it has changed from concave down to concave up. Figure 10.16(b) shows the graph of $y'' = f''(x) = 6x - 8$, and we can see that $y'' < 0$ for $x < \frac{4}{3}$ and $y'' > 0$ for $x > \frac{4}{3}$. Thus the second derivative changes sign at $x = \frac{4}{3}$, so the concavity of the graph of $y = f(x)$ changes at $x = \frac{4}{3}$, $y = -\frac{47}{27}$. The point where concavity changes is called a **point of inflection.**

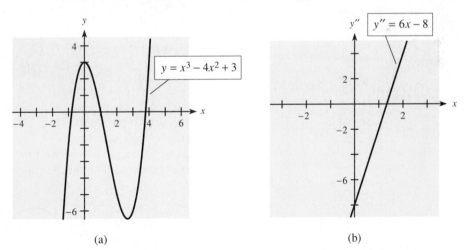

Figure 10.16   (a)   (b)

Point of Inflection	A point (x_0, y_0) on the graph of a function f is called a **point of inflection** if the curve is concave up on one side of the point and concave down on the other side. The second derivative at this point, $f''(x_0)$, will be 0 or undefined.

In general, we can find points of inflection and information about concavity as follows.

Finding Points of Inflection and Concavity

Procedure	Example
To find the point(s) of inflection of a curve and intervals where it is concave upward and where it is concave downward:	Find the points of inflection and concavity of the graph of $y = \dfrac{x^4}{2} - x^3 + 5$.
1. Find the second derivative of the function.	1. $y' = f'(x) = 2x^3 - 3x^2$ $y'' = f''(x) = 6x^2 - 6x$
2. Set the second derivative equal to 0, and solve for x. Potential points of inflection occur at these values of x or at values of x where $f(x)$ is defined and $f''(x)$ is undefined.	2. $0 = 6x^2 - 6x = 6x(x - 1)$ has solutions $x = 0, x = 1$. $f''(x)$ is defined everywhere.
3. Find the potential points of inflection.	3. $(0, 5)$ and $(1, \frac{9}{2})$ are potential points of inflection.
4. If the second derivative has opposite signs on the two sides of one of these values of x, a point of inflection occurs.	4. $\left.\begin{array}{l} f''(-1) = 12 > 0 \\ f''(0) = 0 \\ f''(\frac{1}{2}) = -\frac{3}{2} < 0 \end{array}\right\} \Rightarrow \begin{array}{l}(0, 5) \text{ is a point of} \\ \text{inflection.}\end{array}$ $\left.\begin{array}{l} f''(1) = 0 \\ f''(2) = 12 > 0 \end{array}\right\} \Rightarrow \begin{array}{l}(1, \frac{9}{2}) \text{ is a point of} \\ \text{inflection.}\end{array}$ See the graph in Figure 10.17.

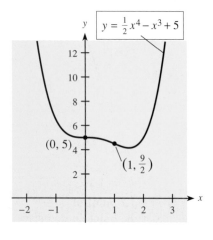

The graph of $y = \frac{1}{2}x^4 - x^3 + 5$ is shown in Figure 10.17. Note the points of inflection at $(0, 5)$ and $(1, \frac{9}{2})$. The point of inflection at $(0, 5)$ is a horizontal point of inflection because $f'(x)$ is also 0 at $x = 0$. A **sign diagram for $f''(x)$,** the second derivative of this function, is shown below. The changes in the sign of $f''(x)$ correspond to changes in concavity and occur at points of inflection.

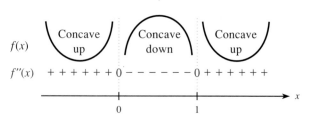

Figure 10.17

EXAMPLE 2

Suppose that a real estate developer wishes to remove pollution from a small lake so he can sell lakefront homes on a "crystal clear" lake. The graph in Figure 10.18 shows the relation between dollars spent on cleaning the lake

Figure 10.18

and the purity of the water. The point of inflection on the graph is called the **point of diminishing returns** on his investment because it is where the *rate* of return on his investment changes from increasing to decreasing. Show that the rate of change in the purity of the lake, $f'(x)$, is maximized at this point, $x = c$. (Assume that $f(c)$, $f'(c)$, and $f''(c)$ are defined.)

Solution

Because $x = c$ is a point of inflection for $f(x)$, we know that the concavity must change at $x = c$. From the figure we see the following.

$$x < c: \quad f(x) \text{ is concave up, so } f''(x) > 0;$$
$$f''(x) > 0 \text{ means that } f'(x) \text{ is increasing.}$$
$$x > c: \quad f(x) \text{ is concave down, so } f''(x) < 0;$$
$$f''(x) < 0 \text{ means that } f'(x) \text{ is decreasing.}$$

Thus $f'(x)$ has $f'(c)$ as its relative maximum.

EXAMPLE 3

Suppose that the daily sales (in thousands of dollars) of a product is given by

$$S = \frac{(-x^3 + 9x^2 + 6)}{6},$$

where x is thousands of dollars spent on advertising. Find the point of diminishing returns for money spent on advertising.

Figure 10.19

(a)

(b)

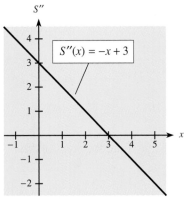

(c)

Solution

We seek the point where the graph of this function changes from concave up to concave down, if such a point exists.

$$\frac{dS}{dx} = S'(x) = \frac{1}{6}(-3x^2 + 18x)$$

$$S''(x) = \frac{1}{6}(-6x + 18) = -x + 3$$

$$S''(x) = 0 \quad \text{when} \quad 0 = -x + 3 \quad \text{or} \quad x = 3.$$

Thus $x = 3$ is a possible point of inflection. We test $S''(x)$ to the left and right of $x = 3$.

$$S''(2) = 1 > 0 \Rightarrow \text{concave up to the left of } x = 3.$$
$$S''(4) = -1 < 0 \Rightarrow \text{concave down to the right of } x = 3.$$

Thus the point of diminishing returns occurs when $x = 3$ (thousand dollars) and $S = 10$ (thousand dollars). Figure 10.19 shows the graphs of S, S', and S''. At $x = 3$, we can see that the point of diminishing returns on the graph of S corresponds to the maximum point of the graph of S' and the zero (or x-intercept) of the graph of S''.

We can use information about points of inflection and concavity to help sketch graphs. For example, if we know that the curve is concave up at a critical point where $f'(x) = 0$, then the point must be a relative minimum because the tangent to the curve is horizontal at the critical point, and only a point at the bottom of a "concave up" curve could have a horizontal tangent (see Figure 10.20(a)).

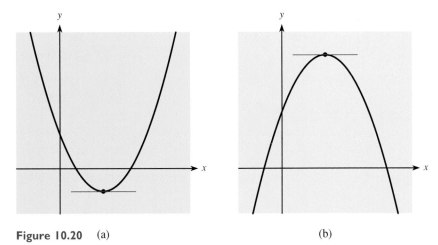

Figure 10.20   (a)   (b)

On the other hand, if the curve is concave down at a critical point where $f'(x) = 0$, then the point is a relative maximum (see Figure 10.20(b)).

Thus we can use the **second-derivative test** to determine if a critical point where $f'(x) = 0$ is a relative maximum or minimum.

Second-Derivative Test	
Procedure	**Example**
To find relative maxima and minima of a function:	Find the relative maxima and minima of $y = f(x) = \frac{1}{3}x^3 - x^2 - 3x + 2$.
1. Find the critical values of the function.	1. $f'(x) = x^2 - 2x - 3$ $0 = (x - 3)(x + 1)$ has solutions $x = -1$ and $x = 3$. No values of x make $x^2 - 2x - 3$ undefined.
2. Substitute the critical values into $f(x)$ to find the critical points.	2. $f(-1) = \frac{11}{3}$ $f(3) = -7$ The critical points are $(-1, \frac{11}{3})$ and $(3, -7)$.
3. Evaluate $f''(x)$ at each critical value for which $f'(x) = 0$. (a) If $f''(x_0) < 0$, a relative maximum occurs at x_0. (b) If $f''(x_0) > 0$, a relative minimum occurs at x_0. (c) If $f''(x_0) = 0$, or $f''(x_0)$ is undefined, the second-derivative test fails; use the first-derivative test.	3. $f''(x) = 2x - 2$ $f''(-1) = 2(-1) - 2 = -4 < 0$, so $(-1, \frac{11}{3})$ is a relative maximum point. $f''(3) = 2(3) - 2 = 4 > 0$, so $(3, -7)$ is a relative minimum point. (The graph is shown in Figure 10.21.)

CHECKPOINT

Figure 10.21

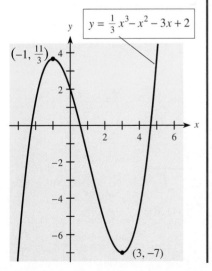

1. If $f''(x) > 0$, then $f(x)$ is concave _____ .
2. Where do possible points of inflection occur?
3. On the graph below, locate any points of inflection (approximately) and label where the curve satisfies $f''(x) > 0$ and $f''(x) < 0$.

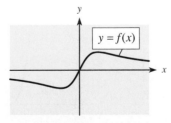

4. Determine whether the following are true or false.
   (a) Let $f(x) = x^4$. Then $f''(x) = 12x^2$ and $f''(0) = 0$ and $f(x)$ has a point of inflection at $x = 0$.
   (b) Let $f(x) = x^5$. Then $f''(x) = 20x^3$ and $f''(0) = 0$ and $f(x)$ has a point of inflection at $x = 0$.

EXAMPLE 4

Find the relative maxima and minima and points of inflection of
$y = 3x^4 - 4x^3$.

Solution

$$y' = f'(x) = 12x^3 - 12x^2$$

Solving $0 = 12x^3 - 12x^2 = 12x^2(x - 1)$ gives $x = 1$ and $x = 0$. Thus
the critical points are $(1, -1)$ and $(0, 0)$.

$$y'' = f''(x) = 36x^2 - 24x$$
$$f''(1) = 12 > 0 \Rightarrow (1, -1) \text{ is a relative minimum point.}$$
$$f''(0) = 0 \Rightarrow \text{the second-derivative test fails.}$$

Since the second-derivative test fails, we must use the first-derivative test
at the critical point $(0, 0)$.

$$\left.\begin{array}{l} f'(-1) = -24 < 0 \\ f'(\tfrac{1}{2}) = -\tfrac{3}{2} < 0 \end{array}\right\} \Rightarrow (0, 0) \text{ is a horizontal point of inflection.}$$

We look for points of inflection by setting $f''(x) = 0$ and solving for x. We
find that $0 = 36x^2 - 24x$ has solutions $x = 0$ and $x = \tfrac{2}{3}$.

$$\left.\begin{array}{l} f''(-1) = 60 > 0 \\ f''(\tfrac{1}{2}) = -3 < 0 \end{array}\right\} \Rightarrow (0, 0) \text{ is a point of inflection.}$$

Thus we see again that $(0, 0)$ is a horizontal point of inflection. This is a
special point, where the curve changes concavity *and* has a horizontal tangent (see Figure 10.22). Testing for concavity on either side of $x = \tfrac{2}{3}$ gives

$$\left.\begin{array}{l} f''(\tfrac{1}{2}) = -3 < 0 \\ f''(1) = 12 > 0 \end{array}\right\} \Rightarrow (\tfrac{2}{3}, -\tfrac{16}{27}) \text{ is a point of inflection.}$$

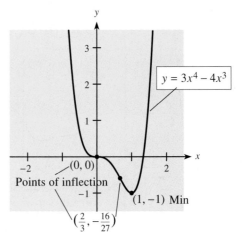

Figure 10.22

Graphing Utilities

We can use graphing utilities to explore the relationship among f, f', and f'', as we did in the previous section for f and f'.

Figure 10.23

Figure 10.24

Figure 10.25

EXAMPLE 5

Figure 10.23 shows the graph of $f(x) = \frac{1}{6}(2x^3 - 3x^2 - 12x + 12)$.
(a) From the graph, identify points where $f''(x) = 0$.
(b) From the graph, observe intervals where $f''(x) > 0$ and where $f''(x) < 0$.
(c) Check the conclusions from parts (a) and (b) by calculating $f''(x)$ and graphing it.

Solution

(a) From Figure 10.23, we can make an initial estimate of the x-value of the point of inflection. It appears to be near $x = \frac{1}{2}$, so we expect $f''(x) = 0$ at (or very near to) $x = \frac{1}{2}$.
(b) We see that the graph is concave downward (so $f''(x) < 0$) to the left of the point of inflection. That is, $f''(x) < 0$ when $x < \frac{1}{2}$. Similarly, $f''(x) > 0$ when $x > \frac{1}{2}$.
(c) $f(x) = \frac{1}{6}(2x^3 - 3x^2 - 12x + 12)$
$f'(x) = \frac{1}{6}(6x^2 - 6x - 12) = x^2 - x - 2$
$f''(x) = 2x - 1$
Thus $f''(x) = 0$ when $x = \frac{1}{2}$.
Figure 10.24 shows the graph of $f''(x) = 2x - 1$. We see that the graph crosses the x-axis ($f''(x) = 0$) when $x = \frac{1}{2}$, is below the x-axis ($f''(x) < 0$) when $x < \frac{1}{2}$, and is above the x-axis ($f''(x) > 0$) when $x > \frac{1}{2}$. This verifies our conclusions from parts (a) and (b).

EXAMPLE 6

Figure 10.25 shows the graph of $f'(x) = -x^2 - 2x$. Use the graph of $f'(x)$ to:
(a) Find intervals where $f(x)$ is concave downward and where it is concave upward.
(b) Find x-values where $f(x)$ has a point of inflection.
(c) Check the conclusions from parts (a) and (b) by finding $f''(x)$ and graphing it.
(d) For $f(x) = \frac{1}{3}(9 - x^3 - 3x^2)$, calculate $f'(x)$ to verify that this could be $f(x)$.

Solution

(a) Concavity for $f(x)$ can be found from the sign of $f''(x)$. Because $f''(x)$ is the first derivative of $f'(x)$, wherever the graph of $f'(x)$ is increasing, it follows that $f''(x) > 0$. Thus, $f''(x) > 0$ and $f(x)$ is concave upward when $x < -1$. Similarly, $f''(x) < 0$, and $f(x)$ is concave downward when $f'(x)$ is decreasing, that is, when $x > -1$.

Figure 10.26

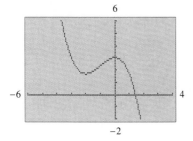

Figure 10.27

(b) From part (a) we know that $f''(x)$ changes sign at $x = -1$, so $f(x)$ has a point of inflection at $x = -1$. Notice that $f'(x)$ has its maximum at the x-value where $f(x)$ has a point of inflection. In fact, points of inflection for $f(x)$ will correspond to relative extrema for $f'(x)$.

(c) For $f'(x) = -x^2 - 2x$, we have $f''(x) = -2x - 2$. Figure 10.26 shows the graph of $y = f''(x)$ and verifies our conclusions from parts (a) and (b).

(d) If $f(x) = \frac{1}{3}(9 - x^3 - 3x^2)$, then $f'(x) = \frac{1}{3}(-3x^2 - 6x) = -x^2 - 2x$. Figure 10.27 shows the graph of $f(x) = \frac{1}{3}(9 - x^3 - 3x^2)$. Notice that the point of inflection and concavity correspond to what we discovered in parts (a) and (b).

The relationships among $f(x)$, $f'(x)$, and $f''(x)$ that we explored in Example 6 can be summarized as follows.

$f(x)$	Concave Upward	Concave Downward	Point of Inflection	
$f'(x)$	increasing	decreasing	maximum	minimum
$f''(x)$	positive (+)	negative (−)	(+) to (−)	(−) to (+)

CHECKPOINT SOLUTIONS

1. up

2. Possible points of inflection occur where $f''(x) = 0$ or $f''(x)$ is undefined.

3. Points of inflection at A, B, and C
   $f''(x) < 0$ to the left of A and between B and C
   $f''(x) > 0$ between A and B and to the right of C

4. (a) False, $f''(x) = 12x^2$ does not change sign from $x < 0$ to $x > 0$.
   (b) True, $f''(x) = 20x^3$ satisfies $f''(x) < 0$ for $x < 0$ and $f''(x) > 0$ for $x > 0$.

EXERCISE 10.2

In problems 1–4, determine whether each function is concave upward or concave downward at the indicated points.

1. $f(x) = x^3 - 3x^2 + 1$ at (a) $x = -2$ (b) $x = 3$

2. $f(x) = x^3 + 6x - 4$ at (a) $x = -5$ (b) $x = 7$

3. $y = 2x^3 + 4x - 8$ at (a) $x = -1$ (b) $x = 4$

4. $y = 4x^3 - 3x^2 + 2$ at (a) $x = 0$ (b) $x = 5$

In problems 5–12, a function and its graph are given. Use the second derivative to determine intervals where the function is concave upward, intervals where it is concave downward, and to locate points of inflection. Check these results against the graph shown.

5. $f(x) = x^3 - 6x^2 + 5x + 6$

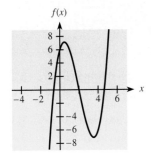

6. $f(x) = 2x^3 - 4x^2 + 5x - 2$

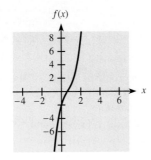

7. $f(x) = \frac{1}{4}x^4 + \frac{1}{2}x^3 - 3x^2 + 3$

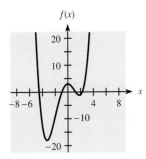

8. $f(x) = \frac{1}{100}(\frac{1}{4}x^4 + 2x^3 - 48x^2 + 8x + 400)$

9. $y = x^3 - 9x^2$

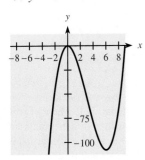

10. $y = 2x^4 - 6x^2 + 4$

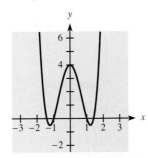

11. $y = \dfrac{4x}{x^2 + 1}$

12. $y = \dfrac{8x}{x^2 + 4}$

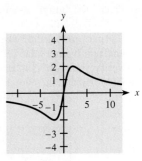

In problems 13–18, a function and its first and second derivatives are given. Use these to find critical values, relative maxima, relative minima, and points of inflection; sketch the graph of each function.

13. $p = 3q^4 - 4q^3 + 1$

$$\frac{dp}{dq} = 12q^2(q - 1)$$

$$\frac{d^2p}{dq^2} = 12q(3q - 2)$$

14. $f(w) = 4w^3 - 6w^2 - 24w$

$f'(w) = 12(w - 2)(w + 1)$

$f''(w) = 12(2w - 1)$

15. $f(x) = 3x^5 - 20x^3$

$f'(x) = 15x^2(x - 2)(x + 2)$

$f''(x) = 60x(x^2 - 2)$

16. $f(x) = x^5 - 5x^4$

$f'(x) = 5x^3(x - 4)$

$f''(x) = 20x^2(x - 3)$

17. $y = x^{1/3}(x - 4)$

$y' = \dfrac{4(x - 1)}{3x^{2/3}}$

$y'' = \dfrac{4(x + 2)}{9x^{5/3}}$

18. $y = x^{4/3}(x - 7)$

$y' = \dfrac{7x^{1/3}(x - 4)}{3}$

$y'' = \dfrac{28(x - 1)}{9x^{2/3}}$

Find the relative maxima, relative minima, and points of inflection, and sketch the graph of the following functions.

19. $y = x^2 - 4x + 2$ 20. $y = x^3 - x^2$

21. $y = x^3 - 3x^2 + 6$ 22. $y = x^4 - 8x^3 + 16x^2$

23. $y = x^4 - 16x^2$ 24. $y = \frac{1}{3}x^3 - 2x^2 + 3x + 2$

In problems 25 and 26, a function and its graph are given.
(a) From the graph estimate where $f''(x) > 0$, where $f''(x) < 0$, and where $f''(x) = 0$.
(b) Use part (a) to decide where $f'(x)$ has its relative maxima and relative minima.
(c) Verify your results in parts (a) and (b) by finding $f'(x)$ and $f''(x)$ and then graphing each with a graphing utility.

25. $f(x) = -\frac{1}{3}x^3 + x^2 + 8x - 12$

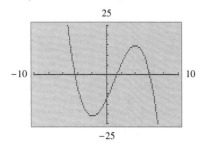

26. $f(x) = \frac{1}{3}x^3 + 2x^2 - 12x - 20$

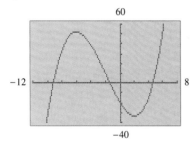

In problems 27 and 28, $f'(x)$ and its graph are given. Use the graph of $f'(x)$ to determine the following.
(a) Where is the graph of $f(x)$ concave upward and concave downward?
(b) Where does $f(x)$ have any points of inflection?
(c) Find $f''(x)$ and graph it, then use that graph to check your conclusions from (a) and (b).
(d) Sketch a possible graph for $f(x)$.

27. $f'(x) = 4x - x^2$

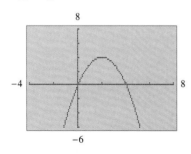

28. $f'(x) = x^2 - x - 2$

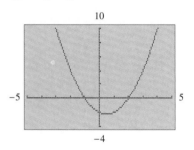

In problems 29 and 30, use the graph shown in Figure 10.28 and identify points from A through I that satisfy the given conditions.

29. (a) $f'(x) > 0$ and $f''(x) > 0$
    (b) $f'(x) < 0$ and $f''(x) < 0$
    (c) $f'(x) = 0$ and $f''(x) > 0$
    (d) $f'(x) > 0$ and $f''(x) = 0$
    (e) $f'(x) = 0$ and $f''(x) = 0$

30. (a) $f'(x) > 0$ and $f''(x) < 0$
    (b) $f'(x) < 0$ and $f''(x) > 0$
    (c) $f'(x) = 0$ and $f''(x) < 0$
    (d) $f'(x) < 0$ and $f''(x) = 0$

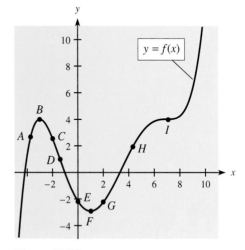

Figure 10.28

In problems 31 and 32 a graph is given. Tell where $f(x)$ is concave upward, concave downward, and has points of inflection on the interval $-2 < x < 2$, if the given graph is the graph of
(a) $f(x)$ (b) $f'(x)$ (c) $f''(x)$.

31.

32.

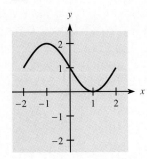

APPLICATIONS

33. **Productivity–diminishing returns** The figure below is a typical graph of worker productivity as a function of time on the job.

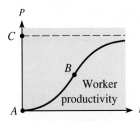

(a) If P represents the productivity and t represents the time, write a symbol that represents the rate of change of productivity with respect to time.
(b) Which of A, B, or C is the critical point for the rate of change found in part (a)? This point actually corresponds to the point at which the rate of production is maximized, or the point for maximum worker efficiency. In economics, this is called the point of diminishing returns.
(c) Which of A, B, or C corresponds to the maximum production?

34. **Population growth** The figure below shows the growth of a population as a function of time.

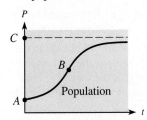

(a) If P represents the population and t represents the time, write a symbol that represents the rate of

change (growth rate) of the population with respect to time.
(b) Which of A, B, or C corresponds to the point at which the growth *rate* attains its maximum?
(c) Which of A, B, or C corresponds to the maximum population?

35. **Advertising and sales** The figure below shows the daily sales volume S as a function of time t since an ad campaign began.

(a) Which of A, B, or C is the point of inflection for the graph?
(b) On which side of C is $d^2S/dt^2 > 0$?
(c) Does the *rate of change* of sales volume attain its minimum at C?

36. **Oxygen purity** The figure below shows the oxygen level P (for purity) in a lake t months after an oil spill.

(a) Which of A, B, or C is the point of inflection for the graph?
(b) On which side of C is $d^2P/dt^2 < 0$?
(c) Does the *rate of change* of purity attain its maximum at C?

37. **Production** Suppose that the total number of units produced by a worker in t hours of an 8-hour shift can be modeled by the **production function** $P(t)$:

$$P(t) = 27t + 12t^2 - t^3.$$

Find the number of hours before
(a) production is maximized.
(b) the rate of production is maximized. That is, find the point of diminishing returns.

38. **Poiseuille's law—velocity of blood** According to Poiseuille's law, the speed S of blood through an artery of radius r at a distance x from the artery wall is given by

$$S = k[r^2 - (r - x)^2],$$

where k is a constant. Find the distance x that maximizes the speed.

39. **Advertising and sales—diminishing returns** Suppose that a company's daily sales volume attributed to an advertising campaign is given by

$$S(t) = \frac{3}{t + 3} - \frac{18}{(t + 3)^2} + 1.$$

How long before
(a) sales volume is maximized?
(b) the rate of change of sales volume is minimized? That is, find the point of diminishing returns.

40. **Oxygen purity—diminishing returns** Suppose that the oxygen level P (for purity) in a body of water t months after an oil spill is given by

$$P(t) = 500\left[1 - \frac{4}{t + 4} + \frac{16}{(t + 4)^2}\right].$$

How long before the
(a) oxygen level reaches its minimum?
(b) rate of change of P is maximized? That is, find the point of diminishing returns.

41. **Poverty in the United States** The following figure shows the number of people in the United States who lived below the poverty level for selected years between 1960 and 1988. The figure also contains the graph of the function used to model these data:

$$p(x) = 1184.01 - 43.136x + 0.5296x^2 - 0.00214x^3,$$

where x is the number of years past 1900.

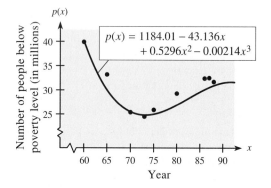

Year	Number of People Below the Poverty Level (in Millions)	Year	Number of People Below the Poverty Level (in Millions)
1960	39.9	1980	29.3
1965	33.2	1986	32.4
1970	25.4	1987	32.5
1973	24.5	1988	31.7
1975	25.9		

SOURCE: *World Almanac,* 1991

The data points show a downturn in the number of people below the poverty line in the late 1980s, but the model does not reflect this. During which year does the model begin to be concave downward, indicating that it may also have a downturn in the future?

42. **Personal savings** The following table gives the personal savings (in billions of dollars) in the United States for selected years.

Year	Personal Savings (Billions)	Year	Personal Savings (Billions)
1980	$153.8	1988	155.7
1985	189.3	1989	152.1
1986	187.5	1990	175.6
1987	142.0	1991	199.6

SOURCE: *Survey of Current Business,* March 1993

Assume that personal savings is modeled by

$$PS(t) = 0.55t^3 - 25.6t^2 + 384t - 1678,$$

where t is the number of years past 1970.
(a) Find a function that models the instantaneous rate of change of personal savings.
(b) Use the model from (a) to find when the rate of personal savings is decreasing, when it is increasing, and when it is maximum.
(c) Find the point of inflection of the function $PS(t)$, and state to what in part (b) it corresponds.

43. **Consumer Price Index** The annual change in the Consumer Price Index (CPI) during certain years is shown in the table below. Assume that the annual change in the Consumer Price Index can be modeled with the function

$$f(t) = -0.0068567913t^4 + 2.32783445t^3 - 295.9642445t^2$$
$$+ 16701.32789t - 352921,$$

where t is the number of years past 1900. A graph of $f(t)$ along with the data in the table is shown in the figure.

Year	Consumer Price Index Annual Percent Change	Year	Consumer Price Index Annual Percent Change
1977	6.5	1985	3.6
1978	7.6	1986	1.9
1979	11.3	1987	3.6
1980	13.5	1988	4.1
1981	10.3	1989	4.8
1982	6.2	1990	5.4
1983	3.2	1991	4.2
1984	4.3		

SOURCE: *The World Almanac and Book of Facts*, 1993

(a) Find $f'(t)$ and $f''(t)$ for the model.
(b) Evaluate $f'(t)$ and $f''(t)$ in 1970, and use those results to describe how the *CPI* is changing (increasing

or decreasing) and also how its rate of change is changing.
(c) Repeat part (b) for 1986.

44. ***Public debt*** The interest paid on the public debt of the United States as a percentage of federal expenditures for selected years is shown in the following table.

Year	Interest Paid as a Percentage of Federal Expenditures	Year	Interest Paid as a Percentage of Federal Expenditures
1930	0	1970	9.9
1940	10.5	1975	9.8
1950	13.4	1980	12.7
1955	9.4	1985	18.9
1960	10.0	1990	21.1
1965	9.6		

SOURCE: Bureau of Public Debt, Department of the Treasury

Assume that the percentage of federal expenditures devoted to payment of interest on the public debt can be modeled with the function

$$d(t) = 0.000593t^3 - 0.0522t^2 + 1.365t + 0.56,$$

where t is the number of years past 1930.
(a) Find $d'(t)$ and $d''(t)$ for the model.
(b) Evaluate $d'(t)$ and $d''(t)$ in 1950, and use the results to describe how interest paid as a percentage of federal expenditures is changing (increasing or decreasing) and how its rate of change is changing.
(c) Repeat part (b) for 1990.

10.3 OPTIMIZATION IN BUSINESS AND ECONOMICS

Objectives

- ❑ To find absolute maxima and minima
- ❑ To maximize revenue, given the total revenue function
- ❑ To minimize the average cost, given the total cost function
- ❑ To find the maximum profit from total cost and total revenue functions, or from a profit function

APPLICATION PREVIEW

Most companies are interested in obtaining the greatest possible profit (instead of just making a profit that is relatively large). Similarly, manufacturers of products are concerned about producing their products for the lowest possible average cost per unit. So rather than just find the relative maxima or relative minima of a function, we will consider where the **absolute maximum** or **absolute minimum** of a function occurs in a given interval. This requires evaluating the function at the endpoints of the given interval as well as finding the relative extrema.

As the name implies, **absolute extrema** are the functional values that are the largest or smallest values over the entire domain of the function (or over the interval of interest).

Absolute Extrema	The value $f(a)$ is the **absolute maximum** for f if $f(a) \geq f(x)$ for all x in the domain of f (or over the interval of interest).
	The value $f(b)$ is the **absolute minimum** for f if $f(b) \leq f(x)$ for all x in the domain of f (or over the interval of interest).

In this section we will discuss how to find the absolute extrema of a function and then use these techniques to solve applications involving revenue, cost, and profit.

Let us begin by considering the graph of $y = (x - 1)^2$, shown in Figure 10.29(a). This graph has a relative minimum at $(1, 0)$. Notice that the relative minimum is the lowest point on the graph. In this case, the point $(1, 0)$ is an **absolute minimum point,** and 0 is the absolute minimum for the function. Similarly, when there is a point that is the highest point on the graph over the domain of the function, we call the point an **absolute maximum point** for the function.

In Figure 10.29(a), we see that there is no relative maximum. However, if the domain of the function is restricted to the interval $[\frac{1}{2}, 2]$, then we get the graph shown in Figure 10.29(b). In this case, there is an absolute maximum of 1 at point $(2, 1)$ and the absolute minimum of 0 is still at $(1, 0)$.

If the domain of $y = (x - 1)^2$ is restricted to the interval $[2, 3]$, the resulting graph is shown in Figure 10.29(c). In this case, the absolute minimum is 1 and occurs at point $(2, 1)$, and its absolute maximum is 4 and occurs at $(3, 4)$.

Figure 10.29

(a)

(b)

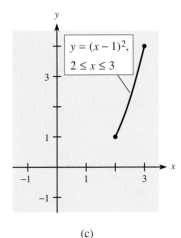

(c)

As the preceding discussion indicates, if the domain of a function is limited, an absolute maximum or minimum may occur at the endpoints of the domain. In testing functions with limited domains for absolute maxima and minima, we must compare the endpoints of the domain with the relative maxima and minima found by taking derivatives. In applications to management, life, and social sciences, a limited domain occurs very often, since many quantities are required to be positive, or at least nonnegative.

MAXIMIZING REVENUE

Since the marginal revenue is the first derivative of the total revenue, it should be obvious that the total revenue function will have a critical point at the point where the marginal revenue equals 0. With the total revenue function $R(x) = 16x - 0.02x^2$, the point where $R'(x) = 0$ is clearly a maximum since $R(x)$ is a parabola that opens downward. But the revenue function may not always be a parabola and the critical point may not always be a maximum, so it is wise to verify that the maximum value occurs at the critical point.

EXAMPLE I

If total revenue for a firm is given by

$$R(x) = 8000x - 40x^2 - x^3,$$

where x is the number of units sold, find the number of units that must be sold to maximize revenue. Find the maximum revenue.

Solution
$R'(x) = 8000 - 80x - 3x^2$, so we must solve $8000 - 80x - 3x^2 = 0$ for x.

$$(40 - x)(200 + 3x) = 0$$
$$40 - x = 0 \qquad 200 + 3x = 0,$$

so

$$x = 40 \quad \text{or} \quad x = -\frac{200}{3}.$$

Now, we reject the negative value for x, but we must verify that $x = 40$ will yield maximum revenue.

$$\left.\begin{array}{l} R'(0) = 8000 > 0 \\ R'(100) = 8000 - 8000 - 30{,}000 < 0 \end{array}\right\} \Rightarrow \text{relative maximum}$$

This test shows that a relative maximum occurs at $x = 40$, giving revenue $R(40) = \$192{,}000$. Because $R(x) < 0$ for all $x > 40$, the revenue function is decreasing to the right of $x = 40$. The revenue at $x = 0$ is $R(0) = 0$, so $R = \$192{,}000$ at $x = 40$ is the (absolute) maximum revenue.

EXAMPLE 2

A travel agency will plan tours for groups of 25 or larger. If the group contains exactly 25 people, the cost is $300 per person. However, each person's cost is reduced by $10 for each additional person above the 25. What size group will produce the largest revenue for the agency?

Solution

The total revenue is

$$R = \text{(number of people)(cost per person)}.$$

If 25 people go, the total revenue will be

$$R = 25 \cdot \$300 = \$7500.$$

But if x additional people go, the number of people will be $25 + x$, and the cost per person will be $(300 - 10x)$ dollars. Then the total revenue will be a function of x,

$$R = R(x) = (25 + x)(300 - 10x),$$

or

$$R(x) = 7500 + 50x - 10x^2.$$

This function will have its maximum where $\overline{MR} = R'(x) = 0$; $R'(x) = 50 - 20x$, and the solution to $0 = 50 - 20x$ is $x = 2.5$. Thus adding 2.5 people to the group should maximize the total revenue. But we cannot add half a person. So we will test the total revenue function for 27 people and 28 people. This will determine the most profitable number because $R(x)$ is concave downward for all x.

For $x = 2$ (giving 27 people) we get $R(2) = 7500 + 50(2) - 10(20)^2 = 7560$. For $x = 3$ (giving 28 people) we get $R(3) = 7500 + 50(3) - 10(3)^2 = 7560$. Note that both 27 and 28 people give the same total revenue and that this revenue is greater than the revenue for 25 people. Thus the revenue is maximized at either 27 or 28 people in the group.

MINIMIZING AVERAGE COST

Since the total cost function is always increasing, we cannot find the number of units that will make the total cost a minimum (except for producing 0 units, which is an absolute minimum). However, we usually can find the number of units that will make the average cost per unit a minimum.

Average Cost   If the total cost is represented by $C = C(x)$, then the **average cost per unit** is

$$\overline{C} = \frac{C(x)}{x}.$$

For example, if $C = 3x^2 + 4x + 2$ is the total cost function for a commodity, the **average cost function** is

$$\overline{C} = \frac{3x^2 + 4x + 2}{x} = 3x + 4 + \frac{2}{x}.$$

Note that the average cost per unit is undefined if no units are produced.

We can use derivatives to find the minimum of the average cost function, as the following example shows.

EXAMPLE 3

If the total cost function for a commodity is given by $C = \frac{1}{4}x^2 + 4x + 100$, where x represents the number of units produced, producing how many units will result in a minimum *average cost* per unit? Find the minimum average cost.

Solution

The average cost function is given by

$$\overline{C} = \frac{\frac{1}{4}x^2 + 4x + 100}{x} = \frac{1}{4}x + 4 + \frac{100}{x}.$$

Then

$$\overline{C}' = \overline{C}'(x) = \frac{1}{4} - \frac{100}{x^2}.$$

Setting $\overline{C}' = 0$ gives

$$0 = \frac{1}{4} - \frac{100}{x^2},$$

$$0 = x^2 - 400, \quad \text{or} \quad x = \pm 20.$$

Since the quantity produced must be positive, 20 units should minimize the average cost per unit. We show it is an absolute minimum by using the second derivative.

$$\overline{C}''(x) = \frac{200}{x^3} \quad \text{so} \quad \overline{C}''(x) > 0 \quad \text{when } x > 0.$$

Thus the minimum average cost per unit occurs if 20 units are produced. The graph of the average cost per unit is shown in Figure 10.30. The minimum average cost per unit is $\overline{C}(20) = \$14$.

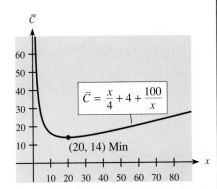

$$\overline{C} = \frac{x}{4} + 4 + \frac{100}{x}$$

(20, 14) Min

Figure 10.30

MAXIMIZING PROFIT

In Chapter 9, we defined the marginal profit function as the derivative of the profit function. That is,

$$\overline{MP} = P'(x).$$

In this chapter we have seen how to use the derivative to find maxima and minima for various functions. Now we can apply those same techniques in the context of **profit maximization.** We can use marginal profit to maximize profit functions.

If there is a physical limitation on the number of units that can be produced in a given period of time, then the endpoints of the interval caused by these limitations should also be checked.

EXAMPLE 4

Suppose that the production capacity for a certain commodity cannot exceed 30. If the total profit function for this commodity is

$$P(x) = 4x^3 - 210x^2 + 3600x - 200,$$

where x is the number of units sold, find the number of items that will maximize profit.

Solution

The restrictions on capacity mean that $P(x)$ is restricted by $0 \le x \le 30$. The marginal profit function is

$$P'(x) = 12x^2 - 420x + 3600.$$

Setting $P'(x)$ equal to 0, we get

$$0 = 12(x - 15)(x - 20),$$

so $P'(x) = 0$ at $x = 15$ *and* $x = 20$. Testing to the right and left of these values (as in the first-derivative test) we get

$$\left. \begin{array}{l} P'(0) = 3600 > 0 \\ P'(18) = -72 < 0 \\ P'(25) = 600 > 0 \end{array} \right\} \begin{array}{l} \Rightarrow \text{relative maximum at } x = 15. \\ \Rightarrow \text{relative minimum at } x = 20. \end{array}$$

Thus, at $(15, 20050)$ the total profit function has a *relative* maximum but we must check the endpoints (0 and 30) before deciding if it is the absolute maximum.

$P(0) = -200$

and $P(30) = 26{,}800$

Thus the absolute maximum profit is \$26,800, and it occurs at the endpoint, $x = 30$. Figure 10.31 shows the graph of the profit function.

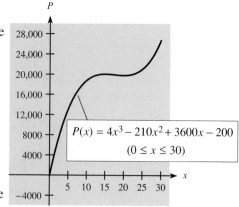

$$P(x) = 4x^3 - 210x^2 + 3600x - 200$$
$$(0 \le x \le 30)$$

Figure 10.31

In a **monopolistic market,** the seller who has a monopoly can control the price by regulating the supply of the product. Since the seller controls the supply, he or she can force the price higher by limiting supply.

If the demand function for the product is $p = f(x)$, total revenue for the sale of x units is $R(x) = px = f(x) \cdot x$. Note that the price p is fixed by the market in a competitive market, but varies with output for the monopolist.

If $\overline{C} = \overline{C}(x)$ represents the average cost per unit sold, then the total cost for the x units sold is $C = \overline{C} \cdot x = \overline{C}x$. Since we have both total cost and total revenue as a function of the quantity, x, we can maximize the profit function, $P(x) = px - \overline{C}x$, where p represents the demand function $p = f(x)$ and $\overline{C}$ represents the average cost function $\overline{C} = \overline{C}(x)$.

EXAMPLE 5

The daily demand function for a product is

$$p = 168 - 0.2x.$$

If a monopolist finds that the average cost is

$$\overline{C} = 120 + x,$$

(a) how many units must be sold to maximize profit?
(b) what is the selling price at this "optimal" level of production?
(c) what is the maximum possible profit?

Solution

(a) The total revenue function for the product is

$$R(x) = px = (168 - 0.2x)x = 168x - 0.2x^2,$$

and the total cost function is

$$C(x) = \overline{C} \cdot x = (120 + x)x = 120x + x^2.$$

Thus the profit function is

$$P(x) = R(x) - C(x) = 168x - 0.2x^2 - (120x + x^2),$$

or $\qquad\qquad\quad P(x) = 48x - 1.2x^2.$

Then, $P'(x) = 48 - 2.4x$, so $P'(x) = 0$ when $x = 20$.
We see that $P''(20) = -2.4$, so by the second-derivative test, $P(x)$ has a maximum at $x = 20$. That is, selling 20 units will maximize profit.

(b) The selling price is determined by $p = 168 - 0.2x$, so the price that will result from supplying 20 units per day is $p = 168 - 0.2(20) = 164$. That is, the "optimal" selling price is $164 per unit.

(c) The profit at $x = 20$ is $P(20) = 48(20) - 1.2(20)^2 = 960 - 480 = 480$. Thus the maximum possible profit is $480 per day.

In a **competitive market,** each firm is so small that its actions in the market cannot affect the price of the product. The price of the product is determined in the market by the intersection of the market demand curve (from all consumers) and market supply curve (from all firms that supply this product). The firm can sell as little or as much as it desires at the given market price, which it cannot change.

Therefore, a firm in a competitive market has a total revenue function given by $R(x) = px$, where p is the market equilibrium price for the product and x is the quantity sold.

EXAMPLE 6

A firm in a competitive market must sell its product for $200 per unit. The average cost per unit (per month) is $\overline{C} = 80 + x$, where x is the number of units sold per month. How many units should be sold to maximize profit?

Solution

If the average cost per unit is $\overline{C} = 80 + x$, then the total cost of x units is given by $C(x) = (80 + x)x = 80x + x^2$. The revenue per unit is $200, so the total revenue is given by $R(x) = 200x$. Thus the profit function is

$$P(x) = R(x) - C(x) = 200x - (80x + x^2), \quad \text{or} \quad P(x) = 120x - x^2.$$

Then $P'(x) = 120 - 2x$. Setting $P'(x) = 0$ and solving for x gives $x = 60$. Since $P''(60) = -2$, the profit is maximized when the firm sells 60 units per month.

CHECKPOINT

1. True or false: If $R(x)$ is the revenue function, we find all possible points where $R(x)$ could be maximized by solving $\overline{MR} = 0$ for x.

2. If $C(x) = \dfrac{x^2}{20} + 10x + 2500$, form $\overline{C}(x)$, the average cost function.

3. (a) If $p = 5000 - x$ gives the demand function in a monopoly market, find $R(x)$, if it is possible with this information.
   (b) If $p = 5000 - x$ gives the demand function in a competitive market, find $R(x)$, if it is possible with this information.

 ### Graphing Utilities

As we have seen, graphing utilities can be used to locate maximum values. In addition, if it is difficult to determine critical values algebraically, we may be able to find them graphically.

 ### EXAMPLE 7

Suppose total revenue and total costs for a company are given by

$$R(x) = 3000 - \frac{3000}{x + 1}$$

$$C(x) = 500 + 12x + x^2,$$

Figure 10.32

(a)

(b)

where x is thousands of units and revenue and costs are in thousands of dollars. Graph $P(x)$ and $P'(x)$ to determine the number of units that yields maximum profit and the amount of the maximum profit.

Solution

$$P(x) = R(x) - C(x)$$

$$P(x) = 3000 - \frac{3000}{x + 1} - (500 + 12x + x^2)$$

$$P(x) = 2500 - \frac{3000}{x + 1} - 12x - x^2$$

$$P'(x) = \frac{3000}{(x + 1)^2} - 12 - 2x$$

Finding the critical values by solving $P'(x) = 0$ is very difficult in this case (try it). This is why we are using a graphing approach. Figure 10.32(a) shows the graph of $P(x)$ and 10.32(b) shows the graph of $P'(x)$. These figures indicate that the maximum profit occurs near $x = 10$.

By adjusting the range for $P'(x)$, we obtain the graph in Figure 10.33. This shows that $P'(x) = 0$ when $x = 9$ (or when 9000 units are sold). The maximum profit is $P(9) = 2500 - 300 - 108 - 81$
$$= 2001 \text{ (thousands of dollars).}$$

Figure 10.33

**CHECKPOINT
SOLUTIONS**

1. False. $\overline{MR} = R'(x)$, but there may also be critical points where $R'(x)$ is undefined or $R(x)$ may be maximized at endpoints of a restricted domain.

2. $\overline{C}(x) = \dfrac{C(x)}{x} = \dfrac{x^2/20 + 10x + 2500}{x} = \dfrac{x}{20} + 10 + \dfrac{2500}{x}$

3. (a) $R(x) = p \cdot x = (5000 - x)x = 5000x - x^2$
   (b) In a competitive market, $R(x) = p \cdot x$, where p is the constant equilibrium price. Thus we need to know the supply function and find the equilibrium price before we can form $R(x)$.

EXERCISE 10.3

MAXIMIZING REVENUE

1. (a) If the total revenue function for a radio is $R = 36x - 0.01x^2$, sale of how many units will maximize the total revenue? Find the maximum revenue.
   (b) Find the maximum revenue if production is limited to at most 1500 radios.
2. (a) If the total revenue function for a blender is $R(x) = 25x - 0.05x^2$, sale of how many units will provide the maximum total revenue? Find the maximum revenue.
   (b) Find the maximum revenue if production is limited to at most 200 blenders.
3. If the total revenue function for a computer is $R(x) = 2000x - 20x^2 - x^3$, find the level of sales that maximizes revenue and find the maximum revenue.
4. A firm has total revenues given by

$$R(x) = 2800x - 8x^2 - x^3$$

   for a product. Find the maximum revenue from sales of that product.
5. An agency charges $10 per person for a trip to a concert if 30 people travel in a group. But for each person above the 30, the charge will be reduced by $.20. How many people will maximize the total revenue for the agency if the trip is limited to at most 50 people?
6. A company handles an apartment building with 50 units. Experience has shown that if the rent for each of the units is $360 per month, all the units will be filled, but one unit will become vacant for each $10 increase in the monthly rate. What rent should be charged to maximize the total revenue from the building if the upper limit on the rent is $450 per month?
7. A cable TV company has 1000 customers paying $20 each month. If each $1 reduction in price attracts 100 new customers, find the price that yields maximum revenue. Find the maximum revenue.
8. A company has determined that its monthly sales revenue S is related to the number of dollars x spent on advertising, with

$$S = 5000 + 8000x - 0.2x^2.$$

   Spending how much money on advertising will maximize their revenue? Find the maximum sales revenue.
9. The function $\bar{R}(x) = R(x)/x$ defines the average revenue for selling x units. For

$$R(x) = 2000x + 20x^2 - x^3,$$

   (a) find the maximum average revenue.
   (b) show that $\bar{R}(x)$ attains its maximum at the x-value where $\bar{R}(x) = \overline{MR}$.
10. For the revenue function given by

$$R(x) = 2800x + 8x^2 - x^3,$$

   (a) find the maximum average revenue.
   (b) show that $\bar{R}(x)$ attains its maximum at the x-value where $\bar{R}(x) = \overline{MR}$.

MINIMIZING AVERAGE COST

11. If the total cost function for a lamp is $C(x) = 25 + 13x + x^2$, producing how many units will result in a minimum average cost per unit? Find the minimum average cost.
12. If the total cost function for a product is $C(x) = 300 + 10x + 0.03x^2$, producing how many units will result in a minimum average cost per unit? Find the minimum average cost.
13. If the total cost function for a product is $C(x) = 100 + x^2$, producing how many units will result in a minimum average cost per unit? Find the minimum average cost.
14. If the total cost function for a product is $C(x) = 250 + 6x + 0.1x^2$, producing how many units will minimize the average cost? Find the minimum average cost.
15. If the total cost function for a product is $C(x) = (x + 4)^3$, where x represents the number of hundreds of units produced, producing how many units will minimize average cost? Find the minimum average cost.
16. If the total cost function for a product is $C(x) = (x + 5)^3$, where x represents the number of hundreds of units produced, producing how many units will minimize average cost? Find the minimum average cost.
17. For the cost function $C(x) = 25 + 13x + x^2$, show that average costs are minimized at the x-value where

$$\bar{C}(x) = \overline{MC}.$$

18. For the cost function $C(x) = 300 + 10x + 0.03x^2$, show that average costs are minimized at the x-value where

$$\bar{C}(x) = \overline{MC}.$$

MAXIMIZING PROFIT

19. If the profit function for a product is $P(x) = 5600x + 85x^2 - x^3 - 200,000$, selling how many items will produce a maximum profit? Find the maximum profit.

20. If the profit function for a commodity is $P = 6400x - 18x^2 - \frac{1}{3}x^3 - 40,000$, selling how many units will result in a maximum profit? Find the maximum profit.

21. A manufacturer estimates that x units of its product can be produced at a total cost of $C(x) = 45,000 + 100x + x^3$. If the manufacturer's total revenue from the sale of x units is $R(x) = 4600x$, determine the level of production x that will maximize the profit. Find the maximum profit.

22. A product can be produced at a total cost $C(x) = 800 + 100x^2 + x^3$, where x is the number produced. If the total revenue is given by $R(x) = 60,000x - 50x^2$, determine the level of production that will maximize the profit. Find the maximum profit.

23. A firm can produce only 1000 units per month. The monthly total cost is given by $C(x) = 300 + 200x$, where x is the number produced. If the total revenue is given by $R(x) = 250x - \frac{1}{100}x^2$, how many items should they produce for maximum profit? Find the maximum profit.

24. A firm can produce 100 units per week. If its total cost function is $C = 500 + 1500x$ and its total revenue function is $R = 1600x - x^2$, how many units should it produce to maximize its profit? Find the maximum profit.

25. A firm has monthly average costs given by

$$\overline{C} = \frac{45,000}{x} + 100 + x,$$

where x is the number of units produced per month. The firm can sell its product in a competitive market for $1600 per unit. If production is limited to 600 units per month, find the number of units that gives maximum profit, and find the maximum profit.

26. A small business has weekly average costs of

$$\overline{C} = \frac{100}{x} + 30 + \frac{x}{10},$$

where x is the number of units produced each week. The competitive market price for this business's product is $46 per unit. If production is limited to 150 units per week, find the level of production that yields maximum profit, and find the maximum profit.

27. The weekly demand function for a product sold by only one firm is $p = 600 - \frac{1}{2}x$, and the average cost of production and sale is $\overline{C} = 300 + 2x$.
    (a) Find the quantity that will maximize profit.
    (b) Find the selling price at this optimal quantity.
    (c) What is the maximum profit?

28. The monthly demand function for a product sold by a monopoly is $p = 8000 - x$, and its average cost is $\overline{C} = 4000 + 5x$. Determine
    (a) the quantity that will maximize profit.
    (b) the selling price at the optimal quantity.
    (c) the maximum profit.

29. The monthly demand function for a product sold by a monopoly is $p = 1960 - \frac{1}{3}x^2$, and the average cost is $\overline{C} = 1000 + 2x + x^2$. If production is limited to 1000 units and x is in hundreds of units, find
    (a) the quantity that will give maximum profit.
    (b) the maximum profit.

30. The monthly demand function for a product sold by a monopoly is $p = 5900 - \frac{1}{2}x^2$, and its average cost is $\overline{C} = 3020 + 2x$. If production is limited to 100 units, find the number of units that maximizes profit. Will the maximum profit result in a profit or loss?

31. An industry having a monopoly on a product has its average weekly costs given by

$$\overline{C} = \frac{10,000}{x} + 60 - 0.03x + 0.00001x^2.$$

The weekly demand for the product is given by $p = 120 - 0.015x$. Find the price the industry should set and the number of units it should produce to obtain maximum profit. Find the maximum profit.

32. A large corporation having monopolistic control in the marketplace has its average daily costs given by

$$\overline{C} = \frac{800}{x} + 100x + x^2.$$

The daily demand for its product is given by $p = 60,000 - 50x$. Find the quantity that gives maximum profit, and find the maximum profit. What selling price should the corporation set for its product?

33. A company handles an apartment building with 50 units. Experience has shown that if the rent for each of the units is $360 per month, all of the units will be filled, but one unit will become vacant for each $10 increase in this monthly rate. If the monthly cost of maintaining the apartment building is $6 per rented unit, what rent should be charged per month to maximize the profit?

34. A travel agency will plan a tour for groups of size 25 or larger. If the group contains exactly 25 people, the cost is $500 per person. However, each person's cost is reduced by $10 for each additional person above the 25. If the travel agency incurs a cost of $125 per person for the tour, what size group will give the agency the maximum profit?

MISCELLANEOUS APPLICATIONS

35. **Sales revenue** The data in the table give sales revenues for Scott Paper Company for various years.

Year	Sales Revenue (in Billions)	Year	Sales Revenue (in Billions)
1983	$2.6155	1989	4.8949
1984	2.7474	1990	5.1686
1985	2.934	1991	4.9593
1986	3.3131	1992	5.0913
1987	3.9769	1993	4.7489
1988	4.5494		

SOURCE: Scott Paper Company, 1993 *Annual Report*

Assume that sales revenues for Scott Paper can be modeled by

$$R(t) = -0.031t^2 + 0.776t + 0.179,$$

where t is the number of years past 1980.
(a) Use the model to find when maximum revenue occurs and the maximum revenue it predicts.
(b) Check your result in (a) against the data in the table.

36. **Revenue** The following table gives the total revenues of AT&T for selected years.

Year	Total revenues (in billions)	Year	Total revenues (in billions)
1985	$63.13	1990	$62.191
1986	$69.906	1991	$63.089
1987	$60.53	1992	$64.904
1989	$61.1	1993	$67.156

SOURCE: *AT&T Annual Report,* 1993

Suppose the data can be modeled by the equation

$$R(t) = 0.253t^2 - 4.03t + 76.84,$$

where t is the number of years past 1980.
(a) Use the model to find the year in which revenue was minimum and find the minimum predicted revenue.
(b) Check your result in (a) against the data in the table.

37. **U.S. trade deficit** The figure shows the U.S. trade deficit in billions of dollars from late 1990 to late 1993. Use the figure to find the following.
(a) Approximately when did the trade deficit reach its absolute minimum, and what was the minimum?

(b) Approximately when did the 12-month moving average reach its absolute minimum, and what was the minimum?
(c) Approximately when did the trade deficit reach its absolute maximum, and what was the maximum?
(d) Approximately when did the 12-month moving average reach its absolute maximum, and what was the maximum?

SOURCE: *Wall Street Journal,* December 17, 1993

38. **Dow-Jones average** The figures show the Dow-Jones average for early 1994 (January through March) as well as the average through the day on March 31, 1994. Use the figures to answer the following.
(a) Approximate the time of day on March 31, 1994, when the Dow-Jones average reached its absolute maximum and when the average reached its absolute minimum.
(b) Find the day of the year in early 1994 when the Dow-Jones reached its absolute maximum, and find its absolute maximum.
(c) Find the approximate day of the year in early 1994 when the Dow-Jones average reached its absolute minimum, and find its absolute minimum.

SOURCE: *Oil City Derrick,* April 1, 1994

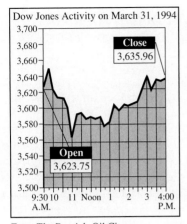

From *The Derrick*, Oil City,
Pennsylvania, April 1, 1994
SOURCE: Telerate Systems, Inc.

39. ***Crime concern*** The graph below shows that the percentage p of people in the United States that believe crime is the most serious domestic problem is a function of time t (in calendar years) and, of course, of the events that occur during those years. Denote this function by $p = f(t)$, and use the graph to answer the following.

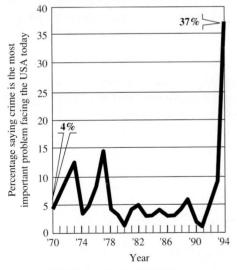

SOURCE: *USA Today*, January 25, 1994

(a) When did $f(t)$ achieve its absolute minimum?
(b) What is the absolute maximum point for $f(t)$?
(c) Is it possible to have $f(t) < 0$? Explain.
(d) What effect do you think a foreign conflict (such as a war) would have on the graph?
(e) Do you think a candidate for public office should have based a campaign extensively on the "law and order" issue in 1980? in 1994? Explain.

40. ***Social Security support*** The graph below shows that the number of workers, W, still in the workforce per Social Security beneficiary as a function of time t, in calendar years. Denote this function by $W = f(t)$, and use the graph to answer the following.

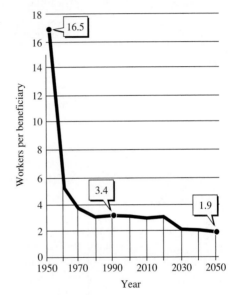

SOURCE: Social Security Administration

(a) What is the absolute maximum for $f(t)$?
(b) What is the absolute minimum for $f(t)$?
(c) Does this graph suggest that Social Security taxes might rise or might fall in the early twenty-first century? Explain.

10.4 APPLICATIONS OF MAXIMA AND MINIMA

Objective

❑ To apply the procedures for finding maxima and minima to solve problems from management, life, and social sciences

APPLICATION PREVIEW

Manufacturers make production runs to restock their inventories. Since there are costs associated with both the production of items and their storage (placement into inventory), a typical question in these **inventory cost models** is "How many items should be produced in each production run to minimize the total costs of production and storage?" This question is a typical example of the kinds of questions and important business applications that require the use of the derivative for finding maxima and minima. As managers, workers, or consumers, we may be interested in such things as maximum revenue, maximum profit, minimum cost, maximum medical dosage, maximum utilization of resources, and so on.

If we have functions that model cost, revenue, or population growth, we can apply the methods of this chapter to find the maxima and minima of the functions.

EXAMPLE 1

Suppose that a new company begins production in 1995 with eight employees and the growth of the company over the next 10 years is predicted by

$$N = N(t) = 8\left(1 + \frac{160t}{t^2 + 16}\right), \qquad 0 \le t \le 10,$$

where N is the number of employees t years after 1995.
(a) In what year will the number of employees be maximized?
(b) What will be the maximum number of employees?

Solution

This function will have a relative maximum when $N'(t) = 0$.

$$N'(t) = 8\left[\frac{(t^2 + 16)(160) - (160t)(2t)}{(t^2 + 16)^2}\right]$$

$$= 8\left[\frac{160t^2 + 2560 - 320t^2}{(t^2 + 16)^2}\right]$$

$$= 8\left[\frac{2560 - 160t^2}{(t^2 + 16)^2}\right]$$

Now, $N'(t) = 0$ when its numerator is 0 (note that the denominator is never 0), so we must solve

$$2560 - 160t^2 = 0$$

$$160(4 + t)(4 - t) = 0,$$

so

$$t = -4 \quad \text{or} \quad t = 4.$$

We are only interested in positive t-values, so we test $t = 4$.

$$\left.\begin{array}{l} N'(0) = 8\left[\dfrac{2560}{256}\right] > 0 \\[4mm] N'(10) = 8\left[\dfrac{-13,440}{(116)^2}\right] < 0 \end{array}\right\} \Rightarrow \quad \text{relative maximum}$$

The relative maximum is

$$N(4) = 8\left(1 + \frac{640}{32}\right) = 168.$$

At $t = 0$, the number of employees is $N(0) = 8$, and it increases to $N(4) = 168$. After $t = 4$, $N(t)$ decreases to $N(10) = 118$ (approximately), so $N(4) = 168$ is the maximum number of employees. Figure 10.34 verifies these conclusions.

Sometimes we must develop the function we need from the statement of the problem. In this case, it is important to understand what is to be maximized or minimized and to express that quantity as a function of *one* variable.

Figure 10.34

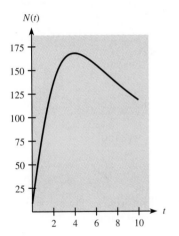

EXAMPLE 2

A farmer needs to enclose a rectangular pasture containing 1,600,000 square feet so that there will be enough grass for his herd. Suppose that along the road adjoining his property he wants to use a more expensive fence and that he needs no fence on one side perpendicular to the road because a river bounds his property on that side. If the fence costs $15 per foot along the road and $10 per foot along the two remaining sides that must be fenced, what dimensions of his rectangular field will minimize his cost? (See Figure 10.35.)

Solution

In Figure 10.35, x represents the length of the pasture along the road (and parallel to the road) and y represents the width. The cost function for the fence used is

$$C = 15x + 10y + 10x = 25x + 10y.$$

We cannot use a derivative to find where C is minimized unless we write C as a function of x or y only. Because the area of the rectangular field must be 1,600,000 square feet, we have

$$A = xy = 1,600,000.$$

Figure 10.35

Figure 10.36

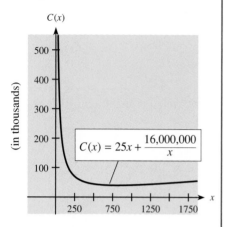

$C(x) = 25x + \dfrac{16,000,000}{x}$

Solving for y in terms of x and substituting gives

$$y = \frac{1,600,000}{x},$$

$$C = 25x + 10\left(\frac{1,600,000}{x}\right) = 25x + \frac{16,000,000}{x}.$$

The derivative of C with respect to x is

$$C'(x) = 25 - \frac{16,000,000}{x^2},$$

and we find the relative minimum of C as follows:

$$0 = 25 - \frac{16,000,000}{x^2}$$

$$0 = 25x^2 - 16,000,000$$

$$25x^2 = 16,000,000$$

$$x^2 = 640,000$$

$$x = 800 \text{ feet.}$$

Testing to see if $x = 800$ gives the minimum cost, we find

$$C''(x) = \frac{32,000,000}{x^3}.$$

$C''(x) > 0$ for $x > 0$, so $C(x)$ is concave up for all positive x. Thus $x = 800$ gives the absolute minimum, and $C(800)$ is the minimum cost. The other dimension of the rectangular field is $x = 1,600,000/800 = 2000$ feet. Figure 10.36 verifies that $C(x)$ reaches its minimum (of 40,000) at $x = 800$.

Figure 10.37

Girth = distance around

(a)

EXAMPLE 3

Postal restrictions limit the size of packages sent through the mail. If the restrictions are that the length plus the girth may not exceed 108 in., find the volume of the largest box with square cross section that can be mailed.

Solution

Let l equal the length of the box and s equal a side of the square end. See Figure 10.37(a). The volume we seek to maximize is given by

$$V = s^2 l.$$

We can use the restriction that girth plus length equals 108,

$$4s + l = 108,$$

to express V as a function of s or l. Since $l = 108 - 4s$, the equation for V becomes

$$V = s^2(108 - 4s)$$

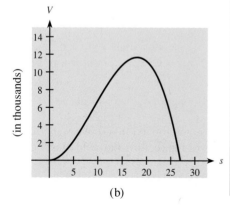

(b)

or

$$V = 108s^2 - 4s^3.$$

Thus we can use dV/ds to find the critical values.

$$\frac{dV}{ds} = 216s - 12s^2$$

$$0 = s(216 - 12s)$$

Critical values are $s = 0$, $s = \frac{216}{12} = 18$. The critical value $s = 0$ will not maximize the volume for, in this case, $V = 0$. Testing to the left and right of $s = 18$ gives

$$V'(17) > 0 \quad \text{and} \quad V'(19) < 0.$$

Thus $s = 18$ in. and $l = 108 - 4(18) = 36$ in. yield a maximum volume of 11,664 cubic inches.

Once again we can graphically verify our results. Figure 10.37(b) shows that $V = s^2(108 - 4s)$ achieves its maximum when $s = 18$.

CHECKPOINT

Suppose that we want to find the minimum value of $C = 5x + 2y$ and we know that x and y must be positive and that $xy = 1000$. Answer the following.

1. What equation do we differentiate to solve this problem?

2. Find the critical values.

3. Find the minimum value of C.

Consider the **inventory cost model** in which x items are produced in each run and items are removed from inventory at a fixed constant rate. Then the number of units in storage changes with time and is illustrated in Figure 10.38. To see how these inventory cost models work, consider the following example.

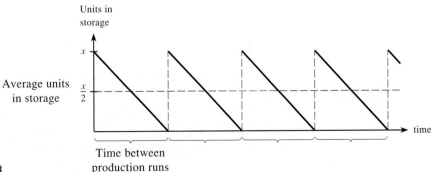

Figure 10.38

EXAMPLE 4

Suppose that a company needs 1,000,000 items during a year and that preparation costs are $800 for each production run. Suppose further that it costs the company $6 to produce each item and $1 to store an item for up to one year. If each production run consists of x items, find x so that the total costs of production and storage are minimized.

Solution

The total production costs are given by

$$\begin{pmatrix} \text{No. of} \\ \text{runs} \end{pmatrix} \begin{pmatrix} \text{cost} \\ \text{per run} \end{pmatrix} + \begin{pmatrix} \text{no. of} \\ \text{items} \end{pmatrix} \begin{pmatrix} \text{cost} \\ \text{per item} \end{pmatrix}$$

$$= \left(\frac{1,000,000}{x} \right) (\$800) + (1,000,000)(\$6).$$

The total storage costs are

$$\begin{pmatrix} \text{Average} \\ \text{no. stored} \end{pmatrix} \begin{pmatrix} \text{storage cost} \\ \text{per item} \end{pmatrix} = \left(\frac{x}{2} \right) (\$1).$$

Thus the total costs of production and storage are

$$C = \left(\frac{1,000,000}{x} \right) (800) + 6,000,000 + \frac{x}{2}.$$

We wish to find x so that C is minimized.

$$C' = \frac{-800,000,000}{x^2} + \frac{1}{2}$$

If $x > 0$, critical values occur when $C' = 0$.

$$0 = \frac{-800,000,000}{x^2} + \frac{1}{2}$$

$$\frac{800,000,000}{x^2} = \frac{1}{2}$$

$$1,600,000,000 = x^2$$

$$x = \pm 40,000$$

Because x must be positive, we could test $x = 40,000$ with the second derivative.

$$C''(x) = \frac{1,600,000,000}{x^3}, \quad \text{so} \quad C''(40,000) > 0.$$

Note that $x = 40,000$ yields an absolute minimum value for C, because $C'' > 0$ for all $x > 0$. That is, production runs of 40,000 items yield minimum total costs for production and storage.

Graphing Utilities

Problems of the types we've studied in this section could also be solved (at least approximately) with a graphing utility. With this approach, our first goal is still to express the quantity to be maximized or minimized as a function of one variable. Then that function can be graphed and the (at least approximate) optimal value can be obtained from the graph.

EXAMPLE 5

A homeowner has to pay for utility line installation to her house from a transformer on the street at the corner of her property. Because of local restrictions the lines must be underground on her property. Suppose that the costs are \$5/foot along the street and \$10/foot underground. How far from the transformer should she enter the property to minimize installation costs? See Figure 10.39(a).

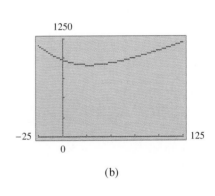

(b)

Figure 10.39
(a)

Solution

If the homeowner has the cable placed underground from the house perpendicular to the street and then to the transformer, the cost would be

$$\$10(48) + \$5(100) = \$980.$$

It may be possible to save some money by placing the cable on a diagonal to the street. By using the Pythagorean Theorem, the length of the underground cable that meets the street x feet closer to the transformer is

$$\sqrt{48^2 + x^2} = \sqrt{2304 + x^2} \quad \text{feet.}$$

Thus the cost C of installation is given by

$$C = 10\sqrt{2304 + x^2} + 5(100 - x) \quad \text{(dollars).}$$

Figure 10.39(b) shows the graph of this function over the possible interval for x ($0 \le x \le 100$). Because any extrema must occur in this interval, we can find the minimum by using TRACE or the MIN command on a calculator or computer. The minimum cost is \$915.70, when $x = 28$ feet (that is, when the cable meets the street 72 feet from the transformer). We can use the derivative of C to verify that $x = 28$ gives the minimum cost. By finding the derivative and graphing it, we can see that the only relative extrema occurs at $x = 28$. (Try this.) Because the cost is lower at $x = 28$ than at either endpoint of the interval, the lowest possible cost occurs there.

CHECKPOINT SOLUTIONS

1. We must differentiate C, but first C must be expressed as a function of one variable: $xy = 1000$ means that $y = 1000/x$. If we substitute $1000/x$ for y in $C = 5x + 2y$, we get

$$C(x) = 5x + 2\left(\frac{1000}{x}\right) = 5x + \frac{2000}{x}.$$

Find $C'(x)$ to solve the problem.

2. $C'(x) = 5 - 2000/x^2$, so $C'(x) = 0$ when

$$5 - \frac{2000}{x^2} = 0 \quad \text{or} \quad x = \pm 20.$$

Since x must be positive, the only critical value is $x = 20$.

3. $C''(x) = 4000/x^3$, so $C''(x) > 0$ for all $x > 0$. Thus $x = 20$ yields the minimum value. Also, when $x = 20$, we have $y = 50$, so the minimum value of C is $C = 5(20) + 2(50) = 200$.

EXERCISE 10.4

APPLICATIONS

1. **Return to sales** The manufacturer of GRIPPER tires modeled its return on sales from television advertising expenditures in two regions, as follows:

Region 1: $S_1 = 30 + 20x_1 - 0.4x_1^2$

Region 2: $S_2 = 20 + 36x_2 - 0.6x_2^2$,

where S_1 and S_2 are the sales in millions of dollars and x_1 and x_2 are millions of dollars of expenditures for television advertising. (a) What advertising expenditures would maximize sales in each district? (b) How much money will be needed to maximize sales in both districts?

2. **Projectiles** A ball thrown into the air from a building 100 ft high travels along a path described by

$$y = \frac{-x^2}{110} + x + 100,$$

where y is its height and x is the horizontal distance from the building. What is the maximum height the ball will reach?

3. **Profit** The profit from a grove of orange trees is given by $x(800 - x)$, where x is the number of orange trees per acre. How many trees per acre will maximize the profit?

4. ***Reaction rates*** The velocity v of an autocatalytic reaction can be represented by the equation

$$v = x(a - x),$$

where a is the amount of material originally present and x is the amount that has been decomposed at any given time. Find the maximum velocity of the reaction.

5. ***Productivity*** Analysis of daily output of a factory during an 8-hour shift shows that the hourly number of units y produced after t hours of production is

$$y = 70t + \tfrac{1}{2}t^2 - t^3, \qquad 0 \le t \le 8.$$

  (a) After how many hours will the hourly number of units be maximized?

  (b) What is the maximum hourly output?

6. ***Productivity*** A time study showed that, on average, the productivity of a worker after t hours on the job can be modeled by

$$P = 27t + 6t^2 - t^3, \qquad 0 \le t \le 8,$$

where P is the number of units produced per hour. After how many hours will productivity be maximized? What is the maximum productivity?

7. ***Consumer expenditure*** Suppose that the demand x for a product is $x = 10,000 - 100p$, where p dollars is the market price per unit. Then the consumer expenditure for the product is

$$E = px = 10,000p - 100p^2.$$

For what market price will expenditure be greatest?

8. ***Production costs*** Suppose that the monthly cost of mining a certain ore is related to the number of pieces of equipment purchased, according to

$$C = 25,000x + \frac{870,000}{x}, \qquad x > 0,$$

where x is the number of pieces of equipment used. Using how many pieces of equipment will minimize the cost?

Medication When medicine is administered, reaction (measured in change of blood pressure or temperature) can be modeled by

$$R = m^2\left(\frac{c}{2} - \frac{m}{3}\right),$$

where c is a positive constant and m is the amount of medicine absorbed into the blood. (Source: Thrall, R. M., et. al., *Some Mathematical Models in Biology,* U.S. Department of Commerce, 1967.)

9. Find the amount of medicine that is being absorbed into the blood when the reaction is maximum.

10. The rate of change of reaction R with respect to the amount of medicine m is defined to be the sensitivity.

  (a) Find the sensitivity, S.

  (b) Find the amount of medicine that is being absorbed into the blood when the sensitivity is maximum.

11. ***Advertising and sales*** An inferior product with a large advertising budget sells well when it is introduced, but sales fall as people discontinue use of the product. Suppose that the weekly sales S are given by

$$S = \frac{200t}{(t + 1)^2}, \qquad t \ge 0,$$

where S is in millions of dollars and t is in weeks. After how many weeks will sales be maximized?

12. ***Revenue*** A newly released film has its weekly revenue given by

$$R(t) = \frac{50t}{t^2 + 36}, \qquad t \ge 0,$$

where R is in millions of dollars and t is in weeks.

  (a) After how many weeks will the weekly revenue be maximized?

  (b) What is the maximum weekly revenue?

13. ***News impact*** Suppose that the percentage p (as a decimal) of people who could correctly identify 2 of 8 defendants in a drug case t days after their trial began is given by

$$p(t) = \frac{6.4t}{t^2 + 64} + 0.05.$$

Find the number of days before the percentage is maximized, and find the maximum percentage.

14. ***Candidate recognition*** Suppose that in an election year the proportion p of voters that recognize a certain candidate's name t months after the campaign started is given by

$$p(t) = \frac{7.2t}{t^2 + 36} + 0.2.$$

After how many months is the proportion maximized?

15. **Minimum fence** Two equal rectangular lots are enclosed by fencing the perimeter of a rectangular lot and then putting a fence across its middle. If each lot is to contain 1200 square feet, what is the minimum amount of fence needed to enclose the lots (include the fence across the middle)?

16. **Minimum fence** The running yard for a dog kennel must contain at least 900 square feet. If a 20-foot side of the kennel is used as part of one side of a rectangular yard with 900 square feet, what dimensions will require the least amount of fencing?

17. **Minimum cost** A rectangular field with one side along a river is to be fenced. Suppose that no fence is needed along the river, the fence on the side opposite the river costs $20 per foot, and the fence on the other sides costs $5 per foot. If the field must contain 45,000 square feet, what dimensions will minimize costs?

18. **Minimum cost** From a tract of land a developer plans to fence a rectangular region and then divide it into two identical rectangular lots by putting a fence down the middle. Suppose that the fence for the outside boundary costs $5 per foot and the fence for the middle costs $2 per foot. If each lot contains 13,500 square feet, find the dimensions of each lot that yield the minimum cost for the fence.

19. **Optimization at a fixed cost** A rectangular area is to be enclosed and divided into thirds. The family has $800 to spend for the fencing material. The outside fence costs $10 per running foot installed, and the dividers costs $20 per running foot installed. What are the dimensions that will maximize the area enclosed? (Answer contains a fraction.)

20. **Minimum cost** A kennel of 640 square feet is to be constructed as shown. The cost is $4 per running foot for the sides and $1 per running foot for the ends and dividers. What are the dimensions of the kennel that will minimize the cost?

21. **Minimum cost** The base of a rectangular box is to be twice as long as it is wide. The volume of the box is 256 cubic inches. Material for the top costs $0.10 per square inch and the material for the sides and bottom costs $0.05 per square inch. Find the dimensions that will make the cost a minimum.

22. **Velocity of air during a cough** According to B. F. Visser, the velocity v of air in the trachea during a cough is related to the radius r of the trachea according to

$$v = ar^2(r_0 - r),$$

where a is a constant and r_0 is the radius of the trachea in a relaxed state. Find the radius r that produces the maximum velocity of air in the trachea during a cough.

23. **Inventory cost model** Suppose that a company needs 1,500,000 items during a year and that preparation for each production run costs $600. Suppose also that it costs $15 to produce each item and $2 per year to store an item. Use the inventory cost model to find the number of items in each production run so that the total costs of production and storage are minimized.

24. **Inventory cost model** Suppose that a company needs 60,000 items during a year and that preparation for each production run costs $400. Suppose further that it costs $4 to produce each item and $0.75 to store each one for one year. Use the inventory cost model to find the number of items in each production run that will minimize the total costs of production and storage.

25. **Inventory cost model** A company needs 150,000 items per year. It costs the company $360 to prepare a production run of these items and $7 to produce each item. If it also costs the company $0.75 per year for each item stored, find the number of items that should be produced in each run so that total costs of production and storage are minimized.

26. **Inventory cost model** A company needs 450,000 items per year. Production costs are $500 to prepare for a production run and $10 for each item produced. Inventory costs are $2 per item per year. Find the number of items that should be produced in each run so that the total costs of production and storage are minimized.

27. **Volume** A rectangular box with a square base is to be formed from a square piece of metal with 12-inch sides. If a square piece with side x is cut from the corners of the metal and the sides are folded up to form an open box, the volume of the box is $V = (12 - 2x)^2 x$. What value of x will maximize the volume of the box?

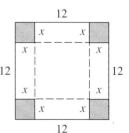

28. **Volume** A square piece of cardboard 36 cm on a side is to be formed into a rectangular box by cutting squares with length x from each corner and folding up the sides. What is the maximum volume possible for the box?

29. **Revenue** The owner of an orange grove must decide when to pick one variety of oranges. She can sell them for $8 a bushel if she sells them now, with each tree yielding an average of 5 bushels. The yield increases by one half bushel per week for the next 5 weeks, but the price per bushel decreases by $0.50 per bushel each week. When should the oranges be picked for maximum return?

30. **Minimum material** A box with an open top and a square base is to be constructed to contain 4000 cubic inches. Find the dimensions that will require the minimum amount of material to construct the box.

31. **Minimum cost** A printer has a contract to print 100,000 posters for a political candidate. He can run the posters by using any number of plates from 1 to 30 on his press. If he uses x metal plates, they will produce x copies of the poster with each impression of the press. The metal plates cost $2.00 to prepare, and it costs $12.50 per hour to run the press. If the press can make 1000 impressions per hour, how many metal plates should he make to minimize costs?

32. **Shortest time** A vacationer on an island 8 miles offshore from a point that is 48 miles from town must travel to town occasionally. (See the figure below.) The vacationer has a boat capable of traveling 30 mph and can go by auto along the coast at 55 mph. At what point should the car be left to minimize the time it takes to get to town?

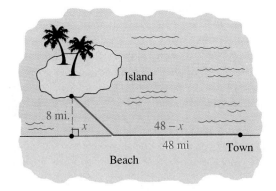

Drug use The percentage of high school seniors who have tried certain drugs is shown in the table for selected years. Use this information in problems 33 and 34.

Year of Graduation	Marijuana	Hallucinogens
1975	47.3	16.3
1978	59.2	14.3
1980	60.3	13.3
1983	57	11.9
1984	54.9	10.7
1985	54.2	10.3
1986	50.9	9.7
1987	50.2	10.3
1988	47.2	8.9
1989	43.7	9.4
1990	40.7	9.4
1991	36.7	9.6

SOURCE: National Institute on Drug Abuse

33. Assume that the percentage of high school seniors who have tried marijuana can be modeled with the function

$$f_m(t) = -0.228596t^2 + 37.0677t - 1444.41,$$

where t is the number of years past 1900.
   (a) Use the model to find the year when the percentage of high school seniors who have tried marijuana was maximum, and find the maximum predicted by the model.
   (b) Compare your results in (a) with the data in the table.

34. The data from the table indicates that the percentage of high school seniors who have tried hallucinogens can be modeled with the function

$$f_h(t) = 0.0241904t^2 - 4.47459t + 216.074,$$

where t is the number of years past 1900.
   (a) Use the model to find the year when the percentage of high school seniors who had tried hallucinogens was minimum and find the minimum predicted by the model.
   (b) Compare your results in (a) with the data in the table.

10.5 ASYMPTOTES; MORE CURVE SKETCHING

Objectives

- ❏ To locate horizontal asymptotes
- ❏ To locate vertical asymptotes
- ❏ To sketch graphs of functions that have vertical and/or horizontal asymptotes

APPLICATION PREVIEW

If the total daily cost of producing plastic rafts for swimming pools is given by

$$C = 500 + 8x + 0.05x^2,$$

where x is the number of units produced per day, then the average cost per unit produced is given by

$$\overline{C} = \frac{500 + 8x + 0.05x^2}{x} = \frac{500}{x} + 8 + 0.05x, \qquad \text{for } x > 0.$$

The graph of this function is very useful in understanding how many rafts should be produced to keep the average daily cost under control. The graph of this function contains a vertical asymptote at $x = 0$. We will discuss graphs and applications involving asymptotes in this section.

The procedures for using the first-derivative test and the second-derivative test are given in previous sections, but none of the graphs discussed in those sections contain vertical asymptotes or horizontal asymptotes. In this section, we consider how to use information about asymptotes along with the first and second derivatives, and we present a unified approach to curve sketching.

ASYMPTOTES

In Section 1.5, we first discussed asymptotes and saw that they are important features of the graphs that have them. Then, in our discussion of limits, we discovered the relationship between certain limits and asymptotes. Limits are used to precisely define and locate asymptotes.

Because a horizontal asymptote tells us the behavior of the functional values (y-coordinates) when x increases or decreases without bound, we use limits at infinity to determine the existence of horizontal asymptotes.

Horizontal Asymptote   The graph of a rational function $y = f(x)$ will have a **horizontal asymptote** at $y = b$, for a constant b, if

$$\lim_{x \to +\infty} f(x) = b \quad \text{or} \quad \lim_{x \to -\infty} f(x) = b.$$

Otherwise, the graph has no horizontal asymptote.

For a rational function f, $\lim_{x \to +\infty} f(x) = b$ if and only if $\lim_{x \to -\infty} f(x) = b$; so we only need to find one of these limits to locate a horizontal asymptote.

Just as with horizontal asymptotes, the formal definition of vertical asymptotes uses limits.

Vertical Asymptote   The line $x = x_0$ is a **vertical asymptote** of the graph of $y = f(x)$ if the values of $f(x)$ approach $+\infty$ or $-\infty$ as x approaches x_0 (from the left or the right).

From our work with limits, recall that a vertical asymptote will occur on the graph of a function at an x-value where the function has its denominator (but not its numerator) equal to zero. These observations allow us to determine where vertical asymptotes occur.

Vertical Asymptote of a   The graph of the rational function
Rational Function

$$h(x) = \frac{f(x)}{g(x)}$$

has a vertical asymptote at $x = c$ if $g(c) = 0$ and $f(c) \neq 0$.

Figure 10.40

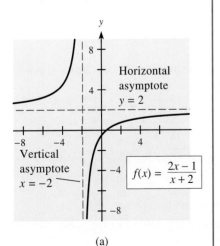

(a)

EXAMPLE 1
Find any vertical and horizontal asymptotes for

(a) $f(x) = \dfrac{2x - 1}{x + 2}$ (b) $f(x) = \dfrac{x^2 + 3}{1 - x}$.

Solution
(a) The denominator of this function is 0 at $x = -2$, and because this value does not make the numerator 0, there is a vertical asymptote at $x = -2$.
    Because the function is rational, we can find horizontal asymptotes by evaluating

$$\lim_{x \to +\infty} \frac{2x - 1}{x + 2} \quad \text{or} \quad \lim_{x \to -\infty} \frac{2x - 1}{x + 2}.$$

We will evaluate both.

$$\lim_{x \to +\infty} \frac{2x - 1}{x + 2} = \lim_{x \to +\infty} \frac{2 - 1/x}{1 + 2/x} = \frac{2 - 0}{1 + 0} = 2$$

$$\lim_{x \to -\infty} \frac{2x - 1}{x + 2} = \lim_{x \to -\infty} \frac{2 - 1/x}{1 + 2/x} = \frac{2 - 0}{1 + 0} = 2$$

Thus there is a horizontal asymptote at $y = 2$. The graph is shown in Figure 10.40(a).

(b) The denominator of this function is 0 at $x = 1$, and because this value does not make the numerator 0, there is a vertical asymptote at $x = 1$.

Figure 10.40 (continued)

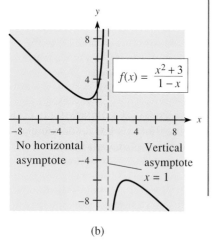

(b)

To find horizontal asymptotes, we evaluate the following.

$$\lim_{x \to +\infty} \frac{x^2 + 3}{1 - x} = \lim_{x \to +\infty} \frac{1 + 3/x^2}{1/x^2 - 1/x} = \frac{1 + 0}{0 - 0} = -\infty$$

This limit is $-\infty$ because the numerator approaches 1 and the denominator approaches 0 through negative values. Thus

$$\lim_{x \to +\infty} \frac{x^2 + 3}{1 - x} \text{ does not exist}$$

and the graph has no horizontal asymptotes. Note also that

$$\lim_{x \to -\infty} \frac{x^2 + 3}{1 - x} \text{ does not exist.}$$

The graph is shown in Figure 10.40(b).

MORE CURVE SKETCHING

We now extend our first- and second-derivative techniques of curve sketching to include functions that have asymptotes.

In general, the following steps are helpful when we sketch the graph of a function.

1. Determine the domain of the function. The domain may be restricted by the nature of the problem or by the equation.

2. Look for vertical asymptotes, especially if the function is a rational function.

3. Look for horizontal asymptotes, especially if the function is a rational function.

4. Find the relative maxima and minima by using the first-derivative test or the second-derivative test.

5. Use the second derivative to find the points of inflection if this derivative is easily found.

6. Use other information (intercepts, for example) and plot additional points to complete the sketch of the graph.

EXAMPLE 2

Sketch the graph of the function $f(x) = \dfrac{x^2}{(x + 1)^2}$.

Solution

1. The domain is the set of all real numbers except $x = -1$.

2. Because $x = -1$ makes the denominator 0 and does not make the numerator 0, there is a vertical asymptote at $x = -1$.

3. Because $\lim\limits_{x \to +\infty} \dfrac{x^2}{(x + 1)^2} = \lim\limits_{x \to +\infty} \dfrac{x^2}{x^2 + 2x + 1}$

$$= \lim\limits_{x \to +\infty} \dfrac{1}{1 + \dfrac{2}{x} + \dfrac{1}{x^2}}$$

$$= \dfrac{1}{1 + 0 + 0} = 1,$$

there is a horizontal asymptote at $y = 1$.

4. To find any maxima and minima, we first find $f'(x)$.

$$f'(x) = \dfrac{(x + 1)^2(2x) - x^2[2(x + 1)]}{(x + 1)^4}$$

$$= \dfrac{2x(x + 1)[(x + 1) - x]}{(x + 1)^4}$$

$$= \dfrac{2x}{(x + 1)^3}$$

Thus $f'(x) = 0$ when $x = 0$ (and $y = 0$), and $f'(x)$ is undefined at $x = -1$ (where the vertical asymptote occurs). Testing $f'(x)$ on either side of $x = 0$ and $x = -1$ gives the following sign diagram. The sign diagram for f' shows that the critical point $(0, 0)$ is a relative minimum and shows how the graph approaches the vertical asymptote at $x = -1$.

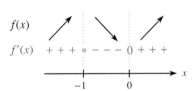

* $x = -1$ is a vertical asymptote.

5. The second derivative is

$$f''(x) = \dfrac{(x + 1)^3(2) - 2x[3(x + 1)^2]}{(x + 1)^6}$$

Factoring $(x + 1)^2$ from the numerator and simplifying gives

$$f''(x) = \dfrac{2 - 4x}{(x + 1)^4}.$$

We can see that $f''(0) = 2 > 0$, so the second-derivative test also shows that $(0, 0)$ is a relative minimum. We see that $f''(x) = 0$ when $x = \frac{1}{2}$. Checking $f''(x)$ between $x = -1$ (where it is undefined) and $x = \frac{1}{2}$ shows the graph is concave up on this interval. Also, note that $f''(x) < 0$ for

Figure 10.41

(a)

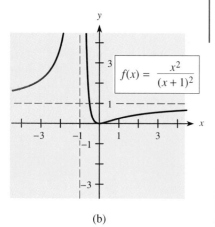

$$f(x) = \frac{x^2}{(x+1)^2}$$

(b)

$x > \frac{1}{2}$, so the point $(\frac{1}{2}, \frac{1}{9})$ is a point of inflection. Also see the sign diagram for $f''(x)$.

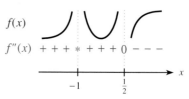

$f(x)$

$f''(x)$ $+ + +$ $*$ $+ + +$ 0 $- - -$

* $x = -1$ is a vertical asymptote.

6. To see how the graph approaches the horizontal asymptote, we check $f(x)$ for large values of $|x|$.

$$f(-100) = \frac{(-100)^2}{(-99)^2} = \frac{10{,}000}{9{,}801} > 1, \quad f(100) = \frac{100^2}{101^2} = \frac{10{,}000}{10{,}201} < 1$$

Thus the graph has the characteristics shown in Figure 10.41(a). The graph is shown in Figure 10.41(b).

When we wish to learn about a function $f(x)$ or sketch its graph, it is important to understand what information we obtain from $f(x)$, from $f'(x)$, and from $f''(x)$. The following summary may be helpful.

Summary

Source	Information Provided
$f(x)$	y-coordinates; horizontal asymptotes, vertical asymptotes; domain restrictions
$f'(x)$	Increasing [$f'(x) > 0$]; decreasing [$f'(x) < 0$]; critical points [$f'(x) = 0$ or $f'(x)$ undefined]; sign-diagram tests for maxima and minima
$f''(x)$	Concave up [$f''(x) > 0$]; concave down [$f''(x) < 0$]; possible points of inflection [$f''(x) = 0$ or $f''(x)$ undefined]; sign-diagram tests for points of inflection; second-derivative test for maxima and minima

CHECKPOINT

1. Let $f(x) = \dfrac{2x + 10}{x - 1}$ and decide whether the following are true or false.

   (a) $f(x)$ has a vertical asymptote at $x = 1$.

   (b) $f(x)$ has $y = 2$ as its horizontal asymptote.

2. Let $f(x) = \dfrac{x^3 - 16}{x} + 1$; then $f'(x) = \dfrac{2x^3 + 16}{x^2}$ and $f''(x) = \dfrac{2x^3 - 32}{x^3}$.

Use these to determine whether the following are true or false.
(a) There are no asymptotes.
(b) $f'(x) = 0$ when $x = -2$
(c) A partial sign diagram for $f'(x)$ is

$f'(x)$ $- - - - 0 + + + + * + + + +$

-2 0

* means $f'(0)$ is undefined.

(d) There is a relative minimum at $x = -2$.
(e) A partial sign diagram for $f''(x)$ is

$f''(x) + + + + * - - - - 0 + + + +$

0 $\sqrt[3]{16}$

* means $f''(0)$ is undefined.

(f) There are points of inflection at $x = 0$ and $x = \sqrt[3]{16}$.

We now consider the problem introduced in the Application Preview.

EXAMPLE 3

If the total daily cost of producing plastic rafts for swimming pools is given by

$$C = 500 + 8x + 0.05x^2,$$

where x is the number of units produced per day, then the average cost per unit produced is given by

$$\overline{C} = \frac{500 + 8x + 0.05x^2}{x} = \frac{500}{x} + 8 + 0.05x, \qquad \text{for } x > 0.$$

Graph this function.
(a) Discuss what happens to average cost as the number of units decreases approaching 0.
(b) Find the level of production that minimizes average cost.

Solution
The domain of $\overline{C}(x)$ does not include 0, and $\lim\limits_{x \to 0^+} \overline{C} = +\infty$, so there is a vertical asymptote at $x = 0$. Thus the average cost per unit increases without bound as the number of units produced approaches zero. Finding the derivative of $\overline{C}(x)$ gives

$$\overline{C}' = -\frac{500}{x^2} + 0.05.$$

Figure 10.42

Setting $\overline{C}' = 0$ and solving for x gives the critical values of x.

$$0 = -\frac{500}{x^2} + 0.05$$
$$0.05x^2 = 500$$
$$x = -100 \quad \text{or} \quad x = 100$$

The second derivative, $\overline{C}'' = 1000x^{-3} = \dfrac{1000}{x^3}$, is positive at $x = 100$, so $\overline{C}(100) = 18$ is the minimum possible average daily cost. The graph of this function is shown in Figure 10.42. The graph verifies that average cost is minimized when 100 units are produced and shows the asymptotic behavior of the function at 0.

Graphing Utilities

If a graphing utility is not available, the procedures previously outlined in this section are necessary to generate a complete and accurate graph. With a graphing utility, the graph of a function is easily generated as long as the viewing window dimensions are appropriate. However, while a graphing utility may reveal the existence of asymptotes, it cannot always precisely locate them. Sometimes we need information provided by derivatives to obtain a window that shows all features of a graph.

Figure 10.43

EXAMPLE 4

Figure 10.43 shows the graph of $f(x) = \dfrac{71x^2}{28(3 - 2x^2)}$.

(a) Determine whether the function has horizontal or vertical asymptotes, and estimate where they occur.
(b) Check your conclusions to part (a) analytically.
(c) Discuss which method is more accurate.

Solution

(a) The graph appears to have a horizontal asymptote somewhere between $y = -1$ and $y = -2$, perhaps near $y = -1.5$. Also, there are two vertical asymptotes located approximately at $x = 1.25$ and $x = -1.25$.
(b) We locate the horizontal asymptote by finding the limit of the function as x approaches infinity.

$$\lim_{x \to \infty} \frac{71x^2}{28(3 - 2x^2)} = \lim_{x \to \infty} \frac{\frac{71x^2}{x^2}}{\frac{84}{x^2} - \frac{56x^2}{x^2}} = \frac{71}{-56} \approx -1.268$$

Vertical asymptotes occur at x-values where $28(3 - 2x^2) = 0$ or

$$3 - 2x^2 = 0$$
$$3 = 2x^2$$
$$\frac{3}{2} = x^2$$
$$\pm\sqrt{\frac{3}{2}} = x \quad \text{or} \quad x \approx \pm1.225$$

(c) The analytical method is, of course, more accurate because asymptotes reveal extreme behavior of the function either vertically or horizontally. An accurate graph shows all features, but not necessarily the details of any feature. Despite this, our estimates from the graph were not too bad.

Even with a graphing utility, sometimes analytic methods are needed to determine an appropriate viewing window.

EXAMPLE 5

The standard viewing window of the graph of $f(x) = \dfrac{x + 10}{x^2 + 300}$ appears blank (check and see). Find any asymptotes, maxima, and minima, and determine an appropriate viewing window. Sketch the graph.

Solution

Because $x^2 + 300 = 0$ has no real solution, there are no vertical asymptotes. We locate the horizontal asymptotes by finding the limit of the function as x approaches infinity.

$$\lim_{x\to\infty} \frac{x + 10}{x^2 + 300} = \lim_{x\to\infty} \frac{\dfrac{x}{x^2} + \dfrac{10}{x^2}}{\dfrac{x^2}{x^2} + \dfrac{300}{x^2}} = \frac{0 + 0}{1 + 0} = 0$$

Hence $y = 0$ (the x-axis) is a horizontal asymptote. We then find an appropriate viewing window by locating the critical points.

$$f'(x) = \frac{(x^2 + 300)(1) - (x + 10)(2x)}{(x^2 + 300)^2}$$
$$= \frac{x^2 + 300 - 2x^2 - 20x}{(x^2 + 300)^2} = \frac{300 - 20x - x^2}{(x^2 + 300)^2}$$

$f'(x) = 0$ when the numerator is zero. Thus

$$300 - 20x - x^2 = 0$$
$$0 = x^2 + 20x - 300$$
$$0 = (x + 30)(x - 10)$$
$$x + 30 = 0 \qquad x - 10 = 0$$
$$x = -30 \qquad\quad x = 10.$$

Figure 10.44

The critical points are $x = -30$, $y = -\frac{1}{60} \approx -0.01666667$ and $x = 10$, $y = \frac{1}{20} = 0.05$. A sign diagram for $f'(x)$ is shown at the right.

Without using the information above, a graphing utility may not give a useful graph. An x-range that includes -30 and 10 is needed. Because $y = 0$ is a horizontal asymptote, these relative extrema are absolute and the y-range must be quite small for the shape of the graph to be seen clearly. Figure 10.44 shows the graph.

EXAMPLE 6

A profit function for a product is given by

$$P(x) = \frac{-x^2 + 16x - 4}{4x^2 + 16}, \qquad \text{for } x \geq 0,$$

where x is in thousands of units and $P(x)$ is in billions of dollars. Because of fixed costs, profit is negative when fewer than 254 units are produced and sold. Will a loss occur at any other level of production and sales?

Figure 10.45

(a)

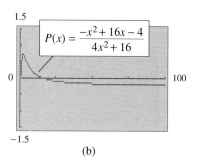

(b)

Solution

Looking at the graph of this function over the range $0 \leq x \leq 10$ (see Figure 10.45(a)), it is not clear whether the graph will eventually cross the x-axis. To see if a loss ever occurs and to see what profit will approach as the number of units produced and sold becomes large, we evaluate the limit of $P(x)$ as $x \to \infty$.

$$\lim_{x \to \infty} \frac{-x^2 + 16x - 4}{4x^2 + 16} = \lim_{x \to \infty} \frac{-1 + \dfrac{16}{x} - \dfrac{4}{x^2}}{4 + \dfrac{16}{x^2}} = -\frac{1}{4}$$

Thus a loss of $\$\frac{1}{4}$ billion is approached as the number of units increases without bound. Figure 10.45(b) shows that the graph does cross the x-axis, where $-x^2 + 16x - 4 = 0$. Using TRACE, SOLVER, the Quadratic Formula, or a program gives $P(x) = 0$ at $x \approx 0.254$ and $x \approx 15.7459$. Thus if 15,746 units or more are produced and sold, the profit is negative.

CHECKPOINT
SOLUTIONS

1. (a) True, $x = 1$ makes the denominator of $f(x)$ equal to zero, while the numerator is non-zero.

   (b) True, $\displaystyle \lim_{x \to \infty} \frac{2x + 10}{x - 1} = 2$ and $\displaystyle \lim_{x \to -\infty} \frac{2x + 10}{x - 1} = 2$.

2. (a) False. There are no horizontal asymptotes, but $x = 0$ is a vertical asymptote.
   (b) True
   (c) True
   (d) True. The relative minimum point is $(-2, f(-2)) = (-2, 13)$.
   (e) True
   (f) False. There is a point of inflection only at $(\sqrt[3]{16}, 1)$. At $x = 0$ the vertical asymptote occurs, so there is no point on the graph, and hence no point of inflection.

EXERCISE 10.5

In problems 1–4, a function and its graph are given. Use the graph to find each of the following, if they exist, then confirm your results analytically.

(a) vertical asymptotes

(b) $\lim\limits_{x \to \infty} f(x)$

(c) horizontal asymptotes

(d) $\lim\limits_{x \to -\infty} f(x)$

1. $f(x) = \dfrac{x - 4}{x - 2}$

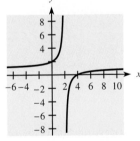

2. $f(x) = \dfrac{8}{x + 2}$

3. $f(x) = \dfrac{x^2}{(x - 2)^2}$

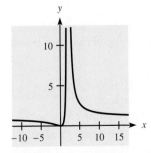

4. $f(x) = \dfrac{3(x^4 + 2x^3 + 6x^2 + 2x + 5)}{(x^2 - 4)^2}$

In problems 5–10, find any horizontal and vertical asymptotes for each function.

5. $y = \dfrac{2x}{x - 3}$

6. $y = \dfrac{3x - 1}{x + 5}$

7. $y = \dfrac{x + 1}{x^2 - 4}$

8. $y = \dfrac{4x}{9 - x^2}$

9. $y = \dfrac{3x^3 - 6}{x^2 + 4}$

10. $y = \dfrac{6x^3}{4x^2 + 9}$

In problems 11–16, a function and its first and second derivatives are given. Use these to find any horizontal and vertical asymptotes, critical points, relative maxima, relative minima, and points of inflection; then sketch the graph of each function.

11. $y = \dfrac{x}{(x - 1)^2}$

$y' = -\dfrac{x + 1}{(x - 1)^3}$

$y'' = \dfrac{2x + 4}{(x - 1)^4}$

12. $y = \dfrac{(x - 1)^2}{x^2}$

$y' = \dfrac{2(x - 1)}{x^3}$

$y'' = \dfrac{6 - 4x}{x^4}$

13. $y = x + \dfrac{3}{\sqrt[3]{x - 3}}$

$y' = \dfrac{(x - 3)^{2/3} - 1}{(x - 3)^{2/3}}$

$y'' = \dfrac{2}{3(x - 3)^{5/3}}$

14. $y = 3\sqrt[3]{x} + \dfrac{1}{x}$

$y' = \dfrac{x^{4/3} - 1}{x^2}$

$y'' = \dfrac{6 - 2x^{4/3}}{3x^3}$

15. $f(x) = \dfrac{9(x - 2)^{2/3}}{x^2}$

$f'(x) = \dfrac{12(3 - x)}{x^3(x - 2)^{1/3}}$

$f''(x) = \dfrac{4(7x^2 - 42x + 54)}{x^4(x - 2)^{4/3}}$

16. $f(x) = \dfrac{3x^{2/3}}{x + 1}$

$f'(x) = \dfrac{2 - x}{x^{1/3}(x + 1)^2}$

$f''(x) = \dfrac{2(2x^2 - 8x - 1)}{3x^{4/3}(x + 1)^3}$

For each function in problems 17–24, find any horizontal and vertical asymptotes, and use information from the derivative to sketch the graph.

17. $f(x) = \dfrac{2x + 2}{x - 3}$ 18. $f(x) = \dfrac{5x - 15}{x + 2}$ 19. $y = \dfrac{x^2 + 4}{x}$

20. $y = \dfrac{x^2 + 4}{x^2}$ 21. $y = \dfrac{27x^2}{(x + 1)^3}$ 22. $y = \left(\dfrac{x + 2}{x - 3}\right)^2$

23. $f(x) = \dfrac{16x}{x^2 + 1}$ 24. $f(x) = \dfrac{4x^2}{x^4 + 1}$

In problems 25–28, a function and its graph are given.
(a) Use the graph to estimate the location of any horizontal or vertical asymptotes.
(b) Use the function to precisely determine the location of any asymptotes.

25. $f(x) = \dfrac{9x}{17 - 4x}$

26. $f(x) = \dfrac{5 - 13x}{3x + 20}$

27. $f(x) = \dfrac{20x^2 + 98}{9x^2 - 49}$

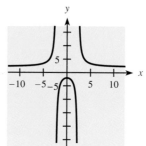

28. $f(x) = \dfrac{15x^2 - x}{7x^2 - 35}$

 For each function in problems 29–34, complete the following steps.
(a) Use a graphing utility to graph the function in the standard viewing window.
(b) Analytically determine the location of any asymptotes and extrema.
(c) Graph the function in a viewing window that shows all features of the graph. State the ranges for x-values and y-values for your viewing window.

29. $f(x) = \dfrac{x + 25}{x^2 + 1400}$ 30. $f(x) = \dfrac{x - 50}{x^2 + 1100}$

31. $f(x) = \dfrac{100(9 - x^2)}{x^2 + 100}$ 32. $f(x) = \dfrac{200x^2}{x^2 + 100}$

33. $f(x) = \dfrac{1000x - 4000}{x^2 - 10x - 2000}$

34. $f(x) = \dfrac{900x + 5400}{x^2 - 30x - 1800}$

APPLICATIONS

35. **Cost-benefit** The percentage p of particulate pollution that can be removed from the smokestacks of an industrial plant by spending C dollars is given by

$$p = \dfrac{100C}{7300 + C}.$$

(a) Find any C-values where the rate of change of p with respect to C does not exist. Make sure that these make sense in the problem.
(b) Find C-values for which p is increasing.
(c) If there is a horizontal asymptote, find it.
(d) Can 100% of the pollution be removed?

36. **Cost-benefit** The percentage p of impurities that can be removed from the waste water of a manufacturing pro-

cess at a cost of C dollars is given by

$$p = \frac{100C}{8100 + C}.$$

(a) Find any C-values where the rate of change of p with respect to C does not exist. Make sure that these make sense in the problem.
(b) Find C-values for which p is increasing.
(c) Find any horizontal asymptotes.
(d) Can 100% of the pollution be removed?

37. **Revenue** A recently released film has its weekly revenue given by

$$R(t) = \frac{50t}{t^2 + 36}, \qquad t \geq 0,$$

where $R(t)$ is in millions of dollars and t is in weeks.
(a) Graph $R(t)$.
(b) When will revenue be maximized?
(c) Suppose that if revenue decreases for 4 consecutive weeks, the film will be removed from theaters and will be released as a video 12 weeks later. When will the video come out?

38. **Production costs** Suppose that the total cost of producing a shipment of a certain product is

$$C = 5000x + \frac{125,000}{x}, \qquad x > 0,$$

where x is the number of machines used in production.
(a) Graph this total cost function.
(b) Using how many machines will minimize the total cost?

39. **Windchill** If x is the wind speed in miles per hour and is greater than or equal to 5, then the windchill (in degrees Fahrenheit) for an air temperature of 0°F can be approximated by the function

$$f(x) = \frac{289.173 - 58.5731x}{x + 1} \qquad x \geq 5.$$

(a) Does $f(x)$ have a vertical asymptote? If so, what is it?
(b) Does $f(x)$ have a vertical asymptote within its domain?
(c) Does $f(x)$ have a horizontal asymptote? If so, what is it?
(d) In the context of windchill, does $\lim_{x \to \infty} f(x)$ have a physical interpretation? If so, what is it, and is it meaningful?

40. **Profit** An entrepreneur starts new companies and sells them when their growth is maximized. Suppose that the

annual profit for a new company is given by

$$P(x) = 22 - \frac{1}{2}x - \frac{18}{x + 1},$$

where P is in thousands of dollars and x is the number of years after the company is formed. If she wants to sell the company before profits begin to decline, after how many years should she sell it?

41. **Productivity** The figure is a typical graph of worker productivity per hour P as a function of time t on the job.
(a) What is the horizontal asymptote?
(b) What is $\lim_{x \to \infty} P(t)$?
(c) What is the horizontal asymptote for $P'(t)$?
(d) What is $\lim_{x \to \infty} P'(t)$?

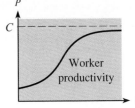

42. **Sales volume** The figure shows a typical curve that gives the volume of sales S as a function of time t since an ad campaign.
(a) What is the horizontal asymptote?
(b) What is $\lim_{t \to \infty} S(t)$?
(c) What is the horizontal asymptote for $S'(t)$?
(d) What is $\lim_{t \to \infty} S'(t)$?

Farm workers The percent of U.S. workers in farm occupations during certain years is shown in the table.

Year	Percent of All Workers in Farm Occupations
1820	71.8
1850	63.7
1870	53
1900	37.5
1920	27
1930	21.2
1940	17.4
1950	11.6
1960	6.1
1970	3.6
1980	2.7
1985	2.8
1990	2.4

Source: *The World Almanac and Book of Facts*, 1993

Assume that the percentage of U.S. workers in farm occupations can be modeled with the function

$$f(t) = 1000 \cdot \frac{-8.0912t + 1558.9}{1.09816t^2 - 122.183t + 21472.6},$$

where t is the number of years past 1800.
Use this model in problems 43 and 44.

43. (a) Find $\lim_{t \to \infty} f(t)$.
    (b) Interpret your answer to part (a).
    (c) Does $f(t)$ have any vertical asymptotes within its domain $t \geq 0$?

44. (a) Use a graphing utility to graph $f(t)$ for $t = 0$ to $t = 220$.
    (b) From the graph, identify t-values where the model is inappropriate, and explain why it is inappropriate.

45. **Barometric pressure** The figure shows a barograph readout of the barometric pressure as recorded by Georgia Southern University's meteorological equipment. The figure shows a tremendous drop in barometric pressure on Saturday morning, March 13, 1993.
    (a) If $B(t)$ is barometric pressure expressed as a function of time, as shown in the figure, does $B(t)$ have a vertical asymptote sometime after 8 A.M. on Saturday, March 13, 1993? Explain why or why not.

(b) Consult your library or some other resource to find out what happened in Georgia (and in the eastern U.S.) on March 13, 1993, to cause such a dramatic drop in barometric pressure.

SOURCE: *Statesboro Herald*, March 14, 1993

KEY TERMS & FORMULAS

Section	Key Term	Formula
10.1	Relative maxima and minima	
	Increasing	$f'(x) > 0$
	Decreasing	$f'(x) < 0$
	Critical points	$f'(x) = 0$ or $f'(x)$ undefined
	Sign diagram for $f'(x)$	
	First-derivative test	
	Horizontal point of inflection	

Section	Key Term	Formula
10.2	Concave up	$f''(x) > 0$
	Concave down	$f''(x) < 0$
	Point of inflection	May occur where $f''(x) = 0$ or $f''(x)$ undefined
	Sign diagram for $f''(x)$	
	Second-derivative test	
10.3	Absolute extrema	
	Average cost	$\bar{C}(x) = C(x)/x$
	Profit maximization	
	Competitive market	$R(x) = p \cdot x$ where $p =$ equilibrium price
	Monopolistic market	$R(x) = p \cdot x$ where $p = f(x)$ is the demand function
10.4	Inventory cost models	
10.5	Asymptotes	
	Horizontal: $y = b$	$\lim\limits_{x \to +\infty} f(x) = b$ or $\lim\limits_{x \to -\infty} f(x) = b$
	Vertical: $x = c$ for rational function $y = f(x)/g(x)$	y unbounded near $x = c$ if $g(c) = 0$ and $f(c) \neq 0$

REVIEW EXERCISES

10.1

In problems 1–4, find all critical points and determine whether they are relative maxima, relative minima, or horizontal points of inflection.

1. $y = -x^2$

2. $p = q^2 - 4q - 5$

3. $f(x) = 1 - 3x + 3x^2 - x^3$

4. $f(x) = \dfrac{3x}{x^2 + 1}$

In problems 5–10,
(a) find all critical values, including those where $f'(x)$ is undefined.
(b) find the relative maxima and minima, if any exist.
(c) find the horizontal points of inflection, if any exist.
(d) sketch the graph.

5. $y = x^3 + x^2 - x - 1$

6. $f(x) = 4x^3 - x^4$

7. $f(x) = x^3 - \dfrac{15}{2}x^2 - 18x + \dfrac{3}{2}$

8. $y = 5x^7 - 7x^5 - 1$

9. $y = x^{2/3} - 1$

10. $y = x^{2/3}(x - 4)^2$

10.2

11. Is the graph of $y = x^4 - 3x^3 + 2x - 1$ concave up or concave down at $x = 2$?

12. Find intervals where the graph of $y = x^4 - 2x^3 - 12x^2 + 6$ is concave upward and where it is concave downward, and find points of inflection.

13. Find the relative maxima, relative minima, and points of inflection of the graph of $y = x^3 - 3x^2 - 9x + 10$.

In problems 14 and 15, find any relative maxima, relative minima, and points of inflection, and sketch each graph.

14. $y = x^3 - 12x$

15. $y = 2 + 5x^3 - 3x^5$

10.3

16. Given $R = 280x - x^2$, find the absolute maximum and minimum for R when (a) $0 \leq x \leq 200$ and (b) $0 \leq x \leq 100$.

17. Given $y = 6400x - 18x^2 - \dfrac{x^3}{3}$, find the absolute maximum and minimum for y when (a) $0 \leq x \leq 50$ and (b) $0 \leq x \leq 100$.

10.5

In problems 18 and 19, use the graphs to find the following:

(a) vertical asymptotes   (b) horizontal asymptotes

(c) $\lim\limits_{x \to +\infty} f(x)$ (d) $\lim\limits_{x \to -\infty} f(x)$

18.

19.

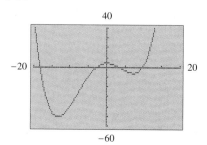

In problems 20 and 21, find any horizontal asymptotes and any vertical asymptotes.

20. $y = \dfrac{3x + 2}{2x - 4}$

21. $y = \dfrac{x^2}{1 - x^2}$

In problems 22–24,

(a) find any horizontal and vertical asymptotes.

(b) find any relative maxima and minima.

(c) sketch each graph.

22. $y = \dfrac{3x}{x + 2}$ 23. $y = \dfrac{8(x - 2)}{x^2}$

24. $y = \dfrac{x^2}{x - 1}$

10.1, 10.2

In problems 25 and 26 a function and its graph are given.

(a) Use the graph to determine (estimate) x-values where $f'(x) > 0$, where $f'(x) < 0$, and where $f'(x) = 0$.

(b) Use the graph to determine x-values where $f''(x) > 0$, where $f''(x) < 0$, and where $f''(x) = 0$.

(c) Check your conclusions to (a) by finding $f'(x)$ and graphing it with a graphing utility.

(d) Check your conclusions to (b) by finding $f''(x)$ and graphing it with a graphing utility.

25. $f(x) = x^3 - 4x^2 + 4x$

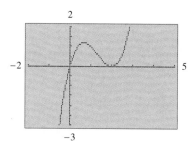

26. $f(x) = 0.0025x^4 + 0.02x^3 - 0.48x^2 + 0.08x + 4$

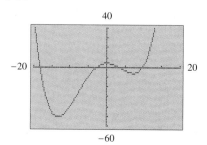

In problems 27 and 28, $f'(x)$ and its graph are given.

(a) Use the graph of $f'(x)$ to determine (estimate) where the graph of $f(x)$ is increasing, where it is decreasing, and where it has relative extrema.

(b) Use the graph of $f'(x)$ to determine where $f''(x) > 0$, where $f''(x) < 0$, and where $f''(x) = 0$.

(c) Verify that the given $f(x)$ has $f'(x)$ as its derivative, and graph $f(x)$ to check your conclusions in (a).

(d) Calculate $f''(x)$ and graph it to check your conclusions in (b).

27. $f'(x) = x^2 + 4x - 5$ $\left(\text{for } f(x) = \dfrac{x^3}{3} + 2x^2 - 5x\right)$

28. $f'(x) = 6x^2 - x^3$ $\left(\text{for } f(x) = 2x^3 - \dfrac{x^4}{4}\right)$

In problems 29 and 30, $f''(x)$ and its graph are given.

(a) Use the graph to determine (estimate) where the graph of $f(x)$ is concave upward, where it is concave downward, and where it has points of inflection.

(b) Verify that the given $f(x)$ has $f''(x)$ as its second derivative, and graph $f(x)$ to check your conclusions in part (a).

29. $f''(x) = 4 - x$ $\left(\text{for } f(x) = 2x^2 - \dfrac{x^3}{6}\right)$

30. $f''(x) = 6 - x - x^2$ $\left(\text{for } f(x) = 3x^2 - \dfrac{x^3}{6} - \dfrac{x^4}{12}\right)$

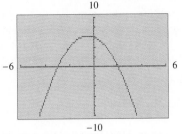

APPLICATIONS

10.3

31. **Cost** If the total cost function for a product is

$$C(x) = 3x^2 + 15x + 75,$$

how many units minimize the average cost? Find the minimum average cost.

32. **Revenue** Suppose the total revenue function for a product is given by

$$R(x) = 32x - 0.01x^2.$$

(a) How many units will maximize the total revenue? Find the maximum revenue.

(b) If production is limited to 2500 units, how many units will maximize the total revenue? Find the maximum revenue.

33. **Profit** Suppose that the profit function for a product is given by

$$P(x) = 1080x + 9.6x^2 - 0.1x^3 - 50{,}000.$$

Find the maximum profit.

34. **Profit** How many units (x) will maximize profit if $R(x) = 46x - x^2$ and $C(x) = 5x^2 + 10x + 3$?

35. **Profit** A product can be produced at a total cost $C(x) = 800 + 4x$, where x is the number produced and is limited to at most 150 units. If the total revenue is given by $R(x) = 80x - \frac{1}{4}x^2$, determine the level of production that will maximize the profit.

36. **Average cost** The total cost function for a product is $C = 2x^2 + 54x + 98$. Producing how many units will minimize average cost?

37. **Revenue** McRobert's TV Shop sells 200 sets per month at a price of $400 per unit. Market research indicates that they can sell one additional set for each $1 they reduce the price. At what selling price will they maximize revenue?

38. **Profit** If, in problem 37, the sets cost the shop $250 each, when will profit be maximized?

39. **Profit** Suppose that for a product in a competitive market the demand function is $p = 1200 - 2x$ and the supply function is $p = 200 + 2x$, where x is the number of units. If a firm's average cost function for this product is

$$\overline{C}(x) = \frac{12,000}{x} + 50 + x,$$

find the maximum profit. Hint: First find the equilibrium price.

40. **Profit** The monthly demand function for a product sold by a monopoly is $p = 800 - x$, and its average cost is $\overline{C} = 200 + x$. Determine
(a) the quantity that will maximize profit.
(b) the selling price at the optimal quantity.

41. **Profit** Suppose that in a monopolistic market the demand function for a commodity is

$$p = 7000 - 10x - \frac{x^2}{3}.$$

If a company's average cost function for this commodity is

$$\overline{C}(x) = \frac{40,000}{x} + 600 + 8x,$$

find the maximum profit.

10.4

42. **Reaction to a drug** The reaction R to an injection of a drug is related to the dosage x according to

$$R(x) = x^2\left(500 - \frac{x}{3}\right).$$

Find the dosage that yields the maximum reaction.

43. **Productivity** The number of parts produced per hour by a worker is given by

$$N = 4 + 3t^2 - t^3,$$

where t is the number of hours on the job without a break. If the worker starts at 8 A.M., when will she be at maximum productivity during the morning?

44. **Population** Population estimates show that the equation $P = 300 + 10t - t^2$ represents the size of the graduating class of a high school, with t representing the

number of years after 1990, $0 \le t \le 10$. What will be the largest graduating class in the decade?

45. **Night brightness** Suppose that an observatory is to be built between cities A and B that are 30 miles apart. For the best viewing, the observatory should be located where the night brightness from these cities is minimum. If the night brightness of city A is 8 times that of city B, then the night brightness b between the two cities and x miles from A is given by

$$b = \frac{8k}{x^2} + \frac{k}{(30 - x)^2},$$

where k is a constant. Find the best location for the observatory; that is, find x that minimizes b.

46. **Product design** A playpen manufacturer wants to make a rectangular enclosure with maximum play area. To remain competitive, he wants the perimeter of the base to be only 16 feet. What dimensions should the playpen have?

47. **Printing design** A printed page is to contain 56 square inches and have a $\frac{3}{4}$-inch margin at the bottom and 1-inch margins at the top and on both sides. Find the dimensions that minimize the size of the page (and hence the costs for paper).

48. **Drug sensitivity** The reaction R to an injection of a drug is related to the dosage x according to

$$R(x) = x^2\left(500 - \frac{x}{3}\right).$$

The sensitivity to the drug is defined by dR/dx. Find the dosage that maximizes sensitivity.

49. **Photosynthesis** The amount of photosynthesis that takes place in a certain plant depends on the intensity of light x according to the equation

$$f(x) = 145x^2 - 30x^3.$$

The rate of change of the amount of photosynthesis with respect to the intensity is $f'(x)$. Find the intensity that maximizes the rate of change.

50. **Inventory cost model** A company needs 288,000 items per year. Production costs are $1500 to prepare for a production run and $30 for each item produced. Inventory costs are $1.50 per year for each item stored. Find the number of items that should be produced in each run so that the total costs of production and storage are minimum.

CONSULTANT'S CORNER

Production Management

Metal Containers, Inc., is reviewing the way in which it submits bids on U.S. Army contracts. The army often requests open-top boxes, with square bases and of specified volumes. The army also specifies the materials for the boxes, with the base usually made of a different material than the sides. The box is put together by riveting a bracket at each of the eight corners. For Metal Containers, the total cost of producing a box is the sum of the cost of the materials for the box and the labor costs associated with affixing each bracket.

Instead of estimating each job separately, the company wants to develop an overall approach that will allow it to cost out proposals more easily. To accomplish this they need you to devise a formula for the total cost of producing each box and determine the dimensions that allow a box of specified volume to be produced at minimum cost. Use the following notation to help you solve this problem.

Cost of the material for the base = A per square unit; Cost of the material for the sides = B per square unit; Cost of each bracket = C; Cost to affix each bracket = D; Length of the sides of the base = x; Height of the box = h; Volume specified by the army = V

1. Write an expression for the company's total cost in terms of these quantities.
2. At the time an order is received for boxes of a specified volume, the costs of the materials and labor will be fixed and only the dimensions will vary. Find a formula for each of the dimensions of the box so that the total cost is a minimum.
3. The army requests bids on boxes of 48 cubic feet with base material costing the container company $12 per square foot and side material costing $8 per square foot. Each bracket costs $5, and the associated labor cost is $1 per bracket. Use your formulas to find the dimensions of the box that meets the army's requirements at a minimum cost. What is this cost?

Metal Containers asks you to determine how best to order the brackets they use on their boxes. You are able to obtain the following information: They use approximately 100,000 brackets a year and the purchase price of each is $5. They buy the same number of brackets (say, n) each time they place an order with their supplier, and it costs $60 to process each order. Metal Containers also has additional costs associated with storing, insuring, and financing their inventory of brackets. These carrying costs amount to 15% of the average value of inventory annually. The brackets are used steadily and deliveries are made just as inventory reaches zero, so that inventory fluctuates between zero and n brackets.

4. If the total annual cost associated with the bracket supply is the sum of the annual purchasing cost and the annual carrying costs, what order size n would minimize the total cost?
5. In the general case of the bracket ordering problem, the order size n that minimizes the total cost of the bracket supply is called the Economic Order Quantity or EOQ. Use the following notations to determine a general formula for the EOQ.

Fixed cost per order = F Unit cost = C
Quantity purchased per year = P Carrying cost (as a decimal rate) = r

Warm-up

Prerequisite problem type	For section	Answer	Section for review
(a) Simplify: $\dfrac{1}{(3x)^{1/2}} \cdot \dfrac{1}{2}(3x)^{-1/2} \cdot 3$ (b) Write with a positive exponent: $\sqrt{x^2 - 1}$	11.1	(a) $\dfrac{1}{2x}$ (b) $(x^2 - 1)^{1/2}$	0.4 Rational exponents
(a) Vertical lines have _____ slopes. (b) Horizontal lines have _____ slopes.	11.3	(a) Undefined (b) 0	1.7 Slopes
Write the equation of the line passing through $(-2, -2)$ with slope 5.	11.3	$y = 5x + 8$	1.7 Equations of lines
Solve: $x^2 + y^2 - 9 = 0$, for y.	11.3	$y = \pm\sqrt{9 - x^2}$	4.1 Quadratic equations
(a) Write $\log_a(x + h) - \log_a x$ as an expression involving one logarithm. (b) Does $\ln x^4 = 4 \ln x$? (c) Does $\dfrac{x}{h} \log_a\left(\dfrac{x + h}{x}\right) = \log_a\left(1 + \dfrac{h}{x}\right)^{x/h}$? (d) Expand $\ln(xy)$ to separate x and y. (e) If $y = a^x$, then $x =$ _____. (f) Simplify $\ln e^x$.	11.1 11.2	(a) $\log_a\left(\dfrac{x + h}{x}\right)$ (b) Yes (c) Yes (d) $\ln x + \ln y$ (e) $\log_a y$ (f) x	5.2 Logarithms
Find the derivative of (a) $y = x^2 - 2x - 2$ (b) $T(q) = 400q - \frac{4}{3}q^2$ (c) $y = \sqrt{9 - x^2}$	11.1– 11.5	(a) $y' = 2x - 2$ (b) $T'(q) =$ $400 - \frac{8}{3}q$ (c) $y' =$ $-x\,(9 - x^2)^{-1/2}$	9.4–9.6 Derivatives
If $\dfrac{dy}{dx} = \dfrac{-x}{\sqrt{9 - x^2}}$, find the slope of the tangent to $y = f(x)$ at $x = \sqrt{5}$.	11.3	Slope $= -\dfrac{\sqrt{5}}{2}$	9.3–9.5 Derivatives

CHAPTER 11

DERIVATIVES CONTINUED

In this chapter we will develop derivative formulas for logarithmic and exponential functions and apply them to problems in the management, life, and social sciences. We will also develop methods for finding derivatives of one variable with respect to another even though the relationship between them may not be functional. This method is called **implicit differentiation.**

We will use implicit derivatives with respect to time to solve problems involving the rates of change of two or more variables. These problems are called **related rate problems.**

The special business and economics applications include elasticity of demand and maximization of taxation revenue.

11.1 DERIVATIVES OF LOGARITHMIC FUNCTIONS

Objective

❑ To find derivatives of logarithmic functions

APPLICATION PREVIEW

The table below shows the expected life span at birth for people born in certain years in the United States. Assume that for these years, the expected life span at birth can be modeled with the function $l(x) = 11.7 + 14.1 \ln x$, where x is the number of years past 1900. The graph of

Year	Life Span (in Years)	Year	Life Span (in Years)
1920	54.1	1975	72.6
1930	59.7	1980	73.7
1940	62.9	1987	75.0
1950	68.2	1988	74.9
1960	69.7	1989	75.2
1970	70.8		

SOURCE: *World Almanac*, 1991

this function is shown in Figure 11.1. If we wanted to use this model to find the rate of change of life span with respect to the number of years past 1900, we would need the derivative of this function and hence the derivative of the logarithmic function $\ln x$.

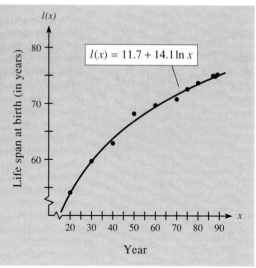

Figure 11.1

In Chapter 5 we defined the logarithmic function $y = \log_a x$ and developed properties of logarithms that will be important in this section. Although logarithmic functions can involve logarithms of any base, including base 10 (called **common logarithms**), most of the problems in calculus and many of the applications to management, life, and social sciences involve logarithms with base e, called **natural logarithms.** We state the properties of logarithms for natural logarithms below.

Properties of Natural Logarithms

I. $\ln e^x = x$ for any real number x

II. $e^{\ln x} = x$ for $x > 0$

III. $\ln(MN) = \ln M + \ln N$, for M and N positive real numbers

IV. $\ln(M/N) = \ln M - \ln N$, for M and N positive real numbers

V. $\ln(M^N) = N \ln M$, for M a positive real number and N any real number

The formula for the derivative of $y = \ln x$ follows.

Derivative of $y = \ln x$

If $y = \ln x$, then

$$\frac{dy}{dx} = \frac{1}{x}.$$

The proof follows.

If $y = f(x) = \ln x$,

$$\frac{dy}{dx} = \lim_{h \to 0} \frac{f(x + h) - f(x)}{h}$$

$$= \lim_{h \to 0} \frac{\ln(x + h) - \ln x}{h}$$

$$= \lim_{h \to 0} \frac{\ln\left(\dfrac{x+h}{x}\right)}{h} \qquad \text{Property IV}$$

$$= \lim_{h \to 0} \frac{x}{x} \cdot \frac{1}{h} \ln\left(\frac{x+h}{x}\right) \qquad \text{Introduce } \frac{x}{x}.$$

$$= \lim_{h \to 0} \frac{1}{x} \cdot \frac{x}{h} \ln\left(1 + \frac{h}{x}\right)$$

$$= \lim_{h \to 0} \frac{1}{x} \ln\left(1 + \frac{h}{x}\right)^{x/h}. \qquad \text{Property V}$$

The natural logarithmic function is continuous when it is defined, so

$$\frac{dy}{dx} = \frac{1}{x} \ln\left[\lim_{h \to 0}\left(1 + \frac{h}{x}\right)^{x/h}\right].$$

We can evaluate

$$\lim_{h \to 0}\left(1 + \frac{h}{x}\right)^{x/h} \qquad (1)$$

by recalling from problem 51 of Exercise 9.1 that

$$\lim_{a \to 0}(1 + a)^{1/a} = e$$

and noting that equation (1) has this form. Thus

$$\lim_{h \to 0}\left(1 + \frac{h}{x}\right)^{x/h} = e$$

and

$$\frac{dy}{dx} = \frac{1}{x} \ln e = \frac{1}{x}.$$

EXAMPLE 1

If $y = x^3 + 3 \ln x$, find dy/dx.

Solution

$$\frac{dy}{dx} = 3x^2 + 3\left(\frac{1}{x}\right) = 3x^2 + \frac{3}{x}$$

EXAMPLE 2

If $y = x^2 \ln x$, find y'.

Solution

By the Product Rule,

$$y' = x^2 \cdot \frac{1}{x} + (\ln x)(2x) = x + 2x \ln x.$$

We can use the Chain Rule to find the formula for the derivative of $y = \ln u$, where $u = f(x)$.

Derivatives of the Natural Logarithmic Functions

If $y = \ln u$, where u is a differentiable function of x, then

$$\frac{dy}{dx} = \frac{1}{u} \cdot \frac{du}{dx}.$$

EXAMPLE 3

Find $f'(x)$ for each of the following.

(a) $f(x) = \ln(x^2)$ (b) $f(x) = \ln(x^4 - 3x + 7)$

Solution

(a) $f'(x) = \dfrac{1}{x^2}(2x) = \dfrac{2x}{x^2} = \dfrac{2}{x}$

(b) $f'(x) = \dfrac{1}{x^4 - 3x + 7}(4x^3 - 3) = \dfrac{4x^3 - 3}{x^4 - 3x + 7}$

EXAMPLE 4

(a) Find $f'(x)$ if $f(x) = \frac{1}{3} \ln(2x^6 - 3x + 2)$.

(b) Find $g'(x)$ if $g(x) = \dfrac{\ln(2x + 1)}{2x + 1}$.

Solution

(a) $f'(x)$ is $\frac{1}{3}$ of the derivative of $\ln(2x^6 - 3x + 2)$.

$$f'(x) = \frac{1}{3} \cdot \frac{1}{2x^6 - 3x + 2}(12x^5 - 3)$$

$$= \frac{4x^5 - 1}{2x^6 - 3x + 2}$$

(b) We begin with the Quotient Rule.

$$g'(x) = \frac{(2x + 1)\dfrac{1}{2x + 1}(2) - [\ln(2x + 1)]2}{(2x + 1)^2}$$

$$= \frac{2 - 2\ln(2x + 1)}{(2x + 1)^2}$$

EXAMPLE 5

Use logarithm properties to find dy/dx when

$$y = \ln[x(x^5 - 2)^{10}].$$

Solution

We use logarithm Properties III and V to rewrite the function.

$$y = \ln x + \ln(x^5 - 2)^{10} \qquad \text{Property III}$$
$$y = \ln x + 10 \ln(x^5 - 2) \qquad \text{Property V}$$

We now take the derivative.

$$\frac{dy}{dx} = \frac{1}{x} + 10 \cdot \frac{1}{x^5 - 2} \cdot 5x^4$$
$$= \frac{1}{x} + \frac{50x^4}{x^5 - 2}.$$

CHECKPOINT

1. If $y = \ln(3x^2 + 2)$, find y'.
2. If $y = \ln x^6$, find y'.
3. If $y = \ln \sqrt[3]{x^2 + 1}$, find y'.

We now return to the life span model in the Application Preview.

EXAMPLE 6

Assume that the average life span (in years) for people born in 1920–1989 can be modeled by

$$l(x) = 11.7 + 14.1 \ln x,$$

where x is the number of years past 1900.
(a) Find the function that models the rate of change of life span.
(b) Does $l(x)$ have a maximum value for $x > 0$?
(c) Evaluate $\lim_{x \to \infty} l'(x)$.
(d) What do (b) and (c) tell us about the average life span?

Solution

(a) The rate of change of life span is given by the derivative.

$$l'(x) = 0 + 14.1 \left(\frac{1}{x}\right) = \frac{14.1}{x}$$

(b) For $x > 0$, we see that $l'(x) > 0$. Hence, $l(x)$ is increasing for all values of $x > 0$, so $l(x)$ never achieves a maximum value. That is, there is no maximum life span.

(c) $\lim_{x \to \infty} l'(x) = \lim_{x \to \infty} \frac{14.1}{x} = 0$

(d) If this model is accurate, life span will continue to increase, but at an ever slower rate.

EXAMPLE 7

Suppose the cost function for x units of a product is given by

$$C(x) = 18{,}250 + 615 \ln(4x + 10),$$

where $C(x)$ is in dollars. Find the marginal cost when 100 units are produced, and explain what it means.

Solution

Marginal cost is given by $C'(x)$.

$$\overline{MC} = C'(x) = 615 \left(\frac{1}{4x + 10}\right)(4) = \frac{2460}{4x + 10}$$

$$\overline{MC}(100) = \frac{2460}{4(100) + 10} = \frac{2460}{410} = 6$$

When 100 units are produced, the marginal cost is 6. This means that the approximate cost of producing the one-hundred-first item is $6.

The **change of base formula,** introduced in Section 5.2, can be used to express logarithms with base a as natural logarithms (that is, logarithms with base e):

$$\log_a x = \frac{\ln x}{\ln a}.$$

We can apply this change of base formula to find the derivative of a logarithm with any base, as the following example illustrates.

EXAMPLE 8

If $y = \log_4(x^3 + 1)$, find dy/dx.

Solution

By using the change of base formula, we have

$$y = \log_4(x^3 + 1) = \frac{\ln(x^3 + 1)}{\ln 4} = \frac{1}{\ln 4} \cdot \ln(x^3 + 1).$$

Thus

$$\frac{dy}{dx} = \frac{1}{\ln 4} \cdot \frac{1}{x^3 + 1} \cdot 3x^2 = \frac{3x^2}{(x^3 + 1)\ln 4}.$$

EXAMPLE 9

Use the graph of $f'(x) = \ln x$ for $x > 0$ to answer the following questions.
(a) At what value a does the graph of $f'(x)$ cross the x-axis (that is, where is $f'(x) = 0$)?
(b) What value a is a critical value for $y = f(x)$?
(c) Does $f(x)$ have a relative maximum or relative minimum at $x = a$?

(d) Graph $y = x \ln x - x$. Does this function appear to satisfy the conditions for $f(x)$ above?

Solution
The graph of $f'(x) = \ln x$ is shown in Figure 11.2.
(a) The graph crosses the x-axis at $x = 1$, so $f'(a) = 0$ if $a = 1$.
(b) $a = 1$ is a critical value of $f(x)$.
(c) Because $f'(x)$ is negative for $x < 1$, $f(x)$ is decreasing for $x < 1$. Because $f'(x)$ is positive for $x > 1$, $f(x)$ is increasing for $x > 1$. Therefore $f(x)$ has a relative minimum at $x = 1$.
(d) The graph of $y = x \ln x - x$ is shown in Figure 11.3. It has a relative minimum at $x = 1$ and may be $f(x)$. To confirm that it is, we take the derivative of $y = x \ln x - x$ and show that $y' = f'(x) = \ln x$.

$$\text{If} \quad y = x \ln x - x, \quad \text{then}$$

$$y' = x\left(\frac{1}{x}\right) + (\ln x)(1) - 1 = \ln x.$$

Figure 11.2

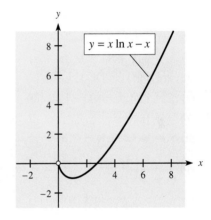

Figure 11.3

CHECKPOINT SOLUTIONS

1. $y' = \dfrac{6x}{3x^2 + 2}$

2. $y' = \dfrac{1}{x^6} \cdot 6x^5 = \dfrac{6}{x}$; or by first using logarithm Property V, we get $y = 6 \ln x$, so
$$y' = 6\left(\frac{1}{x}\right) = \frac{6}{x}.$$

3. By first using logarithm Property V, we get $y = \ln(x^2 + 1)^{1/3} = \frac{1}{3} \ln (x^2 + 1)$. Then
$$y' = \frac{1}{3} \cdot \frac{2x}{x^2 + 1} = \frac{2x}{3(x^2 + 1)}.$$

EXERCISE 11.1

Find the derivatives of the functions in problems 1–10.
1. $f(x) = 4 \ln x$　　　　　　2. $y = 3 \ln x$
3. $y = \ln 8x$　　　　　　　　4. $y = \ln 5x$
5. $y = \ln x^4$　　　　　　　　6. $f(x) = \ln x^3$
7. $f(x) = \ln(4x + 9)$　　　8. $y = \ln(6x + 1)$
9. $y = \ln(2x^2 - x) + 3x$
10. $y = \ln(8x^3 - 2x) - 2x$
11. Find dp/dq if $p = \ln(q^2 + 1)$.

12. Find $\dfrac{ds}{dq}$ if $s = \ln\left(\dfrac{q^2}{4} + 1\right)$.

In each of problems 13–20, find the derivative of the function in (a). Then find the derivative of the function in (b) or show that the function in (b) is the same function as that in (a).
13. (a) $y = \ln x - \ln(x - 1)$

　　(b) $y = \ln \dfrac{x}{x - 1}$

14. (a) $y = \ln(x - 1) + \ln(2x + 1)$
　　(b) $y = \ln[(x - 1)(2x + 1)]$

15. (a) $y = \frac{1}{3} \ln(x^2 - 1)$　　16. (a) $y = 3 \ln(x^4 - 1)$

　　(b) $y = \ln \sqrt[3]{x^2 - 1}$　　　　(b) $y = \ln(x^4 - 1)^3$

17. (a) $y = \ln(4x - 1) - 3 \ln x$

　　(b) $y = \ln\left(\dfrac{4x - 1}{x^3}\right)$

18. (a) $y = 3 \ln x - \ln(x + 1)$

　　(b) $y = \ln\left(\dfrac{x^3}{x + 1}\right)$

19. (a) $y = 3 \ln x + \frac{1}{2} \ln(x + 1)$

　　(b) $y = \ln(x^3 \sqrt{x + 1})$

20. (a) $y = 2 \ln x + \ln(x^4 - x + 1)$

　　(b) $y = \ln[x^2(x^4 - x + 1)]$

21. Find $\dfrac{dp}{dq}$ if $p = \ln\left(\dfrac{q^2 - 1}{q}\right)$.

22. Find $\dfrac{ds}{dt}$ if $s = \ln[t^3(t^2 - 1)]$.

23. Find $\dfrac{dy}{dt}$ if $y = \ln\left(\dfrac{t^2 + 3}{\sqrt{1 - t}}\right)$.

24. Find $\dfrac{dy}{dx}$ if $y = \ln\left(\dfrac{3x + 2}{x^2 - 5}\right)^{1/4}$.

In problems 25–38, find dy/dx.
25. $y = x - \ln x$　　　　　　26. $y = x^2 \ln(2x + 3)$

27. $y = \dfrac{\ln x}{x}$　　　　　　　28. $y = \dfrac{1 + \ln x}{x^2}$

29. $y = \ln(x^4 + 3)^2$　　　　30. $y = \ln(3x + 1)^{1/2}$
31. $y = (\ln x)^4$　　　　　　32. $y = (\ln x)^{-1}$
33. $y = [\ln(x^4 + 3)]^2$　　34. $y = \sqrt{\ln(3x + 1)}$
35. $y = \log_4 x$　　　　　　36. $y = \log_5 x$
37. $y = \log_6(x^4 - 4x^3 + 1)$　　38. $y = \log_2(1 - x - x^2)$

In problems 39–42, find the relative maxima and relative minima and sketch the graph with a graphing utility to check your results.
39. $y = x \ln x$　　　　　　40. $y = x^2 \ln x$
41. $y = x^2 - 8 \ln x$　　　　42. $y = \ln x - x$

APPLICATIONS

43. **Marginal cost** Suppose that the total cost (in dollars) for a product is given by

$$C(x) = 1500 + 200 \ln(2x + 1),$$

where x is the number of units produced.
(a) Find the marginal cost function.
(b) Find the marginal cost when 200 units are produced, and interpret your result.

44. **Investing** The number of years t that it takes to double an investment is a function of the interest rate r compounded continuously, according to

$$t = \frac{\ln 2}{r}.$$

At what rate is the required time changing with respect to r if $r = 10\%$, compounded continuously?

45. **Marginal revenue** The total revenue from the sale of x units of a product is given by

$$R(x) = \frac{2500x}{\ln(10x + 10)}.$$

(a) Find the marginal revenue function.
(b) Find the marginal revenue when 100 units are sold, and interpret your result.

46. **Supply** Suppose that the supply of q units of a product at price x dollars is given by

$$q = 10 + 50 \ln(3x + 1).$$

Find the rate of change of supply with respect to price.

47. **Demand** The demand function for a product is given by $p = 4000/\ln(x + 10)$, where p is the price per unit when x units are demanded.
    (a) Find the rate of change of price with respect to the number of units sold when 40 units are sold.
    (b) Find the rate of change of price with respect to the number of units sold when 90 units are sold.
    (c) Find the second derivative to see if the rate at which the price is changing at 40 units is increasing or decreasing.

48. **pH level** If the pH of a solution is given by

$$pH = -\log[H^+],$$

where $[H^+]$ is the concentration of hydrogen ions (in gram atoms per liter), what is the rate of change of pH with respect to $[H^+]$?

49. **Reynolds number** If the Reynolds number relating to the flow of blood exceeds R, where

$$R = A \ln(r) - Br,$$

and r is the radius of the aorta and A and B are positive constants, the blood flow becomes turbulent. What is the radius r that makes R a maximum?

50. **Decibels** The loudness of sound (L, measured in decibels) perceived by the human ear depends upon intensity levels (I) according to

$$L = 10 \log(I/I_0),$$

where I_0 is the standard threshold of audibility. Using the change of base formula, we get

$$L = \frac{10 \ln(I/I_0)}{\ln 10}.$$

At what rate is the loudness changing with respect to the intensity when the intensity is 100 times the standard threshold of audibility?

51. **Richter scale** The Richter scale reading, R, used for measuring the magnitude of an earthquake with intensity I is determined by

$$R = \frac{\ln(I/I_0)}{\ln 10},$$

where I_0 is a standard minimum threshold of intensity. If $I_0 = 1$, what is the rate of change of the Richter scale reading with respect to intensity?

52. **Violent crime** The following data represents the number of violent crimes (murders, rapes, and armed robberies) per 100,000 people for the years 1987 to 1992.

Year	Violent Crimes (per 100,000)
1987	610
1988	637
1989	663
1990	732
1991	758
1992	765

SOURCE: FBI Crime Report

Letting $t = 0$ in 1980, the above data can be modeled by $y = -22.9 + 321 \ln t$. If this model is accurate, at what rate did violent crimes change in 1990?

11.2 DERIVATIVES OF EXPONENTIAL FUNCTIONS

Objective

☐ To find derivatives of exponential functions

APPLICATION PREVIEW

We have seen in Chapter 6 that the amount that accrues when $100 is invested at 8%, compounded continuously, is

$$S(t) = 100e^{0.08t},$$

where t is the number of years. If we want to find the rate at which the money in this account is growing at the end of one year, then we need to find the derivative of this function, which is an exponential function.

In the previous section we found derivatives of logarithmic functions. In this section we turn our attention to exponential functions. The formula for the derivative of $y = e^x$ is developed as follows.

From Property I of logarithms, we know that

$$\ln e^x = x.$$

Taking the derivative, with respect to x, of both sides of this equation, we have

$$\frac{d}{dx} \ln e^x = \frac{d}{dx} x.$$

Using the Chain Rule for logarithms gives

$$\frac{1}{e^x} \cdot \frac{d}{dx} e^x = 1,$$

and solving for $\frac{d}{dx} e^x$ gives

$$\frac{d}{dx} e^x = e^x.$$

Thus we can conclude the following.

Derivative of   If $y = e^x$, then $\frac{dy}{dx} = e^x$.
$y = e^x$

EXAMPLE 1
If $p = e^q$, find dp/dq.

Solution

$$\frac{dp}{dq} = e^q$$

As with the logarithmic functions, the Chain Rule will permit us to expand our derivative formulas.

Derivatives of   If $y = e^u$, where u is a differentiable function of x, then
Exponential Functions
$$\frac{dy}{dx} = e^u \cdot \frac{du}{dx}.$$

EXAMPLE 2
If $f(x) = e^{4x^3}$, find $f'(x)$.

Solution

$$f'(x) = e^{4x^3} \cdot 12x^2 = 12x^2 e^{4x^3}$$

EXAMPLE 3
If $p = qe^{5q}$, find dp/dq.

Solution
Using the product rule, we get

$$\frac{dp}{dq} = (q \cdot e^{5q} \cdot 5) + (e^{5q} \cdot 1) = 5qe^{5q} + e^{5q}$$

EXAMPLE 4
If $s = 3te^{3t^2+5t}$, find ds/dt.

Solution

$$\frac{ds}{dt} = 3t \cdot e^{3t^2+5t}(6t + 5) + e^{3t^2+5t} \cdot 3$$

$$= (18t^2 + 15t)e^{3t^2+5t} + 3e^{3t^2+5t}$$

CHECKPOINT

1. If $y = 2e^{4x}$, find y'.
2. If $y = e^{x^2+6x}$, find y'.
3. If $s = te^{t^2}$, find ds/dt.

EXAMPLE 5
If $y = e^{\ln x^2}$, find y'.

Solution

$$y' = e^{\ln x^2} \cdot \frac{1}{x^2} \cdot 2x = \frac{2}{x}e^{\ln x^2}$$

By Property II of logarithms (Sections 5.2 and 11.1) $e^{\ln u} = u$, and we can simplify the derivative to

$$y' = \frac{2}{x} \cdot x^2 = 2x.$$

Note that if we had used this property *before* taking the derivative, we would have had

$$y = e^{\ln x^2} = x^2.$$

Then the derivative is $y' = 2x$.

Let us now return to the problem described in the Application Preview.

EXAMPLE 6

The future value when $100 is invested at 8%, compounded continuously, is $S(t) = 100e^{0.08t}$, where t is the number of years. At what rate is the money in this account growing at
(a) the end of one year?
(b) the end of ten years?

Solution

The rate of growth of the money is given by

$$S'(t) = 100e^{0.08t}(.08) = 8e^{0.08t}.$$

(a) The rate of growth of the money at the end of 1 year is

$$S'(1) = 8e^{0.08} = 8.666.$$

Thus, the future value will change by about $8.67 during the next year.

(b) The rate of growth of the money at the end of 10 years is

$$S'(10) = 8e^{0.08(10)} = 17.804.$$

Thus, the future value will change by about $17.80 during the next year.

EXAMPLE 7

If $u = w/e^{3w}$, find u'.

Solution

The function is a quotient, with the denominator equal to e^{3w}. Using the Quotient Rule gives

$$u' = \frac{e^{3w} \cdot 1 - w \cdot e^{3w} \cdot 3}{(e^{3w})^2}$$

$$= \frac{e^{3w} - 3we^{3w}}{e^{6w}}$$

$$= \frac{1 - 3w}{e^{3w}}.$$

EXAMPLE 8

North Forty, Inc., is a wilderness camping equipment manufacturer. The revenue function for their best-selling backpack, the Sierra, can be modeled by the function

$$R(x) = 25xe^{(1-0.01x)},$$

where $R(x)$ is the revenue in thousands of dollars from the sale of x thousand Sierra backpacks. Find the marginal revenue when 75,000 packs are sold, and explain what it means.

Solution

The marginal revenue function is given by $R'(x)$, and to find this derivative we use the Product Rule.

$$R'(x) = \overline{MR} = 25x[e^{(1-0.01x)} \cdot (-0.01)] + e^{(1-0.01x)}(25)$$
$$\overline{MR} = 25e^{(1-0.01x)}(1 - 0.01x)$$

To find the marginal revenue when 75,000 packs are sold, we use $x = 75$.

$$\overline{MR}(75) = 25e^{(1-0.75)}(1 - 0.75) \approx 8.025$$

This means that the sale of one (thousand) more Sierra backpacks will yield approximately \$8.025 (thousand) in additional revenue.

CHECKPOINT

4. If the sales of a product are given by $S = 1000e^{-0.2x}$, where x is the number of days after the end of an advertising campaign, what is the rate of decline in sales 20 days after the end of the campaign?

In a manner similar to that used to find the derivative of $y = e^x$, we can develop a formula for the derivative of $y = a^x$ for any base a.

Derivative of $y = a^u$ If $y = a^x$, then

$$\frac{dy}{dx} = a^x \ln a.$$

If $y = a^u$, where u is a differentiable function of x, then

$$\frac{dy}{dx} = a^u \frac{du}{dx} \ln a.$$

EXAMPLE 9

If $y = 4^x$, find dy/dx.

Solution

$$\frac{dy}{dx} = 4^x \ln 4$$

Graphing Utilities

We can make use of a graphing utility to study the behavior of an exponential function and its first two derivatives.

Figure 11.4

(a)

(b)

(c)

EXAMPLE 10

(a) Graph the three functions $g(x) = 5e^{-x}(x - 2)$, $h(x) = 5xe^{-x}$, and $k(x) = 5e^{-x}(1 - x)$.

(b) These graphs are the graphs of $y = f(x)$, $y = f'(x)$, and $f''(x)$ for some function f. Which of these functions is f, which is f', and which is f''?

(c) Does the graph of $y = h(x)$ cross the x-axis at any point other than the origin?

Solution

(a) The graphs of the three functions are shown in Figure 11.4.

(b) From the graphs, we see that $k(x) > 0$ (that is, the graph is above the x-axis) over intervals where $h(x)$ is increasing and $k(x) < 0$ (that is, the graph is below the x-axis) where $h(x)$ is decreasing, so it seems reasonable that $k(x)$ is the derivative of $h(x)$. In addition, $g(x)$ is negative over intervals where $k(x)$ is decreasing and $g(x)$ is positive where $k(x)$ is increasing, so it seems reasonable that $g(x)$ is the derivative of $k(x)$. (Note also that $g(x) < 0$ over intervals where $h(x)$ is concave down.) Thus we conclude that

$$f(x) = h(x), \quad f'(x) = k(x), \quad \text{and} \quad f''(x) = g(x).$$

We can verify this by finding the derivatives of $h(x)$.

$$f(x) = h(x) = 5xe^{-x}$$
$$f'(x) = 5e^{-x}(1 - x) = k(x)$$
$$f''(x) = 5e^{-x}(x - 2) = g(x)$$

(c) The graph of $y = h(x)$ looks like it might cross the x-axis for some x-value beyond $x = 7$. However, because $5e^{-x} \neq 0$ for any x-value, it follows that

$$h(x) = 5xe^{-x} = (5e^{-x})x = 0 \qquad \text{only when } x = 0.$$

Hence, the graph of $y = h(x)$ only crosses the x-axis once, at the origin. By using a similar argument, we can show that the graphs of the other two functions in Figure 11.4 cross the x-axis only once.

CHECKPOINT SOLUTIONS

1. $y' = 2e^{4x}(4) = 8e^{4x}$

2. $y' = (2x + 6)e^{x^2 + 6x}$

3. By the Product Rule, $\dfrac{ds}{dt} = e^{t^2}(1) + t[e^{t^2}(2t)] = e^{t^2} + 2t^2 e^{t^2}$.

4. The rate of decline is given by dS/dx.

$$\frac{dS}{dx} = 1000e^{-0.2x}(-0.2) = -200e^{-0.2x}$$

$$\frac{dS}{dx}\bigg|_{x=20} = -200e^{(-0.2)(20)}$$

$$= -200e^{-4} \approx -3.663 \text{ sales/day}$$

EXERCISE 11.2

Find the derivatives of the functions in problems 1–32.

1. $y = 5e^x - x$

2. $y = x^2 - 3e^x$

3. $f(x) = e^x - x^e$

4. $f(x) = 4e^x - \ln x$

5. $y = e^{x^3}$

6. $y = e^{x^2-1}$

7. $y = 6e^{3x^2}$

8. $y = 1 - 2e^{-x^3}$

9. $y = 2e^{(x^2+1)^3}$

10. $y = e^{\sqrt{x^2-9}}$

11. $y = e^{\ln x^3}$

12. $y = e^3 + e^{\ln x}$

13. $y = e^{-1/x}$

14. $y = 2e^{\sqrt{x}}$

15. $y = e^{-1/x^2} + e^{-x^2}$

16. $y = \dfrac{2}{e^{2x}} + \dfrac{e^{2x}}{2}$

17. $s = t^2 e^t$

18. $p = 4qe^{q^3}$

19. $y = e^{x^4} - (e^x)^4$

20. $y = 4(e^x)^3 - 4e^{x^3}$

21. $y = \ln(e^{4x} + 2)$

22. $y = \ln(e^{2x} + 1)$

23. $y = e^{-3x} \ln(2x)$

24. $y = e^{2x^2} \ln(4x)$

25. $y = \dfrac{1 + e^{5x}}{e^{3x}}$

26. $y = \dfrac{x}{1 + e^{2x}}$

27. $y = (e^{3x} + 4)^{10}$

28. $y = \dfrac{e^x - e^{-x}}{e^x + e^{-x}}$

29. $y = 6^x$

30. $y = 3^x$

31. $y = 4^{x^2}$

32. $y = 5^{x-1}$

33. (a) What is the slope of the line tangent to $y = xe^{-x}$ at $x = 1$?

(b) Write the equation of the line tangent to the graph of $y = xe^{-x}$ at $x = 1$.

34. (a) What is the slope of the line tangent to $y = e^{-x}/(1 + e^{-x})$ at $x = 0$?

(b) Write the equation of the line tangent to the graph of $y = e^{-x}/(1 + e^{-x})$ at $x = 0$.

35. The equation for the standard normal probability distribution is

$$y = \frac{1}{\sqrt{2\pi}} e^{-z^2/2}.$$

(a) At what value of z will the curve be at its highest point?

(b) Graph this function with a graphing utility to verify your answer.

36. (a) Find the mode of the normal distribution* given by

$$y = \frac{1}{\sqrt{2\pi}} e^{-(x-10)^2/2}.$$

(b) What is the mean of this normal distribution?

(c) Use a graphing utility to verify your answers.

In each of the problems 37–40, find any relative maxima and minima. Use a graphing utility to check your results.

37. $y = \dfrac{e^x}{x}$

38. $y = \dfrac{x}{e^x}$

39. $y = x - e^x$

40. $y = \dfrac{x^2}{e^x}$

APPLICATIONS

41. **Future value** If $\$P$ is invested for n years at 10%, compounded continuously, the future value is given by the function

$$S = Pe^{0.1n}.$$

(a) At what rate is the future value growing at any time (for any n)?

(b) At what rate is the future value growing after 1 year $(n = 1)$?

(c) Is the rate of growth of the future value after 1 year greater than 10%? Why?

42. **Future value** The future value that accrues when $\$700$ is invested at 9%, compounded continuously, is $S(t) = 700e^{0.09t}$, where t is the number of years.

(a) At what rate is the money in this account growing when $t = 4$?

(b) At what rate is it growing when $t = 10$?

*The mode occurs at the highest point on normal curves and equals the mean.

43. **Sales decay** After the end of an advertising campaign, the sales of a product are given by

$$S = 100,000e^{-0.5t},$$

where S is weekly sales and t is the number of weeks since the end of the campaign. Find the rate of change of S (that is, the rate of *sales decay*).

44. **Sales decay** The sales decay for a product is given by

$$S = 50,000e^{-0.8t},$$

where S is the daily sales and t is the number of days since the end of a promotional campaign. Find the rate of sales decay.

45. **Marginal cost** Suppose that the total cost in dollars of producing x units of a product is given by

$$C(x) = 10,000 + 600xe^{x/600}.$$

Find the marginal cost when 600 units are produced.

46. **Marginal revenue** Suppose that the revenue in dollars from the sale of x units of a product is given by

$$R(x) = 1000xe^{-x/50}.$$

Find the marginal revenue function.

47. **Drugs in a bloodstream** The concentration y of a certain drug in the bloodstream at any time t (in hours) is given by $y = 100(1 - e^{-0.462t})$. Find the rate of change of the concentration after one hour. Give your answer to 3 decimal places.

48. **Radioactive decay** The amount of the radioactive isotope thorium-234 present at time t is given by

$$Q(t) = 100e^{-0.02828t}.$$

Find the rate of radioactive decay of the isotope.

49. **Spread of disease** Suppose that the spread of a disease through the student body at an isolated college campus can be modeled by

$$y = \frac{10,000}{1 + 9999e^{-0.99t}},$$

where y is the total number affected at time t (in days). Find the rate of change of y.

50. **Spread of a rumor** The number of people $N(t)$ in a community who are reached by a particular rumor at time t (in days) is given by

$$N(t) = \frac{50,500}{1 + 100e^{-0.7t}}.$$

Find the rate of change of $N(t)$.

51. **Chemical reaction** The number of molecules of a certain substance that have enough energy to activate a reaction is given by $y = 100,000e^{-1/x}$, where y is the number of molecules and x is the (absolute) temperature of the substance. What is the rate of change of y with respect to temperature?

52. **Newton's law of cooling** When a body is moved from one medium to another, its temperature T will change according to the equation

$$T = T_0 + Ce^{kt},$$

where T_0 is the temperature of the new medium, C the temperature difference between the mediums (old–new), t the time in the new medium, and k a constant. If T_0, C, and k are held constant, what is the rate of change of T with respect to time?

53. **World population** Suppose that world population can be considered as growing according to the equation $N = N_0(1 + r)^t$, where N_0 and r are constants. Find the rate of change of N with respect to t.

54. **Blood pressure** Medical research has shown that between heartbeats the pressure in the aorta of a normal adult is a function of time and can be modeled by the equation

$$P = 95e^{-0.491t}.$$

(a) Use the derivative to find the rate at which the pressure changes at any time t.
(b) Use the derivative to find the rate at which the pressure changes after 0.1 seconds.
(c) Is the pressure increasing or decreasing?

55. **Richter scale** The intensity of an earthquake is related to the Richter scale reading R by

$$\frac{I}{I_0} = 10^R,$$

where I_0 is a standard minimum intensity. If $I_0 = 1$, what is the rate of change of the intensity I with respect to the Richter scale reading?

56. **Decibel readings** The intensity level of sound, I, is given by

$$\frac{I}{I_0} = 10^{L/10},$$

where L is the decibel reading and I_0 is the standard

threshold of audibility. At what rate is I/I_0 changing with respect to L when $L = 20$?

57. **National health care** Using data from the Congressional Budget Office (reported in *Newsweek,* October 4, 1993), the national health expenditure H can be modeled by

$$H = 45e^{0.0898t},$$

where t is the number of years past 1960 and H is in billions of dollars. If this model is accurate, at what rate will health care expenditures change in 1997?

58. **Revenue** According to data in *USA Today* (March 1, 1994), the revenue R from wireless technology can be modeled by

$$R = 0.572e^{0.3860t},$$

where t is the number of years past 1985 and R is in billions of dollars. If this model is accurate, at what rate will revenue change in 1999?

59. **U.S. debt** The following table shows the U.S. national debt and percentage of federal expenditures devoted to payment of the interest on this debt for selected years from 1900 to 1989.

Year	U.S. Debt (in Billions)	Percentage to Interest
1900	$1.2	0.0
1910	1.1	0.0
1920	24.2	0.0
1930	16.1	0.0
1940	43.0	10.5
1945	258.7	4.1
1955	272.8	9.4
1965	313.8	9.6
1975	533.2	9.8
1985	1823.1	18.9
1987	2350.3	19.5
1989	2857.4	25.0

SOURCE: *World Almanac,* 1991

Assume that the amount of the national debt (in billions of dollars) can be modeled by the function

$$d(x) = 1956.16 + 29x - 2018.91e^{x/100} + 0.44e^{x/10},$$

where x is the number of years past 1900. This function and the data points are graphed in the following figure.

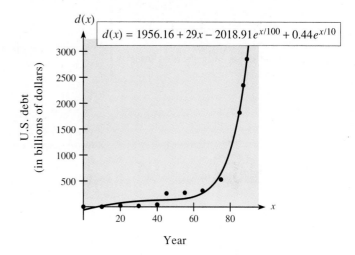

$$d(x) = 1956.16 + 29x - 2018.91e^{x/100} + 0.44e^{x/10}$$

(a) What function describes how fast the national debt is changing?

(b) Find the instantaneous rate of change of the national debt model $d(x)$ in 1940 and 1985.

(c) The Gramm-Rudman Balanced Budget Act (passed in 1985) was designed to limit the growth of the national debt and hence change the function that modeled it. Why was this necessary?

60. **Social Security** The number of workers available to support Social Security beneficiaries is declining, as the following table indicates.

Year	No. of Workers per Retiree
1950	16.5
1960	5.0
1980	3.7
1990	3.4

SOURCE: Social Security Administration

This data can be modeled by the equation

$$y = 16.03e^{-0.0346t},$$

where $t = 0$ in 1940. Using this model, find the rate at which the number of workers was changing in 1990 (when $t = 50$).

11.3 IMPLICIT DIFFERENTIATION

Objectives

❑ To find derivatives by using implicit differentiation
❑ To find slopes of tangents by using implicit differentiation

APPLICATION PREVIEW

In the retail electronics industry, suppose the monthly demand for Precision, Inc., stereo headphones is given by

$$p = \frac{10,000}{(x + 1)^2},$$

where p is the price in dollars per set of headphones and x is demand in hundreds of sets of headphones. If we want to find the rate of change in the quantity demanded with respect to the price, then we need to find dx/dp. While we can solve this equation of x (with some difficulty) so that dx/dp can be found, the resulting equation does not define x as a function of p. In this case, and in other cases where we cannot solve equations for the variable we need, we can find derivatives with a new technique, called **implicit differentiation.**

Figure 11.5

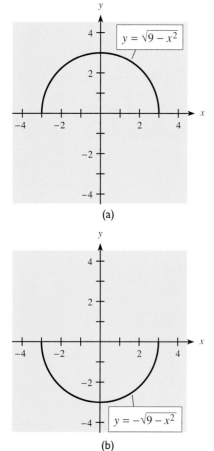

(a)

(b)

When an equation has the form $F(x, y) = 0$, we can say that y is defined implicitly as a function of x whether or not we can solve for y. For example, the equation $xy - 4x + 1 = 0$ is in the form $F(x, y) = 0$, but we can solve for y to write the equation in the form $y = (4x - 1)/x$. We can say that $xy - 4x + 1 = 0$ defines y **implicitly** as a function of x, while $y = (4x - 1)/x$ defines the function explicitly.

The equation $x^2 + y^2 - 9 = 0$ has a circle as its graph. If we solve the equation for y, we get $y = \pm\sqrt{9 - x^2}$, which indicates that y is not a function of x. We can, however, consider the equation as defining *two* functions, $y = \sqrt{9 - x^2}$ and $y = -\sqrt{9 - x^2}$ (see Figure 11.5). We say that the equation $x^2 + y^2 - 9 = 0$ defines the two functions implicitly.

Even though an equation like

$$\ln xy + xe^y + x - 3 = 0$$

may be difficult or even impossible to solve for y, and even though the equation may not represent y as a single function of x, we can use the technique of **implicit differentiation** to find the derivative of y with respect to x. The word *implicit* means we are implying that y is a function of x without verifying it. We simply take the derivative of both sides of $f(x, y) = 0$ and then solve algebraically for dy/dx.

For example, we can find the derivative dy/dx from $x^2 + y^2 - 9 = 0$ by taking the derivative of both sides of the equation.

$$\frac{d}{dx} (x^2 + y^2 - 9) = \frac{d}{dx} (0)$$

$$\frac{d}{dx}(x^2) + \frac{d}{dx}(y^2) + \frac{d}{dx}(-9) = \frac{d}{dx}(0)$$

We have assumed y is a function of x; the derivative of y^2 is treated like the derivative of u^n, where u is a function of x. Thus the derivative is

$$2x + 2y^1 \cdot \frac{dy}{dx} + 0 = 0.$$

Solving for dy/dx gives

$$\frac{dy}{dx} = -\frac{2x}{2y} = -\frac{x}{y}.$$

Let us now compare this derivative with the derivative of the two functions $y = \sqrt{9 - x^2}$ and $y = -\sqrt{9 - x^2}$. The derivative of $y = \sqrt{9 - x^2}$ is

$$\frac{dy}{dx} = \frac{1}{2}(9 - x^2)^{-1/2}(-2x) = \frac{-x}{\sqrt{9 - x^2}},$$

and the derivative of $y = -\sqrt{9 - x^2}$ is

$$\frac{dy}{dx} = -\frac{1}{2}(9 - x^2)^{-1/2}(-2x) = \frac{x}{\sqrt{9 - x^2}}.$$

Note that if we substitute $\pm\sqrt{9 - x^2}$ for y in our "implicit" derivative, we would get the two derivatives that were obtained from the "explicit" functions.

EXAMPLE I

Find the slope of the tangent to the graph of $x^2 + y^2 - 9 = 0$ at $(\sqrt{5}, 2)$.

Solution

The slope of the tangent to the curve is the derivative of the equation, evaluated at the given point. Taking the derivative implicitly gives us $dy/dx = -x/y$. Evaluating the derivative at $(\sqrt{5}, 2)$ gives the slope of the tangent as $-\sqrt{5}/2$.

We also found the derivative of $x^2 + y^2 - 9 = 0$ by solving for y explicitly. The function whose graph contains $(\sqrt{5}, 2)$ is $y = \sqrt{9 - x^2}$, and its derivative is

$$\frac{dy}{dx} = \frac{-x}{\sqrt{9 - x^2}}.$$

Evaluating the derivative at $(\sqrt{5}, 2)$, we get the slope of the tangent: $-\sqrt{5}/2$. Thus we see that both methods give us the same slope for the tangent, but that the implicit method is easier to use.

EXAMPLE 2

Find dy/dx if $x^2 + 4x - 3y^2 + 4y = 0$.

Solution

Taking the derivative implicitly gives

$$\frac{d}{dx}(x^2) + \frac{d}{dx}(4x) + \frac{d}{dx}(-3y^2) + \frac{d}{dx}(4y) = \frac{d}{dx}(0).$$

Now,
$$\frac{d}{dx}(x^2) = 2x \qquad \frac{d}{dx}(4x) = 4$$

$$\frac{d}{dx}(-3y^2) = -3(2y)\frac{dy}{dx} = -6y\frac{dy}{dx}$$

$$\frac{d}{dx}(4y) = 4(1)\frac{dy}{dx} = 4\frac{dy}{dx}$$

and
$$\frac{d}{dx}(0) = 0.$$

So
$$2x + 4 - 6y\frac{dy}{dx} + 4\frac{dy}{dx} = 0$$

$$(-6y + 4)\frac{dy}{dx} = -2x - 4,$$

$$\frac{dy}{dx} = \frac{-2x - 4}{-6y + 4}$$

$$\frac{dy}{dx} = \frac{x + 2}{3y - 2}.$$

EXAMPLE 3

Write the equation of the tangent to the graph of $x^3 + xy + 4 = 0$ at the point $(-2, -2)$.

Solution

Taking the derivative implicitly gives

$$\frac{d}{dx}(x^3) + \frac{d}{dx}(xy) + \frac{d}{dx}(4) = \frac{d}{dx}(0).$$

The $\frac{d}{dx}(xy)$ indicates that we should take the derivative of the *product* of x and y. Since we are assuming y is a function of x and since x is a function of x, we must use the Product Rule to find $\frac{d}{dx}(xy)$.

$$\frac{d}{dx}(xy) = x \cdot 1\frac{dy}{dx} + y \cdot 1 = x\frac{dy}{dx} + y$$

Thus we have $\qquad 3x^2 + \left(x \dfrac{dy}{dx} + y\right) + 0 = 0.$

Solving for dy/dx gives $\qquad \dfrac{dy}{dx} = \dfrac{-3x^2 - y}{x}.$

The slope of the tangent to the curve at $x = -2$, $y = -2$ is

$$m = \frac{-3(-2)^2 - (-2)}{-2} = 5.$$

The equation of the tangent line is

$$y - (-2) = 5[x - (-2)] \quad \text{or} \quad y = 5x + 8.$$

CHECKPOINT

1. Find the following:

   (a) $\dfrac{d}{dx}(x^3)$ \qquad (b) $\dfrac{d}{dx}(y^4)$ \qquad (c) $\dfrac{d}{dx}(x^2y^5)$

2. Find $\dfrac{dy}{dx}$ for $x^3 + y^4 = x^2y^5$.

EXAMPLE 4

At what point(s) does $x^2 + 4y^2 - 2x + 4y - 2 = 0$ have a horizontal tangent? At what point(s) does it have a vertical tangent?

Solution

First we find the derivative implicitly.

$$2x + 8y \cdot y' - 2 + 4y' - 0 = 0$$

$$y' = \frac{2 - 2x}{8y + 4} = \frac{1 - x}{4y + 2}$$

Horizontal tangents will occur where $y' = 0$; that is, where $x = 1$. We can now find the corresponding y-value(s) by substituting 1 for x in the original equation and solving.

$$1 + 4y^2 - 2 + 4y - 2 = 0$$
$$4y^2 + 4y - 3 = 0$$
$$(2y - 1)(2y + 3) = 0$$
$$y = \tfrac{1}{2} \quad \text{or} \quad y = -\tfrac{3}{2}$$

Thus horizontal tangents occur at $(1, \tfrac{1}{2})$, and $(1, -\tfrac{3}{2})$ (see Figure 11.6).

Vertical tangents will occur where the derivative is undefined; that is, where $y = -\tfrac{1}{2}$. To find the corresponding x-value(s), we substitute $-\tfrac{1}{2}$ in

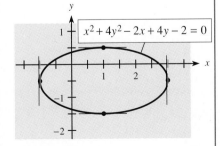

Figure 11.6

the equation for y and solve for x.

$$x^2 + 4(-\tfrac{1}{2})^2 - 2x + 4(-\tfrac{1}{2}) - 2 = 0$$
$$x^2 - 2x - 3 = 0$$
$$(x - 3)(x + 1) = 0$$
$$x = 3 \quad \text{or} \quad x = -1$$

Thus vertical tangents occur at $(3, -\tfrac{1}{2})$ and $(-1, -\tfrac{1}{2})$ (see Figure 11.6).

EXAMPLE 5

If $\ln xy = 6$, find dy/dx.

Solution

Using the properties of logarithms, we have

$$\ln x + \ln y = 6,$$

which leads to the implicit derivative:

$$\frac{1}{x} + \frac{1}{y}\frac{dy}{dx} = 0.$$

Solving gives

$$\frac{dy}{dx} = -\frac{y}{x}.$$

EXAMPLE 6

Find dy/dx if $4x^2 + e^{xy} = 6y$.

Solution

We take the derivative of both sides.

$$\frac{d}{dx}(4x^2) + \frac{d}{dx}(e^{xy}) = \frac{d}{dx}(6y)$$

$$8x + e^{xy} \cdot \frac{d}{dx}(xy) = 6\frac{dy}{dx}$$

$$8x + e^{xy}\left(x\frac{dy}{dx} + y\right) = 6\frac{dy}{dx}$$

$$8x + xe^{xy}\frac{dy}{dx} + ye^{xy} = 6\frac{dy}{dx}$$

$$8x + ye^{xy} = 6\frac{dy}{dx} - xe^{xy}\frac{dy}{dx}$$

$$8x + ye^{xy} = (6 - xe^{xy})\frac{dy}{dx}$$

$$\frac{8x + ye^{xy}}{6 - xe^{xy}} = \frac{dy}{dx}$$

EXAMPLE 7

In the Application Preview, the demand for Precision, Inc., stereo headphones was given by

$$p = \frac{10,000}{(x + 1)^2},$$

where p was the price per set in dollars and x was hundreds of headphone sets demanded. Find the rate of change of demand with respect to price when 19 (hundred) sets are demanded.

Solution

The rate of change of demand with respect to price is dx/dp. Using implicit differentiation, we get the following:

$$\frac{d}{dp}(p) = \frac{d}{dp}\left[\frac{10,000}{(x + 1)^2}\right] = \frac{d}{dp}[10,000(x + 1)^{-2}]$$

$$1 = 10,000\left[-2(x + 1)^{-3}\frac{dx}{dp}\right]$$

$$1 = \frac{-20,000}{(x + 1)^3}\frac{dx}{dp}$$

$$\frac{(x + 1)^3}{-20,000} = \frac{dx}{dp}.$$

When 19 (hundred) headphone sets are demanded we use $x = 19$, and the rate of change of demand with respect to price is

$$\left.\frac{dx}{dp}\right|_{x=19} = \frac{(19 + 1)^3}{-20,000} = \frac{8000}{-20,000} = -0.4.$$

This result means that when 19 (hundred) headphone sets are demanded, if the price per set is increased by \$1, then the expected change in demand would be a decrease of 0.4 hundred, or 40, headphone sets.

Figure 11.7

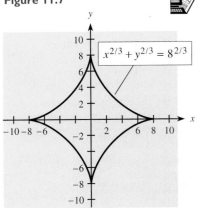

$x^{2/3} + y^{2/3} = 8^{2/3}$

Graphing Utilities

To graph a function with a graphing utility, we need to write y as an *explicit* function of x (such as $y = \sqrt{4 - x^2}$). If an equation defines y as an *implicit* function of x, we have to solve for y in terms of x before we can use the graphing utility. Sometimes we cannot solve for y, and other times, such as in

$$x^{2/3} + y^{2/3} = 8^{2/3},$$

y cannot be written as a single function of x. If this equation is solved for y and a graphing utility is used to graph that function, the resulting graph usually shows only the portion of the graph that lies in quadrants I and II, and sometimes only the part in quadrant I. The complete graph is shown in Figure 11.7. Thus, for the graph of an implicitly defined function, a graphing utility must be used carefully (or sometimes cannot be used).

CHECKPOINT
SOLUTIONS

1. (a) $\dfrac{d}{dx}(x^3) = 3x^2$ (b) $\dfrac{d}{dx}(y^4) = 4y^3 \dfrac{dy}{dx}$

   (c) $\dfrac{d}{dx}(x^2y^5) = x^2\left(5y^4 \dfrac{dy}{dx}\right) + y^5(2x)$ (by the Product Rule)

2. For $x^3 + y^4 = x^2y^5$, we can use the answers to question (1) to obtain

$$3x^2 + 4y^3 \frac{dy}{dx} = x^2\left(5y^4 \frac{dy}{dx}\right) + y^5(2x)$$

$$3x^2 + 4y^3 \frac{dy}{dx} = 5x^2y^4 \frac{dy}{dx} + 2xy^5$$

$$3x^2 - 2xy^5 = (5x^2y^4 - 4y^3) \frac{dy}{dx}$$

$$\frac{3x^2 - 2xy^5}{5x^2y^4 - 4y^3} = \frac{dy}{dx}.$$

EXERCISE 11.3

In problems 1–6, find dy/dx at the given point without first solving for y.

1. $x^2 - 4y - 17 = 0$ at $(1, -4)$
2. $3x^2 - 10y + 400 = 0$ at $(10, 70)$
3. $xy = 8$ at $(2, 4)$
4. $e^y = x$ at $(1, 0)$
5. $x^2 + 3xy - 4 = 0$ at $(1, 1)$
6. $x^2 + 5xy + 4 = 0$ at $(1, -1)$

Find dy/dx for the functions in problems 7–10.

7. $x^2 + 2y^2 - 4 = 0$
8. $x + y^2 - 4y + 6 = 0$
9. $x^2 + 4x + y^2 - 3y + 1 = 0$
10. $x^2 - 5x + y^3 - 3y - 3 = 0$
11. If $x^2 + y^2 = 4$, find y'.
12. If $p^2 + 4p - q = 4$, find dp/dq.
13. If $xy^2 - y^2 = 1$, find y'.
14. If $pq = 4$, find dp/dq.
15. If $pq = 4p - 2$, find dp/dq.
16. If $x^2 - 3y^4 = 2x^5 + 7y^3 - 5$, find dy/dx.
17. If $3x^5 - 5y^3 = 5x^2 + 3y^5$, find dy/dx.
18. If $x^2 + 3x^2y^4 = y + 8$, find dy/dx.

19. If $x^4 + 2x^3y^2 = x - y^3$, find dy/dx.
20. If $(x + y)^2 = 5x^4y^3$, find dy/dx.
21. Find dy/dx for $x^4 + 3x^3y^2 - 2y^5 = (2x + 3y)^2$.
22. Find y' for $2x + 2y = \sqrt{x^2 + y^2}$.

For problems 23–26, find the slope of the tangent to the curve.

23. $x^2 + 4x + y^2 + 2y - 4 = 0$ at $(1, -1)$
24. $x^2 - 4x + 2y^2 - 4 = 0$ at $(2, 2)$
25. $x^2 + 2xy + 3 = 0$ at $(-1, 2)$
26. $y + x^2 = 4$ at $(0, 4)$
27. Write the equation of the tangent to the curve $x^2 - 2y^2 + 4 = 0$ at $(2, 2)$.
28. Write the equation of the tangent to the curve $x^2 + y^2 + 2x - 3 = 0$ at $(-1, 2)$.
29. Write the equation of the tangent to the curve $4x^2 + 3y^2 - 4y - 3 = 0$ at $(-1, 1)$.
30. Write the equation of the tangent to the curve $xy + y = 0$ at $(3, 0)$.
31. If $\ln y = x$, find dy/dx.
32. If $\ln(x + y) = y^2$, find dy/dx.
33. If $x \ln y = 4$, find dy/dx.
34. If $\ln xy = 2$, find dy/dx.

35. Find the slope of the tangent to the curve $x^2 + \ln y = 4$ at the point $(2, 1)$.

36. Write the equation of the tangent to the curve $x \ln y + 2xy = 2$ at the point $(1, 1)$.

37. If $xe^y = 6$, find dy/dx.

38. If $x + e^{xy} = 10$, find dy/dx.

39. If $e^{xy} = 4$, find dy/dx.

40. If $x - xe^y = 3$, find dy/dx.

41. If $ye^x - y = 3$, find dy/dx.

42. If $x^2y = e^{x+y}$, find dy/dx.

43. Find the slope of the line tangent to the graph of $y = xe^{-x}$ at $x = 1$.

44. Find the slope of the line tangent to the curve $ye^x = 4$ at $(0, 4)$.

45. Write the equation of the line tangent to the curve $xe^y = 3$ at $(3, 0)$.

46. Write the equation of the line tangent to the curve $ye^x = 4$ at $(0, 4)$.

47. At what points does the curve defined by $x^2 + 4y^2 - 4x - 4 = 0$ have
    (a) horizontal tangents?
    (b) vertical tangents?

48. At what points does the curve defined by $x^2 + 4y^2 - 4 = 0$ have
    (a) horizontal tangents?
    (b) vertical tangents?

49. In problem 11, the derivative y' was found to be

$$y' = \frac{-x}{y},$$

when $x^2 + y^2 = 4$. Take the implicit derivative of the equation for y' to show

$$y'' = \frac{-y + xy'}{y^2}.$$

50. Find y' implicitly for $x^3 - y^3 = 8$. Then, by taking derivatives implicitly, use it to show

$$y'' = \frac{2x(y - xy')}{y^3}.$$

51. Use the result of problem 49 in parts (a) and (b) below.
    (a) Substitute $-x/y$ for y' in the expression for y'' and simplify to show

$$y'' = -\frac{(x^2 + y^2)}{y^3}.$$

    (b) Does $y'' = -4/y^3$? Why or why not?

52. Use the result of problem 50.
    (a) Substitute x^2/y^2 for y' in the expression for y'' and simplify to show

$$y'' = \frac{2x(y^3 - x^3)}{y^5}.$$

    (b) Does $y'' = -16x/y^5$? Why or why not?

53. Find y'' for $\sqrt{x} + \sqrt{y} = 1$, and simplify.

54. Find y'' for $\dfrac{1}{x} - \dfrac{1}{y} = 1$.

 In each of problems 55 and 56, find the maximum and minimum values of y. Use a graphing utility to verify your conclusion.

55. $x^2 + y^2 - 9 = 0$

56. $4x^2 + y^2 - 8x = 0$

APPLICATIONS

57. **Advertising and sales** Suppose that a company's sales volume y (in thousands of dollars) is related to its advertising expenditures x (in thousands of dollars) according to

$$xy - 20x + 10y = 0.$$

Find the rate of change of sales volume with respect to advertising expenditures when $x = 10$ (thousand dollars).

58. **Insect control** Suppose that the number of mosquitoes N (in thousands) in a certain swampy area near a community is related to the number of pounds of insecticide x sprayed on the nesting areas according to

$$Nx - 10x + N = 300.$$

Find the rate of change of N with respect to x when 49 lb of insecticide are used.

59. **Production** Suppose that a company can produce 12,000 units when the number of hours of skilled labor y and unskilled labor x satisfy

$$384 = (x + 1)^{3/4}(y + 2)^{1/3}.$$

Find the rate of change of skilled labor hours with respect to unskilled labor hours when $x = 255$ and $y = 214$. This can be used to approximate the change in skilled labor hours required to maintain the same production level when unskilled labor hours are increased by one hour.

60. **Production** Suppose that production of 10,000 units of a certain agricultural crop is related to the number of

hours of labor x and the number of acres of the crop y according to

$$300x + 30,000y = 11xy - 0.0002x^2 - 5y.$$

Find the rate of change of the number of hours with respect to the number of acres.

61. **Demand** If the demand function for a product is given by

$$p(q + 1)^2 = 200,000,$$

find the rate of change of quantity with respect to price when $p = \$80$. Interpret this result.

62. **Demand** If the demand function for a commodity is given by

$$p^2(2q + 1) = 100,000,$$

find the rate of change of quantity with respect to price when $p = \$50$. Interpret this result.

63. **Radioactive decay** The number of grams of radium, y, that will remain after t years if 100 grams existed originally can be found using the equation

$$-0.000436t = \ln\left(\frac{y}{100}\right).$$

Use implicit differentiation to find the rate of change of y with respect to t, that is, the rate at which the radium will decay.

64. **Disease control** Suppose the proportion of people affected by a certain disease is described by

$$\ln\left(\frac{P}{1 - P}\right) = 0.5t,$$

where t is the time in months. Find dP/dt, the rate at which P grows.

65. **Temperature-humidity index** The temperature-humidity index (THI) is given by

$$THI = t - 0.55(1 - h)(t - 58),$$

Where t is the air temperature in degrees Fahrenheit and h is the relative humidity. If the THI remains constant, find the rate of change of humidity with respect to temperature if the temperature is 70°F (Source: "Temperature-Humidity Indices," *UMAP Journal*, Fall 1989).

11.4 RELATED RATES

Objective

❑ To use implicit differentiation to solve problems involving related rates

APPLICATION PREVIEW

According to Poiseuille's law, the flow of blood F is related to the radius r of the vessel according to

$$F = kr^4,$$

where k is a constant. When the radius of a blood vessel is reduced, such as by cholesterol deposits, the flow of blood is also restricted. Drugs can be administered that increase the radius of the blood vessel and, hence, the flow of blood. The rate of change of the blood flow and the rate of change of the radius of the blood vessel are time rates of change that are related to each other, so they are called **related rates.** We can use these related rates to find the **percentage rate of change** in the blood flow that corresponds to the percentage rate of change in the radius of the blood vessel caused by the drug.

We have seen that the derivative represents the instantaneous rate of change of one variable with respect to another. When the derivative is taken with respect to time, then it represents the rate at which that variable is changing with respect to time (or the velocity). For example, if distance x is measured in miles and time t in hours, then dx/dt is measured in miles per hour, and would measure how fast x is changing. Similarly, if V represents the volume (in cubic feet) of water in a swimming pool and t is time (in minutes), then dV/dt is measured in cubic feet per minute (ft^3/min) and might measure the rate at which the pool is being filled with water or being emptied.

Sometimes, two (or more) quantities that depend on time are also related to each other. For example, the height of a tree h (in feet) is related to the radius r (in inches) of its trunk and this relationship can be modeled by

$$h = kr^{2/3},$$

where k is a constant.* Of course, both h and r are also related to time, and hence, the rates of change dh/dt and dr/dt are related to each other. Thus they are called **related rates.**

The specific relationship between dh/dt and dr/dt can be found by differentiating $h = kr^{2/3}$ implicitly with respect to time t.

EXAMPLE 1

Suppose for a certain type of tree the height of the tree (in feet) is related to the radius of its trunk (in inches) by

$$h = 15r^{2/3}.$$

Suppose the rate of change of r is $\frac{3}{4}$ inch per year. Find how fast the height is changing when the radius is 8 inches.

Solution

To find how the rates dh/dt and dr/dt are related, we differentiate $h = 15r^{2/3}$ implicitly with respect to time t.

$$\frac{dh}{dt} = 10r^{-1/3}\frac{dr}{dt}$$

Using $r = 8$ inches and $dr/dt = \frac{3}{4}$ inch per year gives

$$\frac{dh}{dt} = 10(8)^{-1/3}(3/4) = \frac{15}{4} = 3\tfrac{3}{4} \text{ feet per year.}$$

While the work in Example 1 shows how to obtain related rates, the different units (feet per year and inches per year) may be somewhat difficult to interpret. For this reason, many applications in the life sciences deal with

*T. McMahon, "Size and Shape in Biology," *Science* 179 (1979): 1201.

percentage rates of change. The percentage rate of change of a quantity is the rate of change of the quantity divided by the quantity.

EXAMPLE 2

As mentioned in the Application Preview, Poiseuille's law expresses the flow of blood F as a function of the radius r of the vessel according to

$$F = kr^4,$$

where k is a constant. When the radius of a blood vessel is restricted, such as by cholesterol deposits, drugs can be administered that will increase the radius of the blood vessel (and hence the blood flow). Find the percentage rate of change in the flow of blood that corresponds to the percentage rate of change in the radius of a blood vessel caused by the drug.

Solution

We seek the percentage rate of change of flow, $(dF/dt)/F$, that results from a given percentage rate of change of the radius $(dr/dt)/r$. We first find the related rates of change by differentiating

$$F = kr^4$$

implicitly with respect to time.

$$\frac{dF}{dt} = k\left(4r^3\,\frac{dr}{dt}\right)$$

Then the percentage rate of change of flow can be found by dividing both sides of the equation by F.

$$\frac{\dfrac{dF}{dt}}{F} = \frac{4kr^3\,\dfrac{dr}{dt}}{F}$$

If we replace F on the right side of the equation with kr^4 and reduce, we get

$$\frac{\dfrac{dF}{dt}}{F} = \frac{4kr^3\,\dfrac{dr}{dt}}{kr^4} = 4\left(\frac{\dfrac{dr}{dt}}{r}\right).$$

Thus we see that the percentage rate of change of the flow of blood is 4 times the corresponding percentage rate of change of the radius of the blood vessel. This means that a drug that would cause a 12% increase in the radius of a blood vessel at a certain time would produce a corresponding 48% increase in blood flow through that vessel at that time.

In the examples considered, the equation relating the time-dependent variables has been given. For some problems, the original equation relating the variables must first be developed from the statement of the problem. These problems can be solved with the aid of the following procedure.

Solving Related Rates Problems

Procedure	Example
To solve related rates problems:	Sand falls at a rate of 5 ft³/min on a conical pile, with the diameter always equal to the height of the pile. At what rate is the height increasing when it is 10 ft?
1. Use geometric and/or physical conditions to write an equation that relates the time-dependent variables.	1. The conical pile has its volume given by $$V = \frac{1}{3}\pi r^2 h.$$
2. Substitute values or relationships that are true *at all times* into the equation.	2. The radius $r = \frac{1}{2}h$ at all times, so $$V = \frac{1}{3}\pi\left[\frac{1}{4}h^2\right]h = \frac{\pi}{12}h^3.$$
3. Differentiate both sides of the equation implicitly with respect to time. This equation is valid for all times.	3. $\dfrac{dV}{dt} = \dfrac{\pi}{12}\left[3h^2\,\dfrac{dh}{dt}\right]$
4. Substitute the values that are known at the instant specified and solve the equation.	4. $\dfrac{dV}{dt} = 5$ at all times, so when $h = 10$, $$5 = \frac{\pi}{4}(10^2)\frac{dh}{dt}.$$
5. Solve for the specified quantity at the given time.	5. $\dfrac{dh}{dt} = \dfrac{20}{100\pi} = \dfrac{1}{5\pi}\,(\text{ft/min})$

Note that you should *not* substitute numerical values for any quantity that varies with time until after the derivative is taken. If values are substituted before the derivative is taken, that quantity will have the constant value resulting from the substitution and hence will have a derivative equal to zero.

EXAMPLE 3

A hot air balloon has a velocity of 50 ft/min and is flying at a constant height of 500 ft. An observer on the ground is watching the balloon approach. How fast is the distance between the balloon and the observer changing when the balloon is 1000 ft from the observer?

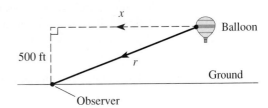

Figure 11.8

Solution

If we let r be the distance between the balloon and the observer and x be the horizontal distance from the balloon to a point directly above the observer, then we see that these quantities are related by the equation

$$x^2 + 500^2 = r^2. \quad \text{(See Figure 11.8)}$$

Because the distance x is decreasing, we know that dx/dt must be negative. Thus we are given that $dx/dt = -50$ at all times, and we need to find dr/dt when $r = 1000$. Taking the derivative with respect to t of both sides of the equation $x^2 + 500^2 = r^2$ gives

$$2x\frac{dx}{dt} + 0 = 2r\frac{dr}{dt}.$$

Using $dx/dt = -50$ and $r = 1000$, we get

$$2x(-50) = 2000\frac{dr}{dt}$$

$$\frac{dr}{dt} = \frac{-100x}{2000} = \frac{-x}{20}.$$

Using $r = 1000$ in $x^2 + 500^2 = r^2$ gives $x^2 = 750{,}000$. So $x = 500\sqrt{3}$, and

$$\frac{dr}{dt} = \frac{-500\sqrt{3}}{20} = -25\sqrt{3} \text{ ft/min.}$$

CHECKPOINT

1. If V represents volume, write a mathematical symbol that represents "the rate of change of volume with respect to time."

2. (a) Differentiate $x^2 + 64 = y^2$ implicitly with respect to time.
   (b) Suppose that we know that y is increasing at 2 units per minute. Use part (a) to find the rate of change of x at the instant when $x = 6$ and $y = 10$.

3. True or false: In solving a related rates problem, we substitute all numerical values into the equation before we take derivatives.

EXAMPLE 4

Suppose that oil is spreading in a circular pattern from a leak at an offshore rig. If the rate at which the radius of the oil slick is growing is 1 ft/min, at what rate is the area of the oil slick growing when the radius is 600 ft?

Solution

The area of the circular oil slick is given by

$$A = \pi r^2,$$

where r is the radius. The rate at which the area is changing is

$$\frac{dA}{dt} = 2\pi r \frac{dr}{dt}.$$

Using $r = 600$ ft and $dr/dt = 1$ ft/min gives

$$\frac{dA}{dt} = 2\pi (600 \text{ ft})(1 \text{ ft/min}) = 1200\pi \text{ ft}^2/\text{min}.$$

Thus when the radius of the oil slick is 600 ft, the area is growing at the rate of 1200π ft²/min, or approximately 3770 ft²/min.

CHECKPOINT SOLUTIONS

1. dV/dt

2. (a) $2x \dfrac{dx}{dt} + 0 = 2y \dfrac{dy}{dt}$

   $x \dfrac{dx}{dt} = y \dfrac{dy}{dt}$

   (b) Use $\dfrac{dy}{dt} = 2$, $x = 6$, and $y = 10$ in part (a) to obtain

   $6 \dfrac{dx}{dt} = 10(2)$ or $\dfrac{dx}{dt} = \dfrac{20}{6} = \dfrac{10}{3}.$

3. False. The numerical values for any quantity that is varying with time should not be substituted until after the derivative is taken.

EXERCISE 11.4

In problems 1–4, find dy/dt using the given values.
1. $y = x^3 - 3x$, for $x = 2$, $dx/dt = 4$
2. $y = 5x^2 + 2x$, for $x = 4$, $dx/dt = 3$
3. $xy = 4$, for $x = 8$, $dx/dt = -2$
4. $xy = x + 3$, for $x = 3$, $dx/dt = -1$

In problems 5–8, assume that x and y are differentiable functions of t. In each case, find dx/dt given that $x = 5$, $y = 12$, and $dy/dt = 2$.

5. $x^2 + y^2 = 169$
6. $y^2 - x^2 = 119$
7. $y^2 = 2xy + 24$
8. $x^2(y - 6) = 12y + 6$

9. If $x^2 + y^2 = z^2$, find dy/dt when $x = 3$, $y = 4$, $dx/dt = 10$, and $dz/dt = 2$.

10. If $s = 2\pi r(r + h)$, find dr/dt when $r = 2$, $h = 8$, $dh/dt = 3$, and $ds/dt = 10\pi$.

11. A point is moving along the graph of the equation $y = -4x^2$. At what rate is y changing when $x = 5$ and is changing at a rate of 2 units/sec?

12. A point is moving along the graph of the equation $y = 5x^3 - 2x$. At what rate is y changing when $x = 4$ and is changing at a rate of 3 units per sec?

13. The radius of a circle is increasing at a rate of 2 ft/min. At what rate is its area changing when the radius is 3 ft? (Recall that for a circle, $A = \pi r^2$.)

14. The area of a circle is changing at a rate of 1 in²/sec. At what rate is its radius changing when the radius is 2 in.?

15. The volume of a cube is increasing at a rate of 64 in³/sec. At what rate is the length of each edge of the cube changing when the edges are 6 in. long? (Recall that for a cube, $V = x^3$.)

16. The lengths of the edges of a cube are increasing at a rate of 8 ft/min. At what rate is the surface area changing when the edges are 24 ft long? (Recall that for a cube, $S = 6x^2$.)

APPLICATIONS

17. **Profit** Suppose that the daily profit (in dollars) from the production and sale of x units of a product is given by

$$P = 180x - \frac{x^2}{1000} - 2000.$$

At what rate is the profit changing when the number of units produced and sold is 100 and is increasing at a rate of 10 units per day?

18. **Profit** Suppose that the monthly revenue and cost (in dollars) for x units of a product are

$$R = 400x - \frac{x^2}{20} \quad \text{and} \quad C = 5000 + 70x.$$

At what rate is the profit changing if the number of units produced and sold is 100 and is increasing at a rate of 10 units per month?

19. **Demand** Suppose that the price p (in dollars) of a product is given by the demand function

$$p = \frac{1000 - 10x}{400 - x},$$

where x represents the quantity demanded. If the daily demand is *decreasing* at a rate of 20 units per day, at what rate is the price changing when the demand is 20 units?

20. **Supply** The supply function for a product is given by $p = 40 + 100\sqrt{2x + 9}$, where x is the number of units supplied and p is the price in dollars. If the price is increasing at a rate of $1 per month, at what rate is the supply changing when $x = 20$?

21. **Capital investment and production** Suppose that the number of units x of a product that is produced per month depends on the number of thousands of dollars y invested, with $x = 30y + 20y^2$. At what rate will production increase if $10,000 is invested and if the investment capital is increasing at a rate of $1000 per month?

22. **Boyle's law** Boyle's law for enclosed gases states that at a constant temperature, the pressure is related to the volume by the equation

$$P = \frac{k}{V},$$

where k is a constant. If the volume is increasing at a rate of 5 cubic inches per hour, at what rate is the pressure changing when the volume is 30 cubic inches and $k = 2$ inch-pounds?

Tumor growth For problems 23 and 24, suppose that a tumor in a person's body has a spherical shape, and that treatment is causing the radius of the tumor to decrease at a rate of 1 millimeter per month.

23. At what rate is the volume decreasing when the radius is 3 millimeters? Recall that $V = \frac{4}{3}\pi r^3$.

24. At what rate is the surface area of the tumor decreasing when the radius is 3 mm? (Recall that for a sphere, $S = 4\pi r^2$.)

25. **Allometric relationships—fish** For many species of fish, the allometric relationship between the weight W and the length L is approximately $W = kL^3$, where k is a constant. Find the percentage rate of change of the weight as a corresponding percentage rate of change of the length.

26. **Blood flow** The resistance R of a blood vessel to the flow of blood is a function of the radius r of the blood vessel and is given by

$$R = \frac{k}{r^4},$$

where k is a constant. Find the percentage rate of change of the resistance of a blood vessel in terms of the percentage rate of change in the radius of the blood vessel.

27. **Allometric relationships—crabs** For fiddler crabs, data gathered by Thompson* show that the allometric relationship between the weight C of the claw and the weight W of the body is given by

$$C = 0.11W^{1.54}.$$

Find the percentage rate of change in the claw weight in terms of the percentage rate of change in the body weight for fiddler crabs.

28. **Body weight and surface area** For human beings, the surface area S of the body is related to the body's weight W according to

$$S = kW^{2/3},$$

---

*d'Arcy Thompson, *On Growth and Form* (Cambridge: Cambridge University Press, 1961).

where k is a constant. Find the percentage rate of change in the body's surface area in terms of the percentage rate of change in the body's weight.

29. **Cell growth** A bacterial cell has a spherical shape. If the volume of the cell is increasing at a rate of 4 cubic micrometers per day, at what rate is the radius of the cell increasing when it is 2 micrometers? (Recall that for a sphere, $V = \frac{4}{3}\pi r^3$.)

30. **Water purification** Assume that water is being purified by flowing through a conical filter that has a height of 15 inches and a radius of 5 inches. If the depth of the water is decreasing at a rate of 1 inch per minute when the depth is 6 inches, at what rate is the (volume of) water flowing out of the filter at this instant?

31. **Volume and radius** Suppose that air is being pumped into a spherical balloon at a rate of 5 in³/min. At what rate is the radius of the balloon increasing when the radius is 5 in.?

32. **Boat docking** Suppose that a boat is being pulled toward a dock by a winch that is 5 ft above the level of the boat deck. If the winch is pulling the cable at a rate of 3 ft/min, at what rate is the boat approaching the dock when it is 12 ft from the dock? Use the figure below.

33. **Ladder safety** A 30-ft ladder is leaning against a wall. If the bottom is pulled away from the wall at a rate of 1 ft/sec, at what rate is the top of the ladder sliding down the wall when the bottom is 18 ft from the wall?

34. **Flight** A kite is 30 ft high and is moving horizontally at of a rate of 10 ft/min. If the kite string is taut, at what rate is the string being played out when 50 ft of string is out?

35. **Flight** A plane is flying at a constant altitude of 1 mile and a speed of 300 mph. If it is flying toward an observer on the ground, how fast is the plane approaching the observer when it is 5 miles from the observer?

36. **Distance** Two boats leave the same port at the same time, with boat A traveling north at 15 knots and boat B traveling east at 20 knots. How fast is the distance between them changing when boat A is 30 nautical miles from port?

37. **Distance** Two cars are approaching an intersection on roads that are perpendicular to each other. Car A is north of the intersection and traveling south at 40 mph. Car B is east of the intersection and traveling west at 55 mph. How fast is the distance between the cars changing when car A is 15 miles from the intersection and car B is 8 miles from the intersection?

38. **Water depth** Water is flowing into a barrel in the shape of a right circular cylinder at the rate of 200 in³/min. If the radius of the barrel is 18 in., at what rate is the depth of the water changing when the water is 30 in. deep?

39. **Water depth** Suppose that water is being pumped into a rectangular swimming pool of uniform depth at 10 ft³/hr. If the pool is 10 ft wide and 25 ft long, at what rate is the water rising when it is 4 ft deep?

11.5 APPLICATIONS IN BUSINESS AND ECONOMICS

Objectives

❑ To find the elasticity of demand
❑ To find the tax per unit that will maximize tax revenue

APPLICATION PREVIEW

In this section, we consider two applications: **elasticity of demand** and **taxation in a competitive market.** Elasticity of demand measures how sensitive the demand for a product is to price changes, and it can be used to measure the effect that price changes have on total revenue. Taxation in a competitive market examines how a tax levied on goods produced shifts in market equilibrium, and it also addresses the associated problem of finding the tax per unit that, despite changes in market equilibrium, maximizes tax revenues.

ELASTICITY OF DEMAND

We know from the law of demand that consumers will respond to changes in prices; if prices increase, the quantity demanded will decrease. But the degree of responsiveness of the consumers to price changes will vary widely for different products. For example, a price increase in insulin will not decrease greatly the demand for it by diabetics, but a price increase in clothes may cause consumers to buy considerably less and wear their old clothes longer. When the response to price changes is considerable, we say the demand is *elastic*. When price changes cause relatively small changes in demand for a product, the demand is said to be *inelastic* for that product.

The **elasticity of demand** on an interval is measured by economists by dividing the rate of change in demand by the rate of change in price. We may write this as

$$E_d = -\frac{\text{change in quantity demanded}}{\text{original quantity demanded}} \div \frac{\text{change in price}}{\text{original price}}$$

or

$$E_d = -\frac{\Delta q}{q} \div \frac{\Delta p}{p}.$$

The demand curve usually has a negative slope, so we have introduced a negative sign into the formula to give us a positive elasticity.

We can write the equation for elasticity as

$$E_d = -\frac{p}{q} \cdot \frac{\Delta q}{\Delta p}.$$

If q is a function of p, then we can write

$$\frac{\Delta q}{\Delta p} = \frac{f(p + \Delta p) - f(p)}{\Delta p},$$

and the limit as Δp approaches 0 gives the **point elasticity of demand:**

$$\eta = \lim_{\Delta p \to 0} \left(-\frac{p}{q} \cdot \frac{\Delta q}{\Delta p} \right) = -\frac{p}{q} \cdot \frac{dq}{dp}.$$

The **point elasticity or elasticity of demand** at the point (q_A, p_A), is

$$\eta = -\frac{p}{q} \cdot \frac{dq}{dp}\bigg|_{(q_A, p_A)}.$$

EXAMPLE I

Find the elasticity of the demand function $p + 5q = 100$ when the
(a) price is \$40,   (b) price is \$60, and   (c) price is \$50.

Solution

Solving the demand function for q gives $q = 20 - \frac{1}{5}p$. Then $dq/dp = -\frac{1}{5}$
and

$$\eta = -\frac{p}{q}\left(-\frac{1}{5}\right).$$

(a) When $p = 40$, $q = 12$ and $\eta = -\frac{p}{q}\left(-\frac{1}{5}\right)\Big|_{(12,\,40)} = -\frac{40}{12}\left(-\frac{1}{5}\right) = \frac{2}{3}.$

(b) When $p = 60$, $q = 8$ and $\eta = -\frac{p}{q}\left(-\frac{1}{5}\right)\Big|_{(8,\,60)} = -\frac{60}{8}\left(-\frac{1}{5}\right) = \frac{3}{2}.$

(c) When $p = 50$, $q = 10$ and $\eta = -\frac{p}{q}\left(-\frac{1}{5}\right)\Big|_{(10,\,50)} = -\frac{50}{10}\left(-\frac{1}{5}\right) = 1.$

Note that in Example 1 the demand equation was $p + 5q = 100$, so the demand "curve" is a straight line, with slope $m = -5$. But the elasticity was $\eta = \frac{2}{3}$ at $(12, 40)$, $\eta = \frac{3}{2}$ at $(8, 60)$, and $\eta = 1$ at $(10, 50)$. This illustrates that the elasticity of demand may be different at different points on the demand curve, even though the slope of the demand "curve" is constant. (See Figure 11.9.) Economists use η to measure how responsive demand is to price at different points on the demand curve for a product.

Figure 11.9

This example shows that the elasticity of demand is more than just the slope of the demand curve, which is the rate at which the demand is changing. Recall that the elasticity measures the consumers' degree of responsiveness to a price change.

Economists classify demand curves according to how responsive demand is to price changes by using elasticity.

ELASTICITY OF DEMAND

- If $\eta > 1$, the demand is **elastic,** and the percent of decrease in demand is greater than the corresponding percent increase in price.
- If $\eta < 1$, the demand is **inelastic,** and the percent of decrease in demand will be less than the corresponding percent increase in price.
- If $\eta = 1$, the demand is **unitary elastic,** and the percent decrease in demand is approximately equal to the corresponding percent increase in price.

We can also use implicit differentiation to find dq/dp in evaluating the point elasticity of demand.

EXAMPLE 2

The demand for a certain product is given by

$$p = \frac{1000}{(q + 1)^2},$$

where p is the price per unit in dollars and q is demand in units of the product. Find the elasticity of demand with respect to price when $q = 19$.

Solution

To find the elasticity, we need to find dq/dp. Using implicit differentiation, we get the following:

$$\frac{d}{dp}(p) = \frac{d}{dp}[1000(q + 1)^{-2}]$$

$$1 = 1000\left[-2(q + 1)^{-3}\frac{dq}{dp}\right]$$

$$1 = \frac{-2000}{(q + 1)^3}\frac{dq}{dp}$$

$$\frac{(q + 1)^3}{-2000} = \frac{dq}{dp}.$$

When $q = 19$, we have $p = 1000/(19 + 1)^2 = 1000/400 = 5/2$ and

$$\frac{dq}{dp}\bigg|_{(q = 19)} = \frac{(19 + 1)^3}{-2000} = \frac{8000}{-2000} = -4.$$

The elasticity of demand when $q = 19$ is

$$\eta = \frac{-p}{q}\cdot\frac{dq}{dp} = -\frac{(5/2)}{19}\cdot(-4) = \frac{10}{19} < 1.$$

Thus the demand for this product is inelastic.

Elasticity is related to revenue in a special way. We can see how by computing the derivative of the revenue function

$$R = pq$$

with respect to p.

$$\frac{dR}{dp} = p \cdot \frac{dq}{dp} + q \cdot 1$$

$$= \frac{q}{q} \cdot p \cdot \frac{dq}{dp} + q = q \cdot \frac{p}{q} \cdot \frac{dq}{dp} + q$$

$$= q(-\eta) + q$$

$$= q(1 - \eta)$$

From this we can summarize the relationship of elasticity and revenue.

Elasticity and Revenue   The rate of change of revenue R with respect to price p is related to elasticity in the following way.

- $\dfrac{dR}{dp} < 0$ if $\eta > 1$, so an increase in price will result in a decrease in revenue when $\eta > 1$ (that is, when demand is elastic).

- $\dfrac{dR}{dp} > 0$ if $\eta < 1$, so an increase in price will result in an increase in revenue when $\eta < 1$ (that is, when demand is inelastic).

- $\dfrac{dR}{dp} = 0$ if $\eta = 1$, so an increase in price will not change revenue when $\eta = 1$ (that is, when demand is unitary elastic). Revenue is optimized at this point.

EXAMPLE 3

The demand for a product is given by

$$p = 10\sqrt{100 - q}, \qquad 0 \le q \le 100.$$

(a) Find the point at which demand is of unitary elasticity, and find intervals in which the demand is inelastic and elastic.
(b) Find where revenue is increasing, decreasing, and maximized.
(c) Use a graphing utility to show the graph of the revenue function $R = pq$, with $0 \le q \le 100$, and confirm the results from part (b).

Solution
The elasticity is

$$\eta = -\frac{10\sqrt{100 - q}}{q} \cdot \frac{dq}{dp}.$$

Finding dq/dp implicitly, we have:

$$1 = 10\left[\frac{1}{2}(100 - q)^{-1/2}\left(-\frac{dq}{dp}\right)\right],$$

so

$$\frac{dq}{dp} = -\frac{1}{5}\sqrt{100 - q}.$$

Thus

$$\eta = -\frac{10\sqrt{100 - q}}{q}\left[-\frac{1}{5}\sqrt{100 - q}\right] = \frac{200 - 2q}{q}.$$

(a) Unitary elasticity occurs where $\eta = 1$.

$$1 = \frac{200 - 2q}{q}$$
$$q = 200 - 2q$$
$$3q = 200$$
$$q = 66\tfrac{2}{3}$$

so unitary elasticity occurs when $66\tfrac{2}{3}$ units are sold at a price of $57.74. For values of q between 0 and $66\tfrac{2}{3}$, $\eta > 1$ and demand is elastic. For values of q between $66\tfrac{2}{3}$ and 100, $\eta < 1$ and demand is inelastic.

(b) When q increases over $0 < q < 66\tfrac{2}{3}$, p decreases, so $\eta > 1$ means R increases. Similarly, when q increases over $66\tfrac{2}{3} < q < 100$, p decreases, so $\eta < 1$ means R decreases. Revenue is maximized where $\eta = 1$, at $q = 66\tfrac{2}{3}$, $p = 57.74$.

(c) The graph of this revenue function,

$$R = 10q\sqrt{1 - q},$$

is shown in Figure 11.10, and confirms our conclusions from (b).

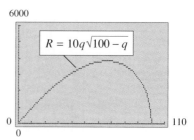

Figure 11.10

CHECKPOINT

1. Write the formula for point elasticity, η.

2. (a) If $\eta > 1$, the demand is called _____ .
   (b) If $\eta < 1$, the demand is called _____ .
   (c) If $\eta = 1$, the demand is called _____ .

3. Find the elasticity of demand for $q = \dfrac{100}{p} - 1$ when $p = 10$ and $q = 9$.

TAXATION IN A COMPETITIVE MARKET

Many taxes imposed by governments are "hidden." That is, the tax is levied on goods produced, and the producers must pay the tax. Of course, the tax becomes a cost to the producers, and they pass the tax on to the consumer in the form of higher prices for goods.

Suppose the government imposes a tax of t dollars on each unit produced and sold by producers. If we are in pure competition in which the consumers' demand depends only on price, the *demand function* will not change. The tax will change the supply function, of course, because at each level of output q, the firm will want to charge a price higher by the amount of the tax.

The graphs of the market demand function, the original market supply function, and the market supply function after taxes are shown in Figure 11.11. Since the tax added to each item is constant, the graph of the supply function is t units above the original supply function. If $p = f(q)$ defines the original supply function, then $p = f(q) + t$ defines the supply function after taxation.

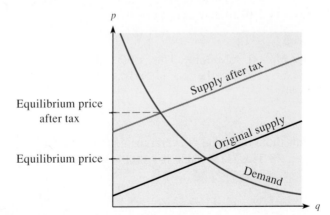

Figure 11.11

Note in this case that after the taxes are imposed, *no* items are supplied at the price that was the equilibrium price before taxation. After the taxes are imposed, the consumers will simply have to pay more for the product. Since taxation does not change the demand curve, the quantity purchased at market equilibrium will be less than it was before taxation. Thus governments planning taxes should recognize that they will not collect taxes on the original equilibrium quantity. They will collect on the *new* equilibrium quantity, a quantity reduced by their taxation. Thus a large tax on each item may reduce the quantity demanded at the new market equilibrium so much that very little revenue would result from the tax.

If the tax revenue is represented by $T = tq$, where t is the tax per unit and q is the equilibrium quantity of the supply and demand functions after taxation, we can use the following procedure for maximizing the total tax revenue in a competitive market.

Maximizing Total Tax Revenue

Procedure	Example
To find the tax per item (under pure competition) that will maximize total tax revenue:	If the demand and supply functions are given by $p = 600 - q$ and $p = 200 + \frac{1}{3}q$, respectively, find the tax rate t that will maximize the total tax revenue T.
1. Write the supply function after taxation.	1. $p = 200 + \frac{1}{3}q + t$
2. Set the demand and (new) supply functions equal, and solve for t.	2. $600 - q = 200 + \frac{1}{3}q + t$ $400 - \frac{4}{3}q = t$
3. Form the total tax revenue function, $T = tq$, and take its derivative with respect to q.	3. $T = tq = 400q - \frac{4}{3}q^2$ $T'(q) = \dfrac{dT}{dq} = 400 - \dfrac{8}{3}q$
4. Set $T' = 0$, solve for q. This is the q that should maximize T. Use the second-derivative test to verify it.	4. $0 = 400 - \frac{8}{3}q$ $q = 150$ $T''(q) = -\frac{8}{3}$. Thus T is maximized at $q = 150$.
5. Substitute the value of q into the equation for t (in step 2). This is the value of t that will maximize T.	5. $t = 400 - \frac{4}{3}(150) = 200$ A tax of $200 per item will maximize the total tax revenue. The total tax revenue for the period would be $200 \cdot (150) = \$30{,}000$.

Note that in the example just given, if a tax of $300 were imposed, the total tax revenue the government would receive would be

$$(\$300)(75) = \$22{,}500.$$

This means that consumers would spend $100 more for each item, suppliers would sell 75 fewer items, and the government would lose $7500 in tax revenue. Thus everyone would suffer if the tax rate were raised above $300.

An infamous example of a tax increase that resulted in decreased tax revenue and economic disaster is the "luxury tax" that went into effect in 1991. This was a 10% excise tax on the sale of more expensive jewelry, furs, airplanes, certain expensive boats, and automobiles costing over $30,000. The Congressional Joint Tax Committee had estimated it would raise $6 million from airplanes alone, but it raised only $53,000 while it destroyed the small airplane market (one company lost $130 million and 480 jobs in one year). It also capsized the boat market and was repealed at the end of 1993 (except for automobiles).*

CHECKPOINT

4. For problems involving taxation in a competitive market, if supply is $p = f(q)$ and demand is $p = g(q)$, is the tax t added to $f(q)$ or to $g(q)$?

EXAMPLE 4

If the demand and supply functions for a product are $p = 900 - 20q - \frac{1}{3}q^2$ and $p = 200 + 10q$, respectively, find the tax per unit that will maximize the tax revenue T.

Solution

After taxation, the supply function is $p = 200 + 10q + t$, where t is the tax per unit. The demand function will meet the new supply function where

$$900 - 20q - \tfrac{1}{3}q^2 = 200 + 10q + t.$$

So

$$t = 700 - 30q - \tfrac{1}{3}q^2.$$

Then the total tax T is $T = tq = 700q - 30q^2 - \frac{1}{3}q^3$, and we maximize T as follows:

$$T'(q) = 700 - 60q - q^2$$
$$0 = -(q + 70)(q - 10)$$
$$q = 10 \quad \text{or} \quad q = -70.$$

Since $q = -70$ is meaningless, we test $q = 10$.

$$\left.\begin{array}{l} T(0) = 0 \\ T'(0) = 700 > 0 \\ T'(q) < 0 \text{ for all } q > 10 \end{array}\right\} \Rightarrow \text{absolute max at } q = 10$$

*Fortune, Sept 6, 1993; *Motor Trend,* December 1993.

The maximum possible tax revenue is

$$T(10) = \$3666.67.$$

The tax per unit that maximizes T is

$$t = 700 - 30(10) - \tfrac{1}{3}(10)^2 = \$366.67.$$

CHECKPOINT
SOLUTIONS

1. $\eta = \dfrac{-p}{q} \cdot \dfrac{dq}{dp}$

2. (a) elastic   (b) inelastic   (c) unitary elastic

3. $\dfrac{dq}{dp} = \dfrac{-100}{p^2}$ and $\dfrac{dq}{dp} = -1$ when $p = 10, q = 9$

     $\eta = \dfrac{-10}{9}(-1) = \dfrac{10}{9}$ (elastic)

4. Tax t is added to supply: $p = f(q) + t$.

EXERCISE 11.5

ELASTICITY OF DEMAND

1. (a) Find the elasticity of the demand function
   $p + 4q = 80$ at $(10, 40)$.
   (b) How will a price increase affect total revenue?

2. (a) Find the elasticity of the demand function
   $2p + 3q = 150$ at the price $p = 15$.
   (b) How will a price increase affect total revenue?

3. (a) Find the elasticity of the demand function
   $p^2 + 2p + q = 49$ at $p = 6$.
   (b) How will a price increase affect total revenue?

4. (a) Find the elasticity of the demand function
   $pq = 81$ at $p = 3$.
   (b) How will a price increase affect total revenue?

5. Suppose that the demand for a product is given by
   $pq + p = 5000$.
   (a) Find the elasticity when $p = \$50$ and $q = 99$.
   (b) Tell what type of elasticity this is: unitary, elastic, or inelastic.
   (c) How would revenue be affected by a price increase?

6. Suppose that the demand for a product is given by
   $2p^2q = 10{,}000 + 9000p^2$.
   (a) Find the elasticity when $p = \$50$ and $q = 4502$.

(b) Tell what type of elasticity this is: unitary, elastic, or inelastic.
   (c) How would revenue be affected by a price increase?

7. Suppose that the demand for a product is given by
   $pq + p + 100q = 50{,}000$.
   (a) Find the elasticity when $p = \$401$.
   (b) Tell what type of elasticity this is.
   (c) How would a price increase affect revenue?

8. Suppose that the demand for a product is given by

   $$(p + 1)\sqrt{q + 1} = 1000.$$

   (a) Find the elasticity when $p = \$39$.
   (b) Tell what type of elasticity this is.
   (c) How would a price increase affect revenue?

9. Suppose the demand function for a product is given by

   $$p = \frac{1}{2} \ln\left(\frac{5000 - q}{q + 1}\right),$$

   where p is in hundreds of dollars and q is the number of tons.

(a) What is the elasticity of demand when the quantity demanded is 2 tons and the price is $371?

(b) Is the demand elastic or inelastic?

10. Suppose the weekly demand function for a product is

$$q = \frac{5000}{1 + e^{2p}} - 1,$$

where p is the price in thousands of dollars and q is the number of units demanded. What is the elasticity of demand when price is $1000 and the quantity demanded is 595?

In problems 11 and 12, the demand functions for specialty steel products are given. For both problems:

(a) Find the elasticity of demand as a function of quantity demanded, q

(b) Find the point at which demand is of unitary elasticity and find intervals in which the demand is inelastic and elastic.

(c) Use information about elasticity in part (b) to decide where the revenue is increasing, is decreasing and is maximized.

(d) Graph the revenue function $R = pq$, and use it to find where revenue is maximized. Is it at the same quantity as that determined in part (c)?

11. $p = 120\sqrt[3]{125 - q}$ 12. $p = 30\sqrt{49 - q}$

TAXATION IN A COMPETITIVE MARKET

13. If the weekly demand function is $p = 30 - q$ and the supply function before taxation is $p = 6 + 2q$, what tax per item will maximize the total tax revenue?

14. If the demand function for a fixed period of time is given by $p = 38 - 2q$ and the supply function before taxation is $p = 8 + 3q$, what tax per item will maximize the total tax revenue?

15. If the demand and supply functions for a product are $p = 800 - 2q$ and $p = 100 + 0.5q$, respectively, find the tax per unit t that will maximize the tax revenue T.

16. If the demand and supply functions for a product are $p = 2100 - 3q$ and $p = 300 + 1.5q$, respectively, find the tax per unit t that will maximize the tax revenue T.

17. If the weekly demand function is $p = 200 - 2q^2$ and the supply function before taxation is $p = 20 + 3q$, what tax per item will maximize the total tax revenue?

18. If the monthly demand function is $p = 7230 - 5q^2$ and the supply function before taxation is $p = 30 + 30q^2$, what tax per item will maximize the total revenue?

19. Suppose the weekly demand for a product is given by $p + 2q = 840$, and the weekly supply before taxation is given by $p = 0.02q^2 + 0.55q + 7.4$. Find the tax per item that produces maximum tax revenue. Find the tax revenue.

20. If the daily demand for a product is given by the function $p + q = 1000$, and the daily supply before taxation is $p = q^2/30 + 2.5q + 920$, find the tax per item that maximizes tax revenue. Find the tax revenue.

21. If the demand and supply functions for a product are $p = 2100 - 10q - 0.5q^2$ and $p = 300 + 5q + 0.5q^2$, respectively, find the tax per unit t that will maximize the tax revenue T.

22. If the demand and supply functions for a product are $p = 5000 - 20q - 0.7q^2$ and $p = 500 + 10q + 0.3q^2$, respectively, find the tax per unit t that will maximize the tax revenue T.

KEY TERMS
& FORMULAS

Section	Key Term	Formula
11.1	Logarithmic function	$y = \log_a x$, defined by $x = a^y$
	Natural logarithm	$\ln x = \log_e x$
	Logarithmic Properties I-V for natural logarithms	$\ln e^x = x;\ e^{\ln x} = x;$
		$\ln(MN) = \ln M + \ln N;$
		$\ln(M/N) = \ln M - \ln N;$
		$\ln(M^N) = N(\ln M)$
	Change of base formula	$\log_a x = \dfrac{\ln x}{\ln a}$

Section	Key Term	Formula
	Derivatives of logarithmic functions	$\dfrac{d}{dx}(\ln x) = \dfrac{1}{x}$
		$\dfrac{d}{dx}(\ln u) = \dfrac{1}{u} \cdot \dfrac{du}{dx}$
11.2	Exponential function	$f(x) = a^x$
	e	$e = \lim\limits_{a \to 0}(1 + a)^{1/a}$
	Derivatives of exponential functions	$\dfrac{d}{dx}(e^x) = e^x$
		$\dfrac{d}{dx}e^u = e^u \dfrac{du}{dx}$
		$\dfrac{d}{dx}a^u = a^u \dfrac{du}{dx} \ln a$
11.3	Implicit differentiation	
11.4	Related rates	
	Percentage rates of change	
11.5	Elasticity of demand	$\eta = \dfrac{-p}{q} \cdot \dfrac{dq}{dp}$
	Elastic	$\eta > 1$
	Inelastic	$\eta < 1$
	Unitary elastic	$\eta = 1$
	Taxation in competitive market	
	Supply function after taxation	$p = f(q) + t$

REVIEW EXERCISES

11.1, 11.2

In problems 1–16, find the indicated derivative.

1. If $y = x^2 e^x$, find dy/dx.

2. If $y = \ln e^{x^2}$, find y'.

3. If $p = \ln\left(\dfrac{q}{q^2 - 1}\right)$, find $\dfrac{dp}{dq}$.

4. If $y = e^{x^2}$, find dy/dx.

5. If $y = 3^{3x-4}$, find dy/dx.

6. If $y = (1 + \ln x)^{10}$, find dy/dx.

7. If $y = \dfrac{\ln x}{x}$, find $\dfrac{dy}{dx}$.

8. If $y = \dfrac{1 + e^{-x}}{1 - e^{-x}}$, find $\dfrac{dy}{dx}$.

9. If $y = \dfrac{\ln(2x + 1)}{2x + 1}$, find $\dfrac{dy}{dx}$.

10. If $x = e^t(t^2 + 4)^2$, find dx/dt.

11.3

11. If $y \ln x = 5$, find dy/dx.

12. Find dy/dx for $e^{xy} = y$.

13. Find dy/dx for $y^2 = 4x - 1$.

14. Find dy/dx if $x^2 + 3y^2 + 2x - 3y + 2 = 0$.

15. Find dy/dx for $3x^2 + 2x^3y^2 - y^5 = 7$.
16. Find the second derivative of $x^2 + y^2 = 1$.
17. Find the slope of the tangent to the curve
    $x^2 + 4x - 3y^2 + 6 = 0$ at (3, 3).
18. Find the points where tangents to the graph of the equation in problem 17 are horizontal.

11.4

19. Suppose $3x^2 - 2y^3 = 10y$, where x and y are differentiable funtions of t. If $dx/dt = 2$, find dy/dt when $x = 10$ and $y = 5$.
20. A right triangle with legs of lengths x and y has its area given by

    $$A = \tfrac{1}{2}xy.$$

    If the rate of change of x is 2 units per minute and the rate of change of y is 5 units per minute, find the rate of change of the area when $x = 4$ and $y = 1$.

APPLICATIONS

11.2

21. **Compound interest** If the future value of $1000 invested for n years at 12%, compounded continuously, is given by

    $$S = 1000e^{0.12n},$$

    find the rate at which the future value is growing after 1 year.

22. **Compound interest**
    (a) In problem 21, find the rate of growth of the future value after 2 years.
    (b) How much faster is the future value growing at the end of 2 years than after 1 year?

23. **Radioactive decay** A breeder reactor converts stable uranium-238 into the isotope plutonium-239. The decay of this isotope is given by

    $$A(t) = A_0 e^{-0.00002876t},$$

    where $A(t)$ is the amount of isotope at time t, in years, and A_0 is the original amount. This isotope has a half-life of 24,101 years (that is, half of it will decay away in 24,101 years).
    (a) At what rate is $A(t)$ decaying at this point in time?
    (b) At what rate is $A(t)$ decaying after 1 year?
    (c) Is the rate of decay at its half-life greater or less than after 1 year?

24. **Marginal Cost** The average cost of producing x units of a product is $\overline{C} = 600e^{x/600}$. What is the marginal cost when 600 units are produced?

25. **Inflation** The impact of inflation on a $20,000 pension can be measured by the purchasing power P of $20,000 after t years. For an inflation rate of 5% per year, compounded annually, P is given by

    $$P = 20{,}000e^{-0.0495t}.$$

    At what rate is purchasing power changing when $t = 10$? (Source: *Viewpoints*, VALIC, Summer 1993)

11.4

26. **Evaporation** A spherical droplet of water evaporates at a rate of 1 mm³/min. Find the rate of change of the radius when the droplet has a radius of 2.5 mm.

27. **Worker Safety** A sign is being lowered over the side of a building at the rate of 2 ft/min. A worker handling a guide line is 7 ft away from a spot directly below the sign. How fast is the worker taking in the guide line at the instant the sign is 25 ft from the worker's hands? See the figure below.

28. **Environment** Suppose that in a study of water birds, the relationship between the square miles of wetlands A and the number of different species S of birds found in the area was found to be

    $$S = kA^{1/3},$$

    where k is constant. Find the percentage rate of change in the number of species in terms of the percentage rate of change of the area.

11.5

29. **Taxes** Can increasing the tax per unit sold actually lead to a decrease in tax revenues?

30. **Taxes** If the demand and supply functions for a product are

$$p = 2800 - 8q - \frac{q^2}{3} \quad \text{and} \quad p = 400 + 2q,$$

respectively, find the tax per unit t that will maximize the tax revenue T.

31. **Taxes** If the supply and demand functions for a product are

$$p = 40 + 20q \quad \text{and} \quad p = \frac{5000}{q + 1},$$

respectively, find the tax t that maximizes tax revenue T.

32. **Elasticity** A demand function is given by

$$pq = 27.$$

(a) Find the elasticity of demand at $(9, 3)$.
(b) How will a price increase affect total revenue?

33. **Elasticity** Suppose the demand for a product is given by

$$p^2(2q + 1) = 10,000.$$

Find the elasticity of demand when $p = \$20$.

34. **Elasticity** Suppose the weekly demand function for a product is given by

$$p = 100e^{-0.1q},$$

where p is the price in dollars and q is the number of tons demanded. What is the elasticity of demand when price is $\$36.79$ and the quantity demanded is 10?

35. **Revenue** A product has the demand function

$$p = 100 - 0.5q.$$

(a) Find the elasticity $\eta(q)$ as a function of q, and graph the function

$$f(q) = \eta(q) - 1.$$

(b) Find the q-intercept of $f(q)$, which gives the quantity for which the product has unitary elasticity.
(c) The revenue function for this product is

$$R(q) = pq = (100 - 0.5q)q.$$

Graph $R(q)$ and find the q-value for which the maximum revenue occurs.
(d) What is the relationship between elasticity and maximum revenue?

CONSULTANT'S CORNER

Inflation

Hollingsworth Pharmaceuticals specializes in manufacturing generic medicines. Recently they developed an antibiotic with outstanding profit potential. The new antibiotic's total costs, sales, and sales growth, as well as projected inflation are described as follows.

Total monthly costs to produce x units (one unit is 100 capsules):

$$C(x) = \begin{cases} 15,000 + 10x & 0 \le x \le 11,000 \\ 15,000 + 10x + 0.001(x - 11,000)^2 & x \ge 11,000 \end{cases}$$

Sales: 10,000 units per month and growing at 1.25% per month
Selling price: $17 per unit
Inflation: Approximately 0.25% per month, affecting both total costs and selling price

Company owners are pleased with the sales growth but are concerned about the projected increase in variable costs when production levels exceed 11,000 units per month. The consensus is that improvements eventually can be made that will reduce costs at higher production levels, thus altering the current cost function model. To properly plan for these changes, Hollingsworth Pharmaceuticals would like you to determine when the company's profits will begin to decrease. To help you determine this, answer the following.

1. If inflation is assumed to be compounded continuously, the selling price and total costs must be multiplied by the factor $e^{0.0025t}$. In addition, if sales growth is assumed to be compounded continuously, then sales must be multiplied by a factor of the form e^{rt}, where r is the monthly sales growth rate (expressed as a decimal) and t is time in months. Use these factors to write each of the following as a function of time t:
   (a) selling price p per unit (including inflation)
   (b) number of units x sold per month (including sales growth)
   (c) total revenue (Recall that $R = px$.)

2. Determine how many months it will be before monthly sales exceed 11,000 units.

3. If you restrict your attention to total costs when $x \ge 11,000$, then, after expanding and collecting like terms, $C(x)$ can be written as follows:

$$C(x) = 136,000 - 12x + 0.001x^2 \quad \text{for } x \ge 11,000.$$

Use this form for $C(x)$ with your result from 1(b) and with the inflationary factor $e^{0.0025t}$ to express these total costs as a function of time.

4. Form the profit function that would be used when monthly sales exceed 11,000 units by using the total revenue function from 1(c) and the total cost function from question 3. This profit function should be a function of time t.

5. Find how long it will be before the profit is maximized. You may have to solve $P'(t) = 0$ by using a graphics calculator or computer to find the t-intercept of the graph of $P'(t)$. In addition, because $P'(t)$ has large numerical coefficients you may want to divide both sides of $P'(t) = 0$ by 1000 before solving or graphing.

803

Warm-up

Prerequisite problem type	For section	Answer	Section for review
Write as a power: (a) $\sqrt{x}$ (b) $\sqrt{x^2 - 9}$	12.1– 12.4	(a) $x^{1/2}$ (b) $(x^2 - 9)^{1/2}$	0.4 Radicals
Expand $(x^2 + 4)^2$.	12.2	$x^4 + 8x^2 + 16$	0.5 Special powers
Divide $x^4 - 2x^3 + 4x^2 - 7x - 1$ by $x^2 - 2x$.	12.3	$x^2 + 4 + \dfrac{x - 1}{x^2 - 2x}$	0.5 Division
Find the derivative of (a) $f(x) = 2x^{1/2}$ (b) $u = x^3 - 3x$	12.1 12.2 12.3	(a) $f'(x) = x^{-1/2}$ (b) $u' = 3x^2 - 3$	9.4 Derivatives
If $y = \dfrac{(x^2 + 4)^6}{6}$, what is y'?	12.2	$(x^2 + 4)^5 2x$	9.6 Derivatives
(a) If $y = \ln u$, what is y'? (b) If $y = e^u$, what is y'?	12.3	(a) $y' = \dfrac{1}{u} \cdot u'$ (b) $y' = e^u \cdot u'$	11.1, 11.2 Derivatives
Solve for y: $\ln y = kt + C$	12.5	$y = e^{kt + C}$	5.2 Logarithmic functions
Solve for k: $0.5 = e^{5600k}$	12.5	$k = -0.00012378$	5.3 Exponential equations

CHAPTER 12

INDEFINITE INTEGRALS

If the marginal cost for a product is $36 at all levels of production, we know that the total cost function is a linear function. In particular $C(x) = 36x + FC$, where FC is the fixed cost. But if the marginal cost changes at different levels of production, the total cost function cannot be linear. In this chapter we will use integration to find total cost functions, given information about marginal costs and fixed costs.

Accountants can use linear regression to translate information about marginal cost into a linear equation defining (approximately) the marginal cost function. By integrating this marginal cost function, it is possible to find an (approximate) function that defines the total cost.

We can also use integration to find total revenue functions from marginal revenue functions, to optimize profit from information about marginal cost and marginal revenue, and to find national consumption functions from information about marginal propensity to consume.

Integration can be used in the social and life sciences to predict growth or decay from expressions giving rates of change. For example, we can determine equations for population size from the rate of growth; we can write equations for the number of radioactive atoms remaining in a substance if we know the rate of disintegration of the substance; and we can determine the volume of blood flow from information about the rate of flow.

12.1 THE INDEFINITE INTEGRAL

Objective

❑ To find certain indefinite integrals

APPLICATION PREVIEW

In our study of the theory of the firm, we have worked with total cost, total revenue, and profit functions and have found their marginal functions. In practice, it is often easier for a company to measure marginal cost, revenue, and profit and use this data to form marginal functions from which it can find total cost, revenue, and profit functions. For example, Jarus Technologies manufactures computer memory boards, and the company's sales records show that the marginal revenue for its 8 MB memory board is given by

$$\overline{MR} = 300 - 0.2x,$$

where x is the number of units sold. If we want to use this function to find the total revenue function for Jarus Technologies' 8 MB memory board, we need to find $R(x)$ from $\overline{MR} = R'(x)$. In this situation, we need to be able to reverse the process of differentiation.

We have studied procedures for and applications of finding derivatives of a given function. We now turn our attention to reversing this process of differentiation. When we know the derivative of a function, the process of finding the function itself is called **antidifferentiation.** For example, if the derivative of a function is $2x$, we know that the function could be $f(x) = x^2$ because

$\dfrac{d}{dx}(x^2) = 2x$. But the function could also be $f(x) = x^2 + 4$ because

$\dfrac{d}{dx}(x^2 + 4) = 2x$. It is clear that any function of the form $f(x) = x^2 + C$, where C is a constant, will have $f'(x) = 2x$ as its derivative. Thus we say that the **general antiderivative** of $f'(x) = 2x$ is $f(x) = x^2 + C$, where C is an arbitrary constant.

Antiderivative   A function $F(x)$ is called an **antiderivative** of a function $f(x)$ if, for every x in the domain of f, $F'(x) = f(x)$. If C is an arbitrary constant, then $F(x) + C$ is called the **general antiderivative** of $f(x)$ because all antiderivatives of $f(x)$ have this form.

EXAMPLE 1

If $f'(x) = 3x^2$, what is $f(x)$?

Solution

The derivative of the function $f(x) = x^3$ is $f'(x) = 3x^2$. But other functions also have this derivative. They will all be of the form $f(x) = x^3 + C$, where C is a constant. Thus we say that $f(x) = x^3 + C$ is the general antiderivative of $f'(x) = 3x^2$.

EXAMPLE 2

If $f'(x) = x^3$, what is $f(x)$?

Solution

We know that the derivative of $f(x) = x^4$ is $4x^3$, so the derivative of $f(x) = \frac{1}{4}x^4$ is $f'(x) = x^3$. Thus any function of the form $f(x) = \frac{1}{4}x^4 + C$ will have the derivative $f'(x) = x^3$.

It is easily seen that

$$\text{if } f'(x) = x^4, \quad \text{then } f(x) = \frac{x^5}{5} + C \quad \text{is the general antiderivative;}$$

$$\text{if } f'(x) = x^5, \quad \text{then } f(x) = \frac{x^6}{6} + C \quad \text{is the general antiderivative.}$$

In general, we have the following.

Antiderivative of $f(x) = x^n$	If $f'(x) = x^n$, then $f(x) = \dfrac{x^{n+1}}{n+1} + C$, for $n \neq -1$.

We can see that this general formula applies for any $n \neq -1$ by noting that the derivative of

$$f(x) = \frac{x^{n+1}}{n+1} + C \quad \text{is} \quad f'(x) = \frac{(n+1)x^n}{n+1} + 0 = x^n.$$

Later, we will discuss the case when $n = -1$.

EXAMPLE 3

What is the general antiderivative of $f'(x) = x^{-1/2}$?

Solution

Using the formula, we get

$$f(x) = \frac{x^{1/2}}{1/2} + C = 2x^{1/2} + C.$$

We can check by noting that the derivative of $2x^{1/2} + C$ is $x^{-1/2}$.

Note that the general antiderivative in Example 3 is a function (actually a number of functions, one for each value of C). Several members of this family of functions are shown in Figure 12.1. Notice that at any given x-value, the tangent line to each curve would have the same slope, indicating all family members have the same derivative.

The process of finding an antiderivative is called **integration.** The function that results when integration takes place is called an **indefinite integral,** or more simply, an **integral.** We can denote the indefinite integral (that is, the general antiderivative) of a function $f(x)$ by $\int f(x)\,dx$. Thus we can write $\int x^2\,dx$ to indicate the general antiderivative of the function $f(x) = x^2$. The expression is read as "the integral of x^2 with respect to x." In this case, x^2 is called the **integrand.** The **integral sign,** $\int$, indicates the process of integration, and the dx indicates that the integral is to be taken with respect to x.

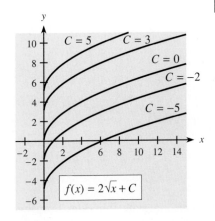

$$f(x) = 2\sqrt{x} + C$$

Figure 12.1

Since the antiderivative of x^2 is $(x^3/3) + C$, we can write

$$\int x^2 \, dx = \frac{x^3}{3} + C.$$

We can now use the integral sign and rewrite the formula for integrating powers of x.

Powers of x Formula

$$\int x^n \, dx = \frac{x^{n+1}}{n+1} + C \quad \text{(for } n \neq -1\text{)}$$

EXAMPLE 4

Find (a) $\displaystyle\int \sqrt[3]{x} \, dx$ and   (b) $\displaystyle\int \frac{1}{x^2} \, dx$.

Solution

(a) $\displaystyle\int \sqrt[3]{x} \, dx = \int x^{1/3} \, dx = \frac{x^{4/3}}{4/3} + C$

$$= \frac{3}{4} x^{4/3} + C = \frac{3}{4} \sqrt[3]{x^4} + C$$

(b) We write the power of x in the numerator so the integral has the form in the formula above.

$$\int \frac{1}{x^2} \, dx = \int x^{-2} \, dx = \frac{x^{-2+1}}{-2+1} + C = \frac{x^{-1}}{-1} + C = \frac{-1}{x} + C$$

Other formulas will be useful in evaluating integrals. The following table shows how some new integration formulas result from differentiation formulas.

Integration Formulas

Derivative	**Resulting Integral**
$\dfrac{d}{dx}(x) = 1$	$\displaystyle\int 1 \, dx = \int dx = x + C$
$\dfrac{d}{dx}[c \cdot u(x)] = c \cdot \dfrac{d}{dx} u(x)$	$\displaystyle\int c\, u(x) \, dx = c \int u(x) \, dx$
$\dfrac{d}{dx}[u(x) \pm v(x)] = \dfrac{d}{dx} u(x) \pm \dfrac{d}{dx} v(x)$	$\displaystyle\int [u(x) \pm v(x)] \, dx = \int u(x) \, dx \pm \int v(x) \, dx$

The formulas above indicate that we can integrate functions term-by-term just as we were able to take derivatives term-by-term.

EXAMPLE 5

Evaluate $\int 4 \, dx$.

Solution

$$\int 4 \, dx = 4\int dx = 4(x + C_1) = 4x + C$$

(Since C_1 is an unknown constant, we can write $4C_1$ as the unknown constant C.)

EXAMPLE 6

Evaluate $\int 8x^5 \, dx$.

Solution

$$\int 8x^5 \, dx = 8\int x^5 \, dx = 8\left(\frac{x^6}{6} + C_1\right) = \frac{4x^6}{3} + C$$

EXAMPLE 7

Evaluate $\int (x^3 + 4x) \, dx$.

Solution

$$\int (x^3 + 4x) \, dx = \int x^3 dx + \int 4x \, dx$$

$$= \left(\frac{x^4}{4} + C_1\right) + \left(4 \cdot \frac{x^2}{2} + C_2\right)$$

$$= \frac{x^4}{4} + 2x^2 + C_1 + C_2$$

$$= \frac{x^4}{4} + 2x^2 + C$$

Note that we need only one constant because the sum of C_1 and C_2 is just a new constant.

EXAMPLE 8

Evaluate $\int (x^2 - 4)^2 \, dx$.

Solution

We expand $(x^2 - 4)^2$ so the integrand is in a form that fits the basic integration formulas.

$$\int (x^2 - 4)^2 \, dx = \int (x^4 - 8x^2 + 16) \, dx = \frac{x^5}{5} - \frac{8x^3}{3} + 16x + C$$

CHECKPOINT

1. True or false:

(a) $\int (4x^3 - 2x)\, dx = \int 4x^3 dx - \int 2x\, dx$
$$= (x^4 + C) - (x^2 + C) = x^4 - x^2$$

(b) $\displaystyle \int \frac{1}{3x^2}\, dx = \frac{1}{3(x^3/3)} + C = \frac{1}{x^3} + C$

2. Evaluate $\int (2x^3 + x^{-1/2} - 4x^{-5})\, dx$.

We now return to the Application Preview and consider how to find total revenue from marginal revenue.

Figure 12.2

(a)

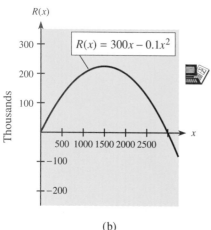

(b)

EXAMPLE 9

Sales records at Jarus Technologies show that the rate of change of the revenue (that is, the marginal revenue) for an 8 MB memory board is $\overline{MR} = 300 - 0.2x$, where x represents the quantity sold. Find the total revenue function for the product.

Solution

We know that the marginal revenue can be found by differentiating the total revenue function. That is,

$$R'(x) = 300 - 0.2x.$$

Thus integrating the marginal revenue function will give the total revenue function.

$$R(x) = \int (300 - 0.2x)\, dx = 300x - 0.1x^2 + K*$$

We can use the fact that there is no revenue when no units are sold to evaluate K. Setting $x = 0$ and $R = 0$ gives $0 = 300(0) - 0.1(0)^2 + K$, so $K = 0$. Thus the total revenue function is

$$R(x) = 300x - 0.1x^2.$$

Graphing Utilities

We can check that the $R(x)$ we found in Example 9 is correct by verifying that $R'(x) = 300 - 0.2x$ and $R(0) = 0$. Also, graphs can help us check the reasonableness of our result. Figure 12.2 shows the graphs of $\overline{MR} = 300 - 0.2x$ and of the $R(x)$ we found. Notice that $R(x)$ passes through the origin, indicating $R(0) = 0$. Also, reading both graphs from left to right, we see that $R(x)$ increases when $\overline{MR} > 0$, attains its maximum when $\overline{MR} = 0$, and decreases when $\overline{MR} < 0$.

*We are using K rather than C here to represent the constant of integration to avoid confusion between the constant C and the cost function $C = C(x)$.

CHECKPOINT SOLUTIONS

1. (a) False, $\int (4x^3 - 2x)\, dx = \int 4x^3\, dx - \int 2x\, dx = (x^4 + C_1) - (x^2 + C_2)$
$$= x^4 - x^2 + C.$$

 (b) False, $\displaystyle\int \frac{1}{3x^2}\, dx = \int \frac{1}{3} \cdot \frac{1}{x^2}\, dx = \frac{1}{3} \int x^{-2}\, dx$
$$= \frac{1}{3} \cdot \frac{x^{-1}}{-1} + C = \frac{-1}{3x} + C.$$

2. $\displaystyle\int (2x^3 + x^{-1/2} - 4x^{-5})\, dx = \frac{2x^4}{4} + \frac{x^{1/2}}{1/2} - \frac{4x^{-4}}{-4} + C$
$$= \frac{x^4}{2} + 2x^{1/2} + x^{-4} + C$$

EXERCISE 12.1

1. If $f'(x) = 4x^3$, what is $f(x)$?
2. If $f'(x) = 5x^4$, what is $f(x)$?
3. If $f'(x) = x^6$, what is $f(x)$?
4. If $g'(x) = x^4$, what is $g(x)$?

Evaluate the following integrals. Check your answers by differentiating.

5. $\int x^7\, dx$

6. $\int x^5\, dx$

7. $\int 8x^5\, dx$

8. $\int 16x^9\, dx$

9. $\int (3^3 + x^{13})\, dx$

10. $\int (5^2 + x^{10})\, dx$

11. $\int (3 - x^{3/2})\, dx$

12. $\int (8 + x^{2/3})\, dx$

13. $\int (x^4 - 9x^2 + 3)\, dx$

14. $\int (3x^2 - 4x - 4)\, dx$

15. $\int (2 + 2\sqrt{x})\, dx$

16. $\int (17 + \sqrt{x^3})\, dx$

17. $\int 6\sqrt[4]{x}\, dx$

18. $\int 3\sqrt[3]{x^2}\, dx$

19. $\displaystyle\int \frac{5}{x^4}\, dx$

20. $\displaystyle\int \frac{6}{x^5}\, dx$

21. $\displaystyle\int \frac{dx}{2\sqrt[3]{x^2}}$

22. $\displaystyle\int \frac{2\, dx}{5\sqrt[3]{x^3}}$

23. $\displaystyle\int \left(x^3 - 4 + \frac{5}{x^6}\right) dx$

24. $\displaystyle\int \left(x^3 - 7 - \frac{3}{x^4}\right) dx$

25. $\displaystyle\int \left(x^9 - \frac{1}{x^3} + \frac{2}{\sqrt[3]{x}}\right) dx$

26. $\displaystyle\int \left(3x^8 + \frac{4}{x^8} - \frac{5}{\sqrt[5]{x}}\right) dx$

In problems 27–32, use algebra to rewrite the integrands, integrate, and simplify.

27. $\int (x + 5)^2 x\, dx$

28. $\int (2x + 1)^2 x\, dx$

29. $\int (4x^2 - 1)^2 x^3\, dx$

30. $\int (x^3 + 1)^2 x\, dx$

31. $\displaystyle\int \frac{x + 1}{x^3}\, dx$

32. $\displaystyle\int \frac{x - 3}{\sqrt{x}}\, dx$

In problems 33 and 34, find the antiderivatives and graph the resulting family members that correspond to $C = 0$, $C = 4$, $C = -4$, $C = 8$, and $C = -8$.

33. $\int (2x - 3)\, dx$

34. $\int (4 - x)\, dx$

In each of problems 35–38, a family of functions is given and graphs of some members of the family are shown. Write the indefinite integral that gives the family.

35. $F(x) = 5x - \dfrac{x^2}{4} + C$

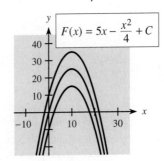

36. $F(x) = \dfrac{x^2}{2} + 3x + C$

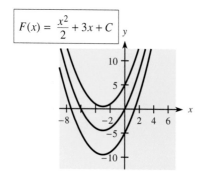

37. $F(x) = x^3 - 3x^2 + C$

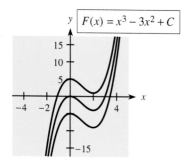

38. $F(x) = 12x - x^3 + C$

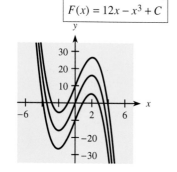

APPLICATIONS

39. **Revenue** If the marginal revenue for a month for a commodity is $\overline{MR} = 3$, what is the total revenue function?

40. **Revenue** If the marginal revenue for a month for a commodity is $\overline{MR} = 5$, what is the total revenue function?

41. **Revenue** If the marginal revenue for a month for a commodity is $\overline{MR} = 0.4x + 3$, find the total revenue function.

42. **Revenue** If the marginal revenue for a month for a commodity is $\overline{MR} = 0.5x + 2$, find the total revenue function.

43. **Revenue** If the marginal revenue for a month is given by $\overline{MR} = 3x + 1$, what is the total revenue from the production and sale of 50 units?

44. **Revenue** If the marginal revenue for a month is given by $\overline{MR} = 5x + 3$, find the total revenue from the sale of 75 units.

45. **Stimulus-response** Suppose that when a sense organ receives a stimulus at time t, the total number of action potentials is $P(t)$. If the rate at which action potentials are produced is $t^3 + 4t^2 + 6$, and if there are 0 action potentials when $t = 0$, find the formula for $P(t)$.

46. **Projectiles** Suppose that a particle has been shot into the air in such a way that the rate at which its height is changing is $v = 320 - 32t$, in feet per second, and suppose that it is 1600 feet high when $t = 10$. Write the equation that describes the height of the particle at any time t.

47. **Pollution** A factory is dumping pollutants into a river at a rate given by $dx/dt = t^{3/4}/600$ tons per week, where t is the time in weeks since the dumping began and x is the number of tons of pollutants.
    (a) Find the equation for total tons of pollutants dumped.
    (b) How many tons were dumped during the first year?

48. **Population growth** The rate of growth of the population of a city is predicted to be

$$\frac{dp}{dt} = 1000t^{1.08},$$

where p is the population at time t, and t is measured in years from the present. Suppose that the current population is 100,000. What is the predicted
    (a) rate of growth 5 years from the present?
    (b) population 5 years from the present?

49. **Average cost** The DeWitt Company has found that the rate of change of its average cost for a product is

$$\overline{C}'(x) = \frac{1}{4} - \frac{100}{x^2},$$

in dollars, where x is the number of units. If the average cost of producing 20 units is $40.00, find
(a) the average cost function for the product.
(b) the average cost of 100 units of the product.

50. **Oil leakage** An oil tanker hits a reef and begins to leak. The efforts of the workers repairing the leak cause the rate at which the oil is leaking to decrease. If the oil was leaking at a rate of 31 barrels per hour at the end of the first hour after the accident, and if the rate is decreasing at a rate of one barrel per hour,
(a) what function describes the rate of loss?
(b) how may barrels of oil will leak in the first 6 hours?
(c) when will the oil leak be stopped? How much will have leaked altogether?

51. **Revenue** Assume that the rate of change of sales revenues for Scott Paper Company can be modeled by

$$\frac{dR}{dt} = -0.062t + 0.776,$$

where t is the number of years past 1980, and sales revenues are in billions of dollars.
(a) If sales revenues in 1987 was $3.9769 billion, find the function that gives sales revenue. Graph this function.
(b) The data in the table show Scott Paper Company's sales revenue in billions of dollars for selected years. Graph the data in the table.
(c) Compare the graphs in parts (a) and (b)

Year	Sales Revenue (in Billions)	Year	Sales Revenue (in Billions)
1983	$2.6155	1989	4.8949
1984	2.7474	1990	5.1686
1985	2.934	1991	4.9593
1986	3.3131	1992	5.0913
1987	3.9769	1993	4.7489
1988	4.5494		

SOURCE: Scott Paper Company, *1993 Annual Report*

52. **Revenue** Assume that the rate of change of AT&T's total revenues can be modeled by

$$\frac{dR}{dt} = 0.506t - 4.03,$$

where t is the number of years past 1980, and total revenues are in billions of dollars.

(a) If AT&T's total revenue in 1985 was $63.13 billion, find the function for total revenue. Graph this function.
(b) The data in the table give AT&T's total revenue for selected years. Graph the data in the table.
(c) Compare the graphs in parts (a) and (b).

Year	Total Revenue (in Billions)	Year	Total Revenue (in Billions)
1985	$63.13	1990	62.191
1986	69.906	1991	63.089
1987	60.53	1992	64.904
1989	61.1	1993	67.156

SOURCE: AT&T *Annual Report, 1993*

53. **Personal Savings** Suppose the rate of personal savings in the United States is given by

$$\frac{dPS}{dt} = 2.1t^2 - 65.4t + 491.6,$$

where t is the number of years past 1970, and personal savings is in billions of dollars.
(a) If personal savings in 1980 was $153.8 billion, find the function that models personal savings. Graph this function.
(b) The data in the table show personal savings in the United States for selected years. Graph the data in the table.
(c) Compare the graphs in parts (a) and (b).

Year	Personal Savings (in Billions)	Year	Personal Savings (in Billions)
1980	$153.8	1988	155.7
1985	189.3	1989	152.1
1986	187.5	1990	175.6
1987	142.0	1991	199.6

SOURCE: *Survey of Current Business,* March 1993

54. **Consumer debt** Assume that the rate of change of consumer debt as a percentage of disposable income can be modeled by

$$\frac{dD}{dt} = -0.002958632t^3 + 0.085011t^2 - 0.621306t + 0.8945,$$

where t is the number of year past 1970.

(a) If consumer debt in 1970 was 18.3% of disposable income, find the function that models consumer debt as a percentage of disposable income. Graph this function.
(b) The data in the table show consumer debt as a percentage of disposable income for selected years. Graph the data in the table.
(c) Compare the graphs in parts (a) and (b).

Year	Consumer Debt as a Percentage of Disposable Income	Year	Consumer Debt as a Percentage of Disposable Income
1970	18.3	1986	21.4
1975	17.9	1987	21.2
1980	18.2	1988	21.0
1982	16.9	1989	20.9
1983	17.7	1990	20.0
1984	19.1	1991	18.9
1985	20.8		

SOURCE: Federal Reserve System

12.2 THE POWER RULE

Objective

❑ To evaluate integrals of the form $\int u^n \cdot u' \, dx$ if $n \neq -1$

APPLICATION PREVIEW

In the previous section, we saw that total revenue could be found by integrating marginal revenue. That is,

$$R(x) = \int \overline{MR} \, dx.$$

For example, if the marginal revenue for a product is given by

$$\overline{MR} = \frac{600}{\sqrt{3x + 1}} + 2,$$

then

$$R(x) = \int \left[\frac{600}{\sqrt{3x + 1}} + 2 \right] dx.$$

To evaluate this integral, however, we need a more general formula than the Powers of x Formula.

In this section, we will extend the Powers of x Formula to a rule for powers of a function of x.

Our goal in this section is to extend the Powers of x Formula

$$\int x^n \, dx = \frac{x^{n+1}}{n + 1} + C \quad (n \neq -1)$$

to powers of a function of x. In order to do this, we must understand the symbol dx.

Recall from Section 9.3 that the derivative of $y = f(x)$ with respect to x can be denoted by dy/dx. As we will see, there are advantages to using dy and dx as separate quantities whose ratio dy/dx equals $f'(x)$.

Differentials | If $y = f(x)$ is a differentiable function with derivative $dy/dx = f'(x)$, then the **differential of x** is dx, and the **differential of y** is dy, where

$$dy = f'(x) \, dx.$$

Although differentials are useful in certain approximation problems, we are interested in the differential notation at this time.

EXAMPLE 1

Find the differential dy if $y = x^3 - 4x^2 + 5$.

Solution

$$dy = f'(x) \, dx = (3x^2 - 8x) \, dx$$

If the dependent variable in a function is u (as in $u = u(x)$), then $du = u'(x) \, dx$.

EXAMPLE 2

If $u = x^2 + 4$, find du.

Solution

Because u is a function of x (that is, $u = u(x)$),

$$du = u'(x) \, dx = 2x \, dx.$$

In terms of our goal of extending the powers of x integration formula, we would suspect that if x is replaced by a function of x, then dx should be replaced by the differential of that function. Let's see if this is true.

Recall that if $y = [u(x)]^n$, the derivative of y is

$$\frac{dy}{dx} = n[u(x)]^{n-1} \cdot u'(x).$$

Using this formula for derivatives, we can see that

$$\int n[u(x)]^{n-1} \cdot u'(x) \, dx = [u(x)]^n + C.$$

It is easy to see that this formula is equivalent to the following formula, called the **Power Rule for Integration.**

Power Rule for Integration | $$\int [u(x)]^n \cdot u'(x) \, dx = \frac{[u(x)]^{n+1}}{n+1} + C, \quad \text{if } n \neq -1$$

Using the fact that

$$du = u'(x) \, dx \quad \text{or} \quad du = u' \, dx,$$

we can write the Power Rule in the alternate form below.

**Power Rule
(Alternate Form)**

$$\int u^n \, du = \frac{u^{n+1}}{n + 1} + C, \qquad n \neq -1$$

Note that this formula has the same form as the formula

$$\int x^n \, dx = \frac{x^{n+1}}{n + 1} + C, \qquad n \neq -1,$$

with the function *u substituted for x and du substituted for dx.*

EXAMPLE 3

Evaluate $\int (x^2 + 4)^5 \cdot 2x \, dx$.

Solution

To use the Power Rule, we must be sure that we have the function $u(x)$, its derivative $u'(x)$, and n.

$$u = x^2 + 4, \quad n = 5$$
$$u' = 2x$$

All required parts are present, so the integral is of the form

$$\int (x^2 + 4)^5 2x \, dx = \int u^5 \cdot u' \, dx = \int u^5 \, du$$

$$= \frac{u^6}{6} + C = \frac{(x^2 + 4)^6}{6} + C.$$

We can check the integration by noting that the derivative of

$$\frac{(x^2 + 4)^6}{6} + C \quad \text{is} \quad (x^2 + 4)^5 \cdot 2x.$$

EXAMPLE 4

Evaluate $\int \sqrt{2x + 3} \cdot 2 \, dx$.

Solution

If we let $u = 2x + 3$, then $u' = 2$, and so we have

$$\int \sqrt{2x + 3} \cdot 2 \, dx = \int \sqrt{u} \, u' \, dx = \int \sqrt{u} \, du$$

$$= \int u^{1/2} \, du = \frac{u^{3/2}}{3/2} + C.$$

Figure 12.3

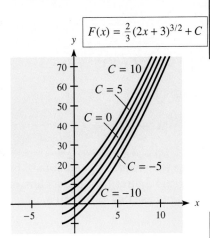

$$F(x) = \frac{2}{3}(2x+3)^{3/2} + C$$

Since $u = 2x + 3$, we have

$$\int \sqrt{2x+3} \cdot 2 \, dx = \frac{2}{3}(2x+3)^{3/2} + C.$$

CHECK: The derivative of $\frac{2}{3}(2x+3)^{3/2} + C$ is $(2x+3)^{1/2} \cdot 2$.

Some members of the family of functions given by

$$\int \sqrt{2x+3} \cdot 2 \, dx = \frac{2}{3}(2x+3)^{3/2} + C$$

are shown in Figure 12.3. Notice from the graphs that the domain of each function is $x \geq -3/2$. This is because $2x + 3$ must be nonnegative so that $(2x+3)^{3/2} = (\sqrt{2x+3})^3$ is a real number.

EXAMPLE 5

Evaluate $\int (x^2 + 4)^4 \cdot x \, dx$.

Solution

If we let $u = x^2 + 4$, then $u' = 2x$. Thus we do not have an integral of the form $\int u^n \cdot u' \, dx$, as we had in Example 3; the factor 2 is not in the integrand. To get the integrand in the correct form, we can multiply by 2 and divide it out as follows:

$$\int (x^2 + 4)^4 \cdot x \, dx = \int (x^2 + 4)^4 \cdot \frac{1}{2}(2x) \, dx.$$

Because $\frac{1}{2}$ is a constant factor, we can factor it outside the integral sign, getting

$$\frac{1}{2} \int (x^2 + 4)^4 \cdot 2x \, dx.$$

Now the integral is in the form $\frac{1}{2} \int u^4 \cdot u' \, dx$. Thus

$$\int (x^2 + 4)^4 \cdot x \, dx = \frac{1}{2} \int (x^2 + 4)^4 \cdot 2x \, dx$$

$$= \frac{1}{2} \frac{(x^2 + 4)^5}{5} + C$$

$$= \frac{1}{10}(x^2 + 4)^5 + C.$$

EXAMPLE 6

Evaluate $\int \sqrt{x^3 - 4} \cdot 5x^2 \, dx$.

Solution

If we let $u = x^3 - 4$, then $u' = 3x^2$. Thus we need the factor 3, rather than 5, in the integrand. If we multiply by the constant factor 3 (and divide it out), we have

$$\int \sqrt{x^3 - 4} \cdot 5x^2 \, dx = \int \sqrt{x^3 - 4} \cdot \frac{5}{3}(3x^2) \, dx$$

$$= \frac{5}{3} \int (x^3 - 4)^{1/2} \cdot 3x^2 \, dx.$$

This integral is of the form $\frac{5}{3} \int u^{1/2} \cdot u' \, dx$, resulting in

$$\frac{5}{3} \cdot \frac{u^{3/2}}{3/2} + C = \frac{5}{3} \cdot \frac{(x^3 - 4)^{3/2}}{3/2} + C = \frac{10}{9}(x^3 - 4)^{3/2} + C.$$

Note that we can factor a constant outside the integral sign to obtain the integrand in the form we seek but if the integral requires the introduction of a variable to obtain the form $u^n \cdot u' \, dx$, we *cannot* use this form, and must try something else.

EXAMPLE 7
Evaluate $\int (x^2 + 4)^2 \, dx$.

Solution
If we let $u = x^2 + 4$, then $u' = 2x$. Since we would have to introduce a variable to get u' in the integral, we cannot solve this problem by using the Power Rule. We must find another method. We can evaluate this integral by squaring and then integrating term-by-term.

$$\int (x^2 + 4)^2 \, dx = \int (x^4 + 8x^2 + 16) \, dx$$

$$= \frac{x^5}{5} + \frac{8x^3}{3} + 16x + C$$

Note that if we had tried to introduce the factor $2x$ into the integral of Example 7, we would get

$$\int (x^2 + 4)^2 \, dx = \int (x^2 + 4)^2 \cdot \frac{1}{2x}(2x) \, dx$$

$$= \frac{1}{2} \int (x^2 + 4)^2 \cdot \frac{1}{x}(2x) \, dx.$$

But we cannot factor the $1/x$ outside the integral, so we do not have the proper form. Again, we can only introduce *a constant factor* to get an integral in the proper form.

EXAMPLE 8
Evalute $\int (2x^2 - 4x)^2 (x - 1) \, dx$.

Solution

If we want to treat this as an integral of the form $\int u^n u'\, dx$, we will have to let $u = 2x^2 - 4x$. Then u' will be $4x - 4$. Multiplying and dividing by 4 will give us this form, as follows.

$$\int (2x^2 - 4x)^2 (x - 1)\, dx = \int (2x^2 - 4x)^2 \cdot \frac{1}{4} \cdot 4(x - 1)\, dx$$

$$= \frac{1}{4} \int (2x^2 - 4x)^2 (4x - 4)\, dx$$

$$= \frac{1}{4} \int u^2 u'\, dx = \frac{1}{4} \cdot \frac{u^3}{3} + C$$

$$= \frac{1}{4} \frac{(2x^2 - 4x)^3}{3} + C$$

$$= \frac{1}{12}(2x^2 - 4x)^3 + C$$

EXAMPLE 9

Evaluate $\int \dfrac{x^2 - 1}{(x^3 - 3x)^3}\, dx$.

Solution

This integral can be treated as $\int u^{-3} u'\, dx$ if we let $u = x^3 - 3x$. Then we can multiply (and divide) by 3 to get $u' = 3(x^2 - 1)$.

$$\int \frac{x^2 - 1}{(x^3 - 3x)^3}\, dx = \int (x^3 - 3x)^{-3} \cdot \frac{1}{3} \cdot 3(x^2 - 1)\, dx$$

$$= \frac{1}{3} \int (x^3 - 3x)^{-3} (3x^2 - 3)\, dx$$

$$= \frac{1}{3} \left[\frac{(x^3 - 3x)^{-2}}{-2} \right] + C$$

$$= \frac{-1}{6(x^3 - 3x)^2} + C$$

CHECKPOINT

1. Which of the following can be evaluated with the Power Rule?
   (a) $\int (4x^2 + 1)^{10} (8x\, dx)$ (b) $\int (4x^2 + 1)^{10} (x\, dx)$
   (c) $\int (4x^2 + 1)^{10} (8\, dx)$ (d) $\int (4x^2 + 1)^{10}\, dx$

2. Which of the following is equal to $\int (2x^3 + 5)^{-2} (6x^2\, dx)$?
   (a) $\dfrac{[(2x^4)/4 + 5x]^{-1}}{-1} \cdot \dfrac{6x^3}{3} + C$ (b) $\dfrac{(2x^3 + 5)^{-1}}{-1} \cdot \dfrac{6x^3}{3} + C$
   (c) $\dfrac{(2x^3 + 5)^{-1}}{-1} + C$

3. True or false: Constants can be factored outside the integral sign.

4. Evaluate:

   (a) $\int (x^3 + 9)^5 (3x^2\, dx)$ (b) $\int (x^3 + 9)^{15} (x^2\, dx)$ (c) $\int (x^3 + 9)^2 (x\ dx)$

We now return to the problem introduced in the Application Preview.

EXAMPLE 10

Suppose that the marginal revenue for a product is given by

$$\overline{MR} = \frac{600}{\sqrt{3x + 1}} + 2.$$

Find the total revenue function.

Solution

$$
\begin{aligned}
R(x) = \int \overline{MR}\ dx &= \int \left[\frac{600}{(3x + 1)^{1/2}} + 2 \right] dx \\
&= \int 600(3x + 1)^{-1/2}\, dx + \int 2\, dx \\
&= 600\left(\frac{1}{3}\right) \int (3x + 1)^{-1/2}(3\ dx) + 2 \int dx \\
&= 200 \frac{(3x + 1)^{1/2}}{1/2} + 2x + K \\
&= 400\sqrt{3x + 1} + 2x + K
\end{aligned}
$$

We know that $R(0) = 0$, so we have

$$0 = 400\sqrt{1} + 0 + K \quad \text{or} \quad K = -400.$$

Thus the total revenue function is

$$R(x) = 400\sqrt{3x + 1} + 2x - 400.$$

Notice in Example 10 that even though $R(0) = 0$, the constant of integration K was *not* 0. This is because $x = 0$ does not necessarily mean that $u(x)$ will also be 0.

CHECKPOINT
SOLUTIONS

1. Expressions (a) and (b) can be evaluated with the Power Rule. For (a), we let $u = 4x^2 + 1$ so that the integral becomes

$$\int u^{10}\, u'\, dx = \int u^{10}\, du.$$

For (b), we let $u = 4x^2 + 1$ again, and the integral becomes

$$\frac{1}{8} \int u^{10} u' \, dx = \frac{1}{8} \int u^{10} \, du.$$

Expressions (c) and (d) do not fit the format of the Power Rule, because neither integral has an x with the dx, outside the power (so they need to be multiplied out before integrating).

2. $u = 2x^3 + 5$, so $u' = 6x^2$

$$\int (2x^3 + 5)^{-2}(6x^2) \, dx = \int u^{-2} u' \, dx = \int u^{-2} \, du$$
$$= -u^{-1} + C = -(2x^3 + 5)^{-1} + C$$

Thus (c) is the correct choice.

3. True

4. (a) $\int (x^3 + 9)^5(3x^2) \, dx = \int u^5 u' \, dx = \dfrac{u^6}{6} + C = \dfrac{(x^3 + 9)^6}{6} + C$

(b) $\int (x^3 + 9)^{15}(x^2 \, dx) = \dfrac{1}{3} \int (x^3 + 9)^{15}(3x^2 \, dx)$

$$= \frac{1}{3} \cdot \frac{(x^3 + 9)^{16}}{16} + C = \frac{(x^3 + 9)^{16}}{48} + C$$

(c) The Power Rule does not fit, so we expand the integrand.

$$\int (x^3 + 9)^2(x \, dx) = \int (x^6 + 18x^3 + 81)x \, dx$$
$$= \int (x^7 + 18x^4 + 81x) \, dx$$
$$= \frac{x^8}{8} + \frac{18x^5}{5} + \frac{81x^2}{2} + C$$

EXERCISE 12.2

Evaluate the following integrals. Check your results by differentiation.

1. $\int (x^2 + 3)^3 2x \, dx$

2. $\int (3x^3 + 1)^{49} 9x^2 \, dx$

3. $\int (15x^2 + 10)^4(30x) \, dx$

4. $\int (8x^4 + 5)^3(32x^3) \, dx$

5. $\int (3x - x^3)^2(3 - 3x^2) \, dx$

6. $\int (4x^2 - 3x)^4(8x - 3) \, dx$

7. $\int (x^2 + 5)^3 x \, dx$

8. $\int (3x^2 - 4)^6 x \, dx$

9. $\int 7(4x - 1)^6 \, dx$

10. $\int 3(5 - x)^{-3} \, dx$

11. $\int (x^2 + 1)^{-3} x \, dx$

12. $\int (3x^4 + 7)^{-4}(5x^3 \, dx)$

13. $\int (x - 1)(x^2 - 2x + 5)^4 \, dx$

14. $\int (2x^3 - x)(x^4 - x^2)^6 \, dx$

15. $\int 2(x^3 - 1)(x^4 - 4x + 3)^{-5} \, dx$

16. $\int 3(x^5 - 2x)(x^6 - 6x^2 + 7)^{-2} \, dx$

17. $\int 7x^3 \sqrt{x^4 + 6} \, dx$

18. $\int 3x \sqrt{5 - x^2} \, dx$

19. $\int (x^3 + 1)^2(3x \, dx)$

20. $\int (x^2 - 5)^2(2x^2 \, dx)$

21. $\int (3x^2 - 1)^2(8x^2 \, dx)$

22. $\int (2x^4 + 3)(8x \, dx)$

23. $\int \sqrt{x^3 - 3x} \, (x^2 - 1) \, dx$

24. $\int \sqrt[3]{x^2 + 2x} \, (x + 1) \, dx$

25. $\displaystyle\int \frac{x^2 \, dx}{(x^3 - 1)^2}$

26. $\displaystyle\int \frac{x \, dx}{(x^2 - 1)^3}$

27. $\displaystyle\int \frac{3x^4 \, dx}{(2x^5 - 5)^4}$

28. $\displaystyle\int \frac{5x^3 \, dx}{(x^4 - 8)^3}$

29. $\displaystyle\int \frac{x^3 - 1}{(x^4 - 4x)^3} \, dx$

30. $\displaystyle\int \frac{3x^5 - 2x^3}{(x^6 - x^4)^5} \, dx$

31. $\displaystyle\int \frac{x^2 - 4x}{\sqrt{x^3 - 6x^2 + 2}} \, dx$

32. $\displaystyle\int \frac{x^2 + 1}{\sqrt{x^3 + 3x + 10}} \, dx$

In problems 33 and 34, (a) evaluate each integral and (b) graph the members of the solution family for $C = -5$, $C = 0$, and $C = 5$.

33. $\int x(x^2 - 1)^3 \, dx$

34. $\int (3x - 11)^{1/3} \, dx$

Each of problems 35 and 36 has the form $\int f(x) \, dx$.

(a) Evaluate each integral to obtain a family of functions.
(b) Find and graph the family member that passes through the point $(0, 2)$. Call that function $F(x)$.
(c) Find any x-values where $f(x)$ is not defined but $F(x)$ is.
(d) At the x-values found in (c), what kind of tangent line does $F(x)$ have?

35. $\displaystyle \int \frac{3dx}{(2x - 1)^{3/5}}$

36. $\displaystyle \int \frac{x^2 \, dx}{(x^3 - 1)^{1/3}}$

In each of problems 37 and 38, a family of functions is given together with the graphs of some functions in the family. Write the indefinite integral that gives the family.

37. $F(x) = (x^2 - 1)^{4/3} + C$

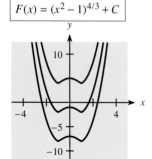

38. $F(x) = 54(4x^2 + 9)^{-1} + C$

39. **Revenue** Suppose that the marginal revenue for a product is given by
$$\overline{MR} = \frac{-30}{(2x + 1)^2} + 30.$$
Find the total revenue.

40. **Revenue** The marginal revenue for a new calculator is given by
$$\overline{MR} = 60,000 - \frac{40,000}{(10 + x)^2},$$
where x represents hundreds of calculators. Find the total revenue function for these calculators.

41. **Physical productivity** The total physical output of a number of machines or workers is called *physical productivity*, and is a function of the number of machines or workers. If $P = f(x)$ is the productivity, dP/dx is the marginal physical productivity. If the marginal physical productivity for bricklayers is $dP/dx = 90(x + 1)^2$, where P is the number of bricks laid per day, find the physical productivity of 4 bricklayers. *Note:* $P = 0$ when $x = 0$.

42. **Production** The rate of production of a new line of products is given by
$$\frac{dx}{dt} = 200\left[1 + \frac{400}{(t + 40)^2}\right],$$
where x is the number of items and t is the number of weeks the product has been in production.
(a) Assuming that $x = 0$ when $t = 0$, find the total number of items produced as a function of time t.
(b) How many items were produced in the 5th week?

43. **Typing speed** The rate of change in typing speed of the average student is $ds/dx = 5(x + 1)^{-1/2}$, where x is the number of typing lessons the student has had.
(a) Find the typing speed as a function of the number of lessons if the average student can type 10 words per minute with no lessons ($x = 0$).
(b) How many words per minute can the average student type after 24 lessons?

44. **Productivity** Because a new employee must learn an assigned task, production will increase with time. Suppose that for the average new employee, the rate of performance is given by
$$\frac{dN}{dt} = \frac{1}{2\sqrt{t + 1}},$$
where N is the number of units completed t hours after

beginning a new task. If 2 units are completed after 3 hours, how many units are completed after 8 hours?

45. **Film attendance** An excellent film with a very small advertising budget must depend largely on word-of-mouth advertising. In this case, the rate at which weekly attendance might grow can be given by

$$\frac{dA}{dt} = \frac{-100}{(t + 10)^2} + \frac{2000}{(t + 10)^3},$$

where t is the time in weeks since release and A is attendance in millions.
  (a) Find the function that describes weekly attendance at this film.
  (b) Find the attendance at this film in the 10th week.

46. **Product quality and advertising** An inferior product with a large advertising budget does well when it is introduced, but sales decline as people discontinue use of the product. Suppose that the rate of weekly sales is given by

$$S'(t) = \frac{400}{(t + 1)^3} - \frac{200}{(t + 1)^2},$$

where S is sales in millions and t is time in weeks.
  (a) Find the function that describes the weekly sales.
  (b) Find the sales for the first week and for the 9th week.

47. **Demographics** Because of the decline of the steel industry, a western Pennsylvania town predicts that its public school population will decrease at the rate

$$\frac{dN}{dx} = \frac{-300}{\sqrt{x + 9}},$$

where x is the number of years and N is the total school population. If the present population ($x = 0$) is 8000, what population size is planned for in 7 years?

48. **Franchise growth** A new fast-food firm predicts that the number of franchises for its products will grow at the rate

$$\frac{dn}{dt} = 9\sqrt{t + 1},$$

where t is the number of years, $0 \le t \le 10$. If there is one franchise ($n = 1$) at present ($t = 0$), how many franchises are predicted for 8 years from now?

49. **Poverty line** Suppose the rate of change of the number of people (in millions) in the United States who lived below the poverty level can be modeled by

$$\frac{dp}{dt} = -0.0061785(t + 60)^2 + 1.0177(t + 60) - 41.3634,$$

where t is the number of years past 1960.

Year	Persons Below the Poverty Level (in Millions)	Year	Persons Below the Poverty Level (in Millions)
1960	39.9	1980	29.3
1965	33.2	1986	32.4
1970	25.4	1989	31.5
1975	25.9	1990	33.6

SOURCE: *World Almanac and Book of Facts,* 1993

  (a) If 33.6 million people lived below the poverty level in 1990, find the function that models the number of people, in millions, in the United States who lived below the poverty level. Graph this function.
  (b) The data in the table show the number of people, in millions, in the United States who lived below the poverty level for selected years. Graph the data in the table.
  (c) Compare the graphs in parts (a) and (b).

50. **U.S. public debt** Suppose the rate of change of the interest paid on the public debt of the United States as a percentage of federal expenditures can be modeled by

$$\frac{dD}{dt} = 0.1779(0.1t + 3)^2 - 2.112(0.1t + 3) + 6.098,$$

where t is the number of years past 1930.

Year	Interest Paid as a Percentage of Federal Expenditures	Year	Interest Paid as a Percentage of Federal Expenditures
1930	0.0	1970	9.9
1940	10.5	1975	9.8
1950	13.4	1980	12.7
1955	9.4	1985	18.9
1960	10.0	1990	21.1
1965	9.6		

SOURCE: Bureau of Public Debt, U.S. Department of Treasury

  (a) If 9.6% of federal expenditures was devoted to interest payments on the public debt in 1965, find the function $D(t)$ that models the interest paid on the public debt of the United States as a percentage of federal expenditures. Graph this function.
  (b) The data in the table give the interest paid on the public debt of the United States as a percentage of

federal expenditures for selected years. Graph the data in the table.
(c) Compare the graphs in parts (a) and (b).

51. **Union membership** Suppose the rate of change of the percentage of United States workers who belonged to unions can be modeled by

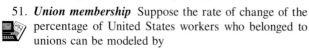

$$\frac{dU}{dt} = 0.1455(0.1t + 2)^2 - 2.18(0.1t + 2) + 7.52,$$

where t is the number of years past 1920.
(a) If 18.0% of United States workers belonged to unions in 1985, find the function $U(t)$ that models the percentage of United States workers who belonged to unions. Graph this function.
(b) The data in the table show the percentage of United States workers who belonged to unions for selected years. Graph the data.
(c) Compare the graphs in parts (a) and (b).

Year	Union Membership as a Percentage of United States Labor Force
1930	11.6
1935	13.2
1940	26.9
1945	35.5
1950	31.5
1955	33.2
1960	31.4
1965	28.4
1970	27.3
1975	25.5
1980	21.9
1985	18.0
1990	16.1

SOURCE: Bureau of Labor Statistics, U.S. Department of Labor

12.3 INTEGRALS INVOLVING LOGARITHMIC AND EXPONENTIAL FUNCTIONS

Objectives

❏ To evaluate integrals of the form
$$\int \frac{u'}{u} \, dx \text{ or equivalently } \int \frac{1}{u} \, du$$
❏ To evaluate integrals of the form
$$\int e^u u' dx \text{ or equivalently } \int e^u \, du$$

APPLICATION PREVIEW

The rate of growth of the market value of a home has typically exceeded the inflation rate. Suppose, for example, that the real estate market has an average annual inflation rate of 8%. Then the rate of change of the value of a house that cost $100,000 can be modeled by

$$\frac{dV}{dt} = 7.7e^{0.077t},$$

where V is the value of the home in hundreds of thousands of dollars and t is the time in years since the home was purchased. To find the market value of such a home 10 years after it was purchased, we would first have to integrate dV/dt. That is, we must be able to integrate an exponential.

In this section, we consider integration formulas that result in natural logarithms and formulas for integrating exponentials.

Recall that the Power Rule for integrals applies only if $n \neq -1$. That is,

$$\int u^n u' \, dx = \frac{u^{n+1}}{n + 1} + C \qquad \text{if } n \neq -1.$$

The following formula applies when $n = -1$.

Logarithmic Formula

If u is a function of x,

$$\int u^{-1} u' \, dx = \int \frac{u'}{u} \, dx = \int \frac{1}{u} \, du = \ln|u| + C$$

This formula is a direct result of the fact that

$$\frac{d}{dx} (\ln|u|) = \frac{1}{u} \cdot u'.$$

Note that we use the absolute value of u because the logarithm is defined only when the quantity is positive. We can see this result by considering the following.

for $u > 0$: $\quad \dfrac{d}{dx} (\ln|u|) = \dfrac{d}{dx} (\ln u) = \dfrac{1}{u} \cdot u'$

for $u < 0$: $\quad \dfrac{d}{dx} (\ln|u|) = \dfrac{d}{dx} [\ln(-u)] = \dfrac{1}{(-u)} \cdot (-u') = \dfrac{1}{u} \cdot u'$

In addition to this verification, we can graphically illustrate the need for the absolute value sign. Figure 12.4(a) shows that $f(x) = 1/x$ is defined for $x \neq 0$, and from Figures 12.4(b) and 12.4(c), we see that $F(x) = \int 1/x \, dx = \ln|x|$ is also defined for $x \neq 0$. But $y = \ln x$ is defined only for $x > 0$.

Figure 12.4

(a)

(b)

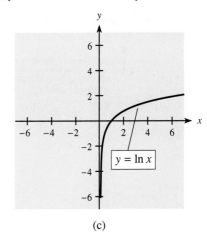

(c)

EXAMPLE 1

Evaluate $\displaystyle\int \frac{4}{4x + 8} \, dx$.

Solution

This integral is of the form

$$\int \frac{u'}{u} \, dx = \ln|u| + C$$

Figure 12.5

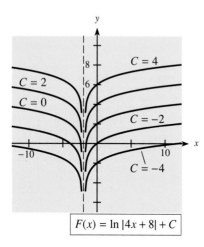

$$F(x) = \ln|4x + 8| + C$$

with $u = 4x + 8$ and $u' = 4$. Thus

$$\int \frac{4}{4x + 8}\, dx = \ln|4x + 8| + C.$$

Figure 12.5 shows several members of the family

$$F(x) = \int \frac{4\, dx}{4x + 8} = \ln|4x + 8| + C.$$

We can choose different values for C and use a graphing utility to easily graph families of curves such as those in Figure 12.5.

EXAMPLE 2

Evaluate $\displaystyle\int \frac{x - 3}{x^2 - 6x + 1}\, dx.$

Solution

This integral is of the form $\int (u'/u)\, dx$, *almost*. If we let $u = x^2 - 6x + 1$, then, $u' = 2x - 6$. If we multiply (and divide) the numerator by 2, we get

$$\int \frac{x - 3}{x^2 - 6x + 1}\, dx = \frac{1}{2} \int \frac{2(x - 3)}{x^2 - 6x + 1}\, dx$$

$$= \frac{1}{2} \int \frac{2x - 6}{x^2 - 6x + 1}\, dx$$

$$= \frac{1}{2} \int \frac{u'}{u}\, dx = \frac{1}{2} \ln|u| + C$$

$$= \frac{1}{2} \ln|x^2 - 6x + 1| + C.$$

If an integral contains a fraction in which the degree of the numerator is equal to or greater than that of the denominator, we should divide the denominator into the numerator as a first step.

EXAMPLE 3

Evaluate $\displaystyle\int \frac{x^4 - 2x^3 + 4x^2 - 7x - 1}{x^2 - 2x}\, dx.$

Solution

Since the numerator is of higher degree than the denominator, we begin by dividing $x^2 - 2x$ into the numerator.

$$
\begin{array}{r}
x^2 + 4 \\
x^2 - 2x \overline{)\, x^4 - 2x^3 + 4x^2 - 7x - 1} \\
\underline{x^4 - 2x^3 } \\
4x^2 - 7x - 1 \\
\underline{4x^2 - 8x } \\
x - 1
\end{array}
$$

Thus

$$\int \frac{x^4 - 2x^3 + 4x^2 - 7x - 1}{x^2 - 2x} \, dx = \int \left(x^2 + 4 + \frac{x - 1}{x^2 - 2x} \right) dx$$

$$= \int (x^2 + 4) \, dx + \frac{1}{2} \int \frac{2(x - 1) \, dx}{x^2 - 2x}$$

$$= \frac{x^3}{3} + 4x + \frac{1}{2} \ln \left| x^2 - 2x \right| + C.$$

CHECKPOINT

1. True or false:

   (a) $\int \frac{3x^2 \, dx}{x^3 + 4} = \ln \left| x^3 + 4 \right| + C$ (b) $\int \frac{2x \, dx}{\sqrt{x^2 + 1}} = \ln \left| \sqrt{x^2 + 1} \right| + C$

   (c) $\int \frac{2}{x} \, dx = 2 \ln \left| x \right| + C$

   (d) $\int \frac{x}{x + 1} \, dx = x \int \frac{1}{x + 1} \, dx = x \ln \left| x + 1 \right| + C$

   (e) To evaluate $\int \frac{4x}{4x + 1} \, dx$, our first step is to divide $4x + 1$ into $4x$.

2. (a) Divide $4x + 1$ into $4x$.   (b) Evaluate $\int \frac{4x}{4x + 1} \, dx$.

We know that

$$\frac{d}{dx} (e^u) = e^u \cdot u'.$$

The corresponding integral is given by the following.

Exponential Formula   If u is a function of x,

$$\int e^u \cdot u' \, dx = \int e^u \, du = e^u + C.$$

EXAMPLE 4
Evaluate $\int 5e^x \, dx$.

Solution
$\int 5e^x \, dx = 5 \int e^x \, dx = 5e^x + C$

EXAMPLE 5
Evaluate $\int 2xe^{x^2} \, dx$.

Solution

Letting $u = x^2$ implies that $u' = 2x$, and the integral is of the form $\int e^u \cdot u' \, dx$. Thus

$$\int 2xe^{x^2} \, dx = \int e^{x^2}(2x) \, dx = \int e^u \cdot u' \, dx = e^u + C = e^{x^2} + C.$$

EXAMPLE 6

Evaluate $\displaystyle\int \frac{x^2 dx}{e^{x^3}}$.

Solution

In order to use $\int e^u \cdot u' \, dx$, we write the exponential in the numerator. Thus

$$\int \frac{x^2 \, dx}{e^{x^3}} = \int e^{-x^3}(x^2 \, dx).$$

This is *almost* of the form $\int e^u \cdot u' \, dx$. Letting $u = -x^3$ gives $u' = -3x^2$. Thus

$$\int e^{-x^3}(x^2 \, dx) = -\frac{1}{3}\int e^{-x^3}(-3x^2 \, dx) = -\frac{1}{3}e^{-x^3} + C = \frac{-1}{3e^{x^3}} + C.$$

CHECKPOINT

3. True or false:

(a) $\int e^{x^2}(2x \, dx) = e^{x^2} \cdot x^2 + C$ (b) $\int e^{-3x} \, dx = -\frac{1}{3}e^{-3x} + C$

(c) $\displaystyle\int \frac{dx}{e^{3x}} = \frac{1}{3}\left(\frac{1}{e^{3x}}\right) + C$ (d) $\displaystyle\int e^{3x+1}(3 \, dx) = \frac{e^{3x+2}}{3x+2} + C$

We now return to the real estate inflation rate problem in the Application Preview.

EXAMPLE 7

Suppose the rate of change of the value of a house that cost $100,000 can be modeled by

$$\frac{dV}{dt} = 7.7e^{0.077t},$$

where V is the market value of the home in hundreds of thousands of dollars and t is the time in years since the home was purchased.
(a) Find the function that expresses the value V in terms of t.
(b) Find the predicted value after 10 years.

Solution

(a) $V = \displaystyle\int \frac{dV}{dt} \, dt = \int 7.7e^{0.077t} \, dt$

$$V = 7.7 \int e^{0.077t} \left(\frac{1}{0.077}\right)(0.077 \, dt)$$

$$V = 7.7 \left(\frac{1}{0.077}\right) \int e^{0.077t}(0.077 \, dt)$$

$$V = 100e^{0.077t} + C$$

Using $V = 100$ (thousand) when $t = 0$, we have

$$100 = 100 + C$$
$$0 = C.$$

Thus, we have the value as a function of time given by

$$V = 100e^{0.077t}.$$

(b) The value after 10 years is found by using $t = 10$.

$$V = 100e^{0.077(10)} = 100e^{0.77} \approx 215.98$$

Thus, after 10 years, the predicted value of the home is \$215,980.

Figure 12.6

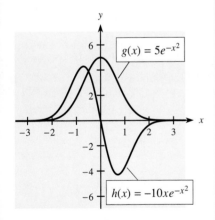

EXAMPLE 8

Figure 12.6 shows the graphs of $g(x) = 5e^{-x^2}$ and $h(x) = -10xe^{-x^2}$. One of these functions is $f(x)$ and the other is $\int f(x) \, dx$ with $C = 0$.

(a) Decide which of $g(x)$ and $h(x)$ is $f(x)$ and which is $\int f(x) \, dx$.
(b) How can the graph of $f(x)$ be used to locate and classify the extrema of $\int f(x) \, dx$?
(c) What feature of the graph of $f(x)$ occurs at the same x-values as the inflection points of the graph of $\int f(x) \, dx$?

Solution

(a) The graph of $h(x)$ looks like the graph of $g'(x)$ because $h(x) > 0$ where $g(x)$ is increasing, $h(x) < 0$ where $g(x)$ is decreasing, and $h(x) = 0$ where $g(x)$ has its maximum. However, if $h(x) = g'(x)$, then equivalently

$$\int h(x) \, dx = g(x) + C$$

so $h(x) = f(x)$ and $g(x) = \int f(x) \, dx$. We can verify this by noting that

$$\int -10xe^{-x^2} \, dx = 5 \int e^{-x^2}(-2x \, dx) = 5e^{-x^2} + C.$$

(b) We know that $f(x)$ is the derivative of $\int f(x) \, dx$, so, as we saw in part (a), the x-intercepts of $f(x)$ locate the critical values and extrema of $\int f(x) \, dx$.
(c) The first derivative of $\int f(x) \, dx$ is $f(x)$, and its second derivative is $f'(x)$. Hence, the inflection points of $\int f(x) \, dx$ occur where $f'(x) = 0$. But $f(x)$ has its extrema where $f'(x) = 0$. Thus, the extrema of $f(x)$ occur at the same x-values as the inflection points of $\int f(x) \, dx$.

CHECKPOINT SOLUTIONS

1. (a) True

   (b) False; $\int \dfrac{2x\,dx}{\sqrt{x^2+1}} = \int (x^2+1)^{-1/2}(2x\,dx)$

   $$= \dfrac{(x^2+1)^{1/2}}{1/2} + C = 2(x^2+1)^{1/2} + C$$

   (c) True

   (d) False. We cannot factor the variable x outside the integral sign.

   (e) True

2. (a) $4x+1\,\overline{\big)\,4x}$
   $$\dfrac{1}{}$$

   $\dfrac{4x+1}{-1}$

   so $\dfrac{4x}{4x+1} = 1 - \dfrac{1}{4x+1}$.

   (b) $\int \dfrac{4x\,dx}{4x+1} = \int\left(1 - \dfrac{1}{4x+1}\right)dx = x - \dfrac{1}{4}\ln|4x+1| + C$

3. (a) False. The correct solution is $e^{x^2} + C$ (see Example 5).

   (b) True

   (c) False; $\int \dfrac{dx}{e^{3x}} = \int e^{-3x}\,dx = -\dfrac{1}{3}e^{-3x} + C = \dfrac{-1}{3e^{3x}} + C$

   (d) False; $\int e^{3x+1}\,(3\,dx) = e^{3x+1} + C$

EXERCISE 12.3

Evaluate the following integrals.

1. $\int \dfrac{3x^2}{x^3+4}\,dx$

2. $\int \dfrac{8x^7}{x^8-1}\,dx$

3. $\int \dfrac{dz}{4z+1}$

4. $\int \dfrac{y}{y^2+1}\,dy$

5. $\int \dfrac{x^3}{x^4+1}\,dx$

6. $\int \dfrac{x^2}{x^3-9}\,dx$

7. $\int \dfrac{4x}{x^2-4}\,dx$

8. $\int \dfrac{5x^2}{x^3-1}\,dx$

9. $\int \dfrac{3x^2-2}{x^3-2x}\,dx$

10. $\int \dfrac{4x^3+2x}{x^4+x^2}\,dx$

11. $\int \dfrac{z^2+1}{z^3+3z+17}\,dz$

12. $\int \dfrac{(x+2)dx}{x^2+4x-9}$

13. $\int \dfrac{x^3-x^2+1}{x-1}\,dx$

14. $\int \dfrac{2x^3+x^2+2x+3}{2x+1}\,dx$

15. $\int \dfrac{x^2+x+3}{x^2+3}\,dx$

16. $\int \dfrac{x^4-2x^2+x}{x^2-2}\,dx$

17. $\int 3e^{3x}\,dx$

18. $\int 4e^{4x}\,dx$

19. $\int e^{-x}\,dx$

20. $\int e^{2x}\,dx$

21. $\int 1000e^{0.1x}\,dx$

22. $\int 1600e^{0.4x}\,dx$

23. $\int 840e^{-0.7x}\,dx$

24. $\int 250e^{-0.5x}\,dx$

25. $\int x^3 e^{3x^4}\,dx$

26. $\int xe^{2x^2}\,dx$

27. $\int \dfrac{3}{e^{2x}}\,dx$

28. $\int \dfrac{4}{e^{1-2x}}\,dx$

29. $\displaystyle\int \frac{x^5}{e^{2-3x^6}}\,dx$ 30. $\displaystyle\int \frac{x^3}{e^{4x^4}}\,dx$

31. $\displaystyle\int \left(e^{4x} - \frac{3}{e^{x/2}}\right) dx$ 32. $\displaystyle\int \left(xe^{3x^2} - \frac{5}{e^{x/3}}\right) dx$

In problems 33 and 34, graphs of two functions (labeled $g(x)$ and $h(x)$) are given. Decide which is the graph of $f(x)$ and which is one member of the family $\int f(x)\,dx$. Check your conclusions by evaluating the integral.

33.

34.

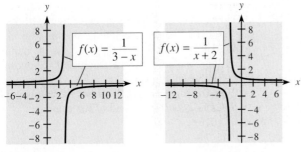

In problems 35 and 36, a function $f(x)$ and its graph are given. Find the family $F(x) = \int f(x)\,dx$ and graph the member that satisfies $F(0) = 0$.

35.   36.

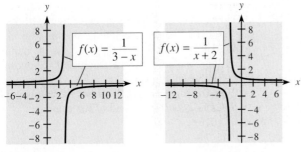

In problems 37–40, a family of functions is given and graphs of some members are shown. Find the function $f(x)$ such that the family is given by $\int f(x)\,dx$.

37. $F(x) = x + \ln|x| + C$

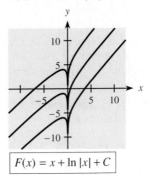

$$\boxed{F(x) = x + \ln|x| + C}$$

38. $F(x) = -\ln(x^2 + 4) + C$

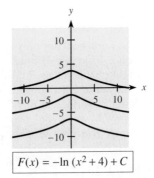

$$\boxed{F(x) = -\ln(x^2 + 4) + C}$$

39. $F(x) = 5xe^{-x} + C$ 40. $F(x) = e^{0.4x} + e^{-0.4x} + C$

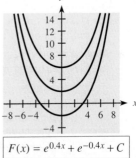

$$\boxed{F(x) = e^{0.4x} + e^{-0.4x} + C}$$

APPLICATIONS

41. **Revenue** Suppose that the marginal revenue from the sale of a product is $\overline{MR} = R'(x) = 6e^{0.01x}$. What is the revenue on the sale of 100 units of the product?

42. **Concentration of a drug** Suppose that the rate at which the concentration of a drug in the blood changes with respect to time t is given by

$$C'(t) = \frac{c}{b - a}(be^{-bt} - ae^{-at}), \qquad t \geq 0,$$

where a, b, and c are constants depending on the drug administered, with $b > a$. Assuming that $C(t) = 0$ when $t = 0$, find the formula for the concentration of the drug in the blood at any time t.

43. **Radioactive decay** The rate of disintegration of a radioactive substance can be described by

$$\frac{dn}{dt} = n_0(-K)e^{-Kt},$$

where n_0 is the number of radioactive atoms present when time t is 0, and K is a positive constant that depends on the substance involved. Using the fact that the constant of integration is 0, integrate dn/dt to find the number of atoms n that are still radioactive after time t.

44. **World population** Since the world contains only about 10 billion acres of arable land, world population is limited. Suppose that the world population is limited to 40 billion people and that the rate of population growth is proportional to how close the world is to this upper limit. Then the rate of growth would be given by $dP/dt = K(40 - P)$, where K is a positive constant. This means that

$$t = \frac{1}{K}\int \frac{1}{40 - P}\, dP.$$

(a) Evaluate this integral to find an expression relating P and t.

(b) Use the properties relating logarithms and exponential functions to write P as a function of t.

45. **Memorization** The rate of vocabulary memorization of the average student in a foreign language is given by

$$\frac{dv}{dt} = \frac{40}{t + 1},$$

where t is the number of continuous hours of study, $0 < t \leq 4$. How many words would the average student memorize in 3 hours?

46. **Population growth** The rate of growth of world population can be modeled by

$$\frac{dn}{dt} = N_0(1 + r)^t \ln(1 + r), \quad r < 1,$$

where t is the time in years from the present and N_0 and r are constants. What function describes world population if the present population is N_0? Use the formula $\int (a^u \ln a)u'\, dx = a^u + C$.

47. **Compound interest** If \$$P$ are invested for n years at 10%, compounded continuously, the rate at which the future value is growing is

$$\frac{dS}{dn} = 0.1Pe^{0.1n}.$$

(a) What function describes the future value at the end of n years?

(b) In how many years will the future value double?

48. **Temperature changes** When an object is moved from one environment to another, its temperature T changes at a rate given by

$$\frac{dT}{dt} = kCe^{kt},$$

where t is the time in the new environment (in hours), C is the temperature difference (old–new) between the two environments, and k is a constant. If the temperature of the body (and the old environment) is 70°F, and $C = -10$°F, what function describes the temperature T of the object t hours after it is moved?

49. **Blood pressure in the aorta** The rate at which blood pressure decreases in the aorta of a normal adult after a heartbeat is

$$\frac{dp}{dt} = -46.645e^{-0.491t},$$

where t is time in seconds.

(a) What function describes the blood pressure in the aorta if $p = 95$ when $t = 0$?

(b) What is the blood pressure 0.1 second after a heartbeat?

50. **Sales and advertising** A store finds that its sales decline after the end of an advertising campaign, with its daily sales for the period declining at the rate $S'(t) = -147.78e^{-0.2t}$, $0 \leq t \leq 100$, where t is the number of days since the end of the campaign. Suppose that $S = 7389$ when $t = 0$.

(a) Find the function that describes the number of daily sales t days after the end of the campaign.

(b) Find the total number of sales 10 days after the end of the advertising campaign.

51. ***Life expectancy*** Suppose the rate of change of the expected life span l at birth of people born in the United States can be modeled by

$$\frac{dl}{dt} = \frac{14.1}{t + 20},$$

where t is the number of years past 1920.
(a) If the expected life span was 72.6 years for people born in 1975, find the function that models the life span. Graph this function.
(b) The data in the table give the expected life span for people born in various years. Graph the data.
(c) Compare the graphs in parts (a) and (b).

Year	Life Span (in Years)
1920	54.1
1930	59.7
1940	62.9
1950	68.2
1960	69.7
1970	70.8
1975	72.6
1980	73.7
1987	75.0
1988	74.9
1989	75.2

SOURCE: *World Almanac,* 1991

52. ***Violent crime*** Suppose the rate of change in the number of violent crimes per 100,000 people in the United States can be modeled by

$$\frac{dc}{dt} = \frac{321}{t + 5},$$

when t is the number of years past 1985.
(a) If there were 758 violent crimes per 100,000 people in 1991, find the function $c(t)$ that models the number of violent crimes. Graph this function.
(b) The data in the table give the number of violent crimes per 100,000 people for selected years. Graph the data in the table.

(c) Compare the graphs in parts (a) and (b).

Year	Violent Crimes (per 100,000)
1987	610
1988	637
1989	663
1990	732
1991	758
1992	765

SOURCE: FBI Crime Report, 1993

53. ***Revenue*** Suppose the rate of change of revenue from wireless technology can be modeled by

$$\frac{dR}{dt} = 0.1994e^{0.3486t},$$

where t is the number of years past 1984 and R is revenue in billions of dollars.
(a) Find a model for revenue, $R(t)$, if revenue from wireless technology was $11 billion in 1993.
(b) Graph the function you found in part (a) and compare it to the graph in the figure.
(c) What does your model predict for revenue from wireless technology in 1995?

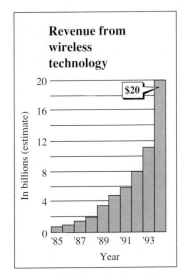

SOURCE: *USA Today,* March 1, 1994

54. **Health care costs** Suppose the rate of change of national health care costs in the United States can be modeled by

$$\frac{dHC}{dt} = 4.0428e^{0.08984t},$$

where t is the number of years past 1960 and HC is the national health care expenditure in billions of dollars.
(a) Use the fact that 1991 health care costs totaled $752 billion to find a function $HC(t)$ that models health care costs.
(b) Graph the function you found in part (a) and compare it to the graph in the figure.
(c) What does your model project for health care costs for the year 2000?

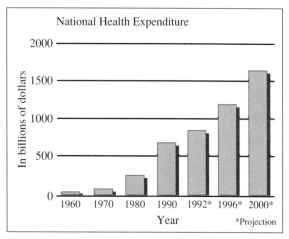

Source: Congressional Budget Office, printed in *Newsweek*, October 4, 1993

12.4 APPLICATIONS OF THE INDEFINITE INTEGRAL IN BUSINESS AND ECONOMICS

Objectives

❑ To use integration to find total cost functions from information involving marginal cost
❑ To optimize profit, given information regarding marginal cost and marginal revenue
❑ To use integration to find national consumption functions from information about marginal propensity to consume and marginal propensity to save

APPLICATION PREVIEW

We turn our attention to applications involving cost, profit, and consumption. A thorough understanding of these concepts is imperative for success in any study of economics and business.

In this section, we will use integration to derive total cost and profit functions from the marginal cost and marginal revenue functions. One of the reasons for the marginal approach in economics is that firms can observe marginal changes in real life. If they know the marginal cost and the total cost when a given quantity is sold, they can develop their total cost function.

In this section, we also use integration to obtain the national consumption function from functions that describe the marginal propensity to consume or the marginal propensity to save.

TOTAL COST AND PROFIT

We know that the marginal cost for a commodity is $\overline{MC} = C'(x)$, where $C(x)$ is the total cost function. Thus if we have the marginal cost function, we can integrate to find the total cost. That is, $C(x) = \int \overline{MC}\, dx$.

If, for example, the marginal cost is $\overline{MC} = 4x + 3$, the total cost is given by

$$C(x) = \int \overline{MC}\, dx$$

$$= \int (4x + 3)\, dx$$

$$= 2x^2 + 3x + K,$$

where K represents the constant of integration. Now, we know that the total revenue is 0 if no items are produced, but the total cost may not be 0 if nothing is produced. The fixed costs accrue whether goods are produced or not. Thus the value for the constant of integration depends on the fixed costs FC of production.

Thus we cannot determine the total cost function from the marginal cost unless additional information is available to help us determine the fixed costs.

EXAMPLE 1

If the marginal cost function for a month for a certain product is $\overline{MC} = 3x + 50$, and if the fixed costs related to the product amount to $100 per month, find the total cost function for the month.

Solution

The total cost function is

$$C(x) = \int (3x + 50)\, dx$$

$$= \frac{3x^2}{2} + 50x + K.$$

But the constant of integration K is found by using the fact that $C(0) = FC = 100$. Thus

$$3(0)^2 + 50(0) + K = 100, \qquad \text{so } K = 100.$$

Thus the total cost for the month is given by

$$C(x) = \frac{3x^2}{2} + 50x + 100.$$

EXAMPLE 2

Suppose monthly records show that the rate of change of the cost (that is, the marginal cost) for a product is $\overline{MC} = 3(2x + 25)^{1/2}$ and that the fixed costs for the month are $11,125. What would be the total cost of producing 300 items per month?

Solution

We can integrate the marginal cost to find the total cost function.

$$C(x) = \int \overline{MC}\, dx = \int 3(2x + 25)^{1/2}\, dx$$

$$= 3 \cdot \left(\frac{1}{2}\right) \int (2x + 25)^{1/2}(2\, dx)$$

$$= \left(\frac{3}{2}\right) \frac{(2x + 25)^{3/2}}{3/2} + K$$

$$= (2x + 25)^{3/2} + K$$

We can find K by using the fact that fixed costs are $11,125.

$$C(0) = 11,125 = (25)^{3/2} + K$$

$$11,125 = 125 + K \qquad \text{or} \qquad K = 11,000$$

Thus the total cost function is

$$C(x) = (2x + 25)^{3/2} + 11,000$$

and the cost of producing 300 items per month is

$$C(300) = (625)^{3/2} + 11,000$$

$$= 26,625 \text{ (dollars)}.$$

It can be shown that the profit is usually maximized when $\overline{MR} = \overline{MC}$. To see that this does not always give us a maximum *positive* profit, consider the following facts concerning the manufacture of widgets over the period of a month:

(a) The marginal revenue is $\overline{MR} = 400 - 30x$.

(b) The marginal cost is $\overline{MC} = 20x + 50$.

(c) When 5 widgets are produced and sold, the total cost is $1750. The profit *should* be maximized when $\overline{MR} = \overline{MC}$, or when $400 - 30x = 20x + 50$. Solving for x gives $x = 7$. To see if our profit is maximized when 7 units are produced and sold, let us examine the profit function.

The profit function is given by $P(x) = R(x) - C(x)$, where

$$R(x) = \int \overline{MR}\, dx \quad \text{and} \quad C(x) = \int \overline{MC}\, dx.$$

Integrating, we get

$$R(x) = \int (400 - 30x)\, dx = 400x - 15x^2 + K,$$

but $K = 0$ for this total revenue function, so

$$R(x) = 400x - 15x^2.$$

The total cost function is

$$C(x) = \int (20x + 50)\, dx = 10x^2 + 50x + K.$$

The value of fixed cost can be determined by using the fact that 5 widgets cost $1750. This tells us $C(5) = 1750 = 250 + 250 + K$, so $K = 1250$.
Thus the total cost is $C(x) = 10x^2 + 50x + 1250$. Now, the profit is

$$P(x) = R(x) - C(x),$$

or

$$P(x) = (400x - 15x^2) - (10x^2 + 50x + 1250).$$

Simplifying gives

$$P(x) = 350x - 25x^2 - 1250.$$

We have found that $\overline{MR} = \overline{MC}$ if $x = 7$, and the graph of $P(x)$ is a parabola that opens downward, so profit is maximized at $x = 7$. But if $x = 7$, profit is

$$P(7) = 2450 - 1225 - 1250 = -25.$$

That is, the production and sale of 7 items result in a loss of $25.

The preceding discussion indicates that, although setting $\overline{MR} = \overline{MC}$ may optimize profit, it does not indicate the level of profit or loss, as forming the profit function does.

If the widget firm is in a competitive market, and its optimal level of production results in a loss, it has two options. It can continue to produce at the optimal level in the short run until it can lower or eliminate its fixed costs, even though it is losing money; or it can take a larger loss (its fixed cost) by stopping production. Producing 7 units causes a loss of $25 per month, and ceasing production results in a loss of $1250 (the fixed cost) per month. If this firm and many others like it cease production, the supply will be reduced, causing an eventual increase in price. The firm can resume production when the price increase indicates that it can make a profit.

EXAMPLE 3

Given that $\overline{MR} = 200 - 4x$, $\overline{MC} = 50 + 2x$, and the total cost of producing 10 Wagbats is $700, at what level should the Wagbat firm hold production in order to maximize the profits?

Solution

Setting $\overline{MR} = \overline{MC}$, we can solve for the production level that maximizes profit.

$$200 - 4x = 50 + 2x$$
$$150 = 6x$$
$$25 = x$$

The level of production that should optimize profit is 25 units. To see whether 25 units maximizes profits or minimizes the losses (in the short run), we must find the total revenue and total cost functions.

$$R(x) = \int (200 - 4x)\, dx = 200x - 2x^2 + K$$

$$= 200x - 2x^2, \qquad \text{since } K = 0$$

$$C(x) = \int (50 + 2x)\, dx = 50x + x^2 + K$$

We find K by noting $C(x) = 700$ when $x = 10$.

$$700 = 50(10) + (10)^2 + K,$$

so $K = 100$.

Thus the cost is given by $C = C(x) = 50x + x^2 + 100$.
At $x = 25$, $R = R(25) = 200(25) - 2(25)^2 = \$3750$ and $C = C(25) = 50(25) + (25)^2 + 100 = \$1975$.

We see that the total revenue is greater than the total cost, so production should be held at 25 units, which results in a maximum profit.

 ## Graphing Utilities

If it is difficult to solve $\overline{MC} = \overline{MR}$ analytically, we could use a graphing utility to solve this equation by finding the point of intersection of the graphs of $\overline{MC}$ and $\overline{MR}$. We may also be able to integrate $\overline{MC}$ and $\overline{MR}$ to find the functions $C(x)$ and $R(x)$ and then use a graphing utility to graph them. From the graphs of $C(x)$ and $R(x)$ we can learn about these functions, and hence about profit.

 ### EXAMPLE 4
Suppose that $\overline{MC} = 1.01(x + 190)^{0.01}$ and $\overline{MR} = (1/\sqrt{2x + 1}) + 2$, where x is the number of thousands of units and both revenue and cost are in thousands of dollars. Suppose further that fixed costs are $\$100,236$ and that production is limited to at most 180 thousand units.
(a) Determine $C(x)$ and $R(x)$ and graph them to determine whether a profit can be made.
(b) Estimate the level of production that yields maximum profit, and find the maximum profit.

Solution

(a) $C(x) = \int \overline{MC}\, dx = \int 1.01\,(x + 190)^{0.01}\, dx$

$$= 1.01\,\frac{(x + 190)^{1.01}}{1.01} + K$$

When we say that fixed costs equal $100,236, we mean $C(0) = 100.236$.

$$100.236 = C(0) = (190)^{1.01} + K$$
$$100.236 = 200.236 + K$$
$$-100 = K$$

Thus $C(x) = (x + 190)^{1.01} - 100$.

$$R(x) = \int \overline{MR}\, dx = \int [(2x + 1)^{-1/2} + 2]\, dx$$

$$= \frac{1}{2} \int (2x + 1)^{-1/2} (2\, dx) + \int 2\, dx$$

$$= \frac{\dfrac{1}{2}(2x + 1)^{1/2}}{1/2} + 2x + K$$

$R(0) = 0$ means

$$0 = R(0) = (1)^{1/2} + 0 + K \qquad \text{or} \qquad K = -1.$$

Thus $R(x) = (2x + 1)^{1/2} + 2x - 1$.

The graphs of $C(x)$ and $R(x)$ are shown in Figure 12.7. (The x-range is chosen to include the production range from 0 to 180 (thousand) units. The y-range is chosen to extend beyond fixed costs of about 100 thousand dollars.)

From the figure we see that a profit can be made as long as the number of units sold exceeds about 95 (thousand). We could locate this value more precisely by using INTERSECT or TRACE and ZOOM.

(b) From the graph we also see that $R(x) - C(x) = P(x)$ is maximum at the right edge of the graph. Because production is limited to at most 180 thousand units, profit will be maximized when $x = 180$ and the maximum profit is

$$P(180) = R(180) - C(180)$$
$$= [(361)^{1/2} + 360 - 1] - [(370)^{1.01} - 100]$$
$$\approx 85.46 \text{ (thousand dollars)}.$$

Figure 12.7

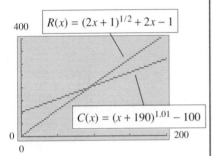

CHECKPOINT

1. True or false:
   (a) If $C(x) = \int \overline{MC}\, dx$, then the constant of integration equals the fixed costs.
   (b) If $R(x) = \int \overline{MR}\, dx$, then the constant of integration equals 0.

2. Find $C(x)$ if $\overline{MC} = \dfrac{100}{\sqrt{x + 1}}$ and fixed costs are $8000.

NATIONAL CONSUMPTION AND SAVINGS

The consumption function is one of the basic ingredients in a larger discussion of how an economy can have persistent high unemployment or persistent high inflation. This study is often called **Keynesian analysis,** after its founder John Maynard Keynes.

If C represents national consumption (in billions of dollars), then a **national consumption function** has the form $C = f(y)$, where y is disposable national income (also in billions of dollars). The **marginal propensity to consume** is the derivative of the national consumption function with respect to y, or $dC/dy = f'(y)$. For example, suppose that

$$C = f(y) = 0.8y + 6$$

is a national consumption function; then the marginal propensity to consume is $f'(y) = 0.8$.

If we know the marginal propensity to consume, we can integrate with respect to y to find national consumption:

$$C = \int f'(y)\, dy = f(y) + K.$$

We can find the unique national consumption function if we have additional information to help us determine the value of K, the constant of integration.

EXAMPLE 5

If consumption is $6 billion when disposable income is 0, and if the marginal propensity to consume is $dC/dy = 0.3 + 0.4/\sqrt{y}$ (in billions of dollars), then find the national consumption function.

Solution

If

$$\frac{dC}{dy} = 0.3 + \frac{0.4}{\sqrt{y}},$$

then

$$C = \int \left(0.3 + \frac{0.4}{\sqrt{y}} \right) dy = 0.3y + 0.8y^{1/2} + K.$$

Now, if $C = 6$ when $y = 0$, then $6 = 0.3(0) + 0.8\sqrt{0} + K$. Thus the constant of integration is $K = 6$, and the consumption function is

$$C = 0.3y + 0.8\sqrt{y} + 6 \qquad \text{(billions of dollars)}.$$

If S represents national savings, we can assume that the disposable national income is given by $y = C + S$, or $S = y - C$. Then the **marginal propensity to save** is $dS/dy = 1 - dC/dy$.

EXAMPLE 6

If the consumption is $9 billion when income is 0, and if the marginal propensity to save is 0.25, find the consumption function.

Solution

If $dS/dy = 0.25$, then $0.25 = 1 - dC/dy$, or $dC/dy = 0.75$. Thus

$$C = \int 0.75 \, dy = 0.75y + K.$$

If $C = 9$ when $y = 0$, then $9 = 0.75(0) + K$, or $K = 9$. Then the consumption function is $C = 0.75y + 9$ (billions of dollars).

CHECKPOINT

3. If the marginal propensity to save is

$$\frac{dS}{dy} = 0.7 - \frac{0.4}{\sqrt{y}},$$

find the marginal propensity to consume.

4. Find the national consumption function if the marginal propensity to consume is

$$\frac{dC}{dy} = \frac{1}{\sqrt{y + 4}} + 0.2$$

and national consumption is $6.8 billion when disposable income is 0.

CHECKPOINT SOLUTIONS

1. (a) False. $C(0)$ equals the fixed costs. It may or may not be the constant of integration. [In Checkpoint problem 2, $C(0) = 8000$, but $K = 7800$.]
   (b) False. We use $R(0) = 0$ to determine the constant of integration, but it may be nonzero. See Example 4.

2. $C(x) = 100 \int (x + 1)^{-1/2} \, dx = 100 \left[\frac{(x + 1)^{1/2}}{1/2} \right] + K$

   $C(x) = 200\sqrt{x + 1} + K$
   When $x = 0$, $C(x) = 8000$, so

   $$8000 = 200\sqrt{1} + K$$

   $$7800 = K$$

   $$C(x) = 200\sqrt{x + 1} + 7800.$$

3. $\dfrac{dC}{dy} = 1 - \dfrac{dS}{dy} = 1 - \left(0.7 - \dfrac{0.4}{\sqrt{y}} \right) = 0.3 + \dfrac{0.4}{\sqrt{y}}$

4. $C(y) = \int \left(\dfrac{1}{\sqrt{y+4}} + 0.2 \right) dy$

$= \int [(y+4)^{-1/2} + 0.2]\, dy$

$= \dfrac{(y+4)^{1/2}}{1/2} + 0.2y + K$

$C(y) = 2\sqrt{y+4} + 0.2y + K$

Using $C(0) = 6.8$ gives $6.8 = 2\sqrt{4} + 0 + K$, or $K = 2.8$.

Thus $C(y) = 2\sqrt{y+4} + 0.2y + 2.8$.

EXERCISE 12.4

TOTAL COST AND PROFIT

1. If the monthly marginal cost for a product is $\overline{MC} = 2x + 100$, with fixed costs amounting to $200, find the total cost function for the month.
2. If the monthly marginal cost for a product is $\overline{MC} = x + 30$, and the related fixed costs are $50, find the total cost function for the month.
3. If the marginal cost for a product is $\overline{MC} = 4x + 2$, and the production of 10 units results in a total cost of $300, find the total cost function.
4. If the marginal cost for a product is $\overline{MC} = 3x + 50$, and the total cost of producing 20 units is $2000, what will be the total cost function?
5. If the marginal cost for a product is $\overline{MC} = 4x + 40$, and the total cost of producing 25 units is $3000, what will be the cost of producing 30 units?
6. If the marginal cost for producing a product is $\overline{MC} = 5x + 10$, with a fixed cost of $800, what will be the cost of producing 20 units?
7. A firm knows that its marginal cost for a product is $\overline{MC} = 3x + 20$, that its marginal revenue is $\overline{MR} = 44 - 5x$, and that the cost of production and sale of 80 units is $11,400. Given this information, find the
   (a) optimal level of production.
   (b) profit function.
   (c) profit or loss at the optimal level.
8. A certain firm's marginal cost for a product is $\overline{MC} = 6x + 60$, its marginal revenue is $\overline{MR} = 180 - 2x$, and its total cost of production of 10 items is $1000. Given this information,
   (a) find the optimal level of production.
   (b) find the profit function.

(c) find the profit or loss at the optimal level of production.
(d) should production be continued for the short run?
(e) should production be continued for the long run?
9. Suppose that the marginal revenue for a product is $\overline{MR} = 900$ and the marginal cost is $\overline{MC} = 30\sqrt{x+4}$, with a fixed cost of $1000.
   (a) Find the profit or loss from the production and sale of 5 units.
   (b) How many units will result in a maximum profit?
10. Suppose that the marginal cost for a product is $\overline{MC} = 60\sqrt{x+1}$ and its fixed cost is $340.00. If the marginal revenue for it is $\overline{MR} = 80x$, find the profit or loss from production and sale of:
    (a) 3 units.
    (b) 8 units.
11. If the average cost of a product changes at the rate

$$\overline{C}'(x) = -6x^{-2} + 1/6,$$

and if the average cost of 10 units is $20.00, find the
(a) average cost function.
(b) average cost of 20 units.
12. If the average cost of a product changes at the rate

$$\overline{C}'(x) = \dfrac{-10}{x^2} + \dfrac{1}{10},$$

and if the average cost of 10 units is $20.00, find the
(a) average cost function.
(b) average cost of 20 units.
13. Suppose for a certain product that marginal cost is given by $\overline{MC} = 1.05(x + 180)^{0.05}$ and marginal revenue is given by $\overline{MR} = (1/\sqrt{0.5x+4}) + 2.8$, where x is in

thousands of units and both revenue and cost are in thousands of dollars. If fixed costs are $200,000 and production is limited to at most 200 thousand units, complete the following.
(a) Find $C(x)$ and $R(x)$.
(b) Graph $C(x)$ and $R(x)$ to determine whether a profit can be made.
(c) Determine the level of production that yields maximum profit, and find the maximum profit (or minimum loss).

14. Suppose for a certain product that the marginal cost is given by $\overline{MC} = 1.02(x + 200)^{0.02}$ and marginal revenue is given by $\overline{MR} = (2/\sqrt{4x + 1}) + 1.75$, where x is in thousands of units and revenue and cost are in thousands of dollars. Suppose further that fixed costs are $150,000 and production is limited to at most 200 thousand units.
(a) Find $C(x)$ and $R(x)$.
(b) Graph $C(x)$ and $R(x)$ to determine whether a profit can be made.
(c) Determine the level of production that yields maximum profit, and find the maximum profit (or minimum loss).

NATIONAL CONSUMPTION AND SAVINGS

15. If consumption is $5 billion when disposable income is 0, and if the marginal propensity to consume is

$$\frac{dC}{dy} = 0.4 + \frac{0.3}{\sqrt{y}} \qquad \text{(in billions of dollars)},$$

find the national consumption function.

16. If consumption is $7 billion when disposable income is 0, and if the marginal propensity to consume is 0.80, find the national consumption function (in billions of dollars).

17. If consumption is $8 billion when income is 0, and if the marginal propensity to consume is

$$\frac{dC}{dy} = 0.3 + \frac{0.2}{\sqrt{y}} \qquad \text{(in billions of dollars)},$$

find the national consumption function.

18. If national consumption is $9 billion when income is 0, and if the marginal propensity to consume is 0.30, what is consumption when disposable income is $20 billion?

19. If consumption is $6 billion when disposable income is 0, and if the marginal propensity to consume is

$$\frac{dC}{dy} = \frac{1}{\sqrt{y + 1}} + 0.4 \qquad \text{(in billions of dollars)},$$

find the national consumption function.

20. If consumption is $5.8 billion when disposable income is 0, and if the marginal propensity to consume is

$$\frac{dC}{dy} = \frac{1}{\sqrt{2y + 9}} + 0.8 \qquad \text{(in billions of dollars)},$$

find the national consumption function.

21. Suppose that the marginal propensity to consume is

$$\frac{dC}{dy} = 0.7 - e^{-2y} \qquad \text{(in billions of dollars)},$$

and that consumption is $5.65 billion when disposable income is 0. Find the national consumption function.

22. Suppose that the marginal propensity to consume is

$$\frac{dC}{dy} = 0.04 + \frac{\ln(y + 1)}{y + 1} \qquad \text{(in billions of dollars)},$$

and that consumption is $6.04 billion when disposable income is 0. Find the national consumption function.

23. Suppose that the marginal propensity to save is

$$\frac{dS}{dy} = 0.15 \qquad \text{(in billions of dollars)}$$

and that consumption is $5.15 billion when disposable income is 0. Find the national consumption function.

24. Suppose that the marginal propensity to save is

$$\frac{dS}{dy} = 0.22 \qquad \text{(in billions of dollars)}$$

and that consumption is $8.6 billion when disposable income is 0. Find the national consumption function.

25. Suppose that the marginal propensity to save is

$$\frac{dS}{dy} = 0.2 - \frac{1}{\sqrt{3y + 7}} \qquad \text{(in billions of dollars)}$$

and that consumption is $6 billion when disposable income is 0. Find the national consumption function.

26. If consumption is $3 billion when disposable income is 0, and if the marginal propensity to save is

$$\frac{dS}{dy} = 0.2 + e^{-1.5y} \qquad \text{(in billions of dollars)},$$

find the national consumption function.

12.5 DIFFERENTIAL EQUATIONS

Objectives

- ❑ To show that a function is the solution to a differential equation
- ❑ To use integration to find the general solution of a differential equation
- ❑ To find particular solutions of differential equations using given conditions
- ❑ To solve separable differential equations
- ❑ To solve applied problems involving separable differential equations

APPLICATION PREVIEW

Carbon-14 dating, used to determine the age of fossils, is based on three facts. First, the half-life of carbon-14 is 5600 years. Second, the amount of carbon-14 in any living organism is essentially constant. Third, when an organism dies, the rate of change of carbon-14 in the organism is proportional to the amount present. If y represents the amount of carbon-14 present in the organism, then we can express the rate of change of carbon-14 by the **differential equation**

$$\frac{dy}{dt} = ky,$$

where k is a constant and t is time in years. In this section, we study methods that allow us to find a function y that satisfies this differential equation and then we use that function to date a fossil.

Recall that we introduced the derivative as an instantaneous rate of change and denoted the instantaneous rate of change of y with respect to time as dy/dt. For many growth or decay processes, such as carbon-14 decay, the rate of change of the amount of a substance with respect to time is proportional to the amount present. As we noted in the Application Preview, this can be represented by the equation

$$\frac{dy}{dt} = ky \qquad (k = \text{constant}).$$

An equation of this type, where y is an unknown function of x or t, is called a **differential equation** because it contains derivatives (or differentials). In this section we restrict ourselves to differential equations where the highest derivative present in the equation is the first derivative. These differential equations are called **first-order differential equations.** Examples of first-order differential equations are

$$f'(x) = \frac{1}{x+1}, \qquad \frac{dy}{dt} = 2t, \quad \text{and} \quad x \, dy = (y+1) \, dx.$$

SOLUTION OF DIFFERENTIAL EQUATIONS

The solution to a differential equation is a function (say $y = f(x)$) that, when used in the differential equation, results in an identify.

EXAMPLE 1

Show that $y = 4e^{-5t}$ is a solution of $dy/dt + 5y = 0$.

Solution

We must show that substituting $y = 4e^{-5t}$ into the equation $dy/dt + 5y = 0$ results in an identity:

$$\frac{d}{dt}(4e^{-5t}) + 5(4e^{-5t}) = 0$$

$$-20e^{-5t} + 20e^{-5t} = 0$$

$$0 = 0$$

Thus $y = 4e^{-5t}$ is a solution.

Now that we know what it means for a function to be a solution to a differential equation, let us consider how to find solutions.

The most elementary differential equations are of the form

$$\frac{dy}{dx} = f(x),$$

where $f(x)$ is a continuous function. These equations are elementary to solve because the solutions are found by integration:

$$y = \int f(x)\, dx.$$

EXAMPLE 2

Find the solution of

$$f'(x) = \frac{1}{x+1}.$$

Solution

The solution is

$$f(x) = \int f'(x)\, dx = \int \frac{1}{x+1}\, dx = \ln|x+1| + C.$$

Figure 12.8

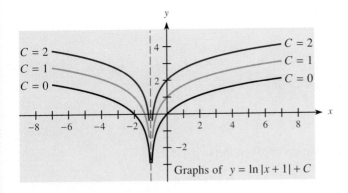

Graphs of $y = \ln|x+1| + C$

The solution in Example 2, $f(x) = \ln|x+1| + C$, is called the **general solution** because every solution to the equation has this form, and different values of C give different **particular solutions.** Figure 12.8 shows the graphs of several members of the family of solutions to this differential equation. (We cannot, of course, show all of them.)

We can find a particular solution to a differential equation when we know that the solution must satisfy additional conditions, such as **initial conditions** or **boundary conditions.** For instance, to find the particular solution to

$$f'(x) = \frac{1}{x + 1} \qquad \text{with the condition that} \qquad f(-2) = 2,$$

we use $f(-2) = 2$ in the general solution, $f(x) = \ln|x + 1| + C$.

$$2 = f(-2) = \ln|-2 + 1| + C$$
$$2 = \ln|-1| + C \qquad \text{so} \qquad C = 2$$

Thus the particular solution is

$$f(x) = \ln|x + 1| + 2$$

and is shown in Figure 12.8 with $C = 2$.

We frequently denote the value of the solution function $y = f(t)$ at the initial time $t = 0$ as $y(0)$ instead of $f(0)$.

CHECKPOINT

1. Given $f'(x) = 2x - [1/(x + 1)]$, $f(0) = 4$,
   (a) find the general solution to the differential equation.
   (b) find the particular solution that satisfies $f(0) = 4$.

Just as we can find the differential of both sides of an equation, we can find the solution to a differential equation of the form

$$G(y)\, dy = F(x)\, dx$$

by integrating both sides.

EXAMPLE 3

Solve $3y^2\, dy = 2x\, dx$, if $y(1) = 2$.

Solution

We find the general solution by integrating both sides.

$$\int 3y^2\, dy = \int 2x\, dx$$
$$y^3 + C_1 = x^2 + C_2$$
$$y^3 = x^2 + C \quad \text{(where } C = C_2 - C_1\text{)}$$

By using $y(1) = 2$, we can find C.

$$2^3 = 1^2 + C$$
$$7 = C$$

Thus the particular solution is given implicitly by

$$y^3 = x^2 + 7.$$

SEPARABLE DIFFERENTIAL EQUATIONS

It is frequently necessary to change the form of a differential equation before it can be solved by integrating both sides.

For example, the equation

$$\frac{dy}{dx} = y^2$$

cannot be solved by simply integrating both sides of the equation with respect to x because we cannot evaluate $\int y^2 \, dx$.

However, we can multiply both sides of $dy/dx = y^2$ by dx/y^2 to obtain an equation that has all terms containing y on one side of the equation and all terms containing x on the other side. That is, we obtain

$$\frac{dy}{y^2} = dx.$$

Separable Differential Equations

When a differential equation can be equivalently expressed in the form

$$g(y) \, dy = f(x) \, dx,$$

we say that the equation is **separable.**

The solution of a separable differential equation is obtained by integrating both sides of the equation after the variables have been separated.

EXAMPLE 4

Solve the differential equation

$$(x^2y + x^2) \, dy = x^3 \, dx.$$

Solution

To write the equation in separable form, we first factor x^2 from the left side and divide both sides by it.

$$x^2(y + 1) \, dy = x^3 \, dx$$

$$(y + 1) \, dy = \frac{x^3}{x^2} \, dx$$

The equation is now separated, so we integrate both sides.

$$\int (y + 1) \, dy = \int x \, dx$$

$$\frac{y^2}{2} + y + C_1 = \frac{x^2}{2} + C_2$$

This equation, as well as the equation

$$y^2 + 2y - x^2 = C, \quad (\text{where } C = 2(C_2 - C_1))$$

gives the solution implicitly.

Note that we need not write both C_1 and C_2 when we integrate because it is always possible to combine the two constants into one.

EXAMPLE 5

Solve the differential equation

$$\frac{dy}{dt} = ky \qquad (k = \text{constant}).$$

Solution

To solve the equation, we write it in separated form as

$$\frac{dy}{y} = k\,dt$$

and integrate both sides as follows:

$$\int \frac{dy}{y} = \int k\,dt$$

$$\ln|y| = kt + C_1.$$

Assuming that $y > 0$ and writing this equation in exponential form gives

$$y = e^{kt+C_1}$$

$$y = e^{kt} \cdot e^{C_1} = Ce^{kt}, \qquad \text{where } C = e^{C_1}.$$

This solution,

$$y = Ce^{kt},$$

is the general solution of the differential equation $dy/dt = ky$ because all solutions have this form, with different values of C giving different particular solutions. The case of $y < 0$ is covered by values of $C < 0$.

CHECKPOINT

2. True or false:
   (a) The general solution to $dy = (x/y)\,dx$ can be found from

   $$\int y\,dy = \int x\,dx.$$

   (b) The first step in solving $dy/dx = -2xy^2$ is to separate it.
   (c) The equation $dy/dx = -2xy^2$ separates as $y^2\,dy = -2x\,dx$.
3. Suppose that $(xy + x)(dy/dx) = x^2y + y$.
   (a) Separate this equation.   (b) Find the general solution.

In many applied problems that can be modeled with differential equations, we know of conditions that allow us to obtain a particular solution.

APPLICATIONS OF DIFFERENTIAL EQUATIONS

We now consider three applications that can be modeled by differential equations. These are radioactive decay (as introduced in the Application Preview), one-container mixture problems (as a model for drugs in an organ), and supply and demand in a competitive market.

EXAMPLE 6

In the Application Preview, we introduced carbon-14 dating and the facts that the process is based on. We said that when an organism dies, the rate of change of the amount of carbon-14 present is proportional to the amount present and is represented by the differential equation

$$\frac{dy}{dt} = ky,$$

where y is the amount present, k is a constant, and t is time in years. If we denote the initial amount of carbon-14 in an organism as y_0, then this represents the amount present at time $t = 0$ (when the organism died). Suppose that anthropologists discover a fossil that contains 1% of the initial amount of carbon-14. Find the age of the fossil. (Recall that the half-life of carbon-14 is 5600 years.)

Solution

We must find a particular solution to

$$\frac{dy}{dt} = ky$$

subject to the fact that when $t = 0$, $y = y_0$, and we must determine the value of k based on the half-life of carbon-14 (i.e., $t = 5600$ years, $y = \frac{1}{2}y_0$ units.) From Example 5, we know that the general solution to the differential equation $dy/dt = ky$ is $y = Ce^{kt}$. Using $t = 0$ and $y = y_0$, we obtain $y_0 = C$, so the equation becomes $y = y_0 e^{kt}$. Using $t = 5600$ and $y = \frac{1}{2}y_0$ in this equation gives

$$\frac{1}{2}y_0 = y_0 e^{5600k} \qquad \text{or} \qquad 0.5 = e^{5600k}.$$

Rewriting this equation in logarithmic form, then solving for k gives

$$\ln(0.5) = 5600k$$
$$-0.69315 = 5600k$$
$$-0.00012378 = k.$$

Thus the equation we seek is

$$y = y_0 e^{-0.00012378t}.$$

Using the fact that $y = 0.01y_0$ when the fossil was discovered, we can find its age t by solving

$$0.01y_0 = y_0 e^{-0.00012378t} \qquad \text{or} \qquad 0.01 = e^{-0.00012378t}.$$

Rewriting this in logarithmic form, then solving gives

$$\ln(0.01) = -0.00012378t$$
$$-4.6051702 = -0.00012378t$$
$$37{,}204 = t.$$

Thus the fossil is approximately 37,200 years old.

Another application of differential equations comes from a group of applications called *one-container mixture problems*. In problems of this type there is a substance whose amount in a container is changing with time, and the goal is to determine the amount of the substance at any time t. The differential equations that model these problems are of the following form:

$$\begin{bmatrix} \text{Rate of change} \\ \text{of the amount} \\ \text{of the substance} \end{bmatrix} = \begin{bmatrix} \text{Rate at which} \\ \text{the substance} \\ \text{enters the container} \end{bmatrix} - \begin{bmatrix} \text{Rate at which} \\ \text{the substance} \\ \text{leaves the container} \end{bmatrix}$$

We consider this application as it applies to the amount of a drug in an organ.

EXAMPLE 7

A liquid carries a drug into an organ of volume 300 cc at a rate of 5 cc/s, and the liquid leaves the organ at the same rate. If the concentration of the drug in the entering liquid is 0.1 g/cc, and if x represents the amount of drug in the organ at any time t, then using the fact that the rate of change of the amount of the drug in the organ, dx/dt, equals the rate in g/s at which the drug enters minus the rate at which it leaves, we have

$$\frac{dx}{dt} = \left(\frac{5 \text{ cc}}{\text{s}}\right)\left(\frac{0.1 \text{ g}}{\text{cc}}\right) - \left(\frac{5\text{cc}}{\text{s}}\right)\left(\frac{x \text{ g}}{300 \text{ cc}}\right)$$

or

$$\frac{dx}{dt} = 0.5 - \frac{x}{60} = \frac{30}{60} - \frac{x}{60} = \frac{30 - x}{60}.$$

Find the amount of the drug in the organ as a function of time t.

Solution

Multiplying both sides of the equation $\dfrac{dx}{dt} = \dfrac{30 - x}{60}$ by $\dfrac{dt}{(30 - x)}$ gives

$$\frac{dx}{30 - x} = \frac{1}{60}\,dt.$$

The equation is now separated, so we can integrate both sides.

$$\int \frac{dx}{30 - x} = \int \frac{1}{60} \, dt$$

$$-\ln(30 - x) = \frac{1}{60}t + C_1 \quad (30 - x > 0)$$

$$\ln(30 - x) = -\frac{1}{60}t - C_1$$

Rewriting this in exponential form gives

$$30 - x = e^{-t/60 - C_1} = e^{-t/60} \cdot e^{-C_1}.$$

Letting $C = e^{-C_1}$ we have

$$30 - x = Ce^{-t/60},$$

so $$x = 30 - Ce^{-t/60},$$

and we have the desired function.

Another application of differential equations involves supply and demand. If p is the price of a given commodity at time t, then we can think of price as a function of time. Similarly the number of units demanded by consumers q_d at any time, and the number of units supplied by producers q_s at any time, may also be considered as functions of time, as well as functions of price.

Both the quantity demanded and the quantity supplied depend not only on the price at the time, but also on the direction and rate of change that consumers and producers feel that prices may be moving. For example, even when prices are high, if consumers feel that prices are rising, the demand may rise. Similarly, if prices are low but producers feel they may go lower, the supply may rise.

If we assume that prices are determined in the marketplace by supply and demand, then the equilibrium price is the one we seek. Equating the number of units supplied with the number demanded gives rise to a first-order differential equation, as the following example shows.

EXAMPLE 8
Suppose the supply and demand functions for a certain commodity in a competitive market are given in hundreds of units by

$$q_s = 30 + p + 5\frac{dp}{dt}$$

$$q_d = 51 - 2p + 4\frac{dp}{dt},$$

where dp/dt denotes the rate of change of the price with respect to time. If, at $t = 0$, the market equilibrium price is 12, express the market equilibrium price as a function of time.

Solution

At market equilibrium $q_s = q_d$, which gives

$$30 + p + 5\frac{dp}{dt} = 51 - 2p + 4\frac{dp}{dt}$$

$$\frac{dp}{dt} = 21 - 3p = 3(7 - p).$$

Separating the variables and solving gives

$$\frac{dp}{7 - p} = 3\,dt$$

$$-\ln(7 - p) = 3t + C_1$$

$$\ln(7 - p) = -3t - C_1$$

$$7 - p = e^{-3t - C_1} = Ce^{-3t}$$

$$7 - Ce^{-3t} = p.$$

Now, using $t = 0$ and $p = 12$ gives $7 - C = 12$, or $C = -5$, so $p = 7 + 5e^{-3t}$ gives the market equilibrium price as a function of time.

Notice in Example 8 that $\lim\limits_{t\to\infty} p = 7$. This means that in this marketplace there is price stability for this item, and we can expect the price of this commodity to approach 7. If $\lim\limits_{t\to\infty} p = \infty$, then prices are increasing as time passes, and we have price instability (inflation) until the equations modeling supply and demand are altered by economic factors.

CHECKPOINT SOLUTIONS

1. (a) $f(x) = \displaystyle\int \left[2x - \frac{1}{x + 1}\right] dx = x^2 - \ln|x + 1| + C$

   (b) If $f(0) = 4$, then $4 = 0^2 - \ln|1| + C$, so $C = 4$.
   Thus $f(x) = x^2 - \ln|x + 1| + 4$ is the particular solution.

2. (a) True, and the solution is

$$\int y\,dy = \int x\,dx \qquad \frac{y^2}{2} = \frac{x^2}{2} + C.$$

   (b) True

   (c) False. It separates as $dy/y^2 = -2x\,dx$, and the solution is

$$\int \frac{-dy}{y^2} = \int 2x\,dx \qquad y^{-1} = x^2 + C$$

$$\frac{1}{y} = x^2 + C$$

$$y = \frac{1}{x^2 + C}.$$

$$3. \quad (a) \ x(y + 1) \frac{dy}{dx} = (x^2 + 1)y \qquad (b) \ \int\left(1 + \frac{1}{y}\right) dy = \int\left(x + \frac{1}{x}\right) dx$$

$$\frac{y + 1}{y} dy = \frac{x^2 + 1}{x} dx \qquad\qquad y + \ln|y| = \frac{x^2}{2} + \ln|x| + C$$

$$\left(1 + \frac{1}{y}\right) dy = \left(x + \frac{1}{x}\right) dx \qquad \text{This is the general solution.}$$

EXERCISE 12.5

In problems 1–4, show that the given function is a solution of the differential equation.

1. $y = x^2$; $\ 4y - 2xy' = 0$

2. $y = x^3$; $\ 3y - xy' = 0$

3. $y = 3x^2 + 1$; $\ 2y \, dx - x \, dy = 2 \, dx$

4. $y = 4x^3 + 2$; $\ 3y \, dx - x \, dy = 6 \, dx$

In problems 5–10, use integration to find the general solution to each differential equation.

5. $dy = xe^{x^2+1} \, dx$

6. $dy = x^2 e^{x^3-1} \, dx$

7. $2y \, dy = 4x \, dx$

8. $4y \, dy = 4x^3 \, dx$

9. $3y^2 \, dy = (2x - 1) \, dx$

10. $4y^3 \, dy = (3x^2 + 2x) \, dx$

In problems 11–14, find the particular solutions.

11. $y' = e^{x-3}$, $\ y(0) = 2$

12. $y' = e^{2x+1}$, $\ y(0) = e$

13. $dy = \left(\frac{1}{x} - x\right) dx$, $\ y(1) = 0$

14. $dy = \left(x^2 - \frac{1}{x + 1}\right) dx$, $\ y(0) = \frac{1}{3}$

In problems 15–28, find the general solution to the given differential equation.

15. $\dfrac{dy}{dx} = \dfrac{x^2}{y}$

16. $y^3 \, dx = \dfrac{dy}{x^3}$

17. $dx = x^3 y \, dy$

18. $dy = x^2 y^3 \, dx$

19. $dx = (x^2 y^2 + x^2) \, dy$

20. $dy = (x^2 y^3 + xy^3) \, dx$

21. $y^2 \, dx = x \, dy$

22. $y \, dx = x \, dy$

23. $\dfrac{dy}{dx} = \dfrac{x}{y}$

24. $\dfrac{dy}{dx} = \dfrac{x^2 + x}{y + 1}$

25. $(x + 1)\dfrac{dy}{dx} = y$

26. $x^2 y \dfrac{dy}{dx} = y^2 + 1$

27. $e^{2x}y \, dy = (y + 1) \, dx$

28. $e^{4x}(y + 1) \, dx + e^{2x}y \, dy = 0$

In problems 29–36, find the particular solution to each differential equation.

29. $\dfrac{dy}{dx} = \dfrac{x^2}{y^3}$, $\ $ when $x = 1$, $y = 1$

30. $\dfrac{dy}{dx} = \dfrac{x + 1}{xy}$, $\ $ when $x = 1$, $y = 3$

31. $2y^2 \, dx = 3x^2 \, dy$, $\ $ when $x = 2$, $y = -1$

32. $(x + 1) \, dy = y^2 \, dx$, $\ $ when $x = 0$, $y = 2$

33. $x^2 e^{2y} \, dy = (x^3 + 1) \, dx$, $\ $ when $x = 1$, $y = 0$

34. $y' = \dfrac{1}{xy}$, $\ $ when $x = 1$, $y = 3$

35. $2xy \dfrac{dy}{dx} = y^2 + 1$, $\ $ when $x = 1$, $y = 2$

36. $xe^y \, dx = (x + 1) \, dy$, $\ $ when $x = 0$, $y = 0$

APPLICATIONS

37. **Allometric growth** If x and y are measurements of certain parts of an organism, then the rate of change in y with respect to x is proportional to the ratio of y to x. That is, these measurements satisfy

$$\frac{dy}{dx} = k\frac{y}{x},$$

which is referred to as an allometric law of growth. Solve this differential equation.

38. **Bimolecular chemical reactions** A bimolecular chemical reaction is one in which two molecules react to give another substance. Suppose that two molecules of a certain substance containing A molecules initially react to form two molecules of a new substance. If x represents

the number of molecules of the new substance, then the rate of change of x is proportional to the square of the number of molecules of the original substance available to be converted. That is,

$$\frac{dx}{dt} = k(A - x)^2.$$

If 40% of the initial amount is converted after 1 hour, how long before 90% is converted?

Compound interest In problems 39 and 40, use the following information.

When interest is compounded continuously, the rate of change of the amount x of the investment is proportional to the amount present. In this case, the proportionality constant is the annual interest rate r (as a decimal); that is,

$$\frac{dx}{dt} = rx.$$

39. (a) If $10,000 is invested at 6% compounded continuously, find an equation for the amount of the investment as a function of time t in years.
    (b) What is the amount of the investment after 1 year? After 5 years?
    (c) How long will it take for this investment to double?

40. (a) If $2000 is invested at 8% compounded continuously, find an equation for the amount of the investment as a function of time t, in years.
    (b) How long will it take for this investment to double?
    (c) What will be the amount of this investment after 35 years?

41. **Bacterial growth** Suppose that the growth of a certain bacteria satisfies

$$\frac{dy}{dt} = ky,$$

where y is the number of organisms and t is the number of hours. If initially there are 10,000 organisms and the number triples after 2 hours, how long before there is 100 times the original population?

42. **Bacterial growth** Suppose that, for a certain bacteria, growth occurs according to

$$\frac{dy}{dt} = ky \quad (t \text{ in hours}).$$

If the doubling rate depends on temperature, find how long it takes for the number of bacteria to reach 50 times the original number at each given temperature in (a) and (b).

(a) At 90°F, the number doubles after 30 minutes ($\frac{1}{2}$ hour).
(b) At 40°F, the number doubles after 3 hours.

43. **Half-life** A breeder reactor converts uranium-238 into an isotope of plutonium-239 at a rate proportional to the amount present at any time. After 10 years, 0.03% of the radioactivity has dissipated (that is, 0.9997 of the initial amount remains). Suppose that initially there is 100 lb of this substance. Find the half-life.

44. **Radioactive decay** A certain radioactive substance has a half-life of 50 hours. How long will it take for 90% of the radioactivity to be dissipated if the amount of material x satisfies

$$\frac{dx}{dt} = kx \quad (t \text{ in hours})?$$

45. **Drug in an organ** Suppose that a liquid carries a drug into a 100-cc organ at a rate of 5 cc/sec and leaves the organ at the same rate. Suppose that the concentration of the drug entering is 0.06 g/cc. If initially there is no drug in the organ, find the amount of drug in the organ as a function of time t.

46. **Drug in an organ** Suppose that a liquid carries a drug into a 250-cc organ at a rate of 10 cc/sec and leaves the organ at the same rate. Suppose that the concentration of the drug entering is 0.15 g/cc. Find the amount of drug in the organ as a function of time t if initially there is none in the organ.

47. **Drug in an organ** Suppose that a liquid carries a drug with concentration 0.1 g/cc into a 200-cc organ at a rate of 5 cc/sec and leaves the organ at the same rate. If initially there is 10 g of the drug in the organ, find the amount of drug in the organ as a function of time t.

48. **Drug in an organ** Suppose that a liquid carries a drug with concentration 0.05 g/cc into a 150-cc organ at a rate of 6 cc/sec and leaves at the same rate. If initially there is 1.5 g of drug in the organ, find the amount of drug in the organ as a function of time t.

49. **Supply and demand** Suppose that for a certain commodity in a competitive market, the supply and demand functions are given by

$$q_d = 60 - p - 3\frac{dp}{dt}$$

$$q_s = 100 + p - 4\frac{dp}{dt},$$

where p is the price. If at $t = 0$, $p = 8$, find the market equilibrium price as a function of time.

50. **Supply and demand** Suppose that for a certain product in a competitive market, the demand function is

$$q_d = 50 - 2p - 3\frac{dp}{dt}$$

and the supply function is

$$q_s = 40 + 2p + \frac{dp}{dt}.$$

If at $t = 0, p = 15$, find the market equilibrium price as a function of time.

51. **Supply and demand** Suppose for a certain product in a competitive market, the supply and demand functions are given by

$$q_s = 120 + p - 4\frac{dp}{dt}$$

$$q_d = 70 - p - 2\frac{dp}{dt}.$$

If $p = 10$ when $t = 0$, find the equilibrium price as a function of time.

52. **Supply and demand** Suppose that the supply and demand functions for a certain commodity in a competitive market are given by

$$q_s = 39 + 4p + \frac{dp}{dt}$$

$$q_d = 84 - 5p - 2\frac{dp}{dt}.$$

If $p = 12$ at $t = 0$, find the equilibrium price as a function of time.

53. **Supply and demand** Suppose that the supply and demand functions for a certain product in a competitive market are given by

$$q_s = 80 + p + 3\frac{dp}{dt}$$

$$q_d = 110 - 2p + 2\frac{dp}{dt}.$$

(a) If $p = 41$ when $t = 0$, find the equilibrium price as a function of time.
(b) Evaluate $\lim_{t \to \infty} p$ to determine whether the equilibrium price is stable or unstable.

54. **Supply and demand** Suppose that the supply and demand functions for a certain product in a competitive

market are given by

$$q_s = 110 + p + 2\frac{dp}{dt}$$

$$q_d = 80 - 2p + 3\frac{dp}{dt}.$$

(a) If $p = 17$ when $t = 0$, find the equilibrium price as a function of time.
(b) Evaluate $\lim_{t \to \infty} p$ to determine whether the equilibrium price is stable or unstable.

55. **Sales and pricing** Suppose that in a certain company the relationship between the price per unit p of their product and the weekly sales volume y in thousands of dollars is given by

$$\frac{dy}{dp} = -\frac{2}{5}\left(\frac{y}{p+8}\right).$$

Solve this differential equation if $y = 8$ when $p = \$24$.

56. **Sales and pricing** Suppose that a chain of auto service stations, Quick-Oil, Inc., has found the relationship between their price p for a oil change and their monthly sales volume y in thousands of dollars to be

$$\frac{dy}{dp} = -\frac{1}{2}\left(\frac{y}{p+5}\right).$$

Solve this differential equation if $y = 18$ when $p = \$20$.

57. **Tumor volume** Let V denote the volume of a tumor, and suppose that the growth rate of the tumor satisfies

$$\frac{dV}{dt} = 0.2Ve^{-0.1t}.$$

If the initial volume of the tumor is 1.86 units, find an equation for V as a function of t.

58. **Gompertz curves** The differential equation

$$\frac{dx}{dt} = x(a - b \ln x),$$

where x represents the number of objects at time t, and a and b are constants, is the model for Gompertz curves. Recall from Section 5.3 that Gompertz curves can be used to study growth or decline of populations, organizations, and revenue from sales of a product, as well as forecast equipment maintenance costs. Solve the differential equation to obtain the Gompertz curve formula

$$x = e^{a/b}e^{-ce^{-bt}}.$$

59. Cell growth If V is the volume of a spherical cell, then, in certain cell growth or for some fetal growth models, the rate of change of V is given by

$$\frac{dV}{dt} = kV^{2/3},$$

where k is a constant depending on the organism. If $V = 0$ when $t = 0$, find V as a function of t.

60. Atmospheric pressure The rate of change of atmospheric pressure P with respect to the altitude above sea level h is proportional to the pressure. That is,

$$\frac{dP}{dh} = kP.$$

Suppose that the pressure at sea level is denoted by P_0, and at 18,000 ft the pressure is half what it is at sea level. Find the pressure, as a percentage of P_0, at 25,000 ft.

61. Newton's law of cooling Newton's law of cooling states that the rate of change of temperature $u = u(t)$ of an object is proportional to the temperature difference between the object and its surroundings, where T is the constant temperature of the surroundings. That is,

$$\frac{du}{dt} = k(u - T).$$

Suppose an object at 0°C is placed in a room where the temperature is 20°C. If the temperature of the object is 8°C after 1 hour, how long will it take for the object to reach 18°C?

62. Newton's law of cooling Newton's law of cooling can be used to estimate time of death. (Actually the estimate may be quite rough because cooling does not begin until metabolic processes have ceased.) Suppose a corpse is discovered at noon in a 70° room, and at that time the body temperature is 96.1°F. If at 1:00 P.M. the temperature is 94.6°F, use Newton's law of cooling to estimate the time of death.

63. Fossil-fuel emissions The amount of carbon in the atmosphere has been estimated to have increased at a rate of about 4.3% per year from 1860 until 1973, except during the Great Depression and the world wars.* This increase is due to carbon emissions from fossil-fuel burning and can be modeled by the differential equation.

$$\frac{dE}{dt} = 0.043E,$$

*American Scientist, Vol. 79, July–August 1990, p. 313

where t is the number of years since 1860 and E is fossil-fuel emissions in gigatons per year.

(a) Solve this differential equation, and find a particular solution that satisfies $E(0) = 0.1$.

(b) Graph your solution and compare it with the graph below, which shows actual emissions since 1860.

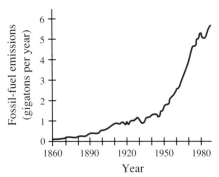

Year

SOURCE: *American Scientist*, Vol 79, July–August 1990, p. 313

64. Mutual fund growth The number of mutual funds established since 1940 can be modeled by

$$\frac{dy}{dt} = 0.662y, \quad y(4) = 68,$$

where t is the number of decades since 1900 and y is the number of mutual funds.

(a) Find the particular solution to this differential equation.

(b) Check your model against the data in the figure for 1960 and 1990.

SOURCE: Investment Company Institute, published in *Investment Digest of Valic Co.*, Vol. 5, No. 2, Summer, 1992.

65. **Impact of inflation** The impact of a 5% inflation rate on a $20,000 per year pension is shown in the accompanying figure. If P represents the purchasing power amount of a $20,000 pension, then the effect of a 5% inflation rate can be modeled by the differential equation

$$\frac{dP}{dt} = -0.05P, \quad P(0) = 20{,}000,$$

where t is in years.

(a) Find the particular solution to this differential equation.
(b) Check your model against the graph in the figure. In particular, find the purchasing power after 30 years. Does your calculation match the graph?

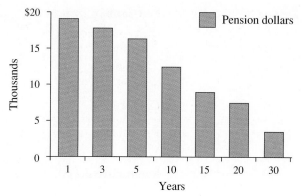

SOURCE: *Viewpoints*, Financial news from Valic Co., Summer 1993

KEY TERMS
& FORMULAS

Section	Key Term	Formula
12.1	General antiderivative of $f'(x)$	$f(x) + C$
	Integral	$\displaystyle\int f(x)\, dx$
	Powers of x Formula	$\displaystyle\int x^n\, dx = \frac{x^{n+1}}{n+1} + C \quad (n \neq -1)$
	Integration Formulas	$\displaystyle\int dx = x + C$
		$\displaystyle\int cu(x)\, dx = c\int u(x)\, dx;\ c = \text{a constant}$
		$\displaystyle\int [u(x) \pm v(x)]\, dx = \int u(x)\, dx \pm \int v(x)\, dx$
12.2	Power Rule	$\displaystyle\int [u(x)]^n u'(x)\, dx = \frac{[u(x)]^{n+1}}{n+1} + C \quad (n \neq -1)$
12.3	Logarithmic Formula	$\displaystyle\int \frac{u'}{u}\, dx = \int \frac{1}{u}\, du = \ln|u| + C$
	Exponential Formula	$\displaystyle\int e^u u'\, dx = \int e^u\, du = e^u + C$
12.4	Total cost	$C(x) = \int \overline{MC}\, dx$
	Total revenue	$R(x) = \int \overline{MR}\, dx$
	Profit	$P(x) = R(x) - C(x)$
	Marginal propensity to consume	$\dfrac{dC}{dy}$
	Marginal propensity to save	$\dfrac{dS}{dy} = 1 - \dfrac{dC}{dy}$
	National consumption	$C = \displaystyle\int f'(y)\, dy = \int \frac{dC}{dy}\, dy$

Section	Key Term	Formula
12.5	Differential equations Solutions General Particular	
	First order	$\dfrac{dy}{dx} = f(x) \Rightarrow y = \int f(x)\,dx$
	Separable	$g(y)\,dy = f(x)\,dx$ $\Rightarrow \int g(y)\,dy = \int f(x)\,dx$
	Radioactive decay	$\dfrac{dy}{dt} = ky$
	Drugs in an organ Supply and demand Equilibrium	Rate = (rate in) − (rate out) $q_s = q_d$

REVIEW EXERCISES

12.1–12.3

Evaluate the following integrals.

1. $\int x^6 dx$

2. $\int x^{1/2}\,dx$

3. $\int (x^3 - 3x^2 + 4x + 5)\,dx$

4. $\int (x^2 - 1)^2\,dx$

5. $\int (x^2 - 1)^2 x\,dx$

6. $\int (x^3 - 3x^2)(x^2 - 2x)\,dx$

7. $\int (x^3 + 4)^2 x\,dx$

8. $\int (x^3 + 4)^6 x^2\,dx$

9. $\displaystyle\int \frac{x^2}{x^3 + 1}\,dx$

10. $\displaystyle\int \frac{x^2}{(x^3 + 1)^2}\,dx$

11. $\displaystyle\int \frac{x^2\,dx}{\sqrt[3]{x^3 - 4}}$

12. $\displaystyle\int \frac{x^2\,dx}{x^3 - 4}$

13. $\displaystyle\int \frac{x^3 + 1}{x^2}\,dx$

14. $\displaystyle\int \frac{x^3 - 3x + 1}{x - 1}\,dx$

15. $\int y^2 e^{y^3}\,dy$

16. $\int (x - 1)^2\,dx$

17. $\displaystyle\int \frac{3x^2}{2x^3 - 7}\,dx$

18. $\displaystyle\int \frac{5\,dx}{e^{4x}}$

19. $\int (x^3 - e^{3x})\,dx$

20. $\int xe^{1+x^2}\,dx$

21. $\displaystyle\int \frac{6x^2}{(x^3 + 1)^2}\,dx$

22. $\displaystyle\int \frac{7x^3}{\sqrt{1 - x^4}}\,dx$

23. $\displaystyle\int \left(\frac{e^{2x}}{2} + \frac{2}{e^{2x}} \right)\,dx$

24. $\displaystyle\int \left[x - \frac{1}{(x + 1)^2} \right]\,dx$

25. (a) $\int (x^2 - 1)^4 x\,dx$ (b) $\int (x^2 - 1)^{10} x\,dx$
    (c) $\int (x^2-1)^7 3x\,dx$ (d) $\int (x^2 - 1)^{-2/3} x\,dx$

26. (a) $\displaystyle\int \frac{2x\,dx}{x^2 - 1}$ (b) $\displaystyle\int \frac{2x\,dx}{(x^2 - 1)^2}$

   (c) $\displaystyle\int \frac{3x\,dx}{\sqrt{x^2 - 1}}$ (d) $\displaystyle\int \frac{3x\,dx}{x^2 - 1}$

12.5

In problems 27–32, find the general solution to each differential equation.

27. $\dfrac{dy}{dt} = 4.6e^{-0.05t}$ 28. $dy = (64 + 76x - 36x^2)\,dx$

29. $\dfrac{dy}{dx} = \dfrac{4x}{y - 3}$

30. $t\,dy = \dfrac{dt}{y + 1}$

31. $\dfrac{dy}{dx} = \dfrac{x}{e^y}$

32. $\dfrac{dy}{dt} = \dfrac{4y}{t}$

In problems 33 and 34, find the particular solution to each differential equation.

33. $y' = \dfrac{x^2}{y + 1}$, $y(0) = 4$

34. $y' = \dfrac{2x}{1 + 2y}$, $y(2) = 0$

APPLICATIONS

12.1

35. **Revenue** If the marginal revenue for a month for a product is $\overline{MR} = 6x + 12$, find the total revenue from the sale of 4 units of the product.

36. **Productivity** Suppose that the rate of change of production of the average worker at a factory is given by

$$\frac{dp}{dt} = 27 + 24t - 3t^2, \quad 0 \le t \le 8,$$

where p is the number of units produced by the worker in t hours. How many units will the average worker produce in an 8-hour shift? (Assume that $p = 0$ when $t = 0$.)

12.2

37. **Oxygen levels in water** The rate of change of the oxygen level in a body of water after an oil spill is given by

$$P'(t) = 400\left[\frac{5}{(t + 5)^2} - \frac{50}{(t + 5)^3}\right],$$

where t is the number of months after the spill. What function gives the oxygen level P at any time t if $P = 400$ when $t = 0$?

38. **Bacterial growth** A population of bacteria grows at the rate

$$r = \frac{100,000}{(t + 100)^2},$$

where t is time. If the population is 1000 when $t = 1$, write the equation that gives the size of the population at any time t.

12.3

39. **Market share** The rate of change of the market share (as a percentage) a firm expects for a new product is

$$\frac{dy}{dt} = 2.4e^{-0.4t},$$

where t is the number of months after the product is introduced.

(a) Write the equation that gives the expected market share y at any time t. (Note that $y = 0$ when $t = 0$.)
(b) What market share does the firm expect after one year?

40. **Revenue** If the marginal revenue for a product is $\overline{MR} = \dfrac{800}{x + 1}$, find the total revenue function.

12.4

41. **Cost** If the marginal cost for a product is $\overline{MC} = 6x + 4$ and the cost of producing 100 items is $31,400,
(a) find the fixed costs.
(b) find the total cost function.

42. **Profit** Suppose a product has a daily marginal revenue $\overline{MR} = 46$ and a daily marginal cost $\overline{MC} = 30 + \frac{1}{5}x$. If the daily fixed cost is $200.00, how many units will give maximum profit and what is the maximum profit?

43. **National consumption** If consumption is $8.5 billion when disposable income is 0, and if the marginal propensity to consume is

$$\frac{dC}{dy} = \frac{1}{\sqrt{2y + 16}} + 0.6 \quad \text{(in billions of dollars),}$$

find the national consumption function.

44. **National consumption** Suppose that the marginal propensity to save is

$$\frac{dS}{dy} = 0.2 - 0.1e^{-2y} \quad \text{(in billions of dollars)}$$

and consumption is $7.8 billion when disposable income is 0. Find the national consumption function.

12.5

45. **Allometric growth** For many species of fish, the length L and weight W of a fish are related by

$$\frac{dW}{dL} = \frac{3W}{L}.$$

The general solution to this differential equation expresses the allometric relationship between the length and weight of a fish. Find the general solution.

46. **Fossil dating** Radioactive beryllium is sometimes used to date fossils found in deep-sea sediment. The amount of radioactive material x satisfies

$$\frac{dx}{dt} = kx.$$

Suppose that 10 units of beryllium are present in a living organism and that the half-life of beryllium is 4.6 million years. Find the age of a fossil if 20% of the original radioactivity is present when the fossil is discovered.

47. **Supply and demand** Suppose that for a certain product the supply and demand functions are given by

$$q_s = 66 + p - \frac{dp}{dt}, \qquad q_d = 186 - 2p - 2\frac{dp}{dt}.$$

Find the equilibrium price, as a function of time, if $p = 15$ when $t = 0$.

48. **Supply and demand** Suppose that for a certain commodity the supply and demand functions are given by

$$q_s = 41 + 7p + 2\frac{dp}{dt}, \qquad q_d = 191 - 3p - 3\frac{dp}{dt}.$$

If $p = 5$ when $t = 0$, find the equilibrium price as a function of time. Evaluate $\lim_{t \to \infty} p$ to determine the stable equilibrium price.

49. **Drug in an organ** Suppose that a liquid carries a drug into a 120-cc organ at a rate of 4 cc/sec and leaves the organ at the same rate. If initially there is no drug in the organ and if the concentration of drug in the liquid is 3 g/cc, find the amount of drug in the organ as a function of time.

50. **Chemical mixture** A 300-gal. tank initially contains a solution with 100 lb of a chemical. A mixture containing 2 lb/gal. of the chemical enters the tank at 3 gal./min, and the well-stirred mixture leaves at the same rate. Find an equation that gives the amount of the chemical in the tank as a function of time. How long will it be before there are 500 lb of chemical in the tank?

CONSULTANT'S CORNER

Employee Production Rate

The manager of a plant has been instructed to hire and train additional employees to manufacture a new product. She must hire a sufficient number of new employees so that within 30 days they will be producing 2500 units of the product each day.

Because a new employee must learn an assigned task, production will increase with training. Suppose that research on similar projects indicates that production increases with training according to the learning curve, so that for the average employee, the rate of production per day is given by

$$\frac{dN}{dt} = be^{-at},$$

where N is the number of units produced per day after t days of training. Because of experience with a very similar project, the manager expects the rate for this project to be

$$\frac{dN}{dt} = 2.5e^{-0.05t}.$$

The manager tested her training program with 5 employees and learned that the average employee could produce 11 units per day after 5 days of training. Based on this information, she must decide how many employees to hire and begin to train so that a month from now they will be producing 2500 units of the product per day. She estimates that it will take her 10 days to hire the employees, and thus she will have 15 days remaining to train them. She also expects a 10% attrition rate during this period.

How many employees would you advise the plant manager to hire? Check your advice by answering the following questions.

1. Use the expected rate of production and the results of the manager's test to find the function relating N and t, that is, $N = N(t)$.
2. Find the number of units the average employee can produce after 15 days of training. How many such employees would be needed to maintain a production rate of 2500 units per day?
3. Explain how you would revise this last result to account for the expected 10% attrition rate. How many new employees should the manager hire?

Warm-up

Prerequisite problem type	For section	Answer	Section for review
Simplify: $\dfrac{1}{n^3}\left[\dfrac{n(n+1)(2n+1)}{6} - \dfrac{2n(n+1)}{2} + n\right]$	13.1	$\dfrac{2n^2 - 3n + 1}{6n^2}$	0.7 Fractions

(a) If $F(x) = \dfrac{x^4}{4} + 4x + C$, what is
    $F(4) - F(2)$?
 	
(b) If $F(x) = -\dfrac{1}{9}\ln\dfrac{9 + \sqrt{81 - 9x^2}}{3x}$,
    what is $F(3) - F(2)$?
	13.2
	13.5
		(a) 68
		
		(b) $\dfrac{1}{9}\ln\left(\dfrac{3 + \sqrt{5}}{2}\right)$
			1.4 Functional notation

Find the limit: (a) $\displaystyle\lim_{n \to +\infty} \dfrac{n^2 + n}{2n^2}$ (b) $\displaystyle\lim_{n \to +\infty} \dfrac{2n^2 - 3n + 1}{6n^2}$ (c) $\displaystyle\lim_{b \to \infty}\left(1 - \dfrac{1}{b}\right)$ (d) $\displaystyle\lim_{b \to \infty}\left(\dfrac{-100{,}000}{e^{0.10b}} + 100{,}000\right)$	13.1 13.7	(a) $\frac{1}{2}$ (b) $\frac{1}{3}$ (c) 1 (d) 100,000	9.2 Limits at infinity
Find the derivative of $y = \ln x$.	13.6	$\dfrac{1}{x}$	11.1 Derivatives of logarithmic functions
Integrate: (a) $\int (x^3 + 4)\, dx$ (b) $\int x\sqrt{x^2 - 9}\, dx$ (c) $\int e^{2x}\, dx$	13.2– 13.7	(a) $\dfrac{x^4}{4} + 4x + C$ (b) $\frac{1}{3}(x^2 - 9)^{3/2} + C$ (c) $\frac{1}{2}e^{2x} + C$	12.1, 12.2, 12.3 Integration

CHAPTER 13

DEFINITE INTEGRALS; TECHNIQUES OF INTEGRATION

We have seen some applications of the indefinite integral in Chapter 12. In this chapter we define the definite integral and discuss a theorem that is useful for evaluating it. We will also see how it can be used to find the areas under certain curves. The definite integral is used to solve many interesting types of problems from economics, finance, and probability.

Definite integrals can be used to approximate the **total value,** the **present value,** and the **future value** of a **continuous income stream.** Improper integrals can be used to find the **capital values** of continuous income streams.

Some consumers are willing to pay more than the market equilibrium price, and some producers are willing to sell for less than this price. The savings are called **consumer's surplus** and **producer's surplus,** respectively, and areas under the demand and supply curves are used to calculate them.

Evaluation of improper integrals is one of three new techniques of integration introduced in this chapter. The other two are use of integral tables and use of integration by parts.

13.1 AREA UNDER A CURVE

Objectives

❑ To use the sum of areas of rectangles to approximate the area under a curve
❑ To use Σ notation to denote sums
❑ To find exact area under a curve

APPLICATION PREVIEW

One way to find the accumulated production (such as the production of ore from a mine) over a period of time is to graph the rate of production as a function of time, and find the area under the resulting curve over a specified time interval. For example, if a coal mine produces at a rate of 30 tons per day, the production over 10 days ($30 \cdot 10 = 300$) could be

863

Figure 13.1

Figure 13.2

(a)

(b)

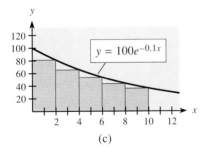

(c)

represented by the area under the line $y = 30$ between $x = 0$ and $x = 10$ (see Figure 13.1).

Using area to determine the accumulated production is very useful when the rate of production function varies at different points in time. For example, if the rate of production is represented by

$$y = 100e^{-0.1x},$$

where x represents the number of days, then the area under the curve (and above the x-axis) from $x = 0$ to $x = 10$ represents the total production over the 10-day period (see Figure 13.2(a)). In order to determine the accumulated production and to solve other types of problems we need a method for finding areas under curves. That is the goal of this section.

To estimate the accumulated production for the example above, we approximate the area under the graph of the production rate function. We can find a rough approximation of the area under this curve by fitting two rectangles to the curve as shown in Figure 13.2(b). The area of the first rectangle is $5 \cdot 100 = 500$ square units, and the area of the second rectangle is $(10 - 5)[100e^{-0.1(5)}] \approx 5(60.65) = 303.25$ square units, so this rough approximation is 803.25 square units. This approximation is clearly larger than the exact area under the curve. Why?

EXAMPLE 1

Fit five rectangles with equal bases inside the area under the curve $y = 100e^{-0.1x}$, and use them to approximate the area under the curve from $x = 0$ to $x = 10$ (see Figure 13.2(c)).

Solution

Each of the five rectangles has base 2, and the height of each rectangle is the value of the function at the right-hand endpoint of the interval forming its base. Thus the areas of the rectangles are as follows:

Rectangle	Base	Height	Area = Base × Height
1	2	$100e^{-0.1(2)} \approx 81.87$	$2(81.87) = 163.74$
2	2	$100e^{-0.1(4)} \approx 67.03$	$2(67.03) = 134.06$
3	2	$100e^{-0.1(6)} \approx 54.88$	$2(54.88) = 109.76$
4	2	$100e^{-0.1(8)} \approx 44.93$	$2(44.93) = 89.86$
5	2	$100e^{-0.1(10)} \approx 36.79$	$2(36.79) = 73.58$

The area under the curve is approximately equal to

$$163.74 + 134.06 + 109.76 + 89.86 + 73.58 = 571.$$

The area is actually 632.12, to two decimal places, so this approximation is much better than the one we obtained with just two rectangles. In general, if we use bases of equal width, the approximation of the area under a curve will improve when more rectangles are used.

Suppose that we wish to find the area between the curve $y = x$ and the x-axis from $x = 0$ to $x = 1$ (see Figure 13.3). One way to approximate this area is to use the areas of rectangles whose bases are on the x-axis and whose heights are the vertical distances from points on their bases to the curve. We can divide the interval $[0, 1]$ into n equal subintervals and use them as the bases of n rectangles whose heights are determined by the curve (see Figure 13.4). The width of each of these rectangles is $1/n$. Using the functional value at the right-hand endpoint of each subinterval as the height of the rectangle, we get n rectangles as shown in Figure 13.4. Because part of each rectangle lies above the curve, the sum of the areas of the rectangles will overestimate the area.

Then, with $y = f(x) = x$ and subinterval width $1/n$, the areas of the rectangles are as follows:

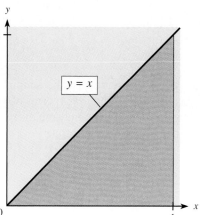

Figure 13.3

Rectangle	Base	Endpoint	Height	Area = Base × Height
1	$\dfrac{1}{n}$	$x_1 = \dfrac{1}{n}$	$f(x_1) = \dfrac{1}{n}$	$\dfrac{1}{n} \cdot \dfrac{1}{n} = \dfrac{1}{n^2}$
2	$\dfrac{1}{n}$	$x_2 = \dfrac{2}{n}$	$f(x_2) = \dfrac{2}{n}$	$\dfrac{1}{n} \cdot \dfrac{2}{n} = \dfrac{2}{n^2}$
3	$\dfrac{1}{n}$	$x_3 = \dfrac{3}{n}$	$f(x_3) = \dfrac{3}{n}$	$\dfrac{1}{n} \cdot \dfrac{3}{n} = \dfrac{3}{n^2}$
$\vdots$				
i	$\dfrac{1}{n}$	$x_i = \dfrac{i}{n}$	$f(x_i) = \dfrac{i}{n}$	$\dfrac{1}{n} \cdot \dfrac{i}{n} = \dfrac{i}{n^2}$
$\vdots$				
n	$\dfrac{1}{n}$	$x_n = \dfrac{n}{n}$	$f(x_n) = \dfrac{n}{n}$	$\dfrac{1}{n} \cdot \dfrac{n}{n} = \dfrac{n}{n^2}$

Note that i/n^2 gives the area of the ith rectangle of *any* value of i. Thus for any value of n, this area can be approximated by the sum

$$A \approx \frac{1}{n^2} + \frac{2}{n^2} + \frac{3}{n^2} + \cdots + \frac{i}{n^2} + \cdots + \frac{n}{n^2}.$$

In particular, we have the following approximations of this area for specific values of n (the number of rectangles).

Figure 13.4

$$n = 5: \quad A \approx \frac{1}{25} + \frac{2}{25} + \frac{3}{25} + \frac{4}{25} + \frac{5}{25} = \frac{15}{25} = 0.60$$

$$n = 10: \quad A \approx \frac{1}{100} + \frac{2}{100} + \frac{3}{100} + \cdots + \frac{10}{100} = \frac{55}{100} = 0.55$$

$$n = 100: \quad A \approx \frac{1}{10{,}000} + \frac{2}{10{,}000} + \frac{3}{10{,}000} + \cdots + \frac{100}{10{,}000}$$

$$= \frac{5050}{10{,}000} = 0.505$$

We can find this sum for any n more easily if we observe that the common denominator is n^2 and that the numerator is the sum of the first n terms of an arithmetic sequence with first term 1 and last term n. As you may recall from Section 6.1, the first n terms of this arithmetic sequence add to $(n + 1)/2$. Thus the area is approximated by

$$A \approx \frac{1 + 2 + 3 + \cdots + n}{n^2} = \frac{n(n + 1)/2}{n^2} = \frac{n + 1}{2n}.$$

Using this formula, we see the following.

$$n = 5: \quad A \approx \frac{5 + 1}{2(5)} = \frac{6}{10} = 0.60$$

$$n = 10: \quad A \approx \frac{10 + 1}{2(10)} = \frac{11}{20} = 0.55$$

$$n = 100: \quad A \approx \frac{100 + 1}{2(100)} = \frac{101}{200} = 0.505$$

Note that as n gets larger, the number of rectangles increases, the area of each rectangle decreases, and the approximation becomes more accurate. If we let n increase without bound, the approximation approaches the exact area.

$$A = \lim_{n \to +\infty} \frac{n + 1}{2n} = \lim_{n \to +\infty} \frac{1 + 1/n}{2} = \frac{1}{2}.$$

We can see that this area is correct, for we are computing the area of a triangle with base 1 and height 1. The formula for the area of a triangle gives

$$A = \frac{1}{2}bh = \frac{1}{2} \cdot 1 \cdot 1 = \frac{1}{2}.$$

A special notation exists that uses the Greek letter Σ (sigma) to express the sum of numbers or expressions. (We used sigma notation informally in Chapter 8.) We may indicate the sum of the n numbers $a_1, a_2, a_3, a_4, \ldots, a_n$ by

$$\sum_{i=1}^{n} a_i = a_1 + a_2 + a_3 + \cdots + a_n.$$

This may be read as "The sum of a_i as i goes from 1 to n." The subscript i in a_i is replaced, first by 1, then by 2, then by 3, . . . , until it reaches the value above the sigma. The i is called the **index of summation,** and it starts with the lower limit, 1, and ends with the upper limit, n. For example, if $x_1 = 2$, $x_2 = 3$, $x_3 = -1$, and $x_4 = -2$, then

$$\sum_{i=1}^{4} x_i = x_1 + x_2 + x_3 + x_4 = 2 + 3 + (-1) + (-2) = 2.$$

The area of the triangle under $y = x$ that we discussed above was approximated by

$$A \approx \frac{1}{n^2} + \frac{2}{n^2} + \frac{3}{n^2} + \cdots + \frac{i}{n^2} + \cdots + \frac{n}{n^2}.$$

Using **sigma notation,** we can write this sum as

$$A \approx \sum_{i=1}^{n} \left(\frac{i}{n^2} \right).$$

Sigma notation allows us to represent the sums of the areas of the rectangles in an abbreviated fashion. Some formulas that simplify computations involving sums follow.

Sum Formulas

I. $\displaystyle\sum_{i=1}^{n} 1 = n$ II. $\displaystyle\sum_{i=1}^{n} cx_i = c \sum_{i=1}^{n} x_i$ $(c = \text{constant})$

III. $\displaystyle\sum_{i=1}^{n} (x_i + y_i) = \sum_{i=1}^{n} x_i + \sum_{i=1}^{n} y_i$ IV. $\displaystyle\sum_{i=1}^{n} i = \frac{n(n+1)}{2}$

V. $\displaystyle\sum_{i=1}^{n} i^2 = \frac{n(n+1)(2n+1)}{6}$

We have found that the area of the triangle discussed above was approximated by

$$A \approx \sum_{i=1}^{n} \frac{i}{n^2}.$$

We can use these formulas to simplify this sum as follows.

$$\sum_{i=1}^{n} \frac{i}{n^2} = \frac{1}{n^2} \sum_{i=1}^{n} i \qquad \text{(Formula II)}$$

$$= \frac{1}{n^2} \left[\frac{n(n+1)}{2} \right] \qquad \text{(Formula IV)}$$

$$= \frac{n+1}{2n}$$

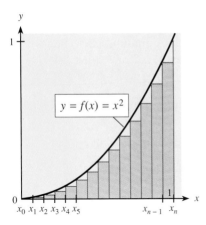

Figure 13.5

Note that this is the same formula we obtained previously using other methods.

The following example shows that we can find the area by evaluating the function at the left-hand endpoints of the subintervals.

EXAMPLE 2

Use rectangles to find the area under $y = x^2$ (and above the x-axis) from $x = 0$ to $x = 1$.

Solution

We again divide the interval $[0, 1]$ into n equal subintervals of length $1/n$. If we evaluate the function at the left-hand endpoints of these subintervals to determine the heights of the rectangles, the sum of the areas of the rectangles will underestimate the area (see Figure 13.5).

Thus we have the following:

Rectangle	Base	Endpoint	Height	Area = Base × Height
1	$\dfrac{1}{n}$	$x_0 = 0$	$f(x_0) = 0$	$\dfrac{1}{n} \cdot 0 = 0$
2	$\dfrac{1}{n}$	$x_1 = \dfrac{1}{n}$	$f(x_1) = \dfrac{1}{n^2}$	$\dfrac{1}{n} \cdot \dfrac{1}{n^2} = \dfrac{1}{n^3}$
3	$\dfrac{1}{n}$	$x_2 = \dfrac{2}{n}$	$f(x_2) = \dfrac{4}{n^2}$	$\dfrac{1}{n} \cdot \dfrac{4}{n^2} = \dfrac{4}{n^3}$
4	$\dfrac{1}{n}$	$x_3 = \dfrac{3}{n}$	$f(x_3) = \dfrac{9}{n^2}$	$\dfrac{1}{n} \cdot \dfrac{9}{n^2} = \dfrac{9}{n^3}$
⋮				
i	$\dfrac{1}{n}$	$x_{i-1} = \dfrac{i-1}{n}$	$\dfrac{(i-1)^2}{n}$	$\dfrac{(i-1)^2}{n^3}$
⋮				
n	$\dfrac{1}{n}$	$x_{n-1} = \dfrac{n-1}{n}$	$\dfrac{(n-1)^2}{n^2}$	$\dfrac{(n-1)^2}{n^3}$

Note that $(i-1)^2/n^3 = (i^2 - 2i + 1)/n^3$ gives the area of the ith rectangle for *any* value of i. The sum of these areas may be written as

$$S = \sum_{i=1}^{n} \frac{i^2 - 2i + 1}{n^3} = \frac{1}{n^3}\left(\sum_{i=1}^{n} i^2 - 2\sum_{i=1}^{n} i + \sum_{i=1}^{n} 1 \right)$$

(Formulas II and III)

$$= \frac{1}{n^3}\left[\frac{n(n+1)(2n+1)}{6} - \frac{2n(n+1)}{2} + n\right]$$

(Formulas V, IV, and I)

$$= \frac{2n^3 + 3n^2 + n}{6n^3} - \frac{n^2 + n}{n^3} + \frac{n}{n^3} = \frac{2n^2 - 3n + 1}{6n^2}.$$

If n is large, there will be a large number of smaller rectangles and the approximation of the area under the curve will be good; the larger n, the better the approximation. For example, if $n = 10$, the area approximation is

$$S(10) = \frac{200 - 30 + 1}{600} = 0.285,$$

whereas if $n = 100$,

$$S(100) = \frac{20{,}000 - 300 + 1}{60{,}000} = 0.328.$$

If we let n increase without bound, we find the exact area.

$$A = \lim_{n\to\infty}\left(\frac{2n^2 - 3n + 1}{6n^2}\right) = \lim_{n\to\infty}\left(\frac{2 - \dfrac{3}{n} + \dfrac{1}{n^2}}{6}\right) = \frac{1}{3}.$$

Note that the approximations with $n = 10$ and $n = 100$ were less than $\frac{1}{3}$. This is because all the rectangles were *under* the curve (see Figure 13.5).

Thus we see that we can determine the area under a curve $y = f(x)$ from $x = a$ to $x = b$ by dividing the interval $[a, b]$ into n equal subintervals of width $(b - a)/n$ and evaluating

$$A = \lim_{n\to\infty} S_R = \lim_{n\to\infty} \sum_{i=1}^{n} f(x_i)\left(\frac{b-a}{n}\right) \qquad \text{(using right-hand endpoints)}$$

or

$$A = \lim_{n\to\infty} S_L = \lim_{n\to\infty} \sum_{i=1}^{n} f(x_{i-1})\left(\frac{b-a}{n}\right) \qquad \text{(using left-hand endpoints)}.$$

CHECKPOINT

1. For the interval $[0, 2]$, determine whether the following are true or false.

   (a) For 4 subintervals, each subinterval has width $\dfrac{1}{2}$.

   (b) For 200 subintervals, each subinterval has width $\dfrac{1}{100}$.

(c) For n subintervals, each subinterval has width $\dfrac{2}{n}$.

(d) For n subintervals, $x_0 = 0$, $x_1 = \dfrac{2}{n}$, $x_2 = 2\left(\dfrac{2}{n}\right)$, . . . , $x_i = i\left(\dfrac{2}{n}\right)$, . . . and $x_n = 2$.

2. If $\dfrac{b - a}{n} = \dfrac{2}{n}$, $x_i = \dfrac{2i}{n}$, and $f(x) = 3x - x^2$, find:

(a) $f(x_i)$

(b) $f(x_i)\dfrac{b - a}{n}$

(c) $\displaystyle\sum_{i=1}^{n} f(x_i)\dfrac{b - a}{n}$, and simplify

(d) $\displaystyle\lim_{n\to\infty}\sum_{i=1}^{n} f(x_i)\dfrac{b - a}{n}$

Figure 13.6

(a)

(b)

EXAMPLE 3

To approximate the area under the graph of $y = \sqrt{x}$ on the interval $[0, 4]$, do the following.

(a) Use a calculator or computer to find S_L and S_R for $n = 8$ on the interval $[0, 4]$.

(b) Predict the area under the curve by evaluating S_L and S_R for larger values of n.

Solution

Figure 13.6 (a) shows the graph of $y = \sqrt{x}$ with 8 rectangles whose heights are determined by evaluating the function at the left-hand endpoint of each interval (the first of these rectangles has height 0). Figure 13.6(b) shows the same graph with 8 rectangles whose heights are determined at the right-hand endpoints.

(a) The formula for the sum of the area of n rectangles over the interval $[0, 4]$, using left-hand endpoints, is

$$S_L = \sum_{i=1}^{n}\sqrt{\frac{4(i - 1)}{n}}\frac{4}{n} = \frac{8}{n^{3/2}}\sum_{i=1}^{n}\sqrt{i - 1}$$

and the formula for the sum of the area of n rectangles over the interval $[0, 4]$, using right-hand endpoints, is

$$S_R = \sum_{i=1}^{n}\sqrt{\frac{4(i)}{n}}\frac{4}{n} = \frac{8}{n^{3/2}}\sum_{i=1}^{n}\sqrt{i}.$$

We have no formula to write either of these summations in a simpler form, but we can use a graphing calculator, computer program, or spreadsheet to compute the sum for any given n. For $n = 8$, the sums are

$$S_L = \frac{1}{\sqrt{8}}\sum_{i=1}^{8}\sqrt{i - 1} \quad \text{and} \quad S_R = \frac{1}{\sqrt{8}}\sum_{i=1}^{8}\sqrt{i}.$$

The following spreadsheet output shows the sum for $n = 8$ rectangles.

1	Area for	$n = 8$		
2		i	S_L	S_R
3		1	0	.35355
4		2	.35355	.5
5		3	.5	.61237
6		4	.61237	.70711
7		5	.70711	.79057
8		6	.79057	.86603
9		7	.86603	.93541
10		8	.93541	1
11	Total		4.76504	5.76504

As line 11 shows, $S_L = 4.76504$ and $S_R = 5.76504$ for $n = 8$.

(b) Larger values of n give better approximations of the area under the curve (try some). For example, $n = 1000$ gives $S_L = 5.3293$ and $S_R = 5.3373$, and since the area under the curve is between these values, the area is approximately 5.33, to two decimal places. By using values of n larger than 1000, we can get better approximations of the area.

CHECKPOINT SOLUTIONS

1. All parts are true.

2. (a) $f(x_i) = f\left(\dfrac{2i}{n}\right) = 3\left(\dfrac{2i}{n}\right) - \left(\dfrac{2i}{n}\right)^2 = \dfrac{6i}{n} - \dfrac{4i^2}{n^2}$

(b) $f(x_i) \dfrac{b - a}{n} = \left(\dfrac{6i}{n} - \dfrac{4i^2}{n^2}\right)\left(\dfrac{2}{n}\right) = \dfrac{12i}{n^2} - \dfrac{8i^2}{n^3}$

(c) $\displaystyle\sum_{i=1}^{n} f(x_i) \dfrac{b - a}{n} = \sum_{i=1}^{n}\left(\dfrac{12i}{n^2} - \dfrac{8i^2}{n^3}\right)$

$\displaystyle = \sum_{i=1}^{n}\dfrac{12i}{n^2} - \sum_{i=1}^{n}\dfrac{8i^2}{n^3}$

$\displaystyle = \dfrac{12}{n^2}\sum_{i=1}^{n}i - \dfrac{8}{n^3}\sum_{i=1}^{n}i^2$

$\displaystyle = \dfrac{12}{n^2}\left[\dfrac{n(n + 1)}{2}\right] - \dfrac{8}{n^3}\left[\dfrac{n(n + 1)(2n + 1)}{6}\right]$

$\displaystyle = \dfrac{6(n + 1)}{n} - \dfrac{4(n + 1)(2n + 1)}{3n^2}$

(d) $\displaystyle\lim_{n\to\infty}\sum_{i=1}^{n} f(x_i)\dfrac{b - a}{n} = \lim_{n\to\infty}\left(\dfrac{6n + 6}{n} - \dfrac{8n^2 + 12n + 4}{3n^2}\right) = 6 - \dfrac{8}{3} = \dfrac{10}{3}$

EXERCISE 13.1

In problems 1–4, approximate the area under each curve over the interval specified by using the indicated number of subintervals (or rectangles) and evaluating the function at the *right-hand* endpoints of the subintervals.

1. $f(x) = 4x - x^2$ from $x = 0$ to $x = 2$; 2 subintervals
2. $f(x) = x^3$ from $x = 0$ to $x = 3$; 3 subintervals
3. $f(x) = 9 - x^2$ from $x = -1$ to $x = 3$; 4 subintervals
4. $f(x) = x^2 + x + 1$ from $x = -1$ to $x = 1$; 4 subintervals

In problems 5–8, approximate the area under each curve by evaluating the function at the *left-hand* endpoints of the subintervals.

5. $f(x) = 4x - x^2$ from $x = 0$ to $x = 2$; 2 subintervals
6. $f(x) = x^3$ from $x = 0$ to $x = 3$; 3 subintervals
7. $f(x) = 9 - x^2$ from $x = -1$ to $x = 3$; 4 subintervals
8. $f(x) = x^2 + x + 1$ from $x = -1$ to $x = 1$; 4 subintervals

When the area under $f(x) = x^2 + x$ from $x = 0$ to $x = 2$ is approximated, the formulas for the sum of n rectangles using *left-hand* endpoints and *right-hand* endpoints are as follows:

Left-hand endpoints: $S_L = \dfrac{14}{3} - \dfrac{6}{n} + \dfrac{4}{3n^2}$

Right-hand endpoints: $S_R = \dfrac{14n^2 + 18n + 4}{3n^2}$

Use these formulas to answer problems 9–13.

9. Find $S_L(10)$ and $S_R(10)$.
10. Find $S_L(100)$ and $S_R(100)$.
11. Find $\lim_{n \to \infty} S_L$ and $\lim_{n \to \infty} S_R$.
12. Compare the right-hand and left-hand values by finding $S_R - S_L$ for $n = 10$, for $n = 100$, and as $n \to \infty$. (Use problems 9–11.)
13. Since $f(x) = x^2 + x$ is increasing over the interval from $x = 0$ to $x = 2$, functional values at the right-hand endpoints are maximum values for each subinterval, and functional values at the left-hand endpoints are minimum values for each subinterval. How would the approximate area using $n = 10$ and *any* other point within each subinterval compare with $S_L(10)$ and $S_R(10)$? What would happen to the area result as $n \to \infty$ if any other point in each subinterval were used?

In problems 14–23, find the value of each sum.

14. $\displaystyle\sum_{k=1}^{3} x_k$, if $x_1 = 1, x_2 = 3, x_3 = -1, x_4 = 5$

15. $\displaystyle\sum_{i=1}^{4} x_i$, if $x_1 = 3, x_2 = -1, x_3 = 3, x_4 = -2$

16. $\displaystyle\sum_{i=3}^{5} (i^2 + 1)$ 17. $\displaystyle\sum_{j=2}^{5} (j^2 - 3)$ 18. $\displaystyle\sum_{i=4}^{7} \left(\dfrac{i-3}{i^2}\right)$

19. $\displaystyle\sum_{j=0}^{4} (j^2 - 4j + 1)$ 20. $\displaystyle\sum_{k=1}^{50} 1$ 21. $\displaystyle\sum_{j=1}^{60} 3$

22. $\displaystyle\sum_{k=1}^{50} (6k^2 + 5)$ 23. $\displaystyle\sum_{k=1}^{30} (k^2 + 4k)$

In problems 24 and 25, use the Sum Formulas I–V to express each of the following without the summation symbol.

24. $\displaystyle\sum_{i=1}^{n} \left(1 - \dfrac{i^2}{n^2}\right)\left(\dfrac{2}{n}\right)$ 25. $\displaystyle\sum_{i=1}^{n} \left(1 - \dfrac{2i}{n} + \dfrac{i^2}{n^2}\right)\left(\dfrac{3}{n}\right)$

Use the function $y = x$ from $x = 0$ to $x = 1$ and n equal subintervals with the function evaluated at the *left-hand* endpoint of each subinterval for problems 26–27.

26. What is the area of the
    (a) first rectangle?   (b) second rectangle?
    (c) ith rectangle?

27. (a) Find a formula for the sum of the areas of the n rectangles (call this S), then find
    (b) $S(10)$. (c) $S(100)$. (d) $S(1000)$. (e) $\lim_{n \to \infty} S$.

28. How do your answers to problems 27(a)–(e) compare with the corresponding calculations in the discussion (after Example 1) of the area under $y = x$ using *right-hand* endpoints?

For problems 29(a)–(e), use the function $y = x^2$ from $x = 0$ to $x = 1$ and n equal subintervals with the function evaluated at the *right-hand* endpoints.

29. (a) Find a formula for the sum of the areas of the n rectangles (call this S), then find
    (b) $S(10)$. (c) $S(100)$. (d) $S(1000)$. (e) $\lim_{n \to \infty} S$.

30. How do your answers to problems 29(a)–(e) compare with the corresponding calculations in Example 2?

31. Use rectangles to find the area between $y = x^2 - 6x + 8$ and the x-axis from $x = 0$ to $x = 2$. Divide the interval $[0, 2]$ into n equal subintervals, so each subinterval has length $2/n$.

32. Use rectangles to find the area between $y = 4x - x^2$ and the x-axis from $x = 0$ to $x = 4$. Divide the interval $[0, 4]$ into n equal subintervals, so each subinterval has length $4/n$.

APPLICATIONS

33. **Speed trials** The graph in the figure gives the times that it takes a 1994 Porsche 911 to reach speeds from 0 mph to 100 mph, in increments of 10 mph, with a smooth

curve connecting them. The area under this curve from $t = 0$ seconds to $t = 14$ seconds represents the total amount of distance traveled over the 14-second period. Count the squares under the curve to estimate this distance. This car will travel 1/4 mile in 14 seconds, to a speed of 100.2 mph. Is your estimate close to this result? (Be careful with time units.)

Source: *Motor Trend*, January 1994

34. **Speed trials** The graph in the figure gives the times that it takes a 1995 Mitsubishi Eclipse GSX to reach speeds from 0 mph to 100 mph, in increments of 10 mph, with a smooth curve connecting them. The area under this curve from $t = 0$ seconds to $t = 21.1$ seconds represents the total amount of distance traveled over the 21.1-second period. Count the squares under the curve to estimate this distance. This car will travel 1/4 mile in 15.4 seconds, to a speed of 89.0 mph, so your estimate should be more than 1/4 mile. Is it? (Be careful with time units.)

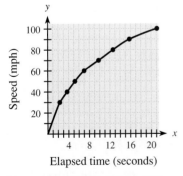

Source: *Road & Track*, June 1994

35. **Pollution monitoring** Suppose the presence of phosphates in certain waste products dumped into a lake promotes the growth of algae. Rampant growth of algae

affects the oxygen supply in the water, so an environmental group wishes to estimate the area of algae growth. They measure the length across the algae growth (see the figure) and obtain the following data (in feet):

x	Length	x	Length
0	0	50	27
10	15	60	24
20	18	70	23
30	18	80	0
40	30		

Use 8 rectangles with bases of 10 feet and lengths measured at the left-hand endpoints to approximate the area of the algae growth.

36. **Drug levels in the blood** The manufacturer of a medicine wants to test how a new 300 mg capsule is released into the bloodstream. After a volunteer is given a capsule, blood samples are drawn every half-hour and the number of milligrams of the drug in the bloodstream is calculated. The results obtained are as follows.

Time t (hr)	$N(t)$ (mg)	Time t (hr)	$N(t)$ (mg)
0	0	2.0	178.3
0.5	247.3	2.5	113.9
1.0	270	3.0	56.2
1.5	236.4	3.5	19.3

Use 7 rectangles, each with height $N(t)$, and with width 0.5 hr, to estimate the area under the graph representing this data. Divide this area by 3.5 hr to estimate the average drug level over this time period.

13.2 THE DEFINITE INTEGRAL; THE FUNDAMENTAL THEOREM OF CALCULUS

Objectives

- ❏ To evaluate definite integrals using the Fundamental Theorem of Calculus
- ❏ To use definite integrals to find the area under a curve

APPLICATION PREVIEW

Suppose that money flows continuously into a slot machine at a casino and grows at a rate given by

$$A' = 100e^{0.1t},$$

where t is the time in hours and $0 \leq t \leq 10$. Then the total amount of money that accumulates over the 10-hour period, if no money is paid out, is given by the **definite integral**

$$\int_0^{10} 100e^{0.1t}\, dt.$$

In Section 13.1, we used the sum of areas of rectangles to approximate the area under curves. In this section, we will see how this sum is related to the definite integral and how to evaluate definite integrals. In addition, we will see how definite integrals will be used to solve several types of applied problems.

In the previous section, we saw that we could determine the area under a curve using equal subintervals and the functional values at either the left-hand endpoints or the right-hand endpoints of the subintervals. In fact, we can use subintervals that are not of equal length, and we can use any point within each subinterval to determine the height of each rectangle. Suppose that we wish to find the area above the x-axis and under the curve $y = f(x)$ over a closed interval $[a, b]$. We can divide the interval into n subintervals (not necessarily equal), with the endpoints of these intervals at $x_0 = a, x_1, x_2 \ldots,$ $x_n = b$. We now choose a point (*any* point) in each subinterval, and denote the points $x_1{}^*, x_2{}^*, \ldots, x_i{}^*, \ldots, x_n{}^*$. Then the ith rectangle (for any i) has height $f(x_i{}^*)$ and width $x_i - x_{i-1}$, so its area is $f(x_i{}^*)(x_i - x_{i-1})$. Then the sum of the areas of the n rectangles is

$$S = \sum_{i=1}^{n} f(x_i{}^*)(x_i - x_{i-1})$$

$$= \sum_{i=1}^{n} f(x_i{}^*)\Delta x_i, \qquad \text{where } \Delta x_i = x_i - x_{i-1}.$$

Because the points in the subinterval may be chosen anywhere in the subinterval, we cannot be sure if the rectangles will underestimate or overestimate the area under the curve. But increasing the number of subintervals (increasing n) and making sure that every interval becomes smaller (just increasing n will not guarantee this if the subintervals are unequal) will in the long run improve the estimation. Thus for any subdivision of $[a, b]$ and any x_i^*, the area is given by

$$A = \lim_{\substack{n \to \infty \\ \max \Delta x_i \to 0}} \sum_{i=1}^{n} f(x_i^*) \, \Delta x_i, \qquad \text{provided that this limit exists.}$$

The preceding discussion takes place in the context of finding the area under a curve. However, if f is any function (not necessarily nonnegative) defined on $[a, b]$, then for each subdivision of $[a, b]$ and each choice of x_i^*, we define the sum S above as the **Riemann sum** of f for the subdivision of $[a, b]$. In addition, the limit of the Riemann sum (as max $\Delta x_i \to 0$) has other important applications, and is called the **definite integral** of $f(x)$ over interval $[a, b]$.

Definite Integral   If f is a function on the interval $[a, b]$, then the *definite integral* of f from a to b is

$$\int_a^b f(x) \, dx = \lim_{\substack{n \to \infty \\ \max \Delta x_i \to 0}} \sum_{i=1}^{n} f(x_i^*) \Delta x_i.$$

If the limit exists, then the definite integral exists and we say that f is integrable on $[a, b]$.

The obvious question is, how is this definite integral related to the indefinite integral (antiderivative) we have been studying? The answer to this question is given by the **Fundamental Theorem of Calculus.**

Fundamental Theorem of Calculus   Let f be a continuous function on the closed interval $[a, b]$; then the definite integral of f exists on this interval, and

$$\int_a^b f(x) \, dx = F(b) - F(a),$$

where F is any function such that $F'(x) = f(x)$ for all x in $[a, b]$.

Stated differently, the theorem says that if the function F is an indefinite integral of a function f that is continuous on the interval $[a, b]$, then

$$\int_a^b f(x) \, dx = F(b) - F(a).$$

We denote $F(b) - F(a)$ by $F(x) \Big|_a^b$.

EXAMPLE 1

Evaluate $\displaystyle\int_2^4 (x^3 + 4) \, dx.$

Solution

$$\int_2^4 (x^3 + 4) \, dx = \frac{x^4}{4} + 4x + C \; \bigg|_2^4$$

$$= \left[\frac{(4)^4}{4} + 4(4) + C \right] - \left[\frac{(2)^4}{4} + 4(2) + C \right]$$

$$= (64 + 16 + C) - (4 + 8 + C)$$

$$= 68 \qquad \text{(Note that the } C\text{'s subtract out)}$$

Note that the Fundamental Theorem states that F can be *any* indefinite integral of f, so we need not add the constant of integration to the integral.

EXAMPLE 2

Evaluate $\displaystyle\int_1^3 (3x^2 + 6x) \, dx$.

Solution

$$\int_1^3 (3x^2 + 6x) \, dx = x^3 + 3x^2 \; \bigg|_1^3$$

$$= (3^3 + 3 \cdot 3^2) - (1^3 + 3 \cdot 1^2)$$

$$= 54 - 4 = 50$$

The properties of definite integrals given below follow from properties of summations.

1. $\displaystyle\int_a^b [f(x) \pm g(x)] \, dx = \int_a^b f(x) \, dx \pm \int_a^b g(x) \, dx$

2. $\displaystyle\int_a^b kf(x) \, dx = k \int_a^b f(x) \, dx, \qquad$ where k is a constant

The following example uses both of these properties.

EXAMPLE 3

Evaluate $\displaystyle\int_3^5 (\sqrt{x^2 - 9} + 2)x \, dx$.

Solution

$$\int_3^5 (\sqrt{x^2 - 9} + 2)x \, dx = \int_3^5 \sqrt{x^2 - 9}(x \, dx) + \int_3^5 2x \, dx$$

$$= \frac{1}{2} \int_3^5 (x^2 - 9)^{1/2}(2x \, dx) + \int_3^5 2x \, dx$$

$$= \frac{1}{2} \left[\frac{2}{3} (x^2 - 9)^{3/2} \right] \bigg|_3^5 + x^2 \bigg|_3^5$$

$$= \frac{1}{3}[(16)^{3/2} - (0)^{3/2}] + (25 - 9)$$

$$= \frac{64}{3} + 16$$

$$= \frac{64}{3} + \frac{48}{3} = \frac{112}{3}$$

In the integral $\int_a^b f(x)\, dx$, we call a the *lower limit* and b the *upper limit* of integration. Although we developed the definte integral with the assumption that the lower limit was less than the upper limit, the following properties permit us to evaluate the definite integral even when that is not the case.

3. $\displaystyle\int_a^a f(x)\, dx = 0$

4. If f is integrable on $[a, b]$, then

$$\int_b^a f(x)\, dx = -\int_a^b f(x)\, dx.$$

The following examples illustrate these properties.

EXAMPLE 4

Evaluate $\displaystyle\int_4^4 x^2\, dx$.

Solution

$$\int_4^4 x^2\, dx = \frac{x^3}{3}\bigg|_4^4 = \frac{4^3}{3} - \frac{4^3}{3} = 0$$

Note that because the limits of integration are the same, it is not necessary to integrate to see that the value of the integral is 0.

EXAMPLE 5

Compare $\displaystyle\int_2^4 3x^2\, dx$ and $\displaystyle\int_4^2 3x^2\, dx$.

Solution

$$\int_2^4 3x^2\, dx = x^3\bigg|_2^4 = 4^3 - 2^3 = 56$$

$$\int_4^2 3x^2\, dx = x^3\bigg|_4^2 = 2^3 - 4^3 = -56$$

Thus

$$\int_4^2 3x^2\, dx = -\int_2^4 3x^2\, dx.$$

Another property of definite integrals is called the additive property.

5. If f is continuous on some interval containing a, b, and c,* then

$$\int_a^b f(x)\ dx = \int_a^c f(x)\ dx + \int_c^b f(x)\ dx.$$

Figure 13.7

(a)

(b)

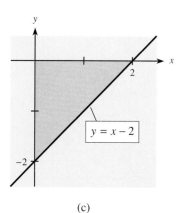

(c)

EXAMPLE 6

Show that $\displaystyle\int_2^3 4x\ dx + \int_3^5 4x\ dx = \int_2^5 4x\ dx.$

Solution

$$\int_2^3 4x\ dx = 2x^2\bigg|_2^3 = 18 - 8 = 10$$

$$\int_3^5 4x\ dx = 2x^2\bigg|_3^5 = 50 - 18 = 32$$

$$\int_2^5 4x\ dx = 2x^2\bigg|_2^5 = 50 - 8 = 42$$

Thus

$$\int_2^3 4x\ dx + \int_3^5 4x\ dx = \int_2^5 4x\ dx.$$

Let us now return to area problems, to see the relationship between the definite integral and the area under a curve. By the formula for the area of a triangle and by summing areas of rectangles, we found the area under the curve (line) $y = x$ from $x = 0$ to $x = 1$ to be $\frac{1}{2}$ (see Figure 13.7(a)). Using the definite integral to find the area gives

$$A = \int_0^1 x\ dx = \frac{x^2}{2}\bigg|_0^1 = \frac{1}{2} - 0 = \frac{1}{2}.$$

In Example 2 of Section 13.1, we used rectangles to find that the area under $y = x^2$ from $x = 0$ to $x = 1$ was $\frac{1}{3}$ (see Figure 13.7(b)). Using the definite integral, we get

$$A = \int_0^1 x^2\ dx = \frac{x^3}{3}\bigg|_0^1 = \frac{1}{3} - 0 = \frac{1}{3},$$

which agrees with the answer obtained in Example 2.

However, not every definite integral represents the area between the curve and the x-axis over an integral. For example,

$$\int_0^2 (x - 2)\ dx = \frac{x^2}{2} - 2x\bigg|_0^2 = (2 - 4) - (0) = -2.$$

*Note that c need not be between a and b.

This would indicate that the area between the curve and the x-axis is negative, but area must be positive. A look at the graph of $y = x - 2$ (see Figure 13.7(c)) shows us what is happening. The region bounded by $y = x - 2$ and x-axis between $x = 0$ and $x = 2$ is a triangle whose base is 2 and height is 2, so its area is $\frac{1}{2}bh = \frac{1}{2}(2)(2) = 2$. The integral has value -2 because $y = x - 2$ lies below the x-axis from $x = 0$ to $x = 2$, and the functional values over the interval $[0, 2]$ are negative. Thus the value of the definite integral over *this* interval does not represent the area between the curve and the x-axis.

In general, the definite integral will give the area under the curve and above the x-axis only when $f(x) \geq 0$ for all x in $[a, b]$.

Area Under a Curve If f is a continuous function on $[a, b]$ and $f(x) \geq 0$ on $[a, b]$, then the exact area between $y = f(x)$ and the x-axis from $x = a$ to $x = b$ is given by

$$\text{Area} \atop \text{(shaded)} = \int_a^b f(x) \, dx.$$

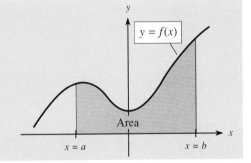

CHECKPOINT

1. True or false:
   (a) For any integral, we can omit the constant of integration (i.e., the $+C$).
   (b) $-\int_{-1}^{3} f(x) \, dx = \int_3^{-1} f(x) \, dx$, if f is integrable on $[-1, 3]$.
   (c) The area between $f(x)$ and the x-axis on the interval $[a, b]$ is given by
   $$\int_a^b f(x) \, dx.$$

2. Evaluate:
   (a) $\displaystyle\int_0^3 (x^2 + 1) \, dx$ (b) $\displaystyle\int_0^3 (x^2 + 1)^4 x \, dx$

If the rate of growth of some function with respect to time t is $f'(t)$, then the total growth of the function during the period from $t = 0$ to $t = k$ can be found by evaluating the definite integral

$$\int_0^k f'(t) \, dt = f(t) \Big|_0^k = f(k) - f(0).$$

For nonnegative rates of growth, this definite integral (and growth) is the same as the area under the graph of $f'(t)$ from $t = 0$ to $t = k$.

We now return to the problem introduced in the Application Preview.

EXAMPLE 7

Suppose that money flows continuously into a slot machine at a casino and grows at a rate given by

$$A' = 100e^{0.1t},$$

where t is time in hours and $0 \leq t \leq 10$. Find the total amount that accumulates in the machine during the 10-hour period, if no money is paid out.

Solution

The total amount is given by

$$A = \int_0^{10} 100e^{0.1t} \, dt = \frac{100}{0.1} \int_0^{10} e^{0.1t} \, (0.1) \, dt$$

$$= 1000e^{0.1t} \Big|_0^{10}$$

$$= 1000e - 1000$$

$$\approx 1718.28 \text{ (dollars)}.$$

In Section 8.5 we stated that the total area under the normal curve is 1 and that the area under the curve from value x_1 to value x_2 represents the probability that a score chosen at random will lie between x_1 and x_2.

The normal distribution is an example of a **continuous distribution** because the values of the random variable are considered over intervals rather than at discrete values. The above statements relating probability and area under the graph apply to other continuous probability distributions determined by **probability density functions.** Thus we can use the definite integral to find the probability that a random variable lies within a given interval. Consider the following example.

EXAMPLE 8

Suppose the probability density function for the life of a computer component is $f(x) = 0.10e^{-0.10x}$, where x is the number of years the component is in use. Find the probability that the component will last between 3 and 5 years.

Solution

The probability that the component will last between 3 and 5 years is the area under the graph of the function between $x = 3$ and $x = 5$. The probability is given by the integral

$$\int_3^5 0.10e^{-0.10x} \, dx = -e^{-0.10x} \Big|_3^5$$

$$= -e^{-0.5} + e^{-0.3}$$

$$\approx -0.6065 + 0.7408$$

$$= 0.1343$$

Graphing Utilities

Some graphics calculators and graphing utilities can be used to get very accurate approximations of definite integrals.

**CHECKPOINT
SOLUTIONS**

1. (a) False. We can omit the constant of integration $(+C)$ only for definite integrals.
   (b) True
   (c) False. Only if $f(x) \geq 0$ on $[a, b]$ is this true.

2. (a) $\displaystyle\int_0^3 (x^2 + 1)\, dx = \frac{x^3}{3} + x \Big|_0^3 = \left(\frac{27}{3} + 3\right) - (0 + 0) = 12$

   (b) $\displaystyle\int_0^3 (x^2 + 1)^4 x\, dx = \frac{1}{2} \int_0^3 (x^2 + 1)^4 (2x\, dx)$

   $\displaystyle = \frac{1}{2} \cdot \frac{(x^2 + 1)^5}{5} \Big|_0^3$

   $\displaystyle = \frac{1}{10}[(3^2 + 1)^5 - (1)^5]$

   $\displaystyle = \frac{1}{10}(10^5 - 1) = 9999.9$

EXERCISE 13.2

Evaluate the following definite integrals.

1. $\displaystyle\int_0^3 4x\, dx$

2. $\displaystyle\int_0^1 8x\, dx$

3. $\displaystyle\int_2^4 dx$

4. $\displaystyle\int_1^5 2\, dy$

5. $\displaystyle\int_2^4 x^3\, dx$

6. $\displaystyle\int_0^5 x^2\, dx$

7. $\displaystyle\int_0^5 4\sqrt[3]{x^2}\, dx$

8. $\displaystyle\int_2^4 3\sqrt{x}\, dx$

9. $\displaystyle\int_2^4 (4x^3 - 6x^2 - 5x)\, dx$

10. $\displaystyle\int_0^2 (x^4 - 5x^3 + 2x)\, dx$

11. $\displaystyle\int_2^3 (x - 4)^2\, dx$

12. $\displaystyle\int_{-1}^3 (x + 2)^3\, dx$

13. $\displaystyle\int_2^4 (x^2 + 2)^3\, x\, dx$

14. $\displaystyle\int_0^3 (2x - x^2)^4 (1 - x)\, dx$

15. $\displaystyle\int_{-1}^2 (x^3 - 3x^2)^3 (x^2 - 2x)\, dx$

16. $\displaystyle\int_0^4 (3x^2 - 2)^4\, x\, dx$

17. $\displaystyle\int_2^3 x\sqrt{x^2 + 3}\, dx$

18. $\displaystyle\int_{-1}^2 x\sqrt[3]{x^2 - 5}\, dx$

19. $\displaystyle\int_1^3 \frac{3}{y^2}\, dy$

20. $\displaystyle\int_1^2 \frac{5}{z^3}\, dz$

21. $\displaystyle\int_0^1 e^{3x}\, dx$

22. $\displaystyle\int_0^2 e^{4x}\, dx$

23. $\int_1^e \dfrac{4}{z} \, dz$

24. $\int_1^e 3y^{-1} \, dy$

25. $\int_4^4 \sqrt{x^2 - 2} \, dx$

26. $\int_2^2 (x^3 - 4x) \, dx$

27. $\int_3^6 \dfrac{x}{3x^2 + 4} \, dx$

28. $\int_0^2 \dfrac{x}{x^2 + 4} \, dx$

29. $\int_1^2 \dfrac{x^2 + 3}{x} \, dx$

30. $\int_1^4 \dfrac{4\sqrt{x} + 5}{\sqrt{x}} \, dx$

31. In the figure, which of the shaded regions A, B, C, or D has the area given by

(a) $\int_a^b f(x) \, dx$?

(b) $-\int_a^b f(x) \, dx$?

A

B

C

D
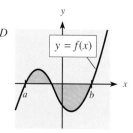

32. For which of the following functions $f(x)$ does

$$\int_0^2 f(x) \, dx$$

give the area between the graph of $f(x)$ and the x-axis from $x = 0$ to $x = 2$?

(a) $f(x) = x^2 + 1$

(b) $f(x) = -x^2$

(c) $f(x) = x - 1$

In problems 33–36,
(a) write the integral that describes the area of the shaded region.
(b) find the area.

33.

34.

35.

36.
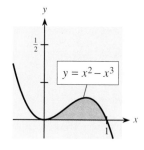

37. Find the area between the curve $y = -x^2 + 3x - 2$ and the x-axis from $x = 1$ to $x = 2$.

38. Find the area between the curve $y = x^2 + 3x + 2$ and the x-axis from $x = -1$ to $x = 3$.

39. Find the area between the curve $y = xe^{x^2}$ and the x-axis from $x = 1$ to $x = 3$.

40. Find the area between the curve $y = e^{-x}$ and the x-axis from $x = -1$ to $x = 1$.

41. How does $\int_{-3}^{-1} x\sqrt{x^2 + 1} \, dx$ compare with $\int_{-3}^{-1} x\sqrt{x^2 + 1} \, dx$?

42. If $\int_{-1}^0 x^3 \, dx = -\frac{1}{4}$ and $\int_0^1 x^3 \, dx = \frac{1}{4}$, what does $\int_{-1}^1 x^3 \, dx$ equal?

43. If $\int_1^2 (2x - x^2) \, dx = \frac{2}{3}$ and $\int_2^4 (2x - x^2) \, dx = -\frac{20}{3}$, what does $\int_1^4 (x^2 - 2x) \, dx$ equal?

44. If $\int_1^2 (2x - x^2) \, dx = \frac{2}{3}$, what does $\int_1^2 (12x - 6x^2) \, dx$ equal?

APPLICATIONS

45. **Total income** The income from an oil change service chain can be considered as flowing continuously at an annual rate given by

$$f(t) = 10{,}000e^{0.02t} \qquad \text{(dollars/year)}.$$

Find the total income for this chain over the first 2 years (from $t = 0$ to $t = 2$).

46. **Total income** Suppose that a vending machine service company models its income by assuming that money flows continuously into the machines, with the annual rate of flow given by

$$f(t) = 120e^{0.01t},$$

in thousands of dollars. Find the total income from the machines over the first 3 years.

Velocity of blood In problems 47 and 48, the velocity of blood through a vessel is given by $v = K(R^2 - r^2)$, where K is the (constant) maximum velocity of the blood, R the (constant) radius of the vessel, and r the distance of the particular corpuscle from the center of the vessel. The rate of flow can be found by measuring the volume of blood that flows past a point in a given time period. This volume, V, is given by

$$V = \int_0^R v(2\pi r \, dr).$$

47. If $R = 0.30$ cm and $v = (0.30 - 3.33r^2)$ cm/sec, find the volume.

48. Develop a general formula for V by evaluating

$$V = \int_0^R v(2\pi r \, dr)$$

using $v = K(R^2 - r^2)$.

Production In problems 49 and 50, the rate of production of a new line of products is given by

$$\frac{dx}{dt} = 200\left[1 + \frac{400}{(t + 40)^2}\right],$$

where x is the number of items produced and t is the number of weeks the products have been in production.

49. How many units were produced in the first 5 weeks?

50. How many units were produced in the 6th week?

51. **Depreciation** If the rate of depreciation of a building is given by $D'(t) = 3000(20 - t)$, $0 \le t \le 20$, what is the total depreciation of the building over the first 10 years ($t = 0$ to $t = 10$)?

52. **Depreciation** What is the total depreciation of the building in problem 51 during the next 10 years ($t = 10$ to $t = 20$)?

53. **Sales and advertising** A store finds that its sales change at a rate given by

$$S'(t) = -3t^2 + 300t,$$

where t is the number of days after an advertising campaign ends and $0 \le t \le 30$. Find
(a) the total sales for the first week after the campaign ends ($t = 0$ to $t = 7$).
(b) the total sales for the second week after the campaign ends ($t = 7$ to $t = 14$).

54. **Health care cost** The health care costs in the United States are given in the table below, with the graph of the function $y = 0.7335x^2 - 3.4997x + 31.9334$ generated from this data. The equation models this data with y in millions of dollars and x in years with $x = 0$ in 1960. Use a definite integral and this model to find the total cost of health care over the period 1960–1989.

Year	Cost ($ Million)
1960	27.1
1965	41.6
1970	74.4
1975	132.9
1980	249.1
1985	420.1
1989	539.9

Source: *World Almanac, 1991*

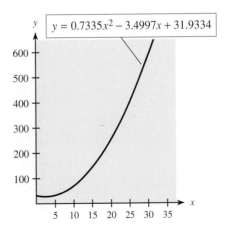

13.3 AREA BETWEEN TWO CURVES

Objectives

❑ To find the area between two curves
❑ To find the average value of a function

APPLICATION PREVIEW

In economics, the **Lorenz curve** is used to represent the inequality of income distribution among different groups in the population of a country. The curve is constructed by plotting the cumulative percentage of families at or below a given income level and the cumulative percentage of total personal income received by these families. For example, the table shows the coordinates of some points on the Lorenz curve $y = L(x)$ that divide the income (for the United States in 1991) into 5 equal income levels (quintiles). The point (0.40, 0.156) is on the Lorenz curve because the families with income in the bottom 40% of the country received 15.6% of the total income in 1991. The graph of the Lorenz curve $y = L(x)$ is shown in Figure 13.8.

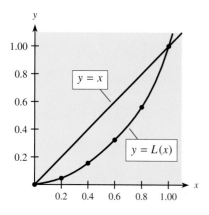

Figure 13.8

U.S. Income Distribution for 1991 (Points on the Lorenz Curve)

x Cumulative Proportion of Families Below Income Level	$y = L(x)$ Cumulative Proportion of Total Income
0	0
0.20	0.047
0.40	0.156
0.60	0.324
0.80	0.561
1	1

SOURCE: *Statistical Abstract of the United States, 1992*

Equality of income would result if each family received an equal proportion of the total income, so that the bottom 20% would receive 20% of the total income, the bottom 40% would receive 40%, and so on. The Lorenz curve representing this would have the equation $y = x$.

The inequality of income distribution is measured by the **Gini coefficient** of income, which measures how far the Lorenz curve falls below $y = x$. It is defined as

$$\frac{\text{Area between } y = x \text{ and } y = L(x)}{\text{Area below } y = x}.$$

Because the area of the triangle below $y = x$ and above the x-axis from $x = 0$ to $x = 1$ is $1/2$, the Gini coefficient of income is

$$\frac{\text{Area between } y = x \text{ and } y = L(x)}{1/2} = 2 \cdot (\text{area between } y = x \text{ and } y = L(x)).$$

In this section we will use the definite integral to find the area between two curves. We will use the area between two curves to find the Gini coefficient of income and to find average cost, average revenue, average profit, and average inventory.

We have used the definite integral to find the area of the region between a curve and the x-axis over an interval where the curve lies above the x-axis. We can easily extend this technique to finding the area between two curves over an interval where one curve lies above the other. (See Figure 13.9.)

Suppose that the graphs of both $y = f(x)$ and $y = g(x)$ lie above the x-axis, and that the graph of $y = f(x)$ lies above $y = g(x)$ throughout the interval from $x = a$ to $x = b$; that is, $f(x) \geq g(x)$ on $[a, b]$.

Then $\displaystyle\int_a^b f(x)\,dx$ gives the area between the graph of $y = f(x)$ and the x-axis (see Figure 13.10(a)), and $\displaystyle\int_a^b g(x)\,dx$ gives the area between the graph of $y = g(x)$ and the x-axis (see Figure 13.10(b)). As Figure 13.10(c) shows,

Figure 13.9

Figure 13.10

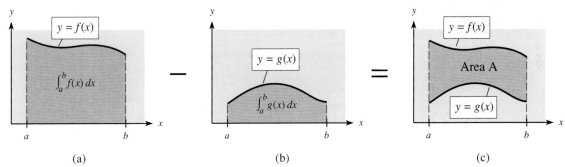

(a)   (b)   (c)

the area of the region between the graphs of $y = f(x)$ and $y = g(x)$ is the difference of these two areas. That is,

$$\text{Area between the curves} = \int_a^b f(x)\,dx - \int_a^b g(x)\,dx.$$

Although Figure 13.10(c) shows the graphs of both $y = f(x)$ and $y = g(x)$ lying above the x-axis, this difference of their integrals will always give the area between their graphs if both functions are continuous and if $f(x) \geq g(x)$ on the interval $[a, b]$. Using the fact that

$$\int_a^b f(x)\,dx - \int_a^b g(x)\,dx = \int_a^b [f(x) - g(x)]\,dx,$$

we have the following result.

| **Area Between Two Curves** | If f and g are continuous functions on $[a, b]$ and if $f(x) \geq g(x)$ on $[a, b]$, then the area of the region bounded by $y = f(x)$, $y = g(x)$, $x = a$, and $x = b$ is $$A = \int_a^b [f(x) - g(x)]\, dx.$$ |

Figure 13.11

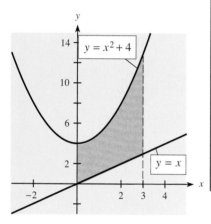

EXAMPLE 1

Find the area of the region bounded by $y = x^2 + 4$, $y = x$, $x = 0$, and $x = 3$.

Solution

We first sketch the graphs of the functions. The graph of the region is shown in Figure 13.11. Since $y = x^2 + 4$ lies above $y = x$ in the interval from $x = 0$ to $x = 3$, the area is

$$A = \int_0^3 [(x^2 + 4) - x]\, dx = \frac{x^3}{3} + 4x - \frac{x^2}{2}\Big|_0^3$$

$$= \left(9 + 12 - \frac{9}{2}\right) - (0 + 0 - 0)$$

$$= 16\tfrac{1}{2} \text{ square units.}$$

We are sometimes asked to find the area enclosed by two curves. In this case, we find the points of intersection of the curves to determine a and b.

EXAMPLE 2

Find the area enclosed by $y = x^2$ and $y = 2x + 3$.

Solution

We first find a and b by finding the x-coordinates of the points of intersection of the graphs. Setting the y-values equal gives

$$x^2 = 2x + 3$$
$$x^2 - 2x - 3 = 0$$
$$(x - 3)(x + 1) = 0$$
$$x = 3, x = -1.$$

Thus $a = -1$ and $b = 3$.

We next sketch the graphs of these functions on the same set of axes. Because the graphs do not intersect on the interval $(-1, 3)$, we can determine which function is larger on this interval by evaluating $2x + 3$ and x^2 at any value c where $-1 < c < 3$. Figure 13.12 shows the region between the graphs, with $2x + 3 \geq x^2$ from $x = -1$ to $x = 3$. The area of the enclosed region is

Figure 13.12

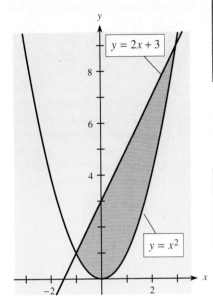

$$A = \int_{-1}^{3} [(2x + 3) - x^2]\, dx = x^2 + 3x - \frac{x^3}{3}\Big|_{-1}^{3}$$

$$= (9 + 9 - 9) - \left(1 - 3 + \frac{1}{3}\right)$$

$$= 10\tfrac{2}{3} \text{ square units.}$$

Some graphs enclose two or more regions because they have more than two points of intersection.

EXAMPLE 3

Find the area of the region enclosed by the graphs of

$$y = f(x) = x^3 - x^2 \text{ and } y = g(x) = 2x.$$

Solution

To find the points of intersection of the graphs, we set the y-values equal and solve for x.

$$x^3 - x^2 = 2x$$
$$x^3 - x^2 - 2x = 0$$
$$x(x - 2)(x + 1) = 0$$
$$x = 0, x = 2, x = -1$$

Graphing these functions between $x = -1$ and $x = 2$, we see that for any x-value in $(-1, 0)$, $f(x) \geq g(x)$, so $f(x) \geq g(x)$ for the region enclosed by the curves from $x = -1$ to $x = 0$. But evaluating the functions for any x-value in $(0, 2)$ shows that $f(x) \leq g(x)$ for the region enclosed by the curves from $x = 0$ to $x = 2$. See Figure 13.13.

Figure 13.13

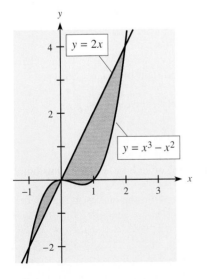

Thus we need one integral to find the area of the region from $x = -1$ to $x = 0$ and a second integral to find the area from $x = 0$ to $x = 2$. The area is found by summing these two integrals.

$$A = \int_{-1}^{0} [(x^3 - x^2) - (2x)]\, dx + \int_{0}^{2} [(2x) - (x^3 - x^2)]\, dx$$

$$= \int_{-1}^{0} (x^3 - x^2 - 2x)\, dx + \int_{0}^{2} (2x - x^3 + x^2)\, dx$$

$$= \left(\frac{x^4}{4} - \frac{x^3}{3} - x^2\right)\Big|_{-1}^{0} + \left(x^2 - \frac{x^4}{4} + \frac{x^3}{3}\right)\Big|_{0}^{2}$$

$$= \left[(0) - \left(\frac{1}{4} - \frac{-1}{3} - 1\right)\right] + \left[\left(4 - \frac{16}{4} + \frac{8}{3}\right) - (0)\right] = \frac{37}{12}$$

Thus the area between the curves is $\frac{37}{12}$ square units.

Graphing Utilities

None of the TI-81, TI-82, or TI-85 calculators will perform symbolic integration, nor does the Casio Fx-7700G. However, the TI-82, TI-85 and Casio Fx-7700G will perform numerical integration to approximate a definite integral. Of course, computer algebra systems will perform both symbolic and numerical integration.

A graphing calculator could be used to find the area enclosed by the graphs of $y = f(x) = x^3 - x^2$ and $y = g(x) = 2x$, found in Example 3. By using SOLVER, INTERSECT, or ZOOM and TRACE, we can find that the curves intersect at $x = -1$, $x = 0$, and $x = 2$. From the graph, shown in Figure 13.12, we see that the curves enclose two regions, with $f(x) > g(x)$ for $-1 < x < 0$, and $g(x) > f(x)$ for $0 < x < 2$. Evaluating the numerical integral of the sum of these functions over these intervals gives the area.

$$\int_{-1}^{0} [(x^3 - x^2) - (2x)]\, dx + \int_{0}^{2} [(2x) - (x^3 - x^2)]\, dx$$

By using a calculator, the integral (and the area) is found to be $0.4166666667 + 2.666666667 = 3.083333333$. This is the same (to 9 decimal places) as the decimal representation of $37/12$, found in Example 3.

CHECKPOINT

1. True or false:
   (a) Over the interval $[a, b]$, the area between the continuous functions $f(x)$ and $g(x)$ is

   $$\int_{a}^{b} [f(x) - g(x)]\, dx.$$

   (b) If $f(x) \geq g(x)$ and the area between $f(x)$ and $g(x)$ is given by

   $$\int_{a}^{b} \left[f(x) - g(x) \right] dx,$$

   then $x = a$ and $x = b$ represent the left and right boundaries, respectively, of the region.
   (c) To find points of intersection of $f(x)$ and $g(x)$, solve $f(x) = g(x)$.

2. Given $f(x) = x^2 + 3x - 9$ and $g(x) = \frac{1}{4}x^2$,
   (a) find the points of intersection of $f(x)$ and $g(x)$.
   (b) determine which function is greater than the other between the points found in part (a).
   (c) set up the integral used to find the area between the curves in the interval between the points found in part (a).
   (d) find the area.

EXAMPLE 4

As we noted in the Application Preview, the inequality of income distribution is measured by the Gini coefficient of income, which is defined as

$$\frac{\text{Area between } y = x \text{ and } y = L(x)}{\text{Area below } y = x} = \frac{\displaystyle\int_0^1 [x - L(x)]\, dx}{1/2}$$

$$= 2\int_0^1 [x - L(x)]\, dx$$

Figure 13.14

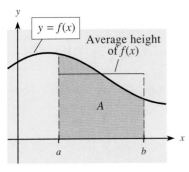

$$A = \int_a^b f(x)\, dx$$

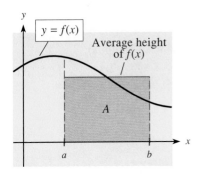

$A = [\text{Average height of } f(x)](b - a)$

Computer software or graphics calculators can be used to model the Lorenz curve for United States income in 1991 from the data in the 1992 *Statistical Abstract of the United States*.

(a) Find the Gini coefficient of income for 1991 if the Lorenz curve is given by $L(x) = 0.897x^{1.864}$.

(b) If the Gini coefficient of income for 1979 is .342, during which year is the distribution of income more equal?

Solution

(a) The Gini coefficient of income for 1991 is

$$2\int_0^1 (x - L(x))\, dx = 2\int_0^1 (x - 0.897x^{1.864})\, dx$$

$$= 2\left[\frac{x^2}{2} - \frac{0.897x^{2.864}}{2.864}\right]\Bigg|_0^1 = 2\left[\frac{1}{2} - 0.313\right] = 0.374.$$

(b) Absolute equality of income would occur if the Gini coefficient of income were 0; and smaller coefficients indicate more equality. Thus the income was more equally distributed in 1979.

If the graph of $y = f(x)$ lies on or above the x-axis from $x = a$ to $x = b$, then the area between the graph and the x-axis is

$$A = \int_a^b f(x)\, dx.$$

The area A is also the area of a rectangle with base equal to $b - a$ and height equal to the *average value* (or average height) of the function $y = f(x)$ (see Figure 13.14).

Thus the average value of the function is

$$\frac{A}{b - a} = \frac{1}{b - a}\int_a^b f(x)\, dx.$$

Even if $f(x) \leq 0$ on all or part of the interval $[a, b]$, we can find the average value by using the integral. Thus we have the following.

Average Value The average value of a continuous function $y = f(x)$ over the interval $[a, b]$ is

$$\text{Average value} = \frac{1}{b - a} \int_a^b f(x) \, dx.$$

EXAMPLE 5

Suppose that the cost function for a product is $C(x) = 400 + x + 0.3x^2$.
(a) What is the average value of $C(x)$ for $x = 10$ to $x = 20$ units?
(b) Find the average cost per unit if 40 units are produced.

Solution

(a) The average value of $C(x)$ is

$$\frac{1}{20 - 10} \int_{10}^{20} (400 + x + 0.3x^2) \, dx = \frac{1}{10} \left(400x + \frac{x^2}{2} + 0.1x^3 \right) \Bigg|_{10}^{20}$$

$$= \frac{1}{10}[(8000 + 200 + 800) - (4000 + 50 + 100)]$$

$$= 485 \text{ (dollars)}.$$

(b) The average cost per unit if 40 units are produced is the average cost function evaluated at $x = 40$.

$$\overline{C}(x) = \frac{C(x)}{x} = \frac{400}{x} + 1 + 0.3x.$$

The average cost per unit if 40 units are produced is

$$\overline{C}(40) = \frac{400}{40} + 1 + 0.3(40) = 23 \text{ (dollars)}.$$

CHECKPOINT

3. Find the average value of $f(x) = x^2 - 4$ over $[-1, 3]$.

EXAMPLE 6

Given the functions $f(x) = x^2 - 4$ and $g(x) = x^3 - 4x$. For each function, do the following.
(a) Graph the function on the interval $[-2, 2]$.
(b) On the graph, "eyeball" the average value (height) of each function on $[-2, 2]$.
(c) Compute the average value of the function over the interval $[-2, 2]$.

Solution

For $f(x) = x^2 - 4$.
(a) The graph of $f(x) = x^2 - 4$ is shown in Figure 13.15(a).

Figure 13.15

(a)

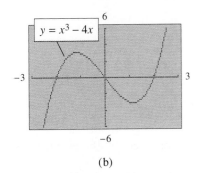

(b)

(b) The average height of $f(x)$ may be near -2.
(c) The average value of $f(x)$ over the integral is given by

$$\frac{1}{2-(-2)}\int_{-2}^{2}(x^2-4)\,dx = \frac{1}{4}\left[\frac{x^3}{3}-4x\right]_{-2}^{2}$$

$$= \left(\frac{8}{12}-2\right)-\left(-\frac{8}{12}+2\right)$$

$$= \frac{4}{3}-4 = -\frac{8}{3} = -2\tfrac{2}{3}.$$

For $g(x) = x^3 - 4x$.
(a) The graph of $g(x) = x^3 - 4x$ is shown in Figure 13.15(b).
(b) The average height of $g(x)$ graph may be approximately 0.
(c) The average value of $g(x)$ is given by

$$\frac{1}{2-(-2)}\int_{-2}^{2}(x^3-4x)\,dx = \frac{1}{4}\left[\frac{x^4}{4}-\frac{4x^2}{2}\right]_{-2}^{2}$$

$$= \left(\frac{16}{16}-\frac{16}{8}\right)-\left(\frac{16}{16}-\frac{16}{8}\right)$$

$$= -1 + 1 = 0.$$

CHECKPOINT SOLUTIONS

1. (a) False. This is true only if $f(x) \geq g(x)$ over $[a, b]$.
   (b) True   (c) True
2. (a) Solve $f(x) = g(x)$, or $x^2 + 3x - 9 = \frac{1}{4}x^2$.

   $$\frac{3}{4}x^2 + 3x - 9 = 0$$

   $$3x^2 + 12x - 36 = 0$$

   $$x^2 + 4x - 12 = 0$$

   $$(x+6)(x-2) = 0$$

   $$x = -6 \mid x = 2$$

   The points of intersection are $(-6, 9)$ and $(2, 1)$.
   (b) Evaluating $g(x)$ and $f(x)$ at any point in the interval $(-6, 2)$ shows $g(x) > f(x)$, so $g(x) \geq f(x)$ on $[-6, 2]$.
   (c) $A = \displaystyle\int_{-6}^{2}\left[\frac{1}{4}x^2 - (x^2 + 3x - 9)\right]dx$
   (d) $A = \left. \dfrac{x^3}{12} - \dfrac{x^3}{3} - \dfrac{3x^2}{2} + 9x \right|_{-6}^{2} = 64$ square units

3. $\dfrac{1}{3-(-1)} \displaystyle\int_{-1}^{3} (x^2 - 4)\, dx = \dfrac{1}{4}\left(\dfrac{x^3}{3} - 4x\right)\Big|_{-1}^{3}$

$$= \dfrac{1}{4}\left[(9 - 12) - \left(-\dfrac{1}{3} + 4\right)\right]$$

$$= \dfrac{1}{4}\left(\dfrac{-20}{3}\right) = -\dfrac{5}{3}$$

EXERCISE 13.3

For each shaded region in problems 1–6, (a) form the integral that represents the area of the shaded region and (b) find the area of the region.

1.

2.

3.

4.

5.

6.

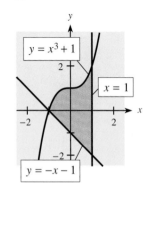

For each shaded region in problems 7–12, (a) find the points of intersection of the curves, (b) form the integral that represents the area of the shaded region, and (c) find the area of the shaded region.

7.

8.

9.

10.

11.

12.

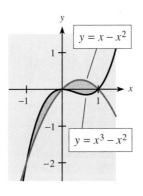

In problems 13–26, equations are given whose graphs enclose a region. In each problem, find the area of the region.

13. $f(x) = x^2 + 2$; $g(x) = -x^2$; $x = 0; x = 2$

14. $f(x) = x^2$; $g(x) = -\frac{1}{10}(10 + x)$; $x = 0; x = 3$

15. $y = x^3 - 1$; $y = x - 1$;   to the right of the y-axis

16. $y = x^2 - 2x + 1$; $y = x^2 - 5x + 4$; $x = 2$

17. $y = \frac{1}{2}x^2$; $y = x^2 - 2x$

18. $y = x^2$; $y = 4x - x^2$

19. $h(x) = x^2$; $k(x) = \sqrt{x}$

20. $g(x) = 1 - x^2$; $h(x) = x^2 + x$

21. $f(x) = x^3$; $g(x) = x^2 + 2x$

22. $f(x) = x^3$; $g(x) = 2x - x^2$

23. $f(x) = \dfrac{3}{x}$; $g(x) = 4 - x$

24. $f(x) = \dfrac{6}{x}$; $g(x) = -x - 5$

25. $y = \sqrt{x + 3}$; $x = -3$; $y = 2$

26. $y = \sqrt{4 - x}$; $x = 4$; $y = 3$

In problems 27–32, find the average value of each function over the given interval.

27. $f(x) = 9 - x^2$ over $[0, 3]$

28. $f(x) = 2x - x^2$ over $[0, 2]$

29. $f(x) = x^2 - x$ over $[-1, 1]$

30. $f(x) = \frac{1}{2}x^3 + 1$ over $[-2, 0]$

31. $f(x) = \sqrt{x} - 2$ over $[1, 4]$

32. $f(x) = \sqrt[3]{x}$ over $[-8, -1]$

33. Use a graphing calculator or computer to find the area between the curves $y = f(x) = x^3 - 4x$ and $y = g(x) = x^2 - 4$.

34. Use a graphing calculator or computer software to find the area between the curves $f(x) = \sqrt[3]{x}$ and $g(x) = x^3 - x$.

APPLICATIONS

35. **Income distribution** Changes in tax laws in the United States during the 1980s were widely thought to help the wealthy at the expense of the poor. The Lorenz curves for the income distribution in 1980 and in 1990 are given below. Find the Gini coefficient of income for both years and determine whether the distribution of income is more or less equal in 1990 than it was in 1970. What was the effect of the tax laws?

1980	$y = 0.916x^{1.821}$
1990	$y = 0.896x^{1.878}$

36. **Income distribution** The Lorenz curves for the income distribution in the United State in 1950 and in 1970 are given below. Find the Gini coefficient of income for both years and compare the distributions of income for these years.

1950	$y = 0.925x^{1.891}$
1970	$y = 0.920x^{1.783}$

37. **Income distribution** In an effort to make the distribution of income more equal, the government of a country passes a tax law that changes the Lorenz curve for one year from $y = 0.998x^{2.6}$ to $y = 0.57x^2 + 0.43x$ for the next year. Find the Gini coefficient of income for both

years and determine whether the distribution of income is more or less equal after the tax law is passed. Interpret the result.

38. **Income distribution** In an effort to make the distribution of income more equal, the government of a country passes a tax law that changes the Lorenz curve for one year from $y = 0.99x^{2.1}$ to $y = 0.32x^2 + 0.68x$ for the next year. Find the Gini coefficient of income for both years and compare the distributions of income before and after the tax law is passed. Interpret the result.

39. **Average profit** For the product whose total cost and total revenue are shown in the figure, represent total revenue by $R(x)$ and total cost by $C(x)$ and write an integral that gives the average profit for the product over the interval from x_0 to x_1.

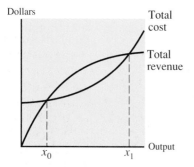

40. **Sales and advertising** The figure shows the sales growth rates under different levels of distribution and advertising from a to b. Set up an integral to determine the extra sales growth if $4 million is used in advertising rather than $2 million.

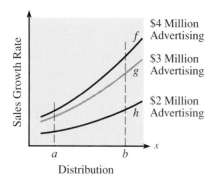

Distribution

41. **Cost** If the cost of producing x units of an item is
$C(x) = x^2 + 400x + 2000,$
(a) use $\overline{C}(x)$ to find the average cost of producing 1000 units.

(b) find the average value of the cost function $C(x)$ over the interval from 0 to 1000.

42. **Inventory management** The figure shows how an inventory of a product is depleted each quarter of a given year. What is the average inventory per month for the first 3 months for this product? (Assume that the graph is a line joining (0, 1300) and (3,100).)

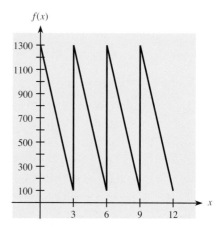

43. **Sales and advertising** The number of daily sales of a product was found to be given by
$$S = 100xe^{-x^2} + 100$$
x days after the start of an advertising campaign for this product. Find the average daily sales during the first 20 days of the campaign, that is, from $x = 0$ to $x = 20$.

44. **Sales and advertising** If no new advertising campaign is begun for the product in problem 43, what is the average number of sales per day for the next 10 days (from $x = 20$ to $x = 30$)?

45. **Demand** If the demand function for a product is given by
$$p = 500 + \frac{1000}{q + 1},$$
where p is the price and q is the number of units demanded, find the average price as demand ranges from 49 to 99 units.

46. **Total income** Suppose that the income from a slot machine in a casino flows continuously at a rate
$$f(t) = 100e^{0.1t},$$

where t is the time in hours since the casino opened. Then the total income during the first 10 hours is given by

$$\int_0^{10} 100e^{0.1t}\, dt.$$

Find the average income over the first 10 hours.

47. **Drug levels in the blood** A drug manufacturer has developed a time-release capsule with the number of milligrams of the drug in the bloodstream given by

$$S = 30x^{18/7} - 240x^{11/7} + 480x^{4/7},$$

where x is in hours and $0 \le x \le 4$. Find the average number of milligrams of the drug in the bloodstream for the first 4 hours after a capsule is taken.

13.4 APPLICATIONS OF DEFINITE INTEGRALS IN BUSINESS AND ECONOMICS

Objectives

❑ To use definite integrals to find total income, present value, and future value of continuous income streams
❑ To use definite integrals to find consumer's surplus
❑ To use definite integrals to find producer's surplus

APPLICATION PREVIEW

The definite integral can be used in a number of applications in business and economics. For example, the definite integral can be used to find the total income over a fixed number of years from a **continuous income stream**. The definite integral can also be used to find the **present value** and the **future value** of a continuous income stream. It can also be used to find the **consumer's surplus** and **producer's surplus** when the demand function and the supply function for a product are known.

CONTINUOUS INCOME STREAMS

An oil company's profits depend upon the amount of oil that can be pumped from a well. Thus we can consider a pump at an oil field as producing a **continuous stream of income** for the owner. Since both the pump and the oil field "wear out" with time, the continuous stream of income is a function of time. Suppose $f(t)$ is the (annual) *rate* of flow of income from this pump; then we can find the total income from the rate of income by using integration. In particular, the total income for k years is given by

$$\text{Total income} = \int_0^k f(t)\, dt.$$

EXAMPLE 1
A small oil company considers the continuous pumping of oil from a well as a continuous income stream with its annual rate of flow at time t given by

$$f(t) = 600e^{-0.2t}$$

in thousands of dollars. Find an estimate of the total income from this well over the next 10 years.

Solution   Total income $= \displaystyle\int_0^{10} f(t)\, dt = \int_0^{10} 600 e^{-0.2t}\, dt$

$$= \left. \frac{600}{-0.2} e^{-0.2t} \right|_0^{10}$$

$$\approx 2594 \text{ (to the nearest integer)}$$

Thus the total income is approximately \$2,594,000.

In addition to the total income from a continuous income stream, the **present value** of the stream is also important. The present value is the value today of a continuous income stream that will be providing income in the future. The present value is useful in deciding when to replace machinery (such as the oil pump in the example) or what new equipment to select.

To find the present value of a continuous stream of income with rate of flow $f(t)$, we first graph the function $f(t)$ and divide the time interval from 0 to k into n subintervals of width Δt_i, $i = 1$ to n.

The total amount of income is the area under this curve between $t = 0$ and $t = k$. We can approximate the amount of income in each subinterval by finding the area of the rectangle in that subinterval. (See Figure 13.16.) We have shown that the future value S that accrues if \$$P$ is invested for t years at an annual rate r, compounded continuously, is

$$S = Pe^{rt}.$$

Thus the present value of the investment that yields the single payment of \$$S$ after t years is

$$P = \frac{S}{e^{rt}} = Se^{-rt}.$$

The contribution to S in the ith subinterval is $S_i = f(t_i)\,\Delta t_i$, and the present value of this amount is

$$P_i = f(t_i)\,\Delta t_i e^{-rt_i}.$$

Thus the total present value of S can be approximated by

$$\sum_{i=1}^{n} f(t_i)\,\Delta t_i e^{-rt_i}.$$

This approximation improves as $\Delta t_i \to 0$ with the present value given by

$$\lim_{\Delta t_i \to 0} \sum_{i=1}^{n} f(t_i)\,\Delta t_i e^{-rt_i}.$$

This limit gives the present value as a definite integral.

Figure 13.16

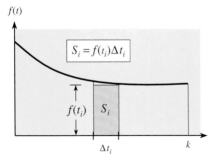

Present Value of a Continuous Income Stream

If $f(t)$ is the rate of continuous income flow earning interest rate r, compounded continuously, then the present value of the continuous income stream is

$$\text{Present value} = \int_0^k f(t)e^{-rt}\,dt,$$

where $t = 0$ to $t = k$ is the time interval.

EXAMPLE 2

Suppose that the oil company in Example 1 is planning to sell the well due to its remote location. Suppose further that the company wants to use the present value of the well over the next 10 years to help establish its selling price. If they determine that the annual rate of flow is

$$f(t) = 600e^{-0.2(t+5)}$$

in thousands of dollars, and if money is worth 10%, compounded continuously, find this present value.

Solution

$$
\begin{aligned}
\text{Present value} &= \int_0^{10} f(t)e^{-rt}\,dt \\[2mm]
&= \int_0^{10} 600e^{-0.2(t+5)}e^{-0.1t}\,dt \\[2mm]
&= \int_0^{10} 600e^{-0.3t-1}\,dt \\[2mm]
&= \frac{600}{-0.3}e^{-0.3t-1}\bigg|_0^{10} \\[2mm]
&= -2000(e^{-4} - e^{-1}) \\[2mm]
&\approx 699 \quad \text{(to the nearest integer)}
\end{aligned}
$$

Thus the present value is $699,000.

Future Value of a Continuous Income Stream

If $f(t)$ is the rate of continuous income flow for k years earning interest rate r, compounded continuously, then the future value of the continuous income stream is

$$FV = e^{rk}\int_0^k f(t)e^{-rt}\,dt.$$

EXAMPLE 3

If the rate of flow of income from an asset is $1000e^{0.02t}$, and if the income is invested at 6%, compounded continuously, find the future value of the asset 4 years from now.

Solution

The future value is given by

$$FV = e^{rk} \int_0^k f(t)e^{-rt}\, dt$$

$$= e^{(0.06)4} \int_0^4 1000e^{0.02t}e^{-0.06t}\, dt = e^{0.24} \int_0^4 1000e^{-0.04t}\, dt$$

$$= e^{0.24}[-25000e^{-0.04t}]_0^4 = -25000e^{0.24}[e^{-0.16}-1]$$

$$\approx 4699.05.$$

CHECKPOINT

1. Suppose that a continuous income stream has an annual rate of flow given by $f(t) = 5000e^{-0.01t}$, and suppose that money is worth 7% compounded continuously. Create the integral used to find the
   (a) total income for the next 5 years.
   (b) present value for the next 5 years.
   (c) future value 5 years from now.

CONSUMER'S SURPLUS

Suppose that the demand for a product is given by $p = f(x)$, while supply of the product is described by $p = g(x)$. The price p_1 where the graphs of these functions intersect is the **equilibrium price** (see Figure 13.17). As the demand curve shows, some consumers (but not all) would be willing to pay more than $\$p_1$ for the product.

For example, some consumers would be willing to buy x_3 units if the price were $\$p_3$. Those consumers willing to pay more than $\$p_1$ are benefiting from the lower price. The total gain for all those consumers willing to pay more than $\$p_1$ is called the **consumer's surplus**, and under proper assumptions the area of the shaded region in Figure 13.17 represents this consumer's surplus.

Looking at Figure 13.18, we see that if the demand curve has equation $p = f(x)$, the consumer's surplus is given by the area between $f(x)$ and the x-axis from 0 to x_1, *minus* the area of the rectangle denoted TR:

$$CS = \int_0^{x_1} f(x)\, dx - p_1 x_1.$$

Note that $p_1 x_1$ is the area of the rectangle that represents the total revenue (see Figure 13.18).

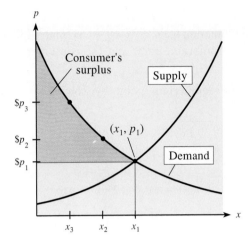

Figure 13.17

Figure 13.18

EXAMPLE 4

The demand function for a product is $p = 100/(x + 1)$. If the equilibrium price is \$20, what is the consumer's surplus?

Solution

We must first find the quantity that will be purchased at this price. Letting $p = 20$, and solving for x, gives

$$20 = \frac{100}{x + 1}$$

$$20(x + 1) = 100$$

$$x + 1 = 5$$

$$x = 4.$$

Thus the equilibrium point is $(4, 20)$. The consumer's surplus is given by the formula

$$CS = \int_0^{x_1} f(x)\, dx - p_1 x_1 = \int_0^4 \frac{100}{x + 1}\, dx - 20 \cdot 4$$

$$= 100 \ln |x + 1| \Big|_0^4 - 80$$

$$= 100(\ln 5 - \ln 1) - 80$$

$$\approx 100(1.6094 - 0) - 80$$

$$= 160.94 - 80$$

$$= 80.94.$$

The consumer's surplus is \$80.94.

EXAMPLE 5

A product's demand function is $p = \sqrt{49 - 6x}$ and supply function is $p = x + 1$. Find the equilibrium point and the consumer's surplus there.

Solution

We can determine the equilibrium point by solving the two equations simultaneously.

$$\sqrt{49 - 6x} = x + 1$$
$$49 - 6x = (x + 1)^2$$
$$0 = x^2 + 8x - 48$$
$$0 = (x + 12)(x - 4)$$
$$x = 4 \quad \text{or} \quad x = -12$$

Thus the equilibrium quantity is 4 and the equilibrium price is \$5 (because $x = -12$ is not a solution). The graphs of the supply and demand functions are shown in Figure 13.19.

The consumer's surplus is given by

$$CS = \int_0^4 f(x)\, dx - p_1 x_1$$

$$= \int_0^4 \sqrt{49 - 6x}\, dx - 5 \cdot 4$$

$$= -\frac{1}{6} \int_0^4 \sqrt{49 - 6x}\, (-6\, dx) - 20$$

$$= -\frac{1}{9} (49 - 6x)^{3/2} \Big|_0^4 - 20$$

$$= -\frac{1}{9} [(25)^{3/2} - (49)^{3/2}] - 20$$

$$= -\frac{1}{9} (125 - 343) - 20$$

$$\approx 24.22 - 20 = 4.22.$$

The consumer's surplus is \$4.22.

Figure 13.19

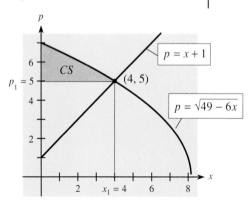

EXAMPLE 6

If a monopoly has a total cost function $C(x) = 60 + 2x^2$ for a product whose demand is given by $p = 30 - x$, find the consumer's surplus at the point where the monopoly has maximum profit.

Solution

We must first find the point where the profit function is maximized. Since the demand for x units is $p = 30 - x$, the total revenue is

$$R(x) = (30 - x)x = 30x - x^2.$$

Thus the profit function is

$$P(x) = R(x) - C(x)$$
$$P(x) = 30x - x^2 - (60 + 2x^2)$$
$$P(x) = 30x - 60 - 3x^2.$$

Then $P'(x) = 30 - 6x$. So $0 = 30 - 6x$ has solution $x = 5$.

Since $P''(5) = -6 < 0$, the profit for the monopolist is maximized when $x = 5$ units are sold at price $p = 25$.

The consumer's surplus at $x = 5$, $p = 25$ is given by

$$CS = \int_0^5 f(x)\, dx - 5 \cdot 25,$$

where $f(x)$ is the demand function.

$$CS = \int_0^5 (30 - x)\, dx - 125$$

$$= 30x - \frac{x^2}{2}\Big|_0^5 - 125$$

$$= \left(150 - \frac{25}{2}\right) - 125$$

$$= \frac{25}{2} = 12.50$$

The consumer's surplus is $12.50.

PRODUCER'S SURPLUS

When a product is sold at the equilibrium price, some producers will also benefit, for they would have sold the product at a lower price. The area between the line $p = p_1$ and the supply curve (from $x = 0$ to $x = x_1$) gives the producer's surplus (see Figure 13.20).

If the supply function is $p = g(x)$, the **producer's surplus** is given by the area between the graph of $p = g(x)$ and the x-axis from 0 to x_1 *subtracted from* the area of the rectangle $0x_1 Ep_1$.

$$PS = p_1 x_1 - \int_0^{x_1} g(x)\, dx$$

Note that $p_1 x_1$ represents the total revenue at the equilibrium point.

Figure 13.20

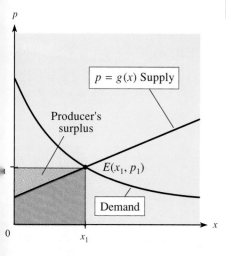

EXAMPLE 7
Suppose that the supply function for a product is $p = x^2 + x$. If the equilibrium price is $20, what is the producer's surplus?

Solution

Since $p = 20$, we can find x as follows:

$$20 = x^2 + x$$
$$0 = x^2 + x - 20$$
$$0 = (x + 5)(x - 4)$$
$$x = -5, x = 4.$$

The equilibrium point is $x = 4$, $p = 20$. The producer's surplus is given by

$$PS = 20 \cdot 4 - \int_0^4 (x^2 + x)\, dx$$

$$= 80 - \left(\frac{x^3}{3} + \frac{x^2}{2} \right) \Big|_0^4$$

$$= 80 - \left(\frac{64}{3} + 8 \right)$$

$$\approx 50.67.$$

The producer's surplus is $50.67. See Figure 13.21.

Figure 13.21

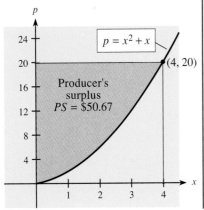

EXAMPLE 8

The demand function for a product is $p = \sqrt{49 - 6x}$ and the supply function is $p = x + 1$. Find the producer's surplus.

Solution

We found the equilibrium point for these functions to be $(4, 5)$ in Example 5 (see Figure 13.19). The producer's surplus is

$$PS = 5 \cdot 4 - \int_0^4 (x + 1)\, dx$$

$$= 20 - \left(\frac{x^2}{2} + x \right) \Big|_0^4$$

$$= 20 - (8 + 4) = 8.$$

The producer's surplus is $8.

CHECKPOINT

2. Suppose that for a certain product the supply function is $p = f(x)$, the demand function is $p = g(x)$, and the equilibrium point is (x_1, p_1). Decide whether the following are true or false.

   (a) $CS = \displaystyle\int_0^{x_1} f(x)\, dx - p_1 x_1$

   (b) $PS = \displaystyle\int_0^{x_1} f(x)\, dx - p_1 x_1$

3. If demand is $p = \dfrac{100}{x + 1}$, supply is $p = x + 1$, and the market equilibrium is (9, 10), create the integral used to find the
(a) consumer's surplus.　(b) producer's surplus.

EXAMPLE 9

Suppose that for a certain product, the demand function is $p = 200e^{-0.01x}$ and the supply function is $p = \sqrt{200x + 49}$.
(a) Use a calculator or computer to find the market equilibrium point.
(b) Find the consumer's surplus.
(c) Find the producer's surplus.

Figure 13.22

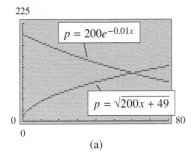

(a)

(b)

Solution

(a) Solving $200e^{-0.01x} = \sqrt{200x + 49}$ requires solving

$$40000e^{-0.02x} = 200x + 49,$$

which is very difficult using algebraic techniques. Using SOLVER or TRACE gives $x = 60$, to the nearest unit, with a price of \$109.76. (See Figure 13.22(a).)

(b) The consumer's surplus is

$$\int_0^{60} 200e^{-0.01x}\, dx - 109.76(60) = \left[-20{,}000e^{-0.01x}\right]_0^{60} - 6585.60$$
$$= -20{,}000e^{-0.6} + 20{,}000 - 6585.60$$
$$= 2438.17.$$

(c) The producer's surplus is

$$60(109.76) - \int_0^{60} \sqrt{200x + 49}\, dx$$
$$= 6585.60 - \frac{1}{200}\left[\frac{(200x + 49)^{3/2}}{3/2}\right]_0^{60}$$
$$= 6585.60 - \frac{1}{300}\left[(12049^{3/2} - 49^{3/2})\right]$$
$$\approx 2178.10.$$

CHECKPOINT SOLUTIONS

1. (a) $\displaystyle\int_0^5 5000e^{-0.01t}\, dt$

(b) $\displaystyle\int_0^5 (5000e^{-0.01t})(e^{-0.07t})\, dt = \int_0^5 5000e^{-0.08t}\, dt$

(c) $\displaystyle e^{(0.07)(5)} \int_0^5 (5000e^{-0.01t})(e^{-0.07t})\, dt = e^{0.35} \int_0^5 5000e^{-0.08t}\, dt$

2. (a) False. Consumer's surplus uses the demand function, so

$$CS = \int_0^{x_1} g(x)\, dx - p_1 x_1.$$

(b) False. Producer's surplus uses the supply function, but the formula is

$$PS = p_1 x_1 - \int_0^{x_1} f(x)\, dx.$$

3. (a) $CS = \displaystyle\int_0^9 \frac{100}{x+1}\, dx - 90$ (b) $PS = 90 - \displaystyle\int_0^9 (x+1)\, dx$

EXERCISE 13.4

CONTINUOUS INCOME STREAMS

1. Find the total income over the next 10 years from a continuous income stream that has an annual rate of flow at time t given by $f(t) = 12{,}000$ (dollars).
2. Find the total income over the next 8 years from a continuous income stream with an annual rate of flow at time t given by $f(t) = 8500$ (dollars).
3. Suppose that a steel company views the production of its continuous caster as a continuous income stream with a monthly rate of flow at time t given by

$$f(t) = 24{,}000e^{0.03t} \quad \text{(dollars)}.$$

Find the total income from this caster in the first year.
4. Suppose that the Quick-Fix Car Service franchise finds that the income generated by its stores can be modeled by assuming that the income is a continuous stream with a monthly rate of flow at time t given by

$$f(t) = 10{,}000e^{0.02t} \quad \text{(dollars)}.$$

Find the total income from a Quick-Fix store for the first 2 years of operation.
5. A small brewery considers the output of its bottling machine as a continuous income stream with an annual rate of flow at time t given by

$$f(t) = 80e^{-0.1t}$$

in thousands of dollars. Find the income from this stream for the next 10 years.
6. A company that services a number of vending machines considers its income as a continuous stream with an

annual rate of flow at time t given by

$$f(t) = 120e^{-0.4t}$$

in thousands of dollars. Find the income from this stream over the next 5 years.

7. A franchise models the profit from its store as a continuous income stream with a monthly rate of flow at time t given by

$$f(t) = 3000e^{0.004t} \quad \text{(dollars)}.$$

When a new store opens, its manager is judged against the model, with special emphasis on the second half of the first year. Find the total profit for the second 6-month period ($t = 6$ to $t = 12$).
8. The Quick-Fix Car Service franchise has a continuous income stream with a monthly rate of flow modeled by $f(t) = 10{,}000e^{0.02t}$ (dollars). Find the total income for years 2 through 5.
9. A continuous income stream has an annual rate of flow at time t given by

$$f(t) = 12{,}000e^{0.04t} \quad \text{(dollars)}.$$

If money is worth 8%, compounded continuously, find the present value of this stream for the next 8 years.
10. A continuous income stream has an annual rate of flow at time t given by

$$f(t) = 9000e^{0.12t} \quad \text{(dollars)}.$$

Find the present value of this stream for the next 10 years, if money is worth 6%, compounded continuously.

11. The income from an established chain of laundromats is a continuous stream with its annual rate of flow at time t given by $f(t) = 63,000$ (dollars). If money is worth 7%, compounded continuously, find the present value and future value of this chain over the next 5 years.

12. The profit from an insurance agency can be considered as a continuous income stream with an annual rate of flow at time t given by $f(t) = 84,000$ (dollars). Find the present value and future value of this agency over the next 12 years, if money is worth 8%, compounded continuously.

13. Suppose that a printing firm considers the production of its presses as a continuous income stream. If the annual rate of flow at time t is given by

$$f(t) = 97.5e^{-0.2(t+3)}$$

in thousands of dollars, and if money is worth 6%, compounded continuously, find the present value and future value of the presses over the next 10 years.

14. Suppose that a vending machine company is considering selling some of its machines. Suppose further that the income from these particular machines is a continuous stream with an annual rate of flow at time t given by

$$f(t) = 12e^{-0.4(t+3)}$$

in thousands of dollars. Find the present value and future value of the machines over the next 5 years if money is worth 10%, compounded continuously.

15. A 58-year-old couple is considering opening a business of their own. They will purchase either an established Gift and Card Shoppe or open a new Video Rental Palace. The Gift Shoppe has a continuous income stream with an annual rate of flow at time t given by

$$G(t) = 30,000 \quad \text{(dollars)},$$

and the Video Palace has a continuous income stream with a projected annual rate of flow at time t given by

$$V(t) = 21,600e^{0.08t} \quad \text{(dollars)}.$$

The initial investment for each business is the same, and money is worth 10%, compounded continuously. Find the present value of each business over the next 7 years (until the couple reaches age 65) to see which is the better buy.

16. If the couple in problem 15 plans to keep the business until age 70 (for the next 12 years), find each present value to see which business is the better buy in this case.

CONSUMER'S SURPLUS

17. The demand function for a product is $p = 34 - x^2$. If the equilibrium price is $9, what is the consumer's surplus?

18. The demand function for a product is $p = 100 - 4x$. If the equilibrium price is $40, what is the consumer's surplus?

19. The demand function for a product is $p = 200/(x + 2)$. If the equilibrium quantity is 8 units, what is the consumer's surplus?

20. The demand function for a certain product is $p = 100/(1 + 2x)$. If the equilibrium quantity is 12 units, what is the consumer's surplus?

21. The demand function for a certain product is $p = 81 - x^2$ and the supply function is $p = x^2 + 4x + 11$. Find the equilibrium point and the consumer's surplus there.

22. The demand function for a product is $p = 49 - x^2$ and the supply function is $p = 4x + 4$. Find the equilibrium point and the consumer's surplus there.

23. If the demand function for a product is $p = 12/(x + 1)$ and the supply function for it is $p = 1 + 0.2x$, find the consumer's surplus under pure competition.

24. If the demand function for a good is $p = 110 - x^2$ and the supply function for it is $p = 2 - \frac{6}{5}x + \frac{1}{5}x^2$, find the consumer's surplus under pure competition.

25. A monopoly has a total cost function $C = 1000 + 120x + 6x^2$ for its product, which has demand function $p = 360 - 3x - 2x^2$. Find the consumer's surplus at the point where the monopoly has maximum profit.

26. A monopoly has a total cost function $C = 500 + 2x^2 + 10x$ for its product, which has demand function $p = -\frac{1}{3}x^2 - 2x + 30$. Find the consumer's surplus at the point where the monopoly has a maximum profit.

PRODUCER'S SURPLUS

27. Suppose that the supply function for a good is $p = 4x^2 + 2x + 2$. If the equilibrium price is $422, what is the producer's surplus there?

28. Suppose that the supply function for a good is $p = 0.1x^2 + 3x + 20$. If the equilibrium price is $36, what is the producer's surplus there?

29. If the supply function for a commodity is $p = 10e^{x/3}$, what is the producer's surplus when 15 units are sold?

30. If the supply function for a commodity is $p = 40 + 100(x + 1)^2$, what is the producer's surplus at $x = 20$?

31. Find the producer's surplus for a product if its demand function is $p = 81 - x^2$ and its supply function is $p = x^2 + 4x + 11$.

32. Find the producer's surplus for a product if its demand function is $p = 49 - x^2$ and its supply function is $p = 4x + 4$.

33. Find the producer's surplus for a product with demand function $p = 12/(x + 1)$ and supply function $p = 1 + 0.2x$.

34. Find the producer's surplus for a product with demand function $p = 110 - x^2$ and supply function $p = 2 - \frac{6}{5}x + \frac{1}{5}x^2$.

35. The demand function for a certain product is $p = 144 - 2x^2$ and the supply function is $p = x^2 + 33x + 48$. Find the producer's surplus at the equilibrium point.

36. The demand function for a product is $p = 280 - 4x - x^2$ and the supply function for it is $p = 160 + 4x + x^2$. Find the producer's surplus at the equilibrium point.

13.5 USING TABLES OF INTEGRALS

Objective

❏ To use tables of integrals to evaluate certain integrals

APPLICATION PREVIEW

In 1982, the rate of increase of new AIDS cases was predicted to be

$$\frac{dN}{dt} = 2^t(700),$$

where t is the number of years after 1982. The number of AIDS cases was 1012 in 1982 ($t = 0$). Finding the predicted number of cases for 1986 ($t = 4$) requires evaluating the integral

$$\int 2^t(700) \, dt.$$

Evaluating this integral is made easier by the existence of a formula such as those given in Table 13.1. The formulas in this table, and others listed in other resources such as the Chemical Rubber Company's *Standard Mathematical Tables*, extend the number of integrals that can be evaluated. Using the formulas is not quite as easy as it may sound because finding the correct formula and using it properly may present problems. The examples in this section illustrate how some of these formulas are used.

Table 13.1   Integration formulas

1. $\displaystyle\int u^n \, du = \frac{u^{n+1}}{n + 1} + C, \quad \text{for } n \neq -1$

2. $\displaystyle\int \frac{du}{u} = \int u^{-1} \, du = \ln |u| + C$

3. $\displaystyle\int a^u \, du = a^u \log_a e + C = \frac{a^u}{\ln a} + C$

4. $\displaystyle\int e^u \, du = e^u + C$

5. $\displaystyle\int \frac{du}{a^2 - u^2} = \frac{1}{2a} \ln \left| \frac{a + u}{a - u} \right| + C$

6. $\displaystyle\int \sqrt{u^2 + a^2}\, du = \frac{1}{2} \left(u \sqrt{u^2 + a^2} + a^2 \ln |u + \sqrt{u^2 + a^2}| \right) + C$

7. $\displaystyle\int \sqrt{u^2 - a^2}\, du = \frac{1}{2} \left(u \sqrt{u^2 - a^2} - a^2 \ln |u + \sqrt{u^2 - a^2}| \right) + C$

8. $\displaystyle\int \frac{du}{\sqrt{u^2 + a^2}} = \ln |u + \sqrt{u^2 + a^2}| + C$

9. $\displaystyle\int \frac{du}{u\sqrt{a^2 - u^2}} = -\frac{1}{a} \ln \left| \frac{a + \sqrt{a^2 - u^2}}{u} \right| + C$

10. $\displaystyle\int \frac{du}{\sqrt{u^2 - a^2}} = \ln |u + \sqrt{u^2 - a^2}| + C$

11. $\displaystyle\int \frac{du}{u\sqrt{a^2 + u^2}} = -\frac{1}{a} \ln \left| \frac{a + \sqrt{a^2 + u^2}}{u} \right| + C$

12. $\displaystyle\int \frac{u\, du}{au + b} = \frac{u}{a} - \frac{b}{a^2} \ln |au + b| + C$

13. $\displaystyle\int \frac{du}{u(au + b)} = \frac{1}{b} \ln \left| \frac{u}{au + b} \right| + C$

14. $\displaystyle\int \ln u\, du = u(\ln u - 1) + C$

15. $\displaystyle\int \frac{u\, du}{(au + b)^2} = \frac{1}{a^2} \left(\ln |au + b| + \frac{b}{au + b} \right) + C$

16. $\displaystyle\int u\sqrt{au + b}\, du = \frac{2(3au - 2b)(au + b)^{3/2}}{15a^2} + C$

17. $\displaystyle\int u\, dv = uv - \int v\, du$

EXAMPLE I

Evaluate $\displaystyle\int \frac{dx}{\sqrt{x^2 + 4}}$.

Solution
We must find a formula in Table 13.1 that is of the same form as this integral. We see that formula 8 has the desired form, *if* we let $u = x$ and $a = 2$. Thus

$$\int \frac{dx}{\sqrt{x^2 + 4}} = \ln |x + \sqrt{x^2 + 4}| + C.$$

EXAMPLE 2

Evaluate $\displaystyle\int_1^2 \frac{dx}{x^2 + 2x}$.

Solution

There does not appear to be any formula having exactly the same form as our integral. But if we rewrite our integral as

$$\int_1^2 \frac{dx}{x(x+2)},$$

we see that formula 13 will work. Letting $u = x$, $a = 1$, and $b = 2$, we get

$$\int_1^2 \frac{dx}{x(x+2)} = \frac{1}{2} \ln \left| \frac{x}{x+2} \right| \Big|_1^2 = \frac{1}{2} \ln \left| \frac{2}{4} \right| - \frac{1}{2} \ln \left| \frac{1}{3} \right|$$

$$= \frac{1}{2} \left(\ln \frac{1}{2} - \ln \frac{1}{3} \right)$$

$$= \frac{1}{2} \ln \frac{3}{2}$$

$$= \frac{1}{2} \ln 1.5.$$

Although the formulas in Table 13.1 are given in terms of the variable u, they may be used with any variable.

EXAMPLE 3

Evaluate $\displaystyle\int \frac{dq}{9-q^2}$.

Solution

The formula that applies in this case is formula 5, with $a = 3$ and $u = q$. Then

$$\int \frac{dq}{9-q^2} = \frac{1}{2 \cdot 3} \ln \left| \frac{3+q}{3-q} \right| + C = \frac{1}{6} \ln \left| \frac{3+q}{3-q} \right| + C.$$

EXAMPLE 4

Evaluate $\displaystyle\int \ln (2x+1) \, dx$.

Solution

This integral has the form of formula 14, with $u = 2x + 1$. But if $u = 2x + 1$, du must be represented by the differential of $2x + 1$ (that is, $2 \, dx$). Thus

$$\int \ln (2x+1) \, dx = \frac{1}{2} \int \ln (2x+1)(2 \, dx)$$

$$= \frac{1}{2}(2x+1)[\ln (2x+1) - 1] + C.$$

CHECKPOINT

1. Can both $\displaystyle\int \frac{dx}{\sqrt{x^2 - 4}}$ and $\displaystyle-\int \frac{dx}{\sqrt{4 - x^2}}$ be evaluated with formula 10 in Table 13.1?

2. Determine the formula used to evaluate $\displaystyle\int \frac{3x}{4x - 5}\,dx$, and show how the formula would be applied.

3. True or false: To use a formula, the given integral must correspond exactly to the formula, including du.

4. True or false: $\displaystyle\int \frac{dx}{x^2(3x^2 - 7)}$ can be evaluated with formula 13.

5. True or false: $\displaystyle\int \frac{dx}{(6x + 1)^2}$ can be evaluated with either formula 1 or formula 15.

6. True or false: $\int\sqrt{x^2 + 4}\,dx$ can be evaluated with any one of formula 1, formula 6, or formula 16.

EXAMPLE 5

Evaluate $\displaystyle\int_1^2 \frac{dx}{x\sqrt{81 - 9x^2}}$.

Solution

This integral is similar to that of formula 9 in Table 13.1. Letting $a = 9$ and $u = 3x$, and multiplying the numerator and denominator by 3, gives the proper form.

$$\int_1^2 \frac{dx}{x\sqrt{81 - 9x^2}} = \int_1^2 \frac{3\,dx}{3x\sqrt{81 - 9x^2}}$$

$$= -\frac{1}{9}\ln\left|\frac{9 + \sqrt{81 - 9x^2}}{3x}\right|\,\Bigg|_1^2$$

$$= \left[-\frac{1}{9}\ln\left(\frac{9 + \sqrt{45}}{6}\right)\right] - \left[-\frac{1}{9}\ln\left(\frac{9 + \sqrt{72}}{3}\right)\right]$$

$$\approx 0.0889$$

It should be pointed out again that the formulas given in Table 13.1 represent only a very small sample of all possible integration formulas. Additional formulas may be found in books of mathematical tables.

EXAMPLE 6

In 1982, the rate of increase of new AIDS cases was predicted to be

$$\frac{dN}{dt} = 2^t(700),$$

where t is the number of years since 1982. If the number of AIDS cases was 1012 in 1982 ($t = 0$), what would be the predicted number of cases for 1986 ($t = 4$), before added attention was given to prevention?

Solution

Finding the predicted number of cases N in 1986 requires evaluating the integral

$$N = \int 2^t(700) \, dt = 700 \int 2^t \, dt$$

and using the fact that $N = 1012$ when $t = 0$ to find the constant C. The integral has the form of formula 3 with $a = 2$ and $u = t$.

$$N = 700 \int 2^t \, dt = 700 \cdot \frac{2^t}{\ln 2} + C$$

Using $N = 1012$ when $t = 0$ gives

$$1012 = 700 \cdot \frac{2^0}{\ln 2} + C$$

$$1012 = 1010 + C$$

$$C = 2,$$

so

$$N = 700 \cdot \frac{2^t}{\ln 2} + 2.$$

The number of cases in 1986 (when $t = 4$) is predicted by this model to be

$$N = 700 \cdot \frac{2^4}{\ln 2} + 2 = 16{,}160.$$

The actual number of cases in 1986 was 13,898. Added attention to prevention has reduced the rate of spread of the disease somewhat, changing the function that models the number of cases. Models using more recent data are discussed in other sections of the text.

Graphing Utilities

Numerical integration with a graphing calculator or computer software is especially useful in evaluating definite integrals when the formulas for the integral are difficult to use or not available. For example, evaluating the definite integral in Example 5 above with a graphing calculator gives 0.08892484. The decimal approximation of the answer in Example 5 is 0.088924836, so the numerical approximation of the answer agrees for the first eight decimal places.

CHECKPOINT
SOLUTIONS

1. No. Although $\int \dfrac{dx}{\sqrt{x^2 - 4}}$ can be evaluated with formula 10 from Table 13.1,

   $-\int \dfrac{dx}{\sqrt{4 - x^2}}$ cannot, because $\sqrt{4 - x^2}$ cannot be rewritten in the form $\sqrt{u^2 - a^2}$ as is needed to use this formula.

2. Use formula 12 with $u = x$, $a = 4$, $b = -5$, and $du = dx$.

$$\int \frac{3x}{4x - 5}\, dx = 3 \int \frac{x\, dx}{4x - 5} = 3 \int \frac{u\, du}{au + b} = 3\left(\frac{u}{a} - \frac{b}{a^2} \ln|au + b| + C\right)$$

$$= \frac{3x}{4} + \frac{15}{16} \ln|4x - 5| + C$$

3. True. An exact correspondence with the formula and du is necessary.

4. False. With $u = x^2$, we must have $du = 2x\, dx$. In this problem there is no x with dx, so the problem cannot correspond to formula 13.

5. False. The integral can be evaluated only with formula 1, not with formula 15. The correspondence is $u = 6x + 1$, $du = 6\, dx$, and $n = -2$.

6. False. The integral can be evaluated only with formula 6, with $u = x$, $du = dx$, and $a = 2$.

EXERCISE 13.5

Evaluate the following integrals.

1. $\displaystyle \int \frac{dx}{16 - x^2}$

2. $\displaystyle \int \frac{dx}{x(3x + 5)}$

3. $\displaystyle \int_1^4 \frac{dx}{x\sqrt{9 + x^2}}$

4. $\displaystyle \int \frac{dx}{x\sqrt{9 - x^2}}$

5. $\displaystyle \int \ln w\, dw$

6. $\displaystyle \int \frac{dv}{v(3v + 8)}$

7. $\displaystyle \int_0^2 \frac{q\, dq}{6q + 9}$

8. $\displaystyle \int_1^5 \frac{dq}{q\sqrt{25 + q^2}}$

9. $\displaystyle \int 3^x\, dx$

10. $\displaystyle \int_0^3 \sqrt{x^2 + 16}\, dx$

11. $\displaystyle \int_5^7 \sqrt{x^2 - 25}\, dx$

12. $\displaystyle \int \frac{x\, dx}{(3x + 2)^2}$

13. $\displaystyle \int w\sqrt{4w + 5}\, dw$

14. $\displaystyle \int \frac{dy}{\sqrt{9 + y^2}}$

15. $\displaystyle \int x\, 5^{x^2}\, dx$

16. $\displaystyle \int \sqrt{9x^2 + 4}\, dx$

17. $\displaystyle \int_0^3 x\sqrt{x^2 + 4}\, dx$

18. $\displaystyle \int x\sqrt{x^4 - 36}\, dx$

19. $\displaystyle \int \frac{5\, dx}{x\sqrt{8 - x^2}}$

20. $\displaystyle \int x\, e^{x^2}\, dx$

21. $\displaystyle \int \frac{dx}{\sqrt{9x^2 - 4}}$

22. $\displaystyle \int \frac{dx}{16 - 4x^2}$

23. $\displaystyle \int_5^6 \frac{dx}{x^2 - 16}$

24. $\displaystyle \int_0^1 \frac{x\, dx}{6 - 5x}$

25. $\displaystyle \int \frac{dx}{\sqrt{(3x + 1)^2 + 1}}$

26. $\displaystyle \int \frac{dx}{9 - (2x + 3)^2}$

27. $\displaystyle \int_0^3 x\sqrt{(x^2 + 1)^2 + 9}\, dx$

28. $\displaystyle \int_1^e x \ln x^2\, dx$

29. $\displaystyle\int \frac{x\,dx}{7 - 3x^2}$

30. $\displaystyle\int_0^1 \frac{e^x}{1 + e^x}\,dx$

31. $\displaystyle\int \frac{dx}{\sqrt{4x^2 + 7}}$

32. $\displaystyle\int e^{2x}\sqrt{3e^x + 1}\,dx$

 Use formulas or numerical integration with a graphing calculator or computer to evaluate the definite integral in problems 33–36.

33. $\displaystyle\int_2^3 \frac{e^{\sqrt{x-1}}}{\sqrt{x - 1}}\,dx$

34. $\displaystyle\int_2^4 \frac{3x}{\sqrt{x^4 - 9}}\,dx$

35. $\displaystyle\int_0^1 \frac{x^3\,dx}{(4x^2 + 5)^2}$

36. $\displaystyle\int_0^1 (e^x + 1)^3\, e^x\,dx$

APPLICATIONS

37. **Producer's surplus** If the supply function for a commodity is $p = 40 + 100 \ln (x + 1)^2$, what is the producer's surplus at $x = 20$?

38. **Consumer's surplus** If the demand function for a good is $p = 5000e^{-x} + 4$, where x is the number of hundreds of bushels of wheat, what is the consumer's surplus at $x = 7$, $p = 9.10$?

39. **Cost** If the marginal cost for a good is $\overline{MC} = \sqrt{x^2 + 9}$ and if the fixed cost is \$300, what is the total cost function of the good?

40. **Total cost** What is the total cost of producing 4 units for the good in problem 39?

41. **Consumer's surplus** Suppose that the demand function for an appliance is

$$p = \frac{400q + 400}{(q + 2)^2}.$$

What is the consumer's surplus if the equilibrium price is \$19 and the equilibrium quantity is 18?

42. **Spread of disease** An isolated community of 1000 people susceptible to a certain disease is exposed when one member returns carrying the disease. If x represents the number infected with the disease at time t (in days), then the rate of change of x is proportional to the product of the number infected, x, and the number still susceptible, $1000 - x$. That is,

$$\frac{dx}{dt} = kx(1000 - x), \quad \text{or} \quad \frac{dx}{x(1000 - x)} = k\,dt.$$

(a) If $k = 0.001$, integrate both sides to solve this differential equation.

(b) Find how long it will be before half the population of the community is affected.

(c) Find the rate of new cases, dx/dt, every other day for the first 13 days.

13.6 INTEGRATION BY PARTS

Objective

❑ To evaluate integrals using the method of integration by parts

APPLICATION PREVIEW

If the value of oil produced by a piece of oil extraction equipment is considered a continuous income stream with an annual rate of flow at time t given by

$$f(t) = 300{,}000 - 2500t, \quad 0 \le t \le 10,$$

and if money can be invested at 8% compounded continuously, then the present value of the piece of equipment is

$$\int_0^{10} (300{,}000 - 2500t)e^{-0.08t}\,dt$$

$$= 300{,}000 \int_0^{10} e^{-0.08t}\,dt - 2500 \int_0^{10} te^{-0.08t}\,dt.$$

The first integral can be evaluated with the formula for the integral of $e^u du$. Evaluating the second integral can be done by using **integration by parts,** which is a special technique that uses formula 17 from Table 13.1. This technique involves rewriting an integral in a form that can be evaluated.

Formula 17 in Table 13.1 follows from the Product Rule for derivatives (actually differentials) as follows.

$$\frac{d}{dx}(uv) = u\frac{dv}{dx} + v\frac{du}{dx} \quad \text{so} \quad d(uv) = u\, dv + v\, du$$

Rearranging the differential form and then integrating both sides gives the following.

$$u\, dv = d(uv) - v\, du$$
$$\int u\, dv = \int d(uv) - \int v\, du$$
$$\int u\, dv = uv - \int v\, du.$$

Integration by Parts

$$\int u\, dv = uv - \int v\, du.$$

Integration by parts is very useful if the integral we seek to evaluate can be treated as the product of one function, u, and the differential dv of a second function, so that the two integrals $\int dv$ and $\int v \cdot du$ can be found. Let us consider an example using this method.

EXAMPLE 1

Evaluate $\int xe^x\, dx$.

Solution

We cannot evaluate this integral using methods we have learned. But we can "split" the integrand into two parts, setting one part equal to u and the other part equal to dv. This "split" must be done in such a way that $\int dv$ and $\int v\, du$ can be evaluated. Letting $u = x$ and $dv = e^x\, dx$ are possible choices. If we make these choices, we have

$$u = x \qquad\qquad dv = e^x\, dx$$

$$du = 1\, dx \qquad v = \int e^x\, dx = e^x$$

Then

$$\int xe^x\, dx = u \cdot v - \int v\, du$$

$$= x \cdot e^x - \int e^x\, dx$$

$$= xe^x - e^x + C.$$

We see that choosing $u = x$ and $dv = e^x\, dx$ worked in evaluating $\int xe^x\, dx$ in Example 1. If we had chosen $u = e^x$ and $dv = x\, dx$, the results would not have been so successful.

How can we select u and dv to make integration by parts work? There are no general rules for separating the integrand into u and dv, but the goal is to select a dv that is integrable and will result in an $\int v\, du$ that is also integrable. There are usually just two reasonable choices, and it may be necessary to try both. Practice will increase your insight and lead to increasingly successful educated guesses. Consider the following examples.

EXAMPLE 2

Evaluate $\int x \ln x\, dx$.

Solution

Let $u = \ln x$ and $dv = x\, dx$. Then

$$du = \frac{1}{x}\, dx \quad \text{and} \quad v = \frac{x^2}{2}.$$

So

$$\int x \ln x\, dx = u \cdot v - \int v\, du$$

$$= (\ln x)\frac{x^2}{2} - \int \frac{x^2}{2} \cdot \frac{1}{x}\, dx$$

$$= \frac{x^2}{2} \ln x - \int \frac{x}{2}\, dx$$

$$= \frac{x^2}{2} \ln x - \frac{x^2}{4} + C.$$

Note that letting $dv = \ln x\, dx$ would lead to great difficulty in evaluating $\int dv$ and $\int v\, du$, so it would not be wise choice.

EXAMPLE 3

Evaluate $\int \ln x^2\, dx$.

Solution

It is frequently good practice to let expressions involving logarithms be part of u in integrating by parts, because the derivatives of logarithmic expressions are usually simple.

In this problem, we can let $u = \ln x^2$. Thus

$$u = \ln x^2 \qquad\qquad dv = dx$$

$$du = \frac{2x}{x^2}\, dx = \frac{2}{x}\, dx \qquad v = x$$

Then,

$$\int \ln x^2\, dx = x \ln x^2 - \int x \cdot \frac{2}{x}\, dx$$

$$= x \ln x^2 - 2x + C.$$

Note that if we write $\ln x^2$ as $2 \ln x$, we can also evaluate this integral using formula 14 in Table 13.1, so integration by parts would not be needed.

CHECKPOINT

1. True or false: In evaluating $\int u\, dv$ by parts,
   (a) the parts u and dv are selected and the parts du and v are calculated.
   (b) the differential (often dx) is always chosen as part of dv.
   (c) the parts du and v are found from u and dv as follows:

   $$du = u'\, dx \qquad \text{and} \qquad v = \int dv.$$

   (d) For $\displaystyle\int \frac{3x}{e^{2x}}\, dx$, we could choose $u = 3x$ and $dv = e^{2x}\, dx$.

2. For $\displaystyle\int \frac{\ln x}{x^4}\, dx$,
   (a) identify u and dv.
   (b) find du and v.
   (c) complete the evaluation of the integral.

Sometimes it is necessary to repeat the integration by parts to complete the evaluation. As before, the goal is to produce a new integral that is simpler.

EXAMPLE 4
Evaluate $\int x^2 e^{2x}\, dx$.

Solution
Let $u = x^2$ and $dv = e^{2x}\, dx$, so $du = 2x\, dx$ and $v = \frac{1}{2}e^{2x}$. Then

$$\int x^2 e^{2x}\, dx = \frac{1}{2}x^2 e^{2x} - \int xe^{2x}\, dx.$$

We cannot evaluate $\int xe^{2x}\, dx$ directly, but this new integral is simpler than the original, and a second integration by parts will be successful. Letting $u = x$

and $dv = e^{2x}\,dx$ gives $du = dx$ and $v = \frac{1}{2}e^{2x}$. So

$$\int x^2 e^{2x}\,dx = \frac{1}{2}x^2 e^{2x} - \left(\frac{1}{2}xe^{2x} - \int \frac{1}{2}e^{2x}\,dx\right)$$

$$= \frac{1}{2}x^2 e^{2x} - \frac{1}{2}xe^{2x} + \frac{1}{4}e^{2x} + C$$

$$= \frac{1}{4}e^{2x}(2x^2 - 2x + 1) + C.$$

The most obvious choices for u and dv are not always the correct ones, as the following example shows. Integration by parts still requires some trial and error.

EXAMPLE 5

Evaluate $\int x^3 \sqrt{x^2 + 1}\,dx$.

Solution

Since x^3 can be integrated easily, it may appear that the new integral would be simplified if we let $u = \sqrt{x^2 + 1}$ and $dv = x^3$. But then du would be $\frac{1}{2}(x^2 + 1)^{-1/2}2x\,dx$, making $\int v\,du$ more complicated than the original integral. However, we can use $\sqrt{x^2 + 1}$ as part of dv, and we can evaluate $\int dv$ if we let $dv = x\sqrt{x^2 + 1}\,dx$. Then

$$u = x^2 \qquad\qquad dv = (x^2 + 1)^{1/2}x\,dx$$

$$du = 2x\,dx \qquad v = \int (x^2 + 1)^{1/2}(x\,dx) = \frac{1}{2}\int (x^2 + 1)^{1/2}(2x\,dx)$$

$$v = \frac{1}{2}\frac{(x^2 + 1)^{3/2}}{3/2} = \frac{1}{3}(x^2 + 1)^{3/2}.$$

Then

$$\int x^3 \sqrt{x^2 + 1}\,dx = \frac{x^2}{3}(x^2 + 1)^{3/2} - \int \frac{1}{3}(x^2 + 1)^{3/2}(2x\,dx)$$

$$= \frac{x^2}{3}(x^2 + 1)^{3/2} - \frac{1}{3}\frac{(x^2 + 1)^{5/2}}{5/2} + C$$

$$= \frac{x^2}{3}(x^2 + 1)^{3/2} - \frac{2}{15}(x^2 + 1)^{5/2} + C$$

$$= \frac{1}{15}(x^2 + 1)^{3/2}[5x^2 - 2(x^2 + 1)] + C$$

$$= \frac{1}{15}(x^2 + 1)^{3/2}(3x^2 - 2) + C.$$

We now consider the problem in the Application Preview.

EXAMPLE 6

Suppose the value of oil produced by a piece of oil extraction equipment is considered a continuous income stream with an annual rate of flow at time t given by

$$f(t) = 300,000 - 2500t, \qquad 0 \le t \le 10,$$

and that money is worth 8%, compounded continuously. Find the present value of the piece of equipment.

Solution

The present value of the piece of equipment is given by

$$\int_0^{10} (300,000 - 2500t)e^{-0.08t}\, dt$$

$$= 300,000 \int_0^{10} e^{-0.08t}\, dt - 2500 \int_0^{10} te^{-0.08t}\, dt$$

$$= \frac{300,000}{-0.08} e^{-0.08t} \Big|_0^{10} - 2500 \int_0^{10} te^{-0.08t}\, dt.$$

The value of the first integral is

$$= \frac{300,000}{-0.08} e^{-0.08t} \Big|_0^{10} = \frac{300,000}{-0.08} e^{-0.8} - \frac{300,000}{-0.08}$$

$$\approx -1,684,983.615 + 3,750,000$$

$$= 2,065,016.385.$$

Evaluating the second of these integrals can be done by using integration by parts, with $u = t$ and $dv = e^{-0.08t}\, dt$. Then $du = 1\, dt$ and $v = \dfrac{e^{-0.08t}}{-0.08}$, and this integral is

$$-2500 \int_0^{10} te^{-0.08t}\, dt = -2500 \frac{te^{-0.08t}}{-0.08} \Big|_0^{10} + 2500 \int_0^{10} \frac{e^{-0.08t}}{-0.08}\, dt$$

$$= \frac{2500}{0.08} te^{-0.08t} \Big|_0^{10} + \frac{2500}{0.0064} e^{-0.08t} \Big|_0^{10}$$

$$= \frac{2500}{0.08} 10e^{-0.8} + \frac{2500}{0.0064} e^{-0.8} - \frac{2500}{0.0064}$$

$$\approx -74,690.572$$

Thus the sum of the integrals is

$$2,065,016.385 + (-74,690.572) = 1,990,325.823$$

and present value of this piece of equipment is $1,990,325.82.

One further note about integration by parts. It can be very useful on certain types of problems, but not on all types. Don't attempt to use integration by parts when easier methods are available.

Graphing Utilities

Using numerical integration with a graphing calculator or computer software to evaluate the integral in Example 6 gives the present value of $1,990,325.80, so this answer is the same as that found in Example 6, to the nearest dollar.

CHECKPOINT SOLUTIONS

1. (a) True   (b) True   (c) True
   (d) False. The product of u and dv must equal the original integrand. Rewrite as

$$\int \frac{3x}{e^{2x}}\, dx = \int 3xe^{-2x}\, dx,$$

then choose $u = 3x$ and $dv = e^{-2x}\, dx$.

2. (a) $u = \ln x$ and $dv = x^{-4}\, dx$

   (b) $du = \dfrac{1}{x}\, dx$ and $v = \displaystyle\int x^{-4}\, dx = \dfrac{x^{-3}}{-3}$

   (c) $\displaystyle\int \frac{\ln x}{x^4}\, dx = uv - \int v\, du$

$$= (\ln x)\left(\frac{x^{-3}}{-3}\right) - \int \frac{x^{-3}}{-3}\cdot\frac{1}{x}\, dx$$

$$= \frac{-\ln x}{3x^3} + \frac{1}{3}\int x^{-4}\, dx$$

$$= -\frac{\ln x}{3x^3} + \frac{1}{3}\left(\frac{x^{-3}}{-3}\right) + C$$

$$= -\frac{\ln x}{3x^3} - \frac{1}{9x^3} + C$$

EXERCISE 13.6

In problems 1–16 use integration by parts to evaluate the integral.

1. $\displaystyle\int xe^{2x}\, dx$

2. $\displaystyle\int xe^{-x}\, dx$

7. $\displaystyle\int \frac{\ln x}{x^2}\, dx$

8. $\displaystyle\int \frac{\ln (x - 1)}{\sqrt{x - 1}}\, dx$

3. $\displaystyle\int x^2 \ln x\, dx$

4. $\displaystyle\int x^3 \ln x\, dx$

9. $\displaystyle\int_1^e \ln x\, dx$

10. $\displaystyle\int \frac{x}{\sqrt{x - 3}}\, dx$

5. $\displaystyle\int_4^6 q\sqrt{q - 4}\, dq$

6. $\displaystyle\int_0^1 y(1 - y)^{3/2}\, dy$

11. $\displaystyle\int x \ln (2x - 3)\, dx$

12. $\displaystyle\int x \ln (4x)\, dx$

13. $\int q^3\sqrt{q^2-3}\,dq$ 14. $\int \dfrac{x^3}{\sqrt{9-x^2}}\,dx$

15. $\int_0^4 x^3\sqrt{x^2+9}\,dx$ 16. $\int \sqrt{x}\,\ln x\,dx$

In problems 17–24, use integration by parts to evaluate the integral. Note that evaluation will require integration by parts more than once.

17. $\int x^2 e^{-x}\,dx$ 18. $\int_0^1 x^2 e^x\,dx$

19. $\int_0^2 x^3 e^{x^2}\,dx$ 20. $\int x^3 e^x\,dx$

21. $\int x^3 \ln^2 x\,dx$ 22. $\int \dfrac{x^2}{\sqrt{x-3}}\,dx$

23. $\int e^{2x}\sqrt{e^x+1}\,dx$ 24. $\int_1^2 (\ln x)^2\,dx$

In problems 25–30, match each of the integrals with the formula or method I–IV that should be used to evaluate it. Then evaluate the integral.

I. Integration by parts   II. $\int e^u\,du$ III. $\int \dfrac{du}{u}$ IV. $\int u^n\,du$

25. $\int xe^{x^2}\,dx$ 26. $\int \dfrac{x}{\sqrt{9-x^2}}\,dx$

27. $\int e^x\sqrt{e^x+1}\,dx$ 28. $\int 4x^2\,e^{x^3}\,dx$

29. $\int_0^4 \dfrac{t}{e^t}\,dt$ 30. $\int x^2\sqrt{x-1}\,dx$

APPLICATIONS

31. **Producer's surplus** If the supply function for a commodity is $p = 30 + 50\ln(2x+1)^2$, what is the producer's surplus at $x = 30$?

32. **Cost** If the marginal cost function for a product is $\overline{MC} = 1 + 3\ln(x+1)$, and if the fixed cost is \$100, find the total cost function.

33. **Present value** Suppose that a machine's production can be considered as a continuous income stream with annual rate of flow at time t given by

$$f(t) = 10{,}000 - 500t \quad \text{(dollars)}.$$

If money is worth 10%, compounded continuously, find the present value of the machine over the next 5 years.

34. **Present value** Suppose that the production of a machine used to mine coal is considered as a continuous income stream with annual rate of flow at time t given by

$$f(t) = 280{,}000 - 14{,}000t \quad \text{(dollars)}.$$

If money is worth 7%, compounded continuously, find the present value of this machine over the next 8 years.

13.7 IMPROPER INTEGRALS AND THEIR APPLICATIONS

Objectives

❏ To evaluate improper integrals
❏ To apply improper integrals to continous income streams and to probability density functions

APPLICATION PREVIEW

We have seen in Section 13.4 that the present value of a continuous income stream over a fixed number of years can be found by using a definite integral. When this notion is extended to an infinite time interval, the result is called the **capital value** of the income stream and is given by

$$\text{Capital value} = \int_0^\infty f(t)e^{-rt}\,dt,$$

where $f(t)$ is the annual rate of flow at time t, and r is the annual interest rate, compounded continuously. This is called an **improper integral.**

Figure 13.23

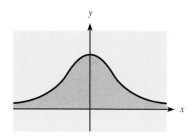

Other applications of calculus to business and to statistics use improper integrals to find the area of a region that extends infinitely to the left or right along the x-axis (see Figure 13.23).

Let us consider how to find the area between the curve $y = 1/x^2$ and x-axis to the right of $x = 1$.

To find the area under this curve from $x = 1$ to $x = b$, where b is any number greater than 1 (see Figure 13.24), we evaluate

$$A = \int_1^b \frac{1}{x^2}\, dx = \frac{-1}{x}\Big|_1^b = \frac{-1}{b} - \left(\frac{-1}{1}\right) = 1 - \frac{1}{b}.$$

Note that the larger b is, the closer the area is to 1. If $b = 100$, $A = 99/100$; if $b = 1000$, $A = 999/1000$; and if $b = 1{,}000{,}000$, $A = 999{,}999/1{,}000{,}000$.

We can represent the area of the region under $1/x^2$ to the right of 1 using the notation

$$\lim_{b \to \infty} \int_1^b \frac{1}{x^2}\, dx = \lim_{b \to \infty} \left(1 - \frac{1}{b}\right),$$

where $\lim\limits_{b \to \infty}$ represents the limit as b gets larger without bound. Clearly

$$\lim_{b \to \infty} \frac{1}{b} = 0,$$

so

$$\lim_{b \to \infty} \left(1 - \frac{1}{b}\right) = 1.$$

Thus the area under the curve $y = 1/x^2$ to the right of $x = 1$ is 1.

In general, we define the area under a curve $y = f(x)$ to the right of $x = a$, with $f(x) \geq 0$, to be

$$\text{Area} = \lim_{b \to \infty} (\text{area from } a \text{ to } b) = \lim_{b \to \infty} \int_a^b f(x)\, dx.$$

This motivates the definition below.

Improper Integral

$$\int_a^\infty f(x)\, dx = \lim_{b \to \infty} \int_a^b f(x)\, dx$$

If the limit defining the improper integral is a finite number, we say that the integral *converges;* if the limit is not a finite number, we say that the integral *diverges.*

Figure 13.24

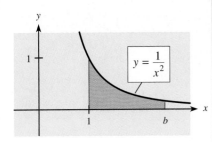

EXAMPLE 1

Evaluate the following improper integrals, if they converge.

(a) $\displaystyle\int_1^\infty \frac{1}{x^3}\,dx$ (b) $\displaystyle\int_1^\infty \frac{1}{x}\,dx$

Solution

(a) $\displaystyle\int_1^\infty \frac{1}{x^3}\,dx = \lim_{b\to\infty}\int_1^b \frac{1}{x^3}\,dx = \lim_{b\to\infty}\left[\frac{x^{-2}}{-2}\right]_1^b$

$$= \lim_{b\to\infty}\left[\frac{-1}{2b^2} - \left(\frac{-1}{2(1)^2}\right)\right]$$

$$= \lim_{b\to\infty}\left(\frac{-1}{2b^2} + \frac{1}{2}\right)$$

Now as $b\to\infty$, $\dfrac{-1}{2b^2}\to 0$, so the limit, and the integral, converge to $\frac{1}{2}$.
That is,

$$\int_1^\infty \frac{1}{x^3}\,dx = \frac{1}{2}.$$

(b) $\displaystyle\int_1^\infty \frac{1}{x}\,dx = \lim_{b\to\infty}\int_1^b \frac{1}{x}\,dx = \lim_{b\to\infty}\Big[\ln|x|\Big]_1^b$

$$= \lim_{b\to\infty}(\ln b - \ln 1)$$

Now $\ln b$ increases without bound as $b\to\infty$, so the limit, and the integral, diverge. We write this as

$$\int_1^\infty \frac{1}{x}\,dx = \infty.$$

Figure 13.25

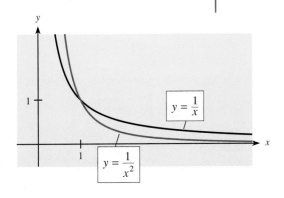

From Example 1 we can conclude that the area under the curve $y = 1/x^3$ to the right of $x = 1$ is $\frac{1}{2}$, while the corresponding area under the curve $y = 1/x$ is infinite. (We have already seen that the corresponding area under $y = 1/x^2$ is 1.)

As Figure 13.25 shows, the graphs of the curves look similar, but the graph of $1/x^2$ gets "close" to the x-axis much more rapidly than the graph of $1/x$. The area under $y = 1/x$ does not converge to a finite number because as $x\to\infty$ the graph of $1/x$ does not approach the x-axis rapidly enough.

EXAMPLE 2

Suppose that an organization wants to establish a trust fund that will provide a continuous income stream with an annual rate of flow at time t given by

$f(t) = 10,000$. If the interest rate remains at 10%, compounded continuously, find the capital value of the fund.

Solution
The capital value of the fund is given by

$$\int_0^\infty f(t) \, e^{-rt} \, dt,$$

where $f(t)$ is the annual rate of flow at time t, and r is the annual interest rate, compounded continuously.

$$\int_0^\infty 10,000e^{-0.10t} \, dt = \lim_{b \to \infty} \int_0^b 10,000e^{-0.10t} \, dt$$

$$= \lim_{b \to \infty} \left[-100,000e^{-0.10t} \right]_0^b$$

$$= \lim_{b \to \infty} \left(\frac{-100,000}{e^{0.10b}} + 100,000 \right)$$

$$= 100,000.$$

Thus the capital value of the fund is $100,000.

Another term for a fund such as the one in Example 2 is a **perpetuity.** Usually the rate of flow of a perpetuity is a constant. If the rate of flow is a constant A, it can be shown that the capital value is given by A/r (see problem 37 in the exercise set).

CHECKPOINT

1. True or false:

(a) $\displaystyle \lim_{b \to +\infty} \frac{1}{b^p} = 0$ if $p > 0$ (b) $\displaystyle \lim_{b \to +\infty} b^p = +\infty$ if $p > 0$

(c) $\displaystyle \lim_{b \to +\infty} e^{-pb} = 0$ if $p > 0$

2. Evaluate the following (if they exist):

(a) $\displaystyle \int_1^\infty \frac{1}{x^{4/3}} \, dx = \lim_{b \to \infty} \int_1^b x^{-4/3} \, dx$ (b) $\displaystyle \int_0^\infty \frac{dx}{\sqrt{x+1}}$

A second improper integral has the form

$$\int_{-\infty}^b f(x) \, dx$$

and is defined by

$$\int_{-\infty}^b f(x) \, dx = \lim_{a \to \infty} \int_{-a}^b f(x) \, dx.$$

The integral converges if the limit is finite. In addition, the improper integral

$$\int_{-\infty}^{\infty} f(x)\, dx$$

is defined by:

$$\int_{-\infty}^{\infty} f(x)\, dx = \lim_{a \to \infty} \int_{-a}^{c} f(x)\, dx + \lim_{b \to \infty} \int_{c}^{b} f(x)\, dx$$

for any finite constant c. (0 is often used for c.) If both limits are finite, the improper integral converges; otherwise, it diverges.

EXAMPLE 3

Evaluate the following integrals.

(a) $\displaystyle\int_{-\infty}^{4} e^{3x}\, dx$ (b) $\displaystyle\int_{-\infty}^{\infty} \frac{x^3}{(x^4 + 3)^2}\, dx$

Solution

(a) $\displaystyle\int_{-\infty}^{4} e^{3x}\, dx = \lim_{a \to \infty} \int_{-a}^{4} e^{3x}\, dx$

$$= \lim_{a \to \infty} \left[\left(\frac{1}{3}\right) e^{3x} \right]_{-a}^{4}$$

$$= \lim_{a \to \infty} \left[\left(\frac{1}{3}\right) e^{12} - \left(\frac{1}{3}\right) e^{-3a} \right]$$

$$= \lim_{a \to \infty} \left[\left(\frac{1}{3}\right) e^{12} - \left(\frac{1}{3}\right)\left(\frac{1}{e^{3a}}\right) \right]$$

$$= \frac{1}{3} e^{12} \qquad \text{(because } 1/e^{3a} \to 0 \text{ as } a \to \infty)$$

(b) $\displaystyle\int_{-\infty}^{\infty} \frac{x^3}{(x^4 + 3)^2}\, dx = \lim_{a \to \infty} \int_{-a}^{0} \frac{x^3}{(x^4 + 3)^2}\, dx + \lim_{b \to \infty} \int_{0}^{b} \frac{x^3}{(x^4 + 3)^2}\, dx$

$$= \lim_{a \to \infty} \left[\frac{1}{4} \frac{(x^4 + 3)^{-1}}{-1} \right]_{-a}^{0} + \lim_{b \to \infty} \left[\frac{1}{4} \frac{(x^4 + 3)^{-1}}{-1} \right]_{0}^{b}$$

$$= \lim_{a \to \infty} \left[-\frac{1}{4}\left(\frac{1}{3} - \frac{1}{a^4 + 3} \right) \right] + \lim_{b \to \infty} \left[-\frac{1}{4}\left(\frac{1}{b^4 + 3} - \frac{1}{3} \right) \right]$$

$$= -\frac{1}{12} + 0 + 0 + \frac{1}{12} = 0$$

$$\left(\text{since } \lim_{a \to \infty} \frac{1}{a^4 + 3} = 0 \quad \text{and} \quad \lim_{b \to \infty} \frac{1}{b^4 + 3} = 0 \right)$$

In Section 13.2, we calculated the probability that a computer component would last between 3 and 5 years when the probability density function for the life span is $f(x) = 0.10e^{-0.10x}$, where x is the number of years, $x \geq 0$. The probability that the component will last more than 3 years is given by the improper integral

$$\int_3^\infty 0.10e^{-0.10x} \, dx,$$

which gives

$$\lim_{b \to \infty} \int_3^b 0.10e^{-0.10x} \, dx = \lim_{b \to \infty} \left[-e^{-0.10x} \right]_3^b$$

$$= \lim_{b \to \infty} \left(-e^{-0.10b} + e^{-0.3} \right)$$

$$= e^{-0.3} = 0.7408.$$

We noted in Chapter 8 that the sum of the probabilities for a probability distribution (i.e., probability density function) equals 1. In particular, we stated that the area under the normal probability curve is 1.

Probability Density Function In general, if $f(x) \geq 0$ for all x, then f is a probability density function for a continuous random variable if and only if

$$\int_{-\infty}^\infty f(x) \, dx = 1.$$

The complete probability density function for the life span of the computer component mentioned above is

$$f(x) = \begin{cases} 0.10e^{-0.10x} & \text{if } x \geq 0 \\ 0 & \text{if } x < 0 \end{cases}.$$

We can verify that

$$\int_{-\infty}^\infty f(x) \, dx = 1$$

for this function.

$$\int_{-\infty}^\infty f(x) \, dx = \int_{-\infty}^0 f(x) \, dx + \int_0^\infty f(x) \, dx = \int_{-\infty}^0 0 \, dx + \int_0^\infty 0.10e^{-0.10x} \, dx$$

$$= 0 + \lim_{b \to \infty} \int_0^b 0.1e^{-0.10x} \, dx$$

$$= \lim_{b \to \infty} \left[-e^{-0.10x} \right]_0^b$$

$$= \lim_{b \to \infty} \left[-e^{-0.10b} + 1 \right]$$

$$= 1$$

In Section 8.2, we found the expected value (mean) of a discrete probability distribution using the formula

$$E(x) = \sum x \Pr(x).$$

For continuous probability distributions, like the normal probability distribution, the expected value, or mean, can be found by evaluating the improper integral

$$\int_{-\infty}^{\infty} x f(x)\, dx.$$

Expected Value (Mean)   If x is a continuous random variable with probability density function f, then the mean of the probability distribution is

$$\mu = \int_{-\infty}^{\infty} x f(x)\, dx.$$

The normal distribution density function, in standard form, is

$$f(x) = \frac{1}{\sqrt{2\pi}} e^{-x^2/2},$$

so the mean of the normal probability distribution is given by

$$\mu = \int_{-\infty}^{\infty} x \left(\frac{1}{\sqrt{2\pi}} e^{-x^2/2} \right) dx$$

$$= \lim_{a \to \infty} \int_{-a}^{0} \frac{1}{\sqrt{2\pi}} x e^{-x^2/2}\, dx + \lim_{b \to \infty} \int_{0}^{b} \frac{1}{\sqrt{2\pi}} x e^{-x^2/2}\, dx$$

$$= \lim_{a \to \infty} \frac{1}{\sqrt{2\pi}} \left[-e^{-x^2/2} \right]_{-a}^{0} + \lim_{b \to \infty} \frac{1}{\sqrt{2\pi}} \left[-e^{-x^2/2} \right]_{0}^{b}$$

$$= \frac{1}{\sqrt{2\pi}} (-1 + 0) + \frac{1}{\sqrt{2\pi}} (0 + 1) = 0.$$

This verifies the statement in Chapter 8 that the mean of the standard normal distribution is 0.

EXAMPLE 4

Find the area of the region below the graph of $f(x) = \dfrac{\ln x}{x^2}$ and above the x-axis. This can be done by evaluating the integral over the interval where $f(x)$ is above the x-axis. It is not possible to find two points where $f(x) = 0$, so we will use graphs to find the interval and to evaluate the integral. To do this,
(a) graph $y = f(x)$ to find the interval over which to integrate.
(b) evaluate the integral of $f(x)$ over this interval.

Solution

(a) The graph of $y = f(x)$ is shown in Figure 13.26(a). Using TRACE and ZOOM shows that $f(1) = 0$ and $f(x)$ approaches 0 as x approaches $+\infty$. Thus the area is found by evaluating

$$\int_1^{+\infty} f(x)\, dx = \int_1^{+\infty} \frac{\ln x}{x^2}\, dx = \lim_{b \to +\infty} \int_1^b \frac{\ln x}{x^2}\, dx$$

(b) Evaluating $\displaystyle\int_1^b \frac{\ln x}{x^2}\, dx$ requires the use of integration by parts, with

$$u = \ln x \quad \text{and} \quad dv = x^{-2}.$$

Thus

$$du = \frac{1}{x}\, dx \quad \text{and} \quad v = \frac{x^{-1}}{-1}$$

$$\int_1^b \frac{\ln x}{x^2}\, dx = (\ln x)\left(\frac{x^{-1}}{-1}\right)\Big|_1^b - \int_1^b (-x^{-1})\frac{1}{x}\, dx = -\frac{\ln x}{x}\Big|_1^b + \frac{x^{-1}}{-1}\Big|_1^b$$

$$= \left(-\frac{\ln b}{b} + \frac{\ln 1}{1}\right) - \left(\frac{1}{b} - 1\right) = 1 - \frac{1}{b} - \frac{\ln b}{b}$$

$$\int_1^\infty f(x)\, dt = \lim_{b \to \infty}\left(1 - \frac{1}{b} - \frac{\ln b}{b}\right) = \lim_{b \to \infty} 1 - \lim_{b \to \infty} \frac{1}{b} - \lim_{b \to \infty} \frac{\ln b}{b}$$

We have not developed a method to evaluate $\displaystyle\lim_{b \to \infty}\left(\frac{\ln b}{b}\right)$, or equivalently, $\displaystyle\lim_{x \to \infty}\left(\frac{\ln x}{x}\right)$. But we can use ZOOM and TRACE to see that the graph of $y = 1 - \frac{1}{x} - \frac{\ln x}{x}$ approaches 1 as x approaches $+\infty$ (see Figure 13.26(b)).

(a)

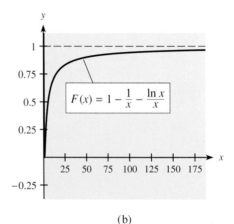

(b)

Figure 13.26

$$\text{Thus } \lim_{b\to\infty}\left(1-\frac{1}{b}-\frac{\ln b}{b}\right)=1, \text{ and the area under } y=f(x) \text{ is}$$

$$\int_1^\infty f(x)\,dx=\lim_{b\to\infty}\left(1-\frac{1}{b}-\frac{\ln b}{b}\right)=1.$$

CHECKPOINT SOLUTIONS

1. (a) True   (b) True   (c) True

2. (a) $\lim\limits_{b\to\infty}\displaystyle\int_1^b x^{-4/3}\,dx=\lim\limits_{b\to\infty}\dfrac{x^{-1/3}}{-1/3}\Big|_1^b=\lim\limits_{b\to\infty}\left(\dfrac{-3}{b^{1/3}}-\dfrac{-3}{1}\right)=0+3=3$

   (b) $\lim\limits_{b\to\infty}\displaystyle\int_0^b (x+1)^{-1/2}\,dx=\lim\limits_{b\to\infty}\dfrac{(x+1)^{1/2}}{1/2}\Big|_0^b=\lim\limits_{b\to\infty}2\sqrt{x+1}\,\Big|_0^b$

   $=\lim\limits_{b\to\infty}(2\sqrt{b+1}-2)=\infty$

   (Integral diverges)

EXERCISE 13.7

In problems 1–20, evaluate the improper integrals that converge.

1. $\displaystyle\int_1^\infty \frac{dx}{x^6}$

2. $\displaystyle\int_1^\infty \frac{1}{x^4}\,dx$

3. $\displaystyle\int_1^\infty \frac{dt}{t^{3/2}}$

4. $\displaystyle\int_5^\infty \frac{dx}{(x-1)^3}$

5. $\displaystyle\int_1^\infty e^{-x}\,dx$

6. $\displaystyle\int_0^\infty x^2 e^{-x^3}\,dx$

7. $\displaystyle\int_1^\infty \frac{dt}{t^{1/3}}$

8. $\displaystyle\int_1^\infty \frac{1}{\sqrt{x}}\,dx$

9. $\displaystyle\int_0^\infty e^{3x}\,dx$

10. $\displaystyle\int_1^\infty xe^{x^2}\,dx$

11. $\displaystyle\int_{-\infty}^{-1} \frac{10}{x^2}\,dx$

12. $\displaystyle\int_{-\infty}^{-2} \frac{x}{\sqrt{x^2-1}}\,dx$

13. $\displaystyle\int_{-\infty}^0 x^2 e^{-x^3}\,dx$

14. $\displaystyle\int_{-\infty}^0 \frac{x}{(x^2+1)^2}\,dx$

15. $\displaystyle\int_{-\infty}^{-1} \frac{6}{x}\,dx$

16. $\displaystyle\int_{-\infty}^{-2} \frac{3x}{x^2+1}\,dx$

17. $\displaystyle\int_{-\infty}^\infty \frac{2x}{x^2+1}\,dx$

18. $\displaystyle\int_{-\infty}^\infty \frac{x}{(x^2+1)^2}\,dx$

19. $\displaystyle\int_{-\infty}^\infty x^3 e^{-x^4}\,dx$

20. $\displaystyle\int_{-\infty}^\infty x^4 e^{-x^5}\,dx$

21. For what value of c does

$$\int_0^\infty \frac{c}{e^{0.5t}}\,dt=1?$$

22. For what value of c does

$$\int_{10}^\infty \frac{c}{x^3}\,dx=1?$$

In problems 23–26, find the area of the region under the graph of $y=f(x)$ and to the right of $x=1$.

23. $f(x)=\dfrac{x}{e^{x^2}}$

24. $f(x)=\dfrac{1}{\sqrt[5]{x^3}}$

25. $f(x)=\dfrac{1}{\sqrt[3]{x^5}}$

26. $f(x)=\dfrac{1}{x\sqrt{x}}$

27. Show that the function

$$f(x)=\begin{cases}\dfrac{200}{x^3} & \text{if } x\geq 10\\[2mm] 0 & \text{otherwise}\end{cases}$$

is a probability density function.

28. Show that

$$f(x) = \begin{cases} 3e^{-3t} & \text{if } t \geq 0 \\ 0 & \text{if } t < 0 \end{cases}$$

is a probability density function.

29. For what value of c is the function

$$f(x) = \begin{cases} c/x^2 & \text{if } x \geq 1 \\ 0 & \text{otherwise} \end{cases}$$

a probability density function?

30. For what value of c is the function

$$f(x) = \begin{cases} c/x^3 & \text{if } x \geq 100 \\ 0 & \text{otherwise} \end{cases}$$

a probability density function?

31. If

$$f(x) = \begin{cases} ce^{-0.5x} & x \geq 0 \\ 0 & x < 0 \end{cases}$$

is a probability density function, what must be the value of c?

32. If

$$f(x) = \begin{cases} ce^{-kx} & \text{if } x \geq 0 \\ 0 & \text{if } x < 0 \end{cases}$$

is a probability density function, what must be the value of c?

33. Find the mean of the probability distribution if the probability density function is

$$f(x) = \begin{cases} \dfrac{200}{x^3} & \text{if } x \geq 10 \\ 0 & \text{otherwise} \end{cases}.$$

34. Find the mean of the probability distribution if the probability density function is

$$f(x) = \begin{cases} c/x^3 & \text{if } x \geq 100 \\ 0 & \text{otherwise} \end{cases}.$$

35. Find the area below the graph of $y = f(x)$ and above the x-axis for $f(x) = 24xe^{-3x}$. Use the graph of $y = f(x)$ to find the interval for which $f(x) \geq 0$ and the graph of the integral of $f(x)$ over this interval to find the area.

36. Find the area below the graph of $y = f(x)$ and above the x-axis for $f(x) = x^2e^{-x}$ and $x \geq 0$. Use the graph of the integral of $f(x)$ over this interval to find the area.

APPLICATIONS

37. **Capital value** Suppose that a continuous income stream has an annual rate of flow at time t given by $f(t) = A$, where A is a constant. If the interest rate is r (as a decimal, $r > 0$), compounded continuously, show that the capital value of the stream is A/r.

38. **Capital value** Suppose that a donor wishes to provide a cash gift to a hospital that will generate a continuous income stream with an annual rate of flow at time t given by $f(t) = \$20,000$. If the annual interest rate is 12%, compounded continuously, find the capital value of this perpetuity.

39. **Capital value** Suppose that a business provides a continuous income stream with an annual rate of flow at time t given by $f(t) = 120e^{0.04t}$ in thousands of dollars. If the interest rate is 9%, compounded continuously, find the capital value of the business.

40. **Capital value** Suppose that the output of the machinery in a factory can be considered as a continuous income stream with annual rate of flow at time t given by $f(t) = 450e^{-0.09t}$ in thousands of dollars. If the annual interest rate is 6%, compounded continuously, find the capital value of the machinery.

41. **Capital value** A business has a continuous income stream with an annual rate of flow at time t given by $f(t) = 56,000e^{0.02t}$ (dollars). If the interest rate is 10%, compounded continuously, find the capital value of the business.

42. **Capital value** Suppose that a business provides a continuous income stream with an annual rate of flow at time t given by $f(t) = 10,800e^{0.06t}$ (dollars). If money is worth 12%, compounded continuously, find the capital value of the business.

43. **Radioactive waste** Suppose that the rate at which a nuclear power plant produces radioactive waste is proportional to the number of years it has been operating, according to $f(t) = 500t$ in pounds per year. Suppose also that the waste decays exponentially at a rate of 3% per year. Then the amount of radioactive waste that will accumulate in b years is given by

$$\int_0^b 500te^{-0.03(b-t)} \, dt.$$

(a) Evaluate this integral.
(b) How much waste will accumulate in the long run? Take the limit as $b \to \infty$ in part (a).

44. **Field intensity** The field intensity around an (infinitely) long straight electrical wire is

$$F = \frac{\mu IM}{10} \int_{-\infty}^{\infty} \frac{dx}{(x^2 + r^2)^{3/2}},$$

where μ, r, I, and M are constants. Evaluate this integral.

45. **Quality control** The probability density function for the life span of an electronics part is $f(t) = 0.08e^{-0.08t}$, where t is the number of months in service. Find the probability that any given part of this type lasts longer than 24 months.

46. **Warranties** A transmission repair firm, desiring to offer a lifetime warranty on its repairs, has determined that the probability density function for transmission failure after repair is $f(t) = 0.3e^{-0.3t}$, where t is the number of months after repair. What is the probability that a transmission chosen at random will last

(a) 3 months or less?

(b) more than 3 months?

KEY TERMS
& FORMULAS

Section	Key Terms	Formula
13.1	Sigma notation	$\displaystyle\sum_{i=1}^{n} 1 = n; \qquad \sum_{i=1}^{n} i = \frac{n(n+1)}{2}$ $\displaystyle\sum_{i=1}^{n} i^2 = \frac{n(n+1)(2n+1)}{6}$
	Area	
	Right-hand endpoints	$\displaystyle\lim_{n \to \infty} \sum_{i=1}^{n} f(x_i)\frac{b-a}{n}$
	Left-hand endpoints	$\displaystyle\lim_{n \to \infty} \sum_{i=1}^{n} f(x_{i-1})\frac{b-a}{n}$
13.2	Riemann sum	$\displaystyle\sum_{i=1}^{n} f(x_i^*)\Delta x_i$
	Definite integral	$\displaystyle\int_{a}^{b} f(x)\, dx = \lim_{\substack{\max \Delta x_i \to 0 \\ (n \to \infty)}} \sum_{i=1}^{n} f(x_i^*)\, \Delta x_i$
	Fundamental Theorem of Calculus	$\displaystyle\int_{a}^{b} f(x)\, dx = F(b) - F(a)$, where $F'(x) = f(x)$
	Definite Integral Properties	$\displaystyle\int_{a}^{a} f(x)\, dx = 0$ $\displaystyle\int_{a}^{b} f(x)\, dx = -\int_{b}^{a} f(x)\, dx$ $\displaystyle\int_{a}^{b} [f(x) \pm g(x)]\, dx = \int_{a}^{b} f(x)\, dx \pm \int_{a}^{b} g(x)\, dx$ $\displaystyle\int_{a}^{b} c \cdot f(x)\, dx = c\int_{a}^{b} f(x)\, dx$

Section	Key Terms	Formula
		$$\int_a^c f(x)\, dx + \int_c^b f(x)\, dx = \int_a^b f(x)\, dx$$
	Area under $f(x)$, where $f(x) \geq 0$	$$A = \int_a^b f(x)\, dx$$
13.3	Area between $f(x)$ and $g(x)$, where $f(x) \geq g(x)$	$$A = \int_a^b [f(x) - g(x)]\, dx$$
	Average value over $[a, b]$	$$\frac{1}{b-a} \int_a^b f(x)\, dx$$
	Lorenz curve	
	Gini coefficient	$$2\int_0^1 [x - L(x)]\, dx$$
13.4	Continuous income streams	
	Total income	$$\int_0^k f(t)\, dt \quad \text{(for } k \text{ years)}$$
	Present value	$$\int_0^k f(t)e^{-rt}\, dt, \quad \text{where } r \text{ is the interest rate}$$
	Future value	$$e^{rk} \int_0^k f(t)e^{-rt}\, dt$$
	Consumer's surplus [demand is $f(x)$]	$$CS = \int_0^{x_1} f(x)\, dx - p_1 x_1$$
	Producer's surplus [supply is $g(x)$]	$$PS = p_1 x_1 - \int_0^{x_1} g(x)\, dx$$
13.5	Integration from tables	See Table 13.1.
13.6	Integration by parts	$$\int u\, dv = uv - \int v\, du$$
13.7	Improper integrals	$$\int_a^\infty f(x)\, dx = \lim_{b \to \infty} \int_a^b f(x)\, dx$$ $$\int_{-\infty}^b f(x)\, dx = \lim_{a \to \infty} \int_{-a}^b f(x)\, dx$$ $$\int_{-\infty}^\infty f(x)\, dx = \int_{-\infty}^c f(x)\, dx + \int_c^\infty f(x)\, dx$$

Section	Key Terms	Formula
	Capital value of a continuous income stream	$\int_{0}^{\infty} f(t) \, e^{-rt} \, dt$
	Probability distribution	$\int_{-\infty}^{\infty} f(x) \, dx = 1$
	Expected value	$\mu = \int_{-\infty}^{\infty} x f(x) \, dx$

REVIEW EXERCISES

13.1

1. Calculate $\displaystyle\sum_{k=1}^{8} (k^2 + 1)$.

2. Use formulas to simplify

$$\sum_{i=1}^{n} \frac{3i}{n^3}.$$

3. Use 6 subintervals of the same size to approximate the area under the graph of $y = 3x^2$ from $x = 0$ to $x = 1$. Use the right-hand endpoints of the subintervals to find the heights of the rectangles.

4. Use rectangles to find the area under the graph of $y = 3x^2$ from $x = 0$ to $x = 1$. Use n equal subintervals.

13.2

5. Use a definite integral to find the area under the graph of $y = 3x^2$ from $x = 0$ to $x = 1$.

6. Find the area between the graph of $y = x^3 - 4x + 5$ and the x-axis from $x = 1$ to $x = 3$.

Evaluate the following integrals.

7. $\displaystyle\int_{1}^{4} 4\sqrt{x^3} \, dx$

8. $\displaystyle\int_{-3}^{2} (x^3 - 3x^2 + 4x + 2) \, dx$

9. $\displaystyle\int_{0}^{5} (x^3 + 4x) \, dx$

10. $\displaystyle\int_{-2}^{3} (x + 2)^2 \, dx$

11. $\displaystyle\int_{-3}^{-1} (x + 1) \, dx$

12. $\displaystyle\int_{2}^{3} \frac{x^2}{2x^3 - 7} \, dx$

13. $\displaystyle\int_{-1}^{2} (x^2 + x) \, dx$

14. $\displaystyle\int_{1}^{4} \left(\frac{1}{x} + \sqrt{x}\right) \, dx$

15. $\displaystyle\int_{0}^{4} (2x + 1)^{1/2} \, dx$

16. $\displaystyle\int_{0}^{1} \frac{x}{x^2 + 1} \, dx$

17. $\displaystyle\int_{0}^{1} e^{-2x} \, dx$

18. $\displaystyle\int_{0}^{1} x e^{x^2} \, dx$

13.3

Find the area between the following curves.

19. $y = x^2 - 3x + 2$ and $y = x^2 + 4$ from $x = 0$ to $x = 5$

20. $y = x^2$ and $y = 4x + 5$

21. $y = x^3$ and $y = x$ from $x = -1$ to $x = 0$

22. $y = x^3 - 1$ and $y = x - 1$

13.5

Evaluate the integrals, using Table 13.1.

23. $\displaystyle\int \sqrt{x^2 - 4} \, dx$

24. $\displaystyle\int_{0}^{1} 3^x \, dx$

25. $\displaystyle\int x \ln x^2 \, dx$

26. $\displaystyle\int \frac{dx}{x(3x + 2)}$

13.6

Use integration by parts to evaluate the following.

27. $\displaystyle\int x^5 \ln x \, dx$

28. $\displaystyle\int x e^{-2x} \, dx$

29. $\displaystyle\int \frac{x \, dx}{\sqrt{x + 5}}$

30. $\displaystyle\int_{1}^{e} \ln x \, dx$

13.7

Evaluate the following improper integrals.

31. $\displaystyle\int_{1}^{\infty} \frac{1}{x} \, dx$

32. $\displaystyle\int_{-\infty}^{-1} \frac{200}{x^3} \, dx$

33. $\displaystyle\int_{0}^{\infty} 5 e^{-3x} \, dx$

34. $\displaystyle\int_{-\infty}^{0} \frac{x}{(x^2 + 1)^2} \, dx$

APPLICATIONS

13.2

35. **Maintenance** Maintenance costs for buildings increase as the buildings age. If the rate of increase in maintenance costs for a building is

$$M'(t) = \frac{14,000}{\sqrt{t + 16}},$$

where M is in dollars and t is time in years, $0 \le t \le 15$, find the total maintenance cost for the first nine years ($t = 0$ to $t = 9$).

36. **Quality control** Suppose the probability density function for the life expectancy of a "disposable" telephone is

$$f(x) = \begin{cases} 1.4e^{-1.4x} & x \ge 0 \\ 0 & x < 0 \end{cases}.$$

Find the probability that the telephone lasts 2 years.

13.3

37. **Savings** The future value of $1000 invested in a savings at 10%, compounded continuously, is $S = 1000e^{0.1t}$, where t is in years. Find the average amount in the savings account during the first five years.

38. **Income stream** Suppose the total income from a video machine is given by

$$I = 50e^{0.2t}, \ 0 \le t \le 4, \ t \text{ in hours.}$$

Find the average income over this four-hour period.

13.4

39. **Consumer's surplus** The demand function for a product under pure competition is $p = \sqrt{64 - 4x}$, and the supply function is $p = x - 1$.
    (a) Find the market equilibrium.
    (b) Find the consumer's surplus at market equilibrium.

40. **Producer's surplus** Find the producer's surplus at market equilibrium for problem 39.

41. **Income streams** Find the total income over the next 10 years from a continuous income stream that has an annual flow rate at time t given by

$$f(t) = 125e^{0.05t},$$

in thousands of dollars.

42. **Income streams** Suppose that a machine's production is considered a continuous income stream with an annual rate of flow at time t given by

$$f(t) = 150e^{-0.2t},$$

in thousands of dollars. If money is worth 8%, compounded continuously, find the present value of the machine over the next 5 years.

13.5

43. **Average cost** Suppose the cost function for a product is given by $C(x) = \sqrt{40,000 + x^2}$. Find the average cost over the first 150 units.

13.6

44. **Income streams** Suppose the present value of a continuous income stream over the next 5 years is given by

$$P = 9000 \int_0^5 te^{-0.08t} \, dt, \quad P \text{ in dollars, } t \text{ in years.}$$

Find the present value.

45. **Cost** If the marginal cost for a product is $\overline{MC} = 3 + x \ln(x + 1)$ and if the fixed cost is $2000, find the total cost function.

13.7

46. **Quality control** Find the probability that a telephone lasts more than 1 year if the probability density function for its life expectancy is given by

$$f(x) = \begin{cases} 1.4e^{-1.4x} & \text{if } x \ge 0 \\ 0 & \text{if } x < 0 \end{cases}.$$

CONSULTANT'S CORNER

Retirement Planning

A 52-year-old client asks an accountant how to plan for his future retirement at age 62. He expects income from Social Security in the amount of $16,000 per year and a retirement pension of $30,000 per year from his employer. He wants to make monthly contributions to an investment plan that pays 8% compounded monthly for 10 years so that he will have a total income of $62,000 per year for 30 years. What will the size of the monthly contributions have to be to accomplish this goal, if it is assumed that money will be worth 8%, compounded continuously throughout the period after he is 62?

To help you answer this question, complete the following.

1. How much money must the client withdraw annually from his investment plan during his retirement so that his total income goal is met?
2. How much money S must the client's account contain when he is 62 so that it will generate this annual amount for 30 years? *Hint:* S can be considered the present value over 30 years of a continuous income stream with the amount found in (1) as its annual rate of flow.
3. The monthly contribution R that would, after 10 years, amount to the present value S found in question 2 can be obtained from the formula

$$R = S\left[\frac{i}{(1 + i)^n - 1}\right],$$

with i representing the monthly interest rate and n the number of months. Find the client's monthly contribution, R.

Warm-up

Prerequisite problem type	For section	Answer	Section for review
If $y = f(x)$, x is the independent variable and y is the _____ variable.	14.1	Dependent	1.4 Functions
What is the domain of $f(x) = \dfrac{3x}{x-1}$?	14.1	All reals except $x = 1$	1.4 Domains
If $C(x) = 5 + 5x$, what is $C(0.20)$?	14.1	6	1.4 Functional notation
(a) Solve for x and y: $\begin{cases} 0 = 50 - 2x - 2y \\ 0 = 60 - 2x - 4y \end{cases}$ (b) Solve for x and y: $\begin{cases} x = 2y \\ x + y - 9 = 0 \end{cases}$	14.4 14.5	(a) $y = 5$, $x = 20$ (b) $x = 6$, $y = 3$	1.8 Systems of equations
If $z = 4x^2 + 5x^3 - 7$, what is $\dfrac{dz}{dx}$?	14.2 14.3 14.4 14.5	$\dfrac{dz}{dx} = 8x + 15x^2$	9.4 Derivatives
If $f(x) = (x^2 - 1)^2$, what is $f'(x)$?	14.2	$f'(x) = 4x(x^2 - 1)$	9.6 Derivatives
If $z = 10y - \ln y$, what is $\dfrac{dz}{dy}$?	14.2	$\dfrac{dz}{dy} = 10 - \dfrac{1}{y}$	11.1 Derivatives of logarithmic functions
If $z = 5x^2 + e^x$, what is $\dfrac{dz}{dx}$?	14.2	$\dfrac{dz}{dx} = 10x + e^x$	11.2 Derivatives of exponential functions
Find the slope of the tangent to $y = 4x^3 - 4e^x$ at $(0, -4)$.	14.2	-4	9.3, 9.4, 11.2 Derivatives

FUNCTIONS OF TWO OR MORE VARIABLES

Although we have been dealing primarily with functions of one variable, many real-life situations involve variables that are functions of two or more variables. For example, the grade you receive in a course is a function of several test grades. The cost of manufacturing a product may involve the cost of labor, the cost of materials, and overhead expenses. The concentration of a substance at any point in a vein after an injection is a function of time since the injection t, the velocity of the blood v, and the distance the point is from the point of injection.

In this chapter we will extend our study to functions of two or more variables. We will extend the derivative concept to functions of several variables by taking partial derivatives, and we will learn how to maximize functions of two variables. We will use these concepts to solve problems in the management, social, and life sciences. In particular, we will discuss joint cost functions, marginal cost, marginal productivity, and marginal demand functions. We will use Lagrange multipliers to optimize functions of two variables subject to a condition that constrains the variables.

14.1 FUNCTIONS OF TWO OR MORE VARIABLES

Objectives

❑ To find the domain and range of a function of two or more variables
❑ To evaluate a function of two or more variables given values for the independent variables

APPLICATION PREVIEW

The relations we have studied up to this point have been limited to two variables, with one of the variables assumed to be a function of the other. But there are many instances where one variable may depend on two or more other variables. For example, the output or production Q (for quantity) of a company can be modeled according to the equation

$$Q = AK^{\alpha}L^{1-\alpha},$$

where A is a constant, K is the company's capital investment, L is the size of the labor force (in work hours), and α is a constant with $0 < \alpha < 1$.

Functions of this type are called **Cobb-Douglas production functions,** and they are frequently used in economics. For example, suppose the Cobb-Douglas production function for a company is given by

$$Q = 4\,K^{0.4}L^{0.6},$$

where Q is dollars of production value. We could use this function to determine the production value for a given amount of capital investment and available work hours of labor. We could also find how production is affected by changes in capital investment or available work hours.

In addition, the demand function for a commodity frequently depends on the price of the commodity, available income, and prices of competing goods. Other examples from economics will be presented later in this chapter.

We write $z = f(x, y)$ to state that z is a function of both x and y. The variables x and y are called the **independent variables** and z is called the **dependent variable.** Thus the function f associates with each pair of possible values for the independent variables (x and y) exactly one value of the dependent variable (z).

The equation $z = x^2 - xy$ defines z as a function of x and y. We can denote this by writing $z = f(x, y) = x^2 - xy$. The domain of the function is the set of all ordered pairs (of real numbers) and the range is the set of all real numbers.

EXAMPLE I

Give the domain of the function

$$g(x, y) = \frac{x^2 - 3y}{x - y}.$$

Solution

The domain of the function is the set of ordered pairs that do not give a 0 denominator. That is, the domain is the set of all ordered pairs where the first and second elements are not equal (that is, where $x \neq y$).

CHECKPOINT

1. Find the domain of the function

$$f(x, y) = \frac{2}{\sqrt{x^2 - y^2}}.$$

We graph the function $z = f(x, y)$ by using three dimensions. We can construct a three-dimensional coordinate space by drawing three mutually perpendicular axes, as in Figure 14.1. By setting up a scale of measurement

Figure 14.1

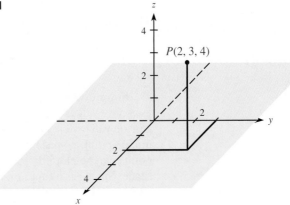

along the three axes from the origin 0, we can determine the three coordinates (x, y, z) for any point P. The point shown in Figure 14.1 is +2 units in the x-direction, +3 units in the y-direction, and +4 units in the z-direction, so the coordinates of the point are $(2, 3, 4)$.

The pairs of axes determine the three **coordinate planes;** the xy-plane, the yz-plane, and the xz-plane. The planes divide the space into eight **octants.** The point $P(2, 3, 4)$ is in the first octant.

If we are given a function $z = f(x, y)$, we can find the z-value corresponding to $x = a$ and $y = b$ by evaluating $f(a, b)$.

EXAMPLE 2
If $z = f(x, y) = x^2 - 4xy + xy^3$, find the following.
(a) $f(1, 2)$ (b) $f(2, 5)$ (c) $f(-1, 3)$

Solution
(a) $f(1, 2) = 1^2 - 4(1)(2) + (1)(2)^3 = 1$
(b) $f(2, 5) = 2^2 - 4(2)(5) + (2)(5)^3 = 214$
(c) $f(-1, 3) = (-1)^2 - 4(-1)(3) + (-1)(3)^3 = -14$

CHECKPOINT

2. If $f(x, y, z) = x^2 + 2y - z$, find $f(2, 3, 4)$.

EXAMPLE 3
The cost of manufacturing a Wosh is given by

$$C(x, y) = 5 + 5x + 2y,$$

where x represents the cost of one ounce of material used and y represents the cost of labor in dollars per hour. If the material cost is \$.20 per ounce and labor costs are \$3.50 per hour, what is the cost of manufacturing one Wosh?

Solution

The cost is

$$C(0.20, 3.50) = 5 + 5(0.20) + 2(3.50)$$
$$= 13 \text{ (dollars)}.$$

For a given function $z = f(x, y)$, we can construct a table of values by assigning values to x and y and finding the corresponding values of z. To each pair of values for x and y there corresponds a unique value of z, and thus a unique point in space. From a table of values such as this, a finite number of points can be plotted. All points that satisfy the equation form a "surface" in space. Since z is a function of x and y, lines parallel to the z-axis will intersect such a surface in at most one point. The graph of the equation $z = 4 - x^2 - y^2$ is a surface like that shown in Figure 14.2. The portion of the surface above the xy-plane is like a bullet and is called a **paraboloid.**

In practical applications of functions of two variables, we will have little need to construct the graphs of the surfaces. For this reason, we will not discuss methods of sketching the graphs. Although you will not be asked to sketch graphs of these surfaces, the fact that the graphs do *exist* will be used in studying relative maxima and minima of functions of two variables.

The properties of functions of one variable can be extended to functions of two variables. The precise definition of continuity for functions of two variables is technical, and may be found in more advanced books. We will limit our study to functions that are continuous and have continuous derivatives in the domain of interest to us. We may think of continuous functions as functions whose graphs consist of surfaces without "holes" or "breaks" in them.

Let the function $U = f(x, y)$ represent the **utility** (that is, satisfaction) derived by a consumer from the consumption of two goods, X and Y, where x and y represent the amounts of X and Y, respectively. Since we will assume

Figure 14.2

$z = 4 - x^2 - y^2$

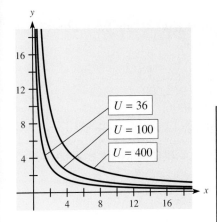

Figure 14.3

that the utility function is continuous, a given level of utility can be derived from an infinite number of combinations of x and y. The graph of all points (x, y) that give the same utility is called an **indifference curve.** A set of indifference curves corresponding to different levels of utility is called an **indifference map** (see Figure 14.3).

EXAMPLE 4

Suppose that the utility function for two goods, X and Y, is $U = x^2y^2$ and a consumer purchases 10 units of X and 2 units of Y.
(a) If the consumer purchases 5 units of X, how many units of Y must be purchased to retain the same level of utility?
(b) Graph the indifference curve for this level of utility.
(c) Graph the indifference curves for this utility function if $U = 100$ and if $U = 36$.

Solution
(a) If $x = 10$ and $y = 2$ satisfy the utility function, then $U = 10^2 \cdot 2^2 = 400$. Thus if x is 5, y must satisfy $400 = 5^2y^2$; so $y = 4$.
(b) The indifference curve for $U = 400$ is $400 = x^2y^2$. The graph for positive x and y is shown in Figure 14.3.
(c) The indifference map in Figure 14.3 contains these indifference curves.

A graphing utility can be used to graph each indifference curve in the indifference map shown in Figure 14.3. To graph the indifference curve for a given value of U, we must recognize that y will be positive and solve for y to express it as a function of x. Try it for $U = 100$. Does your graph agree with the one shown in Figure 14.3?

Sometimes functions of two variables are studied by fixing a value for one variable and graphing the resulting function of a single variable. We'll do this in Section 14.3 with production functions.

EXAMPLE 5

Suppose a company has the Cobb-Douglas production function introduced in the Application Preview,

$$Q = 4K^{0.4}L^{0.6},$$

where Q is thousands of dollars of production value, K is hundreds of dollars of capital investment per week, and L is work hours of labor per week.
(a) If current capital investment is \$72,900 per week and work hours are 3072 per week, find the current weekly production value.
(b) If weekly capital investment is increased to \$97,200 and new employees are hired so there are 4096 total weekly work hours, find the percentage increase in the production value.

Solution

(a) Capital investment of $72,900 means that $K = 729$. We use this value and $L = 3072$ in the production function.

$$Q = 4(729)^{0.4}(3072)^{0.6} = 6912$$

Thus the weekly production value is $6,912,000.

(b) In this case we use $K = 972$ and $L = 4096$.

$$Q = 4(972)^{0.4}(4096)^{0.6} = 9216$$

This is an increase in production value of $9216 - 6912 = 2304$, which is equivalent to a weekly increase of

$$\frac{2304}{6912} = 0.33\tfrac{1}{3} = 33\tfrac{1}{3}\%.$$

CHECKPOINT SOLUTIONS

1. The domain is the set of ordered pairs of real numbers where $x^2 - y^2 > 0$ or $x^2 > y^2$, that is, where $|x| > |y|$.
2. $f(2, 3, 4) = 4 + 6 - 4 = 6$

EXERCISE 14.1

Give the domain of each of the following functions.

1. $z = x^2 + y^2$
2. $z = 4x - 3y$
3. $z = \dfrac{4x - 3}{y}$
4. $z = \dfrac{x + y^2}{\sqrt{x}}$
5. $z = \dfrac{4x^3 y - x}{2x - y}$
6. $z = \sqrt{x - y}$
7. $q = \sqrt{p_1} + 3p_2$
8. $q = \dfrac{p_1 + p_2}{\sqrt{p_1}}$

Evaluate the following functions at the given values of the independent variables.

9. $z = x^3 + 4xy + y^2$; $x = 1, y = -1$
10. $z = 4x^2 - 3xy^3$; $x = 2, y = 2$
11. $z = \dfrac{x - y}{x + y}$; $x = 4, y = -1$
12. $z = \dfrac{x^2 + xy}{x - y}$; $x = 3, y = 2$
13. $C(x_1, x_2) = 600 + 4x_1 + 6x_2$; $x_1 = 400, x_2 = 50$
14. $C(x_1, x_2) = 500 + 5x_1 + 7x_2$;   find $C(200, 300)$.
15. $q_1(p_1, p_2) = \dfrac{p_1 + 4p_2}{p_1 - p_2}$;   find $q_1(40, 35)$.

16. $q_1(p_1, p_2) = \dfrac{5p_1 - p_2}{p_1 + 3p_2}$;   find $q_1(50, 10)$.
17. $z(x, y) = xe^{x+y}$;   find $z(3, -3)$.
18. $f(x, y) = ye^{2x} + y^2$;   find $f(0, 7)$.
19. $f(x, y) = \dfrac{\ln (xy)}{x^2 + y^2}$;   find $f(-3, -4)$.
20. $z(x, y) = x \ln y - y \ln x$;   find $z(1, 1)$.
21. $w = \dfrac{x^2 + 4yz}{xyz}$ at $(1, 3, 1)$
22. $u = f(w, x, y, z) = \dfrac{wx - yz^2}{xy - wz}$ at $(2, 3, 1, -1)$

APPLICATIONS

23. **Investment** The future value S of an investment earning 6%, compounded continuously, is a function of the principal P and the length of time t that the principal has been invested, and is given by

$$S = f(P, t) = Pe^{0.06t}.$$

Find $f(2000, 20)$, and interpret your answer.

24. **Amortization** If $100,000 is borrowed to purchase a home, then the monthly payment R is a function of the interest rate i (expressed as a percent) and the number of years n before the mortgage is paid, and is given by

$$R = f(i, n) = 100,000 \left[\frac{0.01(i/12)}{1 - (1 + 0.01(i/12))^{-12n}} \right].$$

Find $f(7.25, 30)$ and interpret your answer.

25. **Wilson's lot size formula** In economics, the most economical quantity Q of goods (TVs, dresses, gallons of paint, etc.) for a store to order is given by Wilson's lot size formula

$$Q = f(K, M, h) = \sqrt{2KM/h},$$

where K is the cost of placing the order, M is the number of items sold per week, and h is the weekly holding cost for each item (i.e., the cost of storage space, utilities, taxes, security, etc.). Find $f(200, 625, 1)$ and interpret your answer.

26. **Gas law** Suppose that a gas satisfies the universal gas law, $V = nRT/P$, with n equal to 10 moles of the gas and R, the universal gas constant, equal to 0.082054. What is V if $T = 10K$ (degrees kelvin) and $P = 1$ atmosphere?

Temperature-humidity models There are different models for measuring the effects of high temperature and humidity. Two of these are the Summer Simmer Index (S) and the Apparent Temperature (A),* and they are given by

$$S = 1.98T - 1.09(1 - H)(T - 58) - 56.9$$
$$A = 2.70 + 0.885T - 78.7H + 1.20TH,$$

where T is the air temperature (in degrees Fahrenheit) and H is the relative humidity (expressed as a decimal). Use these models in problems 27 and 28.

27. At the Dallas–Fort Worth Airport, the following represent the average daily temperatures and humidities for July.

   Maximum: 97.8°F with 44% humidity
   Minimum: 74.7°F with 80% humidity**

Calculate the Summer Simmer Index S and the Apparent Temperature A for both the average daily maximum and minimum temperatures.

*Bosch, W., and L. G. Cobb, "Temperature-Humidity Indices," UMAP Unit 691, *The UMAP Journal*, 10(3), Fall 1989, 237–256.

**Ruffner, James, and Frank Bair (eds.), *Weather of U.S. Cities*, Gale Research Co., Detroit, MI, 1987.

28. In Orlando, Florida, the following represent the average daily temperatures and humidities for August.

   Maximum: 91.6°F with 60% humidity
   Minimum: 73.4°F with 92% humidity**

Calculate the Summer Simmer Index S and the Apparent Temperature A for both the average daily maximum and minimum temperatures.

29. **Utility** Suppose that the utility function for two goods X and Y is given by $U = xy^2$, and a consumer purchases 9 units of X and 6 units of Y.
   (a) If the consumer purchases 9 units of Y, how many units of X must be purchased to retain the same level of utility?
   (b) If the consumer purchases 81 units of X, how many units of Y must be purchased to retain the same level of utility?
   (c) Graph the indifference curve for the utility level found in parts (a) and (b). Use the graph to confirm your answers to parts (a) and (b).

30. **Utility** Suppose that an indifference curve for two goods, X and Y, has equation $xy = 400$.
   (a) If 25 units of X are purchased, how many units of Y must be purchased to remain on this indifference curve?
   (b) Graph this indifference curve and confirm your results in part (a).

31. **Production** Suppose that a company's production is given by the Cobb-Douglas production function

$$Q = 30K^{1/4}L^{3/4},$$

where K is dollars of capital investment and L is labor hours.
   (a) Find Q if $K = \$10,000$ and $L = 625$ hours.
   (b) Show that if *both* K and L are doubled, then the output is doubled.
   (c) If capital investment is held at $10,000, graph Q as a function of L.

32. **Production** Suppose that the production of a company is given by the Cobb-Douglas production function

$$Q = 70K^{2/3}L^{1/3},$$

where K is dollars of capital investment and L is labor hours.
   (a) Find Q if $K = \$64,000$ and $L = 512$ hours.
   (b) Show that if both K and L are halved, then Q is also halved.
   (c) If capital investment is held at $64,000, graph Q as a function of L.

33. **Production** Suppose that the number of units of a good produced, z, is given by $z = 20xy$, where x is the number of machines working properly and y is the average number of work-hours per machine. Find the production for a week in which
   (a) 12 machines are working properly and the average number of work-hours per machine is 30.
   (b) 10 machines are working properly and the average number of work-hours per machine is 25.

34. **Profit** The Kirk Kelly Kandy Company makes two kinds of candy, Kisses and Kreams. The profit function for the company is

   $$P(x, y) = 100x + 64y - 0.01x^2 - 0.25y^2,$$

   where x is the number of pounds of Kisses sold per week and y is the number of pounds of Kreams. What is their profit if they sell
   (a) 20 pounds of Kisses and 10 pounds of Kreams?
   (b) 100 pounds of Kisses and 16 pounds of Kreams?
   (c) 10,000 pounds of Kisses and 256 pounds of Kreams?

35. **Epidemic** The cost per day to society of an epidemic is

   $$C(x, y) = 20x + 200y,$$

   where C is in dollars, x is the number of people infected on a given day, and y is the number of people who die on a given day. If 14,000 people are infected and 20 people die on a given day, what is the cost to society?

36. **Pesticide** An area of land is to be sprayed with two brands of pesticide: x liters of brand 1 and y liters of brand 2. If the number of insects killed is given by

   $$f(x, y) = 10,000 - 6500e^{-0.01x} - 3500e^{-0.02y},$$

   how many insects would be killed if 80 liters of brand 1 and 120 liters of brand 2 were used?

14.2 PARTIAL DIFFERENTIATION

Objectives

❑ To find partial derivatives of functions of two or more variables
❑ To evaluate partial derivatives of functions of two or more variables at given points
❑ To use partial derivatives to find slopes of tangents to surfaces
❑ To find and evaluate second and higher-order partial derivatives of functions of two variables

APPLICATION PREVIEW

We have used derivatives to find the rate of change of cost with respect to the quantity produced (in Chapter 9). If cost is given as a function of two variables (such as material costs x and labor costs y), we can find the rate of change of cost with respect to *one* of these independent variables. This is done by finding the **partial derivative** of the function with respect to one variable, while holding the other one constant.

For example, suppose that the cost of manufacturing a good is given by

$$C(x, y) = 5 + 5x + 2y,$$

where x represents the cost of one ounce of material used and y represents the labor cost in dollars per hour. To find the rate at which the cost changes with respect to material used x, we treat the y variable as though it were a constant and take the derivative of C with respect to x. This derivative is

$$\frac{\partial C}{\partial x} = 0 + 5 + 0, \quad \text{or} \quad \frac{\partial C}{\partial x} = 5.$$

Note that the partial derivative of $2y$ with respect to x is 0 because y is treated as a constant. Since the partial derivative is a new type of derivative, we use a new symbol to denote it, $\partial C/\partial x$.

The partial derivative $\partial C/\partial x = 5$ tells us that a change of \$1 in the cost of materials will cause an increase of \$5 in total costs, *if* labor costs remain constant. To see the rate at which total cost changes with respect to labor costs, we find $\partial C/\partial y = 0 + 0 + 2$, or $\partial C/\partial y = 2$. Thus if material costs are held constant, an increase of \$1 in labor costs will cause an increase of \$2 in the total cost of the good.

FIRST-ORDER PARTIAL DERIVATIVES

In general, if $z = f(x, y)$ we denote the partial derivative of z with respect to x as $\partial z/\partial x$ and the partial derivative of z with respect to y as $\partial z/\partial y$. Note that dz/dx represents the derivative of a function of one variable, x, and that $\partial z/\partial x$ represents the partial derivative of a function of two or more variables.

As mentioned in the Application Preview, $\partial z/\partial x$ is found by treating y as a constant and taking the derivative of $z = f(x, y)$ with respect to x. Other notations used to represent the partial derivative of $z = f(x, y)$ with respect to x are

$$\frac{\partial z}{\partial x}, \quad \frac{\partial f}{\partial x}, \quad \frac{\partial}{\partial x} f(x, y), \quad f_x(x, y), \quad f_x, \quad \text{and} \quad z_x.$$

If x is held constant in the function $z = f(x, y)$ and the derivative is taken with respect to y, we have the partial derivative of z with respect to y, denoted by

$$\frac{\partial z}{\partial y}, \quad \frac{\partial f}{\partial y}, \quad \frac{\partial}{\partial y} f(x, y), \quad f_y(x, y), \quad f_y, \quad \text{or} \quad z_y.$$

EXAMPLE 1

If $z = 4x^2 + 5x^2y^2 + 6y^3 - 7$, find $\partial z/\partial x$ and $\partial z/\partial y$.

Solution

$$\frac{\partial z}{\partial x} = 8x + 10y^2x$$

$$\frac{\partial z}{\partial y} = 10x^2y + 18y^2$$

EXAMPLE 2

If $z = x^2y + e^x - \ln y$, find z_x and z_y.

Solution

$$z_x = \frac{\partial z}{\partial x} = 2yx + e^x$$

$$z_y = \frac{\partial z}{\partial y} = x^2 - \frac{1}{y}$$

EXAMPLE 3

If $f(x, y) = (x^2 - y^2)^2$, find the following.

(a) f_x (b) f_y

Solution

(a) $f_x = 2(x^2 - y^2)2x = 4x^3 - 4xy^2$

(b) $f_y = 2(x^2 - y^2)(-2y) = -4x^2y + 4y^3$

CHECKPOINT

1. If $z = 100x + 10xy - y^2$, find

(a) z_x.

(b) $\dfrac{\partial z}{\partial y}$.

EXAMPLE 4

If $q = \dfrac{p_1p_2 + 2p_1}{p_1p_2 - 2p_2}$, find $\partial q / \partial p_1$.

Solution

$$\frac{\partial q}{\partial p_1} = \frac{(p_1p_2 - 2p_2)(p_2 + 2) - (p_1p_2 + 2p_1)p_2}{(p_1p_2 - 2p_2)^2}$$

$$= \frac{p_1p_2^2 + 2p_1p_2 - 2p_2^2 - 4p_2 - p_1p_2^2 - 2p_1p_2}{(p_1p_2 - 2p_2)^2}$$

$$= \frac{-2p_2^2 - 4p_2}{p_2^2(p_1 - 2)^2}$$

$$= \frac{-2p_2(p_2 + 2)}{p_2^2(p_1 - 2)^2}$$

$$= \frac{-2(p_2 + 2)}{p_2(p_1 - 2)^2}$$

We may evaluate partial derivatives by substituting values for *x and y* in the same manner we did with derivatives of functions of one variable. For example, if $\partial z/\partial x = 2x - xy$, the value of the partial derivative with respect to *x* at $x = 2, y = 3$ is

$$\left.\frac{\partial z}{\partial x}\right|_{(2,\,3)} = 2(2) - 2 \cdot 3 = -2.$$

Other notations used to denote evaluation of partial derivatives with respect to *x* at (a, b) are

$$\frac{\partial}{\partial x}f(a, b) \quad \text{and} \quad f_x(a, b).$$

We denote the evaluation of partial derivatives with respect to y at (a, b) by

$$\frac{\partial z}{\partial y}\bigg|_{(a, b)}, \quad \frac{\partial}{\partial y} f(a, b), \quad \text{or} \quad f_y(a, b).$$

EXAMPLE 5

Find the partial derivative of $f(x, y) = x^2 + 3xy + 4$ with respect to x at the point $(1, 2, 11)$.

Solution

$$f_x(x, y) = 2x + 3y$$
$$f_x(1, 2) = 2(1) + 3(2) = 8$$

CHECKPOINT

2. If $g(x, y) = 4x^2 - 3xy + 10y^2$, find

   (a) $\dfrac{\partial g}{\partial x}(1, 3)$.

   (b) $g_y(4, 2)$.

EXAMPLE 6

Suppose that a company's sales are related to its television advertising by

$$s = 20{,}000 + 10nt + 20n^2,$$

where n is the number of commercials per day and t is the length of the commercials in seconds. Find the partial derivative of s with respect to n and use the result to find the instantaneous rate of change of sales with respect to the number of commercials per day, if they are currently running ten 30-second commercials.

Solution

The partial derivative of s with respect to n is $\partial s/\partial n = 10t + 40n$. At $n = 10$ and $t = 30$, the rate of change in sales is approximately

$$\frac{\partial s}{\partial n}\bigg|_{\substack{n=10 \\ t=30}} = 10(30) + 40(10) = 700.$$

Thus increasing the number of commercials by one would result in approximately 700 additional sales.

We have seen that the partial derivative $\partial z/\partial x$ is found by holding y constant and taking the derivative of z with respect to x and the partial derivative $\partial z/\partial y$ is found by holding x constant and taking the derivative of z with respect to y. We now give formal definitions of these partial derivatives.

Partial Derivatives

The partial derivative of $z = f(x, y)$ with respect to x at the point (x, y) is

$$\frac{\partial z}{\partial x} = \frac{\partial}{\partial x} f(x, y) = \lim_{h \to 0} \frac{f(x + h, y) - f(x, y)}{h},$$

provided this limit exists.

The partial derivative of $z = f(x, y)$ with respect to y at the point (x, y) is

$$\frac{\partial z}{\partial y} = \frac{\partial}{\partial y} f(x, y) = \lim_{h \to 0} \frac{f(x, y + h) - f(x, y)}{h},$$

provided this limit exists.

We have already stated that the graph of $z = f(x, y)$ is a surface in three dimensions. The partial derivative with respect to x of such a function may be thought of as the slope of the tangent to the surface at a point (x, y, z) on the surface in the *positive direction of the x-axis*. That is, if a plane parallel to the xz-plane cuts the surface, passing through the point (x_0, y_0, z_0), the line in the plane that is tangent to the surface will have a slope equal to $\partial z / \partial x$ evaluated at the point. Thus

$$\left. \frac{\partial z}{\partial x} \right|_{(x_0, y_0)}$$

represents the slope of the tangent to the surface in the positive direction of the x-axis (see Figure 14.4).

Similarly,

$$\left. \frac{\partial z}{\partial y} \right|_{(x_0, y_0)} = \frac{\partial}{\partial y} f(x_0, y_0)$$

represents the slope of the tangent to the surface at (x_0, y_0, z_0) in the positive direction of the y-axis (see Figure 14.5).

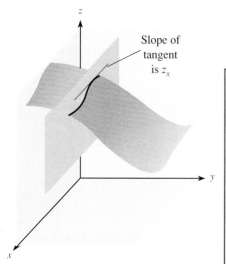

Slope of tangent is z_x

EXAMPLE 7

Find the slope of the tangent in the positive x-direction to the surface

$$z = 4x^3 - 4e^x + 4y^2$$

at the point $(0, 2, 12)$ on the surface.

Solution

$$\frac{\partial z}{\partial x} = 12x^2 - 4e^x$$

Evaluating the partial derivative at the given point, we find the slope of the tangent.

$$\left. \frac{\partial z}{\partial x} \right|_{(0, 2)} = 12(0)^2 - 4e^0 = -4$$

Figure 14.4

This tells us that z is *decreasing* at a rate of approximately 4 units for each increase of 1 unit in x at this point. That is, the slope of the tangent to the surface in the positive x-direction is negative at $(0, 2, 12)$.

EXAMPLE 8

Using the same function and point as that of Example 7, find the slope of the tangent in the positive y-direction to the surface at the point.

Solution

$$\frac{\partial z}{\partial y} = 8y$$

$$\left.\frac{\partial z}{\partial y}\right|_{(0,2)} = 8(2) = 16$$

So at this point in the positive y-direction the function is increasing at a rate of 16 units in the z-value for each unit increase in y.

Up to this point we have considered derivatives of functions of two variables. We can easily extend the concept to functions of three or more variables. We can find the partial derivative with respect to any one independent variable by taking the derivative of the function with respect to that variable while holding all other independent variables constant.

EXAMPLE 9

If $u = f(w, x, y, z) = 3x^2y + w^3 - 4xyz$, find the following.

(a) $\dfrac{\partial u}{\partial w}$ (b) $\dfrac{\partial u}{\partial x}$ (c) $\dfrac{\partial u}{\partial y}$ (d) $\dfrac{\partial u}{\partial z}$

Solution

(a) $\dfrac{\partial u}{\partial w} = 3w^2$ (b) $\dfrac{\partial u}{\partial x} = 6xy - 4yz$

(c) $\dfrac{\partial u}{\partial y} = 3x^2 - 4xz$ (d) $\dfrac{\partial u}{\partial z} = -4xy$

EXAMPLE 10

If $C = 4x_1 + 2x_1^2 + 3x_2 - x_1x_2 + x_3^2$, find the following.

(a) $\dfrac{\partial C}{\partial x_1}$ (b) $\dfrac{\partial C}{\partial x_2}$ (c) $\dfrac{\partial C}{\partial x_3}$

Solution

(a) $\dfrac{\partial C}{\partial x_1} = 4 + 4x_1 - x_2$ (b) $\dfrac{\partial C}{\partial x_2} = 3 - x_1$ (c) $\dfrac{\partial C}{\partial x_3} = 2x_3$

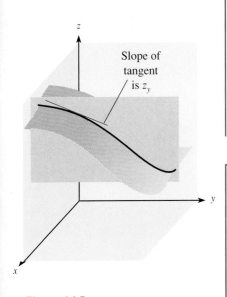

Slope of tangent is z_y

Figure 14.5

CHECKPOINT

3. If $f(w, x, y, z) = 8xy^2 + 4yz - xw^2$, find

(a) $\dfrac{\partial f}{\partial x}$.   (b) $\dfrac{\partial f}{\partial w}$.   (c) $\dfrac{\partial f}{\partial y}(1, 2, 1, 3)$.   (d) $\dfrac{\partial f}{\partial z}(0, 2, 1, 3)$.

HIGHER-ORDER PARTIAL DERIVATIVES

Just as we have taken derivatives of derivatives to obtain higher-order derivatives of functions of one variable, we may also take partial derivatives of partial derivatives to obtain higher-order partial derivatives of a function of more than one variable. If $z = f(x, y)$, then the partial derivative functions z_x and z_y are called *first partials*. Partial derivatives of z_x and z_y are called *second partials*, so $z = f(x,y)$ has *four* **second partial derivatives.** The notations for these second partial derivatives follow.

Second Partial Derivatives

$z_{xx} = \dfrac{\partial^2 z}{\partial x^2} = \dfrac{\partial}{\partial x}\left(\dfrac{\partial z}{\partial x}\right)$: both derivatives taken with respect to x.

$z_{yy} = \dfrac{\partial^2 z}{\partial y^2} = \dfrac{\partial}{\partial y}\left(\dfrac{\partial z}{\partial y}\right)$: both derivatives taken with respect to y.

$z_{xy} = \dfrac{\partial^2 z}{\partial y\, \partial x} = \dfrac{\partial}{\partial y}\left(\dfrac{\partial z}{\partial x}\right)$: first derivative taken with respect to x, second with respect to y.

$z_{yx} = \dfrac{\partial^2 z}{\partial x\, \partial y} = \dfrac{\partial}{\partial x}\left(\dfrac{\partial z}{\partial y}\right)$: first derivative taken with respect to y, second with respect to x.

EXAMPLE 11

If $z = x^3 y - 3xy^2 + 4$, find each of the second partial derivatives of the function.

Solution

Since

$$z_x = 3x^2 y - 3y^2 \quad \text{and} \quad z_y = x^3 - 6xy,$$

$$z_{xx} = \frac{\partial}{\partial x}(3x^2 y - 3y^2) = 6xy$$

$$z_{xy} = \frac{\partial}{\partial y}(3x^2 y - 3y^2) = 3x^2 - 6y$$

$$z_{yy} = \frac{\partial}{\partial y}(x^3 - 6xy) = -6x$$

$$z_{yx} = \frac{\partial}{\partial x}(x^3 - 6xy) = 3x^2 - 6y.$$

Note that z_{xy} and z_{yx} are equal for the function in Example 11. This will always occur if the derivatives of this function are continuous.

▶ **$z_{xy} = z_{yx}$** If the second partial derivatives z_{xy} and z_{yx} of a function $z = f(x, y)$ are continuous at a point, they are equal there.

EXAMPLE 12
Find each of the second partial derivatives of $z = x^2y + e^{xy}$.

Solution
Since $z_x = 2xy + e^{xy} \cdot y$,

$$z_{xx} = 2y + e^{xy} \cdot y^2 = 2y + y^2e^{xy}$$
$$z_{xy} = 2x + (e^{xy} \cdot 1 + ye^{xy} \cdot x)$$
$$= 2x + e^{xy} + xye^{xy}.$$

Since $z_y = x^2 + e^{xy} \cdot x$,

$$z_{yx} = 2x + (e^{xy} \cdot 1 + xe^{xy} \cdot y)$$
$$= 2x + e^{xy} + xye^{xy}$$
$$z_{yy} = 0 + xe^{xy} \cdot x = x^2e^{xy}.$$

CHECKPOINT

4. If $z = 4x^3y^4 + 4xy$, find:
   (a) z_{xx}.   (b) z_{yy}.   (c) z_{xy}.   (d) z_{yx}.
5. If $z = x^2 + 4e^{xy}$, find z_{xy}.

We can find partial derivatives of order higher than the second. For example, we can find the third-order partial derivatives z_{xyx} and z_{xyy} for the function in Example 11 from the second derivative $z_{xy} = 3x^2 - 6y$:

$$z_{xyx} = 6x$$
$$z_{xyy} = -6.$$

EXAMPLE 13
If $z = x^3y^2 + 4 \ln x$, find z_{xyy}.

Solution

$$z_x = 3x^2y^2 + 4 \cdot \frac{1}{x}$$
$$z_{xy} = 3x^2(2y) + 0 = 6x^2y$$
$$z_{xyy} = 6x^2$$

CHECKPOINT SOLUTIONS

1. (a) $z_x = 100 + 10y$

   (b) $\dfrac{\partial z}{\partial y} = 10x - 2y$

2. (a) $\dfrac{\partial g}{\partial x} = 8x - 3y$ and $\dfrac{\partial g}{\partial x}(1, 3) = 8(1) - 3(3) = -1$

   (b) $g_y = -3x + 20y$ and $g_y(4, 2) = -3(4) + 20(2) = 28$

3. (a) $\dfrac{\partial f}{\partial x} = 8y^2 - w^2$ (b) $\dfrac{\partial f}{\partial w} = -2xw$

   (c) $\dfrac{\partial f}{\partial y} = 16xy + 4z$ and $\dfrac{\partial f}{\partial y}(1, 2, 1, 3) = 16(2)(1) + 4(3) = 44$

   (d) $\dfrac{\partial f}{\partial z} = 4y$ and $\dfrac{\partial f}{\partial z}(0, 2, 1, 3) = 4(1) = 4$

4. $z_x = 12x^2y^4 + 4y$ and $z_y = 16x^3y^3 + 4x$

   (a) $z_{xx} = 24xy^4$ (b) $z_{yy} = 48x^3y^2$

   (c) $z_{xy} = 48x^2y^3 + 4$ (d) $z_{yx} = 48x^2y^3 + 4$

5. $z_x = 2x + 4e^{xy}(y) = 2x + 4ye^{xy}$

   Calculation of z_{xy} requires the Product Rule.

   $$z_{xy} = 0 + (4y)(e^{xy}x) + (e^{xy})(4) = 4xye^{xy} + 4e^{xy}$$

EXERCISE 14.2

1. If $z = x^4 - 5x^2 + 6x + 3y^3 - 5y + 7$, find $\dfrac{\partial z}{\partial x}$ and $\dfrac{\partial z}{\partial y}$.

2. If $z = x^5 - 6x + 4y^4 - y^2$, find $\dfrac{\partial z}{\partial x}$ and $\dfrac{\partial z}{\partial y}$.

3. If $z = x^3 + 4x^2y + 6y^2$, find z_x and z_y.

4. If $z = 3xy + y^2$, find z_x and z_y.

5. If $f(x, y) = (x^3 + 2y^2)^3$, find $\dfrac{\partial f}{\partial x}$ and $\dfrac{\partial f}{\partial y}$.

6. If $f(x, y) = (xy^3 + y)^2$, find $\dfrac{\partial f}{\partial x}$ and $\dfrac{\partial f}{\partial y}$.

7. If $f(x, y) = \sqrt{2x^2 - 5y^2}$, find f_x and f_y.

8. If $g(x, y) = x\sqrt{y - x}$, find g_x and g_y.

9. If $C(x, y) = 600 - 4xy + 10x^2y$, find $\dfrac{\partial C}{\partial x}$ and $\dfrac{\partial C}{\partial y}$.

10. If $C(x, y) = 1000 - 4x + xy^2$, find $\dfrac{\partial C}{\partial x}$ and $\dfrac{\partial C}{\partial y}$.

11. If $Q(s, t) = \dfrac{2s - 3t}{s^2 + t^2}$, find $\dfrac{\partial Q}{\partial s}$ and $\dfrac{\partial Q}{\partial t}$.

12. If $q = \dfrac{5p_1 + 4p_2}{p_1 + p_2}$, find $\dfrac{\partial q}{\partial p_1}$ and $\dfrac{\partial q}{\partial p_2}$.

13. If $z = e^{2x} + y \ln x$, find z_x and z_y.

14. If $z = \ln(1 + x^2y) - ye^{-x}$, find z_x and z_y.

15. If $f(x, y) = 100e^{xy}$, find $\dfrac{\partial f}{\partial x}$ and $\dfrac{\partial f}{\partial y}$.

16. If $f(x, y) = \ln(xy + 1)$, find $\dfrac{\partial f}{\partial x}$ and $\dfrac{\partial f}{\partial y}$.

17. Find the partial derivative of

    $$f(x, y) = 4x^3 - 5xy + y^2$$

    with respect to x at the point $(1, 2, -2)$.

18. Find the partial derivative of

    $$f(x, y) = 3x^2 + 4x + 6xy$$

    with respect to y at $x = 2$, $y = -1$.

19. Find the slope of the tangent in the positive x-direction to the surface $z = 5x^3 - 4xy$ at the point $(1, 2, -3)$.

20. Find the slope of the tangent in the positive y-direction to the surface $z = x^3 - 5xy$ at $(2, 1, -2)$.

21. Find the slope of the tangent in the positive y-direction to the surface $z = e^{xy}$ at $(0, 1, 1)$.

22. Find the slope of the tangent in the positive x-direction to the surface $z = \ln(xy)$ at $(1, 1, 0)$.

23. If $u = f(w, x, y, z) = y^2 - x^2z + 4x$, find the following.

    (a) $\dfrac{\partial u}{\partial w}$ (b) $\dfrac{\partial u}{\partial x}$ (c) $\dfrac{\partial u}{\partial y}$ (d) $\dfrac{\partial u}{\partial z}$

24. If $u = x^2 + 3xy + xz$, find the following.

    (a) u_x (b) u_y (c) u_z

25. If $C(x_1, x_2, x_3) = 4x_1^2 + 5x_1x_2 + 6x_2^2 + x_3$, find the following.

    (a) $\dfrac{\partial C}{\partial x_1}$ (b) $\dfrac{\partial C}{\partial x_2}$ (c) $\dfrac{\partial C}{\partial x_3}$

26. If $f(x, y, z) = 2x\sqrt{yz - 1} + x^2z^3$, find the following.

    (a) $\dfrac{\partial f}{\partial x}$ (b) $\dfrac{\partial f}{\partial y}$ (c) $\dfrac{\partial f}{\partial z}$

27. If $z = x^2 + 4x - 5y^3$, find the following.

    (a) z_{xx} (b) z_{xy} (c) z_{yx} (d) z_{yy}

28. If $z = x^3 - 5y^2 + 4y + 1$, find the following.

    (a) z_{xx} (b) z_{xy} (c) z_{yx} (d) z_{yy}

29. If $z = x^2y - 4xy^2$, find the following.

    (a) z_{xx} (b) z_{xy} (c) z_{yx} (d) z_{yy}

30. If $z = xy^2 + 4xy - 5$, find the following.

    (a) z_{xx} (b) z_{xy} (c) z_{yx} (d) z_{yy}

31. If $z = x^2 - xy + 4y^3$, find z_{xyx}.

32. If $z = x^3 - 4x^2y + 5y^3$, find z_{yyx}.

33. If $f(x, y) = x^3y + 4xy^4$, find $\left. \dfrac{\partial^2}{\partial x^2} f(x, y) \right|_{(1, -1)}$.

34. If $f(x, y) = x^4y^2 + 4xy$, find $\left. \dfrac{\partial^2}{\partial y^2} f(x, y) \right|_{(1, 2)}$.

35. If $f(x_1, x_2) = 3x_1^2 - 2x_1x_2$, find the following.

    (a) $\left. \dfrac{\partial^2 f(x_1, x_2)}{\partial x_1^2} \right|_{(3, 2)}$ (b) $\left. \dfrac{\partial^2 f(x_1, x_2)}{\partial x_2^2} \right|_{(3, 2)}$

36. If $f(x, y) = x^3y^2 + 4xy^3$, find the following.

    (a) $\left. \dfrac{\partial^2 f(x, y)}{\partial x^2} \right|_{(1, 3)}$ (b) $\left. \dfrac{\partial^2 f(x, y)}{\partial y^2} \right|_{(1, 3)}$

37. If $z = x^2y + ye^{x^2}$, find $z_{yx}|_{(1, 2)}$.

38. If $z = xy^3 + x \ln y^2$, find $z_{xy}|_{(2, 1)}$.

39. If $f(x, y) = x^2 + e^{xy}$, find the following.

    (a) $\dfrac{\partial^2 f}{\partial x^2}$ (b) $\dfrac{\partial^2 f}{\partial y\, \partial x}$ (c) $\dfrac{\partial^2 f}{\partial x\, \partial y}$ (d) $\dfrac{\partial^2 f}{\partial y^2}$

40. If $z = xe^{xy}$, find the following.

    (a) z_{xx} (b) z_{yy} (c) z_{xy} (d) z_{yx}

41. If $f(x, y) = y^2 - \ln xy$, find the following.

    (a) $\dfrac{\partial^2 f}{\partial x^2}$ (b) $\dfrac{\partial^2 f}{\partial y\, \partial x}$

    (c) $\dfrac{\partial^2 f}{\partial x\, \partial y}$ (d) $\dfrac{\partial^2 f}{\partial y^2}$

42. If $f(x, y) = x^3 + \ln(xy - 1)$, find the following.

    (a) $\dfrac{\partial^2 f}{\partial x^2}$ (b) $\dfrac{\partial^2 f}{\partial y^2}$

    (c) $\dfrac{\partial^2 f}{\partial x\, \partial y}$ (d) $\dfrac{\partial^2 f}{\partial y\, \partial x}$

43. If $w = 4x^3y + y^2z + z^3$, find the following.

    (a) w_{xxy} (b) w_{xyx} (c) w_{xyz}

44. If $w = 4xyz + x^3y^2z + x^3$, find the following.

    (a) w_{xyz} (b) w_{xzz} (c) w_{yyz}

APPLICATIONS

45. **Mortgage** When a homeowner has a 25-year variable rate mortgage loan, the monthly payment R is a function of the amount of the loan A and the current interest rate i (as a percent) that is, $R = f(A, i)$. Interpret each of the following.

    (a) $f(100{,}000, 8) = 1289$

    (b) $\dfrac{\partial f}{\partial i}(100{,}000, 8) = 62.51$

46. **Mass transportation ridership** Suppose in a certain city the number of people N using the mass transportation system is a function of the fare f and the daily cost of downtown parking p, so that $N = N(f, p)$. Interpret each of the following.

    (a) $N(5, 10) = 6500$

    (b) $\dfrac{\partial N}{\partial f}(5, 10) = -400$

    (c) $\dfrac{\partial N}{\partial p}(5, 10) = 250$

47. **Wilson's lot size formula** In economics, the most economical quantity Q of goods (TVs, dresses, gallons of paint, etc.) for a store to order is given by Wilson's lot size formula

$$Q = \sqrt{2KM/h},$$

where K is the cost of placing the order, M is the number of items sold per week, and h is the weekly holding costs for each item (i.e., the cost of storage space, utilities, taxes, security, etc.).

(a) Explain why $\dfrac{\partial Q}{\partial M}$ will be positive.

(b) Explain why $\dfrac{\partial Q}{\partial h}$ will be negative.

48. **Cost** Suppose that the total cost of producing a product is $C(x, y) = 25 + 2x^2 + 3y^2$, where x is the cost per pound for material and y is cost per hour for labor.
   (a) If material costs are held constant, at what rate will the total cost increase for each $1 per hour increase in labor?
   (b) If the labor costs are held constant, at what rate will the total cost increase for each increase of $1 in material cost?

49. **Pesticide** Suppose that the number of insects killed by two brands of pesticide is given by

   $$f(x, y) = 10{,}000 - 6500e^{-0.01x} - 3500e^{-0.02y},$$

   where x is the number of liters of brand 1 and y is the number of liters of brand 2.
   (a) What is the rate of change of insect deaths with respect to the number of liters of brand 1?
   (b) What is the rate of change of insect deaths with respect to the number of liters of brand 2?

50. **Profit** Suppose that the profit from the sale of Kisses and Kreams is given by

   $$P(x, y) = 100x + 64y - 0.01x^2 - 0.25y^2,$$

   where x is the number of pounds of Kisses and y is the number of pounds of Kreams.
   (a) Find $\partial P/\partial x$, and give the approximate rate of change of profit with respect to the number of pounds of Kisses if present sales are 20 pounds of Kisses and 10 pounds of Kreams.
   (b) Find $\partial P/\partial y$, and give the approximate rate of change of profit with respect to the number of pounds of Kreams that are sold if 100 pounds of Kisses and 16 pounds of Kreams are currently being sold.

51. **Utility** If $U = f(x, y)$ is the utility function for goods X and Y, the *marginal utility* of X is $\partial U/\partial x$ and the *marginal utility* of Y is $\partial U/\partial y$. If $U = x^2y^2$, find the marginal utility of
   (a) X.   (b) Y.

52. **Utility** If the utility function for goods X and Y is $U = xy + y^2$, find the marginal utility of
   (a) X.   (b) Y.

53. **Production** Suppose that the output Q (in thousands of units) of a certain company is $Q = 75K^{1/3}L^{2/3}$, where K is the capital expenditures in thousands of dollars and L is the number of labor hours. Find $\partial Q/\partial K$ and $\partial Q/\partial L$ when capital expenditures are $729,000 and the labor hours total 1728. Interpret each answer.

54. **Production** Suppose that the production Q (in hundreds of gallons of paint) of a paint manufacturer can be modeled by $Q = 140K^{1/2}L^{1/2}$, where K is the company's capital expenditures in thousands of dollars and L is the size of the labor force (in hours worked). Find $\partial Q/\partial K$ and $\partial Q/\partial L$ when capital expenditures are $250,000 and the labor hours are 1225. Interpret each answer.

Windchill factor Dr. Paul Siple conducted studies testing the effect of wind on the formation of ice at various temperatures, and developed the concept of the windchill factor, which we hear reported during winter weather reports. The windchill temperatures for selected air temperatures and wind speeds are shown in the table below. For example, the table shows that an air temperature of 15°F together with a wind speed of 35 mph feels the same as when the air temperature is −27°F and there is no wind.

	Air Temperature (°F)						
Wind Speed (mph)	35	25	15	5	−5	−15	−25
5	33	21	12	0	−10	−21	−31
15	16	2	−11	−25	−38	−51	−65
25	8	−7	−22	−36	−51	−66	−81
35	4	−12	−27	−43	−58	−74	−89
45	2	−14	−30	−46	−62	−78	−93

SOURCE: *World Almanac,* 1991

One form of the formula that meteorologists use to calculate windchill temperatures is

$$WC = 48.064 + 0.474t - 0.020ts - 1.85s + 0.304t\sqrt{s} - 27.74\sqrt{s},$$

where s is wind speed and t is the actual air temperature. Use this equation to answer problems 55 and 56.

55. (a) To see how the windchill temperature changes with wind speed, find $\partial WC/\partial s$.
   (b) Find $\partial WC/\partial s$ when the temperature is 10°F and the wind speed is 25 mph. What does this mean?

56. (a) To see how windchill temperature changes with temperature, find $\partial WC/\partial t$.
   (b) Find $\partial WC/\partial t$ when the temperature is 10°F and the wind speed is 25 mph. What does this mean?

14.3 APPLICATIONS OF FUNCTIONS OF TWO VARIABLES IN BUSINESS AND ECONOMICS

Objectives

- ❏ To evaluate cost functions at given levels of production
- ❏ To find marginal costs from total cost and joint cost functions
- ❏ To find marginal productivity for given production functions
- ❏ To find marginal demand functions from demand functions for a pair of related products

APPLICATION PREVIEW

In this section, we consider three classes of applications of functions of two variables and their partial derivatives. We begin with joint cost functions and their marginals. Next, we consider production functions and revisit Cobb-Douglas production functions. Marginal productivity is introduced and its meaning is explained. Finally, we consider demand functions for two products in a competitive market. Partial derivatives are used to define marginal demands, and these marginals are used to classify the products as competitive or complementary.

JOINT COST AND MARGINAL COST

Suppose that a firm produces two commodities using the same inputs in different proportions. In such a case the **joint cost function** is of the form $C = Q(x, y)$, where x and y represent the quantities of each commodity and C represents the total cost for the two commodities. Then $\partial C/\partial x$ is the **marginal cost** with respect to product x, and $\partial C/\partial y$ is the **marginal cost** with respect to product y.

EXAMPLE I

If the joint cost function for two products is

$$C = Q(x, y) = 50 + x^2 + 8xy + y^3,$$

find the marginal cost with respect to

(a) x.   (b) y.   (c) x at $(5, 3)$.   (d) y at $(5, 3)$.

Solution

(a) The marginal cost with respect to x is $\partial C/\partial x = 2x + 8y$.

(b) The marginal cost with respect to y is $\dfrac{\partial C}{\partial y} = 8x + 3y^2$.

(c) $\left.\dfrac{\partial C}{\partial x}\right|_{(5,\,3)} = 2(5) + 8(3) = 34$

Thus if 5 units of product x and 3 units of product y are produced, the total cost will increase approximately \$34 for each unit increase in product x if y is held constant.

(d) $\left.\dfrac{\partial C}{\partial y}\right|_{(5,\,3)} = 8(5) + 3(3)^2 = 67$

Thus if 5 units of product x and 3 units of product y are produced, the total cost will increase approximately $67 for each unit increase in product y if x is held constant.

PRODUCTION FUNCTIONS

An important problem in economics concerns how the factors necessary for production determine the output of a product. For example, the output of a product depends on available labor, land, capital, material, and machines. If the amount of output z of a product depends on the amounts of two inputs x and y, then the quantity z is given by the **production function** $z = f(x, y)$.

EXAMPLE 2

Suppose that it is known that z bushels of a crop can be harvested according to the function

$$z = (21)\frac{6xy - 4x^2 - 3y}{2x + 0.01y}$$

when $100x$ work-hours of labor are employed on y acres of land. What would be the output (in bushels) if 200 work-hours were used on 300 acres?

Solution
Since $z = f(x, y)$,

$$f(2, 300) = (21)\frac{6(2)(300) + 4(2)^2 - 3(300)}{2(2) + 3}$$

$$= (21)\frac{3600 + 16 - 900}{7} = 8148 \text{ (bushels)}.$$

If $z = f(x, y)$ is a production function, $\partial z/\partial x$ represents the change in the output z with respect to input x while input y remains constant. This partial derivative is called the **marginal productivity of x.** The partial derivative $\partial z/\partial y$ is the **marginal productivity of y**, and measures the rate of change of z with respect to input y.

Marginal productivity (for either input) will be positive over a wide range of inputs, but it increases at a decreasing rate, and may eventually reach a point where it no longer increases, and begins to decrease.

EXAMPLE 3

If a production function is given by $z = 5x^{1/2}y^{1/4}$, find the marginal productivity of
(a) x.   (b) y.

Solution

(a) $\dfrac{\partial z}{\partial x} = \dfrac{5}{2}x^{-1/2}y^{1/4}$ (b) $\dfrac{\partial z}{\partial y} = \dfrac{5}{4}x^{1/2}y^{-3/4}$

Note that the marginal productivity of x is positive for all values of x, but that it decreases as x gets larger (because of the negative exponent). The same is true for the marginal productivity of y.

 Graphing Utilities

If we have a production function and fix a value for one variable, then we can use a graphing utility to analyze the marginal productivity with respect to the other variable.

EXAMPLE 4

Suppose the Cobb-Douglas production function for a company is given by

$$z = 100x^{1/4}y^{3/4},$$

where x is the company's capital investment and y is the size of the labor force (in work-hours).
(a) Find the marginal productivity of x.
(b) If the current labor force is 625 work-hours, substitute $y = 625$ in your answer to part (a) and graph the result.
(c) From the graph in (b), what can be said about the effect on production of additional capital investment when the work-hours remain at 625?
(d) Find the marginal productivity of y.
(e) If current capital investment is $10,000, substitute $x = 10,000$ in your answer to part (d) and graph the result.
(f) From the graph in (e), what can be said about the effect on production of additional work-hours when capital investment remains at $10,000?

Solution

(a) $z_x = 25x^{-3/4}y^{3/4}$
(b) If $y = 625$, then z_x becomes

$$z_x = 25x^{-3/4}(625)^{3/4} = 25\left(\frac{1}{x^{3/4}}\right)(125) = \frac{3125}{x^{3/4}}.$$

The graph of z_x can be limited to quadrant I because the capital investment is $x > 0$, and hence $z_x > 0$. Knowledge of asymptotes can help us determine range values for x and z_x that give an accurate graph. See Figure 14.6.
(c) Figure 14.6 shows that $z_x > 0$ for $x > 0$. This means that any increases in capital investment will result in increases in productivity. However, Figure 14.6 also shows that z_x is decreasing for $x > 0$, which means that increases in capital investment have a diminishing impact on productivity.

Figure 14.6

Figure 14.7

(d) $z_y = 75x^{1/4}y^{-1/4}$

(e) If $x = 10,000$, then z_y becomes

$$z_y = 75(10,000)^{1/4}\left(\frac{1}{y^{1/4}}\right) = \frac{750}{y^{1/4}}.$$

The graph is shown in Figure 14.7.

(f) Figure 14.7 also shows that $z_y > 0$ when $y > 0$, so that increasing work-hours increases productivity. Note that z_y is decreasing for $y > 0$, but it does so more slowly than z_x. This indicates that increases in work-hours have a diminishing impact on productivity, but still a more significant one than increases in capital expenditures (as seen in Figure 14.6).

DEMAND FUNCTIONS

Suppose that two products are sold at prices p_1 and p_2, respectively, in a competitive market consisting of a fixed number of consumers with given tastes and incomes. Then the amount of each *one* of the products demanded by the consumers is dependent on the prices of *both* products on the market. If q_1 represents the demand for the first product, then $q_1 = f(p_1, p_2)$ is the **demand function** for that product. The graph of such a function is called a **demand surface**. An example of a demand function in two variables is $q_1 = 400 - 2p_1 - 4p_2$. Here q_1 is a function of two variables p_1 and p_2. If $p_1 = \$10$ and $p_2 = \$20$, the demand would equal $400 - 2(10) - 4(20) = 300$.

EXAMPLE 5

The demand functions for two products are

$$q_1 = 50 - 5p_1 - 2p_2$$
$$q_2 = 100 - 3p_1 - 8p_2.$$

What is the demand for each of the products if the price of the first is $p_1 = \$5$ and the price of the second is $p_2 = \$8$?

Solution

$$q_1 = 50 - 5(5) - 2(8) = 9$$
$$q_2 = 100 - 3(5) - 8(8) = 21$$

Thus if these are the prices, the demand for product 2 is higher than the demand for product 1.

EXAMPLE 6

Using the demand equations in Example 5, find a pair of prices p_1 and p_2 such that the demands for product 1 and product 2 are equal.

Solution
We want q_1 to equal q_2. Setting $q_1 = q_2$, we see that

$$50 - 5p_1 - 2p_2 = 100 - 3p_1 - 8p_2$$
$$6p_2 - 50 = 2p_1$$
$$p_1 = 3p_2 - 25$$

Now, any pair of positive values that satisfies this equation will make the demands equal. Letting $p_2 = 10$, we see that $p_1 = 5$ will satisfy the equation. Thus the prices $p_1 = 5$ and $p_2 = 10$ will make the demands equal. The prices $p_1 = 2$ and $p_2 = 9$ will also make the demands equal. Many pairs of values (that is, all those satisfying $p_1 = 3p_2 - 25$) will equalize the demands.

If the demand functions for a pair of related products, product 1 and product 2, are $q_1 = f(p_1, p_2)$ and $q_2 = g(p_1, p_2)$, respectively, then the partial derivatives of q_1 and q_2 are called **marginal demand functions.**

$\dfrac{\partial q_1}{\partial p_1}$ is the marginal demand of q_1 with respect to p_1.

$\dfrac{\partial q_1}{\partial p_2}$ is the marginal demand of q_1 with resepct to p_2.

$\dfrac{\partial q_2}{\partial p_1}$ is the marginal demand of q_2 with resepct to p_1.

$\dfrac{\partial q_2}{\partial p_2}$ is the marginal demand of q_2 with respect to p_2.

For typical demand functions, if the price of product 2 is fixed, the demand for product 1 will decrease as its price p_1 increases. In this case the marginal demand of q_1 with respect to p_1 will be negative; that is, $\partial q_1/\partial p_1 < 0$. Similarly, $\partial q_2/\partial p_2 < 0$.

But what about $\partial q_2/\partial p_1$ and $\partial q_1/\partial p_2$? If $\partial q_2/\partial p_1$ and $\partial q_1/\partial p_2$ are both positive, the two products are **competitive** because an increase in price p_1 will result in an increase in demand for product 2 (q_2) if the price p_2 is held constant, while an increase in price p_2 will increase the demand for product 1 (q_1) if p_1 is held constant. Stated more simply, an increase in the price of one of the two products will result in an increased demand for the other, so the products are in competition. For example, an increase in the price of a Japanese automobile will result in an increase in demand for an American automobile if the price of the American automobile is held constant.

If $\partial q_2/\partial p_1$ and $\partial q_1/\partial p_2$ are both negative, the products are **complementary** because an increase in the price of one product will cause a decrease in demand for the other product if the price of the second product doesn't change. Under these conditions, a *decrease* in the price of product 1 would result in an *increase* in the demand for product 2, and a decrease in the price of product 2 would result in an increase in the demand for product 1. For example, a decrease in the price of gasoline would result in an increase in the demand for large automobiles.

If the signs of $\partial q_2/\partial p_1$ and $\partial q_1/\partial p_2$ are different, the products are neither competitive nor complementary. This situation rarely occurs, but is possible.

EXAMPLE 7

The demand functions for two related products, product 1 and product 2, are given by

$$q_1 = 400 - 5p_1 + 6p_2 \qquad q_2 = 250 + 4p_1 - 5p_2$$

(a) Determine the four marginal demands.
(b) Are product 1 and product 2 complementary or competitive?

Solution

(a) $\dfrac{\partial q_1}{\partial p_1} = -5 \qquad \dfrac{\partial q_2}{\partial p_2} = -5 \qquad \dfrac{\partial q_1}{\partial p_2} = 6 \qquad \dfrac{\partial q_2}{\partial p_1} = 4$

(b) Since $\partial q_1/\partial p_2$ and $\partial q_2/\partial p_1$ are both positive, products 1 and 2 are competitive.

CHECKPOINT

1. If the joint cost function for two product is

$$C = 100 + 3x + 10xy + y^2,$$

find the marginal cost with respect to
(a) x.   (b) y at $(7, 3)$.

2. If the production function for a product is

$$P = 10x^{1/4}y^{1/2},$$

find the marginal productivity of x.

3. If the demand functions for two products are

$$q_1 = 200 - 3p_1 - 4p_2 \quad \text{and} \quad q_2 = 50 - 6p_1 - 5p_2,$$

what is the marginal demand of
(a) q_1 with respect to p_1?   (b) q_2 with respect to p_2?

CHECKPOINT SOLUTIONS

1. (a) $\dfrac{\partial C}{\partial x} = 3 + 10y$ (b) $\dfrac{\partial C}{\partial y} = 10x + 2y$

$\dfrac{\partial C}{\partial y}(7, 3) = 10(7) + 2(3) = 76$

2. $\dfrac{\partial P}{\partial x} = \dfrac{2.5y^{1/2}}{x^{3/4}}$

3. (a) $\dfrac{\partial q_1}{\partial p_1} = -3$ (b) $\dfrac{\partial q_2}{\partial p_2} = -5$

EXERCISE 14.3

JOINT COST AND MARGINAL COST

1. The cost of manufacturing one item is given by

$$C(x, y) = 30 + 3x + 5y,$$

where x is the cost of one hour of labor and y is the cost of one pound of material. If the hourly cost of labor is $4, and the material costs $3 per pound, what is the cost of manufacturing one of these items?

2. The manufacture of one unit of a product has its cost given by

$$C(x, y, z) = 10 + 8x + 3y + z,$$

where x is the cost of one pound of one raw material, y is the cost of one pound of a second material, and z is the cost of one work-hour of labor. If the cost of the first raw material is $16 per pound, the cost of the second raw material is $8 per pound, and labor costs $8 per work-hour, what will it cost to produce one unit of the product?

3. The total cost of producing one unit of a product is

$$C(x, y) = 30 + 2x + 4y + \frac{xy}{50},$$

where x is the cost per pound of raw materials and y is the cost per hour of labor.
   (a) If labor costs are held constant, at what rate will the total cost increase for each increase of $1 per pound in material cost?
   (b) If material costs are held constant, at what rate will the total cost increase for each $1 per hour increase in labor costs?

4. The total cost of producing an item is

$$C(x, y) = 40 + 4x + 6y + \frac{x^2 y}{100},$$

where x is the cost per pound of raw materials and y is the cost per hour for labor. How will an increase of
   (a) $1 per pound of raw materials affect the total cost?
   (b) $1 per hour in labor costs affect the total cost?

5. The total cost of producing one unit of a product is given by

$$C(x, y) = 20x + 70y + \frac{x^2}{1000} + \frac{xy^2}{100},$$

where x represents the cost per pound of raw materials and y represents the hourly rate for labor. The present cost for raw materials is $10 per pound and the present hourly rate for labor is $4. How will an increase of
   (a) $1 per pound for raw materials affect the total cost?
   (b) $1 per hour in labor costs affect the total cost?

6. The total cost of producing one unit of a product is given by

$$C(x, y) = 30 + 10x^2 + 20y - xy,$$

where x is the hourly labor rate and y is the cost per pound of raw materials. The current hourly rate is $5, and the raw materials cost $6 per pound. How will an increase of
   (a) $1 per pound for the raw materials affect the total cost?
   (b) $1 in the hourly labor rate affect the total cost?

7. The joint cost function for two products is

$$C(x, y) = 30 + x^2 + 3y + 2xy,$$

where x represents the quantity of product X produced and y represents the quantity of product Y produced. Find the marginal cost with respect to
   (a) x if 8 units of product X and 10 units of product Y are produced.
   (b) y if 8 units of product X and 10 units of product Y are produced.

8. The joint cost function for products X and Y is

$$C(x, y) = 40 + 3x^2 + y^2 + xy,$$

where x represents the quantity of X and y represents the quantity of Y. Find the marginal cost with respect to
   (a) x if 20 units of product X and 15 units of product Y are produced.
   (b) y if 20 units of X and 15 units of Y are produced.

9. If the joint cost function for two products is

$$C(x, y) = x\sqrt{y^2 + 1},$$

   (a) Find the marginal cost (function) with respect to x.
   (b) Find the marginal cost with respect to y.
10. Suppose the joint cost function for x units of product X and y units of product Y is given by

$$C(x, y) = 2500\sqrt{xy + 1}.$$

   Find the marginal cost with respect to
   (a) x.   (b) y.
11. Suppose that the joint cost function for two products is

$$C(x, y) = 1200 \ln(xy + 1) + 10{,}000.$$

   Find the marginal cost with respect to
   (a) x.   (b) y.
12. Suppose that the joint cost function for two products is

$$C(x, y) = y \ln(x + 1).$$

   Find the marginal cost with respect to
   (a) x.   (b) y.

PRODUCTION FUNCTIONS

13. Suppose that the production function for a product is $z = \sqrt{4xy}$, where x represents the number of work-hours per month and y is the number of available machines. Determine the marginal productivity of
   (a) x.   (b) y.
14. Suppose the production function for a product is

$$z = 60x^{2/5}y^{3/5},$$

   where x is the capital expenditures and y is the number of work-hours. Find the marginal productivity of
   (a) x.   (b) y.
15. Suppose that the production function for a product is $z = \sqrt{x} \ln(y + 1)$, where x represents the number of work-hours and y represents the available capital (per week). Find the marginal productivity of
   (a) x.   (b) y.
16. Suppose that a company's production function for a certain product is

$$z = (x + 1)^{1/2} \ln(y^2 + 1),$$

   where x is the number of work-hours of unskilled labor and y is the number of work-hours of skilled labor. Find the marginal productivity of
   (a) x.   (b) y.

For problems 17–19, suppose that the production function for an agricultural product is given by

$$z = \frac{11xy - 0.0002x^2 - 5y}{0.03x + 3y},$$

where x is the number of hours of labor and y is the number of acres of the crop.
17. Find the output when $x = 300$ and $y = 500$.
18. Find the marginal productivity of the number of acres of the crop (y) when $x = 300$ and $y = 500$.
19. Find the marginal productivity of the number of hours of labor (x) when $x = 300$ and $y = 500$.
20. If a production function is given by $z = 12x^{3/4}y^{1/3}$, find the marginal productivity of
   (a) x.   (b) y.
21. Suppose the Cobb-Douglas production function for a company is given by

$$z = 400x^{3/5}y^{2/5},$$

   where x is the company's capital investment and y is the size of the labor force (in work-hours).
   (a) Find the marginal productivity of x.
   (b) If the current labor force is 1024 work-hours, substitute $y = 1024$ in your answer to part (a) and graph the result.
   (c) Find the marginal productivity of y.
   (d) If current capital investment is $59,049, substitute $x = 59{,}049$ in your answer to part (c) and graph the result.
   (e) Interpret the graphs in parts (b) and (d) with regard to what they say about the effects on productivity of increased capital investment (part (b)) and of an increased labor force (part (d)).

22. Suppose the Cobb-Douglas production function for a company is given by

$$z = 300x^{2/3}y^{1/3},$$

   where x is the company's capital investment and y is the size of the labor force (in work-hours).
   (a) Find the marginal productivity of x.
   (b) If the current labor force is 729 work-hours, substitute $y = 729$ in your answer to part (a) and graph the result.
   (c) Find the marginal productivity of y.
   (d) If the current capital investment is $27,000, substitute $x = 27{,}000$ in your answer to part (c) and graph the result.
   (e) Interpret the graphs in parts (b) and (d) with regard to what they say about the effects on productivity of increased capital investment (part (b)) and of an increased labor force (part (d)).

DEMAND FUNCTIONS

23. The demand functions for two products are given by

$$q_1 = 300 - 8p_1 - 4p_2$$
$$q_2 = 400 - 5p_1 - 10p_2.$$

Find the demand for each of the products if the price of the first is $p_1 = 10$ and the price of the second is $p_2 = 8$.

24. The demand functions for two products are given by

$$q_1 = 900 - 9p_1 + 2p_2$$
$$q_2 = 1200 + 6p_1 - 10p_2.$$

Find the demands q_1 and q_2 if $p_1 = \$10$ and $p_2 = \$12$.

25. Find a pair of prices p_1 and p_2 such that the demands for the two products in problem 23 will be equal.

26. Find a pair of prices p_1 and p_2 such that the demands for the two products in problem 24 will be equal.

In problems 27–30, the demand functions for two related products, A and B, are given. Complete parts (a)–(e) for each problem.

(a) Find the marginal demand of q_A with respect to p_A.
(b) Find the marginal demand of q_A with respect to p_B.
(c) Find the marginal demand of q_B with respect to p_B.
(d) Find the marginal demand of q_B with respect to p_A.
(e) Are the two goods competitive or complementary?

27. $\begin{cases} q_A = 400 - 3p_A - 2p_B \\ q_B = 250 - 5p_A - 6p_B \end{cases}$

28. $\begin{cases} q_A = 600 - 4p_A + 6p_B \\ q_B = 1200 + 8p_A - 4p_B \end{cases}$

29. $\begin{cases} q_A = 5000 - 50p_A - \dfrac{600}{p_B + 1} \\ q_B = 10{,}000 - \dfrac{400}{p_A + 4} + \dfrac{400}{p_B + 4} \end{cases}$

30. $\begin{cases} q_A = 2500 + \dfrac{600}{p_A + 2} - 40p_B \\ q_B = 3000 - 100p_A + \dfrac{400}{p_B + 5} \end{cases}$

14.4 MAXIMA AND MINIMA

Objective

❑ To find relative maxima, minima, and saddle points of functions of two variables

APPLICATION PREVIEW

Adele Lighting manufactures 20-inch lamps and 31-inch lamps. Suppose x is the number of thousands of 20-inch lamps, and the demand for these is given by $p_1 = 50 - x$, where p_1 is in dollars. Similarly, suppose y is the number of thousands of 31-inch lamps, and the demand for these is given by $p_2 = 60 - 2y$, where p_2 is also in dollars. Adele Lighting's joint cost function for these lamps is $C = 2xy$ (in thousands of dollars). Therefore, Adele Lighting's profit (in thousands of dollars) is a function of the two variables x and y. In order to determine Adele's maximum profit, we need to develop methods for finding maximum values for a function of two variables.

In our study of differentiable functions of one variable, we saw that for a relative maximum or minimum to occur at a point, the tangent line to the curve had to be horizontal at that point. The function $z = f(x, y)$ describes a surface in three dimensions. If all partial derivatives of $f(x, y)$ exist, then the surface described by $z = f(x, y)$ must have a horizontal plane tangent to the surface at a point in order to have a relative maximum or minimum at that

Figure 14.8   (a)   (b)

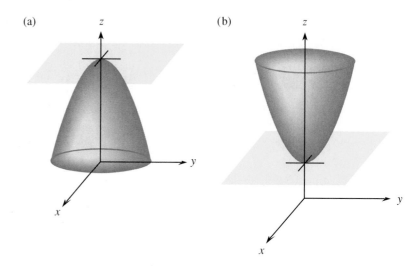

point (see Figure 14.8). But if the plane tangent to the surface at the point is horizontal, then all the tangent lines to the surface at that point must also be horizontal, for they lie in the tangent plane. In particular, the tangent line in the direction of the x-axis will be horizontal, so $\partial z/\partial x = 0$ at the point; and the tangent line in the direction of the y-axis will be horizontal, so $\partial z/\partial y = 0$ at the point. Thus we can determine the *critical points* for a surface by finding those points where *both* $\partial z/\partial x = 0$ and $\partial z/\partial y = 0$.

How can we determine whether a critical point is a relative maximum or relative minimum, or neither of these? Finding that $\partial^2 z/\partial x^2 < 0$ and $\partial^2 z/\partial y^2 < 0$ is not enough to tell us that we have a relative maximum. The "second derivative" test we must use involves the values of the second partial derivatives and the value of D at the critical point (a, b), where D is defined as follows:

$$D = \frac{\partial^2 z}{\partial x^2} \cdot \frac{\partial^2 z}{\partial y^2} - \left(\frac{\partial^2 z}{\partial x\, \partial y}\right)^2.$$

We shall state, without proof, the result that determines whether there is a relative maximum or minimum, or neither, at a point (a, b).

Test for Maxima and Minima   Let $z = f(x, y)$ be a function for which

$$\frac{\partial z}{\partial x} = \frac{\partial z}{\partial y} = 0 \quad \text{at a point } (a, b),$$

and suppose that all second partial derivatives are continuous there. For

$$D = \frac{\partial^2 z}{\partial x^2} \cdot \frac{\partial^2 z}{\partial y^2} - \left(\frac{\partial^2 z}{\partial x\, \partial y}\right)^2$$

a. If $D > 0$ and $\partial^2 z/\partial x^2 > 0$ at (a, b), then a relative minimum occurs at (a, b). In this case, $\partial^2 z/\partial y^2 > 0$ at (a, b) also.

b. If $D > 0$ and $\partial^2 z/\partial x^2 < 0$ at (a, b), then a relative maximum occurs at (a, b). In this case, $\partial^2 z/\partial y^2 < 0$ at (a, b) also.

c. If $D < 0$ at (a, b), there is neither a relative maximum nor minimum at (a, b).

d. If $D = 0$ at (a, b), then the test fails; the function must be investigated near the point.

We can test for relative maxima and minima by using the following procedure.

Maxima and Minima of $z = f(x, y)$

Procedure	Example
To find relative maxima and minima of $z = f(x, y)$:	Test $z = 4 - 4x^2 - y^2$ for relative maxima and minima.
1. Find $\partial z/\partial x$ and $\partial z/\partial y$.	1. $\dfrac{\partial z}{\partial x} = -8x$; $\quad \dfrac{\partial z}{\partial y} = -2y$
2. Find the point(s) that satisfy *both* $\partial z/\partial x = 0$ and $\partial z/\partial y = 0$. These are the critical points.	2. $\dfrac{\partial z}{\partial x} = 0$ if $x = 0$. $\quad \dfrac{\partial z}{\partial y} = 0$ if $y = 0$. The critical point is $(0, 0, 4)$.
3. Find all second partial derivatives.	3. $\dfrac{\partial^2 z}{\partial x^2} = -8$; $\quad \dfrac{\partial^2 z}{\partial y^2} = -2$ $\dfrac{\partial^2 z}{\partial x \partial y} = \dfrac{\partial^2 z}{\partial y \partial x} = 0$
4. Evaluate D at the critical point(s).	4. At $(0, 0)$, $D = (-8)(-2) - 0^2 = 16$.
5. Use the test for maxima and minima to determine if relative maxima or minima occur.	5. $D > 0$, $\partial^2 z/\partial x^2 < 0$, and $\partial^2 z/\partial y^2 < 0$. A relative maximum occurs at $(0, 0)$. The graph is shown in Figure 14.9.

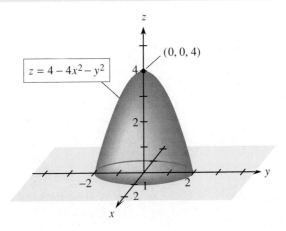

$z = 4 - 4x^2 - y^2$

$(0, 0, 4)$

Figure 14.9

EXAMPLE 1

Test $x^2 + y^2 - 2x + 1$ for relative maxima and minima.

Solution

1. $\dfrac{\partial z}{\partial x} = 2x - 2; \quad \dfrac{\partial z}{\partial y} = 2y$

2. $\dfrac{\partial z}{\partial x} = 0$ if $x = 1.$ $\quad \dfrac{\partial z}{\partial y} = 0$ if $y = 0.$

   Both are 0 if $x = 1$ *and* $y = 0$, so the critical point is $(1, 0, 0)$.

3. $\dfrac{\partial^2 z}{\partial x^2} = 2; \quad \dfrac{\partial^2 z}{\partial y^2} = 2 \quad \dfrac{\partial^2 z}{\partial x \, \partial y} = \dfrac{\partial^2 z}{\partial y \, \partial x} = 0$

4. At $(1, 0)$, $D = 2 \cdot 2 - 0^2 = 4.$

5. $D > 0$, $\partial^2 z / \partial x^2 > 0$,
   and $\partial^2 z / \partial y^2 > 0.$
   A relative minimum
   occurs at $(1, 0)$.
   (See Figure 14.10.)

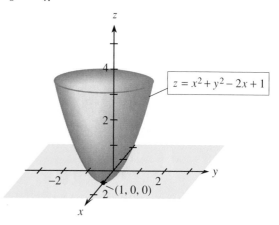

$z = x^2 + y^2 - 2x + 1$

Figure 14.10

EXAMPLE 2

Test $z = y^2 - x^2$ for relative maxima and minima.

Solution

1. $\dfrac{\partial z}{\partial x} = -2x; \quad \dfrac{\partial z}{\partial y} = 2y$

2. $\dfrac{\partial z}{\partial x} = 0$ if $x = 0; \quad \dfrac{\partial z}{\partial y} = 0$ if $y = 0.$

   So both equal 0 if $x = 0$, $y = 0$. The critical point is $(0, 0, 0)$.

3. $\dfrac{\partial^2 z}{\partial x^2} = -2; \quad \dfrac{\partial^2 z}{\partial y^2} = 2; \quad \dfrac{\partial^2 z}{\partial x \, \partial y} = \dfrac{\partial^2 z}{\partial y \, \partial x} = 0$

4. $D = (-2)(2) - 0 = -4$

5. $D < 0$, so the critical point is neither a relative maximum nor minimum.
   As Figure 14.11 shows, the surface formed has the shape of a saddle. For

this reason, critical points that are neither relative maxima nor minima are called **saddle points.**

Figure 14.11

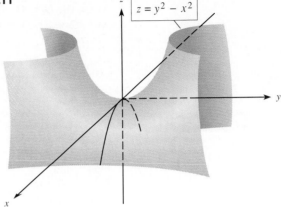

$z = y^2 - x^2$

The following example involves a surface with two critical points.

EXAMPLE 3

Test $z = x^3 + y^3 + 6xy$ for relative maxima and minima.

Solution

1. $\dfrac{\partial z}{\partial x} = 3x^2 + 6y; \quad \dfrac{\partial z}{\partial y} = 3y^2 + 6x$

2. $\dfrac{\partial z}{\partial x} = 0$ if $0 = 3x^2 + 6y$; that is, if $y = -\frac{1}{2}x^2$.

$\dfrac{\partial z}{\partial y} = 0$ if $0 = 3y^2 + 6x$; that is, if $x = -\frac{1}{2}y^2$.

Since *both* conditions must be satisfied, we can substitute $-\frac{1}{2}y^2$ for x in $y = -\frac{1}{2}x^2$, and obtain

$$y = -\frac{1}{2}\left(-\frac{1}{2}y^2\right)^2$$

$$= -\frac{1}{8}y^4$$

$$y + \frac{1}{8}y^4 = 0$$

$$y(8 + y^3) = 0.$$

So $y = 0$ or $y^3 = -8$; thus, $y = 0$ or $y = -2$. If $y = 0$, $x = -\frac{1}{2}(0)^2 = 0$, so one critical point is $(0, 0, 0)$. If $y = -2$, $x = -\frac{1}{2}(-2)^2 = -2$, so the second critical point is $(-2, -2, 8)$.

3. $\dfrac{\partial^2 z}{\partial x^2} = 6x;$ $\dfrac{\partial^2 z}{\partial y^2} = 6y;$ $\dfrac{\partial^2 z}{\partial x\, \partial y} = \dfrac{\partial^2 z}{\partial y\, \partial x} = 6$

Thus $D = (6x)(6y) - (6)^2.$

4. At $(0, 0)$, $D = 0 \cdot 0 - (6)^2 = -36 < 0.$
   At $(-2, -2)$, $D = (-12)(-12) - 36 = 108 > 0.$

5. At $(0, 0)$, $D < 0$, so a saddle point occurs at $(0, 0, 0)$.
   At $(-2, -2)$, $D > 0$, $\partial^2 z/\partial x^2 = 6(-2) = -12$, and $\partial^2 z/\partial y^2 = 6(-2) = -12$, so a relative maximum occurs at $(-2, -2)$. It is 8.

CHECKPOINT

Suppose that $z = 4 - x^2 - y^2 + 2x - 4y.$
1. Find z_x and z_y.
2. Solve $z_x = 0$ and $z_y = 0$ simultaneously to find the critical point(s) for the graph of this function.
3. Test the point(s) for relative maxima and minima.

Let us now solve the problem introduced in the Application Preview.

EXAMPLE 4

Maximize Adele Lighting's profit if the demand function is $p_1 = 50 - x$ for 20-inch lamps and $p_2 = 60 - 2y$ for 31-inch lamps and the joint cost function is $C = 2xy$. Recall that x and y are thousands of lamps, p_1 and p_2 are in dollars, and C is in thousands of dollars.

Solution

The profit function is $P(x, y) = p_1 x + p_2 y - C(x, y)$. Thus,

$$P(x, y) = (50 - x)x + (60 - 2y)y - 2xy$$
$$= 50x - x^2 + 60y - 2y^2 - 2xy$$

gives the profit in thousands of dollars. To maximize the profit, we proceed as follows.

$$P_x = 50 - 2x - 2y \quad \text{and} \quad P_y = 60 - 4y - 2x.$$

Solving simultaneously $P_x = 0$ and $P_y = 0$, We have

$$\begin{cases} 0 = 50 - 2x - 2y \\ 0 = 60 - 2x - 4y. \end{cases}$$

Subtraction gives $-10 + 2y = 0$, so $y = 5$. Thus $0 = 40 - 2x$, so $x = 20$. Now

$$P_{xx} = -2, P_{yy} = -4, \text{ and } P_{xy} = -2, \quad \text{and}$$
$$D = (P_{xx})(P_{yy}) - (P_{xy})^2 = (-2)(-4) - (-2)^2 = +4.$$

Since $P_{xx} < 0$, $P_{yy} < 0$, and $D > 0$, the values $x = 20$ and $y = 5$ yield maximum profit. Therefore, when $x = 20$ and $y = 5$, $p_1 = 30$, $p_2 = 50$, and the maximum profit is

$$P(20, 5) = 600 + 250 - 200 = 650.$$

That is, Adele Lighting's maximum profit is \$650,000 when the company sells 20,000 of the 20-inch lamps at \$30 each and 5000 of the 31-inch lamps at \$50 each.

In Chapter 8, we used linear regression to find the equation of the straight line that is the best fit for a set of data. If the relationship between x and y is approximately linear, then we can find a straight line

$$\hat{y} = a + bx,$$

where the values of $\hat{y}$ approximate the y-values of the points we know. For any given x-values, the sum of the squares of the deviations (see Figure 14.12) between the y-values of the points and the $\hat{y}$-values from the equation gives rise to a function of the two variables, a (the y-intercept) and b (the slope) of the regression line. We can use maxima and minima techniques for this function of a and b to derive the regression formulas used in Chapter 8. These techniques can also be adapted to develop the formulas that form the basis of curve-fitting programs that graphics calculators and computers use to generate an equation that fits a set of data points.

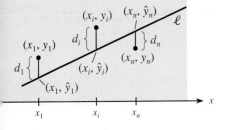

Figure 14.12

CHECKPOINT SOLUTIONS

1. $z_x = -2x + 2$, $z_y = -2y - 4$
2. $-2x + 2 = 0$ gives $x = 1$.
   $-2y - 4 = 0$ gives $y = -2$.
   Thus the critical point is $(1, -2, 9)$.
3. $z_{xx} = -2$, $z_{yy} = -2$, and $z_{xy} = 0$, so

$$D(x, y) = (-2)(-2) - (0)^2 = 4.$$

   Hence, at $(1, -2, 9)$ we have $D > 0$ and $z_{xx} < 0$, so $(1, -2, 9)$ is a relative minimum.

EXERCISE 14.4

In the following problems, test for relative maxima and minima.

1. $z = 9 - x^2 - y^2$
2. $z = 16 - 4x^2 - 9y^2$
3. $z = x^2 + y^2 + 4$
4. $z = x^2 + y^2 - 4$
5. $z = x^2 + y^2 - 2x + 4y + 5$
6. $z = 4x^2 + y^2 + 4x + 1$
7. $z = x^2 + 6xy + y^2 + 16x$
8. $z = x^2 - 4xy + y^2 - 6y$

9. $z = \dfrac{x^2 - y^2}{9}$

10. $z = \dfrac{y^2}{4} - \dfrac{x^2}{9}$

11. $z = 24 - x^2 + xy - y^2 + 36y$

12. $z = 46 - x^2 + 2xy - 4y^2$

13. $z = x^2 + xy + y^2 - 4y + 10x$

14. $z = x^2 + 5xy + 10y^2 + 8x - 40y$

15. $z = x^3 + y^3 - 6xy$

16. $z = x^3 + y^3 + 3xy$

APPLICATIONS

17. **Profit** Suppose that the profit from the sale of Kisses and Kreams is given by

$$P(x, y) = 100x + 64y - 0.01x^2 - 0.25y^2,$$

where x is the number of pounds of Kisses and y is the number of pounds of Kreams. Selling how many pounds of Kisses and Kreams will maximize profit?

18. **Profit** The profit from the sales of two products is given by

$$P(x, y) = 20x + 70y - x^2 - y^2,$$

where x is the number of units of product 1 sold and y is the number of units of product 2. Selling how much of each product will maximize profit?

19. **Nutrition** A new food is designed to add weight to mature beef cattle. The increase in weight is given by $W = xy(20 - x - 2y)$, where x is the number of units of the first ingredient and y is the number of units of the second ingredient. How many units of each ingredient will maximize the weight gain?

20. **Profit** The profit for a grain crop is related to fertilizer and labor. If the profit per acre is

$$P = 100x + 40y - 5x^2 - 2y^2,$$

where x is the number of units of fertilizer and y is the number of work-hours, what values of x and y will maximize the profit?

21. **Production** Suppose that

$$P = 3.78x^2 + 1.5y^2 - 0.09x^3 - 0.01y^3$$

is the production function for a product with x units of one input and y units of a second input. Find the values of x and y that will maximize production.

22. **Production** Suppose that x units of one input and y units of a second input result in

$$P = 40x + 50y - x^2 - y^2 - xy$$

units of a product. Determine the inputs x and y that will maximize P.

23. **Production** Suppose that a manufacturer produces two brands of a product, brand 1 and brand 2. If the demand for brand 1 is $p_1 = 10 - x$, for brand 2, $p_2 = 40 - 2y$, and if the joint cost function is $C = xy$, how many of each brand should be produced to maximize profit?

24. **Production** Suppose that a firm produces two products, A and B, that sell for \$$a$ and \$$b$, respectively, with the total cost of producing x units of A and y units of B equal to $C(x, y)$. Show that profit from these products is maximized when

$$\frac{\partial C}{\partial x}(x, y) = a \quad \text{and} \quad \frac{\partial C}{\partial y}(x, y) = b.$$

25. **Manufacturing** Find the values for each of the dimensions of an open-top box of length x, width y, and height $500,000/(xy)$ such that the box requires the least amount of material to make.

26. **Manufacturing** Find the values for each of the dimensions of a closed-top box of length x, width y, and height z if the volume equals 27,000 cubic inches and the box requires the least amount of material to make. *Hint:* First write z in terms of x and y as in problem 25.

27. **Profit** A company manufactures two products, A and B. If x is the number of units of A and y is the number of units of B, then the cost and revenue functions are

$$C(x, y) = 2x^2 - 2xy + y^2 - 7x + 10y + 11$$
$$R(x, y) = 5x + 4y.$$

Find the number of each type of product that should be manufactured to maximize profit.

28. **Production** Let x be the number of work-hours required and y be the amount of capital required to produce z units of a product. Show that the average production per work-hour, z/x, is maximized when

$$\frac{\partial z}{\partial x} = \frac{z}{x}.$$

Use $z = f(x, y)$ and assume a maximum exists.

14.5 MAXIMA AND MINIMA OF FUNCTIONS SUBJECT TO CONSTRAINTS; LAGRANGE MULTIPLIERS

Objective

☐ To find the maximum or minimum value of a function of two or more variables subject to a condition that constrains the variables

APPLICATION PREVIEW

Many practical problems require that a function of two or more variables be maximized or minimized subject to certain conditions, or constraints, that limit the variables involved. For example, a firm will want to maximize its profits within the limits (constraints) imposed by its production capacity. Similarly, a city planner may want to locate a new building to maximize access to public transportation, yet may be constrained by the availability and cost of building sites.

Specifically, suppose that the utility function for commodities X and Y is given by $U = x^2y^2$, where x and y are the amounts of X and Y, respectively. If p_1 and p_2 represent the prices of X and Y, respectively, and I represents the consumer's income available to purchase these two commodities, the equation $p_1x + p_2y = I$ is called the *budget constraint*. If the price of X is \$2, the price of Y is \$4, and the income available is \$40, then the budget constraint is $2x + 4y = 40$. Thus, we seek to maximize the consumer's utility $U = x^2y^2$ subject to the budget constraint $2x + 4y = 40$. In this section we develop methods to solve this type of constrained maximum or minimum.

We can obtain maxima and minima for a function $z = f(x, y)$ subject to the constraint $g(x, y) = 0$ by using the method of **Lagrange multipliers,** named for the famous eighteenth-century mathematician Joseph Louis Lagrange. Lagrange multipliers can be used with functions of two or more variables when the constraints are given by an equation.

In order to find the critical values of a function $f(x, y)$ subject to the constraint $g(x, y) = 0$, we will use the new variable λ to form the **objective function**

$$F(x, y, \lambda) = f(x, y) + \lambda g(x, y).$$

It can be shown that the critical values of $F(x, y, \lambda)$ will satisfy the constraint $g(x, y)$ and will also be critical points of $f(x, y)$. Thus we need only find the critical points of $F(x, y, \lambda)$ to find the required critical points.

To find the critical points of $F(x, y, \lambda)$, we must find the points that make all the partial derivatives equal to 0. That is, the points must satisfy

$$\partial F/\partial x = 0, \quad \partial F/\partial y = 0, \quad \text{and} \quad \partial F/\partial \lambda = 0.$$

Since $F(x, y, \lambda) = f(x, y) + \lambda g(x, y)$, these equations may be written as

$$\frac{\partial f}{\partial x} + \lambda \frac{\partial g}{\partial x} = 0$$

$$\frac{\partial f}{\partial y} + \lambda \frac{\partial g}{\partial y} = 0$$

$$g(x, y) = 0.$$

Finding the values of x and y that satisfy these three equations simultaneously gives the critical values.

   This method will not tell us if the critical points correspond to maxima or minima, but this can be determined either from the physical setting for the problem or by testing according to a procedure similar to that used for unconstrained maxima and minima. The following examples will illustrate the use of Lagrange multipliers.

EXAMPLE I
Find the maximum value of $z = x^2y$ subject to $x + y = 9$, $x \geq 0$, $y \geq 0$.

Solution
The function to be maximized is $f(x, y) = x^2y$.
The constraint is $g(x, y) = 0$, where $g(x, y) = x + y - 9$.
The objective function is

$$F(x, y, \lambda) = f(x, y) + \lambda g(x, y),$$

or

$$F(x, y, \lambda) = x^2y + \lambda(x + y - 9).$$

Thus

$$\frac{\partial F}{\partial x} = 2xy + \lambda(1) = 0, \quad \text{or} \quad 2xy + \lambda = 0$$

$$\frac{\partial F}{\partial y} = x^2 + \lambda(1) = 0, \quad \text{or} \quad x^2 + \lambda = 0$$

$$\frac{\partial F}{\partial \lambda} = 0 + 1(x + y - 9) = 0, \quad \text{or} \quad x + y - 9 = 0.$$

Solving the first two equations for λ and substituting gives

$$\lambda = -2xy$$
$$\lambda = -x^2$$
$$2xy = x^2$$
$$2xy - x^2 = 0$$
$$x(2y - x) = 0.$$

So

$$x = 0 \quad \text{or} \quad x = 2y.$$

Since $x = 0$ could not make $z = x^2 y$ a maximum, we substitute $x = 2y$ into $x + y - 9 = 0$.

$$2y + y = 9$$
$$y = 3$$
$$x = 6$$

Thus the function $z = x^2 y$ is maximized at 108 when $x = 6$, $y = 3$, if the constraint is $x + y = 9$. Testing values near $x = 6$, $y = 3$, and satisfying the constraint shows that the function is maximized there. (Try $x = 5.5$, $y = 3.5$, $x = 7$, $y = 2$, and so on.)

EXAMPLE 2
Find the minimum value of the function $z = x^3 + y^3 + xy$ subject to the constraint $x + y - 4 = 0$.

Solution
The function to be minimized is $f(x, y) = x^3 + y^3 + xy$.
The constraint function is $g(x, y) = x + y - 4$.
The objective function is

$$F(x, y, \lambda) = f(x, y) + \lambda g(x, y),$$

or

$$F(x, y, \lambda) = x^3 + y^3 + xy + \lambda(x + y - 4).$$

Then

$$\frac{\partial F}{\partial x} = 3x^2 + y + \lambda = 0$$

$$\frac{\partial F}{\partial y} = 3y^2 + x + \lambda = 0$$

$$\frac{\partial F}{\partial \lambda} = x + y - 4 = 0.$$

Solving the first two equations for λ and substituting gives the following.

$$\lambda = -(3x^2 + y)$$
$$\lambda = -(3y^2 + x)$$
$$3x^2 + y = 3y^2 + x$$

Solving $x + y - 4 = 0$ for y gives $y = 4 - x$. Substituting for y in the equation above, we get

$$3x^2 + (4 - x) = 3(4 - x)^2 + x$$
$$3x^2 + 4 - x = 48 - 24x + 3x^2 + x$$
$$22x = 44 \quad \text{or} \quad x = 2.$$

Thus when $x + y - 4 = 0$, $x = 2$ and $y = 2$ give the minimum value $z = 20$.

EXAMPLE 3

Find the maximum value of the function $z = 2xy - 2x^2 - 2y^2$ subject to the constraint $2x + y = 7$.

Solution
The constraint function is $g(x, y) = 2x + y - 7$, so the objective function is

$$F(x, y, \lambda) = 2xy - 2x^2 - 2y^2 + \lambda(2x + y - 7).$$

Then

$$\frac{\partial F}{\partial x} = 2y - 4x + 2\lambda = 0$$

$$\frac{\partial F}{\partial y} = 2x - 4y + \lambda = 0$$

$$\frac{\partial F}{\partial \lambda} = 2x + y - 7 = 0.$$

Solving the first two equations for λ gives

$$\lambda = -y + 2x$$
$$\lambda = -2x + 4y$$
$$-y + 2x = -2x + 4y$$
$$4x = 5y$$
$$x = \frac{5y}{4}.$$

Substituting for x in $2x + y - 7 = 0$ gives $\frac{5}{2}y + y - 7 = 0$, or $y = 2$. Thus the critical values are $y = 2$, $x = \frac{5}{2}$. Testing values near $x = \frac{5}{2}$, $y = 2$, shows that the function has a relative maximum there. The maximum value is $-21/2$.

CHECKPOINT

Find the minimum value of $f(x, y) = x^2 + y^2 - 4xy$, subject to the constraint $x + y = 10$, by:

1. forming the objective function $F(x, y, \lambda)$;

2. finding $\dfrac{\partial F}{\partial x}$, $\dfrac{\partial F}{\partial y}$, and $\dfrac{\partial F}{\partial \lambda}$;

3. setting the three partial derivatives (from question 2) equal to 0, and solving the equations simultaneously for x and y;

4. finding the value of $f(x, y)$ at the critical values of x and y.

We can also use Lagrange multipliers to find the maxima and minima of functions of three (or more) variables, subject to two (or more) constraints. The method involves using two multipliers, one for each constraint, to form an objective function $F = f + \lambda g_1 + \mu g_2$. We leave further discussion for more advanced courses.

We can easily extend the method to functions of three or more variables, as the following example shows.

EXAMPLE 4

Find the minimum value of the function $w = x + y^2 + z^2$, subject to the constraint $x + y + z = 1$.

Solution

The function to be minimized is $f(x, y, z) = x + y^2 + z^2$. The constraint is $g(x, y, z) = 0$, where $g(x, y, z) = x + y + z - 1$.
The objective function is

$$F(x, y, z, \lambda) = f(x, y, z) + \lambda g(x, y, z),$$

or

$$F(x, y, z, \lambda) = x + y^2 + z^2 + \lambda(x + y + z - 1).$$

Then

$$\frac{\partial F}{\partial x} = 1 + \lambda = 0$$

$$\frac{\partial F}{\partial y} = 2y + \lambda = 0$$

$$\frac{\partial F}{\partial z} = 2z + \lambda = 0$$

$$\frac{\partial F}{\partial \lambda} = x + y + z - 1 = 0.$$

Solving the first three equations simultaneously gives

$$\lambda = -1 \qquad y = \frac{1}{2} \qquad z = \frac{1}{2}.$$

Substituting these values in the fourth equation (which is the constraint), we get $x + \frac{1}{2} + \frac{1}{2} - 1 = 0$, so $x = 0, y = \frac{1}{2}, z = \frac{1}{2}$. Thus $w = \frac{1}{2}$ is the minimum value since other values of x, y, and z satisfying $x + y + z = 1$ give larger values for w.

Let us now return to the utility problem posed in the Application Preview.

EXAMPLE 5

Find x and y that maximize the utility function $U = x^2y^2$ subject to the budget constraint $2x + 4y = 40$.

Solution

First we rewrite the constraint as $2x + 4y - 40 = 0$. Then the objective function is

$$F(x, y, \lambda) = x^2y^2 + \lambda(2x + 4y - 40).$$

$$\frac{\partial F}{\partial x} = 2xy^2 + 2\lambda, \qquad \frac{\partial F}{\partial y} = 2x^2y + 4\lambda, \qquad \frac{\partial F}{\partial \lambda} = 2x + 4y - 40$$

Setting these partial derivatives equal to 0 and solving gives

$$-\lambda = xy^2 = x^2y/2, \quad \text{or} \quad xy^2 - x^2y/2 = 0,$$

so

$$xy(y - x/2) = 0$$

yields $x = 0$, $y = 0$, or $x = 2y$. Neither $x = 0$ nor $y = 0$ maximizes utility. If $x = 2y$, we have

$$0 = 4y + 4y - 40.$$

Thus $y = 5$ and $x = 10$.

Testing values near $x = 10$, $y = 5$ shows that these values maximize utility at $U = 2500$.

Figure 14.13 shows the budget constraint $2x + 4y = 40$ from Example 5 graphed with the indifference curves for $U = x^2y^2$ that correspond to $U = 500$, $U = 2500$, and $U = 5000$.

Whenever an indifference curve intersects the budget constraint, that utility level is attainable within the budget. Notice that the highest attainable utility (such as $U = 2500$, found in Example 5) corresponds to the indifference curve that touches the budget constraint at exactly one point; that is, that has the budget constraint as a tangent line. Notice also that utility levels greater than $U = 2500$ are not attainable within the budget because the indifference curve "misses" the budget constraint line (as for $U = 5000$).

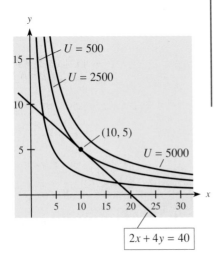

Figure 14.13

EXAMPLE 6

Suppose that the Cobb-Douglas production function for a certain manufacturer gives the number of units of production z according to

$$z = f(x, y) = 100x^{4/5}y^{1/5},$$

where x is the number of units of labor and y is the number of units of capital. Suppose further that labor costs \$160 per unit, capital costs \$200 per unit, and the total cost for capital and labor is limited to \$100,000, so that production is constrained by

$$160x + 200y = 100,000.$$

Find the number of units of labor and the number of units of capital that maximize production.

Solution

The objective function is

$$F(x, y, \lambda) = 100x^{4/5}y^{1/5} + \lambda(160x + 200y - 100,000).$$

$$\frac{\partial F}{\partial x} = 80x^{-1/5}y^{1/5} + 160\lambda, \qquad \frac{\partial F}{\partial y} = 20x^{4/5}y^{-4/5} + 200\lambda,$$

$$\frac{\partial F}{\partial \lambda} = 160x + 200y - 100,000$$

Setting these partial derivatives equal to 0 and solving gives

$$\lambda = \frac{-80x^{-1/5}y^{1/5}}{160} = \frac{-20x^{4/5}y^{-4/5}}{200} \qquad \text{or} \qquad \frac{y^{1/5}}{2x^{1/5}} = \frac{x^{4/5}}{10y^{4/5}}.$$

This means $5y = x$. Using this in $\dfrac{\partial F}{\partial \lambda} = 0$ gives

$$160(5y) + 200y - 100,000 = 0$$
$$1000y = 100,000$$
$$y = 100$$
$$x = 5y = 500.$$

Thus, production is maximized at $z = 100(500)^{4/5}(100)^{1/5} \approx 36,239$ when $x = 500$ (units of labor) and $y = 100$ (units of capital). See Figure 14.14.

In problems of this type, economists call the value of $-\lambda$ the **marginal productivity of money**. In this case

$$-\lambda = \frac{y^{1/5}}{2x^{1/5}} = \frac{(100)^{0.2}}{2(500)^{0.2}} \approx 0.362.$$

This means that each additional dollar spent on production results in approximately 0.362 additional units produced.

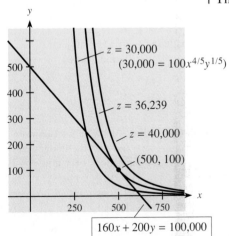

$160x + 200y = 100,000$

Figure 14.14

Finally, Figure 14.14 shows the graph of the constraint together with some production function curves that correspond to different production levels.

CHECKPOINT SOLUTIONS

1. $F(x, y, \lambda) = x^2 + y^2 - 4xy + \lambda(10 - x - y)$

2. $\dfrac{\partial F}{\partial x} = 2x - 4y - \lambda, \quad \dfrac{\partial F}{\partial y} = 2y - 4x - \lambda, \quad \dfrac{\partial F}{\partial \lambda} = 10 - x - y$

3. $0 = 2x - 4y - \lambda$ (1)
   $0 = 2y - 4x - \lambda$ (2)
   $0 = 10 - x - y$ (3)
   From equations (1) and (2) we have the following:

$$\lambda = 2x - 4y \quad \text{and} \quad \lambda = 2y - 4x \quad \text{so}$$

$$2x - 4y = 2y - 4x$$

$$6x = 6y \quad \text{or} \quad x = y.$$

   Using $x = y$ in equation (3) gives $0 = 10 - x - x$, or $2x = 10$. Thus $x = 5$ and $y = 5$.

4. $f(5, 5) = 25 + 25 - 100 = -50$ is the minimum because other values satisfying the constraint give larger z-values.

EXERCISE 14.5

1. Find the minimum value of $z = x^2 + y^2$ subject to the condition $x + y = 6$.
2. Find the minimum value of $z = 4x^2 + y^2$ subject to the constraint $x + y = 5$.
3. Find the minimum value of $z = 3x^2 + 5y^2 - 2xy$ subject to the constraint $x + y = 5$.
4. Find the maximum value of $z = 2xy - 3x^2 - 5y^2$ subject to the constraint $x + y = 5$.
5. Find the maximum value of $z = x^2y$ subject to $x + y = 6$, $x \geq 0$, $y \geq 0$.
6. Find the maximum value of the function $z = x^3y^2$ subject to $x + y = 10$, $x \geq 0$, $y \geq 0$.
7. Find the maximum value of the function $z = 2xy - 2x^2 - 4y^2$ subject to the condition $x + 2y = 8$.
8. Find the minimum value of $z = 2x^2 + y^2 - xy$ subject to the constraint $2x + y = 8$.
9. Find the minimum value of $z = x^2 + y^2$ subject to the condition $2x + y + 1 = 0$.
10. Find the minimum value of $z = x^2 + y^2$ subject to the condition $xy = 1$.

11. Find the minimum value of the function $w = x^2 + y^2 + z^2$ subject to the constraint $x + y + z = 3$.
12. Find the minimum value of the function $w = x^2 + y^2 + z^2$ subject to the condition $2x - 4y + z = 21$.
13. Find the maximum value of $w = xz + y$ subject to the constraint $x^2 + y^2 + z^2 = 1$.
14. Find the maximum value of $w = x^2yz$ subject to $4x + y + z = 4$, $x \geq 0$, $y \geq 0$, and $z \geq 0$.

APPLICATIONS

15. **Utility** Suppose that the utility function for two commodities is given by $U = x^2y$ and that the budget constraint is $3x + 6y = 18$. What values of x and y will maximize utility?
16. **Utility** Suppose that the budget constraint in problem 15 is $5x + 20y = 80$. What values of x and y will maximize $U = x^2y$?
17. **Utility** Suppose that the utility function for two products is given by $U = x^2y$, and the budget constraint is $2x + 3y = 120$. Find the values of x and y that maxi-

mize utility. Check by graphing the budget constraint with the indifference curve for maximum utility and with two other indifference curves.

18. **Utility** Suppose that the utility function for two commodities is given by $U = x^2y^3$, and the budget constraint is $10x + 15y = 250$. Find the values of x and y that maximize utility. Check by graphing the budget constraint with the indifference curve for maximum utility and with two other indifference curves.

19. **Production** A company has the Cobb-Douglas production function
$$z = 400x^{0.6}y^{0.4},$$
where x is the number of units of labor and y is the number of units of capital. Suppose labor costs $150 per unit, capital costs $100 per unit, and the total cost of labor and capital is limited to $100,000.
   (a) Find the number of units of labor and the number of units of capital that maximize production.
   (b) Find the marginal productivity of money and interpret it.
   (c) Graph the constraint with the optimal value or production and with two other z-values (one smaller than the optimal value and one larger).

20. **Production** Suppose a company has the Cobb-Douglas production function
$$z = 100^{0.75}y^{0.25},$$
where x is the number of units of labor and y is the number of units of capital. Suppose further that labor costs $90 per unit, capital costs $150 per unit, and the total costs of labor and capital are limited to $90,000.
   (a) Find the number of units of labor and the number of units of capital that maximize production.
   (b) Find the marginal productivity of money and interpret it.

(c) Graph the constraint with the optimal value for production and with two other z-values (one smaller than the optimal value and one larger).

21. **Cost** A firm has two plants, X and Y. Suppose that the cost of producing x units at plant X is $x^2 + 1200$ and the cost of producing y units of the same product at plant Y is given by $3y^2 + 800$. If they have an order for 1200 units, how many should they produce at each plant to fill this order and minimize the cost of production?

22. **Cost** Suppose that the cost of producing x units at plant X is $(3x + 4)x$ and that the cost of producing y units of the same product at plant Y is $(2y + 8)y$. If the firm owning the plants has an order for 149 units, how many should it produce at each plant to fill this order and minimize its cost of production?

23. **Revenue** On the basis of past experience a company has determined that its sales revenue is related to its advertising according to the formula $s = 20x + y^2 + 4xy$, where x is the amount spent in radio advertising and y is the amount spent in television advertising. If the company plans to spend $30,000 on these two means of advertising, how much should it spend on each method to maximize its sales revenue?

24. **Manufacturing** Find the dimensions x, y, and z of the rectangular box with the largest volume that satisfies
$$3x + 4y + 12z = 12.$$

25. **Manufacturing** Find the dimensions of the box with square base, open top, and volume 500,000 cc that requires the least materials.

26. **Manufacturing** Show that a box with a square base, an open top, and a fixed volume requires the least material to build if it has a height equal to one-half the length of one side of the base.

KEY TERMS
& FORMULAS

Section	Key Term	Formula
14.1	Function of two variables	
	Variables: independent, dependent	
	Domain	
	Coordinate planes	
	Utility	
	Indifference curve	
	Indifference map	

Section	Key Term	Formula
14.2	First-order partial derivative	
	With respect to x	$z_x = \dfrac{\partial z}{\partial x}$
	With respect to y	$z_y = \dfrac{\partial z}{\partial y}$
	Higher-order partial derivatives	
	Second partial derivatives	z_{xx}, z_{yy}, z_{xy}, and z_{yx}
14.3	Joint cost function	$C = Q(x, y)$
	Marginal cost	
	Marginal productivity	
	Demand function	
	Marginal demand function	
	Competitive products	
	Complementary products	
14.4	Critical values for maxima and minima	Solve simultaneously $\begin{cases} z_x = 0 \\ z_y = 0 \end{cases}$.
	Test for critical values	Use $D(x, y) = (z_{xx})(z_{yy}) - (z_{xy})^2$.
14.5	Maxima and minima subject to constraints	
	Lagrange multipliers	
	Objective function	$F(x, y, \lambda) = f(x, y) + \lambda g(x, y)$

REVIEW EXERCISES

14.1

1. What is the domain of $z = \dfrac{3}{2x - y}$?

2. What is the domain of $z = \dfrac{3x + 2\sqrt{y}}{x^2 + y^2}$?

3. If $w(x, y, z) = x^2 - 3yz$, find $w(2, 3, 1)$.

4. If $Q(K, L) = 70K^{2/3}L^{1/3}$, find $Q(64{,}000, 512)$.

14.2

5. Find $\dfrac{\partial z}{\partial x}$ if $z = 5x^3 + 6xy + y^2$.

6. Find $\dfrac{\partial z}{\partial y}$ if $z = 12x^5 - 14x^3y^3 + 6y^4 - 1$.

In problems 7–12, find z_x and z_y.

7. $z = 4x^2y^3 + \dfrac{x}{y}$

8. $z = \sqrt{x^2 + 2y^2}$

9. $z = (xy + 1)^{-2}$

10. $z = e^{x^2y^3}$

11. $z = e^{xy} + y \ln x$

12. $z = e^{\ln xy}$

13. Find the partial derivative of $f(x, y) = 4x^3 - 5xy^2 + y^3$ with respect to x at the point $(1, 2, -8)$.

14. Find the slope of the tangent in the x-direction to the surface $z = 5x^4 - 3xy^2 + y^2$ at $(1, 2, -3)$.

In problems 15–18, find the second partials.
(a) z_{xx}　(b) z_{yy}　(c) z_{xy}　(d) z_{yx}

15. $z = x^2y - 3xy$

16. $z = 3x^3y^4 - \dfrac{x^2}{y^2}$

17. $z = x^2e^{y^2}$

18. $z = \ln(xy + 1)$

14.4

19. Test $z = 16 - x^2 - xy - y^2 + 24y$ for maxima and minima.

20. Test $z = x^3 + y^3 - 3xy$ for maxima and minima.

14.5

21. Find the minimum value of $z = 4x^2 + y^2$ subject to the constraint $x + y = 10$.

22. Find the maximum value of $z = x^4y^2$ subject to $x + y = 9$, $x \geq 0$, $y \geq 0$.

APPLICATIONS

14.1

23. **Utility** Suppose that the utility function for two goods X and Y is given by $U = x^2y$, and a consumer purchases 6 units of X and 15 units of Y. If the consumer purchases 60 units of Y, how many units of X must be purchased to retain the same level of utility?

24. **Utility** Suppose that an indifference curve for two products, X and Y, has the equation $xy = 1600$. If 80 units of X are purchased, how many units of Y must be purchased?

14.2

25. **Concorde sonic booms** The width of the region on the ground on either side of the path of France's Concorde jet in which people hear the sonic boom is given by

$$w = f(T, h, d) = 2\sqrt{Th/d},$$

where T is the air temperature at ground level in degrees Kelvin, h is the Concorde's altitude in kilometers, and d is the vertical temperature gradient (the temperature drop in degrees Kelvin per kilometer).*
(a) Suppose the Concorde approaches Washington, D.C., from Europe on a course that takes it south of Nantucket Island at an altitude 16.8 km. If the surface temperature is 293°K and the vertical temperature gradient is 5°K/km, how far south of Nantucket must the plane pass to keep the sonic boom off the island?
(b) Interpret $f(287, 17.1, 4.9) \approx 63.3$.
(c) Find $\dfrac{\partial f}{\partial h}(293, 16.8, 5)$ and interpret the result.
(d) Find $\dfrac{\partial f}{\partial d}(293, 16.8, 5)$ and interpret the result.

14.3

26. **Cost** The joint cost function for two products is $C(x, y) = x^2\sqrt{y^2 + 13}$. Find the marginal cost with respect to
(a) x if 20 units of x and 6 units of y are produced.
(b) y if 20 units of x and 6 units of y are produced.

27. **Production** Suppose that the production function for a company is given by

$$Q = 80K^{1/4}L^{3/4},$$

where Q is the output (in hundreds of units), K is the capital expenditures (in thousands of dollars), and L is

the labor-hours. Find $\partial Q/\partial K$ and $\partial Q/\partial L$ when expenditures are \$625,000 and total labor hours are 4096. Interpret the results.

28. **Marginal demand** The demand functions for two related products, product A and product B, are given by

$$q_A = 400 - 2p_A - 3p_B$$
$$q_B = 300 - 5p_A - 6p_B.$$

(a) Find the marginal demand of q_A with respect to p_A.
(b) Find the marginal demand of q_B with respect to p_B.
(c) Are the products complementary or competitive?

29. **Marginal demand** Suppose that the demand functions for two related products, A and B, are given by

$$q_A = 800 - 40p_A - \frac{2}{p_B + 1}$$

$$q_B = 1000 - \frac{10}{p_A + 4} - 30p_B.$$

Determine whether the products are competitive or complementary.

14.4

30. **Profit** The profit from the sale of two products is given by $P(x, y) = 40x + 80y - x^2 - y^2$, where x is the number of units of product 1 and y is the number of units of product 2. Selling how much of each product will maximize profit?

14.5

31. **Utility** If the utility function for two commodities is $U = x^2y$, and the budget constraint is $4x + 5y = 60$, find the values of x and y that maximize utility.

32. **Production** Suppose a company has the Cobb-Douglas production function

$$z = 300x^{2/3}y^{1/3},$$

where x is the number of units of labor and y is the number of units of capital. Suppose labor costs and capital costs are each \$50 per unit and total costs are limited to \$75,000.
(a) Find the number of units of labor and the number of units of capital that maximize production.
(b) Find the marginal productivity of money and interpret your result.
(c) Graph the constraint with the production function when $z = 180,000$, $z = 300,000$, and the optimal z-value.

---

* Balachandra, N. K., W. L. Donn, and D. H. Rind, "Concorde Sonic Booms as an Atmospheric Probe," *Science,* 1 July 1977, Vol. 197, p. 47.

Advertising

To model sales of its tires, the manufacturer of GRIPPER tires used the quadratic equation $S = a_0 + a_1x + a_2x^2 + b_1y$, where S is regional sales in millions of dollars, x is TV advertising expenditures in millions of dollars, and y is other promotional expenditures in millions of dollars. (See the Consultant's Corner, Marginal Return to Sales, on page 681.)

Although this model represents the relationship between advertising and sales dollars for small changes in advertising expenditures, it is clear to the vice president of advertising that it does not apply to large expenditures for TV advertising on a national level. He knows from experience that increased expenditures for TV advertising do result in more sales, but at a decreasing rate of return for the product.

The vice president is aware that some advertising agencies model the relationship between advertising and sales by the function

$$S = b_0 + b_1(1 - e^{-ax}) + c_1y,$$

where $a > 0$, S is sales in millions of dollars, x is TV advertising expenditures in millions of dollars, and y is other promotional expenditures in millions of dollars.* The equation

$$S_n = 24.58 + 325.18(1 - e^{-x/14}) + b_1y$$

has the form mentioned previously as being used by some advertising agencies. For TV advertising expenditures up to \$20 million, this equation closely approximates

$$S_1 = 30 + 20x - 0.4x^2 + b_1y$$

which, in the Consultant's Corner on page 681, was used with fixed promotional expenses to describe advertising and sales in Region 1.

To help the vice president decide if this is a better model for large expenditures, answer the following questions.

1. What is $\partial S_1/\partial x$? Does this indicate that sales might actually decline after some amount is spent on TV advertising? If so, what is this amount?
2. Does the quadratic model $S_1(x, y)$ indicate that sales will become negative after some amount is spent on TV advertising? Does this model cease to be useful in predicting sales after a certain point?
3. What is $\partial S_n/\partial x$? Does this indicate that sales will continue to rise if additional money is devoted to TV advertising? Is S_n growing at a rate that is increasing or decreasing when promotional sales are held constant? Is S_n a better model for large expenditures?
4. If this model does describe the relationship between advertising and sales, and if promotional expenditures are held constant at y_0, is there an upper limit to the sales, even if an unlimited amount of money is spent on TV advertising? If so, what is it?

*Mansfield, Edwin, *Managerial Economics*, W. W. Norton and Co., 1990.

APPENDIX

Table I Areas Under the Standard Normal Curve

The value of A is the area under the standard normal curve between $z = 0$ and $z = z_0$, for $z_0 \geq 0$. Areas for negative values of z_0 are obtained by symmetry.

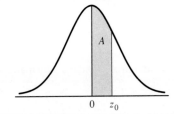

z_0	A	z_0	A	z_0	A	z_0	A
.00	.0000	.41	.1591	.82	.2939	1.23	.3907
.01	.0040	.42	.1628	.83	.2967	1.24	.3925
.02	.0080	.43	.1664	.84	.2996	1.25	.3944
.03	.0120	.44	.1700	.85	.3023	1.26	.3962
.04	.0160	.45	.1736	.86	.3051	1.27	.3980
.05	.0199	.46	.1772	.87	.3079	1.28	.3997
.06	.0239	.47	.1808	.88	.3106	1.29	.4015
.07	.0279	.48	.1844	.89	.3133	1.30	.4032
.08	.0319	.49	.1879	.90	.3159	1.31	.4049
.09	.0359	.50	.1915	.91	.3186	1.32	.4066
.10	.0398	.51	.1950	.92	.3212	1.33	.4082
.11	.0438	.52	.1985	.93	.3238	1.34	.4099
.12	.0478	.53	.2019	.94	.3264	1.35	.4115
.13	.0517	.54	.2054	.95	.3289	1.36	.4131
.14	.0557	.55	.2088	.96	.3315	1.37	.4147
.15	.0596	.56	.2123	.97	.3340	1.38	.4162
.16	.0636	.57	.2157	.98	.3365	1.39	.4177
.17	.0675	.58	.2190	.99	.3389	1.40	.4192
.18	.0714	.59	.2224	1.00	.3413	1.41	.4207
.19	.0754	.60	.2258	1.01	.3438	1.42	.4222
.20	.0793	.61	.2291	1.02	.3461	1.43	.4236
.21	.0832	.62	.2324	1.03	.3485	1.44	.4251
.22	.0871	.63	.2357	1.04	.3508	1.45	.4265
.23	.0910	.64	.2389	1.05	.3531	1.46	.4279
.24	.0948	.65	.2422	1.06	.3554	1.47	.4292
.25	.0987	.66	.2454	1.07	.3577	1.48	.4306
.26	.1026	.67	.2486	1.08	.3599	1.49	.4319
.27	.1064	.68	.2518	1.09	.3621	1.50	.4332
.28	.1103	.69	.2549	1.10	.3643	1.51	.4345
.29	.1141	.70	.2580	1.11	.3665	1.52	.4357
.30	.1179	.71	.2612	1.12	.3686	1.53	.4370
.31	.1217	.72	.2642	1.13	.3708	1.54	.4382
.32	.1255	.73	.2673	1.14	.3729	1.55	.4394
.33	.1293	.74	.2704	1.15	.3749	1.56	.4406
.34	.1331	.75	.2734	1.16	.3770	1.57	.4418
.35	.1368	.76	.2764	1.17	.3790	1.58	.4430
.36	.1406	.77	.2794	1.18	.3810	1.59	.4441
.37	.1443	.78	.2823	1.19	.3830	1.60	.4452
.38	.1480	.79	.2852	1.20	.3849	1.61	.4463
.39	.1517	.80	.2881	1.21	.3869	1.62	.4474
.40	.1554	.81	.2910	1.22	.3888	1.63	.4485

Table I Areas Under the Standard Normal Curve (*Continued*)

z_0	A	z_0	A	z_0	A	z_0	A
1.64	.4495	2.08	.4812	2.52	.4941	2.96	.4985
1.65	.4505	2.09	.4817	2.53	.4943	2.97	.4985
1.66	.4515	2.10	.4821	2.54	.4945	2.98	.4986
1.67	.4525	2.11	.4826	2.55	.4946	2.99	.4986
1.68	.4535	2.12	.4830	2.56	.4948	3.00	.4987
1.69	.4545	2.13	.4834	2.57	.4949	3.01	.4987
1.70	.4554	2.14	.4838	2.58	.4951	3.02	.4987
1.71	.4564	2.15	.4842	2.59	.4952	3.03	.4988
1.72	.4573	2.16	.4846	2.60	.4953	3.04	.4988
1.73	.4582	2.17	.4850	2.61	.4955	3.05	.4989
1.74	.4591	2.18	.4854	2.62	.4956	3.06	.4989
1.75	.4599	2.19	.4857	2.63	.4957	3.07	.4989
1.76	.4608	2.20	.4861	2.64	.4959	3.08	.4990
1.77	.4616	2.21	.4865	2.65	.4960	3.09	.4990
1.78	.4625	2.22	.4868	2.66	.4961	3.10	.4990
1.79	.4633	2.23	.4871	2.67	.4962	3.11	.4991
1.80	.4641	2.24	.4875	2.68	.4963	3.12	.4991
1.81	.4649	2.25	.4878	2.69	.4964	3.13	.4991
1.82	.4656	2.26	.4881	2.70	.4965	3.14	.4992
1.83	.4664	2.27	.4884	2.71	.4966	3.15	.4992
1.84	.4671	2.28	.4887	2.72	.4967	3.16	.4992
1.85	.4678	2.29	.4890	2.73	.4968	3.17	.4992
1.86	.4686	2.30	.4893	2.74	.4969	3.18	.4993
1.87	.4693	2.31	.4896	2.75	.4970	3.19	.4993
1.88	.4700	2.32	.4898	2.76	.4971	3.20	.4993
1.89	.4706	2.33	.4901	2.77	.4972	3.21	.4993
1.90	.4713	2.34	.4904	2.78	.4973	3.22	.4994
1.91	.4719	2.35	.4906	2.79	.4974	3.23	.4994
1.92	.4726	2.36	.4909	2.80	.4974	3.24	.4994
1.93	.4732	2.37	.4911	2.81	.4975	3.25	.4994
1.94	.4738	2.38	.4913	2.82	.4976	3.26	.4994
1.95	.4744	2.39	.4916	2.83	.4977	3.27	.4995
1.96	.4750	2.40	.4918	2.84	.4977	3.28	.4995
1.97	.4756	2.41	.4920	2.85	.4978	3.29	.4995
1.98	.4762	2.42	.4922	2.86	.4979	3.30	.4995
1.99	.4767	2.43	.4925	2.87	.4980	3.31	.4995
2.00	.4773	2.44	.4927	2.88	.4980	3.32	.4996
2.01	.4778	2.45	.4929	2.89	.4981	3.33	.4996
2.02	.4783	2.46	.4931	2.90	.4981	3.34	.4996
2.03	.4788	2.47	.4932	2.91	.4982	3.35	.4996
2.04	.4793	2.48	.4934	2.92	.4983	3.36	.4996
2.05	.4798	2.49	.4936	2.93	.4983	3.37	.4996
2.06	.4803	2.50	.4938	2.94	.4984	3.38	.4996
2.07	.4808	2.51	.4940	2.95	.4984	3.39	.4997

Table I Areas Under the Standard Normal Curve (*Continued*)

z_0	A	z_0	A	z_0	A	z_0	A
3.40	.4997	3.52	.4998	3.64	.4999	3.76	.4999
3.41	.4997	3.53	.4998	3.65	.4999	3.77	.4999
3.42	.4997	3.54	.4998	3.66	.4999	3.78	.4999
3.43	.4997	3.55	.4998	3.67	.4999	3.79	.4999
3.44	.4997	3.56	.4998	3.68	.4999	3.80	.4999
3.45	.4997	3.57	.4998	3.69	.4999	3.81	.4999
3.46	.4997	3.58	.4998	3.70	.4999	3.82	.4999
3.47	.4997	3.59	.4998	3.71	.4999	3.83	.4999
3.48	.4998	3.60	.4998	3.72	.4999	3.84	.4999
3.49	.4998	3.61	.4999	3.73	.4999	3.85	.4999
3.50	.4998	3.62	.4999	3.74	.4999	3.86	.4999
3.51	.4998	3.63	.4999	3.75	.4999		

Table II Amount of \$1 at Compound Interest $\left(1 + \dfrac{j}{k}\right)^{kn}$

Periods	$\frac{1}{2}\%$	$\frac{3}{4}\%$	1%	$1\frac{1}{2}\%$	2%
1	1.005 000	1.007 500	1.010 000	1.015 000	1.020 000
2	1.010 025	1.015 056	1.020 100	1.030 225	1.040 400
3	1.015 075	1.022 669	1.030 301	1.045 678	1.061 208
4	1.020 151	1.030 339	1.040 604	1.061 364	1.082 432
5	1.025 251	1.038 067	1.051 010	1.077 284	1.104 081
6	1.030 378	1.045 852	1.061 520	1.093 443	1.126 162
7	1.035 529	1.053 696	1.072 135	1.109 845	1.148 686
8	1.040 707	1.061 599	1.082 857	1.126 493	1.171 659
9	1.045 911	1.069 561	1.093 685	1.143 390	1.195 093
10	1.051 140	1.077 583	1.104 622	1.160 541	1.218 994
11	1.056 396	1.085 664	1.115 668	1.177 949	1.243 374
12	1.061 678	1.093 807	1.126 825	1.195 618	1.268 242
13	1.066 986	1.102 010	1.138 093	1.213 552	1.293 607
14	1.072 321	1.110 276	1.149 474	1.231 756	1.319 479
15	1.077 683	1.118 603	1.160 969	1.250 232	1.345 868
16	1.083 071	1.126 992	1.172 579	1.268 986	1.372 786
17	1.088 487	1.135 445	1.184 304	1.288 020	1.400 241
18	1.093 929	1.143 960	1.196 147	1.307 341	1.428 246
19	1.099 399	1.152 540	1.208 109	1.326 951	1.456 811
20	1.104 896	1.161 184	1.220 190	1.346 855	1.485 947
21	1.110 420	1.169 893	1.232 392	1.367 058	1.515 666
22	1.115 972	1.178 667	1.244 716	1.387 564	1.545 980
23	1.121 552	1.187 507	1.257 163	1.408 377	1.576 899
24	1.127 160	1.196 414	1.269 735	1.429 503	1.608 437
25	1.132 796	1.205 387	1.282 432	1.450 945	1.640 606
26	1.138 460	1.214 427	1.295 256	1.472 710	1.673 418
27	1.144 152	1.223 535	1.308 209	1.494 800	1.706 886
28	1.149 873	1.232 712	1.321 291	1.517 222	1.741 024
29	1.155 622	1.241 957	1.354 504	1.539 981	1.775 845
30	1.161 400	1.251 272	1.347 849	1.563 080	1.811 362
31	1.167 207	1.260 656	1.361 327	1.586 526	1.847 589
32	1.173 043	1.270 111	1.374 941	1.610 324	1.884 541
33	1.178 908	1.279 637	1.388 690	1.634 479	1.922 231
34	1.184 803	1.289 234	1.402 577	1.658 996	1.960 676
35	1.190 727	1.298 904	1.416 603	1.683 881	1.999 890
36	1.196 681	1.308 645	1.430 769	1.709 140	2.039 887
37	1.202 664	1.318 460	1.445 076	1.734 777	2.080 685
38	1.208 677	1.328 349	1.459 527	1.760 798	2.122 299
39	1.214 721	1.338 311	1.474 123	1.787 210	2.164 745
40	1.220 794	1.348 349	1.488 864	1.814 018	2.208 040

Table II Amount of $1 at Compound Interest $\left(1 + \dfrac{j}{k}\right)^{kn}$ (*Continued*)

Periods	3%	4%	5%	6%	7%
1	1.030 000	1.040 000	1.050 000	1.060 000	1.070 000
2	1.060 900	1.081 600	1.102 500	1.123 600	1.114 900
3	1.092 727	1.124 864	1.157 625	1.191 016	1.225 043
4	1.125 509	1.169 859	1.215 506	1.262 477	1.310 796
5	1.159 274	1.216 653	1.276 282	1.338 226	1.402 552
6	1.194 052	1.265 319	1.340 096	1.418 519	1.500 730
7	1.229 874	1.315 932	1.407 100	1.503 630	1.605 781
8	1.266 770	1.368 569	1.477 455	1.593 848	1.718 186
9	1.304 773	1.423 312	1.551 328	1.689 479	1.838 459
10	1.343 916	1.480 244	1.628 895	1.790 848	1.967 151
11	1.384 234	1.539 454	1.710 339	1.898 299	2.104 852
12	1.425 761	1.601 032	1.795 856	2.012 196	2.252 192
13	1.468 534	1.665 074	1.885 649	2.132 928	2.409 845
14	1.512 590	1.731 676	1.979 932	2.260 904	2.578 534
15	1.557 967	1.800 944	2.078 928	2.396 558	2.759 032
16	1.604 706	1.872 981	2.182 875	2.540 352	2.952 164
17	1.652 848	1.947 901	2.292 018	2.692 773	3.158 815
18	1.702 433	2.025 817	2.406 619	2.854 339	3.379 932
19	1.753 506	2.106 849	2.526 950	3.025 600	3.616 528
20	1.806 111	2.191 123	2.653 298	3.207 135	3.869 684
21	1.860 295	2.278 768	2.785 963	3.399 564	4.140 562
22	1.916 103	2.369 919	2.925 261	3.603 537	4.430 402
23	1.973 587	2.464 716	3.071 524	3.819 750	4.740 530
24	2.032 794	2.563 304	3.225 100	4.048 935	5.072 367
25	2.093 778	2.665 836	3.386 355	4.291 871	5.427 433
26	2.156 591	2.772 470	3.555 673	4.549 383	5.807 353
27	2.221 289	2.883 369	3.733 456	4.822 346	6.213 868
28	2.287 928	2.998 703	3.920 129	5.111 687	6.648 838
29	2.356 566	3.118 651	4.116 136	5.418 388	7.114 257
30	2.427 262	3.243 398	4.321 942	5.743 491	7.612 255
31	2.500 080	3.373 133	4.538 039	6.088 101	8.145 113
32	2.575 083	3.508 059	4.764 941	6.453 387	8.715 271
33	2.652 335	3.648 381	5.003 189	6.840 590	9.325 340
34	2.731 905	3.794 316	5.253 348	7.251 025	9.978 114
35	2.813 862	3.946 089	5.516 015	7.686 087	10.676 582
36	2.898 278	4.103 933	5.791 816	8.147 252	11.423 942
37	2.985 227	4.268 090	6.081 407	8.636 087	12.223 618
38	3.074 783	4.438 813	6.385 477	9.154 252	13.079 272
39	3.167 027	4.616 366	6.704 751	9.703 507	13.994 821
40	3.262 038	4.801 021	7.039 989	10.285 718	14.974 458

Table II Amount of \$1 at Compound Interest $\left(1 + \dfrac{j}{k}\right)^{kn}$ *(Continued)*

Periods	8%	9%	10%	11%	12%
1	1.080 000	1.090 000	1.100 000	1.110 000	1.120 000
2	1.166 400	1.188 100	1.210 000	1.232 100	1.254 400
3	1.259 712	1.295 029	1.331 000	1.367 631	1.404 928
4	1.360 489	1.411 582	1.464 100	1.518 070	1.573 519
5	1.469 328	1.538 624	1.610 510	1.685 058	1.762 342
6	1.586 874	1.677 100	1.771 561	1.870 415	1.973 823
7	1.713 824	1.828 039	1.948 717	2.076 160	2.210 681
8	1.850 930	1.992 563	2.143 589	2.304 538	2.475 963
9	1.999 005	2.171 893	2.357 948	2.558 037	2.773 079
10	2.158 925	2.367 364	2.593 742	2.839 421	3.105 848
11	2.331 639	2.580 426	2.853 117	3.151 757	3.478 550
12	2.518 170	2.812 665	3.138 428	3.498 451	3.895 976
13	2.719 624	3.065 805	3.452 271	3.883 280	4.363 493
14	2.937 194	3.341 727	3.797 498	4.310 441	4.887 112
15	3.172 169	3.642 482	4.177 248	4.784 590	5.473 566
16	3.425 943	3.970 306	4.594 973	5.310 894	6.130 394
17	3.700 018	4.327 633	5.054 470	5.895 073	6.866 041
18	3.996 020	4.717 120	5.559 917	6.543 553	7.689 966
19	4.315 701	5.141 661	6.115 909	7.623 344	8.612 762
20	4.660 957	5.604 411	6.727 500	8.062 312	9.646 293
21	5.033 834	6.108 808	7.400 250	8.949 166	10.803 848
22	5.436 540	6.658 600	8.140 275	9.933 574	12.100 310
23	5.871 464	7.257 874	8.954 303	11.026 267	13.552 347
24	6.341 181	7.911 083	9.849 733	12.239 157	15.178 629
25	6.848 475	8.623 081	10.834 706	13.585 464	17.000 064
26	7.396 353	9.399 158	11.918 177	15.079 865	19.040 072
27	7.988 062	10.245 082	13.109 994	16.738 650	21.324 881
28	8.627 106	11.167 140	14.420 994	18.579 902	23.883 867
29	9.317 275	12.172 182	15.863 093	20.623 691	26.749 931
30	10.062 657	13.267 678	17.449 403	22.892 297	29.959 922
31	10.867 670	14.461 770	19.194 343	25.410 450	33.555 113
32	11.737 083	15.763 329	21.113 777	28.205 599	37.581 726
33	12.676 050	17.182 028	23.225 155	31.308 215	42.091 534
34	13.690 134	18.728 411	25.547 670	34.752 119	47.142 518
35	14.785 344	20.413 968	28.102 437	38.574 852	52.799 620
36	15.968 172	22.251 225	30.912 681	42.818 085	59.135 574
37	17.245 626	24.253 835	34.003 948	47.528 075	66.231 843
38	18.625 276	26.436 680	37.404 344	52.756 163	74.179 664
39	20.115 298	28.815 982	41.144 779	58.559 341	83.081 224
40	21.724 522	31.409 420	45.259 257	65.000 868	93.050 971

Table III   Amount of an Ordinary Annuity of $1 ($S_{\overline{n}|i}$)

Periods	1%	2%	3%	4%	5%	6%
1	1.000 000	1.000 000	1.000 000	1.000 000	1.000 000	1.000 000
2	2.010 000	2.020 000	2.030 000	2.040 000	2.050 000	2.060 000
3	3.030 100	3.060 400	3.090 900	3.121 600	3.152 500	3.183 600
4	4.060 401	4.121 608	4.183 627	4.246 464	4.310 125	4.374 616
5	5.101 005	5.204 040	5.309 136	5.416 323	5.525 631	5.637 093
6	6.152 015	6.308 121	6.468 410	6.632 975	6.801 913	6.975 319
7	7.213 535	7.434 284	7.662 462	7.898 294	8.142 008	8.393 838
8	8.285 670	8.582 969	8.892 336	9.214 226	9.549 109	9.897 468
9	9.368 527	9.754 629	10.159 106	10.582 795	11.026 564	11.491 316
10	10.462 212	10.949 721	11.463 879	12.006 107	12.577 893	13.180 795
11	11.566 834	12.168 716	12.807 796	13.486 351	14.206 787	14.971 643
12	12.682 503	13.412 090	14.192 030	15.025 805	15.917 126	16.869 941
13	13.809 328	14.680 332	15.617 790	16.626 838	17.712 983	18.882 138
14	14.947 421	15.973 939	17.086 324	18.291 911	19.598 632	21.015 066
15	16.096 895	17.293 418	18.598 914	20.023 588	21.578 564	23.275 970
16	17.257 864	18.639 286	20.156 881	21.824 531	23.657 492	25.672 529
17	18.430 443	20.012 072	21.761 588	23.697 512	25.840 366	28.212 880
18	19.614 747	21.412 313	23.414 435	26.645 413	28.132 385	30.905 653
19	20.810 895	22.840 559	25.116 869	27.671 229	30.539 004	33.759 992
20	22.019 004	24.297 371	26.870 375	29.778 079	33.065 954	36.787 592
21	23.239 194	25.783 318	28.676 486	31.969 202	35.719 252	39.992 727
22	24.471 586	27.298 985	30.536.780	34.247 970	38.505 214	43.392 291
23	25.716 301	28.844 964	32.452 884	36.617 889	41.430 475	46.995 829
24	26.973 464	30.421 864	34.426 470	39.082 604	44.501 999	50.815 578
25	28.243 199	32.030 301	36.459 264	41.645 908	47.727 099	54.864 513
26	29.525 631	33.670 907	38.553 042	44.311 745	51.113 454	59.156 384
27	30.820 887	35.344 325	40.709 634	47.084 214	54.669 126	63.705 767
28	32.129 096	37.051 212	42.930 923	49.967 583	58.402 583	68.528 113
29	33.450 387	38.792 236	45.218 850	52.966 286	62.322 712	73.639 800
30	34.784 891	40.568 081	47.575 416	56.084 938	66.438 847	79.058 188
31	36.132 740	42.379 443	50.002 678	59.328 335	70.760 790	84.801 679
32	37.494 067	44.227 031	52.502 759	62.701 469	75.298 829	90.889 780
33	38.869 008	46.111 572	55.077 842	66.209 528	80.063 771	97.343 167
34	40.257 698	48.033 804	57.730 177	69.857 909	85.066 959	104.183 757
35	41.660 275	49.994 480	60.462 082	73.652 225	90.320 307	111.434 783
36	43.076 878	51.994 369	63.275 945	77.598 314	95.836 323	119.120 870
37	44.507 646	54.034 257	66.174 223	81.702 247	101.628 139	127.268 122
38	45.952 723	56.114 942	69.159 450	85.970 336	107.709 546	135.904 209
39	47.412 250	58.237 241	72.234 233	90.409 150	114.095 023	145.058 462
40	48.886 373	60.401 986	75.401 260	95.025 516	120.799 774	154.761 970

Appendix

Table III Amount of an Ordinary Annuity of $1 ($S_{\overline{n}|i}$) (*Continued*)

Periods	7%	8%	9%	10%	11%	12%
1	1.000 000	1.000 000	1.000 000	1.000 000	1.000 000	1.000 000
2	2.070 000	2.080 000	2.090 000	2.100 000	2.110 000	2.120 000
3	3.214 900	3.246 400	3.278 100	3.310 000	3.342 100	3.374 400
4	4.439 943	4.506 112	4.573 129	4.641 000	4.709 731	4.779 328
5	5.750 739	5.866 601	5.984 711	6.105 100	6.227 801	6.352 847
6	7.153 291	7.335 929	7.523 335	7.715 610	7.912 860	8.115 189
7	8.654 021	8.922 803	9.200 435	9.487 171	9.783 274	10.089 012
8	10.259 803	10.636 628	11.028 474	11.435 888	11.859 434	12.299 693
9	11.977 989	12.487 558	13.021 036	13.579 477	14.136 972	14.775 656
10	13.816 448	14.486 563	15.192 930	15.937 425	16.722 009	17.548 735
11	15.783 599	16.645 488	17.560 293	18.531 167	19.561 430	20.654 583
12	17.888 451	18.977 127	20.140 720	21.384 284	22.713 187	24.133 133
13	20.140 643	21.495 297	22.953 385	24.522 712	26.211 638	28.029 109
14	22.550 488	24.214 920	26.019 189	27.974 984	30.094 918	32.392 602
15	25.129 022	27.152 114	29.360 916	31.772 482	34.405 359	37.279 715
16	27.888 054	30.324 283	33.003 399	35.949 730	39.189 949	42.753 281
17	30.840 218	33.750 226	36.973 705	40.544 703	44.500 843	48.883 674
18	33.999 033	37.450 244	41.301 338	45.599 174	50.395 936	55.749 715
19	37.378 965	41.446 263	46.018 458	51.159 091	56.939 489	63.439 681
20	40.995 493	45.761 965	51.160 120	57.275 000	64.202 833	72.052 443
21	44.865 177	50.422 922	56.764 530	64.002 501	72.265 145	81.698 736
22	49.005 740	55.456 756	62.873 338	71.402 750	81.214 310	92.502 584
23	53.436 142	60.893 296	69.531 939	79.543 025	91.147 885	104.602 894
24	58.176 672	66.764 760	76.789 813	88.497 328	102.174 152	118.155 242
25	63.249 039	73.105 940	84.700 896	98.347 061	114.413 309	133.333 871
26	68.676 471	79.954 416	93.323 977	109.181 767	127.998 773	150.333 935
27	74.483 824	87.350 769	102.723 135	121.099 944	143.078 638	169.374 007
28	80.697 692	95.338 831	112.968 217	134.209 938	159.817 288	190.698 888
29	87.346 531	103.965 937	124.135 357	148.630 932	178.397 190	214.582 755
30	94.460 788	113.283 212	136.307 539	164.494 026	199.020 881	241.322 686
31	102.073 043	123.345 869	149.575 217	181.943 428	221.913 178	271.292 608
32	110.218 156	134.213 539	164.036 987	201.137 771	247.323 628	304.847 721
33	118.933 427	145.950 622	179.800 316	222.251 548	275.529 227	342.429 447
34	128.258 767	158.626 671	196.982 344	245.476 703	306.837 442	384.520 981
35	138.236 881	172.316 805	215.710 755	271.024 374	341.589 561	431.663 499
36	148.913 462	187.102 150	236.124 723	299.126 811	380.164 413	484.463 119
37	160.337 405	203.070 322	258.375 948	330.039 493	422.982 498	543.598 693
38	172.561 023	220.315 948	282.629 783	364.043 442	470.510 573	609.830 536
39	185.640 295	238.941 223	309.066 464	401.447 787	523.266 737	686.010 201
40	199.635 116	259.056 521	337.882 446	442.592 566	581.826 078	767.091 425

Table IV Present Value of an Annuity of $1 ($a_{\overline{n}|i}$)

Periods	1%	2%	3%	4%	5%	6%
1	0.990 099	0.980 392	0.970 874	0.961 538	0.952 381	0.943 396
2	1.970 395	1.941 561	1.913 470	1.886 095	1.859 410	1.833 393
3	2.940 985	2.883 883	2.828 611	2.775 091	2.723 248	2.673 012
4	3.901 965	3.807 729	3.717 098	3.629 895	3.545 951	3.465 106
5	4.853 431	4.713 460	4.579 707	4.451 822	4.329 477	4.212 364
6	5.795 476	5.601 431	5.417 191	5.242 137	5.075 692	4.917 324
7	6.728 194	6.471 991	6.230 283	6.002 055	5.786 373	5.582 381
8	7.651 678	7.325 482	7.019 692	6.732 745	6.463 213	6.209 794
9	8.566 017	8.162 237	7.786 109	7.435 332	7.107 822	6.801 692
10	9.471 304	8.982 585	8.530 203	8.110 896	7.721 735	7.360 087
11	10.367 628	9.786 848	9.252 624	8.760 477	8.306 414	7.886 875
12	11.255 077	10.575 342	9.954 004	9.385 074	8.863 252	8.383 844
13	12.133 740	11.348 374	10.634 955	9.985 648	9.393 573	8.852 683
14	13.003 703	12.106 249	11.296 073	10.563 123	9.898 641	9.294 984
15	13.865 052	12.849 264	11.937 935	11.118 387	10.379 658	9.712 249
16	14.717 874	13.577 710	12.561 102	11.652 296	10.837 770	10.105 895
17	15.562 251	14.291 872	13.166 118	12.165 669	11.274 066	10.477 260
18	16.398 268	14.992 032	13.753 513	12.659 297	11.689 587	10.827 604
19	17.226 008	15.678 462	14.323 799	13.133 939	12.085 321	11.158 117
20	18.045 553	16.351 434	14.877 475	13.590 326	12.462 210	11.469 921
21	18.856 983	17.011 210	15.415 024	14.029 160	12.821 153	11.764 077
22	18.660 379	17.658 049	15.936 917	14.451 115	13.163 003	12.041 582
23	20.455 821	18.292 205	16.443 608	14.856 842	13.488 574	12.303 379
24	21.243 387	18.913 926	16.935 542	15.246 963	13.798 642	12.550 358
25	22.023 155	19.523 457	17.413 148	15.622 080	14.093 945	12.783 356
26	22.795 203	20.121 036	17.876 842	15.982 769	14.375 185	13.003 166
27	23.559 607	20.706 898	18.327 032	16.329 586	14.643 034	13.210 534
28	24.316 443	21.281 273	18.764 108	16.663 063	14.898 127	13.406 164
29	25.065 785	21.844 385	19.188 455	16.983 715	15.141 074	13.590 721
30	25.807 708	22.396 456	19.600 441	17.292 033	15.372 451	13.764 831
31	26.542 285	22.937 702	20.000 429	17.588 494	15.592 810	13.929 086
32	27.269 589	23.468 335	20.388 766	17.873 552	15.802 677	14.084 043
33	27.989 692	23.988 564	20.765 792	18.147 646	16.002 549	14.230 230
34	28.702 666	24.498 592	21.131 837	18.411 198	16.192 904	14.368 141
35	28.408 580	24.998 620	21.487 220	18.664 613	16.374 194	14.498 246
36	30.107 505	25.488 843	21.832 253	18.908 282	16.546 852	14.620 987
37	30.799 510	25.969 454	22.167 235	19.142 579	16.711 287	14.736 780
38	31.484 663	26.440 641	22.492 462	19.367 864	16.867 893	14.846 019
39	32.163 033	26.902 589	22.808 215	19.584 485	17.017 041	14.949 075
40	32.834 686	27.355 480	23.114 772	19.792 774	17.159 086	15.046 297

Appendix

Table IV Present Value of an Annuity of $1 ($a_{\overline{n}|i}$) (*Continued*)

Periods	7%	8%	9%	10%	11%	12%
1	0.934 579	0.925 926	0.917 431	0.909 091	0.900 901	0.892 857
2	1.808 018	1.783 265	1.759 111	1.735 537	1.712 523	1.690 051
3	2.624 316	2.577 097	2.531 295	2.486 852	2.443 715	2.401 831
4	3.387 211	3.312 127	3.239 720	3.169 865	3.102 446	3.037 349
5	4.100 197	3.992 710	3.889 651	3.790 787	3.695 897	3.604 776
6	4.766 540	4.622 880	4.485 919	4.355 261	4.230 538	4.111 407
7	5.389 289	5.206 370	5.032 953	4.868 419	4.712 196	4.563 757
8	5.971 299	5.746 639	5.534 819	5.334 926	5.146 123	4.967 640
9	6.515 232	6.246 888	5.995 247	5.759 024	5.537 048	5.328 250
10	7.023 582	6.710 081	6.417 658	6.144 567	5.889 232	5.650 223
11	7.498 674	7.138 964	6.805 191	6.495 061	6.206 515	5.937 699
12	7.942 686	7.536 078	7.160 725	6.813 692	6.492 356	6.194 374
13	8.357 651	7.903 776	7.486 904	7.103 356	6.749 870	6.423 548
14	8.745 468	8.244 237	7.786 150	7.366 687	6.981 865	6.628 168
15	9.107 914	8.559 479	8.060 688	7.606 080	7.190 870	6.810 864
16	9.446 649	8.851 369	8.312 558	7.823 709	7.379 162	6.973 986
17	9.763 223	9.121 638	8.543 631	8.021 553	7.548 794	7.119 630
18	10.059 087	9.371 887	8.755 625	8.201 412	7.701 617	7.249 670
19	10.335 595	9.603 599	8.950 115	8.364 920	7.839 294	7.365 777
20	10.594 014	9.818 147	9.128 546	8.513 564	7.963 328	7.469 444
21	10.835 527	10.016 803	9.292 244	8.648 694	8.075 070	7.562 003
22	11.061 241	10.200 744	9.442 425	8.771 540	8.175 739	7.644 646
23	11.272 187	10.371 059	9.580 207	8.883 218	8.266 432	7.718 434
24	11.469 334	10.528 758	9.706 612	8.984 744	8.348 137	7.784 316
25	11.653 583	10.674 776	9.822 580	9.077 040	8.421 745	7.843 139
26	11.825 779	10.809 978	9.928 972	9.160 945	8.488 058	7.895 660
27	11.986 709	10.935 165	10.026 580	9.237 223	8.547 800	7.942 554
28	12.137 111	11.051 079	10.116 128	9.306 567	8.601 622	7.984 423
29	12.277 674	11.158 406	10.198 283	9.369 606	8.650 110	8.021 806
30	12.409 041	11.257 783	10.273 654	9.429 914	8.693 793	8.055 184
31	12.531 814	11.349 799	10.342 802	9.479 013	8.733 146	8.084 986
32	12.646 555	11.434 999	10.406 240	9.526 376	8.768 600	8.111 594
33	12.753 790	11.513 888	10.464 441	9.569 432	8.800 541	8.135 352
34	12.854 009	11.586 934	10.517 835	9.608 575	8.829 316	8.156 564
35	12.947 672	11.654 568	10.566 821	9.644 159	8.855 240	8.175 504
36	13.035 208	11.717 193	10.611 763	9.676 508	8.878 594	8.192 414
37	13.117 017	11.775 179	10.652 993	9.705 917	8.899 635	8.207 513
38	13.193 473	11.828 869	10.690 820	9.732 651	8.918 590	8.220 993
39	13.264 928	11.878 582	10.725 523	9.756 956	8.935 666	8.233 030
40	13.331 709	11.924 613	10.757 360	9.779 051	8.951 051	8.243 777

Table V   Periodic Payment to Amortize \$1 at the End of n Periods $\left(\dfrac{1}{a_{\,\overline{n}|i}}\right)$

Periods	1%	2%	3%	4%	5%	6%
1	1.010 000	1.020 000	1.030 000	1.040 000	1.050 000	1.060 000
2	0.507 512	0.515 049	0.522 611	0.530 196	0.537 805	0.545 437
3	0.340 022	0.346 755	0.353 530	0.360 349	0.367 209	0.374 110
4	0.256 281	0.262 624	0.269 027	0.275 490	0.282 012	0.288 591
5	0.206 040	0.212 158	0.218 355	0.224 627	0.230 975	0.237 396
6	0.172 548	0.178 526	0.184 598	0.190 762	0.197 017	0.203 363
7	0.148 628	0.154 512	0.160 506	0.166 610	0.172 820	0.179 135
8	0.130 690	0.136 510	0.142 456	0.148 528	0.154 722	0.161 036
9	0.116 749	0.122 515	0.128 434	0.134 493	0.140 690	0.147 022
10	0.105 592	0.111 327	0.117 231	0.123 291	0.129 505	0.135 868
11	0.096 454	0.102 178	0.108 077	0.114 149	0.120 389	0.126 793
12	0.088 849	0.094 560	0.100 462	0.106 552	0.112 825	0.119 277
13	0.082 415	0.088 118	0.094 030	0.100 144	0.106 456	0.112 960
14	0.076 901	0.082 602	0.088 526	0.094 669	0.101 024	0.107 585
15	0.072 124	0.077 825	0.083 767	0.089 941	0.096 342	0.102 963
16	0.067 945	0.073 650	0.079 611	0.085 820	0.092 270	0.098 952
17	0.064 258	0.069 970	0.075 953	0.082 199	0.088 699	0.095 445
18	0.060 982	0.066 702	0.072 709	0.078 993	0.085 546	0.092 357
19	0.058 052	0.063 782	0.069 814	0.076 139	0.082 745	0.089 621
20	0.055 415	0.061 157	0.067 216	0.073 582	0.080 243	0.087 185
21	0.053 031	0.058 785	0.064 872	0.071 280	0.077 996	0.085 005
22	0.050 864	0.056 631	0.062 747	0.069 199	0.075 971	0.083 046
23	0.048 886	0.054 668	0.060 814	0.067 309	0.074 137	0.081 278
24	0.047 073	0.052 871	0.059 047	0.065 587	0.072 471	0.079 679
25	0.045 407	0.051 220	0.057 428	0.064 012	0.070 952	0.078 227
26	0.043 869	0.049 699	0.055 938	0.062 567	0.069 564	0.076 904
27	0.042 446	0.048 293	0.054 564	0.061 239	0.068 292	0.075 697
28	0.041 124	0.046 990	0.053 293	0.060 013	0.067 123	0.074 593
29	0.039 895	0.045 778	0.052 115	0.058 880	0.066 046	0.073 580
30	0.038 748	0.044 650	0.051 019	0.057 830	0.065 051	0.072 649
31	0.037 676	0.043 596	0.049 999	0.056 855	0.064 132	0.071 792
32	0.036 671	0.042 611	0.049 047	0.055 949	0.063 280	0.071 002
33	0.035 727	0.041 687	0.048 156	0.055 104	0.062 490	0.070 273
34	0.034 840	0.040 819	0.047 322	0.054 315	0.061 755	0.069 598
35	0.034 004	0.040 002	0.046 539	0.053 577	0.061 072	0.068 974
36	0.033 214	0.039 233	0.045 804	0.052 887	0.060 434	0.068 395
37	0.032 468	0.038 507	0.045 112	0.052 240	0.059 840	0.067 857
38	0.031 761	0.037 821	0.044 459	0.051 632	0.059 284	0.067 358
39	0.031 092	0.037 171	0.043 844	0.051 061	0.058 765	0.066 894
40	0.030 456	0.036 556	0.043 262	0.050 523	0.058 278	0.066 462

Table V Periodic Payment to Amortize \$1 at the End of *n* Periods $\left(\dfrac{1}{a_{\overline{n}|i}}\right)$ (*Continued*)

Periods	*7%*	*8%*	*9%*	*10%*	*11%*	*12%*
1	1.070 000	1.080 000	1.090 000	1.100 000	1.110 000	1.120 000
2	0.553 092	0.560 769	0.568 469	0.576 190	0.583 934	0.591 698
3	0.381 052	0.388 034	0.395 055	0.402 115	0.409 213	0.416 349
4	0.295 228	0.301 921	0.308 669	0.315 471	0.322 326	0.329 234
5	0.243 891	0.250 456	0.257 092	0.263 797	0.270 570	0.277 410
6	0.209 796	0.216 315	0.222 920	0.229 607	0.236 377	0.243 226
7	0.185 553	0.192 072	0.198 691	0.205 405	0.212 215	0.219 118
8	0.167 468	0.174 015	0.180 674	0.187 444	0.194 321	0.201 303
9	0.153 486	0.160 080	0.166 799	0.173 641	0.180 602	0.187 679
10	0.142 378	0.149 029	0.155 820	0.162 745	0.169 801	0.176 984
11	0.133 357	0.140 076	0.146 947	0.153 963	0.161 121	0.168 415
12	0.125 902	0.132 695	0.139 651	0.146 763	0.154 027	0.161 437
13	0.119 651	0.126 522	0.133 567	0.140 779	0.148 151	0.155 677
14	0.114 345	0.121 297	0.128 433	0.135 746	0.143.228	0.150 871
15	0.109 795	0.116 830	0.124 059	0.131 474	0.139 065	0.146 824
16	0.105 858	0.112 977	0.120 300	0.127 817	0.135 517	0.143 390
17	0.102 425	0.109 629	0.117 046	0.124 664	0.132 471	0.140 457
18	0.099 413	0.106 702	0.114 212	0.121 930	0.129 843	0.137 937
19	0.096 753	0.104 128	0.111 730	0.119 547	0.127 563	0.135 763
20	0.094 393	0.101 852	0.109 546	0.117 460	0.125 576	0.133 879
21	0.092 289	0.099 832	0.107 617	0.115 624	0.123 838	0.132 240
22	0.090 406	0.098 032	0.105 905	0.114 005	0.122 313	0.130 811
23	0.088 714	0.096 422	0.104 382	0.112 572	0.120 971	0.129 560
24	0.087 189	0.094 978	0.103 023	0.111 300	0.119 787	0.128 463
25	0.085 811	0.093 679	0.101 806	0.110 168	0.118 740	0.127 500
26	0.084 561	0.092 507	0.100 715	0.109 159	0.117 813	0.126 652
27	0.083 426	0.091 448	0.099 735	0.108 258	0.116 989	0.125 904
28	0.082 392	0.090 489	0.098 852	0.107 451	0.116 257	0.125 244
29	0.081 449	0.089 619	0.098 056	0.106 728	0.115 605	0.124 660
30	0.080 586	0.088 827	0.097 336	0.106 079	0.115 025	0.124 144
31	0.079 797	0.088 107	0.096 686	0.105 496	0.114 506	0.123 686
32	0.079 073	0.087 451	0.096 096	0.104 972	0.114 043	0.123 280
33	0.078 408	0.086 852	0.095 562	0.104 499	0.113 629	0.122 920
34	0.077 797	0.086 304	0.095 077	0.104 074	0.113 259	0.122 601
35	0.077 234	0.085 803	0.094 636	0.103 690	0.112 927	0.122 317
36	0.076 715	0.085 345	0.094 235	0.103 343	0.112 630	0.122 064
37	0.076 237	0.084 924	0.093 870	0.103 030	0.112 364	0.121 640
38	0.075 795	0.084 539	0.093 538	0.102 747	0.112 125	0.121 840
39	0.075 387	0.084 185	0.093 236	0.102 491	0.111 911	0.121 462
40	0.075 009	0.083 860	0.092 960	0.102 259	0.111 719	0.121 304

Table VI Periodic Payment for a Sinking Fund of $1 at the End of *n* Periods $\left(\dfrac{1}{s_{\overline{n}|i}}\right)$

Periods	1%	2%	3%	4%	5%	6%
1	1.000 000	1.000 000	1.000 000	1.000 000	1.000 000	1.000 000
2	0.497 512	0.495 049	0.492 611	0.490 196	0.487 805	0.485 437
3	0.330 022	0.326 755	0.323 530	0.320 349	0.317 209	0.314 110
4	0.246 281	0.242 624	0.239 027	0.235 490	0.232 012	0.228 591
5	0.196 040	0.192 158	0.188 355	0.184 627	0.180 975	0.177 396
6	0.162 548	0.158 526	0.154 598	0.150 762	0.147 017	0.143 363
7	0.138 628	0.134 512	0.130 506	0.126 610	0.122 820	0.119 135
8	0.120 690	0.116 510	0.112 456	0.108 528	0.104 722	0.101 036
9	0.106 740	0.102 515	0.098 434	0.094 493	0.090 690	0.087 022
10	0.095 582	0.091 327	0.087 231	0.083 291	0.079 505	0.075 868
11	0.086 454	0.082 178	0.078 077	0.074 149	0.070 389	0.066 793
12	0.078 849	0.074 560	0.070 462	0.066 552	0.062 825	0.059 277
13	0.072 415	0.068 118	0.064 030	0.060 144	0.056 456	0.052 960
14	0.066 901	0.062 602	0.058 526	0.054 669	0.051 024	0.047 585
15	0.062 124	0.057 825	0.053 767	0.049 941	0.046 342	0.042 963
16	0.057 945	0.053 650	0.049 611	0.045 820	0.042 270	0.038 952
17	0.054 258	0.049 970	0.045 953	0.042 199	0.038 699	0.035 445
18	0.050 982	0.046 702	0.042 709	0.038 993	0.035 546	0.032 357
19	0.048 052	0.043 782	0.039 814	0.036 139	0.032 745	0.029 621
20	0.045 415	0.041 157	0.037 216	0.033 582	0.030 243	0.027 185
21	0.043 031	0.038 785	0.034 872	0.031 280	0.027 996	0.025 005
22	0.040 864	0.036 631	0.032 747	0.029 199	0.025 971	0.023 046
23	0.038 886	0.034 668	0.030 814	0.027 309	0.024 137	0.021 278
24	0.037 073	0.032 871	0.029 047	0.025 587	0.022 471	0.019 679
25	0.035 407	0.031 220	0.027 428	0.024 012	0.020 952	0.018 227
26	0.033 869	0.029 699	0.025 938	0.022 567	0.019 564	0.016 904
27	0.032 446	0.028 293	0.024 564	0.021 239	0.018 292	0.015 697
28	0.031 124	0.026 990	0.023 293	0.020 013	0.017 123	0.014 593
29	0.029 895	0.025 778	0.022 115	0.018 880	0.016 046	0.013 580
30	0.028 748	0.024 650	0.021 019	0.017 830	0.015 051	0.012 649
31	0.027 676	0.023 596	0.019 999	0.016 855	0.014 132	0.011 792
32	0.026 671	0.022 611	0.019 047	0.015 949	0.013 280	0.011 002
33	0.025 727	0.021 687	0.018 156	0.015 104	0.012 490	0.010 273
34	0.024 840	0.020 819	0.017 322	0.014 315	0.011 755	0.009 598
35	0.024 004	0.020 002	0.016 539	0.013 577	0.011 072	0.008 974
36	0.023 214	0.019 233	0.015 804	0.012 887	0.010 434	0.008 395
37	0.022 468	0.018 507	0.015 112	0.012 240	0.009 840	0.007 857
38	0.021 761	0.017 821	0.014 459	0.011 632	0.009 284	0.007 358
39	0.021 092	0.017 171	0.013 844	0.011 061	0.008 764	0.006 894
40	0.020 456	0.016 556	0.013 262	0.010 523	0.008 278	0.006 462

Appendix

Table VI   Periodic Payment for a Sinking Fund of $1 at the End of *n* Periods $\left(\dfrac{1}{s_{\overline{n}|i}}\right)$ *(Continued)*

Periods	7%	8%	9%	10%	11%	12%
1	1.000 000	1.000 000	1.000 000	1.000 000	1.000 000	1.000 000
2	0.483 092	0.480 769	0.478 469	0.476 190	0.473 934	0.471 698
3	0.311 052	0.308 034	0.305 055	0.302 115	0.299 213	0.296 349
4	0.225 228	0.221 921	0.218 669	0.215 471	0.212 326	0.209 234
5	0.173 891	0.170 456	0.167 092	0.163 797	0.160 570	0.157 410
6	0.139 796	0.136 315	0.132 920	0.129 607	0.126 377	0.123 226
7	0.115 553	0.112 072	0.108 691	0.105 405	0.102 215	0.099 118
8	0.097 468	0.094 015	0.090 674	0.087 444	0.084 321	0.081 303
9	0.083 486	0.080 080	0.076 799	0.073 641	0.070 602	0.067 679
10	0.072 378	0.069 029	0.065 820	0.062 745	0.059 801	0.056 984
11	0.063 357	0.060 076	0.056 947	0.053 963	0.051 121	0.048 415
12	0.055 902	0.052 695	0.049 651	0.046 763	0.044 027	0.041 437
13	0.049 651	0.046 522	0.043 567	0.040 779	0.038 151	0.035 677
14	0.044 345	0.041 297	0.038 433	0.035 746	0.033 228	0.030 871
15	0.039 795	0.036 830	0.034 059	0.031 474	0.029 065	0.026 824
16	0.035 858	0.032 977	0.030 300	0.027 817	0.025 517	0.023 390
17	0.032 425	0.029 629	0.027 046	0.024 664	0.022 471	0.020 457
18	0.029 413	0.026 702	0.024 212	0.021 930	0.019 843	0.017 937
19	0.026 753	0.024 128	0.021 730	0.019 547	0.017 563	0.015 763
20	0.024 393	0.021 852	0.019 546	0.017 460	0.015 576	0.013 879
21	0.022 289	0.019 832	0.017 617	0.015 624	0.013 838	0.012 240
22	0.020 406	0.018 032	0.015 905	0.014 005	0.012 313	0.010 811
23	0.018 714	0.016 422	0.014 382	0.012 572	0.010 971	0.009 560
24	0.017 189	0.014 978	0.013 023	0.011 300	0.009 787	0.008 463
25	0.015 811	0.013 679	0.011 806	0.010 168	0.008 740	0.007 500
26	0.014 561	0.012 507	0.010 715	0.009 159	0.007 813	0.006 652
27	0.013 426	0.011 448	0.009 735	0.008 258	0.006 989	0.005 904
28	0.012 392	0.010 489	0.008 852	0.007 451	0.006 257	0.005 244
29	0.011 449	0.009 619	0.008 056	0.006 728	0.005 605	0.004 660
30	0.010 586	0.008 827	0.007 336	0.006 079	0.005 025	0.004 144
31	0.009 797	0.008 107	0.006 686	0.005 496	0.004 506	0.003 686
32	0.009 073	0.007 451	0.006 096	0.004 972	0.004 043	0.003 280
33	0.008 408	0.006 852	0.005 562	0.004 499	0.003 629	0.002 920
34	0.007 797	0.006 304	0.005 077	0.004 074	0.003 259	0.002 601
35	0.007 234	0.005 803	0.004 636	0.003 690	0.002 927	0.002 317
36	0.006 715	0.005 345	0.004 235	0.003 343	0.002 630	0.002 064
37	0.006 237	0.004 924	0.003 870	0.003 030	0.002 364	0.001 840
38	0.005 795	0.004 539	0.003 538	0.002 747	0.002 125	0.001 640
39	0.005 387	0.004 185	0.003 236	0.002 491	0.001 911	0.001 462
40	0.005 009	0.003 860	0.002 960	0.002 259	0.001 719	0.001 304

Answers

Following are the answers to all Chapter Review Exercises and to odd-numbered Section Exercises.

Exercise 0.1 *(page 9)*

1. $\in$ **3.** $\in$ **5.** $\notin$ **7.** $\{1, 2, 3, 4, 5, 6, 7\}$
9. $\{x: x$ is a natural number greater than 2 and less than 8$\}$
11. yes   **13.** no   **15.** $D \subseteq C$
17. $D \subseteq A$
19. $A \subseteq B$ or $B \subseteq A$ **21.** yes   **23.** no
25. A and B, B and D, C and D
27. $A \cap B = \{4, 6\}$ **29.** $A \cap B = \varnothing$
31. $A \cup B = \{1, 2, 3, 4, 5\}$
33. $A \cup B = \{1, 2, 3, 4\} = B$
35. $A' = \{4, 6, 9, 10\}$
37. $(A \cup B)' = \{6, 9\}$ **39.** $A \cap B' = \{1, 2, 5, 7\}$
41. $A' \cup B' = \{1, 2, 4, 5, 6, 7, 9, 10\}$
43. $\{1, 2, 3, 5, 7, 9\}$ **45.** $\{2, 6, 8\}$
47. $A - B = \{1, 7\}$ **49.** $A - B = \varnothing$ or $\{\,\}$
51. (a) $A = \{$Cadillac, Corvette, RX-7, Maxima$\}$
   $B = \{$Cadillac, Corvette, RX-7, Maxima$\}$
   $C = \{$RX-7, Maxima, Bonneville, Celica, Corolla$\}$
   (b) $A \subseteq B$ and $B \subseteq A$, in fact $A = B$
   (c) $\{$RX-7, Maxima$\}$; the set of cars with a base price of at least \$20,000 in any year and that had a base price increase of at least 75% over 10 years
   (d) the set of cars whose 1993 base price was less than \$20,000
   (e) $\{$Camaro, Escort, Mustang, Accord, Civic$\}$
53. (a) 130   **(b)** 840   **(c)** 520
55. (a)

U, E 25, 40, F 20, 15

   (b) 40   **(c)** 85
   (d) 25
57.

Math, U, 17, 6, 32, Fine Arts, 12, 3, 1, 25, 4, Economics

   (a) 25   **(b)** 43   **(c)** 53

59. (a) and (b)

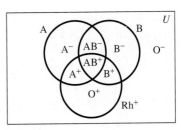

Exercise 0.2 *(page 16)*

1. (a) irrational   **(b)** rational, integer
   (c) rational, integer, natural   **(d)** meaningless
3. (a) Commutative   **(b)** Distributive
   (c) Multiplicative identity
5. $<$ **7.** $<$ **9.** $>$ **11.** -4 **13.** 4
15. 2   **17.** $\frac{-4}{3}$ **19.** 3   **21.** $\frac{17}{11}$ **23.** entire line
25. $(1, 3]$ **27.** $(2, 10)$ **29.** $x \leq 5$ **31.** $x > 4$
33. $(-3, 4)$ ⟶ -3 -1 1 4
35. $(4, +\infty)$ ⟶ 0 2 4 6
37. $[-1, +\infty)$ ⟶ -1 1 3 5
39. $(-\infty, 0) \cup (7, +\infty)$ ⟵⟶ -2 0 2 4 6 7 8
41. -0.000038585
43. 9122.387471   **45.** 398.73507
47. (a) 68.38%   **(b)** 64.85%   61.31%
   (c) It's decreasing.   **(d)** The percentage of those who have tried cigarettes cannot be less than zero.
   (e) $t = 125,\ 2100$

Exercise 0.3 *(page 22)*

1. -64 **3.** -16 **5.** $\frac{1}{9}$ **7.** $-\frac{9}{4}$ **9.** 6^8
11. $\frac{1}{10}$ **13.** 3^0 **15.** $(\frac{3}{2})^2 = \frac{9}{4}$ **17.** $1/x^6$
19. x/y^2 **21.** x^7 **23.** $x^{-2} = 1/x^2$ **25.** x^4
27. y^{12} **29.** x^{12} **31.** x^2y^2 **33.** $16/x^4$
35. $x^8/(16y^4)$ **37.** $-16a^2/b^2$ **39.** $2/(xy^2)$
41. $1/(x^9y^6)$ **43.** $(a^{18}c^{12})/b^6$
45. (a) $1/(2x^4)$ **(b)** $1/(16x^4)$ **(c)** $1/x^4$ **(d)** 8
47. x^{-1} **49.** $8x^3$ **51.** $\frac{1}{4}x^{-2}$ **53.** $-\frac{1}{8}x^3$

55. 2.0736   **57.** -7776
59. $A = \$2114.81; I = \$914.81$
61. $A = \$9607.70; I = \$4607.70$
63. (a) 5   **(b)** 132   **(c)** 1862   **(d)** 3611

Exercise 0.4   *(page 30)*

1. $\frac{16}{3}$ **3.** -8 **5.** 8   **7.** not real   **9.** $\frac{9}{4}$
11. (a) 4   **(b)** $\frac{1}{4}$ **13.** $m^{3/2}$ **15.** $(m^2 n^5)^{1/4}$
17. $\sqrt[4]{x^7}$ **19.** $-1/(4\sqrt[4]{x^5})$ **21.** $y^{3/4}$ **23.** $z^{19/4}$
25. $1/y^{5/2}$ **27.** x **29.** $1/y^{21/10}$ **31.** $x^{1/2}$ **33.** $1/x$
35. $8x^2$ **37.** $8x^2 y^2 \sqrt{2y}$ **39.** $2x^2 y \sqrt[3]{5x^2 y^2}$
41. $6x^2 y \sqrt{x}$ **43.** $42x^3 y^2 \sqrt{x}$ **45.** $2xy^5/3$
47. $2b\sqrt[4]{b}/(3a^2)$ **49.** $\sqrt{6}/3$ **51.** $\sqrt{mx}/x$
53. $\sqrt[3]{mx^2}/x^2$ **55.** $-\frac{2}{3}x^{-2/3}$ **57.** $3x^{3/2}$
59. $(3\sqrt{x})/2$ **61.** $1/(2\sqrt{x})$ **63.** 74 kg
65. 39,491   **67.** 10

Exercise 0.5   *(page 38)*

1. (a) 2   **(b)** -1 **(c)** 10   **(d)** one
3. (a) 5   **(b)** -14 **(c)** 0   **(d)** several
5. (a) 5   **(b)** 0   **(c)** 2   **(d)** -5
7. -12 **9.** $\frac{-7}{31}$ **11.** $21pq - 2p^2$
13. $m^2 - 7n^2 - 3$ **15.** $3q + 12$
17. $x^2 - 1$ **19.** $35x^5$ **21.** $3rs$
23. $2ax^4 + a^2 x^3 + a^2 bx^2$ **25.** $6y^2 - y - 12$
27. $2 - 5x^2 + 2x^4$ **29.** $16x^2 + 24x + 9$
31. $x^4 - x^2 + \frac{1}{4}$ **33.** $4x^2 - 1$ **35.** $0.01 - 16x^2$
37. $x^3 - 8$ **39.** $x^8 + 3x^6 - 10x^4 + 5x^3 + 25x$
41. $3 + m + 2m^2 n$ **43.** $8x^3 y^2/3 + 5/(3y) - 2x^2/(3y)$
45. $x^3 + 3x^2 + 3x + 1$ **47.** $8x^3 - 36x^2 + 54x - 27$
49. $0.1x^2 - 1.995x - 0.1$
51. $x^2 - 2x + 5 - 11/(x + 2)$
53. $x^2 + 3x - 1 + (-4x + 2)/(x^2 + 1)$
55. $x + 2x^2$ **57.** $x - x^{1/2} - 2$
59. $x - 9$ **61.** $4x^2 + 4x$
63. (a) $9x^2 - 21x + 13$ **(b)** 5   **65.** $55x$
67. (a) $4000 - x$ **(b)** $0.10x$
   (c) $0.08(4000 - x)$
   (d) $0.10x + 0.08(4000 - x)$
69. $(15 - 2x)(10 - 2x)x$

Exercise 0.6   *(page 45)*

1. $3b(3a - 4a^2 + 6b)$ **3.** $2x(2x + 4y^2 + y^3)$
5. $(5 - x^2)(y - 4)$ **7.** $(6 + y)(x - m)$

9. $(x + 2)(x + 6)$ **11.** $(x - 3)(x + 2)$
13. $2(x - 7)(x + 3)$ **15.** $(7x + 4)(x - 2)$
17. $(x - 5)^2$ **19.** $(7a + 12b)(7a - 12b)$
21. $2x(x - 2)^2$ **23.** $(2x - 3)(x + 2)$
25. $3(x + 4)(x - 3)$ **27.** $2x(x + 2)(x - 2)$
29. $(5x + 2)(2x + 3)$ **31.** $(5x - 1)(2x - 9)$
33. $(3x - 1)(3x + 8)$
35. $(y^2 + 4x^2)(y + 2x)(y - 2x)$
37. $(x + 2)^2 (x - 2)^2$
39. $(2x + 1)(2x - 1)(x + 1)(x - 1)$
41. $(2x - 21)(5x + 3)$
43. $(10x - 9)(x + 7)$ **45.** $(x + 1)^3$
47. $(x - 4)^3$ **49.** $(x - 4)(x^2 + 4x + 16)$
51. $(3 + 2x)(9 - 6x + 4x^2)$ **53.** $x + 1$
55. $1 + x$ **57.** $7x - 3x^3$
59. $m^2(c/2 - m/3)$ **61.** $4x(4 - x)^2$

Exercise 0.7   *(page 54)*

1. $2y^3/z$ **3.** $\frac{1}{3}$ **5.** $(x - 1)/(x - 3)$
7. $(2xy^2 - 5)/(y + 3)$ **9.** $20x/y$ **11.** $\frac{32}{3}$
13. $2x^2 - 7x + 6$
15. $-(x + 1)(x + 3)/[(x - 1)(x - 3)]$
17. $15bc^2/2$ **19.** $5y/(y - 3)$
21. $\dfrac{-x(x - 3)(x + 2)}{x + 3}$ **23.** $\dfrac{1}{x + 1}$ **25.** $\dfrac{4a - 4}{a(a - 2)}$
27. $\dfrac{-x^2 + x + 1}{x + 1}$ **29.** $\dfrac{16a + 15a^2}{12(x + 2)}$ **31.** $\dfrac{79x + 9}{30(x - 2)}$
33. $\dfrac{-4y}{(x - 2y)^2(x + 2y)}$ **35.** $\dfrac{9x + 4}{(x - 2)(x + 2)(x + 1)}$
37. $(7x - 3x^3)/\sqrt{3 - x^2}$ **39.** $\frac{1}{6}$ **41.** xy
43. $\dfrac{x + 1}{x^2}$ **45.** $\dfrac{1}{\sqrt{a}} = \dfrac{\sqrt{a}}{a}$ **47.** $\dfrac{x - 2}{(x - 3)\sqrt{x^2 + 9}}$
49. (a) -12 **(b)** $\frac{25}{36}$ **51.** $2b - a$
53. $\dfrac{x^3 + y^3}{x^2 y^2(x + y)} = \dfrac{x^2 - xy + y^2}{x^2 y^2}$
55. $(1 - 2\sqrt{x} + x)/(1 - x)$
57. $1/(\sqrt{x + h} + \sqrt{x})$ **59.** $(bc + ac + ab)/abc$
61. (a) $\dfrac{0.1x^2 + 55x + 4000}{x}$
   (b) $0.1x^2 + 55x + 4000$
63. $\dfrac{t^2 + 9t + 10}{(t + 3)^2}$

Chapter 0 Review Exercises *(page 57)*

1. yes   **2.** no   **3.** no   **4.** {1, 2, 3, 4, 9}
5. {5, 6, 7, 8, 10}   **6.** {1, 2, 3, 4, 9}
7. yes, $(A' \cup B')' = \{1, 3\} = A \cap B$
8. **(a)** commutative property of addition
   (b) associative property of multiplication
   (c) distributive law
9. **(a)** irrational   **(b)** rational, integer
   (c) meaningless
10. **(a)** >   **(b)** <   **(c)** >   **11.** 6
12. 142   **13.** 10   **14.** 5/4   **15.** 9   **16.** -29
17. 13/4   **18.** -10.62857888
19. **(a)** [0, 5], closed

   (b) $[-3, 7)$, half-open

   (c) $(-4, 0)$, open

20. **(a)** $-1 < x < 16$ **(b)** $-12 \le x \le$ **(c)** $x < -1$
21. **(a)** 1   **(b)** $2^7 = 128$ **(c)** 4^6 **(d)** 7
22. **(a)** $1/x^2$ **(b)** x^{10} **(c)** x^9 **(d)** $1/y^8$ **(e)** y^6
23. $-x^2y^2/36$ **24.** $9y^8/(4x^4)$ **25.** $y^2/(4x^4)$
26. $-x^8z^4/y^4$ **27.** $3x/(y^7z)$ **28.** $x^5/(2y^3)$
29. **(a)** 4   **(b)** 2/7
30. **(a)** $x^{1/2}$ **(b)** $x^{2/3}$ **(c)** $x^{-1/4}$
31. **(a)** $\sqrt[3]{x^2}$ **(b)** $1/\sqrt{x} = \sqrt{x}/x$ **(c)** $-x\sqrt{x}$
32. **(a)** $5y\sqrt{2x}/2$ **(b)** $\sqrt[3]{x^2y}/x^2$ **33.** $x^{5/6}$ **34.** y
35. $x^{17/4}$ **36.** $x^{11/3}$ **37.** $x^{2/5}$ **38.** x^2y^8
39. $2xy^2\sqrt{3xy}$ **40.** $25x^3y^4\sqrt{2y}$ **41.** $6x^2y^4\sqrt[3]{5x^2y^2}$
42. $8a^2b^4\sqrt{2a}$ **43.** $2xy$ **44.** $4x\sqrt{3xy}/(3y^4)$
45. $-x - 2$ **46.** $-x^2 - x$
47. $4x^3 + xy + 4y - 4$ **48.** $24x^5y^5$
49. $3x^2 - 7x + 4$ **50.** $3x^2 + 5x - 2$
51. $4x^2 - 7x - 2$ **52.** $6x^2 - 11x - 7$
53. $4x^2 - 12x + 9$ **54.** $16x^2 - 9$
55. $2x^4 + 2x^3 - 5x^2 + x - 3$
56. $8x^3 - 12x^2 + 6x - 1$
57. $x^3 - y^3$ **58.** $(2/y) - (3xy/2) - 3x^2$
59. $3x^2 + 2x - 3 + (-3x + 7)/(x^2 + 1)$
60. $x^3 - x^2 + 2x + 7 + 21/(x - 3)$
61. $x^2 - x$ **62.** $2x - a$ **63.** $x^3(2x - 1)$
64. $2(x^2 + 1)^2(1 + x)(1 - x)$ **65.** $(2x - 1)^2$
66. $(4 + 3x)(4 - 3x)$ **67.** $2x^2(x + 2)(x - 2)$
68. $(x - 7)(x + 3)$ **69.** $(3x + 2)(x - 1)$
70. $(4x + 3)(3x - 8)$ **71.** $(2x + 3)^2(2x - 3)^2$

72. $x^{2/3} + 1$ **73.** **(a)** $\dfrac{x}{(x + 2)}$ **(b)** $\dfrac{2xy(2 - 3xy)}{2x - 3y}$
74. $\dfrac{x^2 - 4}{x(x + 4)}$ **75.** $\dfrac{(x + 3)}{(x - 3)}$ **76.** $\dfrac{x^2(3x - 2)}{(x - 1)(x + 2)}$
77. $(6x^2 + 9x - 1)/(6x^2)$ **78.** $\dfrac{4x - x^2}{4(x - 2)}$
79. $-\dfrac{x^2 + 2x + 2}{x(x - 1)^2}$ **80.** $\dfrac{x(x - 4)}{(x - 2)(x + 1)(x - 3)}$
81. $\dfrac{(x - 1)^3}{x^2}$ **82.** $\dfrac{1 - x}{1 + x}$ **83.** $3(\sqrt{x} + 1)$
84. $2/(\sqrt{x} + \sqrt{x - 4})$
85. **(a)**

R: recognized
E: exercise
C: community involvement

   (b) 10   **(c)** 100

Exercise 1.1 *(page 69)*

1. $x = 1$ **3.** $x = -9/4$ **5.** $x = 0$ **7.** $x = 0$
9. $x = 5/2$ **11.** $x = 17/13$ **13.** $x = 13/5$
15. $x = 40$ **17.** $x = 1.7$ **19.** $x = -29/2$
21. $x = -1$ **23.** $x = -1/3$ **25.** $x = 5/3$
27. $x = 0$ **29.** $x \approx -0.279$ **31.** $x \approx -1147.362$
33. $y = -3x/2 + 20$ **35.** $y = x - 1$
37. **(a)** 3333 approx.   **(b)** 15,873   **39.** $-40°$
41. **(a)** 59.2% (approx.)   **(b)** $t \approx 107$, in 2082
43. $t \approx 26$ in 2014   **45.** 9000
47. **(a)** T $= (7n - 52)/12$ **(b)** $61°$
49. \$4000   **51.** \$307   **53.** \$246   **55.** \$300
57. 96   **59.** lost \$40   **61.** \$8800

Exercise 1.2 *(page 78)*

1.

$y = 6x - 2$

3.

$y = \frac{1}{3}x + \frac{2}{3}$

5.

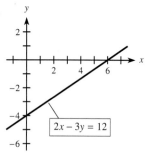

$2x - 3y = 12$

7.

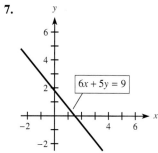

$6x + 5y = 9$

9.

$y = 3$

11.

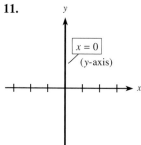

$x = 0$
(y-axis)

13.

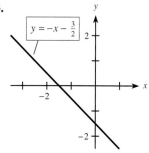

$y = -x - \frac{3}{2}$

15. (e)   **17.** (c)

19.

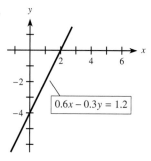

$0.6x - 0.3y = 1.2$

21.

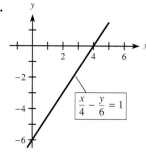

$\dfrac{x}{4} - \dfrac{y}{6} = 1$

23.

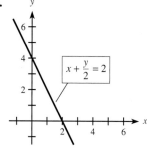

$x + \dfrac{y}{2} = 2$

25. See Exercise 19. $x/2 - y/4 = 1$ is equivalent to $0.6x - 0.3y = 1.2$.

27. (a)   **(b)** 73,750

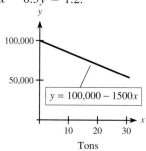

$y = 100,000 - 1500x$

Tons

(c) (0, 100,000) means when there are 0 tons of pollutants, the stream has 100,000 fish.
(50, 25,000) means when there are 50 tons of pollutants, the stream has 25,000 fish.

29. (a)

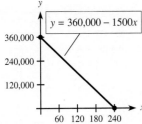

(b) 240 months

(c) After 60 months of use, the value of the building is $270,000.

31. (a) (20, 57.05)

(b) In the year 2000 ($t = 20$), the per capita tax burden is expected to be $57.05 (hundred).

(c)

33. (a)

(b)

(c) At $t = 0$, $R_{FS} = 88.381$ and $R_{SP} = 104.238$. These are different from 100 because these equations are a good fit to the data but are not perfect; the data are not exactly linear.

(d) These equations show past performance; future sales, market conditions, profits, and confidence cannot be measured and are not part of these equations.

35. (a)

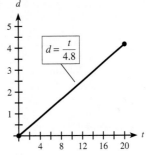

(b) When the time between lightning flashes is 9.6 sec, the storm is 2 miles away.

37.

39.

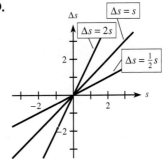

Exercise 1.3   (page 86)

1. Linear

3. Linear

5. Linear

$y = \frac{1}{2}x - 4$

7. Not linear

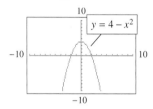
$y = 4 - x^2$

9. Not linear

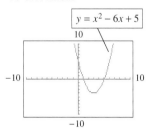
$y = x^2 - 6x + 5$

11. Not linear

$y = x^3 - 3x$

13. Not linear

$y = \frac{1}{3}x^3 - x^2 - 3x + 2$

15. Not linear

$y = \frac{12x}{x^2 + 1}$

17.

$y = x^3 - 12x - 1$

19.

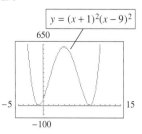
$y = (x + 1)^2(x - 9)^2$

21. (a) $y = 0.01\,x^3 + 0.3x^2 - 72x + 150$

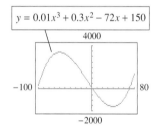
$y = 0.01x^3 + 0.3x^2 - 72x + 150$

(b) Standard window

23. (a) $y = \dfrac{x + 15}{x^2 + 400}$ $y = \dfrac{x^3 + 19x^2 - 62x - 840}{20}$

(b) Standard window

25. (a) The equation is linear, so the graph will be a line. Use the intercepts to determine a window.

(b) Window: x-min $= -5$, x-max $= 35$, y-min $= -0.06$, y-max $= 0.02$

(c) $y = 0.001x - 0.03$

$y = 0.001x - 0.03$

27.

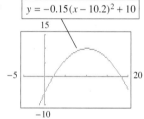
$y = -0.15(x - 10.2)^2 + 10$

29. $y = \dfrac{(x^3 + 19x^2 - 62x - 840)}{20}$

31.

33. (a)

(b) The coordinates mean that when a male's salary is $50 thousand, a female's is $40.536 thousand.

35. (a)

(b) $E \geq 0$ when $0 \leq p \leq 100$

37. (a)　　　　　　　　　　　　**(b)**

(c) From 1980–1995 revenues were rising to what appears to be a maximum. From 1980–2020 revenues rose then declined below 0. This model cannot be valid for this period because revenue cannot be negative.

39. (a)

(b) Near $p = 0$, cost grows without bound.
(c) The coordinates of the point mean that the cost of obtaining stream water with 1% of the current pollution levels would cost $282,150.
(d) The p-intercept means that the cost of stream water with 100% of the current pollution levels would cost $0.

41. (a)

(b)

(c) predicts $325 (nearest dollar), actual is $305.

Exercise 1.4　*(page 98)*

1. yes　　**3.** no　　**5.** no
7. (a) -10　　**(b)** 6　　**(c)** -34　　**(d)** 2.8
9. (a) -3　　**(b)** 1　　**(c)** 13　　**(d)** 6
11. (a) $63/8$　　**(b)** 6　　**(c)** -6
13. (a) no, $f(2 + 1) = f(3) = 13$ but $f(2) + f(1) = 10$
　　(b) $1 + x + h + x^2 + 2xh + h^2$
　　(c) no, $f(x) + f(h) = 2 + x + h + x^2 + h^2$
　　(d) no, $f(x) + h = 1 + x + x^2 + h$
15. (a) $-2x^2 - 4xh - 2h^2 + x + h$
　　(b) $-4xh - 2h^2 + h$　　**(c)** $-4x - 2h + 1$
17. The vertical line test shows that graph (a) represents a function of x, but graph (b) does not.
19. (a) 10　　**(b)** 6
21. (a) $b = a^2 - 4a$　　**(b)** $(1, -3)$, yes
　　(c) $(3, -3)$, yes　　**(d)** $x = 0, x = 4$, yes

23. (a) $3x + x^3$ **(b)** $3x - x^3$ **(c)** $3x^4$ **(d)** $\dfrac{3}{x^2}$

25. (a) $\sqrt{2x} + x^2$ **(b)** $\sqrt{2x} - x^2$
   (c) $x^2\sqrt{2x}$ **(d)** $\dfrac{\sqrt{2x}}{x^2}$

27. (a) $-8x^3$ **(b)** $1 - 2(x-1)^3$
   (c) $[(x-1)^3 - 1]^3$ **(d)** $(x-1)^6$

29. (a) $2\sqrt{x^4 + 5}$ **(b)** $16x^2 + 5$
   (c) $2\sqrt{2\sqrt{x}}$ **(d)** $4x$

31. D: all reals; R: reals $y \geq 4$

33. D: reals $x \geq -4$: R: reals $y \geq 0$

35. $x \geq 1, x \neq 2$ **37.** $-7 \leq x \leq 7$

39. (a) $f(20) = 103{,}000$ means if \$103,000 is borrowed
   it can be repaid in 20 years (of \$800 per month
   payments).
   (b) no; $f(5 + 5) = f(10) = 69{,}000$, but
   $f(5) + f(5) = 80{,}000$
   (c) $t \approx 26$ years. If someone has found a home to
   buy and it requires a loan of \$115,000, then this
   question answers how long that loan would have
   to be so payments fit the budget.

41. (a) $f(1950) = 16.5$ means that in 1950 there were
   16.5 workers supporting each Social Security
   beneficiary.
   (b) 3.4
   (c) The parts of the graph that correspond to data
   prior to 1995 would be the same, but data on this
   graph beyond 1995 are predictions for the future
   and only might be accurate.
   (d) Domain: $1950 \leq t \leq 2050$
   Range: $1.9 \leq n \leq 16.5$

43. (a) $s \geq 0$
   (b) $f(10) \approx -29.33$ means if the air temperature is
   $-5°F$ and there is a 10 mph wind, then the
   temperature feels like $-29.33°F$.
   (c) $f(0)$ should equal the air temperature, $-5°F$.

45. (a) yes **(b)** $v \geq -\frac{1}{4}$ **(c)** $v \geq 0$

47. (a) yes **(b)** all reals
   (c) D: $32° \leq F \leq 212°$; R: $0° \leq C \leq 100°$
   (d) $4.44°C$

49. (a) 6 **(b)** 20 **(c)** 42

51. (a) $N(2) = 48\frac{2}{3}$; the number is 48. This means 2
   seconds after a stimulus, 48 action potentials are
   produced.
   (b) $N(10) = 248\frac{2}{51}$; the number is 248. This means
   that 10 seconds after a stimulus, 248 action
   potentials are produced.

53. (a) $0 \leq p < 100$
   (b) \$5972.73; to remove 45% of the particulate
   pollution would cost \$5972.73.
   (c) \$65,700; to remove 90% of the particulate
   pollution would cost \$65,700.
   (d) \$722,700; to remove 99% of the particulate
   pollution would cost \$722,700.
   (e) \$1,817,700; to remove 99.6% of the particulate
   pollution would cost \$1,817,700.

55. (a) yes **(b)** $A(2) = 96; A(30) = 600$
   (c) $0 < x < 50$

57. (a) $(p\circ x)(t) = 180(1000 + 10t)$
   $$-\frac{(1000 + 10t)^2}{100} - 200$$
   (b) $x = 1150, p = \$193{,}575$

59. $L = 2x + 3200/x$ **61.** $R = (30 + x)(10 - 0.20x)$

Exercise 1.5 *(page 110)*

1. k **3.** l **5.** r **7.** o **9.** a **11.** d
13. g **15.** j **17.** m
19. (a) -1 **(b)** 1 **(c)** 1 **(d)** no
21. (a) 7 **(b)** undefined **(c)** 2.99 **(d)** 1.9
23. (a) 37 **(b)** undefined **(c)** 2.99 **(d)** 1.81
25. (a) 2 **(b)** 4 **(c)** 0 **(d)** 2
27.

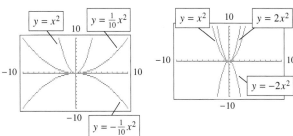

(a) For $0 < |a| < 1$, the graph opens wider than
   $f(x) = x^2$.
(b) For $|a| > 1$, the graph opens narrower than
   $f(x) = x^2$.
(c) For $a < 0$, the graph is reflected over the x-axis,
   i.e., it is upside down from when $a > 0$.

29.

(a) and **(b)** Each graph is the graph of $f(x) = |x|$ shifted a units on the x-axis, to the right if $a > 0$ and to the left if $a < 0$.

31.

(a) and **(b)** Each graph is the graph of $f(x) = 1/x$ shifted a units on the y-axis, upward if $a > 0$ and downward if $a < 0$.

33. (a)

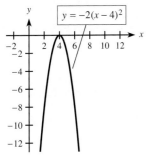

(b) polynomial   **(c)** no asymptotes   **(d)** turning point at $x = 4$

35. (a)

(b) polynomial   **(c)** no asymptotes   **(d)** turning points at $x = 3$ and $x = \frac{1}{3}$

37. (a)

(b) polynomial

(c) no asymptotes

(d) turning points at $x = 0$ and approximately $x = -2.8$ and $x = 2.8$

39. (a)

(b) rational

(c) vertical: $x = -1$, horizontal: $y = 2$

(d) no turning points

41. (a)

$$f(x) = \begin{cases} -x & \text{if } x < 0 \\ 5x & \text{if } x \geq 0 \end{cases}$$

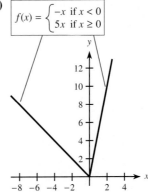

(b) piecewise

(c) no asymptotes

(d) turning point at $x = 0$

43.

45. (a) $x = 0.30, f(0.30) = 18$
(b) $x = 30, f(30) = 19$
(c) $x = 40, f(40) = 20$
(d)

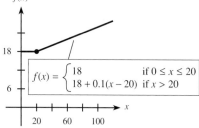

47. (a) $C(780) = 48.92$ means the monthly charge for 780 kWh of electricity is $48.92.
(b) $89.33
49. (a) $T(58,676) = 11,632.28$ means if taxable income is $58,676, then tax due is $11,632.28.
(b) No. They pay 13% more only on the income above $36,900. Tax on $36,900 is $5535.00 and on $36,901 is $5535.28.
51. (a) $f(15) = -31$ means when the air temperature is 0°F and the wind speed is 15 mph, the air temperature feels like −31°F.
(b) −55°F
53. (a)

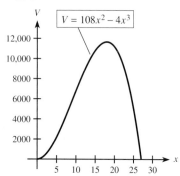

(b) The x-intercepts correspond to boxes with no volume but that satisfy the equation.

(c) The turning point x-coordinate gives the length of the side of the base of the box whose V-coordinate is the maximum volume.
(d) No, $x < 0$ makes no physical sense and $x > 27$ gives a negative volume, which is impossible.
(e) Maximum volume is approximately 12,000 cubic inches. (Actually it is 11,664 when $x = 18$.)
55.

57. (a)

(b) Turning point at (4, 24) means costs will be $24 when 4 plates are used.
(c) No, the problem statement says the number of plates is from 1 to 10.
59.

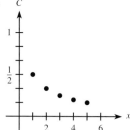

Exercise 1.6   *(page 122)*

1. (a) $C(x) = 17x + 3400$ **(b)** $6800
3. (a) $C(x) = 43x + 1850$
   (b) $1850; fixed costs   **(c)** $3441
5. (a) $R(x) = 34x$ **(b)** $10,200
7. (a) $R(x) = 80x$ **(b)** $6000
9. (a) $P(x) = 17x - 3400$ **(b)** $1700   **(c)** 200

11. (a) $P(x) = 37x - 1850$ **(b)** $-\$740$, loss of $740
(c) 50
13. (a) 50 and 550 units   **(b)** 300 units   **(c)** $5000
15.   **17.**

19. demand decreases
21. (a) 650 (approx)   **(b)** 300   **(c)** shortage
23. 16 demanded, 25 supplied; surplus
25. (a) $3 billion   **(b)** $7.2 billion
(c)

27. (a) $5.6 billion   **(b)** $14.6 billion
(c)

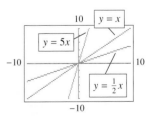

29. (a) $16 billion   **(b)** $16.5 billion   **(c)** 47%

Exercise 1.7   *(page 133)*

1. $m = 2$ **3.** $m = -5$ **5.** $m = 0$
7. $m = 7/3, b = -1/4$ **9.** $m = 0, b = 3$
11. no slope, no y-intercept   **13.** $m = 1/4, b = 0$
15. $m = -2/3, b = 2$ **17.** $m \approx -0.831, b \approx 29$.

19.   **21.**

23.   **25.**

27.

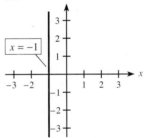

29. $y = 2x - 4$ **31.** $-x + 13y = 32$ **33.** $y = 0$
35. $y = -\frac{3}{5}x - \frac{41}{5}$ **37.** $y = -\frac{6}{5}x + \frac{23}{5}$
39. $y = -\frac{8}{5}x + \frac{128}{5}$
41. (a)

(b) x-intercept: $x = 0$ y-intercept: $y = 0$
(c) (i)   line rises left to right when $m > 0$
    (ii) line falls left to right when $m < 0$
    (iii) line close to horizontal when m is close to 0
    (iv) line close to vertical when $|m|$ is large

43. (a) $m = 0.1369\ b = -5.091255$
   (b) The y-intercept indicates that when there were 0 terminals, the amount transacted was negative. This is impossible. The model must be restricted to when both $x > 0$ and $y \geq 0$.
   (c) The slope means that the transaction amount increases by $0.1369 (billion) when the number of ATMs increases by 1 (thousand).
45. $y = 4.95 + 0.0838x$
47. (a) $f = 0.838m - 1387.4$ **(b)** $23,752.60
49. $p = 85,000 - 1700x$
51. (a) $m = 5, b = 250$
   (b) $\overline{MC} = 5$ means each additional unit produced costs $5.   **(c)** $250
   (d) slope = marginal cost; c-intercept = fixed costs   **(e)** 5, 5
53. (a) 27   **(b)** $\overline{MR} = 27$ means each additional unit sold brings in $27.   **(c)** 27, 27
55. (a) $P(x) = 22x - 250$ **(b)** 22   **(c)** $\overline{MP} = 22$
   (d) Each unit sold adds $22 to profits at all levels of production, so produce and sell as much as possible.
57. $C(x) = 27x + 3000$

59. $p = -2q/3 + 1060$ **61.** $p = 0.0001q + 0.5$
63. $R = 3.2t - 0.2$ **65.** $P = 58x - 8500$
67. $y = 0.48x - 71$ **69. (a)** 2055   **(b)** 2008

Exercise 1.8   *(page 146)*

1. $x = 2, y = 2$ **3.** no solution   **5.** $x = 2, y = 5$
7. $x = 14/11, y = 6/11$ **9.** $x = 10/3, y = 2$
11. $x = 2, y = 1$ **13.** $x = 1, y = 1$ **15.** no solution
17. $x = -52/7, y = -128/7$ **19.** dependent
21. $x = 1, y = 7$ **23.** $x = -17, y = 7, z = 5$
25. $x = 4, y = 12, z = -1$
27. $x = 44, y = -9, z = -1/2$
29. $68,000 at 18%; $77,600 at 10%
31. $13,500 at 10%; $10,000 at 12%

33. $A = 4$ oz, $B = 6\frac{2}{3}$ oz   **35.** $A = 4550, B = 1500$
37. 7 cc of 20%; 3 cc of 5%
39. 10,000 at $20; 6000 at $30   **41.** 80 cc
43. 5 oz of A, 1 oz of B, 5 oz of C
45. $A = 200, B = 100, C = 200$

Exercise 1.9   *(page 156)*

1. (a) Revenue passes through the origin   **(b)** $2000
   (c) 400 units   **(d)** $\overline{MC} = 2.5; \overline{MR} = 7.5$ **3.** 33
5. (a) $R(x) = 12x; C(x) = 8x + 1600$ **(b)** 400 units
7. (a) $P(x) = 4x - 1600$ **(b)** $x = 400$ units to break even   **9.** 3600 units
11. (a) $R(x) = 54.90x$ **(b)** $C(x) = 14.90x + 20,200$
   (c) 505
13. 20,000 units
15. (a) demand falls; supply rises   **(b)** (30, $25)
17. (a) $q = 20$ **(b)** $q = 40$
   (c) shortage, 20 units short
19. shortage
21.

$$p = -\tfrac{1}{2}q + 28$$

23.

$$p = \tfrac{1}{3}q + \tfrac{34}{3}$$

25. $q = 20, p = 18

27. $q = 10, p = 180 **29.** $q = 100, p = 325
31. (a) $15   **(b)** $q = 100, p = 100
   (c) $q = 50, p = 110 **(d)** yes
33. $q = 8; p = 188 **35.** $q = 500, p = 40$
37. $q = 1200, p = 15
39. (a) $C(y) = 0.6y + 8$ **(b)** 0.6

41.

$C = 0.6y + 8$

43. **(a)** 0.65
(b) $5 billion

12.

$2x - 3y = 6$

13.

$5x + 3y = 6$

45.
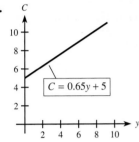
$C = 0.65y + 5$

47. They are equal.

49. **(a)** .84 **(b)** $C = .84y + 7.2$

14.

$y = \dfrac{x-3}{4}$

15.

$y = -x^2 + 2x + 3$

Chapter 1 Review Exercises *(page 161)*

1. $x = 7$ **2.** $x = \frac{31}{3}$ **3.** $x = -13$ **4.** $-\frac{29}{8}$
5. $x = -\frac{1}{9}$ **6.** $x = 10.05$ **7.** $x = \frac{2}{5}$
8. $x \approx 1.11$ **9.**

$y = 3x$

16.

$y = \dfrac{x^3 - 27x + 54}{15}$

10.

$y = 5$

11.

$x = -2$

17. (a)

$y = (x + 6)(x - 3)(x - 15)$

(b)

Standard window view
$y = (x + 6)(x - 3)(x - 15)$

(c) The graph in (a) shows the complete graph. The graph in (b) shows a piece that rises toward the high point and a piece between the high and low points.

18. (a)

(b)

(c)

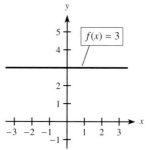

(c) The graph in (a) is a complete graph. The one in (b) shows pieces that fall toward the minimum point and rise from it.

19. yes   **20.** no   **21.** yes
22. D: reals $x \le 9$; R: reals $y \ge 0$
23. reals $x \ge -3$ except $x \ne 0$
24. (a) 2   **(b)** 37   **(c)** 29/4
25. (a) 0   **(b)** 9/4   **(c)** 10.01
26. (a) -36 **(b)** $-x^2 + 7x + 8$
    (c) $-x^2 - 2xh - h^2 + 9x + 9h$
27. $9 - 2x - h$ **28.** yes   **29.** no   **30.** 4
31. $x = 0, x = 4$ **32.** $y = 1, y = -1$
33. (a) 7   **(b)** $x = -1, x = 3$
    (c)

34. (a)

(b)

35. (a) 0   **(b)** 10,000   **(c)** -25 **(d)** 0.1
36. (a) -2 **(b)** 0   **(c)** 1   **(d)** 4
37.

38.

39.

40.

41.

42.

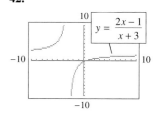

43. $m = -\frac{3}{4}$ **44.** $m = 1$ **45.** undefined

46. $m = -\frac{2}{5}, b = 2$ **47.** $m = -\frac{4}{3}, b = 2$

48. $y = 4x + 2$ **49.** $y = -\frac{1}{2}x + 3$

50. $y = \frac{2}{5}x + \frac{9}{5}$ **51.** $y = -x - 5$

52. $y = -\frac{11}{8}x + \frac{17}{4}$ **53.** $x = -1$ **54.** $y = 4x + 2$

55. $y = \frac{4}{3}x + \frac{10}{3}$ **56.** $x = 2, y = 1$

57. $x = 10, y = -1$ **58.** $x = 3, y = -2$

59. no solution **60.** $x = 10, y = -71$

61. $x = 2, y = 1$ **62.** $x = 1, y = -1, z = 2$

63. $x = 11, y = 10, z = 9$ **64.** 95%

65. 40,000 mi. He would normally drive more than 40,000 miles in 5 years so he should buy diesel.

66. **(a)** yes **(b)** no **(c)** 4

67. **(a)** $565.44
   (b) The monthly payment on a $70,000 loan is $494.75.
   (c) For this particular function, $f(120)$ equals $f(40) + f(80)$, or $2f(60)$. $f(120) = $848.14.

68. **(a)** $x = 12; C(12) = $18.68
   (b) $x = 825; C(825) = $909.70

69. **(a)** **(b)** $0 \le x \le 6$

70. **(a)** $C(x) = 2x + 500$ **(b)** $700

71. **(a)** $R(x) = 4x$ **(b)** $400

72. **(a)** $P(x) = 2x - 500$ **(b)** loss of $300 **(c)** 250

73. **(a)** 12 supplied; 14 demanded **(b)** shortage
   (c) increase

74. **(a)** $2 billion **(b)** $6 billion

75.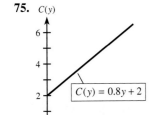

76. $100,000 at 9.5%; $50,000 at 11%

77. 2.8 liters of 20%; 1.2 liters of 70%

78.

79. **(a)** $4 billion **(b)** 0.6
   (c) $C(y) = 4 + 0.6y$

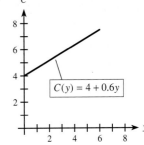

80. **(a)** 38.80 **(b)** 61.30 **(c)** 22.50 **(d)** 200

81. **(a)** $C(x) = 22x + 1500$ **(b)** $R(x) = 52x$
   (c) $P(x) = 30x - 1500$ **(d)** $\overline{MC} = 22$
   (e) $\overline{MR} = 52$ **(f)** $\overline{MP} = 30$ **(g)** $x = 50$

82. $q = 300, p = $150 **83.** $q = 700, p = 80$

Exercise 2.1 *(page 176)*

1. 3 **3.** $\begin{bmatrix} -1 & -2 & -3 \\ 1 & 0 & -1 \\ -2 & 3 & 4 \end{bmatrix}$ **5.** A, C, D, F, Z

7. A, F, and Z are 3×3; C and D are 2×2

9. 1 **11.** no **13.** $\begin{bmatrix} 0 & 0 & 0 \\ 0 & 0 & 0 \\ 0 & 0 & 0 \end{bmatrix}$

15. $\begin{bmatrix} 9 & 5 \\ 4 & 7 \end{bmatrix}$ **17.** $\begin{bmatrix} 0 & -2 & -1 \\ 4 & 2 & 0 \\ 2 & 3 & 7 \end{bmatrix}$ **19.** impossible

21. impossible **23.** A **25.** impossible

27. $x = 3, y = 2, z = 3, w = 4$

29. $x = 4, y = 1, w = 3, z = 3$

31. $x = 2, y = 2, z = -3$ **33.** $\begin{bmatrix} 825 & 580 & 1560 \\ 810 & 650 & 350 \end{bmatrix}$

35. (a) $A = \begin{bmatrix} 0 & 1 & 1 & 0 \\ 1 & 0 & 1 & 1 \\ 0 & 0 & 0 & 1 \\ 0 & 1 & 1 & 0 \end{bmatrix}$ **(b)** $B = \begin{bmatrix} 0 & 1 & 1 & 1 \\ 1 & 0 & 1 & 1 \\ 0 & 1 & 0 & 1 \\ 1 & 1 & 1 & 0 \end{bmatrix}$

(c) person 2

37. (a) $\begin{array}{c} \\ \text{I} \\ \text{II} \\ \text{III} \end{array} \begin{bmatrix} \text{wt} & 1 \\ 140 & 5.5 \\ 151 & 5.7 \\ 141 & 5.5 \end{bmatrix}$ **(b)** $\begin{array}{c} \\ \text{I} \\ \text{II} \\ \text{III} \end{array} \begin{bmatrix} \text{wt} & 1 \\ 250 & 12.5 \\ 215 & 11.8 \\ 190 & 9.8 \end{bmatrix}$

39.

$A = \begin{array}{c} \text{M} \quad\;\; \text{F} \\ \begin{bmatrix} 54.4 & 55.6 \\ 62.1 & 66.6 \\ 67.4 & 74.1 \\ 70.7 & 78.1 \end{bmatrix} \end{array}$ $B = \begin{array}{c} \text{M} \quad\;\; \text{F} \\ \begin{bmatrix} 45.5 & 45.2 \\ 51.5 & 54.9 \\ 61.1 & 67.4 \\ 65.3 & 73.6 \end{bmatrix} \end{array}$

$C = \begin{bmatrix} 8.9 & 10.4 \\ 10.6 & 11.7 \\ 6.3 & 6.7 \\ 5.4 & 4.5 \end{bmatrix}$

41. (a) 80 **(b)** 80 **(c)** 86 **(d)** 74

(e) $\begin{bmatrix} 80 & 75 \\ 58 & 106 \end{bmatrix}$ **(f)** $\begin{bmatrix} 176 & 127 \\ 139 & 143 \end{bmatrix}$

43.

	1	2	3	4	5	6
WHN1	23	23	15	26	20	34
WHN2	24	23	16	28	21	33
WHN3	18	17	15	18	17	22

45.

	1	2	3	4	5	6
WHN1	0	−2	−1	−2	−3	4
WHN2	0	−4	−1	4	−2	8
WHN3	0	−1	−1	−2	−2	−2

The flow rate is the same under rule 1.

47. $\begin{bmatrix} 1.00 & 0.94 & 0.95 & 0.95 & 0.94 & 0.85 & 0.83 & 0.85 & 0.85 \\ 0.85 & 0.99 & 0.95 & 0.95 & 0.94 & 0.95 & 0.83 & 0.85 & 0.85 \\ 0.85 & 0.84 & 1.00 & 0.95 & 0.94 & 0.95 & 0.93 & 0.85 & 0.85 \\ 0.85 & 0.84 & 0.85 & 1.00 & 0.94 & 0.95 & 0.93 & 0.95 & 0.85 \\ 0.85 & 0.84 & 0.85 & 0.85 & 0.99 & 0.95 & 0.93 & 0.95 & 0.95 \\ 0.95 & 0.84 & 0.85 & 0.85 & 0.84 & 1.00 & 0.93 & 0.95 & 0.95 \\ 0.95 & 0.94 & 0.85 & 0.85 & 0.84 & 0.85 & 0.98 & 0.95 & 0.95 \\ 0.95 & 0.94 & 0.95 & 0.85 & 0.84 & 0.85 & 0.83 & 1.00 & 0.95 \\ 0.95 & 0.94 & 0.95 & 0.95 & 0.84 & 0.85 & 0.83 & 0.85 & 1.00 \end{bmatrix}$

49. Work Centers

1	2	3	4	5	6	7	8	9
[0.88	0.90	0.92	0.95	0.96	0.93	0.91	0.89	0.87]

Number 5; Number 9

Exercise 2.2 *(page 191)*

1. [32] **3.** [11 17] **5.** $\begin{bmatrix} 3 & 0 & 6 \\ 9 & 6 & 3 \\ 12 & 0 & 9 \end{bmatrix}$

7. $\begin{bmatrix} 28 & 16 \\ 10 & 18 \end{bmatrix}$ **9.** impossible **11.** $\begin{bmatrix} 29 & 25 \\ 10 & 12 \end{bmatrix}$

13. $\begin{bmatrix} 22 & 16 \\ 20 & 19 \end{bmatrix}$ **15.** $\begin{bmatrix} 8 & -2 & 2 & 4 \\ 13 & -5 & 12 & -11 \end{bmatrix}$

17. $\begin{bmatrix} 7 & 5 & 3 & 2 \\ 14 & 9 & 11 & 3 \\ 13 & 10 & 12 & 3 \end{bmatrix}$

19. impossible **21.** $\begin{bmatrix} 13 & 9 & 3 & 4 \\ 9 & 7 & 16 & 1 \end{bmatrix}$

23. $\begin{bmatrix} 17 & 0 & 14 \\ 8 & 2 & 11 \end{bmatrix}$ **25.** $\begin{bmatrix} 9 & 0 & 8 \\ 13 & 4 & 11 \\ 16 & 0 & 17 \end{bmatrix}$

27. $\begin{bmatrix} 161 & 126 \\ 42 & 35 \end{bmatrix}$ **29.** yes **31.** no

33. $\begin{bmatrix} -55 & 88 & 0 \\ -42 & 67 & 0 \\ 28 & -44 & 1 \end{bmatrix}$ **35.** $\begin{bmatrix} 0 & 0 & 0 \\ 0 & 0 & 0 \\ 0 & 0 & 0 \end{bmatrix}$

37. $\begin{bmatrix} 0 & 0 & 0 \\ 0 & 0 & 0 \\ 0 & 0 & 0 \end{bmatrix}$ **39.** A **41.** Z

43. no (see problem 35)

45. $AB = \begin{bmatrix} 1 & 0 \\ 0 & 1 \end{bmatrix}$, $BA = \begin{bmatrix} 1 & 0 \\ 0 & 1 \end{bmatrix}$

47. $\begin{cases} x + 2y = 0 \\ 3x + 4y = 2 \end{cases}$ **49.** $\begin{cases} x + \quad\;\; 2z = 0 \\ 3x + y \quad\;\; = 7 \\ x + 2y + z = 3 \end{cases}$

51. $\begin{bmatrix} 2 - 2 + 2 \\ 6 - 4 - 4 \\ 4 + 0 - 2 \end{bmatrix} = \begin{bmatrix} 2 \\ -2 \\ 2 \end{bmatrix}$; is solution

53. $\begin{bmatrix} 1+2+2 \\ 4+0+1 \\ 2+2+1 \end{bmatrix} = \begin{bmatrix} 5 \\ 5 \\ 5 \end{bmatrix}$; is solution

55. $\begin{bmatrix} 1 & 0 & 0 & 0 & 0 \\ 0 & 1 & 0 & 0 & 0 \\ 0 & 0 & 1 & 0 & 0 \\ 0 & 0 & 0 & 1 & 0 \\ 0 & 0 & 0 & 0 & 1 \end{bmatrix}$ (Some entries may appear as decimal approximations of 0.)

57. (a) $\begin{bmatrix} 26{,}000 & 39{,}000 \\ 16{,}250 & 9750 \\ 22{,}750 & 37{,}700 \end{bmatrix}$ **(b)** $\begin{bmatrix} 8000 & 72{,}000 \\ 5000 & 18{,}000 \\ 7000 & 69{,}600 \end{bmatrix}$

59. (a) $\begin{bmatrix} 20{,}000 & 20{,}800 & 9600 \\ 12{,}000 & 41{,}600 & 4000 \\ 6400 & 16{,}000 & 96{,}000 \end{bmatrix}$

(b) paid half due in 60 days and all due in 30 days

61. 172, 208, 268, 327, 101, 123, 268, 327, 216, 263, 162, 195, 176, 215, 343, 417

63. $\begin{bmatrix} 9{,}371{,}700 & 276{,}573{,}000 \\ 6{,}884{,}150 & 455{,}532{,}000 \\ 3{,}818{,}100 & 650{,}760{,}000 \\ 17{,}239{,}300 & 3{,}188{,}724{,}000 \\ 11{,}685{,}700 & 818{,}873{,}000 \end{bmatrix}$

65. (a) $\begin{bmatrix} .08 & .22 & .12 \\ .10 & .08 & .19 \\ .05 & .07 & .09 \\ .10 & .26 & .15 \\ .12 & .04 & .24 \end{bmatrix}$

(b) 0.08, 0.22, 0.12 consumed by carnivores 1, 2, 3
(c) Plant 5 by 1, Plant 4 by 2, Plant 5 by 3

67. $\begin{array}{c} \\ A \\ B \end{array}\begin{array}{cc} X & Y \\ \begin{bmatrix} 59 & 60 \\ 79 & 77 \end{bmatrix} \end{array}$

69. (a) $[1000 \quad 4000 \quad 2000 \quad 1000]$
(b) $[45{,}000 \quad 55{,}000 \quad 90{,}000 \quad 70{,}000]$

(c) $\begin{bmatrix} 5 \\ 3 \\ 4 \\ 4 \end{bmatrix}$ **(d)** $[\$1{,}030{,}000]$ **(e)** $\begin{array}{c} \\ A \\ B \\ C \\ D \end{array}\begin{array}{c} \$ \\ \begin{bmatrix} 65 \\ 145 \\ 125 \\ 135 \end{bmatrix} \end{array}$

71. $\begin{bmatrix} 0.84 & 10.2 & 12.24 & 1.32 & 6.72 & 4.32 \\ 0.515 & 0.206 & 6.283 & 1.339 & 0.206 & 1.03 \\ 2.31 & 0.42 & 9.24 & 1.26 & 1.26 & 5.04 \\ 302.16 & 76.08 & 97.92 & 42.24 & 65.16 & 173.04 \\ 31.5 & 1.05 & 1.05 & 1.05 & 1.05 & 1.05 \\ 765.233 & 0 & 0 & 0 & 0 & 0 \end{bmatrix}$

Exercise 2.3 *(page 207)*

1. $\left[\begin{array}{ccc|c} -1 & 2 & 1 & 7 \\ 0 & 7 & 5 & 21 \\ 4 & 2 & 2 & 1 \end{array}\right]$ **3.** $\left[\begin{array}{ccc|c} -3 & 2 & 4 & 0 \\ 2 & -1 & 2 & 0 \\ 1 & -2 & -4 & 0 \end{array}\right]$

5. $\left[\begin{array}{ccc|c} 1 & -3 & 4 & 2 \\ 2 & 0 & 2 & 1 \\ 1 & 2 & 1 & 1 \end{array}\right]$ **7.** $x = 2, y = 1/2, z = -5$

9. $x = 18, y = 10, z = 0$ **11.** $x = -5, y = 2, z = 1$
13. $x = 4, y = 1, z = -2$ **15.** $x = \frac{1}{3}, y = \frac{5}{3}$
17. $x = 15, y = -13, z = 2$
19. $x = 15, y = 0, z = 2$ **21.** $x = 1, y = -1, z = 1$
23. $x_1 = 1, x_2 = 0, x_3 = 1, x_4 = 0$
25. $x = 1, y = 3, z = 1, w = 0$
27. no solution **29.** $x = (11 + 2z)/3, y = (-1 - z)/3$
31. $x = 0, y = -z$ **33.** $x = 0, y = -1/3, z = 2/3$
35. $x = 1 - z, y = \frac{1}{2}z$ **37.** $x = 2z - 2, y = 1 + z$
39. $x = \frac{9}{2}, y = -\frac{1}{2}, z = -1$ **41.** $x_1 = \frac{23}{38}x_3, x_2 = \frac{12}{19}x_3$
43. (a) $-4 \times (\text{Row } 2) + \text{Row } 5 \rightarrow \text{Row } 5$

(b) $\left[\begin{array}{ccccc|c} 1 & 0 & 0 & -1 & -3 & -18 \\ 0 & 1 & 1.5 & 1.5 & -1 & -1.5 \\ 0 & 0 & 3.5 & 0.5 & 4 & 32.5 \\ 0 & 0 & -2.5 & 4.5 & 0 & 10.5 \\ 0 & 0 & 0 & 6 & -5 & -1 \end{array}\right]$

45. no solution
47. $x_1 = \frac{41}{26} - \frac{6}{13}x_4; x_2 = \frac{7}{26} + \frac{5}{13}x_4; x_3 = -\frac{9}{26} + \frac{1}{13}x_4$
49. $x = (b_2 c_1 - b_1 c_2)/(a_1 b_2 - a_2 b_1)$
51. I = 80g, II = 50g, III = 50 g
53. 3 passenger, 4 transport, 4 jumbo
55. 2 of Portfolio I, 2 of Portfolio II
57. $\frac{3}{8}$ lb of red meat, 6 slices of bread, 4 glasses of milk
59. $ 13,500 @ 15% and $ 10,000 @ 16%
61. Type I = 3 Type IV, Type II = 1000 − 2(Type IV), Type III = 500 − Type IV, Type IV = any amount (less than 500 bags)
63. Oil = 6138 − 3.3C − 0.6R, Bank = 7.5C + R − 12,787.50 R = any amount (less than 10,230), C = any amount (less than 1705)

65. Bacteria III = any amount (between 1800 and 2300)
　Bacteria I　= 6900 − 3 (Bacteria III)
　Bacteria II　= $\frac{1}{2}$ (Bacteria III) − 900
67. There are 3 possibilities:
　　(1) 4 of I and 2 of II
　　(2) 5 of I, 1 of II, and 1 of III
　　(3) 6 of I and 2 of III

Exercise 2.4　*(page 223)*

1. $\begin{bmatrix} 1 & 0 & 0 \\ 0 & 1 & 0 \\ 0 & 0 & 1 \end{bmatrix}$　**3.** yes　**5.** $\begin{bmatrix} \frac{1}{3} & 0 & 0 \\ 0 & \frac{1}{3} & 0 \\ 0 & 0 & \frac{1}{3} \end{bmatrix}$

7. $\begin{bmatrix} 2 & -7 \\ -1 & 4 \end{bmatrix}$　**9.** no inverse　**11.** $\begin{bmatrix} \frac{5}{2} & -1 \\ -2 & 1 \end{bmatrix}$

13. $\begin{bmatrix} -\frac{1}{10} & \frac{7}{10} \\ \frac{1}{5} & -\frac{2}{5} \end{bmatrix}$　**15.** $\begin{bmatrix} -1 & 1 & 0 \\ 1 & 0 & 0 \\ -1 & 0 & 1 \end{bmatrix}$

17. $\begin{bmatrix} \frac{1}{3} & -\frac{1}{3} & \frac{1}{3} \\ -\frac{2}{3} & -\frac{1}{3} & \frac{7}{3} \\ \frac{1}{3} & \frac{2}{3} & -\frac{5}{3} \end{bmatrix}$　**19.** no inverse　**21.** no inverse

23. $\begin{bmatrix} 2 & 2 & 0 & 2 & 2 \\ 1 & 0 & 2 & 2 & 1 \\ 0 & 1 & 0 & 2 & 1 \\ 2 & 0 & 2 & 2 & 1 \\ 1 & 0 & 0 & 0 & 2 \end{bmatrix}$

25. $x = 1, y = 1, z = 1$　**27.** $\begin{bmatrix} x \\ y \\ z \end{bmatrix} = \begin{bmatrix} 1 \\ 1 \\ 2 \end{bmatrix}$

29. $x = 2, y = 1$　**31.** $x = 1, y = 2$
33. $x = 1, y = 1, z = 1$　**35.** $x = 1, y = 3, z = 2$
37. $x_1 = 5.6, x_2 = 5.4, x_3 = 3.25, x_4 = 6.1, x_5 = .4$
39. −2　**41.** 14　**43.** −5　**45.** −19　**47.** yes
49. no　**51.** Hang on　**53.** Answers in back
55. $x_0 = 2400, y_0 = 1200$
57. A = 5.5 mg and B = 8.8 mg for Patient I; A = 10 mg and B = 16 mg for Patient II
59. $68,000 at 18%, $77,600 at 10%
61. 7cc of 20%, 3cc of 5%
63. (a) $\begin{bmatrix} 0 & 1 & 0 \\ 0 & 0 & 1 \\ 1 & 1 & 1 \end{bmatrix}$　**(b)** 108

Exercise 2.5　*(page 236)*

1. (a) 15　**(b)** 4　**3.** 8　**5.** 40
7. most: raw materials; least: fuels
9. raw materials, manufacturing, service
11. 5 units of manufacturing; 10 units of ag; 105 units of utilities
13. farm products = 200; machinery = 40
15. agricultural products = 100; oil = 700
17. utilities = 200; manufacturing = 400
19. mining = 310; manufacturing = 530
21. electronic components = 1200; computers = 320
23. fishing = 100; oil = 1250
25. development = $21,000; promotional = $12,000
27. engineering = $15,000; computer = $13,000
29. agricultural goods = 400; manufactured goods = 500; fuels = 400
31. electronics = 1240; steel = 1260; autos = 720
33. products = $\frac{7}{17}$ households; machinery = $\frac{1}{17}$ household
35. government = $\frac{10}{19}$ households; industry = $\frac{11}{19}$ households
37. manufacturing = 3 households; utilities = 3 households

39. $\begin{bmatrix} 24 \\ 96 \\ 24 \\ 120 \\ 492 \\ 3456 \end{bmatrix}$ 3456 bolts, 492 braces, 120 sheets

41. $\begin{bmatrix} 10 \\ 10 \\ 20 \\ 56 \\ 20 \\ 26 \\ 300 \end{bmatrix}$ 56　2 × 4's, 20 braces, 26 clamps, 300 nails

Chapter 2 Review Exercises　*(page 240)*

1. 4　**2.** 0　**3.** A, B　**4.** none

5. D, F, G, I　**6.** $\begin{bmatrix} -2 & 5 & 11 & -8 \\ -4 & 0 & 0 & -4 \\ 2 & 2 & -1 & -9 \end{bmatrix}$

7. Zero matrix　**8.** Order

9. $\begin{bmatrix} 6 & -1 & -9 & 3 \\ 10 & 3 & -1 & 4 \\ -2 & -2 & -2 & 14 \end{bmatrix}$ **10.** $\begin{bmatrix} 3 & -3 \\ 4 & -1 \\ 2 & -6 \\ 1 & -2 \end{bmatrix}$

11. $\begin{bmatrix} 2 & 5 \\ 1 & 1 \end{bmatrix}$ **12.** $\begin{bmatrix} 12 & -6 \\ 15 & 0 \\ 18 & 0 \\ 3 & 9 \end{bmatrix}$ **13.** $\begin{bmatrix} 4 & 0 \\ 0 & 4 \end{bmatrix}$

14. $\begin{bmatrix} 2 & -12 \\ -8 & -22 \end{bmatrix}$ **15.** $\begin{bmatrix} 9 & 20 \\ 4 & 5 \end{bmatrix}$ **16.** $\begin{bmatrix} 5 & 16 \\ 6 & 15 \end{bmatrix}$

17. $\begin{bmatrix} 2 & 37 & 61 & -55 \\ -2 & 9 & -3 & -20 \\ 10 & 10 & -14 & -30 \end{bmatrix}$ **18.** $\begin{bmatrix} 43 & -23 \\ 33 & -12 \\ -13 & 15 \end{bmatrix}$

19. $\begin{bmatrix} 10 & 16 \\ 15 & 25 \\ 18 & 30 \\ 6 & 11 \end{bmatrix}$ **20.** $\begin{bmatrix} 17 & 73 \\ 7 & 28 \end{bmatrix}$ **21.** $\begin{bmatrix} 3 & 7 \\ 23 & 42 \end{bmatrix}$

22. *F* **23.** *F* **24.** $\begin{bmatrix} 1 & 0 \\ 0 & 1 \end{bmatrix}$ **25.** *F* **26.** $(1, 2, 1)$

27. $x = 22, y = 9$ **28.** $x = -3, y = 3, z = 4$

29. $x = -\frac{3}{2}, y = 7, z = -\frac{11}{2}$ **30.** no solution

31. $x = 2 - 2z; y = -1 - 2z$

32. $x_1 = 1, x_2 = 11, x_3 = -4, x_4 = -5$

33. yes **34.** $\begin{bmatrix} \frac{1}{2} & \frac{1}{4} \\ \frac{5}{2} & \frac{7}{4} \end{bmatrix}$ **35.** $\begin{bmatrix} -1 & -2 & 8 \\ 1 & 2 & -7 \\ 1 & 1 & -4 \end{bmatrix}$

36. $\begin{bmatrix} 2 & 1 & -2 \\ 7 & 5 & -8 \\ -13 & -9 & 15 \end{bmatrix}$ **37.** $\begin{bmatrix} -33 \\ 30 \\ 19 \end{bmatrix}$ **38.** $\begin{bmatrix} 4 \\ 5 \\ -13 \end{bmatrix}$

39. $A^{-1} = \begin{bmatrix} -41 & 32 & 5 \\ 17 & -13 & -2 \\ -9 & 7 & 1 \end{bmatrix}; x = 4, y = -2, z = 2$

40. No **41.** $\begin{bmatrix} 250 & 140 \\ 480 & 700 \end{bmatrix}$ **42.** $\begin{bmatrix} 1030 & 800 \\ 700 & 1200 \end{bmatrix}$

43. **(a)** higher in June **(b)** higher in July

44. $\begin{array}{l} \\ \end{array}\begin{bmatrix} 865 & 885 \\ 210 & 270 \end{bmatrix}\begin{array}{l} \text{Robes} \\ \text{Hoods} \end{array}$ men women

45. $\begin{bmatrix} 1750 \\ 480 \end{bmatrix}\begin{array}{l} \text{Robes} \\ \text{Hoods} \end{array}$

46. $[2230]$

47. **(a)** $[0.20 \ 0.30 \ 0.50]$ **(b)** $\begin{bmatrix} 0.013469 \\ 0.013543 \\ 0.006504 \end{bmatrix}$

(c) $[0.20 \ 0.30 \ 0.50] \begin{bmatrix} 0.013469 \\ 0.013543 \\ 0.006504 \end{bmatrix} =$

$0.20(0.013469) + 0.30(0.013543) +$
$0.50(0.006504) = 0.0100087$

(d) The historical return of the portfolio, .0100087, is the estimated expected monthly return of the portfolio. This is roughly 1% per month.

48. 400 fast food, 700 computer, 200 pharmaceutical

49. $A = 2C, B = 2000 - 4C$ **50.** $S = 6000; A = 800$

51. $S = 1000; C = 500$ **52.** $X = \begin{bmatrix} 366 \\ 322 \\ 402 \end{bmatrix}$, approx.

53. $G = \frac{64}{93}H; A = \frac{59}{93}H; M = \frac{40}{93}H$

Exercise 3.1 *(page 250)*

1. $-\frac{1}{2} < x \le 3$ **3.** $(1, 3]$, half-open

5. $(2, 10)$, open **7.** $[-3, 2]$, closed

9. $(-4, 3)$, open **11.** $[4, 6]$, closed

13. $[4, 9)$, half-open **15.** $x < 2$ **17.** $x \le -4$

19. $x < 2$ **21.** $x < -3$ **23.** $x < \frac{3}{2}$ **25.** $x \le -1$

27. $x \le \frac{5}{28}$ **29.** $x \ge 1.949$

31. $x < -4$

33. $x < 3$

35. $x \ge -\frac{7}{2}$

37. $x < 2$

39. $x < \frac{20}{17}$

41. $1 < x < 4$ **43.** $x > 4$ **45.** $-50 \le x < -22$

47. $x > 80$

49. $695 + 5.75x \le 900$; 35 or fewer

51. **(a)** $0 \le I \le 22,750$
$22,751 \le I \le 55,100$
$55,101 \le I \le 115,000$
$115,001 \le I \le 250,000$
$I > 250,000$

(b) $0 \leq T \leq 3412.50$
$3412.50 < T \leq 12{,}470.50$
$12{,}470.50 < T \leq 31{,}039.50$
$31{,}039.50 < T \leq 79{,}639.50$
$T > 79{,}639.50$

53. $C \geq 37$

55. (a) (1) $x_4 \leq 1100$, (2) $x_4 \leq 1000$,
(3) $x_4 \leq 900$, (4) $x_4 \geq 0$
(b) $0 \leq x_4 \leq 900$
(c) $x_1 \in [100, 1000]$, $x_2 \in [200, 1100]$,
$x_3 \in [300, 1200]$

11. (0, 0), (20, 10), (0, 15), (25, 0)
13. (0, 0), (7, 0), (5, 5), (2, 6), (0, 4)
15. (0, 5), (1, 2), (3, 1), (6, 0)
17.

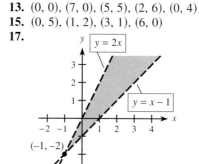

Exercise 3.2   *(page 257)*

1.

3.

5.

7.

9.

19.

21.

23.

25.

27.

29.

31.

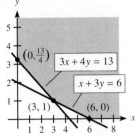

$(0, \frac{13}{4})$ $3x + 4y = 13$

$x + 3y = 6$

$(3, 1)$ $(6, 0)$

33. (a) Let x = Deluxe model
and y = Economy model
$3x + 2y \leq 24$
$\frac{1}{2}x + y \leq 8$
$x \geq 0, y \geq 0$

(b)

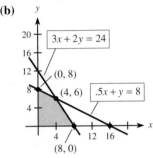

$3x + 2y = 24$

$(0, 8)$

$(4, 6)$ $.5x + y = 8$

$(8, 0)$

Corners: $(0, 0), (8, 0), (4, 6), (0, 8)$

35. Let x = cord-type trimmer
and y = cordless trimmer.
(a) Constraints are:
$x + y \leq 300$
$2x + 4y \leq 800$
$x \geq 0, y \geq 0$

(b)

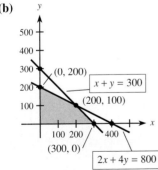

$(0, 200)$

$x + y = 300$

$(200, 100)$

$(300, 0)$

$2x + 4y = 800$

37. Let x = minutes of radio
and y = minutes of television

(a) Constraints are:
$x + y \geq 80$
$0.006x + 0.09y \geq 2.16$
$x \geq 0, y \geq 0$

(b)

250 $0.006x + 0.09y = 2.16$

$(0, 80)$

50 $(60, 20)$ $(360, 0)$

-50 50 250 350

$x + y = 80$

39. (a) x = lbs of regular
y = lbs of all beef
$0.18x + 0.75y \leq 1020$
$0.2x + 0.2y \geq 500$
$0.3x \qquad \leq 600$

(b)

$0.2x + 0.2y = 500$

$0.3x = 600$

$0.18x + 0.75y = 1020$

B

C

A

1000 2000 3000

$A = (2000, 500)$
$B = (2000, 880)$
$C = (1500, 1000)$

Exercise 3.3   *(page 267)*

1. $(12, 0)$
3. max = 20 at $(4, 4)$; min = 0 at $(0, 0)$
5. max = 68 at $(12, 4)$; min = 14 at $(2, 2)$
7. no max; min = 10 at $(2, 1)$
9. $(0, 0), (0, 20), (10, 18), (15, 10), (20, 0)$; max = 66
at $(10, 18)$
11. $(0, 60), (10, 30), (20, 20), (70, 0)$; min = 90 at
$(10, 30)$
13. max = 382 at $x = 10, y = 38$
15. max = 80 at $x = 20, y = 10$
17. min = 115 at $x = 5, y = 7$
19. min = 36 at $x = 3, y = 0$
21. max = 6 at $(0, 2)$
23. max = 22 at $(2, 4)$

25. max = 30 on line between (0, 5) and (3, 4)
27. min = 32 at (2, 3)   **29.** min = 9 at (2, 3)
31. min = 75 at (15, 15)
33. max = 10 at (2, 4)
35. min = 3100 at (40, 60)
37. max = $132 at (4, 6)
39. max = $10,000 at 200 cord-type, 100 cordless
41. 250 fish: 150 of I, 100 of II
43. R = $366,000 with 6 satellite and 17 full-service branches
45. radio = 60, TV = 20, C = $16,000
47. 30 days for Factory 1 and 20 days for Factory 2; cost = $700,000
49. 60 days for Location I and 70 days for Location II; cost = $86,000
51. reg = 2000 lb; beef = 880 lb; profit = $1328
53. (0, 0), (0, 200), (270, 0), (210, 60) $11,400,000 with 210 Carolina and 60 Savannah models

Exercise 3.4   *(page 286)*

1. $3x + 5y + s_1 = 15, 3x + 6y + s_2 = 20$

3.
$$\begin{bmatrix} 2 & 5 & 1 & 0 & 0 & | & 30 \\ 1 & 5 & 0 & 1 & 0 & | & 25 \\ -2 & -4 & 0 & 0 & 1 & | & 0 \end{bmatrix}$$

5.
$$\begin{bmatrix} 1 & 5 & 1 & 0 & 0 & | & 200 \\ 2 & 3 & 0 & 1 & 0 & | & 134 \\ -4 & -9 & 0 & 0 & 1 & | & 0 \end{bmatrix}$$

7.
$$\begin{bmatrix} 2 & 7 & 9 & 1 & 0 & 0 & 0 & | & 100 \\ 6 & 5 & 1 & 0 & 1 & 0 & 0 & | & 145 \\ 1 & 2 & 7 & 0 & 0 & 1 & 0 & | & 90 \\ -2 & -5 & -2 & 0 & 0 & 0 & 1 & | & 0 \end{bmatrix}$$

9.
$$\begin{bmatrix} 2 & 4 & 1 & 0 & 0 & | & 24 \\ 1 & ① & 0 & 1 & 0 & | & 5 \\ -4 & -11 & 0 & 0 & 1 & | & 0 \end{bmatrix}$$

11.
$$\begin{bmatrix} 10 & ㉗ & 1 & 0 & 0 & 0 & | & 200 \\ 4 & 51 & 0 & 1 & 0 & 0 & | & 400 \\ 15 & 27 & 0 & 0 & 1 & 0 & | & 350 \\ -6 & -7 & 0 & 0 & 0 & 1 & | & 0 \end{bmatrix}$$

13.
$$\begin{bmatrix} 2 & 0 & 1 & -\frac{3}{4} & 0 & | & 12 \\ ③ & 1 & 0 & \frac{1}{3} & 0 & | & 15 \\ -4 & 0 & 0 & 3 & 1 & | & 15 \end{bmatrix}$$

15. Solution is complete

17.
$$\begin{bmatrix} 4 & 4 & 1 & 0 & 0 & 2 & 0 & | & 12 \\ ② & ④ & 0 & 1 & 0 & -1 & 0 & | & 4 \\ -3 & -11 & 0 & 0 & 1 & -1 & 0 & | & 6 \\ -3 & -3 & 0 & 0 & 0 & 4 & 1 & | & 150 \end{bmatrix}$$

Either circled number may act as the next pivot, but only one of them.

19. No solution is possible   **21.** $x = 11, y = 9; f = 20$
23. $x = 0, y = 14, z = 11; f = 525$
25. $x = 50, y = 10, z = 0, f = 100$. Multiple solutions are possible.
27. $x = 0, y = 5, f = 50$ **29.** $x = 4, y = 3, f = 17$
31. $x = 2, y = 5, f = 17$ **33.** $x = 4, y = 3, f = 11$
35. $x = 0, y = 2, z = 5, f = 40$
37. $x = 15, y = 15, z = 25, f = 780$
39. $x = 6, y = 2, z = 26, f = 206$
41. $x = 8, y = 16, f = 32$ **43.** no solution
45. $x = 0, y = 50$ or $x = 40, y = 40, f = 600$

47. (a)
$$\begin{bmatrix} 1 & 1 & 1 & 0 & 0 & | & 60 \\ 1 & 3 & 0 & 1 & 0 & | & 120 \\ -40 & -60 & 0 & 0 & 1 & | & 0 \end{bmatrix}$$
(b) Max profit is $3000 with 30 ink jet and 30 laser printers

49. (a)
$$\begin{bmatrix} 4 & 3 & 1 & 0 & 0 & | & 48 \\ 3 & 5 & 0 & 1 & 0 & | & 40 \\ -30 & -30 & 0 & 0 & 1 & | & 0 \end{bmatrix}$$
(b) Max profit is $3900 with 11 axles and 2 wheels

51. 500 tomatoes, 1800 peaches; $P = 4100$
53. 21 newspaper, 13 radio
55. Medium 1 = 10, Medium 2 = 10, Medium 3 = 12
57. $1650 profit with 46A, 20B, 6C
59. 8000 Regular, 0 Special, and 1000 Kitchen Magic; $32,000

Exercise 3.5   *(page 298)*

1.
$$\begin{bmatrix} 5 & 2 & | & 10 \\ 1 & 2 & | & 6 \\ 3 & 1 & | & g \end{bmatrix} \text{transpose} = \begin{bmatrix} 5 & 1 & | & 3 \\ 2 & 2 & | & 1 \\ 10 & 6 & | & g \end{bmatrix}$$

3. Maximize $f = 10x + 6y$ subject to $5x + y \le 3$, $2x + 2y \le 1, x \ge 0, y \ge 0$

5.
$$\begin{bmatrix} 5 & 1 & 1 & 0 & 0 & | & 3 \\ 2 & 2 & 0 & 1 & 0 & | & 1 \\ -10 & -6 & 0 & 0 & 1 & | & 0 \end{bmatrix}$$

7. $g = 5$ at $x = 0, y = 5$

9. $\begin{bmatrix} 1 & 1 & | & 9 \\ 1 & 3 & | & 15 \\ \hline 5 & 2 & | & g \end{bmatrix}$ transpose $= \begin{bmatrix} 1 & 1 & | & 5 \\ 1 & 3 & | & 2 \\ \hline 9 & 15 & | & g \end{bmatrix}$

11. $\begin{bmatrix} 4 & 1 & | & 11 \\ 3 & 2 & | & 12 \\ 3 & 1 & | & 6 \\ \hline 3 & 1 & | & g \end{bmatrix}$ transpose $= \begin{bmatrix} 4 & 3 & 3 & | & 3 \\ 1 & 2 & 1 & | & 1 \\ \hline 11 & 12 & 6 & | & g \end{bmatrix}$

13. $\begin{bmatrix} 1 & 3 & 0 & | & 1 \\ 4 & 6 & 1 & | & 3 \\ 0 & 4 & 1 & | & 1 \\ \hline 12 & 48 & 8 & | & g \end{bmatrix}$ transpose $= \begin{bmatrix} 1 & 4 & 0 & | & 12 \\ 3 & 6 & 4 & | & 48 \\ 0 & 1 & 1 & | & 8 \\ \hline 1 & 3 & 1 & | & g \end{bmatrix}$

15. Maximize $f = 9x_1 + 15x_2$ subject to
$$x_1 + x_2 \le 5$$
$$x_1 + 3x_2 \le 2$$

17. Maximize $f = 11x_1 + 12x_2 + 6x_3$
subject to $4x_1 + 3x_2 + 3x_3 \le 3$
$$x_1 + 2x_2 + x_3 \le 1$$

19. Maximize $f = x_1 + 3x_2 + x_3$
subject to $\qquad x_1 + 4x_2 \le 12$
$$3x_1 + 6x_2 + 4x_3 \le 48$$
$$x_2 + x_3 \le 8$$

21. $y_1 = 0,\ y_2 = 9;\ g = 18$ (min)

23. $y_1 = 2,\ y_2 = 3;\ g = 9$ (min)

25. $y_1 = \frac{2}{5},\ y_2 = \frac{1}{5},\ y_3 = \frac{1}{5};\ g = 16$ (min)

27. min $= 32$ at $x = 16,\ y = 0$

29. min $= 28$ at $x = 2,\ y = 0,\ z = 1$

31. Line 1 for 77 hours, Line 2 for 7 hours; $17,080

33. $A = 12$ weeks, $B = 0$ weeks, $C = 0$ weeks;
cost $= \$12,000$

35. Factory 1: 30 days, Factory 2: 20 days. Minimum cost $700,000.

37. 105 minutes on radio, nothing on TV. Minimum cost $10,500.

39. 16 oz. of Food I, 0 oz. of Food II, 0 oz. of Food III,
Minimum cost $= \$16$.

Exercise 3.6 *(page 309)*

1. $-3x + y \le -5$ **3.** $-6x - y \le -40$

5. $\begin{bmatrix} 1 & 2 & 1 & 0 & 0 & | & 6 \\ -4 & -2 & 0 & 1 & 0 & | & -12 \\ \hline -4 & -5 & 0 & 0 & 1 & | & 0 \end{bmatrix}$

7. $\begin{bmatrix} 6 & 4 & 1 & 0 & 0 & | & 24 \\ -5 & -2 & 0 & 1 & 0 & | & -16 \\ \hline -2 & -2 & 0 & 0 & 1 & | & 0 \end{bmatrix}$

9. (a) Maximize $f = 2x + 3y$ subject to
$$7x + 4y \le 28$$
$$3x - y \le -2$$
$$x \ge 0,\ y \ge 0$$

(b) $\begin{bmatrix} 7 & 4 & 1 & 0 & 0 & | & 28 \\ 3 & \boxed{-1} & 0 & 1 & 0 & | & -2 \\ \hline -2 & -3 & 0 & 0 & 1 & | & 0 \end{bmatrix}$

11. (a) Maximize $-g = -3x - 8y$ subject to
$$4x - 5y \le 50$$
$$x + y \le 80$$
$$x - 2y \le -4$$
$$x \ge 0,\ y \ge 0$$

(b) $\begin{bmatrix} 4 & -5 & 1 & 0 & 0 & 0 & | & 50 \\ 1 & 1 & 0 & 1 & 0 & 0 & | & 80 \\ 1 & \boxed{-2} & 0 & 0 & 1 & 0 & | & -4 \\ \hline 3 & 8 & 0 & 0 & 0 & 1 & | & 0 \end{bmatrix}$

13. $x = 6,\ y = 8,\ z = 12;\ f = 120$

15. $x = 10,\ y = 17;\ f = 57$ **17.** $x = 5,\ y = 7;\ f = 31$

19. $x = 4,\ y = 8;\ f = 16$ **21.** $x = 12,\ y = 8;\ f = 76$

23. $x = 5,\ y = 15;\ f = 45$

25. $x = 10,\ y = 20;\ f = 120$

27. $x = 20,\ y = 10,\ z = 0;\ f = 40$

29. $x = 5,\ y = 0,\ z = 3;\ f = 22$

31. $x = 0,\ y = 44,\ z = 98;\ f = 34$

33. $x = 70,\ y = 0,\ z = 40;\ f = 2100$

35. Produce 200 of each at Monaca; produce 300 commercial components and 550 domestic furnaces at Hamburg; profit $= \$355,250$

37. Produce 200 of each at Monaca; produce 300 commercial components and 550 domestic furnaces at Hamburg; cost $= \$337,750$

39. regular $= 2000$ lb; beef $= 880$ lb; profit $= \$1328$

41. I $= 3$ million, II $= 0$, III $= 3$ million;
cost $= \$180,000$

Chapter 3 Review Exercises   *(page 312)*

1. $x \le 3$

2. $x \ge -20/3$

3. $x \ge -15/13$

4. $x \le -3/2$

5. (a) closed; $[0, 5]$ **(b)** half-open; $[3, 7)$
(c) open; $(-3, 2)$
6. (a) $-1 < x < 16$ **(b)** $-12 < x \le 8$
(c) $x \le -1$

7.

8.

9.

10.

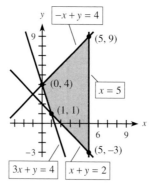

11. $f = 66$ at $(6, 6)$ **12.** $f = 91$ at $(9, 2)$
13. $f = 43$ at $(7, 9)$ **14.** $f = 31$ at $(12, 7)$
15. $g = 24$ at $(3, 3)$ **16.** $g = 20$ at $(5, 1)$
17. $g = \frac{247}{3}$ at $\left(\frac{212}{3}, \frac{7}{3}\right)$ **18.** $g = 80$ at $(5, 45)$
19. $f = 168$ at $(12, 7)$ **20.** $f = 260$ at $(60, 20)$
21. $f = 360$ at $(40, 30)$ **22.** $f = 270$ at $(5, 3, 2)$
23. $f = 640$ on the line between $(160, 0)$
and $(90, 70)$
24. no solution **25.** $g = 32$ at $y_1 = 2$, $y_2 = 3$
26. $g = 20$ at $y_1 = 4$, $y_2 = 2$
27. $g = 7$ at $y_1 = 1$, $y_2 = 5$
28. $g = 16$ at $y_1 = \frac{2}{5}$, $y_2 = \frac{1}{5}$, $y_3 = \frac{1}{5}$
29. $f = 165$ at $x = 20$, $y = 21$
30. $f = 54$ at $x = 6$, $y = 5$
31. $f = 156$ at $x = 15$, $y = 2$
32. $f = 31$ at $x = 4$, $y = 5$
33. $p = \$14{,}750$ when 110 large and 75 small swingsets
are made
34. $C = \$300{,}000$ when #1 operates 30 days, #2
operates 25 days

35. $P = \$320$; I = 40, II = 20
36. food I = 0 oz, food II = 3 oz; $C = \$0.60$ (min)
37. Cost = \$5.60; $A = 40$ lb, $B = 0$ lb
38. Cost = \$8500; $A = 20$ days, $B = 15$ days, $C = 0$
days
39. pancake mix = 8000 lb; cake mix = 3000 lb;
profit = \$3550
40. Texas: 55 desks, 65 computer tables; Louisiana: 75
desks, 65 computer tables; cost = \$4245

Exercise 4.1 *(page 329)*

1. $x^2 + 5x - 3 = 0$ **3.** $x^2 + 2x - 1 = 0$
5. $y^2 + 3y - 2 = 0$ **7.** $\frac{3}{2}, -\frac{3}{2}$ **9.** 0, 1
11. 0, 140 **13.** $-1, 3$ **15.** $-7, 3$ **17.** $\frac{1}{2}$
19. $-6, 2$ **21. (a)** $-\frac{1}{3}, \frac{1}{2}$ **(b)** $-0.33, 0.50$
23. (a) $2 + 2\sqrt{2}, 2 - 2\sqrt{2}$ **(b)** $4.83, -0.83$
25. no real solutions **27. (a)** $-\frac{7}{4}, \frac{3}{4}$ **(b)** $-1.75, 0.75$
29. (a) $(1 \pm \sqrt{31})/5$ **(b)** $1.31, -0.91$
31. $\sqrt{7}, -\sqrt{7}$
33. no real solutions **35.** $1, -9$
37. $(2 + \sqrt{5})/3 \approx 1.41, (2 - \sqrt{5})/3 \approx -0.08$
39. $-\frac{2}{7} \approx -0.29$ **41.** $-2, 5$
43. $\frac{2}{3} \approx 0.67, 3$ **45.** $0.69, -0.06$
47. $(5 + \sqrt{37})/2 \approx 5.54, (5 - \sqrt{37})/2 \approx -3.08$
49. $x = 20$ or $x = 70$ **51.** $x = 10$ or $x = 345\frac{5}{9}$
53. (a) $v = 28\sqrt{30}$ **(b)** $v = 28\sqrt{70}$
(c) $h = v^2/1960 - 10$
55. (a) $r = R/\sqrt{2}$ **(b)** $r = R\sqrt{3}/2$ **(c)** $v_r = 0$
57. 59.7 mph **59. (a)** $1.93, 13.99$ **(b)** 1994
61. 1997 **63.** \$80

Exercise 4.2 *(page 338)*

1. $(-1, -\frac{1}{2})$; min **3.** $(1, 9)$; max
5. (a) $x = 3$ **(b)** $f(3) = 9$
7. (a) $x = -1$ **(b)** $f(-1) = -4$
9. zeros $(-2, 0), (0, 0)$

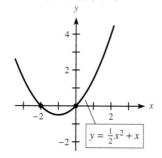

11. zeros $(-2, 0)(4, 0)$

13. vertex $(0, -4)$; zeros $(-2, 0)$, $(2, 0)$

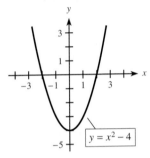

15. vertex $(2, 1)$; zeros $(0, 0)$, $(4, 0)$

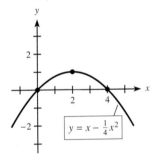

17. vertex $(-2, 0)$; zero $(-2, 0)$

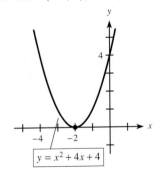

19. vertex $(-1, -3\frac{1}{2})$; zeros $(-1 + \sqrt{7}, 0)$, $(-1 - \sqrt{7}, 0)$

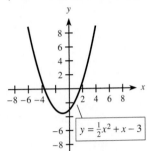

21. vertex $(1, -8)$; zeros $(-3, 0)$, $(5, 0)$

23. vertex $(-\frac{5}{2}, \frac{25}{4})$; zeros $(-5, 0)$, $(0, 0)$

25. vertex $(-6, 3)$; no real zeros

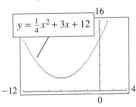

27. vertex $(10, 64)$; zeros $(90, 0)$, $(-70, 0)$

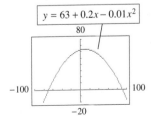

29. vertex $(0, -0.01)$; zeros $(10, 0)$, $(-10, 0)$

31. (a)

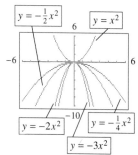

(b) When $a > 0$, the graph of $y = ax^2$ has the same basic shape as $y = x^2$. When $a < 0$, the graph is turned upside down.

(c) When $0 < a < 1$, the graph of $y = ax^2$ opens more gradually than $y = x^2$. When $a > 1$, the opening is narrower than $y = x^2$.

33. (a)

(b) The graph of $y = x^2 + k$ is the same graph as $y = x^2$ shifted k units on the y-axis.

35. (a) 80 units   **(b)** \$540

37.

39. 400

41.

43.

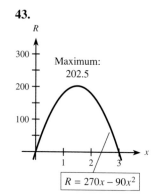

45. Equation **(a)** $(384.62, 202.31)$ **(b)** $(54, 46)$
Projectile **(a)** goes higher.

47. (a) quadratic
(b) $a > 0$ because the graph opens upward

49. (a)

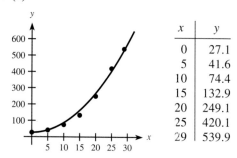

x	y
0	27.1
5	41.6
10	74.4
15	132.9
20	249.1
25	420.1
29	539.9

(b) The shape appears to be quadratic.
(c) $y = 0.6x^2 + 27.1$ fits fairly well.
(d) The model predicts 567.1.
(e) Yes, \$666.2 million is significantly larger than the predicted \$567.1 million. Yes.

51. (a) and **(b)**

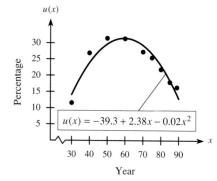

53. (a) 1999 and 1919
(b) after 1999 and before 1919, when the percentage is less than 0

Exercise 4.3 (page 347)

1. (a) and **(b)**

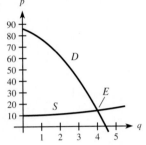

(c) See E on graph. **(d)** $q = 4, p = \$14$

3. (a) and **(b)**

(c) See E on graph. **(d)** $q = 4, p = \$6.60$

5. $q = 10, p = \$196$ **7.** $p = \$27.08, q = 216\frac{2}{3}$

9. $p = \$40, q = 30$ **11.** $q = 90, p = \$50$

13. $q = 70, p = \$62$ **15.** $x = 40$ units, $x = 50$ units

17. $x = 50, x = 300$

19. $x = 15$ units; reject $x = 100$ **21.** $\$41,173.61$

23. $\$87.50$ **25.** $x = 55, P(55) = 2025$

27. (a) $P(x) = -x^2 + 350x - 15,000$; max is $\$15,625$

(b) no **(c)** x-values agree

29. (a) $x = 28$ units, $x = 1000$ units **(b)** $\$651,041.67$

(c) $P(x) = -x^2 + 1028x - 28,000$; max is $\$236,196$

(d) $\$941$

31. (a) $t \approx 8$; 1988 $R = \$60.792$ billion

(b) The data shows a smaller revenue, $R = \$60.53$ billion in 1987

(c)

$R(t) = 0.253t^2 - 4.03t + 76.84$

(d) Except for 1986 ($t = 6$) the model fits the data quite well.

33. (a) $P(t) = -0.019t^2 + 0.284t - 0.546$

(b) 1987

(c)

$P(t) = -0.019t^2 + 0.284t - 0.546$

(d) The model projects decreasing profits, and, except for 1992, the data supports this.

(e) Management would be interested in increasing revenues or reducing costs (or both) to improve profits.

Chapter 4 Review Exercises (page 351)

1. $x = 0, x = -\frac{5}{3}$ **2.** $x = 0, x = \frac{4}{3}$

3. $x = -2, x = -3$

4. $x = (-5 + \sqrt{47})/2, x = (-5 - \sqrt{47})/2$

5. no real solutions **6.** $x = \frac{\sqrt{3}}{2}, x = -\frac{\sqrt{3}}{2}$ **7.** $\frac{5}{7}, -\frac{4}{5}$

8. $(-1 + \sqrt{2})/4, (-1 - \sqrt{2})/4$

9. $7/2, 100$ **10.** $13/5, 90$

11. no real solutions **12.** $z = -9, z = 3$

13. $x = 8, x = -2$ **14.** $x = 3, x = -1$

15. $x = (-a \pm \sqrt{a^2 - 4b})/2$

16. $r = (2a \pm \sqrt{4a^2 + x^3c})/x$

17. $1.64, -7051.64$ **18.** $0.41, -2.38$

19. vertex $(-2, -2)$; zeros $(0, 0), (-4, 0)$

20. vertex $(0, 4)$; zeros $(4, 0), (-4, 0)$

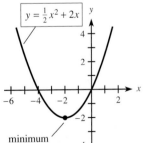

$y = \frac{1}{2}x^2 + 2x$

minimum

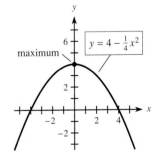

$y = 4 - \frac{1}{4}x^2$

maximum

21. vertex $(\frac{1}{2}, \frac{25}{4})$; zeros $(-2, 0)$, $(3, 0)$

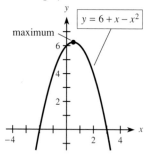

22. vertex $(2, 1)$; no real zeros

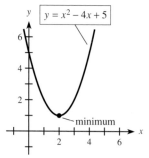

23. vertex $(-3, 0)$; zero $(-3, 0)$

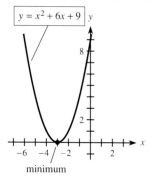

24. vertex $(\frac{3}{2}, 0)$; zero $(\frac{3}{2}, 0)$

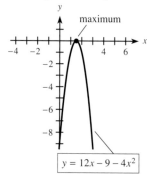

25. vertex $(0, -3)$; zeros $(-3, 0)$, $(3, 0)$

26. vertex $(0, 2)$; no real zeros

27. vertex $(-1, 4)$; no real zeros

28. vertex $(\frac{7}{2}, \frac{9}{4})$; zeros $(2, 0)$, $(5, 0)$

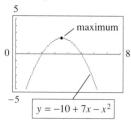

29. vertex $(100, 1000)$; zeros $(0, 0)$, $(200, 0)$

30. vertex $(75, -6.35)$, zeros $(50, 0)$, $(100, 0)$

$y = 50 - 1.5x + 0.01x^2$

minimum

31. (a) $(1, -4\frac{1}{2})$ (b) $x = -2, x = 4$ (c) B
32. (a) $(0, 49)$ (b) $x = -7, x = 7$ (c) D
33. (a) $(7, 25)$ approximately, actual is $(7, 24\frac{1}{2})$
   (b) $x = 0, x = 14$ (c) A
34. (a) $(-1, 9)$ (b) $x = -4, x = 2$ (c) C
35. (a) $x = 200$ ft   (b) $A = 30,000$ sq ft
36.   **37.**

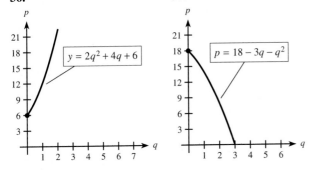

$y = 2q^2 + 4q + 6$

$p = 18 - 3q - q^2$

38. (a)

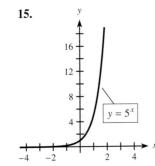

$p = 85 - 0.2q - 0.1q^2$

Market Equilibrium

$p = 0.1q^2 + 1$

   (b) $p = 41, q = 20$
39. $p = 400, q = 10$
40. $p = 10, q = 20$
41. $x = 46 + 2\sqrt{89}, x = 46 - 2\sqrt{89}$
42. $(15, 1275), (60, 2400)$
43. max revenue = \$2500; max profit = \$506.25

44. max profit = 12.25; break-even $x = 100, x = 30$

$(65, 12.25)$

$(30, 0)$

$(100,0)$

$P(x) = 1.3x - 0.01x^2 - 30$

45. $x = 50, P(50) = 640$
46. (a) $C = 15,000 + 140x + 0.04x^2$;
   $R = 300x - 0.06x^2$
   (b) $100, 1500$ (c) 2500
   (d) $P = 160x - 15,000 - 0.1x^2$; max at 800
   (e) at 2500: $P = -240,000$; at 800: $P = 49,000$

Exercise 5.1   (page 362)

1. 3.162278 **3.** 0.01296525 **5.** 1.44225
7. 7.3891 **9.** 0.04979
11.

$y = 4^x$

13.

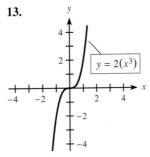

$y = 2(x^3)$

15.

$y = 5^x$

17.

$y = e^x$

19.

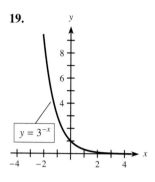

$y = 3^{-x}$

27. (a)

$f(x) = 100(1 + e^{-x})$

$f(x) = 50(1 + e^{-x})$

$f(x) = 10(1 + e^{-x})$

(b) As c changes, the y-intercept and the asymptote change.

21.

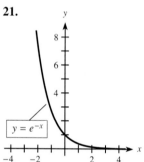

$y = e^{-x}$

29. (a)

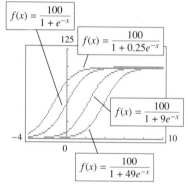

$f(x) = \dfrac{100}{1 + e^{-x}}$

$f(x) = \dfrac{100}{1 + 0.25e^{-x}}$

$f(x) = \dfrac{100}{1 + 9e^{-x}}$

$f(x) = \dfrac{100}{1 + 49e^{-x}}$

(b) Different c values change the y-intercept and how the graph approaches the asymptote.

31. Decay exponential   **33.** $1884.54

23. All graphs have the same basic shape. For larger positive k, the graphs fall more sharply. For positive k nearer 0, the graphs fall more slowly.

$f(x) = e^{-0.5x}$

$f(x) = e^{-x}$

$f(x) = e^{-0.1x}$

$f(x) = e^{-x}$

$f(x) = e^{-5x}$

$f(x) = e^{-2x}$

35.

$N = 4.1(1.02)^t$

25. $y = f(x) + C$ is the same graph as $y = f(x)$ but shifted C units on the y-axis.

$f(x) = 4^x + 6$

$f(x) = 4^x$

$f(x) = 4^x + 3$

$f(x) = 4^x - 5$

$f(x) = 4^x - 2$

37.

$N = 4.1(1.05)^t$

39.

$y = 100,000e^{-1/x}$

41.

43. $1972.17

45.

47.

49.

51.

Exercise 5.2   *(page 376)*

1. $2^4 = 16$ **3.** $4^{1/2} = 2$ **5.** $x = 8$
7. $x = \frac{1}{2}$ **9.** $\log_2 32 = 5$ **11.** $\log_4(\frac{1}{4}) = -1$
13.

15.

17.

19. (a) 3   **(b)** -1 **21. (a)** $\frac{2}{3}$ **(b)** 5
23. (a) x **(b)** 2   **25. (a)** x **(b)** 3
27. (a) 4.9   **(b)** 0.4   **(c)** 12.4   **(d)** 0.9
29. $\log x - \log(x + 1)$ **31.** $\log_7 x + \frac{1}{3}\log_7(x + 4)$
33. $\ln(x/y)$ **35.** $\log_5[x^{1/2}(x + 1)]$
37. (a)

(b) For each c, the domain is $x > c$ and the vertical asymptote is at $x = c$.
(c) Each x-intercept is at $x = c + 1$
(d) The graph of $y = f(x - c)$ is the graph of $y = f(x)$ shifted c units on the x-axis.
39. (a) 4.0875   **(b)** -0.1544
41.   **43.**

45. If $\log_a M = u$ and $\log_a N = v$, then $a^u = M$ and $a^v = N$. Therefore $\log_a(M/N) = \log_a(a^u/a^v) = \log_a(a^{u-v}) = u - v = \log_a M - \log_a N$.
47. (a) $\log_{1.005}(2) = t$ **(b)** 139 months (approx.)
49. (a) $230.26   **(b)** 31 (nearest unit)
51. 5.903   **53.** 15.8   **55.** 4:1   **57.** 140 decibels
59. 2512:1   **61. (a)** 7.4   **(b)** 4.2   **(c)** 2.2

63. (a) 0.001   **(b)** 1.62×10^{-8} **(c)** 0.0000001

65.

67. (a)

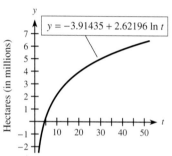

(b) 6.34 million hectares

Exercise 5.3   *(page 386)*

1. (a) 2038   **(b)** 4.9 months
3. (a) 69%   **(b)** 37,204 yrs (approx.)
5. 24.5 yrs   **7.** 128,402
9. (a) 600   **(b)** 2119   **(c)** 3000
   (d)

11. 1979
13. (a) 37   **(b)** 1.5 hours   **15. (a)** 52   **(b)** tenth
17. (a) $4.98   **(b)** 8   **19.** $502   **21.** $420.09
23. $2707   **25. (a)** $10,100.31   **(b)** 6.03 yrs
27. (a) $5469.03   **(b)** 7 yrs, 9 mos (approx.)
31. (a) 44.7 yrs (approx.)
   (b) The intent of such a plan would be to reduce
       future increases in health care expenditures. A
       new model may not be exponential, or, if it were,
       would be one that rose more gently.
33. (a) 18.3 yrs (approx.)   **(b)** $42,340.00
35. (a) 0.23 km³   **(b)** 5.9 yrs
37. (a) 0 lb   **(b)** 29.95 lb   **(c)** 0.21 min
39. (a) $x = 16$; 0.2%   **(b)** 10.56 min   **(c)** 0.06%

Chapter 5 Review Exercises *(page 390)*

1. (a) $\log_2 y = x$ **(b)** $\log_3 2x = y$
2. (a) $7^{-2} = \frac{1}{49}$ **(b)** $4^{-1} = x$

3.

4.

5.

6.

7.

8.

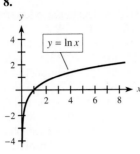

15. 0   **16.** 2   **17.** $\frac{1}{2}$ **18.** -1
19. 8   **20.** 1   **21.** 5   **22.** 3.15
23. -2.7 **24.** 0.6   **25.** 5.1
26. 15.6   **27.** $\log y + \log z$
28. $\frac{1}{2}\ln(x+1) - \frac{1}{2}\ln x$ **29.** no
30. -2 **31.** 5   **32.** 1   **33.** 0
34. 3.4939   **35.** -1.5845

36.

37.

9.

10.

38. Growth exponential because the general outline has the same shape as a growth exponential.

39. (a)

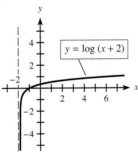

(b) The function $y = 0.8356\,(1.454)^t$ is an exponential because the variable is the exponent.

40. Exponential because the general shape is similar to the graph of a growth exponential.

41. (a) -3.9 (b) $0.14B_0$ (c) $0.004B_0$ (d) yes

42. (a) 3000   (b) 8603   (c) 10,000

43. (a) $27,440.58   (b) 12 weeks

44. $1366.19   **45.** 5.8 years

46. (a) $5532.77   (b) 5.13 years

11.

12.

13.

14.

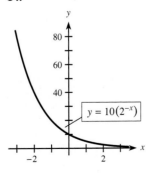

Exercise 6.1   *(page 402)*

1. (a) $9600   (b) $19,600
3. (a) $30   (b) $1030
5. $864   **7.** $3850   **9.** 13%
11. (a) 6.5%   (b) 5.65%
13. $1631.07   **15.** pay on time
17. 3, 6, 9, 12, 15, 18, 21, 24, 27, 30
19. $\frac{1}{3}, \frac{2}{3}, 1, \frac{4}{3}, \frac{5}{3}, 2, \frac{7}{3}, \frac{8}{3}$
21. $-\frac{1}{4}, \frac{1}{4}, -\frac{1}{4}, \frac{1}{4}, -\frac{1}{4}, \frac{1}{4}$

23. $-\frac{1}{2}, \frac{1}{4}, -\frac{1}{6}, \frac{1}{8}, -\frac{1}{10}, \frac{1}{12}$

25. $-1, -\frac{1}{4}, -\frac{1}{15}, 0; a_{10} = \frac{1}{20}$

27. **(a)** $d = 3, a_1 = 2$ **(b)** 11, 14, 17

29. **(a)** $d = \frac{3}{2}, a_1 = 3$ **(b)** $\frac{15}{2}, 9, \frac{21}{2}$

31. 105   **33.** -35 **35.** 203   **37.** 2185

39. 1907.5   **41.** 2550   **43.** $-15,862.5$

45. 21, 34, 55   **47.** $2400

49. the job starting at $20,000

51. **(a)** $3000   **(b)** $4500   **(c)** Plan II, by $1500

      (d) $10,000   **(e)** $13,500   **(f)** Plan II, by $3500

      (g) Plan II

Exercise 6.2   *(page 415)*

(Minor differences may occur due to rounding)

1. **(a)** 6%   **(b)** 6   **(c)** 3% = 0.03   **(d)** 12

3. **(a)** 8%   **(b)** 7   **(c)** 2% = 0.02   **(d)** 28

5. **(a)** 9%   **(b)** 5   **(c)** $(\frac{9}{12})$% = .0075   **(d)** 60

7. $24,846.78   **9.** $3086.45   **11.** $4755.03

13. $6844.36   **15.** $7309.98   **17.** $502.47

19. $50.26 more at 8%   **21.** 8.24%   **23.** 6.18%

25. 8% compounded monthly, 8% compounded quarterly, 8% compounded annually

27. The higher graph is for continuous compounding because its yield is higher (i.e., its effective annual rate)

29. 26.8%   **31.** 3 years   **33.** 4%

35. $3996.02   **37.** **(a)** $63,128.75

      (b) $14,263.10

39. 5.12 years (approx.)

41. **(a)** 24, 48, 96   **(b)** 24, 16, $\frac{32}{3}$

43. 40,960   **45.** $4 \cdot (\frac{3}{2})^{15}$

47. $\dfrac{6(1 - 3^{17})}{-2}$ **49.** $\frac{8}{3}[1 + (\frac{1}{2})^{21}]$

51. $\dfrac{3^{35} - 1}{2}$ **53.** $18[1 - (\frac{2}{3})^{18}]$

55. $350,580 (approx.)   **57.** 24.4 million (approx.)

59. 35 years   **61.** 40.5 ft   **63.** $4096

65. 320,000   **67.** $7,231,366   **69.** $1801.14

71. 15,625   **73.** 305,175,780

Exercise 6.3   *(page 429)*

1. $3\frac{1}{2}$% = 0.035 for 22 periods

3. $\frac{1}{4}$% = 0.0025 for 60 periods

5. **(a)** The higher graph is $1120 per year.

      (b) 22 years (approx.) Contributions: smaller = $22,000; larger = $24,640

7. $14.49   **9.** $7328.22   **11.** $1072.97

13. $1074.82   **15.** $6042.22   **17.** $4774.55

19. **(a)** $103,213.29   **(b)** $90,350.81   **(c)** Patty

21. $6.153291 or $6.15   **23.** $1180.78

25. $4152.32   **27.** $26,517.13   **29.** $976.32

31. $1165.23   **33.** $2115.07   **35.** $11,689.59

37. $2,128,391

39. **(a)** The higher graph corresponds to 8%.

      (b) $1500 (approx.)

      (c) With an interest rate of 10%, a present value of about $9000 is needed to purchase an annuity of $1000 for 25 years. If the interest rate is 8%, about $10,500 is needed.

41. $7957.86   **43.** $74,993.20

45. $19,922.97   **47.** $1317.98

Exercise 6.4   *(page 440)*

1. **(a)** The 10-year loan because the loan must be paid more quickly.

      (b) The 25-year loan because the loan is paid more slowly.

3. $2504.56   **5.** $1288.29

7.

Period	Payment	Interest
1	$39,505.50	$9000.00
2	39,505.50	6254.51
3	39,505.43	3261.92
	118,516.43	18,516.43

Period	Balance Reduction	Unpaid Balance
		$100,000.00
1	$30,505.50	69,494.50
2	33,250.99	36,243.51
3	36,243.51	0.00
	100,000.00	

9.

Period	Payment	Interest
1	$5380.54	$600.00
2	5380.54	456.58
3	5380.54	308.87
4	5380.54	156.71
	21,522.16	1522.16

Period	Balance Reduction	Unpaid Balance
		$20,000.00
1	$4780.54	15,219.46
2	4923.96	10,295.50
3	5071.67	5,223.83
4	5223.83	0.00
	20,000.00	

11. $4466.29   **13.** $3443.61

15. (a) $4359.23   (b) $87,184.60   (c) $37,184.60

17. (a) $1237.78   (b) $9902.24   (c) $1902.24

19. $2967.75   **21.** $62,473.28   **23.** $36,547.10

25. (a) $89,120.53   (b) $6451.45

27. The line is the total amount paid ($644.30 per month × the number of months). The other curve is the total amount paid toward principal.

29.

length ≈ $60,000 amount paid to interest

31. A sinking fund is a savings plan so the 10% rate (a) is better.

33. $6274.50   **35.** $4372.20   **37.** $2921.40

39. $265.25

41. Amortization costs = $17,698.42/year
Interest + sinking fund costs = $17,274.54/year

Chapter 6 Review Exercises *(page 443)*

1. $1, \frac{1}{4}, \frac{1}{9}, \frac{1}{16}$ **2.** 25   **3.** 19   **4.** $4\frac{1}{2}$

5. 8   **6.** $40\frac{4}{9}$ **7.** $10,880   **8.** $6\frac{2}{3}\%$

9. $4650   **10.** the $20,000 job ($245,000 vs $234,000)

11. (a) 40   (b) 2% = 0.02

12. (a) $S = P(1 + i)^n$ (b) $S = Pe^{jt}$

13. $11,239.42   **14.** $372.79   **15.** $1601.03

16. $1616.07   **17.** $11,466.64   **18.** $3466.64

19. 14.5 years   **20.** 34.3 months   **21.** 16.32%

22. 13.29%   **23.** 7.44%   **24.** (b) monthly

25. 2^{63} **26.** $2^{32} - 1$

27. $29,428.47   **28.** $31,194.18

29. $1863.93   **30.** $64,176.58   **31.** $19,972.79

32. $12,007.09   **33.** $88.85

34. (a) $592.76   (b) $177,828   (c) $85,828

35. $3443.61   **36.** $34,597.40   **37.** $6069.44

38. $32,834.69

Exercise 7.1   *(page 458)*

1. $\frac{1}{4}$ **3.** (a) $\frac{2}{5}$ (b) 0   (c) 1

5. (a) $\frac{1}{6}$ (b) 0   (c) $\frac{1}{2}$

7. (a) $\frac{3}{10}$ (b) $\frac{1}{2}$ (c) $\frac{1}{5}$ (d) $\frac{3}{5}$ (e) $\frac{7}{10}$

9. (a) $\frac{1}{13}$ (b) $\frac{1}{2}$ (c) $\frac{1}{4}$

11. {HH, HT, TH, TT};   (a) $\frac{1}{4}$ (b) $\frac{1}{2}$ (c) $\frac{1}{4}$

13. (a) $\frac{1}{12}$ (b) $\frac{1}{12}$ (c) $\frac{1}{36}$

15. (a) $\frac{1}{2}$ (b) $\frac{5}{12}$

17. (a) 431/1200
(b) If fair, $\Pr(6) = \frac{1}{6}$; 431/1200 not close to $\frac{1}{6}$, so not a fair die

19. (a) 2:3   (b) 3:2   **21.** (a) 8:3   (b) 3:8

23. (a) $\frac{1}{21}$ (b) $\frac{20}{21}$ (c) $\frac{1}{21}$

25. (a) $\frac{57}{80}$ (b) $\frac{23}{80}$

27. (a) 1/3601   (b) 100/3601   (c) 3500/3601

29. S = {A+, A−, B+, B−, AB+, AB−, O+, O−}

31. .46   **33.** (a) .04   (b) .96

35. (a) .627   (b) .373   **37.** (a) $\frac{1}{5}$ (b) $\frac{1}{14}$

39. $\frac{1}{3}$ **41.** .20   **43.** $\frac{1}{3}$

45. .22; yes, .39 is much higher than .22

47. $\frac{1}{4}$ **49.** $\frac{3}{8}$

51. (a) no   (b) {BB, BG, GB, GG}
(c) $\frac{1}{2}$ **53.** $\frac{1}{5}$ **55.** $\frac{3}{125}$

57. $\Pr(A) = 0.000019554$ or approx. 1.9 accidents per 100,000
$\Pr(B) = 0.000035919$ or approx. 3.6 accidents per 100,000
$\Pr(C) = 0.000037679$ or approx. 3.8 accidents per 100,000   Intersection C is the most dangerous

59. (a) 557/1200   (b) 11/120

61. (a) boy: 1/5; girl: 4/5
(b) boy: .4946; girl: .5054   (c) part b

Exercise 7.2   *(page 468)*

1. $\frac{1}{2}$ **3.** $\frac{1}{6}$ **5.** $\frac{2}{5}$

7. (a) $\frac{1}{7}$ (b) $\frac{5}{7}$ **9.** $\frac{3}{4}$ **11.** $\frac{2}{3}$

13. $\frac{10}{17}$ **15.** $\frac{2}{3}$

17. (a) $\frac{1}{2}$ (b) $\frac{1}{3}$ (c) $\frac{8}{9}$ (d) $\frac{1}{9}$

19. .54　　**21. (a)** 362/425　　**(b)** $\frac{66}{85}$
23. $\frac{17}{50}$　　**25. (a)** $\frac{8}{15}$　　**(b)** $\frac{7}{15}$
27. (a) $\frac{11}{12}$　　**(b)** $\frac{5}{6}$　　**29. (a)** $\frac{1}{2}$　　**(b)** $\frac{7}{8}$　　**(c)** $\frac{3}{4}$
31. .56　　**33.** .965　　**35. (a)** .72　　**(b)** .84　　**(c)** $\frac{61}{100}$
37. $\frac{51}{65}$　　**39.** .13

Exercise 7.3　　*(page 478)*

1. (a) $\frac{1}{2}$　　**(b)** $\frac{1}{13}$　　**3. (a)** $\frac{2}{3}$　　**(b)** $\frac{1}{3}$　　**5.** $\frac{4}{9}$
7. (a) $\frac{2}{3}$　　**(b)** $\frac{4}{9}$　　**(c)** $\frac{3}{5}$　　**9. (a)** $\frac{1}{4}$　　**(b)** $\frac{1}{2}$
11. $\frac{1}{36}$　　**13. (a)** $\frac{1}{8}$　　**(b)** $\frac{7}{8}$　　**15. (a)** $\frac{3}{50}$　　**(b)** $\frac{1}{15}$
17. (a) $\frac{4}{25}$　　**(b)** $\frac{9}{25}$　　**(c)** $\frac{6}{25}$　　**(d)** 0
19. $\frac{5}{68}$　　**21. (a)** $\frac{1}{5}$　　**(b)** $\frac{3}{5}$　　**23. (a)** $\frac{1}{17}$　　**(b)** 13/204
25. (a) $\frac{13}{17}$　　**(b)** $\frac{4}{17}$　　**(c)** $\frac{8}{51}$　　**27.** $\frac{31}{52}$　　**29.** $\frac{25}{96}$
31. $\frac{43}{50}$　　**33.** $\frac{65}{87}$　　**35.** 35/435 $= \frac{7}{87}$　　**37.** $\frac{1}{10}$
39. $\frac{3}{14}$　　**41.** 0.004292　　**43.** 0.06　　**45.** 0.45
47. $(.95)^5 = 0.774$　　**49.** 0.06
51. (a) 0.366　　**(b)** 0.634
53. (a) 0.4565　　**(b)** 0.5435
55. (a) $\left(\frac{1}{3}\right)^3\left(\frac{1}{5}\right)^4 = 1/16{,}875$
　　(b) $\left(\frac{2}{3}\right)^3\left(\frac{4}{5}\right)^4 = 2048/16{,}875$　　**(c)** 14,827/16,875
57. (a) 364/365　　**(b)** $\frac{1}{365}$　　**59. (a)** .59　　**(b)** .41

Exercise 7.4　　*(page 489)*

1. (a) $\frac{4}{17}$　　**(b)** $\frac{4}{17}$　　**3. (a)** $\left(\frac{5}{9}\right)\cdot\left(\frac{4}{8}\right)\cdot\left(\frac{3}{7}\right) = \frac{5}{42}$
　　(b) $\frac{10}{21}$　　**(c)** $\frac{4}{9}$
5. (a) $\left(\frac{5}{15}\right)\cdot\left(\frac{5}{15}\right) = \frac{1}{9}$　　**(b)** $\left(\frac{5}{15}\right)\cdot\left(\frac{4}{15}\right) + \left(\frac{4}{15}\right)\cdot\left(\frac{5}{15}\right) = \frac{8}{45}$
　　(c) $\frac{16}{25}$　　**7. (a)** $\left(\frac{7}{11}\right)\left(\frac{4}{10}\right) = \frac{14}{55}$　　**(b)** $\frac{28}{55}$
9. (a) $\frac{1}{30}$　　**(b)** $\frac{1}{2}$　　**(c)** $\frac{5}{6}$　　**11.** $\frac{3}{5}$
13. (a) $\frac{6}{25}$　　**(b)** $\frac{9}{25}$　　**(c)** $\frac{12}{25}$　　**(d)** $\frac{19}{25}$
15. $\frac{2}{3}$　　**17.** $\frac{2}{3}$　　**17.** 0.3095
21. (a) 81/10,000　　**(b)** 1323/5000
23. (a) $\frac{6}{35}$　　**(b)** $\frac{6}{35}$　　**(c)** $\frac{12}{35}$
25. (a) $\frac{4}{7}$　　**(b)** $\frac{5}{14}$　　**(c)** $\frac{7}{10}$　　**(d)** $\frac{16}{25}$
27. $\frac{17}{45}$　　**29.** 0.079　　**31. (a)** 49/100　　**(b)** $\frac{12}{49}$

Exercise 7.5　　*(page 497)*

1. 360　　**3.** 151,200　　**5.** 5! = 120　　**7.** 1
9. (a) $6 \cdot 5 \cdot 4 \cdot 3 = 360$　　**(b)** $6^4 = 1296$
11. 1　　**13.** $n + 1$　　**15.** 16　　**17.** 15　　**19.** 21
21. 1　　**23.** 1　　**25.** 1
27. $_{13}C_{10} = 286$; $_{12}C_9 + {}_{12}C_{10} = 220 + 66 = 286$
29. 10　　**31.** 604,800　　**33.** 120　　**35.** 24　　**37.** 64
39. 720　　**41.** $2^{10} = 1024$　　**43.** 10! = 3,628,800
45. 252　　**47.** 30,045,015　　**49.** 792　　**51.** 210
53. 2,891,999,880

Exercise 7.6　　*(page 503)*

1. $\frac{1}{120}$　　**3. (a)** 120　　**(b)** $\frac{1}{120}$　　**5.** .639
7. (a) 1/10,000　　**(b)** 1/5040　　**9.** $1/10^6$
11. (a) $\left(\frac{1}{5}\right)^{10} = 1/9{,}765{,}625$　　**(b)** $\left(\frac{4}{5}\right)^{10} = .107$
　　(c) .893　　**13.** $1/10! = 1/3{,}628{,}800$
15. (a) $\frac{1}{22}$　　**(b)** $\frac{6}{11}$　　**(c)** $\frac{9}{22}$
17. 0.098　　**19.** $\dfrac{_{90}C_{28} \cdot {}_{10}C_2}{_{100}C_{30}}$
21. (a) 0.119　　**(b)** 0.0476　　**(c)** 0.476
23. 0.0238　　**25. (a)** 0.721　　**(b)** 0.262　　**(c)** 0.279
27. (a) $\frac{1}{3}$　　**(b)** $\frac{1}{6}$　　**29.** $\dfrac{_{20}C_{10}}{_{80}C_{10}} = .00000011$
31. (a) 0.033　　**(b)** 0.633　　**33. (a)** 0.0005　　**(b)** 0.002
35. 0.00748

Exercise 7.7　　*(page 512)*

1. can　　**3.** cannot, sum $\neq 1$　　**5.** cannot, not square
7. can　　**9.** [0.248 0.752]
11. [0.228　0.236　0.536]　　**13.** $\left[\frac{1}{4}\ \frac{3}{4}\right]$　　**15.** $\left[\frac{1}{4}\ \frac{1}{4}\ \frac{1}{2}\right]$
17. [0.5　0.4　0.1]; [0.44　0.43　0.13];
　　[0.431　0.43　0.139]; [0.4292　0.4291　0.1417]
19.
$$\begin{array}{c} \\ R \\ N \end{array}\begin{array}{cc} R & N \\ \left[0.8\right. & \left.0.2\right] \\ \left[0.3\right. & \left.0.7\right] \end{array}$$
21. 0.45
23.
$$\begin{array}{c} \\ A \\ F \\ V \end{array}\begin{bmatrix} A & F & V \\ 0 & 0.7 & 0.3 \\ 0.6 & 0 & 0.4 \\ 0.8 & 0.2 & 0 \end{bmatrix}$$
25. [0.3928　0.37　0.2372]
27. [46/113　38/113　29/113]
29.
$$\begin{array}{c} \\ r \\ u \end{array}\begin{bmatrix} r & u \\ 0.7 & 0.3 \\ 0.1 & 0.9 \end{bmatrix};\ [1/4\ \ 3/4]$$
31. $\left[\frac{1}{14}\ \frac{3}{14}\ \frac{5}{7}\right]$　　**33.** $\left[\frac{4}{7}\ \frac{2}{7}\ \frac{1}{7}\right]$
35. [49/100　42/100　9/100]

Chapter 7 Review Exercises *(page 515)*

1. $\frac{1}{3}$　　**2.** $\frac{1}{13}$　　**3.** $\frac{1}{3}$
4. (a) $\frac{1}{4}$　　**(b)** $\frac{1}{2}$　　**(c)** $\frac{1}{4}$
5. (a) $\frac{3}{8}$　　**(b)** $\frac{1}{8}$　　**(c)** $\frac{3}{8}$　　**6.** $\frac{2}{13}$　　**7.** 16/169
8. $\frac{3}{4}$　　**9.** $\frac{2}{13}$　　**10.** $\frac{7}{13}$
11. (a) $\frac{2}{9}$　　**(b)** $\frac{2}{3}$　　**(c)** $\frac{7}{9}$
12. $\frac{2}{7}$　　**13.** $\frac{1}{2}$　　**14.** $\frac{8}{15}$
15. (a) $\frac{3}{14}$　　**(b)** $\frac{4}{7}$　　**(c)** $\frac{3}{8}$

16. 30   **17.** 35   **18.** 26^3 **19.** $\frac{5}{8}$ **20.** $\frac{29}{50}$

21. $\frac{5}{56}$ **22.** $\frac{33}{56}$ **23.** $\frac{15}{22}$ **24.** $4! = 24$

25. $_8P_4 = 1680$ **26.** $_{12}C_4 = 495$ **27.** $_8C_4 = 70$

28. **(a)** $_{12}C_2 = 66$ **(b)** $_{12}C_3 = 220$ **29.** 62,193,780

30. If her assumption about blood groups as accurate, there would be $4 \cdot 2 \cdot 4 \cdot 8 = 256$, not 288 unique groups.

31. **(a)** 63/2000   **(b)** $\frac{60}{63}$ **32.** $\frac{1}{24}$ **33.** $\frac{3}{2500}$

34. $\frac{3}{500}$ **35.** **(a)** .3398   **(b)** .1975   **36.** $\frac{1}{10}$

37. **(a)** $(_{10}C_5)(_2C_1)/_{12}C_6$

(b) $\dfrac{(_{10}C_5)(_2C_1) + (_{10}C_4)(_2C_2)}{_{12}C_6}$

38. [.135   .51   .355], [.09675   .3305   .57275], [.0640875   .288275   .6476375]

39. [12/265   68/265   37/53]

Exercise 8.1   *(page 523)*

1. .0595   **3.** **(a)** $\frac{1}{64}$ **(b)** $\frac{5}{16}$ **(c)** $\frac{15}{64}$ **5.** .0284

7. **(a)** .2304   **(b)** .0102   **(c)** .3174

9. .0585   **11.** .2759   **13.** **(a)** .375   **(b)** .0625

15. **(a)** .1157   **(b)** .4823

17. **(a)** $\frac{27}{64}$ **(b)** $\frac{27}{128}$ **(c)** $\frac{81}{256}$

19. **(a)** .0729   **(b)** .5905   **(c)** .9914

21. .2457   **23.** .0007

25. **(a)** .1323   **(b)** .0308

27. **(a)** .9044   **(b)** .0914   **(c)** .0043

29. **(a)** .8683   **(b)** .2099   **31.** .740

Exercise 8.2   *(page 531)*

1. no; $\text{Pr}(x) \not\geq 0$ **3.** yes   **5.** yes   **7.** no; $\Sigma \text{Pr}(x) > 1$

9. $\frac{15}{8}$ **11.** 5   **13.** $\frac{8}{3}$ **15.** 2

17. $\mu = \frac{13}{8}$, $\sigma^2 = 1.48$, $\sigma = 1.22$

19. $\mu = 4.2$, $\sigma^2 = 9.56$, $\sigma = 3.09$

21. $\mu = \frac{13}{3}$, $\sigma^2 = 2.22$, $\sigma = 1.49$

23. 3   **25.** 2   **27.** 1.85

29. TV, 37,500; P.A., 35,300

31. 2   **33.** 0   **35.** $-\$0.39$

37. Expect to lose $2 each time.

39. If he buys 0, his profit is 0. If the buys 100, his expected profit is $3(100)(.25) + \$3(50)(.20) + \$3(10)(.55)+$ $\$(-1)\,50(.20) + \$(-1)90(.55) = \$62$. If he buys 200, his expected profit is $3(180)(.25) + \$3(50)(.20) +$ $\$3(10)(.55) + (-\$1)(20)(.25) + (-\$1)(150)(.20) +$ $(-\$1)(190)(.55) = \$42$. He should buy 100.

41. TV: 37,500; newspapers: 39,000;   newspapers

43. $10

Exercise 8.3   *(page 538)*

1. **(a)**

x	$\text{Pr}(x)$
0	125/216
1	25/72
2	5/72
3	1/216

(b) $3(\frac{1}{6}) = \frac{1}{2}$ **(c)** $\sqrt{3(\frac{1}{6})(\frac{5}{6})} = (\frac{1}{6})\sqrt{15}$

3. **(a)** 42   **(b)** 3.55   **5.** 2, 1.29

7. 4   **9.** 2   **11.** 2   **13.** 15

15. $a^6 + 6a^5b + 15a^4b^2 + 20a^3b^3 + 15a^2b^4 + 6ab^5 + b^6$

17. $x^4 + 4x^3h + 6x^2h^2 + 4xh^3 + h^4$

19. **(a)** $100(.10) = 10$ **(b)** $\sqrt{100(.10)(90)} = 3$

21. **(a)** 60,000   **(b)** $\sqrt{24,000} = 154.919$

23. 59,690   **25.** **(a)** 4   **(b)** 1.79

27. 2, 1.41   **29.** 300

Exercise 8.4   *(page 549)*

1.

3.

5.

7.

9. 3   **11.** 13   **13.** 2   **15.** 1

17. mode = 2, median = 4.5, mean = 6

19. mode = 17, median = 18.5, mean = 23.5

21. mode = 5.3, median = 5.3, mean = 5.32   **23.** 9

25. 14   **27.** 4, 8.5714, 2.9277   **29.** 14, 4.6667, 2.1602

31. 2.73, 1.35   **33.** 6.75, 2.96

35.

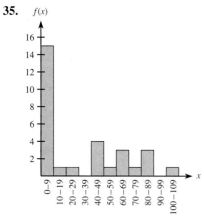

Total Executions in U.S. 1950–79

37. Michael Jordan's extremely high salary caused the mean for his major to be unusually high.

39. The mode will be the lowest measure.

41. (a) $\bar{x} = \dfrac{\Sigma x}{n} = \$480$ (b) $\bar{x} = \dfrac{\Sigma xf}{\Sigma f} = \$400$

43. $\bar{x} = 17{,}800, \; s = 262.467$

45.

Gold Prod.	1249.1	thousand oz
Gold Sales	1233.9	thousand oz
Revenue	423	\$/oz
Aver. price	355.7	\$/oz
Oper. cost	179.7	\$/oz
Net income	138.3	\$/oz

47. $\bar{x} = 159.8, \; s = 17.123$ **49.** $\bar{x} = 9.5, \; s = 1.07$

Exercise 8.5   *(page 559)*

1. 0.4641   **3.** 0.4641   **5.** 0.9153   **7.** 0.1070
9. 0.0166   **11.** 0.0227   **13.** 0.8849   **15.** 0.1915
17. 0.3944   **19.** 0.5381   **21.** 0.2957   **23.** 0.3446
25. 0.7258   **27.** (a) 0.3413   (b) 0.3944   **29.** 0.9876
31. (a) 0.4192   (b) 0.0227   (c) 0.0581   (d) 0.8965
33. (a) 0.0668   (b) 0.3085   (c) 0.3830
35. (a) 0.0475   (b) 0.2033   (c) 0.5934
37. (a) 0.0227   (b) 0.1587   (c) 0.8186

Exercise 8.6   *(page 566)*

1. $\hat{y} = 5.7x - 1.4$ **3.** $\hat{y} = .11x + 2$
5. $\hat{y} = -2.7x + 5$ **7.** $r = .993$

9. $r = .750$ **11.** $r = -.414$
13. (a) $\hat{p} = -2.9q + 317$ (b) $r = -.978$
15. (a) $\hat{y} = .57x - .0287$ (b) $r = .999$
17. (a) $\hat{y} = .91x - .24$ (b) $r = .689$
19. (a) $\hat{y} = .016x + 17$ (b) $r = 0.583$
21. (a) $y = 27.0 + 0.631x$
  (b) \$325 (to the nearest dollar)   (c) $r = 0.983$
23. $y = 2.70 + 0.051x$ **25.** $\hat{q} = 34{,}726 - 55.09p$
27. (a) $\hat{y} = .959x - 1.126, \; r = 0.998$
  (b)

29. (a) $\hat{y} = 25.09x + 399.09$
  (b) $r = 0.991$; is reasonable
  (c) \$700

Chapter 8 Review Exercises   *(page 571)*

1. $\left(\frac{4}{5}\right)^5$ **2.** (a) $(99{,}999/100{,}000)^{99{,}999} \approx 0.37$
  (b) $1 - (99{,}999/100{,}000)^{100{,}000} \approx 0.63$
3. $\frac{33}{81}$ **4.** 2.4   **5.** yes   **6.** no; $\Sigma\Pr(x) \neq 1$
7. yes   **8.** no; $\Pr(x) \not\geq 0$ **9.** 2
10. (a) 4.125   (b) 2.7344   (c) 1.654
11. (a) $\frac{37}{12}$ (b) 0.9097   (c) 0.9538
12. $\mu = 4, \; \sigma = (2\sqrt{3})/3$ **13.** (a) 1   (b) $\left(\frac{4}{5}\right)^4$
14. $x^5 + 5x^4y + 10x^3y^2 + 10x^2y^3 + 5xy^4 + y^5$
15.

16. 3   **17.** $\frac{77}{26} = 2.96$ **18.** 3

19.

20. 15   **21.** 14   **22.** 15
23. $\bar{x} = 3.86$; $s^2 = 6.81$; $s = 2.61$
24. $\bar{x} = 2$; $s^2 = 2.44$; $s = 1.56$
25. 0.9165   **26.** 0.1498   **27.** 0.1039   **28.** 0.3413
29. 0.6826   **30.** .1360   **31.** 0.297   **32.** 0.16308
33. $\$-.50$ **34.** 455
35.

Class Marks

36. 30.3%   **37.** 8.35%
38. (a) 0.4773   (b) 0.1360   (c) 0.0227
39. (a) $\hat{y} = 0.838x - 1.364$, $r = 0.999$
    (b)

40. (a) $\hat{y} = 0.143x + 9.997$, $r = 0.996$
    (b) $\hat{y} = 319.6$ is the prediction for $x = 2165.3$

Exercise 9.1   *(page 590)*

1. (a) 1   (b) 1   **3.** (a) -8 (b) -8
5. (a) 10   (b) does not exist   **7.** (a) 0   (b) -6
9. (a) does not exist $(+\infty)$ (b) does not exist $(+\infty)$
    (c) does not exist $(+\infty)$ (d) does not exist
11. (a) 3   (b) -6 (c) does not exist   (d) -6

13.

x	$f(x)$
1.9	-0.0526
1.99	-0.0050
1.999	-0.0005
$\downarrow$	$\downarrow$
2	0
$\uparrow$	$\uparrow$
2.001	0.0005
2.01	0.0050
2.1	0.0476

$\lim_{x \to 2} f(x) = 0$

15.

x	$f(x)$
0.9	-1.9
0.99	-1.99
0.999	-1.999
$\downarrow$	$\downarrow$
1	-2
$\uparrow$	$\uparrow$
1.001	-2.001
1.01	-2.01
1.1	-2.1

$\lim_{x \to 1} f(x) = -2$

17.

x	$f(x)$
.9	3.5
.99	3.95
.999	3.995
$\downarrow$	$\downarrow$
1	
$\uparrow$	
1.001	4.99599
1.01	4.9599
1.1	4.59

$\left. \begin{array}{c} 4 \\ 5 \end{array} \right\}$ Different

$\lim_{x \to 1} f(x)$ does not exist

19. 1   **21.** -4 **23.** 0   **25.** -2
27. 6   **29.** 0   **31.** does not exist
33. -3 **35.** does not exist   **37.** does not exist
39. $3x^2$ **41.** -2 **43.** $\frac{1}{30}$ **45.** does not exist
47. 9   **49.** does not exist

51.

a	$(1 + a)^{1/a}$
0.1	2.5937
0.01	2.7048
0.001	2.7169
0.0001	2.7181
0.00001	2.71827
$\downarrow$	$\downarrow$
0	≈ 2.718

53. (a) 2   (b) 6   (c) -8 (d) $-\frac{1}{2}$
55. $150,000
57. (a) $32 (thousands)   (b) $55.04 (thousands)
59. (a) $2800   (b) $700   (c) $560
61. (a) 1.52 units/hr   (b) 0.85 units /hr   (c) lunch
63. 0; $p \to 100^-$ means the water approaches not being treated (containing 100% or all of its impurities); the associated costs of non-treatment approach zero.
65. (a) $3315.00   (b) $3315.00   (c) $3315.00

67. $1091.70

69. 3867 (approx.) This corresponds to the Dow-Jones opening average on January 17, 1994.

71. (a) -1.449 (approx.)
(b) This predicts the percentage of US workers in farm occupations as the year approaches 2000.
(c) The model is inaccurate because the percentage must be 0 or positive, and cannot be negative.

Exercise 9.2   *(page 604)*

1. (a) continuous   **(b)** discontinuous; $f(1)$ does not exist
(c) discontinuous; $\lim_{x \to 3} f(x)$ does not exist
(d) discontinuous; $f(0)$ does not exist and $\lim_{x \to 0} f(x)$ does not exist

3. continuous   **5.** continuous

7. discontinuous; $f(-3)$ does not exist

9. discontinuous; $f(-1)$ and $\lim_{x \to -1} f(x)$ do not exist

11. continuous   **13.** discontinuous; $\lim_{x \to 1} f(x)$ does not exist

15. continuous   **17.** discontinuity at $x = -2$; $g(-2)$ does not exist

19. discontinuity at $x = -2$; $g(-2)$ and $\lim_{x \to -2} g(x)$ do not exist

21. continuous   **23.** continuous

25. discontinuity at $x = 3$; $\lim_{x \to 3} f(x)$ does not exist

27. vertical asymptote: $x = -2$; $\lim_{x \to -\infty} f(x) = 0$; $\lim_{x \to -\infty} f(x) = 0$

29. vertical asymptotes: $x = -2$, $x = 3$; $\lim_{x \to -\infty} f(x) = 2$; $\lim_{x \to -\infty} f(x) = 2$

31. 0   **33.** 1   **35.** $\frac{5}{3}$ **37.** 0

39. does not exist $(+\infty)$

41. (a) no, not at $p = -8$ **(b)** yes   **(c)** yes
(d) $p > 0$

43. (a) yes, $q = -1$ **(b)** yes

45. (a) R/i **(b)** $10,000   **47.** yes, $0 \le p \le 100$

49. 100%; No, for p to approach 100% (as a limit) requires spending to increase without bound, which is impossible.

51. $R(x)$ is discontinuous at $x = 36{,}900$, $x = 89{,}150$, $x = 140{,}000$, and $x = 250{,}000$.

53. (a) $79.40   **(b)** $\lim_{x \to 100} C(x) = 19.40$; $\lim_{x \to 500} C(x) = 49.40$ **(c)** yes

55. (a) 31.3% (approx.)   **(b)** does not exist $(+\infty)$
(c) No, because values of $d(t)$ only make sense if they are between 0 and 100, as $d(t)$ is a percentage of Federal expenditures.

57. (a) no   **(b)** From the early 1930s to the mid 1950s the rate was continuous at 3¢ for 1 oz.

(c) Exponential. The general shape of the curve looks similar to an exponential growth function.

Exercise 9.3   *(page 619)*

1. (a) 32   **(b)** 32   **(c)** $(4, 64)$

3. (a) verification   **(b)** -5 **(c)** -5 **(d)** $(-1, 3)$

5. (a) $P(1, 1)$, $A(3, 0)$ **(b)** $-\frac{1}{2}$ **(c)** $-\frac{1}{2}$ **(d)** $-\frac{1}{2}$

7. (a) $P(1, 3)$, $A(0, 3)$ **(b)** 0   **(c)** 0   **(d)** 0

9. $f'(x) = -6$

11. $f'(x) = 8x - 2$

13. (a) $p'(q) = 2q + 4$ **(b)** $p'(5) = 14$

15. (a) $C'(x) = 28x + 5$ **(b)** $C'(4) = 117$

17. $f'(-3) = \frac{5}{2}$; $f(-3) = -9$

19. $y = 3x - 4$

21. (a) a, b, d **(b)** c **(c)** A, C, E

23. (a) A, B, C, D **(b)** A, D

25. (a) $f'(x) = 2x + 1$ **(b)** $f'(2) = 5$ **(c)** $y = 5x - 4$
(d)

27. (a) $f'(x) = 3x^2$ **(b)** $f'(1) = 3$ **(c)** $y = 3x + 1$
(d)

29. (a) -32 ft/sec   **(b)** the ball is falling, so S is decreasing

31. (a) $\frac{-100}{3}$ **(b)** $\frac{-4}{3}$ **33.** 68 mph

35. (a) $R'(x) = \overline{MR} = 300 - 2x$ **(b)** 200
(c) -100 **(d)** 0   **(e)** It changes from increasing to decreasing

37. 200

39. (a) 100; the expected profit from the sale of the 201st car is $100.
(b) -100; the expected profit from the sale of the 301st car is a loss of $100.

41. (a) 1.039   **(b)** If humidity changes by 1%, the heat index will change by about 1.039°F.

43. (a) 1.164 million infections per year (approx.)
(b) Low: 3.6 million infections per year
High: 20 million infections per year

Exercise 9.4 (page 631)

1. $y' = 0$ **3.** $y' = 1$ **5.** $f'(x) = 6x^2 - 5x^4$

7. $y' = 24x^3 - 10x + 1$

9. $g'(x) = 90x^8 - 25x^4 + 21x^2 + 5$

11. (a) 19 **(b)** 19 **13. (a)** 0 **(b)** 0

15. $y' = -5x^{-6} - 8x^{-9}$

17. $y' = 11x^{8/3} - \frac{7}{2}x^{3/4} - \frac{1}{2}x^{-1/2}$

19. $f'(x) = -4x^{-9/5} - \frac{8}{3}x^{-7/3}$

21. $g'(x) = -\dfrac{12}{x^5} - \dfrac{10}{x^6} + \dfrac{5}{3\sqrt[3]{x^2}}$

23. $h'(x) = \dfrac{-30}{x^4} + \dfrac{4}{\sqrt[5]{x^7}} + 2x$ **25.** $y = -3x + 6$

27. $y = 3$ **29.** $y = -\frac{1}{2}x + 3$

31. (a) $y = 8x - 3$

(b)

$y = 3x^2 + 2x$ $y = 8x - 3$

(c) $x{:}0.7 \to 1.6$; $y{:}3.0 \to 7.9$

33. (a) $f'(x) = -2 - 2x$

(b)

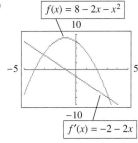

$f(x) = 8 - 2x - x^2$

$f'(x) = -2 - 2x$

(c) $f'(x) = 0$ at $x = -1$; $f'(x) > 0$ for $x < -1$
$f'(x) < 0$ for $x > -1$

(d) $f(x)$ has max when $x = -1$
$f(x)$ rises for $x < -1$
$f(x)$ falls for $x > -1$

35. (a) $f'(x) = 3x^2 - 12$

(b)

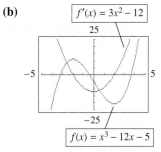

$f'(x) = 3x^2 - 12$

$f(x) = x^3 - 12x - 5$

(c) $f'(x) = 0$ at $x = -2$ and $x = 2$
$f'(x) > 0$ for $x < -2$ and $x > 2$
$f'(x) < 0$ for $-2 < x < 2$

(d) $f(x)$ has max when $x = -2$, min when $x = 2$
$f(x)$ rises when $x < -2$ and when $x > 2$
$f(x)$ falls when $-2 < x < 2$

37. (a) $y' = 3x^2$; $u'(x) = 1$; $v'(x) = 2x$ **(b)** no
(c) No, it fails for u and v given in part (a).

39. (a) 40, revenue increasing
(b) -20, revenue decreasing

41. (a) 920 **(b)** 926

43. (a) -4; if the price changes to \$26, the quantity demanded will change by approximately -4 units
(b) $-\frac{1}{2}$; if the price changes to \$101, the quantity demanded will change by approximately $-\frac{1}{2}$ unit

45. (a) $\overline{C}(x)' = (-4000/x^2) + 0.1$ **(b)** 200

47. (a) $-120{,}000$
(b) If the impurities change from 1% to 2%, then the expected change in cost is $-120{,}000$ (dollars).

49. (a) $WC = 55.174 - 2.15s - 23.18\sqrt{s}$
(b) -4.468
(c) If the wind speed changes by $+1$ mph (to 26 mph), then the wind chill will change by approximately $-4.468°$F.

51. (a) -0.78% per year **(b)** -0.82% per year

53. (a) $f'(t) = -0.0274271652t^3 + 6.98350335\,t^2 - 591.9284890t + 16701.32789$
(b) $f'(81) \approx -2.0$; $f'(88) \approx 1.0$
(c) $f'(81) \approx -2.0$ means that from 1981 to 1982 the consumer price index dropped 2.0%. $f'(88) \approx 1.0$ means that from 1988 to 1989 the consumer price index rose by 1.0%.

55. (a) $f'_m(t) = -0.457192t + 37.0677$
(b) $f'_m(75) \approx 2.78$ and $f'_m(85) \approx -1.79$

(c)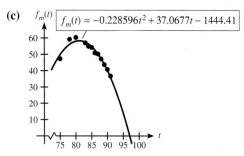

$f_m(t) = -0.228596t^2 + 37.0677t - 1444.41$

(d) In those years when $f_m(t) < 0$; i.e., below the
t-axis, the model canot be valid. Thus, after about
1997 the model cannot be used.

Exercise 9.5   *(page 642)*

1. $y' = 3x^2 + 2x - 6$ **3.** $dp/dq = 9q^2 - 2q + 6$
5. $f'(x) = (x^{12} + 3x^4 + 4)(12x^2) +$
    $(4x^3 - 1)(12x^{11} + 12x^3)$
7. $y' = (7x^6 - 5x^4 + 2x^2 - 1)(36x^8 + 21x^6 - 10x + 3) +$
    $(4x^9 + 3x^7 - 5x^2 + 3x)(42x^5 - 20x^3 + 4x)$
9. $y' = (x^2 + x + 1)(\frac{1}{3}x^{-2/3} - x^{-1/2}) +$
    $(x^{1/3} - 2x^{1/2} + 5)(2x + 1)$
11. (a) 40   **(b)** 40   **13. (a)** 104   **(b)** 104
15. $y' = (-x^2 - 1)/(x^2 - 1)^2$
17. $\dfrac{dp}{dq} = (q^2 - 4q - 1)/(q - 2)^2$
19. $\dfrac{dy}{dx} = \dfrac{4x^5 - 4x^3 - 16x}{(x^4 - 2x^2 + 5)^2}$
21. $\dfrac{dz}{dx} = 2x + \dfrac{2x - x^2}{(1 - x - 2x^2)^2}$
23. $\dfrac{dp}{dq} = \dfrac{2q + 1}{\sqrt[3]{q^2}\,(1 - q)^2}$
25. $y' = \dfrac{2x^3 - 6x^2 - 8}{(x - 2)^2}$
27. (a) $\frac{3}{5}$ **(b)** $\frac{3}{5}$
29. $y = 44x - 32$ **31.** $y = \frac{10}{3}x - \frac{10}{3}$
33. $(-2, 0)$ $(4, -108)$

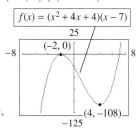

$f(x) = (x^2 + 4x + 4)(x - 7)$

35. $(0, 0)$ $(4, 8)$

$f(x) = \dfrac{x^2}{x - 2}$

37. (a) $f'(x) = \dfrac{20x}{(x^2 + 1)^2}$

(b)

$f(x) = \dfrac{10x^2}{x^2 + 1}$

$f'(x) = \dfrac{20x}{(x^2 + 1)^2}$

(c) $f' = 0$ at $x = 0$ **(d)** f has a min at $x = 0$
$f' > 0$ for $x > 0$ f increasing for $x > 0$
$f' < 0$ for $x < 0$ f decreasing for $x < 0$

39. $f'(x) = \displaystyle\lim_{h \to 0} \dfrac{\dfrac{u(x + h)}{v(x + h)} - \dfrac{u(x)}{v(x)}}{h}$

$= \displaystyle\lim_{h \to 0} \dfrac{u(x + h)\,v(x) - u(x)\,v(x + h)}{h \cdot v(x)\,v(x + h)}$

$= \displaystyle\lim_{h \to 0} \dfrac{u(x + h)\,v(x) - u(x)\,v(x) + u(x)v(x) - u(x)v(x + h)}{h \cdot v(x)\,v(x + h)}$

$= \displaystyle\lim_{h \to 0} \dfrac{v(x)\left[\dfrac{u(x + h) - u(x)}{h}\right] - u(x)\left[\dfrac{v(x + h) - v(x)}{h}\right]}{v(x)\,v(x + h)}$

$= \dfrac{v(x)\,u'(x) - u(x)\,v'(x)}{[v(x)]^2}$

41. $C'(p) = 810{,}000/(100 - p)^2$
43. $R'(49) \approx 30.00$ The expected revenue from the sale
of the next unit (the 50th) is about \$30.00.
45. $R'(5) = -50$ As the group changes by 1 person (to
31), the revenue will drop by about \$50.
47. $S = 1000x - x^2$ **49.** $\dfrac{dR}{dn} = \dfrac{r(1 - r)}{[1 + (n - 1)r]^2}$
51. (a) $P'(6) \approx 0.045$ During the next (7th) month of
the campaign, the proportion of voters who

recognize the candidate will change by about 0.045, or 4.5%.

(b) $P'(12) \approx -0.010$ During the next (13th) month of the campaign, the proportion of voters who recognize the candidate will drop by about 0.010, or 1%.

(c) It is better for $P'(t)$ to be positive, that is to have increasing recognition.

53. (a) $f'(20) \approx -0.79$

(b) At 0°F, if the wind speed changes by 1mph (to 21 mph), the wind chill will change by about -0.79°F.

55. (a) $C'(5) = 0.3$

(b) From 1990 to 1991, the model predicts a change of 0.3% in the cost of living.

(c)

$C(t) = (t+5)(-0.218t + 3.57) - 20.13$

(d) The model fails to be valid when $C(t) < 0$, below the t-axis.

57. (a) $f'(t) =$
$$1000 \left[\frac{26.65629658t^2 - 3423.843248t + 16{,}731.97758}{(1.09816t^2 - 122.183t + 21{,}472.6)^2} \right]$$

(b) $f'(70) \approx -0.276$ $f'(170) \approx 0.195$

(c) $f'(70)$ means that from 1870 to 1871, the model predicts a change of about -0.276% in U.S. workers in farm occupations. $f'(170)$ means that from 1970 to 1971, the model predicts a change of about 0.195% in U.S. workers in farm occupations.

Exercise 9.6   *(page 651)*

1. $6x(x^2 + 1)^2$

3. $4(8x - 4)(4x^2 - 4x + 1)^3 = 16(2x - 1)^7$

5. $-6x/(x^2 + 2)^4$ **7.** $-4(x + 2)(x^2 + 4x)^{-3}$

9. $-3(2x + 3)(x^2 + 3x + 4)^{-4}$

11. $\dfrac{-3(6x^2 + 3)}{4(2x^3 + 3x + 5)^{7/4}}$ **13.** $(x + 2)/\sqrt{x^2 + 4x + 5}$

15. $(6 - 4x)/\sqrt{3x - x^2}$ **17.** $16x(x^2 - 3)^4$

19. $\dfrac{15(3x + 1)^4 - 3}{7}$ **21.** 96,768   **23.** 2

25. $y = -3x + 4$ **27.** $9x - 5y = 2$

29. (a) $x = 0, x = 2, x = -2$

(b) $(0, -52), (2, 12), (-2, 12)$

(c) $f(x) = (x^2 - 4)^3 + 12$

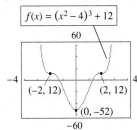

$f(x) = (x^2 - 4)^3 + 12$

60

$(-2, 12)$ $(2, 12)$

$(0, -52)$

-60

31. (a) $f'(x) = 8x(1 - x^2)^{1/3}$

(b)

$f(x) = 5 - 3(1 - x^2)^{4/3}$

15

-15

$f'(x) = 8x(1 - x^2)^{1/3}$

(c) $f'(x) = 0$ at $x = -1, x = 0, x = 1$
$f'(x) > 0$ for $x < -1$ and $0 < x < 1$
$f'(x) < 0$ for $-1 < x < 0$ and $x > 1$

(d) $f(x)$ has a maximum at $x = -1$ and $x = 1$, a minimum at $x = 0$
$f(x)$ increasing for $x < -1$ and $0 < x < 1$
$f(x)$ decreasing for $-1 < x < 0$ and $x > 1$

33. (a) $y' = 2x^2$ **(b)** $y' = -2/x^4$

(c) $y' = 2(2x)^2$ **(d)** $y' = \dfrac{-18}{(3x)^4}$

35. 10 ft/sec   **37.** \$1499.85 (approx.)

39. (a) -0.114 (approx.)

(b) If the price changes by \$1, to \$22, the weekly sales volume will change by approximately -0.114 thousand units.

41. (a) $-$\$3.20 per unit

(b) If the quantity demanded changes from 49 to 50 units, the change in price will be about $-$\$3.20.

43. $\dfrac{dy}{dx} = \left(\dfrac{8k}{5} \right)(x - x_0)^{3/5}$

45. $\dfrac{dp}{dq} = \dfrac{-100}{(2q + 1)^{3/2}}$ **47.** $\dfrac{dK_c}{dv} = \dfrac{8}{\sqrt{4v + 1}}$

49. (a) $658.75. If the interest changed from 6% to 7%, the amount of the investment would change by about $658.75

(b) $2156.94. If the interest rate changed from 12% to 13%, the amount of the investment would change by about $2156.94.

51. $d'(30) \approx -0.170$ means from 1960 to 1961 the model predicts the interest paid as a percent of federal expenditures would change by -0.17%. $d'(60) \approx 1.500$ means from 1990 to 1991 the model predicts the interest paid as a percent of federal expenditures would change by 1.500%.

53. $p'(10) \approx -0.399$ means from 1970 to 1971, the model predicts the number of persons below the poverty level would change by -0.399 million (a decrease). $p'(30) \approx 0.184$ means from 1990 to 1991, the model predicts the number of persons below the poverty level would change by 0.184 million (an increase).

Exercise 9.7 *(page 659)*

1. 0   **3.** $4(-4x^{-5}); -16/x^5$

5. $15x^2 + 4(-x^{-2}); 15x^2 - 4/x^2$

7. $(x^2 - 2)1 + (x + 4)(2x); 3x^2 + 8x - 2$

9. $\dfrac{x^2(3x^2) - (x^3 + 1)(2x)}{(x^2)^2}; (x^3 - 2)/x^3$

11. $(3x^2 - 4)(x^3 - 4x)^9$

13. $\frac{5}{3}x^3[3(4x^5 - 5)^2(20x^4)] + (4x^5 - 5)^3(5x^2);$ $5x^2(4x^5 - 5)^2(24x^5 - 5)$

15. $(x - 1)^2(2x) + (x^2 + 1)2(x - 1);$ $2(x - 1)(2x^2 - x + 1)$

17. $\dfrac{(x^2 + 1)\, 3(x^2 - 4)^2(2x) - (x^2 - 4)^3(2x)}{(x^2 + 1)^2};$ $\dfrac{2x(x^2 - 4)^2(2x^2 + 7)}{(x^2 + 1)^2}$

19. $3[(q + 1)(q^3 - 3)]^2[(q + 1)3q^2 + (q^3 - 3)1];$ $3(4q^3 + 3q^2 - 3)[(q + 1)(q^3 - 3)]^2$

21. $4[x^2(x^2 + 3x)]^3[x^2(2x + 3) + (x^2 + 3x)(2x)];$ $4x^2(4x + 9)[x^2(x^2 + 3x)]^3$

23. $4\left(\dfrac{2x - 1}{x^2 + x}\right)^3\left[\dfrac{(x^2 + x)2 - (2x - 1)(2x + 1)}{(x^2 + x)^2}\right];$ $\dfrac{4(-2x^2 + 2x + 1)(2x - 1)^3}{(x^2 + x)^5}$

25. $(8x^4 + 3)^2 3(x^3 - 4x)^2(3x^2 - 4) +$ $(x^3 - 4x)^3 2(8x^4 + 3)(32x^3);$ $(8x^4 + 3)(x^3 - 4x)^2(136x^6 - 352x^4 + 27x^2 - 36)$

27. $\dfrac{(4 - x^2)\frac{1}{3}(x^2 + 5)^{-2/3}(2x) - (x^2 + 5)^{1/3}(-2x)}{(4 - x^2)^2};$ $\dfrac{2x(2x^2 + 19)}{3\sqrt[3]{(x^2 + 5)^2}\,(4 - x^2)^2}$

29. $(x^2)\frac{1}{4}(4x - 3)^{-3/4}(4) + (4x - 3)^{1/4}(2x);$ $(9x^2 - 6x)/\sqrt[4]{(4x - 3)^3}$

31. $(2x)\frac{1}{2}(x^3 + 1)^{-1/2}(3x^2) + (x^3 + 1)^{1/2}(2);$ $(5x^3 + 2)/\sqrt{x^3 + 1}$

33. (a) $F_1'(x) = 12x^3(x^4 + 1)^4$ **(b)** $F_2'(x) = \dfrac{-12x^3}{(x^4 + 1)^6}$

(c) $F_3'(x) = 12x^3(3x^4 + 1)^4$ **(d)** $F_4'(x) = \dfrac{-300x^3}{(5x^4 + 1)^6}$

35. $dP/dx = 90(3x + 1)^2$ **37. (a)** $59,900

(b) it is increasing

39. $C'(y) = 1/\sqrt{y + 1} + 0.4$

41. $dV/dx = 144 - 96x + 12x^2$

43. (a) -1.6 This means that from the 9th to the 10th week, sales are expected to change by -1.6 (decrease).

45. (a) $f'(t) =$

$1000\left[\dfrac{8.885432188\,(t + 20)^2 - 1446.629069\,(t + 20) - 364,210.17}{[1.09816\,(t + 20)^2 - 122.183\,(t + 20) + 21472.6]^2}\right]$

(b) 1850: $f'(30) \approx -1.263$
1950: $f'(130) \approx -0.494$

(c) $f'(30)$ means that from 1850 to 1851, the model predicts a change of -1.263% in U.S. workers in farm occupations.
$f'(130)$ means that from 1950 to 1951, the model predicts a change of -0.491% in U.S. workers in farm occupations.

Exercise 9.8 *(page 664)*

1. $24x - 30$ **3.** $60x - 2$ **5.** $6x - 2x^{-3}$

7. $6x + \frac{1}{4}x^{-3/2}$ **9.** $60x^2 - 96$

11. $1008x^6 - 720x^3$ **13.** $-6/x^4$ **15.** $\frac{3}{8}x^{-5/2}$

17. $20x^3 + \frac{1}{4}x^{-3/2}$ **19.** $\frac{3}{8}(x + 1)^{-5/2}$ **21.** 0

23. $-15/(16x^{7/2})$ **25.** $\frac{3}{8}(x - 1)^{-5/2}$ **27.** $\dfrac{4}{(x^2 + 4)^{3/2}}$

29. $-2(x + 1)^{-3}$ **31.** 26

33. (a) $f'(x) = 3x^2 - 6x$ $f''(x) = 6x - 6$

(b)

$f'(x) = 3x^2 - 6x$

$f''(x) = 6x - 6$

$f(x) = x^3 - 3x^2 + 5$

(c) $f''(x) = 0$ at $x = 1$
$f''(x) > 0$ for $x > 1$
$f''(x) < 0$ for $x < 1$
(d) $f'(x)$ has a min at $x = 1$
$f'(x)$ is increasing for $x > 1$
$f'(x)$ is decreasing for $x < 1$
(e) $f''(x) < 0$ **(f)** $f''(x) > 0$
35. (a) $f'(x) = -x^2 - 2x + 3$ $f''(x) = -2x - 2$
(b)

$f(x) = -\frac{1}{3}x^3 - x^2 + 3x + 7$

$f''(x) = -2x - 2$

$f'(x) = -x^2 - 2x + 3$

(c) $f''(x) = 0$ at $x = -1$
$f''(x) > 0$ for $x < -1$
$f''(x) < 0$ for $x > -1$
(d) $f'(x)$ has a max at $x = -1$
$f'(x)$ is increasing for $x < -1$
$f'(x)$ is decreasing for $x > -1$
(e) $f''(x) < 0$ **(f)** $f''(x) > 0$
37. $a = 0.12$ **39.** -0.02
41. (a) $\dfrac{dR}{dm} = mc - m^2$ **(b)** $\dfrac{d^2R}{dm^2} = c - 2m$
(c) Second
43. (a) 0.0009 (approx.)
(b) When one more unit is sold (beyond 25), the marginal revenue will change by about 0.0009 thousand dollars per unit, or $0.90 per unit.
45. (a) $S' = \dfrac{-3}{(t+3)^2} + \dfrac{36}{(t+3)^3}$ **(b)** $S''(15) = 0$

(c) After 15 weeks, the rate of change of the rate of Sales is zero because the rate of Sales reaches a minimum value.
47. (a) $p'(t) = -0.006t^2 + 0.276t - 2.54$
(b) 1970: $p''(10) = 0.156$
1990: $p''(30) = -0.084$
(c) $p'(10) = 0.82$ means that from 1970 to 1971, the number of people who lived below the poverty level was expected to change by about 0.82 million.
$p''(10) = 0.156$ means that from 1970 to 1971, the rate of change of the number of people who lived below the poverty level was expected to change by about 0.156 million per year.
Thus, from 1970 to 1971 the number of people who lived below the poverty level was increasing at an increasing rate.
49. (a) $f'(t) = -0.0274272(t+70)^3 +$ $6.98349\,(t+70)^2 - 591.928(t+70) +$ 16701.327
(b) 1980: $f''(10) \approx -1.172$
1991: $f''(21) \approx -2.307$
(c) $f'(10) \approx -1.303$ means that from 1980 to 1981 the expected change in the CPI was about -1.303%.
$f''(10) \approx -1.172$ means that from 1980 to 1981 the expected change in the rate of change of the CPI was about -1.172% per year.

Exercise 9.9 *(page 673)*

1. $\overline{MC} = 8$ **3.** $\overline{MC} = 13 + 2x$
5. $\overline{MC} = 3x^2 - 12x + 24$ **7.** $\overline{MC} = 27 + 3x^2$
9. (a) $10; the cost will increase by $10. **(b)** $11
11. $46; the cost will increase by $46.
13. $\overline{MC}$

$\overline{MC} = 2x + 4$

15. (a) $\overline{MR} = 4$ **(b)** The sale of each additional item brings in $4 revenue at all levels or production.
17. (a) $3500; this is revenue from the sale of 100 units.
(b) $\overline{MR} = 36 - 0.02x$ **(c)** $34; Revenue will increase by $34.

(d) Actual revenue from the sale of the 101st item is $33.99.

19. (a) $\overline{MR} = 36 - 0.02x$ (b) $x = 1800$
(c) $32,400

21. $\overline{MP} = 5$
23. (a) $0   (b) $\overline{MP} = 30 - 2x$ (c) $-$10; profit will decrease by $10 if one additional unit is sold.
(d) The sale of the 21st item results in a loss of $11.
25. (a)

(b) 15   (c) 15   (d) $25
27. 70   29. $200
31. (a) Data: $R(7) = 60.53$ Model: $R(7) = 61.027$
$R(7)$ represents AT&T's total revenues (in billions of dollars) for 1987.
(b) $R'(t) = 0.506\,t - 4.03$
(c) $R'(12) = 2.042$
(d) $R'(12) = 2.042$ means that from 1992 to 1993, the model predicts that AT&T's revenue rose by about $2.042 billion.
33. (a) $C'(10) = 0.252$
(b) This means that from 1990 to 1991, the model predicts that Scott Paper Company's costs and expenses will change by about $0.252 billion. (Actual change for that period was $0.0319 billion.)

Chapter 9 Review Exercises (page 677)

1. (a) 2   (b) 2   2. (a) 0   (b) 0
3. (a) 2   (b) 1   4. (a) 2   (b) does not exist

5. (a) does not exist   (b) 2
6. (a) does not exist   (b) does not exist
7. 55   8. 0   9. -2 10. 6   11. $\frac{1}{2}$ 12. $\frac{1}{5}$
13. no limit   14. 0   15. 4   16. no limit
17. 3   18. no limit   19. $6x$ 20. $1 - 4x$
21. (a) yes   (b) no   22. (a) yes   (b) no
23. 2   24. no limit   25. 1   26. no
27. yes   28. yes   29. discontinuity at $x = 5$
30. discontinuity at $x = 2$ 31. continuous
32. discontinuity at $x = 1$
33. (a) $x = 0, x = 1$ (b) 0   (c) 0
34. (a) $x = -1, x = 0$ (b) $\frac{1}{2}$ (c) $\frac{1}{2}$
35. -2 36. 0   37. true   38. false
39. $f'(x) = 6x + 2$ 40. $f'(x) = 1 - 2x$
41. (a) no   (b) no   42. (a) yes   (b) no
43. $20x^4 - 18x^2$ 44. $8x$ 45. 3   46. $1/(2\sqrt{x})$
47. 0   48. $-4/(3\sqrt[3]{x^4})$ 49. $\dfrac{-1}{x^2} + \dfrac{1}{2\sqrt{x^3}}$
50. $\dfrac{-3}{x^3} - \dfrac{1}{3\sqrt[3]{x^2}}$ 51. $y = 15x - 18$
52. $y = 34x - 48$
53. (a) $x = 0, x = 2$ (b) $(0, 1)$ $(2, -3)$

54. (a) $x = 0, x = 2, x = -2$
(b) $(0, 8)$ $(2, -24)$ $(-2, -24)$

55. $9x^2 - 26x + 4$ 56. $15x^4 + 9x^2 + 2x$
57. $\dfrac{2(1 - q)}{q^3}$ 58. $\dfrac{1 - 3t}{[2\sqrt{t}(3t + 1)^2]}$ 59. $\dfrac{9x + 2}{2\sqrt{x}}$
60. $\dfrac{5x^6 + 2x^4 + 20x^3 - 3x^2 - 4x}{(x^3 + 1)^2}$

61. $(9x^2 - 24x)(x^3 - 4x^2)^2$

62. $6(30x^5 + 24x^3)(5x^6 + 6x^4 + 5)^5$

63. $72x^3(2x^4 - 9)^8$ **64.** $\dfrac{-(3x^2 - 4)}{2\sqrt{(x^3 - 4x)^3}}$

65. $2x(2x^4 + 5)^7(34x^4 + 5)$ **66.** $\dfrac{-2(3x + 1)(x + 12)}{(x^2 - 4)^2}$

67. $36[(3x + 1)(2x^3 - 1)]^{11}(8x^3 + 2x^2 - 1)$

68. $\dfrac{3}{(1 - x)^4}$ **69.** $\dfrac{(2x^2 - 4)}{\sqrt{x^2 - 4}}$ **70.** $\dfrac{2x - 1}{(3x - 1)^{4/3}}$

71. $y'' = \frac{-1}{4}x^{-3/2} - 2$ **72.** $y'' = 12x^2 - 2/x^3$

73. $\dfrac{d^5y}{dx^5} = 0$ **74.** $\dfrac{d^5y}{dx^5} = -30(1 - x)$

75. $\dfrac{d^3y}{dx^3} = -4/[(x^2 - 4)^{3/2}]$ **76.** $\dfrac{d^4y}{dx^4} = \dfrac{2x(x^2 - 3)}{(x^2 + 1)^3}$

77. **(a)** $x'(10) = -1$ means if price changes from \$10 to \$11, the number of units demanded will change by about -1.

   (b) $x'(20) = -\frac{1}{4}$ means if price changes from \$20 to \$21, the number of units demanded will change by about $-\frac{1}{4}$.

78. $\dfrac{dq}{dp} = \dfrac{-p}{\sqrt{0.02p^2 + 500}}$

79. $x'(10) = \frac{1}{6}$ means if price changes from \$10 to \$11, the number of units supplied will change by about $\frac{1}{6}$.

80. **(a)** $\overline{MC} = 6x + 6$ **(b)** 186

   (c) If a 31st unit is produced, costs will change by about \$186.

81. $C'(4) = 53$ means that a 5th unit produced would change total costs by about \$53.

82. **(a)** $\overline{MR} = 40 - 0.04x$ **(b)** $x = 1000$ units

83. $\overline{MP}(10) = 48$ means if an 11th unit is sold, profit would change by about \$48.

84. **(a)** $\overline{MR} = 80 - 0.08x$ **(b)** 72

   (c) If a 101st unit is sold, revenue would change by about \$72.

85. $\dfrac{120x(x + 1)}{(2x + 1)^2}$ **86.** $\overline{MP} = 4500 - 3x^2$

87. $\overline{MP} = 16 - 0.2x$

Exercise 10.1 *(page 697)*

1. **(a)** $dy/dx = 3x^2 - 3$ **(b)** $x = 1, x = -1$

   (c) $(1, 2), (-1, 6)$

   (d) relative max at $(-1, 6)$, relative min at $(1, 2)$

3. **(a)** $dy/dx = 3x^2 + 6x + 3$ **(b)** $x = -1$

   (c) $(-1, -3)$ **(d)** Horizontal point of inflection at $(-1, -3)$.

5. **(a)** $\dfrac{dy}{dx} = x - 1$ **(b)** $x = 1$ **(c)** $(1, -\frac{1}{2})$

   (d) decreasing: $x < 1$
   increasing: $x > 1$

   (e)

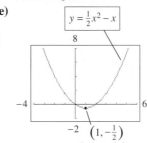

relative minimum

7. **(a)** $dy/dx = x^2 + x - 2$

   (b) $x = -2, x = 1$ **(c)** $(-2, \frac{13}{3}), (1, -\frac{1}{6})$

   (d) increasing: $x < -2$ and $x > 1$
   decreasing: $-2 < x < 1$

   (e)

9. **(a)** $\dfrac{dy}{dx} = \dfrac{2}{3x^{1/3}}$ **(b)** $x = 0$ **(c)** $(0, 0)$

   (d) decreasing: $x < 0$
   increasing: $x > 0$

   (e)

11. (1, 1) rel max;
(0, 0), (2, 0) rel min

13. $(\frac{1}{2}, -\frac{27}{16})$ min;
HPI (2, 0)

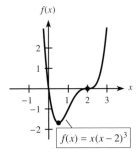

15. (3, 4) rel max;
(5, 0) rel min;
HPI (0, 0)

17. (0, 0) rel max;
(2, −4.8) rel min

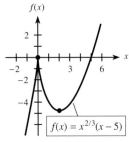

19. HPI $(1, \frac{4}{3})$
no max or min

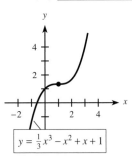

21. (−6, 128) rel max
(4, $-38\frac{2}{3}$) rel min

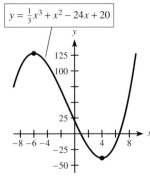

23. HPI (0, 0);
(3, −27) min

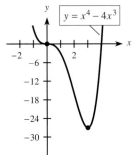

25. (−1, 3) rel max;
(1, −1) rel min;
HPI (0, 1)

27. **(a)** $f'(x) = 0$ at $x = -\frac{1}{2}$
$f'(x) > 0$ for $x < -\frac{1}{2}$
$f'(x) < 0$ for $x > -\frac{1}{2}$
(b) $f'(x) = -1 - 2x$ verifies these conclusions

29. **(a)** $f'(x) = 0$ at $x = 0$, $x = -3$, $x = 3$
(b) $f'(x) > 0$ for $-3 < x < 3$, $x \neq 0$
(c) $f'(x) < 0$ for $x < -3$ and $x > 3$

31. (50, 300.5) rel max
(100, 238) rel min

33. (0, −10) rel min
(−60, 1070) HPI

$f(x) = 0.00025x^4 + 0.04x^3 + 1.8x^2 − 10$

35. $f(x) = 0.6\left[\dfrac{10x^3}{3} − \dfrac{19x^2}{2} + 9x − \dfrac{20}{3}\right]$

(a)

(b) (0.9, −2.299) rel max
(1, −2.3) rel min

(c) x: 0.7 to 1.2
y: −2.305 to −2.295

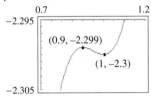

37. $f(x) = 7.5x^4 − x^3 + 2$

(a)

(b) (0, 2) HPI
(0.1, 1.99975) rel min

(c) x: −0.1 to 0.2
y: 1.9997 to 2.0007

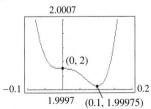

39. critical values: $x = −1$, $x = 2$
$f(x)$ increasing for $x < −1$ and $x > 2$
$f(x)$ decreasing for $−1 < x < 2$
rel max at $x = −1$; rel min at $x = 2$

possible graph for $f(x)$

41. critical values: $x = 0$, $x = 3$
$f(x)$ increasing for $x > 3$
$f(x)$ decreasing for $x < 3$, $x \neq 0$
rel min at $x = 3$; HPI at $x = 0$

possible graph of $f(x)$

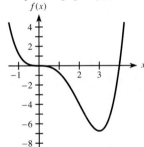

43. Graph on left is $f(x)$; on right is $f'(x)$ because $f(x)$ is
increasing when $f'(x) > 0$ (i.e., above the x-axis) and
$f(x)$ is decreasing when $f'(x) < 0$ (i.e., below the
x-axis).

45. decreasing for $t \geq 0$

47. (a) $2 \pm \sqrt{13}$
(b) $2 + \sqrt{13} \approx 5.6$
(c) $0 \leq t < 2 + \sqrt{13}$

$P(t) = 27t + 6t^2 − t^3$

49. (a) $x = 5$ **(b)** decreasing for $0 < x < 5$
(c) $x > 5$

51. (a) at $x = 150$, increasing; at $x = 250$, changing from
increasing to decreasing; at $x = 350$, decreasing
(b) increasing for $x < 250$ **(c)** 250 units

53. (a) $t = 6$ **(b)** 6 weeks
55. (a) 10   **(b)** Jan 1
57. (a) 1973   **(b)** yes
59. (a) and **(b)** critical points: (19.7, 11.7) rel max and
    (39, 9.6), rel min
    (c) These reflect the data quite well. The relative
      maximum corresponds to about 1950, when the
      data show a relative maximum. The relative
      minimum corresponds to 1969, also close to a
      relative minimum in the data.

Exercise 10.2   *(page 710)*

1. (a) concave down   **(b)** concave up
3. (a) concave down   **(b)** concave up
5. concave up when $x > 2$; concave down when $x < 2$;
   POI at $x = 2$
7. concave up when $x < -2$ and $x > 1$
   concave down when $-2 < x < 1$
   points of inflection at $x = -2$ and $x = 1$
9. concave up when $x > 3$; concave down when $x < 3$;
   POI at $x = 3$
11. concave up when $-\sqrt{3} < x < 0$ and $x > \sqrt{3}$
   concave down when $x < -\sqrt{3}$ and $0 < x < \sqrt{3}$
   points of inflection at $x = -\sqrt{3}, x = 0, x = \sqrt{3}$
13. points of inflection: $(0, 1)$ and $(\frac{2}{3}, \frac{11}{27})$; min: $(1, 0)$

15. $(-2, 64)$ rel max; $(2, -64)$ rel min; and points of
   inflection: $(-\sqrt{2}, 39.6)$, $(0, 0)$ and $(\sqrt{2}, -39.6)$

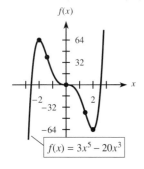

17. $(1, -3)$ min; points of inflection: $(-2, 7.6)$ and $(0, 0)$

19. no points of inflection; $(2, -2)$ min

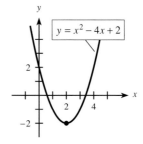

21. $(0, 6)$ rel max; $(2, 2)$ rel min; point of inflection:
   $(1, 4)$

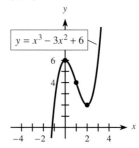

23. $(0, 0)$ rel max; $(2\sqrt{2}, -64)$, $(-2\sqrt{2}, -64)$ min;
   points of inflection: $(2\sqrt{6}/3, -320/9)$ and
   $(-2\sqrt{6}/3, -320/9)$

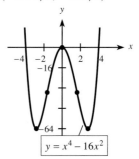

25. **(a)** $f''(x) = 0$ when $x = 1$
$f''(x) > 0$ when $x < 1$
$f''(x) < 0$ when $x > 1$
(b) rel max for $f'(x)$ at $x = 1$
no rel min
(c)

27. **(a)** concave up when $x < 2$ concave down when $x > 2$
(b) point of inflection at $x = 2$
(c)

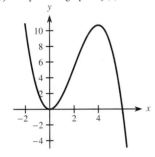

(d) possible graph of $f(x)$

29. **(a)** *G*   **(b)** *C*   **(c)** *F*   **(d)** *H*   **(e)** *I*
31. **(a)** concave up when $x < 0$
concave down when $x > 0$
point of inflection at $x = 0$
(b) concave up when $-1 < x < 1$
concave down when $x < -1$ and $x > 1$
POI at $x = -1$ and $x = 1$
(c) concave up when $x > 0$
concave down when $x < 0$
point of inflection at $x = 0$
33. **(a)** $P'(t)$ **(b)** *B*   **(c)** *C*

35. **(a)** *C*   **(b)** right   **(c)** yes
37. **(a)** in an 8-hour shift, max when $t = 8$ **(b)** 4 hr
39. **(a)** 9 days   **(b)** 15 days
41. $x \approx 82.5$
43. **(a)** $f'(t) = -0.0274271652t^3 + 6.98350335t^2 - 591.9284890t + 16701.32789$
$f''(t) = -0.0822814956t^2 + 13.9670067t - 591.928489$
(b) $f'(70) \approx 78.0$; $f''(70) \approx -17.4$ The CPI is increasing at a decreasing rate.
(c) $f'(86) \approx 0.26$; $f''(86) \approx 0.68$ The CPI is increasing at an increasing rate (though the increase and its rate are small).

Exercise 10.3 *(page 724)*

1. **(a)** $x = 1800$ units, $R = \$32,400$
(b) $x = 1500$ units, $R = \$31,500$
3. $x = 20$ units, $R = \$24,000$ **5.** 40 people
7. $p = \$15$, $R = \$22,500$
9. **(a)** max $= \$2100$ **(b)** $x = 10$
11. $x = 5$ units, $\overline{C} = \$23$
13. $x = 10$ units, $\overline{C} = \$20$
15. 200 units ($x = 2$), $\overline{C} = \$108$ **17.** $x = 5$
19. $x = 80$ units, $P = \$280,000$
21. $x = 10\sqrt{15} \approx 39$ units, $P \approx \$71,181$ (using $x = 39$)
23. $x = 1000$ units, $P = \$39,700$
25. $x = 600$ units, $P = \$495,000$
27. **(a)** 60   **(b)** $\$570$ **(c)** $\$9000$
29. **(a)** 1000 units   **(b)** $\$8066.67$ (approx.)
31. 2000 units priced at $\$90$/unit; max profit is $\$90,000$/wk
33. rent $= \$430$
35. **(a)** $t \approx 12.5$ (in 1993), $R = \$5.0353$ billion
(b) the data show a max in 1990 of $\$5.1686$ billion
37. **(a)** Aug 1991; approx. $\$3$ billion
(b) Aug or Sept 1991; approx. $\$5.5$ billion
(c) Dec 1992 or Jan 1993; approx. $\$12$ billion
(d) May 1993; approx. $\$9.5$ billion
39. **(a)** About Jan 1980 or Jan 1991
(b) (Jan 25, 1994, 37%)
(c) No, either some citizens think crime is the most important problem or no citizens think that.
(d) Crime would be a less important issue, and the percent would drop.
(e) Not 1980 because crime was a low priority. Yes in 1994, as seen from the high concern about crime then.

Exercise 10.4   *(page 734)*

1. **(a)** $x_1 = \$25$ million, $x_2 = \$30$ million
   (b) \$55 million
3. 400 trees   **5. (a)** 5   **(b)** 237.5   **7.** \$50
9. $m = c$ **11.** 1 week
13. $t = 8$, $p = 45\%$ **15.** 240 ft   **17.** $300' \times 150'$
19. 20 ft long, $6\frac{2}{3}$ ft across (dividers run across)
21. $4'' \times 8'' \times 8''$ high   **23.** 30,000   **25.** 12,000
27. $x = 2$ **29.** 3 weeks from now   **31.** 25 plates
33. **(a)** in 1981; max $\approx 58.3\%$
    (b) The data show a max about that time, but it is
        higher.

Exercise 10.5   *(page 747)*

1. **(a)** $x = 2$ **(b)** 1   **(c)** 1   **(d)** $y = 1$
3. **(a)** $x = 2$ **(b)** 1   **(c)** 1   **(d)** $y = 1$
5. HA: $y = 2$; VA: $x = 3$
7. HA: $y = 0$; VA: $x = -2$, $x = 2$
9. HA: none; VA: none
11. HA: $y = 0$; VA: $x = 1$; $(-1, -\frac{1}{4})$ rel min; point of
    inflection: $(-2, -\frac{2}{9})$

13. VA: $x = 3$; $(2, -1)$ rel max; $(4, 7)$ rel min

15. HA: $y = 0$; VA: $x = 0$
    $(2, 0)$ rel min; $(3, 1)$ rel max
    points of inflection: $(1.87, 0.66)$, $(4.13, 0.87)$

17. HA: $y = 2$; VA: $x = 3$
    no max, min, nor points of inflection

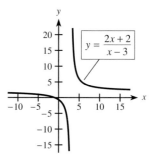

19. VA: $x = 0$; $(-2, -4)$ rel max; $(2, 4)$ rel min

21. VA: $x = -1$; HA: $y = 0$; $(0, 0)$ rel min; $(2, 4)$
rel max; points of inflection: $(2 + \sqrt{3}, 3.55)$
and $(2 - \sqrt{3}, 0.95)$

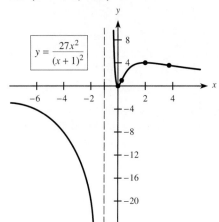

$$y = \frac{27x^2}{(x+1)^2}$$

23. HA: $y = 0$; $(1, 8)$ rel max; $(-1, -8)$ rel min; points
of inflection: $(0, 0)$, $(-\sqrt{3}, -4\sqrt{3})$, and $(\sqrt{3}, 4\sqrt{3})$

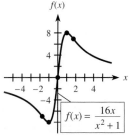

$$f(x) = \frac{16x}{x^2 + 1}$$

25. (a) HA: approx. $y = -2$; VA: approx. $x = 4$
   (b) HA: $y = -\frac{9}{4}$; VA: $x = \frac{17}{4}$

27. (a) HA: approx. $y = 2$; VA: approx. $x = 2.5$,
   $x = -2.5$
   (b) HA: $y = \frac{20}{9}$; VA: $x = \frac{7}{3}$, $x = -\frac{7}{3}$

29. $f(x) = \dfrac{x + 25}{x^2 + 1400}$

   (a)

   (b) HA: $y = 0$; rel min $(-70, -0.0071)$
   rel max $(20, 0.025)$

(c) x: -500 to 400
   y: $-.01$ to 0.03

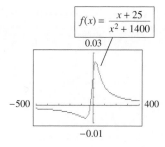

$$f(x) = \frac{x + 25}{x^2 + 1400}$$

31. $f(x) = \dfrac{100(9 - x^2)}{x^2 + 100}$

   (a)

   (b) HA: $y = -100$; rel max $(0, 9)$
   (c) x: -75 to 75
   y: -120 to 20

$$f(x) = \frac{100(9 - x^2)}{x^2 + 100}$$

33. $f(x) = \dfrac{1000x - 4000}{x^2 - 10x - 2000}$

   (a)

   (b) HA: $y = 0$
   VA: $x = -40$, $x = 50$
   no max nor min
   (c) x: -200 to 200
   y: -200 to 200

$$f(x) = \frac{1000x - 4000}{x^2 - 10x - 2000}$$

35. (a) none   **(b)** $C \geq 0$ **(c)** $p = 100$ **(d)** no

37. (a)

$$R(t) = \frac{50t}{t^2 + 36}$$

(6, 4.17) Max

(b) 6 weeks   **(c)** 22 weeks after its release

39. (a) yes, $x = -1$ **(b)** no; domain is $x \geq 5$
   (c) yes, $y = -58.5731$
   (d) At 0°F, as the wind speed increases, there is a limiting wind chill of about -58.6°F. This is meaningful because at high wind speeds, additional wind probably has little noticeable effect.

41. (a) $P = C$ **(b)** C **(c)** $P' = 0$ **(d)** 0

43. (a) 0
   (b) As the years past 1800 increase, the percentage of workers in farm occupations approaches 0.
   (c) no

45. (a) No. Barometric pressure can drop off the scale (as shown) but it cannot decrease without bound. In fact, it must always be positive.
   (b) See your library with regard to a significant weather event in March 1993.

Chapter 10 Review Exercises   *(page 751)*

1. $(0, 0)$ max   **2.** $(2, -9)$ min   **3.** HPI $(1, 0)$

4. $(1, \frac{3}{2})$ max, $(-1, -\frac{3}{2})$ min

5. (a) $\frac{1}{3}, -1$ **(b)** $(-1, 0)$ rel max, $(\frac{1}{3}, -\frac{32}{27})$ rel min
   (c) none
   (d)

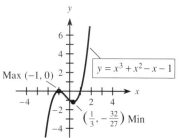

$y = x^3 + x^2 - x - 1$

Max $(-1, 0)$

$(\frac{1}{3}, -\frac{32}{27})$ Min

6. (a) $3, 0$ **(b)** $(3, 27)$ max   **(c)** $(0, 0)$

(d)

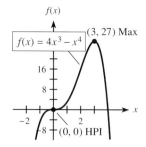

$f(x) = 4x^3 - x^4$

$(3, 27)$ Max

$(0, 0)$ HPI

7. (a) $-1, 6$
   (b) $(-1, 11)$ rel max, $(6, -160.5)$ rel min
   (c) none
   (d)

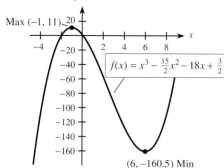

Max $(-1, 11)$

$$f(x) = x^3 - \frac{15}{2}x^2 - 18x + \frac{3}{2}$$

$(6, -160.5)$ Min

8. (a) $0, \pm 1$ **(b)** $(-1, 1)$ rel max, $(1, -3)$ rel min
   (c) $(0, -1)$ **(d)**

Max $(-1, 1)$

$(0, -1)$ HPI

$(1, -3)$ Min

$y = 5x^7 - 7x^5 - 1$

9. (a) 0   **(b)** $(0, -1)$ min   **(c)** none
   (d)

$y = x^{2/3} - 1$

$(0, -1)$ Min

Vertical tangent

10. (a) 0, 1, 4

(b) (0, 0) rel min, (1, 9) rel max, (4, 0) rel min

(c) none   **(d)**

11. concave up

12. concave up when $x < -1$ and $x > 2$
concave down when $-1 < x < 2$
points of inflection at $(-1, -3)$ and $(2, -42)$

13. $(-1, 15)$ rel max; $(3, -17)$ rel min; point of inflection $(1, -1)$

14. $(-2, 16)$ rel max; $(2, -16)$ rel min;
point of inflection $(0, 0)$

15. $(1, 4)$ rel max; $(-1, 0)$ rel min; points of inflection:
$$\left(\frac{1}{\sqrt{2}}, 2 + \frac{7}{4\sqrt{2}}\right), (0, 2), \text{ and } \left(-\frac{1}{\sqrt{2}}, 2 - \frac{7}{4\sqrt{2}}\right)$$

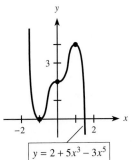

16. (a) (0, 0) absolute min; (140, 19,600) absolute max

(b) (0, 0) absolute min; (100, 18,000) absolute max

17. (a) (50, 233,333) absolute max; (0, 0) absolute min

(b) (64, 248,491) absolute max; (0, 0) absolute min

18. (a) $x = 1$ **(b)** $y = 0$ **(c)** 0   **(d)** 0

19. (a) $x = -1$ **(b)** $y = \frac{1}{2}$ **(c)** $\frac{1}{2}$ **(d)** $\frac{1}{2}$

20. HA: $y = \frac{3}{2}$; VA: $x = 2$

21. HA: $y = -1$; VA: $x = 1, x = -1$

22. (a) HA: $y = 3$; VA: $x = -2$

(b) no max nor min   **(c)**

23. (a) HA: $y = 0$; VA: $x = 0$

(b) (4, 1) max   **(c)**

24. (a) HA: none; VA: $x = 1$

(b) (0, 0) rel max; (2, 4) rel min

(c)

25. (a) $f'(x) > 0$ for $x < \frac{2}{3}$ (approx.) and $x > 2$
$f'(x) < 0$ for about $\frac{2}{3} < x < 2$
$f'(x) = 0$ about $x = \frac{2}{3}$ and $x = 2$

(b) $f''(x) > 0$ for $x > \frac{4}{3}$
$f''(x) < 0$ for $x < \frac{4}{3}$
$f''(x) = 0$ at $x = \frac{4}{3}$

(c)

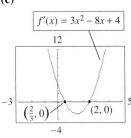

$f'(x) = 3x^2 - 8x + 4$

(d)

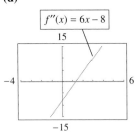

$f''(x) = 6x - 8$

(d)

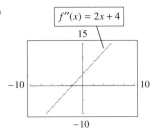

$f''(x) = 2x + 4$

26. (a) $f'(x) > 0$ for about $-13 < x < 0$ and $x > 7$
$f'(x) < 0$ for about $x < -13$ and $0 < x < 7$
$f'(x) = 0$ for about $x = 0$, $x = -13$, $x = 7$
(b) $f''(x) > 0$ for about $x < -8$ and $x > 4$
$f''(x) < 0$ for about $-8 < x < 4$
$f''(x) = 0$ for about $x = -8$ and $x = 4$
(c)

$f'(x) = 0.01x^3 + 0.06x^2 - 0.96x + 0.08$

(d)

$f''(x) = 0.03x^2 + 0.12x - 0.96$

27. (a) $f(x)$ increasing for $x < -5$ and $x > 1$
$f(x)$ decreasing for $-5 < x < 1$
$f(x)$ has rel max at $x = -5$, rel min at $x = 1$
(b) $f''(x) > 0$ for $x > -2$ (where $f'(x)$ increases)
$f''(x) < 0$ for $x < -2$ (where $f'(x)$ decreases)
$f''(x) = 0$ for $x = -2$
(c)

$f(x) = \dfrac{x^3}{3} + 2x^2 - 5x$

28. (a) $f(x)$ increasing for $x < 6$, $x \neq 0$
$f(x)$ decreasing for $x > 6$
$f(x)$ has max at $x = 6$, point of inflection at $x = 0$
(b) $f''(x) > 0$ for $0 < x < 4$
$f''(x) < 0$ for $x < 0$ and $x > 4$
$f''(x) = 0$ at $x = 0$ and $x = 4$
(c)

$f(x) = 2x^3 - \dfrac{x^4}{4}$

(d)

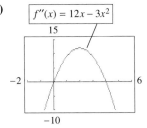

$f''(x) = 12x - 3x^2$

29. (a) $f(x)$ is concave up for $x < 4$
$f(x)$ is concave down for $x > 4$
$f(x)$ has point of inflection at $x = 4$
(b) $f''(x) = 4 - x$

$f(x) = 2x^2 - \dfrac{x^3}{6}$

30. (a) $f(x)$ is concave up for $-3 < x < 2$

$f(x)$ is concave down for $x < -3$ and $x > 2$

$f(x)$ has points of inflection at $x = -3$ and $x = 2$

(b) $f''(x) = 6 - x - x^2$

$$f(x) = 3x^2 - \frac{x^3}{6} - \frac{x^4}{12}$$

31. $x = 5$ units, $\overline{C} = \$45$

32. (a) $x = 1600$ units, $R = \$25,600$

(b) $x = 1600$ units, $R = \$25,600$

33. $P = \$54,000$ at $x = 100$ units   **34.** $x = 3$ units

35. $x = 152$ units   **36.** $x = 7$ units   **37.** \$300

38. selling 175 sets at \$425 each

39. \$93,625 at 325 units   **40. (a)** 150   **(b)** \$650

41. \$208,490.67 at 64 units   **42.** $x = 1000$ units

43. 10:00 A.M.   **44.** 325 in 1995

45. 20 mi from A, 10 mi from B **46.** 4 ft $\times$ 4 ft

47. $8\frac{3}{4}'' \times 10''$ **48.** 500   **49.** $\frac{29}{18}$ **50.** 24,000

Exercise 11.1   (page 764)

1. $f'(x) = 4/x$ **3.** $y' = 1/x$

5. $y' = 4/x$ **7.** $f'(x) = \dfrac{4}{4x + 9}$

9. $y' = \dfrac{4x - 1}{2x^2 - x} + 3$ **11.** $dp/dq = 2q/(q^2 + 1)$

13. (a) $y' = \dfrac{1}{x} - \dfrac{1}{x - 1}$ **(b)** $y' = \dfrac{-1}{x(x - 1)}$

15. (a) $y' = \dfrac{2x}{3(x^2 - 1)}$ **(b)** $y' = \dfrac{2x}{3(x^2 - 1)}$

17. (a) $y' = \dfrac{4}{4x - 1} - \dfrac{3}{x}$ **(b)** $y' = \dfrac{-8x + 3}{x(4x - 1)}$

19. (a) $y' = \dfrac{3}{x} + \dfrac{1}{2(x + 1)}$ **(b)** $y' = \dfrac{7x + 6}{2x(x + 1)}$

21. $\dfrac{dp}{dq} = \dfrac{(q^2 + 1)}{q(q^2 - 1)}$ **23.** $\dfrac{dy}{dt} = -\dfrac{(3t^2 - 4t - 3)}{2(1 - t)(t^2 + 3)}$

25. $\dfrac{dy}{dx} = 1 - \dfrac{1}{x}$ **27.** $y' = (1 - \ln x)/x^2$

29. $y' = 8x^3/(x^4 + 3)$ **31.** $y' = \dfrac{4(\ln x)^3}{x}$

33. $y' = \dfrac{8x^3 \ln(x^4 + 3)}{x^4 + 3}$ **35.** $y' = \dfrac{1}{x \ln 4}$

37. $y' = \dfrac{4x^3 - 12x^2}{(x^4 - 4x^3 + 1) \ln 6}$

39. rel min$(e^{-1}, -e^{-1})$

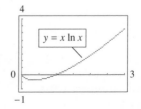

$y = x \ln x$

41. rel min$(2, 4 - 8 \ln 2)$

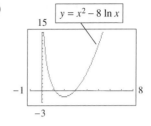

$y = x^2 - 8 \ln x$

43. (a) $\overline{MC} = \dfrac{400}{2x + 1}$

(b) $\overline{MC} = \dfrac{400}{401} \approx 1.0$; the approximate cost of the 201st unit is \$1.00

45. (a) $\overline{MR} = \dfrac{2500[(x + 1) \ln(10x + 10) - x]}{(x + 1) \ln^2(10x + 10)}$

(b) 309.67; at 100 units, selling one additional unit yields \$309.67.

47. (a) -5.23 **(b)** -1.89 **(c)** increasing

49. A/B **51.** $dR/dI = 1/(I \ln 10)$

Exercise 11.2   (page 771)

1. $y' = 5e^x - 1$ **3.** $f'(x) = e^x - exe^{-1}$

5. $y' = 3x^2 e^{x^3}$ **7.** $y' = 36xe^{3x^2}$

9. $y' = 12x(x^2 + 1)^2 e^{(x^2 + 1)^3}$ **11.** $y' = 3x^2$

13. $y' = e^{-1/x}/x^2$ **15.** $y' = \dfrac{2}{x^3}e^{-1/x^2} - 2xe^{-x^2}$

17. $ds/dt = te^t(t + 2)$ **19.** $4x^3 e^{x^4} - 4 e^{4x}$

21. $\dfrac{4e^{4x}}{e^{4x} + 2}$ **23.** $y' = e^{-3x}/x - 3e^{-3x} \ln(2x)$

25. $y' = (2e^{5x} - 3)/e^{3x} = 2e^{2x} - 3e^{-3x}$

27. $y' = 30e^{3x}(e^{3x} + 4)^9$ **29.** $y' = 6^x \ln 6$

31. $y' = 4^{x^2}(2x \ln 4)$

33. (a) $y'(1) = 0$ **(b)** $y = e^{-1}$

35. (a) $z = 0$ **(b)**

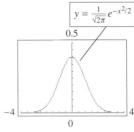

$y = \frac{1}{\sqrt{2\pi}} e^{-x^2/2}$

37. rel min at $x = 1$, $y = e$

39. rel max at $x = 0$, $y = -1$

41. (a) $(0.1)\, Pe^{0.1n}$ **(b)** $(0.1)\, Pe^{0.1}$ **(c)** Yes, since $e^{0.1n} > 1$ for any $n \geq 1$.

43. $\dfrac{dS}{dt} = -50{,}000e^{-0.5t}$ **45.** $1200e$

47. 29.107 **49.** $y' = \dfrac{98{,}990{,}100e^{-0.99t}}{(1 + 9999e^{-0.99t})^2}$

51. $y' = 100{,}000e^{-1/x}/x^2$

53. $\dfrac{dN}{dt} = N_0(1 + r)^t \cdot \ln(1 + r)$

55. $\dfrac{dI}{dR} = 10^R \ln 10$ **57.** 112.066 ($\$billion/yr$)

59. (a) $d'(x) = 29 - 20.1891e^{x/100} + 0.044e^{x/10}$
   (b) $\$1.28$ (billion/yr), $\$198.01$ (billion/yr)
   (c) U.S. could not tolerate exponential growth in its debt. The percent of federal expenditures devoted to payment of interest on the debt was too high.

Exercise 11.3   (page 780)

1. $\frac{1}{2}$ **3.** 2 **5.** $-\frac{5}{3}$

7. $-x/(2y)$ **9.** $-(2x + 4)/(2y - 3)$ **11.** $-x/y$

13. $y' = \dfrac{-y}{2(x - 1)}$ **15.** $\dfrac{dp}{dq} = \dfrac{p}{4 - q}$

17. $\dfrac{dy}{dx} = \dfrac{x(3x^3 - 2)}{3y^2(1 + y^2)}$ **19.** $\dfrac{dy}{dx} = \dfrac{4x^3 + 6x^2y - 1}{-2y - 3y^2}$

21. $\dfrac{(4x^3 + 9x^2y^2 - 8x - 12y)}{(18y + 12x - 6x^3y + 10y^4)}$ **23.** undefined

25. 1 **27.** $y = \frac{1}{2}x + 1$ **29.** $y = 4x + 5$

31. y **33.** $\dfrac{-y \ln y}{x}$, or $-\dfrac{4y}{x^2}$ **35.** -4

37. $-1/x$ **39.** $-y/x$

41. $ye^x/(1 - e^x)$ **43.** 0 **45.** $y = -\frac{1}{3}x + 1$

47. horizontal: $(2, \sqrt{2})$, $(2, -\sqrt{2})$; vertical: $(2 + 2\sqrt{2}, 0)$, $(2 - 2\sqrt{2}, 0)$

51. (b) yes, because $x^2 + y^2 = 4$ **53.** $1/(2x\sqrt{x})$

55. max at $(0, 3)$; min at $(0, -3)$ **57.** $\frac{1}{2}$ **59.** $-\frac{243}{128}$

61. At $p = \$80$, $q = 49$ and $dq/dp = -\frac{5}{16}$, which means that if the price is increased to $\$81$, quantity demanded will decrease by approximately $\frac{5}{16}$ units.

63. $-0.000436\, y$ **65.** $\dfrac{dh}{dt} = -\dfrac{3}{44} - \dfrac{h}{12}$

Exercise 11.4   (page 787)

1. 36 **3.** $\frac{1}{8}$ **5.** $-\frac{24}{5}$

7. $\frac{7}{6}$ **9.** -5 if $z = 5$, -10 if $z = -5$

11. -80 units/sec   **13.** 12π ft^2/min   **15.** $\frac{16}{27}$ in/sec

17. $\$1798$/day   **19.** $\$0.42$/day

21. 430 units/month   **23.** 36π mm^3/month

25. $\dfrac{\frac{dW}{dt}}{W} = 3\left(\dfrac{\frac{dL}{dt}}{L}\right)$ **27.** $\dfrac{\frac{dC}{dt}}{C} = 1.54\left(\dfrac{\frac{dW}{dt}}{W}\right)$

29. $\dfrac{1}{4\pi}$ micrometers/day

31. $1/(20\pi)$ in/min   **33.** $-.75$ ft/sec   **35.** $-120\sqrt{6}$

37. 61.18 mph   **39.** $\frac{1}{25}$ ft/hr

Exercise 11.5   (page 798)

1. (a) 1 **(b)** no change   **3. (a)** 84
   (b) Revenue will decrease

5. (a) $\frac{100}{99}$ **(b)** elastic   **(c)** decrease

7. (a) 0.81 **(b)** inelastic   **(c)** increase

9. (a) $\eta = 11.1$ (approx.)   **(b)** elastic

11. (a) $\eta = \dfrac{375 - 3q}{q}$
   (b) unitary: $q = 93.75$; inelastic: $q > 93.75$; elastic: $q < 93.75$
   (c) As q increases over $0 < q < 93.75$, p decreases, so elastic demand means R increases. Similarly, R decreases for $q > 93.75$.
   (d) Maximum for R when $q = 93.75$; yes.

13. $\$12$/item   **15.** $t = \$350$ **17.** $\$115$/item

19. $\$483$ per item; $\$40{,}100$ **21.** $\$1100$/item

Chapter 11   Review Exercises   (page 800)

1. $x^2e^x + 2xe^x$ **2.** $2x$ **3.** $1/q - \left[\dfrac{2q}{(q^2 - 1)}\right]$

4. $dy/dx = 2xe^{x^2}$ **5.** $dy/dx = 3^{3x-3}/\ln 3$

6. $y' = \dfrac{10(1 + \ln x)^9}{x}$ **7.** $y' = \dfrac{1 - \ln x}{x^2}$

8. $dy/dx = -2e^{-x}/(1 - e^{-x})^2$

9. $dy/dx = \dfrac{2[1 - \ln(2x + 1)]}{(2x + 1)^2}$

10. $dx/dt = e^t(t^2 + 4)(t + 2)^2$ **11.** $y' = \dfrac{-y}{x \ln x}$

12. $dy/dx = ye^{xy}/(1 - xe^{xy})$ **13.** $dy/dx = 2/y$

14. $\dfrac{dy}{dx} = \dfrac{2(x + 1)}{3(1 - 2y)}$ **15.** $y' = \dfrac{6x(1 + xy^2)}{y(5y^3 - 4x^3)}$

16. $d^2y/dx^2 = -(x^2 + y^2)/y^3$ **17.** 5/9

18. $(-2, \pm \sqrt{\tfrac{2}{3}})$ **19.** 3/4   **20.** 11 sq. units/min

21. 135.3   **22. (a)** 152.5   **(b)** 1.13 times faster

23. (a) $-0.00002876\, A_0$ **(b)** $-0.00002876\, A_0$
   (c) less   **24.** $1200e$

25. $603.48   **26.** $1/(25\pi)$ mm/min   **27.** $\frac{48}{25}$ ft/min

28. $\dfrac{dS/dt}{S} = \dfrac{1}{3}\left(\dfrac{dA/dt}{A}\right)$ **29.** yes   **30.** $t = 1446.67$

31. $880   **32. (a)** 1   **(b)** no change   **33.** $\frac{25}{12}$, elastic

34. 1

35. (a)   **(b)** $q = 100$

(c) max revenue at $q = 100$

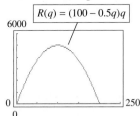

(d) Revenue is maximized where elasticity is unitary.

Exercise 12.1   *(page 811)*

1. $x^4 + C$ **3.** $\frac{1}{7}x^7 + C$ **5.** $\frac{1}{8}x^8 + C$
7. $\frac{4}{5}x^6 + C$ **9.** $27x + \frac{1}{14}x^{14} + C$

11. $3x - \frac{2}{5}x^{5/2} + C$ **13.** $\frac{1}{5}x^5 - 3x^3 + 3x + C$

15. $2x + \frac{4}{3}x\sqrt{x} + C$ **17.** $\frac{24}{5}x\sqrt[4]{x} + C$

19. $-5/(3x^3) + C$ **21.** $\frac{3}{2}\sqrt[3]{x} + C$

23. $\dfrac{1}{4}x^4 - 4x - \dfrac{1}{x^5} + C$

25. $\dfrac{1}{10}x^{10} + \dfrac{1}{2x^2} + 3x^{2/3} + C$

27. $\frac{1}{4}x^4 + \frac{10}{3}x^3 + \frac{25}{2}x^2 + C$ **29.** $2x^8 - \frac{4}{3}x^6 + \frac{1}{4}x^4 + C$

31. $-1/x - 1/(2x^2) + C$

33.

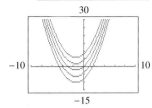

35. $\int(5 - \frac{1}{2}x)\,dx$ **39.** $R(x) = 3x$
41. $R(x) = 0.2x^2 + 3x$ **43.** $3800
45. $P(t) = \frac{1}{4}t^4 + \frac{4}{3}t^3 + 6t$
47. (a) $x = t^{7/4}/1050$ **(b)** 0.96 tons
49. (a) $x/4 + 100/x + 30$ **(b)** $56
51. (a) $R = -0.031t^2 + 0.776t + 0.0639$

(b)   Graph of $R(t)$ with data points

53. (a) $PS = 0.7t^3 - 32.7t^2 + 491.6t - 2192.2$

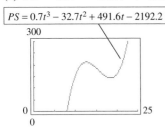

(b)

Graph of $PS(t)$ with data points

Exercise 12.2   *(page 821)*

1. $\frac{1}{4}(x^2 + 3)^4 + C$ **3.** $(15x^2 + 10)^5/5 + C$

5. $\frac{1}{3}(3x - x^3)^3 + C$ **7.** $\frac{1}{8}(x^2 + 5)^4 + C$

9. $\frac{1}{4}(4x - 1)^7 + C$ **11.** $-\frac{1}{4}(x^2 + 1)^{-2} + C$

13. $\frac{1}{10}(x^2 - 2x + 5)^5 + C$ **15.** $-\frac{1}{8}(x^4 - 4x + 3)^{-4} + C$

17. $\frac{7}{6}(x^4 + 6)^{3/2} + C$ **19.** $\frac{3}{8}x^8 + \frac{6}{5}x^5 + \frac{3}{2}x^2 + C$

21. $\frac{72}{7}x^7 - \frac{48}{5}x^5 + \frac{8}{3}x^3 + C$ **23.** $\frac{2}{9}(x^3 - 3x)^{3/2} + C$

25. $\dfrac{-1}{[3(x^3 - 1)]} + C$ **27.** $\dfrac{-1}{[10(2x^5 - 5)^3]} + C$

29. $\dfrac{-1}{[8(x^4 - 4x)^2]} + C$ **31.** $\frac{2}{3}\sqrt{x^3 - 6x^2 + 2} + C$

33. (a) $f(x) = \frac{1}{8}(x^2 - 1)^4 + C$

(b)

35. (a) $F(x) = \frac{15}{4}(2x - 1)^{2/5} + C$

(b)

(c) $x = \frac{1}{2}$ **(d)** vertical

37. $\displaystyle\int \dfrac{8x(x^2 - 1)^{1/3}}{3}\, dx$

39. $R(x) = \dfrac{15}{2x + 1} + 30x - 15$

41. 3720   **43. (a)** $s = 10\sqrt{x + 1}$ **(b)** 50

45. (a) $100/(t + 10) - 1000/(t + 10)^2$

(b) 2.5 million

47. 7400

49. (a) $P = -0.0020595\ (t + 60)^3 +$
$0.50885(t + 60)^2 - 41.3634t - 1345.8075$

(b)   Graph of P with data points

(c) The data seem to fit the model quite well.

51. (a) $U = 0.485(0.1t + 2)^3 - 10.9(0.1t + 2)^2$
$+ 7.52t + 18.874$

(b)

Graph of U with data points

$$U = 0.485(0.1t + 2)^3 - 10.9(0.1t + 2)^2 + 7.52t + 18.874$$

(c) Overall the data fit the model quite well, but the maximum points for each do not match.

Exercise 12.3 *(page 830)*

1. $\ln|x^3 + 4| + C$ 3. $\frac{1}{4}\ln|4z + 1| + C$
5. $\frac{1}{4}\ln|x^4 + 1| + C$ 7. $2\ln|x^2 - 4| + C$
9. $\ln|x^3 - 2x| + C$ 11. $\frac{1}{3}\ln|z^3 + 3z + 17| + C$
13. $\frac{1}{3}x^3 + \ln|x - 1| + C$
15. $x + \frac{1}{2}\ln|x^2 + 3| + C$
17. $e^{3x} + C$ 19. $-e^{-x} + C$
21. $10{,}000\, e^{0.1x} + C$
23. $-1200\, e^{-0.7x} + C$
25. $\frac{1}{12}e^{3x^4} + C$ 27. $-\frac{3}{2}e^{-2x} + C$ 29. $\frac{1}{18}e^{3x^6 - 2} + C$
31. $\frac{1}{4}e^{4x} + 6/e^{x/2} + C$
33. $f(x) = h(x),\ \int f(x)\, dx = g(x)$
35. $F(x) = -\ln|3 - x| + C$

$$F(x) = \ln 3 - \ln |3 - x|$$

37. $\displaystyle\int \left(1 + \frac{1}{x}\right) dx$ 39. $5e^{-x} - 5xe^{-x}$
41. $\$1030.97$ 43. $n = n_0 e^{-Kt}$
45. 55 47. **(a)** $Pe^{0.1n}$ **(b)** approx. 7 yrs
49. **(a)** $p = 95e^{-0.491t}$ **(b)** ≈ 90.45
51. **(a)**

$$\ell = 14.1 \ln |t + 20| + 11.7$$

(b)

Graph of ℓ with data points

$$\ell = 14.1 \ln |t + 20| + 11.7$$

(c) The model is a very good fit to these data.
53. **(a)** $R = 0.572\, e^{0.3486t} + C$
(b)

$$R = 0.572 e^{0.3486t} - 2.181$$

(c) $\$24.28$ billion

Exercise 12.4 *(page 842)*

1. $C(x) = x^2 + 100x + 200$
3. $C(x) = 2x^2 + 2x + 80$ 5. $\$3750$
7. **(a)** $x = 3$ units is optimal level
   (b) $P(x) = -4x^2 + 24x - 200$ **(c)** loss of $\$164$
9. **(a)** $\$2960$ **(b)** 896
11. **(a)** $\overline{C}(x) = 6/x + x/6 + 266/15$ **(b)** $\$21.37$
13. **(a)** and **(b)**

$$R(x) = 4\sqrt{0.5x + 4} + 2.8x - 8$$
$$C(x) = (x + 180)^{1.05} - 33.365$$

hundreds of thousands

(c) Max profit is $\$114.743$ thousand at $x = 200$ thousand units
15. $C(y) = 0.4y + 0.6\sqrt{y} + 5$
17. $C(y) = 0.3y + 0.4\sqrt{y} + 8$
19. $C = 2\sqrt{y + 1} + 0.4y + 4$
21. $C = 0.7y + 0.5e^{-2y} + 5.15$

23. $C = 0.85y + 5.15$

25. $C = 0.8y + \dfrac{2\sqrt{3y + 7}}{3} + 4.24$

Exercise 12.5   (page 853)

1. $4y - 2xy' = 4x^2 - 2x(2x) = 0$
3. $2y\,dx - x\,dy = 2(3x^2 + 1)\,dx - x(6x\,dx) = 2\,dx$ ✔
5. $y = \frac{1}{2}e^{x^2+1} + C$ **7.** $y^2 = 2x^2 + C$
9. $y^3 = x^2 - x + C$ **11.** $y = e^{x-3} - e^{-3} + 2$

13. $y = \ln x - \dfrac{x^2}{2} + \dfrac{1}{2}$ **15.** $\dfrac{y^2}{2} = \dfrac{x^3}{3} + C$

17. $\dfrac{1}{2x^2} + \dfrac{y^2}{2} = C$ **19.** $\dfrac{1}{x} + y + \dfrac{y^3}{3} = C$

21. $\dfrac{1}{y} + \ln|x| = C$ **23.** $x^2 - y^2 = C$

25. $y = C(x + 1)$ **27.** $y - \ln|y + 1| = -\frac{1}{2}e^{-2x} + C$

29. $3y^4 = 4x^3 - 1$ **31.** $2y = 3x + 4xy$ or $y = \dfrac{3x}{2 - 4x}$

33. $e^{2y} = x^2 - \dfrac{2}{x} + 2$ **35.** $y^2 + 1 = 5x$

37. $y = Cx^k$
39. (a) $x = 10{,}000e^{0.06t}$ **(b)** \$10,618,37; \$13,498.59
   (c) 11.55 years   **41.** ≈ 8.4 hrs
43. $\approx 23{,}100$ yrs   **45.** $x = 6(1 - e^{-0.05t})$
47. $x = 20 - 10e^{-0.025t}$ **49.** $p = 28e^{2t} - 20$
51. $p = 35e^t - 25$
53. (a) $p = 10 + 31e^{-3t}$ **(b)** stable; $p \to 10$

55. $y = \dfrac{32}{(p + 8)^{2/5}}$ **57.** $V = 1.86e^{2-2e^{-0.1t}}$

59. $V = \dfrac{k^3t^3}{27}$ **61.** $t \approx 4.5$ hr

63. (a) $E(t) = 0.1e^{0.043t}$
   (b) The graph is an accurate, but smooth,
   representation of the given graph.

65. (a) $P(t) = 20{,}000e^{-0.05t}$
   (b) $P(30) \approx \$4463$. This result corresponds quite
   well to the graph.

Chapter 12 Review Exercises   (page 858)

1. $\frac{1}{7}x^7 + C$ **2.** $\frac{2}{3}x^{3/2} + C$
3. $\frac{1}{4}x^4 - x^3 + 2x^2 + 5x + C$
4. $\frac{1}{5}x^5 - \frac{2}{3}x^3 + x + C$
5. $\frac{1}{6}(x^2 - 1)^3 + C$ **6.** $\frac{1}{6}(x^3 - 3x^2)^2 + C$
7. $\frac{1}{8}x^8 + \frac{8}{5}x^5 + 8x^2 + C$ **8.** $\frac{1}{21}(x^3 + 4)^7 + C$

9. $\frac{1}{3}\ln|x^3 + 1| + C$ **10.** $\dfrac{-1}{3(x^3 + 1)} + C$

11. $\frac{1}{2}(x^3 - 4)^{2/3} + C$ **12.** $\frac{1}{3}\ln|x^3 - 4| + C$

13. $\dfrac{1}{2}x^2 - \dfrac{1}{x} + C$

14. $\frac{1}{3}x^3 + \frac{1}{2}x^2 - 2x - \ln|x - 1| + C$
15. $\frac{1}{3}e^{y^3} + C$ **16.** $x^3/3 - x^2 + x + C$

17. $\frac{1}{2}\ln|2x^3 - 7| + C$ **18.** $\dfrac{-5}{4e^{4x}} + C$

19. $x^4/4 - e^{3x}/3 + C$ **20.** $\frac{1}{2}e^{x^2+1} + C$

21. $\dfrac{-2}{x^3 + 1} + C$ **22.** $-\frac{7}{2}\sqrt{1 - x^4} + C$

23. $\frac{1}{4}e^{2x} - e^{-2x} + C$ **24.** $x^2/2 + 1/(x + 1) + C$
25. (a) $\frac{1}{10}(x^2 - 1)^5 + C$ **(b)** $\frac{1}{22}(x^2 - 1)^{11} + C$
   (c) $\frac{3}{16}(x^2 - 1)^8 + C$ **(d)** $\frac{3}{2}(x^2 - 1)^{1/3} + C$

26. (a) $\ln|x^2 - 1| + C$ **(b)** $\dfrac{-1}{x^2 - 1} + C$

   (c) $3\sqrt{x^2 - 1} + C$ **(d)** $\frac{3}{2}\ln|x^2 - 1| + C$
27. $y = C - 92e^{-0.05t}$
28. $y = 64x + 38x^2 - 12x^3 + C$
29. $(y - 3)^2 = 4x^2 + C$
30. $(y + 1)^2 = 2 \ln|t| + C$

31. $e^y = \dfrac{x^2}{2} + C$ **32.** $y = Ct^4$

33. $3(y + 1)^2 = 2x^3 + 75$ **34.** $(1 + 2y)^2 = 4x^2 - 15$
35. 96   **36.** 472
37. $400[1 - 5/(t + 5) + 25/(t + 5)^2]$
38. $p = 1990.099 - 100{,}000/(t + 100)$
39. (a) $y = -60e^{-0.04t} + 60$ **(b)** 23%
40. $R = 800 \ln(x + 1)$
41. (a) \$1000   **(b)** $C(x) = 3x^2 + 4x + 1000$
42. 80 units, \$440
43. $C = \sqrt{2y + 16} + 0.6y + 4.5$
44. $C = 0.8y - 0.05e^{-2y} + 7.85$ **45.** $W = CL^3$
46. ≈ 10.7 million years   **47.** $p = 40 - 25e^{-3t}$
48. $p = 15 - 10e^{-2t}; p \to 15$ **49.** $x = 360(1 - e^{-t/30})$
50. $x = 600 - 500e^{-0.01t};\ \approx 161$ min

Exercise 13.1 (page 872)

1. 7 sq units **3.** 22 sq units **5.** 3 sq units
7. 30 sq units **9.** $S_L(10) = 4.08$; $S_R(10) = 5.28$
11. both equal 14/3
13. It would lie between $S_L(10)$ and $S_R(10)$. It would equal 14/3.
15. 3 **17.** 42 **19.** -5 **21.** 180 **23.** 11,315
25. $3 - \dfrac{3(n+1)}{n} + \dfrac{(n+1)(2n+1)}{2n^2} = \dfrac{2n^2 - 3n + 1}{2n^2}$
27. (a) $S = (n-1)/2n$ **(b)** 9/20 **(c)** 99/200
(d) 999/2000 **(e)** $\frac{1}{2}$
29. (a) $S = \dfrac{(n+1)(2n+1)}{6n^2}$
(b) $77/200 = 0.385$ **(c)** $6767/20{,}000 \approx 0.3384$
(d) $667{,}667/2{,}000{,}000 \approx 0.3338$ **(e)** $\frac{1}{3}$ **31.** $\frac{20}{3}$
33. There are approximately 90 squares under the curve, each representing 1 second by
$$10 \text{ mph} = 10 \times \frac{1 \text{ hr}}{3600} \times \frac{1 \text{ mile}}{\text{hour}} = \frac{1}{360} \text{ mile. The area}$$
under the curve is approximately $90\frac{1}{360} = \frac{1}{4}$ mile.
35. 1550 sq ft

Exercise 13.2 (page 881)

1. 18 **3.** 2 **5.** 60 **7.** $12\sqrt[3]{25}$ **9.** 98
11. $\frac{7}{3}$ **13.** 12,960 **15.** 0 **17.** $8\sqrt{3} - \frac{7}{3}\sqrt{7}$
19. 2 **21.** $e^3/3 - 1/3$ **23.** 4 **25.** 0
27. $\frac{1}{6}\ln(112/31)$ **29.** $\frac{3}{2} + 3\ln 2$
31. (a) A, C **(b)** B
33. $\int_0^4 (2x - \frac{1}{2}x^2)\, dx$ **(b)** 16/3
35. (a) $\int_{-1}^0 (x^3 + 1)\, dx$ **(b)** 3/4
37. $\frac{1}{6}$ **39.** $\frac{1}{2}(e^9 - e)$
41. same absolute values, opposite signs
43. 6 **45.** 20,405.39 **47.** 0.04 cm^3
49. 1222 (approx.) **51.** $450,000
53. (a) $7007 **(b)** $19,649

Exercise 13.3 (page 892)

1. (a) $\int_0^2 (4 - x^2)\, dx$ **(b)** $\frac{16}{3}$
3. (a) $\int_0^1 [(2 - x) - \sqrt[3]{x}]\, dx$ **(b)** $\frac{3}{4}$
5. (a) $\int_0^2 [(4 - x^2) - (\frac{1}{4}x^3 - 2)]\, dx$ **(b)** $\frac{25}{3}$
7. (a) $(-1, 1), (2, 4)$ **(b)** $\int_{-1}^2 [(x + 2) - x^2]\, dx$
(c) 9/2

9. (a) $(0, 0), (\frac{5}{2}, -\frac{15}{4})$
(b) $\int_0^{5/2}[(x - x^2) - (x^2 - 4x)]\, dx$ **(c)** $\frac{125}{24}$
11. (a) $(-2, -4), (0, 0), (2, 4)$
(b) $\int_{-2}^0 [(x^3 - 2x) - 2x]\, dx + \int_0^2 [2x - (x^3 - 2x)]\, dx$
(c) 8
13. $\frac{28}{3}$ **15.** $\frac{1}{4}$ **17.** $\frac{16}{3}$ **19.** $\frac{1}{3}$ **21.** $\frac{37}{12}$
23. $4 - 3\ln 3$ **25.** $\frac{8}{3}$ **27.** 6 **29.** 0 **31.** $-\frac{4}{9}$
33. $11.8\overline{3}$
35. 1980:0.351; 1990: 0.377. The difference in incomes widened.
37. Before: 0.446; After: 0.19. The change decreases the difference in income.
39. ave. profit $= \dfrac{1}{x_1 - x_0} \displaystyle\int_{x_0}^{x_1} (R(x) - C(x))\, dx$
41. (a) $1402 **(b)** $535,333.33
43. 102.5 units **45.** $513.86 **47.** 147 mg

Exercise 13.4 (page 904)

1. $120,000 **3.** $346,664 (nearest dollar)
5. $506,000 (nearest thousand)
7. $18,660 (nearest dollar)
9. $82,155 (nearest dollar)
11. $PV = \$265{,}781$ (nearest dollar), $FV = \$377{,}161$ (nearest dollar)
13. $PV = \$190{,}519$ (nearest dollar), $FV = \$347{,}148$ (nearest dollar)
15. Gift shop: $151,024; Video rental: $141,093; Gift shop is a better buy.
17. $83.33 **19.** $161.89 **21.** (5, 56); $83.33
23. $11.50 **25.** $204.17 **27.** $2766.67
29. $17,839.58 **31.** $133.33 **33.** $2.50
35. $103.35

Exercise 13.5 (page 911)

1. $\frac{1}{8}\ln |(4 + x)/(4 - x)| + C$
3. $\frac{1}{3}\ln[(3 + \sqrt{10})/2]$ **5.** $w(\ln w - 1) + C$
7. $\frac{1}{3} + \frac{1}{4}\ln(\frac{3}{7})$ **9.** $3^x \log_3 e + C$ or $3^x/\ln 3 + C$
11. $\frac{1}{2}[7\sqrt{24} - 25\ln(7 + \sqrt{24}) + 25\ln 5]$
13. $\dfrac{(6w - 5)(4w + 5)^{3/2}}{60} + C$ **15.** $\frac{1}{2}(5^{x^2})\log_5 e + C$
17. $\frac{1}{3}(13^{3/2} - 8)$ **19.** $\dfrac{-5}{\sqrt{8}}\ln\left|\dfrac{\sqrt{8} + \sqrt{8 - x^2}}{x}\right| + C$
21. $\frac{1}{3}\ln|3x + \sqrt{9x^2 - 4}| + C$ **23.** $\frac{1}{8}\ln(\frac{9}{5})$
25. $\frac{1}{3}\ln|3x + 1 + \sqrt{(3x + 1)^2 + 1}| + C$

27. $\frac{1}{4}[10\sqrt{109} - \sqrt{10} + 9\ln(10 + \sqrt{109}) - 9\ln(1 + \sqrt{10})]$

29. $-\frac{1}{6}\ln|7 - 3x^2| + C$

31. $\frac{1}{2}\ln|2x + \sqrt{4x^2 + 7}| + C$ **33.** $2(e^{\sqrt{2}} - e) \approx 2.7899$

35. $\frac{1}{32}[\ln(9/5) - 4/9] \approx .004479$ **37.** \$3391.10

39. $C = \frac{1}{2}x\sqrt{x^2 + 9} + \frac{9}{2}\ln\left|\dfrac{x + \sqrt{x^2 + 9}}{3}\right| + 300$

41. $CS = \$399$ approx.

Exercise 13.6   *(page 918)*

1. $\frac{1}{2}xe^{2x} - \frac{1}{4}e^{2x} + C$ **3.** $\frac{1}{5}x^3\ln x - \frac{1}{9}x^3 + C$

5. $\dfrac{104\sqrt{2}}{15}$ **7.** $-(1 + \ln x)/x + C$ **9.** 1

11. $\dfrac{x^2}{2}\ln(2x - 3) - \frac{1}{4}x^2 - \frac{3}{4}x - \frac{9}{8}\ln(2x - 3) + C$

13. $\frac{1}{5}(q^2 - 3)^{3/2}(q^2 + 2) + C$ **15.** 282.4

17. $-e^{-x}(x^2 + 2x + 2) + C$

19. $(3e^4 + 1)/2$

21. $\frac{1}{4}x^4\ln^2 x - \frac{1}{8}x^4\ln x + \frac{1}{32}x^4 + C$

23. $\frac{2}{15}(e^x + 1)^{3/2}(3e^x - 2) + C$

25. $\frac{1}{2}e^{x^2} + C$ **27.** $\frac{2}{3}(e^x + 1)^{3/2}$ **29.** $-5e^{-4} + 1$

31. \$2794.46   **33.** \$34,836.73

Exercise 13.7   *(page 927)*

1. 1/5   **3.** 2   **5.** 1/e   **7.** diverges   **9.** diverges

11. 10   **13.** diverges   **15.** diverges   **17.** diverges

19. 0   **21.** 0.5   **23.** 1/(2e)   **25.** $\frac{3}{2}$

27. $\int_{-\infty}^{\infty} f(x)\,dx = 1$ **29.** $c = 1$ **31.** $c = \frac{1}{2}$

33. 20   **35.** area $= \frac{8}{3}$ **37.** $\int_0^{\infty} Ae^{-rt}\,dt = A/r$

39. \$2,400,000   **41.** \$700,000

43. (a) $500\left[\dfrac{e^{-0.03b} + 0.03b - 1}{0.0009}\right]$ **(b)** limit $= \infty$

45. 0.147

Chapter 13 Review Exercises   *(page 931)*

1. 212   **2.** $\dfrac{3(n + 1)}{2n^2}$ **3.** $\frac{91}{72}$ **4.** 1

5. 1   **6.** 14   **7.** $\frac{248}{5}$ **8.** $-\frac{205}{4}$

9. $\frac{825}{4}$ **10.** $\frac{125}{3}$ **11.** -2

12. $\frac{1}{6}\ln 47 - \frac{1}{6}\ln 9$ **13.** $\frac{9}{2}$ **14.** $\ln 4 + \frac{14}{3}$ **15.** $\frac{26}{3}$

16. $\frac{1}{2}\ln 2$ **17.** $(1 - e^{-2})/2$

18. $(e - 1)/2$ **19.** 95/2   **20.** 36   **21.** $\frac{1}{4}$ **22.** $\frac{1}{2}$

23. $\frac{1}{2}x\sqrt{x^2 - 4} - 2\ln|x + \sqrt{x^2 - 4}| + C$

24. $2\log_3 e$ **25.** $\frac{1}{2}x^2(\ln x^2 - 1) + C$

26. $\frac{1}{2}\ln|x| - \frac{1}{2}\ln|3x + 2| + C$

27. $\frac{1}{6}x^6\ln x - \frac{1}{36}x^6 + C$

28. $(-xe^{-2x}/2) - (e^{-2x}/4) + C$

29. $2x\sqrt{x + 5} - \frac{4}{3}(x + 5)^{3/2} + C$ **30.** 1   **31.** ∞

32. -100 **33.** $\frac{5}{3}$ **34.** $-\frac{1}{2}$ **35.** \$28,000

36. $e^{-2.8} \approx 0.061$ **37.** \$1297.44   **38.** \$76.60

39. (a) $(7, 6)$ **(b)** \$7.33   **40.** \$24.50

41. \$1,621,803   **42.** \$403,609   **43.** \$217.42

44. \$86,557.41

45. $-\dfrac{x^2}{4} + \dfrac{7}{2}x + \dfrac{x^2 - 1}{2}\ln(x + 1) + 2000$

46. $e^{-1.4} \approx 0.247$

Exercise 14.1   *(page 940)*

1. $\{(x, y): x \text{ and } y \text{ are real numbers}\}$

3. $\{(x, y): x \text{ and } y \text{ are real numbers and } y \neq 0\}$

5. $\{(x, y): x \text{ and } y \text{ are real numbers and } 2x - y \neq 0\}$

7. $\{(p_1, p_2): p_1 \text{ and } p_2 \text{ are real numbers and } p_1 \geq 0\}$

9. -2 **11.** $\frac{5}{3}$ **13.** 2500   **15.** 36   **17.** 3

19. $\frac{1}{25}\ln(12)$ **21.** $\frac{13}{3}$

23. \$6640.23; the amount that results when \$2000 is invested for 20 years

25. 500; if the cost of placing an order is \$400, the number of items ordered is 625, and the weekly holding cost per item is \$1, then the most economical order size is 500.

27. Max: $S \approx 112.5°F$; $A \approx 106.3°F$
Min: $S \approx 87.4°F$; $A \approx 77.6°F$

29. (a) $x = 4$ **(b)** $y = 2$ **(c)**

31. (a) 37,500

(b) $30(2K)^{1/4}(2L)^{3/4} = 30(2^{1/4})(2^{3/4})K^{1/4}L^{3/4} = 2[30\ K^{1/4}L^{3/4}]$

(c)

$$Q = 300L^{3/4}$$

33. (a) 7200   **(b)** 5000   **35.** $284,000

Exercise 14.2   *(page 950)*

1. $\dfrac{\partial z}{\partial x} = 4x^3 - 10x + 6$ $\dfrac{\partial z}{\partial y} = 9y^2 - 5$

3. $z_x = 3x^2 + 8xy$ $z_y = 4x^2 + 12y$

5. $\dfrac{\partial f}{\partial x} = 9x^2(x^3 + 2y^2)^2$ $\dfrac{\partial f}{\partial y} = 12y(x^3 + 2y^2)^2$

7. $f_x = 2x(2x^2 - 5y^2)^{-1/2}$ $f_y = -5y(2x^2 - 5y^2)^{-1/2}$

9. $\dfrac{\partial C}{\partial x} = -4y + 20xy$ $\dfrac{\partial C}{\partial y} = -4x + 10x^2$

11. $\dfrac{\partial Q}{\partial s} = \dfrac{2(t^2 + 3st - s^2)}{(s^2 + t^2)^2}$ $\dfrac{\partial Q}{\partial t} = \dfrac{3t^2 - 4st - 3s^2}{(s^2 + t^2)^2}$

13. $z_x = 2e^{2x} + \dfrac{y}{x}$ $z_y = \ln x$

15. $\dfrac{\partial f}{\partial x} = 100ye^{xy}$ $\dfrac{\partial f}{\partial y} = 100xe^{xy}$

17. 2   **19.** 7   **21.** 0

23. (a) 0   **(b)** $-2xz + 4$ **(c)** $2y$ **(d)** $-x^2$

25. (a) $8x_1 + 5x_2$ **(b)** $5x_1 + 12x_2$ **(c)** 1

27. (a) 2   **(b)** 0   **(c)** 0   **(d)** $-30y$

29. (a) $2y$ **(b)** $2x - 8y$ **(c)** $2x - 8y$ **(d)** $-8x$

31. 0   **33.** -6 **35. (a)** 6   **(b)** 0

37. $2 + 2e$

39. (a) $2 + y^2e^{xy}$ **(b)** $xye^{xy} + e^{xy}$
   (c) $xye^{xy} + e^{xy}$ **(d)** x^2e^{xy}

41. (a) $1/x^2$ **(b)** 0   **(c)** 0   **(d)** $2 + 1/y^2$

43. (a) $24x$ **(b)** $24x$ **(c)** 0

45. (a) For a mortgage of $100,000 and an 8% interest rate, the monthly payment is $1289.

(b) The rate of change of the payment with respect to the interest rate is $62.51. That is, if the rate goes from 8% to 9% on a $100,000 mortage, the approximate increase in the monthly payment is $62.51.

47. (a) If the number of items sold per week changes by one, the most economical order quantity should

also increase. $\dfrac{\partial Q}{\partial M} = \sqrt{\dfrac{K}{Mh}} > 0$

(b) If the weekly storage costs change by one, the most economical order quantity should decrease.

$$\dfrac{\partial Q}{\partial h} = -\sqrt{\dfrac{KM}{\partial h}} < 0$$

49. (a) $65e^{-0.01x}$ **(b)** $70e^{-0.02y}$

51. (a) $2xy^2$ **(b)** $2x^2y$

53. $\dfrac{\partial Q}{\partial K} = 44\frac{4}{9}$; If labor hours are held constant at 1728

and K changes by $1 (thousand) to $730,000, Q will

change by about $44\frac{4}{9}$ thousand units. $\dfrac{\partial Q}{\partial L} = 37\frac{1}{2}$; If

capital expenditures are held constant at $729,000 and L changes by 1 hour (to 1729), Q will change by about $37\frac{1}{2}$ thousand units.

55. (a) $\dfrac{\partial WC}{\partial s} = -0.020t - 1.85 + \dfrac{0.152t}{\sqrt{s}} - \dfrac{13.87}{\sqrt{s}}$

(b) at $t = 10$, $s = 25$, $\dfrac{\partial WC}{\partial s} = -0.20 - 1.85 +$

$0.304 - 2.774 = -4.52$

This means that the rate of change of WC with respect to s is -4.52 if $t = 10°$F, $s = 25$ mph. An increase in speed causes a decrease in wind chill temperature.

Exercise 14.3   *(page 959)*

1. 57   **3. (a)** $2 + y/50$ **(b)** $4 + x/50$

5. (a) 20.18   **(b)** 70.80   **7. (a)** 36   **(b)** 19

9. (a) $\sqrt{y^2 + 1}$ **(b)** $xy/\sqrt{y^2 + 1}$

11. (a) $1200y/(xy + 1)$ **(b)** $1200x/(xy + 1)$

13. (a) $\sqrt{y/x}$ **(b)** $\sqrt{x/y}$

15. (a) $\ln(y + 1)/(2\sqrt{x})$ **(b)** $\sqrt{x}/(y + 1)$

17. $z = 1092$ approx.   **19.** $z_x = 3.6$

21. **(a)** $z_x = \dfrac{240y^{2/5}}{x^{2/5}}$ **(b)** 5000

(c) $z_y = \dfrac{160x^{3/5}}{y^{3/5}}$ **(d)** 80,000

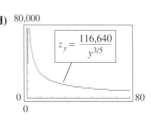

(e) Both z_x and z_y are positive, so that increases in both capital investment and in work hours result in increases in productivity. However, both are decreasing, so that such increases have a diminishing effect on productivity. Also, z_y decreases more slowly than z_x so that increases in work hours have a more significant impact on productivity than increases in capital investment.

23. $q_1 = 188$; $q_2 = 270$

25. any values for p_1 and p_2 that satisfy $6p_2 - 3p_1 = 100$ and that make q_1 and q_2 nonnegative, such as $p_1 = 10$, $p_2 = 21\frac{2}{3}$

27. **(a)** -3 **(b)** -2 **(c)** -6 **(d)** -5
    (e) complementary

29. **(a)** -50 **(b)** $600/(p_B + 1)^2$
    (c) $-400/(p_B + 4)^2$
    (d) $400/(p_A + 4)^2$ **(e)** competitive

Exercise 14.4   *(page 967)*

1. max $(0, 0, 9)$ **3.** min$(0, 0, 4)$ **5.** min$(1, -2, 0)$
7. saddle $(1, -3, 8)$ **9.** saddle $(0, 0, 0)$
11. max$(12, 24, 456)$ **13.** min$(-8, 6, -52)$
15. saddle $(0, 0, 0)$; min$(2, 2, -8)$
17. $x = 5000, y = 128$ **19.** $x = \frac{20}{3}, y = \frac{10}{3}$
21. $x = 28, y = 100$ **23.** $x = 0, y = 10$
25. length $= 100$, width $= 100$, height $= 50$
27. $x = 3, y = 0$

Exercise 14.5   *(page 976)*

1. 18 at $(3, 3)$ **3.** 35 at $(3, 2)$ **5.** 32 at $(4, 2)$
7. -28 at $(3, \frac{5}{2})$ **9.** $\frac{1}{5}$ at $(-\frac{2}{5}, -\frac{1}{5})$

11. 3 at $(1, 1, 1)$ **13.** 1 at $(0, 1, 0)$
15. $x = 4, y = 1$ **17.** $x = 40, y = \frac{40}{3}$

19. **(a)** $x = 400, y = 400$
    (b) $-\lambda = 1.6$ means that each additional dollar spent on production results in approximately 1.6 additional units produced.
    (c) 1250

21. $x = 900, y = 300$
23. $x = \$10,003.33, y = \$19,996.67$
25. length $= 100$ cm, width $= 100$ cm, height $= 50$ cm

Chapter 14 Review Exercises   *(page 978)*

1. $\{(x, y): x \text{ and } y \text{ are real numbers and } y \neq 2x\}$
2. $\{(x, y): x \text{ and } y \text{ are real numbers with } y \geq 0 \text{ and } (x, y) \neq (0, 0)\}$
3. -5 **4.** 896,000
5. $15x^2 + 6y$ **6.** $24y^3 - 42x^3y^2$
7. $z_x = 8xy^3 + 1/y$; $z_y = 12x^2y^2 - x/y^2$
8. $z_x = x/\sqrt{x^2 + 2y^2}$; $z_y = 2y/\sqrt{x^2 + 2y^2}$
9. $z_x = -2y/(xy + 1)^3$; $z_y = -2x/(xy + 1)^3$
10. $z_x = 2xy^3e^{x^2y^3}$; $z_y = 3x^2y^2e^{x^2y^3}$
11. $z_x = ye^{xy} + y/x$; $z_y = xe^{xy} + \ln x$
12. $z_x = y$; $z_y = x$ **13.** -8 **14.** 8
15. **(a)** $2y$ **(b)** 0   **(c)** $2x - 3$ **(d)** $2x - 3$
16. **(a)** $18xy^4 - 2/y^2$ **(b)** $36x^3y^2 - 6x^2/y^4$
    (c) $36x^2y^3 + 4x/y^3$ **(d)** $36x^2y^3 + 4x/y^3$
17. **(a)** $2e^{y^2}$ **(b)** $4x^2y^2e^{y^2} + 2x^2e^{y^2}$
    (c) $4xye^{y^2}$ **(d)** $4xye^{y^2}$
18. **(a)** $-y^2/(xy + 1)^2$ **(b)** $-x^2/(xy + 1)^2$
    (c) $1/(xy + 1)^2$ **(d)** $1/(xy + 1)^2$
19. max$(-8, 16, 208)$
20. saddle$(0, 0, 0)$; min$(1, 1, -1)$
21. 80 at $(2, 8)$ **22.** 11,664 at $(6, 3)$

23. 3 units   **24.** 20

25. (a) 62.8 km (approx.)

 (b) If the surface temperature is 287°K, the Concorde's altitude is 17.1 km, and the vertical temperature gradient is 4.9°K/km, then the width of the sonic boom is about 63.3 km.

 (c) $\dfrac{\partial f}{\partial h} \approx 1.87$ If the surface temperature is 293°K and the temperature gradient is 5°K/km, then a change of 1 km in the Concorde's altitude would result in a change in the width of the sonic boom of about 1.87 km.

 (d) $\dfrac{\partial f}{\partial d} \approx -6.28$ If the surface temperature is 293°K and the Concorde's altitude is 16.8 km, then a change of 1°K/km in the temperature gradient would result in a change in the width of the sonic boom of about −6.28 km.

26. (a) 280   **(b)** 2400/7

27. $\partial Q/\partial K = 81.92$; $\partial Q/\partial L = 37.5$

28. (a) −2   **(b)** −6   **(c)** complementary

29. competitive   **30.** $x = 20$, $y = 40$

31. $x = 10$, $y = 4$

32. (a) $x = 1000$, $y = 500$

 (b) $-\lambda \approx 159$ means each additional dollar spent on production results in approximately 159 additional units.

 (c)

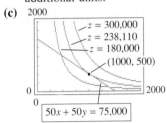

Index